Lewin's GENES X

Jones and Bartlett Titles in Biological Science

LEWIN'S
GENES X

JOCELYN E. KREBS
University of Alaska, Anchorage

ELLIOTT S. GOLDSTEIN
Arizona State University

STEPHEN T. KILPATRICK
University of Pittsburgh at Johnstown

JONES AND BARTLETT PUBLISHERS
Sudbury, Massachusetts
BOSTON   TORONTO   LONDON   SINGAPORE

World Headquarters

Jones and Bartlett Publishers
40 Tall Pine Drive
Sudbury, MA 01776
978-443-5000
info@jbpub.com
www.jbpub.com

Jones and Bartlett Publishers Canada
6339 Ormindale Way
Mississauga, Ontario L5V 1J2
Canada

Jones and Bartlett Publishers International
Barb House, Barb Mews
London W6 7PA
United Kingdom

Jones and Bartlett's books and products are available through most bookstores and online booksellers. To contact Jones and Bartlett Publishers directly, call 800-832-0034, fax 978-443-8000, or visit our website, www.jbpub.com.

Substantial discounts on bulk quantities of Jones and Bartlett's publications are available to corporations, professional associations, and other qualified organizations. For details and specific discount information, contact the special sales department at Jones and Bartlett via the above contact information or send an email to specialsales@jbpub.com.

Production Credits

Chief Executive Officer: Clayton Jones
Chief Operating Officer: Don W. Jones, Jr.
President, Higher Education and Professional
   Publishing: Robert W. Holland, Jr.
V.P., Sales: William J. Kane
V.P., Design and Production: Anne Spencer
V.P., Manufacturing and Inventory Control:
   Therese Connell
Publisher, Higher Education: Cathleen Sether
Acquisitions Editor, Science: Molly Steinbach
Senior Editorial Assistant, Science: Jessica S. Acox

Editorial Assistant, Science: Caroline Perry
Production Manager: Louis C. Bruno, Jr.
Associate Production Editor: Leah Corrigan
Senior Marketing Manager: Andrea DeFronzo
Composition: Shepherd, Inc.
Cover Design: Kristin E. Parker
Photo Research and Permissions Manager:
   Kimberly Potvin
Cover Image: © Science VU/DOE/Visuals Unlimited
Printing and Binding: Courier Kendallville
Cover Printing: Courier Kendallville

About the Cover: A false color SEM image of a human X chromosome.

Library of Congress Cataloging-in-Publication Data

Lewin's genes X. — 10th ed. / [edited by] Jocelyn E. Krebs, Stephen T. Kilpatrick, Elliott S. Goldstein.
   p. ; cm.
   Rev. ed. of: Genes IX / Benjamin Lewin. c2008.
   Includes bibliographical references and index.
   ISBN 978-0-7637-6632-0 (alk. paper)
   1. Genetics. 2. Genes. I. Lewin, Benjamin. II. Krebs, Jocelyn E. III. Kilpatrick, Stephen T. IV. Goldstein, Elliott S. V.
Lewin, Benjamin. Genes IX. VI. Title: Genes X. VII. Title: Lewin's genes 10. VIII. Title: Lewin's genes ten.
   [DNLM: 1. Genes. 2. DNA—genetics. 3. Genetic Processes. 4. Genome. 5. Proteins—genetics. 6. RNA—genetics.
QU 470 L6723 2010]
   QH430.L4 2010
   576.5—dc22
                                                                                    2009042681

6048

Printed in the United States of America
13 12 11 10 09   10 9 8 7 6 5 4 3 2 1

Dedication

To Benjamin Lewin, for setting the bar high.

To my mother, Ellen Baker, for raising me with a love of science; to the memory of my stepfather, Barry Kiefer, for convincing me science would stay fun; and to my partner, Susannah Morgan, for always pretending my biology jokes are funny. Finally, to my son, Rhys, who may someday read a future edition of this book.

Jocelyn Krebs

To my family: my wife, Suzanne, whose patience, understanding, and confidence in me are amazing; my children, Andy, Hyla, and Gary, who have taught me so much about using the computer; and my grandchildren, Seth and Elena, whose smiles and giggles inspire me. And to the memory of my mentor and dear friend, Lee A. Snyder, whose professionalism, guidance, and insight demonstrated the skills necessary to be a scientist and teacher. I have tried to live up to his expectations. This is for you, Doc.

Elliott Goldstein

To my wife, Lori, for our many years of love, support, and sometimes tolerance; to my daughter, Jennifer, who will actually read this book; to my son, Andrew, who continually renews my faith in humanity; and to my daughter, Sarah, who brings me joy daily.

Stephen Kilpatrick

Brief Table of Contents

Contents

Chapter 4. The Interrupted Gene 79

Edited by Donald Forsdyke

Chapter 5. The Content of the Genome 98

Chapter 6. Genome Sequences and Gene Numbers 118

Chapter 18. Somatic Recombination and Hypermutation in the Immune System 458

Edited by Paolo Casali

PART 3. TRANSCRIPTION AND POSTTRANSCRIPTIONAL MECHANISMS 503

Chapter 19. Prokaryotic Transcription 504

Edited by Richard Gourse

Contents **xvii**

Preface

Of the diverse ways to study the living world, molecular biology has been most remarkable in the speed and breadth of its expansion. New data are acquired daily, and new insights into well-studied processes come on a scale measured in weeks or months rather than years. It's difficult to believe that the first complete organismal genome sequence was obtained less than fifteen years ago. The structure and function of genes and genomes and their associated cellular processes are sometimes elegantly and deceptively simple but frequently amazingly complex, and no single book can do justice to the realities and diversities of natural genetic systems.

This book is aimed at advanced students in molecular genetics and molecular biology. In order to provide the most current understanding of the rapidly-changing subjects in molecular biology, we have enlisted twenty-one scientists to provide revisions and content updates in their individual fields of expertise. Their expert knowledge has been incorporated throughout the text. Much of the revision and reorganization of this edition follows that of the second edition of *Lewin's Essential GENES*, but there are many updates and features that are new to this book. Most notably, there are two new chapters: Chapter 3 ("Methods in Molecular Biology and Genetic Engineering") provides an introduction to the concepts and practice of laboratory techniques in molecular biology early on in the book, and Chapter 8 ("Genome Evolution") combines, expands, and updates material that had been scattered among various chapters in previous editions, as well as introducing a number of topics new to this book. This edition is generally updated and reorganized for a more logical flow of topics, and many chapters have been renamed to better indicate their contents. In particular, discussion of chromatin organization and nucleosome structure now precedes the discussion of eukaryotic transcription, because chromosome organization is critical to all DNA transactions in the cell, and current research in the field of transcriptional regulation is heavily biased toward the study of the role of chromatin in this process. The discussion of transcriptional activation and chromatin remodeling has accordingly been combined into one chapter (Chapter 28). Two chapters on transposons and retroposons have been combined into one (Chapter 17). In addition, some chapters have been revised to contain extensive new material. The original introductory chapter on messenger RNA has been entirely rewritten to cover more advanced topics (Chapter 22, "mRNA Stability and Localization"), and the regulatory RNA chapter has been dramatically expanded to include material on RNAi pathways (Chapter 30, "Regulatory RNA"). Many new figures are included in this book, some reflecting new developments in the field, particularly in the topics of chromatin structure and function, epigenetics, and regulation by noncoding and microRNAs in eukaryotes.

This book is organized into four parts. **Part 1 (Genes and Chromosomes)** comprises Chapters 1 through 10. Chapters 1 and 2 serve as an introduction to the structure and function of DNA and contain basic coverage of DNA replication and gene expression. Chapter 3 provides information on molecular laboratory techniques. Chapter 4 introduces the interrupted structures of eukaryotic genes, and Chapters 5 through 8 discuss genome structure and evolution. Chapters 9 and 10 discuss the structure of eukaryotic chromosomes.

Part 2 (DNA Replication and Recombination) comprises Chapters 11 through 18. Chapters 11 to 14 provide detailed discussions of DNA replication in plasmids, viruses, and prokaryotic and eukaryotic cells. Chapters 15 through 18 cover recombination and its roles in DNA repair and the human immune system, with Chapter 16 discussing DNA repair pathways in detail and Chapter 17 focusing on different types of transposable elements.

Part 3 (Transcription and Posttranscriptional Mechanisms) includes Chapters 19 through 25.

Chapters 19 and 20 provide more in-depth coverage of bacterial and eukaryotic transcription. Chapters 21 through 23 are concerned with RNA, discussing messenger RNA, RNA stability and localization, RNA processing, and the catalytic roles of RNA. Chapters 24 and 25 discuss translation and the genetic code.

Part 4 (Gene Regulation) comprises Chapters 26 through 30. In Chapter 26, the regulation of bacterial gene expression via operons is discussed. Chapter 27 covers the regulation of expression of genes during phage development as they infect bacterial cells. Chapters 28 and 29 cover eukaryotic gene regulation, including epigenetic modifications. Finally, Chapter 30 covers RNA-based control of gene expression in prokaryotes and eukaryotes.

For instructors who prefer to order topics with the essentials of DNA replication and gene expression followed by more advanced topics, the following chapter sequence is suggested:

Introduction: Chapters 1–2
Gene and Genome Structure: Chapters 5–7
DNA Replication: Chapters 11–14
Transcription: Chapters 19–22
Translation: Chapters 24–25
Regulation of Gene Expression: Chapters 9–10 and 26–30

Other chapters can be covered at the instructor's discretion.

Pedagogical Features

This edition contains several features to help students learn as they read. Each chapter begins with a *Chapter Outline*, and each section is summarized with a bulleted list of *Key Concepts*. *Key Terms* are highlighted in bold type in the text and compiled in the *Glossary* at the end of the book. Finally, each chapter concludes with an expanded and updated list of *References*, which provides both primary literature and current reviews to supplement and reinforce the chapter content. Additional instructional tools are available online and on the Instructor's media CD-ROM.

Ancillaries

Jones and Bartlett Publishers offers an impressive array of traditional and interactive multimedia supplements to assist instructors and aid students in mastering molecular biology. Additional information and review copies of any of the following items are available through your Jones and Bartlett sales representative or by visiting http://www.jbpub.com/biology.

For the Student
Interactive Student Study Guide
Jones and Bartlett Publishers and Brent Nielsen of Brigham Young University have developed an interactive, electronic study guide dedicated exclusively to this title. Students will find a variety of study aids and resources at http://biology.jbpub.com/lewin/genesx, all designed to explore the concepts of molecular biology in more depth and to help students master the material in the book. A variety of activities are available to help students review class material, such as chapter summaries, Web-based learning exercises, study quizzes, a searchable glossary, and links to animations, videos, and podcasts, all to help students master important terms and concepts.

For Instructors
Instructor's ToolKit CD-ROM
The *Instructor's Media CD-ROM* provides the instructor with the following resources:

- The **PowerPoint® Image Bank** provides all of the illustrations, photographs, and tables (to which Jones and Bartlett Publishers holds the copyright or has permission to reprint digitally) inserted into PowerPoint slides. With the Microsoft® PowerPoint program, you can quickly and easily copy individual image slides into your existing lecture slides.
- A set of **PowerPoint Lecture Outline Slides**, created by author Stephen Kilpatrick, of the University of Pittsburgh at Johnstown, provides outline summaries and relevant images for each chapter of *Lewin's GENES X*. A PowerPoint viewer is provided on the CD, and instructors with the Microsoft PowerPoint software can customize the outlines, figures, and order of presentation.

Online Instructor Resources
The **Test Bank**, updated and expanded by author Stephen Kilpatrick, is provided as a text file with 750 questions in a variety of formats. The Test Bank is easily compatible with most course management software.

Acknowledgments

The authors would like to thank the following individuals for their assistance in the preparation of this book: The editorial, production, marketing, and sales teams at Jones and Bartlett have been exemplary in all aspects of this project. Cathy Sether, Caroline Perry, Megan Turner, Kimberly Potvin, Leah Corrigan, and Lou Bruno deserve special mention. Cathy brought us together on this project and in doing so launched an efficient and amiable partnership. She has provided able leadership and has been an excellent resource as we ventured into new territories. Caroline, Lou and Leah have handled the daily responsibilities of the writing and production phases with friendly professionalism and helpful guidance. Megan and Kimberly have made the process of choosing and revising figures very smooth.

We thank the editors of individual chapters, whose expertise, enthusiasm, and careful judgment brought the manuscript up to date in many critical areas. We also thank Brent Nielsen of Brigham Young University for an early version of Section 8.3, and David Rand of Brown University for suggestions for improvement to Chapter 8.

Jocelyn E. Krebs
Elliott S. Goldstein
Stephen T. Kilpatrick

About the Authors

Benjamin Lewin founded the journal *Cell* in 1974 and was Editor until 1999. He founded the Cell Press journals *Neuron*, *Immunity*, and *Molecular Cell*. In 2000, he founded Virtual Text, which was acquired by Jones and Bartlett Publishers in 2005. He is also the author of *Essential GENES* and *CELLS*.

Jocelyn E. Krebs received a B.A. in Biology from Bard College, Annandale-on-Hudson, NY, and a Ph.D. in Molecular and Cell Biology from the University of California, Berkeley. For her Ph.D. thesis, she studied the roles of DNA topology and insulator elements in transcriptional regulation.

She performed her postdoctoral training as an American Cancer Society Fellow at the University of Massachusetts Medical School in the laboratory of Dr. Craig Peterson, where she focused on the roles of histone acetylation and chromatin remodeling in transcription. In 2000, Dr. Krebs joined the faculty in the Department of Biological Sciences at the University of Alaska, Anchorage, where she is now an Associate Professor. She directs a research group studying chromatin structure and function in transcription and DNA repair in the yeast *Saccharomyces cerevisiae* and the role of chromatin remodeling in embryonic development in the frog *Xenopus*. She teaches courses in molecular biology for undergraduates, graduate students, and first-year medical students. She also teaches a Molecular Biology of Cancer course and has taught Genetics and Introductory Biology. She lives in Eagle River, AK, with her partner and son, and a house full of dogs and cats. Her non-work passions include hiking, camping, and snowshoeing.

Elliott S. Goldstein earned his B.S. in Biology from the University of Hartford (Connecticut) and his Ph.D. in Genetics from the University of Minnesota, Department of Genetics and Cell Biology. Following this, he was awarded an N.I.H. Postdoctoral Fellowship to work with Dr. Sheldon Penman at the Massachusetts Institute of Technology. After leaving Boston, he joined the faculty at Arizona State University in Tempe, where he is an Associate Professor in the Cellular, Molecular, and Biosciences program in the School of Life Sciences and in the Honors Disciplinary Program. His research interests are in the area of molecular and developmental genetics of early embryogenesis in *Drosophila melanogaster*. In recent years, he has focused on the *Drosophila* counterparts of the human proto-oncogenes *jun* and *fos*. His primary teaching responsibilities are in the undergraduate General Genetics course as well as the graduate level Molecular Genetics course. Dr. Goldstein lives in Tempe with his wife, his high school sweetheart. They have three children and two grandchildren. He is a bookworm who loves reading as well as underwater photography. His pictures can be found at http://www.public.asu.edu/~elliotg/.

Stephen T. Kilpatrick received a B.S. in Biology from Eastern College (now Eastern University) in St. Davids, PA, and a Ph.D. from the Program in Ecology and Evolutionary Biology at Brown University. His thesis research was an investigation of the population genetics of interactions between the mitochondrial and nuclear genomes of *Drosophila melanogaster*. Since 1995, Dr. Kilpatrick has taught at the University of Pittsburgh at Johnstown in Johnstown, PA. His regular teaching duties include undergraduate courses in nonmajors biology, introductory majors biology, and advanced undergraduate courses in genetics, evolution, molecular genetics, and biostatistics. He has also supervised a number of undergraduate research projects in evolutionary genetics. Dr. Kilpatrick's major professional focus has been in biology education. He has participated in the development and authoring of ancillary materials for several introductory biology, genetics, and molecular genetics texts as well as writing articles for educational reference publications. For his classes at Pitt-Johnstown, Dr. Kilpatrick has developed many active learning exercises in introductory biology, genetics, and evolution. Dr. Kilpatrick resides in Johnstown, PA, with his wife and three children. Outside of scientific interests, he enjoys music, literature, and theater and occasionally performs in local community theater groups.

Chapter Editors

Esther Siegfried completed her work on this book while teaching at the University of Pittsburgh at Johnstown. She is now an Assistant Professor of Biology at Pennsylvania State University, Altoona. Her research interests include signal transduction pathways in *Drosophila* development.

John Brunstein is a Clinical Assistant Professor with the Department of Pathology and Laboratory Medicine at the University of British Columbia. His research interests focus on the development of, validation strategies for, and implementation of novel molecular diagnostic technologies.

Donald Forsdyke, emeritus professor of biochemistry at Queen's University in Canada, studied lymphocyte activation/inactivation and the associated genes. In the 1990s he obtained evidence supporting his 1981 hypothesis on the origin of introns, and immunologists in Australia shared a Nobel Prize for work that supported his 1975 hypothesis on the positive selection of the lymphocyte repertoire. His books include *The Origin of Species, Revisited* (2001), *Evolutionary Bioinformatics* (2006), and *"Treasure Your Exceptions": The Science and Life of William Bateson* (2008).

Hank W. Bass is an Associate Professor of Biological Science at Florida State University. His laboratory works on the structure and function of meiotic chromosomes and telomeres in maize using molecular cytology and genetics.

Stephen D. Bell is the Professor of Microbiology in the Sir William Dunn School of Pathology, Oxford University. His research group is studying gene transcription, DNA replication, and cell division in the Archaeal domain of life.

Søren Johannes Sørensen is a Professor in the Department of Biology and Head of the Section of Microbiology at the University of Copenhagen. The main objective of his studies is to evaluate the extent of genetic flow within natural communities and the responses to environmental perturbations. Molecular techniques such as DGGE and high-throughput sequencing are used to investigate resilience and resistance of microbial community structure. Dr. Sørensen has more than twenty years' experience in teaching molecular microbiology at both the bachelor and graduate levels.

Lars Hestbjerg Hansen is an Associate Professor in the Section of Microbiology, Department of Biology, at the University of Copenhagen. His research interests include the bacterial maintenance and interchange of plasmid DNA, especially focused on plasmidborne mechanisms of bacterial resistance to antibiotics. Dr. Hansen's laboratory has developed and is currently working with new flow-cytometric methods for estimating plasmid transfer and stability. Dr. Hansen is the Science Director of Prokaryotic Genomics at Copenhagen High-Throughput Sequencing Facility, focusing on using high-throughput sequencing to describe bacterial and plasmid diversity in natural environments.

Barbara Funnell is a Professor of Molecular Genetics at the University of Toronto. Her laboratory studies chromosome dynamics in bacterial cells, and in particular the mechanisms of action of proteins involved in plasmid and chromosome segregation.

Peter Burgers is Professor of Biochemistry and Molecular Biophysics at Washington University School of Medicine. His laboratory has a long-standing interest in the biochemistry and genetics of DNA replication in eukaryotic cells, and in the study of responses to DNA damage and replication stress that result in mutagenesis and in cell cycle checkpoints.

Hannah L. Klein is a Professor of Biochemistry, Medicine, and Pathology at New York University Langone Medical Center. She studies pathways of DNA damage repair and recombination and genome stability.

Samantha Hoot is a postdoctoral researcher in the laboratory of Dr. Hannah Klein at New York University Langone Medical Center. She received her PhD from the University of Washington. Her interests include the role of recombination in genome stability in yeast and the molecular mechanisms of drug resistance in pathogenic fungi.

Damon Lisch is an Associate Research Professional at the University of California at Berkeley. He is interested in the regulation of transposable elements in plants and the ways in which transposon activity has shaped plant genome evolution. His laboratory investigates the complex behavior and epigenetic regulation of the *Mutator* system of transposons in maize and related species.

Paolo Casali, MD, is the Donald L. Bren Professor of Medicine, Molecular Biology & Biochemistry, and Director of the Institute for Immunology at the University of California, Irvine. He works on B lymphocyte differentiation and regulation of antibody gene expression as well as molecular mechanisms of generation of autoantibodies. He has been Editor-in-Chief of *Autoimmunity* since 2002. He is a member of the American Association of Immunologists, an elected "Young Turk" of the American Society for Clinical Investigation, and an elected Fellow of the American Association for Advancement of Science. He has served on several NIH immunology study sections and scientific panels.

Richard Gourse is a Professor in the Department of Bacteriology at the University of Wisconsin, Madison, and an Editor of the *Journal of Bacteriology*. His primary interests lie in transcription initiation and the regulation of gene expression in bacteria. His laboratory has long focused on rRNA promoters and the control of ribosome synthesis as a means of uncovering fundamental mechanisms responsible for regulation of transcription and translation.

Xiang-Dong Fu is a Professor of Cellular and Molecular Medicine at the University of California, San Diego. His laboratory studies mechanisms underlying constitutive and regulated pre-mRNA splicing in mammalian cells, coupling between transcription and RNA processing, RNA genomics, and roles of RNA processing in development and diseases.

Ellen Baker is an Associate Professor of Biology at the University of Nevada, Reno. Her research interests have focused on the role of polyadenylation in mRNA stability and translation.

Douglas J. Briant teaches at the University of Victoria in British Columbia. His research has investigated bacterial RNA processing and the role of ubiquitin in cell signalling pathways.

Cheryl Keller Capone is an Instructor in the Department of Biology at The Pennsylvania State University and teaches cell and molecular biology. Her research interests include embryonic muscle development in *Drosophila melanogaster* and the molecular mechanisms involved in the clustering and postsynaptic targeting of $GABA_A$ receptors.

John Perona is a Professor of Biochemistry in the Department of Chemistry and Biochemistry, and the Interdepartmental Program in Biomolecular Science and Engineering, at the University of California, Santa Barbara. His laboratory studies structure-function relationships and catalytic mechanisms in aminoacyl-tRNA synthetases, tRNA-dependent amino acid modification enzymes, and tRNA-modifying enzymes.

Liskin Swint-Kruse is an Assistant Professor in Biochemistry and Molecular Biology at the University of Kansas School of Medicine. Her research utilizes bacterial transcription regulators in studies that bridge biophysics of DNA-binding and bioinformatics analyses of transcription repressor families to advance the principles of protein engineering.

Trygve Tollefsbol is a Professor of Biology at the University of Alabama at Birmingham and a Senior Scientist of the Center for Aging, Comprehensive Cancer Center, and Clinical Nutrition Research Center. He has long been involved with elucidating epigenetic mechanisms, especially as they pertain to cancer, aging, and differentiation. He has been the editor and primary contributor of numerous books including *Epigenetic Protocols*, *Cancer Epigenetics*, and *Epigenetics of Aging*.

Genes and Chromosomes

PART

I

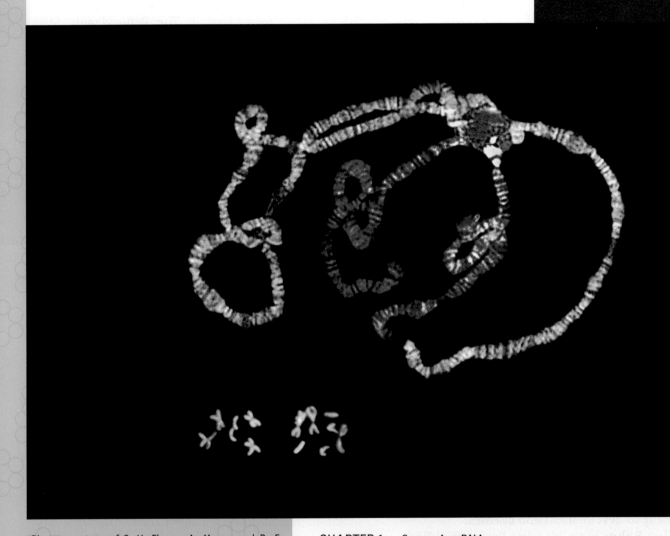

Photo courtesy of S. V. Flores, A. Mena, and B. F. McAllister. Used with permission of Bryant McAllister, Department of Biology, University of Iowa.

© Mopic/ShutterStock, Inc.

Genes Are DNA

CHAPTER OUTLINE

- The nitrogenous bases of each chain are flat purine or pyrimidine rings that face inward and pair with one another by hydrogen bonding to form only A-T or G-C pairs.
- The diameter of the double helix is 20 Å, and there is a complete turn every 34 Å, with ten base pairs per turn (~10.4 base pairs per turn in solution).
- The double helix has a major (wide) groove and a minor (narrow) groove.

1.7 DNA Replication Is Semiconservative

- The Meselson–Stahl experiment used "heavy" isotope labeling to show that the single polynucleotide strand is the unit of DNA that is conserved during replication.
- Each strand of a DNA duplex acts as a template for synthesis of a daughter strand.
- The sequences of the daughter strands are determined by complementary base pairing with the separated parental strands.

1.8 Polymerases Act on Separated DNA Strands at the Replication Fork

- Replication of DNA is undertaken by a complex of enzymes that separate the parental strands and synthesize the daughter strands.
- The replication fork is the point at which the parental strands are separated.
- The enzymes that synthesize DNA are called DNA polymerases.
- Nucleases are enzymes that degrade nucleic acids; they include DNases and RNases and can be categorized as endonucleases or exonucleases.

1.9 Genetic Information Can Be Provided by DNA or RNA

- Cellular genes are DNA, but viruses may have genomes of RNA.
- DNA is converted into RNA by transcription, and RNA may be converted into DNA by reverse transcription.
- The translation of RNA into protein is unidirectional.

1.10 Nucleic Acids Hybridize by Base Pairing

- Heating causes the two strands of a DNA duplex to separate.
- The T_m is the midpoint of the temperature range for denaturation.
- Complementary single strands can renature when the temperature is reduced.
- Denaturation and renaturation/hybridization can occur with DNA–DNA, DNA–RNA, or RNA–RNA combinations and can be intermolecular or intramolecular.

- The ability of two single-stranded nucleic acids to hybridize is a measure of their complementarity.

1.11 Mutations Change the Sequence of DNA

- All mutations are changes in the sequence of DNA.
- Mutations may occur spontaneously or may be induced by mutagens.

1.12 Mutations May Affect Single Base Pairs or Longer Sequences

- A point mutation changes a single base pair.
- Point mutations can be caused by the chemical conversion of one base into another or by errors that occur during replication.
- A transition replaces a G-C base pair with an A-T base pair or vice versa.
- A transversion replaces a purine with a pyrimidine, such as changing A-T to T-A.
- Insertions and/or deletions can result from the movement of transposable elements.

1.13 The Effects of Mutations Can Be Reversed

- Forward mutations alter the function of a gene, and back mutations (or revertants) reverse their effects.
- Insertions can revert by deletion of the inserted material, but deletions cannot revert.
- Suppression occurs when a mutation in a second gene bypasses the effect of mutation in the first gene.

1.14 Mutations Are Concentrated at Hotspots

- The frequency of mutation at any particular base pair is statistically equivalent, except for hotspots, where the frequency is increased by at least an order of magnitude.

1.15 Many Hotspots Result from Modified Bases

- A common cause of hotspots is the modified base 5-methylcytosine, which is spontaneously deaminated to thymine.
- A hotspot can result from the high frequency of change in copy number of a short, tandemly repeated sequence.

1.16 Some Hereditary Agents Are Extremely Small

- Some very small hereditary agents do not code for polypeptide, but consist of RNA or protein with heritable properties.

1.17 Summary

1.1 Introduction

The hereditary basis of every living organism is its **genome**, a long sequence of DNA that provides the complete set of hereditary information carried by the organism. The genome includes chromosomal DNA as well as DNA in plasmids and (in eukaryotes) organellar DNA as found in mitochondria and chloroplasts. We use the term *information* because the genome does not itself perform an active role in the development of the organism. It is the sequence of the individual subunits, or bases, of the DNA that determines development. By a complex series of interactions, the DNA sequence produces all of the proteins of the organism at the appropriate time and place.

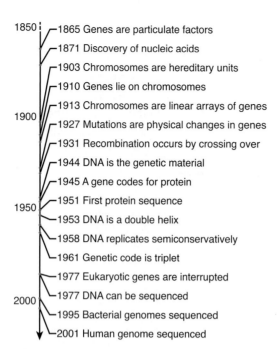

1850
—1865 Genes are particulate factors
—1871 Discovery of nucleic acids
—1903 Chromosomes are hereditary units
—1910 Genes lie on chromosomes
1900
—1913 Chromosomes are linear arrays of genes
—1927 Mutations are physical changes in genes
—1931 Recombination occurs by crossing over
—1944 DNA is the genetic material
—1945 A gene codes for protein
—1951 First protein sequence
1950
—1953 DNA is a double helix
—1958 DNA replicates semiconservatively
—1961 Genetic code is triplet
—1977 Eukaryotic genes are interrupted
—1977 DNA can be sequenced
2000
—1995 Bacterial genomes sequenced
—2001 Human genome sequenced

FIGURE 1.1 A brief history of genetics.

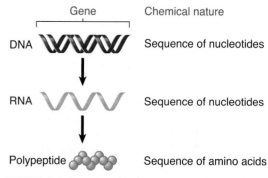

FIGURE 1.2 A gene codes for an RNA, which may code for protein.

Proteins serve a diverse series of roles in the development and functioning of an organism; they can form part of the structure of the organism, have the capacity to build the structures, perform the metabolic reactions necessary for life, and participate in regulation as transcription factors, receptors, key players in signal transduction pathways, and other molecules.

Physically, the genome may be divided into a number of different DNA molecules, or **chromosomes**. The ultimate definition of a genome is the sequence of the DNA of each chromosome. Functionally, the genome is divided into genes. Each gene is a sequence of DNA that encodes a single type of RNA or polypeptide. Each of the discrete chromosomes comprising the genome may contain a large number of genes. Genomes for living organisms may contain as few as ~500 genes (for a mycoplasma, a type of bacterium), ~20,000 to 25,000 for a human being, or as many as ~50,000 to 60,000 for rice.

In this chapter, we explore the gene in terms of its basic molecular construction. **FIGURE 1.1** summarizes the stages in the transition from the historical concept of the gene to the modern definition of the genome.

The first definition of the gene as a functional unit followed from the discovery that individual genes are responsible for the production of specific proteins. Later, the chemical differences between the DNA of the gene and its protein product led to the suggestion that a gene codes for a protein. This in turn led to the discovery of the complex apparatus by which the DNA sequence of a gene determines the amino acid sequence of a polypeptide.

Understanding the process by which a gene is expressed allows us to make a more rigorous definition of its nature. **FIGURE 1.2** shows the basic theme of this book. A gene is a sequence of DNA that directly produces a single strand of another nucleic acid, RNA, with a sequence that is identical to one of the two polynucleotide strands of DNA. In many cases, the RNA is in turn used to direct production of a polypeptide. In other cases, such as rRNA and tRNA genes, the RNA transcribed from the gene is the functional end product. Thus a gene is a sequence of DNA that codes for an RNA, and in protein-coding (or **structural**) genes, the RNA in turn codes for a polypeptide.

From the demonstration that a gene consists of DNA, and that a chromosome consists of a long stretch of DNA representing many genes, we will move to the overall organization of the genome. In Chapter 4, *The Interrupted Gene*, we take up in more detail the organization of the gene and its representation in proteins. In Chapter 5, *The Content of the Genome*, we consider the total number of genes, and in Chapter 7, *Clusters and Repeats*, we discuss other components of the genome and the maintenance of its organization.

1.2 DNA Is the Genetic Material of Bacteria and Viruses

Key concepts

- Bacterial transformation provided the first support that DNA is the genetic material of bacteria. Genetic properties can be transferred from one bacterial strain to another by extracting DNA from the first strain and adding it to the second strain.
- Phage infection showed that DNA is the genetic material of viruses. When the DNA and protein components of bacteriophages are labeled with different radioactive isotopes, only the DNA is transmitted to the progeny phages produced by infecting bacteria.

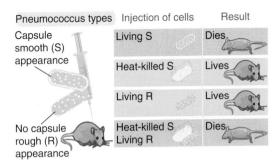

FIGURE 1.3 Neither heat-killed S-type nor live R-type bacteria can kill mice, but simultaneous injection of both can kill mice just as effectively as the live S-type.

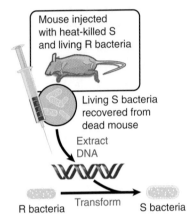

FIGURE 1.4 The DNA of S-type bacteria can transform R-type bacteria into the same S-type.

The idea that the genetic material of organisms is DNA has its roots in the discovery of **transformation** by Frederick Griffith in 1928. The bacterium *Streptococcus* (formerly *Pneumococcus*) *pneumoniae* kills mice by causing pneumonia. The virulence of the bacterium is determined by its capsular polysaccharide, which allows the bacterium to escape destruction by its host. Several types of *S. pneumoniae* have different capsular polysaccharides, but they all have a smooth (S) appearance. Each of the S types can give rise to variants that fail to produce the capsular polysaccharide and therefore have a rough (R) surface (consisting of the material that was beneath the capsular polysaccharide). The R types are avirulent and do not kill the mice, because the absence of the polysaccharide capsule allows the animal to destroy the bacteria.

When S bacteria are killed by heat treatment, they can no longer harm the animal. **FIGURE 1.3**, however, shows that when heat-killed S bacteria and avirulent R bacteria are jointly injected into a mouse, it dies as the result of a pneumonia infection. Virulent S bacteria can be recovered from the mouse's blood.

In this experiment, the dead S bacteria were of type III. The live R bacteria had been derived from type II. The virulent bacteria recovered from the mixed infection had the smooth coat of type III. So, some property of the dead IIIS bacteria can transform the live IIR bacteria so that they make the capsular polysaccharide and become virulent. **FIGURE 1.4** shows the identification of the component of the dead bacteria responsible for trans-

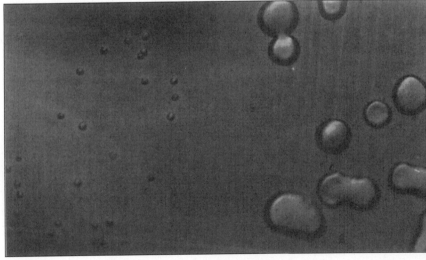

FIGURE 1.5 Rough (left) and smooth (right) colonies of *S. pneumoniae*. © Avery, et al., 1944. Originally published in **The Journal of Experimental Medicine**, 79: 137–158. Used with permission of The Rockefeller University Press.

formation. This was called the **transforming principle**. It was purified in a cell-free system in which extracts from the dead IIIS bacteria were added to the live IIR bacteria before being plated on agar and assayed for transformation (**FIGURE 1.5**). Purification of the transforming

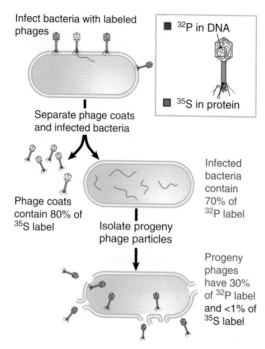

Infect bacteria with labeled phages

■ ^{32}P in DNA

■ ^{35}S in protein

Separate phage coats and infected bacteria

Phage coats contain 80% of ^{35}S label

Isolate progeny phage particles

Infected bacteria contain 70% of ^{32}P label

Progeny phages have 30% of ^{32}P label and <1% of ^{35}S label

FIGURE 1.6 The genetic material of phage T2 is DNA.

principle in 1944 by Avery, MacLeod, and McCarty showed that it is deoxyribonucleic acid (DNA).

Having shown that DNA is the genetic material of bacteria, the next step was to demonstrate that DNA is the genetic material in a quite different system. Phage T2 is a virus that infects the bacterium *Escherichia coli*. When phage particles are added to bacteria, they attach to the outside surface, some material enters the cell, and then ~20 minutes later each cell bursts open, or lyses, to release a large number of progeny phage.

FIGURE 1.6 illustrates the results of an experiment in 1952 by Alfred Hershey and Martha Chase in which bacteria were infected with T2 phages that had been radioactively labeled either in their DNA component (with ^{32}P) or in their protein component (with ^{35}S). The infected bacteria were agitated in a blender, and two fractions were separated by centrifugation. One fraction contained the empty phage "ghosts" that were released from the surface of the bacteria, and the other consisted of the infected bacteria themselves. Previously, it had been shown that phage replication occurs intracellularly, so that the genetic material of the phage would have to enter the cell during infection.

Most of the ^{32}P label was present in the fraction containing infected bacteria. The progeny phage particles produced by the infection

contained ~30% of the original ^{32}P label. The progeny received less than 1% of the protein contained in the original phage population. The phage ghosts consist of protein and therefore carried the ^{35}S radioactive label. This experiment directly showed that only the DNA of the parent phages enters the bacteria and becomes part of the progeny phages, which is exactly the pattern expected of genetic material.

A phage reproduces by commandeering the machinery of an infected host cell to manufacture more copies of itself. The phage possesses genetic material with properties analogous to those of cellular genomes: its traits are faithfully expressed and are subject to the same rules that govern inheritance of cellular traits. The case of T2 reinforces the general conclusion that DNA is the genetic material of the genome of a cell or a virus.

1.3 DNA Is the Genetic Material of Eukaryotic Cells

Key concepts

- DNA can be used to introduce new genetic traits into animal cells or whole animals.
- In some viruses, the genetic material is RNA.

When DNA is added to eukaryotic cells growing in culture, it enters the cells, and in some of them this results in the production of new proteins. When an isolated gene is used, its incorporation leads to the production of a particular protein, as depicted in **FIGURE 1.7**. Although for historical reasons these experiments are described as **transfection** when performed with animal cells, they are a direct counterpart to bacterial transformation. The DNA that is introduced into the recipient cell becomes part of its genome and is inherited with it, and expression of the new DNA results in a new trait upon the cells (synthesis of thymidine kinase in the example of Figure 1.7). At first, these experiments were successful only with individual cells growing in culture, but in later experiments DNA was introduced into mouse eggs by microinjection and became a stable part of the genome of the mouse. Such experiments show directly that DNA is the genetic material in eukaryotes, and that it can be transferred between different species and remain functional.

The genetic material of all known organisms and many viruses is DNA. Some viruses,

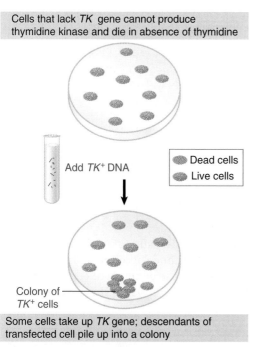

Cells that lack *TK* gene cannot produce thymidine kinase and die in absence of thymidine

Add *TK*+ DNA

Dead cells
Live cells

Colony of *TK*+ cells

Some cells take up *TK* gene; descendants of transfected cell pile up into a colony

FIGURE 1.7 Eukaryotic cells can acquire a new phenotype as the result of transfection by added DNA.

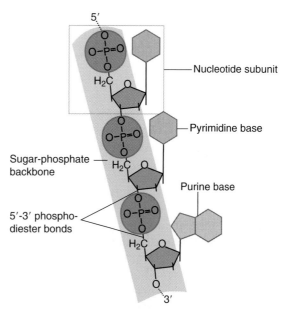

Nucleotide subunit

Pyrimidine base

Sugar-phosphate backbone

Purine base

5'-3' phospho-diester bonds

FIGURE 1.8 A polynucleotide chain consists of a series of 5'–3' sugar–phosphate links that form a backbone from which the bases protrude.

though, use RNA as the genetic material. As a result, the general nature of the genetic material is that it is always nucleic acid; specifically, it is DNA, except in the RNA viruses.

1.4 Polynucleotide Chains Have Nitrogenous Bases Linked to a Sugar–Phosphate Backbone

Key concepts

- A nucleoside consists of a purine or pyrimidine base linked to the 1' carbon of a pentose sugar.
- The difference between DNA and RNA is in the group at the 2' position of the sugar. DNA has a deoxyribose sugar (2'–H); RNA has a ribose sugar (2'-OH).
- A nucleotide consists of a nucleoside linked to a phosphate group on either the 5' or 3' carbon of the (deoxy)ribose.
- Successive (deoxy)ribose residues of a polynucleotide chain are joined by a phosphate group between the 3' carbon of one sugar and the 5' carbon of the next sugar.
- One end of the chain (conventionally written on the left) has a free 5' end and the other end of the chain has a free 3' end.
- DNA contains the four bases adenine, guanine, cytosine, and thymine; RNA has uracil instead of thymine.

The basic building block of nucleic acids (DNA and RNA) is the nucleotide, which has three components:

- a nitrogenous base,
- a sugar, and
- one or more phosphates.

The nitrogenous base is a **purine** or **pyrimidine** ring. The base is linked to the 1' ("one prime") carbon on a pentose sugar by a glycosidic bond from the N_1 of pyrimidines or the N_9 of purines. The pentose sugar linked to a nitrogenous base is called a **nucleoside**. To avoid ambiguity between the numbering systems of the heterocyclic rings and the sugar, positions on the pentose are given a prime (').

Nucleic acids are named for the type of sugar: DNA has 2"–deoxyribose, whereas RNA has ribose. The difference is that the sugar in RNA has a hydroxyl (—OH) group on the 2' carbon of the pentose ring. The sugar can be linked by its 5' or 3' carbon to a phosphate group. A nucleoside linked to a phosphate is a **nucleotide**.

A **polynucleotide** is a long chain of nucleotides. **FIGURE 1.8** shows that the backbone of the polynucleotide chain consists of an alternating series of pentose (sugar) and phosphate residues. The chain is formed by linking the 5' carbon of one pentose ring to the 3' carbon of the next pentose ring via a phosphate group; thus the sugar–phosphate backbone is said to consist of 5'–3' phosphodiester linkages. Specifically, the 3′ carbon of one pentose is bonded to one oxygen of the phosphate, while the 5′ carbon of the other pentose is bonded to the opposite oxygen of the phosphate. The nitrogenous bases "stick out" from the backbone.

Each nucleic acid contains four types of nitrogenous bases. The same two purines, adenine (A) and guanine (G), are present in both DNA and RNA. The two pyrimidines in DNA are cytosine (C) and thymine (T); in RNA uracil (U) is found instead of thymine. The only difference between uracil and thymine is the presence of a methyl group at position C_5.

The terminal nucleotide at one end of the chain has a free 5′ phosphate group, whereas the terminal nucleotide at the other end has a free 3′ hydroxyl group. It is conventional to write nucleic acid sequences in the 5′ to 3′ direction—that is, from the 5′ terminus at the left to the 3′ terminus at the right.

1.5 Supercoiling Affects the Structure of DNA

Key concepts

- Supercoiling occurs only in "closed" DNA with no free ends.
- Closed DNA is either circular DNA or linear DNA in which the ends are anchored so that they are not free to rotate.
- A closed DNA molecule has a linking number (L), which is the sum of twist (T) and writhe (W).
- The linking number can be changed only by breaking and reforming bonds in the DNA backbone.

The two strands of DNA are wound around each other to form a double helical structure (described in detail in the next section); the double helix can also wind around itself to change the overall conformation, or *topology*, of the DNA molecule in space. This is called **supercoiling**. The effect can be imagined like a rubber band twisted around itself. Supercoiling creates tension in the DNA, and thus can only occur if the DNA has no free ends (otherwise the free ends can rotate to relieve the tension) or in linear DNA (**FIGURE 1.9**, top) if it is anchored to a protein scaffold, as in eukaryotic chromosomes. The simplest example of a DNA with no free ends is a circular molecule. The effect of supercoiling can be seen by comparing the nonsupercoiled circular DNA lying flat in Figure 1.9 (center) with the supercoiled circular molecule that forms a twisted (and therefore more condensed) shape (Figure 1.9, bottom).

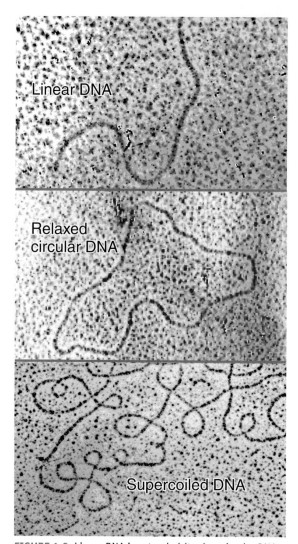

FIGURE 1.9 Linear DNA is extended (top); a circular DNA remains extended if it is relaxed (nonsupercoiled) (center); but a supercoiled DNA has a twisted and condensed form (bottom). Photos courtesy of Nirupam Roy Choudhury, International Centre for Genetic Engineering and Biotechnology (ICGEB).

The consequences of supercoiling depend on whether the DNA is twisted around itself in the same direction as the two strands within the double helix (clockwise) or in the opposite direction. Twisting in the same direction produces *positive supercoiling*, which overwinds the DNA so that there are more base pairs per turn. Twisting in the opposite direction produces *negative supercoiling*, or underwinding, so there are fewer base pairs per turn. Both types of supercoiling of the double helix in space are tensions in the DNA (which is why DNA molecules with no supercoiling are called "relaxed"). Negative supercoiling can be thought of as creating tension in the DNA that is relieved by the unwinding of the double helix. The effect of severe negative supercoiling is to generate a region in

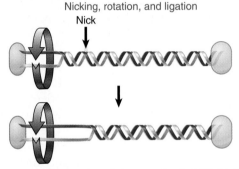

Rotation about a free end

Rotation at fixed ends

Strand separation compensated by positive supercoiling

Nicking, rotation, and ligation

Nick

FIGURE 1.10 Separation of the strands of a DNA double helix can be achieved in several ways.

which the two strands of DNA have separated (technically, zero base pairs per turn).

Topological manipulation of DNA is a central aspect of all its functional activities (recombination, replication, and transcription) as well as of the organization of its higher-order structure. All synthetic activities involving double-stranded DNA require the strands to separate. The strands do not simply lie side by side, though; they are intertwined. Their separation therefore requires the strands to rotate about each other in space. Some possibilities for the unwinding reaction are illustrated in **FIGURE 1.10**.

Unwinding a short linear DNA presents no problems, as the DNA ends are free to spin around the axis of the double helix to relieve any tension. DNA in a typical chromosome, however, is not only extremely long, but is also coated with proteins that serve to anchor the DNA at numerous points. As a result, even a linear eukaryotic chromosome does not functionally possess free ends.

Consider the effects of separating the two strands in a molecule whose ends are not free to rotate. When two intertwined strands are pulled apart from one end, the result is to *increase* their winding about each other far-

ther along the molecule, resulting in positive supercoiling elsewhere in the molecule to balance the underwinding generated in the single-stranded region. The problem can be overcome by introducing a transient nick in one strand. An internal free end allows the nicked strand to rotate about the intact strand, after which the nick can be sealed. Each repetition of the nicking and sealing reaction releases one superhelical turn.

A closed molecule of DNA can be characterized by its **linking number (L)**, which is the number of times one strand crosses over the other in space. Closed DNA molecules of identical sequence may have different linking numbers, reflecting different degrees of supercoiling. Molecules of DNA that are the same except for their linking numbers are called *topological isomers*.

The linking number is made up of two components: the **writhing number (W)** and the **twisting number (T)**.

The twisting number, T, is a property of the double helical structure itself, representing the rotation of one strand about the other. It represents the total number of turns of the duplex and is determined by the number of base pairs per turn. For a relaxed closed circular DNA lying flat in a plane, the twist is the total number of base pairs divided by the number of base pairs per turn.

The writhing number, W, represents the turning of the axis of the duplex in space. It corresponds to the intuitive concept of supercoiling, but does not have exactly the same quantitative definition or measurement. For a relaxed molecule, W = 0, and the linking number equals the twist.

We are often concerned with the change in linking number, ΔL, given by the equation

$$\Delta L = \Delta W + \Delta T.$$

The equation states that any change in the total number of revolutions of one DNA strand about the other can be expressed as the sum of the changes of the coiling of the duplex axis in space (ΔW) and changes in the helical repeat of the double helix itself (ΔT). In the absence of protein binding or other constraints, the twist of DNA does not tend to vary—in other words, the 10.5 bp/turn helical repeat is a very stable conformation for DNA in solution. Thus, any ΔL (change in linking number) is mostly likely to be expressed by a change in W; that is, by a change in supercoiling.

A decrease in linking number (that is, a change of $-\Delta L$) corresponds to the introduction of some combination of negative supercoiling (ΔW) and/or underwinding (ΔT). An increase in linking number, measured as a change of $+\Delta L$, corresponds to an increase in positive supercoiling and/or overwinding.

We can describe the change in state of any DNA by the specific linking difference, $\sigma = \Delta L/L0$, for which L0 is the linking number when the DNA is relaxed. If all of the change in linking number is due to change in W (that is, $\Delta T = 0$), the specific linking difference equals the supercoiling density. In effect, σ as defined in terms of $\Delta L/L0$ can be assumed to correspond to superhelix density so long as the structure of the double helix itself remains constant.

The critical feature about the use of the linking number is that this parameter is an invariant property of any individual *closed* DNA molecule. The linking number cannot be changed by any deformation short of one that involves the breaking and rejoining of strands. A circular molecule with a particular linking number can express the number in terms of different combinations of T and W, but it cannot change their sum so long as the strands are unbroken. (In fact, the partition of L between T and W prevents the assignment of fixed values for the latter parameters for a DNA molecule in solution.)

The linking number is related to the actual enzymatic events by which changes are made in the topology of DNA. The linking number of a particular closed molecule can be changed only by breaking one or both strands, using the free end to rotate one strand about the other, and rejoining the broken ends. When an enzyme performs such an action, it must change the linking number by an integer; this value can be determined as a characteristic of the reaction. The reactions to control supercoiling in the cell are performed by topoisomerase enzymes (see Chapter 14, *DNA Replication*).

1.6 DNA Is a Double Helix

Key concepts

- The B-form of DNA is a double helix consisting of two polynucleotide chains that run antiparallel.
- The nitrogenous bases of each chain are flat purine or pyrimidine rings that face inward and pair with one another by hydrogen bonding to form only A-T or G-C pairs.
- The diameter of the double helix is 20 Å, and there is a complete turn every 34 Å, with ten base pairs per turn (~10.4 base pairs per turn in solution).
- The double helix has a major (wide) groove and a minor (narrow) groove.

By the 1950s, the observation by Erwin Chargaff that the bases are present in different amounts in the DNAs of different species led to the concept that the sequence of bases is the form in which genetic information is carried. Given this concept, there were two remaining challenges: working out the structure of DNA, and explaining how a sequence of bases in DNA could determine the sequence of amino acids in a protein.

Three pieces of evidence contributed to the construction of the double helix model for DNA by James Watson and Francis Crick in 1953:

- X-ray diffraction data collected by Rosalind Franklin and Maurice Wilkins showed that the B-form of DNA (which is more hydrated than the A-form) is a regular helix, making a complete turn every 34 Å (3.4 nm), with a diameter of ~20 Å (2 nm). The distance between adjacent nucleotides is 3.4 Å (0.34 nm), thus there must be 10 nucleotides per turn. (In aqueous solution, the structure averages 10.4 nucleotides per turn.)
- The density of DNA suggests that the helix must contain two polynucleotide chains. The constant diameter of the helix can be explained if the bases in each chain face inward and are restricted so that a purine is always paired with a pyrimidine, avoiding partnerships of purine–purine (which would be too wide) or pyrimidine–pyrimidine (which would be too narrow).
- Chargaff also observed that regardless of the absolute amounts of each base, the proportion of G is always the same as the proportion of C in DNA, and the proportion of A is always the same as that of T. Consequently, the composition of any DNA can be described by its G-C content, or the sum of the proportions of G and C bases. (The proportions of A and T bases can be determined by subtracting the G-C content from 1.) G-C content ranges from 0.26 to 0.74 for different species.

Watson and Crick proposed that the two polynucleotide chains in the double helix associate by hydrogen bonding between the nitrogenous bases. Normally, G can hydrogen bond specifically only with C, whereas A can bond specifically only with T. This hydrogen bonding between bases is described as base pairing, and the paired bases (G forming three hydrogen bonds with C, or A forming two hydrogen bonds with T) are said to be **complementary**.

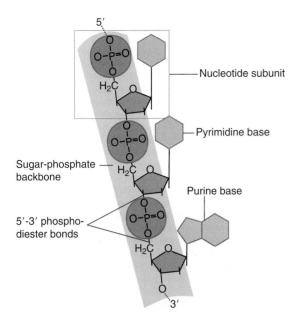

FIGURE 1.11 The double helix maintains a constant width because purines always face pyrimidines in the complementary A-T and G-C base pairs. The sequence in the figure is T-A, C-G, A-T, G-C.

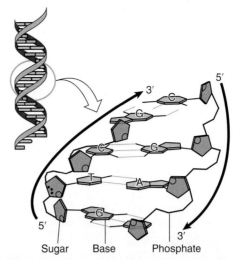

FIGURE 1.12 Flat base pairs lie perpendicular to the sugar–phosphate backbone.

Complementary base pairing occurs because of complementary shapes of the complementary bases at the interfaces of where they pair, along with the location of just the right functional groups in just the right geometry along those interfaces so that hydrogen bonds can form.

The Watson–Crick model has the two polynucleotide chains running in opposite directions, so they are said to be **antiparallel**, as illustrated in **FIGURE 1.11**. Looking in one direction along the helix, one strand runs in the 5′ to 3′ direction, whereas its complement runs 3′ to 5′.

The sugar–phosphate backbones are on the outside of the double helix and carry negative charges on the phosphate groups. When DNA is in solution *in vitro*, the charges are neutralized by the binding of metal ions, typically Na^+. In the cell, positively charged proteins provide some of the neutralizing force. These proteins play important roles in determining the organization of DNA in the cell.

The base pairs are on the inside of the double helix. They are flat and lie perpendicular to the axis of the helix. Using the analogy of the double helix as a spiral staircase, the base pairs form the steps, as illustrated schematically in **FIGURE 1.12**. Proceeding up the helix, bases are stacked above one another like a pile of plates.

Each base pair is rotated ~36° around the axis of the helix relative to the next base pair, so ~10 base pairs make a complete turn of 360°. The twisting of the two strands around one another forms a double helix with a **minor groove** that is ~12 Å (1.2 nm) across and a **major groove** that is ~22 Å (2.2 nm) across, as can be seen from the scale model of **FIGURE 1.13**. In B-DNA, the double helix is said to be "right-handed"; the turns run clockwise as viewed along the helical axis. (The A-form of DNA, observed when DNA is dehydrated, is also a right-handed helix and is shorter and thicker than the B-form. A third DNA structure, Z-DNA, is longer and narrower than the B-form, and is a left-handed helix.)

It is important to realize that the Watson–Crick model of the B-form represents an average structure, and that there can be local variations in the precise structure. If it has more base pairs per turn it is said to be **overwound**; if it has fewer base pairs per turn it is **underwound**. The degree of local winding can be affected by the overall conformation of the DNA double helix or by the binding of proteins to specific sites on the DNA.

Another structural variant is *bent DNA*. A series of eight to ten adenine residues on one strand can result in intrinsic bending of the double helix. This structure allows tighter packing with consequences for nucleosome assembly (see Chapter 10, *Chromatin*) and gene regulation.

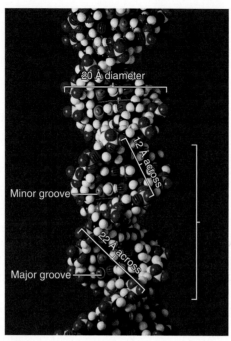

FIGURE 1.13 The two strands of DNA form a double helix. © Photodisc.

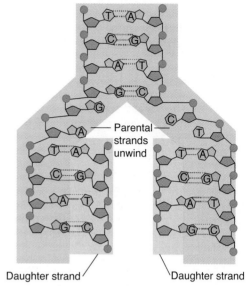

FIGURE 1.14 Base pairing provides the mechanism for replicating DNA.

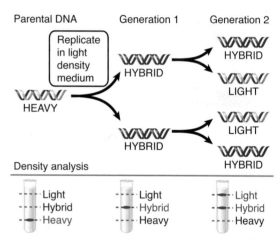

FIGURE 1.15 Replication of DNA is semiconservative.

1.7 DNA Replication Is Semiconservative

Key concepts

- The Meselson–Stahl experiment used "heavy" isotope labeling to show that the single polynucleotide strand is the unit of DNA that is conserved during replication.
- Each strand of a DNA duplex acts as a template for synthesis of a daughter strand.
- The sequences of the daughter strands are determined by complementary base pairing with the separated parental strands.

It is crucial that DNA is reproduced accurately. The two polynucleotide strands are joined only by hydrogen bonds, so they are able to separate without the breakage of covalent bonds. The specificity of base pairing suggests that both of the separated parental strands could act as template strands for the synthesis of complementary daughter strands. **FIGURE 1.14** shows the principle that a new daughter strand is assembled from each parental strand. The sequence of the daughter strand is determined by the parental strand: an A in the parental strand causes a T to be placed in the daughter strand, a parental G directs incorporation of a daughter C, and so on.

The top part of Figure 1.14 shows an unreplicated parental duplex with the original two parental strands. The lower part shows the two daughter duplexes produced by complementary base pairing. Each of the daughter duplexes is identical in sequence to the original parent duplex, containing one parental strand and one newly synthesized strand. The structure of DNA carries the information needed for its own replication. The consequences of this mode of replication, called **semiconservative replication**, are illustrated in **FIGURE 1.15**. The parental duplex is replicated to form two daughter duplexes, each of which consists of one parental strand and one newly synthesized daughter strand. The unit conserved from one generation to the next is one of the two individual strands comprising the parental duplex.

Figure 1.15 illustrates a prediction of this model. If the parental DNA carries a "heavy" density label because the organism has been grown in medium containing a suitable isotope (such as ^{15}N), its strands can be distinguished from those that are synthesized when the organism is transferred to a medium containing "light" isotopes. The parental DNA is a duplex of two "heavy" strands (red). After one generation of growth in "light" medium, the duplex DNA is "hybrid" in density—it consists of one "heavy" parental strand (red) and one "light" daughter strand (blue). After a second generation, the two strands of each hybrid duplex have separated. Each strand gains a "light" partner, so that now one half of the duplex DNA remains hybrid and the other half is entirely "light" (both strands are blue).

The individual strands of these duplexes are entirely "heavy" or entirely "light." This pattern was confirmed experimentally by Matthew Meselson and Franklin Stahl in 1958. Meselson and Stahl followed the semiconservative replication of DNA through three generations of growth of *E. coli*. When DNA was extracted from bacteria and separated in a density gradient by centrifugation, the DNA formed bands corresponding to its density—heavy for parental, hybrid for the first generation, and half hybrid and half light in the second generation.

1.8 Polymerases Act on Separated DNA Strands at the Replication Fork

Key concepts

- Replication of DNA is undertaken by a complex of enzymes that separate the parental strands and synthesize the daughter strands.
- The replication fork is the point at which the parental strands are separated.
- The enzymes that synthesize DNA are called DNA polymerases.
- Nucleases are enzymes that degrade nucleic acids; they include DNases and RNases and can be categorized as endonucleases or exonucleases.

Replication requires the two strands of the parental duplex to undergo separation, or **denaturation**. The disruption of the duplex, however, is only transient and is reversed, or undergoes **renaturation**, as the daughter duplex is formed. Only a small stretch of the duplex DNA is denatured at any moment during replication. ("Denaturation" is also used to describe the loss of functional protein structure; it is a general term implying that the natural

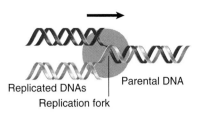

FIGURE 1.16 The replication fork is the region of DNA in which there is a transition from the unwound parental duplex to the newly replicated daughter duplexes.

conformation of a macromolecule has been converted to some nonfunctional form.)

The helical structure of a molecule of DNA during replication is illustrated in **FIGURE 1.16**. The unreplicated region consists of the parental duplex opening into the replicated region where the two daughter duplexes have formed. The duplex is disrupted at the junction between the two regions, which is called the **replication fork**. Replication involves movement of the replication fork along the parental DNA, so that there is continuous denaturation of the parental strands and formation of daughter duplexes.

The synthesis of DNA is aided by specific enzymes (**DNA polymerases**) that recognize the template strand and catalyze the addition of nucleotide subunits to the polynucleotide chain that is being synthesized. They are accompanied in DNA replication by ancillary enzymes such as helicases that unwind the DNA duplex, a primase that synthesizes an RNA primer required by DNA polymerase, and ligase that connects discontinuous DNA strands. Degradation of nucleic acids also requires specific enzymes: deoxyribonucleases (**DNases**) degrade DNA, and ribonucleases (**RNases**) degrade RNA. The nucleases fall into the general classes of **exonucleases** and **endonucleases**:

- Endonucleases break individual phosphodiester linkages within RNA or DNA molecules, generating discrete fragments. Some DNases cleave both strands of a duplex DNA at the target site, whereas others cleave only one of the two strands. Endonucleases are involved in cutting reactions, as shown in **FIGURE 1.17**.
- Exonucleases remove nucleotide residues one at a time from the end of the molecule, generating mononucleotides. They always function on a single nucleic acid strand, and each exonuclease proceeds in a specific direction, that is, starting either at a 5' or a 3' end and

FIGURE 1.17 An endonuclease cleaves a bond within a nucleic acid. This example shows an enzyme that attacks one strand of a DNA duplex.

FIGURE 1.18 An exonuclease removes bases one at a time by cleaving the last bond in a polynucleotide chain.

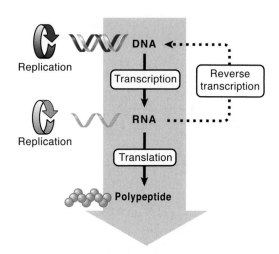

FIGURE 1.19 The central dogma states that information in nucleic acid can be perpetuated or transferred, but the transfer of information into protein is irreversible.

proceeding toward the other end. They are involved in trimming reactions, as shown in **FIGURE 1.18**.

1.9 Genetic Information Can Be Provided by DNA or RNA

Key concepts
- Cellular genes are DNA, but viruses may have genomes of RNA.
- DNA is converted into RNA by transcription, and RNA may be converted into DNA by reverse transcription.
- The translation of RNA into protein is unidirectional.

The **central dogma** is the dominant paradigm of molecular biology. Structural genes exist as sequences of nucleic acid, but function by being expressed in the form of polypeptides. Replication makes possible the inheritance of genetic information, whereas transcription and translation are responsible for its expression to another form.

FIGURE 1.19 illustrates the roles of replication, transcription, and translation in the context of the central dogma:

- Transcription of DNA by a DNA-dependent **RNA polymerase** generates RNA molecules. Messenger RNAs (mRNAs) are translated to polypeptides. Other types of RNA, such as rRNAs and tRNAs, are functional themselves and are not translated.
- A genetic system may involve either DNA or RNA as the genetic material. Cells use only DNA. Some viruses use RNA, and replication of viral RNA by an RNA-dependent RNA polymerase occurs in the infected cell.
- The expression of cellular genetic information is usually unidirectional. Transcription of DNA generates RNA molecules; the exception is the reverse transcription of retroviral RNA to DNA that occurs when retroviruses infect cells (see below). Generally, polypeptides cannot be retrieved for use as genetic information; translation of RNA into polypeptide is always irreversible.

These mechanisms are equally effective for the cellular genetic information of prokaryotes or eukaryotes and for the information carried by viruses. The genomes of all living organisms consist of duplex DNA. Viruses have genomes that consist of DNA or RNA, and there are examples of each type that are double-stranded (dsDNA or dsRNA) or single-stranded (ssDNA or ssRNA). Details of the mechanism used to replicate the nucleic acid vary among viruses, but the principle of replication via synthesis of complementary strands remains the same, as illustrated in **FIGURE 1.20**.

Cellular genomes reproduce DNA by the mechanism of semiconservative replication. Double-stranded viral genomes, whether DNA or RNA, also replicate by using the individual strands of the duplex as templates to synthesize complementary strands.

Viruses with single-stranded genomes use the single strand as a template to synthesize a complementary strand; this complementary strand in turn is used to synthesize its complement (which is, of course, identical to the original strand). Replication may involve the formation of stable double-stranded intermedi-

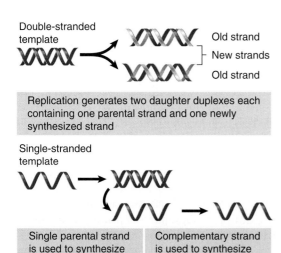

Double-stranded template

Old strand

New strands

Old strand

Replication generates two daughter duplexes each containing one parental strand and one newly synthesized strand

Single-stranded template

Single parental strand is used to synthesize complementary strand

Complementary strand is used to synthesize copy of parental strand

FIGURE 1.20 Double-stranded and single-stranded nucleic acids both replicate by synthesis of complementary strands governed by the rules of base pairing.

Genome	Gene Number	Base Pairs
Organisms		
Plants	<50,000	$<10^{11}$
Mammals	30,000	$\sim 3 \times 10^9$
Worms	14,000	$\sim 10^8$
Flies	12,000	1.6×10^8
Fungi	6,000	1.3×10^7
Bacteria	2–4,000	$<10^7$
Mycoplasma	500	$<10^6$
dsDNA Viruses		
Vaccinia	<300	187,000
Papova (SV40)	~6	5,226
Phage T4	~200	165,000
ssDNA Viruses		
Parvovirus	5	5,000
Phage fX174	11	5,387
dsRNA Viruses		
Reovirus	22	23,000
ssRNA Viruses		
Coronavirus	7	20,000
Influenza	12	13,500
TMV	4	6,400
Phage MS2	4	3,569
STNV	1	1,300
Viroids		
PSTV RNA	0	359

FIGURE 1.21 The amount of nucleic acid in the genome varies over an enormous range.

ates or use double-stranded nucleic acid only as a transient stage.

The restriction of a unidirectional transfer of information from DNA to RNA in cells is not absolute. It is broken by the retroviruses, which have genomes consisting of a single-stranded RNA molecule. During the retroviral cycle of infection, the RNA is converted into a single-stranded DNA by the process of **reverse transcription**, which is accomplished by the enzyme *reverse transcriptase*, an RNA-dependent DNA polymerase. The resulting ssDNA is in turn converted into a double-stranded DNA. This duplex DNA becomes part of the genome of the host cell and is inherited like any other gene. Thus reverse transcription allows a sequence of RNA to be retrieved and used as DNA in a cell.

The existence of RNA replication and reverse transcription establishes the general principle that information in the form of either type of nucleic acid sequence can be converted into the other type. In the usual course of events, however, the cell relies on the processes of DNA replication, transcription, and translation. On rare occasions, though (possibly mediated by an RNA virus), information from a cellular RNA is converted into DNA and inserted into the genome. Although retroviral reverse transcription is not necessary for the regular operations of the cell, it becomes a mechanism of potential importance when we consider the evolution of the genome.

The same principles for the perpetuation of genetic information apply to the massive genomes of plants or amphibians as well as the tiny genomes of mycoplasma and the even smaller genomes of DNA or RNA viruses. **FIGURE 1.21** presents some examples that illustrate the range of genome types and sizes. The reasons for such variation in genome size and gene number will be explored in Chapters 5 and 6.

Among the various living organisms, with genomes varying in size over a 100,000-fold range, a common principle prevails: the DNA codes for all the proteins that the cell(s) of the organism must synthesize, and the proteins in turn (directly or indirectly) provide the functions needed for survival. A similar principle describes the function of the genetic information of viruses, whether DNA or RNA: the nucleic acid codes for the protein(s) needed to package the genome and for any other functions in addition to those provided by the host cell that are needed to reproduce the virus. (The smallest virus—the satellite tobacco necrosis virus [STNV]—cannot replicate independently. It requires the presence of a "helper" virus—the tobacco necrosis virus [TNV], which is itself a normally infectious virus.)

1.10 Nucleic Acids Hybridize by Base Pairing

Key concepts

- Heating causes the two strands of a DNA duplex to separate.
- The T_m is the midpoint of the temperature range for denaturation.
- Complementary single strands can renature when the temperature is reduced.
- Denaturation and renaturation/hybridization can occur with DNA–DNA, DNA–RNA, or RNA–RNA combinations and can be intermolecular or intramolecular.
- The ability of two single-stranded nucleic acids to hybridize is a measure of their complementarity.

A crucial property of the double helix is the capacity to separate the two strands without disrupting the covalent bonds that form the polynucleotides and at the (very rapid) rates needed to sustain genetic functions. The specificity of the processes of denaturation and renaturation is determined by complementary base pairing.

The concept of base pairing is central to all processes involving nucleic acids. Disruption of the base pairs is crucial to the function of a double-stranded nucleic acid, whereas the ability to form base pairs is essential for the activity of a single-stranded nucleic acid. **FIGURE 1.22** shows that base pairing enables complementary single-stranded nucleic acids to form a duplex.

- An intramolecular duplex region can form by base pairing between two complementary sequences that are part of a single-stranded nucleic acid.
- A single-stranded nucleic acid may base pair with an independent, complementary single-stranded nucleic acid to form an intermolecular duplex.

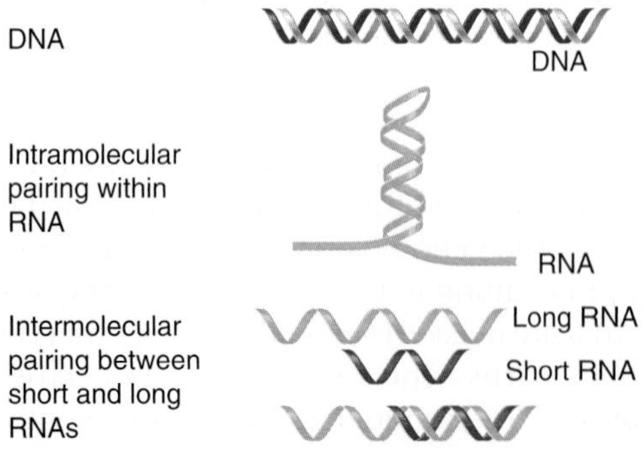

Formation of duplex regions from single-stranded nucleic acids is most important for RNA, but is also important for single-stranded viral DNA genomes. Base pairing between independent complementary single strands is not restricted to DNA–DNA or RNA–RNA, but can also occur between DNA and RNA.

The lack of covalent bonds between complementary strands makes it possible to manipulate DNA *in vitro*. The hydrogen bonds that stabilize the double helix are disrupted by heating or low salt concentration. The two strands of a double helix separate entirely when all the hydrogen bonds between them are broken.

Denaturation of DNA occurs over a narrow temperature range and results in striking changes in many of its physical properties. The midpoint of the temperature range over which the strands of DNA separate is called the **melting temperature** (T_m), and it depends on the G-C content of the duplex. Each G-C base pair has three hydrogen bonds; as a result it is more stable than an A-T base pair, which has only two hydrogen bonds. The more G-C base pairs in a DNA, the greater the energy that is needed to separate the two strands. In solution under physiological conditions, a DNA that is 40% G-C (a value typical of mammalian genomes) denatures with a T_m of about 87° C, so duplex DNA is stable at the temperature of the cell.

The denaturation of DNA is reversible under appropriate conditions. Renaturation depends on specific base pairing between the complementary strands. **FIGURE 1.23** shows that the reaction takes place in two stages. First, single strands of DNA in the solution encounter one another by chance; if their sequences are complementary, the two strands base pair to generate a short double-stranded region.

DNA · · · · · DNA

Intramolecular pairing within RNA · · · · · RNA

Intermolecular pairing between short and long RNAs · · · · · Long RNA / Short RNA

FIGURE 1.22 Base pairing occurs in duplex DNA and also in intra- and intermolecular interactions in single-stranded RNA (or DNA).

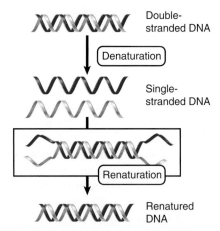

Double-stranded DNA

Denaturation

Single-stranded DNA

Renaturation

Renatured DNA

FIGURE 1.23 Denatured single strands of DNA can renature to give the duplex form.

This region of base pairing then extends along the molecule, much like a zipper, to form a lengthy duplex. Complete renaturation restores the properties of the original double helix. The property of renaturation applies to any two complementary nucleic acid sequences. This is sometimes called **annealing**, but the reaction is more generally called **hybridization** whenever nucleic acids from different sources are involved, as in the case when DNA hybridizes to RNA. The ability of two nucleic acids to hybridize constitutes a precise test for their complementarity because only complementary sequences can form a duplex.

The purpose of the hybridization reaction is to combine two single-stranded nucleic acids in solution and then to measure the amount of double-stranded material that forms. **FIGURE 1.24** illustrates a procedure in which a DNA preparation is denatured and the single strands are attached to a filter. A second denatured DNA (or RNA) preparation is then added. The filter is treated so that the second preparation can attach to it only if it is able to base pair with the DNA that was originally attached. Usually the second preparation is labeled so that the hybridization reaction can be measured as the amount of label retained by the filter. Alternatively, hybridization in solution can be measured as the change in UV-absorbance of

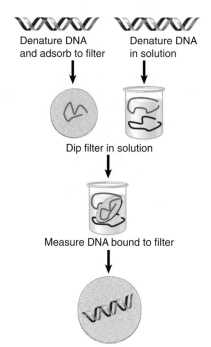

FIGURE 1.24 Filter hybridization establishes whether a solution of denatured DNA (or RNA) contains sequences complementary to the strands immobilized on the filter.

a nucleic acid solution at 260 nm as detected via spectrophotometry. As DNA denatures to single strands with increasing temperature, UV-absorbance of the DNA solution increases; UV-absorbance consequently decreases as ssDNA hybridizes to complementary DNA or RNA with decreasing temperature.

The extent of hybridization between two single-stranded nucleic acids is determined by their complementarity. Two sequences need not be perfectly complementary to hybridize. If they are similar but not identical, an imperfect duplex is formed in which base pairing is interrupted at positions where the two single strands are not complementary.

1.11 Mutations Change the Sequence of DNA

Key concepts
- All mutations are changes in the sequence of DNA.
- Mutations may occur spontaneously or may be induced by mutagens.

Mutations provide decisive evidence that DNA is the genetic material. When a change in the sequence of DNA causes an alteration in the sequence of a protein, we may conclude that the DNA codes for that protein. Furthermore, a corresponding change in the phenotype of the organism may allow us to identify the function of that protein. The existence of many mutations in a gene may allow many variant forms of a protein to be compared, and a detailed analysis can be used to identify regions of the protein responsible for individual enzymatic or other functions.

All organisms suffer a certain number of mutations as the result of normal cellular operations or random interactions with the environment. These are called **spontaneous mutations**, and the rate at which they occur (the "background level") is characteristic for any particular organism. Mutations are rare events, and of course those that have deleterious effects are selected against during evolution. It is therefore difficult to observe large numbers of spontaneous mutants from natural populations.

The occurrence of mutations can be increased by treatment with certain compounds. These are called **mutagens**, and the changes they cause are called **induced mutations**. Most mutagens either modify a particular base of DNA or become incorporated into the nucleic acid. The potency of a mutagen is judged by how much it increases the rate of mutation above background. By using mutagens, it becomes possible to induce many changes in any gene.

Mutation rates can be measured at several levels of resolution: mutation across the whole genome (as the rate per genome per generation), mutation in a gene (as the rate per locus per generation), or mutation at a specific nucleotide site (as the rate per base pair per generation). These rates correspondingly decrease as a smaller unit is observed.

Spontaneous mutations that inactivate gene function occur in bacteriophages and bacteria at a relatively constant rate of $3–4 \times 10^{-3}$ per genome per generation. Given the large variation in genome sizes between bacteriophages and bacteria, this corresponds to great differences in the mutation rate per base pair. This suggests that the overall rate of mutation has been subject to selective forces that have balanced the deleterious effects of most mutations against the advantageous effects of some mutations. This conclusion is strengthened by the observation that an archaean that lives under harsh conditions of high temperature and acidity (which are expected to damage DNA) does not show an elevated mutation rate, but in fact has an overall mutation rate just below the average range. **FIGURE 1.25** shows that in bacteria, the mutation rate corresponds to $\sim 10^{-6}$ events per locus per generation or to an average rate of change per base pair of $10^{-9}–10^{-10}$ per generation. The rate at individual base pairs

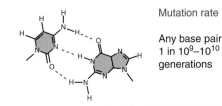

| | Mutation rate |
| | Any base pair
1 in 10^9–10^{10}
generations |

....ATCGGACTTACCGGTTA....
....TAGCCTGAATGGCCAAT....

Any gene
1 in 10^5–10^6
generations

The genome
1 in 300
generations

FIGURE 1.25 A base pair is mutated at a rate of 10^{-9}–10^{-10} per generation, a gene of 1000 bp is mutated at $\sim 10^{-6}$ per generation, and a bacterial genome is mutated at 3×10^{-3} per generation.

varies very widely, over a 10,000-fold range. We have no accurate measurement of the rate of mutation in eukaryotes, although usually it is thought to be somewhat similar to that of bacteria on a per-locus per-generation basis.

1.12 Mutations May Affect Single Base Pairs or Longer Sequences

Key concepts
- A point mutation changes a single base pair.
- Point mutations can be caused by the chemical conversion of one base into another or by errors that occur during replication.
- A transition replaces a G-C base pair with an A-T base pair or vice versa.
- A transversion replaces a purine with a pyrimidine, such as changing A-T to T-A.
- Insertions and/or deletions can result from the movement of transposable elements.

Any base pair of DNA can be mutated. A **point mutation** changes only a single base pair and can be caused by either of two types of event:

- Chemical modification of DNA directly changes one base into a different base.
- An error during the replication of DNA causes the wrong base to be inserted into a polynucleotide.

Point mutations can be divided into two types, depending on the nature of the base substitution:

- The most common class is the **transition**, which results from the substitution of one pyrimidine by the other, or of one purine by the other. This replaces a G-C pair with an A-T pair or vice versa.
- The less common class is the **transversion**, in which a purine is replaced by a pyrimidine or vice versa, so that an A-T pair becomes a T-A or C-G pair.

As shown in **FIGURE 1.26**, the mutagen nitrous acid performs an oxidative deamination that converts cytosine into uracil, resulting in a transition. In the replication cycle following the transition, the U pairs with an A, instead of the G with which the original C would have paired. So the C-G pair is replaced by a T-A pair when the A pairs with the T in the next replication cycle. (Nitrous acid can also deaminate adenine, causing the reverse transition from A-T to G-C.)

Transitions are also caused by base mispairing, which occurs when noncomplementary bases pair instead of the usual Watson–Crick pairs. Base mispairing usually occurs as an aberration resulting from the incorporation into DNA of an abnormal base that has flexible pairing properties. FIGURE 1.27 shows the example of the mutagen bromouracil (BrdU), an analog of thymine that contains a bromine atom in place of thymine's methyl group and can be incorporated into DNA in place of thymine. BrdU has flexible pairing properties, though, because the presence of the bromine atom allows a tautomeric shift from a keto (=O) form to an enol (–OH) form. The enol form of BrdU can pair with guanine, which after replication leads to substitution of the original A-T pair by a G-C pair.

The mistaken pairing can occur either during the original incorporation of the base or in a subsequent replication cycle. The transition is induced with a certain probability in each replication cycle, so the incorporation of BrdU has continuing effects on the sequence of DNA.

Point mutations were thought for a long time to be the principal means of change in individual genes. We now know, though, that insertions of short sequences are quite frequent. Often, the insertions are the result of transposable elements, which are sequences of DNA with the ability to move from one site to another (see Chapter 17, *Transposable Elements and Retroviruses*). An insertion within a coding region usually abolishes the activity of the gene

because it may alter the reading frame; such an insertion is a *frameshift mutation*. (Similarly, a deletion within a coding region is usually a frameshift mutation.) Where such insertions have occurred, deletions of part or all of the inserted material, and sometimes of the adjacent regions, may subsequently occur.

A significant difference between point mutations and insertions is that mutagens can increase the frequency of point mutations, but do not affect the frequency of transposition. Both insertions and deletions of short sequences (often called *indels*) can occur by other mechanisms, though—for example, those involving errors during replication or recombination. In addition, a class of mutagens called the acridines introduce very small insertions and deletions.

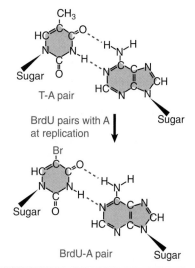

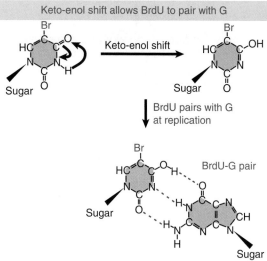

FIGURE 1.26 Mutations can be induced by chemical modification of a base.

FIGURE 1.27 Mutations can be induced by the incorporation of base analogs into DNA.

1.13 The Effects of Mutations Can Be Reversed

Key concepts

- Forward mutations alter the function of a gene, and back mutations (or revertants) reverse their effects.
- Insertions can revert by deletion of the inserted material, but deletions cannot revert.
- Suppression occurs when a mutation in a second gene bypasses the effect of mutation in the first gene.

FIGURE 1.28 shows that the possibility of reversion mutations, or **revertants**, is an important characteristic that distinguishes point mutations and insertions from deletions:

- A point mutation can revert either by restoring the original sequence or by gaining a compensatory mutation elsewhere in the gene.
- An insertion can revert by deletion of the inserted sequence.

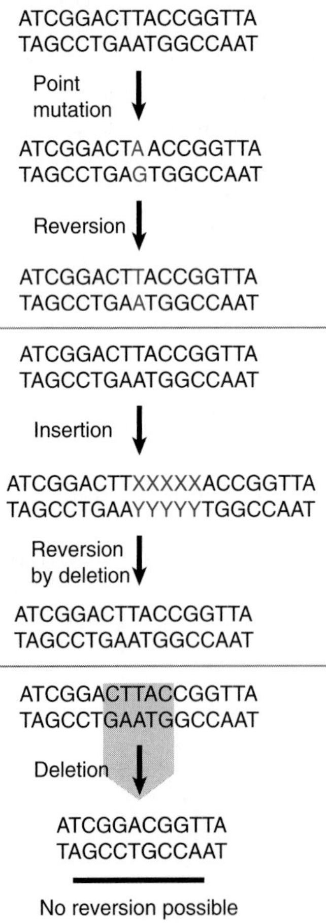

FIGURE 1.28 Point mutations and insertions can revert, but deletions cannot revert.

- A deletion of a sequence cannot revert in the absence of some mechanism to restore the lost sequence.

Mutations that inactivate a gene are called **forward mutations**. Their effects are reversed by **back mutations**, which are of two types: true reversions and second-site reversions.

An exact reversal of the original mutation is called a **true reversion**. So if an A-T pair has been replaced by a G-C pair, another mutation to restore the A-T pair will exactly regenerate the original sequence. The exact removal of a transposable element following its insertion is another example of a true reversion.

The second type of back mutation, **second-site reversion**, may occur elsewhere in the gene, and its effects compensate for the first mutation. For example, one amino acid change in a protein may abolish gene function, but a second alteration may compensate for the first and restore protein activity.

A forward mutation results from any change that alters the function of a gene product, whereas a back mutation must restore the original function to the altered gene product. The possibilities for back mutations are thus much more restricted than those for forward mutations. The rate of back mutations is correspondingly lower than that of forward mutations, typically by a factor of ~10.

Mutations in other genes can also occur to circumvent the effects of mutation in the original gene. This is called a **suppression mutation**. A locus in which a mutation suppresses the effect of a mutation in another locus is called a suppressor. For example, a point mutation may cause an amino acid substitution in a polypeptide, while a second mutation in a tRNA gene may cause it to recognize the mutated codon, and as a result insert the original amino acid during translation. (Note that this suppresses the original mutation but causes errors during translation of other mRNAs.)

1.14 Mutations Are Concentrated at Hotspots

Key concept

- The frequency of mutation at any particular base pair is statistically equivalent, except for hotspots, where the frequency is increased by at least an order of magnitude.

So far we have dealt with mutations in terms of individual changes in the sequence of DNA that influence the activity of the DNA in which they occur. When we consider mutations in terms of the alteration of function of the gene, most genes within a species show more or less similar rates of mutation relative to their size. This suggests that the gene can be regarded as a target for mutation, and that damage to any part of it can alter its function. As a result, susceptibility to mutation is roughly proportional to the size of the gene. Are all base pairs in a gene equally susceptible, though, or are some more likely to be mutated than others?

What happens when we isolate a large number of independent mutations in the same gene? Each is the result of an individual mutational event. Most mutations will occur at different sites, but some will occur at the same position. Two independently isolated mutations at the same site may constitute exactly the same change in DNA (in which case the same mutation has happened more than once), or they may constitute different changes (three different point mutations are possible at each base pair).

The histogram of **FIGURE 1.29** shows the frequency with which mutations are found at each base pair in the *lacI* gene of *E. coli*. The statistical probability that more than one mutation occurs at a particular site is given by random-hit kinetics (as seen in the Poisson distribution). Some sites will gain one, two, or three mutations, whereas others will not gain any. Some sites gain far more than the number of mutations expected from a random distribution; they may have 10× or even 100× more mutations than predicted by random hits. These sites are called **hotspots**. Spontaneous mutations may occur at hotspots, and different mutagens may have different hotspots.

1.15 Many Hotspots Result from Modified Bases

Key concepts

- A common cause of hotspots is the modified base 5-methylcytosine, which is spontaneously deaminated to thymine.
- A hotspot can result from the high frequency of change in copy number of a short, tandemly repeated sequence.

A major cause of spontaneous mutation is the presence of an unusual base in the DNA. In addition to the four standard bases of DNA, modified bases are sometimes found. The name reflects their origin; they are produced by chemical modification of one of the four standard bases. The most common modified base is 5-methylcytosine, which is generated when a methylase enzyme adds a methyl group to cytosine residues at specific sites in the DNA. Sites containing 5-methylcytosine are hotspots for spontaneous point mutation in *E. coli*. In each case, the mutation is a G-C to A-T transition. The hotspots are not found in mutant strains of *E. coli* that cannot methylate cytosine.

The reason for the existence of these hotspots is that cytosine bases suffer a higher frequency of spontaneous deamination. In this reaction, the amino group is replaced by a keto group. Recall that deamination of cytosine generates uracil (see Figure 1.26). **FIGURE 1.30** compares this reaction with the deamination of 5-methylcytosine where deamination generates thymine. The effect is to generate the mismatched base pairs G-U and G-T, respectively.

All organisms have repair systems that correct mismatched base pairs by removing and replacing one of the bases. The operation of

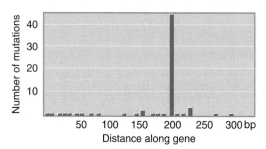

FIGURE 1.29 Spontaneous mutations occur throughout the *lacI* gene of *E. coli*, but are concentrated at a hotspot.

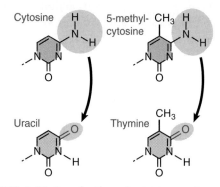

FIGURE 1.30 Deamination of cytosine produces uracil, whereas deamination of 5-methylcytosine produces thymine.

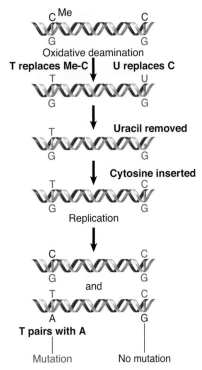

FIGURE 1.31 The deamination of 5-methylcytosine produces thymine (by C-G to T-A transitions), whereas the deamination of cytosine produces uracil (which usually is removed and then replaced by cytosine).

these systems determines whether mismatched pairs such as G-U and G-T result in mutations.

FIGURE 1.31 shows that the consequences of deamination are different for 5-methylcytosine and cytosine. Deaminating the (rare) 5-methylcytosine causes a mutation, whereas deaminating cytosine does not have this effect. This happens because the DNA repair systems are much more effective in accurately repairing G-U than G-T.

E. coli contain an enzyme, uracil-DNA-glycosidase, that removes uracil residues from DNA (see *Section 16.5, Base Excision Repair Systems Require Glycosylases*). This action leaves an unpaired G residue, and a repair system then inserts a complementary C base. The net result of these reactions is to restore the original sequence of the DNA. Thus, this system protects DNA against the consequences of spontaneous deamination of cytosine. (This system is not, however, efficient enough to prevent the effects of the increased deamination caused by nitrous acid; see Figure 1.26.)

Note that the deamination of 5-methylcytosine creates thymine and results in a mismatched base pair, G-T. If the mismatch is not corrected before the next replication cycle a mutation results. The bases in the mispaired

G-T first separate and then pair with the correct complements to produce the wild-type G-C in one daughter DNA and the mutant A-T in the other.

Deamination of 5-methylcytosine is the most common cause of mismatched G-T pairs in DNA. Repair systems that act on G-T mismatches have a bias toward replacing the T with a C (rather than the alternative of replacing the G with an A), which helps to reduce the rate of mutation (see *Section 16.7, Controlling the Direction of Mismatch Repair*). These systems are not, however, as effective as those that remove U from G-U mismatches. As a result, deamination of 5-methylcytosine leads to mutation much more often than does deamination of cytosine.

5-methylcytosine also creates hotspots in eukaryotic DNA. It is common at CpG dinucleotides that are concentrated in regions called CpG islands (see *Section 20.13, CpG Islands Are Regulatory Targets*). Although 5-methylcytosine accounts for ~1% of the bases in human DNA, sites containing the modified base account for ~30% of all point mutations.

The importance of repair systems in reducing the rate of mutation is emphasized by the effects of eliminating the mouse enzyme MBD4, a glycosylase that can remove T (or U) from mismatches with G. The result is to increase the mutation rate at CpG sites by a factor of 3. (The reason the effect is not greater is that MBD4 is only one of several systems that act on G-T mismatches; most likely the elimination of all the systems would increase the mutation rate much more.)

The operation of these systems casts an interesting light on the use of T in DNA as compared to U in RNA. It may relate to the need for stability of DNA sequences; the use of T means that any deaminations of C are immediately recognized because they generate a base (U) that is not usually present in the DNA. This greatly increases the efficiency with which repair systems can function (compared with the situation when they have to recognize G-T mismatches, which can be produced also by situations where removing the T would not be the appropriate correction). In addition, the phosphodiester bond of the backbone is more easily broken when the base is U.

Another type of hotspot, though not often found in coding regions, is the "slippery sequence"—a homopolymer run, or region where a very short sequence (one or a few nucleotides) is repeated many times in tandem. During replication, a DNA polymerase may skip one repeat or replicate the same repeat twice, leading to a decrease or increase in repeat number.

1.16 Some Hereditary Agents Are Extremely Small

Key concept
- Some very small hereditary agents do not code for polypeptide, but consist of RNA or protein with heritable properties.

Viroids (or subviral pathogens) are infectious agents that cause diseases in higher plants. They are very small circular molecules of RNA. Unlike viruses—for which the infectious agent consists of a virion, a genome encapsulated in a protein coat—the viroid RNA is itself the infectious agent. The viroid consists solely of the RNA molecule, which is extensively folded by imperfect base pairing, forming a characteristic rod as shown in FIGURE 1.32. Mutations that interfere with the structure of this rod reduce the infectivity of the viroid.

A viroid RNA consists of a single molecule that is replicated autonomously and accurately in infected cells. Viroids are categorized into several groups. A given viroid is assigned to a group according to sequence similarity with other members of the group. For example, four viroids in the PSTV (potato spindle tuber viroid) group have 70%–83% sequence similarity with PSTV. Different isolates of a particular viroid strain vary from one another in sequence, which may result in phenotypic differences among infected cells. For example, the "mild" and "severe" strains of PSTV differ by three nucleotide substitutions.

Viroids are similar to viruses in having heritable nucleic acid genomes, but differ from viruses in both structure and function. Viroid RNA does not appear to be translated into poly-peptide, so it cannot itself code for the functions needed for its survival. This situation poses two as yet unanswered questions: How does viroid RNA replicate, and how does it affect the phenotype of the infected plant cell?

Replication must be carried out by enzymes of the host cell. The heritability of the viroid sequence indicates that viroid RNA is the template for replication.

Viroids are presumably pathogenic because they interfere with normal cellular processes. They might do this in a relatively random way—for example, by taking control of an essential enzyme for their own replication or by interfering with the production of necessary cellular RNAs. Alternatively, they might behave as abnormal regulatory molecules, with particular effects upon the expression of individual genes.

An even more unusual agent is the cause of scrapie, a degenerative neurological disease of sheep and goats. The disease is similar to the human diseases of kuru and Creutzfeldt–Jakob disease, which affect brain function. The infectious agent of scrapie does not contain nucleic acid. This extraordinary agent is called a **prion** (proteinaceous infectious agent). It is a 28 kD hydrophobic glycoprotein, PrP. PrP is coded by a cellular gene (conserved among the mammals) that is expressed in normal brain cells. The protein exists in two forms: the version found in normal brain cells is called PrP^c and is entirely degraded by proteases. The version found in infected brains is called PrP^{sc} and is extremely resistant to degradation by proteases. PrP^c is converted to PrP^{sc} by a conformational change that confers protease-resistance, and which has yet to be fully defined.

As the infectious agent of scrapie, PrP^{sc} must in some way modify the synthesis of its normal cellular counterpart so that it becomes infectious instead of harmless (see *Section 29.12, Prions Cause Diseases in Mammals*). Mice that lack

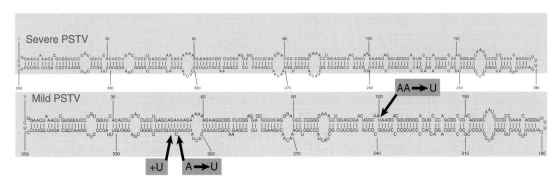

FIGURE 1.32 PSTV RNA is a circular molecule that forms an extensive double-stranded structure, interrupted by many interior loops. The severe and mild forms differ at three sites.

a PrP gene cannot develop scrapie, which demonstrates that PrP is essential for development of the disease.

1.17 Summary

Two classic experiments provided strong evidence that DNA is the genetic material of bacteria, viruses, and eukaryotic cells. DNA isolated from one strain of *Pneumococcus* bacteria can confer properties of that strain upon another strain. In addition, DNA is the only component that is inherited by progeny phages from parental phages. DNA can be used to transfect new properties into eukaryotic cells.

DNA is a double helix consisting of antiparallel strands in which the nucleotide units are linked by 5' to 3' phosphodiester bonds. The backbone is on the exterior; purine and pyrimidine bases are stacked in the interior in pairs in which A is complementary to T and G is complementary to C. In semiconservative replication, the two strands separate and daughter strands are assembled by complementary base pairing. Complementary base pairing is also used to transcribe an RNA from one strand of a DNA duplex.

A stretch of DNA may encode a polypeptide. The genetic code describes the relationship between the sequence of DNA and the sequence of the polypeptide. In general, only one of the two strands of DNA encodes a polypeptide. A codon consists of three nucleotides that encode a single amino acid. A coding sequence of DNA consists of a series of codons, which are read from a fixed starting point. In most cases only one of the three possible reading frames can be translated into polypeptide.

A mutation consists of a change in the sequence of A-T and G-C base pairs in DNA. A mutation in a coding sequence may change the sequence of amino acids in the corresponding polypeptide. A frameshift mutation alters the subsequent reading frame by inserting or deleting a base; this causes an entirely new series of amino acids to be coded after the site of mutation. A point mutation changes only the amino acid represented by the codon in which the mutation occurs. Point mutations may be reverted by back mutation of the original mutation. Insertions may revert by loss of the inserted material, but deletions cannot revert. Mutations may also be suppressed indirectly when a mutation in a different gene counters the original defect.

The natural incidence of mutations is increased by mutagens. Mutations may be concentrated at hotspots. A type of hotspot responsible for some point mutations is caused by deamination of the modified base 5-methylcytosine. Forward mutations occur at a rate of $\sim 10^{-6}$ per locus per generation; back mutations are rarer. Not all mutations have an effect on the phenotype.

Although all genetic information in cells is carried by DNA, viruses have genomes of double-stranded or single-stranded DNA or RNA. Viroids are subviral pathogens that consist solely of small molecules of RNA with no protective packaging. The RNA does not code for protein and its mode of perpetuation and of pathogenesis is unknown. Scrapie results from a proteinaceous infectious agent, or prion.

References

1.2 Introduction

Review

Cairns, J., Stent, G., and Watson, J. D. (1966). *Phage and the Origins of Molecular Biology*. Cold Spring Harbor Symp. Quant. Biol.

Judson, H. (1978). *The Eighth Day of Creation*. Knopf, New York.

Olby, R. (1974). *The Path to the Double Helix*. MacMillan, London.

1.2 DNA Is the Genetic Material of Bacteria and Viruses

Research

Avery, O. T., MacLeod, C. M., and McCarty, M. (1944). Studies on the chemical nature of the substance inducing transformation of pneumococcal types. *J. Exp. Med.* 98, 451–460.

Griffith, F. (1928). The significance of pneumococcal types. *J. Hyg.* 27, 113–159.

Hershey, A. D. and Chase, M. (1952). Independent functions of viral protein and nucleic acid in growth of bacteriophage. *J. Gen. Physiol.* 36, 39–56.

1.3 DNA Is the Genetic Material of Eukaryotic Cells

Research

Pellicer, A., Wigler, M., Axel, R., and Silverstein, S. (1978). The transfer and stable integration of the HSV thymidine kinase gene into mouse cells. *Cell* 14, 133–141.

1.6 DNA Is a Double Helix

Review

Watson, J. D. (1981). *The Double Helix: A Personal Account of the Discovery of the Structure of DNA* (Norton Critical Editions). W. W. Norton, New York, NY.

Research

Watson, J. D. and Crick, F. H. C. (1953). A structure for DNA. *Nature* 171, 737–738.

Watson, J. D., and Crick, F. H. C. (1953). Genetic implications of the structure of DNA. *Nature* 171, 964–967.

Wilkins, M. F. H., Stokes, A. R., and Wilson, H. R. (1953). Molecular structure of DNA. *Nature* 171, 738–740.

1.7 DNA Replication Is Semiconservative

Review

Holmes, F. (2001). *Meselson, Stahl, and the Replication of DNA: A History of the Most Beautiful Experiment in Biology.* Yale University Press, New Haven, CT.

Research

Meselson, M. and Stahl, F. W. (1958). The replication of DNA in *E. coli. Proc. Natl. Acad. Sci. USA* 44, 671–682.

1.11 Mutations Change the Sequence of DNA

Reviews

Drake, J. W., Charlesworth, B., Charlesworth, D., and Crow, J. F. (1998). Rates of spontaneous mutation. *Genetics* 148, 1667–1686.

Drake, J. W. and Balz, R. H. (1976). The biochemistry of mutagenesis. *Annu. Rev. Biochem.* 45, 11–37.

Research

Drake, J. W. (1991). A constant rate of spontaneous mutation in DNA-based microbes. *Proc. Natl. Acad. Sci. USA* 88, 7160–7164.

Grogan, D. W., Carver, G. T., and Drake, J. W. (2001). Genetic fidelity under harsh conditions: analysis of spontaneous mutation in the thermoacidophilic archaeon *Sulfolobus acidocaldarius. Proc. Natl. Acad. Sci. USA* 98, 7928–7933.

1.12 Mutations May Affect Single Base Pairs or Longer Sequences

Review

Maki, H. (2002). Origins of spontaneous mutations: specificity and directionality of base-substitution, frameshift, and sequence-substitution mutageneses. *Annu. Rev. Genet.* 36, 279–303.

1.14 Mutations Are Concentrated at Hotspots

Research

Coulondre, C. et al. (1978). Molecular basis of base substitution hotspots in *E. coli. Nature* 274, 775–780.

Millar, C. B., Guy, J., Sansom, O. J., Selfridge, J., MacDougall, E., Hendrich, B., Keightley, P. D., Bishop, S. M., Clarke, A. R., and Bird, A. (2002). Enhanced CpG mutability and tumorigenesis in MBD4-deficient mice. *Science* 297, 403–405.

1.16 Some Hereditary Agents Are Extremely Small

Reviews

Diener, T. O. (1986). Viroid processing: a model involving the central conserved region and hairpin. *Proc. Natl. Acad. Sci. USA* 83, 58–62.

Diener, T. O. (1999). Viroids and the nature of viroid diseases. *Arch. Virol. Suppl.* 15, 203–220.

Prusiner, S. B. (1998). Prions. *Proc. Natl. Acad. Sci. USA* 95, 13363–13383.

Research

Bueler, H. et al. (1993). Mice devoid of PrP are resistant to scrapie. *Cell* 73, 1339–1347.

McKinley, M. P., Bolton, D. C., and Prusiner, S. B. (1983). A protease-resistant protein is a structural component of the scrapie prion. *Cell* 35, 57–62.

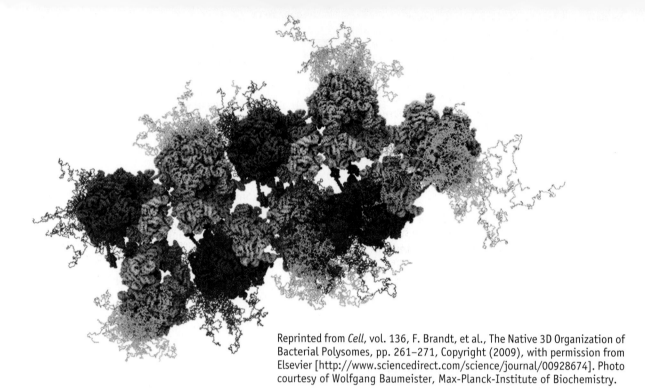

Reprinted from *Cell*, vol. 136, F. Brandt, et al., The Native 3D Organization of Bacterial Polysomes, pp. 261–271, Copyright (2009), with permission from Elsevier [http://www.sciencedirect.com/science/journal/00928674]. Photo courtesy of Wolfgang Baumeister, Max-Planck-Institute of Biochemistry.

Genes Code for Proteins
Edited by Esther Siegfried

CHAPTER OUTLINE

2.1 Introduction

The gene is the functional unit of heredity. Each gene is a sequence within the genome that functions by giving rise to a discrete product (which may be a polypeptide or an RNA). The basic behavior of the gene was defined by Mendel more than a century ago. Summarized in his two laws (segregation and independent assortment), the gene was recognized as a "particulate factor" that passes largely unchanged from parent to progeny. A gene may exist in alternative forms. These forms are called **alleles**.

In diploid organisms with two sets of chromosomes, one of each chromosome pair is inherited from each parent. This is also true for genes. One of the two copies of each gene is the paternal allele (inherited from the father), the other is the maternal allele (inherited from the mother). This common pattern of inheritance led to the discovery that chromosomes in fact carry the genes.

Each chromosome consists of a linear array of genes. Each gene resides at a particular location on the chromosome. The location is more formally called a genetic **locus**. The alleles of a gene are the different forms that are found at its locus.

The key to understanding the organization of genes into chromosomes was the discovery of genetic linkage—the tendency for genes on the same chromosome to remain together in the progeny instead of assorting independently as predicted by Mendel's laws. Once the unit of **genetic recombination** (reassortment) was introduced as the measure of linkage, the construction of genetic maps became possible.

The resolution of the recombination map of a multicellular eukaryote is restricted by the small number of progeny that can be obtained from each mating. Recombination occurs so infrequently between nearby points that it is rarely observed between different mutations in the same gene. As a result, classical linkage maps of eukaryotes can place the genes in order, but cannot determine relationships within a gene. By moving to a microbial system in which a very large number of progeny can be obtained from each genetic cross, researchers could demonstrate that recombination occurs within genes. It follows the same rules that

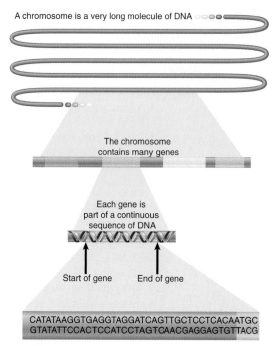

A chromosome is a very long molecule of DNA

The chromosome contains many genes

Each gene is part of a continuous sequence of DNA

Start of gene   End of gene

CATATAAGGTGAGGTAGGATCAGTTGCTCCTCACAATGC
GTATATTCCACTCCATCCTAGTCAACGAGGAGTGTTACG

FIGURE 2.1 Each chromosome has a single long molecule of DNA within which are the sequences of individual genes.

were previously deduced for recombination between genes.

Mutations within a gene can be arranged into a linear order, showing that the gene itself has the same linear construction as the array of genes on a chromosome. Thus the genetic map is linear within as well as between loci: it consists of an unbroken sequence within which the genes reside. This conclusion leads naturally into the modern view summarized in **FIGURE 2.1** that the genetic material of a chromosome consists of an uninterrupted length of DNA representing many genes. Having defined the gene as an uninterrupted length of DNA, it should be noted that in eukaryotes many genes are interrupted by sequences in the DNA that are then excised from the mRNA (see Chapter 4, *The Interrupted Gene*).

2.2 A Gene Codes for a Single Polypeptide

Key concepts

- The one gene: one enzyme hypothesis summarizes the basis of modern genetics: that a gene is a stretch of DNA coding for one or more isoforms of a single polypeptide.
- Some genes do not encode polypeptides, but encode structural or regulatory RNAs.
- Most mutations damage gene function and are recessive to the wild-type allele.

The first systematic attempt to associate genes with enzymes, carried out by Beadle and Tatum in the 1940s, showed that each stage in a metabolic pathway is catalyzed by a single enzyme and can be blocked by mutation in a different gene. This led to the **one gene : one enzyme hypothesis**. Each metabolic step is catalyzed by a particular enzyme, whose production is the responsibility of a single gene. A mutation in the gene alters the activity of the protein for which it is responsible.

A modification in the hypothesis is needed to accommodate proteins that consist of more than one subunit. If the subunits are all the same, the protein is a **homomultimer** and is represented by a single gene. If the subunits are different, the protein is a **heteromultimer**. Stated as a more general rule applicable to any heteromultimeric protein, the one gene: one enzyme hypothesis becomes more precisely expressed as the **one gene : one polypeptide hypothesis**. Even this general rule needs to be refined because many genes encode multiple, related polypeptides through alternative splicing of the mRNA (see Chapter 21, *RNA Splicing and Processing*).

Identifying which protein represents a particular gene can be a protracted task. The mutation responsible for Mendel's wrinkled-pea mutant was identified only in 1990 as an alteration that inactivates the gene for a starch-branching enzyme!

It is important to remember that a gene does not directly generate a polypeptide. As shown previously in Figure 1.2, a gene codes for an RNA, which may in turn code for a polypeptide. Many genes code for polypeptides, but some genes code for RNAs that do not give rise to polypeptides. These RNAs may be structural components of the apparatus responsible for synthesizing proteins or, as has become evident in recent years, have roles in regulating gene expression (see Chapter 30, *Regulatory RNA*). The basic principle is that the gene is a sequence of DNA that specifies the sequence of an independent product. The process of gene expression may terminate in a product that is either RNA or polypeptide.

A mutation is a random event with regard to the structure of the gene, so the greatest probability is that it will damage or even abolish gene function. Most mutations that affect gene function are recessive: *they represent an absence of function, because the mutant gene has been prevented from producing its usual product.* **FIGURE 2.2** illustrates the relationship between recessive

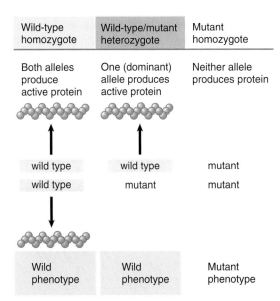

Wild-type homozygote	Wild-type/mutant heterozygote	Mutant homozygote
Both alleles produce active protein	One (dominant) allele produces active protein	Neither allele produces protein
wild type	wild type	mutant
wild type	mutant	mutant
Wild phenotype	Wild phenotype	Mutant phenotype

FIGURE 2.2 Genes code for proteins; dominance is explained by the properties of mutant proteins. A recessive allele does not contribute to the phenotype in the wild-type/mutant heterozygote because it produces no protein (or protein that is nonfunctional). If both alleles are the recessive mutant allele, no active protein is produced.

and wild-type alleles. When a heterozygote contains one wild-type allele and one mutant allele, the wild-type allele is able to direct production of the normal gene product. The wild-type allele is therefore dominant. (This assumes that an adequate *amount* of product is made by the single wild-type allele. When this is not true, the smaller amount made by one allele as compared to two alleles results in the intermediate phenotype of a partially dominant allele in a heterozygote.)

- A mutation in a gene affects only the product (protein or RNA) coded by the mutant copy of the gene and does not affect the product coded by any other allele.
- Failure of two mutations to complement (produce wild phenotype) when they are present in *trans* configuration in a heterozygote means that they are part of the same gene.

How do we determine whether two mutations that cause a similar phenotype lie in the same gene? If they map close together, they may be alleles. They could, however, also represent mutations in two *different* genes whose pro-

teins are involved in the same function. The **complementation test** is used to determine whether two mutations lie in the same gene or in different genes. The test consists of making a heterozygote for the two mutations.

If the mutations lie in the same gene, the parental genotypes can be represented as:

$$\frac{m_1}{m_1} \text{ and } \frac{m_2}{m_2}$$

The first parent provides an m_1 mutant allele and the second parent provides an m_2 allele, so that the heterozygote has the constitution:

$$\frac{m_1}{m_2}$$

No wild-type gene is present, so the heterozygote has mutant phenotype and the alleles fail to complement. If the mutations lie in different genes, the parental genotypes can be represented as:

$$\frac{m_1+}{m_1+} \text{ and } \frac{+m_2}{+m_2}$$

Each chromosome has a wild-type copy of one gene (represented by the plus sign) and a mutant copy of the other. Then the heterozygote has the constitution:

$$\frac{m_1+}{+m_2}$$

in which the two parents between them have provided a wild-type copy of each gene. The heterozygote has wild phenotype, and thus the two genes are said to *complement*.

The complementation test is shown in more detail in **FIGURE 2.3**. The basic test consists of the comparison shown in the top part of the figure. If two mutations lie in the same gene, we see a difference in the phenotypes of the *trans* configuration and the *cis* configuration. The *trans* configuration is mutant because each allele has a (different) mutation, whereas the *cis* configuration is wild-type because one allele has two mutations and the other allele has no mutations. The lower part of the figure shows that if the two mutations lie in different genes, we always see a wild phenotype. There is always one wild-type and one mutant allele of each gene, and the configuration is irrelevant. Failure to complement means that two mutations are part of the *same* genetic unit. Mutations that do not complement one another are said to comprise part of the same *complementation group*. Another term used to describe the

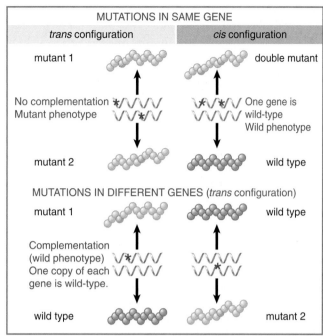

FIGURE 2.3 The cistron is defined by the complementation test. Genes are represented by spirals; red stars identify sites of mutation.

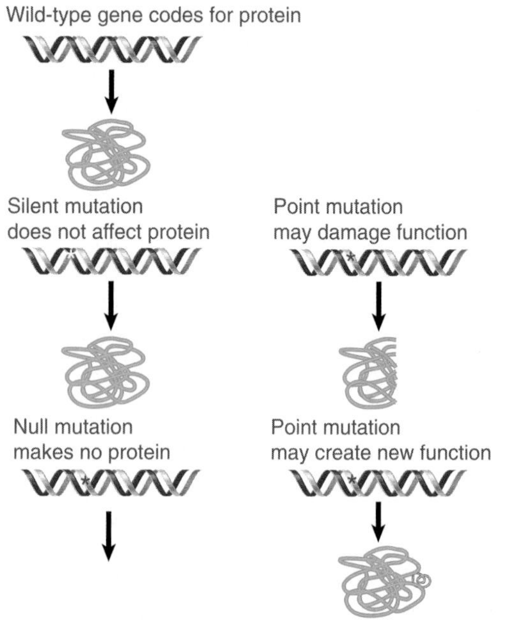

FIGURE 2.4 Mutations that do not affect protein sequence or function are silent. Mutations that abolish all protein activity are null. Point mutations that cause loss-of-function are recessive; those that cause gain-of-function are dominant.

unit defined by the complementation test is the **cistron**. This is the same as the gene. Basically these three terms all describe a stretch of DNA that functions as a unit to give rise to an RNA or protein product. The properties of the gene with regard to complementation are explained by the fact that this product is a single molecule that behaves as a functional unit.

2.4 Mutations May Cause Loss-of-Function or Gain-of-Function

Key concepts

- Recessive mutations are due to loss-of-function by the protein product.
- Dominant mutations result from a gain-of-function.
- Testing whether a gene is essential requires a null mutation (one that completely eliminates its function).
- Silent mutations have no effect, either because the base change does not change the sequence or amount of protein, or because the change in protein sequence has no effect.

The various possible effects of mutation in a gene are summarized in **FIGURE 2.4**.

When a gene has been identified, insight into its function in principle can be gained by generating a mutant organism that entirely lacks the gene. A mutation that completely eliminates gene function—usually because the gene has been deleted—is called a **null mutation**. If a gene is essential, a null mutation is lethal when homozygous or hemizygous.

To determine what effect a gene has upon the phenotype, it is essential to characterize a null mutant. Generally, if a null mutant fails to affect a phenotype, we may safely conclude that the gene function is not necessary. Some genes have overlapping functions, though, and removal of one gene is not sufficient to significantly affect the phenotype. Null mutations, or other mutations that impede gene function (but do not necessarily abolish it entirely), are called **loss-of-function mutations**. A loss-of-function mutation is recessive (as in the example of Figure 2.2). Loss-of-function mutations that affect protein activity but retain sufficient activity so that the phenotype is not altered are referred to as *leaky mutations*. Sometimes a mutation has the opposite effect and causes a protein to acquire a new function or expression pattern; such a change is called a **gain-of-function mutation**. A gain-of-function mutation is dominant.

Not all mutations in protein-coding genes lead to a detectable change in the phenotype. Mutations without apparent effect are called **silent mutations**. They comprise two types: One type involves base changes in DNA that do not cause any change in the amino acid present in the corresponding protein. The second type changes the amino acid, but the replacement in the protein does not affect its activity; these are called **neutral substitutions**.

2.5 A Locus May Have Many Different Mutant Alleles

Key concept
- The existence of multiple alleles allows heterozygotes that represent any pairwise combination of alleles to exist.

If a recessive mutation is produced by every change in a gene that prevents the production of an active protein, there should be a large number of such mutations in any one gene. Many amino acid replacements may change the structure of the protein sufficiently to impede its function.

Different variants of the same gene are called *multiple alleles*, and their existence makes possible a heterozygote with two mutant alleles. The relationship between these multiple alleles takes various forms.

In the simplest case, a wild-type allele codes for a product that is functional. Mutant allele(s) code for products that are nonfunctional. There are often cases, though, in which a series of mutant alleles affect the same phenotype to differing extents. For example, wild-type function of the *white* locus of *Drosophila melanogaster* is required for development of the normal red color of the eye. The locus is named for the effect of extreme (null) mutations, which cause the fly to have white eyes in mutant homozygotes.

To denote wild-type and mutant alleles, the wild-type genotype is indicated by a plus superscript after the name of the locus (w^+ is the wild-type allele for [red] eye color in *D. melanogaster*). Sometimes + is used by itself to describe the wild-type allele, and only the mutant alleles are indicated by the name of the locus.

An entirely defective form of the gene (or absence of phenotype) may be indicated by a minus superscript. To distinguish among a variety of mutant alleles with different effects, other superscripts may be introduced, such as w^i or w^a.

The w^+ allele is dominant over any other allele in heterozygotes. There are many different mutant alleles. **FIGURE 2.5** shows a (small) sample. Although some alleles produce no visible pigment, and therefore the eyes are white, many alleles produce some color. Each of these

Allele	Phenotype of homozygote
w^+	red eye (wild type)
w^{bl}	blood
w^{ch}	cherry
w^{bf}	buff
w^h	honey
w^a	apricot
w^e	eosin
w^l	ivory
w^z	zeste (lemon-yellow)
w^{sp}	mottled, color varies
w^1	white (no color)

FIGURE 2.5 The *w* locus has an extensive series of alleles whose phenotypes extend from wild-type (red) color to complete lack of pigment.

mutant alleles must therefore represent a different mutation of the gene, which does not eliminate its function entirely, but leaves a residual activity that produces a characteristic phenotype. These alleles are named for the color of the eye in a homozygote. (Most *w* alleles affect the quantity of pigment in the eye. The examples in the figure are arranged in [roughly] declining amount of color, but others, such as w^{sp}, affect the pattern in which it is deposited.)

When multiple alleles exist, an organism may be a heterozygote that carries two different mutant alleles. The phenotype of such a heterozygote depends on the nature of the residual activity of each allele. The relationship between two mutant alleles is in principle no different from that between wild-type and mutant alleles: one allele may be dominant, there may be partial dominance, or there may be codominance.

2.6 A Locus May Have More Than One Wild-type Allele

Key concept
- A locus may have a polymorphic distribution of alleles with no individual allele that can be considered to be the sole wild-type.

There is not necessarily a unique wild-type allele at any particular locus. Control of the human blood group system provides an example. Lack of function is represented by the null type *O* group. The functional alleles *A* and *B*, however, provide activities that are codominant with one another and dominant over *O* group. The basis for this relationship is illustrated in **FIGURE 2.6**.

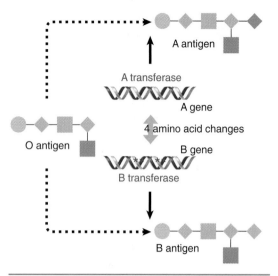

Phenotype	Genotype	Transferase Activity
O	OO	None
A	AO or AA	N-Ac-gal transferase
B	BO or BB	Gal transferase
AB	AB	GalN-Ac-Gal-transferase

FIGURE 2.6 The ABO blood group locus codes for a galactosyltransferase whose specificity determines the blood group.

The O (or H) antigen is generated in all individuals and consists of a particular carbohydrate group that is added to proteins. The *ABO* locus codes for a galactosyltransferase enzyme that adds a further sugar group to the O antigen. The specificity of this enzyme determines the blood group. The *A* allele produces an enzyme that uses the cofactor UDP-N-acetylgalactose, creating the A antigen. The *B* allele produces an enzyme that uses the cofactor UDP-galactose, creating the B antigen. The A and B versions of the transferase protein differ in four amino acids that presumably affect its recognition of the type of cofactor. The *O* allele has a transferase mutation (a small deletion) that eliminates activity, so no modification of the O antigen occurs.

This explains why *A* and *B* alleles are dominant in the *AO* and *BO* heterozygotes: the corresponding transferase activity creates the A or B antigen. The *A* and *B* alleles are codominant in *AB* heterozygotes, because both transferase activities are expressed. The *OO* homozygote is a null that has neither activity and therefore lacks both antigens.

Neither *A* nor *B* can be regarded as uniquely wild type, because they represent alternative activities rather than loss or gain of function. A situation such as this, in which there are multiple functional alleles in a population, is described as a **polymorphism** (see *Section 5.3, Individual Genomes Show Extensive Variation*).

2.7 Recombination Occurs by Physical Exchange of DNA

Key concepts

- Recombination is the result of crossing-over that occurs at chiasmata and involves two of the four chromatids.
- Recombination occurs by a breakage and reunion that proceeds via an intermediate of hybrid DNA that depends on the complementarity of the two strands of DNA.
- The frequency of recombination between two genes is proportional to their physical distance; recombination between genes that are very closely linked is rare.
- For genes that are very far apart on a single chromosome, the frequency of recombination is not proportional to their physical distance because recombination happens so frequently.

Genetic recombination describes the generation of new combinations of alleles that occurs at each generation in diploid organisms. The two copies of each chromosome may have different alleles at some loci. By exchanging corresponding parts between the chromosomes, recombinant chromosomes that are different from the parental chromosomes can be generated.

Recombination results from a physical exchange of chromosomal material. This is visible in the form of the *crossing-over* that occurs during meiosis (the specialized division that produces haploid germ cells). Meiosis starts with a cell that has duplicated its chromosomes, so that it has four copies of each chromosome. Early in meiosis, all four copies are closely associated (synapsed) in a structure called a *bivalent*. Each individual chromosomal unit is called a *chromatid* at this stage. Pairwise exchanges of material occur between two nonidentical (nonsister) chromatids.

The visible result of a crossing-over event is called a **chiasma** and is illustrated diagrammatically in **FIGURE 2.7**. A chiasma represents a site at which two of the chromatids in a bivalent have been broken at corresponding points. The broken ends have been rejoined crosswise, generating new chromatids. Each new chromatid consists of material derived from one chromatid on one side of the junction point, with material from the other chromatid on the opposite side.

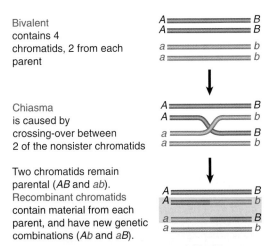

Bivalent contains 4 chromatids, 2 from each parent

Chiasma is caused by crossing-over between 2 of the nonsister chromatids

Two chromatids remain parental (*AB* and *ab*). Recombinant chromatids contain material from each parent, and have new genetic combinations (*Ab* and *aB*).

FIGURE 2.7 Chiasma formation is responsible for generating recombinants.

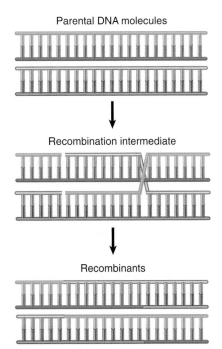

Parental DNA molecules

Recombination intermediate

Recombinants

FIGURE 2.8 Recombination involves pairing between complementary strands of the two parental duplex DNAs.

The two recombinant chromatids have reciprocal structures. The event is described as a *breakage and reunion*. Its nature explains why a single recombination event can produce only 50% recombinants: each individual recombination event involves only two of the four associated chromatids.

The complementarity of the two strands of DNA is essential for the recombination process. Each of the chromatids shown in Figure 2.7 consists of a very long duplex of DNA. For them to be broken and reconnected without any loss of material requires a mechanism to recognize exactly corresponding positions through complementary base pairing.

Recombination involves a process in which the single strands in the region of the crossover exchange their partners. **FIGURE 2.8** shows that this creates a stretch of *hybrid DNA*, in which the single strand of one duplex is paired with its complement from the other duplex. Each duplex DNA corresponds to one of the chromatids involved in recombination in Figure 2.7. The mechanism, of course, involves other stages (strands must be broken and resealed), which we discuss in more detail in Chapter 15 (*Homologous and Site-Specific Recombination*), but the crucial feature that makes precise recombination possible is the complementarity of DNA strands. The figure shows only some stages of the reaction, but we see that a stretch of hybrid DNA forms in the recombination intermediate when a single strand crosses over from one duplex to the other. Each recombinant consists of one parental duplex DNA at the left, which is connected by a stretch of hybrid DNA to the other parental duplex at the right.

The formation of hybrid DNA requires the sequences of the two recombining duplexes to be close enough to allow pairing between the complementary strands. If there are no differences between the two parental genomes in this region, formation of hybrid DNA will be perfect. The reaction can be tolerated, however, even when there are small differences. In this case, the hybrid DNA has points of mismatch, at which a base in one strand faces a base in the other strand that is not complementary to it. The correction of such mismatches is another feature of genetic recombination (see Chapter 16, *Repair Systems Handle Damage to DNA*).

Over chromosomal distances, recombination events occur more or less at random with a characteristic frequency. The probability that a crossover will occur within any specific region of the chromosome is more or less proportional to the length of the region, up to a saturation point. For example, a large human chromosome usually has three or four crossover events per meiosis, whereas a small chromosome has only one on average.

FIGURE 2.9 compares three situations: two genes on different chromosomes, two genes that are far apart on the same chromosome, and two genes that are close together on the same chromosome. Genes on different chromosomes segregate independently according to Mendel's laws, resulting in the production

FIGURE 2.9 Genes on different chromosomes segregate independently so that all possible combinations of alleles are produced in equal proportions. Recombination occurs so frequently between genes that are far apart on the same chromosome that they effectively segregate independently. Recombination is reduced, however, when genes are closer together, and for adjacent genes may hardly ever occur.

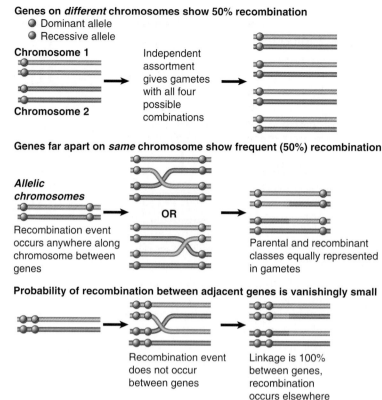

Genes on *different* chromosomes show 50% recombination
- Dominant allele
- Recessive allele

Chromosome 1

Independent assortment gives gametes with all four possible combinations

Chromosome 2

Genes far apart on *same* chromosome show frequent (50%) recombination

Allelic chromosomes

OR

Recombination event occurs anywhere along chromosome between genes

Parental and recombinant classes equally represented in gametes

Probability of recombination between adjacent genes is vanishingly small

Recombination event does not occur between genes

Linkage is 100% between genes, recombination occurs elsewhere

of 50% parental types and 50% recombinant types during meiosis. When genes are sufficiently far apart on the same chromosome, the probability of one or more recombination events in the region between them becomes so high that they behave in the same way as genes on different chromosomes and show 50% recombination.

When genes are close together, though, the probability of a recombination event between them is reduced, and occurs only in some proportion of meioses. For example, if it occurs in one quarter of the meioses, the overall rate of recombination is 12.5% (because a single recombination event produces 50% recombination, and this occurs in 25% of meioses). When genes are very close together, as shown in the bottom panel of Figure 2.9, recombination between them may never be observed in phenotypes of higher eukaryotes.

This leads us to the view that a chromosome contains an array of many genes. Each protein-coding gene is an independent unit of expression, and is represented in one or more polypeptide chains. The properties of a gene can be changed by mutation. The allelic combina-

tions present on a chromosome can be changed by recombination. We can now ask, "what is the relationship between the sequence of a gene and the sequence of the polypeptide chain it represents?"

2.8 The Genetic Code Is Triplet

Key concepts
- The genetic code is read in triplet nucleotides called codons.
- The triplets are nonoverlapping and are read from a fixed starting point.
- Mutations that insert or delete individual bases cause a shift in the triplet sets after the site of mutation.
- Combinations of mutations that together insert or delete three bases (or multiples of three) insert or delete amino acids, but do not change the reading of the triplets beyond the last site of mutation.

Each gene represents a particular polypeptide chain. The concept that each protein consists of a particular series of amino acids dates from Sanger's characterization of insulin in the 1950s. The discovery that a gene consists of DNA presents us with the issue of how a sequence of nucleotides in DNA represents a sequence of amino acids in protein.

The sequence of nucleotides in DNA is important not because of its structure *per se*, but because it *codes* for the sequence of amino acids that constitutes the corresponding polypeptide. The relationship between a sequence of DNA and the sequence of the corresponding protein is called the **genetic code.**

The structure and/or enzymatic activity of each protein follows from its primary sequence of amino acids and its overall conformation, which is determined by interactions between the amino acids. By determining the sequence of amino acids in each protein, the gene is able to carry all the information needed to specify an active polypeptide chain. In this way, a single type of structure—the gene—is able to represent itself in innumerable polypeptide forms.

Together the various protein products of a cell undertake the catalytic and structural activities that are responsible for establishing its phenotype. Of course, in addition to sequences that code for proteins, DNA also contains certain sequences whose function is to be recognized by regulator molecules, usually proteins. Here the function of the DNA is determined by its sequence directly, not via any intermediary code. Both types of region—genes expressed as proteins and sequences recognized as such—constitute genetic information.

The genetic code is deciphered by a complex apparatus that interprets the nucleic acid sequence. This apparatus is essential if the information carried in DNA is to have meaning. In any given region, only one of the two strands of DNA codes for protein, so we write the genetic code as a sequence of bases (rather than base pairs).

The genetic code is read in groups of three nucleotides, each group representing one amino acid. Each trinucleotide sequence is called a **codon**. A gene includes a series of codons that is read sequentially from a starting point at one end to a termination point at the other end. Written in the conventional 5′ to 3′ direction, the nucleotide sequence of the DNA strand that codes for protein corresponds to the amino acid sequence of the protein written in the direction from N-terminus to C-terminus.

The genetic code is read in *nonoverlapping triplets from a fixed starting point*:

- The use of a *fixed starting point* means that assembly of a protein must start at one end and work to the other, so that different parts of the coding sequence cannot be read independently.

The nature of the code predicts that two types of mutations, base substitution and base insertion/deletion, will have different effects. If a particular sequence is read sequentially, such as:

UUU AAA GGG CCC (codons)
aa1   aa2   aa3   aa4 (amino acids)

then a base substitution, or point mutation, will affect only one amino acid. For example, the substitution of an A by some other base (X) causes aa2 to be replaced by aa5:

UUU AAX GGG CCC
aa1   aa5   aa3   aa4

because only the second codon has been changed.

A mutation that inserts or deletes a single base, though, will change the triplet sets for the entire subsequent sequence. A change of this sort is called a **frameshift**. An insertion might take the form:

UUU AAX AGG GCC C
aa1   aa5   aa6   aa7

The new sequence of triplets is completely different from the old one, and as a result the entire amino acid sequence of the protein is altered beyond the site of mutation. Thus the function of the protein is likely to be lost completely.

Frameshift mutations are induced by the **acridines**. The acridines are compounds that bind to DNA and distort the structure of the double helix, causing additional bases to be incorporated or omitted during replication. Each mutagenic event sponsored by an acridine results in the addition or removal of a single base pair.

If an acridine mutant is produced by, say, addition of a nucleotide, it should revert to wild-type by deletion of the nucleotide. Reversion also can be caused by deletion of a different base, though, at a site close to the first. Combinations of such mutations provided revealing evidence about the nature of the genetic code.

FIGURE 2.10 illustrates the properties of frameshift mutations. An insertion or deletion changes the entire protein sequence following the site of mutation. The combination of an insertion *and* a deletion, though, causes the code to be read incorrectly only between the

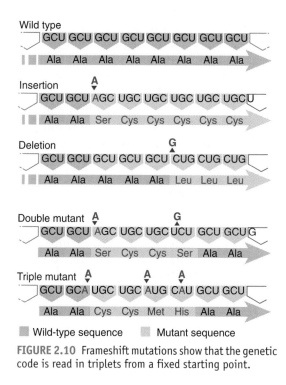

FIGURE 2.10 Frameshift mutations show that the genetic code is read in triplets from a fixed starting point.

two sites of mutation; correct reading resumes after the second site.

In 1961, genetic analysis of acridine mutations in the *rII* region of the phage T4 showed that all the mutations could be classified into one of two sets, described as (+) and (−). Either type of mutation by itself causes a frameshift: the (+) type by virtue of a base addition, and the (−) type by virtue of a base deletion. Double mutant combinations of the types (+ +) and (− −) continue to show mutant behavior. Combinations of the types (+ −) or (− +), however, suppress one another, giving rise to a description in which one mutation is described as a *frameshift suppressor* of the other. (In the context of this work, "suppressor" is used in an unusual sense because the second mutation is in the same gene as the first.)

These results show that the genetic code must be read as a sequence that is fixed by the starting point. Thus additions or deletions compensate for each other, whereas double additions or double deletions remain mutant. This does not, however, reveal how many nucleotides make up each codon.

When triple mutants are constructed, only (+ + +) and (− − −) combinations show the wild phenotype, whereas other combinations remain mutant. If we take three additions or three deletions to correspond respectively to the

addition or omission overall of a single amino acid, this implies that the code is read in triplets. An incorrect amino acid sequence is found between the two outside sites of mutation and the sequence on either side remains wild-type, as indicated in Figure 2.10.

2.9 Every Sequence Has Three Possible Reading Frames

Key concept

• In general, only one reading frame is translated, and the other two are blocked by frequent termination signals.

If the genetic code is read in nonoverlapping triplets, there are three possible ways of translating any nucleotide sequence into protein, depending on the starting point. These are called **reading frames**. For the sequence

A C G A C G A C G A C G A C G A C G

the three possible reading frames are

ACG ACG ACG ACG ACG ACG ACG
CGA CGA CGA CGA CGA CGA CGA
GAC GAC GAC GAC GAC GAC GAC

A reading frame that consists exclusively of triplets representing amino acids is called an **open reading frame** or **ORF**. A sequence that is translated into protein has a reading frame that starts with a special **initiation codon** (**AUG**) and then extends through a series of triplets representing amino acids until it ends at one of three types of **termination codon** (see Chapter 25, *Using the Genetic Code*).

A reading frame that cannot be read into protein because termination codons occur frequently is said to be **closed** or **blocked**. If a sequence is blocked in all three reading frames, it cannot have the function of coding for protein.

When the sequence of a DNA region of unknown function is obtained, each possible reading frame is analyzed to determine whether it is open or blocked. Usually no more than one of the three possible frames of reading is open in any single stretch of DNA. **FIGURE 2.11** shows an example of a sequence that can be read in only one reading frame because the alternative reading frames are blocked by frequent termination codons. A long open reading frame is unlikely to exist by chance; if it were not translated into protein, there would have been no

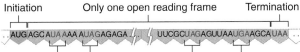

Second reading frame is closed   Third reading frame is closed

FIGURE 2.11 An open reading frame starts with AUG and continues in triplets to a termination codon. Blocked reading frames may be interrupted frequently by termination codons.

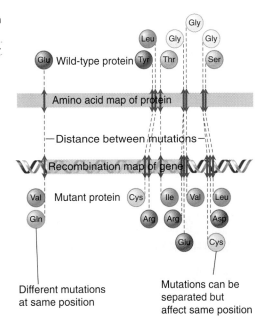

Different mutations at same position

Mutations can be separated but affect same position

FIGURE 2.12 The recombination map of the tryptophan synthetase gene corresponds with the amino acid sequence of the protein.

selective pressure to prevent the accumulation of termination codons. Thus the identification of a lengthy open reading frame is taken to be *prima facie* evidence that the sequence is translated into protein in that frame. An ORF for which no protein product has been identified is sometimes called an **unidentified reading frame** (**URF**).

2.10 Prokaryotic Genes Are Colinear with Their Proteins

Key concepts

- A prokaryotic gene consists of a continuous length of 3N nucleotides that encodes N amino acids.
- The gene, mRNA, and protein are all colinear.

By comparing the nucleotide sequence of a gene with the amino acid sequence of a protein, we can determine directly whether the gene and the protein are *colinear*; that is, whether the sequence of nucleotides in the gene corresponds exactly with the sequence of amino acids in the protein. In bacteria and their viruses, there is an exact equivalence. Each gene contains a continuous stretch of DNA whose length is directly related to the number of amino acids in the protein that it represents. A gene with an open reading frame of 3N bp is required to code for a protein of N amino acids, according to the genetic code.

The equivalence of the bacterial gene and its product means that a physical map of DNA will exactly match an amino acid map of the protein. How well do these maps fit with the recombination map?

The **colinearity** of gene and protein was originally investigated in the tryptophan synthetase gene of *E. coli.* Genetic distance was measured by the percent recombination

between mutations; protein distance was measured by the number of amino acids separating sites of replacement. **FIGURE 2.12** compares the two maps. The order of seven sites of mutation is the same as the order of the corresponding sites of amino acid replacement, and the recombination distances are relatively similar to the actual distances in the protein. The recombination map expands the distances between some mutations, but otherwise there is little distortion of the recombination map relative to the physical map.

The recombination map makes two further general points about the organization of the gene. Different mutations may cause a wild-type amino acid to be replaced with different substituents. If two such mutations cannot recombine, they must involve different point mutations at the same position in DNA. If the mutations can be separated on the genetic map, but affect the same amino acid on the upper map (the connecting lines converge in the figure), they must involve point mutations at different positions that affect the same amino acid. This happens because the unit of genetic recombination (1 bp) is smaller than the unit coding for the amino acid (3 bp).

2.11 Several Processes Are Required to Express the Protein Product of a Gene

Key concepts

- A prokaryotic gene is expressed by transcription into mRNA and then translation of the mRNA into protein.
- In eukaryotes, a gene may contain internal regions that are not represented in protein.
- Internal regions are removed from the mRNA transcript by RNA splicing to give an mRNA that is colinear with the protein product.
- Each mRNA consists of an untranslated 5′ region, a coding region, and an untranslated 3′ trailer.

In comparing gene and protein, we are restricted to dealing with the sequence of DNA stretching between the points corresponding to the ends of the protein. A gene is not directly translated into protein, though, but instead is expressed via the production of a **messenger RNA** (abbreviated to **mRNA**), a nucleic acid intermediate actually used to synthesize a protein (as we see in detail in Chapter 22, *mRNA Stability and Localization*).

Messenger RNA is synthesized by the same process of complementary base pairing used to replicate DNA, with the important difference that it corresponds to only one strand of the DNA double helix. **FIGURE 2.13** shows that the sequence of mRNA is complementary with the sequence of one strand of DNA and is identical (apart from the replacement of T with U) with the other strand of DNA. The convention for writing DNA sequences is that the top strand runs 5′→3′, with the sequence that is the same as RNA.

The process by which a gene gives rise to a protein is called **gene expression**. In bacteria, it consists of two stages. The first stage is **tran-scription**, when an mRNA copy of one strand of the DNA is produced. The second stage is **translation** of the mRNA into protein. This is the process by which the sequence of an mRNA is read in triplets to give the series of amino acids that make the corresponding protein.

An mRNA includes a sequence of nucleotides that corresponds with the sequence of amino acids in the protein. This part of the nucleic acid is called the **coding region**. Note, however, that the mRNA includes additional sequences on either end; these sequences do not directly encode polypeptide. The 5′ untranslated region is called the **leader** or **5′ UTR**, and the 3′ untranslated region is called the **trailer** or **3′ UTR**.

The *gene* includes the entire sequence represented in messenger RNA. Sometimes mutations impeding gene function are found in the additional, noncoding regions, confirming the view that these comprise a legitimate part of the genetic unit.

FIGURE 2.14 illustrates this situation, in which the gene is considered to comprise a continuous stretch of DNA that is needed to produce a particular protein. It includes the sequence coding for that protein, but also includes sequences on either side of the coding region.

A bacterium consists of only a single compartment, so transcription and translation occur in the same place, as illustrated in **FIGURE 2.15**.

In eukaryotes transcription occurs in the nucleus, but the mRNA product must be *transported* to the cytoplasm in order to be translated. For the simplest eukaryotic genes (just like in bacteria) the translated RNA is in fact the transcribed copy of the gene. For more complex genes, however, the immediate transcript of the gene is a **pre-mRNA** that requires **RNA processing** to generate the mature mRNA. The basic stages of gene expression in a eukaryote are outlined in **FIGURE 2.16**. This results in a spa-

---

DNA consists of two base-paired strands

top strand
5′ ATGCCGTTAGACCGTTAGCGGACCTGAC
3′ TACGGCAATCTGGCAATCGCCTGGACTG
bottom strand

↓ RNA synthesis

5′ AUGCCGUUAGACCGUUAGCGGACCUGAC 3

RNA has same sequence as DNA top strand; is complementary to DNA bottom strand

FIGURE 2.13 RNA is synthesized by using one strand of DNA as a template for complementary base pairing.

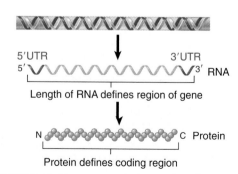

5′UTR   3′UTR
5′ ⌇⌇⌇⌇⌇⌇⌇⌇⌇⌇⌇ 3′ RNA

Length of RNA defines region of gene

↓

N ⦷⦷⦷⦷⦷⦷⦷⦷⦷⦷ C Protein

Protein defines coding region

FIGURE 2.14 The gene may be longer than the sequence coding for protein.

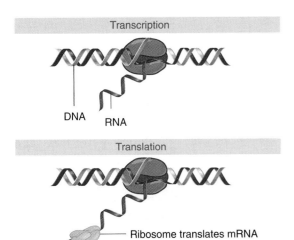

FIGURE 2.15 Transcription and translation take place in the same compartment in bacteria.

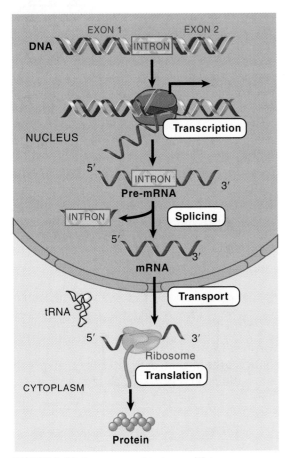

FIGURE 2.16 Gene expression is a multistage process.

tial separation between transcription (in the nucleus) and translation (in the cytoplasm).

The most important stage in processing is **splicing**. Many genes in eukaryotes (and a majority in multicellular eukaryotes) contain internal regions called **introns** that do

not code for protein. The process of splicing removes these regions from the pre-mRNA to generate an RNA that has a continuous open reading frame (see Figure 4.1). (The remaining, expressed regions of the mRNA are called **exons**.) Other processing events that occur at this stage involve the modification of the 5′ and 3′ ends of the pre-mRNA (see Figure 21.1).

Translation is accomplished by a complex apparatus that includes both protein and RNA components. The actual "machine" that undertakes the process is the **ribosome**, a large complex that includes some large RNAs (**ribosomal RNAs**, abbreviated to **rRNAs**) and many small proteins. The process of recognizing which amino acid corresponds to a particular nucleotide triplet requires an intermediate **transfer RNA** (abbreviated to **tRNA**); there is at least one tRNA species for every amino acid. Many ancillary proteins are involved. We describe translation in Chapter 24, *Translation*, but note for now that the ribosomes are the large structures in Figure 2.14 that move along the mRNA.

The important point to note at this stage is that the process of gene expression involves RNA not only as the essential substrate, but also in providing components of the apparatus. The rRNA and tRNA components are coded by genes and are generated by the process of transcription (just like mRNA, except that there is no subsequent stage of translation).

2.12 Proteins Are *trans*-acting, but Sites on DNA Are *cis*-acting

Key concepts

- All gene products (RNA or proteins) are *trans*-acting. They can act on any copy of a gene in the cell.
- *cis*-acting mutations identify sequences of DNA that are targets for recognition by *trans*-acting products. They are not expressed as RNA or protein and affect only the contiguous stretch of DNA.

A crucial step in the definition of the gene was the realization that all its parts must be present on one contiguous stretch of DNA. In genetic terminology, sites that are located on the same DNA are said to be in *cis*. Sites that are located on two different molecules of DNA are described as being in *trans*. So two mutations may be in *cis* (on the same DNA) or in *trans* (on different DNAs). The complementation test uses this concept to determine whether two

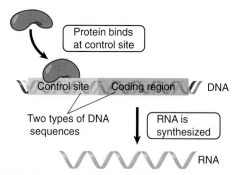

FIGURE 2.17 Control sites in DNA provide binding sites for proteins; coding regions are expressed via the synthesis of RNA.

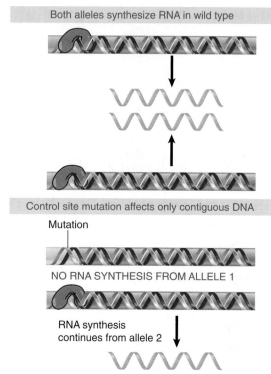

FIGURE 2.18 A *cis*-acting site controls expression of the adjacent DNA but does not influence the other allele.

mutations are in the same gene (see Figure 2.3). We may now extend the concept of the difference between *cis* and *trans* effects from defining the coding region of a gene to describing the interaction between and a gene and its regulatory elements.

Suppose that the ability of a gene to be expressed is controlled by a protein that binds to the DNA close to the coding region. In the example depicted in **FIGURE 2.17**, mRNA can be synthesized only when the protein is bound to the DNA. Now suppose that a mutation occurs in the DNA sequence to which this protein binds, so that the protein can no longer recognize the DNA. As a result, the DNA can no longer be expressed.

So a gene can be inactivated either by a mutation in a control site or by a mutation in a coding region. The mutations cannot be distinguished genetically, because both have the property of acting only on the DNA sequence of the single allele in which they occur. They have identical properties in the complementation test, and a mutation in a control region is therefore defined as comprising part of the gene in the same way as a mutation in the coding region.

FIGURE 2.18 shows that a deficiency in the control site *affects only the coding region to which it is connected; it does not affect the ability of the other allele to be expressed.* A mutation that acts solely by affecting the properties of the contiguous sequence of DNA is called a ***cis*-acting sequence**. It should be noted that in many eukaryotes the control region can influence the expression of DNA at some distance, but nonetheless the control region resides in the same DNA molecule as the coding sequence.

We may contrast the behavior of the *cis*-acting mutation shown in Figure 2.17 with the result of a mutation in the gene coding for the regulator protein. **FIGURE 2.19** shows that the

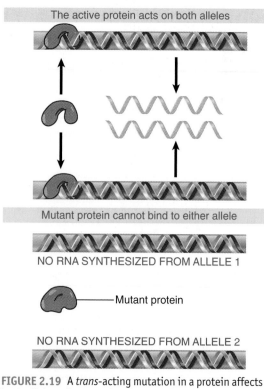

FIGURE 2.19 A *trans*-acting mutation in a protein affects both alleles of a gene that it controls.

absence of regulator protein would prevent *both* alleles from being expressed. A mutation of this sort is said to be in a ***trans*-acting sequence**.

Reversing the argument, if a mutation is *trans*-acting, we know that its effects must be exerted through some diffusible product (either a protein or a regulatory RNA) that acts on multiple targets within a cell. If a mutation is *cis*-acting, though, it must function via affecting directly the properties of the contiguous DNA, which means that it is *not expressed in the form of RNA or protein*.

2.13 Summary

A chromosome consists of an uninterrupted length of duplex DNA that contains many genes. Each gene (or cistron) is transcribed into an RNA product, which in turn is translated into a polypeptide sequence if the gene codes for protein. An RNA or protein product of a gene is said to be *trans*-acting. A gene is defined as a unit of a single stretch of DNA by the complementation test. A site on DNA that regulates the activity of an adjacent gene is said to be *cis*-acting.

When a gene codes for protein, the relationship between the sequence of DNA and sequence of the protein is given by the genetic code. Only one of the two strands of DNA codes for protein. A codon consists of three nucleotides that represent a single amino acid. A coding sequence of DNA consists of a series of codons, read from a fixed starting point and nonoverlapping. Usually one of the three possible reading frames can be translated into protein.

A gene may have multiple alleles. Recessive alleles are caused by loss-of-function mutations that interfere with the function of the protein. A null allele has total loss-of-function. Dominant alleles are caused by gain-of-function mutations that create a new property in the protein.

References

2.8 The Genetic Code Is Triplet

Review

Roth, J. R. (1974). Frameshift mutations. *Annu. Rev. Genet.* 8, 319–346.

Research

Benzer, S. and Champe, S. P. (1961). Ambivalent rII mutants of phage T4. *Proc. Natl. Acad. Sci. USA* 47, 403–416.

Crick, F. H. C., Barnett, L., Brenner, S., and Watts-Tobin, R. J. (1961). General nature of the genetic code for proteins. *Nature* 192, 1227–1232.

2.10 Prokaryotic Genes Are Colinear with Their Proteins

Research

Yanofsky, C., Drapeau, G. R., Guest, J. R., and Carlton, B. C. (1967). The complete amino acid sequence of the tryptophan synthetase A protein (μ subunit) and its colinear relationship with the genetic map of the A gene. *Proc. Natl. Acad. Sci. USA* 57, 2966–2968.

Yanofsky, C. et al. (1964). On the colinearity of gene structure and protein structure. *Proc. Natl. Acad. Sci. USA*, 51, 266–272.

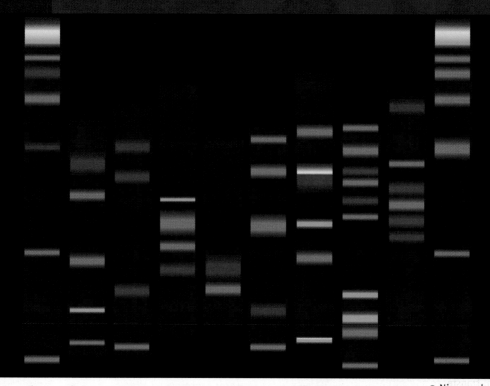

© Nicemonkey/Dreamstime.com

Methods in Molecular Biology and Genetic Engineering

Edited by John Brunstein

CHAPTER OUTLINE

- The next generation of sequencing techniques aims to increase automation and decrease time and cost of sequencing.

3.8 PCR and RT-PCR

- PCR permits the exponential amplification of a desired sequence, using primers that anneal to the sequence of interest.
- RT-PCR uses reverse transcriptase to convert RNA to DNA for use in a PCR reaction.
- Real-time, or quantitative, PCR detects the products of PCR amplification during their synthesis, and is more sensitive and quantitative than conventional PCR.
- PCR depends on the use of thermostable DNA polymerases that can withstand multiple cycles of template denaturation.

3.9 Blotting Methods

- Southern blotting involves the transfer of DNA from a gel to a membrane, followed by detection of specific sequences by hybridization with a labeled probe.
- Northern blotting is similar to Southern blotting, but involves the transfer of RNA from a gel to a membrane.
- Western blotting entails separation of proteins on an SDS gel, transfer to a nitrocellulose membrane, and detection proteins of interest using antibodies.

3.10 DNA Microarrays

- DNA microarrays comprise known DNA sequences spotted or synthesized on a small chip.
- Genome-wide transcription analysis is performed using labeled cDNA from experimental samples hybridized to a microarray containing sequences from all ORFs of the organism being used.

- SNP arrays permit genome-wide genotyping of single nucleotide polymorphisms.
- Array comparative genome hybridization (array-CGH) allow the detection of copy number changes in any DNA sequence compared between two samples.

3.11 Chromatin Immunoprecipitation

- Chromatin immunoprecipitation allows detection of specific protein–DNA interactions in vivo.
- "ChIP on chip" allows mapping of all the protein-binding sites for a given protein across the entire genome.

3.12 Gene Knockouts and Transgenics

- ES (embryonic stem) cells that are injected into a mouse blastocyst generate descendant cells that become part of a chimeric adult mouse.
- When the ES cells contribute to the germline, the next generation of mice may be derived from the ES cell.
- Genes can be added to the mouse germline by transfecting them into ES cells before the cells are added to the blastocyst.
- An endogenous gene can be replaced by a transfected gene using homologous recombination.
- The occurrence of successful homologous recombination can be detected by using two selectable markers, one of which is incorporated with the integrated gene, the other of which is lost when recombination occurs.
- The Cre/*lox* system is widely used to make inducible knockouts and knock-ins.

3.13 Summary

3.1 Introduction

Today, the field of molecular biology focuses on the mechanisms by which cellular processes are carried out by the various biological macromolecules in the cell, with a particular emphasis on the structure and function of genes and genomes. Molecular biology as a field, however, was originally born from the development of tools and methods that allow the direct manipulation of DNA both *in vitro* and *in vivo* in numerous organisms.

Two essential items in the molecular biologist's toolkit are **restriction endonucleases**, which allow DNA to be cut into precise pieces, and **cloning vectors**, such as plasmids or phages used to "carry" inserted foreign DNA fragments for the purposes of producing more material or a protein product. The term *genetic engineering* was originally used to describe the range of manipulations of DNA that become possible with the ability to clone a gene by placing its DNA into another context in which it could be propagated. From this beginning, when recombinant DNA was used as a tool to analyze gene structure and expression, we moved to the ability to change the DNA content of bacteria and eukaryotic cells by directly introducing cloned DNA that could become part of the genome. Then, by changing the genetic content in conjunction with the ability to develop an animal from an embryonic cell, it became possible to generate multicellular eukaryotes with deletions or additions of specific genes that are inherited via the germline. We now use genetic engineering to describe a range of activities including the manipulation of DNA, the introduction of changes into specific somatic cells within an animal or plant, and even changes in the germline itself.

As research has advanced, more and more sensitive methods for detecting and amplifying DNA have been developed. Now that we have entered the era of routine whole-genome

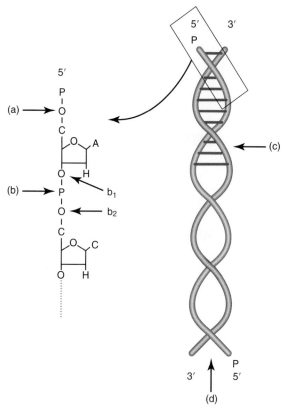

FIGURE 3.1 The target of a phosphatase is shown in (a), a terminal phosphomonoester bond. The target of a nuclease is shown in (b), the phosphodiester bond between two adjacent nucleotides. Note that the nuclease can cleave either the first ester bond from the 3′ end of the terminal nucleotide (b1) or the second ester bond from the 5′ end of the next nucleotide (b2). Nucleases can cleave internal bonds (c) as an endonuclease, or start from an end and progress into the fragment (d) as an exonuclease.

sequencing, methods to assess the content, function and expression of entire genomes have become commonplace. This chapter will discuss some of the most common methods used in molecular biology, ranging from the very first tools developed by molecular biologists, to some of the most recently developed methods now in use.

3.2 Nucleases

Key concepts

- Nucleases hydrolyze an ester bond within a phosphodiester bond.
- Phosphatases hydrolyze the ester bond in a phosphomonoester bond.
- Nucleases have a multiplicity of specificities.
- Restriction endonucleases can be used to cleave DNA into defined fragments.
- A map can be generated by using the overlaps between the fragments generated by different restriction enzymes.

Nucleases are one of the most valuable tools in a molecular biology laboratory. One class of enzymes, the restriction endonucleases that we will discuss below, was critical for the cloning revolution. **Nucleases** are enzymes that degrade nucleic acids, the opposite function of polymerases. They hydrolyze, or break an ester bond in a phosphodiester linkage between adjacent nucleotides in a polynucleotide chain as shown in **FIGURE 3.1**.

There is another, related class of enzymes that can hydrolyze an ester bond in a nucleotide chain (a monoesterase, usually called a **phosphatase**). The critical difference between a phosphatase and a nuclease is shown in Figure 3.1. A phosphatase can only hydrolyze a terminal ester bond linking a phosphate (or di- or tri-phosphate) to a terminal nucleotide at the 3′ or 5′ end, while a nuclease can hydrolyze an internal ester bond in a diester link, between adjacent bases.

Phosphatases are important enzymes in the laboratory because they allow the removal of a terminal phosphate from a polynucleotide chain. This is often required for a subsequent step of connecting or **ligating** chains together. This also allows one to replace the phosphate with a radioactive ^{32}P molecule.

We can divide nucleases into different groups based on a number of different features. First, we can distinguish between **endonucleases** and **exonucleases** as shown in Figure 3.1. An endonuclease can hydrolyze internal bonds within a polynucleotide chain, whereas an exonuclease must start at the end of a chain and hydrolyze from that end position.

The specificity of nucleases ranges from none to extreme. Nucleases may be specific for DNA, as DNases, or RNA, as RNases, or even be specific for a DNA/RNA hybrid, as RNaseH (which cleaves the RNA strand of a hybrid duplex). Nucleases may be specific for either single-strand nucleotide chains, duplex chains, or both.

When a nuclease, either endo- or exo-, hydrolyzes an ester bond in a phosphodiester linkage, it will have specificity for either of the two ester bonds, generating either 5′ nucleotides or 3′ nucleotides, as seen in Figure 3.1. An exonuclease may attack a polynucleotide chain from either the 5′ end and hydrolyze 5′ to 3′ or attack from the 3′ end and hydrolyze 3′ to 5′, as shown in Figure 3.1.

Nucleases may have a sequence preference, such as pancreatic RNase A, which preferentially cuts after a pyrimidine, or T1 RNase, which cuts single-stranded RNA chains after a

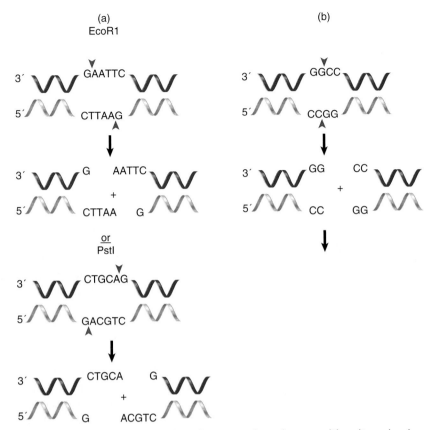

(a)
EcoR1

or
PstI

(b)

FIGURE 3.2 (a) A restriction endonuclease may cleave its recognition site and make a staggered cut, leaving a 5′ overhang or a 3′ overhang. (b) A restriction endonuclease may cleave its recognition site and make a blunt end cut.

G. At the extreme end of sequence specificity lie the **restriction endonucleases**, usually called *restriction enzymes*. These are endonucleases from eubacteria and archaea that recognize a specific DNA sequence. Their name typically derives from the bacteria in which they were discovered. For example, EcoR1 is the first restriction enzyme from an *E. coli R* strain.

Broadly speaking, there are three different classes of restriction enzymes and several subclasses. In 1978, the Nobel Prize in Medicine was awarded to Daniel Nathans, Werner Arber, and Hamilton Smith for the discovery of restriction endonucleases. It was this discovery that enabled scientists to develop the methods to clone DNA, as we will see in the next section. Thousands of restriction enzymes are known, many of which are now commercially available. Restriction enzymes have to do two things: (1) recognize a specific sequence, and (2) cut, or restrict, that sequence.

The type II restriction enzymes (with several subgroups) are the most common. Type II enzymes are distinguished because the recognition site and cleavage site are the same. These sites range in length from 4 to 8 bp. The sites are typically *inversely palindromic*, that is, reading the same forward and backward on complementary strands as shown in **FIGURE 3.2**. Restriction enzymes can cut the DNA in two different ways as shown in Figure 3.2. The first and more common is a staggered cut, which leaves single-stranded overhangs, or "sticky ends." The overhang may be a 3′ or a 5′ overhang. The second way is a blunt double stranded cut, which does not leave an overhang. An additional level of specificity determines whether or not the enzyme will cut DNA containing a methylated base. The degree of specificity in the site also varies. Most enzymes are very specific, while some will allow multiple bases at one or two positions within the site.

Restriction enzymes from different bacteria may have the same recognition site but cut the DNA differently. One may make a blunt cut and the other may make a staggered cut, or one may leave a 3′ overhang while the second may

A B   A B   A

1000 200   1900   600 800 500

FIGURE 3.3 A restriction map is a linear sequence of sites separated by defined distances on DNA. The map identifies the three sites cleaved by enzyme A and the two sites cleaved by enzyme B. Thus A produces four fragments, which overlap those of B, and B produces three fragments, which overlap those of A.

leave a 5′ overhang. These different enzymes are called *isoschizomers*.

Types I and III enzymes differ from type II enzymes in that the recognition site and cleavage site are different and are usually not palindromes. With a type I enzyme, the cleavage site can be up to 1000 bp away from the recognition site. Type III enzymes have closer cleavage sites, usually 20 to 30 bp away.

A *restriction map* represents a linear sequence of the sites at which particular restriction enzymes find their targets. When a DNA molecule is cut with a suitable restriction enzyme, it is cleaved into distinct negatively charged fragments. These fragments can be separated on the basis of their size by gel electrophoresis (described later, in *Section 3.6, DNA Separation Techniques*; see Figure 3.14). By analyzing the restriction fragments of DNA, we can generate a map of the original molecule in the form shown in **FIGURE 3.3**. The map shows the positions at which particular restriction enzymes cut DNA. *So the DNA is divided into a series of regions of defined lengths that lie between sites recognized by the restriction enzymes.* A restriction map can be obtained for any sequence of DNA, irrespective of whether we have any knowledge of its function. If the sequence of the DNA is known, a restriction map can be generated *in silico* by simply searching for the recognition sites of known enzymes. Knowing the restriction map of a DNA sequence of interest is extremely valuable in DNA cloning, which is described in the next section.

3.3 Cloning

Key concepts

- Cloning a fragment of DNA requires a specially engineered vector.
- Blue/white selection allows the identification of bacteria that contain the vector plasmid and vector plasmids that contain an insert.

Cloning has a very simple definition: to **clone** is to make identical copies, whether it is done

by a copy machine for a piece of paper, cloning Dolly the sheep, or cloning DNA, which is what we will discuss here. Cloning can also be considered an amplification process, in which we currently have one copy and we want many identical copies. Cloning DNA typically involves **recombinant DNA**. This also has a very simple definition: a DNA molecule from two (or more) different sources.

In order to clone a fragment of DNA, a recombinant DNA molecule must be created and copied many times. There are two different DNAs needed: a **vector**, or cloning vehicle, and an **insert**, or the molecule to be cloned. The two most popular classes of vectors are derived from plasmids and viruses, respectively.

Over the years, vectors have been specifically engineered for safety, selection ability, and high growth rate. "Safety" means that the vector will not integrate into a genome (unless engineered specifically for that purpose) and the recombinant vector will not autotransfer to another cell. (We will discuss selection below.) In general, about a microgram of vector DNA will be ligated with about a microgram of the insert DNA that we wish to clone. Both the vector and insert should be restricted with the same restriction endonuclease to create compatible DNA ends. Let us now examine the details and the variables that will affect the process.

We will start with the insert, the DNA fragment that you want to amplify. The insert could come from one of many different sources, such as restricted genomic DNA, either size selected on an agarose gel, or unselected; a larger fragment from another clone to be **subcloned** (meaning taking a smaller part of the larger fragment); a PCR fragment (see *Section 3.8, PCR and RT-PCR*); or even a DNA fragment synthesized *in vitro*. The size and the nature of the fragment ends must be known. Are the ends blunt or do they have overhanging single strands (recall *Section 3.2, Nucleases*), and if so, what are their sequences? The answer to this question comes from how the fragments were created (what restriction enzyme(s) were used to cut the DNA, or what PCR primers were used to amplify the DNA).

The vector is selected based on the answers to these questions. For this exercise, we will use a common type of plasmid cloning vector called a *blue/white selection vector*, as shown in **FIGURE 3.4**. This vector has been constructed with a number of important elements. It has an *ori*, or origin of replication (see Chapter 14, *DNA Replication*), to allow plasmid replication, which will provide

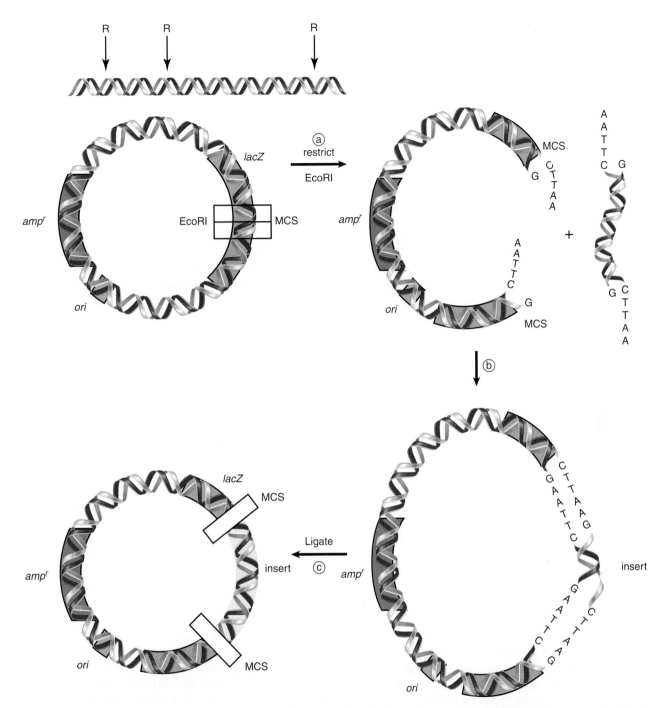

FIGURE 3.4 (a) A plasmid that contains three key sites (an origin of replication, *ori*; a gene for ampicillin resistance, *amp^r*; and *lacZ* with an MCS), together with the insert DNA to be cloned, is restricted with EcoR1. (b) Restricted insert fragments and vector will be combined, and (c) ligated together. The final pool of this DNA will be transformed into *E. coli*.

the actual amplification step, in a bacterial cell. It contains a gene that codes for resistance to the antibiotic ampicillin, *amp^r*, which will allow selection of bacteria that contain the vector. It also contains the *E. coli lacZ* gene (see Chapter 26, *The Operon*), which will allow selection of an insert DNA fragment in the vector.

The *lacZ* gene has been engineered to contain an **MCS**, or **multiple cloning site**. This

is an oligonucleotide sequence with a series of different restriction endonuclease recognition sites arranged in tandem in the same reading frame as the *lacZ* gene itself. This is the heart of blue/white selection. The *lacZ* gene codes for the β-galactosidase (β-gal) enzyme, which cleaves the galactoside bond in lactose. It will also cleave the galactoside bond in an artificial substrate called X-gal (5-bromo-4-chloro-3-

indolyl-beta-D-galactopyranoside), which can be added to bacterial growth media and has a blue color when cleaved by the intact enzyme. *If a fragment of DNA is cloned (inserted) into the MCS, the lacZ gene will be disrupted, inactivating it, and the resulting β-gal will no longer be able to cleave X-gal, resulting in white bacterial colonies rather than blue colonies.* This is the blue/white selection mechanism.

Let us now begin the cloning experiment. Following along in Figure 3.4, both the vector and the insert are cut with the same restriction enzyme in order to generate compatible single-stranded sticky ends. The variables here are the ability to select different enzymes that recognize different restriction sites as long as they generate the same overhang sequence. An enzyme that makes a blunt cut can also be used, although that will make the next step, ligation, less efficient. Two completely different ends with different overhangs can also be used if an exonuclease is used to trim the ends and produce blunt ends. (Continuing with the same reasoning, randomly sheared DNA can also be used if the ends are then blunted for ligation.) If we are forced to use a type I or type III restriction enzyme, the ends must also be blunted. An important alternative is to use two different restriction enzymes that leave different overhangs on each end. The advantages to this are that neither the vector nor the insert will self-circularize, and the orientation of how the insert goes into the vector can be controlled; this is called *directional cloning*. We will select the vector that has the appropriate restriction endonuclease sites.

The next step is to combine the two pools of DNA fragments, vector and insert, in order to connect or ligate them. A 5- or 10-to-1 molar ratio of insert to vector is usually used. Too much vector and vector–vector dimers will be produced. Too much insert and multiple inserts per vector will be produced. The size of the insert is important; too large (over ~10 kb) an insert will not be efficiently cloned in a plasmid vector, which will necessitate using an alternative virus-based vector. Ligation is often performed overnight on ice to slow the ligation reaction down and generate fewer multimers.

The pool of randomly generated ligated DNA molecules is now used to "transform" *E. coli*. **Transformation** is the process by which DNA is introduced into a host cell. *E. coli* does not normally undergo physiological transformation. As a result, DNA must be forced into the cell. There are two common methods of transformation: washing the bacteria in a high

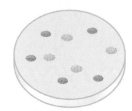

FIGURE 3.5 After transformation into *E. coli* of restricted and ligated vector plus insert DNA, the bacterial cells are plated onto agar plates containing ampicillin, IPTG, and the color indicator, X-gal. Overnight incubation at 37°C will yield both blue and white colonies. The white colonies will be used to prepare DNA for further analysis.

salt wash of $CaCl_2$, or *electroporation*, in which an electric current is applied. Both methods create small pores or holes in the cell wall. Even with these methods, only a tiny fraction of bacterial cells will be transformed. The strain of *E. coli* is important. It should not have a restriction system or a modification system to methylate the incoming DNA. The strain should also be compatible with the blue/white system, which means that it should contain the α-complementing fragment of LacZ (the *lacZ* gene contained in most plasmids does not function without this fragment). DH5α is a commonly used strain.

Transformation results in a pool of multiple types of bacteria, most of which are not wanted because they either contain vector with no insert or have not taken up any DNA at all. We must select the handful of bacteria that contain recombinant plasmid from the millions that do not. The transformed bacterial cells are plated on an agar plate containing both the antibiotic ampicillin and an artificial β-gal inducer called IPTG (Isopropyl Thiogalactoside). The ampicillin in the plate will kill the vast majority of bacterial cells, namely all of those that have not been transformed with the *amp^r* plasmid. The remaining bacteria can now grow and form visible colonies. As shown in **FIGURE 3.5**, two different types of colonies—blue ones that contain a vector without an insert (because β-gal cleaved X-gal into a blue compound) and white ones, for which the inactivated β-gal did not cleave X-gal and so remained colorless, are seen.

This is not quite the end of the story. False positive clones, such as those that were formed as vector-only dimers, must be identified and removed. In order to do so, plasmid DNA must be at least partly purified from each candidate colony, and restricted and run on a gel to check for the insert size. Sequencing the fragment to be absolutely certain a random contaminant has not been cloned is also suggested. Sequencing is described in *Section 3.7, DNA Sequencing*.

3.4 Cloning Vectors Can Be Specialized for Different Purposes

Key concepts

- Cloning vectors may be bacterial plasmids, phages, cosmids, or yeast artificial chromosomes.
- Shuttle vectors can be propagated in more than one type of host cell.
- Expression vectors contain promoters that allow transcription of any cloned gene.
- Reporter genes can be used to measure promoter activity or tissue-specific expression.
- Numerous methods exist to introduce DNA into different target cells.

Vector	Features	Isolation of DNA	DNA limit
Plasmid	High copy number	Physical	10 kb
Phage	Infects bacteria	Via phage packaging	20 kb
Cosmid	High copy number	Via phage packaging	48 kb
BAC	Based on F plasmid	Physical	300 kb
YAC	Origin + centromere + telomere	Physical	>1 Mb

FIGURE 3.6 Cloning vectors may be based on plasmids or phages or may mimic eukaryotic chromosomes.

In the example in the previous section, we described the use of a vector that is designed simply for amplifying insert DNA, with inserts up to ~10 kb. It is often desirable to clone larger inserts, though, and sometimes the goal is not just to amplify the DNA, but also to express cloned genes in cells, investigate properties of a promoter, or create various fusion proteins (defined below). FIGURE 3.6 summarizes the properties of the most common classes of cloning vectors. These include vectors based on bacteriophage genomes, which can be used in bacteria but have the disadvantage that only a limited amount of DNA can be packaged into the viral coat (although more than can be carried in a plasmid). The advantages of plasmids and phages are combined in the **cosmid**, which propagates like a plasmid but uses the packaging mechanism of phage lambda to deliver the DNA to the bacterial cells. Cosmids can carry inserts of up to 47 kb (the maximum length of DNA that can be packaged into the phage head).

The vector used for cloning the largest possible DNA inserts is the **yeast artificial chromosome (YAC)**. A YAC has a yeast origin to support replication, a centromere to ensure proper segregation, and telomeres to afford stability. In effect, it is propagated just like a yeast chromosome. YACs have the largest capacity of any cloning vector, and can propagate with inserts measured in the Mb length range.

An extremely useful class of vectors known as **shuttle vectors** can be used in more than one species of host cell. The example shown in FIGURE 3.7 contains origins of replication and selectable markers for both *E. coli* and the yeast *S. cerevisiae*. It can replicate as a circular multicopy plasmid in *E. coli*. It has a yeast centromere, and also has yeast telomeres adjacent to *Bam*HI restriction sites, so that cleavage with *Bam*HI generates a YAC that can be propagated in yeast.

Other vectors, such as **expression vectors**, may contain promoters to drive expression of genes. Any open reading frame can be inserted into the vector and expressed without further modification. These promoters can be continuously active, or may be *inducible* so that they are only expressed under specific conditions.

Alternatively, the goal may be to study the function of a cloned promoter of interest in order to understand the normal regulation of a gene. In this case, rather than using the actual gene, we can use an easily detected **reporter gene** under control of the promoter of interest.

The type of reporter gene that is most appropriate depends on whether we are interested in quantitating the efficiency of the promoter (and, for example, determining the effects of mutations in it or the activities of transcription factors that bind to it), or determining its tissue-specific pattern of expression. FIGURE 3.8 summarizes a common system for assaying promoter activity. A cloning vector is created that has a eukaryotic promoter linked to the coding region of *luciferase*, a gene that encodes the enzyme responsible for bioluminescence in the firefly. In general, a transcription termination signal is added to ensure the proper generation of the mRNA. The hybrid vector is introduced into target cells, and the cells are grown and subjected to any appropriate experimental treatments. The level of luciferase activity is measured by addition of its substrate luciferin. Luciferase activity results in light emission that can be measured at 562 nanometers (nm), and is directly proportional to the amount of enzyme that was made, which in turn depends upon the activity of the promoter.

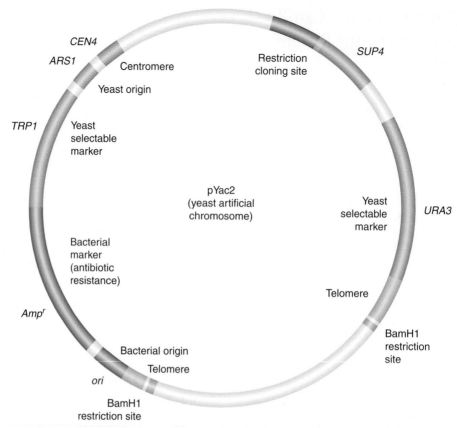

FIGURE 3.7 pYac2 is a cloning vector with features to allow replication and selection in both bacteria and yeast. Bacterial features (described in blue) include an origin of replication and antibiotic resistance gene. Yeast features (described in red and yellow) include an origin, centromere, two selectable markers, and telomeres.

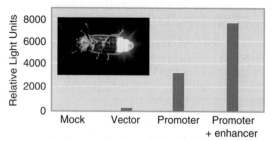

FIGURE 3.8 Luciferase (derived from fireflies such as the one shown here) is a popular reporter gene. The graph shows the results from mammalian cells transfected with a luciferase vector driven by a minimal promoter or the promoter plus a putative enhancer. The levels of luciferase activity correlate with the activities of the promoters. Photo © Cathy Keifer/Dreamstime.com.

Some very striking reporters are now available for visualizing gene expression. The *lacZ* gene, described in the blue-white selection strategy above, also serves as a very useful reporter gene. **FIGURE 3.9** shows what happens when the *lacZ* gene is placed under the control of a promoter that regulates the expression of a gene in the mouse nervous system. The tissues in which this promoter is normally active can

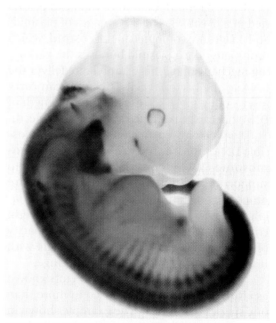

FIGURE 3.9 Expression of a *lacZ* gene can be followed in the mouse by staining for β-galactosidase (in blue). In this example, *lacZ* was expressed under the control of a promoter of a mouse gene that is expressed in the nervous system. The corresponding tissues can be visualized by blue staining. Photo courtesy of Robb Krumlauf, Stowers Institute for Medical Research.

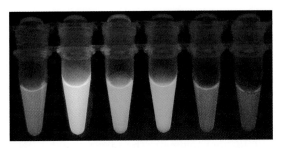

(a)

(b)

FIGURE 3.10 (a) Since the discovery of GFP, derivatives that fluoresce in different colors have been engineered. Photo courtesy of Joachim Goedhart, Molecular Cytology, SILS, University of Amsterdam. (b) A live transgenic mouse expressing human rhodopsin (a protein expressed in the retina of the eye) fused to GFP. Reprinted from *Vision Res.*, vol. 45, T. G. Wensel, et al., Rhodopsin-EGFP knock-ins . . . , pp. 3445–3453. Copyright 2005, with permission from Elsevier [http://www.sciencedirect .com/science/journal/00426989]. Photo courtesy of Theodore G. Wensel, Baylor College of Medicine.

A viral vector introduces DNA by infection

Liposomes may fuse with the membrane

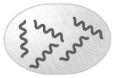

Microinjection introduces DNA directly into the cytoplasm or nucleus

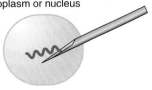

Nanospheres can be shot into the cell by a gene gun

FIGURE 3.11 DNA can be released into target cells by methods that pass it across the membrane naturally, such as by means of a viral vector (in the same way as a viral infection) or by encapsulating it in a liposome (which fuses with the membrane). Alternatively, it can be passed manually, by microinjection, or by coating it on the exterior of nanoparticles that are shot into the cell by a gene gun that punctures the membrane at very high velocity.

be visualized by providing the X-gal substrate to stain the embryo.

One of the most popular reporters that can be used to visualize patterns of gene expression is GFP (green fluorescent protein), which is obtained from jellyfish. GFP is a naturally fluorescent protein that, when excited with one wavelength of light, emits fluorescence in another wavelength. In addition to the original GFP, numerous variants that fluoresce in different colors, such as yellow (YFP), cyan (CFP), and blue (BFP), have been developed. GFP and its variants can be used as reporter genes on their own, or they can be used to generate *fusion proteins* in which a protein of interest is fused to GFP and can thus be visualized in living tissues, as is shown in the example in **FIGURE 3.10**.

Vectors are introduced into different species in a variety of different ways. Bacteria and simple eukaryotes like yeast can be transformed easily, using chemical treatments that permeabilize the cell membranes, as discussed above in *Section 3.3, Cloning*. Many types of cells cannot be

transformed so easily, though, and other methods must be used, as summarized in **FIGURE 3.11**. Some types of cloning vectors use natural methods of infection to pass the DNA into the cell, such as a viral vector that uses the viral infective process to enter the cell. *Liposomes* are small spheres made from artificial membranes, which can contain DNA or other biological materials. Liposomes can fuse with plasma membranes and release their contents into the cell. *Microinjection* uses a very fine needle to puncture the cell membrane. A solution containing DNA can be introduced into the cytoplasm, or directly into the nucleus in the case where the nucleus is large enough to be chosen as a target (such as an egg). The thick cell walls of plants are an impediment to many transfer methods, and the "gene gun" was invented as a means for overcoming this obstacle. A gene gun shoots very small particles into the cell by propelling them through the wall at high velocity. The particles can consist of gold or nanospheres coated with DNA. This method now has been adapted for use with a variety of species, including mammalian cells.

3.5 Nucleic Acid Detection

Key concept

- Hybridization of a labeled nucleic acid to complementary sequences can identify specific nucleic acids.

There are a number of different ways to detect DNA and RNA. The classical method relies on ability of nucleic acids to absorb light at 260 nanometers. The amount of light absorbed is proportional to the amount of nucleic acid present. There is a slight difference in the amount of absorption by single-stranded as compared to double-stranded nucleic acids, but not DNA versus RNA. Protein contamination can affect the outcome, but because proteins absorb maximally at 280 nm, tables have been published of 260/280 ratios that allow quantitation of the amount of nucleic acid present.

DNA and RNA can be nonspecifically stained with ethidium bromide (EtBr) to make visualization more sensitive. EtBr is an organic tricyclic compound that binds strongly to double-stranded DNA (and RNA) by intercalating into the double helix between the stacked base pairs. It binds to DNA, and as a result it is a strong mutagen and care must be taken when using it. EtBr fluoresces when exposed to UV light, which increases the sensitivity. SYBR green is a safer alternate DNA stain.

We will focus on the detection of *specific* sequences of nucleic acids. The ability to identify a specific sequence relies on hybridization of a **probe** with a known sequence to a target. The probe will detect and bind to a sequence to which it is **complementary**. The percent of match does not have to be perfect, but as the match percentage decreases, the stability of the nucleic acid hybrid decreases. G-C base pairs are more stable than A-T base pairs so that base composition (usually referred to as % G-C) is an important variable. The second set of variables that affects hybrid stability is extrinsic; it includes the buffer conditions (concentration and composition) and the temperature at which hybridization occurs. This is called the **stringency**, under which the hybridization is carried out.

The probe functions as a single-stranded molecule (if it is double stranded, it must be melted). The target may be single stranded or double stranded. If the target is double stranded, it also must be melted to single strands to begin the hybridization process. The reaction can take place in solution (for example, during sequencing or PCR; see *Sections 3.7, DNA Sequencing* and *3.8, PCR and RT-PCR*), or can be performed when the target has been bound to a membrane support such as a nitrocellulose filter (see *Section 3.9, Blotting Methods*). The target may be DNA (called a Southern blot) or RNA (called a Northern blot); the probe is usually DNA.

For this exercise, let's use a Southern blot from an experiment in which we have restricted a large DNA fragment into smaller fragments and subcloned the individual fragments (see *Section 3.2, Cloning*). Starting with the clones on the plate from Figure 3.5, we will isolate plasmid DNA from each white clone and restrict the DNA with the same restriction enzymes that we used to clone the fragments. The DNA fragments will be separated on an agarose gel and blotted onto nitrocellulose (see *Section 3.6, DNA Separation Techniques*).

In order to increase the sensitivity from the optical range, the probe must be labeled. We will begin with radiolabeling and then describe alternate labeling without radioactivity. For most reactions, ^{32}P is used, but ^{33}P (with a longer half-life but less penetrating ability) and ^3H (for special purposes described below) are also used. Probes can be radiolabeled in several different ways. One is *end labeling*, in which a strand of DNA (which has no 5' phosphate) is labeled using a kinase and ^{32}P. Alternatively, a probe can be generated by *nick-translation* or *random priming* with ^{32}P using the Klenow DNA polymerase fragment and labeled nucleotides (see *Section 14.4, DNA Polymerases Have Various Nuclease Activities*) or during a PCR reaction (see *Section 3.8, PCR and RT-PCR*).

In performing nucleic acid hybridization studies, standard procedures are typically used that allow hybridization over a large range of G-C content. Hybridization experiments are performed in a standardized buffer called SSC (standard sodium citrate), which is usually prepared as a 20× concentrated stock solution. Hybridization is typically carried out within a standard temperature range of 45°C to 65°C, depending upon the required stringency.

The actual hybridization between a labeled probe and a target DNA bound to a membrane usually takes place in a closed (or sealed) container in a buffer that contains a set of molecules to reduce background hybridization of probe to the filter. Hybridization experiments typically are performed overnight to ensure maximum probe-to-target hybridization. The hybridization reaction is stochastic and depends upon the abundance of each different sequence. The more copies of a sequence, the greater the chance of a given probe molecule encountering its complementary sequence.

The next step is to wash the filter to remove all of the probe that is not specifically bound to a complementary sequence of nucleic acid. Depending on the type of experiment, the stringency of the wash is usually set quite high to avoid spurious results. Higher stringency conditions include higher temperature (closer to the melting temperature of the probe) and lower concentration of cations. (Lower salt concentrations result in less shielding of the negative phosphate groups of the DNA backbone, which in turn inhibits strand annealing.) In some experiments, however, where one is looking specifically for hybridization to targets with a lower percent match (such as finding a copy of species X DNA using a probe from species Y), hybridization would be performed at lower stringency.

The last step is the identification of which target DNA band on the gel (and thus the filter) has been bound by the radiolabeled probe. The washed nitrocellulose filter is subjected to **autoradiography**. The dried filter will be placed against a sheet of X-ray film. To amplify the radioactive signal, intensifying screens can be used. These are special screens placed on either side of the filter/film pair that act to bounce the radiation back through the film. Alternatively, a *phosphorimaging* screen (a solid-state liquid scintillation device) can be used. This is more sensitive and faster than X-ray film, but results in somewhat lower resolution. The length of time for autoradiography is empirical. An estimate of the total radioactivity can be made with a handheld radiation monitor. Sample results are seen in FIGURE 3.12. One band on the filter has blackened the X-ray film. The film can be aligned to the filter to determine which band corresponds to the probe.

A simple modification of the autoradiography procedure called *in situ* **hybridization** allows one to peer into a cell and determine the location, at a microscopic level, of specific nucleic acid sequences. We simply modify a few steps in the above process to perform the hybridization between our probe, usually labeled with ³H, and complementary nucleic acids in an intact cell or tissue. The goal is to determine exactly where the target is located. The cell or tissue slice is mounted on a microscope slide. Following hybridization, a photographic emulsion instead of film is applied to the slide, covering it. The emulsion, when developed, is transpar-

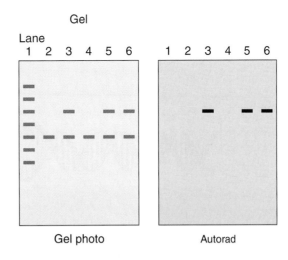

FIGURE 3.12 An autoradiogram of a gel prepared from the colonies described in Figure 3.5. The gel was blotted onto nitrocellulose and probed with a radioactive gene fragment. Lane 1 contains a set of standard DNA size markers. Lane 2 is the original vector cleaved with EcoR1. Lanes 3 to 6 each contain plasmid DNA from one of the white clones from Figure 3.4 that was restricted with EcoR1. The inset shows a photograph of the gel; the radioactive bands are marked with an asterisk.

ent to visible light so that it is possible to see the exact location in the cell where the grains in the emulsion blackened by the radioactivity are located. Development time can be weeks to months because ³H has less energetic radiation and its longer half-life results in lower activity.

There are nonradioactive alternatives to the procedures described above that use either colorimetric or fluorescence labeling. Digoxygenin-labeled probe is a commonly used colorimetric procedure. Probe bound to target is localized with an anti-digoxygenin antibody coupled to alkaline phosphatase to develop color. The advantage is the time required to see the results. It is typically a single day, but sensitivity is usually less than with radioactivity. FISH, or Fluorescence *in situ* hybridization, is another very common nonradioactive procedure that uses a fluorescently labeled probe. This method is illustrated in FIGURE 3.13. Multiple fluorophores in different colors are available—about a dozen now—but ratios of different probe colors combinations can be used to create additional colors.

These procedures are more picturesque but less quantitative than traditional scintillation counting. At best, these procedures can be called semiquantitative. It is possible to use an optical scanner to quantitate the amount of signal produced on film, but care must be taken to ensure the time of exposure during the experiment is within a linear range.

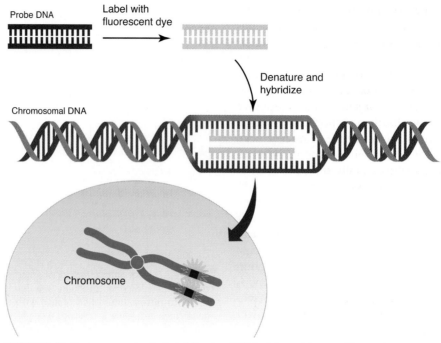

FIGURE 3.13 Fluorescence *in situ* hybridization (FISH). Adapted from an illustration by Darryl Leja, National Human Genome Research Institute (www.genome.gov).

3.6 DNA Separation Techniques

Key concepts

- Gel electrophoresis separates DNA fragments by size, using an electric current to cause the DNA to migrate toward a positive charge.
- DNA can also be isolated using density gradient centrifugation.

With a few exceptions, the individual pieces of DNA (chromosomes) making up a living organism's genome are on the order of megabases in length, making them too physically large to be manipulated easily in the laboratory. Individual genes or chromosomal regions of interest by contrast are often quite small and readily manageable, on the order of hundreds or a few thousand base pairs in length. A necessary first step, therefore, in many experimental processes investigating a specific gene or region, is to break the large original chromosomal DNA molecule down into smaller manageable pieces and then begin isolation and selection of the particular relevant fragment or fragments of interest. This breakage can be done by mechanical shearing of chromosomes, in a process that produces breakages randomly to produce a uniform size distribution of assorted molecules. This approach is useful if a random-ness in breakpoints is required, such as to create a library of short DNA molecules that "tile" or partially overlap each other while together representing a much larger genomic region, such as an entire chromosome or genome. Alternatively, restriction endonucleases (see *Section 3.2, Nucleases*) may be employed to cut large DNA molecules into defined shorter segments in a way that is reproducible. This reproducibility is frequently useful, in that a DNA section of interest can be identified in part by its size. Consider a hypothetical gene *genX* on a bacterial chromosome, with the entire gene lying between two EcoRI sites spaced 2.3 kb apart. Digestion of the bacterial DNA with EcoRI will yield a range of small DNA molecules, but *genX* will always occur on the same 2.3 kb fragment. Depending on the size and complexity of the starting genome, there may be several other DNA segments of similar size produced, or in a simple enough system, this 2.3 kb size may be unique to the *genX* fragment. In this latter case, detection or visualization of a 2.3 kb fragment is enough to definitively identify the presence of *genX*. Many of the earliest laboratory techniques developed in working with DNA relate to separating and concentrating DNA molecules based on size expressly to take advantage of these concepts. An ability to separate DNA molecules based on size allows for taking a complex mixture of many fragment sizes and selecting a much smaller, less complex subset of interest for further study.

The simplest method for separation and visualization of DNA molecules based on size is gel electrophoresis. In neutral agarose gel electrophoresis, the most basic type of gel, this is done by preparing a small slab of gel in an electrically conductive, mildly basic buffer. While similar to the gelatins used to make dessert dishes, this type of gel is made from agarose, a polysaccharide that is derived from seaweed and has very uniform molecular sizes. Preparation of agarose gels of a specific percentage of agarose by mass (usually in the range of 0.8% to 3%) creates, in effect, a molecular sieve, with a "mesh" pore size being determined by the percentage of agarose (higher percentages yielding smaller pores). The gel is poured in a molten state into a rectangular container, with discrete wells being formed near one end of the product. After cooling and solidifying, the slab is submerged in the same conductive, mildly alkaline, buffer, and samples of mixed DNA fragments are placed in the preformed wells. A DC electric current is then applied to the gel, with the positive charge being at the opposite end of the gel from the wells. The alkalinity of the solution ensures that the DNA molecules have a uniform negative charge from their backbone phosphates, and the DNA fragments begin to be drawn electrostatically toward the positive electrode. Shorter DNA fragments are able to move through the agarose pores with less resistance than longer fragments, and so over time the smallest DNA molecules move the furthest from the wells and the largest move the least. All fragments of a given size will move at about the same rate, effectively concentrating any population of equal-sized molecules into a discrete band at the same distance from the well. Addition of a DNA-binding fluorescent dye, such as ethidium bromide or SYBR green, to the gel stains these DNA bands such that they can be directly seen by eye when the gel is exposed to fluorescence-exciting light. In practice, a standard sample consisting of a set of DNA molecules of a known size is run in one of the wells, with sizes of bands in other wells estimated in comparison to the standard, as shown in FIGURE 3.14. DNA molecules of roughly 50 to 10,000 base pairs can be quickly separated, identified, and sized to within about 10% accuracy by this simple method, which remains a common laboratory technique. DNAs can be separated not only by size, but also by shape. Supercoiled DNA, which is compact compared

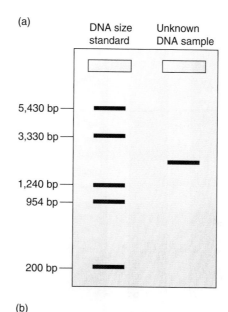

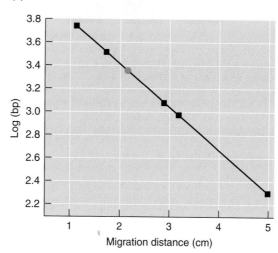

FIGURE 3.14 DNA sizes can be determined by gel electrophoresis. A DNA size standard and a DNA of unknown size are run in two lanes of a gel, depicted schematically. The migration of the DNAs of known size in the standard is graphed to create a standard curve (migration distance in cm vs. log bp). The point shown in green is for the DNA of unknown size. Adapted from an illustration by Michael Blaber, Florida State University.

to relaxed or linear DNA, migrates more rapidly on a gel, and the more supercoiling, the faster the migration, as seen in FIGURE 3.15.

Variations on this method primarily relate to changing the gel matrix from agarose to other molecules such as synthetic polyacrylamides, which can have even more precisely controlled pore sizes. These can offer finer size resolution of DNA molecules from roughly 10 to 1500 base pairs in size. Both resolution and sensitivity are further improved by making these types of gels as thin as possible, normally requiring they be formed between glass plates for mechanical strength. When chemical denaturants such as

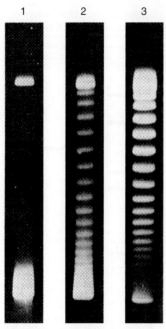

FIGURE 3.15 Supercoiled DNAs separated by agarose gel electrophoresis. Lane 1 contains untreated negatively supercoiled DNA (lower band). Lanes 2 and 3 contain the same DNA that was treated with a type 1 topoisomerase for 5 and 30 minutes, respectively. The topoisomerase makes a single strand break in the DNA and relaxes negative supercoils in single steps (one supercoil relaxed per strand broken and reformed). Reproduced from W. Keller, *Proc. Natl. Acad. Sci. USA* 72 (1975): 2550–2554. Photo courtesy of Walter Keller, University of Basel.

urea are added to the buffer system, the DNA molecules are forced to unfold (losing any secondary structures) and take on hydrodynamic properties related only to molecule length. This approach can clearly resolve DNA molecules differing in length by only a single nucleotide. Denaturing polyacrylamide electrophoresis is a key component of common DNA sequencing techniques whereby the separation and detection of a series of single nucleotide length difference DNA products allows for the reading of the underlying order of nucleotide bases (see *Section 3.7, DNA Sequencing*).

The next level of refinement to this technique is to place the gel matrix in a very fine capillary, which can be even thinner than a glass plate-supported gel and thus still further improve on sensitivity and resolution capacity. Unlike a glass-supported slab gel where multiple lanes can be run side by side, a capillary can only handle one sample at a time; however, a capillary can be run clean of sample and reused, making it ideal for system automation and high throughput applications. Instruments with multiple parallel capillaries allow for par-

allel analysis of multiple samples to further increase throughput. Technologies of this form mark the apparent apex of chain termination-based sequencing methods.

Further miniaturization of capillaries onto the surfaces of inert "chips" with etched-in microfluidic reservoirs, valves, pumps, and mixing chambers can be employed to create entire "lab-on-a-chip" disposable nucleic acid sample analysis cartridges. These cartridges can process, separate, perform size analysis, and quantitate DNA or RNA in a small input sample. Frequently, these devices are controlled and have data output processed by a computer, which in turn will manipulate the data output in order to present it as a traditional stained agarose or polyacrylamide gel—in effect bringing the technology full circle.

Another method for separating DNA molecules from other contaminating biomolecules, or in some cases for fractionation of specific small DNA molecules from other DNAs, is through the use of gradients, as depicted in **FIGURE 3.16**. The most frequent implementation of this is *isopycnic banding*, which is based on the fact that specific DNA molecules have unique densities based on their G-C content. Under the influence of extreme g-forces, such as through ultracentrifugation, a high concentration solution of a salt (such as cesium chloride) will form a stable density gradient from low density (near top of tube/ center of rotor) to high density (near bottom of tube or outside of rotor). When placed on top of this gradient (or even mixed uniformly within the gradient) and subjected to continued centrifugation, individual DNA molecules will migrate to a position in the gradient where their density matches that of the surrounding medium. Individual DNA bands can then be either visualized (for example, through the incorporation of DNA-binding fluorescent dyes in the gradient matrix and exposure to fluorescence excitation), or recovered by careful puncture of the centrifuge tube and fractional collection of the tube contents. This method can also be used to separate double-stranded from single-stranded molecules and RNA from DNA molecules, again based solely of density differences.

Choice of the gradient matrix material, its concentration, and the centrifugation conditions can influence the total density range separated by the process, with very narrow ranges being used to fractionate one particular type of DNA molecule from others, and wider

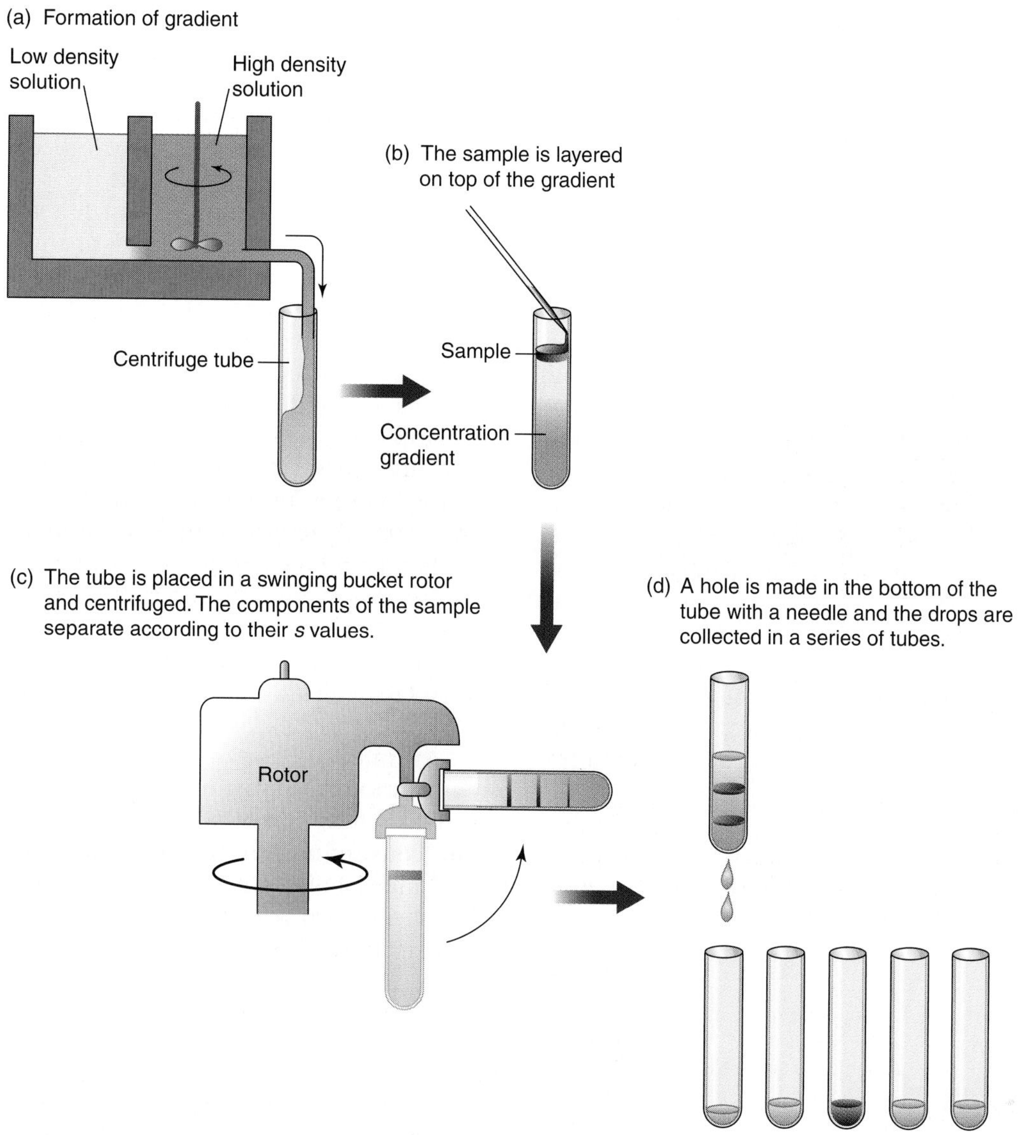

(a) Formation of gradient

Low density solution

High density solution

Centrifuge tube

(b) The sample is layered on top of the gradient

Sample

Concentration gradient

(c) The tube is placed in a swinging bucket rotor and centrifuged. The components of the sample separate according to their *s* values.

Rotor

(d) A hole is made in the bottom of the tube with a needle and the drops are collected in a series of tubes.

FIGURE 3.16 Gradient centrifugation separates samples based on their density.

ranges being used to separate DNAs in general from other biomolecules. Historically, one of the best-known uses of this technique was in the Meselson–Stahl experiment of 1958 (introduced in *Section 1.7, DNA Replication Is Semiconservative*), in which the stepwise density changes in the DNA genomes of bacteria shifted from growth in "heavy" nitrogen (^{15}N) to "regular" nitrogen (^{14}N) were observed. The method's capacity to differentially band DNA with pure ^{15}N, half ^{15}N/half ^{14}N, and pure ^{14}N conclusively demonstrated the semiconservative nature of DNA replication. Today, the method is most frequently employed as a large-scale preparative purification technique with wider density ranges to purify DNAs as a group away from proteins and RNAs.

3.7 DNA Sequencing

Key concepts

- Chain termination sequencing uses dideoxynucleotides to terminate DNA synthesis at particular nucleotides.
- Fluorescently tagged ddNTPs and capillary gel electrophoresis allow automated, high-throughput DNA sequencing.
- The next generation of sequencing techniques aim to increase automation and decrease time and cost of sequencing.

The most commonly used method of DNA sequencing hasn't changed much since Frederick Sanger and colleagues developed a technique in 1977 called *dideoxy sequencing*. This method requires many identical copies of the DNA, an oligonucleotide **primer** that is complementary to a short stretch of the DNA, DNA polymerase, deoxynucleotides (dNTPS: dATP, dCTP, dGTP, and dTTP), and **dideoxynucleotides (ddNTPS)**. Dideoxynucleotides are modified nucleotides that can be incorporated into the growing DNA strand but lack the 3′ hydroxyl group needed to attach the next nucleotide. Thus, their incorporation terminates the synthesis reaction. The ddNTPs are added at much lower concentrations than the normal nucleotides so that they are incorporated at low rate, randomly, often only after synthesis has proceeded normally for a strand length of up to several hundred nucleotides.

Originally, four separate reactions were necessary, with a single different ddNTP added to each one. The reason for this was that the strands were labeled with radioisotopes and could not be distinguished from each other on the basis of the label. Thus, the reactions were loaded into adjacent lanes on a denaturing acrylamide gel and separated by electrophoresis at a resolution that distinguished between strands differing by a length of one nucleotide. The gel was transferred to a solid support, dried, and exposed to a film. The results were read from top to bottom, with a band appearing in the ddATP lane indicating that the strand terminated with an adenine, the next band appearing in the ddTTP lane indicating that the next base was a thymine, and so on.

Two recent modifications have aided in the automation and scaling up of the procedure. The incorporation of a different fluorescent label for each ddNTP allows a single reaction to be run that is read as the strands are hit with a laser and pass by an optical sensor. The information as to which ddNTP terminated the fragment is fed directly into a computer. The second modification is the replacement of large slabs of polyacrylamide gels with very thin, long, glass capillary tubes filled with gel, as described previously in *Section 3.6, DNA Separation Techniques*. These tubes can dissipate heat more rapidly, allowing the electrophoresis to be run at a higher voltage, greatly reducing the time required for separation. A schematic illustrating this process is shown in **FIGURE 3.17**. These modifications, with their resulting automation and increased throughput, ushered in the era of whole-genome sequencing.

A number of "second-generation" sequencing technologies are currently under development. These aim to eliminate the need for time-consuming gel separation and reliance on human labor. Sequencing-by-synthesis and sequencing through nanopores are two of the many technologies currently being explored.

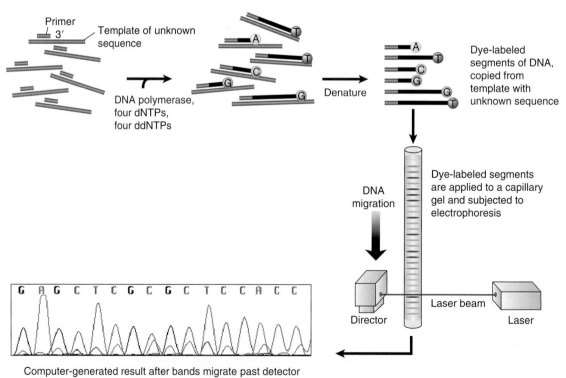

FIGURE 3.17 DideoxyNTP sequencing using fluorescent tags. Inset photo courtesy of Jan Kieleczawa.

Sequencing-by-synthesis relies on the detection and identification of each nucleotide as it is added to a growing strand. In one such application, the primer is tethered to a glass surface and the complementary DNA to be sequenced anneals to the primer. Sequencing proceeds by adding polymerase and fluorescently labeled nucleotides individually, washing away any unused dNTPs. After illuminating with a laser, the nucleotide that has been incorporated into the DNA strand can be detected. Other versions use nucleotides with reversible termination, so that only one nucleotide can be incorporated at a time even if there is a stretch of homopolymeric DNA (such as a run of adenines). Still another version, called *pyrosequencing*, detects the release of pyrophosphate from the newly added base. Although these technologies are still under development, they have the advantage that many parallel reactions can be run.

A completely different approach aims to detect individual nucleotides as a DNA sequence is run through a silicone nanopore. Tiny transistors are used to control a current passing through the pore. As a nucleotide passes through the pore, it disturbs the current in a manner unique to its chemical structure. If successful, this technology has the advantage of reading DNA by simply using electronics, with no chemistry or optical detection required. Nevertheless, there are many kinks to work out of the process before it becomes feasible.

3.8 PCR and RT-PCR

Key concepts

- PCR permits the exponential amplification of a desired sequence, using primers that anneal to the sequence of interest.
- RT-PCR uses reverse transcriptase to convert RNA to DNA for use in a PCR reaction.
- Real-time, or quantitative, PCR detects the products of PCR amplification during their synthesis, and is more sensitive and quantitative than conventional PCR.
- PCR depends on the use of thermostable DNA polymerases that can withstand multiple cycles of template denaturation.

Few advances in the life sciences have had the broad-reaching and even paradigm-shifting impact of the **polymerase chain reaction** (**PCR**). While evidence exists that the underlying core principles of the method were understood and in fact used in practice by a

few isolated people prior to 1983, credit for independent conceptualization of the mature technology and foresight of its applications must go to Kary Mullis, who was awarded the 1993 Nobel Prize in Chemistry for his insight.

The underlying concepts are simple and based on the knowledge that DNA polymerases require a template strand with an annealed primer containing a 3' hydroxyl to commence strand extension. The steps of PCR are illustrated in **FIGURE 3.18**. While in the context of normal cellular DNA replication (see Chapter 14, *DNA Replication*) this primer is in the form of a short RNA molecule provided by DNA primase, it can equally well be provided in the form of a short, single-stranded synthetic DNA oligonucleotide having a defined sequence complementary to the 3' end of any known sequence of interest. Heating of the double-stranded target sequence of interest (known as the "template molecule," or just "template" for short) to near 100°C in appropriate buffer causes thermal denaturation as the template strands melt apart from each other (Figure 3.18a and b). Rapid cooling to the annealing temperature (or "T_m") of the primer/template pair and a vast molar excess of the short, kinetically active synthetic primer ensures that a primer molecule finds and appropriately anneals to its complementary target sequence more rapidly than the original opposing strand can do so (Figure 3.18c). If presented to a polymerase, this annealed primer presents a defined location from which to commence primer extension (Figure 3.18d). In general, this extension will occur until either the polymerase is forced off the template or it reaches the 5' end of the template molecule and effectively runs out of template to copy.

The ingenuity of PCR arises from simultaneously incorporating a nearby second primer of opposing polarity (that is, complementary to the opposite strand the first primer anneals to) and then subjecting the mixture of template, two primers (at high concentrations), thermostable DNA polymerase, and dNTP containing polymerase buffer to repeated cycles of thermal denaturation, annealing, and primer extension. Consider just the first cycle of the process: denaturation and annealing occur as described above, but with both primers, creating the situation depicted in **FIGURE 3.19**. If polymerase extension is allowed to proceed for a short period of time (on the order of 1 minute per 1000 base pairs), each of the primers will be extended out and past the location of the other, thus creating a new complementary

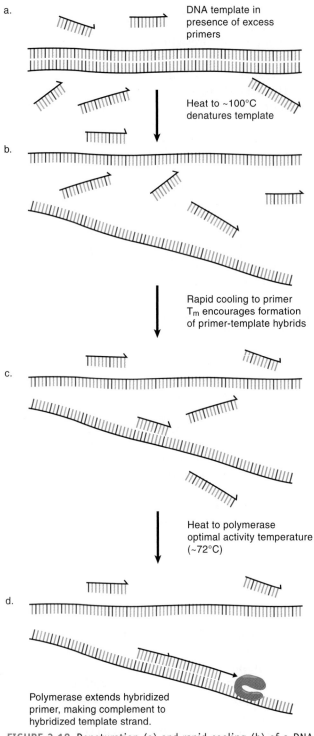

a. DNA template in presence of excess primers

Heat to ~100°C denatures template

b.

Rapid cooling to primer T$_m$ encourages formation of primer-template hybrids

c.

Heat to polymerase optimal activity temperature (~72°C)

d.

Polymerase extends hybridized primer, making complement to hybridized template strand.

FIGURE 3.18 Denaturation (a) and rapid cooling (b) of a DNA template molecule in the presence of excess primer allows the primer to hybridize to any complementary sequence region of the template (c). This provides a substrate for polymerase action and primer extension (d), creating a complementary copy of one template strand downstream from the primer.

annealing site for the opposing primer. Raising the temperature back to denaturation stops the primer elongation process and displaces the polymerases and newly created strands. As the system is cooled once more to the annealing temperature, each of the newly formed short, single DNA strands serves as an annealing site for its opposite polarity primer. In this second thermal cycle, extension of the primers proceeds only as far as the template exists—that is, the 5' end of the opposing primer sequence. The process has now made both strands of the short, defined, precisely primer-to-primer DNA sequence. Repeating the thermal steps of denaturation, annealing, and primer extension lead to an exponential (2^N, where N is number of thermal cycles) increase in the number of this defined product, allowing for phenomenal levels of "sequence amplification." Close consideration of the process reveals that while this also creates uncertain length products from the extension of each primer off the original template molecule with each cycle, these products accrue in a linear fashion and are quickly vastly outnumbered by the primer-to-primer defined product (known as the **amplicon**). In fact, within 40 thermal cycles of an idealized PCR reaction, a single template DNA molecule generates approximately 10^{12} amplicons—more than enough to go from an invisible target to a clearly visible fluorescent dye stained product.

Perhaps not surprisingly, there are many technical complexities underlying this deceptively simple description. Primer design must take into account issues such as DNA secondary structures, uniqueness of sequence, and similarity of T$_M$ between primers. Use of a thermostable polymerase (that is, one that is not inactivated by the high temperatures used in the denaturation steps) is an essential concept identified by Mullis and coworkers. Within this constraint, though, different enzyme sources with differing properties (such as exonuclease activities for increased accuracy) can be exploited to meet individual application needs. Buffer composition (including agents such as DMSO to help reduce secondary structural barriers to effective amplification, and inclusion of divalent cations such as Mg^{2+} at sufficient concentration to not be depleted by chelation to nucleotides) often needs some optimization for effective reactions. In general, the PCR process works best when the primers are within short distances of each other (100 to 500 base pairs), but well-optimized reactions have been successful at distances into the tens of kilobases. "Hot start" techniques—frequently through covalent modification of the polymerase—can be employed to ensure that no inappropriate primer annealing and extension can occur prior

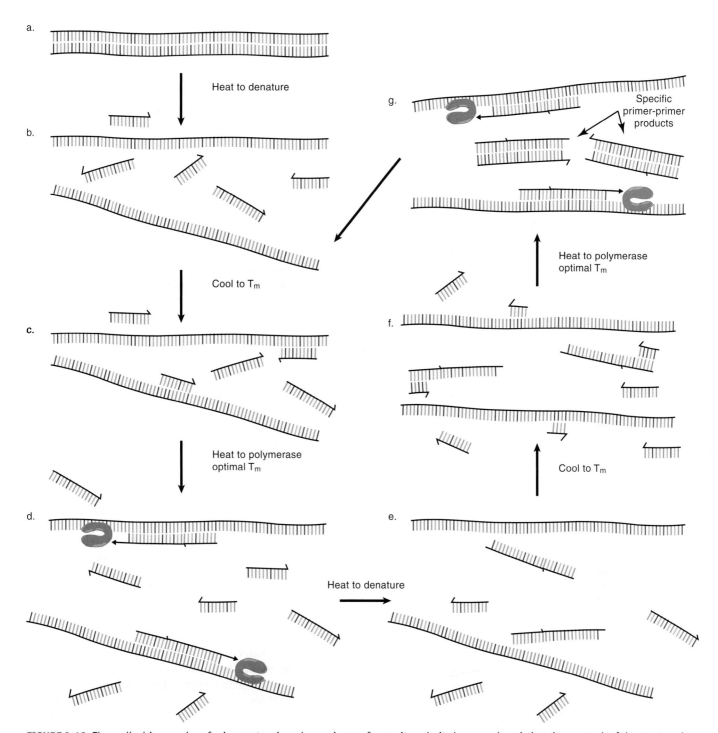

FIGURE 3.19 Thermally driven cycles of primer extension where primers of opposite polarity have nearby priming sites on each of the two template strands leads to the exponential production of the short, primer to primer-defined sequence (the "amplicon").

to the first denaturation step, thereby avoiding the production of incorrect products. Generally, somewhere around 40 thermal cycles marks an effective limit for a PCR reaction with good kinetics in the presence of appropriate template, as depletion of dNTPs into amplicons effectively occurs around this point and a "plateau phase" occurs wherein no more product is made. Con-versely, if the appropriate template was not present in the reaction, proceeding beyond 40 cycles primarily increases the likelihood of pro-duction of rare, incorrect products.

Pairing PCR with a preliminary reverse transcription step (either random-primed or using one of the PCR primers to direct activity of the RNA-dependent DNA polymerase [reverse transcriptase]) allows for RNA templates to be converted to cDNA and then subject to regular PCR, in a variation known as **reverse transcription PCR** (**RT-PCR**). In general, the subsequent discussion uses the term *PCR* to refer to both PCR and RT-PCR.

Detection of PCR products can be done in a number of ways. Post-reaction "endpoint techniques" include gel electrophoresis and DNA-specific dye staining. Long a staple of molecular biological techniques (described above in *Section 3.6, DNA Separation Techniques*), this is a simple but effective technique to rapidly visualize both that an amplicon was produced and that it is of an expected size. If the particular application requires exact, to-the-nucleotide product sizing, capillary electrophoresis can be used instead. Hybridization of PCR products to microarrays or suspension bead arrays can be used to detect specific amplicons when more than one product sequence may come out of an assay. These in turn use a variety of methods for amplicon labeling, including chemiluminescence, fluorescence, and electrochemical techniques. Alternatively, **real-time PCR** methodologies employ some way of directly detecting the ongoing production of amplicons in the reaction vessel, most commonly through monitoring a direct or indirect fluorescence change linked to amplicon production by optical methods. These methods allow the reaction vessel to stay sealed throughout the process. In contrast to endpoint methods where final amplicon concentration bears little relationship to starting template concentration, real-time methods show good correlations between the thermocycle number at which clear signals are measurable (usually referred to as the **threshold cycle** or C_T) and the starting template concentration. Thus, real-time methods are effective template quantification approaches. As a result, these methods are often referred to as **quantitative PCR** (**qPCR**) methods.

Conceptually, the simplest method for real-time PCR detection is based on the use of dyes that selectively bind and become fluorescent in the presence of double-stranded DNA, such as SYBR green. Production of a PCR product during thermocycling leads to an exponential increase in the amount of double-stranded product present at the annealing and extension thermal steps of each cycle. The real-time instrument monitors fluorescence in each reaction tube during these thermal steps of each cycle and calculates the change in fluorescence per cycle to generate a sigmoidal amplification curve. A cutoff threshold value placed approximately midrange in the exponential phase of this curve is used for calculating the C_T of each sample and can be used for quantitation if appropriate controls are present.

A potential issue with this approach is that the reporter dyes are not sequence specific, so any spurious products produced by the reaction can lead to false positive signals. In practice, this is usually controlled for by performance of a melt point analysis at the end of regular thermocycling. The reaction is cooled to the annealing temperature, and then the temperature is slowly raised while fluorescence is constantly monitored. Specific amplicons will have a characteristic melt point at which fluorescence is lost, while nonspecific amplicons will demonstrate a broad range of melt points, giving a gradual loss in sample fluorescence.

A number of alternate approaches use probe-based fluorescence reporters, which avoid this potential nonspecific signal. Probe-based approaches work through the application of a process called **fluorescence resonant energy transfer** (**FRET**). In simple terms, FRET occurs when two fluorophores are in close proximity and the emission wavelength of one (the reporter) matches the excitation wavelength of the other (the quencher). Photons emitted at the reporter dye emission wavelength are effectively captured by the nearby quencher dye and reemitted at the quencher emission wavelength. In the simplest form of this approach, two short oligonucleotide probes with homology to adjoining sequences within the expected amplicon are included in the assay reaction; one probe carries the reporter dye, and the other the quencher. If specific PCR product is formed in the reaction, then at each annealing step these two probes can anneal to the single-stranded product and thereby place the reporter and quencher molecules close to each other. Illumination of the reaction with the excitation wavelength of the reporter dye will lead to FRET and fluorescence at the quencher dye's characteristic emission frequency. By contrast, if the homologous template for the probe molecules is not present (that is, the expected PCR product), the two dyes will not be colocalized and excitation of the reporter dye will lead to fluorescence at its emission frequency. This is illustrated in FIGURE 3.20. As with the DNA-binding dye approach, the real-time instrument monitors the quencher emission wavelength during each cycle and generates

a.

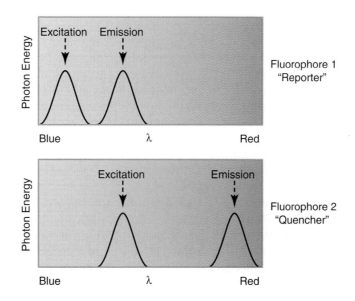

b. When Reporter and Quencher are not in very close proximity:

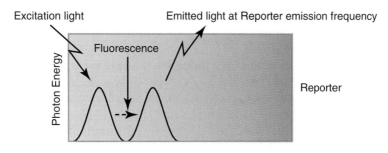

c. When Reporter and Quencher are in close proximity:

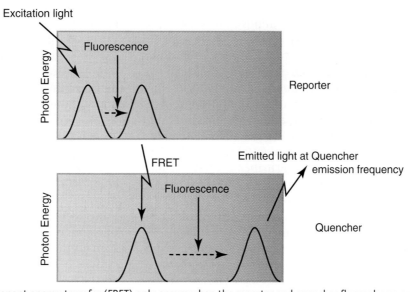

FIGURE 3.20 Fluorescence resonant energy transfer (FRET) only occurs when the reporter and quencher fluorophores are very close to each other, leading to the detection of light at the quencher emission frequency when the reporter is stimulated by light of its excitation frequency. If the reporter and quencher are not colocalized, stimulation of the reporter instead leads to detection of light at the reporter emission frequency. By placing the reporter and quencher fluorophores on single-stranded nucleic acid probes complementary to the expected amplicon, different variations on this method can be designed such that the occurrence of FRET can be used to monitor the production of sequence-specific amplicons.

a similar sigmoidal amplification curve. Multiple alternate ways of exploiting FRET for this process exist, including 5′ fluorogenic nuclease assays, molecular beacons, and molecular scorpions. Although the details of these differ, the underlying concept is similar and all generate data in a similar fashion.

The applications of the PCR process are incredibly diverse. The simple appearance or nonappearance of an amplicon in a properly controlled reaction can be taken as evidence for the presence or absence, respectively, of the assay target template. This leads to medical applications such as the detection of infectious disease agents at sensitivities, specificities, and speeds much greater than alternate methods. The fact that while the two primer sites must be of known sequence, the internal section may be any sequence of a general length, leads directly to applications where a PCR product for a region known to vary between species (or even between individuals) can be produced and subject to sequence analysis to identify the species (or individual identity, in the latter case) of the sample template. Coupled with single molecule sensitivity, this has provided criminal forensics with tools powerful enough to identify individuals from residual DNA on crime scene traces as simple as cigarette butts, smudged fingerprints, or a single hair. Evolutionary biologists have made use of PCR to amplify DNA from well-preserved samples, such as insects in amber millions of years old, with subsequent sequencing and phylogenetic analysis yielding fascinating results on the continuity and evolution of life on Earth. Quantitative real-time approaches have applications in medicine (for example, monitoring viral loads in transplant patients), research (such as examining transcriptional activation of a specific target gene in a single cell), or environmental monitoring (for instance, water purification quality control).

In general, PCR reactions are run with carefully optimized T_M values that maximize sensitivity and amplification kinetics while ensuring that primers will only anneal to their exact hybridization matches. Lowering the T_M of a PCR reaction—in effect, relaxing the reaction stringency and allowing primers to anneal to not quite perfect hybridization partners—has useful applications as well, such as in searching a sample for an unknown sequence suspected to be similar to a known one. This technique has been successfully employed for the discovery of new virus species, when primers matching a similar virus species are employed. Similarly, during a PCR-directed cloning of a gene or region of interest, use of planned mismatches in the primer sequence and slightly lowered T_Ms can be used to introduce wanted mutations in a process called *site-directed mutagenesis*. Differential detection of single nucleotide polymorphisms (SNPs) (see *Section 5.3, Individual Genomes Show Extensive Variation*), which can be directly indicative of particular genotypes or serve as surrogate linked markers for nearby genetic targets of interest, can be done through design of PCR primers with a 3′ terminal nucleotide specific to the expected polymorphism. At the optimal T_M, this final crucial nucleotide can only hybridize and provide a 3′ hydroxyl to the waiting polymerase if the matching SNP occurs, in a process known by several names, including Amplification Refractory Mutation Selection (ARMS) or Allele-SPEcific PCR (ASPE).

The PCR process described thus far has been restricted to amplification of a single target per reaction, or "simplex" PCR. Although this is the most common application, it is possible to combine multiple, independent PCR reactions into a single reaction, allowing for an experiment to query a single minute specimen for the presence, absence, or possibly the amount of multiple unrelated sequences. This *multiplex PCR* is particularly useful in forensics applications and medical diagnostic situations, but entails rapidly increasing levels of complexity in ensuring that multiple primer sets do not have unwanted interactions that lead to undesired false products. At best, multiplexing tends to result in loss of some sensitivity for each individual PCR due to effective competition between them for limited polymerase and nucleotides.

A final point of interest to many students with regard to PCR is its consideration from a philosophical perspective. In practice, performance of this now incredibly pervasive method requires the use of a thermostable polymerase, as previously indicated. These polymerases (of which there are a number of varieties) primarily derive from bacterial DNA polymerases originally identified in extremophiles living in boiling hot springs and deep-sea volcanic thermal vents. Few people would have been likely to suspect that studying deep-sea thermal vent microbes would be of such direct importance in so many other aspects of science, including ones with impact on their daily lives. These unexpected links between topics serve to highlight the importance of basic research on all manner of subjects; critical discoveries can come from the least expected avenues of research.

Blotting Methods

Key concepts

- Southern blotting involves the transfer of DNA from a gel to a membrane, followed by detection of specific sequences by hybridization with a labeled probe.
- Northern blotting is similar to Southern blotting, but involves the transfer of RNA from a gel to a membrane.
- Western blotting entails separation of proteins on an SDS gel, transfer to a nitrocellulose membrane, and detection proteins of interest using antibodies.

After nucleic acids are separated by size in a gel matrix, they can be detected using dyes that are nonsequence specific, or specific sequences can be detected using a method generically referred to as *blotting*. Although slower and more involved than direct visualization by fluorescent dye staining, blotting techniques have two major advantages: they have a greatly increased sensitivity relative to dye staining, and they allow for the specific detection of defined sequences of interest among many similarly sized bands on a gel.

The method was first developed for application to DNA agarose gels, and was briefly introduced in *Section 3.5, Nucleic Acid Detection*. In this form, the method is referred to as **Southern blotting** (after the method's inventor, Dr. Edwin Southern). A schematic of this process is shown in **FIGURE 3.21**. A regular agarose gel is made, run (and if desired, stained) as described

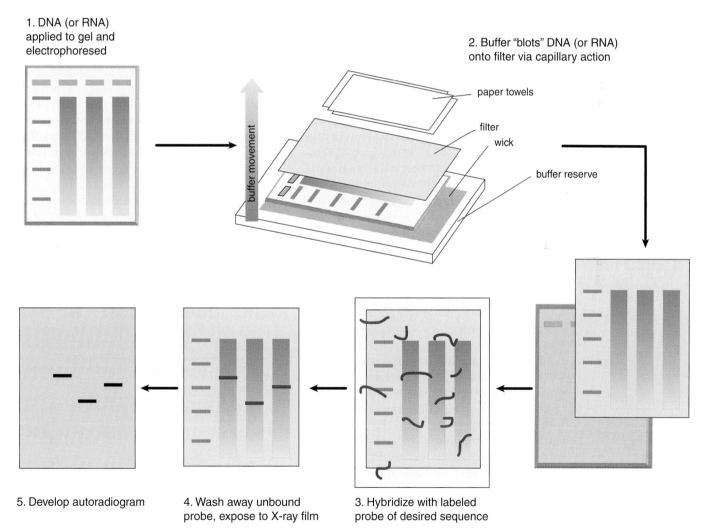

1. DNA (or RNA) applied to gel and electrophoresed

2. Buffer "blots" DNA (or RNA) onto filter via capillary action

paper towels

filter

wick

buffer reserve

buffer movement

5. Develop autoradiogram

4. Wash away unbound probe, expose to X-ray film

3. Hybridize with labeled probe of desired sequence

FIGURE 3.21 To perform a Southern blot, DNA digested with restriction enzymes is electrophoresed to separate fragments by size. Double-stranded DNA is denatured in an alkali solution either before or during blotting. The gel is placed on a wick (such as a sponge) in a container of transfer buffer and a membrane (nylon or nitrocellulose) is placed on top of the gel. Absorbent materials such as paper towels are placed on top. Buffer is drawn from the reservoir through the gel by capillary action, transferring the DNA to the membrane. The membrane is then incubated with a labeled probe (usually DNA). The unbound probe is washed away, and the bound probe is detected by autoradiography or phosphorimaging. In Northern blotting, RNA is run on a gel rather than DNA.

previously. Following this, the gel is soaked in alkali buffer to denature the DNA, then placed in contact with a sheet of porous membrane (commonly nitrocellulose or nylon) and a buffer is drawn through the gel and then the membrane either by capillary action (for instance, by wicking into a stack of dry paper towel) or by a gentle vacuum pressure. This slow flow of buffer in turn draws each nucleic acid band in the gel out of the gel matrix and onto the membrane surface. Nucleic acids bind to the membrane, which in many cases is positively charged to increase efficiency of DNA binding. This in effect creates a "contact print" of the order and position of all nucleic acid bands as size resolved in the gel. To make the elution of large DNA molecules from the gel matrix more efficient, the gel is sometimes treated with a mild acid after electrophoresis but before transfer. This induces acid depurination and creates random strand breaks in the DNA within the gel, such that large molecules are broken into smaller subsections that elute more readily, but remain in the same physical location as their original gel band.

Following transfer, the nucleic acids are fixed to the membrane either through drying or through exposure to UV light, which can create physical crosslinks between the membrane and the nucleic acids (primarily pyrimidines). The blot is now ready for blocking, where it is immersed in a warmed, low-salt buffer containing materials that will bind to and block areas of the blot that may bind organic compounds nonspecifically. Following blocking, a probe molecule is introduced. The probe consists of a labeled (isotopically or chemically, such as through incorporation of biotinylated nucleotides) copy of the target sequence of interest, which has been heat denatured and rapidly cooled to place it in a single-stranded form. When this is added to the warmed buffer and allowed to incubate with the blocked membrane, the probe will attempt to hybridize to homologous sequences on the membrane surface. Following this hybridization step, the membrane is generally washed in warm buffer without probe or blocking agent to remove nonspecifically associated probe molecules, and then visualized; in the case of isotopically labeled probes, this can be done by simply exposing the membrane to a piece of film or a phosphorimager screen. Decay of the label (usually ^{32}P or ^{35}S) leads to the production of an image in which any hybridized DNA bands become visible on the developed film or scanned phosphor screen. For chemically labeled probes, chemi-

luminescent or fluorescent detection strategies are used in an analogous manner.

A final benefit of the Southern blotting technique is that the observed band intensity is related to the amount of target on the membrane—in other words, it is a quantitative method. If a suitable standard (such as a dilution series of unlabelled probe sequence) is included in the gel, then comparison of this standard to target band intensities allows for determination of target quantity in the starting sample. This information can be useful for applications such as determining viral copy number in a host cell sample.

Numerous variations on the Southern blot approach exist, such as use of a denaturing gel matrix for an otherwise analogous process on RNA molecules (referred to as "northern blotting"). In this case, there is no initial digestion step, so intact RNAs are separated by size, usually on a formaldehyde or other denaturing gel, which eliminates RNA secondary structures. This allows measurement of actual RNA sizes, and like Southern blotting, provides a similarly quantitative method for detection of any type of RNA. If mRNA is the target of interest, it is possible to separate mRNA from all the other classes of RNA in the cell. mRNA (and some noncoding RNA) differs from other RNAs in that it is polyadenylated (it has a string of adenine residues added to the 3′ end; see *Section 21.15, The 3′ Ends of mRNAs Are Generated by Cleavage and Polyadenylation*). Poly(A)+ mRNA can therefore be enriched by use of an oligo(dT) column, in which oligomers of oligo(dT) are immobilized on a solid support and used to capture mRNA from the total RNA in a sample. This is illustrated in **FIGURE 3.22**.

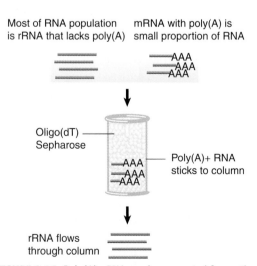

Most of RNA population is rRNA that lacks poly(A)

mRNA with poly(A) is small proportion of RNA

Oligo(dT) Sepharose

Poly(A)+ RNA sticks to column

rRNA flows through column

FIGURE 3.22 Poly(A)+ RNA can be separated from other RNAs by fractionation on an oligo(dT) column.

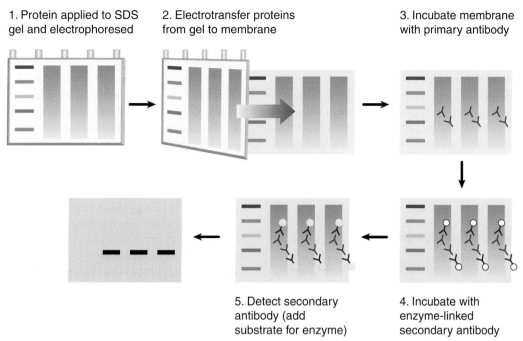

1. Protein applied to SDS gel and electrophoresed

2. Electrotransfer proteins from gel to membrane

3. Incubate membrane with primary antibody

5. Detect secondary antibody (add substrate for enzyme)

4. Incubate with enzyme-linked secondary antibody

FIGURE 3.23 In a western blot, proteins are separated by size on an SDS gel, transferred to a nitrocellulose membrane, and detected using an antibody. The primary antibody detects the protein and the enzyme-linked secondary antibody detects the primary antibody. The secondary antibody is detected in this example via addition of a chemiluminescent substrate, which results in emission of light that can be detected on X-ray film.

A conceptually similar process for proteins based on protein-separation gels and blotting to membrane is known as "western blotting." This method is depicted in **FIGURE 3.23**. There are some key differences between the procedures for blotting proteins compared to nucleic acids. First, protein-separation gels typically contain the detergent SDS, which both serves to unfold the proteins so that they will migrate according to size rather than shape, and also provides a uniform negative charge to all proteins so that they will migrate toward the positive pole of the gel. (In the absence of SDS, each protein has a specific individual charge at a given pH; it is possible to separate proteins based on these charges, rather than size, in a technique called *isoelectric focusing*.)

Once the proteins are separated on the gel, they are transferred to a nitrocellulose membrane using an electric current to effect the transfer, rather than the capillary or vacuum methods used for nucleic acids. The most significant difference in western blotting is the method of detecting proteins on the membrane. Complementary base pairing can't be used to detect a protein, so westerns use *antibodies* to recognize the protein of interest. The antibody can either recognize the protein itself, if such an antibody is available, or can recognize an **epitope tag** that has been fused to the pro-

tein sequence. An epitope tag is a short peptide sequence that is recognized by a commercially available antibody; the DNA encoding the tag can be cloned in-frame to a gene of interest, resulting in a product containing the epitope (typically at the N- or C-terminus of the protein). Sequences for the most commonly used epitope tags (such as the HA, FLAG, and myc tags) are often available in expression vectors for ease of fusion (see *Section 3.4, Cloning Vectors Can Be Specialized for Different Purposes*).

The antibody that recognizes the target on the membrane is known as the *primary antibody*. The final stage of western blotting is detection of the primary antibody with a *secondary antibody*, which is the antibody that can be visualized. Secondary antibodies are raised in a different species than the primary antibody used and recognize the constant region of the primary antibody. (For example, a "goat antirabbit" antibody will recognize a primary antibody raised in a rabbit; see Chapter 18, *Recombination in the Immune System*, for a review of antibody structure.) The secondary antibody is typically linked to a moiety that allows its visualization—for example, a fluorescent dye or an enzyme such as alkaline phosphatase or horseradish peroxidase. These enzymes serve as visualization tools because they can convert added substrates to a colored product (*colorimetric detection*), or can release light as a reaction product (*chemiluminescent detection*). Use of primary and secondary antibodies (rather than linking a visualizer to the primary antibody) increases the sensitivity of western blotting. The result is semiquantitative detection of the protein of interest.

Continuing in the same vein, techniques used to identify interactions between DNA and proteins (through protein gel separation and blotting followed by probing with a DNA) are "southwestern blotting." When an RNA probe is used the technique is "northwestern blotting."

3.10 DNA Microarrays

Key concepts

- DNA microarrays comprise known DNA sequences spotted or synthesized on a small chip.
- Genome-wide transcription analysis is performed using labeled cDNA from experimental samples hybridized to a microarray containing sequences from all ORFs of the organism being used.
- SNP arrays permit genome-wide genotyping of single nucleotide polymorphisms.
- Array comparative genome hybridization (array-CGH) allow the detection of copy number changes in any DNA sequence compared between two samples.

A logical technical progression from Southern and northern blotting is the microarray. Instead of having the unknown sample on the membrane and the probe in solution, this effectively reverses the two. These originated in the form of "slot-blots" or "dot-blots," where a researcher would spot individual DNA sequences of interest directly onto a hybridization membrane, in an ordered pattern, with each spot consisting of a different, single-known sequence. Drying of the membrane immobilized these spots, creating a premade blotting array. In use, the researcher would then take a nucleic acid sample of interest, such as total cellular DNA, and then fragment and randomly and uniformly label this DNA (originally with a radioisotopic label). This labeled mix of sample DNA could then be used exactly as in a Southern blot as a probe to hybridize to the premade blot. Labeled DNA sequences homologous to any of the array spots would hybridize and be retained in the known, fixed location of that spot and be visualized by autoradiography. By viewing the autoradiogram and knowing the physical location of each specific probe spot, the pattern of hybridized versus nonhybridized spots could be read out to indicate the presence or absence of each of the corresponding known sequences in the unknown sample.

Technological improvements to this approach followed rapidly through miniaturization of the size and physical density of the immobilized spots, going from membranes with 30 to 100 spots to glass microscope slides with up to 1000 spots. Today, silicon chip substrates have hundreds of thousands (and now up to a million or more) of individual spots in an area about the size of a postage stamp.

In order to visualize the distinct spots in such a high-density array, automated optical microscopy is used and fluorescence has replaced radiolabeling both to allow for increased spatial resolution (higher spot density), as well as easier quantification of each hybridization signal. In parallel with the increased total number of spots per array, the length of each unique probe has generally become shorter, allowing for each spot in the array to be specific to a smaller target area—in effect, giving greater "resolution" on a molecular scale. Although the potential applications of microarrays are really only limited by the user's imagination, there are a number of particular applications where they have become standard tools.

The first of these is in gene expression profiling, where a total mRNA sample from a specimen of interest (such as tissue in a disease state or under a particular environmental challenge) is collected and converted *en masse* to cDNA by a random primed reverse transcription. A label is incorporated into the cDNA during its synthesis (either through use of labeled nucleotides or having the primers themselves with a label); this can either be a fluorophore ("direct labeling") or another hapten (such as biotin), which can at a later stage be exposed to a fluorophore conjugate that will bind the hapten (in the present example, streptavidin–phycoerythrin conjugate might be used) in what is called "indirect labeling." This labeled cDNA is then hybridized to an array where the immobilized spots consist of complementary strands to a number of known mRNAs from the target organism. Hybridization, washing, and visualization allows for the detection of those spots that have bound their complimentary labeled cDNA, and thus the readout of which genes are being expressed in the original sample. This process is depicted in **FIGURE 3.24**. As with Southern blotting,

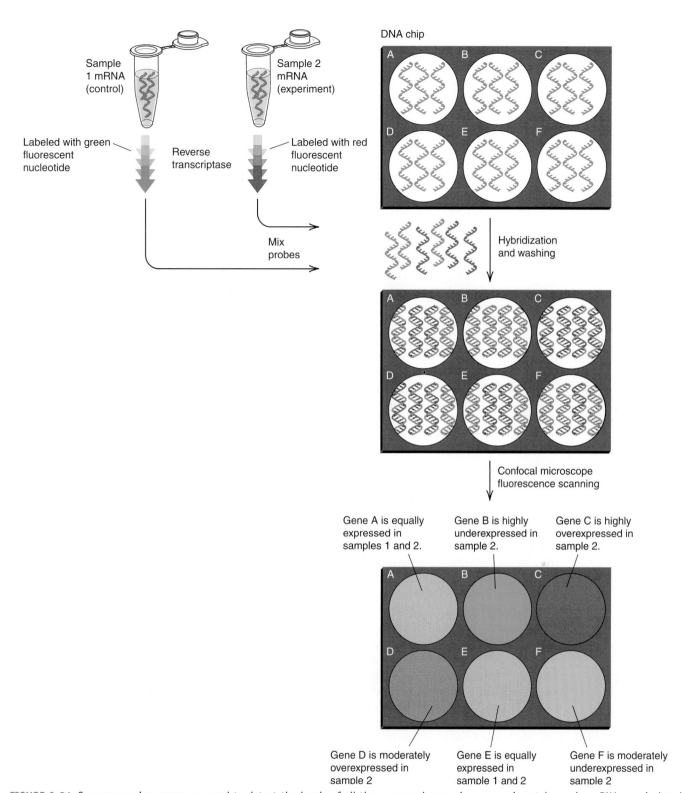

FIGURE 3.24 Gene expression arrays are used to detect the levels of all the expressed genes in an experimental sample. mRNAs are isolated from control and experimental cells or tissues and reverse transcribed in the presence of fluorescently labeled nucleotides (or primers), resulting in labeled cDNAs with different fluorophores (red and green strands) for each sample. Competitive hybridization of the red and green cDNAs to the microarray is proportional to the relative abundance of each mRNA in the two samples. The relative levels of red and green fluorescence are measured by microscopic scanning and displayed as a single color. Red or orange indicates increased expression in the red (experimental) sample, green or yellow-green indicates lower expression, and yellow indicates equal levels of expression in the control and experiment.

the method is quantitative, meaning that the observed signal on each spot corresponds to the original level of its particular mRNA. Clever selection of the sequence of each of the immobilized spots, such as choosing short probe sequences that are complementary to particular alternate exons of a gene, can even allow the method to differentiate and quantitate the relative levels of alternate splicing products from a single gene. By comparison of the data from such experiments performed in parallel on experimental tissue and control tissue, an experiment can collect a snapshot of the total cellular "global" changes in gene expression patterns, often with useful insight to the state or condition of the experimental tissue.

A second major application is in genotyping. Analysis of the human genome (and other organisms) has led to the identification of large numbers of **single nucleotide polymorphisms** (**SNPs**), which are single nucleotide substitutions at a specific genetic locus (see *Section 5.4, RFLPs and SNPs Can Be Used for Genetic Mapping*). Individual SNPs occur at known frequencies, which often differ between populations. The most straightforward examples are where the SNP creates a missense mutation within a gene of interest, such as one involved in metabolism of a drug. People carrying one allele of the SNP may clear a drug from circulation at a very different rate from those with an alternate allele, and thus determination of a patient's allele at this SNP may be an important consideration in choosing an appropriate drug dosage. An example of this that has come all the way from theory into everyday use is CYP450 SNP genotyping to determine appropriate dosage of the anticoagulant warfarin. Another is in SNP genotyping of the K-Ras oncogene in some types of cancer patients, in order to determine whether EGFR-inhibitory drugs will be of therapeutic value. Other SNPs may be of no direct biological consequence but can become a valuable genetic marker if found to be closely associated to a particular allele of interest—that is, if in genetic terms it is closely linked. Hundreds of thousands of SNPs have been mapped in the human genome, and arrays that can be probed with a subject's DNA allow for the genotype at each of these to be simultaneously determined, with concurrent determination of what the linked genetic alleles are. In effect, this allows for much of the genotype of the subject to be inferred from a single experiment at vastly less time and expense than actually sequencing the entire subject genome. With a view toward the future, however, it should be noted that SNP genotyping, in the common case of linked alleles as opposed to direct missense mutation alleles, is indirect inference and has at least some potential for being inaccurate.

Sequencing, on the other hand, is definitive. If emerging sequencing technologies improve to the point of offering an entire human genome in 24 hours for a competitive cost to SNP genotyping, it may move to become the dominant approach for genotyping.

A third major application of DNA microarrays is *array comparative genome hybridization* (*array-CGH*). This is a technique that is augmenting, and in some cases replacing, cytogenetics for the detection and localization of chromosomal abnormalities that change the copy number of a given sequence—that is, deletions or duplications. In this technique, the array chip (known as a **tiling array**) is spotted with an organism's genomic sequences that together represent the entire genome; the higher the density of the array, the smaller genetic region each spot represents and thus the higher resolution the assay can provide. Two DNA samples (one from normal control tissue and one from the tissue of interest) are each randomly labeled with a different fluorophore, such that one sample, for example, is green and the other is red (similar to the mRNA labeling described earlier for the expression arrays). These two differentially labeled specimens are mixed at exactly equal ratios for total DNA, and then hybridized to the chip. Regions of DNA that occur equally in the two samples will hybridize equally to their complementary array spots, giving a "mixed" color signal. By comparison, any DNA regions that occur more in one sample than the other will outcompete and thus show a stronger color on its complementary probe spot than the deficient sample will. Computer-assisted image analysis can read out and quantitate small color changes on each array spot and thus detect hemizygous loss of even very small regions in a test sample. The resolution and facility for automation provided by this technique compared to conventional cytogenetics is leading to its increasing adoption in diagnostic settings for the detection of chromosomal copy number changes associated with a range of hereditary diseases.

Tiling arrays are also often used for chromatin immunoprecipitation (ChIP) studies, which can identify sequences interacting with a DNA-binding protein or complex on a genome-wide scale; this is described in *Section 3.11, Chromatin Immunoprecipitation*.

In addition to the chiplike solid phase arrays described, lower density arrays for focused applications (with up to a few hundred targets, as opposed to millions) can be made in microbead-based formats. In these approaches, each microscopic bead has a distinct optical signal or code, and its surface can be coated with the target DNA sequence. Different bead codes can be mixed and matched into a single sample of labeled sample DNA or cDNA, and then sorted, detected, and quantitated by optical and/or flow sorting methods. Although of much lower density than chip-type arrays, bead arrays can be modified and adapted much more readily to suit a particular focused biological question, and in practice show faster three-dimensional hybridization kinetics than chips, which effectively have two-dimensional kinetics.

3.11 Chromatin Immunoprecipitation

Key concepts

- Chromatin immunoprecipitation allows detection of specific protein–DNA interactions *in vivo*.
- "ChIP on chip" allows mapping of all the protein-binding sites for a given protein across the entire genome.

Most of the methods discussed thus far in this chapter are *in vitro* methods that allow the detection or manipulation of nucleic acids or proteins that have been isolated from cells (or produced synthetically). Many other powerful molecular techniques have been developed, though, that allow either direct visualization of the *in vivo* behavior of macromolecules (such as imaging of GFP fusions in live cells), or that allow researchers to take a "snapshot" of the *in vivo* localization or interactions of macromolecules at a particular condition or point in time.

Throughout this book, we will discuss numerous proteins that function by interacting directly with DNA, such as chromatin proteins, or the factors that perform replication, repair, and transcription. While much of our understanding of these processes is derived from *in vitro* reconstitution experiments, it is critical to map the dynamics of protein–DNA interactions in living cells in order to fully understand these complex functions. The powerful technique of **chromatin immunoprecipitation (ChIP)**

was developed to capture such interactions. (*Chromatin* refers to the native state of eukaryotic DNA *in vivo*, in which it is packaged extensively with proteins; this is discussed in Chapter 10, *Chromatin*.) ChIP allows researchers to detect the presence of any protein of interest at a specific DNA sequence *in vivo*.

FIGURE 3.25 shows the process of chromatin immunoprecipitation. This method depends on the use of an antibody to detect the protein of interest. As was discussed earlier for western blots (see *Section 3.9, Blotting Methods*), this antibody can be against the protein itself, or against an epitope-tagged target.

The first step in ChIP is typically the crosslinking of the cell (or tissue or organism) of interest by fixing it with formaldehyde. This serves two purposes: (1) it kills the cell and arrests all ongoing processes at the time of fixation, providing the snapshot of cellular activity; and (2) it covalently links any protein and DNA that are in very close proximity, thus preserving protein–DNA interactions through the subsequent analysis. ChIP can be performed on cells or tissues under different experimental conditions (such as different phases of the cell cycle, or after specific treatments) to look for changes in protein–DNA interactions under different conditions.

After crosslinking, the chromatin is then isolated from the fixed material and cleaved into small chromatin fragments, usually 200 to 1000 bp each. This can be achieved by sonication, which uses high intensity sound waves to nonspecifically shear the chromatin. Nucleases (either sequence-specific or nonsequence-specific) can be used to fragment the DNA. These small chromatin fragments are then incubated with the antibody against the protein target of interest. These antibodies can then be used to immunoprecipitate the protein by pulling the antibodies out of the solution using heavy beads coated with a protein (such as Protein A) that binds to the antibodies.

After washing away unbound material, the remaining material contains the protein of interest still crosslinked to any DNA it was associated with *in vivo*. This is sometimes called a "guilt by association" assay, because the DNA target is only isolated due to its interaction with the protein of interest. The final stages of ChIP entail reversal of the crosslinks so that the DNA can be purified, and detection of specific DNA sequences using PCR or blotting methods. Quantitative (real-time) PCR

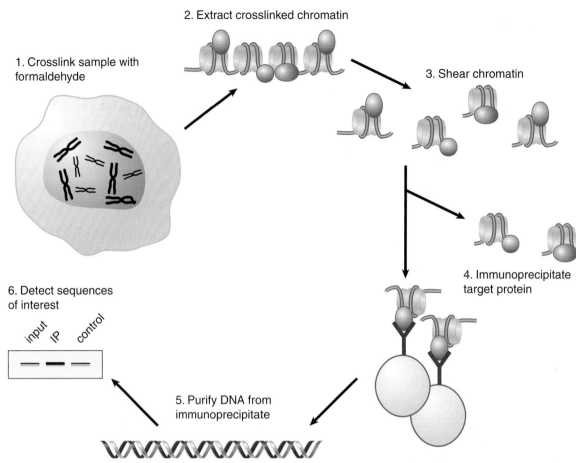

1. Crosslink sample with formaldehyde

2. Extract crosslinked chromatin

3. Shear chromatin

4. Immunoprecipitate target protein

5. Purify DNA from immunoprecipitate

6. Detect sequences of interest

input   IP   control

FIGURE 3.25 Chromatin immunoprecipitation detects protein–DNA interactions in the native chromatin context *in vivo*. Proteins and DNA are cross-linked, chromatin is broken into small fragments, and an antibody is used to immunoprecipitate the protein of interest. Associated DNA is then purified and analyzed by either identifying specific sequences by PCR (as shown), or by labeling the DNA and applying to a tiling array to detect genome-wide interactions.

is usually the method of choice for detecting the DNA.

In addition to revealing the presence of a specific protein at a given DNA sequence (such as a transcription factor bound to the promoter of a gene of interest), highly specialized antibodies can provide even more detailed information. For example, antibodies can be developed that distinguish between different posttranslational modifications of the same protein. As a result, ChIP can distinguish the difference between RNA polymerase II engaged in initiation at the promoter of a gene from pol II that has entered the elongation phase of transcription, because pol II is differentially phosphorylated in these two states (see *Section 20.8, Initiation Is Followed by Promoter Clearance and Elongation*), and antibodies exist that recognize these phosphorylation events.

A variation on the ChIP procedure allows researchers to query the localization of a given protein (or modified version of a protein) across large genomic regions—or even entire genomes. In this variation, known as "ChIP on chip," the only difference is the fate of the DNA that is purified from the immunoprecipitated material. Rather than querying specific sequences in this DNA via PCR, the DNA is labeled in bulk and hybridized to a DNA microarray (usually a genome tiling array, such as described in the previous section). This allows a researcher to obtain a genome-wide footprint of all of the binding sites of the protein of interest. For example, putative origins of replication (which are difficult to identify in multicellular eukaryotes) can be detected *en masse* by performing a ChIP against proteins in the origin recognition complex (ORC).

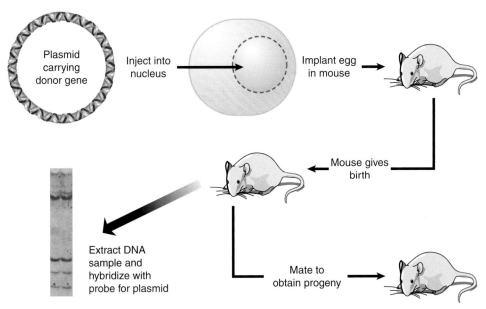

FIGURE 3.26 Transfection can introduce DNA directly into the germline of animals. Photo reproduced from P. Chambon, *Sci. Am.* 244 (1981): 60–71. Used with permission of Pierre Chambon, Institute of Genetics and Molecular and Cellular Biology, College of France.

3.12 Gene Knockouts and Transgenics

Key concepts

- ES (embryonic stem) cells that are injected into a mouse blastocyst generate descendant cells that become part of a chimeric adult mouse.
- When the ES cells contribute to the germline, the next generation of mice may be derived from the ES cell.
- Genes can be added to the mouse germline by transfecting them into ES cells before the cells are added to the blastocyst.
- An endogenous gene can be replaced by a transfected gene using homologous recombination.
- The occurrence of successful homologous recombination can be detected by using two selectable markers, one of which is incorporated with the integrated gene, the other of which is lost when recombination occurs.
- The Cre/*lox* system is widely used to make inducible knockouts and knock-ins.

An organism that gains new genetic information from the addition of foreign DNA is described as **transgenic**. For simple organisms, such as bacteria or yeast, it is easy to generate transgenics by transformation with DNA constructs containing sequences of interest. Transgenesis in multicellular organisms, however, can be much more challenging.

The approach of directly injecting DNA can be used with mouse eggs, as shown in **FIG-**URE 3.26. Plasmids carrying the gene of interest are injected into the nucleus of the oocyte or into the pronucleus of the fertilized egg. The egg is implanted into a pseudopregnant mouse (a mouse that has mated with a vasectomized male to trigger a receptive state). After birth, the recipient mouse can be examined to see whether it has gained the foreign DNA, and, if so, whether it is expressed. Typically, a minority (~15%) of the injected mice carry the transfected sequence. In general, multiple copies of the plasmid appear to have been integrated in a tandem array into a single chromosomal site. The number of copies varies from 1 to 150, and they are inherited by the progeny of the injected mouse. The levels of gene expression from *transgenes* introduced in this way is highly variable, both due to copy number and the site of integration. A gene may be highly expressed if it integrates within an active chromatin domain, but not if it integrates in or near a silenced region of the chromosome.

Transgenesis with novel or mutated genes can be used to study genes of interest in the whole animal. In addition, defective genes can be replaced by functional genes using transgenic techniques. One example is the cure of the defect in the *hypogonadal* mouse. The *hpg* mouse has a deletion that removes the distal part of the gene coding for the precursor to GnRH (gonadotropin-releasing hormone) and GnRH-associated peptide (GAP). As a result, the

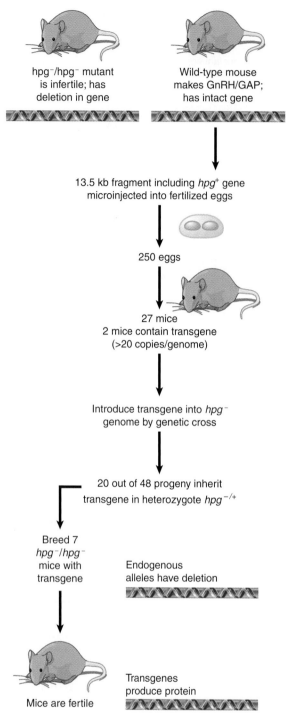

FIGURE 3.27 Hypogonadism can be averted in the progeny of *hpg* mice by introducing a transgene that has the wild-type sequence.

mouse is infertile. When an intact *hpg* gene is introduced into the mouse by transgenic techniques, it is expressed in the appropriate tissues. **FIGURE 3.27** summarizes experiments to introduce a transgene into a line of *hpg*–/– homozygous mutant mice. The resulting progeny are normal. This provides a striking demonstration that expression of a transgene under normal regulatory control can be indistinguishable from the behavior of the normal allele.

Although promising, there are impediments to using such techniques to cure human genetic defects. The transgene must be introduced into the germline of the *preceding* generation, the ability to express a transgene is not predictable, and an adequate level of expression of a transgene may be obtained in only a small minority of the transgenic individuals. In addition, the large number of transgenes that may be introduced into the germline, and their erratic expression, could pose problems in cases in which overexpression of the transgene is harmful. In other cases, the transgene can integrate near an oncogene and activate it, promoting carcinogenesis.

A more versatile approach for studying the functions of genes is to eliminate the gene of interest. Transgenesis methods allow DNA to be *added* to cells or animals, but in order to understand the function of a gene, it is most useful to be able to *remove* the gene or its function and observe the resulting phenotype. The most powerful techniques for changing the genome use *gene targeting* to delete or replace genes by homologous recombination. Gene deletions are usually referred to as **knockouts**, whereas replacement of a gene with an alternative mutated version is called a **knock-in**.

In simple organisms such as yeast, this is again a very simple process in which DNA encoding a selectable marker flanked by short regions of homology to a target gene is transformed into the yeast. As little as 40 bp or so of homology will result in extremely efficient replacement of the target gene by the introduced marker gene, via homologous recombination using the short regions of homology.

In some organisms, and in mammalian cells in culture, there is no good method for deleting endogenous genes. Instead, researchers use **knockdown** approaches, which reduce the amount of a gene product (RNA or protein) produced, even while the endogenous gene is intact. There are several different knockdown methods, but one of the most powerful is the use of RNA interference (RNAi) to selectively target specific mRNAs for destruction. (RNAi is described in *Section 30.5, MicroRNAs Are Widespread Regulators in Eukaryotes* and *Section 30.6, How Does RNA Interference Work*?) Briefly, introduction of double-stranded RNA into most eukaryotic cells triggers a response in which these RNAs are cleaved by a nuclease called Dicer

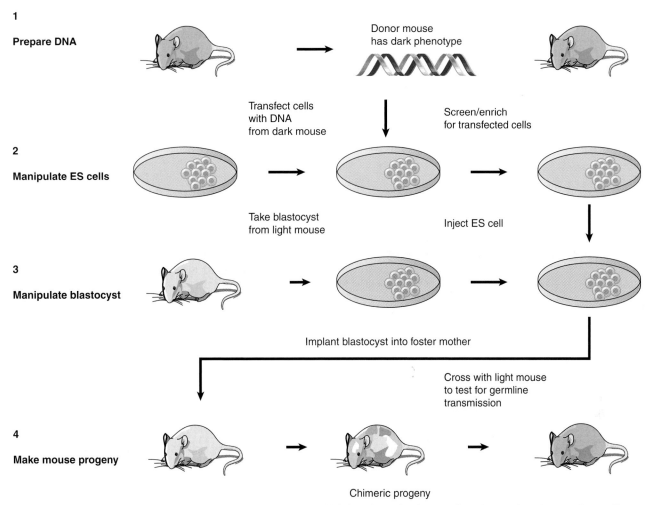

1

Prepare DNA

Donor mouse
has dark phenotype

Transfect cells
with DNA
from dark mouse

Screen/enrich
for transfected cells

2

Manipulate ES cells

Take blastocyst
from light mouse

Inject ES cell

3

Manipulate blastocyst

Implant blastocyst into foster mother

Cross with light mouse
to test for germline
transmission

4

Make mouse progeny

Chimeric progeny

FIGURE 3.28 ES cells can be used to generate mouse chimeras, which breed true for the transfected DNA when the ES cell contributes to the germline.

into 21 bp dsRNA fragments, unwound into single strands, then used by another enzyme, RISC, to find and anneal to mRNAs containing complementary sequence. When a complementary mRNA is found, it is cleaved and destroyed. In practice, this means that the mRNA for any gene can be targeted for destruction by introduction of a dsRNA designed to anneal to the target of interest. The means of introducing the dsRNA depends on the species being targeted; in mammalian cells one method is transfection with DNA encoding a self-annealing RNA that forms a hairpin contain the targeting sequence.

In some multicellular organisms gene deletion is possible, but the process is more complicated than in organisms like yeast. In mammals, the target is usually the genome of an embryonic stem (ES) cell, which is then used to generate a mouse with the knockout. ES cells are derived from the mouse blastocyst (an early stage of development, which precedes implan-

tation of the egg in the uterus). **FIGURE 3.28** illustrates the general approach.

ES cells are transfected with DNA in the usual way (most often by microinjection or electroporation). By using a donor that carries an additional sequence, such as a drug-resistance marker or some particular enzyme, it is possible to select ES cells that have obtained an integrated transgene carrying any particular donor trait. This results in a population of ES cells in which there is a high proportion carrying the marker.

These ES cells are then injected into a recipient blastocyst. The ability of the ES cells to participate in normal development of the blastocyst forms the basis of the technique. The blastocyst is implanted into a foster mother, and in due course develops into a *chimeric* mouse. Some of the tissues of the chimeric mice are derived from the cells of the recipient blastocyst; other tissues are derived from the injected ES cells. The proportions of tissues in the adult

Wild-type gene is modified to provide donor

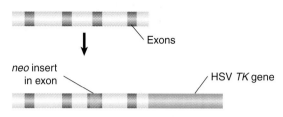

Nonhomologous recombination inserts whole donor unit
at random location

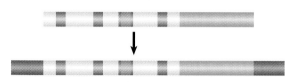

Homologous recombination inserts *neo* into target and
separates *TK* gene

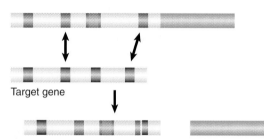

FIGURE 3.29 A transgene containing *neo* within an exon and *TK* downstream can be selected by resistance to G418 and loss of *TK* activity.

FIGURE 3.29 illustrates the knockout technique that is used to disrupt endogenous genes. The basis for the technique is the design of a knockout construct with two different markers that are designed to allow nonhomologous and homologous recombination events in the ES cells to be distinguished. The donor DNA is homologous to a target gene, but has two key modifications. First, the gene is inactivated by interrupting or replacing an exon with a gene encoding a selectable marker (most often the *neo^R* gene that confers resistance to the drug G418 is used). Second, a *counterselectable* marker (a gene that can be selected *against*) is added on one side of the gene; for example, the *TK* gene of the herpes virus.

When this knockout construct is introduced into an ES cell, homologous and nonhomologous recombinations will result in different outcomes. Nonhomologous recombination inserts the entire construct, including the flanking *TK* gene. These cells are resistant to neomycin, and they also express thymidine kinase, which makes them *sensitive* to the drug gancyclovir (thymidine kinase phosphorylates gancyclovir, which makes it toxic). In contrast, homologous recombination involves two exchanges within the sequence of the donor gene, resulting in the loss of the flanking *TK* gene. Cells in which homologous recombination has occurred therefore gain neomycin resistance in the same way as cells that have nonhomologous recombination, but they do *not* have thymidine kinase activity, and so are resistant to gancyclovir. Thus plating the cells in the presence of neomycin plus gancyclovir specifically selects those in which homologous recombination has replaced the endogenous gene with the donor gene.

The presence of the *neo^R* gene in an exon of the donor gene disrupts translation, and thereby creates a null allele. A particular target gene can therefore be knocked out by this means; once a mouse with one null allele has been obtained, it can be bred to generate the homozygote. This is a powerful technique for investigating whether a particular gene is essential, and what functions in the animal are perturbed by its loss. Sometimes phenotypes can even be observed in the heterozygote.

A major extension of ability to manipulate a target genome has been made possible by using the phage *Cre/lox* system to engineer site-specific recombination in a eukaryotic cell. The Cre enzyme catalyzes a site-specific recom-

mouse that are derived from cells in the recipient blastocyst and from injected ES cells varies widely in individual progeny; if a visible marker (such as coat-color gene) is used, areas of tissue representing each type of cell can be seen.

To determine whether the ES cells contributed to the germline, the chimeric mouse is crossed with a mouse that lacks the donor trait. Any progeny that have the trait must be derived from germ cells that have descended from the injected ES cells. By this means, it is known that an entire mouse has been generated from an original ES cell!

When a donor DNA is introduced into the cell, it may insert into the genome by either nonhomologous or homologous recombination. Homologous recombination is relatively rare, probably representing <1% of all recombination events, and thus occurring at a frequency of ~10^{-7}. By designing the donor DNA appropriately, though, we can use selective techniques to identify those cells in which homologous recombination has occurred.

bination reaction between two *lox* sites, which are identical 34-bp sequences (see *Section 15.18, Site-Specific Recombination Resembles Topoisomerase Activity*). **FIGURE 3.30** shows that the consequence of the reaction is to excise the stretch of DNA between the two *lox* sites.

The great utility of the Cre/*lox* system is that it requires no additional components and works when the Cre enzyme is produced in any cell that has a pair of *lox* sites. **FIGURE 3.31** shows that we can control the reaction to make it work in a particular cell by placing the *cre* gene under the control of a regulated promoter. The procedure starts with two mice. One mouse has the *cre* gene, typically controlled by a promoter that can be turned on specifically in a certain cell or under certain conditions. The other mouse has a target sequence flanked by *lox* sites. When we cross the two mice, the progeny have both elements of the system; and the system can be turned on by controlling the promoter of the *cre* gene. This allows the sequence between the *lox* sites to be excised in a controlled way.

The Cre/*lox* system can be combined with the knockout technology to give us even more control over the genome. Inducible knockouts can be made by flanking the *neo^R* gene (or any other gene that is used similarly in a selective procedure) with *lox* sites. After the knockout has been made, the target gene can be reactivated by causing Cre to excise the *neo^R* gene in some particular circumstance (such as in a specific tissue).

FIGURE 3.32 shows a modification of this procedure that allows a knock-in to be created. Basically, we use a construct in which some mutant version of the target gene is used to replace the endogenous gene, replying on the usual selective procedures. Then, when the inserted gene is reactivated by excising the *neo^R* sequence, we have in effect replaced the original gene with a different version.

A useful variant of this method is to introduce a wild-type copy of the gene of interest in which the gene itself (or one of its exons) is flanked by *lox* sites. This results in a normal animal that can be crossed to a mouse containing *Cre* under control of a tissue-specific or otherwise regulated promoter. The offspring of this cross are *conditional knockouts*, in which the function of the gene is lost only in cells that express *Cre*. This is particularly useful for studying genes that are essential for embryonic development; genes in this class would be lethal in homozygous embryos and thus are very difficult to study.

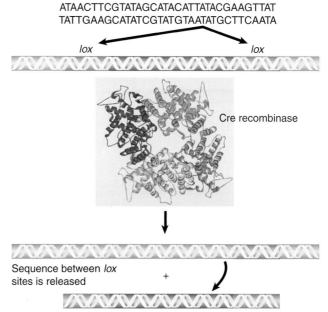

FIGURE 3.30 The Cre recombinase catalyzes a site-specific recombination between two identical *lox* sites, releasing the DNA between them. Structure from Protein Data Bank: 1OUQ. E. Ennifar, et al., *Nucleic Acids Res.* 31 (2003): 5449–5460.

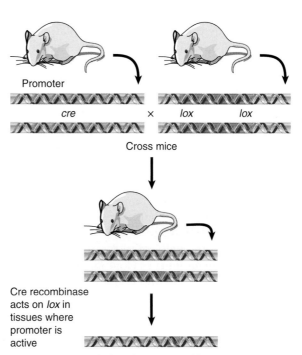

FIGURE 3.31 By placing the Cre recombinase under the control of a regulated promoter, it is possible to activate the excision system only in specific cells. One mouse is created that has a promoter-*cre* construct, and another that has a target sequence flanked by *lox* sites. The mice are crossed to generate progeny that have both constructs. Then excision of the target sequence can be triggered by activating the promoter.

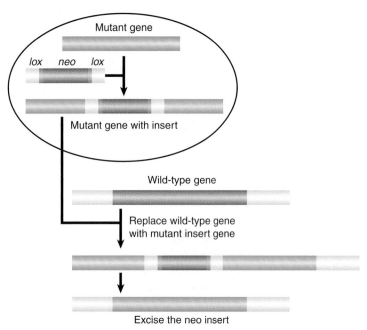

FIGURE 3.32 An endogenous gene is replaced in the same way as when a knockout is made (see Figure 3.29), but the neomycin gene is flanked by *lox* sites. After the gene replacement has been made using the selective procedure, the neomycin gene can be removed by activating Cre, leaving an active insert.

With these techniques, we are able to investigate the functions and regulatory features of genes in whole animals. The ability to introduce DNA into the genome allows us to make changes in it, to add new genes that have had particular modifications introduced *in vitro*, or to inactivate existing genes. Thus it becomes possible to delineate the features responsible for tissue-specific gene expression. Ultimately, we may expect routinely to replace defective genes in the genome in a targeted manner.

3.13 Summary

DNA can be manipulated and propagated using the techniques of cloning. These include digestion by restriction endonucleases, which cut DNA at specific sequences, and insertion into cloning vectors, which permit DNA to be maintained and amplified in host cells such as bacteria. Cloning vectors can have specialized functions as well, such as allowing expression of the product of a gene of interest, or fusion of a promoter of interest to an easily assayed reporter gene.

DNA (and RNA) can be detected nonspecifically by the use of dyes that bind independent of sequence. Specific nucleic acid sequences can be detected using base complementarity. Specific primers can be used to detect and amplify particular DNA targets via PCR. RNA can be reverse transcribed into DNA to be used in PCR; this is known as reverse transcription (RT)-PCR. Labeled probes can be used to detect DNA or RNA on Southern or northern blots, respectively. Proteins are detected on western blots using antibodies.

DNA microarrays are solid supports (usually silicon chips or glass slides) on which DNA sequences corresponding to ORFs or complete genomic sequences are arrayed. Microarrays are used to detect gene expression, for SNP genotyping, and to detect changes in DNA copy number, as well as many other applications.

Protein-DNA interactions can be detected *in vivo* using chromatin immunoprecipitation. The DNA obtained in a chromatin immunoprecipitation experiment can be used as a probe on a genome tiling array to map all localization sites for a given protein in the genome.

New sequences of DNA may be introduced into a cultured cell by transfection or into an animal egg by microinjection. The foreign sequences may become integrated into the genome, often as large tandem arrays. The array appears to be inherited as a unit in a cultured cell. The sites of integration appear to be random. A transgenic animal arises when the integration event occurs into a genome that enters the germ cell lineage. Often a transgene responds to tissue and temporal regulation in a manner that resembles the endogenous gene. Under conditions that promote homologous recombination, an inactive sequence can be used to replace a functional gene, thus creating a knockout, or deletion, of the target locus. Extensions of this technique can be used to make conditional knockouts, where the activity of the gene can be turned on or off (such as by Cre-dependent recombination), and knock-ins, where a donor gene specifically replaces a target gene. Transgenic mice can be obtained by injecting recipient blastocysts with ES cells that carry transfected DNA. Knockdowns, mostly commonly achieved using RNA interference, can be used to eliminate gene products in cell types for which knockout technologies are not available.

4

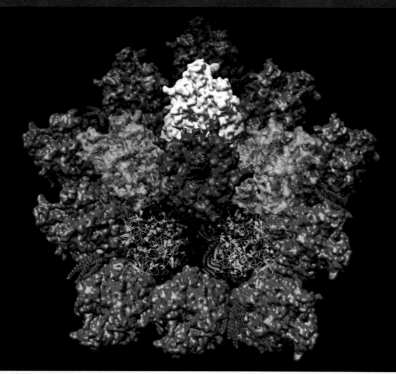

Reproduced from *J. Virol.*, 2006, vol. 80, pp. 12049–12059, DOI and reproduced with permission from the American Society of Microbiology. Photo courtesy of Phoebe L. Stewart, Vanderbilt University Medical Center.

The Interrupted Gene

Edited by Donald Forsdyke

CHAPTER OUTLINE

4.1 Introduction

4.2 An Interrupted Gene Consists of Exons and Introns

- Introns are removed by RNA splicing, which occurs in *cis* in individual RNA molecules.
- Mutations in exons can affect polypeptide sequence; mutations in introns can affect RNA processing and hence may influence the sequence and/or production of a polypeptide.

4.3 Exon and Intron Base Compositions Differ

- The four "rules" for DNA base composition are the first and second parity rules, the cluster rule, and the GC rule. Exons and introns can be distinguished on the basis of all rules except the first.
- The second parity rule suggests an extrusion of structured stem-loop segments from duplex DNA, which would be greater in introns.
- The rules relate to genomic characteristics, or "pressures," that constitute the genome phenotype.

4.4 Organization of Interrupted Genes May Be Conserved

- Introns can be detected when genes are compared with their RNA transcription products by either restriction mapping, electron microscopy, or sequencing.

- The positions of introns are usually conserved when homologous genes are compared between different organisms. The lengths of the corresponding introns may vary greatly, though.
- Introns usually do not encode proteins.

4.5 Exon Sequences under Negative Selection Are Conserved but Introns Vary

- Comparisons of related genes in different species show that the sequences of the corresponding exons are usually conserved, but the sequences of the introns much less so.
- Introns evolve more rapidly than their neighboring exons (in other words, they vary more between species) when the exons are under selective pressure to retain the capacity to encode useful proteins.

4.6 Exon Sequences under Positive Selection Vary but Introns Are Conserved

- Under positive selection an individual happening to have an advantageous mutation survives (in other words, is able to produce more fertile progeny) relative to others without the mutation.
- Due to intrinsic genomic pressures, such as that which conserves the potential to extrude stem-loops from duplex DNA, introns evolve more slowly than exons that are under positive selection pressure.

4.1 Introduction

The simplest form of a gene is a length of DNA that is colinear with a protein. Bacterial genes are almost always of this type, in which a continuous sequence of $3N$ bases encodes a protein of N amino acids. In the 1960s and 1970s it was found in eukaryotes that both ribosomal RNAs (rRNAs) and messenger RNAs (mRNAs) are first synthesized as long precursor transcripts that are subsequently shortened (see *Section 21.20, Production of rRNA Requires Cleavage Events and Involves Small RNAs*). Thus eukaryotic genes are much longer than the functional transcripts they produce. It is natural to think that the shortening involved a trimming of additional, perhaps regulatory, sequences at the 5′ and/or 3′ ends of transcripts, leaving the rRNA or protein-encoding sequence of the precursor intact.

A eukaryotic gene, however, can include additional sequences that lie both *within* and outside the region that is operational with respect to phenotype. Protein-encoding sequences can be interrupted, as can the

5′ and 3′ sequences (UTRs) that flank the protein-encoding sequences within mRNA. The interrupting sequences are removed from the **primary (RNA) transcript** during gene expression, generating an mRNA that includes a continuous base sequence corresponding to the polypeptide product as determined by the genetic code. The sequences of DNA comprising an interrupted protein-encoding gene are divided into the two categories depicted in **FIGURE 4.1**:

- **Exons** are the sequences represented in the mature RNA. A **mature transcript** starts and ends with exons that correspond to the 5′ and 3′ ends of the RNA.
- **Introns** are the sequences that are removed when the primary RNA transcript is processed to give the mature RNA.

The exon sequences are in the same order in the gene and in the RNA, but an **interrupted gene** is longer than its mature RNA product because of the presence of the introns.

The expression of interrupted genes requires an additional process that is not needed for uninterrupted genes. The DNA of an interrupted gene gives rise to an RNA *transcript* corresponding exactly to the sequence of that DNA. This RNA is only a precursor; it is not used for producing protein. Introns are removed from the RNA to give a messenger RNA that consists only of a series of exons. This process, **RNA splicing** (see *Section 2.11, Several Processes Are Required to Express the Protein Product of a Gene*), involves precisely deleting introns from the primary transcript and joining the ends of the RNA on either side of each intron to form a covalently intact molecule (see Chapter 21, *RNA Splicing and Processing*).

In simple terms, a eukaryotic gene comprises the region in the genome between points corresponding to the 5′ and 3′ terminal bases of mature RNA. We know that transcription starts at the 5′ end of the mRNA and usually extends beyond the 3′ end, which is then generated by cleavage of the 3′ extension (see *Section 21.15, The 3′ Ends of mRNAs Are Generated by Cleavage and Polyadenylation*). In some contexts the gene can also be considered to include regions on both sides of the segment corresponding to the primary transcript. These regions may be involved in regulation of the initiation and termination of transcription (see *Section 2.12, Proteins Are* trans-*acting, but Sites on DNA Are* cis-*acting*).

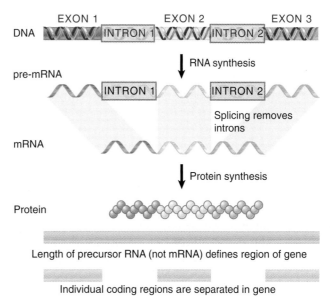

FIGURE 4.1 Interrupted genes are expressed via a precursor RNA. Introns are removed when the exons are spliced together. The mRNA has only the sequences of the exons.

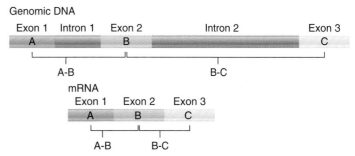

FIGURE 4.2 Exons remain in the same order in mRNA as in DNA, but distances along the gene do not correspond to distances along the mRNA or protein products. The distance from A–B in the gene is smaller than the distance from B–C; but the distance from A–B in the mRNA (and protein) is greater than the distance from B–C.

4.2 An Interrupted Gene Consists of Exons and Introns

Key concepts

- Introns are removed by RNA splicing, which occurs in *cis* in individual RNA molecules.
- Mutations in exons can affect polypeptide sequence; mutations in introns can affect RNA processing and hence may influence the sequence and/or production of a polypeptide.

How does the existence of introns change our view of the gene? Through splicing, exons are joined together in the same order as they occur in DNA, so the colinearity of gene and protein is maintained. From **FIGURE 4.2** we may deduce that the *order* of mutations in a gene remains the same as the order of amino acid replacements (mutations) in the corresponding polypeptide. The *distances* in the gene, however, do not correspond with the distances in the polypeptide. Genetic distances, as seen on a recombination map, bear little relationship to the distances between the corresponding points in the polypeptide. The length of a gene is the length of the initial (precursor) RNA, not the length of the resulting mature mRNA. All exons of a gene are on one RNA molecule, and their splicing together is an *intra*molecular reaction. There is usually no joining of exons carried by different RNA molecules, so there is rarely cross-splicing of sequences.

Mutations that directly affect the sequence of a protein lie in exons. What are the effects of mutations in the introns? The introns are not part of the mature messenger RNA, so mutations in them do not directly affect protein structure. They can affect mRNA production, though—for example, by inhibiting the splicing of exons. A mutation of this sort acts only on the allele that carries it.

Mutations that affect splicing are usually deleterious. The majority are single-base substitutions at the junctions between introns and exons. They may cause an exon to be left out, cause an intron to be included, or cause splicing to occur at an aberrant site. The most common result is to introduce a termination codon that shortens the polypeptide sequence. Thus, intron mutations may affect not only the production of a polypeptide, but also its sequence. About 15% of the point mutations that cause human diseases are caused by disruption of splicing.

Eukaryotic genes are not necessarily interrupted. Some correspond directly with the polypeptide product in the same manner as prokaryotic genes. In the yeast *Saccharomyces cerevisiae*, most genes are uninterrupted. In multicellular eukaryotes most genes are interrupted, and the introns are usually much longer than exons.

4.3 Exon and Intron Base Compositions Differ

Key concepts

- The four "rules" for DNA base composition are the first and second parity rules, the cluster rule, and the GC rule. Exons and introns can be distinguished on the basis of all rules except the first.
- The second parity rule suggests an extrusion of structured stem-loop segments from duplex DNA, which would be greater in introns.
- The rules relate to genomic characteristics, or "pressures," that constitute the genome phenotype.

In the 1940s Erwin Chargaff initiated studies of DNA base composition that led to four "rules," beginning with the *first parity rule* for duplex DNA (see Chapter 1, *Genes Are DNA*). This rule applied to most regions of DNA, including both exons and introns. Base A in one strand of the duplex is matched by a complementary base (T) in the other strand, and base G in one strand of the duplex is matched by a complementary base (C) in the other strand. By extension, the rule applied not only to single bases, but also to dinucleotides, trinucleotides, and oligonucleotides. Thus, GT would pair with its reverse complement AC, and ATG would pair with its reverse complement CAT. There is also

a *second parity rule*, which is that, to a close approximation, the first parity rule, including its extension, applies to single-stranded DNA. The second parity rule applies more closely to introns than to exons, partly due to a further rule—that purines and pyrimidines tend to cluster separately together in DNA. This *cluster rule* as applied to exons is that the purines, A and G, tended to be clustered in one DNA strand of the DNA duplex (usually the non-template strand) and these are complemented by clusters of the pyrimidines, T and C, in the template strand.

The fact that in single-stranded DNA an oligonucleotide is accompanied, in series, by equal quantities of its reverse complementary oligonucleotide, suggests that duplex DNA has the potential to extrude folded stem-loop structures, the stems of which can display base parity and the loops of which can display some degree of base clustering. Indeed, the potential for such secondary structure was found to be greater in introns than exons, especially in exons under positive selection pressure (see *Section 4.6, Exon Sequences under Positive Selection Vary but Introns Are Conserved*).

Finally, there is the *GC rule*, that the overall proportion of G+C in a genome (GC content) tends to be a species-specific character (although individual genes within that genome tend to have distinctive values). The GC content tends to be greater in exons than in introns. Chargaff's four rules are seen to relate to characters or "pressures" that are *intrinsic* to the genome, contributing to what was termed the "genome phenotype" (see *Section 4.11, Genetic Information Is Not Completely Contained in DNA*).

4.4 Organization of Interrupted Genes May Be Conserved

Key concepts

- Introns can be detected when genes are compared with their RNA transcription products by either restriction mapping, electron microscopy, or sequencing.
- The positions of introns are usually conserved when homologous genes are compared between different organisms. The lengths of the corresponding introns may vary greatly, though.
- Introns usually do not encode proteins.

When a gene is uninterrupted, the restriction map of its DNA corresponds with the map of its mRNA. When a gene possesses an intron, the map at each end of the gene corresponds to the map at each end of the message sequence. Within the gene, however, the maps diverge because additional regions that are found in the gene are not represented in the message. Each such region corresponds to an intron. The example of **FIGURE 4.3** compares the restriction maps of a β-globin gene and its mRNA. There are two introns, each of which contains a series of restriction sites that are absent from the **cDNA**. The pattern of restriction sites in the exons is the same in both the cDNA and the gene. Better than mapping, comparison of the base sequences of a gene and its mRNA permits precise definition of introns. An intron usually has no open reading frame. An intact reading frame is created in an mRNA sequence by the removal of the introns from the primary transcript.

The structures of eukaryotic genes show extensive variation. Some genes are uninterrupted and their sequences are colinear with those of the corresponding mRNAs. Most multicellular eukaryotic genes are interrupted, but the introns vary enormously in both number and size.

All major classes of genes may be interrupted: nuclear genes encoding proteins, nucleolar genes encoding rRNA, and genes encoding tRNA. Thus the interruptions are not confined to protein-encoding sequences. Interruptions also are found in mitochondrial genes in unicellular/oligocellular eukaryotes and in chloroplast genes. Interrupted genes do not appear to be excluded from any class of eukaryote (though rare in some individual species) and have even been found in bacteria and bacteriophages. They are, however, extremely rare in prokaryotic genomes.

Some interrupted genes possess only one or a few introns. The globin genes provide a much-studied example (see *Section 4.10, Members of a Gene Family Have a Common Organization*). The two general types of globin gene, α and β, share a common type of structure. The consistency of the organization of mammalian globin genes is evident from the structure of a "generic" globin gene summarized in **FIGURE 4.4**.

Interruptions occur at homologous positions (relative to the coding sequence) in all known active globin genes, including those of mammals, birds, and frogs. The first intron is always fairly short, and the second usually is longer, but the actual lengths can vary. Most of the variation in overall lengths between different globin genes results from the variation in the second intron. In mice, the second intron in the α-globin gene is only 150 bp long, so the overall length of the gene is 850 bp, compared with the major β-globin gene for which the intron length of 585 bp gives the gene a total length of 1382 bp. The variation in length of the genes is much greater than the range of lengths of the mRNAs (α-globin mRNA = 585 bases; β-globin mRNA = 620 bases).

The example of DHFR (dihydrofolate reductase), a somewhat

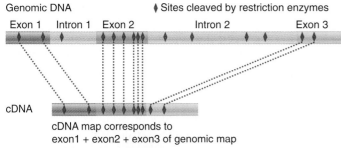

FIGURE 4.3 Comparison of the restriction maps of cDNA and genomic DNA for mouse β-globin shows that the gene has two introns that are not present in the cDNA. The exons can be aligned exactly between cDNA and gene.

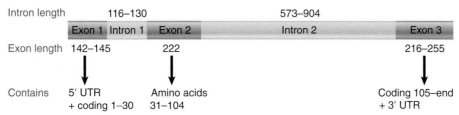

FIGURE 4.4 All functional globin genes have an interrupted structure with three exons. The lengths indicated in the figure apply to the mammalian β-globin genes. Note that exon 2, which is entirely protein-encoding, can be 31 bases in length. This is not divisible by 3, so here an intron boundary intersects a codon.

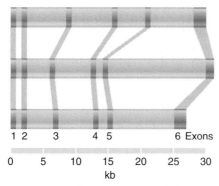

FIGURE 4.5 Mammalian genes for DHFR have the same relative organization of rather short exons and very long introns, but vary extensively in the lengths of introns.

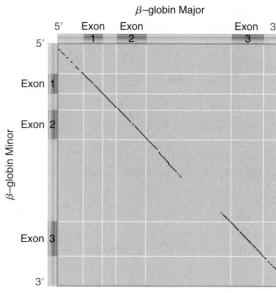

FIGURE 4.6 The sequences of the mouse β^{maj}- and β^{min}-globin genes are closely related in coding regions but differ in the flanking regions and long intron. Data provided by Philip Leder, Harvard Medical School.

larger gene, is shown in **FIGURE 4.5**. The mammalian DHFR gene is organized into six exons that correspond to a 2000-base mRNA. They extend over a great length of DNA because the introns are very long. In three mammals the exons remain essentially the same, and the relative positions of the introns are unaltered. The lengths of individual introns vary extensively, though, resulting in a variation in the length of the gene from 25 to 31 kb.

The globin and DHFR genes present examples of a general phenomenon: *Genes that are related by evolution have related organizations with conservation of the positions of (at least some) of the introns.*

4.5 Exon Sequences under Negative Selection Are Conserved but Introns Vary

Key concepts

- Comparisons of related genes in different species show that the sequences of the corresponding exons are usually conserved, but the sequences of the introns much less so.
- Introns evolve more rapidly than their neighboring exons (in other words, they vary more between species) when the exons are under selective pressure to retain the capacity to encode useful proteins.

Is a gene unique in its genome? The answer can be ambiguous. The entire length of the gene is unique as such, but its exons often are related to those of other genes. As a general rule,

when two genes are related, the relationship between their exons is closer than the relationship between their introns. In an extreme case, the exons of two genes may code for the same polypeptide sequence, whereas the introns are different. This implies that the two genes originated by a duplication of some common ancestral gene. Then differences accumulated between the copies, but they were less in the exons because of the need to encode common protein functions.

As we will see later, exons can be considered as basic building blocks that are assembled in various combinations. Some genes may arise by duplication and translocation of individual exons. A gene may have some exons that are related to exons of another gene, but the other exons may be unrelated. Usually the introns are not related in such cases.

The relationship between two genes can be plotted in the form of a dot matrix, as in **FIGURE 4.6**. A dot is placed to indicate each position at which the same base is found in each gene. The dots form a line at an angle of 45° if two sequences are identical. The line is broken by regions that lack similarity and is displaced laterally or vertically by deletions or insertions in one sequence relative to the other.

When the two β-globin genes of the mouse are compared, such a line extends through the three exons and through the small intron. The line peters out in the flanking regions and in the large intron. This is a typical pattern, in which coding sequences are well related and the relationship can sometimes extend beyond the boundaries of the exons. The pattern is usually lost, though, in longer introns and in the regions on either side of the gene.

The overall degree of divergence between two exons is related to the differences between the proteins. It is caused mostly by base substitutions. In the translated regions, the exons are under the constraint of encoding amino acids, so they are limited in their potential to change. In other words, the exon sequences are conserved by the *negative selection* of individuals in which the sequences have changed (have not been conserved) to result in a phenotype that is less able to survive and produce fertile progeny. Many of the changes do not affect codon meanings because they change a codon into one that represents the same amino acid. In this case, the polypeptide will not change and negative selection will not operate on the phenotype conferred by the polypeptide. Similarly, changes can occur in untranslated regions (corresponding to the 5′ and 3′ UTRs of the mRNA) without affecting the nature of the polypeptide.

In the corresponding introns, the pattern of divergence involves both changes in size (due to deletions and insertions) and base substitutions. Introns evolve much more rapidly than exons when the exons are under negative selection pressure. When a gene is compared in different species, there are times when its exons are homologous, but its introns have diverged so much that a correspondence between their sequences cannot be recognized.

Mutations generally occur at the same rate in both exons and introns, but are removed more effectively from the exons in the population by adverse selection of individuals with mutated exons. In the absence of this constraint imposed by a protein-encoding function, though, an intron appears freely able to accumulate point substitutions and other changes. Indeed, it is sometimes possible to locate exons in uncharted sequences by virtue of their conservation relative to introns (see *Section 5.6, Eukaryotic Protein-Coding Genes Can Be Identified by the Conservation of Exons*). From this description it is all too easy to conclude that introns do not have a sequence-specific function. Genes under positive selection, however, cast a different light on the problem.

4.6 Exon Sequences under Positive Selection Vary but Introns Are Conserved

Key concepts

- Under positive selection an individual happening to have an advantageous mutation survives (in other words, is able to produce more fertile progeny) relative to others without the mutation.
- Due to intrinsic genomic pressures, such as that which conserves the potential to extrude stem-loops from duplex DNA, introns evolve more slowly than exons that are under positive selection pressure.

A mutation that confers a more advantageous phenotype on an organism, relative to its unmutated fellow organisms, may result in the preferential survival (*positive selection*) of that organism. Pathogenic bacteria are killed by an antibiotic, but a bacterium with a mutation that confers antibiotic resistance survives (in other words, is positively selected). Mutations conferring venom-resistance on prey of venomous snakes can result in the positive selection of that prey relative to its fellows that succumb to the poison (in other words, are negatively selected). Likewise, a snake that, when confronted by a venom-resistant prey population, has a mutation that enhances the power of its venom, will be positively selected. This can trigger an attack–defense cycle—an "arm's race" between two protagonist species.

In such situations the pattern of exon conservation and intron variation seen in genes under negative selection can be reversed because exons evolve faster than introns. Thus, a plot similar to Figure 4.6 will have lines in introns and gaps in exons. Another way of showing this is to plot base substitutions along the length of a gene. **FIGURE 4.7** shows a plot of the substitutions observed when two snake venom alkaline phosphatase genes are compared. The protein-encoding parts of exons (2, 3, and the first half of exon 4) have many base substitutions (in other words, they are varying), whereas the three introns have relatively few (in other words, they are conserved).

What is being conserved in introns? First, intron sequences needed for RNA splicing—the 5′ and 3′ splice sites and the branch site—are conserved (see Chapter 21, *RNA Splicing and Processing*). In addition to these,

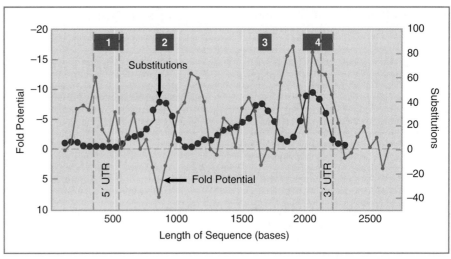

FIGURE 4.7 The sequences of snake venom phospholipase genes differ in coding regions, but are closely related in introns and flanking regions. Fold potential (here the contribution of base order to the potential to extrude stem–loop structures) is low (more positive) in the protein-encoding exons and high (more negative) in introns. The positions of the four exons are shown as numbered boxes. Modified from D. R. Forsdyke, Conservation of Stem-Loop Potential in Introns of Snake Venom Phospholipase A2 Genes: An Application of FORS-D Analysis, *Mol. Biol. Evol.*, vol. 12 (6), pp. 1157–1165, by permission of Oxford University Press.

base order has been adapted to promote the potential of the duplex DNA in the region to extrude stem–loop structures (fold potential). Thus, a plot of base order-dependent fold potential along the length of the gene shows that fold potential (measured in negative units) is high (more negative) in introns, and low (more positive) in exons (Figure 4.7). This reciprocal relationship between substitution frequency and the contribution of base order to fold potential is a character-istic of DNA sequences under positive selection. Indeed, the low (more positive) value of fold potential in an exon provides evaluation of the extent to which it has been under positive selection, without the need to compare two sequences (the classical way of determining if selection is positive or negative).

4.7 Genes Show a Wide Distribution of Sizes

Key concepts

- Most genes are uninterrupted in *S. cerevisiae*, but are interrupted in multicellular eukaryotes.
- Exons are usually short, typically coding for <100 amino acids.
- Introns are short in unicellular/oligocellular eukaryotes, but can be many kb in multicellular eukaryotes.
- The overall length of a gene is determined largely by its introns.

FIGURE 4.8 shows the relative extents of interruption of genes in a yeast, an insect, and mammals. In *Saccharomyces cerevisiae*, the great majority of genes (>96%) are uninterrupted, and those that have exons are usually compact. There are no *S. cerevisiae* genes with more than four exons.

In insects and mammals the situation is reversed. Only a few genes have uninterrupted coding sequences (6% in mammals). Insect genes tend to have a small number of exons—typically fewer than 10. Mammalian genes are split into more pieces. Approximately

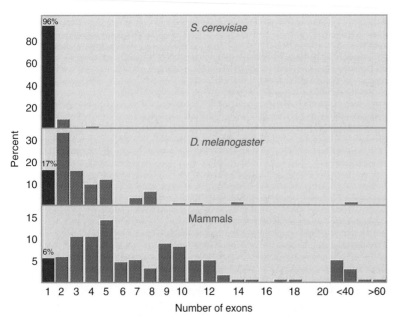

FIGURE 4.8 Most genes are uninterrupted in yeast, but most genes are interrupted in flies and mammals. (Uninterrupted genes have only one exon and are totaled in the leftmost column.)

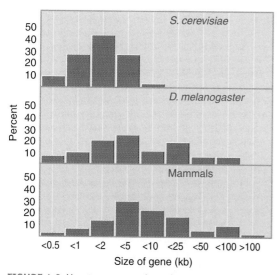

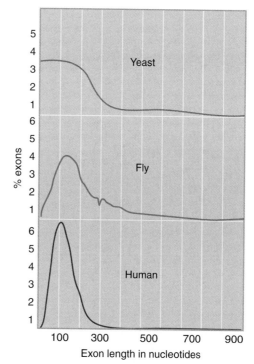

FIGURE 4.9 Yeast genes are short, but genes in flies and mammals have a dispersed bimodal distribution extending to very long sizes.

FIGURE 4.10 Exons coding for proteins usually are short.

50% of mammalian genes have >10 introns. Examining how this impacts gene size, we see in **FIGURE 4.9** that there is a striking difference between yeast and multicellular eukaryotes. The average yeast gene is 1.4 kb long, and very few are longer than 5 kb. The predominance of interrupted genes in multicellular eukaryotes means that the gene can be much larger than the unit that codes for polypeptide. Relatively few genes in flies or mammals are shorter than 2 kb, and many have lengths between 5 kb and 100 kb. The average human gene is 27 kb long (see Figure 6.12). The longest known is dystrophin, at 2000 kb.

In fungi (except some yeasts, such as *S. cerevisiae*), the majority of genes are interrupted, but they have a relatively small number of exons (<6) and are fairly short (<5 kb). The switch to long genes occurs within multicellular eukaryotes. The fruit fly has a markedly bimodal distribution of gene length—many short and some long. With increase in gene length, the relationship between genome size and organism complexity is lost (see Figure 8.7).

FIGURE 4.10 shows that exons encoding stretches of protein tend to be fairly small. In multicellular eukaryotes, the average exon codes for ~50 amino acids, and the general distribution fits well with the idea that genes have evolved by the slow addition of units that code for small, individual domains of proteins (see *Section 8.6, How Did Interrupted Genes Evolve?*). There are no significant differences in the sizes of exons in different types of multicellular eukaryotes, although the distribution is more compact in vertebrates, for which there are few exons longer than 200 bp. In yeast, there

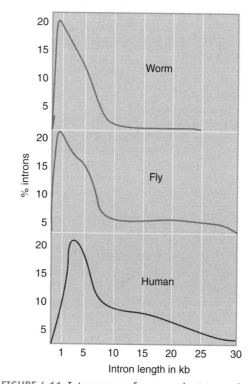

FIGURE 4.11 Introns range from very short to very long.

are some longer exons that represent uninterrupted genes for which the coding sequence is intact. There is a tendency for exons coding for untranslated 5′ and 3′ regions to be longer than those that encode proteins.

FIGURE 4.11 shows that introns vary widely in size. In worms and flies, the average intron

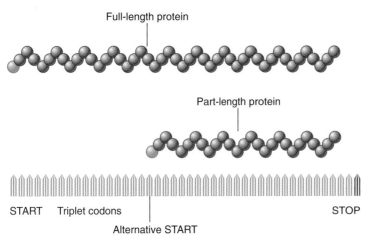

FIGURE 4.12 Two proteins can be generated from a single gene by starting (or terminating) expression at different points.

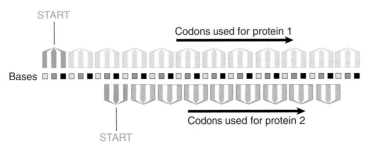

FIGURE 4.13 Two genes may share the same sequence by reading codons in different frames.

4.8 Some DNA Sequences Code for More Than One Polypeptide

Key concepts

- Usage of alternative translation initiation or termination codons allows one polypeptide to be equivalent to a fragment of another.
- Different polypeptides can be produced from the same sequence of DNA when the mRNA is read in different reading frames (as two overlapping genes).
- Otherwise identical polypeptides, differing by the presence or absence of certain regions, can be generated by differential (alternative) splicing when certain exons are included or excluded. This may take the form of including or excluding individual exons, or of choosing between alternative exons.

Most genes consist of a sequence that, when appropriately spliced at the RNA level, codes for one polypeptide. Sometimes, though, the sequence codes for more than one polypeptide.

Overlapping genes occur in the relatively simple situation in which one gene is part of another. The first half (or second half) of a gene is used independently to specify a polypeptide that represents the first (or second) half of the polypeptide specified by the full gene (see **FIGURE 4.12**). It is as if cleavage had taken place in the polypeptide product to generate a part-length as well as a full-length form.

Two genes overlap in a more subtle manner when the same sequence of DNA is shared between two different polypeptides. This situation arises when a sequence is translated in more than one reading frame. An mRNA sequence usually is read in only one of the three potential reading frames. In some cases (found especially in viruses and mitochondria), though, there is an overlap between two adjacent genes so that codons are read in different reading frames (see **FIGURE 4.13**). The distance of overlap is usually relatively short, so that most of the sequence representing the polypeptide retains a unique coding function.

In some cases, genes can be *nested*. This occurs when a complete gene is found within the intron of a larger "host" gene. Nested genes often lie on the strand opposite to that of the host gene.

is not much longer than the exons. There are no very long introns in worms, but flies contain many. In vertebrates, the size distribution is much wider, extending from approximately the same length as the exons (<200 bp) up to 60 kb in extreme cases. The distribution curves for worms tend to be unimodal, and for humans tend to be bimodal, whereas flies combine both patterns (trimodal).

There is no correlation between gene size and mRNA size in multicellular eukaryotes, nor is there a good correlation between gene size and number of exons. The size of a gene therefore depends primarily on the lengths of its individual introns. In mammals, insects, and birds, the "average" gene is approximately 5× the length of its mRNA.

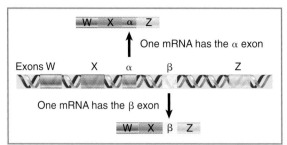

FIGURE 4.14 Alternative splicing generates the α and β variants of troponin T.

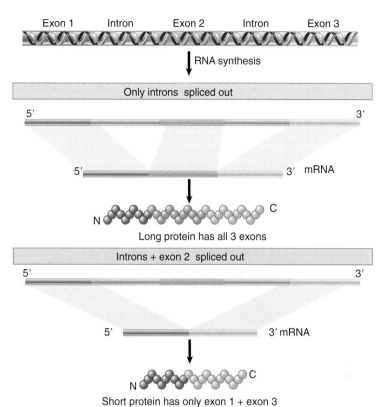

FIGURE 4.15 Alternative splicing uses the same pre-mRNA to generate mRNAs that have different combinations of exons.

In some genes there are switches in the pathway for connecting the exons that result in *alternative* patterns of gene expression. A single gene may generate a variety of mRNA products that differ in their content of exons. Certain exons may be optional; in other words, they may be included or spliced out. There also may be exons that are treated as mutually exclusive—one or the other is included, but not both. The alternative mRNA forms that result from differential splicing produce proteins in which one part is common and the other part varies.

In some cases, the alternative means of expression do not affect the sequence of the polypeptide. For example, changes that affect the 5′ UTR or the 3′ UTR may have regulatory consequences, but the same polypeptide is made. In other cases, one exon is substituted for another, as in **FIGURE 4.14**.

In this example, the polypeptides produced by the two mRNAs contain sequences that overlap extensively, but are different within the alternatively spliced region. The 3′ half of the troponin T gene of rat muscle contains five exons, but only four are used to construct an individual mRNA. Three exons, *WXZ*, are the same in both expression patterns. In one pattern, however, the α exon is spliced between *X* and *Z*; in the other pattern, the β exon is used. The α and β forms of troponin T differ in the sequence of the amino acids between sequences *W* and *Z*, depending on which of the alternative exons—α or β—is used. Either one of the α and β exons can be used to form an individual mRNA, but both cannot be used in the same mRNA.

FIGURE 4.15 shows that **alternative splicing** can lead to the inclusion of an exon in some mRNAs while leaving it out of others. An ini-tial single transcript can be spliced in either of two ways. In the first pathway, two introns are spliced out and the three exons are joined together. In the second pathway, the second exon is not recognized. As a result, a single large intron is spliced out. This intron consists of intron 1 + exon 2 + intron 2. In effect, exon 2 has been treated in this pathway as if it were part of a single intron. The pathways produce two polypeptides that are the same at their ends, but one has an additional sequence in the middle. So the region of DNA codes for more than one polypeptide. (Other types of combinations that are produced by alternative splicing are discussed in *Section 21.12, Alternative Splicing Is a Rule, Rather Than an Exception, in Multicellular Eukaryotes*).

Sometimes two pathways operate simultaneously, with a certain proportion of the RNAs being spliced in each way. Sometimes the pathways are alternatives that are expressed under different conditions—one in one cell type and another in another cell type.

So, displaying an apparent need for economy of space, alternative (differential) splicing can generate polypeptides with overlapping sequences from a single stretch of DNA. Multiple products can be made from an individual locus. Yet the genomes of multicellular eukaryotes appear to be extremely spacious and large genes are often quite dispersed. Alternative splicing expands the number of polypeptides relative to the number of genes by ~15% in flies and worms, but has much bigger effects in humans, for which ~60% of genes may have alternative modes of expression (see *Section 6.5, The Human Genome Has Fewer Genes Than Originally Expected*). About 80% of the alternative splicing events result in a change in the polypeptide sequence.

4.9 Some Exons Can Be Equated with Protein Functional Domains

Key concepts

- Proteins can consist of independent functional modules the boundaries of which, in some cases, can be equated with those of exons.
- The exons of some genes appear homologous to the exons of others, suggesting a common exon ancestry.

The issue of the evolution of interrupted genes will be more fully considered in *Section 8.6, How Did Interrupted Genes Evolve?*. If current proteins evolved by combining ancestral protein units that were originally separate, though, the accretion of units is likely to have occurred sequentially, with one exon being added at a time. For an organism so endowed to be positively selected, each accretion would have to improve upon the advantages of prior accretions. Are the different function-encoding segments from which these genes may have originally been pieced together reflected in their present structures? If a protein sequence were randomly interrupted, sometimes the interruption would intersect a domain and sometimes it would lie between domains. If we can equate the functional domains of current proteins with the individual exons of the corresponding genes, then this would suggest selective interdomain interruptions rather than random ones.

In some cases there is a clear relationship between the structures of a gene and its protein product, but these may be special cases. The example *par excellence* is provided by the immunoglobulin (antibody) proteins—an extracellular system for self/not-self discrimination that aids the elimination of foreign pathogens. Immunoglobulins are encoded by genes in which every exon corresponds exactly with a known functional protein domain. Banks of variable sequence domains are tapped so that each cell acquires the ability to secrete a cell-specific immunoglobulin with distinctive binding capacity for a foreign antigen that, perchance, may one day penetrate the organism's extracellular space (see Chapter 18, *Recombination in the Immune System*). FIGURE 4.16 compares the structure of an immunoglobulin with its gene.

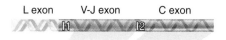
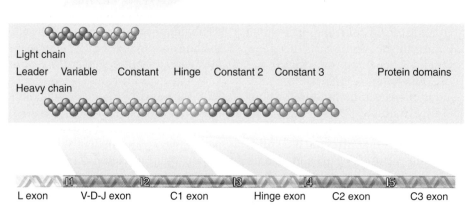

FIGURE 4.16 Immunoglobulin light chains and heavy chains are coded by genes whose structures (in their expressed forms) correspond with the distinct domains in the protein. Each protein domain corresponds to an exon; introns are numbered 1 to 5.

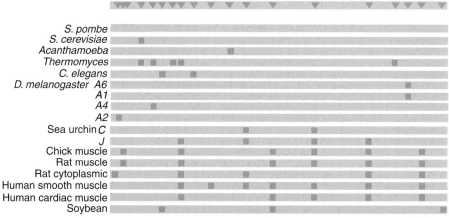

FIGURE 4.20 Actin genes vary widely in their organization. The sites of introns are indicated by dark boxes.

process of intron insertion continued independently in the different lines of evolution.

Irrespective of whether introns were present in actin genes early or late, there appears to have been no consistent guidance from actin protein domains or subdomains as to where introns should locate. On the other hand, when exons are under negative selection (resulting in homology conservation), in-series recombination between members of an expanding gene family (that could cause a contraction in family size) would be decreased by intron diversification (resulting in homology interruption), and introns would come to reside where this could best be achieved. Alleles would have similar exons and introns, so in-parallel interallelic recombination (as in meiosis) would be unimpaired until divergence into species occurred—a process that could be accompanied by intron relocations. The relationships between the intron locations found in different species could then be used to construct a tree for the evolution of the actin gene.

Thus, the relationship between exons and protein domains is somewhat erratic. In some cases there is a clear 1:1 relationship; in others no pattern can be discerned, as if the protein had been randomly interrupted irrespective of the underlying domain structure. One possibility is that the removal of introns has fused the adjacent exons. This means that the intron must have been precisely removed, without changing the integrity of the coding region. An alternative is that some introns arose by insertion into a coherent domain. Together with the variations that we see in exon placement in cases such as the actin genes, this argues that intron positions can be adjusted in the course of evolution.

The equation of at least some exons with protein domains, and the appearance of related exons in different proteins, supports the view that the duplication and juxtaposition of exons has played an important role in evolution. It is possible that the number of ancestral exons—from which all proteins have been derived by duplication, variation, and recombination—could be relatively small (a few thousand or tens of thousands). By taking exons as the building blocks of evolution, this view is consistent with the "introns early" model (the hypothesis that introns have always been part of gene structures and that some lineages have subsequently lost them; see *Section 8.6, How Did Interrupted Genes Evolve?*) for the origin of genes encoding proteins.

4.11 Genetic Information Is Not Completely Contained in DNA

Key concepts

- Genetic information includes not only that related to characters corresponding to the conventional phenotype, but also that related to characters (pressures) corresponding to the genome phenotype.
- In certain contexts, the definition of the gene can be seen as reversed from "one gene:one protein" to "one protein:one gene."
- Positional information may be important in development.
- Sequences transferred "horizontally" from other species to the germline could locate within introns or intergenic DNA and thence transfer "vertically" through the generations. Some of these may be involved in intracellular nonself recognition.

By genetic information we mean all information that passes "vertically" through the germline, not just genic information. The word "gene" and its adjective "genic" have different meanings in different contexts, but in most circumstances there is little confusion when context is considered. In situations in which a stretch of DNA is responsible for production of one particular polypeptide, current usage regards the entire sequence of DNA—from the first point represented in the messenger RNA to the last point corresponding to its end—as comprising the "gene," exons, introns, and all.

When the sequences representing polypeptides overlap or have alternative forms of expression, we may reverse the usual description of the gene. Instead of saying "one gene–one polypeptide," we may describe the relationship as "one polypeptide–one gene." So we regard the sequence involved in production of the polypeptide (including introns as well as exons) as constituting the gene, while recognizing that from the perspective of another polypeptide, part of this same sequence also belongs to *its* gene. This allows the use of descriptions such as "overlapping" or "alternative" genes.

We can now see how far we have come from the one gene:one enzyme hypothesis of the early part of the twentieth century. The driving question was then the nature of the gene. It was thought that genes represented "ferments" (enzymes), but what was the fundamental nature of ferments? Once it was discovered that most genes represent proteins, the paradigm became fixed as the concept that every genetic unit functions through the synthesis of a particular protein. Either directly or indirectly, protein-encoding pressure was responsible for what we can now refer to as the *conventional phenotype*. We now recognize that genetic units corresponding to polypeptides may also include information corresponding to the *genome phenotype*—manifestations of which include fold pressure, purine-loading (AG) pressure and GC pressure. There may be conflict between different pressures, such as competition for space in the gamete that will transfer genomic information to the next generation. Thus, a protein might function most efficiently with the basic amino acid lysine (codon AAA) in a certain position. GC pressure might require the substitution of another basic amino acid, though—arginine (codon CGG). Alternatively, fold pressure might require the corresponding nucleic acid to fold into a stem-loop structure

where CCG would pair, in antiparallel configuration, with the arginine codon. A lysine codon in this position would disrupt the structure, so again a less efficient polypeptide would have to suffice.

The conventional phenotype, however, remains the central paradigm of molecular biology: A genic DNA sequence either directly encodes a particular polypeptide or is adjacent to the segment that actually codes for that polypeptide. How far does this paradigm take us beyond explaining the basic relationship between genes and proteins?

The development of multicellular organisms rests on the use of different genes to generate the different cell phenotypes of each tissue. The expression of genes is determined by a regulatory network that takes the form of a cascade. Expression of the first set of genes at the start of embryonic development leads to expression of the genes involved in the next stage of development, which in turn leads to a further stage, and so on until all the tissues of the adult are functioning. The molecular nature of this regulatory network is still somewhat unknown, but we assume that it consists of genes that code for products (often protein, but sometimes RNA) that can act on other genes.

Although such a series of interactions is almost certainly the means by which the developmental program is executed, we can ask whether it is entirely sufficient. One specific question concerns the nature and role of *positional information*. We know that all parts of a fertilized egg are not equal; one of the features responsible for development of different tissue parts from different regions of the egg is location of information (presumably specific macromolecules) within the cell.

We do not fully understand how these particular regions are formed, though particular examples have been well studied (see *Section 22.10, Some Eukaryotic mRNA Are Localized to Specific Regions of a Cell*). We assume, however, that the existence of positional information in the egg leads to the differential expression of genes in the cells subsequently formed in these regions. This leads to the development of the adult organism, which in turn leads to the development of an egg with the appropriate positional information.

This possibility prompts us to ask whether some information needed for development of the organism is contained in a form that we cannot directly attribute to a sequence of DNA (although the expression of particular sequences may be needed to perpetuate the positional information). Put in a more general way, we might ask the following: when we read out the entire sequence of DNA comprising the genome of some organism and interpret it in terms of proteins and regulatory regions, could we in principle construct an organism (or even a single living cell) by controlled expression of the proper genes?

Once tissues and organs have developed they not only have to be maintained, but also protected against potential pathogens. Banks of variable region genes have diversified in the germline, and continue to diversify somatically, to endow multicellular organisms with the ability to (i) respond extracellularly by the synthesis of immunoglobulin antibodies directed against pathogens (see Figure 4.16), and (ii) remember past pathogens so that future responses will be faster and greater (immunological memory; see Chapter 18, *Recombination in the Immune System*). Should it escape such *extracellular* defenses, though, the nucleic acid of a pathogenic virus could gain entry to cells and *intracellular* defenses, perhaps functioning like restriction enzymes (see *Section 3.2, Nucleases*) would be needed.

While so far only resting on evidence from bacteria infected by bacteriophages (see Chapter 27, *Phage Strategies*), host defenses could include rapid local or genome-wide transcription of DNA in the hope that the transcripts would happen to be "antisense" (capable of base-pairing with pathogen "sense" transcripts) to form double-stranded RNAs, which could then act as an alarm signal triggering secondary defenses (see the example of bacterial CRISPRs discussed in *Section 30.4, Bacteria Contain Regulator RNAs*). The host could create a memory for previous intracellular invaders by converting some pathogen transcripts into DNA by reverse transcription and inserting them into its genome in inactive form for future rapid transcription, as antisense, in times of stress. Thus, some pathogen nucleic acid might enter the germline "horizontally" (within a generation) so that the parental memory of the pathogen could subsequently be transferred "vertically" to offspring. The diversity of some elements found within introns and extragenic DNA (see Chapter 17, *Transposable Elements and Retroviruses*) could in part reflect such past pathogen attacks.

▣ 4.12 Summary

Eukaryotic genomes contain genes that are interrupted by intron sequences. The proportion of interrupted genes is low in yeasts and increases in the lower eukaryotes; few genes are uninterrupted in higher eukaryotes. The size of a gene is determined primarily by the lengths of its introns. The range of gene sizes in mammals is generally from 1 to 100 kb, but there are some that are even larger.

Introns are found in all classes of eukaryotic genes, both those encoding protein products and those encoding independently functioning RNAs. The structure of an interrupted gene is the same in all tissues. Exons are spliced together at the RNA level in the order of their locations in DNA, and the introns, which usually have no protein-encoding function, are removed. Some genes are expressed with alternative splicing patterns, so a particular sequence is removed as an intron in some situations, but retained as an exon in others.

Positions of introns often are found to be conserved when the organization of homologous genes is compared between species. In genes under negative selection pressure, intron sequences vary—and may even appear unrelated—although exon sequences remain closely related. This conservation of exons, which relates to conventional phenotypic characters, can be used to identify related genes in different species. In genes under positive selection pressure, however, exon sequences vary, although intron sequences can remain more similar. This conservation of introns relates to characters corresponding to the genome phenotype, such as fold pressure, which may relate to error correction in DNA.

Some genes share some of their exons with other genes, suggesting that they have been assembled by addition of exons representing individual polypeptide modules of the protein. Such modules may have been incorporated into a variety of different proteins and sometimes correspond to functional domains. The idea that genes have been assembled by accretion of exons is consistent with the hypothesis that introns were present in the genes of ancestral organisms, and thus facilitating the assembly process. Some of the relationships between homologous genes can be explained by loss of introns from the primordial genes, with different introns being lost in different lines of descent.

References

4.1   Introduction

Reviews

Crick, F. (1979) Split genes and RNA splicing. *Science* 204, 264–271.

Harris, H. (1994) An RNA heresy in the fifties. *Trends Biochem. Sci.* 19, 303–305.

Hong, X., Schofield, D. G., and Lynch, M. (2006). Intron size, abundance, and distribution within untranslated regions of genes. *Mol. Biol. Evol.* 23, 2392–2404.

Research

Glover, D. M. and Hogness, D. S. (1977). A novel arrangement of the 8S and 28S sequences in a repeating unit of *D. melanogaster* rDNA. *Cell* 10, 167–176.

Scherrer, K. et al. (1970). Nuclear and cytoplasmic messenger-like RNAs and their relation to the active messenger RNA in polyribosomes of HeLa cells. *Cold Spring Harb. Symp. Quant. Biol.* 35, 539–554.

4.2   An Interrupted Gene Consists of Exons and Introns

Review

Forsdyke, D. R. (2006). Exons and introns. In: *Evolutionary Bioinformatics*. Springer, New York, pp. 207–224. *See also* http://post.queensu.ca/~forsdyke/introns.htm

4.3   Exon and Intron Base Compositions Differ

Reviews

Forsdyke, D. R. and Mortimer, J. R. (2000). Chargaff's legacy. *Gene* 261, 127–137. (*See* http://post.queensu.ca/~forsdyke/bioinfo2.htm)

Forsdyke, D. R. and Bell, S. J. (2004). Purine-loading, stem-loops, and Chargaff's second parity rule: a discussion of the application of elementary principles to early chemical observations. *Applied Bioinformatics* 3, 3–8. (*See* http://post.queensu.ca/~forsdyke/bioinfo5.htm.)

Research

Babak, T., Blencowe, B. J., and Hughes, T. R. (2007). Considerations in the identification of functional RNA structural elements in genomic alignments. *BMC Bioinf.* 8, article number 33.

Bechtel, J. M. et al. (2008). Genomic mid-range inhomogeneity correlates with an abundance of RNA secondary structure. *BMC Genomics* 9, article number 284.

Bultrini, E. et al. (2003). Pentamer vocabularies characterizing introns and intron-like inter-genic tracts from *Caenorhabditis elegans* and *Drosophila melanogaster*. *Gene* 304, 183–192.

Ko, C. H. et al. (1998). U-richness is a defining feature of plant introns and may function as an intron recognition signal in maize. *Plant Mol. Biol.* 36, 573–583.

Zhang, C., Li, W.-H., Krainer, A. R., and Zhang, M. Q. (2008). RNA landscape of evolution for optimal exon and intron discrimination. *Proc. Natl. Acad. Sci. USA* 105, 5797–5802.

4.4   Organization of Interrupted Genes May Be Conserved

Review

Fedoroff, N. V. (1979). On spacers. *Cell* 16, 697–710.

Research

Berget, S. M., Moore, C., and Sharp, P. (1977). Spliced segments at the 5′ terminus of adenovirus 2 late mRNA. *Proc. Natl. Acad. Sci. USA* 74, 3171–3175.

Chow, L. T., Gelinas, R. E., Broker, T. R., and Roberts, R. J. (1977). An amazing sequence arrangement at the 5′ ends of adenovirus 2 mRNA. *Cell* 12, 1–8.

Jeffreys, A. J. and Flavell, R. A. (1977). The rabbit β-globin gene contains a large insert in the coding sequence. *Cell* 12, 1097–1108.

4.6   Exon Sequences under Positive Selection Vary but Introns Are Conserved

Forsdyke, D. R. (1995). Conservation of stem-loop potential in introns of snake venom phospholipase A_2 genes: an application of FORS-D analysis. *Mol. Biol. Evol.* 12, 1157–1165.

Forsdyke, D. R. (1995). Reciprocal relationship between stem-loop potential and substitution density in retroviral quasispecies under positive Darwinian selection. *J. Mol. Evol.* 41, 1022–1037. (*See* http://post.queensu.ca/~forsdyke/hiv01.htm.)

Forsdyke, D. R. (1996). Stem-loop potential in MHC genes: a new way of evaluating positive Darwinian selection. *Immunogenetics* 43, 182–189.

4.7   Genes Show a Wide Distribution of Sizes

Hawkins, J. D. (1988). A survey of intron and exon lengths. *Nucleic Acids Res.* 16, 9893–9905.

Naora, H. and Deacon, N. J. (1982). Relationship between the total size of exons and introns in protein-coding genes of higher eukaryotes. *Proc. Natl. Acad. Sci. USA* 79, 6196–6200.

4.8   Some DNA Sequences Code for More Than One Polypeptide

Research

Pan, Q. et al. (2008). Deep surveying of alternative splicing complexity in the human transcriptome by high-throughput sequencing. *Nature Genetics* 40, 1413–1415.

Sultan, M. et al. (2008). A global view of gene activity and alternative splicing by deep sequencing of the human transcriptome. *Science* 321, 956–960.

4.9 Some Exons Can Be Equated with Protein Functional Domains

Reviews

Blake, C. C. (1985). Exons and the evolution of proteins. *Int. Rev. Cytol.* 93, 149–185.

Doolittle, R. F. (1985). The genealogy of some recently evolved vertebrate proteins. *Trends Biochem. Sci.* 10, 233–237.

4.10 Members of a Gene Family Have a Common Organization

Review

Dixon, B. and Pohajdek, B. (1992). Did the ancestral globin gene of plants and animals contain only two introns? *Trends Biochem. Sci.* 17, 486–488.

Research

Matsuo, K. et al. (1994). Short introns interrupting the Oct-2 POU domain may prevent recombi-nation between POU family members without interfering with potential POU domain "shuffling" in evolution. *Biol. Chem. Hopp-Seyler* 375, 675–683.

Weber, K. and Kabsch, W. (1994). Intron positions in actin genes seem unrelated to the secondary structure of the protein. *EMBO. J.* 13, 1280–1286.

4.11 Genetic Information Is Not Completely Contained in DNA

Reviews

Barrangou, R. et al. (2007). CRISPR provides acquired resistance against viruses in prokaryotes. *Science* 315, 1709–1712.

Bernardi, G. and Bernardi, G. (1986). Compositional constraints and genome evolution. *J. Mol. Evol.* 24, 1–11.

Forsdyke, D. R., Madill, C. A., and Smith, S. D. (2002). Immunity as a function of the unicellular state: implications of emerging genomic data. *Trends Immunol.* 23, 575–579. (*See* http://post.queensu.ca/~forsdyke/pfalcip01.htm.)

Jeffares, D. C., Penkett, C. J., and Bähler, J. (2008). Rapidly regulated genes are intron poor. *Trends in Genetics* 24, 375–378.

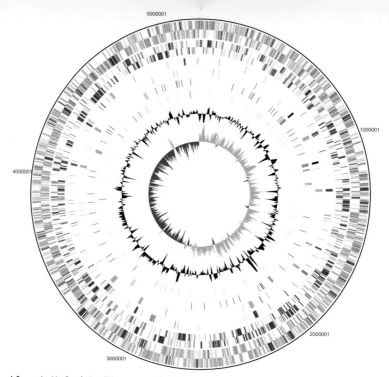

5000001
1000001
2000001
3000001
4000001

The Content of the Genome

CHAPTER OUTLINE

5.1 Introduction

5.2 Genomes Can Be Mapped at Several Levels of Resolution

- Linkage maps are based on the frequency of recombination between genetic markers; restriction maps are based on the physical distances between markers.
- Molecular characterization of mutations can be used to reconcile linkage maps with physical maps.

5.3 Individual Genomes Show Extensive Variation

- Polymorphism may be detected at the phenotypic level when a sequence affects gene function, at the restriction fragment level when it affects a restriction enzyme target site, and at the sequence level by direct analysis of DNA.
- The alleles of a gene show extensive polymorphism at the sequence level, but many sequence changes do not affect function.

5.4 RFLPs and SNPs Can Be Used for Genetic Mapping

- RFLPs and SNPs can be the basis for linkage maps and are useful for establishing parent–offspring relationships.

5.5 Eukaryotic Genomes Contain Both Nonrepetitive and Repetitive DNA Sequences

- The kinetics of DNA reassociation after a genome has been denatured distinguish sequences by their frequency of repetition in the genome.
- Polypeptides are generally coded by sequences in nonrepetitive DNA.
- Larger genomes within a taxonomic group do not contain more genes, but have large amounts of repetitive DNA.
- A large part of repetitive DNA may be made up of transposons.

5.6 Eukaryotic Protein-Coding Genes Can Be Identified by the Conservation of Exons

- Conservation of exons can be used as the basis for identifying coding regions by identifying fragments whose sequences are present in multiple organisms.
- Human disease genes are identified by mapping and sequencing DNA of patients to find differences from normal DNA that are genetically linked to the disease.

5.1 Introduction

One key question about the genome is how many genes it contains. An even more fundamental question, however, is "what is a gene?" Clearly, genes cannot solely be defined as a sequence of DNA that codes for polypeptide, because many genes codes for multiple polypeptides, and many code for RNAs that serve other functions. Given the variety of RNA functions and the complexities of gene expression, it seems prudent to focus on the gene as a unit of transcription. Large areas of chromosomes previously thought to be devoid of genes now appear to be extensively transcribed, though, so at present the definition of a "gene" is a moving target.

We can attempt to characterize both the total number of genes and the number of protein-coding genes at four levels, which correspond to successive stages in gene expression:

- The **genome** is the complete set of genes of an organism. Ultimately it is defined by the complete DNA sequence, although as a practical matter it may not be possible to identify every gene unequivocally solely on the basis of sequence.
- The **transcriptome** is the complete set of genes expressed under particu-

lar conditions. It is defined in terms of the set of RNA molecules that is present and can refer to a single cell type or to any more complex assembly of cells, up to the complete organism. Some genes generate multiple mRNAs, so the transcriptome is likely to be larger than the number of genes defined directly in the genome. The transcriptome includes noncoding RNAs (such as tRNAs, rRNAs, and miRNAs [see Chapter 30, *Regulatory RNA*], and a host of other RNAs with as-yet-unknown functions), as well as mRNAs.

- The **proteome** is the complete set of polypeptides encoded by the whole genome or produced in any particular cell or tissue. It should correspond to the mRNAs in the transcriptome, although there can be differences of detail reflecting changes in the relative abundance or stabilities of mRNAs and proteins. There may also be posttranslational modifications to proteins that allow more than one protein to be produced from a single transcript. (This is called *protein splicing*; see *Section 23.12, Protein Splicing Is Autocatalytic.*)
- Proteins may function independently or as part of multiprotein or

multimolecular complexes, such as holoenzymes and metabolic pathways where enzymes are clustered together. The RNA polymerase holoenzyme (see *Section 19.5, RNA Polymerase Holoenzyme Consists of the Core Enzyme and Sigma Factor*) and the spliceosome (see *Section 21.8, The Spliceosome Assembly Pathway*) are two examples. If we could identify all protein–protein interactions, we could define the total number of independent complexes of proteins. This is sometimes referred to as the **interactome**.

The maximum number of protein-coding genes in the genome can be identified directly by characterizing open reading frames (ORFs). Large-scale mapping of this nature is complicated by the fact that interrupted genes may consist of many separated open reading frames, and alternative splicing can result in the use of subsets or variously combined portions of these ORFs. We do not necessarily have information about the functions of the protein products—or indeed proof that they are expressed at all—so this approach is restricted to defining the *potential* of the genome. A strong presumption exists, however, that any conserved ORF is likely to be expressed.

Another approach is to define the number of genes directly in terms of the transcriptome (by directly identifying all the mRNAs) or proteome (by directly identifying all the polypeptides). This gives an assurance that we are dealing with *bona fide* genes that are expressed under known circumstances. It allows us to ask how many genes are expressed in a particular tissue or cell type, what variation exists in the relative levels of expression, and how many of the genes expressed in one particular cell are unique to that cell or are also expressed elsewhere. In addition, analysis of the transcriptome can reveal how many different mRNAs (e.g., mRNAs containing different combinations of exons) are generated from a given gene.

Concerning the types of genes, we may ask whether a particular gene is *essential*: what is the phenotypic effect of a null mutant? If a null mutation is lethal, or the organism has a visible defect, we may conclude that the gene is essential or at least conveys a selective advantage. Some genes, however, can be deleted without apparent effect on the phenotype. Are these genes really dispensable, or does a selective disadvantage result from the absence of the gene, perhaps in other circumstances, or over longer periods of time?

In some cases, the absence of these genes could be compensated for by a redundant mechanism, such as a gene duplication, providing a backup for an essential function.

5.2 Genomes Can Be Mapped at Several Levels of Resolution

Key concepts

- Linkage maps are based on the frequency of recombination between genetic markers; restriction maps are based on the physical distances between markers.
- Molecular characterization of mutations can be used to reconcile linkage maps with physical maps.

Defining the contents of a genome essentially means making a map. We can think about mapping genes and genomes at several levels of resolution:

- A **genetic** (or **linkage**) **map** identifies the distance between loci in terms of recombination frequencies. It is limited by its reliance on the occurrence of recombination of variable markers that are either visible (such as phenotypic traits) or can be visualized (such as by electrophoresis). For example, a linkage map can be constructed by measuring recombination between sites in genomic DNA that have sequence variations generating differences in the susceptibility to cleavage by certain restriction enzymes. These variations are common, and as a result such a map can be prepared for any organism irrespective of the occurrence of mutants. Recombination frequencies can be distorted relative to the physical distance between sites, and thus a linkage map does not accurately represent physical distances along a chromosome.

- In **restriction mapping**, a restriction map is constructed by cleaving DNA into fragments with restriction enzymes and measuring the physical distances, in terms of the length of DNA (determined by migration on an electrophoretic gel), between the sites of cleavage. A restriction map does not intrinsically identify sites of interest, such as a gene. For it to be related to the genetic map, mutations have to be characterized in terms of their effects upon the restriction sites. Large changes in the genome can be recognized because they affect the sizes or numbers of restriction fragments. Point mutations are more difficult to detect because they change only a single restriction site or lie between restriction sites and are undetectable.

- The ultimate genomic map is the sequence of the DNA. From the sequence, we can identify genes and the distances between them. By analyzing the protein-coding potential of a sequence of the DNA, we can hypothesize about its function. The basic assumption here is that natural selection prevents the accumulation of damaging mutations in sequences that code for proteins. Reversing the argument, we may assume that an intact coding sequence with accompanying transcription signals is likely to be used to generate a protein.

By comparing a wild-type DNA sequence with that of a mutant allele, we can determine the nature of a mutation and its exact site of occurrence. This provides a way to determine the relationship between the genetic map (based entirely on sites of mutation) and the physical map (based on, or even comprising, the sequence of DNA).

Similar techniques are used to identify and sequence genes and to map the genome, although there is of course a difference of scale. In each case, the principle is to characterize a series of overlapping fragments of DNA that can be connected into a continuous map. The crucial feature is that each segment is identified as adjacent to the next segment on the map by the overlap between them, so that we can be sure no segments are missing. This principle is applied both at the level of ordering large fragments into a map and in connecting the sequences that make up the fragments.

5.3 Individual Genomes Show Extensive Variation

Key concepts

- Polymorphism may be detected at the phenotypic level when a sequence affects gene function, at the restriction fragment level when it affects a restriction enzyme target site, and at the sequence level by direct analysis of DNA.
- The alleles of a gene show extensive polymorphism at the sequence level, but many sequence changes do not affect function.

The original Mendelian view of the genome classified alleles as either wild-type or mutant. Subsequently we recognized the existence of multiple alleles, each with a different effect on the phenotype. In some cases it may not even be appropriate to define any one allele as "wild-type."

The coexistence of multiple alleles at a locus is called genetic **polymorphism**. Any site at which multiple alleles exist as stable components of the population is by definition polymorphic. A locus is usually defined as polymorphic if two or more alleles are present at a frequency of >1% in the population.

What is the basis for the polymorphism among the mutant alleles? They possess different mutations that may alter the protein function, thus producing changes in phenotype. The population dynamics of these different alleles are partly determined by their selective effects on phenotype. If we compare the restriction maps or the DNA sequences of these alleles they, too, will be polymorphic in the sense that each map or sequence will be different from the others.

Although not evident from the phenotype, the wild type may itself be polymorphic. Multiple versions of the wild-type allele may be distinguished by differences in sequence that do not affect their function, and which therefore do not produce phenotypic variants. A population may have extensive polymorphism at the level of genotype. Many different sequence variants may exist at a given locus; some of them are evident because they affect the phenotype, whereas others are hidden because they have no visible effect. These mutant alleles are selectively neutral, with their population dynamics mainly a result of random genetic drift (see Chapter 8, *Genome Evolution*).

So there may be a continuum of changes at a locus, including those that change DNA sequence but do not change protein sequence, those that change protein sequence without changing function, those that result in proteins with different activities, and those that result in mutant proteins that are nonfunctional.

A change in a single nucleotide when alleles are compared is called a **single nucleotide polymorphism (SNP)**. On average, one occurs every ~1330 bases in the human genome. Defined by their SNPs, every human being is unique. SNPs can be detected by various means, ranging from direct comparisons of sequence to mass spectroscopy or biochemical methods that produce differences based on sequence variations in a defined region.

One aim of genetic mapping is to obtain a catalog of common variants. The observed frequency of SNPs per genome predicts that, over the human population as a whole (taking the sum of all human genomes of all living individuals), there should be >10 million SNPs that occur at a frequency of >1%. More than six million have already been identified.

Some polymorphisms in the genome can be detected by comparing the restriction maps of different individuals. The criterion is a change in the pattern of fragments produced by cleavage with a restriction enzyme. **FIGURE 5.1** shows that when a target site is present in the genome of one individual and absent from another, the extra cleavage in the first genome will generate two fragments corresponding to the single fragment in the second genome. A difference

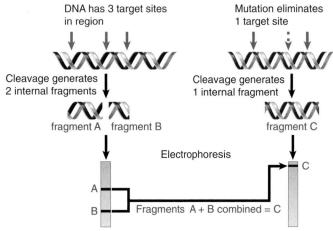

FIGURE 5.1 A point mutation that affects a restriction site is detected by a difference in restriction fragments.

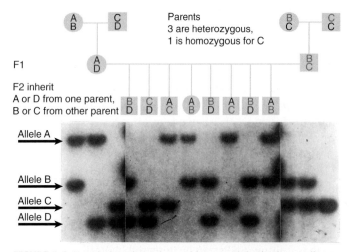

FIGURE 5.2 Restriction site polymorphisms are inherited according to Mendelian rules. Four alleles for a restriction marker are found in all possible pairwise combinations and segregate independently at each generation. Photo courtesy of Ray L. White, Ernest Gallo Clinic and Research Center, University of California, San Francisco.

in restriction maps between two individuals is called a **restriction fragment length polymorphism (RFLP)** or "riflip." Basically, an RFLP is an SNP that is located in the target site for a restriction enzyme. It can be used as a genetic marker in exactly the same way as any other marker. Instead of examining some feature of the phenotype, we directly assess the genotype, as revealed by the restriction map. **FIGURE 5.2** shows a pedigree of a restriction polymorphism followed through three generations. It displays Mendelian segregation at the level of DNA marker fragments.

The restriction map is independent of gene function; as a result, an RFLP at this level can be detected irrespective of whether the sequence change affects the phenotype. Probably very few of the RFLPs in a genome actually affect the phenotype. Most involve sequence changes that have no effect on the production of proteins (e.g., because they lie between genes).

The sequencing of complete individual genomes is now possible and allows the assessment of individual DNA-level variations, both neutral SNPs and those linked to diseases or disease susceptibilities. Although the sequencing of "celebrity" genomes (such as those of James Watson and Craig Venter) receive more press coverage, rapid genome sequencing of anonymous individuals is potentially more informative.

5.4 RFLPs and SNPs Can Be Used for Genetic Mapping

> **Key concept**
> • RFLPs and SNPs can be the basis for linkage maps and are useful for establishing parent–offspring relationships.

Recombination frequency between a restriction marker and a visible phenotypic marker can be measured, as illustrated in **FIGURE 5.3**. Thus a genetic map can include both genotypic and phenotypic markers.

Restriction markers are not limited to those genome changes that affect the phenotype; as a result, they provide the basis for an extremely powerful technique for identifying genetic variants at the molecular level. A typical problem concerns a mutation with known effects on the phenotype, where the relevant genetic locus can be placed on a genetic map, but for which we have no knowledge about the corresponding gene or protein. Many damaging or fatal human diseases fall into this category. For

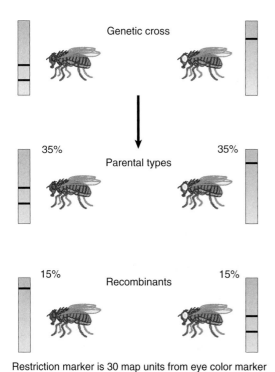

Restriction marker is 30 map units from eye color marker

FIGURE 5.3 A restriction polymorphism can be used as a genetic marker to measure recombination distance from a phenotypic marker (such as eye color). The figure simplifies the situation by showing only the DNA bands corresponding to the allele of one genome in a diploid.

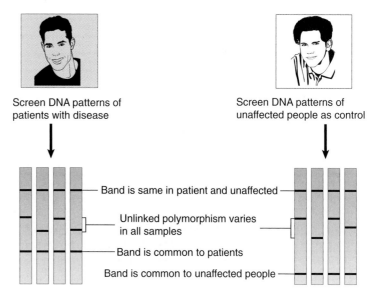

FIGURE 5.4 If a restriction marker is associated with a phenotypic characteristic, the restriction site must be located near the gene responsible for the phenotype. The mutation changing the band that is common in healthy people into the band that is common in patients is very closely linked to the disease gene.

example, cystic fibrosis shows recessive Mendelian inheritance, but the molecular nature of the mutant function was unknown until it could be identified as a result of characterizing the gene.

If restriction polymorphisms occur at random in the genome, some should occur near or within any particular target gene. We can identify such restriction markers by virtue of their tight association with the mutant phenotype. If we compare the restriction map of DNA from patients suffering from a disease with the DNA of healthy people, we may find that a particular restriction site is always present (or always absent) from the patients.

A hypothetical example is shown in **FIGURE 5.4**. This situation corresponds to finding 100% linkage between the restriction marker and the locus producing the phenotype. It would imply that the restriction marker lies so close to the mutant gene that it is never separated from it by recombination; it may in fact be the same mutation.

The identification of such a marker has two important consequences:

- It may offer a diagnostic procedure for detecting the disease. Some of the human diseases with a known inheritance pattern but ill-defined in molecular terms cannot be easily diagnosed. If a restriction marker is closely linked to the phenotype, its presence can be used to diagnose the probability of carrying the disease allele.

- It may lead to isolation of the gene. The restriction marker must lie relatively near the gene on the genetic map if the two loci rarely or never recombine. "Relatively near" in genetic terms can be a substantial distance in terms of base pairs of DNA, but it provides a starting point from which we can proceed along the DNA to the gene itself.

The frequent occurrence of SNPs in the human genome makes them useful for genetic mapping. From the several million SNPs that have already been identified, there is on average an SNP every ~1 kb. This should allow rapid localization of new disease genes by locating them using the nearest SNPs.

On the same principle, RFLP mapping has been in use for some time. Once an RFLP has been assigned to a linkage group (i.e., a chromosome), it can be placed on the genetic map. RFLP mapping of both the human and mouse genomes has led to the construction of linkage maps for both. Any site with an unknown position can be tested for linkage to these sites, and by this means can be rapidly placed on the map. There are fewer RFLPs than SNPs, so the resolution of the RFLP map is in principle more limited.

The large proportion of polymorphic sites means that every individual has a unique constellation of SNPs and RFLPs. The particular combination of sites found in a specific region is called a **haplotype**, a genotype in miniature. Haplotype was originally introduced as a concept to describe the genetic constitution of the major histocompatibility locus, a region specifying proteins of importance in the immune system (see Chapter 18, *Recombination in the Immune System*). The term has now been extended to describe the particular combination of alleles, restriction sites, or any other genetic markers present in some defined area of the genome. Using SNPs, a detailed haplotype map of the human genome has been made; this enables disease-causing genes to be mapped more easily.

The existence of RFLPs provides the basis for a technique to establish unequivocal parent–offspring relationships. In cases for which parentage is in doubt, a comparison of the RFLP map in a suitable chromosome region between potential parents and child allows absolute assignment of the relationship. The use of DNA restriction analysis to identify individuals has been called **DNA fingerprinting**. Analysis of especially variable "minisatellite" sequences is used in mapping the human genome (see *Section 7.8, Minisatellites Are Useful for Genetic Mapping*).

5.5 Eukaryotic Genomes Contain Both Nonrepetitive and Repetitive DNA Sequences

Key concepts

- The kinetics of DNA reassociation after a genome has been denatured distinguish sequences by their frequency of repetition in the genome.
- Polypeptides are generally coded by sequences in nonrepetitive DNA.
- Larger genomes within a taxonomic group do not contain more genes, but have large amounts of repetitive DNA.
- A large part of repetitive DNA may be made up of transposons.

The general nature of the eukaryotic genome can be assessed by the kinetics of reassociation of denatured DNA. This technique was used

extensively before large-scale DNA sequencing became possible.

Reassociation kinetics identifies two general types of genomic sequences:

- **Nonrepetitive DNA** consists of sequences that are unique: there is only one copy in a haploid genome.
- **Repetitive DNA** consists of sequences that are present in more than one copy in each genome.

Repetitive DNA often is divided into two general types:

- Moderately repetitive DNA consists of relatively short sequences that are repeated typically 10–1000× in the genome. The sequences are dispersed throughout the genome and are responsible for the high degree of secondary structure formation in pre-mRNA, when inverted repeats in the introns pair to form duplex regions.
- Highly repetitive DNA consists of very short sequences (typically <100 bp) that are present many thousands of times in the genome, often organized as long regions of tandem repeats (see *Section 7.5, Satellite DNAs Often Lie in Heterochromatin*). Neither class is found in coding regions.

The proportion of the genome occupied by nonrepetitive DNA varies widely among taxonomic groups. **FIGURE 5.5** summarizes the genome organization of some representative organisms. Prokaryotes contain nonrepeti-

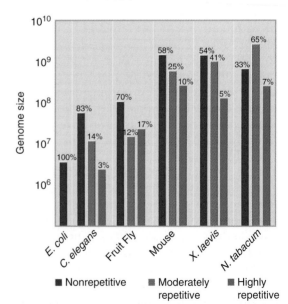

FIGURE 5.5 The proportions of different sequence components vary in eukaryotic genomes. The absolute content of nonrepetitive DNA increases with genome size but reaches a plateau at ~2 × 10⁹ bp.

tive DNA almost exclusively. For unicellular eukaryotes, most of the DNA is nonrepetitive; <20% falls into one or more moderately repetitive components. In animal cells, up to half of the DNA often is occupied by moderately and highly repetitive components. In plants and amphibians, the moderately and highly repetitive components may account for up to 80% of the genome, so that the nonrepetitive DNA is reduced to a minority component.

A significant part of the moderately repetitive DNA consists of **transposons**, short sequences of DNA (~1 kb) that have the ability to move to new locations in the genome and/or to make additional copies of themselves (see Chapter 17, *Transposable Elements and Retrotransposons*). In some multicellular eukaryotic genomes they may even occupy more than half of the genome (see *Section 6.5, The Human Genome Has Fewer Genes Than Originally Expected*).

Transposons are sometimes viewed as **selfish DNA**, which is defined as sequences that propagate themselves within a genome without contributing to the development and functioning of the organism. Transposons may cause genome rearrangements, which could confer selective advantages. It is fair to say, though, that we do not really understand why selective forces do not act against transposons becoming such a large proportion of the eukaryotic genome. It may be that they are selectively neutral as long as they do not interrupt or delete coding or regulatory regions. Many organisms actively suppress transposition, perhaps because in some cases deleterious chromosome breakages result (see Figure 17.6). Another term used to describe the apparent excess of DNA in some genomes is *"junk" DNA*, meaning genomic sequences without any apparent function, though this name may simply affect our failure to understand the functions of many of these sequences. Of course, it is likely that there is a balance in the genome between the generation of new sequences and the elimination of unwanted sequences, and some proportion of DNA that apparently lacks function may be in the process of being eliminated.

The length of the nonrepetitive DNA component tends to increase with overall genome size as we proceed up to a total genome size ~3×10^9 (characteristic of mammals). Further increases in genome size, however, generally reflect an increase in the amount and proportion of the repetitive components, so that it is rare for an organism to have a nonrepetitive DNA component >2×10^9. The nonrepetitive

DNA content of genomes therefore accords better with our sense of the relative complexity of the organism. *E. coli* has 4.2×10^6 bp of nonrepetitive DNA, *C. elegans* has an order of magnitude more (6.6×10^7 bp), *D. melanogaster* has ~10^8 bp, and mammals have yet another order of magnitude more, at ~2×10^9 bp.

What type of DNA corresponds to protein-coding genes? Reassociation kinetics typically shows that mRNA is derived from nonrepetitive DNA. The amount of nonrepetitive DNA is therefore a better indication of the coding potential than is the C-value. (More detailed analysis based on genomic sequences, however, shows that many exons have related sequences in other exons [see *Section 4.5, Exon Sequences under Negative Selection Are Conserved but Introns Vary*]. Such exons evolve by a duplication to give copies that initially are identical, but which then diverge in sequence during evolution.)

5.6 Eukaryotic Protein-Coding Genes Can Be Identified by the Conservation of Exons

Key concepts

- Conservation of exons can be used as the basis for identifying coding regions by identifying fragments whose sequences are present in multiple organisms.
- Human disease genes are identified by mapping and sequencing DNA of patients to find differences from normal DNA that are genetically linked to the disease.

Some major approaches to identifying eukaryotic protein-coding genes are based on the contrast between the conservation of exons and the variation of introns. In a region containing a gene whose function has been conserved among a range of species, the sequence representing the polypeptide should have two distinctive properties:

- It must have an open reading frame, and
- it is likely to have a related (orthologous) sequence in other species.

These features can be used to identify functional genes.

Suppose we know by linkage analysis that a particular genetic trait is located in a given chromosomal region. If we lack knowledge about the nature of the gene product, how are we to identify the gene in a region that may be, for example, >1 Mb in size?

An approach that has proved successful with some genes of medical importance is to screen relatively short fragments from the region for the two properties expected of a conserved gene. First we seek to identify fragments that cross-hybridize with the genomes of other species, then we examine these fragments for open reading frames.

The first criterion can be applied by performing a **zoo blot**. We use short fragments from the region as labeled probes to test for homologous DNA from a variety of species by Southern blotting (a technique for transferring DNA fragments from an electrophoretic gel to a filter membrane, followed by hybridization of a probe to detect the complementary or near-complementary sequence). If we find hybridizing fragments in several species related to that of the probe (which is usually prepared from human DNA), the probe becomes a candidate for an exon of the gene.

The candidates are sequenced, and if they contain open reading frames they are used to isolate surrounding genomic regions. If these appear to be part of an exon, they can then be used to identify the entire gene, to isolate the corresponding cDNA (DNA reverse transcribed from the mRNA) or mRNA itself, and ultimately to identify the protein. In these days of whole genome sequencing, however, much of this analysis can be performed *in silico*, searching databases of complete genomes for homologs of the putative gene of interest.

When a human disease is caused by a change in a known protein, the gene that is responsible can be identified because it encodes the protein, and its responsibility for the disease can be confirmed by showing that it has inactivating mutations in the DNA of patients but not in normal DNA. In many cases, though, we do not know the cause of a disease at the molecular level, and it is necessary to identify the gene without any information about its protein product.

The basic criterion for identifying a gene involved in a human disease is to show that in every patient with the disease the gene has a mutation that is not present in normal DNA. The extensive polymorphism between individual genomes, though, means that we may find many changes when we compare patient DNA with normal DNA. Before the sequencing of the human genome, genetic linkage could be used to identify a region containing a disease gene, but the region could contain many candidate genes. For a very large gene, with introns spread over a long distance of the genome, it was difficult to identify the critical mutations in patients. The availability of high-resolution SNP maps and of the genome sequence now makes it much easier to pinpoint a smaller region containing the gene in which sequences of normal and patient DNA can be directly compared.

An example of the process by which a disease gene can be tracked down is provided by the gene responsible for Duchenne muscular dystrophy (DMD), a degenerative disorder of muscle that is X-linked and affects 1 in 3500 human males. The steps in identifying the gene are summarized in **FIGURE 5.6**.

Linkage analysis localized the DMD locus to chromosomal band Xp21. Patients with the disease often have chromosomal rearrangements involving this band. By comparing the ability of X-linked DNA probes to hybridize with DNA from patients with normal DNA, cloned fragments were obtained that correspond to the region that was rearranged or deleted in patients' DNA.

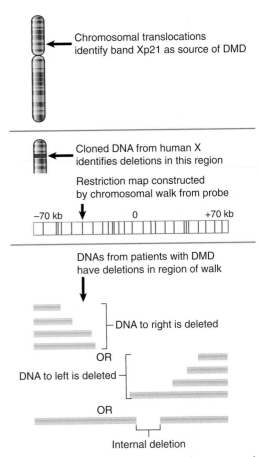

FIGURE 5.6 The gene involved in Duchenne muscular dystrophy was tracked down by chromosome mapping and "walking" to a region in which deletions can be identified with the occurrence of the disease.

Once some DNA in the general vicinity of the target gene has been obtained, it is possible to "walk" along the chromosome until the gene is reached. A **chromosomal walk** was used to construct a restriction map of the region on either side of the probe, which covered a region of >100 kb. Analysis of the DNA from a series of patients identified large deletions in this region that extended in either direction. The most telling deletion is one that is contained entirely within the region, because this delineates a segment that must be important in gene function and indicates that the gene—or at least part of it—lies in this region.

After identifying the region of the gene, its exons and introns needed to be identified. A zoo blot identified fragments that cross-hybridize with the mouse X chromosome and with other mammalian DNAs. As summarized in FIGURE 5.7, these were scrutinized for open reading frames and the sequences typically found at exon–intron junctions. Fragments that met these criteria were used as probes to identify homologous sequences in a cDNA library prepared from muscle mRNA.

The cDNA corresponding to the gene identifies an unusually large (14 kb) mRNA. Hybridization back to the genome shows that the mRNA is encoded by >60 exons, which are spread over ~2000 kb of DNA. This makes DMD one of the longest identified genes.

The gene codes for a protein of ~500 kD called *dystrophin*, which is a component of muscle and is present in rather low amounts. All patients with the disease have deletions at this locus and lack (or have defective) dystrophin.

Muscle also has the distinction of having the largest known protein, titin, with almost 27,000 amino acids. The *titin* gene has the largest number of exons (178) and the longest single exon in the human genome (17,000 bp).

Another technique that allows genomic fragments to be scanned rapidly for the presence of exons is called **exon trapping**. FIGURE 5.8 shows that it starts with a vector that contains a strong promoter and has a single intron between two exons. When this vector is transfected into cells, its transcription generates large amounts of an RNA containing the sequences of the two exons. A restriction site lies within the intron and is used to insert genomic fragments from a region of interest. If a fragment does not contain an exon, there is no change in the splicing pattern, and the RNA contains only the same sequences as the parental vector. If the genomic fragment contains an

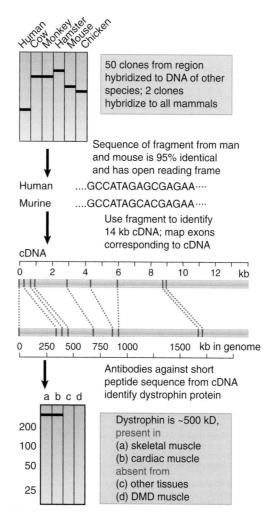

FIGURE 5.7 The Duchenne muscular dystrophy gene was characterized by zoo blotting, cDNA hybridization, genomic hybridization, and identification of the protein.

exon flanked by two partial intron sequences, though, the splicing sites on either side of this exon are recognized and the sequence of the exon is inserted into the RNA between the two exons of the vector. This can be detected readily by reverse transcribing the cytoplasmic RNA into cDNA and using PCR (called *RT-PCR*, which will be described in the next section; see also *Section 3.8, PCR and RT-PCR*) to amplify the sequences between the two exons of the vector. So the appearance in the amplified population of sequences from the genomic fragment indicates that an exon has been "trapped." In mammalian protein-coding genes introns are usually large and exons are small; thus there is a high probability that a random piece of genomic DNA will contain the required structure of an exon surrounded by partial introns. In fact, exon trapping may mimic the events that have occurred naturally

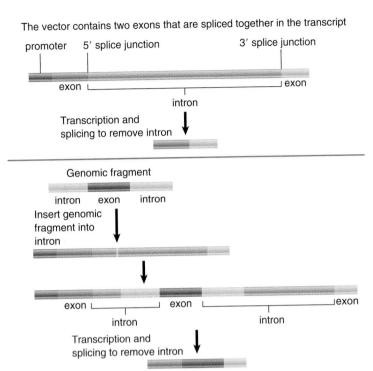

The vector contains two exons that are spliced together in the transcript

promoter   5′ splice junction   3′ splice junction

exon   exon
                    intron

Transcription and
splicing to remove intron

Genomic fragment

intron   exon   intron

Insert genomic
fragment into
intron

exon   exon   exon
    intron   intron

Transcription and
splicing to remove intron

FIGURE 5.8 A special splicing vector is used for exon trapping. If an exon is present in the genomic fragment, its sequence will be recovered in the cytoplasmic RNA. If the genomic fragment consists solely of sequences from within an intron, though, splicing does not occur, and the mRNA is not exported to the cytoplasm.

during evolution of genes (see *Section 8.6, How Did Interrupted Genes Evolve?*).

Ultimately, exons can be identified by the large-scale sequencing of cellular mRNAs that is now feasible.

<table><tr><td>5.7</td><td></td></tr></table>

The Conservation of Genome Organization Helps to Identify Genes

Key concepts

- Methods for identifying active genes are not perfect and many corrections must be made to preliminary estimates.
- Pseudogenes must be distinguished from active genes.
- There are extensive syntenic relationships between the mouse and human genomes, and most active genes are in a syntenic region.

Once we have determined the sequence of a genome, we still have to identify the genes within it. Coding sequences represent a very small fraction of the total genome. Potential exons can be identified as uninterrupted open reading frames flanked by appropriate sequences. What criteria need to be satisfied to identify a functional (intact) gene from a series of exons?

FIGURE 5.9 shows that a functional gene should consist of a series of exons for which the first exon immediately follows a promoter, the internal exons are flanked by appropriate splicing junctions, the last exon is followed by 3′ processing signals, and a single ORF starting with an initiation codon and ending with a termination codon can be deduced by joining the exons together. Internal exons can be identified as open reading frames flanked by splicing junctions. In the simplest cases, the first and last exons contain the start and end of the coding region, respectively (as well as the 5′ and 3′ untranslated regions). In more complex cases, the first or last exons may have only untranslated regions and may therefore be more difficult to identify.

The algorithms that are used to connect exons are not completely effective when the genome is very large and the exons may be separated by very large distances. For example, the initial analysis of the human genome mapped 170,000 exons into 32,000 genes. This is unlikely to be correct because it gives an average of 5.3 exons per gene, whereas the average of individual genes that have been fully characterized is 10.2. Either we have missed many exons, or they should be connected differently into a smaller number of genes in the whole genome sequence.

Even when the organization of a gene is correctly identified, there is the problem of distinguishing functional genes from pseudogenes. Many pseudogenes can be recognized by obvious defects in the form of multiple mutations that create a nonfunctional coding sequence. Pseudogenes that have arisen more recently have not accumulated as many mutations and thus may be more difficult to recognize. In an extreme example, the mouse has only one functional *Gapdh* gene (coding for glyceraldehyde phosphate dehydrogenase), but has ~400 pseudogenes. Approximately 100 of these pseudogenes initially appeared to be active in the mouse genome sequence, and individual examination was necessary to exclude them from the list of functional genes. Pseudogenes with relatively intact coding sequences but mutated transcription signals are more difficult to identify. (Some pseudogenes do generate RNAs that play a role in gene regulation; see *Section 30.5, MicroRNAs Are Regulators in Eukaryotes*.)

How can putative protein-coding genes be verified? If it can be shown that a DNA sequence is transcribed and processed into a translatable mRNA, it is assumed that it is func-

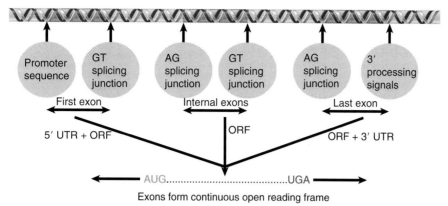

FIGURE 5.9 Exons of protein-coding genes are identified as coding sequences flanked by appropriate signals (with untranslated regions at both ends). The series of exons must generate an open reading frame with appropriate initiation and termination codons.

tional. One technique for doing this is **reverse transcription polymerase chain reaction (RT-PCR)** (see *Section 3.8, PCR and RT-PCR*) in which RNA isolated from cells is reverse-transcribed to DNA and subsequently amplified to many copies using the polymerase chain reaction. The amplified DNA products can then be sequenced or otherwise analyzed to determine if they have the appropriate structural features of a mature transcript. RT-PCR can also be used as a quantitative assessment of gene expression.

Confidence that a gene is functional can be increased by comparing regions of the genomes of different species. There has been extensive overall reorganization of sequences between the mouse and human genomes, as seen in the simple fact that there are 23 chromosomes in the human haploid genome and 20 chromosomes in the mouse haploid genome. At the local level, though, the order of genes is generally the same: when pairs of human and mouse homologs are compared, the genes located on either side also tend to be homologs. This relationship is called **synteny**.

FIGURE 5.10 shows the relationship between mouse chromosome 1 and the human chromosomal set. We can recognize 21 segments in this mouse chromosome that have syntenic counterparts in human chromosomes. The extent of reshuffling that has occurred between the genomes is shown by the fact that the segments are spread among six different human chromosomes. The same types of relationships are found in all mouse chromosomes except for the X chromosome, which is syntenic only with the human X chromosome. This is explained by the fact that the X is a special case, subject to dosage compensation to adjust for the difference between the one copy of males and the

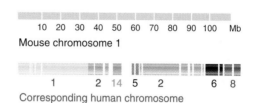

FIGURE 5.10 Mouse chromosome 1 has 21 segments of 1 to 25 Mb that are syntenic with regions corresponding to parts of six human chromosomes.

two copies of females (see *Section 29.5, X Chromosomes Undergo Global Changes*). This restriction may apply selective pressure against the translocation of genes to and from the X chromosome.

Comparison of the mouse and human genome sequences shows that >90% of each genome lies in syntenic blocks that range widely in size from 300 kb to 65 Mb. There are a total of 342 syntenic segments, with an average length of 7 Mb (0.3% of the genome). Ninety-nine percent of mouse genes have a homolog in the human genome; for 96% that homolog is in a syntenic region.

Comparison of genomes provides interesting information about the evolution of species. The number of gene families in the mouse and human genomes is the same, and a major difference between the species is the differential expansion of particular families in the mouse genome. This is especially noticeable in genes that affect phenotypic features that are unique to the species. Of 25 families for which the size has been expanded in mouse, 14 contain genes specifically involved in rodent reproduction and five contain genes specific to the immune system.

A validation of the importance of the identification of syntenic blocks comes from pairwise comparisons of the genes within them. For example, a gene that is not in a syntenic location (that is, its context is different in the two species being compared) is twice as likely to be a pseudogene. Put another way, translocation away from the original locus tends to be associated with the formation of pseudogenes. The lack of a related gene in a syntenic position is therefore grounds for suspecting that an apparent gene may really be a pseudogene. Overall, >10% of the genes that are initially identified by analysis of the genome are likely to turn out to be pseudogenes.

As a general rule, comparisons between genomes add significantly to the effectiveness of gene prediction. When sequence features indicating active genes are conserved—for example, between human and mouse genomes—there is an increased probability that they identify active orthologs.

Identifying genes coding for RNAs other than mRNA is more difficult because we cannot use the criterion of the open reading frame. It is certainly true that the comparative genome analysis described above has increased the rigor of the analysis. For example, analysis of either the human or the mouse genome alone identifies ~500 genes coding for tRNA, but comparison of features suggests that <350 of these genes are in fact functional in each genome.

An active gene can be located through the use of an **expressed sequence tag (EST)**, a short portion of a transcribed sequence usually obtained from sequencing one or both ends of a cloned fragment from a cDNA library. An EST can confirm that a suspected gene is actually transcribed or help identify genes that influence particular disorders. Through the use of a physical mapping technique such as *in situ* hybridization (see *Section 7.5, Satellite DNAs Often Lie in Heterochromatin*), the chromosomal location of an EST can be determined.

5.8 Some Organelles Have DNA

Key concepts

- Mitochondria and chloroplasts have genomes that show non-Mendelian inheritance. Typically they are maternally inherited.
- Organelle genomes may undergo somatic segregation in plants.
- Comparisons of mitochondrial DNA suggest that it is descended from a single population that existed 200,000 years ago in Africa.

The first evidence for the presence of genes outside the nucleus was provided by **non-**

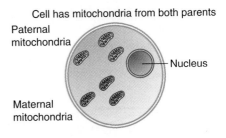

Cell has mitochondria from both parents

Paternal mitochondria

Nucleus

Maternal mitochondria

Possible outcomes of stochastic segregation

Cells usually have both types of mitochondria

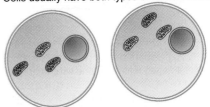

Uneven distribution gives cells with only one type

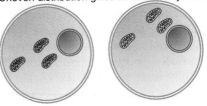

FIGURE 5.11 When paternal and maternal mitochondrial alleles differ, a cell has two sets of mitochondrial DNAs. Mitosis usually generates daughter cells with both sets. Somatic variation may result if unequal segregation generates daughter cells with only one set.

Mendelian inheritance in plants (observed in the early years of the twentieth century, just after the rediscovery of Mendelian inheritance). Non-Mendelian inheritance is defined by the failure of the offspring of a mating to display Mendelian segregation for parental characters, and is therefore taken to indicate the presence of genes that reside outside the nucleus and do not utilize segregation on the meiotic and mitotic spindles to distribute copies to gametes or to daughter cells, respectively. **FIGURE 5.11** shows that this happens when the mitochondria inherited from the male and female parents have different alleles, and a daughter cell receives an unbalanced distribution of mitochondria from only one parent (see *Section 13.11, How Do Mitochondria Replicate and Segregate?*). This is also true of chloroplasts in plants; both mitochondria and chloroplasts contain genomes with functional genes (see below).

The extreme form of non-Mendelian inheritance is uniparental inheritance, which occurs when the genotype of only one parent is inherited and that of the other parent is not passed to the offspring. In less extreme examples, one parental genotype exceeds the other genotype in the offspring. In animals and most plants it is the mother whose genotype is preferentially (or solely) inherited. This effect is sometimes described as **maternal inheritance**. The important point is that the genotype contributed by the parent of one particular sex predominates, as seen in abnormal segregation ratios when a cross is made between mutant and wild type. This contrasts with the behavior of Mendelian genetics, which occurs when reciprocal crosses show the contributions of both parents to be equally inherited.

The bias in parental genotypes is established at, or soon after, the formation of a zygote. There are various possible causes. The contribution of maternal or paternal information to the organelles of the zygote may be unequal; in the most extreme case, only one parent contributes. In other cases the contributions are equal, but the information provided by one parent does not survive. Combinations of both effects are possible. Whatever the cause, the unequal representation of the information from the two parents contrasts with nuclear genetic information, which derives equally from each parent.

Some non-Mendelian inheritance results from the presence in mitochondria and chloroplasts of DNA genomes that are inherited independently of nuclear genes. In effect, the organelle genome comprises a length of DNA that has been physically sequestered in a defined part of the cell and is subject to its own form of expression and regulation. An organelle genome can code for some or all of the tRNAs and rRNAs, but codes for only some of the polypeptides needed to perpetuate the organelle. The other polypeptides are encoded in the nucleus, expressed via the cytoplasmic protein synthetic apparatus, and imported into the organelle.

Genes not residing within the nucleus are generally described as **extranuclear genes**; they are transcribed and translated in the same organelle compartment (mitochondrion or chloroplast) in which they reside. By contrast, nuclear genes are expressed by means of cytoplasmic protein synthesis. (The term "cytoplasmic inheritance" sometimes is used to describe the behavior of genes in organelles. We shall not use this description, though, because it is important to be able to distinguish between events in the general cytosol and those in specific organelles.)

Animals show maternal inheritance of mitochondria, which can be explained if the mitochondria are contributed entirely by the ovum and not at all by the sperm. **FIGURE 5.12** shows that the sperm contributes only a copy

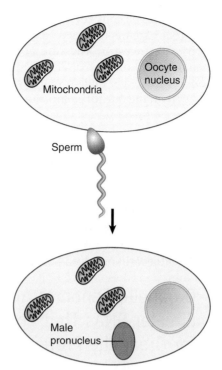

FIGURE 5.12 DNA from the sperm enters the oocyte to form the male pronucleus in the fertilized egg, but all the mitochondria are provided by the oocyte.

of the nuclear DNA. Thus the mitochondrial genes are derived exclusively from the mother, and in males they are discarded each generation. Chloroplasts are generally also maternally inherited, though some plant taxonomic groups show paternal or biparental inheritance of chloroplasts.

The chemical environment of organelles is different from that of the nucleus, and organelle DNA therefore evolves at its own distinct rate. If inheritance is uniparental, there can be no recombination between parental genomes. In fact, recombination usually does not occur in those cases for which organelle genomes are inherited from both parents. Organelle DNA has a different replication system from that of the nucleus; as a result, the error rate during replication may be different. Mitochondrial DNA accumulates mutations more rapidly than nuclear DNA in mammals, but in plants the accumulation in the mitochondrion is slower than in the nucleus; chloroplast DNA has an intermediate mutation rate.

One consequence of maternal inheritance is that the sequence of mitochondrial DNA is more sensitive than nuclear DNA to reductions in the size of the breeding population. Comparisons of mitochondrial DNA sequences in a range of human populations allow an evolutionary tree to be constructed. The divergence among human mitochondrial DNAs spans

0.57%. A tree can be constructed in which the mitochondrial variants diverged from a common (African) ancestor. The rate at which mammalian mitochondrial DNA accumulates mutations is 2 to 4% per million years, which is >10× faster than the rate for globin. Such a rate would generate the observed divergence over an evolutionary period of 140,000 to 280,000 years. This implies that human mitochondrial DNA is descended from a single population that lived in Africa ~200,000 years ago. This cannot be interpreted as evidence that there was only a single population at that time, however; there may have been many populations, and some or all of them may have contributed to modern human *nuclear* genetic variation.

5.9 Organelle Genomes Are Circular DNAs That Code for Organelle Proteins

Key concepts

- Organelle genomes are usually (but not always) circular molecules of DNA.
- Organelle genomes code for some, but not all, of the proteins found in the organelle.
- Animal cell mitochondrial DNA is extremely compact and typically codes for 13 proteins, 2 rRNAs, and 22 tRNAs.
- Yeast mitochondrial DNA is 5× longer than animal cell mtDNA because of the presence of long introns.

Most organelle genomes take the form of a single circular molecule of DNA of unique sequence (denoted **mtDNA** in the mitochondrion and **ctDNA** or **cpDNA** in the chloroplast). There are a few exceptions in unicellular eukaryotes for which mitochondrial DNA is a linear molecule.

Usually there are several copies of the genome in the individual organelle. There are multiple organelles per cell; therefore there are many organelle genomes per cell, so the organelle genome can be considered a repetitive sequence.

Chloroplast genomes are relatively large, usually ~140 kb in higher plants and <200 kb in unicellular eukaryotes. This is comparable to the size of a large bacteriophage genome, such as that of T4 at ~165 kb. There are multiple copies of the genome per organelle, typically 20 to 40 in a higher plant, and multiple copies of the organelle per cell, typically 20 to 40.

Mitochondrial genomes vary in total size by more than an order of magnitude. Animal cells have small mitochondrial genomes (approximately 16.5 kb in mammals). There are several hundred mitochondria per cell, and each mitochondrion has multiple copies of the DNA. The total amount of mitochondrial DNA relative to nuclear DNA is small; it is estimated to be <1%.

In yeast, the mitochondrial genome is much larger. In *Saccharomyces cerevisiae*, the exact size varies among different strains but averages ~80 kb. There are ~22 mitochondria per cell, which corresponds to ~4 genomes per organelle. In growing cells, the proportion of mitochondrial DNA can be as high as 18%.

Plants show an extremely wide range of variation in mitochondrial DNA size, with a minimum of ~100 kb. The size of the genome makes it difficult to isolate intact, but restriction mapping in several plants suggests that the mitochondrial genome is usually a single sequence that is organized as a circle. Within this circle there are short homologous sequences. Recombination between these elements generates smaller, subgenomic circular molecules that coexist with the complete, "master" genome—a good example of the apparent complexity of plant mitochondrial DNAs.

With mitochondrial genomes sequenced from many organisms, we can now see some general patterns in the representation of functions in mitochondrial DNA. **FIGURE 5.13** summarizes the distribution of genes in mitochondrial genomes. The total number of protein-coding genes is rather small and does not correlate with the size of the genome. The 16 kb mammalian mitochondrial genomes encode 13 proteins, whereas the 60 to 80 kb yeast mitochondrial genomes encode as few as eight proteins. The much larger plant mitochondrial genomes encode more proteins. Introns are found in most mitochondrial genes, although not in the very small mammalian genomes.

The two major rRNAs are always encoded by the mitochondrial genome. The number of tRNAs encoded by the mitochondrial genome varies from none to the full complement (25 to 26 in mitochondria). This accounts for the variation in Figure 5.13.

Species	Size (kb)	Protein-coding genes	RNA-coding genes
Fungi	19–100	8–14	10–28
Protists	6–100	3–62	2–29
Plants	186–366	27–34	21–30
Animals	16–17	13	4–24

FIGURE 5.13 Mitochondrial genomes have genes coding for (mostly complex I–IV) proteins, rRNAs, and tRNAs.

The major part of the protein-coding activity is devoted to the components of the multisubunit assemblies of respiration complexes I–IV. Many ribosomal proteins are encoded in protist and plant mitochondrial genomes, but there are few or none in fungi and animal genomes. There are genes encoding proteins involved in cytoplasm-to-mitochondrion import in many protist mitochondrial genomes.

Animal mitochondrial DNA is extremely compact. There are extensive differences in the detailed gene organization found in different animal taxonomic groups, but the general principle of a small genome encoding a restricted number of functions is maintained. In mammalian mitochondria, the genome is extremely compact. There are no introns, some genes actually overlap, and almost every base pair can be assigned to a gene. With the exception of the **D loop**, a region involved with the initiation of DNA replication, no more than 87 of the 16,569 bp of the human mitochondrial genome lie in intercistronic regions.

The complete nucleotide sequences of animal mitochondrial genomes show extensive homology in organization. The map of the human mitochondrial genome is summarized in **FIGURE 5.14**. There are 13 protein-coding regions. All of the proteins are components of the electron transfer system of cellular respiration. These include cytochrome b, three subunits of cytochrome oxidase, one of the subunits of ATPase, and seven subunits (or associated proteins) of NADH dehydrogenase.

The fivefold discrepancy in size between the *S. cerevisiae* (84 kb) and mammalian (16 kb) mitochondrial genomes alone alerts us to the fact that there must be a great difference in their genetic organization in spite of their common function. The number of endogenously synthesized products concerned with mitochondrial enzymatic functions appears to be similar. Does the additional genetic material in yeast mitochondria represent other proteins, perhaps concerned with regulation, or is it unexpressed?

The map in **FIGURE 5.15** accounts for the major RNA and protein products of the yeast mitochondrion. The most notable feature is the dispersion of loci on the map.

The two most prominent loci are the interrupted genes *box* (coding for cytochrome b) and *oxi3* (coding for subunit 1 of cytochrome oxidase). Together these two genes are almost as long as the entire mitochondrial genome in mammals! Many of the long introns in these genes have open reading frames in register with the preceding exon (see *Section 23.5, Some Group I Introns Code for Endonucleases That Sponsor*

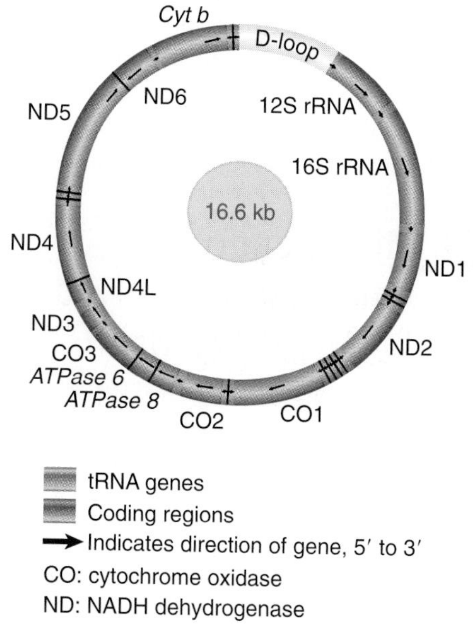

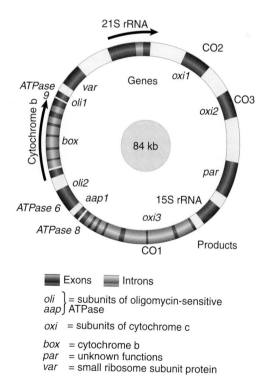

tRNA genes
Coding regions
→ Indicates direction of gene, 5′ to 3′
CO: cytochrome oxidase
ND: NADH dehydrogenase

FIGURE 5.14 Human mitochondrial DNA has 22 tRNA genes, two rRNA genes, and 13 protein-coding regions. Fourteen of the 15 protein-coding or rRNA-coding regions are transcribed in the same direction. Fourteen of the tRNA genes are expressed in the clockwise direction and eight are read counter-clockwise.

Exons   Introns

$\left.\begin{array}{l} oli \\ aap \end{array}\right\}$ = subunits of oligomycin-sensitive ATPase

oxi = subunits of cytochrome c

box = cytochrome b
par = unknown functions
var = small ribosome subunit protein

FIGURE 5.15 The mitochondrial genome of *S. cerevisiae* contains both interrupted and uninterrupted protein-coding genes, rRNA genes, and tRNA genes (positions not indicated). Arrows indicate direction of transcription.

Mobility). This adds several proteins, all synthesized in low amounts, to the complement of the yeast mitochondrion.

The remaining genes are uninterrupted. They correspond to the other two subunits of cytochrome oxidase coded by the mitochondrion, to the subunit(s) of the ATPase, and (in the case of *var1*) to a mitochondrial ribosomal protein. The total number of yeast mitochondrial genes is unlikely to exceed ~25.

5.10 The Chloroplast Genome Codes for Many Proteins and RNAs

Key concept

- Chloroplast genomes vary in size, but are large enough to code for 50 to 100 proteins as well as rRNAs and tRNAs.

What genes are carried by chloroplasts? Chloroplast DNAs vary in length from ~120 to 217 kb (the largest in geranium). The sequenced chloroplast genomes (>100 in total) have 87 to 183 genes. FIGURE 5.16 summarizes the functions coded by the chloroplast genome in land plants. There is more variation in the chloroplast genomes of algae.

The chloroplast genome is generally similar to that of mitochondria, except that more genes are involved. The chloroplast genome encodes all the rRNA and tRNA species needed for protein synthesis. The ribosome includes two small rRNAs in addition to the major species. The tRNA set may include all of the necessary genes. The chloroplast genome codes for ~50 proteins, including RNA polymerase and ribosomal proteins. Again, the rule is that organelle genes are transcribed and translated within the organelle. About half of the chloroplast genes encode proteins involved in protein synthesis.

Introns in chloroplasts fall into two general classes. Those in tRNA genes are usually (although not inevitably) located in the anticodon loop, like the introns found in yeast nuclear tRNA genes (see *Section 21.18, tRNA Splicing Involves Cutting and Rejoining in Separate Reactions*). Those in protein-coding genes resemble the introns of mitochondrial genes (see Chapter 23, *Catalytic RNA*). This places the endosymbiotic event at a time in evolution before the separation of prokaryotes with uninterrupted genes.

The role of the chloroplast is to be the site of photosynthesis. Many of its genes encode proteins of photosynthetic complexes located in the thylakoid membranes. The constitution of these complexes shows a different balance from that of mitochondrial complexes. Although some complexes are like mitochondrial complexes in having some subunits encoded by the organelle genome and some by the nuclear genome, other chloroplast complexes are coded entirely by one genome. For example, the gene for the large subunit of ribulose bisphosphate carboxylase (RuBisCO, which catalyzes the carbon fixation reaction of the Calvin cycle), *rbcL*, is contained in the chloroplast genome; variation in this gene is frequently used as a basis for reconstructing plant phylogenies. The gene for the small rubisco subunit, *rbcS*, is, however, usually carried in the nuclear genome. On the other hand, genes for photosystem protein complexes are found on the chloroplast genome, while those for the LHC (light-harvesting complex) proteins are nuclear-encoded.

5.11 Mitochondria and Chloroplasts Evolved by Endosymbiosis

Key concepts

- Both mitochondria and chloroplasts are descended from bacterial ancestors.
- Most of the genes of the mitochondrial and chloroplast genomes have been transferred to the nucleus during the organelle's evolution.

How is it that an organelle evolved so that it contains genetic information for some of its functions, whereas the information for other functions is encoded in the nucleus? FIGURE 5.17 shows the endosymbiotic hypothesis for mitochondrial evolution, in which primitive cells captured bacteria that provided the function of

Genes	Types
RNA-coding	
16S rRNA	1
23S rRNA	1
4.5S rRNA	1
5S rRNA	1
tRNA	30–32
Gene Expression	
r-proteins	20–21
RNA polymerase	3
Others	2
Chloroplast functions	
Rubisco and thylakoids	31–32
NADH dehydrogenase	11
Total	105–113

FIGURE 5.16 The chloroplast genome in land plants codes for four rRNAs, 30 tRNAs, and ~60 proteins.

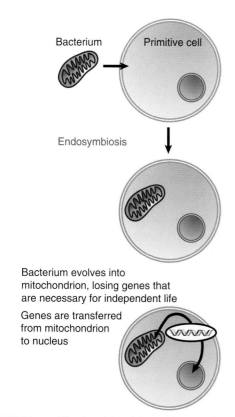

Bacterium   Primitive cell

Endosymbiosis

Bacterium evolves into
mitochondrion, losing genes that
are necessary for independent life

Genes are transferred
from mitochondrion
to nucleus

FIGURE 5.17 Mitochondria originated by an endosymbiotic event when a bacterium was captured by a eukaryotic cell.

cellular respiration and over time evolved into mitochondria. At this point, the proto-organelle must have contained all of the genes needed to specify its functions. A similar mechanism has been proposed for the origin of chloroplasts.

Sequence homologies suggest that mitochondria and chloroplasts evolved separately, from lineages that are common with different eubacteria, with mitochondria sharing an origin with α-purple bacteria and chloroplasts sharing an origin with cyanobacteria. The closest known relative of mitochondria among the bacteria is *Rickettsia* (the causative agent of typhus), which is an obligate intracellular parasite that is probably descended from free-living bacteria. This reinforces the idea that mitochondria originated in an endosymbiotic event involving an ancestor that is also common to *Rickettsia*.

The endosymbiotic origin of the chloroplast is emphasized by the relationships between its genes and their counterparts in bacteria. The organization of the rRNA genes in particular is closely related to that of a cyanobacterium, which pins down more precisely the last common ancestor between chloroplasts and bacteria. Not surprisingly, cyanobacteria are photosynthetic.

Two changes must have occurred as the bacterium became integrated into the recipient cell and evolved into the mitochondrion (or chloroplast). The organelles have far fewer genes than an independent bacterium and have lost many of the gene functions that are necessary for independent life (such as metabolic pathways). The majority of genes encoding organelle functions are in fact now located in the nucleus, so these genes must have been transferred there from the organelle.

Transfer of DNA between an organelle and the nucleus has occurred over evolutionary history and still continues. The rate of transfer can be measured directly by introducing into an organelle a gene that can function only in the nucleus, for example, because it contains a nuclear intron, or because the protein must function in the cytosol. In terms of providing the material for evolution, the transfer rates from organelle to nucleus are roughly equivalent to the rate of single gene mutation. DNA introduced into mitochondria is transferred to the nucleus at a rate of 2×10^{-5} per generation. Experiments to measure transfer in the reverse direction, from nucleus to mitochondrion, suggest that the rate is much lower, $<10^{-10}$. When a nuclear-specific antibiotic resistance gene is introduced into chloroplasts, its transfer to the nucleus and successful expression can be followed by screening seedlings for resistance to the antibiotic. This shows that transfer occurs at a rate of 1 in 16,000 seedlings, or 6×10^{-5} per generation.

Transfer of a gene from an organelle to the nucleus requires physical movement of the DNA, of course, but successful expression also requires changes in the coding sequence. Organelle proteins that are encoded by nuclear genes have special sequences that allow them to be imported into the organelle after they have been synthesized in the cytoplasm. These sequences are not required by proteins that are synthesized within the organelle. Perhaps the process of effective gene transfer occurred at a period when compartments were less rigidly defined, so that it was easier both for the DNA to be relocated and for the proteins to be incorporated into the organelle irrespective of the site of synthesis.

Phylogenetic maps show that gene transfers have occurred independently in many different lineages. It appears that transfers of mitochondrial genes to the nucleus occurred only early in animal cell evolution, but it is possible that the process is still continuing in plant cells. The number of transfers can be large; there are >800 nuclear genes in *Arabidopsis,* whose sequences are related to genes in the chloroplasts of other plants. These genes are candidates for evolution from genes that originated in the chloroplast.

5.12 Summary

The DNA sequences composing a eukaryotic genome can be classified into three groups:

- nonrepetitive sequences that are unique;
- moderately repetitive sequences that are dispersed and repeated a small number of times, with some copies not being identical; and
- highly repetitive sequences that are short and usually repeated as tandem arrays.

The proportions of these types of sequences are characteristic for each genome, although larger genomes tend to have a smaller proportion of nonrepetitive DNA. Almost 50% of the human genome consists of repetitive sequences, the vast majority corresponding to transposon sequences. Most structural genes are located in nonrepetitive DNA. The amount of nonrepetitive DNA is a better reflection of the complexity of the organism than the total genome size; the greatest amount of nonrepetitive DNA in genomes is ~2×10^9 bp.

Non-Mendelian inheritance is explained by the presence of DNA in organelles in the cytoplasm. Mitochondria and chloroplasts are membrane-bounded systems in which some proteins are synthesized within the organelle, whereas others are imported. The organelle genome is usually a circular DNA that codes for all the RNAs and some of the proteins required by the organelle.

Mitochondrial genomes vary greatly in size from the small 16 kb mammalian genome to the 570 kb genome of higher plants. The larger genomes may code for additional functions. Chloroplast genomes range in size from ~120 to 217 kb. Those that have been sequenced have similar organizations and coding functions. In both mitochondria and chloroplasts, many of the major proteins contain some subunits synthesized in the organelle and some subunits imported from the cytosol. Transfers of DNA have occurred between chloroplasts or mitochondria and nuclear genomes.

References

5.3 Individual Genomes Show Extensive Variation

Review

Levy, S. and Strausberg, R. L. (2008). Human genetics: individual genomes diversify. *Nature* 456, 49–51.

Research

Altshuler, D., Brooks, L. D., Chakravarti, A., Collins, F. S., Daly, M. J., and Donnelly, P. (2005). A haplotype map of the human genome. *Nature* 437, 1299–1320.

Altshuler, D., Pollara, V. J., Cowles, C. R., Van Etten, W. J., Baldwin, J., Linton, L., and Lander, E. S. (2000). An SNP map of the human genome generated by reduced representation shotgun sequencing. *Nature* 407, 513–516.

Mullikin, J. C., Hunt, S. E., Cole, C. G., Mortimore, B. J., Rice, C. M., Burton, J., Matthews, L. H., Pavitt, R., Plumb, R. W., Sims, S. K., Ainscough, R. M., Attwood, J., Bailey, J. M., Barlow, K., Bruskiewich, R. M., Butcher, P. N., Carter, N. P., Chen, Y., and Clee, C. M. (2000). An SNP map of human chromosome 22. *Nature* 407, 516–520.

5.4 RFLPs and SNPs Can Be Used for Genetic Mapping

Reviews

Gusella, J. F. (1986). DNA polymorphism and human disease. *Annu. Rev. Biochem.* 55, 831–854.

White, R., Leppert, M., Bishop, D. T., et al. (1985). Construction of linkage maps with DNA markers for human chromosomes. *Nature* 313, 101–105.

Research

Altshuler, D., Brooks, L. D., Chakravarti, A., Collins, F. S., Daly, M. J., and Donnelly, P. (2005). A haplotype map of the human genome. *Nature* 437, 1299–1320.

Dib, C., Faure, S., Fizames, C., et al. (1996). A comprehensive genetic map of the human genome based on 5,264 microsatellites. *Nature* 380, 152–154.

Dietrich, W. F., Miller, J., Steen, R., et al. (1996). A comprehensive genetic map of the mouse genome. *Nature* 380, 149–152.

Donis-Keller, J., Green, P., Helms, C., et al. (1987). A genetic linkage map of the human genome. *Cell* 51, 319–337.

Hinds, D. A., Stuve, L. L., Nilsen, G. B., Halperin, E., Eskin, E., Ballinger, D. G., Frazer, K. A., and Cox, D. R. (2005). Whole-genome patterns of common DNA variation in three human populations. *Science* 307, 1072–1079.

Sachidanandam, R., Weissman, D., Schmidt, S., et al. (2001). A map of human genome sequence variation containing 1.42 million single nucleotide polymorphisms. The International SNP Map Working Group. *Nature* 409, 928–933.

5.5 Eukaryotic Genomes Contain Both Nonrepetitive and Repetitive DNA Sequences

Reviews

Britten, R. J. and Davidson, E. H. (1971). Repetitive and nonrepetitive DNA sequences and a speculation on the origins of evolutionary novelty. *Q. Rev. Biol.* 46, 111–133.

Davidson, E. H. and Britten, R. J. (1973). Organization, transcription, and regulation in the animal genome. *Q. Rev. Biol.* 48, 565–613.

5.6 Eukaryotic Protein-Coding Genes Can Be Identified by the Conservation of Exons

Research

Buckler, A. J., Chang, D. D., Graw, S. L., Brook, J. D., Haber, D. A., Sharp, P. A., and Housman, D. E. (1991). Exon amplification: a strategy to isolate mammalian genes based on RNA splicing. *Proc. Natl. Acad. Sci. USA* 88, 4005–4009.

Kunkel, L. M., Monaco, A. P., Middlesworth, W., Ochs, H. D., and Latt, S. A. (1985). Specific cloning of DNA fragments absent from the DNA of a male patient with an X chromosome deletion. *Proc. Natl. Acad. Sci. USA* 82, 4778–4782.

Monaco, A. P., Bertelson, C. J., Middlesworth, W., Colletti, C. A., Aldridge, J., Fischbeck, K. H., Bartlett, R., Pericak-Vance, M. A., Roses, A. D., and Kunkel, L. M. (1985). Detection of deletions spanning the Duchenne muscular dystrophy locus using a tightly linked DNA segment. *Nature* 316, 842–845.

5.8 Some Organelles Have DNA

Research

Cann, R. L., Stoneking, M., and Wilson, A. C. (1987). Mitochondrial DNA and human evolution. *Nature* 325, 31–36.

5.9 Organelle Genomes Are Circular DNAs That Code for Organelle Proteins

Reviews

Attardi, G. (1985). Animal mitochondrial DNA: an extreme example of economy. *Int. Rev. Cytol.* 93, 93–146.

Boore, J. L. (1999). Animal mitochondrial genomes. *Nucleic Acids Res.* 27, 1767–1780.

Clayton, D. A. (1984). Transcription of the mammalian mitochondrial genome. *Annu. Rev. Biochem.* 53, 573–594.

Gray, M. W. (1989). Origin and evolution of mitochondrial DNA. *Annu. Rev. Cell Biol.* 5, 25–50.

Lang, B. F., Gray, M. W., and Burger, G. (1999). Mitochondrial genome evolution and the origin of eukaryotes. *Annu. Rev. Genet.* 33, 351–397.

Research

Anderson, S., Bankier, A. T., Barrell, B. G., et al. (1981). Sequence and organization of the human mitochondrial genome. *Nature* 290, 457–465.

5.10 The Chloroplast Genome Codes for Many Proteins and RNAs

Reviews

Palmer, J. D. (1985). Comparative organization of chloroplast genomes. *Annu. Rev. Genet.* 19, 325–354.

Shimada, H. and Sugiura, M. (1991). Fine structural features of the chloroplast genome: comparison of the sequenced chloroplast genomes. *Nucleic Acids Res.* 11, 983–995.

Sugiura, M., Hirose, T., and Sugita, M. (1998). Evolution and mechanism of translation in chloroplasts. *Annu. Rev. Genet.* 32, 437–459.

5.11 Mitochondria and Chloroplasts Evolved by Endosymbiosis

Review

Lang, B. F., Gray, M. W., and Burger, G. (1999). Mitochondrial genome evolution and the origin of eukaryotes. *Annu. Rev. Genet.* 33, 351–397.

Research

Adams, K. L., Daley, D. O., Qiu, Y. L., Whelan, J., and Palmer, J. D. (2000). Repeated, recent and diverse transfers of a mitochondrial gene to the nucleus in flowering plants. *Nature* 408, 354–357.

Arabidopsis Initiative (2000). Analysis of the genome sequence of the flowering plant *Arabidopsis thaliana*. *Nature* 408, 796–815.

Huang, C. Y., Ayliffe, M. A., and Timmis, J. N. (2003). Direct measurement of the transfer rate of chloroplast DNA into the nucleus. *Nature* 422, 72–76.

Thorsness, P. E. and Fox, T. D. (1990). Escape of DNA from mitochondria to the nucleus in *S. cerevisiae*. *Nature* 346, 376–379.

Photo courtesy of Massimiliano Di Ventra, University of California, San Diego. More information at J. Lagerqvist, M. Zwolak, and M. Di Ventra, *Nano Lett.* 6 (2006): 779–782.

Genome Sequences and Gene Numbers

CHAPTER OUTLINE

6.1 Introduction

6.2 Prokaryotic Gene Numbers Range Over an Order of Magnitude
- The minimum number of genes for a parasitic prokaryote is about 500; for a free-living nonparasitic prokaryote it is about 1500.

6.3 Total Gene Number Is Known for Several Eukaryotes
- There are 6000 genes in yeast; 18,500 in a worm; 13,600 in a fly; 25,000 in the small plant *Arabidopsis*; and probably 20,000 to 25,000 in mice and humans.

6.4 How Many Different Types of Genes Are There?
- The sum of the number of unique genes and the number of gene families is an estimate of the number of types of genes.
- The minimum size of the proteome can be estimated from the number of types of genes.

6.5 The Human Genome Has Fewer Genes Than Originally Expected
- Only 1% of the human genome consists of exons.
- The exons comprise ~5% of each gene, so genes (exons plus introns) comprise ~25% of the genome.
- The human genome has 20,000 to 25,000 genes.

- ~60% of human genes are alternatively spliced.
- Up to 80% of the alternative splices change protein sequence, so the proteome has ~50,000 to 60,000 members.

6.6 How Are Genes and Other Sequences Distributed in the Genome?
- Repeated sequences (present in more than one copy) account for >50% of the human genome.
- The great bulk of repeated sequences consist of copies of nonfunctional transposons.
- There are many duplications of large chromosome regions.

6.7 The Y Chromosome Has Several Male-Specific Genes
- The Y chromosome has ~60 genes that are expressed specifically in the testis.
- The male-specific genes are present in multiple copies in repeated chromosomal segments.
- Gene conversion between multiple copies allows the active genes to be maintained during evolution.

6.8 How Many Genes Are Essential?
- Not all genes are essential. In yeast and flies, deletions of <50% of the genes have detectable effects.

6.1 Introduction

Since the first complete organismal genomes were sequenced in 1995, both the speed and range of sequencing have improved greatly. The first genomes to be sequenced were small bacterial genomes, <2 Mb in size. By 2002, the human genome of >3000 Mb had been sequenced. Genomes have now been sequenced from a wide range of organisms, including bacteria, archaeans, yeasts and other unicellular eukaryotes, plants, and animals, including worms, flies, and mammals

Perhaps the single most important piece of information provided by a genome sequence is the number of genes. (See *Section 5.1* for a discussion about the difficulties of defining a gene; for our purposes, the term "gene" refers to a DNA sequence transcribed to mRNA, tRNA, or rRNA.) *Mycoplasma genitalium*, a free-living parasitic bacterium, has the smallest known genome of any organism, with only ~470 genes. The genomes of free-living bacteria have from 1700 to 7500 genes. Archaean genomes have a smaller range, from 1500 to 2700 genes. Unicellular eukaryotic genomes start with about 5300 genes. Worms and flies have roughly 18,500 and 13,500 genes, respectively, but the number rises only to ~25,000 for the mouse and human genomes.

FIGURE 6.1 summarizes the minimum number of genes found in six groups of organisms. A cell requires ~500 genes, a free-living cell requires ~1500 genes, a cell with a nucleus requires >5000 genes, a multicellular organism requires >10,000 genes, and an organism with a nervous system requires >13,000 genes.

500 genes
Intracellular (parasitic) bacterium

1,500 genes
Free-living bacterium

5,000 genes
Unicellular eukaryote

13,000 genes
Multicellular eukaryote

25,000 genes
Higher plants

25,000 genes
Mammals

FIGURE 6.1 The minimum gene number required for any type of organism increases with its complexity. Photo of intracellular bacterium courtesy of Gregory P. Henderson and Grant J. Jensen, California Institute of Technology. Photo of free-living bacterium courtesy of Karl O. Stetter, Universität Regensburg. Photo of unicellular eukaryote courtesy of Eishi Noguchi, Drexel University College of Medicine. Photo of multicellular eukaryote courtesy of Carolyn B. Marks and David H. Hall, Albert Einstein College of Medicine, Bronx, NY. Photo of higher plant courtesy of Keith Weller/USDA. Photo of mammal © Photodisc.

Many species may have more than the minimum number of genes required, so the number of genes can vary widely even among closely related species.

Within bacteria and unicellular eukaryotes, most genes are unique. Within multicellular eukaryotic genomes, however, some genes are arranged into families of related members. Of course, some genes are unique (meaning the family has only one member), but many belong to families with ten or more members. The number of different families may be a better indication of the overall complexity of the organism than the number of genes.

Some of the most insightful information comes from comparing genome sequences. With the sequences now available for both the human and chimpanzee genomes, it is possible to begin to address some of the questions about what makes humans unique.

6.2 Prokaryotic Gene Numbers Range Over an Order of Magnitude

Key concept

- The minimum number of genes for a parasitic prokaryote is about 500; for a free-living nonparasitic prokaryote it is about 1500.

Large-scale efforts have now led to the sequencing of many genomes. The range of genome sizes for organisms with completely sequenced genomes is summarized in **FIGURE 6.2**. They extend from the 0.6×10^6 bp of a mycoplasma to the 3.3×10^9 bp of the human genome, and include several important experimental animals, such as yeasts, the fruit fly, and a nematode worm. Although not yet completely sequenced, many plant genomes are much larger; the genome of bread wheat (*Triticum aestivum* L.) is 17 Gb (five times the size of the human genome), though it should be noted that the species is hexaploid.

The sequences of the genomes of prokaryotes show that virtually all of the DNA (typically 85%–90%) codes for RNA or polypeptide. **FIGURE 6.3** shows that the range of genome sizes is about an order of magnitude, and that the genome size is proportional to the number of genes. The typical gene averages about 1000 bp in length.

All of the prokaryotes with genome sizes below 1.5 Mb are parasites—they can live within a eukaryotic host that provides them with small molecules. Their genome sizes

Species	Genomes (Mb)	Genes	Lethal loci
Mycoplasma genitalium	0.58	470	~300
Rickettsia prowazekii	1.11	834	
Haemophilus influenzae	1.83	1,743	
Methanococcus jannaschi	1.66	1,738	
B. subtilis	4.2	4,100	
E. coli	4.6	4,288	1,800
S. cerevisiae	13.5	6,034	1,090
S. pombe	12.5	4,929	
A. thaliana	119	25,498	
O. sativa (rice)	466	~30,000	
D. melanogaster	165	13,601	3,100
C. elegans	97	18,424	
H. sapiens	3,300	~25,000	

FIGURE 6.2 Genome sizes and gene numbers are known from complete sequences for several organisms. Lethal loci are estimated from genetic data.

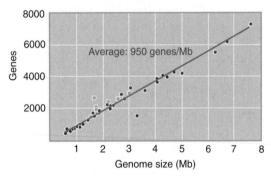

- Obligate parasitic bacteria
- Other bacteria
- Archaea

FIGURE 6.3 The number of genes in bacterial and archaeal genomes is proportional to genome size.

suggest the minimum number of functions required for a cellular organism. All classes of genes are reduced in number compared to prokaryotes with larger genomes, but the most significant reduction is in loci coding for enzymes concerned with metabolic functions (which are largely provided by the host cell) and with regulation of gene expression. *Mycoplasma genitalium* has the smallest genome, with ~470 genes.

Archaeans have biological properties that are intermediate between those of other prokaryotes and those of eukaryotes, but their genome sizes and gene numbers fall in the same range as those of bacteria. Their genome sizes vary from 1.5 to 3 Mb, corresponding to 1500 to 2700 genes. *Methanococcus jannaschii* is a methane-producing species that lives under high pressure and temperature. Its total gene number is similar to that of *Haemophilus influenzae*, but fewer of its genes can be identified on the basis of comparison with genes known in other organisms. Its apparatus for gene expression resembles that of eukaryotes more than of prokaryotes, but its apparatus for cell division better resembles that of prokaryotes.

The genomes of archaea and the smallest free-living bacteria suggest the minimum number of genes required to make a cell able to function independently in its environment. The smallest archaeal genome has ~1500 genes. The free-living nonparasitic bacterium with the smallest known genome is the thermophile *Aquifex aeolicus*, with a 1.5 Mb genome and 1512 genes. A "typical" gram-negative bacterium, *H. influenzae*, has 1743 genes, each of which is ~900 bp. So we can conclude that ~1500 genes are required by an exclusively free-living organism.

Prokaryotic genome sizes extend over about an order of magnitude, from 0.6 Mb to <8 Mb. As expected, the larger genomes have more genes. The prokaryotes with the largest genomes, *Sinorhizobium meliloti* and *Mesorhizobium loti*, are nitrogen-fixing bacteria that live on plant roots. Their genome sizes (~7 Mb) and total gene numbers (>7500) are similar to those of yeasts.

The size of the genome of *E. coli* is in the middle of the range for prokaryotes. The common laboratory strain has 4288 genes, with an average length of ~950 bp, and an average separation between genes of 118 bp. There can be quite significant differences between strains, though. The known extremes among strains of *E. coli* are from 4.6 Mb with 4249 genes to 5.5 Mb with 5361 genes.

We still do not know the functions of all of these genes. In most of these genomes, ~60% of the genes can be identified on the basis of homology with known genes in other species. These genes fall approximately equally into classes whose products function in metabolism, cell structure or transport of components, and gene expression and its regulation. In virtually every genome, >25% of the genes cannot yet be ascribed any function. Many of these genes can be found in related organisms, implying that they have a conserved function.

There has been some emphasis on sequencing the genomes of pathogenic bacteria, given their medical significance. An important insight into the nature of pathogenicity has been provided by the demonstration that **pathogenicity islands** are a characteristic feature of their genomes. These are large regions (~10 to 200 kb) that are present in the genomes of pathogenic species but absent from the genomes of nonpathogenic variants of the same or related species. Their G-C content often differs from that of the rest of the genome, and it is likely that they migrate between bacteria by a process of **horizontal transfer**. For example, the bacterium that causes anthrax (*Bacillus anthracis*) has two large plasmids (extrachromosomal DNA), one of which has a pathogenicity island that includes the gene coding for the anthrax toxin.

6.3 Total Gene Number Is Known for Several Eukaryotes

Key concept

- There are 6000 genes in yeast; 18,500 in a worm; 13,600 in a fly; 25,000 in the small plant *Arabidopsis*; and probably 20,000 to 25,000 in mice and humans.

As soon as we look at eukaryotic genomes, the relationship between genome size and gene number is weakened. The genomes of unicellular eukaryotes fall in the same size range as the largest bacterial genomes. Multicellular eukaryotes have more genes, but the number does not correlate with genome size, as can be seen from **FIGURE 6.4**.

The most extensive data for unicellular eukaryotes are available from the sequences of the genomes of the yeasts *Saccharomyces cerevisiae*

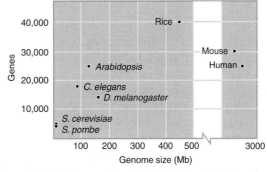

FIGURE 6.4 The number of genes in a eukaryote varies from 6000 to 40,000 but does not correlate with the genome size or the complexity of the organism.

5% of *S. cerevisiae* genes have 1 intron on average

1400 bp — Gene
200 bp — Intergenic space
Introns

1426 bp
500 bp

43% of *S. pombe* genes have introns
Average interrupted gene has 2 introns

FIGURE 6.5 The *S. cerevisiae* genome of 13.5 Mb has 6000 genes, almost all uninterrupted. The *S. pombe* genome of 12.5 Mb has 5000 genes, almost half having introns. Gene sizes and spacing are fairly similar.

and *Schizosaccharomyces pombe*. **FIGURE 6.5** summarizes the most important features. The yeast genomes of 13.5 Mb and 12.5 Mb have ~6000 and ~5000 genes, respectively. The average open reading frame (ORF) is ~1.4 kb, so that ~70% of the genome is occupied by coding regions. The major difference between them is that only 5% of *S. cerevisiae* genes have introns, compared to 43% in *S. pombe*. The density of genes is high; organization is generally similar, although the spaces between genes are a bit shorter in *S. cerevisiae*. About half of the genes identified by sequence were either known previously or related to known genes. The remaining genes were previously unknown, which gives some indication of the number of new types of genes that may be discovered.

The identification of long reading frames on the basis of sequence is quite accurate. ORFs coding for <100 amino acids, however, cannot be identified solely by sequence because of the high occurrence of false positives. Analysis of gene expression suggests that only ~300 of 600 such ORFs in *S. cerevisiae* are likely to be active genes.

A powerful way to validate gene structure is to compare sequences in closely related species—if a gene is active, it is likely to be conserved. Comparisons between the sequences of four closely related yeast species suggest that 503 of the genes originally identified in *S. cerevisiae* do not have counterparts in the other species and therefore should be deleted from the catalog. This reduces the total estimated gene number for *S. cerevisiae* to 5726.

The genome of *Caenorhabditis elegans* DNA varies between regions rich in genes and regions in which genes are more sparsely distributed. The total sequence contains ~18,500 genes. Only ~42% of the genes have putative counterparts outside Nematoda.

The fly genome is larger than the worm genome, but there are fewer genes in some species (~14,000 in *D. melanogaster*) and more in others (e.g., ~23,000 in *D. persimilis*). The number of different transcripts is somewhat larger as the result of alternative splicing. We do not understand why *C. elegans*—arguably, a less complex organism—has 30% more genes than the fly, but it may be because *C. elegans* has a larger average number of genes per gene family than does *D. melanogaster*, so the number of *unique* genes of the two species is more similar. A comparison of twelve *Drosophila* genomes reveals that there can be a fairly large range of gene number among closely related species. In some cases, there are several thousand genes that are species-specific. This emphasizes forcefully the lack of an exact relationship between gene number and complexity of the organism.

The plant *Arabidopsis thaliana* has a genome size intermediate between the worm and the fly, but has a larger gene number (25,000) than either. This again shows the lack of a clear relationship and also emphasizes the special quality of plants, which may have more genes (due to ancestral duplications) than animal cells. A majority of the *Arabidopsis* genome is found in duplicated segments, suggesting that there was an ancient doubling of the genome (to result in a tetraploid). Only 35% of *Arabidopsis* genes are present as single copies.

The genome of rice (*Oryza sativa*) is ~4× larger than that of *Arabidopsis*, but the number of genes is only ~50% larger, probably ~40,000. Repetitive DNA occupies 42%–45% of the genome. More than 80% of the genes found in *Arabidopsis* are also found in rice. Of these common genes, ~8000 are found in *Arabidopsis* and rice but not in any of the bacterial or animal genomes that have been sequenced. This is probably the set of genes that codes for plant-specific functions, such as photosynthesis.

From the twelve sequenced *Drosophila* genomes, we can form an impression of how many genes are devoted to each type of function. **FIGURE 6.6** breaks down the functions into different categories. Among the genes that are identified, we find over 3000 enzymes, ~900 transcription factors, and ~700 transporters and ion channels. About a quarter of the genes encode products of unknown function.

Polypeptide size increases from prokaryotes to eukaryotes. The archaean *M. jannaschi* and bacterium *E. coli* have average

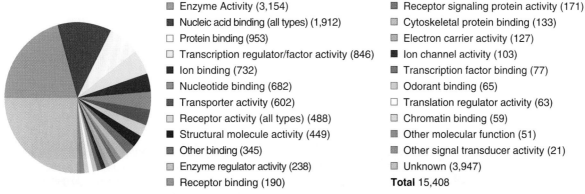

☐ Enzyme Activity (3,154)	☐ Receptor signaling protein activity (171)
■ Nucleic acid binding (all types) (1,912)	☐ Cytoskeletal protein binding (133)
☐ Protein binding (953)	☐ Electron carrier activity (127)
☐ Transcription regulator/factor activity (846)	■ Ion channel activity (103)
■ Ion binding (732)	☐ Transcription factor binding (77)
■ Nucleotide binding (682)	☐ Odorant binding (65)
■ Transporter activity (602)	☐ Translation regulator activity (63)
☐ Receptor activity (all types) (488)	☐ Chromatin binding (59)
■ Structural molecule activity (449)	☐ Other molecular function (51)
■ Other binding (345)	☐ Other signal transducer activity (21)
☐ Enzyme regulator activity (238)	☐ Unknown (3,947)
☐ Receptor binding (190)	**Total** 15,408

FIGURE 6.6 Functions of *Drosophila* genes based on comparative genomics of twelve species. The functions of about a quarter of the genes of *Drosophila* are unknown. Adapted from *Drosophila* 12 Genomes Consortium, *Nature* 450 (2007): 203–218.

polypeptide lengths of 287 and 317 amino acids, respectively, whereas *S. cerevisiae* and *C. elegans* have average lengths of 484 and 442 amino acids, respectively. Large polypeptides (>500 amino acids) are rare in bacteria, but comprise a significant component (~1/3) in eukaryotes. The increase in length is due to the addition of extra domains, with each domain typically constituting 100–300 amino acids. The increase in polypeptide size, however, is responsible for only a very small part of the increase in genome size.

Another insight into gene number is obtained by counting the number of expressed protein-coding genes. If we relied upon the estimates of the number of different mRNA species that can be counted in a cell, we would conclude that the average vertebrate cell expresses ~10,000 to 20,000 genes. The existence of significant overlaps between the mRNA populations in different cell types would suggest that the total expressed gene number for the organism should be within the same order of magnitude. The estimate for the total human gene number of 20,000 to 25,000 (see *Section 6.5, The Human Genome Has Fewer Genes Than Originally Expected*) would imply that a significant proportion of the total gene number is actually expressed in any given cell.

Eukaryotic genes are transcribed individually, with each gene producing a **monocistronic mRNA**. There is only one general exception to this rule: in the genome of *C. elegans*, ~15% of the genes are organized into units transcribed to **polycistronic mRNAs** (which are associated with the use of *trans*-splicing to allow expression of the downstream genes in these units; see *Section 21.14, trans-splicing Reactions Use Small RNAs*).

6.4 How Many Different Types of Genes Are There?

Key concepts
- The sum of the number of unique genes and the number of gene families is an estimate of the number of types of genes.
- The minimum size of the proteome can be estimated from the number of types of genes.

Some genes are unique; others belong to families in which the other members are related (but not usually identical). The proportion of unique genes declines with genome size and the proportion of genes in families increases.

Some genes are present in more than one copy or are related to one another, so the number of different types of genes is less than the total number of genes. We can divide the total number of genes into sets that have related members, as defined by comparing their exons. (A gene family arises by duplication of an ancestral gene followed by accumulation of changes in sequence between the copies. Most often the members of a family are related but not identical.) The number of types of genes is calculated by adding the number of unique genes (for which there is no other related gene at all) to the numbers of families that have two or more members.

FIGURE 6.7 compares the total number of genes with the number of distinct families in each of six genomes. In bacteria, most genes are unique, so the number of distinct families is close to the total gene number. The situation is different even in the unicellular eukaryote *S. cerevisiae*, for which there is a significant proportion of repeated genes. The most striking

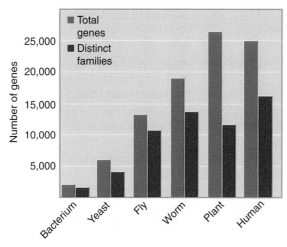

FIGURE 6.7 Many genes are duplicated, and as a result the number of different gene families is much less than the total number of genes. The histogram compares the total number of genes with the number of distinct gene families.

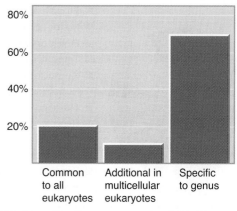

FIGURE 6.9 The fly genome can be divided into genes that are (probably) present in all eukaryotes, additional genes that are (probably) present in all multicellular eukaryotes, and genes that are more specific to subgroups of species that include flies.

	Unique genes	Families with 2–4 members	Families with >4 members
H. influenzae	89%	10%	1%
S. cerevisiae	72%	19%	9%
D. melanogaster	72%	14%	14%
C. elegans	55%	20%	26%
A. thaliana	35%	24%	41%

FIGURE 6.8 The proportion of genes that are present in multiple copies increases with genome size in higher eukaryotes.

effect is that the number of genes increases quite sharply in the higher eukaryotes, but the number of gene families does not change much.

FIGURE 6.8 shows that the proportion of unique genes drops sharply with genome size. When genes are present in families, the number of members in a family is small in bacteria and unicellular eukaryotes, but is large in multicellular eukaryotes. Much of the extra genome size of *Arabidopsis* is accounted for by families with more than four members.

If every gene is expressed, the total number of genes will account for the total number of polypeptides required to make the organism (the proteome). There are two conditions, however, that cause the proteome to be different from the total gene number. First, genes can be duplicated, and as a result some of them code for the same polypeptide (although it may be expressed in a different time or place) and others may code for related polypeptides that again play the same role in different times or

places. Second, the proteome can be larger than the number of genes because some genes can produce more than one polypeptide by means of alternative splicing.

What is the core proteome—the basic number of the different types of polypeptides in the organism? Although difficult to estimate because of the possibility of alternative splicing, a minimum estimate is given by the number of gene families, ranging from 1400 in the bacterium, to ~4000 in the yeast, and 11,000 to 14,000 for the fly and the worm.

What is the distribution of the proteome by type of protein? The 6000 proteins of the yeast proteome include 5000 soluble proteins and 1000 transmembrane proteins. About half of the proteins are cytoplasmic, a quarter are in the nucleus, and the remainder are split between the mitochondrion and the endoplasmic reticulum (ER)/Golgi system.

How many genes are common to all organisms (or to groups such as bacteria or multicellular eukaryotes), and how many are specific to lower-level taxonomic groups? **FIGURE 6.9** shows the comparison of fly genes to those of the worm (another multicellular eukaryote) and yeast (a unicellular eukaryote). Genes that code for corresponding polypeptides in different organisms are called **orthologous genes**, or **orthologs** (see *Section 4.10, The Members of a Gene Family Have a Common Organization*). Operationally, we usually consider that two genes in different organisms are orthologs if their sequences are similar over >80% of the length. By this criterion, ~20% of the fly genes have orthologs in both yeast and worm. These genes are probably required by all eukaryotes.

The proportion increases to 30% when fly and worm are compared, probably representing the addition of gene functions that are common to multicellular eukaryotes. This still leaves a major proportion of genes as coding for proteins that are required specifically by either flies or worms, respectively.

A minimum estimate of the size of an organismal proteome can be deduced from the number and structures of genes, and a cellular or organismal proteome size can also be directly measured by analyzing the total polypeptide content of a cell or organism. By such approaches, some proteins have been identified that were not suspected on the basis of genome analysis; this has led to the identification of new genes. Several methods are used for large-scale analysis of proteins. Mass spectrometry can be used for separating and identifying proteins in a mixture obtained directly from cells or tissues. Hybrid proteins bearing tags can be obtained by expression of cDNAs made by linking the sequences of ORFs to appropriate expression vectors that incorporate the sequences for affinity tags. This allows array analysis to be used to analyze the products. These methods also can be effective in comparing the proteins of two tissues—for example, a tissue from a healthy individual and one from a patient with a disease—to pinpoint the differences.

Once we know the total number of proteins, we can ask how they interact. By definition, proteins in structural multiprotein assemblies must form stable interactions with one another. Proteins in signaling pathways interact with one another transiently. In both cases, such interactions can be detected in test systems where essentially a readout system magnifies the effect of the interaction. One popular such system is the two-hybrid assay discussed in *Section 28.4, The Two-Hybrid Assay Detects Protein-Protein Interactions*. Such assays cannot detect all interactions: for example, if one enzyme in a metabolic pathway releases a soluble metabolite that then interacts with the next enzyme, the proteins may not interact directly.

As a practical matter, assays of pairwise interactions can give us an indication of the minimum number of independent structures or pathways. An analysis of the ability of all 6000 (predicted) yeast proteins to interact in pairwise combinations shows that ~1000 proteins can bind to at least one other protein. Direct analyses of complex formation have identified 1440 different proteins in 232 multiprotein complexes. This is the beginning of an analysis that will lead to definition of the number of functional assemblies or pathways. A comparable analysis of 8100 human proteins identified 2800 interactions, but is more difficult to interpret in the context of the larger proteome.

In addition to functional genes, there are also copies of genes that have become nonfunctional (identified as such by interruptions in their protein-coding sequences). These are called pseudogenes (see *Section 8.11, Pseudogenes Are Nonfunctional Gene Copies*). The number of pseudogenes can be large. In the mouse and human genomes, the number of pseudogenes is ~10% of the number of (potentially) active genes (see *Section 5.7, The Conservation of Genome Organization Helps to Identify Genes*). Some of these pseudogenes may serve some function by producing regulatory microRNAs; see Chapter 30, *Regulatory RNA*.

6.5 The Human Genome Has Fewer Genes Than Originally Expected

Key concepts

- Only 1% of the human genome consists of exons.
- The exons comprise ~5% of each gene, so genes (exons plus introns) comprise ~25% of the genome.
- The human genome has 20,000 to 25,000 genes.
- ~60% of human genes are alternatively spliced.
- Up to 80% of the alternative splices change protein sequence, so the proteome has ~50,000 to 60,000 members.

The human genome was the first vertebrate genome to be sequenced. This massive task has revealed a wealth of information about the genetic makeup of our species and about the evolution of genomes in general. Our understanding is deepened further by the ability to compare the human genome sequence with other sequenced vertebrate genomes.

Mammal genomes generally fall into a narrow size range, ~3×10^9 bp (see *Section 8.7, Why Are Some Genomes So Large?*). The mouse genome is ~14% smaller than the human genome, probably because it has had a higher rate of deletion. The genomes contain similar gene families and genes, with most genes having an ortholog in the other genome, but with differences in the number of members of a family, especially in those cases for which the functions are specific to the species (see *Section 5.7, The Conservation of Genome Organization Helps to Identify Genes*). Originally estimated to have ~30,000 genes, the mouse genome is now thought to have about the same number of genes as the human genome, 20,000 to 25,000. **FIGURE 6.10** plots the distribution of the mouse genes. The 23,000 protein-coding genes are accompanied

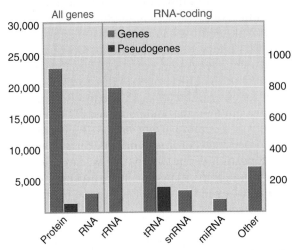

FIGURE 6.10 The mouse genome has ~23,000 protein-coding genes, which have ~1200 pseudogenes. There are ~3000 RNA-coding genes.

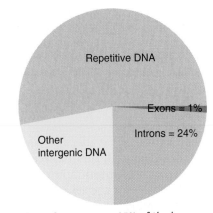

FIGURE 6.11 Genes occupy 25% of the human genome, but protein-coding sequences are only a tiny part of this fraction.

by ~3000 genes representing RNAs that do not code for proteins; these are generally small (aside from the ribosomal RNAs). Almost half of these genes code for transfer RNAs. In addition to the active genes, ~1200 pseudogenes have been identified.

The human (haploid) genome contains 22 autosomes plus the X and Y chromosomes. The chromosomes range in size from 45 to 279 Mb of DNA, making a total genome size of 3286 Mb (~3.3 × 10⁹ bp). On the basis of chromosome structure, the genome can be divided into regions of euchromatin (containing many active genes) and heterochromatin, with a much lower density of active genes (see *Section 9.7, Chromatin Is Divided into Euchromatin and Heterochromatin*). The euchromatin comprises the majority of the genome, ~2.9 × 10⁹ bp. The identified genome sequence represents ~90% of the euchromatin. In addition to providing information on the genetic content of the genome, the sequence also identifies features that may be of structural importance (see *Section 9.8, Chromosomes Have Banding Patterns*).

FIGURE 6.11 shows that a tiny proportion (~1%) of the human genome is accounted for by the exons that actually code for polypeptides. The introns that constitute the remaining sequences of protein-coding genes bring the total of DNA concerned with producing proteins to ~25%. As shown in **FIGURE 6.12**, the average human gene is 27 kb long, with nine exons that include a total coding sequence of 1340 bp. The average coding sequence is therefore only 5% of the length of an average protein-coding gene.

Two independent sequencing efforts for the human genome produced estimates of ~30,000 and ~40,000 genes, respectively. One measure of the accuracy of the analyses is whether they identify the same genes. The surprising answer is that the overlap between the two sets of genes

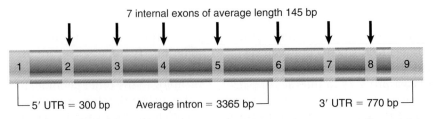

FIGURE 6.12 The average human gene is 27 kb long and has nine exons, usually comprising two longer exons at each end and seven internal exons. The UTRs in the terminal exons are the untranslated (noncoding) regions at each end of the gene. (This is based on the average. Some genes are extremely long, which makes the median length 14 kb with seven exons.)

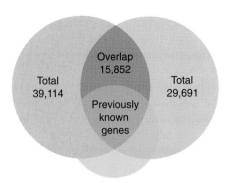

FIGURE 6.13 The two sets of genes identified in the human genome overlap only partially, as shown in the two large upper circles. They include, however, almost all previously known genes, as shown by the overlap with the smaller, lower circle.

is only ~50%, as summarized in **FIGURE 6.13**. An earlier analysis of the human gene set based on RNA transcripts had identified ~11,000 genes, almost all of which are present in both the large human gene sets, and which account for the major part of the overlap between them. So there is no question about the authenticity of half of each human gene set, but we have yet to establish the relationship between the other half of each set. The discrepancies illustrate the pitfalls of large-scale sequence analysis! As the sequence is analyzed further (and as other genomes are sequenced with which it can be compared), the number of valid genes seems to decline, and is now generally thought to be ~20,000 to 25,000.

By any measure, the total human gene number is much less than was originally expected—most estimates before the genome was sequenced were ~100,000. It shows a relatively small increase over flies and worms (13,600 and 18,500, respectively), not to mention the plants *Arabidopsis* (25,000) (see Figure 6.2) and rice (40,000). We should not, however, be particularly surprised by the notion that it does not take a great number of additional genes to make a more complex organism. The difference in DNA sequences between the human and chimpanzee genomes is extremely small (there is >99% similarity), so it is clear that the functions and interactions between a similar set of genes can produce very different results. The functions of specific groups of genes may be especially important, because detailed comparisons of orthologous genes in humans and chimpanzees suggest that there has been rapid evolution of certain classes of genes, including some involved in early development, olfaction, and hearing—all functions that are relatively specialized in these species.

The number of protein-coding genes is less than the number of potential polypeptides because of mechanisms such as alternative splicing, alternate promoter selection, and alternate poly(A) site selection that can result in several polypeptides from the same gene (see *Section 21.12, Alternative Splicing Is a Rule Rather Than Exception in Higher Eukaryotic Cells*). The extent of alternative splicing is greater in humans than in flies or worms; it may affect as many as 60% of the genes, so the increase in size of the human proteome relative to that of the other eukaryotes may be larger than the increase in the number of genes. A sample of genes from two chromosomes suggests that the proportion of the alternative splices that actually result in changes in the polypeptide sequence may be as high as 80%. This could increase the size of the proteome to 50,000 to 60,000 members.

In terms of the diversity of the number of gene families, however, the discrepancy between humans and the other eukaryotes may not be so great. Many of the human genes belong to gene families. An analysis of ~25,000 genes identified 3500 unique genes and 10,300 gene pairs. As can be seen from Figure 6.7, this extrapolates to a number of gene families only slightly larger than that of worms or flies.

6.6 How Are Genes and Other Sequences Distributed in the Genome?

Key concepts

- Repeated sequences (present in more than one copy) account for >50% of the human genome.
- The great bulk of repeated sequences consist of copies of nonfunctional transposons.
- There are many duplications of large chromosome regions.

Are genes uniformly distributed in the genome? Some chromosomes are relatively poor in genes and have >25% of their sequences as "deserts"—regions longer than 500 kb where there are no ORFs. Even the most gene-rich chromosomes have >10% of their sequences as deserts. So overall, ~20% of the human genome consists of deserts that have no protein-coding genes.

Repetitive sequences account for ~50% of the human genome, as seen in

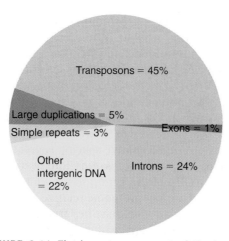

FIGURE 6.14 The largest component of the human genome consists of transposons. Other repetitive sequences include large duplications and simple repeats.

FIGURE 6.14. The repetitive sequences fall into five classes:

- Transposons (either active or inactive) account for the vast majority (45% of the genome). All transposons are found in multiple copies.
- Processed pseudogenes (~3000 in all, account for ~0.1% of total DNA). (These are sequences that arise by insertion of a reverse transcribed DNA copy of an mRNA sequence into the genome; see *Section 8.11, Pseudogenes Are Nonfunctional Gene Copies*.)
- Simple sequence repeats (highly repetitive DNA such as [CA] account for ~3%).
- Segmental duplications (blocks of 10 to 300 kb that have been duplicated into a new region) account for ~5%. For a small percentage of cases, these duplications are found on the same chromosome; in the other cases, the duplicates are on different chromosomes.
- Tandem repeats form blocks of one type of sequence (especially found at centromeres and telomeres).

The sequence of the human genome emphasizes the importance of transposons. (Many transposons have the capacity to replicate themselves and insert into new locations. They may function exclusively as DNA elements or may have *an active form that is RNA* [see *Chapter 17, Transposable Elements and Retroviruses*]. Their distribution in the human genome is summarized in Figure 17.39.) Most of the transposons

in the human genome are nonfunctional; very few are currently active. The high proportion of the genome occupied by these elements, however, indicates that they have played an active role in shaping the genome. One interesting feature is that some present genes originated as transposons and evolved into their present condition after losing the ability to transpose. At least 50 genes appear to have originated in this manner.

Segmental duplication at its simplest involves the tandem duplication of some region within a chromosome (typically because of an aberrant recombination event at meiosis; see *Section 7.2, Unequal Crossing-over Rearranges Gene Clusters*). In many cases, however, the duplicated regions are on different chromosomes, implying that either there was originally a tandem duplication followed by a translocation of one copy to a new site, or that the duplication arose by some different mechanism altogether. The extreme case of a segmental duplication is when a whole genome is duplicated, in which case the diploid genome initially becomes tetraploid. As the duplicated copies develop differences from one another, the genome may gradually become effectively a diploid again, although homologies between the diverged copies leave evidence of the event. This is especially common in plant genomes. The present state of analysis of the human genome identifies many individual duplicated regions, and there is evidence for a whole-genome duplication in the vertebrate lineage (see *Section 8.12, Genome Duplication Has Played a Role in Plant and Vertebrate Evolution*).

One curious feature of the human genome is the presence of sequences that do not appear to have coding functions, but that nonetheless show an evolutionary conservation higher than the background level. As detected by comparison with other genomes (such as the mouse genome), these represent about 5% of the total genome. Are these sequences associated with protein-coding sequences in some functional way? Their density on chromosome 18 is the same as elsewhere in the genome, although chromosome 18 has a significantly lower concentration of protein-coding genes. This suggests indirectly that their function is not connected with structure or expression of protein-coding genes.

The Y Chromosome Has Several Male-Specific Genes

Key concepts

- The Y chromosome has ~60 genes that are expressed specifically in the testis.
- The male-specific genes are present in multiple copies in repeated chromosomal segments.
- Gene conversion between multiple copies allows the active genes to be maintained during evolution.

The sequence of the human genome has significantly extended our understanding of the role of the sex chromosomes. It is generally thought that the X and Y chromosomes have descended from a common, very ancient autosome pair. Their evolution has involved a process in which the X chromosome has retained most of the original genes, whereas the Y chromosome has lost most of them.

The X chromosome is like the autosomes insofar as females have two copies and recombination can take place between them. The density of genes on the X chromosome is comparable to the density of genes on other chromosomes.

The Y chromosome is much smaller than the X chromosome and has many fewer genes. Its unique role results from the fact that only males have the Y chromosome, of which there is only one copy, so Y-linked loci are effectively haploid instead of diploid like all other human genes.

For many years, the Y chromosome was thought to carry almost no genes except for one or a few genes that determine maleness. The vast majority of the Y chromosome (>95% of its sequence) does not undergo crossing-over with the X chromosome, which led to the view that it could not contain active genes because there would be no means to prevent the accumulation of deleterious mutations. This region is flanked by short *pseudoautosomal regions* that exchange frequently with the X chromosome during male meiosis. It was originally called the nonrecombining region, but now has been renamed the *male-specific region*.

Detailed sequencing of the Y chromosome shows that the male-specific region contains three types of sequences, as illustrated in **FIGURE 6.15**:

- The *X-transposed sequences* consist of a total of 3.4 Mb comprising some large blocks resulting from a transposition from band q21 in the X chromosome about three or four million years ago. This is specific to the human lineage. These sequences do not recombine with the X chromosome and have become largely inactive. They now contain only two active genes.

- The *X-degenerate segments* of the Y are sequences that have a common origin with the X chromosome (going back to the common autosome from which both X and Y have descended) and contain genes or pseudogenes related to X-linked genes. There are 14 active genes and 13 pseudogenes. The active genes have, in a sense, thus far defied the trend for genes to be eliminated

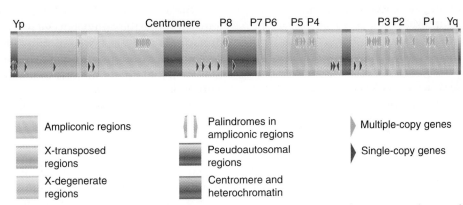

FIGURE 6.15 The Y chromosome consists of X-transposed regions, X-degenerate regions, and amplicons. The X-transposed X-degenerate regions have two and fourteen single-copy genes, respectively. The amplicons have eight large palindromes (P1–P8), which contain nine gene families. Each family contains at least two copies.

from chromosomal regions that cannot recombine at meiosis.

- The *ampliconic segments* have a total length of 10.2 Mb and are internally repeated on the Y chromosome. There are eight large palindromic blocks. They include nine protein-coding gene families, with copy numbers per family ranging from 2 to 35. The name "amplicon" reflects the fact that the sequences have been internally amplified on the Y chromosome.

Totaling the genes in these three regions, the Y chromosome contains 156 transcription units, of which half represent protein-coding genes and half represent pseudogenes.

The presence of the active genes is explained by the fact that the existence of closely related gene copies in the ampliconic segments allows gene conversion between multiple copies of a gene to be used to regenerate active copies. The most common needs for multiple copies of a gene are quantitative (to provide more protein product) or qualitative (to code for proteins with slightly different properties or that are expressed in different times or places). In this case, though, the essential function is evolutionary. In effect, the existence of multiple copies allows recombination within the Y chromosome itself to substitute for the evolutionary diversity that is usually provided by recombination between allelic chromosomes.

Most of the protein-coding genes in the ampliconic segments are expressed specifically in testis and are likely to be involved in male development. If there are ~60 such genes out of a total human gene set of ~25,000, then the genetic difference between male and female humans is ~0.2%.

6.8 How Many Genes Are Essential?

Key concepts

- Not all genes are essential. In yeast and flies, deletions of <50% of the genes have detectable effects.
- When two or more genes are redundant, a mutation in any one of them may not have detectable effects.
- We do not fully understand the persistence of genes that are apparently dispensable in the genome.

The force of natural selection ensures that functional genes are retained in the genome. Mutations occur at random, and a common mutational effect in an ORF will be to damage the protein product. An organism with a damaging mutation will be at a disadvantage in competition, and ultimately the mutation may be eliminated. The frequency of a disadvantageous allele in the population is balanced, however, between the generation of new mutants and the elimination of the allele by selection. Reversing this argument, whenever we see an intact, expressed ORF in the genome, we assume that its product plays a useful role in the organism. Natural selection must have prevented mutations from accumulating in the gene. The ultimate fate of a gene that ceases to be functional is to accumulate mutations until it is no longer recognizable.

The maintenance of a gene implies that it does not confer a selective disadvantage to the organism. In the course of evolution, though, even a small relative advantage may be the subject of natural selection, and a phenotypic defect may not necessarily be immediately detectable as the result of a mutation. Also, in diploid organisms, a new recessive mutation may be "hidden" in heterozygous form for many generations. We should like to know, however, how many genes are actually essential—meaning that their absence is lethal to the organism. In the case of diploid organisms, it means of course that the homozygous null mutation is lethal.

We might assume that the proportion of essential genes will decline with increase in genome size, given that larger genomes may have multiple related copies of particular gene functions. So far this expectation has not been borne out by the data (see Figure 6.2).

One approach to the issue of gene number is to determine the number of essential genes by mutational analysis. If we saturate some specified region of the chromosome with mutations that are lethal, the mutations should map into a number of complementation groups that correspond to the number of lethal loci in that region. By extrapolating to the genome as a whole, we may estimate the total essential gene number.

In the organism with the smallest known genome (*M. genitalium*), random insertions have detectable effects only in about two-thirds of the genes. Similarly, fewer than half of the genes of *E. coli* appear to be essential. The proportion is even lower in the yeast *S. cerevisiae*. When insertions were introduced at random into the genome in one early analysis, only 12% were lethal, and another 14% impeded growth.

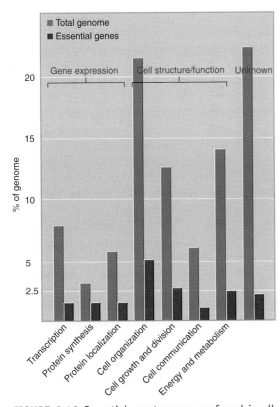

FIGURE 6.16 Essential yeast genes are found in all classes. Blue bars show total proportion of each class of genes; red bars show those that are essential.

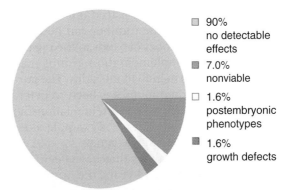

☐ 90%
no detectable
effects

☐ 7.0%
nonviable

☐ 1.6%
postembryonic
phenotypes

☐ 1.6%
growth defects

FIGURE 6.17 A systematic analysis of loss of function for 86% of worm genes shows that only 10% have detectable effects on the phenotype.

The majority (70%) of the insertions had no effect. A more systematic survey based on completely deleting each of 5916 genes (>96% of the identified genes) shows that only 18.7% are essential for growth on a rich medium (that is, when nutrients are fully provided). FIGURE 6.16 shows that these include genes in all categories. The only notable concentration of defects is in genes coding for products involved in protein synthesis, where ~50% are essential. Of course, this approach underestimates the number of genes that are essential for the yeast to live in the wild, when it is not so well provided with nutrients.

FIGURE 6.17 summarizes the results of a systematic analysis of the effects of loss of gene function in the worm *C. elegans*. The sequences of individual genes were predicted from the genome sequence, and by targeting an inhibitory RNA against these sequences (see *Section 30.3, Noncoding RNAs Can Be Used to Regulate Gene Expression*) a large collection of worms was made in which one (predicted) gene was prevented from functioning in each worm. Detectable effects on the phenotype were only observed for 10% of these knockdowns, suggesting that most genes do not play essential roles.

There is a greater proportion of essential genes (21%) among those worm genes that have counterparts in other eukaryotes, suggesting that highly conserved genes tend to have more basic functions. There is also an increased proportion of essential genes among those that are present in only one copy per haploid genome, compared with those for which there are multiple copies of related or identical genes. This suggests that many of the multiple genes might be relatively recent duplications that can substitute for one another's functions.

Extensive analyses of essential gene number in a multicellular eukaryote have been made in *Drosophila* through attempts to correlate visible aspects of chromosome structure with the number of functional genetic units. The notion that this might be possible originated from the presence of bands in the polytene chromosomes of *D. melanogaster*. (These chromosomes are found at certain developmental stages and represent an unusually extended physical form, in which a series of bands [more formally called chromomeres] are evident; see *Section 9.10, Polytene Chromosomes Form Bands*.) From the time of the early concept that the bands might represent a linear order of genes, there has been an attempt to correlate the organization of genes with the organization of bands. There are ~5000 bands in the *D. melanogaster* haploid set; they vary in size over an order of magnitude, but on average there is ~20 kb of DNA per band.

The basic approach is to saturate a chromosomal region with mutations. Usually the mutations are simply collected as lethals, without analyzing the cause of the lethality. *Any mutation that is lethal is taken to identify a locus that is essential for the organism. Sometimes mutations cause visible deleterious effects short of lethality, in which case we also define them as essential loci.*

When the mutations are placed into complementation groups, the number can be compared with the number of bands in the region, or individual complementation groups may even be assigned to individual bands. The purpose of these experiments has been to determine whether there is a consistent relationship between bands and genes. For example, does every band contain a single gene?

Totaling the analyses that have been carried out over the past 35 years, the number of essential complementation groups is ~70% of the number of bands. It is an open question whether there is any functional significance to this relationship. Irrespective of the cause, the equivalence gives us a reasonable estimate for the essential gene number of ~3600. By any measure, the number of essential loci in *Drosophila* is significantly less than the total number of genes.

If the proportion of essential human genes is similar to that of other eukaryotes, we would predict a range of ~4000 to 8000 genes in which mutations would be lethal or produce evidently damaging effects. At present, 1300 genes in which mutations cause evident defects have been identified. This is a substantial proportion of the expected total, especially in view of the

fact that many lethal genes may act so early in development that we never see their effects. This sort of bias may also explain the results in **FIGURE 6.18**, which show that the majority of known genetic defects are due to point mutations (where there is more likely to be at least some residual function of the gene).

How do we explain the persistence of genes whose deletion appears to have no effect? The most likely explanation is that the organism has alternative ways of fulfilling the same function. The simplest possibility is that there is **redundancy**, and that some genes are present in multiple copies. This is certainly true in some cases, in which multiple (related) genes must be knocked out in order to produce an effect. In a slightly more complex scenario, an organism might have two separate biochemical pathways capable of providing some activity. Inactivation of either pathway by itself would not be damaging, but the simultaneous occurrence of mutations in genes from both pathways would be deleterious.

Situations such as these can be tested by combining mutations. In this approach, deletions in two genes, neither of which is lethal by itself, are introduced into the same strain. If the double mutant dies, the strain is called a **synthetic lethal**. This technique has been used to great effect with yeast, where the isolation of double mutants can be automated. The procedure is called **synthetic genetic array analysis (SGA)**. **FIGURE 6.19** summarizes the results of an analysis in which an SGA screen was made for each of 132 viable deletions by testing whether it could survive in combination with any one of 4700 viable deletions. Every one of the tested genes had at least one partner with which the combination was lethal, and most of the tested genes had many such partners; the median is ~25 partners, and the greatest number is shown by one tested gene that had 146 lethal partners. A small proportion (~10%) of the interacting mutant pairs code for polypeptides that interact physically.

This result goes some way toward explaining the apparent lack of effect of so many deletions. Natural selection will act against these deletions when they are found in lethal pairwise combinations. To some degree, the organism is protected against the damaging effects of mutations by built-in redundancy. There is, however, a price in the form of accumulating the "genetic load" of mutations that are not deleterious in themselves, but that may cause serious problems when combined with other

Missense/nonsense	58%
Splicing	10%
Regulatory	<1%
Small deletions	16%
Small insertions	6%
Large deletions	5%
Large rearrangements	2%

FIGURE 6.18 Most known genetic defects in human genes are due to point mutations. The majority directly affect the protein sequence. The remainder is due to insertions, deletions, or rearrangements of varying sizes.

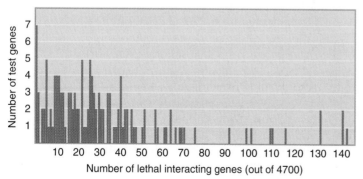

FIGURE 6.19 All 132 mutant test genes have some combinations that are lethal when they are combined with each of 4700 nonlethal mutations. The chart shows how many lethal interacting genes there are for each test gene.

such mutations in future generations. Presumably, the loss of the individual genes in such circumstances produces a sufficient disadvantage to maintain the functional gene during the course of evolution.

About 10,000 Genes Are Expressed at Widely Differing Levels in a Eukaryotic Cell

Key concepts

- In any given cell, most genes are expressed at a low level.
- Only a small number of genes, whose products are specialized for the cell type, are highly expressed.
- mRNAs expressed at low levels overlap extensively when different cell types are compared.
- The abundantly expressed mRNAs are usually specific for the cell type.
- ~10,000 expressed genes may be common to most cell types of a higher eukaryote.

The proportion of DNA containing protein-coding genes being expressed in a specific cell at a specific time can be determined by the amount of the DNA that can hybridize with the mRNAs isolated from that cell. Such a saturation analysis conducted for many cell types at various times typically identifies ~1% of the DNA being expressed as mRNA. From this we can calculate the number of protein-coding genes, so long as we know the average length of an mRNA. For a unicellular eukaryote such as yeast, the total number of expressed protein-coding genes is ~4000. For somatic tissues of multicellular eukaryotes, including both plants and vertebrates, the number usually is 10,000 to 15,000. (The only consistent exception to this type of value is presented by mammalian brain cells, for which much larger numbers of genes appear to be expressed, although the exact number is not certain.)

Kinetic analysis of the reassociation of an RNA population can be used to determine its sequence complexity. This type of analysis typically identifies three components in a eukaryotic cell. Just as with a DNA reassociation curve, a single component hybridizes over about two decades of Rot (RNA concentration × time) values, and a reaction extending over a greater range must be resolved by computer curve-fitting into individual components. Again, this represents what is really a continuous spectrum of sequences.

An example of an excess mRNA × cDNA reaction that generates three components is given in **FIGURE 6.20**:

- The first component has the same characteristics as a control reaction of ovalbumin mRNA with its DNA copy. This suggests that the first component is in fact just ovalbumin mRNA (which indeed occupies about half of the messenger mass in oviduct tissue).
- The next component provides 15% of the reaction, with a total complexity of 15 kb. This corresponds to seven to eight mRNA species of average length 2000 bases.
- The last component provides 35% of the reaction, which corresponds to a complexity of 26 Mb. This corresponds to ~13,000 mRNA species of average length 2000 bases.

From this analysis, we can see that about half of the mass of mRNA in the cell represents a single mRNA, ~15% of the mass is provided by a mere seven to eight mRNAs, and ~35% of the mass is divided into the large number of 13,000 mRNA species. It is therefore obvious that the mRNAs comprising each component must be present in very different amounts.

The average number of molecules of each mRNA per cell is called its **abundance**. It can

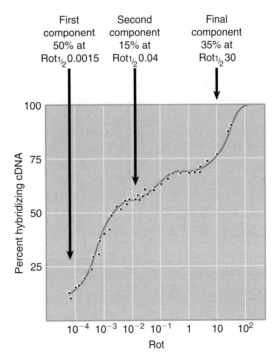

FIGURE 6.20 Hybridization between excess mRNA and cDNA identifies several components in chick oviduct cells, each characterized by the $Rot_{1/2}$ of reaction.

be calculated quite simply if the total mass of a specific mRNA species in the cell is known. In the example of chick oviduct cells shown in Figure 6.20, the total mRNA can be accounted for as 100,000 copies of the first component (ovalbumin mRNA), 4000 copies of each of seven or eight other mRNAs in the second component, and only ~5 copies of each of the 13,000 remaining mRNAs that constitute the last component.

We can divide the mRNA population into two general classes, according to their abundance:

- The oviduct is an extreme case, with so much of the mRNA represented by only one species, but most cells do contain a small number of RNAs present in many copies each. This **abundant mRNA** component typically consists of <100 different mRNAs present in 1000 to 10,000 copies per cell. It often corresponds to a major part of the mass, approaching 50% of the total mRNA.

- About half of the mass of the mRNA consists of a large number of sequences, of the order of 10,000, each represented by only a small number of copies in the mRNA—say, <10. This is the **scarce mRNA** (or **complex mRNA**) class. It is this class that drives a saturation reaction.

Many somatic tissues of multicellular eukaryotes have an expressed gene number in the range of 10,000 to 20,000. How much overlap is there between the genes expressed in different tissues? For example, the expressed gene number of chick liver is ~11,000 to 17,000, compared with the value for oviduct of ~13,000 to 15,000. How many of these two sets of genes are identical? How many are specific for each tissue? These questions are usually addressed by analyzing the transcriptome—the set of sequences represented in RNA.

We see immediately that there are likely to be substantial differences among the genes expressed in the abundant class. Ovalbumin, for example, is synthesized only in the oviduct, and not at all in the liver. This means that 50% of the mass of mRNA in the oviduct is specific to that tissue.

The abundant mRNAs represent only a small proportion of the number of expressed genes, though. In terms of the total number of genes of the organism, and of the number of changes in transcription that must be made between different cell types, we need to know the extent of overlap between the genes represented in the scarce mRNA classes of different cell phenotypes.

Comparisons between different tissues show that, for example, ~75% of the sequences expressed in liver and oviduct are the same. In other words, ~12,000 genes are expressed in both liver and oviduct, ~5000 additional genes are expressed only in liver, and ~3000 additional genes are expressed only in oviduct.

The scarce mRNAs overlap extensively. Between mouse liver and kidney, ~90% of the scarce mRNAs are identical, leaving a difference between the tissues of only 1000 to 2000 in terms of the number of expressed genes. The general result obtained in several comparisons of this sort is that only ~10% of the mRNA sequences of a cell are unique to it. The majority of sequences are common to many—perhaps even all—cell types.

This suggests that the common set of expressed gene functions, numbering perhaps ~10,000 in mammals, comprise functions that are needed in all cell types. Sometimes this type of function is referred to as a **housekeeping gene** or **constitutive gene**. It contrasts with the activities represented by specialized functions (such as ovalbumin or globin) needed only for particular cell phenotypes. These are sometimes called **luxury genes**.

6.10 Expressed Gene Number Can Be Measured *en masse*

Key concepts

- DNA microarray technology allows a snapshot to be taken of the expression of the entire genome in a yeast cell.
- ~75% (~4500 genes) of the yeast genome is expressed under normal growth conditions.
- DNA microarray technology allows detailed comparisons of related animal cells to determine (for example) the differences in expression between a normal cell and a cancer cell.

Recent technology allows more systematic and accurate estimates of the number of expressed protein-coding genes. One approach (serial analysis of gene expression, or SAGE) allows a unique sequence tag to be used to identify each mRNA. The technology then allows the abundance of each tag to be measured. This approach identifies 4665 expressed genes in *S. cerevisiae* growing under normal conditions, with abundances varying from 0.3 to >200 transcripts/

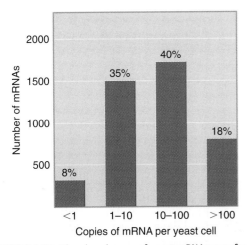

FIGURE 6.21 The abundances of yeast mRNAs vary from <1 per cell (meaning that not every cell has a copy of the mRNA) to >100 per cell (coding for the more abundant proteins).

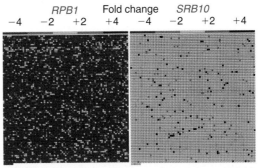

FIGURE 6.22 DNA microarray analysis allows change in expression of each gene to be measured. Each square represents one gene (top left is first gene on chromosome I, bottom right is last gene on chromosome XVI). Change in expression relative to wild type is indicated by red (reduction), white (no change), or blue (increase). Photos courtesy of Rich A. Young, Whitehead Institute, Massachusetts Institute of Technology.

cell. This means that ~75% of the total gene number (~6000) is expressed under these conditions. FIGURE 6.21 summarizes the number of different mRNAs that is found at each different abundance level.

The most powerful new technology uses chips that contain **microarrays**, arrays of many tiny DNA oligonucleotide samples. Their construction is made possible by knowledge of the sequence of the entire genome. In the case of *S. cerevisiae*, each of 6181 ORFs is represented on the microarray by twenty 25-mer oligonucleotides that perfectly match the sequence of the message and twenty mismatch oligonucleotides that differ at one base position. The expression level of any gene is calculated by subtracting the average signal of a mismatch from its perfect match partner. The entire yeast genome can be represented on four chips. This technology is sensitive enough to detect transcripts of 5460 genes (~90% of the genome), and shows that many genes are expressed at low levels, with abundances of 0.1 to 0.2 transcript/cell. An abundance of <1 transcript/cell means that not all cells have a copy of the transcript at any given moment.

The technology allows not only measurement of levels of gene expression, but also detection of differences in expression in mutant cells compared to wild-type cells growing under different growth conditions, and so on. The results of comparing two states are expressed in the form of a grid, in which each square represents a particular gene and the relative change in expression is indicated by color. The left part of FIGURE 6.22 shows the effect of a mutation (*RPB1*) in RNA polymerase II, the enzyme that

produces mRNA, which as might be expected causes the expression of most genes to be heavily reduced. By contrast, the right part shows that a mutation in an ancillary component of the transcription apparatus (*SRB10*) has much more restricted effects, causing increases in expression of some genes.

The extension of this technology to animal cells will allow the general descriptions based on RNA hybridization analysis to be replaced by exact descriptions of the genes that are expressed, and the abundances of their products, in any given cell type. A gene expression map of *D. melanogaster* detects transcriptional activity in some stage of the life cycle in almost all (93%) of predicted genes and shows that 40% have alternatively spliced forms.

6.11 Summary

Genomes that have been sequenced include those of many bacteria and archaea, yeasts, a worm, a fly, a mouse, and a human. The minimum number of genes required to make a living cell (a parasite) is ~470. The minimum number required to make a free-living cell is ~1700. A typical gram-negative bacterium has ~1500 genes. Genomes of strains of *E. coli* vary from 4300 to 5400 genes. The average bacterial gene is ~1000 bp long and is separated from the next gene by a space of ~100 bp. The yeasts *S. pombe* and *S. cerevisiae* have 5000 and 6000 genes, respectively.

Although the fly *D. melanogaster* has a larger genome than the worm *C. elegans*, the fly has fewer genes (13,600) than the worm (18,500). The plant *Arabidopsis* has 25,000 genes, and the lack of a clear relationship between genome size and gene number is shown by the fact that the rice genome is 4× larger but contains only 50% more genes (~40,000). Mice and humans each have 20,000 to 25,000 genes, which is much less than had been originally expected. The complexity of development of an organism may depend on the nature of the interactions between genes as well as their total number.

About 8000 genes are common to prokaryotes and eukaryotes and are likely to be involved in basic functions. A further 12,000 genes are found in multicellular organisms. Another 8000 genes are found in animals, and an additional 8000 (largely involved with the immune and nervous systems) are found in vertebrates. In each organismal genome that has been sequenced, only ~50% of the genes have defined functions. Analysis of lethal genes suggests that only a minority of genes is essential in each organism.

The sequences comprising a eukaryotic genome can be classified in three groups: nonrepetitive sequences are unique; moderately repetitive sequences are dispersed and repeated a small number of times in the form of related, but not identical, copies; and highly repetitive sequences are short and usually repeated as tandem arrays. The proportions of the types of sequence are characteristic for each genome, although larger genomes tend to have a smaller proportion of nonrepetitive DNA. Almost 50% of the human genome consists of repetitive sequences, the vast majority corresponding to transposon sequences. Most structural genes are located in nonrepetitive DNA. The complexity of nonrepetitive DNA is a better reflection of the complexity of the organism than the total genome complexity.

Genes are expressed at widely varying levels. There may be 10^5 copies of mRNA for an abundant gene whose protein is the principal product of the cell, 10^3 copies of each mRNA for <10 moderately abundant messages, and <10 copies of each mRNA for >10,000 scarcely expressed genes. Overlaps between the mRNA populations of cells of different phenotypes are extensive; the majority of mRNAs are present in most cells.

References

6.2 Prokaryotic Gene Numbers Range Over an Order of Magnitude

Reviews

Bentley, S. D. and Parkhill, J. (2004). Comparative genomic structure of prokaryotes. *Annu. Rev. Genet.* 38, 771–792.

Hacker, J. and Kaper, J. B. (2000). Pathogenicity islands and the evolution of microbes. *Annu. Rev. Microbiol.* 54, 641–679.

Research

Blattner, F. R. et al. (1997). The complete genome sequence of *Escherichia coli* K-12. *Science* 277, 1453–1474.

Deckert, G. et al. (1998). The complete genome of the hyperthermophilic bacterium *Aquifex aeolicus*. *Nature* 392, 353–358.

Galibert, F. et al. (2001). The composite genome of the legume symbiont *Sinorhizobium meliloti*. *Science* 293, 668–672.

6.3 Total Gene Number Is Known for Several Eukaryotes

Research

Adams, M. D. et al. (2000). The genome sequence of *D. melanogaster*. *Science* 287, 2185–2195.

Arabidopsis Initiative (2000). Analysis of the genome sequence of the flowering plant *Arabidopsis thaliana*. *Nature* 408, 796–815.

C. elegans Sequencing Consortium (1998). Genome sequence of the nematode *C. elegans*: a platform for investigating biology. *Science* 282, 2012–2022.

Duffy, A., and Grof, P. (2001). Psychiatric diagnoses in the context of genetic studies of bipolar disorder. *Bipolar Disord.* 3, 270–275.

Dujon, B. et al. (1994). Complete DNA sequence of yeast chromosome XI. *Nature* 369, 371–378.

Goff, S. A. et al. (2002). A draft sequence of the rice genome (*Oryza sativa* L. ssp. *japonica*). *Science* 296, 92–114.

Johnston, M. et al. (1994). Complete nucleotide sequence of *S. cerevisiae* chromosome VIII. *Science* 265, 2077–2082.

Kellis, M., Patterson, N., Endrizzi, M., Birren, B., and Lander, E. S. (2003). Sequencing and comparison of yeast species to identify genes and regulatory elements. *Nature* 423, 241–254.

Oliver, S. G. et al. (1992). The complete DNA sequence of yeast chromosome III. *Nature* 357, 38–46.

Wilson, R. et al. (1994). 22 Mb of contiguous nucleotide sequence from chromosome III of *C. elegans*. *Nature* 368, 32–38.

Wood, V. et al. (2002). The genome sequence of *S. pombe*. *Nature* 415, 871–880.

6.4 How Many Different Types of Genes Are There?

Reference

Rual, J. F., Venkatesan, K., Hao, T., Hirozane-Kishikawa, T., Dricot, A., Li, N., Berriz, G. F., Gibbons, F. D., Dreze, M., Ayivi-Guedehoussou, N., et al. (2005). Towards a proteome-scale map of the human protein–protein interaction network. *Nature* 437, 1173–1178.

Reviews

Aebersold, R. and Mann, M. (2003). Mass spectrometry-based proteomics. *Nature* 422, 198–207.

Hanash, S. (2003). Disease proteomics. *Nature* 422, 226–232.

Phizicky, E., Bastiaens, P. I., Zhu, H., Snyder, M., and Fields, S. (2003). Protein analysis on a proteomic scale. *Nature* 422, 208–215.

Sali, A., Glaeser, R., Earnest, T., and Baumeister, W. (2003). From words to literature in structural proteomics. *Nature* 422, 216–225.

Research

Agarwal, S., Heyman, J. A., Matson, S., Heidtman, M., Piccirillo, S., Umansky, L., Drawid, A., Jansen, R., Liu, Y., Miller, P., Gerstein, M., Roeder, G. S., and Snyder, M. (2002). Subcellular localization of the yeast proteome. *Genes Dev.* 16, 707–719.

Arabidopsis Initiative (2000). Analysis of the genome sequence of the flowering plant *Arabidopsis thaliana*. *Nature* 408, 796–815.

Gavin, A. C. et al. (2002). Functional organization of the yeast proteome by systematic analysis of protein complexes. *Nature* 415, 141–147.

Ho, Y. et al. (2002). Systematic identification of protein complexes in *S. cerevisiae* by mass spectrometry. *Nature* 415, 180–183.

Rubin, G. M. et al. (2000). Comparative genomics of the eukaryotes. *Science* 287, 2204–2215.

Uetz, P. et al. (2000). A comprehensive analysis of protein–protein interactions in *S. cerevisiae*. *Nature* 403, 623–630.

Venter, J. C. et al. (2001). The sequence of the human genome. *Science* 291, 1304–1350.

6.5   The Human Genome Has Fewer Genes Than Originally Expected

Research

Clark, A. G. et al. (2003). Inferring nonneutral evolution from human–chimp–mouse orthologous gene trios. *Science* 302, 1960–1963.

Hogenesch, J. B., Ching, K. A., Batalov, S., Su, A. I., Walker, J. R., Zhou, Y., Kay, S. A., Schultz, P. G., and Cooke, M. P. (2001). A comparison of the Celera and Ensembl predicted gene sets reveals little overlap in novel genes. *Cell* 106, 413–415.

International Human Genome Sequencing Consortium (2001). Initial sequencing and analysis of the human genome. *Nature* 409, 860–921.

International Human Genome Sequencing Consortium (2004). Finishing the euchromatic sequence of the human genome. *Nature* 431, 931–945.

Venter, J. C. et al. (2001). The sequence of the human genome. *Science* 291, 1304–1350.

Waterston et al. (2002). Initial sequencing and comparative analysis of the mouse genome. *Nature* 420, 520–562.

6.6   How Are Genes and Other Sequences Distributed in the Genome?

Reference

Nusbaum, C., Cody, M. C., Borowsky, M. L., Kamal, M., Kodira, C. D., Taylor, T. D., Whittaker, C. A., Chang, J. L., Cuomo, C. A., Dewar, K., et al. (2005). DNA sequence and analysis of human chromosome 18. *Nature* 437, 551–555.

6.7   The Y Chromosome Has Several Male-Specific Genes

Research

Skaletsky, H. et al. (2003). The male-specific region of the human Y chromosome is a mosaic of discrete sequence classes. *Nature* 423, 825–837.

6.8   How Many Genes Are Essential?

Research

Giaever et al. (2002). Functional profiling of the *S. cerevisiae* genome. *Nature* 418, 387–391.

Goebl, M. G. and Petes, T. D. (1986). Most of the yeast genomic sequences are not essential for cell growth and division. *Cell* 46, 983–992.

Hutchison, C. A. et al. (1999). Global transposon mutagenesis and a minimal mycoplasma genome. *Science* 286, 2165–2169.

Kamath, R. S., Fraser, A. G., Dong, Y., Poulin, G., Durbin, R., Gotta, M., Kanapin, A., Le Bot, N., Moreno, S., Sohrmann, M., Welchman, D. P., Zipperlen, P., and Ahringer, J. (2003). Systematic functional analysis of the *C. elegans* genome using RNAi. *Nature* 421, 231–237.

Tong, A. H. et al. (2004). Global mapping of the yeast genetic interaction network. *Science* 303, 808–813.

6.9   About 10,000 Genes Are Expressed at Widely Differing Levels in a Eukaryotic Cell

Research

Hastie, N. B. and Bishop, J. O. (1976). The expression of three abundance classes of mRNA in mouse tissues. *Cell* 9, 761–774.

6.10   Expressed Gene Number Can Be Measured *en masse*

Reviews

Mikos, G. L. G. and Rubin, G. M. (1996). The role of the genome project in determining gene function: insights from model organisms. *Cell* 86, 521–529.

Young, R. A. (2000). Biomedical discovery with DNA arrays. *Cell* 102, 9–15.

Research

Holstege, F. C. P. et al. (1998). Dissecting the regulatory circuitry of a eukaryotic genome. *Cell* 95, 717–728.

Hughes, T. R., Marton, M. J., Jones, A. R., Roberts, C. J., Stoughton, R., Armour, C. D., Bennett, H. A., Coffey, E., Dai, H., He, Y. D., Kidd, M. J., King, A. M., Meyer, M. R., Slade, D., Lum, P. Y., Stepaniants, S. B., Shoemaker, D. D. et al. (2000). Functional discovery via a compendium of expression profiles. *Cell* 102, 109–126.

Stolc, V. et al. (2004). A gene expression map for the euchromatic genome of *Drosophila melanogaster*. *Science* 306, 655–660.

Velculescu, V. E. et al. (1997). Characterization of the yeast transcriptosome. *Cell* 88, 243–251.

© MedicalRF.com/Visuals Unlimited

Clusters and Repeats

7.1 Introduction

A set of genes descended by duplication and variation from a single ancestral gene is called a **gene family**. Its members may be clustered together or dispersed on different chromosomes (or a combination of both). Genome analysis to identify paralogous sequences shows that many genes belong to families; the 20,000 to 25,000 genes identified in the human genome fall into ~15,000 families, so the average gene has ~2 relatives in the genome (see Figure 6.7). Gene families vary enormously in the degree of relatedness between members, from those consisting of multiple identical members to those for which the relationship is quite distant. Genes are usually related only by their exons, with introns having diverged (see *Section 4.5, Exon Sequences Are Conserved but Introns Vary*). Genes may also be related by only some of their exons, whereas others are unique (see *Section 4.9, Some Exons Can Be Equated with Protein Functional Domains*).

Some members of the gene family may evolve to become **pseudogenes**. Pseudogenes (ψ) are defined by their possession of sequences that are related to those of the functional genes, but that cannot be translated into a functional polypeptide. (See *Section 8.11, Pseudogenes Are Nonfunctional Gene Copies*, for further discussion.)

Some pseudogenes have the same general structure as functional genes, with sequences corresponding to exons and introns in the usual locations. They may have been rendered inactive by mutations that prevent any or all of the stages of gene expression. The changes can take the form of abolishing the signals for initiating transcription, preventing splicing at the exon–intron junctions, or prematurely terminating translation.

The initial event that allows related exons or genes to develop is a duplication, when a copy is generated of some sequence within the genome. *Tandem duplication* (when the duplicates remain together) may arise through errors in replication or recombination. Separation of the duplicates can occur by a **translocation** that transfers material from one chromosome

to another. A duplicate at a new location may also be produced directly by a transposition event that is associated with copying a region of DNA from the vicinity of the transposon. Duplications of intact genes, collections of exons, or even individual exons may occur. When an intact gene is involved, duplication generates two copies of a gene whose activities are indistinguishable, but then usually the copies diverge as each accumulates different substitutions.

The members of a structural gene family usually have related or even identical functions, although they may be expressed at different times or in different cell types. For example, different globin proteins are expressed in embryonic and adult red blood cells, whereas different actins are utilized in muscle and nonmuscle cells. When genes have diverged significantly, or when only some exons are related, the proteins may have different functions.

Some gene families consist of identical members. Clustering is a prerequisite for maintaining identity between genes, although clustered genes are not necessarily identical. **Gene clusters** range from extremes in which a duplication has generated two adjacent related genes to cases where hundreds of identical genes lie in a tandem array. Extensive tandem repetition of a gene may occur when the product is needed in unusually large amounts. Examples are the genes for rRNA or histone proteins. This creates a special situation with regard to the maintenance of identity and the effects of selective pressure.

Gene clusters offer us an opportunity to examine the forces involved in evolution of the genome over larger regions than single genes. Duplicated sequences, especially those that remain in the same vicinity, provide the substrate for further evolution by recombination. A population evolves by the classical homologous recombination illustrated in **FIGURES 7.1** and **7.2**, in which an exact crossing-over occurs (see Chapter 15, *Homologous and Site-Specific Recombination*). The recombinant chromosomes have the same organization as the parental chromosome; they contain precisely the same loci in

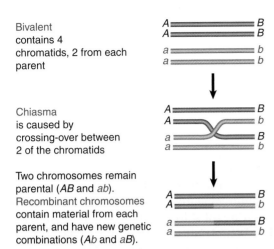

Bivalent
contains 4
chromatids, 2 from each
parent

Chiasma
is caused by
crossing-over between
2 of the chromatids

Two chromosomes remain
parental (*AB* and *ab*).
Recombinant chromosomes
contain material from each
parent, and have new genetic
combinations (*Ab* and *aB*).

FIGURE 7.1 Chiasma formation represents the generation of recombinants.

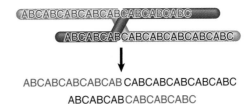

ABCABCABCABCAB CABCABCABCABCABC

ABCABCAB CABCABCABC

FIGURE 7.3 Unequal crossing-over results from pairing between nonequivalent repeats in regions of DNA consisting of repeating units. Here the repeating unit is the sequence ABC, and the third repeat of the blue chromosome has aligned with the first repeat of the black chromosome. Throughout the region of pairing, ABC units of one chromosome are aligned with ABC units of the other chromosome. Crossing-over generates chromosomes with ten and six repeats each, instead of the eight repeats of each parent.

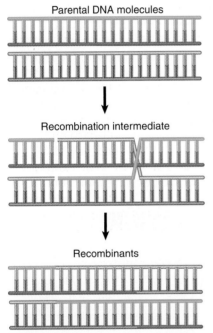

Parental DNA molecules

Recombination intermediate

Recombinants

FIGURE 7.2 Recombination involves pairing between complementary strands of the two parental duplex DNAs.

the same order, but contain different combinations of alleles, providing the raw material for natural selection. The existence of duplicated sequences, however, allows aberrant events to occur occasionally, which changes the number of copies of genes and not just the combination of alleles.

Unequal crossing-over (also known as **nonreciprocal recombination**) describes a recombination event occurring between two sites that are similar or identical, but not precisely homologous in position. The feature that makes such events possible is the existence of repeated sequences. **FIGURE 7.3** shows that this allows one copy of a repeat in one chromosome to misalign for recombination with a different copy of the repeat in the homologous chromosome, instead of with the corresponding copy. When recombination occurs, this increases the number of repeats in one chromosome and decreases it in the other. In effect, one recombinant chromosome has a deletion and the other has an insertion. This mechanism is responsible for the evolution of clusters of related sequences. We can trace its operation in expanding or contracting the size of an array in both gene clusters and regions of highly repeated DNA.

The highly repetitive fraction of the genome consists of multiple tandem copies of very short repeating units. These often have unusual properties. One is that they may be identified as a separate peak on a density gradient analysis of DNA (see *Section 3.6, DNA Separation Techniques*), which gave rise to the name **satellite DNA**. They often are associated with heterochromatic regions of the chromosomes and in particular with centromeres (which contain the points of attachment for segregation on a mitotic or meiotic spindle). As a result of their repetitive organization, they show some of the same behavior with regard to evolution as the tandem gene clusters. In addition to the satellite sequences, there are shorter stretches of DNA called **minisatellites**, tandem repeats in which each repeat is less than 10 base pairs in length, and they have similar properties. They are useful in showing a high degree of divergence between individual genomes that can be used for mapping or identification purposes.

All of these events that change the constitution of the genome are rare, but they are significant over the course of evolution.

7.2 Unequal Crossing-over Rearranges Gene Clusters

Key concepts

- When a genome contains a cluster of genes with related sequences, mispairing between nonallelic loci can cause unequal crossing-over. This produces a deletion in one recombinant chromosome and a corresponding duplication in the other.
- Different thalassemias are caused by various deletions that eliminate α- or β-globin genes. The severity of the disease depends on the individual deletion.

There are frequent opportunities for rearrangement in a cluster of related or identical genes. We can see the results by comparing the mammalian β-globin clusters (see *Section 8.10, Globin Clusters Arise by Duplication and Divergence,* for discussion of the evolution of the globin gene family). Although all β-globin clusters serve the same function, and all have the same general organization, each is different in size, there is variation in the total number and types of β-globin genes, and the numbers and structures of pseudogenes are different. All of these changes must have occurred since the mam-

malian radiation ~85 million years ago (the time of the common ancestor to all the mammals).

The comparison makes the general point that gene duplication, rearrangement, and variation is as important a factor in evolution as the slow accumulation of point mutations in individual genes (see Chapter 8, *Genome Evolution*). What types of mechanisms are responsible for gene reorganization?

As described in the introduction to this chapter, unequal crossing-over can occur as the result of pairing between two sites that are not homologous in position. Usually, recombination involves corresponding sequences of DNA held in exact alignment between the two homologous chromosomes. When there are two copies of a gene on each chromosome, though, an occasional misalignment allows pairing between them. (This requires some of the adjacent regions to go unpaired.) This can happen in a region of short repeats (see Figure 7.3) or in a gene cluster. **FIGURE 7.4** shows that unequal crossing-over in a gene cluster can have two consequences—quantitative and qualitative:

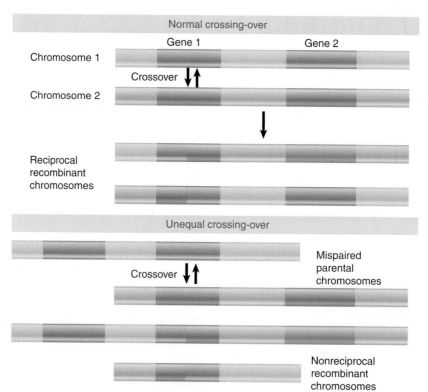

FIGURE 7.4 Gene number can be changed by unequal crossing-over. If gene 1 of one chromosome pairs with gene 2 of the other chromosome, the other gene copies are excluded from pairing. Recombination between the mispaired genes produces one chromosome with a single (recombinant) copy of the gene and one chromosome with three copies of the gene (one from each parent and one recombinant).

- In the quantitative scenario the number of repeats increases in one chromosome and decreases in the other. In effect, one recombinant chromosome has a deletion and the other has an insertion. This happens irrespective of the exact location of the crossover. In the figure, the first recombinant has an increase in the number of gene copies from two to three, whereas the second has a decrease from two to one.
- If the recombination event occurs within a gene (as opposed to between genes), the qualitative result depends on whether the recombining genes are identical or only related. If the nonhomologous gene copies 1 and 2 are identical in sequence, there is no change in the sequence of either gene. Unequal crossing-over, however, also can occur when the sequences of adjacent genes are very similar (although the probability is less than when they are identical). In this case, each of the recombinant genes has a sequence that is different from either parent.

The determination of whether the chromosome has a selective advantage or disadvantage will depend on the consequence of any change in the sequence of the gene product, as well as on the change in the number of gene copies.

An obstacle to unequal crossing-over is presented by the interrupted structure of the genes. In a case such as the globins, the corresponding exons of adjacent gene copies are likely to be well-enough related to support pairing; however, the sequences of the introns have diverged appreciably. The restriction of pairing to the exons considerably reduces the continuous length of DNA that can be involved. This lowers the chance of unequal crossing-over. So divergence between introns could enhance the stability of gene clusters by hindering the occurrence of unequal crossing-over.

Thalassemias result from mutations that reduce or prevent synthesis of either α- or β-globin. The occurrence of unequal crossing-over in the human globin gene clusters is revealed by the nature of certain thalassemias.

Many of the most severe thalassemias result from deletions of part of a cluster. In at least some cases, the ends of the deletion lie in regions that are homologous, which is exactly what would be expected if it had been generated by unequal crossing-over.

FIGURE 7.5 summarizes the deletions that cause the α-thalassemias. α-thal-1 deletions are

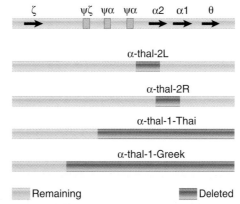

FIGURE 7.5 α-thalassemias result from various deletions in the α-globin gene cluster.

long, varying in the location of the left end, with the positions of the right ends located beyond the known genes. They eliminate both the α genes. The α-thal-2 deletions are short and eliminate only one of the two α genes. The L deletion removes 4.2 kb of DNA, including the α2 gene. It probably results from unequal crossing-over, because the ends of the deletion lie in homologous regions, just to the right of the ψα and α2 genes, respectively. The R deletion results from the removal of exactly 3.7 kb of DNA, the precise distance between the α1 and α2 genes. It appears to have been generated by unequal crossing-over between the α1 and α2 genes themselves. This is precisely the situation depicted in Figure 7.4.

Depending on the diploid combination of thalassemic alleles, an affected individual may have any number of α chains from zero to three. There are few differences from the wild type (four α genes) in individuals with three or two α genes. If an individual has only one α gene, though, the excess β chains form the unusual tetramer β_4, which causes **HbH** (hemoglobin H) **disease**. The complete absence of α genes results in **hydrops fetalis**, which is fatal at or before birth.

The same unequal crossing-over that generated the thalassemic chromosome should also have generated a chromosome with three α genes. Individuals with such chromosomes have been identified in several populations. In some populations, the frequency of the triple α locus is about the same as that of the single α locus; in others, the triple α genes are much less common than single α genes. This suggests that (unknown) selective factors operate in different populations to adjust the gene numbers.

Variations in the number of α genes are found relatively frequently, which suggests that unequal crossing-over in the cluster must be

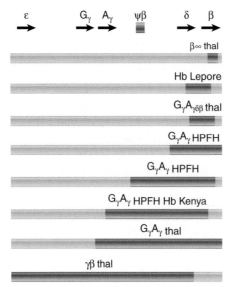

FIGURE 7.6 Deletions in the β-globin gene cluster cause several types of thalassemia.

fairly common. It occurs more often in the α cluster than in the β cluster, possibly because the introns in α genes are much shorter and therefore present less of an impediment to mispairing between nonhomologous loci.

The deletions that cause β-thalassemias are summarized in **FIGURE 7.6**. In some (rare) cases, only the β gene is affected. These have a deletion of 600 bp, extending from the second intron through the 3′ flanking regions. In the other cases, more than one gene of the cluster is affected. Many of the deletions are very long, extending from the 5′ end indicated on the map for >50 kb toward the right.

The **Hb Lepore** type provides the classic evidence that deletion can result from unequal crossing-over between linked genes. The β and δ genes differ by only ~7% in sequence. Unequal crossing-over deletes the material between the genes, thus fusing them together (Figure 7.4). The fused gene produces a single β-like chain that consists of the N-terminal sequence of δ joined to the C-terminal sequence of β.

Several types of Hb Lepore are known, the difference between them lying in the point of transition from δ to β sequences. Thus when the δ and β genes pair for unequal crossing-over, the exact point of recombination determines the position at which the switch from δ to β sequence occurs in the amino acid chain.

The reciprocal of this event has been found in the form of **Hb anti-Lepore**, which is produced by a gene that has the N-terminal part of β and the C-terminal part of δ. The fusion gene lies between normal δ and β genes. Although heterozygotes for this mutation are phenotypically normal, those that also carry a β deletion in *trans* show a mild β-thalassemia.

Evidence that unequal crossing-over can occur between more distantly related genes is provided by the identification of **Hb Kenya**, another fused hemoglobin. This contains the N-terminal sequence of the ^Aγ gene and the C-terminal sequence of the β gene. The fusion must have resulted from unequal crossing-over between ^Aγ and α, which differ ~20% in sequence.

From the differences between the globin gene clusters of various mammals, we see that duplication followed (sometimes) by variation has been an important feature in the evolution of each cluster. The human thalassemic deletions demonstrate that unequal crossing-over continues to occur in both globin gene clusters. Each such event generates a duplication as well as the deletion, and we must account for the fate of both recombinant loci in the population. Deletions can also occur (in principle) by recombination between homologous sequences lying on the same chromosome. This does not generate a corresponding duplication.

It is difficult to estimate the natural frequency of these events, because evolutionary forces rapidly adjust the levels of the variant clusters in the population. Generally a contraction in gene number is likely to be deleterious and selected against. In some populations, though, there may be a balancing advantage that maintains the deleted form at a low frequency. In small populations, genetic drift is likely to play a role in eliminating effectively neutral new duplications.

The structures of the present human clusters show several duplications that attest to the importance of such mechanisms. The functional sequences include two α genes encoding the same polypeptide, fairly similar β and δ genes, and two almost identical γ genes. These comparatively recent independent duplications have persisted in the population, not to mention the more distant duplications that originally generated the various types of globin genes. Other duplications may have given rise to pseudogenes or have been lost. We expect continual duplication and deletion to be a feature of all gene clusters.

Genes for rRNA Form Tandem Repeats Including an Invariant Transcription Unit

Key concepts

- Ribosomal RNA is coded by a large number of identical genes that are tandemly repeated to form one or more clusters.
- Each rDNA cluster is organized so that transcription units giving a joint precursor to the major rRNAs alternate with nontranscribed spacers.
- The genes in an rDNA cluster all have an identical sequence.
- The nontranscribed spacers consist of shorter repeating units whose number varies so that the lengths of individual spacers are different.

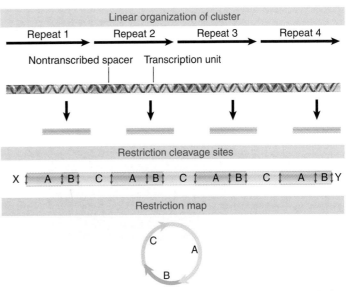

FIGURE 7.7 A tandem gene cluster has an alternation of transcription unit and nontranscribed spacer and generates a circular restriction map.

In the case of the globin genes discussed above, there are differences between the individual members of the cluster that allow selective pressure to act somewhat differently (though because of linkage, not independently) upon each gene. A contrast is provided by two cases of large gene clusters that contain many identical copies of the same gene or genes. Most eukaryotic organisms contain multiple copies of the genes for the histone proteins that are a major component of the chromosomes, and in most organismal genomes there are multiple copies of the genes that code for the ribosomal RNAs. These situations pose some interesting evolutionary questions.

Ribosomal RNA is the predominant product of transcription, constituting some 80%–90% of the total mass of cellular RNA in both eukaryotes and prokaryotes. The number of major rRNA genes varies from one (e.g., in *Coxiella burnetii*, an obligate intracellular bacterium, and *Mycoplasma pneumoniae*), to seven in *E. coli*, to 100 to 200 in unicellular/oligocellular eukaryotes, to several hundred in multicellular eukaryotes. The genes for the large and small rRNA (found in the large and small subunits of the ribosome, respectively) usually form a tandem pair. (The sole exception is the yeast mitochondrion.)

The lack of any detectable variation in the sequences of the rRNA molecules implies that all of the copies of each gene must be identical, or at least must have differences below the level of detection in rRNA (~1%). A point of major interest is what mechanism(s) are used to pre-

vent variations from accruing in the individual sequences.

In bacteria, the multiple rRNA genes are dispersed. In most eukaryotic genomes, the rRNA genes are contained in a tandem cluster or clusters. Sometimes these regions are called **rDNA**. (In some cases, the proportion of rDNA in the total DNA, together with its atypical base composition, is great enough to allow its isolation as a separate fraction directly from sheared genomic DNA.) An important diagnostic feature of a tandem cluster is that it generates a circular restriction map (see *Section 3.2, Nucleases*, for a description of restriction mapping), as shown in **FIGURE 7.7**.

Suppose that each repeat unit has three restriction sites. When we map these fragments by conventional means, we find that A is next to B, which is next to C, which is next to A, generating the circular map. If the cluster is large, the internal fragments (A, B, and C) will be present in much greater quantities than the terminal fragments (X and Y), which connect the cluster to adjacent DNA. In a cluster of 100 repeats, X and Y would be present at 1% of the level of A, B, and C. This can make it difficult to obtain the ends of a gene cluster for mapping purposes.

The region of the nucleus where 18S and 28S rRNA synthesis occurs has a characteristic appearance, with a core of fibrillar nature surrounded by a granular cortex. The fibrillar core

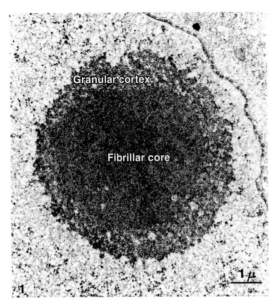

FIGURE 7.8 The nucleolar core identifies rDNA under transcription, and the surrounding granular cortex consists of assembling ribosomal subunits. This thin section shows the nucleolus of the newt *Notopthalmus viridescens*. Photo courtesy of Oscar Miller.

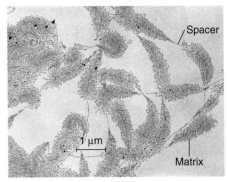

FIGURE 7.9 Transcription of rDNA clusters generates a series of matrices, each corresponding to one transcription unit and separated from the next by the nontranscribed spacer. Photo courtesy of Oscar Miller.

is where the rRNA is transcribed from the DNA template, and the granular cortex is formed by the ribonucleoprotein particles into which the rRNA is assembled. The whole area is called the **nucleolus**. Its characteristic morphology is evident in **FIGURE 7.8**.

The particular chromosomal regions associated with a nucleolus are called **nucleolar organizers**. Each nucleolar organizer corresponds to a cluster of tandemly repeated 18/28S rRNA genes on one chromosome. The concentration of the tandemly repeated rRNA genes, together with their very intensive transcription, is responsible for creating the characteristic morphology of the nucleoli.

The pair of major rRNAs is transcribed as a single precursor in both bacteria (where 5S and 16/23S rRNAs are cotranscribed) and the eukaryotic nucleolus (where the 18S and 28S rRNAs are transcribed). In eukaryotes, 5S genes are also typically found in tandem clusters transcribed as a precursor with transcribed spacers. Following transcription, the precursor is cleaved to release the individual rRNA molecules. The transcription unit is shortest in bacteria and is longest in mammals (where it is known as 45S RNA, according to its rate of sedimentation). An rDNA cluster contains many transcription units, each separated from the next by a nontranscribed spacer. The alternation of transcription unit and **nontranscribed spacer** can be seen directly in electron micro-

graphs. The example shown in **FIGURE 7.9** is taken from the newt *Notopthalmus viridescens*, in which each transcription unit is intensively expressed, so that many RNA polymerases are simultaneously engaged in transcription on one repeating unit. The polymerases are so closely packed that the RNA transcripts form a characteristic matrix displaying increasing length along the transcription unit.

The length of the nontranscribed spacer varies a great deal between and (sometimes) within species. In yeast there is a short nontranscribed spacer that is relatively constant in length. In the fly *D. melanogaster* there is almost a twofold variation in the length of the nontranscribed spacer between different copies of the repeating unit. A similar situation is seen in the amphibian *X. laevis*. In each of these cases, all of the repeating units are present as a single tandem cluster on one particular chromosome. (In the example of *D. melanogaster*, this happens to be the sex chromosome. The cluster on the X chromosome is larger than the one on the Y chromosome, so female flies have more copies of the rRNA genes than male flies do.)

In mammals the repeating unit is much larger, comprising the transcription unit of ~13 kb and a nontranscribed spacer of ~30 kb. Usually, the genes lie in several dispersed clusters—in the cases of humans and mice the clusters reside on five and six chromosomes, respectively. One interesting (but unanswered) question is how the corrective mechanisms that presumably function within a single cluster to ensure constancy of rRNA sequence are able to work when there are several clusters.

The variation in length of the nontranscribed spacer in a single gene cluster contrasts with the conservation of sequence of the transcription unit. In spite of this variation, the sequences of longer nontranscribed spacers remain homologous with those of the shorter nontranscribed spacers. This implies that each nontranscribed spacer is internally repetitious, so that the variation in length results from changes in the number of repeats of some subunit.

The general nature of the nontranscribed spacer is illustrated by the example of *X. laevis* (FIGURE 7.10). Regions that are fixed in length alternate with regions that vary in length. Each of the three repetitive regions comprises a variable number of repeats of a rather short sequence. One type of repetitious region has repeats of a 97-bp sequence; the other, which occurs in two locations, has a repeating unit found in two forms, 60 and 81 bp long. The variation in the number of repeating units in the repetitive regions accounts for the overall variation in spacer length. The repetitive regions are separated by shorter constant sequences called **Bam islands**. (This description takes its name from their isolation via the use of the BamHI restriction enzyme.) From this type of organization, we see that the cluster has evolved by duplications involving the promoter region.

We need to explain the lack of variation in the expressed copies of the repeated genes. One hypothesis would be that there is a quantitative demand for a certain number of "good" sequences. This would, however, enable mutated sequences to accumulate up to a point at which their proportion of the cluster is great enough for selection to act against them. We can exclude this hypothesis because of the lack of such variation in the cluster.

The lack of variation implies that there is purifying selection against individual variations. Another hypothesis would be that the entire cluster is regenerated periodically from one or a very few members. As a practical matter, any mechanism would need to involve regeneration every generation. We can exclude this hypothesis because a regenerated cluster would not show variation in the nontranscribed regions of the individual repeats.

We are left with a dilemma. Variation in the nontranscribed regions suggests that there is frequent unequal crossing-over. This will change the size of the cluster, but will not otherwise change the properties of the individual repeats. So how are mutations prevented from accumulating? We'll see in the next section that continuous contraction and expansion of a cluster may provide a mechanism for homogenizing its copies.

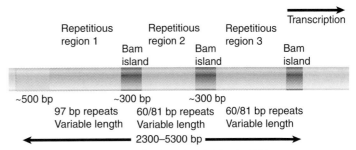

FIGURE 7.10 The nontranscribed spacer of *X. laevis* rDNA has an internally repetitive structure that is responsible for its variation in length. The Bam islands are short constant sequences that separate the repetitious regions.

7.4 Crossover Fixation Could Maintain Identical Repeats

Key concepts

- Unequal crossing-over changes the size of a cluster of tandem repeats.
- Individual repeating units can be eliminated or can spread through the cluster.

Not all duplicated copies of genes become pseudogenes. How can selection prevent the accumulation of deleterious mutations?

The duplication of a gene is likely to result in an immediate relaxation of the selection pressure on the sequence of one of the two copies. Now that there are two identical copies, a change in the sequence of one will not deprive the organism of a functional product, because the original product can continue to be encoded by the other copy. Then the selective pressure on the two genes is diffused, until one of them mutates sufficiently away from its original function to refocus all the selective pressure on the other.

Immediately following a gene duplication, changes might accumulate more rapidly in one of the copies, eventually leading to a new function (or to its disuse in the form of a pseudogene). If a new function develops, the gene then evolves at the same, slower rate characteristic of the original function. Probably this is the sort of mechanism responsible for the separation of functions between embryonic and adult globin genes.

Yet there are instances in which duplicated genes retain the same function, coding for identical or nearly identical products. Identical polypeptides are encoded by the two human α-globin genes, and there is only a single amino acid difference between the two γ-globin polypeptides. How does selection maintain their sequence identities?

The most obvious possibility is that the two genes do not actually have identical functions, but instead differ in some (undetected) property, such as time or place of expression. Another possibility is that the need for two copies is quantitative, because neither by itself produces a sufficient amount of product.

In more extreme cases of repetition, however, it is impossible to avoid the conclusion that no single copy of the gene is essential. When there are many copies of a gene, the immediate effects of mutation in any one copy must be very slight. The consequences of an individual mutation are diluted by the large number of copies of the gene that retain the wild-type sequence. Many mutant copies could accumulate before a lethal effect is generated.

Lethality becomes quantitative, a conclusion reinforced by the observation that half of the units of the rDNA cluster of *X. laevis* or *D. melanogaster* can be deleted without ill effect. So how are these units prevented from gradually accumulating deleterious mutations? What chance is there for the rare favorable mutation to display its advantages in the cluster?

The basic principle of hypotheses to explain the maintenance of identity among repeated copies is to suppose that nonallelic genes are continually regenerated from *one* of the copies of a preceding generation. In the simplest case of two identical genes, when a mutation occurs in one copy, either it is by chance eliminated (because the sequence of the other copy takes over), or it is spread to both duplicates. Spreading exposes a mutation to selection. The result is that the two genes evolve together as though only a single locus existed. This is called **concerted evolution** or **coincidental evolution**. It can be applied to a pair of identical genes or (with further assumptions) to a cluster containing many genes.

One mechanism for this concerted evolution is that the sequences of the nonallelic genes are directly compared with one another and homogenized by enzymes that recognize any differences. This can be done by exchanging single strands between them to form genes, one of whose strands derives from one copy, and one from the other copy. Any differences are revealed as improperly paired bases, which attract attention from enzymes able to excise and replace a base, so that only A-T and G-C pairs survive. This type of event is called **gene conversion** and is associated with genetic recombination. We should be able to ascertain the scope of such events by comparing the sequences of duplicate genes. If they are subject to concerted evolution, we should not see the accumulation of silent substitutions (those that do not change the amino acid sequence; see *Section 8.5, The Rate of Neutral Substitution Can Be Measured from Divergence of Repeated Sequences*) between them because the homogenization process applies to these as well as to the replacement sites (those that, if mutated, will change the amino acid sequence). We know that the extent of the maintenance mechanism need not extend beyond the gene itself, as there are cases of duplicate genes whose flanking sequences are entirely different. Indeed, we may see abrupt boundaries that mark the ends of the sequences that were homogenized.

We must remember that the existence of such mechanisms can invalidate the determination of the history of such genes via their divergence, because the divergence reflects only the time since the last homogenization/regeneration event, not the original duplication.

The **crossover fixation** model suggests that an entire cluster is subject to continual rearrangement by the mechanism of unequal crossing-over. Such events can explain the concerted evolution of multiple genes if unequal crossing-over causes all the copies to be regenerated physically from one copy.

Following the sort of event depicted in Figure 7.4, for example, the chromosome carrying a triple locus could suffer deletion of one of the genes. Of the two remaining genes, 1½ represent the sequence of one of the original copies; only ½ of the sequence of the other original copy has survived. Any mutation in the first region now exists in both genes and is subject to selection.

Tandem clustering provides frequent opportunities for "mispairing" of loci whose sequences are the same, but that lie in different positions in their clusters. By continually expanding and contracting the number of units via unequal crossing-over, it is possible for all the units in one cluster to be derived from rather a small proportion of those in an ancestral cluster. The variable lengths of the spacers are consistent with the idea that unequal crossing-over events take place in spacers that are internally mispaired. This can explain the homogeneity of the genes compared with the variability of the spacers. The genes are exposed to selection when individual repeating units are amplified within the cluster; however, the spacers are functionally irrelevant and can accumulate changes.

In a region of nonrepetitive DNA, recombination occurs between precisely matching points on the two homologous chromosomes, thus generating reciprocal recombinants. The basis for this precision is the ability of two duplex DNA sequences to align exactly. We know that unequal recombination can occur when there are multiple copies of genes whose exons are related, even though their flanking and intervening sequences may differ. This happens because of the mispairing between corresponding exons in nonallelic genes.

Imagine how much more frequently misalignment must occur in a tandem cluster of identical or nearly identical repeats. Except at the very ends of the cluster, the close relationship between successive repeats makes it impossible even to define the exactly corresponding repeats! This has two consequences: there is continual adjustment of the size of the cluster; and there is homogenization of the repeating unit.

Consider a sequence consisting of a repeating unit "ab" with ends "x" and "y." If we represent one chromosome in black and the other in red, the exact alignment between "allelic" sequences would be:

xababababababababababababababababy
xababababababababababababababababy

It is likely, however, that *any* sequence ab in one chromosome could pair with *any* sequence ab in the other chromosome. In a misalignment such as:

xababababababababababababababababy
  xababababababababababababababababy,

the region of pairing is no less stable than in the perfectly aligned pair, although it is shorter. We do not know very much about how pairing is initiated prior to recombination, but very likely it starts between short corresponding regions and then spreads. If it starts within highly repetitive satellite DNA, it is more likely than not to involve repeating units that do not have exactly corresponding locations in their clusters.

Now suppose that a recombination event occurs within the unevenly paired region. The recombinants will have different numbers of repeating units. In one case, the cluster has become longer; in the other, it has become shorter,

xababababababababababababababababy
×
  xababababababababababababababababy
↓
xababababababababababababababababababy
+
  xababababababababababababy,

where "×" indicates the site of the crossover.

If this type of event is common, clusters of tandem repeats will undergo continual expansion and contraction. This can cause a particular repeating unit to spread through the cluster, as illustrated in **FIGURE 7.11**. Suppose that the cluster consists initially of a sequence abcde, where each letter represents a repeating unit. The different repeating units are closely enough related to one another to mispair for recombination. Then by a series of unequal recombination events, the size of the repetitive region increases or decreases, and one unit spreads to replace all the others.

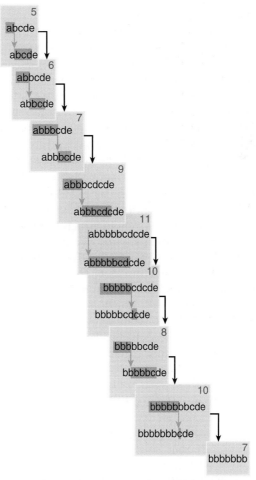

FIGURE 7.11 Unequal recombination allows one particular repeating unit to occupy the entire cluster. The numbers indicate the length of the repeating unit at each stage.

The crossover fixation model predicts that *any sequence of DNA that is not under selective pressure will be taken over by a series of identical tandem repeats generated in this way*. The critical assumption is that the process of crossover fixation is fairly rapid relative to mutation, so that new mutations either are eliminated (their repeats are lost) or come to take over the entire cluster. In the case of the rDNA cluster, of course, a further factor is imposed by selection for a functional transcribed sequence.

7.5 Satellite DNAs Often Lie in Heterochromatin

Key concepts

- Highly repetitive DNA (or satellite DNA) has a very short repeating sequence and no coding function.
- Satellite DNA occurs in large blocks that can have distinct physical properties.
- Satellite DNA is often the major constituent of centromeric heterochromatin.

Repetitive DNA is characterized by its (relatively) rapid rate of renaturation. The component that renatures most rapidly in a eukaryotic genome is called *highly repetitive* DNA and consists of very short sequences repeated many times in tandem in large clusters. As a result of its short repeating unit, it is sometimes described as **simple sequence DNA**. This type of component is present in almost all multicellular eukaryotic genomes, but its overall amount is extremely variable. In mammalian genomes it is typically <10%, but in (for example) the fly *Drosophila virilis*, it amounts to ~50%. In addition to the large clusters in which this type of sequence was originally discovered, there are smaller clusters interspersed with nonrepetitive DNA. It typically consists of short sequences that are repeated in identical or related copies in the genome.

In addition to simple sequence DNA, multicellular eukaryotes have *complex satellites* with longer repeat units, usually in heterochromatin but sometimes in euchromatic regions (see below for a discussion of heterochromatin and euchromatin). For example, *Drosophila* species have the 1.688 g-cm^{-3} class of satellite DNA that consists of a 359-bp repeat unit. In humans, the α satellite family, found in centromeric regions, has a repeat unit length of 171 bp. The human β satellite family has ±68-bp repeat units interspersed with a longer 3.3 kb repeat unit that includes pseudogenes.

The tandem repetition of a short sequence often has distinctive physical properties that can be used to isolate it. In some cases, the repetitive sequence has a base composition distinct from the genome average, which allows it to form a separate fraction by virtue of its distinct buoyant density. A fraction of this sort is called *satellite DNA*. The term satellite DNA is essentially synonymous with simple sequence DNA. Consistent with its simple sequence, this DNA may or may not be transcribed, but it is not translated. (In some species there is evidence that short RNAs are required for heterochromatin formation, suggesting that there is transcription of sequences in heterochromatic regions of chromosomes, which contain satellite DNA; see *Section 30.7, Heterochromatin Formation Requires microRNAs.*)

Tandemly repeated sequences are especially liable to undergo misalignments during chromosome pairing, and thus the sizes of tandem clusters tend to be highly polymorphic, with wide variations between individuals. In fact, the smaller clusters of such sequences can be used to characterize individual genomes in the technique of "DNA fingerprinting" (see *Section 7.8, Minisatellites Are Useful for Genetic Mapping*).

The buoyant density of a duplex DNA depends on its G-C content according to the empirical formula

$$\rho = 1.660 + 0.00098 \, (\%\text{G-C}) \, \text{g-cm}^{-3}$$

Buoyant density usually is determined by centrifuging DNA through a density gradient of CsCl. The DNA forms a band at the position corresponding to its own density. Fractions of DNA differing in G-C content by >5% can usually be separated on a density gradient.

When eukaryotic DNA is centrifuged on a density gradient, two types of material may be distinguished:

- Most of the genome forms a continuum of fragments that appear as a rather broad peak centered on the buoyant density corresponding to the average G-C content of the genome. This is called the *main band*.
- Sometimes an additional, smaller peak (or peaks) is seen at a different value. This material is the *satellite DNA*.

Satellites are present in many eukaryotic genomes. They may be either heavier or lighter than the main band, but it is uncommon for them to represent >5% of the total DNA. A clear example is provided by mouse DNA, as shown

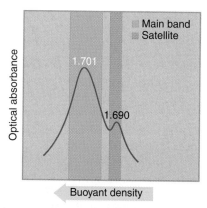

FIGURE 7.12 Mouse DNA is separated into a main band and a satellite by centrifugation through a density gradient of CsCl.

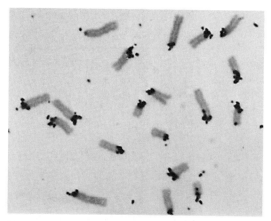

FIGURE 7.13 Cytological hybridization shows that mouse satellite DNA is located at the centromeres. Photo courtesy of Mary Lou Pardue and Joseph G. Gall, Carnegie Institution.

in FIGURE 7.12. The graph is a quantitative scan of the bands formed when mouse DNA is centrifuged through a CsCl density gradient. The main band contains 92% of the genome and is centered on a buoyant density of 1.701 g-cm^{-3} (corresponding to its average G-C of 42%, typical for a mammal). The smaller peak represents 8% of the genome and has a distinct buoyant density of 1.690 g-cm^{-3}. It contains the mouse satellite DNA, whose G-C content (30%) is much lower than any other part of the genome.

The behavior of satellite DNA on density gradients is often anomalous. When the actual base composition of a satellite is determined, it is different from the prediction based on its buoyant density. The reason is that ρ is a function not just of base composition, but also of the constitution in terms of nearest neighbor pairs. For simple sequences, these are likely to deviate from the random pairwise relationships needed to obey the equation for buoyant density. In addition, satellite DNA may be methylated, which changes its density.

Often, most of the highly repetitive DNA of a genome can be isolated in the form of satellites. When a highly repetitive DNA component does not separate as a satellite, on isolation its properties often prove to be similar to those of satellite DNA. That is to say, highly repetitive DNA consists of multiple tandem repeats with anomalous centrifugation. Material isolated in this manner is sometimes referred to as a **cryptic satellite**. Together the cryptic and apparent satellites usually account for all the large, tandemly repeated blocks of highly repetitive DNA. When a genome has more than one type of highly repetitive DNA, each exists in its own satellite block (although sometimes different blocks are adjacent).

Where in the genome are the blocks of highly repetitive DNA located? An extension of nucleic acid hybridization techniques allows the location of satellite sequences to be determined directly in the chromosome complement. In the technique of *in situ* **hybridization**, the chromosomal DNA is denatured by treating cells that have been squashed on a cover slip. Next, a solution containing a labeled single-stranded DNA or RNA probe is added. The probe hybridizes with its complementary sequences in the denatured genome. The location of the sites of hybridization can be determined by a technique to detect the label, such as autoradiography or fluorescence (see Figure 3.13).

Satellite DNA is found in regions of **heterochromatin**. Heterochromatin is the term used to describe regions of chromosomes that are permanently tightly coiled up and inert, in contrast with the **euchromatin** that represents most of the genome (see *Section 9.7, Chromatin Is Divided into Euchromatin and Heterochromatin*). Heterochromatin is commonly found at centromeres (the regions where the kinetochores are formed at mitosis and meiosis for controlling chromosome segregation). The centromeric location of satellite DNA suggests that it has some structural function in the chromosome. This function could be connected with the process of chromosome segregation.

An example of the localization of satellite DNA for the mouse chromosomal complement is shown in FIGURE 7.13. In this case, one end of each chromosome is labeled, because this is where the centromeres are located in *Mus musculus* chromosomes.

7.6 Arthropod Satellites Have Very Short Identical Repeats

Key concept
- The repeating units of arthropod satellite DNAs are only a few nucleotides long. Most of the copies of the sequence are identical.

In the arthropods, as typified by insects and crustaceans, each satellite DNA appears to be rather homogeneous. Usually, a single, very short repeating unit accounts for >90% of the satellite. This makes it relatively straightforward to determine the sequence.

The fly *Drosophila virilis* has three major satellites and a cryptic satellite; together they represent >40% of the genome. The sequences of the satellites are summarized in **FIGURE 7.14**. The three major satellites have closely related sequences. A single base substitution is sufficient to generate either satellite II or III from the sequence of satellite I.

The satellite I sequence is present in other species of *Drosophila* related to *D. virilis* and so its presence probably preceded speciation. The sequences of satellites II and III seem to be specific to *D. virilis*, and so may have evolved from satellite I following speciation.

The main feature of these satellites is their very short repeating unit: only 7 bp. Similar satellites are found in other species. *D. melanogaster* has a variety of satellites, several of which have very short repeating units (5, 7, 10, or 12 bp). Comparable satellites are found in crustaceans.

The close sequence relationship found among the *D. virilis* satellites is not necessarily a feature of other genomes, for which the satellites may have unrelated sequences. *Each satellite has arisen by a lateral amplification of a very short sequence.* This sequence may represent a variant of a previously existing satellite (as in *D. virilis*), or could have some other origin.

Satellites are continually generated and lost from genomes. This makes it difficult to ascertain evolutionary relationships, because a current satellite could have evolved from some previous satellite that has since been lost. The important feature of these satellites is that *they represent very long stretches of DNA of very low sequence complexity, within which constancy of sequence can be maintained.*

One feature of many of these satellites is a pronounced asymmetry in the orientation of base pairs on the two strands. In the example of the *D. virilis* satellites shown in Figure 7.14, in each of the major satellites one of the strands is much richer in T and G bases. This increases its buoyant density, so that upon denaturation this heavy strand (H) can be separated from the complementary light strand (L). This can be useful in sequencing the satellite.

7.7 Mammalian Satellites Consist of Hierarchical Repeats

Key concept
- Mouse satellite DNA has evolved by duplication and mutation of a short repeating unit to give a basic repeating unit of 234 bp in which the original half-, quarter-, and eighth-repeats can be recognized.

In the mammals, as typified by various rodents, the sequences comprising each satellite show appreciable divergence between tandem repeats. Common short sequences can be recognized by their preponderance among the oligonucleotide fragments released by chemical or enzymatic treatment. The predominant short sequence, however, usually accounts for only a small minority of the copies. The other short sequences are related to the predominant sequence by a variety of substitutions, deletions, and insertions.

A series of these variants of the short unit, however, can constitute a longer repeating unit that is itself repeated in tandem with some variation. Thus mammalian satellite DNAs are constructed from a hierarchy of repeating units. These longer repeating units constitute the sequences that renature in reassociation analyses. They also can be recognized by digestion with restriction enzymes.

Satellite	Predominant Sequence	Total Length	Genome Proportion
I	A C A A A C T T G T T T G A	1.1×10^7	25%
II	A T A A A C T T A T T T G A	3.6×10^6	8%
III	A C A A A T T T G T T T A A	3.6×10^6	8%
Cryptic	A A T A T A G T T A T A T C		

FIGURE 7.14 Satellite DNAs of *D. virilis* are related. More than 95% of each satellite consists of a tandem repetition of the predominant sequence.

When any satellite DNA is digested with an enzyme that has a recognition site in its repeating unit, one fragment will be obtained for every repeating unit in which the site occurs. In fact, when the DNA of a eukaryotic genome is digested with a restriction enzyme, most of it gives a general smear due to the random distribution of cleavage sites. Satellite DNA generates sharp bands, though, because a large number of fragments of identical or almost identical size are created by cleavage at restriction sites that lie a regular distance apart.

Determining the sequence of satellite DNA can be difficult. For example, we can cut the region into fragments with restriction endonucleases and attempt to obtain a sequence directly. If, however, there is appreciable divergence between individual repeating units, different nucleotides will be present at the same position in different repeats, so the sequencing gels will be obscure. If the divergence is not too great—say, within ~2%—it may be possible to determine an average repeating sequence.

Individual segments of the satellite can be inserted into plasmids for cloning. A difficulty is that the satellite sequences tend to be excised from the chimeric plasmid by recombination in the bacterial host. When the cloning succeeds, though, it is possible to determine the sequence of the cloned segment unambiguously. Although this gives the actual sequence of a repeating unit or units, we should need to have many individual such sequences to reconstruct the type of divergence typical of the satellite as a whole.

Using either sequencing approach, the information we can gain is limited to the distance that can be analyzed on one set of sequence gels. The repetition of divergent tandem copies makes it difficult to reconstruct longer sequences by obtaining overlaps between individual restriction fragments.

The satellite DNA of the mouse *M. musculus* is cleaved by the enzyme EcoRII into a series of bands, including a predominant monomeric fragment of 234 bp. This sequence must be repeated with few variations throughout the 60%–70% of the satellite that is cleaved into the monomeric band. We may analyze this sequence in terms of its successively smaller constituent repeating units.

FIGURE 7.15 depicts the sequence in terms of two half-repeats. By writing the 234 bp sequence so that the first 117 bp are aligned with the second 117 bp, we see that the two halves are quite closely related. They differ at 22 positions, corresponding to 19% divergence. This means that the current 234 bp repeating unit must have been generated at some time in the past by duplicating a 117 bp repeating unit, after which differences accumulated between the duplicates.

Within the 117 bp unit we can recognize two further subunits. Each of these is a quarter-repeat relative to the whole satellite. The four quarter-repeats are aligned in FIGURE 7.16. The upper two lines represent the first half-repeat of Figure 7.15; the lower two lines represent the second half-repeat. We see that the divergence between the four quarter-repeats has increased

FIGURE 7.15 The repeating unit of mouse satellite DNA contains two half-repeats, which are aligned to show the identities (in blue).

FIGURE 7.16 The alignment of quarter-repeats identifies homologies between the first and second half of each half-repeat. Positions that are the same in all four quarter-repeats are shown in green. Identities that extend only through 3/4 of the quarter repeats are in black, with the divergent sequences in red.

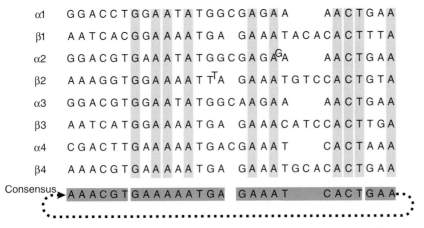

α1   G G A C C T G G A A T A T G G C G A G A A   A A C T G A A

β1   A A T C A C G G A A A A T G A G A A A T A C A C A C T T T A

α2   G G A C G T G A A A T A T G G C G A G A^{G}A   A A C T G A A

β2   A A A G G T G G A A A A T T^{T}A G A A A T G T C C A C T G T A

α3   G G A C G T G G A A T A T G G C A A G A A   A A C T G A A

β3   A A T C A T G G A A A A T G A G A A A C A T C C A C T T G A

α4   C G A C T T G A A A A A T G A C G A A A T   C A C T A A A

β4   A A A C G T G A A A A A T G A G A A A T G C A C A C T G A A

Consensus → A A A C G T G A A A A A T G A   G A A A T   C A C T G A A

Ancestral?   A A A C G T G A A A A A T G A   G A A A T G C A C A C T G A A

FIGURE 7.17 The alignment of eighth-repeats shows that each quarter-repeat consists of an α and a β half. The consensus sequence gives the most common base at each position. The "ancestral" sequence shows a sequence very closely related to the consensus sequence, which could have been the predecessor to the α and β units. (The satellite sequence is continuous, so that for the purposes of deducing the consensus sequence we can treat it as a circular permutation, as indicated by joining the last GAA triplet to the first 6 bp.)

to 23 out of 58 positions, or 40%. The first three quarter-repeats are somewhat more similar, and a large proportion of the divergence is due to changes in the fourth quarter-repeat.

Looking within the quarter-repeats, we find that each consists of two related subunits (one-eighth-repeats), shown as the α and β sequences in **FIGURE 7.17**. The α sequences all have an insertion of a C, and the β sequences all have an insertion of a trinucleotide sequence, relative to a common consensus sequence. This suggests that the quarter-repeat originated by the duplication of a sequence like the consensus sequence, after which changes occurred to generate the components we now see as α and β. Further changes then took place between tandemly repeated αβ sequences to generate the individual quarter- and half-repeats that exist today. Among the one-eighth-repeats, the present divergence is 19/31 = 61%.

The consensus sequence is analyzed directly in **FIGURE 7.18**, which demonstrates that the current satellite sequence can be treated as derivatives of a 9 bp sequence. We can recognize three variants of this sequence in the satellite, as indicated at the bottom of the figure. If in one of the repeats we take the next most frequent base at two positions instead of the most frequent, we obtain three closely related 9 bp sequences:

G A A A A A C G T
G A A A A A T G A
G A A A A A A C T

```
              G G A C C T
        G G A A T A T G G C
        G A G A A A A C T
        G A A A A T C A C
        G G A A A A T G A
        G A A A T C A C T
        T T A G G A C G T
        G A A A T A T G G C
        G A G A A A A C T
        G A A A A A G G T
        G G A A A A T T A
        G A A A T* C A C T
        G T A G G A C G T
        G G A A T A T G G C
        A A G A A A A C T
        G A A A A T C A T
        G G A A A A T G A
        G A A A C* C A C T
        T G A C G A C T T
        G A A A A A T G A C
        G A A A T C A C T
        A A A A A A C G T
        G A A A A A T G A
        G A A A T* C A C T
        G A A
```

$G_{20} A_{16} A_{21} A_{20} A_{12} A_{17} T_8 G_{11} A_5$

$T_7 C_5 A_8 C_9 T_{15}$

C_7

* indicates inserted triplet in β sequence
C in position 10 is extra base in α sequence

FIGURE 7.18 The existence of an overall consensus sequence is shown by writing the satellite sequence in terms of a 9 bp repeat.

The origin of the satellite could well lie in an amplification of one of these three nonamers (9 bp units). The overall consensus sequence of the present satellite is GAAAAA$^{AG}_{TC}$T, which is effectively an amalgam of the three 9 bp repeats.

The average sequence of the monomeric fragment of the mouse satellite DNA explains its properties. The longest repeating unit of 234 bp is identified by the restriction cleavage. The unit of reassociation between single strands of denatured satellite DNA is probably the 117 bp half-repeat, because the 234 bp fragments can anneal both in register and in half-register (in the latter case, the first half-repeat of one strand renatures with the second half-repeat of the other).

So far, we have treated the present satellite as though it consisted of identical copies of the 234 bp repeating unit. Although this unit accounts for the majority of the satellite, vari-

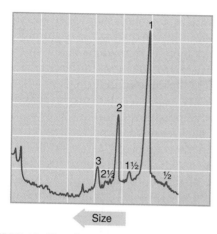

FIGURE 7.19 Digestion of mouse satellite DNA with the restriction enzyme EcoRII identifies a series of repeating units (1, 2, 3) that are multimers of 234 bp and also a minor series (½, 1½, 2½) that includes half-repeats (see text this page). The band at the far left is a fraction resistant to digestion.

ants of it also are present. Some of them are scattered at random throughout the satellite, whereas others are clustered.

The existence of variants is implied by our description of the starting material for the sequence analysis as the "monomeric" fragment. When the satellite is digested by an enzyme that has one cleavage site in the 234 bp sequence, it also generates dimers, trimers, and tetramers relative to the 234 bp length. They arise when a repeating unit has lost the enzyme cleavage site as the result of mutation.

The monomeric 234 bp unit is generated when two adjacent repeats each have the recognition site. A dimer occurs when one unit has lost the site, a trimer is generated when two adjacent units have lost the site, and so on. With some restriction enzymes, most of the satellite is cleaved into a member of this repeating series, as shown in the example of FIGURE 7.19. The declining number of dimers, trimers, and so forth shows that there is a random distribution of the repeats in which the enzyme's recognition site has been eliminated by mutation.

Other restriction enzymes show a different type of behavior with the satellite DNA. They continue to generate the same series of bands. They cleave, however, only a small proportion of the DNA, say 5%–10%. This implies that a certain region of the satellite contains a concentration of the repeating units with this particular restriction site. Presumably the series of repeats in this domain all are derived from an ancestral variant that possessed this recognition site (although in the usual way, some members since have lost it by mutation).

A satellite DNA suffers unequal recombination. This has additional consequences when there is internal repetition in the repeating unit. Let us return to our cluster consisting of "ab" repeats. Suppose that the "a" and "b" components of the repeating unit are themselves sufficiently well related to pair. Then the two clusters can align in *half-register*, with the "a" sequence of one aligned with the "b" sequence of the other. How frequently this occurs will depend on the closeness of the relationship between the two halves of the repeating unit. In mouse satellite DNA, reassociation between the denatured satellite DNA strands *in vitro* commonly occurs in the half-register.

When a recombination event occurs out of register, it changes the length of the repeating units that are involved in the reaction:

xababababababababababababababababaaby
×
xababababababababababababababababababy
↓
xababababababababababaabababababababababy
+
xababababababababababbbabababababababababy

In the upper recombinant cluster, an "ab" unit has been replaced by an "aab" unit. In the lower cluster, the "ab" unit has been replaced by a "b" unit.

This type of event explains a feature of the restriction digest of mouse satellite DNA. Figure 7.19 shows a fainter series of bands at lengths of ½, 1½, 2½, and 3½ repeating units, in addition to the stronger integral length repeats. Suppose that in the preceding example, "ab" represents the 234 bp repeat of mouse satellite DNA, generated by cleavage at a site in the "b" segment. The "a" and "b" segments correspond to the 117 bp half-repeats.

Then, in the upper recombinant cluster, the "aab" unit generates a fragment of 1½ times the usual repeating length. In the lower recombinant cluster, the "b" unit generates a fragment of half of the usual length. (The multiple fragments in the half-repeat series are generated in the same way as longer fragments in the integral series, when some repeating units have lost the restriction site by mutation.)

Turning the argument the other way around, the identification of the half-repeat series on the gel shows that the 234 bp repeating unit consists of two half-repeats closely related enough to pair sometimes for recombination. Also visible in Figure 7.19 are some rather faint bands corresponding to ¼- and ¾-spacings. These will be generated in the same way as the ½-spacings, when recombination occurs between clusters aligned in a quarter-register. The decreased relationship between quarter-repeats compared with half-repeats explains the reduction in frequency of the ¼- and ¾-bands compared with the ½-bands.

7.8 Minisatellites Are Useful for Genetic Mapping

Key concept

- The variation between microsatellites or minisatellites in individual genomes can be used to identify heredity unequivocally by showing that 50% of the bands in an individual are derived from a particular parent.

Sequences that resemble satellites in consisting of tandem repeats of a short unit, but that overall are much shorter—consisting of (for example) 5 to 50 repeats—are common in mammalian genomes. They were discovered by chance as fragments whose size is extremely variable in genomic libraries of human DNA. The variability is seen when a population contains fragments of many different sizes that represent the same genomic region; when individuals are examined, it turns out that there is extensive polymorphism, and that many different alleles can be found.

Whether a repeat cluster is called a minisatellite or a microsatellite depends on both the length of the repeat unit and the number of repeats in the cluster. The name **microsatellite** is usually used when the length of the repeating unit is <10 bp; the number of repeats is smaller than that of minisatellites. The name minisatellite is used when the length of the repeating unit is ~10 to 100 bp and there is a greater number of repeats. The terminology is not, however, precisely defined. These types of sequences are also called **variable number tandem repeat (VNTR)** regions. VNTRs used in human forensics are microsatellites that generally have <20 copies of a 2 to 6 bp repeat.

The cause of the variation between individual genomes at microsatellites or minisatellites is that individual alleles have different numbers of the repeating unit. For example, one minisatellite has a repeat length of 64 bp and is found in the population with the following approximate distribution:

7%	18 repeats
11%	16 repeats
43%	14 repeats
36%	13 repeats
4%	10 repeats

The rate of genetic exchange at minisatellite sequences is high, ~10^{-4} per kb of DNA. (The frequency of exchanges per actual locus is assumed to be proportional to the length of the minisatellite.) This rate is ~10× greater than the rate of homologous recombination at meiosis for any random DNA sequence.

The high variability of minisatellites makes them especially useful for genomic mapping, because there is a high probability that individuals will vary in their alleles at such a locus. An example of mapping by minisatellites is illustrated in **FIGURE 7.20**. This shows an extreme case in which two individuals both are heterozygous at a minisatellite locus, and in fact all four alleles are different. All progeny gain one allele from each parent in the usual way, and it is possible unambiguously to determine the source of every allele in the progeny. In the terminology of human genetics, the meioses

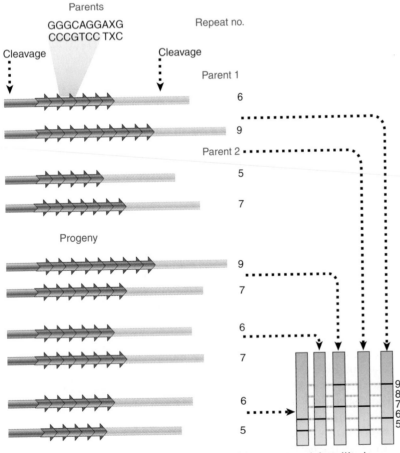

FIGURE 7.20 Alleles may differ in the number of repeats at a minisatellite locus, so that cleavage on either side generates restriction fragments that differ in length. By using a minisatellite with alleles that differ between parents, the pattern of inheritance can be followed.

described in this figure are highly informative because of the variation between alleles.

One family of minisatellites in the human genome share a common "core" sequence. The core is a G-C-rich sequence of 10 to 15 bp, showing an asymmetry of purine/pyrimidine distribution on the two strands. Each individual minisatellite has a variant of the core sequence, but ~1000 minisatellites can be detected on Southern blot (see *Section 3.9, Blotting Methods*) by a probe consisting of the core sequence.

Consider the situation shown in Figure 7.20, but multiplied many times by the existence of many such sequences. The effect of the variation at individual loci is to create a unique pattern for every individual. This makes it possible to assign heredity unambiguously between parents and progeny by showing that 50% of the bands in any individual are derived from a particular parent. This is the basis of the technique known as **DNA fingerprinting**.

Both microsatellites and minisatellites are unstable, although for different reasons. Microsatellites undergo intrastrand mispairing, when slippage during replication leads to expansion of the repeat, as shown in **FIGURE 7.21**. Systems that repair damage to DNA—in particular those that recognize mismatched base pairs—are important in reversing such changes, as shown by a large increase in frequency when repair genes are inactivated. Mutations in repair systems are an important contributory factor in the development of cancer, thus tumor cells often display variations in microsatellite sequences. Minisatellites undergo the same sort of unequal crossing-over between repeats that we have discussed for satellites (see Figure 7.3). One telling case is that increased variation is associated with a recombination hotspot. The recombination event is not usually associated with recombination between flanking markers, but has a complex form in which the new mutant allele gains information from both the sister chromatid and the other (homologous) chromosome.

It is not clear at what repeating length the cause of the variation shifts from replication slippage to unequal crossing-over.

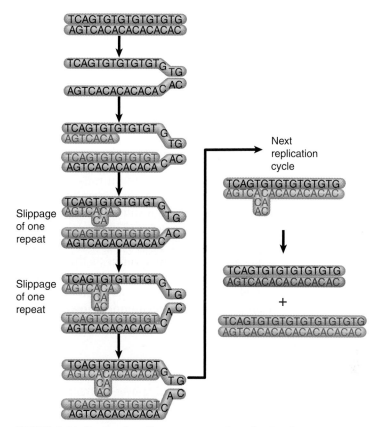

FIGURE 7.21 Replication slippage occurs when the daughter strand slips back one repeating unit in pairing with the template strand. Each slippage event adds one repeating unit to the daughter strand. The extra repeats are extruded as a single-strand loop. Replication of this daughter strand in the next cycle generates a duplex DNA with an increased number of repeats.

7.9 Summary

Most genes belong to families, which are defined by the possession of related sequences in the exons of individual members. Families evolve by the duplication of a gene (or genes), followed by divergence between the copies.

Some copies suffer inactivating mutations and become pseudogenes that no longer have any function.

A tandem cluster consists of many copies of a repeating unit that includes the transcribed sequence(s) and a nontranscribed spacer(s). rRNA gene clusters code only for a single rRNA precursor. Maintenance of active genes in clusters depends on mechanisms such as gene conversion or unequal crossing-over, which cause mutations to spread through the cluster so that they become exposed to evolutionary pressure.

Satellite DNA often consists of very short sequences repeated many times in tandem. Its distinct centrifugation properties reflect its biased base composition. Satellite DNA is concentrated in centromeric heterochromatin, but its function (if any) is unknown. The individual repeating units of arthropod satellites are identical. Those of mammalian satellites are related and can be organized into a hierarchy reflecting the evolution of the satellite by the amplification and divergence of randomly chosen sequences.

Unequal crossing-over appears to have been a major determinant of satellite DNA organization. Crossover fixation explains the ability of variants to spread through a cluster.

Minisatellites and microsatellites consist of even shorter repeating sequences than satellites, generally <10 bp for microsatellites and 10 to 50 bp for minisatellites, with a shorter cluster length than satellites have. The number of repeating units is usually 5 to 50. There is high variation in the repeat number between individual genomes. A microsatellite repeat number varies as the result of slippage during replication; the frequency is affected by systems that recognize and repair damage in DNA. Minisatellite repeat number varies as the result of recombination-like events. Variations in repeat number can be used to determine hereditary relationships by the technique known as DNA fingerprinting.

References

7.2 Unequal Crossing-over Rearranges Gene Clusters

Research

Bailey, J. A., Gu, Z., Clark, R. A., Reinert, K., Samonte, R. V., Schwartz, S., Adams, M. D., Myers, E. W., Li, P. W., and Eichler, E. E. (2002). Recent segmental duplications in the human genome. *Science* 297, 1003–1007.

7.3 Genes for rRNA Form Tandem Repeats Including an Invariant Transcription Unit

Research

Afseth, G., and Mallavia, L. P. (1997). Copy number of the 16S rRNA gene in *Coxiella burnetii*. *Eur. J. Epidemiol.* 13, 729–731.

7.4 Crossover Fixation Could Maintain Identical Repeats

Research

Charlesworth, B., Sniegowski, P., and Stephan, W. (1994). The evolutionary dynamics of repetitive DNA in eukaryotes. *Nature* 371, 215–220.

7.6 Arthropod Satellites Have Very Short Identical Repeats

Research

Smith, C. D., Shu, S.-Q., Mungall, C. J., and Karpen, G. H. (2007). The release 5.1 annotation of *Drosophila melanogaster* heterochromatin. *Science* 316, 1586–1591.

7.7 Mammalian Satellites Consist of Hierarchical Repeats

Review

Waterston, R. H. et al. (2002). Initial sequencing and comparative analysis of the mouse genome. *Nature* 420, 520–562.

7.8 Minisatellites Are Useful for Genetic Mapping

Research

Jeffreys, A. J., Murray, J., and Neumann, R. (1998). High-resolution mapping of crossovers in human sperm defines a minisatellite-associated recombination hotspot. *Mol. Cell* 2, 267–273.

Jeffreys, A. J., Royle, N. J., Wilson, V., and Wong, Z. (1988). Spontaneous mutation rates to new length alleles at tandem-repetitive hypervariable loci in human DNA. *Nature* 332, 278–281.

Jeffreys, A. J., Tamaki, K., MacLeod, A., Monckton, D. G., Neil, D. L., and Armour, J. A. (1994). Complex gene conversion events in germline mutation at human minisatellites. *Nat. Genet.* 6, 136–145.

Jeffreys, A. J., Wilson, V., and Thein, S. L. (1985). Hypervariable minisatellite regions in human DNA. *Nature* 314, 67–73.

Strand, M., Prolla, T. A., Liskay, R. M., and Petes, T. D. (1993). Destabilization of tracts of simple repetitive DNA in yeast by mutations affecting DNA mismatch repair. *Nature* 365, 274–276.

Muller Element

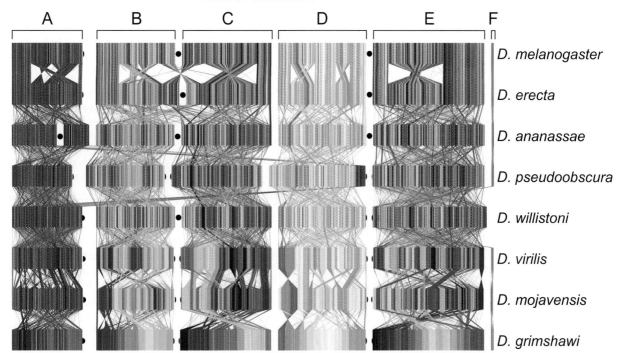

A   B   C   D   E   F

D. melanogaster
D. erecta
D. ananassae
D. pseudoobscura
D. willistoni
D. virilis
D. mojavensis
D. grimshawi

Reproduced from A. Bhutkar, S. W. Schaeffer, et al., *Genetics* 179 (2008): 1657–1680. Used with permission of the Genetics Society of America. Photo courtesy of Arjun Bhutkar, Koch Institute for Integrative Cancer Research, Massachusetts Institute of Technology, and Stephen W. Schaeffer, Department of Biology, The Pennsylvania State University.

Genome Evolution

8.1 Introduction

The growing number of complete genome sequences has provided valuable opportunities to study genome structure and organization. As genome sequences of related species become available, though, there are now opportunities to compare not only individual gene differences but also large-scale genomic differences in such aspects as gene distribution, the proportions of nonrepetitive and repetitive DNA and their functional potentials, and the number of copies of repetitive sequences. By making these comparisons, we can gain insight into the historical genetic events that have shaped the genomes of individual species and of the adaptive and non-adaptive forces at work following these events.

The availability of the genome sequences of genetic "model organisms" (e.g., *E. coli*, yeast, *Drosophila*, *Arabidopsis*, and human) in the early part of this decade allowed comparisons between major taxonomic groups such as prokaryote vs. eukaryote, animal vs. plant, or vertebrate vs. invertebrate. More recently, however, data from multiple genomes within lower-level taxonomic groups (classes down to genera) have allowed closer examination of genome evolution. Such comparisons have the advantage of highlighting changes that have occurred much more recently and are unobscured by additional changes, such as multiple mutations at the same site. In addition, evolutionary events specific to a taxonomic group can be explored. For example, human–chimpanzee comparisons can provide information about primate-specific genome evolution, particularly when compared with an *outgroup* (a species that is less closely related, but close enough to show substantial similarity) such as the mouse. One recent milestone in this field of *comparative genomics* is the completion of genome sequences of twelve species of the genus *Drosophila*. These types of fine-scale comparisons will continue as more genomes from the same species become available.

What questions can be addressed by comparative genomics? First, the evolution of individual genes can be explored by comparing genes descended from a common ancestor. To some extent, the evolution of a genome is a result of the evolution of a collection of individual genes, so comparisons of homologous sequences within and between genomes can help to answer questions about the adaptive (i.e., naturally selective) and nonadaptive changes that occur to these sequences. The forces that shape coding sequences are usually quite different than those that affect noncoding regions (such as introns, untranslated regions, or regulatory regions) of the same gene: coding and regulatory regions more directly influence phenotype (though in different ways), making selection a more important aspect of their evolution than for noncoding regions. Second, one can also explore the mechanisms that result in changes in the structure of the genome, such as gene duplication, expansion and contraction of repetitive arrays, transposition, and polyploidization.

8.2

DNA Sequences Evolve by Mutation and a Sorting Mechanism

Key concepts

- The probability of a mutation is influenced by the likelihood that the particular error will occur and the likelihood that it will be repaired.

- In small populations, the frequency of a mutation will change randomly and new mutations are likely to be eliminated by chance.

- The frequency of a neutral mutation largely depends on genetic drift, the strength of which depends on the size of the population.

- The frequency of a mutation that affects phenotype will be influenced by negative or positive selection.

Biological evolution is based on two sets of processes: the generation of genetic variation and the sorting of that variation in subsequent generations. Variation among chromosomes can be generated by recombination (see Chapter 15, *Homologous and Site-Specific Recombination*) and variation among sexually reproducing organisms results from the combined processes of meiosis and fertilization. Ultimately, however, variation among DNA sequences is a result of mutation.

Mutation occurs when DNA is altered by replication error or chemical changes to nucleotides, or when electromagnetic radiation breaks or forms chemical bonds, and the damage remains unrepaired at the time of the next DNA replication event (see Chapter 16, *Repair Systems*). Regardless of the cause, the initial damage can be considered an "error." In principle, a base can mutate to any of the other three standard bases, though the three possible mutations are not equally likely due to biases incurred by the mechanisms of damage (see *Section 8.14, There May Be Biases in Mutation, Gene Conversion, and Codon Usage*) and differences in the likelihood of repair of the damage.

For example, if one assumes that mutation from one base to any of the other three is equally probable, then transversion mutations (from a pyrimidine to a purine, or vice versa) would be twice as frequent as transition mutations (from one pyrimidine to another, or one purine to another; see *Section 1.12, Mutations May Affect Single Base Pairs or Longer Sequences*). The observation is usually the opposite, though: transitions occur roughly twice as frequently as transversions. This may be because (1) spontaneous transitional errors occur more frequently than transversional errors; (2) transversional errors are more likely to be detected and corrected by DNA repair mechanisms; or (3) both of these are true. Given that transversional errors result in distortion of the DNA duplex as either pyrimidines or purines are paired together, and that base-pair geometry is used as a fidelity mechanism (see *Section 14.5, DNA Polymerases Control the Fidelity of Replication*), it is less likely for a DNA polymerase to make a transversional error. The distortion also makes it easier for transversional errors to be detected by postreplication repair mechanisms. As shown in **FIGURE 8.1**, a basic model of mutation would be that the probabilities of transitions are equal (α), as are those of transversions (β), and that $\alpha > \beta$. More complex models could have different probabilities for the individual substitution mutations, and could be tailored to

	A	T	C	G
A	–	β	β	α
T	β	–	α	β
C	β	α	–	β
G	α	β	β	–

FIGURE 8.1 A simple model of mutational change in which α is the probability of a transition and β is the probability of a transversion. Reproduced from MEGA (Molecular Evolutionary Genetics Analysis) by S. Kumar, K. Tamura, and J. Dudley. Used with permission of Masatoshi Nei, Pennsylvania State University.

individual taxonomic groups from actual data on mutation rates in those groups.

If a mutation occurs in the coding region of a protein-coding gene, it can be characterized by its effect on the polypeptide product of the gene. A substitution mutation that does not change the amino acid sequence of the polypeptide product is a **synonymous mutation**; this is a specific type of *silent mutation*. (Silent mutations include those that occur in noncoding regions.) A **nonsynonymous mutation** in a coding region does alter the amino acid sequence of the polypeptide product, creating either a missense codon (for a different amino acid) or a nonsense (termination) codon. The effect of the mutation on the phenotype of the organism will influence the fate of the mutation in subsequent generations.

Mutations in genes other than those encoding polypeptides and mutations in noncoding sequences may of course also be subject to selection. In noncoding regions, a mutational change may alter the regulation of a gene by directly changing a regulatory sequence or by changing the secondary structure of the DNA in such a way that some aspect of the gene's expression (transcription rate, RNA processing, mRNA structure influencing translation rate) is affected. Many changes in noncoding regions, though, may be selectively **neutral mutations**, having no effect on the phenotype of the organism.

If a mutation is selectively neutral or near-neutral, then its fate is predictable only in terms of probability. The random changes in the frequency of a mutational variant in a population are called **genetic drift**; this is a type of "sampling error" in which, by chance, the offspring genotypes of a particular set of parents do not precisely match those predicted by Mendelian inheritance. In a very large population, the random effects of genetic drift tend to average

out, so there is little change in the frequency of each variant. In a small population, however, these random changes can be quite significant and genetic drift can have a major effect on the genetic variation of the population. **FIGURE 8.2** shows a simulation comparing the random changes in allele frequency for seven populations of ten individuals each with those of seven populations of one hundred individuals each. Each population begins with two alleles, each with a frequency of 0.5. After fifty generations, most of the small populations have lost one or the other allele, while the large populations have retained both alleles (though their allele frequencies have randomly drifted from the original 0.5).

Genetic drift is a random process; thus the eventual fate of a particular variant is not strictly predictable, but a probability can be assigned according to the current frequency of the variant. In other words, a new mutation (with a low frequency in a population) is very likely to be lost from the population by chance. If by chance it becomes more frequent, though, it has a greater probability of being retained in the population. Over the long term, a variant may either be lost from the population or *fixed*, replacing all other variants, but in the short term there may be randomly fluctuating variation for a given locus, particularly in smaller populations where **fixation** or loss occurs more quickly.

On the other hand, if a new mutation is not selectively neutral and does affect phenotype, natural selection will play a role in its increase or decrease in frequency in the population. The speed of its frequency change will partly depend on how much of an advantage or disadvantage the mutation confers to the organisms that carry it. It will also depend on whether it is dominant or recessive; in general, because dominant mutations are "exposed" to natural selection when they first appear, they are affected by selection more rapidly.

Mutations are random with regard to their effects, and thus the common result of a non-neutral mutation is for the phenotype to be negatively affected, so selection often acts primarily to eliminate new mutations (though this may be somewhat delayed in the likely event that the mutation is recessive). This is called *negative* (or *purifying*) selection (see *Section 4.5, Exon Sequences under Negative Selection Are Conserved but Introns Vary*). The overall result of negative selection is for there to be little variation within a population as new variants are

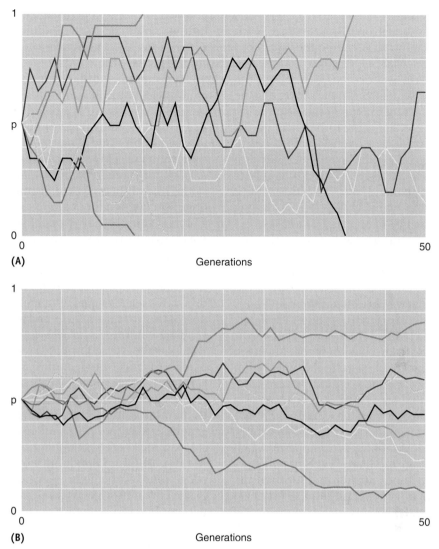

FIGURE 8.2 The fixation or loss of alleles by random genetic drift occurs more rapidly in (A) populations of 10 than in (B) populations of 100. p is the frequency of one of two alleles at a locus in the population. Data courtesy of Kent E. Holsinger, University of Connecticut (http://darwin.eeb.uconn.edu).

generally eliminated. More rarely, a new mutation may be subject to positive selection (see *Section 4.6, Exon Sequences under Positive Selection Vary but Introns Are Conserved*) if it happens to confer an advantageous phenotype. This type of selection will also tend to reduce variation within a population, as the new mutation eventually replaces the original sequence, but may result in greater variation *between* populations, provided they are isolated from one another, as different mutations occur in these different populations.

The question of how much observed genetic variation in a population or species (or the lack of such variation) is due to selection and how much is due to genetic drift is a long-standing one in population genetics. In the next section, we will look at some ways that selection

on DNA sequences may be detected by testing for significant differences from the expectations of evolution of neutral mutations.

8.3 Selection Can Be Detected by Measuring Sequence Variation

Key concepts

- The ratio of nonsynonymous to synonymous substitutions in the evolutionary history of a gene is a measure of positive or negative selection.

- Low heterozygosity of a gene may indicate recent selective events.

- Comparing the rates of substitution among related species can indicate whether selection on the gene has occurred.

Many methods have been used over the years for analyzing selection on DNA sequences. With the development of DNA sequencing techniques in the 1970s (see Chapter 3, *Methods in Molecular Biology and Genetic Engineering*), the automation of sequencing in the 1990s, and the development of high-throughput sequencing over the past decade, large numbers of partial or complete genome sequences are becoming available. Coupled with the polymerase chain reaction (PCR) to amplify specific genomic regions, DNA sequence analysis has become a valuable tool in many applications, including the study of selection on genetic variants.

There is now an abundance of DNA sequence data from a wide range of organisms in various publicly available databases. Homologous gene sequences have been obtained from many species as well as from different individuals of the same species. This allows for determination of genetic changes across species lineages as compared to changes within a species. These comparisons have led to the observation that some species (such as *Drosophila melanogaster*) have high levels of DNA sequence polymorphism among individuals, most likely as a result of neutral mutations and random genetic drift within populations. (Other species, such as humans, have moderate levels of polymorphism, and without further investigation the relative roles of genetic drift and selection in keeping these levels low is not immediately clear. This is one use for techniques to detect selection on sequences.) By conducting both interspecific and intraspecific DNA sequence analysis, the level of divergence due to species differences can be determined.

Some neutral mutations are synonymous mutations (see *Section 8.2, DNA Sequences Evolve by Mutation and a Sorting Mechanism*), but not all synonymous mutations are neutral. While this may at first seem contradictory, the levels of individual tRNAs for a given amino acid are not the same in a cell. Some cognate tRNAs (different tRNAs that carry the same amino acid) are more abundant than others, and a specific codon may lack sufficient tRNAs, whereas a different codon for the same amino acid may have a sufficient number. In the case of a codon that requires a rare tRNA in that organism, ribosomal frameshifting or other alterations in translation may occur (see *Section 25.16, Frameshifting Occurs at Slippery Sequences*). It also may be that a particular codon is necessary to maintain mRNA structure. Alternatively, there may be a nonsynonymous mutation to an amino acid with the same general characteristics, with little or no effect on the folding and activity of the protein. In either case neutral sequence changes have little effect on the organism. A nonsynonymous mutation may result in an amino acid with different properties, however, such as a change from a polar to a nonpolar amino acid, or from a hydrophobic amino acid to a hydrophilic one in a protein embedded in a phospholipid bilayer. Such changes are likely to have functional effects that are deleterious to the role of the polypeptide and thus to the organism. Depending on the location of the amino acid in the polypeptide, such a change may cause only slight disruption of protein folding and activity. Only in rare cases is an amino acid change advantageous; in this case the mutational change may become subjected to positive selection and ultimately lead to fixation of this variant in the population.

One common approach for determining selection is to use codon-based sequence information to study the evolutionary history of a gene. This can be done by counting the number of synonymous (K_s) and nonsynonymous (K_a) amino acid substitutions in orthologous genes (see *Section 6.4, How Many Different Types of Genes Are There?*), and determining the K_a/K_s ratio. This ratio is indicative of the selective constraints on the gene. A K_a/K_s ratio of 1 is expected for those genes that evolve neutrally, with amino acid sequence changes being neither favored nor disfavored. In this case the changes that occur do not usually affect the activity of the polypeptide, and this serves as a suitable control. A K_a/K_s ratio <1 is most commonly observed, and indicates negative selection where amino acid replacements are disfavored because they affect the activity of the polypeptide. Thus there is selective pressure to retain the original functional amino acid at these sites in order to maintain proper protein function.

Positive selection occurs when the K_a/K_s ratio is >1, but is rarely observed. This indicates that the amino acid changes are advantageous and may become fixed in the population. Some examples of this are antigenic proteins of some pathogens, such as viral coat proteins, which are under strong selection pressure to evade the immune response of the host, and some reproductive proteins that are under sexual selection. For example, the K_a/K_s ratios for the peptide-binding regions of mammalian MHC genes, the products of which function in immunological self-recognition by displaying both "self" and "nonself" antigens, are typically in the range of 2 to 10, indicating strong selection for new variants. This is expected since these proteins represent the cellular uniqueness of individual organisms. The detection of a positive K_a/K_s ratio may be rare in part because the average value must be greater than one over a length of sequence. If a single substitution in a gene is being positively selected, but flanking regions are under negative selection, the average ratio across the sequence may actually be negative. In contrast, the K_a/K_s ratios for histone genes are typically much less than one, suggesting strong purifying selection on these genes. Histones are DNA-binding proteins that make up the basic structure of chromatin (see Chapter 10, *Chromatin*) and alterations to their structures are likely to result in deleterious effects on chromosome integrity and gene expression.

In addition to the difficulty of detecting strong selection on a single substitution variant when K_a/K_s is averaged over a stretch of DNA, mutational hotspots may also affect this measure. There have been reports of unusually highly mutable regions of some protein-coding genes that encode a high proportion of polar amino acids; such a bias may influence the interpretation of the K_a/K_s ratio because a higher point mutation rate may be incorrectly interpreted as a higher substitution rate. Although codon-based methods of detecting selection can be useful, their limitations must be taken into account.

Intraspecific DNA sequence analysis can be used to detect positive selection by comparing the nucleotide sequence between two alleles, or two individuals of the same species. Nucleotide sequences evolve neutrally at a certain rate; variation in this rate at specific nucleotides affects the heterozygosity (the proportion of heterozygotes at a locus). If a variant sequence is favored the site will show a reduction in nucleotide heterozygosity, and the variant will increase in frequency and eventually become fixed in the population. Nearby linked neutral variants may also become fixed, a phenomenon termed **genetic hitchhiking**. These regions are characterized by having a lower level of DNA sequence polymorphism. (It is important to remember, though, that reduced polymorphism can have other causes, such as purifying selection or genetic drift.)

In practice it is more reliable to carry out both interspecific and intraspecific DNA sequence comparisons to assess deviations from neutral evolution. By including sequence information from at least one closely related species, species-specific DNA polymorphisms can be distinguished from ancestral polymorphisms, and more accurate information can be obtained regarding the link between the polymorphisms and between-species differences. With this combined analysis the degree of nonsynonymous changes between species can be determined. If evolution is primarily neutral, the ratio of nonsynonymous to synonymous changes *within* species is expected to be the same as the ratio *between* species. An excess of nonsynonymous changes may be evidence for positive selection on these amino acids, whereas a lower ratio may indicate that negative selection is conserving sequences.

One example is the comparison of twelve sequences of the *Adh* gene in *D. melanogaster* to each other and to *Adh* sequences from *D. simulans* and *D. yakuba*, as shown in **FIGURE 8.3**. A simple contingency chi-square test on these data shows that there are significantly more fixed nonsynonymous changes between species than similar changes polymorphic in *D. melanogaster*. The high proportion of nonsynonymous differences among species suggests positive selection on *Adh* variants in these species, as does the lower proportion of such differences in one species given that nonneutral variation would not be expected to persist for very long within a species.

Relative rate tests can also be used for determining selection. This involves (at a minimum) three related species, two closely related and one outgroup representative. The substitution rate is compared between the close relatives, and each is compared to the outgroup species to see if the substitution rates are similar. This removes the dependence of the analysis on time, as long as the phylogenetic relationship between the species is certain. If the rate of substitutions between relatives compared to the rate between these and the outgroup species is different, this may be an indication of selection on the sequence. For example, the protein lysozyme, which functions to digest bacterial cell

	Nonsynonymous	Synonymous
Fixed	7	17
Polymorphic	2	42

FIGURE 8.3 Nonsynonymous and synonymous variation in the *Adh* locus in *Drosophila melanogaster* ("polymorphic") and between *D. melanogaster*, *D. simulans*, and *D. yakuba* ("fixed"). Adapted from J. H. McDonald and M. Kreitman, *Nature* 351 (1991): 652–654.

walls and is a general antibiotic in many species, has evolved to be active at low pH in ruminating mammals, where it functions to digest dead bacteria in the gut. **FIGURE 8.4** shows that the number of amino acid (i.e., nonsynonymous) substitutions for lysozyme in the cow/deer (ruminant) lineage is higher than that of the nonruminant pig outgroup.

This method must take into account that some genes accumulate nucleotide or amino acid substitutions more rapidly (these are said to be *fast-clock*; see *Section 8.4, A Constant Rate of Sequence Divergence Is a Molecular Clock*) in some species than in others, possibly due to differences in metabolic rate, generation time, DNA replication time, or DNA repair efficiency. To deal with this difference, additional related species need to be examined in order to identify and eliminate fast-clock effects. The reliability of this approach is improved if larger numbers of distantly related species are included. It is difficult, though, to make accurate comparisons between taxonomic groups due to the inherent rate differences. As more work in this area has been done, corrections have been developed to adjust for differences in substitution rates.

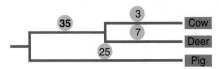

FIGURE 8.4 A higher number of nonsynonymous substitutions in lysozyme sequences in the cow/deer lineage as compared to the pig lineage is a result of adaptation of the protein for digestion in ruminant stomachs. Adapted from N. H. Barton, et al. *Evolution*. Cold Spring Harbor Laboratory Press, 1991. Original figure appeared in J. H. Gillespie, *The Causes of Molecular Evolution*. Oxford University Press, 2007.

Another method for detecting selection utilizes estimates of polymorphism at specific genetic loci. For example, sequence analysis of the *Teosinte branched 1* (*tb1*) locus, an important gene in domesticated maize, has been used to characterize the nucleotide substitution rate in domesticated and native maize (teosinte) varieties, with an estimate of 2.9×10^{-8} to 3.3×10^{-8} base substitutions per year. **FIGURE 8.5** shows the ratio of a measure of nucleotide diversity (π) of the *tb1* region in domesticated maize to π in wild teosinte. For a neutrally evolving gene in these two species this ratio is ~0.75, but is <0.1 in this region. The interpretation is that strong selection in domesticated maize has severely reduced variation for this gene.

As genome-wide data on nucleotide diversity become available, regions of low diversity may be used to detect selection. Millions of single nucleotide polymorphisms (SNPs) are being characterized in humans, nonhuman animals, and plants, along with other species. One approach that has been applied to the human genome is to look for an association between an allele's frequency and its **linkage disequilibrium** with other genetic markers surrounding it. Linkage disequilibrium is a measure of an association between an allele at one locus and an allele at a different locus. A new mutation occurs on one chromosome; thus it initially has high linkage disequilibrium with alleles at other polymorphic loci on the same chromosome. In a large population, a neutral allele is expected to rise to fixation slowly, so recombination and mutation will break up associations between loci as reflected by a decay in linkage disequilibrium. On the other hand, an allele under positive selection will rise to fixation quickly

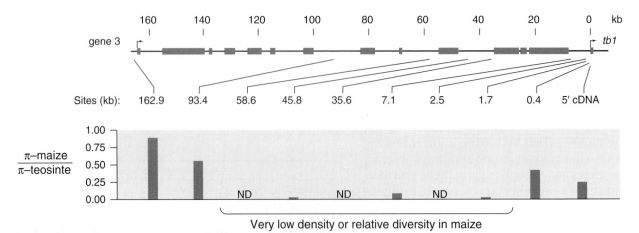

FIGURE 8.5 Nucleotide diversity (π) of the *tb1* region in domesticated maize is much lower than in wild teosinte, indicating strong selection on this locus in maize. Reproduced from R. M. Clark, et al., *Proc. Natl. Acad. Sci. USA* 101 (2004): 700–707. © 2004 National Academy of Sciences, U.S.A. Courtesy of John F. Doebley, University of Wisconsin, Madison.

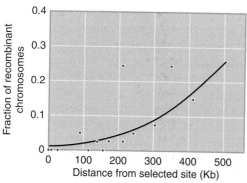

FIGURE 8.6 The fraction of recombinants between an allele of *G6PD* and alleles at nearby loci on a human chromosome remains low, suggesting that the allele has rapidly increased in frequency by positive selection. The allele confers resistance to malaria. Adapted from E. T. Wang, et al., *Proc. Natl. Acad. Sci. USA* 103 (2006): 135–140.

and linkage disequilibrium will be maintained. By sampling SNPs across the genome, a general background level of linkage disequilibrium that accounts for local variations in rates of recombination can be established and significantly high measures of linkage disequilibrium can be detected. **FIGURE 8.6** shows the slowly decreasing linkage disequilibrium (measured by the increasing fraction of recombinant chromosomes) with increasing chromosomal distance from a variant of the *G6PD* locus that confers resistance to malaria in African human populations. This pattern suggests that this allele has been under strong recent selection—carrying along with it linked alleles at other loci—and that recombination has not yet had time to break up these interlocus associations.

8.4 A Constant Rate of Sequence Divergence Is a Molecular Clock

Key concepts

- The sequences of orthologous genes in different species vary at nonsynonymous sites (where mutations have caused amino acid substitutions) and synonymous sites (where mutation has not affected the amino acid sequence).
- Synonymous substitutions accumulate ~10× faster than nonsynonymous substitutions.
- The evolutionary divergence between two DNA sequences is measured by the corrected percent of positions at which the corresponding nucleotides differ.
- Substitutions may accumulate at a more or less constant rate after genes separate, so that the divergence between any pair of globin sequences is proportional to the time since they shared common ancestry.

Most changes in gene sequences occur by mutations that accumulate slowly over time. Point mutations and small insertions and deletions occur by chance, probably with more or less equal probability in all regions of the genome. The exceptions to this are *hotspots*, where mutations occur much more frequently. Recall from *Section 8.2* that most nonsynonymous mutations are deleterious and will be eliminated by negative selection, whereas the rare advantageous substitution will spread through the population and eventually replace the original sequence (fixation). Neutral variants are expected to be lost or fixed in the population due to random genetic drift. What proportion of mutational changes in a protein-coding gene sequence are selectively neutral is a historically contentious issue.

The rate at which substitutions accumulate is a characteristic of each gene, presumably depending at least in part on its functional flexibility with regard to change. Within a species, a gene evolves by mutation, followed by fixation within the single population. Recall that when we scrutinize the gene pool of a species, we see only the variants that have been maintained, whether by selection or genetic drift. When multiple variants are present they may be stable, or they may in fact be transient because they are in the process of being displaced.

When a single species separates into two new species, each of the resulting species now constitutes an independent pool for evolution. By comparing orthologous genes in two species, we see the differences that have accumulated between them since the time when their ancestors ceased to interbreed. Some genes are highly conserved, showing little or no change from species to species. This indicates that most changes are deleterious and therefore eliminated.

The difference between two genes is expressed as their **divergence**, the percent of positions at which the nucleotides are different, corrected for the possibility of convergent mutations (the same mutation at the same site in two separate lineages) and true revertants. There is usually a difference in the rate of evolution among the three codon positions within genes, because mutations at the third base position often are synonymous.

In addition to the coding sequence, a gene contains untranslated regions. Here again, most mutations are potentially neutral, apart from their effects on either secondary structure or (usually rather short) regulatory signals.

Although synonymous mutations are expected to be neutral with regard to the polypeptide, they could affect gene expression via the sequence change in RNA (see *Section 8.2, DNA Sequences Evolve by Mutation and a Sorting Mechanism*). Another possibility is that a change in synonymous codons calls for a different tRNA to respond, influencing the efficiency of translation. Species generally show a **codon bias**; when there are multiple codons for the amino acid, one codon is found in protein-coding genes in a high percentage, whereas the remaining codons are found in low percentages. There is a corresponding percentage difference in the tRNA species that recognize these codons. Consequently, a change from a common to a rare synonymous codon may reduce the rate of translation due to a lower concentration of appropriate tRNAs. (Alternatively, there may be a nonadaptive explanation for codon bias; see *Section 8.14, There May Be Biases in Mutation, Gene Conversion, and Codon Usage*.)

FIGURE 8.7 shows the divergence of three types of proteins (representing nonsynony-

mous changes) over time by comparing species for which there is paleontological evidence for the time of divergence. There are two striking features of these data. First, the three types of proteins evolve at different rates: fibrinopeptides evolve quickly, cytochrome *c* evolves slowly, and hemoglobin evolves at an intermediate rate. Second, for each protein type the rate of evolution is approximately constant over millions of years. In other words, for a given type of protein, the divergence between any pair of sequences is (more or less) proportional to the time since they separated. This provides a **molecular clock** that measures the accumulation of substitutions at an approximately constant rate during the evolution of a given protein-coding gene.

There can also be molecular clocks for paralogous proteins diverging within a species lineage. To take the example of the human β- and δ-globin chains (see *Section 7.2, Unequal Crossing-over Rearranges Gene Clusters* and *Section 8.10, Globin Clusters Arise by Duplication and Divergence*), there are ten differences in 146 residues,

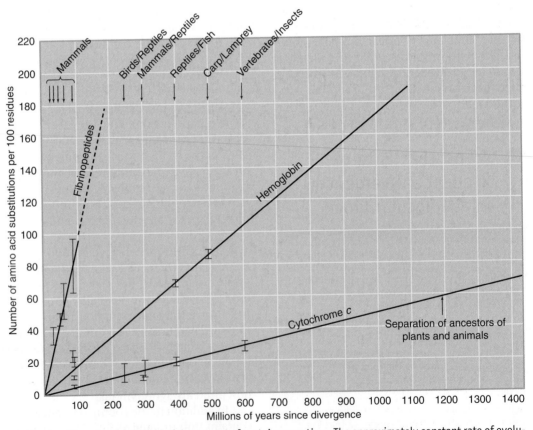

FIGURE 8.7 The rate of evolution of three types of proteins over time. The approximately constant rate of evolution of each protein type is a molecular clock. Reproduced with kind permission from Springer Science+Business Media: *J. Mol. Evol.*, The structure of cytochrome and the rates of molecular evolution, vol. 1, 1971, pp. 26–45, R. E. Dickerson, fig. 3. Courtesy of Richard Dickerson, University of California, Los Angeles.

a divergence of 6.9%. The DNA sequence has 31 changes in 441 residues. These changes are distributed very differently, however, in the nonsynonymous and synonymous sites. There are 11 changes in the 330 nonsynonymous sites, but 20 changes in only 111 synonymous sites. This gives corrected rates of divergence of 3.7% in the nonsynonymous sites and 32% in the synonymous sites, an order of magnitude in difference.

The striking difference in the divergence of nonsynonymous and synonymous sites demonstrates the existence of much greater constraints on nucleotide changes that influence protein constitution relative to those that do not. Many fewer amino acid changes are neutral.

Suppose we take the rate of synonymous substitutions to indicate the underlying rate of mutational fixation (assuming there is no selection at all at the synonymous sites). Then over the period since the β and δ genes diverged, there should have been changes at 32% of the 330 nonsynonymous sites, for a total of 105. All but 11 of them have been eliminated, which means that ~90% of the mutations were not maintained.

The rate of divergence can be measured as the percent difference per million years, or as its reciprocal, the unit evolutionary period (UEP), the time in millions of years that it takes for 1% divergence to accrue. Once the rate of the molecular clock has been established by pairwise comparisons between species (remembering the practical difficulties in establishing the actual time of speciation), it can be applied to paralogous genes within a species. From their divergence, we can calculate how much time has passed since the duplication that generated them.

By comparing the sequences of orthologous genes in different species, the rate of divergence at both nonsynonymous and synonymous sites can be determined, as plotted in **FIGURE 8.8**.

In pairwise comparisons, there is an average divergence of 10% in the nonsynonymous sites of either the α- or β-globin genes of mammal lineages that have been separated since the mammalian radiation occurred ~85 million years ago. This corresponds to a nonsynonymous divergence rate of 0.12% per million years.

The rate is approximately constant when the comparison is extended to genes that diverged in the more distant past. For example, the average nonsynonymous divergence

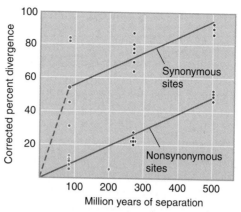

FIGURE 8.8 Divergence of DNA sequences depends on evolutionary separation. Each point on the graph represents a pairwise comparison.

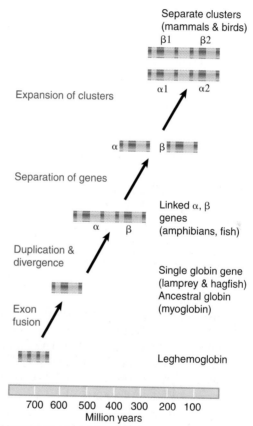

FIGURE 8.9 All globin genes have evolved by a series of duplications, transpositions, and mutations from a single ancestral gene.

between orthologous mammalian and chicken globin genes is 23%. Relative to a common ancestor at ~270 million years ago, this gives a rate of 0.09% per million years.

Going back further, we can compare the α- with the β-globin genes within a species. They have been diverging since the original duplication event 500 million years ago (see **FIGURE 8.9**). They have an average nonsynonymous divergence of ~50%, which gives a rate of 0.1% per million years.

The summary of these data in Figure 8.8 shows that nonsynonymous divergence in the globin genes has an average rate of ~0.096% per million years (for a UEP of 10.4). Considering the uncertainties in estimating the times at which the species diverged, the results lend good support to the idea that there is a constant molecular clock.

The data on synonymous site divergence are much less clear. In every case, it is evident that the synonymous site divergence is much greater than the nonsynonymous site divergence, by a factor that varies from 2 to 10. The range of synonymous site divergences in pairwise comparisons, though, is too great to establish a molecular clock, so we must base temporal comparisons on the nonsynonymous sites.

From Figure 8.8, it is clear that the rate at synonymous sites is not constant over time. If we assume that there must be zero divergence at zero years of separation, we see that the rate of synonymous site divergence is much greater for the first ~100 million years of separation. One interpretation is that a fraction of roughly half of the synonymous sites is rapidly (within 100 million years) saturated by mutations; this fraction behaves as neutral sites. The other fraction accumulates mutations more slowly, at a rate approximately the same as that of the nonsynonymous sites; this fraction represents sites that are synonymous with regard to the polypeptide, but that are under selective constraint for some other reason.

Now we can reverse the calculation of divergence rates to estimate the times since paralogous genes were duplicated. The difference between the human β and δ genes is 3.7% for nonsynonymous sites. At a UEP of 10.4, these genes must have diverged $10.4 \times 3.7 = 40$ million years ago—about the time of the separation of the major primate lineages: New World monkeys, Old World monkeys, and great apes (including humans). All of these taxonomic groups have both β and δ genes, which suggests that the gene divergence commenced just before this point in evolution.

Proceeding further back, the divergence between the nonsynonymous sites of γ and ε genes is 10%, which corresponds to a duplication event ~100 million years ago. The separation between embryonic and fetal globin genes therefore may have just preceded or accompanied the mammalian radiation.

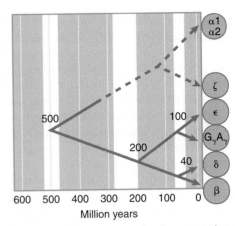

FIGURE 8.10 Nonsynonymous site divergences between pairs of β-globin genes allow the history of the human cluster to be reconstructed. This tree accounts for the separation of classes of globin genes.

An evolutionary tree for the human globin genes is presented in **FIGURE 8.10**. Paralogous groups that evolved before the mammalian radiation—such as the separation of β/δ from γ—should be found in all mammals. Paralogous groups that evolved afterward—such as the separation of β- and δ-globin genes—should be found in individual lineages of mammals.

In each species, there have been comparatively recent changes in the structures of the clusters. We know this because we see differences in gene number (one adult β-globin gene in humans, two in the mouse) or in type (most often concerning whether there are separate embryonic and fetal genes).

When sufficient data have been collected on the sequences of a particular gene or gene family, the analysis can be reversed, and comparisons between orthologous genes can be used to assess taxonomic relationships. If a molecular clock has been established, the time to common ancestry between the previously analyzed species and a species newly introduced to the analysis can be estimated.

8.5 The Rate of Neutral Substitution Can Be Measured from Divergence of Repeated Sequences

Key concept

- The rate of substitution per year at neutral sites is greater in the mouse genome than in the human genome, probably because of a higher mutation rate.

We can make the best estimate of the rate of substitution at neutral sites by examining sequences that do not code for polypeptide. (We use the term "neutral" here rather than "synonymous," because there is no coding potential.) An informative comparison can be made by comparing the members of a common repetitive family in the human and mouse genomes.

The principle of the analysis is summarized in **FIGURE 8.11**. We start with a family of related sequences that have evolved by duplication and substitution from an original ancestral sequence. We assume that the ancestral sequence can be deduced by taking the base that is most common at each position. Then we can calculate the divergence of each individual family member as the proportion of bases that differ from the deduced ancestral sequence. In this example, individual members vary from 0.13 to 0.18 divergence and the average is 0.16.

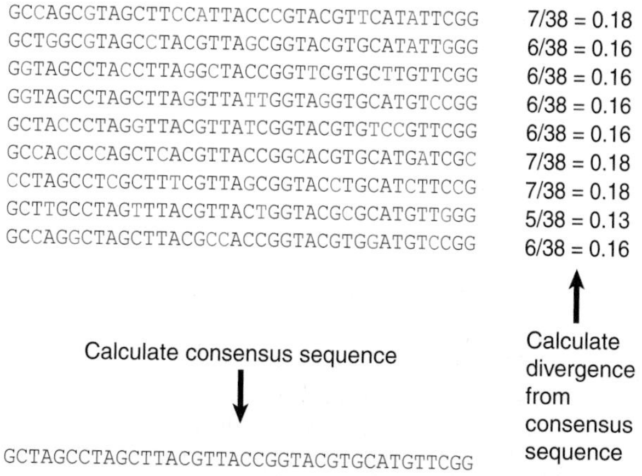

One family used for this analysis in the human and mouse genomes derives from a sequence that is thought to have ceased to be active at about the time of the common ancestor between humans and rodents (the LINEs family; see *Section 17.18, Retroelements Fall into Three Classes*). This means that it has been diverging under limited selective pressure for the same length of time in both species. Its average divergence in humans is ~0.17 substitutions per site, corresponding to a rate of 2.2×10^{-9} substitutions per base per year over the 75 million years since the separation. In the mouse genome, however, neutral substitutions have occurred at twice this rate, corresponding to 0.34 substitutions per site in the family, or a rate of 4.5×10^{-9}. Note, however, that if we calculated the rate per generation instead of per year, it would be greater in humans than in the mouse (~2.2×10^{-8} as opposed to ~10^{-9}).

These figures probably underestimate the rate of substitution in the mouse; at the time of divergence the rates in both lineages would have been the same, and the difference must have evolved since then. The current rate of neutral substitution per year in the mouse is probably two to three times greater than the historical average. At first glance, these rates would seem to reflect the balance between the occurrence of mutations (which may be higher in species with higher metabolic rates, like the

```
GCCAGCGTAGCTTCCATTACCCGTACGTTCATATTCGG     7/38 = 0.18
GCTGGCGTAGCCTACGTTAGCGGTACGTGCATATTGGG     6/38 = 0.16
GGTAGCCTACCTTAGGCTACCGGTTCGTGCTTGTTCGG     6/38 = 0.16
GGTAGCCTAGCTTAGGTTATTGGTAGGTGCATGTCCGG     6/38 = 0.16
GCTACCCTAGGTTACGTTATCGGTACGTGTCCGTTCGG     6/38 = 0.16
GCCACCCCAGCTCACGTTACCGGCACGTGCATGATCGC     7/38 = 0.18
CCTAGCCTCGCTTTCGTTAGCGGTACCTGCATCTTCCG     7/38 = 0.18
GCTTGCCTAGTTTACGTTACTGGTACGCGCATGTTGGG     5/38 = 0.13
GCCAGGCTAGCTTACGCCACCGGTACGTGGATGTCCGG     6/38 = 0.16
```

Calculate consensus sequence ↓

```
GCTAGCCTAGCTTACGTTACCGGTACGTGCATGTTCGG
```

↑ Calculate divergence from consensus sequence

FIGURE 8.11 An ancestral consensus sequence for a family is calculated by taking the most common base at each position. The divergence of each existing current member of the family is calculated as the proportion of bases at which it differs from the ancestral sequence.

mouse) and the loss of them due to genetic drift, which is largely a function of population size; genetic drift is a type of "sampling error" where allele frequencies fluctuate more widely in smaller populations. In addition to eliminating neutral alleles more quickly, smaller population sizes also allow faster fixation and loss of neutral alleles. Rodent species tend to have short generation times (allowing more opportunities for substitutions per year), but species with short generation times also tend to have larger population sizes; thus the effects of more substitutions per year but less fixation of neutral alleles would cancel each other out. The higher substitution rate in mice is probably due primarily to a higher mutation rate.

Comparing the mouse and human genomes allows us to assess whether syntenic (homologous) regions show signs of conservation or have differed at the rate predicted from accumulation of neutral substitutions. The proportion of sites that show signs of selection is ~5%. This is much higher than the proportion found in exons (~1%). This observation implies that the genome includes many more stretches whose sequence is important for functions other than coding for RNA. Known regulatory elements are likely to comprise only a small part of this proportion. This number also suggests that most (i.e., the rest) of the genome sequences do not have any function that depends on the exact sequence.

8.6 How Did Interrupted Genes Evolve?

Key concepts

- A major evolutionary question is whether genes originated with introns or whether they were originally uninterrupted.
- Interrupted genes that correspond either to proteins or to independently functioning non-protein-encoding RNAs probably originated in an interrupted form (the "introns early" hypothesis).
- The interruption allowed base order to better satisfy the potential for stem–loop extrusion from duplex DNA, perhaps to facilitate recombination repair of errors.
- A special class of introns is mobile and can insert themselves into genes.

The structure of many eukaryotic genes suggests a concept of the eukaryotic genome as a sea of mostly unique DNA sequences in which exon "islands" separated by intron "shallows" are strung out in individual gene "archipelagoes." What was the original form of genes?

- The **"introns early" model** is the proposal that introns have always been an integral part of the gene. Genes originated as interrupted structures, and those now without introns have lost them in the course of evolution.
- The **"introns late" model** is the proposal that the ancestral protein-coding sequences were uninterrupted and that introns were subsequently inserted into them.

In simple terms, can the difference between eukaryotic and prokaryotic gene organizations be accounted for by the acquisition of introns in the eukaryotes or by the loss of introns from the prokaryotes?

One advantage of the "introns early" model is that the mosaic structure of genes suggests an ancient combinatorial approach to the construction of genes to encode novel proteins. Suppose that an early cell had a number of separate protein-coding sequences: it is likely to have evolved by reshuffling different polypeptide units to construct new proteins. Inasmuch as we may recognize the advantages of this mechanism for gene evolution, though, it does not follow that it constituted—or contributed to—the selective pressure for the *initial* evolution of the mosaic structure. Introns may have greatly assisted, but might not have been critical for, the recombination of protein-coding gene segments. Thus, a disproof of the combinatorial hypothesis would neither disprove the introns early hypothesis nor support the introns late hypothesis.

If a protein-coding unit (now known as an exon) must be a continuous series of codons, every such reshuffling event would require a precise recombination of DNA to place separate protein-coding units in sequence and in the same reading frame (a ⅓ probability in any one random joining event). If, however, this combination doesn't produce a functional protein, the cell might be damaged because the original sequence of protein-coding units might have been lost.

The cell might survive, though, if some of the experimental recombination occurs in RNA transcripts, leaving the DNA intact. If a translocation event could place two protein-coding units in the same transcription unit, various RNA splicing experiments to combine the two proteins into a single polypeptide chain could be explored. If some combinations are not successful, the original protein-coding units remain available for further trials. In addition, this scenario does not require the two protein-coding units to be recombined precisely into a continuous coding sequence. There is evidence supporting this scenario: different genes have related exons, as if each gene had been assembled by a process of **exon shuffling** (see *Section 4.9, Some Exons Can Be Equated with Protein Functional Domains*).

FIGURE 8.12 illustrates the result of a translocation of a random sequence that includes an exon into a gene. In some organisms, exons are

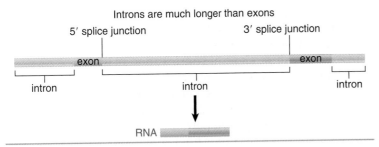

Introns are much longer than exons

5' splice junction   3' splice junction

exon   exon

intron   intron   intron

RNA

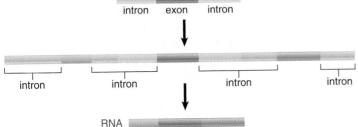

Sequence including an exon translocates into random target site

intron exon intron

intron   intron   intron   intron

RNA

FIGURE 8.12 An exon surrounded by flanking sequences that is translocated into an intron may be spliced into the RNA product.

very small compared to introns, so it is likely that the exon will insert within an intron and be flanked by functional 5′ and 3′ splice junctions. Splicing junctions are recognized in sequential pairs, so the splicing mechanism should recognize the 5′ splicing junction of the original intron and the 3′ splicing junction of the introduced exon, instead of the 3′ splice junction of the original intron. Similarly, the 5′ splicing junction of the new exon and the 3′ splicing junction of the original intron may be recognized as a pair, so the new exon will remain between the original two exons in the mature RNA transcript. As long as the new exon is in the same reading frame as the original exons (a ⅓ probability at each end), a new, longer protein will be produced. Exon shuffling events could have been responsible for generating new combinations of exons during evolution. (Note that the mechanism of this process is mimicked by the technique of *exon trapping* that is used to screen for functional exons [see Figure 5.8]).

Given that it is difficult to envision (1) the assembly of long chains of amino acids by some template-independent process, and (2) that such assembled chains would be able to self-replicate, it is widely believed that the most successful early self-replicating life forms were nucleic acids—probably RNA. Indeed, RNA molecules can act both as coding templates and as catalysts (i.e., *ribozymes*; see Chapter 23, *Catalytic RNA*). It was probably by virtue of their catalytic activities that prototypic molecules in the early "RNA world" were able to self-replicate; the templating property would have emerged later.

Many functions mediated by nucleic acid could have competed for genome space in the RNA world. As set out earlier (see *Section 4.3, Exon and Intron Base Compositions Differ*), these functions can be seen as exerting pressures: AG pressure (the pressure for purine-enrichment in exons); GC pressure (the genome-wide pressure for a distinctive balance between the proportions of the two sets of Watson–Crick pairing bases); single-strand parity pressure (the genome-wide pressure for parity between A and T, and between G and C, in single-stranded nucleic acids) and, probably related to the latter, fold pressure (the genome-wide pressure for single-stranded nucleic acid, whether in free form or extruded from duplex forms, to adopt secondary and higher order stem–loop structures; see Figure 4.7). For present purposes, the functions served by these pressures need not concern us. The fact that the pressures are

so widely spread among organisms suggests important roles in the economy of life (survival and reproduction), rather than mere neutrality.

To these pressures competing for genome space would have been added pressures for increased catalytic activities, ribozyme pressure being supplemented or superseded by protein pressure (the pressure to encode a sequence of amino acids with potential enzymatic activity) once a translation system had evolved. Mutation that happened to generate protein-coding potential would have been favored, but would also be competing against preexisting nucleic acid level pressures. In other words, exons may have been latecomers to an evolving molecular system. Given the redundancy of the genetic code, especially at the third base positions of codons, accommodations could have been explored in the course of evolution, so that a protein-encoding region would, to a degree, have been subject to selection by nucleic acid pressures *within itself*. Thus, coding sequences could be selected for both their protein-coding potential and their effects on DNA structure.

Constellations of exons that were *slowly* evolving under negative selection (see *Section 4.5, Exon Sequences under Negative Selection Are Conserved but Introns Vary*) would have been able to adapt to accommodate nucleic acid pressures. Exon sequences that could accommodate both protein and nucleic acid pressures would have been conserved. However, those evolving more *rapidly* under positive selection (see *Section 4.6, Exon Sequences under Positive Selection Vary but Introns Are Conserved*) would not have been able to afford this luxury. Thus, some nucleic acid level pressures (e.g., fold pressure) would have been diverted to neighboring introns—hence the conservation of the latter (see Figure 4.7).

Some RNA transcripts perform functions by virtue of their secondary and higher-order structures, not by acting as templates for translation. These RNAs, which often interact with proteins, include *Xist* that is involved in X-chromosome inactivation (see *Section 29.5, X Chromosomes Undergo Global Changes*) and the tRNAs and rRNAs that facilitate the translation of mRNAs. Generally, these single-stranded RNAs have the same sequence of bases as one strand (the RNA-synonymous strand) of the corresponding DNA.

It is important to note that because these RNAs have structures that serve their distinctive functions (often cytoplasmic), it does not follow that the *same* structures will serve the functions (nuclear) of the corresponding DNAs equally well. Thus, we should not be surprised that, even though there is no ultimate protein product, RNA genes are interrupted and the transcripts are spliced to generate mature RNA products (see Figure 21.40). Similarly, there are sometimes introns in the 5′ and 3′ untranslated regions of pre-mRNAs that must be spliced out.

Information for the overtly functional parts of genes can be seen, then, as having had to intrude into genomes that were already adapted to numerous preexisting pressures operating at the nucleic acid level. A reconfiguration of pressures usually could not have occurred if the genic function-encoding parts existed as contiguous sequences. The outcome was that DNA segments corresponding to the genic function-encoding parts were often interrupted by other DNA segments catering to the basic needs of the genome. A further fortuitous outcome would have been a facilitation of the intermixing of functional parts to allow the evolutionary testing of new combinations.

Apart from the above pressures on genome space, there are selection pressures acting at the organismal level. For example, birds tend to have shorter introns than mammals, which has led to the controversial hypothesis that there has been selection pressure for compaction of the genome because of the metabolic demands of flight. For many microorganisms (such as bacteria and yeast), evolutionary success can be equated with the ability to rapidly replicate DNA. Smaller genomes can be more rapidly replicated than larger, so it may be the pressure for compaction of genomes that led to uninterrupted genes in most microorganisms. Long protein-encoding sequences had to accommodate numerous genomic pressures in addition to protein-pressure.

Some species have alternative forms of rRNA and tRNA genes, both with and without introns. For tRNAs, which all have the same general conformation, it seems unlikely that the two regions of the gene evolved independently because the two regions base pair to fold the molecule into a functional shape. In this case, the intron must have been inserted into a continuous gene.

There is evidence that introns have been lost from some members of gene families. See *Section 4.10, Members of a Gene Family Have a Common Organization*, for examples from the insulin and actin gene families. In the case of the actin gene family, it is sometimes not clear whether the presence of an intron in a member of the family indicates the ancestral state or an inser-

tion event. Overall, current evidence suggests that genes originally had sequences now called introns, but can evolve with both the loss and gain of introns.

Organelle genomes show the evolutionary connections between prokaryotes and eukaryotes. There are many general similarities between mitochondria or chloroplasts and certain bacteria because those organelles originated by endosymbiosis, in which a bacterial cell dwelled within the cytoplasm of a eukaryotic prototype. Although there are similarities to bacterial genetic processes—such as protein and RNA synthesis—some organelle genes possess introns and therefore resemble eukaryotic nuclear genes. Introns are found in several chloroplast genes, including some that are homologous to *E. coli* genes. This suggests that the endosymbiotic event occurred before introns were lost from the prokaryotic lineage.

Mitochondrial genome comparisons are particularly striking. The genes of yeast and mammalian mitochondria encode virtually identical proteins, in spite of a considerable difference in gene organization. Vertebrate mitochondrial genomes are very small and extremely compact, whereas yeast mitochondrial genomes are larger and have some complex interrupted genes. Which is the ancestral form? Yeast mitochondrial introns (and certain other introns) can be mobile—they are independent sequences that can splice out of the RNA and insert DNA copies elsewhere—which suggests that they may have arisen by insertions into the genome (see *Section 23.5, Some Group I Introns Code for Endonucleases That Sponsor Mobility* and *Section 23.6, Group II Introns May Code for Multifunction Proteins*). While most evidence supports "introns early," there is reason to believe that, in addition to the introduction of mobile elements, ongoing accommodations to various extrinsic and intrinsic (genomic) pressures might result, from time to time, in the emergence of new introns ("introns late").

As for the role of introns, it is easy to dismiss intronic characteristics such as an enhanced potential to extrude stem–loop structures, as an adaptation to assist accurate splicing. An analogy has been drawn between the transmission of genic messages and the transmission of electronic messages, though, in which a message sequence is normally interrupted by error-correcting codes. While there is no evidence that similar types of code operate in genomes, it is possible that fold pressure arose to aid the detection and correction of sequence errors by recombination repair. So important would be the latter that in many circumstances fold pressure might trump protein pressure (see *Section 4.11, Genetic Information Is Not Completely Contained in DNA*, and Chapter 16, *Repair Systems*).

8.7 Why Are Some Genomes So Large?

Key concepts

- There is no clear correlation between genome size and genetic complexity.
- There is an increase in the minimum genome size associated with organisms of increasing complexity.
- There are wide variations in the genome sizes of organisms within many taxonomic groups.

The total amount of DNA in the (haploid) genome is a characteristic of each living species known as its **C-value**. There is enormous variation in the range of C-values, from $<10^6$ bp for a mycoplasma to $>10^{11}$ bp for some plants and amphibians.

FIGURE 8.13 summarizes the range of C-values found in different taxa (groups of organisms classified together). There is an increase in the minimum genome size found in each group as the complexity increases. Although C-values are greater in the multicellular eukaryotes, we do see some wide variations in the genome sizes within some taxa.

Plotting the minimum amount of DNA required for a member of each group suggests in **FIGURE 8.14** that an increase in genome size is required for increased complexity in prokaryotes and lower eukaryotes.

Mycoplasma are the smallest prokaryotes and have genomes only ~3× the size of a large bacteriophage and smaller than those of some megaviruses. More typical bacterial genome sizes start at $~2 \times 10^6$ bp. Unicellular eukaryotes (whose lifestyles may resemble those of prokaryotes) get by with genomes that are small, too, although they are larger than those of most bacteria. Being eukaryotic *per se* does not imply a vast increase in genome size; a yeast may have a genome size of $~1.3 \times 10^7$ bp, which is only about twice the size of an average bacterial genome.

A further twofold increase in genome size is adequate to support the slime mold *Dictyostelium discoideum*, which is able to live in either unicellular or multicellular modes. Another increase in complexity is necessary to produce the first fully multicellular organisms; the nematode worm *Caenorhabditis elegans* has a DNA content of 8×10^7 bp.

We also can see the steady increase in genome size with complexity in the listing in **FIGURE 8.15** of some of the most commonly analyzed organisms. It is necessary for insects, birds, amphibians, and mammals to have larger genomes than those of unicellular eukaryotes. After this point, though, there is no clear relationship between genome size and morphological complexity of the organism.

We know that genes are much larger than the sequences needed to code for polypeptides,

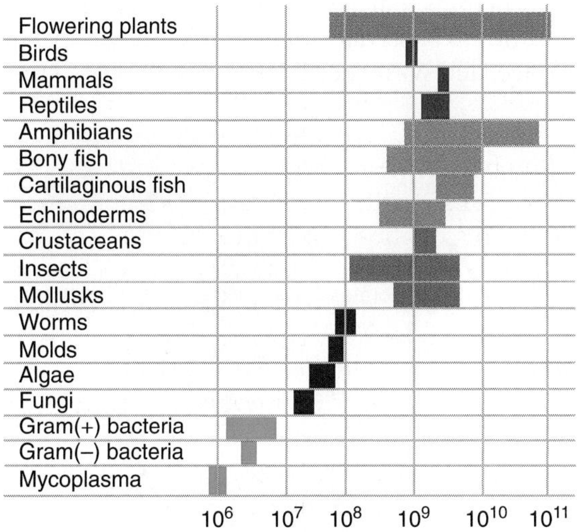

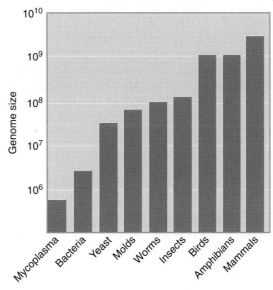

FIGURE 8.13 DNA content of the haploid genome increases with morphological complexity of lower eukaryotes, but varies extensively within some groups of higher eukaryotes. The range of DNA values within each group is indicated by the shaded area.

FIGURE 8.14 The minimum genome size found in each taxon increases from prokaryotes to mammals.

because exons may comprise only a small part of the total length of a gene. This explains why there is much more DNA than is needed to provide reading frames for all the proteins of the organism. Large parts of an interrupted gene may not code for polypeptide. In addition, in multicellular organisms there also may be significant lengths of DNA between genes, some of which functions in gene regulation. So it is not possible to deduce from the overall size of the genome anything about the number of genes or the complexity of the organism.

The **C-value paradox** refers to the lack of correlation between genome size and genetic and morphological complexity (such as the number of different cell types). There are some extremely curious observations about relative genome size, such as that the toad *Xenopus* and humans have genomes of essentially the same size. In some taxa there are extremely large variations in DNA content between organisms that do not vary much in complexity, as will be shown in the next section in Figure 8.16. (This is especially marked in insects, amphibians, and plants, but does not occur in birds, reptiles, and mammals, which all show little variation within the group—with an ~2× range of genome sizes.) A cricket has a genome 11× the size of that of a fruit fly. In amphibians, the smallest genomes are <10^9 bp, whereas the largest are ~10^{11} bp. There is unlikely to be a large difference in the number of genes needed to specify these amphibians. Fish have about the same number of genes as mammals have, but some fish genomes (such as that of fugu) are more compact, with smaller introns and shorter intergenic spaces, while others are tetraploid. The extent to which this variation is selectively neutral or subject to natural selection is not yet fully understood.

In mammals, additional complexity is also a consequence of the alternative splicing of genes that allows two or more protein variants to be produced from the same gene (see Chapter 21, *RNA Splicing and Processing*). With such mechanisms, increased complexity need not be accompanied by an increased number of genes.

8.8 Morphological Complexity Evolves by Adding New Gene Functions

Key concepts

- In general, comparisons of eukaryotes to prokaryotes, multicellular to unicellular eukaryotes, and vertebrate to invertebrate animals show a positive correlation between gene number and morphological complexity as additional genes are needed with generally increased complexity.
- Most of the genes that are unique to vertebrates are concerned with the immune or nervous systems.

Comparison of the human genome sequence with sequences found in other species is revealing about the process of evolution. **FIGURE 8.16** shows an analysis of human genes according to the breadth of their distribution among all cellular organisms. Starting with the most generally distributed (top right corner of the figure), ~21% of genes are common to eukaryotes and prokaryotes. These tend to code for proteins that are essential for all living forms—typically basic metabolism, replication, transcription, and translation. Moving clockwise, another ~32% of genes are found in eukaryotes in general—for example, they may be found in yeast. These tend to code for proteins involved in functions that are general to eukaryotic cells but not to bacteria—for example, they may be concerned with specifying organelles or cyto-

Phylum	Species	Genome (bp)
Algae	*Pyrenomas salina*	6.6×10^5
Mycoplasma	*M. pneumoniae*	1.0×10^6
Bacterium	*E. coli*	4.2×10^6
Yeast	*S. cerevisiae*	1.3×10^7
Slime mold	*D. discoideum*	5.4×10^7
Nematode	*C. elegans*	8.0×10^7
Insect	*D. melanogaster*	1.8×10^8
Bird	*G. domesticus*	1.2×10^9
Amphibian	*X. laevis*	3.1×10^9
Mammal	*H. sapiens*	3.3×10^9

FIGURE 8.15 The genome sizes of some common experimental organisms.

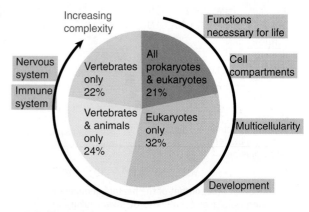

FIGURE 8.16 Human genes can be classified according to how widely their homologs are distributed in other species.

skeletal components. Another ~24% of genes are generally found in animals. These include genes necessary for multicellularity and for development of different tissue types. Approximately twenty-two percent of genes are unique to vertebrates. These mostly encode proteins of the immune and nervous systems; they encode very few enzymes, consistent with the idea that enzymes have ancient origins, and that metabolic pathways originated early in evolution. We see, therefore, that the evolution of more complex morphology and specialization requires the addition of groups of genes representing the necessary new functions.

One way to define essential proteins is to identify the proteins present in all proteomes. Comparing the human proteome in more detail with the proteomes of other organisms, 46% of the yeast proteome, 43% of the worm proteome, and 61% of the fly proteome is represented in the human proteome. A key group of ~1300 proteins is present in all four proteomes. The common proteins are basic "housekeeping" proteins required for essential functions, falling into the types summarized in FIGURE 8.17. The main functions are concerned with transcription and translation (35%), metabolism (22%), transport (12%), DNA replication and modification (10%), protein folding and degradation (8%), and cellular processes (6%), with the remaining 7% dedicated to various other functions.

One of the striking features of the human proteome is that it has many unique proteins compared with other eukaryotes, but has relatively few unique protein domains (portions of proteins having a specific function). Most protein domains appear to be common to the animal kingdom. There are many unique protein architectures, however, defined as unique combinations of domains. FIGURE 8.18 shows that the greatest proportion of unique proteins consists of transmembrane and extracellular proteins. In yeast, the vast majority of architectures are concerned with intracellular proteins. About twice as many intracellular architectures are found in flies (or worms), but there is a strikingly higher proportion of transmembrane and extracellular proteins, as might be expected from the additional functions required for the interactions between the cells of a multicellular organism. The additions in intracellular architectures required in a vertebrate (human) are relatively small, but there is, again, a higher proportion of transmembrane and extracellular architectures.

It has long been known that the genetic difference between humans and chimpanzees (our nearest relative) is very small, with ~99% identity between genomes. The sequence of the chimpanzee genome now allows us to investigate the 1% of differences in more detail to see whether features responsible for "humanity" can be identified. The comparison shows 35×10^6 nucleotide substitutions (1.2% sequence difference overall), 5×10^6 deletions or insertions (making ~1.5% of the euchromatic sequence specific to each species), and many chromosomal rearrangements. Homologous proteins are usually very similar: 29% are identical, and in most cases there are only one or two amino acid differences in the protein between the species. In fact, nucleotide

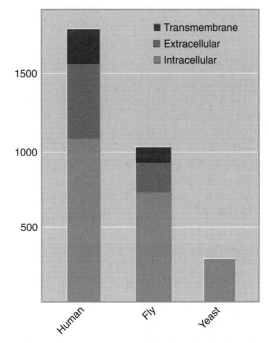

FIGURE 8.18 Increasing complexity in eukaryotes is accompanied by accumulation of new proteins for transmembrane and extracellular functions.

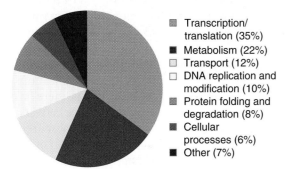

- Transcription/ translation (35%)
- Metabolism (22%)
- Transport (12%)
- DNA replication and modification (10%)
- Protein folding and degradation (8%)
- Cellular processes (6%)
- Other (7%)

FIGURE 8.17 Common eukaryotic proteins are concerned with essential cellular functions.

substitutions occur less often in genes coding for polypeptides than are likely to be involved in specifically human traits, suggesting that protein evolution is not a major factor in human–chimpanzee differences. This leaves larger-scale changes in gene structure and/or changes in gene regulation as the major candidates. Some 25% of nucleotide substitutions occur in CpG dinucleotides (among which are many potential regulator sites).

8.9 Gene Duplication Contributes to Genome Evolution

Key concept

- Duplicated genes may diverge to generate different genes, or one copy may become an inactive pseudogene.

Exons act as modules for building genes that are tried out in the course of evolution in various combinations (see *Section 4.9, Some Exons Can Be Equated with Protein Functional Domains*). At one extreme, an individual exon from one gene may be copied and used in another gene. At the other extreme, an entire gene, including both exons and introns, may be duplicated. In such a case, mutations can accumulate in one copy without elimination by natural selection as long as the other copy is under selection to remain functional. The selectively neutral copy may then evolve to a new function, become expressed at a different time or in a different cell type from the first copy, or become a nonfunctional pseudogene.

FIGURE 8.19 summarizes our present view of the rates at which these processes occur. There is a ~1% probability that a given gene will be included in a duplication in a period of one million years. After the gene has duplicated, differences evolve as the result of the occurrence of different mutations in each copy. These accumulate at a rate of ~0.1% per million years (see *Section 8.4, A Constant Rate of Sequence Divergence Is a Molecular Clock*).

Unless the gene encodes a product that is required in high concentration in the cell, the organism is not likely to need to retain two identical copies of the gene. As differences evolve between the duplicated genes, one of two types of event is likely to occur:

- Both of the gene copies remain necessary. This can happen either because the differences between them gener-

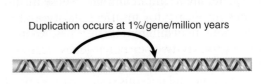

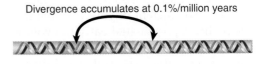

Duplication occurs at 1%/gene/million years

Divergence accumulates at 0.1%/million years

Silencing of one copy takes ~4 million years

Active   Pseudogene

FIGURE 8.19 After a globin gene has been duplicated, differences may accumulate between the copies. The genes may acquire different functions or one of the copies may become inactive.

ate proteins with different functions, or because they are expressed specifically at different times or in different cell types.

- If this does not happen, one of the genes is likely to become a pseudogene because it will by chance gain a deleterious mutation, and there will be no purifying selection to eliminate this copy so by genetic drift the mutant version may increase in frequency and fix in the species. Typically this takes ~4 million years for globin genes; in general, the time to fixation of a neutral mutant depends on the generation time and the effective population size, with genetic drift being a stronger force in smaller populations. In such a situation, it is purely a matter of chance which of the two copies becomes inactive. (This can contribute to incompatibility between different individuals, and ultimately to speciation, if different copies become inactive in different populations.)

Analysis of the human genome sequence shows that ~5% of the genome comprises duplications of identifiable segments ranging in length from 10 to 300 kb. These duplications have arisen relatively recently; that is, there has not been sufficient time for divergence between them for their homology to become obscured. They include a proportional share (~6%) of the expressed exons, which shows that the duplications are occurring more or less irrespective of genetic content. The genes in these duplications may be especially interesting because of the implication that they have evolved recently and therefore could be important for recent evolutionary developments (such as the separation of the human lineage from that of other primates).

8.10 Globin Clusters Arise by Duplication and Divergence

Key concepts

- All globin genes are descended by duplication and mutation from an ancestral gene that had three exons.
- The ancestral gene gave rise to myoglobin, leghemoglobin, and α and β globins.
- The α- and β-globin genes separated in the period of early vertebrate evolution, after which duplications generated the individual clusters of separate α- and β-like genes.
- Once a gene has been inactivated by mutation, it may accumulate further mutations and become a pseudogene (ψ), which is homologous to the active gene(s) but has no functional role.

The most common type of gene duplication generates a second copy of the gene close to the first copy. In some cases, the copies remain associated and further duplication may generate a cluster of related genes. The best-characterized example of a gene cluster is that of the globin genes, which constitute an ancient gene family fulfilling a function that is central to animals: the transport of oxygen.

The major constituent of the vertebrate red blood cell is the globin tetramer, which is associated with its heme (iron-binding) group in the form of hemoglobin. Functional globin genes in all species have the same general structure: they are divided into three exons, as shown previously in Figure 4.4. We conclude that all globin genes have evolved from a single ancestral gene, and by tracing the history of individual globin genes within and between species we may learn about the mechanisms involved in the evolution of gene families.

In red blood cells of adult mammals, the globin tetramer consists of two identical α chains and two identical β chains. Embryonic red blood cells contain hemoglobin tetramers that are different from the adult form. Each tetramer contains two identical α-like chains and two identical β-like chains, each of which is related to the adult polypeptide and is later replaced by it in the adult form of the protein.

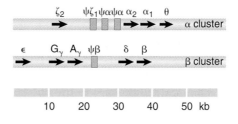

FIGURE 8.20 Each of the α-like and β-like globin gene families is organized into a single cluster that includes functional genes and pseudogenes (ψ).

This is an example of developmental control, in which different genes are successively switched on and off to provide alternative products that fulfill the same function at different times.

The division of globin chains into α-like and β-like reflects the organization of the genes. Each type of globin is encoded by genes organized into a single cluster. The structures of the two clusters in the primate genome are illustrated in FIGURE 8.20. Pseudogenes are indicated by the symbol ψ.

Stretching over 50 kb, the β cluster contains five functional genes (ε, two γ, δ, and β) and one nonfunctional pseudogene (ψβ). The two γ genes differ in their coding sequence in only one amino acid: the G variant has glycine at position 136, whereas the A variant has alanine.

The more compact α cluster extends over 28 kb and includes one active ζ gene, one nonfunctional ζ pseudogene, two α genes, two nonfunctional α pseudogenes, and the θ gene of unknown function. The two α genes code for the same protein. Two (or more) identical genes present on the same chromosome are described as **nonallelic** copies.

The details of the relationship between embryonic and adult hemoglobins vary with the species. The human pathway has three stages: embryonic, fetal, and adult. The distinction between embryonic and adult is common to mammals, but the number of preadult stages varies. In humans, ξ and α are the two α-like chains. ε, γ, δ, and β are the β-like chains. FIGURE 8.21 shows how the chains are expressed at different stages of development. There is also tissue-specific expression associated with the developmental expression: embryonic hemoglobin genes are expressed in the yolk sac, fetal genes are expressed in the liver, and adult genes are expressed in bone marrow.

In the human pathway, ζ is the first α-like chain to be expressed, but it is soon replaced by α. In the β-pathway, ε and γ are expressed first, with δ and β replacing them later. In adults, the $\alpha_2\beta_2$ form provides 97% of the hemoglobin, $\alpha_2\delta_2$ provides ~2%, and ~1% is provided by persistence of the fetal form $\alpha_2\gamma_2$.

What is the significance of the differences between embryonic and adult globins? The embryonic and fetal forms have a higher affinity for oxygen, which is necessary in order to obtain oxygen from the mother's blood. This helps to explain why there is no direct equivalent (although there is temporal expression of globins) in, for example, the chicken, for which the embryonic stages occur outside the mother's body (that is, within the egg).

Functional genes are defined by their expression to RNA and ultimately by the polypeptides they encode. Pseudogenes are defined as such by their inability to produce functional polypeptides; the reasons for their inactivity vary, and the deficiencies may be in transcrip-

tion or translation (or both). A similar general organization is found in other vertebrate globin gene clusters, but details of the types, numbers, and order of genes all vary, as illustrated in **FIGURE 8.22**. Each cluster contains both embryonic and adult genes. The total lengths of the clusters vary widely. The longest known cluster is found in the goat genome, where a basic cluster of four genes has been duplicated twice. The distribution of active genes and pseudogenes differs in each case, illustrating the random nature of the evolution of one copy of a duplicated gene to a pseudogene.

The characterization of these gene clusters provides an important general point. There may be more members of a gene family, both functional and nonfunctional, than we would suspect on the basis of protein analysis. The extra functional genes may represent duplicates that code for identical polypeptides, or they may be related to—but different from—known proteins (and presumably expressed only briefly or in low amounts).

With regard to the question of how much DNA is needed to code for a particular function, we see that coding for the β-like globins requires a range of 20 to 120 kb in different mammals. This is much greater than we would expect just from scrutinizing the known β-globin proteins or from even considering the individual genes. Clusters of this type are not common, though; most genes are found as individual loci.

From the organization of globin genes in a variety of species, we should be able to trace the evolution of present globin gene clusters from a single ancestral globin gene. Our present view of the evolutionary history was pictured earlier, in Figure 8.9.

The leghemoglobin gene of plants, which is related to the globin genes, may provide some clues about the ancestral form, though of course the modern leghemoglobin gene has evolved for just as long as the animal globin genes. (Leghemoglobin is an oxygen carrier found in the nitrogen-fixing root nodules of legumes.) The furthest back that we can trace a true globin gene is to the sequence of the single chain of mammalian myoglobin, which diverged from the globin lineage ~800 million years ago in the ancestors of mammals. The myoglobin gene has the same organization as globin genes, so we may take the three-exon structure to represent that of their common ancestor.

Some members of the class *Chondrichthyes* (cartilaginous fish) have only a single type of globin chain, so they must have diverged from

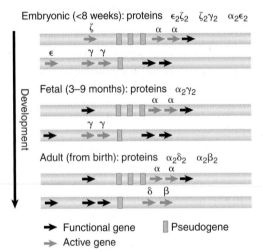

FIGURE 8.21 Different hemoglobin genes are expressed during embryonic, fetal, and adult periods of human development.

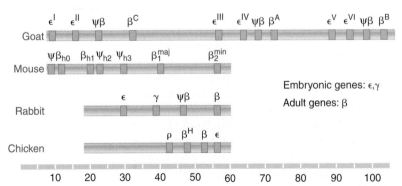

FIGURE 8.22 Clusters of β-globin genes and pseudogenes are found in vertebrates. Seven mouse genes include two early embryonic genes, one late embryonic gene, two adult genes, and two pseudogenes. Rabbit and chick each have four genes.

the lineage of other vertebrates before the ancestral globin gene was duplicated to give rise to the α and β variants. This appears to have occurred ~500 million years ago, during the evolution of the *Osteichthyes* (bony fish).

The next stage of globin evolution is represented by the state of the globin genes in the amphibian *Xenopus laevis*, which has two globin clusters. Each cluster, though, contains both α and β genes, of both larval and adult types. The cluster must therefore have evolved by duplication of a linked α–β pair, followed by divergence between the individual copies. Later the entire cluster was duplicated.

The amphibians separated from the reptilian/mammalian/avian line ~350 million years ago, so the separation of the α- and β-globin genes must have resulted from a transposition in the reptilian/mammalian/avian forerunner after this time. This probably occurred in the period of early tetrapod evolution. There are separate clusters for α and β globins in both birds and mammals; so the α and β genes must have been physically separated before the mammals and birds diverged from their common ancestor, an event estimated to have occurred ~270 million years ago.

Evolutionary changes have taken place within the separate α and β clusters in more recent times, as we saw from the description of the divergence of the individual genes in Section 8.4.

8.11 Pseudogenes Are Nonfunctional Gene Copies

Key concepts

- Processed pseudogenes result from reverse transcription and integration of mRNA transcripts.
- Nonprocessed pseudogenes result from incomplete duplication or second-copy mutation of functional genes.
- Some pseudogenes may gain functions different from those of their parent genes, such as regulation of gene expression, and take on different names.

As discussed earlier in this chapter, pseudogenes are copies of functional genes that have altered or missing regions such that they presumably do not produce functional protein or RNA products. For example, as compared to their functional counterparts, many pseudogenes have frameshift or nonsense mutations

that disable their protein-coding functionality. There are two types of pseudogenes characterized by their modes of origin.

Processed pseudogenes result from the reverse transcription of mature mRNA transcripts into cDNA copies, followed by their integration into the genome. This may occur at a time when active reverse transcriptase is present in the cell, such as during active retroviral infection or retroposon activity (see Chapter 17, *Transposable Elements and Retroviruses*). The transcript has undergone processing (see Chapter 21, *RNA Splicing and Processing*); as a result, a processed pseudogene usually lacks the regulatory regions necessary for normal expression. So, while it initially contains the coding sequence of a functional polypeptide, it is inactive as soon as it is formed. Such pseudogenes also lack introns and may contain the remnant of the mRNA's poly(A) tail (see *Section 21.15, The 3' Ends of mRNAs Are Generated by Cleavage and Polyadenylation*) as well as the flanking direct repeats characteristic of insertion of retroelements (see *Section 17.14, Viral DNA Integrates into the Chromosome*).

The second type, **nonprocessed pseudogenes**, arise from inactivating mutations in one copy of a multiple-copy or single-copy gene or from incomplete duplication of an active gene. Often, these are created by mechanisms that result in tandem duplications. An example of a β-globin pseudogene is shown in **FIGURE 8.23**. If a gene is duplicated in its entirety with intact regulatory regions, there may be two active copies for a time, but inactivating mutations in one copy would not necessarily be subject to negative selection. Thus, gene families are ripe for the origin of nonprocessed pseudogenes as

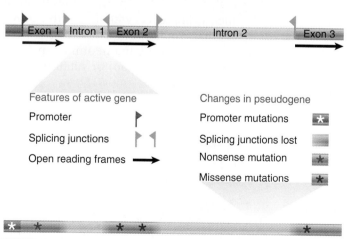

FIGURE 8.23 Many changes have occurred in a β-globin gene since it became a pseudogene.

evidenced by existence of several pseudogenes in the globin gene family (see *Section 8.10, Globin Clusters Arise by Duplication and Divergence*). Alternatively, an incomplete duplication of an active gene, resulting in a copy missing regulatory regions and/or coding sequence, would be "dead on arrival" as an instant pseudogene.

There are approximately 20,000 pseudogenes in the human genome. Ribosomal protein (RP) pseudogenes comprise a large family of pseudogenes, with approximately 2000 copies. These are processed pseudogenes; presumably the high copy number is a function of the high expression rate of the approximately 80 copies of active ribosomal protein genes. Their insertion into the genome is apparently mediated by the L1 retrotransposon (see *Section 17.18, Retroelements Fall into Three Classes*). RP genes are highly conserved among species; thus it is possible to identify RP pseudogene orthologs in species with a long history of separate evolution and for which whole-genome sequences are available. For example, as shown in **FIGURE 8.24**, more than two-thirds of human RP pseudogenes are found in the chimpanzee genome, whereas less than a dozen are shared between humans and rodents. This suggests that most RP pseudogenes are of more recent origin in both primates and rodents, and that most ancestral RP pseudogenes have been lost by deletion or mutational decay beyond recognition.

Interestingly, the rate of evolution of RP pseudogenes is slower than that of the neutral rate (as determined by the rate of substitution in ancient repeats across the genome), suggesting negative selection and implying a functional role for RP pseudogenes. Although by definition pseudogenes are nonfunctional, there are clear examples of former pseudogenes (originally identified as pseudogenes because of sequence differences with their active counterparts that would presumably render them nonfunctional) becoming neofunctionalized (taking on a new function) or subfunctional-

ized (taking on a subfunction or complementary function of the parent gene). Once functional, they would be subject to selection and thus evolve more slowly than expected under a neutral model.

How might a pseudogene gain a new function? One possibility is that translation, but not transcription, of the pseudogene has been disabled. The pseudogene encodes an RNA transcript that is no longer translatable but can affect expression or regulation of the still-functional "parent" gene. In the mouse, the processed pseudogene *Makorin1-p1* stabilizes transcripts of the functional *Makorin1* gene. Several endogenous siRNAs (see *Section 30.5, MicroRNAs Are Widespread Regulators in Eukaryotes*) are encoded by pseudogenes. A second possibility is that a processed pseudogene may be inserted in a location that provides them with new regulatory regions, such as transcription factor binding sites, which allow them to be expressed in a tissue-specific manner unlike that of the parent genes.

8.12 Genome Duplication Has Played a Role in Plant and Vertebrate Evolution

Key concepts

- Genome duplication occurs when polyploidization increases the chromosome number by a multiple of two.
- Genome duplication events can be obscured by the evolution and/or loss of duplicates as well as by chromosome rearrangements.
- Genome duplication has been detected in the evolutionary history of many flowering plants and of vertebrate animals.

As discussed in Section 8.9 (*Gene Duplication Contributes to Genome Evolution*), genomes can evolve via duplication and divergence of individual genes or chromosomal segments carrying blocks of genes. It appears, though, that some of the major metazoan lineages have had genome duplications in their evolutionary histories. Genome duplication is accomplished by **polyploidization**, as when a tetraploid (4N) variety arises from a diploid (2N) ancestral lineage.

There are two major mechanisms of polyploidization. **Autopolyploidy** occurs when a species endogenously gives rise to a polyploid variety; this usually involves fertilization by unreduced gametes. **Allopolyploidy** is a

Number RP pseudogenes shared between species	
Human-chimpanzee	1,282
Human-mouse	6
Human-rat	11
Mouse-rat	394

FIGURE 8.24 Most human RP pseudogenes are of recent origin; many are shared with the chimpanzee but absent from rodents. Adapted from S. Balasubramanian, et al., *Genome Biol.* 20 (2009): R2.

result of hybridization between two reproductively compatible species such that diploid sets of chromosomes from both parental species are retained in the hybrid offspring. As with autopolyploids, the process generally involves the accidental production of unreduced gametes. In both cases, new tetraploids are usually reproductively isolated from the diploid parental species because backcrossed hybrids are triploid and sterile, as some chromosomes are without homologs during meiosis.

Following the successful establishment of a polyploidy species, many mutations may be essentially neutral. As with gene duplications, nonsynonymous substitutions are "covered" by the redundant functional copy of the same gene. In the case of a genome duplication, the deletion of a gene or chromosomal segment, or the loss of a chromosome pair, may have little phenotypic effect. In addition to the loss of chromosomal segments, chromosomal rearrangements such as inversions and translocations will shuffle the locations and orders of blocks of genes. Over a long period of time, such events can obscure ancestral polyploidization. There may, however, still be evidence of polyploidization in the presence of redundant chromosomes or chromosomal segments within a genome.

One successful approach to detecting ancient polyploidization is to compare many pairs of paralogous (duplicated) genes within a species and establish an age distribution of gene duplication events. Many events of approximately the same age can be taken as evidence of polyploidization. As seen in **FIGURE 8.25**, genome duplication events will appear as peaks above the general pattern of random events of gene duplication and copy loss. This approach, along with an analysis of chromosomal locations of gene duplications, suggests that the evolutionary histories of the unicellular yeast *Saccharomyces cerevisiae* and many flowering plants include one or more genome duplication events. The genetic model land plant *Arabidopsis thaliana*, for example, has a history of two, possibly three, polyploidization events.

As polyploidization is more common in plants than in animals, it is not surprising that most detected examples of genome duplication are in plant species. Genome duplication appears to have played an important role in vertebrate evolution, though, specifically in ray-finned fishes. As evidence, the zebrafish genome contains seven *Hox* clusters as compared to four clusters in tetrapod genomes,

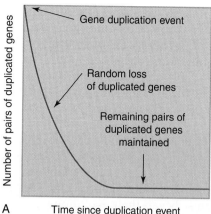

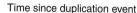

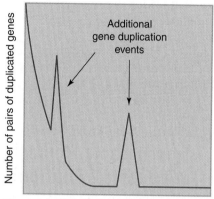

FIGURE 8.25 (A) A constant rate of gene duplication and loss shows an exponentially decreasing age distribution of duplicated gene pairs. (B) A genome duplication event shows a secondary peak in the age distribution as many genes are duplicated at the same time. Adapted from G. Blanc and K. H. Wolfe, *Plant Cell* 16 (2004): 1667–1678.

suggesting that there was a tetraploidization event followed by secondary loss of one cluster. The analysis of other fish genomes suggests that this event occurred before the diversification of this taxonomic group. The presence of four *Hox* clusters in tetrapods (and at least four in other vertebrates), together with the observation of other shared gene duplications as compared to invertebrate animal genomes, itself suggests that there may have been two major polyploidization events prior to the evolution of vertebrates. In reference to "two rounds of polyploidization," this has been termed the **2R hypothesis**.

This hypothesis leads to the prediction that many vertebrate genes, like the *Hox* clusters, will be found in 4× the copy number as compared to their orthologs in invertebrate species. The subsequent observation that less than 5% of vertebrate genes show this 4:1 ratio seems weak support for the hypothesis at best. It is to be expected, though, that after nearly

500 million years of evolution many of the additional copies of genes would have been deleted, evolved significantly to take on new functions, or become pseudogenes and decayed beyond recognition. Stronger support, however, comes from analyses that take into account the map position of duplications that date to the time of the common ancestor of vertebrates. The ancient gene duplications that do show the 4:1 pattern tend to be found in clusters, even after half a billion years of chromosomal rearrangements. The vertebrates evidently began their evolutionary history as octoploids. The 2R hypothesis is tempting as an explanation for the burst of morphological complexity that accompanied the evolution of vertebrates, though as yet there is little evidence of a direct correlation between the genomic and morphological changes in this taxonomic group.

8.13 What Is The Role of Transposable Elements in Genome Evolution?

Key concept

- Transposable elements tend to increase in copy number when introduced to a genome but are kept in check by negative selection and transposition regulation mechanisms.

Transposable elements (TEs) are mobile genetic elements that can be integrated into the genome at multiple sites and (for some elements) also excised from an integration site. (See Chapter 17, *Transposable Elements and Retroviruses*, for an extensive discussion of the types and mechanisms of TEs.) The insertion of a TE at a new site in the genome is called **transposition**. One type of TE, the retrotransposon, transposes via an RNA intermediate; a new copy of the element is created by transcription, followed by reverse transcription to DNA and subsequent integration at a new site.

Most TEs integrate at sequences that are random (at least with respect to their functions). As such, they are a major source of the problems associated with insertion mutations: frameshifts if inserted into coding regions and altered gene expression if inserted into regulatory regions. The number of copies of a given TE

in a species' genome therefore depends on several factors: the rate of integration of the TE; its rate of excision (if any); selection on individuals with phenotypes altered by TE integration; and regulation of transposition.

TEs effectively act as intracellular parasites and, like other parasites, may need to strike an evolutionary balance between their own proliferation and the detrimental effects on the "host" organism. Studies on *Drosophila* TEs confirm that the mutational integration of TEs generally have deleterious, sometimes lethal, phenotypic effects. This suggests that negative selection plays an important role in the regulation of transposition; individuals with high levels of transposition are less likely to survive and reproduce. One might, however, expect that both TEs and their hosts may evolve mechanisms to limit transposition, and in fact both are observed. In one example of TE self-regulation, the *Drosophila* P element encodes a transposition repressor protein that is active in somatic tissue (see *Section 17.10, P Elements Are Activated in the Germline*). In addition, there are two major cellular mechanisms for transposition regulation:

- In an RNA interference-like mechanism (see *Section 30.6, How Does RNA Interference Work?*) involving piRNAs (see *Section 30.5, MicroRNAs Are Widespread Regulators in Eukaryotes*), the RNA intermediates of retrotransposons can be selectively degraded.
- In mammals, plants, and fungi, a DNA methyltransferase methylates cytosines within TEs, resulting in transcriptional silencing (see *Section 29.8, DNA Methylation Is Responsible for Imprinting*).

In any case, it is rare for TE proliferation to continue unchecked but rather to be limited by negative selection and/or regulation of transposition. Following introduction of a TE to a genome, though, the copy number may increase to many thousands or millions before some equilibrium is achieved, particularly if TEs are integrated into introns or intergenic DNA where phenotypic effects will be absent or minimal. As a result, genomes may contain a high proportion of moderately or highly repetitive sequences (see *Section 5.5, Eukaryotic Genomes Contain Both Nonrepetitive and Repetitive DNA Sequences*).

8.14 There May Be Biases in Mutation, Gene Conversion, and Codon Usage

Key concepts

- Mutational bias may account for a high AT content in organismal genomes.
- Gene conversion bias, which tends to increase GC content, may act in partial opposition to the mutational bias.
- Codon bias may be a result of adaptive mechanisms that favor particular sequences, and of gene conversion bias.

As discussed earlier in this chapter (see *Section 8.2, DNA Sequences Evolve by Mutation and a Sorting Mechanism*), the probability of a particular mutation is a function of the probability that a particular replication error or DNA-damaging event will occur and the probability that the error will be detected and repaired before the next DNA replication. To the extent that there is bias in these two events, there is bias in the types of mutations that occur (e.g., a bias for transition mutations over transversion mutations despite the greater number of possible transversions).

Observations of the distributions of types of mutations over a taxonomically wide range of species (including prokaryotes and unicellular and multicellular eukaryotes), assessed by direct observation of mutational variants or by comparing sequence differences in pseudogenes, show a consistent pattern of a bias toward a high AT genomic content. The reasons for this are complex, and different mechanisms may be more or less important in different taxonomic groups, but there are two likely mechanisms. First, the common mutational source of spontaneous deamination of cytosine to uracil, or of 5-methylcytosine to thymine (see Figure 1.30), promotes the transition mutation of C-G to T-A. Uracil in DNA is more likely to be repaired than thymine (see *Section 1.15, Many Hotspots Result from Modified Bases*), so methylated cytosines (often found in CG doublets) are not only mutation hotspots but specifically biased toward producing a T-A pair. Second, oxidation of guanine to 8-oxoguanine can result in a C-G to A-T transversion because 8-oxoguanine pairs more stably with adenine than with cytosine.

Despite this *mutational bias*, in analyses in which the expected equilibrium base composition is predicted from the observed rates of specific types of mutations, the observed AT content is generally lower than expected. This suggests that some mechanism or mechanisms are working to counteract the mutational bias toward AT. One possibility is that this is adaptive; a highly biased base composition limits the mutational possibilities and consequently limits evolutionary potential. As discussed below, though, there may be a nonadaptive explanation.

A second possible source of bias in genomic base composition is *gene conversion*, which occurs when heteroduplex DNA containing mismatched base pairs, often resulting from the resolution of a Holliday junction during recombination or double-strand break repair, is repaired using the mutated strand as template (see *Section 7.4, Crossover Fixation Could Maintain Identical Repeats*, and *Section 15.3, Double-Strand Breaks Initiate Recombination*). Interestingly, observations of gene conversion events in animals and fungi show a clear bias toward G-C, though the mechanism is unclear. In support of this observation, chromosomal regions of high recombinational activity show more mutations to G-C, and regions with low recombinational activity tend to be AT-rich. The observed rates of gene conversion per site tend to be of the same order of magnitude or higher than mutation rates; thus gene conversion bias alone may account for the lower-than-expected AT content being driven higher by mutational bias. *Gene conversion bias* may also be partly responsible for another universally observed bias in genome composition, *codon bias* (see *Section 8.4, A Constant Rate of Sequence Divergence Is a Molecular Clock*).

Due to the degeneracy of the genetic code, most of the amino acids found in polypeptides are represented by more than one codon in a genetic message. The alternate codons are not generally found in equal frequencies in genes, though; particularly in highly expressed genes, one codon of the two, four, or six that call for a particular amino acid is often used at a much higher frequency than the others. As discussed in *Section 8.4*, one explanation for this bias is that a particular codon may be more efficient at recruiting an abundant tRNA species, such that the rate or accuracy of translation is greater with higher usage of that codon. There may be additional adaptive consequences of particular exon sequences: some may contribute to splicing efficiency, form secondary structures that affect mRNA stability, or be less subject to frameshift mutations than others (e.g., mononucleotide repeats that promote slippage). Biased gene conversion remains a (nonadaptive) possibility as well, though. Intriguingly, the synonymous site for most codons is the 3' end, and high-usage codons in eukaryotes almost always end in G or C, as is consistent with the hypothesis that biased gene conversion drives codon bias. Clearly, the causes of codon bias are complex and may involve both adaptive and nonadaptive mechanisms.

8.15 Summary

New variation in a genome is introduced by mutation. Although mutation is random with respect to function, the types of mutations that actually occur are biased by the probabilities of various changes to DNA and of types of DNA repair. This variation is sorted by random genetic drift (if variation is selectively neutral and/or populations are small) and negative or positive selection (if the variation affects phenotype).

The past influence of selection on a gene sequence can be detected from comparing homologous sequences among and within species. The K_a/K_s ratio compares nonsynonymous with synonymous changes; either an excess or a deficiency of nonsynonymous mutations may indicate positive or negative selection, respectively. Comparing the rates of evolution or the amount of variation for a locus among different species can also be used to assess past selection on DNA sequences.

Synonymous substitutions accumulate more rapidly than nonsynonymous substitutions (which affect the amino acid sequence). The rate of divergence at nonsynonymous sites can sometimes be used to establish a molecular clock, which can be calibrated in percent divergence per million years. The clock can then be used to calculate the time of divergence between any two members of the family.

Certain genes share only some of their exons with other genes, suggesting that they have been assembled by addition of exons representing functional "modular units" of the protein. Such modular exons may have been incorporated into a variety of different proteins. The hypothesis that genes have been assembled by accumulation of exons implies that introns were present in the genes of proto-eukaryotes. Some of the relationships between orthologous genes can be explained by loss of introns from the primordial genes, with different introns being lost in different lines of descent.

The proportions of repetitive and nonrepetitive DNA are characteristic for each genome, although larger genomes tend to have a smaller proportion of unique sequence DNA. The amount of nonrepetitive DNA is a better reflection of the complexity of the organism than the total genome size; the greatest amount of nonrepetitive DNA in genomes is ~2×10^9 bp.

About 5000 genes are common to prokaryotes and eukaryotes (though individual species may not carry all of these genes) and most are likely to be involved in basic functions. A further 8000 genes are found in multicellular organisms. Another 5000 genes are found in animals, and an additional 5000 (largely involved with the immune and nervous systems) are found in vertebrates.

An evolving set of genes may remain together in a cluster or may be dispersed to new locations by chromosomal rearrangement. The organization of existing clusters can sometimes be used to infer the series of events that has occurred. These events act with regard to sequence rather than function, and therefore include pseudogenes as well as active genes. Pseudogenes that arise by gene duplication and inactivation are nonprocessed, whereas those that arise via an RNA intermediate are processed. Pseudogenes may become secondarily functional due to gain-of-function mutations or via their untranslatable RNA products.

In some taxonomic groups, genome duplication (or polyploidization) can provide raw material for subsequent genome evolution. This process has shaped many flowering plant genomes and appears to have been a factor in early vertebrate evolution.

Copies of transposable elements can propagate within genomes and sometimes result in a large proportion of repetitive sequences in genomes. The number of copies of an element is kept in check by selection, self-regulation, and host regulatory mechanisms.

There are several sources of bias affecting the base composition of a genome. Mutational bias tends to result in higher AT content, whereas gene conversion bias acts to lower it somewhat. The universally observed codon biases of protein-coding sequences in genomes may be influenced by selection as well as gene conversion bias.

References

8.1   Introduction

Review

Lynch, M. (2007). *The Origins of Genome Architecture*. Sinauer Associates Inc., Sunderland, MA.

8.3   Selection Can Be Detected by Measuring Sequence Variation

Research

Clark, R. M., Linton, E., Messing, J., and Doebley, J. F. (2004). Pattern of diversity in the genomic region near the maize domestication gene *tb1*. *Proc. Natl. Acad. Sci. USA*. 101, 700–707.

Clark, R. M., Tavaré, S., and Doebley, J. (2005). Estimating a nucleotide substitution rate for maize from polymorphism at a major domestication locus. *Mol. Biol. Evol.* 22, 2304–2312.

Geetha, V., Di Francesco, V., Garnier, J., and Munson, P. J. (1999). Comparing protein sequence-based and predicted secondary structure-based methods for identification of remote homologs. *Protein Eng.* 12, 527–534.

McDonald, J. H. and Kreitman, M. (1991). Adaptive protein evolution at the *Adh* locus in *Drosophila*. *Nature* 351, 652–654.

Robinson, M., Gouy, M., Gautier, C., and Mouchiroud, D. (1998). Sensitivity of the relative-rate test to taxonomic sampling. *Mol. Biol. Evol.* 15, 1091–1098.

Wang, E. T., Kodama, G., Baldi, P., and Moyzis, R. K. (2006). Global landscape of recent inferred Darwinian selection for *Homo sapiens*. *Proc. Natl. Acad. Sci. USA*. 103, 135–140.

8.4   A Constant Rate of Sequence Divergence Is a Molecular Clock

Research

Dickerson, R. E. (1971). The structure of cytochrome *c* and the rates of molecular evolution. *J. Mol. Evol.* 1, 26–45.

8.5   The Rate of Neutral Substitution Can Be Measured from Divergence of Repeated Sequences

Research

Waterston, R. H. et al. (2002). Initial sequencing and comparative analysis of the mouse genome. *Nature* 420, 520–562.

8.6   How Did Interrupted Genes Evolve?

Review

Belshaw, R. and Bensasson, D. (2005). The rise and fall of introns. *Heredity* 96, 208–213.

Joyce, G. F. and Orgel, L. E. (2006). Progress toward understanding the origin of the RNA world. In: *The RNA World: The Nature of Modern RNA Suggests a Prebiotic RNA World*. 3rd ed. Cold Spring Harbor, New York: Cold Spring Harbor Laboratory Press.

Research

Barrette, I. H., McKenna, S., Taylor, D. R., and Forsdyke, D. R. (2001). Introns resolve the conflict between base order-dependent stem-loop potential and the encoding of RNA or protein: further evidence from overlapping genes. *Gene* 270, 181–189. (See http://post.queensu.ca/~forsdyke/introns1.htm.)

Coulombe-Huntington, J. and Majewski, J. (2007). Characterization of intron-loss events in mammals. *Genome Research* 17, 23–32.

Forsdyke, D. R. (1981). Are introns in-series error detecting sequences? *J. Theoret. Biol.* 93, 861–866.

Forsdyke, D.R. (1995). A stem-loop "kissing" model for the initiation of recombination and the origin of introns. *Mol. Biol. Evol.* 12, 949–958.

Hughes, A. L. and Friedman, R. (2008). Genome size reduction in the chicken has involved massive loss of ancestral protein-coding genes. *Mol. Biol. Evol.* 25, 2681–2688.

Raible, F. et al. (2005). Vertebrate-type intron-rich genes in the marine annelid *Platynereis dumerilii*. *Science* 310, 1325–1326.

Roy, S. W. and Gilbert, W. (2006). Complex early genes. *Proc. Natl. Acad. Sci. USA*. 102, 1986–1991.

8.7   Why Are Some Genomes So Large?

Review

Gall, J. G. (1981). Chromosome structure and the C-value paradox. *J. Cell Biol.* 91, 3s–14s.

Gregory, T. R. (2001). Coincidence, coevolution, or causation? DNA content, cell size, and the C-value enigma. *Biol. Rev. Camb. Philos. Soc.* 76, 65–101.

8.8   Morphological Complexity Evolves by Adding New Gene Functions

Reference

The Chimpanzee Sequencing and Analysis Consortium (2005). Initial sequence of the chimpanzee genome and comparison with the human genome. *Nature* 437, 69–87.

Research

Giaever et al. (2002). Functional profiling of the *S. cerevisiae* genome. *Nature* 418, 387–391.

Goebl, M. G. and Petes, T. D. (1986). Most of the yeast genomic sequences are not essential for cell growth and division. *Cell* 46, 983–992.

Hutchison, C. A. et al. (1999). Global transposon mutagenesis and a minimal mycoplasma genome. *Science* 286, 2165–2169.

Kamath, R. S., Fraser, A. G., Dong, Y., Poulin, G., Durbin, R., Gotta, M., Kanapin, A., Le Bot, N., Moreno, S., Sohrmann, M., Welchman, D. P., Zipperlen, P., and Ahringer, J. (2003). Systematic functional analysis of the *C. elegans* genome using RNAi. *Nature* 421, 231–237.

Tong, A. H. et al. (2004). Global mapping of the yeast genetic interaction network. *Science* 303, 808–813.

8.9   Gene Duplication Contributes to Genome Evolution

Research
Bailey, J. A., Gu, Z., Clark, R. A., Reinert, K., Samonte, R. V., Schwartz, S., Adams, M. D., Myers, E. W., Li, P. W., and Eichler, E. E. (2002). Recent segmental duplications in the human genome. *Science* 297, 1003–1007.

8.10   Globin Clusters Arise by Duplication and Divergence

Review
Hardison, R. (1998). Hemoglobins from bacteria to man: evolution of different patterns of gene expression. *J. Exp. Biol.* 201, 1099–1117.

8.11   Pseudogenes Are Nonfunctional Gene Copies

Research
Balasubramanian, S., Zheng, D., Liu, Y.-J., Fang, G., Frankish, A., Carriero, N., Robilotto, R., Cayting, P., and Gerstein, M. (2009). Comparative analysis of processed ribosomal protein pseudogenes in four mammalian genomes. *Genome Biol.* 10, R2.

Esnault, C., Maestre, J., and Heidmann, T. (2000). Human LINE retrotransposons generate processed pseudogenes. *Nat. Genet.* 24, 363–367.

Kaneko, S., Aki, I., Tsuda, K., Mekada, K., Moriwaki, K., Takahata, N., and Satta, Y. (2006). Origin and evolution of processed pseudogenes that stabilize functional Makorin1 mRNAs in mice, primates and other mammals. *Genetics* 172, 2421–2429.

Review
Balakirev, E. S. and Ayala, F. J. (2003). Pseudogenes: are they "junk" or functional DNA? *Ann. Rev. Genet.* 37, 123–151.

8.12   Genome Duplication Has Played a Role in Plant and Vertebrate Evolution

Research
Abbasi, A. A. (2008). Are we degenerate tetraploids? More genomes, new facts. *Biol. Direct* 3, 50.

Blanc, G. and Wolfe, K. H. (2004). Widespread paleopolyploidy in model plant species inferred from age distributions of duplicate genes. *Plant Cell* 16, 1667–1678.

Dehal, P. and Boore, J. L. (2005). Two rounds of whole genome duplication in the ancestral vertebrate. *PLoS Biol.* 3, e314.

Review
Furlong, R. F. and Holland, P. W. (2002). Were vertebrates octoploid? *Phil. Trans. R. Soc. Lond. B* 357, 531–544.

Kasahara, M. (2007). The 2R hypothesis: an update. *Curr. Opin. Immunol.* 19, 547–552.

8.14   There May Be Biases in Mutation, Gene Conversion, and Codon Usage

Research
Rocha, E. P. C. (2004). Codon usage bias from tRNA's point of view: Redundancy, specialization, and efficient decoding for translation optimization. *Genome Res.* 14, 2279–2286.

Photo courtesy of Venigalla Rao, The Catholic University of America, Washington, DC, and Steven McQuinn, Independent Science Artist. Capsid volume data by Siyang Sun, Andrei Fokine, and Michael Rossmann, Purdue University. Additional information at *Cell* 135 (2008): 1251–1262 and *Ann. Rev. Gent.* 42 (2008): 642–681.

Chromosomes

Edited by Hank W. Bass

CHAPTER OUTLINE

9.1 Introduction

A general principle is evident in the organization of all cellular genetic material. It exists as a compact mass that is confined to a limited volume, and its various activities, such as replication and transcription, must be accomplished within this space. The organization of this material must accommodate local transitions between inactive and active states.

The condensed state of nucleic acid results from its binding to basic proteins. The positive charges of these proteins neutralize the negative charges of the nucleic acid. The structure of the nucleoprotein complex is determined by the interactions of the proteins with the DNA (or RNA).

A common problem is presented by the packaging of DNA into phages, viruses, bacterial cells, and eukaryotic nuclei. The length of the DNA as an extended molecule would vastly exceed the dimensions of the compartment that contains it. The DNA (or in the case of some viruses, the RNA) must be compressed exceedingly tightly to fit into the space available. *Thus in contrast with the customary picture of DNA as an extended double helix, structural deformation of DNA to bend or fold it into a more compact form is the rule rather than exception.*

The magnitude of the discrepancy between the length of the nucleic acid and the size of its compartment is evident from the examples summarized in **FIGURE 9.1**. For bacteriophages

Compartment	Shape	Dimensions	Type of Nucleic Acid	Length
TMV	filament	0.008 x 0.3 μm	One single-stranded RNA	2 μm = 6.4 kb
Phage fd	filament	0.006 x 0.85 μm	One single-stranded DNA	2 μm = 6.0 kb
Adenovirus	icosahedron	0.07 μm diameter	One double-stranded DNA	11 μm = 35.0 kb
Phage T4	icosahedron	0.065 x 0.10 μm	One double-stranded DNA	55 μm = 170.0 kb
E. coli	cylinder	1.7 x 0.65 μm	One double-stranded DNA	1.3 mm = 4.2×10^3 kb
Mitochondrion (human)	oblate spheroid	3.0 x 0.5 μm	~10 identical double-stranded DNAs	50 μm = 16.0 kb
Nucleus (human)	spheroid	6 μm diameter	46 chromosomes of double-stranded DNA	1.8 m = 6×10^6 kb

FIGURE 9.1 The length of nucleic acid is much greater than the dimensions of the surrounding compartment.

and for eukaryotic viruses, the nucleic acid genome, whether single-stranded or double-stranded DNA or RNA, effectively fills the container (which can be rodlike or spherical).

For bacteria or for eukaryotic cell compartments, the discrepancy is hard to calculate exactly, because the DNA is contained in a compact area that occupies only part of the compartment. The genetic material is seen in the form of the **nucleoid** in bacteria, and as the mass of **chromatin** in eukaryotic nuclei at interphase (between divisions), or as maximally condensed chromosomes during mitosis.

The density of DNA in these compartments is high. In a bacterium it is ~10 mg/ml, in a eukaryotic nucleus it is ~100 mg/ml, and in the phage T4 head it is >500mg/ml. Such a concentration in solution would be equivalent to a gel of great viscosity. We do not entirely understand the physiological implications of such high concentrations of DNA, such as the effect this has upon the ability of proteins to find their binding sites on DNA.

The packaging of chromatin is flexible; it changes during the eukaryotic cell cycle. At the time of division (mitosis or meiosis), the genetic material becomes even more tightly packaged, and individual **chromosomes** become recognizable.

The overall compression of the DNA can be described by the **packing ratio**, which is the length of the DNA divided by the length of the unit that contains it. For example, the smallest human chromosome contains ~4.6×10^7 bp of DNA (~10 times the genome size of the bacterium *E. coli*). This is equivalent to 14,000 μm (= 1.4 cm) of extended DNA. At the most condensed moment of mitosis, the chromosome is ~2 μm long. Thus the packing ratio of DNA in the chromosome can be as great as 7000.

Packing ratios cannot be established with such certainty for the more amorphous overall structures of the bacterial nucleoid or eukaryotic chromatin. The usual reckoning, however, is that mitotic chromosomes are likely to be five to ten times more tightly packaged than interphase chromatin, which indicates a typical packing ratio of 1000 to 2000.

A major unanswered question concerns the *specificity* of higher-order packaging. Is the DNA folded into a *particular* pattern, or is it different in each individual copy of the genome? How does the pattern of packaging change when a segment of DNA is replicated or transcribed?

9.2 Viral Genomes Are Packaged into Their Coats

Key concepts

- The length of DNA that can be incorporated into a virus is limited by the structure of the headshell.
- Nucleic acid within the headshell is extremely condensed.
- Filamentous RNA viruses condense the RNA genome as they assemble the headshell around it.
- Spherical DNA viruses insert the DNA into a preassembled protein shell.

From the perspective of packaging the *individual* sequence, there is an important difference between a cellular genome and a virus. The cellular genome is essentially indefinite in size; the number and location of individual sequences can be changed by duplication, deletion, and rearrangement. Thus it requires a *generalized* method for packaging its DNA, one that is insensitive to the total content or distribution of sequences. By contrast, two restrictions define the needs of a virus. The amount of nucleic acid to be packaged is *predetermined* by the size of the genome, and it must all fit within a coat assembled from a protein or proteins coded by the viral genes.

A virus particle is deceptively simple in its superficial appearance. The nucleic acid genome is contained within a **capsid**, which is a symmetrical or quasisymmetrical structure assembled from one or only a few proteins. Attached to the capsid (or incorporated into it) are other structures; these structures are assembled from distinct proteins and are necessary for infection of the host cell.

The virus particle is tightly constructed. The internal volume of the capsid is rarely much greater than the volume of the nucleic acid it must hold. The difference is usually less than twofold, and often the internal volume is barely larger than the nucleic acid.

In its most extreme form, the restriction that the capsid must be assembled from proteins encoded by the virus means that the entire shell is constructed from a single type of subunit. The rules for assembly of identical subunits into closed structures restrict the capsid to one of two types. For the first type, the protein subunits stack sequentially in a helical array to form a *filamentous* or rodlike shape. For the second type, they form a pseudospherical shell—a type of structure that conforms to a polyhedron with **icosahedral symmetry**. Some viral capsids are assembled from more than a single type of protein subunit, but although this extends the exact types of structures that can be formed, viral capsids still all conform to the general classes of quasicrystalline filaments or icosahedrons.

There are two types of solution to the problem of how to construct a capsid that contains nucleic acid:

- The protein shell can be assembled around the nucleic acid, thereby condensing the DNA or RNA by protein–nucleic acid interactions during the process of assembly.
- The capsid can be constructed from its component(s) in the form of an empty shell, into which the nucleic acid must be inserted, being condensed as it enters.

The capsid is assembled around the genome for single-stranded RNA viruses. The principle of assembly is that *the position of the RNA within the capsid is determined directly by its binding to the proteins of the shell*. The best-characterized example is TMV (tobacco mosaic virus). Assembly starts at a duplex hairpin that lies within the RNA sequence. From this **nucleation center**,

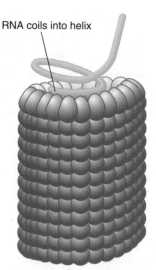

RNA coils into helix

FIGURE 9.2 A helical path for TMV RNA is created by the stacking of protein subunits in the virion.

assembly proceeds bidirectionally along the RNA until it reaches the ends. The unit of the capsid is a two-layer disk, with each layer containing 17 identical protein subunits. The disk is a circular structure, which forms a helix as it interacts with the RNA. At the nucleation center, the RNA hairpin inserts into the central hole in the disk, and the disk changes conformation into a helical structure that surrounds the RNA. Additional disks are added, with each new disk pulling a new stretch of RNA into its central hole. The RNA becomes coiled in a helical array on the inside of the protein shell, as illustrated in **FIGURE 9.2**.

The spherical capsids of DNA viruses are assembled in a different way, as best characterized for the phages lambda and T4. In each case, an empty headshell is assembled from a small set of proteins. *The duplex genome then is inserted into the head*, accompanied by a structural change in the capsid.

FIGURE 9.3 summarizes the assembly of lambda. It starts with a small headshell that contains a protein "core." This is converted to an empty headshell of more distinct shape. At this point the DNA packaging begins, the headshell expands in size though remaining the same shape, and finally the full head is sealed by the addition of the tail.

A double-stranded DNA that spans short distances is a fairly rigid rod, yet it must be compressed into a compact structure to fit within the capsid. We should like to know whether packaging involves a smooth coiling of the DNA into the head or whether it requires abrupt bends.

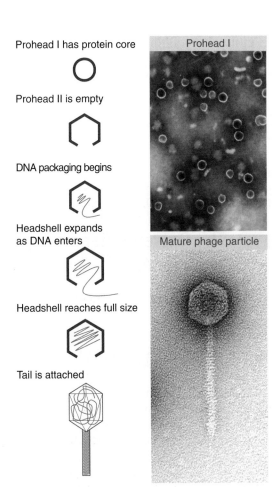

Prohead I has protein core

Prohead II is empty

DNA packaging begins

Headshell expands as DNA enters

Headshell reaches full size

Tail is attached

FIGURE 9.3 Maturation of phage lambda passes through several stages. The empty head changes shape and expands when it becomes filled with DNA. The electron micrographs show the particles at the start and the end of the maturation pathway. Top photo reproduced from D. Cue and M. Feiss, *Proc. Natl. Acad. Sci. USA* 90 (1993): 9240–9294. Copyright © 2004 National Academy of Sciences, U.S.A. Photo courtesy of Michael G. Feiss, University of Iowa. Bottom photo courtesy of Robert Duda, University of Pittsburgh.

Inserting DNA into a phage head involves two types of reaction: *translocation* and *condensation*. Both are energetically unfavorable.

Translocation is an active process in which the DNA is driven into the head by an ATP-dependent mechanism. A common mechanism is used for many viruses that replicate by a rolling circle mechanism to generate long tails that contain multimers of the viral genome. The best-characterized example is phage lambda. The genome is packaged into the empty capsid by the **terminase** enzyme. **FIGURE 9.4** summarizes the process.

The terminase was first recognized for its role in generating the ends of the linear phage DNA by cleaving at *cos* sites. (The name *cos* reflects the fact that it generates cohesive ends that have complementary single-stranded tails.)

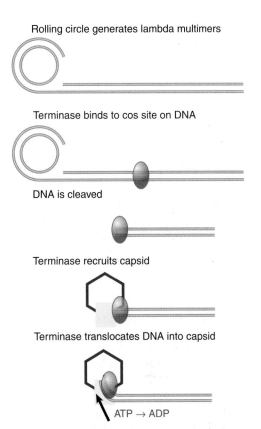

Rolling circle generates lambda multimers

Terminase binds to cos site on DNA

DNA is cleaved

Terminase recruits capsid

Terminase translocates DNA into capsid

ATP → ADP

FIGURE 9.4 Terminase protein binds to specific sites on a multimer of virus genomes generated by rolling circle replication. It cuts the DNA and binds to an empty virus capsid, and then uses energy from hydrolysis of ATP to insert the DNA into the capsid.

The phage genome codes two subunits that make up the terminase. One subunit binds to a *cos* site; at this point it is joined by the other subunit, which cuts the DNA. The terminase assembles into a hetero-oligomer in a complex that also includes IHF (integration host factor, a dimer that is coded by the bacterial genome). It then binds to an empty capsid and uses ATP hydrolysis to power translocation along the DNA. The translocation drives the DNA into the empty capsid.

Another method of packaging uses a structural component of the phage. In the *Bacillus subtilis* phage φ29, the motor that inserts the DNA into the phage head is the structure that connects the head to the tail. It functions as a rotary motor, where the motor action effects the linear translocation of the DNA into the phage head. The same motor is used to eject the DNA from the phage head when it infects a bacterium.

Little is known about the mechanism of condensation into an empty capsid, except that the capsid contains "internal proteins" as well as DNA. One possibility is that they provide some sort of "scaffolding" onto which the DNA condenses. (This would be a counterpart to the use of the proteins of the shell in the plant RNA viruses.)

How specific is the packaging? It cannot depend on particular sequences, because deletions, insertions, and substitutions all fail to interfere with the assembly process. The relationship between DNA and the headshell has been investigated directly by determining which regions of the DNA can be chemically crosslinked to the proteins of the capsid. The surprising answer is that all regions of the DNA are more or less equally susceptible. This probably means that when DNA is inserted into the head it follows a general rule for condensing, but the pattern is not determined by particular sequences.

These varying mechanisms of virus assembly all accomplish the same end: packaging a single DNA or RNA molecule into the capsid. Some viruses, though, have genomes that consist of multiple nucleic acid molecules. Reovirus contains ten double-stranded RNA segments, all of which must be packaged into the capsid. Specific sorting sequences in the segments may be required to ensure that the assembly process selects one copy of each different molecule in order to collect a complete set of genetic information. In the simpler case of phage φ6, which packages three different segments of double-stranded RNA into one capsid, the RNA segments must bind in a specific order: as each is incorporated into the capsid, it triggers a change in the conformation of the capsid that creates binding sites for the next segment.

Some plant viruses are multipartite: their genomes consist of segments, each of which is packaged into a *different* capsid. An example is alfalfa mosaic virus (AMV), which has four different single-stranded RNAs, each of which is packaged independently into a coat comprising the same protein subunit. A successful infection depends on the entry of one of each type into the cell.

The four components of AMV exist as particles of different sizes. This means that the same capsid protein can package each RNA into its own characteristic particle. This is a departure from the packaging of a unique length of nucleic acid into a capsid of fixed shape.

The assembly pathway of viruses whose capsids have only one authentic form may be diverted by mutations that cause the formation of aberrant *monster* particles in which the head is longer than usual. These mutations show that a capsid protein(s) has an intrinsic ability to assemble into a particular type of structure, but the exact size and shape may vary.

Some of the mutations occur in genes that code for assembly factors, which are needed for head formation, but are not themselves part of the headshell. Such ancillary proteins limit the options of the capsid protein, reducing variation in the assembly pathway. Comparable proteins are employed in the assembly of cellular chromatin (see Chapter 10, *Chromatin*).

9.3 The Bacterial Genome Is a Nucleoid

Key concepts
- The bacterial nucleoid is ~80% DNA by mass and can be unfolded by agents that act on RNA or protein.
- The proteins that are responsible for condensing the DNA have not been identified.

Although bacteria do not display structures with the distinct morphological features of eukaryotic chromosomes, their genomes nonetheless are organized into definite bodies. The genetic material can be seen as a fairly compact clump (or series of clumps) that occupies about a third of the volume of the cell. **FIGURE 9.5** displays a thin section through a bacterium in which this nucleoid is evident.

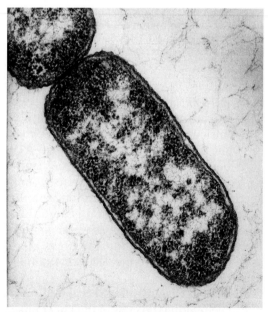

FIGURE 9.5 A thin section shows the bacterial nucleoid as a compact mass in the center of the cell. Photo courtesy of the Molecular and Cell Biology Instructional Laboratory Program, University of California, Berkeley.

When *E. coli* cells are lysed, fibers are released in the form of loops attached to the broken envelope of the cell. As can be seen from **FIGURE 9.6**, the DNA of these loops is not found in the extended form of a free duplex, but instead is compacted by association with proteins.

Several DNA-binding proteins with a superficial resemblance to eukaryotic chromosomal proteins have been isolated in *E. coli*. What criteria should we apply for deciding whether a DNA-binding protein plays a structural role in the nucleoid? It should be present in sufficient quantities to bind throughout the genome, and mutations in its gene should cause some disruption of structure or of functions associated with genome survival (for example, segregation to daughter cells). None of the known candidate proteins fully satisfies these genetic conditions.

Protein HU is a dimer that condenses DNA, possibly wrapping it into a beadlike structure. It is related to IHF, which has a structural role in building a protein complex in specialized recombination reactions. Null mutations in either of the genes coding for the subunits of HU (*hupA* and *-B*) have little effect, but loss of both functions causes a cold-sensitive phenotype and some loss of superhelicity in DNA. These results raise the possibility that HU plays some general role in nucleoid condensation.

Protein H1 (also known as H-NS) binds DNA, interacting preferentially with sequences that are bent. Mutations in its gene have turned up in a variety of guises (*osmZ, bglY, pilG*), each

of which is identified as an apparent regulator of a different system. These results probably reflect the effect that H1 has on the local topology of DNA, with effects upon gene expression that depend upon the particular promoter.

We might expect that the absence of a protein required for nucleoid structure would have serious effects upon viability. Why, then, are the effects of deletions in the genes for proteins HU and H1 relatively restricted? One explanation is that these proteins are *redundant*, and that any one can substitute for the others so that deletions of *all* of them would be necessary to interfere seriously with nucleoid structure. Another possibility is that we have yet to identify the proteins responsible for the major features of nucleoid integrity. Yet another possibility is that we have underestimated their contribution to fitness by using laboratory tests that evaluate some, but not all, of the conditions in which these proteins contribute to reproduction or survival.

The nucleoid can be isolated directly in the form of a very rapidly sedimenting complex, which consists of ~80% DNA by mass. (The analogous complexes in eukaryotes contain ~50% DNA by mass; see *Section 9.4, The Bacterial Genome Is Supercoiled*.) The bacterial nucleoid can be unfolded by treatment with reagents that destroy RNA or protein. The possible role of proteins in stabilizing its structure is evident. The role of RNA has been quite refractory to analysis.

9.4 The Bacterial Genome Is Supercoiled

Key concepts

- The nucleoid has ~400 independent negatively supercoiled domains.
- The average density of supercoiling is ~1 turn/100bp.

The DNA of the bacterial nucleoid isolated *in vitro* behaves as a closed duplex structure, as judged by its response to ethidium bromide. This small molecule intercalates between base pairs to generate *positive* superhelical turns in "closed" circular DNA molecules; that is, molecules in which both strands have covalent integrity. (In "open" circular molecules, which contain a nick in one strand, or with linear molecules, the DNA can rotate freely in response to the intercalation, thus relieving the tension.)

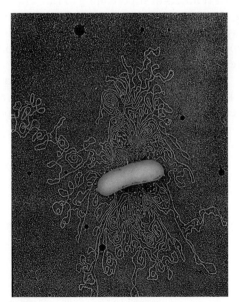

FIGURE 9.6 The nucleoid spills out of a lysed *E. coli* cell in the form of loops of a fiber. © G. Murti/Photo Researchers, Inc.

In a natural closed DNA that is *negatively* supercoiled, the intercalation of ethidium bromide first removes the negative supercoils and then introduces positive supercoils. The amount of ethidium bromide needed to achieve zero supercoiling is a measure of the original density of negative supercoils.

Some nicks occur in the compact nucleoid during its isolation; they can also be generated by limited treatment with DNAase. This does not, however, abolish the ability of ethidium bromide to introduce positive supercoils. This capacity of the genome to retain its response to ethidium bromide in the face of nicking means that it must have many independent chromosomal **domains**, and that *the supercoiling in each domain is not affected by events in the other domains.*

This autonomy suggests that the structure of the bacterial chromosome has the general organization depicted diagrammatically in **FIGURE 9.7**. Each domain consists of a loop of DNA, the ends of which are secured in some (unknown) way that does not allow rotational events to propagate from one domain to another.

Early data suggested that each domain consists of ~40 kb of DNA, but more recent analysis suggests that the domains may be smaller, at ~10 kb each. This would correspond to ~400 domains in the *E. coli* genome. The ends of the domains appear to be randomly distributed instead of located at predetermined sites on the chromosome.

The existence of separate domains could permit different degrees of supercoiling to be maintained in different regions of the genome. This could be relevant in considering the different susceptibilities of particular bacterial promoters to supercoiling (see *Section 19.17, Supercoiling Is an Important Feature of Transcription*).

As shown in **FIGURE 9.8**, supercoiling in the genome can in principle take either of two forms:

- If a supercoiled DNA is free its path is *unconstrained*, and negative supercoils generate a state of torsional tension that is transmitted freely along the DNA within a domain. It can be relieved by unwinding the double helix, as described in *Section 1.5, Supercoiling Affects the Structure of DNA*. The DNA is in a dynamic equilibrium between the states of tension and unwinding.

- Supercoiling can be *constrained* if proteins are bound to the DNA to hold it in a particular three-dimensional configuration. In this case, the supercoils are represented by the path the DNA follows in its fixed association with the proteins. The energy of interaction between the proteins and the supercoiled DNA stabilizes the nucleic acid, so that no tension is transmitted along the molecule.

Are the supercoils in *E. coli* DNA constrained *in vivo* or is the double helix subject to the torsional tension characteristic of free DNA? Measurements of supercoiling *in vitro* encounter the difficulty that constraining proteins may have been lost during isolation. Various approaches suggest that DNA is under torsional stress *in vivo*.

One approach is to measure the effect of nicking the DNA. Unconstrained supercoils are released by nicking, whereas constrained super-

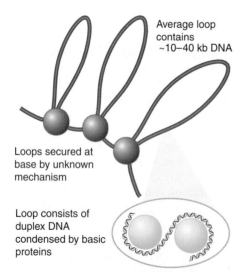

FIGURE 9.7 The bacterial genome consists of a large number of loops of duplex DNA (in the form of a fiber), each of which is secured at the base to form an independent structural domain.

Average loop contains ~10–40 kb DNA

Loops secured at base by unknown mechanism

Loop consists of duplex DNA condensed by basic proteins

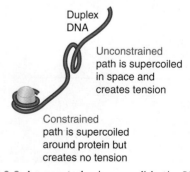

FIGURE 9.8 An unrestrained supercoil in the DNA path creates tension, but no tension is transmitted along DNA when a supercoil is restrained by protein binding.

Duplex DNA

Unconstrained path is supercoiled in space and creates tension

Constrained path is supercoiled around protein but creates no tension

coils are unaffected. Nicking releases ~50% of the overall supercoiling. This suggests that about half of the supercoiling is transmitted as tension along DNA, with the other half being absorbed by protein binding.

Another approach uses the crosslinking reagent psoralen, which binds more readily to DNA when it is under torsional tension. The reaction of psoralen with *E. coli* DNA *in vivo* corresponds to an average density of one negative superhelical turn/200 bp ($\sigma = -0.05$).

We can also examine the ability of cells to form alternative DNA structures; for example, to generate cruciforms at palindromic sequences. From the change in linking number that is required to drive such reactions, it is possible to calculate the original supercoiling density. This approach suggests an average density of $\sigma = -0.025$, or one negative superhelical turn/100 base pairs.

Thus supercoils *do* create torsional tension *in vivo*. There may be variation about an average level, and the precise range of densities is difficult to measure. It is, however, clear that the level is sufficient to exert significant effects on DNA structure—for example, in assisting melting in particular regions such as origins or promoters.

Many of the important features of the structure of the compact nucleoid remain to be established. What is the specificity with which domains are constructed? Do the same sequences always lie at the same relative locations, or can the contents of individual domains shift? How is the integrity of the domain maintained? Biochemical analysis by itself is unable to answer these questions fully, but if it is possible to devise suitable selective techniques, the properties of structural mutants should lead to a molecular analysis of nucleoid construction.

9.5 Eukaryotic DNA Has Loops and Domains Attached to a Scaffold

Key concepts

- DNA of interphase chromatin is negatively supercoiled into independent domains of ~85 kb.
- Metaphase chromosomes have a protein scaffold to which the loops of supercoiled DNA are attached.

Interphase chromatin is a tangled mass occupying a large part of the nuclear volume, in contrast with the highly organized and repro-

ducible ultrastructure of mitotic chromosomes. What controls the distribution of interphase chromatin within the nucleus?

Some indirect evidence on its nature is provided by the isolation of the genome as a single, compact body. Using the same technique that was developed for isolating the bacterial nucleoid (see *Section 9.4, The Bacterial Genome Is Supercoiled*), nuclei can be lysed on top of a sucrose gradient. This releases the genome in a form that can be collected by centrifugation. As isolated from *Drosophila melanogaster*, it can be visualized as a compactly folded fiber (10 nm in diameter) consisting of DNA bound to proteins.

Supercoiling measured by the response to ethidium bromide corresponds to about one negative supercoil/200 bp. These supercoils can be removed by nicking with DNase, although the DNA remains in the form of the 10 nm fiber. This suggests that the supercoiling is caused by the arrangement of the fiber in space, and that it represents the existing torsion.

Full relaxation of the supercoils requires one nick/85 kb, thus identifying the average length of "closed" DNA. This region could comprise a loop or domain similar in nature to those identified in the bacterial genome. Loops can be seen directly when the majority of proteins are extracted from mitotic chromosomes. The resulting complex consists of the DNA associated with ~8% of the original protein content. As seen in FIGURE 9.9, the protein-depleted chromosomes take the form of a protein-depleted metaphase **scaffold** that still resembles the general form of a mitotic chromosome, surrounded by a halo of DNA.

The metaphase scaffold consists of a dense network of fibers. Threads of DNA emanate from the scaffold, apparently as loops of average length 10 to 30 μm (30 to 90 kb). The DNA can be digested without affecting the integrity of the scaffold, which consists of a set of specific proteins. This suggests a form of organization in which loops of DNA of ~60 kb are anchored in a central proteinaceous scaffold. In interphase nuclei, this underlying proteinaceous structure changes its organization to occupy the entire nucleus; during interphase this structure is referred to as the *matrix* rather than the scaffold.

The appearance of the scaffold resembles a mitotic pair of sister chromatids. The sister scaffolds usually are tightly connected (but sometimes are separate), and are joined only by a few fibers. Could this be the structure responsible for maintaining the shape of the mitotic chromosomes? Could it be generated by

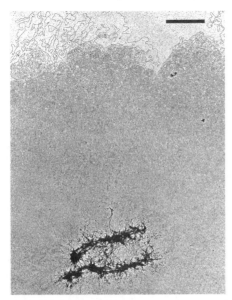

FIGURE 9.9 Histone-depleted chromosomes consist of a protein scaffold to which loops of DNA are anchored. Reprinted from *Cell*, vol. 12, J. R. Paulson and U. K. Laemmli, The structure of histone-depleted metaphase chromosomes, pp. 817–828. Copyright 1977, with permission from Elsevier [http://www.sciencedirect.com/science/journal/00928674]. Photo courtesy of Ulrich K. Laemmli, University of Geneva, Switzerland.

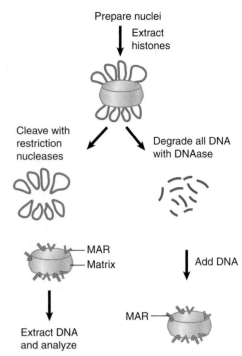

FIGURE 9.10 Matrix-associated regions may be identified by characterizing the DNA retained by the matrix isolated *in vivo* or by identifying the fragments that can bind to the matrix from which all DNA has been removed.

bringing together the protein components that usually secure the bases of loops in interphase chromatin?

9.6 Specific Sequences Attach DNA to an Interphase Matrix

Key concepts

- DNA is attached to the nuclear matrix at specific sequences called MARs or SARs.
- The MARs are A-T-rich but do not have any specific consensus sequence.

Is DNA attached to the scaffold via specific sequences? DNA sites attached to proteinaceous structures in interphase nuclei are called **MARs (matrix attachment regions)**; they are sometimes also called *SARs* (scaffold attachment regions), as the same sequences appear to attach to the protein substructure in both metaphase and interphase cells. (The nature of the structure in interphase cells to which they are connected is not clear.) Chromatin often appears to be attached to a matrix, and there have been many suggestions that this attachment is necessary for transcription or replication. When nuclei are depleted of proteins, the DNA extrudes as loops from a residual protein-aceous structure, as occurs in scaffold preparations. Attempts to relate the proteins found in this preparation to structural elements of intact cells have not been successful, though.

Are particular DNA regions associated with this matrix? *In vivo* and *in vitro* approaches are summarized in **FIGURE 9.10**. Both start by isolating the matrix as a crude nuclear preparation containing chromatin and nuclear proteins. Different treatments can then be used to characterize DNA in the matrix or to identify DNA able to attach to it.

To analyze the existing MARs, the chromosomal loops can be decondensed by extracting the proteins. Removal of the DNA loops by treatment with restriction nucleases leaves only the (presumptive) *in vivo* MAR sequences attached to the matrix.

The complementary approach is to remove *all* the DNA from the matrix by treatment with DNase, at which point isolated fragments of DNA can be tested for their ability to bind to the matrix *in vitro*.

The same sequences should be associated with the matrix *in vivo* or *in vitro*. Once a potential MAR has been identified, the size of the minimal region needed for association *in vitro* can be determined by deletions. This enables us to identify proteins that bind to the MAR sequences.

A surprising feature is the lack of conservation of sequence in MAR fragments. They are usually ~70% A-T-rich, but otherwise lack any consensus sequences. Other interesting sequences, however, often are in the DNA stretch containing the MAR. *cis*-acting sites that regulate transcription are common, and a recognition site for topoisomerase II is usually present in the MAR. It is therefore possible that a MAR serves more than one function by providing a site for attachment to the matrix and containing other sites at which topological changes in DNA are effected.

What is the relationship between the chromosome scaffold of dividing cells and the matrix of interphase cells? Are the same DNA sequences attached to both structures? In several cases, the same DNA fragments that are found with the nuclear matrix *in vivo* can be retrieved from the metaphase scaffold. Fragments that contain MAR sequences can bind to a metaphase scaffold, so it therefore seems likely that DNA contains a single type of attachment site. In interphase cells the attachment site is connected to the nuclear matrix, whereas in mitotic cells it is connected to the chromosome scaffold.

The nuclear matrix and chromosome scaffold consist of different proteins, although there are some common components. Topoisomerase II is a prominent component of the chromosome scaffold, and is a constituent of the nuclear matrix. This suggests that the control of topology is important in both cases.

FIGURE 9.11 The sister chromatids of a mitotic pair each consist of a fiber (~30 nm in diameter) compactly folded into the chromosome. © Biophoto Associates/Photo Researchers, Inc.

molecules of DNA.) Thus in accounting for interphase chromatin and mitotic chromosome structure, we have to explain the packaging of a single, exceedingly long molecule of DNA into a form in which it can be transcribed and replicated, and can become cyclically more and less compressed.

Individual eukaryotic chromosomes come into the limelight for a brief period, during the act of cell division. Only then can each be seen as a compact unit. **FIGURE 9.11** is an electron micrograph of a replicated chromosome isolated and photographed at metaphase. The sister chromatids are evident at this stage, and will give rise to the daughter chromosomes upon their separation during the anaphase stage of mitosis. Each chromatid consists of a fiber with a diameter of ~30 nm and a nubbly appearance. The DNA is five to ten times more condensed in chromosomes than in interphase chromatin.

During most of the life cycle of the eukaryotic cell, however, its genetic material occupies an area of the nucleus in which individual chromosomes cannot be distinguished by conventional microscopy. The global structure of the interphase chromatin does not change visibly between divisions. No disruption is evident during the period of replication, when the amount of chromatin doubles. Chromatin is fibrillar, although the overall configuration of the fiber in space is hard to discern in detail. The fiber itself, however, is similar or identical to that of the mitotic chromosomes.

9.7 Chromatin Is Divided into Euchromatin and Heterochromatin

Key concepts

- Individual chromosomes can be seen only during mitosis.
- During interphase, the general mass of chromatin is in the form of euchromatin, which is slightly less tightly packed than mitotic chromosomes.
- Regions of heterochromatin remain densely packed throughout interphase.

Each chromosome contains a single, very long duplex of DNA, folded into a fiber that runs continuously throughout the chromosome. This explains why chromosome replication is semiconservative like the individual DNA molecule. (This would not necessarily be the case if a chromosome carried many independent

As can be seen in the nuclear section of **FIGURE 9.12**, chromatin can be divided into two types of material:

- In most regions, the fibers are much less densely packed than in the mitotic chromosome. This material is called **euchromatin**. It has a relatively dispersed appearance in the nucleus and occupies most of the nuclear region in Figure 9.12.

- Some regions of chromatin are very densely packed with fibers, displaying a condition comparable to that of the chromosome at mitosis. This material is called **heterochromatin**. It is typically found at centromeres, but occurs at other locations as well. It passes through the cell cycle with relatively little change in its degree of condensation. It forms a series of discrete clumps in Figure 9.12, but often the various heterochromatic regions, especially those associated with centromeres, aggregate into a densely staining **chromocenter**. The common form of heterochromatin that always remains heterochromatic is called *constitutive heterochromatin*. In contrast, there is another sort of heterochromatin, called *facultative heterochromatin*, in which regions of euchromatin are converted to a heterochromatic state.

The same fibers run continuously between euchromatin and heterochromatin, which implies that these states represent different degrees of condensation of the genetic material. In the same way, euchromatic regions exist in different states of condensation during interphase and during mitosis. Thus the genetic material is organized in a manner that permits alternative states to be maintained side by side in chromatin, and allows cyclical changes to occur in the packaging of euchromatin between interphase and division. We discuss the molecular basis for these states in Chapter 10, *Chromatin*, and Chapter 29, *Epigenetic Effects Are Inherited*.

The structural condition of the genetic material is correlated with its activity. The common features of constitutive heterochromatin are:

- It is permanently condensed.
- It often consists of multiple repeats of a few sequences of DNA that are not transcribed or are transcribed at very low levels. (Genes that reside in heterochromatic regions are generally less transcriptionally active than their euchromatic counterparts, but there are exceptions to this general rule.)
- The density of genes in this region is very much reduced compared with euchromatin, and genes that are translocated into or near it are often inactivated. The one dramatic exception to this is the ribosomal DNA in the nucleolus, which has the general compacted appearance and behavior of heterochromatin (such as late replication), yet is engaged in very active transcription.
- It replicates late in S phase and has a reduced frequency of genetic recombination relative to euchromatic gene-rich areas of the genome.

We have some molecular markers for changes in the properties of the DNA and protein components (see *Section 29.3, Heterochromatin Depends on Interactions with Histones*). They include reduced acetylation of histone proteins, increased methylation at particular sites on histones, and hypermethylation of cytidine bases in DNA. These molecular changes result in the condensation of the chromatin, which is responsible for its inactivity.

Although active genes are contained within euchromatin, only a small minority of the sequences in euchromatin are transcribed at any time. Thus location in euchromatin is *necessary* for gene expression, but is not *sufficient* for it.

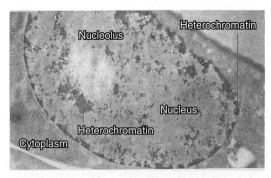

FIGURE 9.12 A thin section through a nucleus stained with Feulgen shows heterochromatin as compact regions clustered near the nucleolus and nuclear membrane. Photo courtesy of Edmund Puvion, Centre National de la Recherche Scientifique.

Chromosomes Have Banding Patterns

Key concepts

- Certain staining techniques cause the chromosomes to have the appearance of a series of striations, which are called G-bands.
- The bands are lower in G-C content than the interbands.
- Genes are concentrated in the G-C-rich interbands.

As a result of the diffuse state of chromatin, we cannot directly determine the specificity of its organization. We can, however, ask whether the structure of the (mitotic) chromosome is ordered. Do particular sequences always lie at particular sites, or is the folding of the fiber into the overall structure a more random event?

At the level of the chromosome, each member of the complement has a different and reproducible ultrastructure. When mitotic chromosomes are subjected to proteolytic enzyme (trypsin) treatment followed by staining with the chemical dye Giemsa, they generate distinct chromosome-specific patterns called **G-bands**. FIGURE 9.13 presents an example of the human set.

Until the development of this technique, human chromosomes could be distinguished only by their overall size and the relative location of the centromere. G-banding allows each chromosome to be identified by its characteristic banding pattern. This pattern allows transloca-

tions from one chromosome to another to be identified by comparison with the original diploid set. FIGURE 9.14 shows a diagram of the **bands** of the human X chromosome. The bands are large structures, each ~10^7 bp of DNA, and each of which could include many hundreds of genes.

The banding technique is of enormous practical use, but the mechanism of banding remains a mystery. All that is certain is that the dye stains untreated chromosomes more or less uniformly. Thus the generation of bands depends on a variety of treatments that change the response of the chromosome (presumably by extracting the component that binds the stain from the nonbanded regions). Similar bands can be generated by an assortment of other treatments.

The only known feature that distinguishes bands from **interbands** is that the bands have a lower G-C content than the interbands. This is a peculiar result. If there are ~10 bands on a large chromosome with a total content of ~100 Mb, this means that the chromosome is divided into regions of ~5 Mb in length that alternate between low G-C (band) and high G-C (interband) content. There is a tendency for genes (as identified by hybridization with mRNAs) to

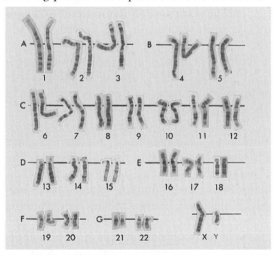

FIGURE 9.13 G-banding generates a characteristic lateral series of bands in each member of the chromosome set. Photo courtesy of Lisa Shaffer, Washington State University–Spokane.

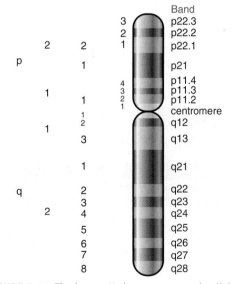

FIGURE 9.14 The human X chromosome can be divided into distinct regions by its banding pattern. The short arm is *p* and the long arm is *q*; each arm is divided into larger regions that are further subdivided. This map shows a low resolution structure; at higher resolution, some bands are further subdivided into smaller bands and interbands, e.g., *p21* is divided into *p21.1*, *p21.2*, and *p21.3*.

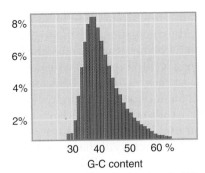

FIGURE 9.15 There are large fluctuations in G-C content over short distances. Each bar shows the percent of 20 kb fragments with the given G-C content.

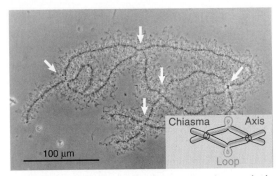

FIGURE 9.16 A lampbrush chromosome is a meiotic bivalent in which the two pairs of sister chromatids are held together at chiasmata (indicated by arrows). Photo courtesy of Joseph G. Gall, Carnegie Institution.

be located in the interband regions. All of this argues for some long-range sequence-dependent organization.

The human genome sequence confirms the basic observation. **FIGURE 9.15** shows that there are distinct fluctuations in G-C content when the genome is divided into small tranches (DNA segments or lengths). The average of 41% G-C is common to mammalian genomes. There are regions as low as 30% or as high as 65%. The average length of regions with >43% G-C is 200 to 250 kb. This makes it clear that the band/interband structure does not correspond directly with the more numerous homogeneous segments that alternate in G-C content, although the bands do tend to contain a higher content of low G-C segments. Genes are concentrated in regions of higher G-C content. We have yet to understand how the G-C content affects chromosome structure.

9.9 Lampbrush Chromosomes Are Extended

Key concept

- Sites of gene expression on lampbrush chromosomes show loops that are extended from the chromosomal axis.

It would be extremely useful to visualize gene expression in its natural state in order to see what structural changes are associated with transcription. The compression of DNA in chromatin, coupled with the difficulty of identifying particular genes within it, makes it impossible to visualize the transcription of individual active genes.

Gene expression can be visualized directly in certain unusual situations in which the chromosomes are found in a highly extended form that allows individual loci (or groups of loci) to be distinguished. Lateral differentiation of structure is evident in many chromosomes

when they first appear for meiosis. At this stage, the chromosomes resemble a series of beads on a string. The beads are densely staining granules, properly known as **chromomeres**. Chromomeres are larger and distinct from individual nucleosomes, which are also sometimes referred to as beads on a string (see Chapter 10, *Chromatin*). In general, though, there is little gene expression at meiosis, and it is not practical to use this material to identify the activities of individual genes. An exceptional situation that allows the material to be examined is presented by **lampbrush chromosomes**, which have been best characterized in certain amphibians.

Lampbrush chromosomes are formed during an unusually extended meiosis, which can last up to several months. During this period, the chromosomes are structured as a stretched-out form in which they can be visualized in the light microscope. At a later point during meiosis, the chromosomes revert to their usual compact size. Thus the extended state provides unique visual accessibility to the structure of the chromosome.

The lampbrush chromosomes are meiotic bivalents, each consisting of paired homologous chromosomes that have been replicated. The sister chromatids remain connected along their lengths and each homolog appears, therefore, as a single fiber. **FIGURE 9.16** shows an example in which the homologs have desynapsed, and are held together only by chiasmata that indicate points of chromosome crossover. Each sister chromatid pair forms a series of ellipsoidal chromomeres, ~1 to 2 μm in diameter, which are connected by a very fine thread. This thread contains the two sister duplexes of DNA and runs continuously along the chromosome, through the chromomeres.

The lengths of the individual lampbrush chromosomes in the newt *Notophthalmus viride-*

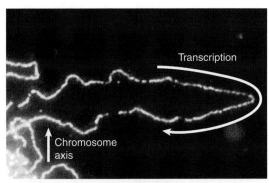

FIGURE 9.17 A lampbrush chromosome loop is surrounded by a matrix of ribonucleoprotein. Reproduced from J. G. Gall, et al., *Molecular Biology of the Cell* 10 (1999): 4385–4402. Copyright 1999 by American Society of Cell Biology. Reproduced with permission of the American Society of Cell Biology in the format of Textbook via Copyright Clearance Center. Photo courtesy of Joseph G. Gall, Carnegie Institution.

scens range from 400 to 800 μm, compared with the range of 15 to 20 μm seen later in meiosis. Thus the lampbrush chromosomes are ~30 times less compacted along their axes than their somatic counterparts. The total length of the entire lampbrush chromosome set is 5 to 6 mm and is organized into ~5000 chromomeres.

The lampbrush chromosomes take their name from the lateral loops that extrude from the chromomeres at certain positions. The arrangement of fibers around the chromosome axis resemble the cleaning fibers of a lampbrush. The loops extend in pairs, one from each sister chromatid. The loops are continuous with the axial thread, which suggests that they represent chromosomal material extruded from its more compact organization in the chromomere.

The loops are surrounded by a matrix of ribonucleoproteins that contain nascent RNA chains. Often, a transcription unit can be defined by the increase in the length of the RNP moving around the loop. An example is shown in FIGURE 9.17.

Thus the loop is an extruded segment of DNA that is being actively transcribed. In some cases, loops corresponding to particular genes have been identified. For these cases, the structure of the transcribed gene—and the nature of the product—can be scrutinized *in situ*.

9.10 Polytene Chromosomes Form Bands

Key concept

- Polytene chromosomes of dipterans have a series of bands that can be used as a cytological map.

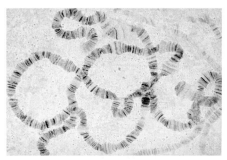

FIGURE 9.18 The polytene chromosomes of *D. melanogaster* form an alternating series of bands and interbands. Photo courtesy of José Bonner, Indiana University.

The interphase nuclei of some tissues of the larvae of dipteran flies contain chromosomes that are greatly enlarged relative to their usual condition. They possess both increased diameter and greater length. FIGURE 9.18 shows an example of a chromosome set from the salivary gland of *D. melanogaster*. The members of this set are called **polytene chromosomes**.

Each member of the polytene set consists of a visible series of bands (more properly, but rarely, described as chromomeres). The bands range in size from the largest, with a breadth of ~0.5 μm, to the smallest, at ~0.05 μm. (The smallest can be distinguished only under an electron microscope.) The bands contain most of the mass of DNA and stain intensely with appropriate reagents. The regions between them stain more lightly and are called interbands. There are ~5000 bands in the *D. melanogaster* set.

The centromeres of all four chromosomes of *D. melanogaster* aggregate to form a chromocenter that consists largely of heterochromatin. (In the male it includes the entire Y chromosome.) The remaining ~75% of the genome is organized into alternating bands and interbands in the polytene chromosomes. The length of the chromosome set is ~2000 μm. The DNA in extended form would stretch for ~40,000 μm, so the packing ratio is ~20. This demonstrates vividly the extension of the genetic material relative to the usual states of interphase chromatin or mitotic chromosomes.

What is the structure of these giant chromosomes? Each is produced by the successive replications of a synapsed diploid pair of chromosomes. The replicas do not separate, but instead remain attached to each other in their extended state. At the start of the process, each synapsed pair has a DNA content of 2C (where C represents the DNA content of the individual chromosome). This amount then doubles up to nine times, at its maximum giving a content of 1024C. The number of doublings is different in the various tissues of the *D. melanogaster* larva.

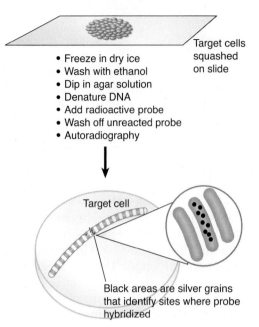

- Freeze in dry ice
- Wash with ethanol
- Dip in agar solution
- Denature DNA
- Add radioactive probe
- Wash off unreacted probe
- Autoradiography

Target cells squashed on slide

Target cell

Black areas are silver grains that identify sites where probe hybridized

FIGURE 9.19 Individual bands containing particular genes can be identified by *in situ* hybridization.

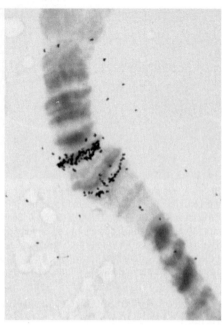

FIGURE 9.20 A magnified view of bands 87A and 87C shows their hybridization *in situ* with labeled RNA extracted from heat-shocked cells. Photo courtesy of José Bonner, Indiana University.

Each chromosome can be visualized as a large number of parallel fibers running longitudinally that are tightly condensed in the bands and less so in the interbands. It is likely that each fiber represents a single (C) haploid chromosome. This gives rise to the name polytene ("many threads"). The degree of polyteny is the number of haploid chromosomes contained in the giant chromosome.

The banding pattern is characteristic for each strain of *Drosophila*. The constant number and linear arrangement of the bands was first noted in the 1930s, when it was realized that they form a *cytological map* of the chromosomes. Rearrangements—such as deletions, inversions, or duplications—result in alterations of the order of bands.

The linear array of bands can be equated with the linear array of genes. Thus genetic rearrangements, as seen in a linkage map, can be correlated with structural rearrangements of the cytological map. Ultimately, a particular mutation can be located in a particular band. The total number of genes in *D. melanogaster* exceeds the number of bands, so there are probably multiple genes in most or all bands.

The positions of particular genes on the cytological map can be determined directly by the technique of ***in situ* hybridization**. A modern version of this protocol using fluorescent probes was described in *Section 3.5, Nucleic Acid Detection* (see Figure 3.13). Although fluores-

cent probes are currently preferred, when the method was originally developed a radioactive probe representing the gene of interest was used; the protocol is summarized in **FIGURE 9.19**. A probe representing a gene (most often a labeled cDNA clone derived from the mRNA) is hybridized with the denatured DNA of the polytene chromosomes *in situ*. Autoradiography identifies the position or positions of the corresponding genes by the superimposition of grains at a particular band or bands. An example is shown in **FIGURE 9.20**. Using *in situ* hybridization, it is possible to determine directly the band within which a particular sequence lies.

9.11 Polytene Chromosomes Expand at Sites of Gene Expression

Key concept

- Bands that are sites of gene expression on polytene chromosomes expand to give "puffs."

One of the intriguing features of the polytene chromosomes is that transcriptionally active sites can be visualized. Some of the bands pass transiently through an expanded state in which they appear like a **puff** on the chromosome, when chromosomal material is extruded from

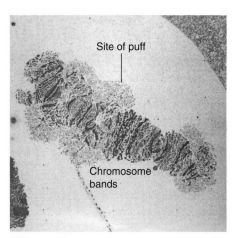

Site of puff

Chromosome bands

FIGURE 9.21 Chromosome IV of the insect *C. tentans* has three Balbiani rings in the salivary gland. Reprinted from *Cell*, vol. 4, B. Daneholt, Transcription in polytene chromosomes, pp. 1–9. Copyright 1975, with permission from Elsevier [http://www.sciencedirect.com/science/journal/00928674]. Photo courtesy of Bertil Daneholt, Karolinska Institutet.

the axis. Examples of some very large puffs (called Balbiani rings) are shown in **FIGURE 9.21**.

What is the nature of the puff? It consists of a region in which the chromosome fibers unwind from their usual state of packing in the band. The fibers remain continuous with those in the chromosome axis. Puffs usually emanate from single bands, although when they are very large, as typified by the Balbiani rings, the swelling may be so extensive as to obscure the underlying array of bands.

The pattern of puffs is related to gene expression. During larval development, puffs appear and regress in temporal and tissue-specific patterns. A characteristic pattern of puffs is found in each tissue at any given time. Many puffs are induced by the hormone ecdysone that controls *Drosophila* development. Some puffs are induced directly by the hormone; others are induced indirectly by the products of earlier puffs.

The puffs are sites where RNA is being synthesized. The accepted view of puffing has been that expansion of the band is a consequence of the need to relax its structure in order to synthesize RNA. Puffing has therefore been viewed as a consequence of transcription. A puff can be generated by a single active gene. The sites of puffing differ from ordinary bands in accumulating additional proteins, which include RNA polymerase II and other proteins associated with transcription.

The features displayed by lampbrush and polytene chromosomes suggest a general con-

clusion. In order to be transcribed, the genetic material is dispersed from its usual, more tightly packed state. The question to keep in mind is whether this dispersion at the gross level of the chromosome mimics the events that occur at the molecular level within the mass of ordinary interphase euchromatin.

Do the bands of a polytene chromosome have a functional significance? That is, does each band correspond to some type of genetic unit? You might think that the answer would be immediately evident from the sequence of the fly genome, because by mapping interbands to the sequence it should be possible to determine whether a band has any fixed type of identity. Thus far, though, no pattern has been found that identifies a functional significance for the bands.

9.12 The Eukaryotic Chromosome Is a Segregation Device

Key concept

- A eukaryotic chromosome is held on the mitotic spindle by the attachment of microtubules to the kinetochore that forms in its centromeric region.

During mitosis, the sister chromatids move to opposite poles of the cell. Their movement depends on the attachment of the chromosome to microtubules, which are connected at their other end to the poles. The microtubules comprise a cellular filamentous system, which is reorganized at mitosis so that they connect the chromosomes to the poles of the cell. The sites in the two regions where microtubule ends are organized—in the vicinity of the centrioles at the poles and at the chromosomes—are called **microtubule organizing centers**, or **MTOCs**.

FIGURE 9.22 illustrates the separation of sister chromatids as mitosis proceeds from metaphase to telophase. The region of the chromosome that is responsible for its segregation at mitosis and meiosis is called the **centromere**. The centromeric region on each sister chromatid is moved along microtubules to the opposite pole. Opposing this motive force, "glue" proteins called cohesins hold the sister chromatids together. Initially the sister chromatids separate at their centromeres, and then they are released completely from one another during anaphase when the cohesins are degraded. The centromere is moved toward the pole during mitosis, and the attached chromosome appears to be

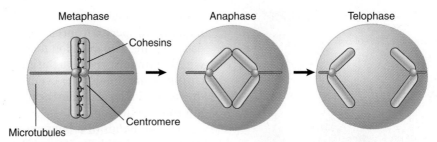

FIGURE 9.22 Chromosomes are pulled to the poles via microtubules that attach at the centromeres. The sister chromatids are held together until anaphase by glue proteins (cohesins). The centromere is shown here in the middle of the chromosome (metacentric), but can be located anywhere along its length, including close to the end (acrocentric) and at the end (telocentric).

"dragged along" behind it. The chromosome therefore provides a device for attaching a large number of genes to the apparatus for division. The centromere essentially acts as the luggage handle for the entire chromosome and its location typically appears as a constricted region connecting all four chromosome arms, as can be seen in the photo of Figure 9.11, which shows the sister chromatids at the metaphase stage of mitosis.

The centromere is essential for segregation, as shown by the behavior of chromosomes that have been broken. A single break generates one piece that retains the centromere, and another, an **acentric fragment**, that lacks it. The acentric fragment does not become attached to the mitotic **spindle**, and as a result it fails to be included in either of the daughter nuclei. When chromosome movement relies on discrete centromeres, there can be *only* one centromere per chromosome. When translocations generate chromosomes with more than one centromere, aberrant structures form at mitosis. This is because the two centromeres on the *same* sister chromatid can be pulled toward different poles, thus breaking the chromosome. In some species, though, the centromeres are holocentric, being diffuse and spread along the entire length of the chromosome. Species with holocentric chromosomes still make spindle fiber attachments for mitotic chromosome separation, but do not require one and only one regional or point centromere per chromosome. Most of the molecular analysis of centromeres has been done on canonical point (budding yeast) or regional (fly, mammalian, rice) centromeres.

The regions flanking the centromere often are rich in satellite DNA sequences and display

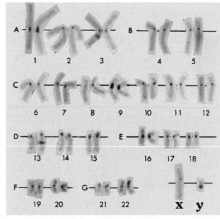

FIGURE 9.23 C-banding generates intense staining at the centromeres of all chromosomes. Photo courtesy of Lisa Shaffer, Washington State University–Spokane.

a considerable amount of heterochromatin. The entire chromosome is condensed, though, so centromeric heterochromatin is not immediately evident in mitotic chromosomes. It can, however, be visualized by a technique that generates "C-bands." In the example of **FIGURE 9.23**, all the centromeres show as darkly staining regions. Although it is common, heterochromatin cannot be identified around *every* known centromere, which suggests that it is unlikely to be essential for the division mechanism.

The centromeric chromatin comprises DNA sequences, specialized centromeric variants, and a group of specific proteins that are responsible for establishing the structure that attaches the chromosome to the microtubules. This structure is called the **kinetochore**. It is a darkly staining fibrous object of diameter or length ~400 nm. The kinetochore provides a microtubule attachment point on the chromosome.

9.13 Regional Centromeres Contain a Centromeric Histone H3 Variant and Repetitive DNA

Key concepts

- Centromeres are characterized by a centromere-specific histone H3 variant, and often have heterochromatin that is rich in satellite DNA sequences.
- Centromeres in higher eukaryotic chromosomes contain large amounts of repetitive DNA and unique histone variants.
- The function of repetitive DNA is not known.

The region of the chromosome at which the centromere forms was originally thought to be defined by DNA sequences, yet recent studies in plants, animals, and fungi have shown that centromeres are more likely to be specified epigenetically by chromatin structure. Centromere-specific histone H3 (CENP-A/CenH3; see *Section 10.5, Histone Variants Produce Alternative Nucleosomes*) appears to be a primary determinant in establishing functional centromeres and kinetochore assembly sites. This finding explains the old puzzle of why specific DNA sequences could not be identified as "the centromeric DNA" and why there is so much variation in centromere-associated DNA sequences among closely related species. **FIGURE 9.24** shows a model for the epigenetic specification of centromeres, with the kinetochore connecting to the clusters of CenH3 nucleosomes, which protrude from the bulk chromatin. This model explains how centromeres can reposition themselves without concomitant transposition of satellite DNA sequences. New questions of centromere function include what determines or

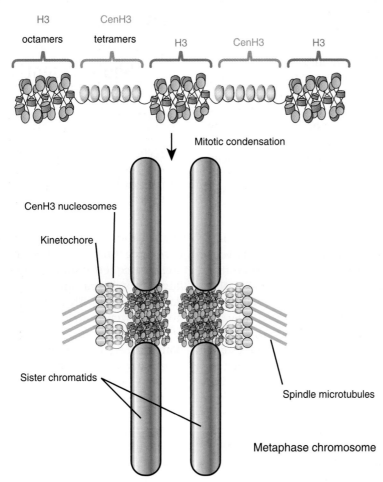

FIGURE 9.24 A model of the overall structure of a regional centromere. The CenH3-containing nucleosomes (orange) occur in clusters that protrude from the chromosome and bind to kinetochore proteins that in turn connect to spindle microtubules. Adapted from Y. Datal, et al., *Proc. Natl. Acad. Sci. USA* 104 (2007): 15974–15981.

restricts the sites of CenH3 installation, and how do chromosomes maintain one such region per chromosome?

Centromeres are highly specialized chromatin structures that occupy the same site for many generations, despite the fact that they can be repositioned without DNA transposition. In eukaryotic chromosomes, the centromere-specific histone H3 variant CenH3 (CENP-A in humans) replaces the normal H3 histone at sites where centromeres reside and kinetochores attach chromosomes to spindle fibers. This specialized centromeric chromatin is the foundation for binding of other centromere-associated proteins. This view represents a paradigm shift in how we understand centromere formation, identity, and function. CenH3 is a nucleosomal protein and not a DNA sequence *per se*; thus the centromere is now regarded as being primarily epigenetic in its specification. The role of satellite DNA sequences, which are also characteristic of centromeres, remains difficult to ascertain, despite their prevalence and conservation.

The length of DNA required for centromeric function is often quite long. The short, discrete elements of *Saccharomyces cerevisiae* may be an exception to the general rule. *S. cerevisiae* is the only case so far in which centromeric DNA can be identified by its ability to confer stability on plasmids. A related approach, though, has been used with the yeast *Schizosaccharomyces pombe*. *S. pombe* has only three chromosomes, and the region containing each centromere has been identified by deleting most of the sequences of each chromosome to create a stable minichromosome. This approach locates the centromeres within regions of 40 to 100 kb that consist largely or entirely of repetitious DNA. Attempts to localize centromeric functions in *Drosophila* chromosomes suggest that they are dispersed in a large region of 200 to 600 kb. The large size of this type of centromere may reflect multiple specialized functions, including kinetochore assembly and sister chromatid pairing.

The size of the centromere in *Arabidopsis* is comparable. Each of the five chromosomes has a centromeric region in which recombination is very largely suppressed. This region occupies >500 kb. The primary motif comprising the heterochromatin of primate centromeres is the α satellite DNA, which consists of tandem arrays of a 171 bp repeating unit (see *Section 7.5, Satellite DNAs Often Lie in Heterochromatin*). There is significant variation between individual repeats, although those at any centromere tend to be better related to one another than to members of the family in other locations.

Current models for regional centromere organization and function invoke alternating chromatin domains, with clusters of CenH3 nucleosomes interspersed among clusters of nucleosomes with H3 and H2A.Z. The CenH3 nucleosomes form the chromatin foundation for recruitment and assembly of the other proteins that eventually comprise a functional kinetochore. The formation of neocentromeres that contain CenH3, but not α-satellite DNA, provide important evidence for the idea of centromeres being epigenetically determined. Key questions remain as to the role of repetitive DNA and alternating chromatin domains in forming the large bipartite kinetochore structure on replicated sister centromeres.

9.14 Point Centromeres in *S. cerevisiae* Contain Short, Essential DNA Sequences

Key concepts

- *CEN* elements are identified in *S. cerevisiae* by the ability to allow a plasmid to segregate accurately at mitosis.
- *CEN* elements consist of the short, conserved sequences *CDE-I* and *CDE-III* that flank the A-T-rich region *CDE-II*.

If a centromeric sequence of DNA is responsible for segregation, any molecule of DNA possessing this sequence should move properly at cell division, whereas any DNA lacking it should fail to segregate. This prediction has been used to isolate centromeric DNA in the yeast *S. cerevisiae*. Yeast chromosomes do not display visible kinetochores comparable to those of multicellular eukaryotes, but otherwise divide at mitosis and segregate at meiosis by the same mechanisms.

Genetic engineering has produced plasmids of yeast that are replicated like chromosomal sequences (see *Section 11.8, Replication Origins Can Be Isolated in Yeast*). They are unstable at mitosis and meiosis, though, and disappear from a majority of the cells because they segregate erratically. Fragments of chromosomal DNA containing centromeres have been isolated by their ability to confer mitotic stability on these plasmids.

TCACATGATGATATTTGATTTTATTATATTTTTAAAAAAAGTAAAAAATAAAAAGTAGTTTATTTTTAAAAAATAAAATTTAAAATATTTCACAAAATGATTTCCGAA
AGTGTACTACTATAAACTAAAATAATATAAAAATTTTTTCATTTTTTATTTTTCATCAAATAAAAATTTTTTATTTTAAATTTTATAAAGTGTTTTACTAAAGGCTT

CDE-I   CDE-II 80–90 bp, >90% A + T   CDE-III

FIGURE 9.25 Three conserved regions can be identified by the sequence homologies between yeast *CEN* elements.

A centromeric DNA region (*CEN*) fragment is identified as the minimal sequence that can confer stability upon such a plasmid. Another way to characterize the function of such sequences is to modify them *in vitro* and then reintroduce them into the yeast cell, where they replace the corresponding centromere on the chromosome. This allows the sequences required for *CEN* function to be defined directly in the context of the chromosome.

A *CEN* fragment derived from one chromosome can replace the centromere of another chromosome with no apparent consequence. This result suggests that centromeres are interchangeable. They are used simply to attach the chromosome to the spindle, and play no role in distinguishing one chromosome from another.

The sequences required for centromeric function fall within a stretch of ~120 bp. The centromeric region is packaged into a nuclease-resistant structure and binds a single microtubule. We may therefore look to the *S. cerevisiae* centromeric region to identify proteins that bind centromeric DNA and proteins that connect the chromosome to the spindle.

As summarized in **FIGURE 9.25**, three types of sequence element can be distinguished in the *CEN* region:

- Cell cycle-dependent element (*CDE*)-*I* is a sequence of 9 bp that is conserved with minor variations at the left boundary of all centromeres.

- *CDE-II* is a >90% A-T-rich sequence of 80 to 90 bp found in all centromeres; its function could depend on its length rather than exact sequence. Its constitution is reminiscent of some short, tandemly repeated (satellite) DNAs (see *Section 7.6, Arthropod Satellites Have Very Short Identical Repeats*). Its base composition may cause some characteristic distortions of the DNA double helical structure.

- *CDE-III* is an 11 bp sequence highly conserved at the right boundary of all centromeres. Sequences on either side of the element are less well conserved, and may also be needed for centromeric function. (*CDE-III* could be longer than 11 bp if it turns out that the flanking sequences are essential.)

Mutations in *CDE-I* or *CDE-II* reduce, but do not inactivate, centromere function, but point mutations in the central CCG of *CDE-III* completely inactivate the centromere.

9.15 The *S. cerevisiae* Centromere Binds a Protein Complex

Key concepts

- A specialized protein complex that is an alternative to the usual chromatin structure is formed at *CDE-II*.
- The CBF3 protein complex that binds to *CDE-III* is essential for centromeric function.
- The proteins that bind *CEN* serve as an assembly platform for the kinetochore and provide the connection to microtubules.

Can we identify proteins that are necessary for the function of *CEN* sequences? There are several genes in which mutations affect chromosome segregation, and whose proteins are localized at centromeres. The contributions of these proteins to the centromeric structure are summarized in **FIGURE 9.26**.

The *CEN* region recruits three DNA-binding factors: Cbf1, CBF3 (an essential four-protein complex), and Mif2 (CENP-C in multicellular eukaryotes). In addition, a specialized chromatin structure is built by binding the *CDE-II*

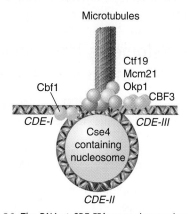

FIGURE 9.26 The DNA at *CDE-II* is wound around an alternative nucleosome including Cse4, *CDE-III* is bound by the CBF3 complex, and *CDE-I* is bound by a Cbf1 homodimer. These proteins are connected by the group of Ctf19, Mcm21, and Okp1 proteins, and numerous other factors serve to link this complex to a microtubule.

region to a protein called Cse4, a variant of the histone proteins that comprise the basic subunits of chromatin (CENP-A in multicellular eukaryotes; see *Section 10.5, Histone Variants Produce Alternative Nucleosomes*). A protein called Scm3 is required for proper association of Cse4 with *CEN*. Inclusion of histone variants related to Cse4 are a universal aspect of centromere construction in all species. The basic interaction consists of bending the DNA of the *CDE-II* region around a protein aggregate; the reaction is probably assisted by the occurrence of intrinsic bending in the *CDE-II* sequence.

CDE-I is bound by a homodimer of Cbf1; this interaction is not essential for centromere function, but in its absence the fidelity of chromosome segregation is reduced ~10×. The 240 kD heterotetramer CBF3, binds to *CDE-III*. This interaction is essential for centromeric function.

The proteins bound at *CDE-I*, *CDE-II*, and *CDE-III* also interact with another group of proteins (Ctf19, Mcm21, and Okp1), which in turn link the centromeric complex to the kinetochore proteins (~70 individual kinetochore proteins have been identified in yeast) and to the microtubule.

The overall model suggests that the complex is localized at the centromere by a protein structure that resembles the normal building block of chromatin (the nucleosome). The bending of DNA at this structure allows proteins bound to the flanking elements to become part of a single complex. The DNA-binding components of the complex form a scaffold for assembly of the kinetochore, linking the centromere to the microtubule. The construction of kinetochores follows a similar pattern, and uses related components, in a wide variety of organisms.

9.16 Telomeres Have Simple Repeating Sequences

Key concepts

- The telomere is required for the stability of the chromosome end.
- A telomere consists of a simple repeat where a C+A-rich strand has the sequence $C_{>1}(A/T)_{1-4}$.

Another essential feature in all chromosomes is the **telomere**, which "seals" the chromosome ends. We know that the telomere must be a special structure, because chromosome ends generated by breakage are "sticky" and tend to react with other chromosomes, whereas natural ends are stable.

We can apply two criteria in identifying a telomeric sequence:

- It must lie at the end of a chromosome (or, at least at the end of an authentic linear DNA molecule).
- It must confer stability on a linear molecule.

The problem of finding a system that offers an assay for function again has been brought to the molecular level by using yeast. All of the plasmids that survive in yeast (by virtue of possessing *ARS* and *CEN* elements) are circular DNA molecules. Linear plasmids are unstable (because they are degraded). Could an authentic telomeric DNA sequence confer stability on a linear plasmid? Fragments from yeast DNA that prove to be located at chromosome ends can be identified by such an assay, and a region from the end of a known natural linear DNA molecule—the extrachromosomal rDNA of *Tetrahymena*—is able to render a yeast plasmid stable in linear form.

Telomeric sequences have been characterized from a wide range of eukaryotes. The same type of sequence is found in plants and humans, so the construction of the telomere seems to follow a nearly universal principle. Each telomere consists of a long series of short, tandemly repeated sequences. There may be 100 to 1000 repeats, depending on the organism.

All telomeric sequences can be written in the general form $C_n(A/T)_m$, where $n > 1$ and m is 1 to 4. **FIGURE 9.27** shows a generic example. One unusual property of the telomeric sequence is the extension of the G-T-rich strand, which for 14 to 16 bases is usually a single strand. The G-tail is probably generated because there is a specific limited degradation of the C-A-rich strand.

Some indications about how a telomere functions are given by some unusual properties of the ends of linear DNA molecules. In a trypanosome population, the ends vary in length. When an individual cell clone is followed, the telomere grows longer by 7 to 10 bp (one to two repeats) per generation. Even more revealing

CCCCAACCCCAACCCCAACCCCAACCCCAACCCCAA
GGGGTTGGGGTTGGGGTTGGGGTTGGGGTTGGGGTT

CCCCAACCCCAACCCCAA 5′
GGGGTTGGGGTTGGGGTTGGGGTTGGGGTTGGGGTT3′

FIGURE 9.27 A typical telomere has a simple repeating structure with a G-T-rich strand that extends beyond the C-A-rich strand. The G-tail is generated by a limited degradation of the C-A-rich strand.

is the fate of ciliate telomeres introduced into yeast. After replication in yeast, yeast telomeric repeats are added onto the ends of the *Tetrahymena* repeats.

Addition of telomeric repeats to the end of the chromosome in every replication cycle could solve the difficulty of replicating linear DNA molecules discussed in *Section 12.2, The Ends of Linear DNA Are a Problem for Replication*. The addition of repeats by *de novo* synthesis would counteract the loss of repeats resulting from failure to replicate up to the end of the chromosome. Extension and shortening would be in dynamic equilibrium.

If telomeres are continually being lengthened (and shortened), their exact sequence may be irrelevant. All that is required is for the end to be recognized as a suitable substrate for addition. This explains how the ciliate telomere functions in yeast.

9.17 Telomeres Seal the Chromosome Ends and Function in Meiotic Chromosome Pairing

Key concept

- The protein TRF2 catalyzes a reaction in which the 3′ repeating unit of the G+T-rich strand forms a loop by displacing its homolog in an upstream region of the telomere.

Isolated telomeric fragments do not behave as though they contain single-stranded DNA; instead, they show aberrant electrophoretic mobility and other properties.

Guanine bases have an unusual capacity to associate with one another. The single-stranded G-rich tail of the telomere can form "quartets" of G residues. Each quartet contains four guanines that hydrogen bond with one another to form a planar structure. Each guanine comes from the corresponding position in a successive TTAGGG repeating unit. **FIGURE 9.28** shows an organization based on a recent crystal structure. The quartet that is illustrated represents an association between the first guanine in each repeating unit. It is stacked on top of another quartet that has the same organization, but is formed from the second guanine in each repeating unit. A series of quartets could be stacked like this in a helical manner. Although the formation of this structure attests to the unusual properties of the G-rich sequence *in vitro*, it does not of course demonstrate whether the quartet forms *in vivo*.

What feature of the telomere is responsible for the stability of the chromosome end? **FIGURE 9.29** shows that a loop of DNA forms at the telomere. The absence of any free end may be the crucial feature that stabilizes the end of the chromosome. The average length of the loop in animal cells is 5 to 10 kb. The loop is formed when the 3′ single-stranded end of the telomere (TTAGGG)$_n$ displaces the same sequence

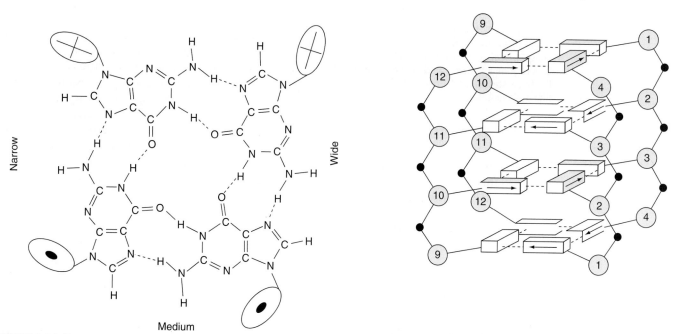

FIGURE 9.28 The crystal structure of a short repeating sequence from the human telomere forms three stacked G quartets. The top quartet contains the first G from each repeating unit. This is stacked above quartets that contains the second G (G3, G9, G15, G21) and the third G (G4, G10, G16, G22).

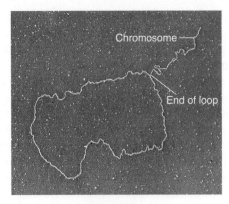

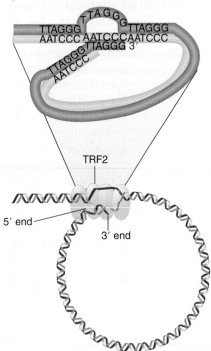

FIGURE 9.29 A loop forms at the end of chromosomal DNA. The 3' single-stranded end of the telomere (TTAGGG) *n* displaces the homologous repeats from duplex DNA to form a t-loop. The reaction is catalyzed by TRF2. Photo courtesy of Jack Griffith, University of North Carolina at Chapel Hill.

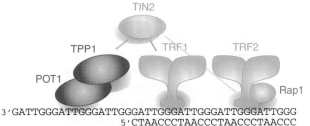

3' GATTGGGATTGGGATTGGGATTGGGATTGGGATTGGGATTGGG
5' CTAACCCTAACCCTAACCCTAACCC

FIGURE 9.30 A schematic of how shelterin might be positioned on telomeric DNA, highlighting the duplex telomeric DNA interactions of TRF1 and TRF2 and the binding of POT1 to the single-stranded TTAGGG repeats. Although one of the shelterin complexes may have the depicted structure, telomeres contain numerous copies of the complex bound along the ds TTAGGG repeat array. It is not known whether all (or even most) shelterin is present as a six-protein complex. Nucleosomes are omitted for simplicity. Reprinted, with permission, from the *Annual Review of Genetics*, Volume 42 © 2008 by Annual Reviews www.annualreviews.org. Courtesy of Titia de Lange, The Rockefeller University.

in an upstream region of the telomere. This converts the duplex region into a structure like a D-loop, where a series of TTAGGG repeats are displaced to form a single-stranded region, and the tail of the telomere is paired with the homologous strand.

The reaction is catalyzed by the telomere-binding protein TRF2, which together with other proteins forms a complex that stabilizes the chromosome ends. Its importance in protecting the ends is indicated by the fact the deletion of TRF2 causes chromosome rearrangements to occur.

In mammals, six telomeric proteins (TRF1, TRF2, Rap1, TIN2, TPP1, and POT1) comprise a complex called shelterin, depicted in FIGURE 9.30. Shelterin functions to protect telomeres from DNA damage repair pathways and to regulate telomere length control by telomerase. Increasing roles for telomeres in aging, cancer, and cell differentiation reveal that telomeres are more than static caps at the ends of linear chromosomes.

Besides their role in capping the ends of linear chromosomes, telomeres also have an ancient and conserved function in meiosis, where they cluster on the nuclear envelope just prior to homologous chromosome synapsis. This clustering defines the "bouquet" stage of meiosis, as shown in FIGURE 9.31, and represents a once-in-a-life cycle configuration. The telomere clustering involves motility forces that act across the nuclear envelope via microtubules, actin, or other filamentous systems. Genetic disruption of meiotic telomere clustering results in chromosome recombination and segregation defects, including the production of aneuploid daughter cells or sterility. Interestingly, fruit flies, which lack canonical telomerase-based telomeres, do not exhibit meiotic telomere clustering, but have evolved other mechanisms to ensure homologous chromosome pairing.

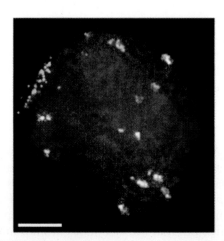

FIGURE 9.31 The meiotic telomere cluster is visualized by telomere FISH. Microscopic image of a maize nucleus fixed at meiotic prophase (zygotene stage), subjected to telomere (green) and centromere (white) FISH, and counterstained for total DNA with DAPI (red). This pseudo-colored image is a 2-D projection of a 3D, multi-color image dataset. Photo courtesy of S. P. Murphy and H. W. Bass, Florida State University.

9.18 Telomeres Are Synthesized by a Ribonucleoprotein Enzyme

Key concepts

- Telomerase uses the 3'-OH of the G+T telomeric strand to prime synthesis of tandem TTGGGG repeats.
- The RNA component of telomerase has a sequence that pairs with the C+A-rich repeats.
- One of the protein subunits is a reverse transcriptase that uses the RNA as template to synthesize the G+T-rich sequence.

The telomere has three widely conserved functions:

- The first is to protect the chromosome end. Any other DNA end—for example, the end generated by a double-strand break—becomes a target for repair systems. The cell has to be able to distinguish the telomere.
- The second is to allow the telomere to be extended. If it is not extended, it becomes shorter with each replication cycle (because replication cannot start at the very end).
- The third is to facilitate meiotic chromosome reorganization for efficient pair-ing and recombination of homologous chromosomes.

Proteins that bind to the telomeres contribute to the solution of all of these. In yeast, different sets of proteins solve the first two problems, but both are bound to the telomere via the same protein, Cdc13:

- The Stn1 protein protects against degradation (specifically, against any extension of the degradation of the C-A-strand that generates the G-tail).
- A **telomerase** enzyme extends the C-A-rich strand. Its activity is influenced by two proteins that have ancillary roles, such as controlling the length of the extension.

The telomerase uses the 3'−OH of the G+T telomeric strand as a primer for synthesis of tandem TTGGGG repeats. Only dGTP and dTTP are needed for the activity. The telomerase is a large ribonucleoprotein that consists of a templating RNA (coded by *TLC1*) and a protein with catalytic activity (*EST2*). The short RNA component (159 bases long in *Tetrahymena*, and 192 bases long in *Euplotes*) includes a sequence of 15 to 22 bases that is identical to two repeats of the C-rich repeating sequence. This RNA provides the template for synthesizing the G-rich repeating sequence. The protein component of the telomerase is a catalytic subunit that can act only upon the RNA template provided by the nucleic acid component.

FIGURE 9.32 shows the action of telomerase. The enzyme progresses discontinuously: the template RNA is positioned on the DNA primer, several nucleotides are added to the primer, and then the enzyme translocates to begin again. The telomerase is a specialized example of a reverse transcriptase, an enzyme that synthesizes a DNA sequence using an RNA template (see *Section 17.13, Viral DNA Is Generated by Reverse Transcription*). We do not know how the complementary (C-A-rich) strand of the telomere is assembled, but we may speculate that it could be synthesized by using the 3'−OH of a terminal G-T hairpin as a primer for DNA synthesis.

Telomerase synthesizes the individual repeats that are added to the chromosome ends, but does not itself control the number of repeats. Other proteins are involved in determining the length of the telomere. They can be identified by the *EST1* and *EST3* mutants in yeast that have altered telomere lengths. These proteins may bind telomerase and influence the length of the telomere by controlling the access of telomerase to its substrate. Proteins that bind telomeres in mammalian cells have been found, but less is known about their functions.

Each organism has a characteristic range of telomere lengths. They are long in mammals (typically 5 to 15 kb in humans) and short in yeast (typically ~300 bp in *S. cerevisiae*). The basic control mechanism is that the probability that a telomere will be a substrate for telomerase increases as the length of the telomere shortens; we do not know if this is a continuous effect or if it depends on the length falling below some critical value. When telomerase acts on a telomere, it may add several repeating units. The enzyme's intrinsic mode of action is to dissociate after adding one repeat; addition of several repeating units depends on other proteins that cause telomerase to undertake more than one round of extension. The number of repeats that is added is not influenced by the length of the telomere itself, but instead is controlled by ancillary proteins that associate with telomerase.

The minimum features required for existence as a chromosome are:

- Telomeres to ensure survival.
- A centromere to support segregation.
- An origin to initiate replication.

All of these elements have been put together to construct a yeast artificial chromosome (YAC; see *Section 3.4, Cloning Vectors Can Be Specialized for Different Purposes*). This is a useful method for perpetuating foreign sequences. It turns out that the synthetic chromosome is stable only if it is longer than 20 to 50 kb. We do not know the basis for this effect, but the ability to construct a synthetic chromosome allows us to investigate the nature of the segregation device in a controlled environment.

9.19 Telomeres Are Essential for Survival

Key concepts

- Telomerase is expressed in actively dividing cells and is not expressed in quiescent cells.
- Loss of telomeres results in senescence.
- Escape from senescence can occur if telomerase is reactivated, or via unequal homologous recombination to restore telomeres.

Telomerase activity is found in all dividing cells and is generally turned off in terminally differentiated cells that do not divide. **FIGURE 9.33** shows that if telomerase is mutated in a dividing cell, the telomeres become gradually shorter with each cell division. An example of the effects of such a mutation in yeast is shown in

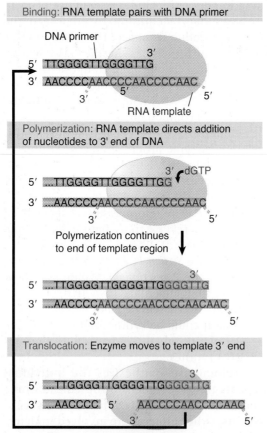

FIGURE 9.32 Telomerase positions itself by base pairing between the RNA template and the protruding single-stranded DNA primer. It adds G and T bases one at a time to the primer, as directed by the template. The cycle starts again when one repeating unit has been added.

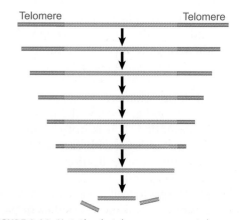

FIGURE 9.33 Mutation in telomerase causes telomeres to shorten in each cell division. Eventual loss of the telomere causes chromosome breaks and rearrangements.

FIGURE 9.34, where the telomere length shortens over ~120 generations from 400 bp to zero.

Loss of telomeres has dire effects. When the telomere length reaches zero, it becomes difficult for the cells to divide successfully. Attempts to divide typically generate chromosome breaks and translocations. This causes an increased rate of mutation. In yeast this is associated with a loss of viability and the culture becomes predominantly occupied by senescent cells (which are elongated and nondividing, and eventually die).

Some cells grow out of the senescing culture. They have acquired the ability to extend their telomeres by an alternative to telomerase activity. The survivors fall into two groups. The members of one group have circularized their chromosomes: they now have no telomeres, and as a result they have become independent of telomerase. The other group uses unequal crossing-over to extend their telomeres (**FIGURE 9.35**). The telomere is a repeating structure, so it is possible for two telomeres to misalign when chromosomes pair. Recombination between the mispaired regions generates an unequal crossing-over, as shown previously in Figure 7.3: when the length of one recombinant chromosome increases, the length of the other decreases.

Cells usually suppress unequal crossing-over because of its potentially deleterious consequences. Two systems are responsible for suppressing crossing-over between telomeres. One is provided by telomere-binding proteins. In yeast, the frequency of recombination between telomeres is increased by deletion of the gene *TAZ1*, which codes for a protein that regulates telomerase activity. The second is a general system that undertakes mismatch repair. In addition to correcting mismatched base pairs that may arise in DNA, this system suppresses recombination between mispaired regions. As shown in Figure 9.35, this includes telomeres. When it is mutated, a greater proportion of telomerase-deficient yeast survives the loss of telomeres because recombination between telomeres generates some chromosomes with longer telomeres.

When eukaryotic cells are placed in culture, they usually divide for a fixed number of generations and then enter senescence. The reason appears to be a decline in telomere length because of the absence of telomerase expression. Cells enter a crisis from which some emerge, but typically the cells that emerge have chromosome rearrangements that have resulted from lack of protection of chromosome ends. These rearrangements may cause mutations that contribute to the tumorigenic state. The absence of telomerase expression in this situation is due to failure to express the gene, and reactivation of telomerase is one of the mechanisms by which these cells then survive continued culture.

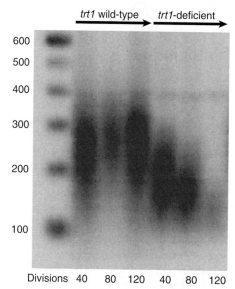

FIGURE 9.34 Telomere length is maintained at ~350 bp in wild-type yeast, but a mutant in the *trt1* gene coding for the RNA component of telomerase rapidly shortens its telomeres to zero length. Reproduced from T. M. Nakamura, et al., *Science* 277 (1997): 955–959 [http://www.sciencemag.org]. Reprinted with permission from AAAS. Photo courtesy of Thomas R. Cech, Howard Hughes Medical Institute.

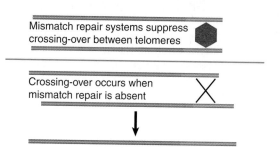

FIGURE 9.35 Crossing-over in telomeric regions is usually suppressed by mismatch-repair systems, but can occur when they are mutated. An unequal crossing-over event extends the telomere of one of the products, allowing the chromosome to survive in the absence of telomerase.

9.20 Summary

The genetic material of all organisms and viruses takes the form of tightly packaged nucleoprotein. Some virus genomes are inserted into preformed virions, whereas others assemble a protein coat around the nucleic acid. The bacterial genome forms a dense nucleoid, with ~20% protein by mass, but details of the interaction of the proteins with DNA are not known. The DNA is organized into ~100 domains that maintain independent supercoiling, with a density of unrestrained supercoils corresponding to ~1/100 to 200 bp. In eukaryotes, interphase chromatin and metaphase chromosomes both appear to be organized into large loops. Each loop may be an independently supercoiled domain. The bases of the loops are connected to a metaphase scaffold or to the nuclear matrix by specific DNA sites.

Transcriptionally active sequences reside within the euchromatin that comprises the majority of interphase chromatin. The regions of heterochromatin are packaged ~5 to 10× more compactly, and are transcriptionally inert. All chromatin becomes densely packaged during cell division, when the individual chromosomes can be distinguished. The existence of a reproducible ultrastructure in mammalian chromosomes is indicated by the production of G-bands by treatment with Giemsa stain. The bands are very large regions (~10^7 bp) that can be used to map chromosomal translocations or other large changes in structure.

Lampbrush chromosomes of amphibians and polytene chromosomes of insects have unusually extended structures, with packing ratios <100. Polytene chromosomes of *D. melanogaster* are divided into ~5000 bands. These bands vary in size by an order of magnitude, with an average of ~25 kb. Transcriptionally active regions can be visualized in even more unfolded ("puffed") structures, in which material is extruded from the axis of the chromosome. This may resemble the changes that occur on a smaller scale when a sequence in euchromatin is transcribed.

The centromeric region contains the kinetochore, which is responsible for attaching a chromosome to the mitotic spindle. The centromere often is surrounded by heterochromatin. Centromeric sequences have been identified only in yeast *S. cerevisiae*, where they consist of short, conserved elements. These elements, *CDE-I* and *CDE-III*, bind Cbf1 and the CBF3 complex, respectively, and a long A-T-rich region called *CDE-II* binds Cse4 to form a specialized structure in chromatin. Another group of proteins that binds to this assembly provides the connection to microtubules.

Telomeres make the ends of chromosomes stable. Almost all known telomeres consist of multiple repeats in which one strand has the general sequence $C_n(A/T)_m$, where $n > 1$ and $m = 1$ to 4. The other strand, $G_n(T/A)_m$, has a single protruding end that provides a template for addition of individual bases in defined order. The enzyme telomerase is a ribonucleoprotein whose RNA component provides the template for synthesizing the G-rich strand. This overcomes the problem of the inability to replicate at the very end of a duplex. The telomere stabilizes the chromosome end because the overhanging single strand $G_n(T/A)_m$ displaces its homolog in earlier repeating units in the telomere to form a loop, so there are no free ends that resemble double-strand breaks.

References

9.2 Viral Genomes Are Packaged into Their Coats

Reviews

Black, L. W. (1989). DNA packaging in dsDNA bacteriophages. *Annu. Rev. Immunol.* 43, 267–292.

Butler, P. J. (1999). Self-assembly of tobacco mosaic virus: the role of an intermediate aggregate in generating both specificity and speed. *Philos. Trans. R. Soc. Lond. B Biol. Sci.* 354, 537–550.

Klug, A. (1999). The tobacco mosaic virus particle: structure and assembly. *Philos. Trans. R. Soc. Lond. B Biol. Sci.* 354, 531–535.

Mindich, L. (2000). Precise packaging of the three genomic segments of the double-stranded-RNA bacteriophage phi6. *Microbiol. Mol. Biol. Rev.* 63, 149–160.

Research

Caspar, D. L. D. and Klug, A. (1962). Physical principles in the construction of regular viruses. *Cold Spring Harbor Symp. Quant. Biol.* 27, 1–24.

de Beer, T., Fang, J., Ortega, M., Yang, Q., Maes, L., Duffy, C., Berton, N., Sippy, J., Overduin, M., Feiss, M., and Catalano, C. E. (2002). Insights into specific DNA recognition during the assembly of a viral genome packaging machine. *Mol. Cell* 9, 981–991.

Dube, P., Tavares, P., Lurz, R., and van Heel, M. (1993). The portal protein of bacteriophage SPP1: a DNA pump with 13-fold symmetry. *EMBO J.* 12, 1303–1309.

Fraenkel-Conrat, H. and Williams, R. C. (1955). Reconstitution of active tobacco mosaic virus from its inactive protein and nucleic acid components. *Proc. Natl. Acad. Sci. USA.* 41, 690–698.

Jiang, Y. J., Aerne, B. L., Smithers, L., Haddon, C., Ish-Horowicz, D., and Lewis, J. (2000). Notch signalling and the synchronization of the somite segmentation clock. *Nature* 408, 475–479.

Zimmern, D. (1977). The nucleotide sequence at the origin for assembly on tobacco mosaic virus RNA. *Cell* 11, 463–482.

Zimmern, D. and Butler, P. J. (1977). The isolation of tobacco mosaic virus RNA fragments containing the origin for viral assembly. *Cell* 11, 455–462.

9.3 The Bacterial Genome Is a Nucleoid

Reviews

Brock, T. D. (1988). The bacterial nucleus: a history. *Microbiol. Rev.* 52, 397–411.

Drlica, K. and Rouviere-Yaniv, J. (1987). Histone-like proteins of bacteria. *Microbiol. Rev.* 51, 301–319.

9.4 The Bacterial Genome Is Supercoiled

Review

Hatfield, G. W. and Benham, C. J. (2002). DNA topology-mediated control of global gene expression in *Escherichia coli. Annu. Rev. Genet.* 36, 175–203.

Research

Pettijohn, D. E. and Pfenninger, O. (1980). Supercoils in prokaryotic DNA restrained *in vitro. Proc. Natl. Acad. Sci. USA.* 77, 1331–1335.

Postow, L., Hardy, C. D., Arsuaga, J., and Cozzarelli, N. R. (2004). Topological domain structure of the *Escherichia coli* chromosome. *Genes Dev.* 18, 1766–1779.

9.5 Eukaryotic DNA Has Loops and Domains Attached to a Scaffold

Research

International Human Genome Sequencing Consortium. (2001). Initial sequencing and analysis of the human genome. *Nature* 409, 860–921.

Saccone, S., De Sario, A., Wiegant, J., Raap, A. K., Della Valle, G., and Bernardi, G. (1993). Correlations between isochores and chromosomal bands in the human genome. *Proc. Natl. Acad. Sci. USA.* 90, 11929–11933.

Venter, J. C. et al. (2001). The sequence of the human genome. *Science* 291, 1304–1350.

9.6 Specific Sequences Attach DNA to an Interphase Matrix

Review

Chattopadhyay, S. and Pavithra, L. (2007). MARs and MARBPs: key modulators of gene regulation and disease manifestation. *Subcell. Bio.* 41, 213–230.

9.12 The Eukaryotic Chromosome Is a Segregation Device

Review

Hyman, A. A. and Sorger, P. K. (1995). Structure and function of kinetochores in budding yeast. *Annu. Rev. Cell Dev. Biol.* 11, 471–495.

9.13 Regional Centromeres Contain a Centromeric Histone H3 Variant and Repetitive DNA

Review

Dalal, Y. (2009). Epigenetic specification of centromeres. *Biochem. Cell Biol.* 87, 273–282.

Ekwall, K. (2007). Epigenetic control of centromere behavior. *Annu. Rev. Genet.* 41, 63–81.

Research

Black, B. E., Foltz, D. R., Chakravarthy, S., Luger, K., Woods, Jr., V. L., and Cleveland, D. W. (2004). Structural determinants for generating centromeric chromatin. *Nature* 430, 578–582.

Depinet, T. W., Zackowski, J. L., Earnshaw, W. C., Kaffe, S., Sekhon, G. S., Stallard, R., Sullivan, B. A., Vance, G. H., Van Dyke, D. L., Willard, H. F, Zinn, A. B., and Schwartz, S. (1997). Characterization of neo-centromeres in marker chromosomes lacking detectable alpha-satellite DNA. *Hum. Mol. Genet.* 6, 1195–1204.

Foltz, D. R., Jansen, L. E., Black, B. E., Bailey, A. O., Yates, 3rd, J. R., and Cleveland, D. W. (2006). The human CENP-A centromeric nucleosome-associated complex. *Nat. Cell Biol.* 8, 458–469.

Sun, X., Wahlstrom, J., and Karpen, G. (1997). Molecular structure of a functional *Drosophila* centromere. *Cell* 91, 1007–1019.

9.14 Point Centromeres in *S. cerevisiae* Contain Short, Essential DNA Sequences

Reviews

Blackburn, E. H. and Szostak, J. W. (1984). The molecular structure of centromeres and telomeres. *Annu. Rev. Biochem.* 53, 163–194.

Clarke, L. and Carbon, J. (1985). The structure and function of yeast centromeres. *Annu. Rev. Genet.* 19, 29–56.

Research

Fitzgerald-Hayes, M., Clarke, L., and Carbon, J. (1982). Nucleotide sequence comparisons and functional analysis of yeast centromere DNAs. *Cell* 29, 235–244.

9.15 The *S. cerevisiae* Centromere Binds a Protein Complex

Review

Bloom, K. (2007). Centromere dynamics. *Curr. Opin. Genet. Dev.* 17, 151–156.

Kitagawa, K. and Hieter, P. (2001). Evolutionary conservation between budding yeast and human kinetochores. *Nat. Rev. Mol. Cell Biol.* 2, 678–687.

Research

Lechner, J. and Carbon, J. (1991). A 240 kd multisubunit protein complex, CBF3, is a major component of the budding yeast centromere. *Cell* 64, 717–725.

Meluh, P. B. and Koshland, D. (1997). Budding yeast centromere composition and assembly as revealed by *in vitro* cross-linking. *Genes Dev.* 11, 3401–3412.

Meluh, P. B. et al. (1998). Cse4p is a component of the core centromere of *S. cerevisiae*. *Cell* 94, 607–613.

Ortiz, J., Stemmann, O., Rank, S., and Lechner, J. (1999). A putative protein complex consisting of Ctf19, Mcm21, and Okp1 represents a missing link in the budding yeast kinetochore. *Genes Dev.* 13, 1140–1155.

9.16 Telomeres Have Simple Repeating Sequences

Reviews

Blackburn, E. H. and Szostak, J. W. (1984). The molecular structure of centromeres and telomeres. *Annu. Rev. Biochem.* 53, 163–194.

Zakian, V. A. (1989). Structure and function of telomeres. *Annu. Rev. Genet.* 23, 579–604.

Research

Dejardin, J. and Kingston, R. E. (2009). Purification of proteins associated with specific genomic loci. *Cell* 136, 175–186.

Wellinger, R. J., Ethier, K., Labrecque, P., and Zakian, V. A. (1996). Evidence for a new step in telomere maintenance. *Cell* 85, 423–433.

9.17 Telomeres Seal the Chromosome Ends and Function In Meiotic Chromosome Pairing

Reviews

Palm, W. and de Lange, T. (2008). How shelterin protects mammalian telomeres. *Annu. Rev. Genet.* 42, 301–334.

Scherthan, H. (2007). Telomere attachment and clustering during meiosis. *Cell Mol. Life Sci.* 64, 117–124.

Research

Bass, H. W., Marshall, W. F., Sedat, J. W., Agard, D. A., and Cande, W. Z. (1997). Telomeres cluster de novo before the initiation of synapsis: a three-dimensional spatial analysis of telomere positions before and during meiotic prophase. *J. Cell Biol.* 137, 5–18.

Chikashige, Y., Tsutsumi, C., Yamane, M., Okamasa, K., Haraguchi, T., and Hiraoka, Y. (2006). Meiotic proteins bqt1 and bqt2 tether telomeres to form the bouquet arrangement of chromosomes. *Cell* 125, 59–69.

Griffith, J. D. et al. (1999). Mammalian telomeres end in a large duplex loop. *Cell* 97, 503–514.

Henderson, E., Hardin, C. H., Walk, S. K., Tinoco, I., and Blackburn, E. H. (1987). Telomeric oligonucleotides form novel intramolecular structures containing guanine-guanine base pairs. *Cell* 51, 899–908.

Karlseder, J., Broccoli, D., Dai, Y., Hardy, S., and de Lange, T. (1999). p53- and ATM-dependent apoptosis induced by telomeres lacking TRF2. *Science* 283, 1321–1325.

Parkinson, G. N., Lee, M. P., and Neidle, S. (2002). Crystal structure of parallel quadruplexes from human telomeric DNA. *Nature* 417, 876–880.

van Steensel, B., Smogorzewska, A., and de Lange, T. (1998). TRF2 protects human telomeres from end-to-end fusions. *Cell* 92, 401–413.

Williamson, J. R., Raghuraman, K. R., and Cech, T. R. (1989). Monovalent cation-induced structure of telomeric DNA: the G-quartet model. *Cell* 59, 871–880.

9.18 Telomeres Are Synthesized by a Ribonucleoprotein Enzyme

Reviews

Blackburn, E. H. (1991). Structure and function of telomeres. *Nature* 350, 569–573.

Blackburn, E. H. (1992). Telomerases. *Annu. Rev. Biochem.* 61, 113–129.

Blackburn, E. H., Greider, C. W., and Szostak, J. W. (2006). Telomeres and telomerase: the path from maize, *Tetrahymena* and yeast to human cancer and aging. *Nat. Med.* 12, 1133–1138.

Collins, K. (1999). Ciliate telomerase biochemistry. *Annu. Rev. Biochem.* 68, 187–218.

Smogorzewska, A. and de Lange, T. (2004). Regulation of telomerase by telomeric proteins. *Annu. Rev. Biochem.* 73, 177–208.

Zakian, V. A. (1995). Telomeres: beginning to understand the end. *Science* 270, 1601–1607.

Zakian, V. A. (1996). Structure, function, and replication of *S. cerevisiae* telomeres. *Annu. Rev. Genet.* 30, 141–172.

Research

Greider, C. and Blackburn, E. H. (1987). The telomere terminal transferase of *Tetrahymena* is a ribonucleoprotein enzyme with two kinds of primer specificity. *Cell* 51, 887–898.

Murray, A., and Szostak, J. W. (1983). Construction of artificial chromosomes in yeast. *Nature* 305, 189–193.

Pennock, E., Buckley, K., and Lundblad, V. (2001). Cdc13 delivers separate complexes to the telomere for end protection and replication. *Cell* 104, 387–396.

Shippen-Lentz, D. and Blackburn, E. H. (1990). Functional evidence for an RNA template in telomerase. *Science* 247, 546–552.

Teixeira, M. T., Arneric, M., Sperisen, P., and Lingner, J. (2004). Telomere length homeostasis is achieved via a switch between telomerase-extendible and -nonextendible states. *Cell* 117, 323–335.

9.19 Telomeres Are Essential for Survival

Review

Bailey, S. M. and Murname, J. P. (2006). Telomeres, chromosome instability and cancer. *Nucleic Acids Res.* 34, 2408–2417.

Research

Hackett, J. A., Feldser, D. M., and Greider, C. W. (2001). Telomere dysfunction increases mutation rate and genomic instability. *Cell* 106, 275–286.

Nakamura, T. M., Cooper, J. P., and Cech, T. R. (1998). Two modes of survival of fission yeast without telomerase. *Science* 282, 493–496.

Nakamura, T. M., Morin, G. B., Chapman, K. B., Weinrich, S. L., Andrews, W. H., Lingner, J., Harley, C. B., and Cech, T. R. (1997). Telomerase catalytic subunit homologs from fission yeast and human. *Science* 277, 955–959.

Rizki, A. and Lundblad, V. (2001). Defects in mismatch repair promote telomerase-independent proliferation. *Nature* 411, 713–716.

10

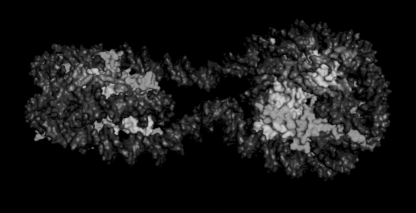

Structure from Protein Data Base 1ZBB. T. Schalch, et al., *Nature* 436 (2005): 138–141. Photo courtesy of Chris Nelson, University of Victoria.

Chromatin

CHAPTER OUTLINE

10.6 DNA Structure Varies on the Nucleosomal Surface

- DNA is wrapped 1.67 times around the histone octamer.
- DNA on the nucleosome shows regions of smooth curvature and regions of abrupt kinks.
- The structure of the DNA is altered so that it has an increased number of base pairs/turn in the middle, but a decreased number at the ends.
- ~0.6 negative turns of DNA are absorbed by the change in bp/turn from 10.5 in solution to an average of 10.2 on the nucleosomal surface, which explains the linking-number paradox.

10.7 The Path of Nucleosomes in the Chromatin Fiber

- 10 nm chromatin fibers consist of a string of nucleosomes.
- 30 nm fibers have six nucleosomes/turn, which are organized into a two-start helix.
- Histone H1, histone tails, and increased ionic strength all promote the formation of the 30 nm fiber.

10.8 Replication of Chromatin Requires Assembly of Nucleosomes

- Histone octamers are not conserved during replication, but H2A-H2B dimers and $H3_2$-$H4_2$ tetramers are conserved.
- There are different pathways for the assembly of nucleosomes during replication and independently of replication.
- Accessory proteins are required to assist the assembly of nucleosomes.
- CAF-1 and ASF1 are histone assembly proteins that are linked to the replication machinery.
- A different assembly protein, HIRA, and the histone H3.3 variant are used for replication-independent assembly.

10.9 Do Nucleosomes Lie at Specific Positions?

- Nucleosomes may form at specific positions as the result of either the local structure of DNA or proteins that interact with specific sequences.
- A common cause of nucleosome positioning is when proteins binding to DNA establish a boundary.
- Positioning may affect which regions of DNA are in the linker and which face of DNA is exposed on the nucleosome surface.
- DNA sequence determinants (exclusion or preferential binding) may be responsible for half of the *in vivo* nucleosome positions.

10.10 Nucleosomes Are Displaced and Reassembled During Transcription

- Most transcribed genes retain a nucleosomal structure, though the organization of the chromatin changes during transcription.

- Some heavily transcribed genes appear to be exceptional cases that are devoid of nucleosomes.
- RNA polymerase displaces histone octamers during transcription *in vitro*, but octamers reassociate with DNA as soon as the polymerase has passed.
- Nucleosomes are reorganized when transcription passes through a gene.
- Additional factors are required both for RNA polymerase to displace octamers during transcription and for the histones to reassemble into nucleosomes after transcription.

10.11 DNase Sensitivity Detects Changes in Chromatin Structure

- Hypersensitive sites are found at the promoters of expressed genes, as well as other important sites such as origins of replication and centromeres.
- Hypersensitive sites are generated by the binding of factors that exclude histone octamers.
- A domain containing a transcribed gene is defined by increased sensitivity to degradation by DNase I.

10.12 Insulators Define Transcriptionally Independent Domains

- Insulators are able to block passage of any activating or inactivating effects from enhancers, silencers, and other control elements.
- Insulators can provide barriers against the spread of heterochromatin.
- Insulators are specialized chromatin structures that typically contain hypersensitive sites.
- In most cases, two insulators can protect the region between them from all external effects.
- Different insulators are bound by different factors, and may use alternative mechanisms for enhancer blocking and/or heterochromatin barrier formation.

10.13 An LCR May Control a Domain

- LCRs are located at the 5′ end of a chromosomal domain and typically consist of multiple DNAse hypersensitive sites.
- LCRs regulate gene clusters.
- LCRs usually regulate loci that show complex developmental or cell-type specific patterns of gene expression.
- LCRs control the transcription of target genes in the locus by direct interactions, forming looped structures.

10.14 Summary

10.1 Introduction

Chromatin has a compact organization in which most DNA sequences are structurally inaccessible and functionally inactive. Within this mass is the minority of active sequences. What is the general structure of chromatin, and what is the difference between active and inactive sequences? The high overall packing ratio of the genetic material immediately suggests that DNA cannot be directly packaged into the final structure of chromatin. There must be hierarchies of organization.

The fundamental subunit of chromatin has the same type of design in all eukaryotes. The **nucleosome** contains ~200 bp of DNA, organized by an octamer of small, basic proteins into a beadlike structure. The protein components are **histones**. They form an interior core; the DNA lies on the surface of the particle. Additional regions of the histones, known as the **histone tails**, extend from the surface. Nucleosomes are an invariant component of euchromatin and heterochromatin in the interphase nucleus and of mitotic chromosomes. The nucleosome provides the first level of organization, compacting the DNA ~6-fold over the length of naked DNA, resulting in a fiber ~10 nm in diameter. Its components and structure are well characterized.

The second level of organization is the coiling of the **10 nm fiber** of nucleosomes into a helical array to constitute the fiber of diameter ~30 nm that is found in both interphase chromatin and mitotic chromosomes. This compacts the DNA ~40-fold. The structure of this fiber requires the histone tails and is stabilized by **linker histones**.

This **30 nm fiber** is then further folded and compacted into interphase chromatin or into mitotic chromosomes. This results in ~1000-fold compaction in euchromatin, cyclically interchangeable with packing into mitotic chromosomes to achieve an overall ~10,000-fold compaction. Heterochromatin generally maintains ~10,000-fold compaction in both interphase and mitosis.

In this chapter, we will describe the structure and relationships between these levels of organization to characterize the events involved in cyclical packaging, replication, and transcription. Association with additional proteins, as well as modifications of existing chromosomal proteins, are involved in changing the structure of chromatin. Both replication and transcription require unwinding of DNA, and thus first involve an unfolding of the structure that allows the relevant enzymes to manipulate the DNA. This is likely to involve changes in all levels of organization.

When chromatin is replicated, the nucleosomes must be reproduced on both daughter duplex molecules. In addition to asking how the nucleosome itself is assembled, we must inquire what happens to other proteins present in chromatin. Replication disrupts the structure of chromatin, which indicates that it both poses a problem for maintaining regions with specific structure and offers an opportunity to change the structure.

The mass of chromatin contains up to twice as much protein as DNA. Approximately half of the protein mass is accounted for by the nucleosomes. The mass of RNA is <10% of the mass of DNA. Much of the RNA consists of nascent transcripts still associated with the template DNA.

The **nonhistones** include all the proteins found in chromatin except the histones. They are more variable between tissues and species, and they comprise a smaller proportion of the mass than the histones. They also comprise a much larger number of proteins, so that any individual protein is present in amounts much smaller than any histone. The functions of nonhistone proteins include control of gene expression and higher-order structure. Thus RNA polymerase may be considered to be a prominent nonhistone. The HMG (high-mobility group) proteins comprise a discrete and well-defined subclass of nonhistones (at least some of which are transcription factors).

10.2 DNA Is Organized in Arrays of Nucleosomes

Key concepts

- MNase cleaves linker DNA and releases individual nucleosomes from chromatin.
- >95% of the DNA is recovered in nucleosomes or multimers when MNase cleaves DNA in chromatin.
- The length of DNA per nucleosome varies for individual tissues or species in a range from 154 to 260 bp.
- Nucleosomal DNA is divided into the core DNA and linker DNA depending on its susceptibility to MNase.
- The core DNA is the length of 146 bp that is found on the core particles produced by prolonged digestion with MNase.
- Linker DNA is the region of 8 to 114 bp that is susceptible to early cleavage by nucleases.

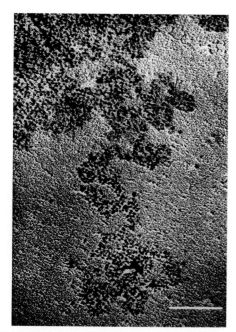

FIGURE 10.1 Chromatin spilling out of lysed nuclei consists of a compactly organized series of particles. The bar is 100 nm. Reprinted from *Cell*, vol. 4, P. Oudet, M. Gross-Bellard, and P. Chambon, Electron microscopic and biochemical evidence . . . , pp. 281–300. Copyright 1975, with permission from Elsevier [http://www.sciencedirect.com/science/journal/00928674]. Photo courtesy of Pierre Chambon, College of France.

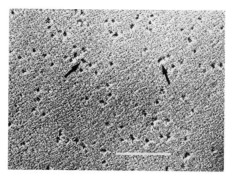

FIGURE 10.2 Individual nucleosomes are released by digestion of chromatin with micrococcal nuclease. The bar is 100 nm. Reprinted from *Cell*, vol. 4, P. Oudet, M. Gross-Bellard, and P. Chambon, Electron microscopic and biochemical evidence . . . , pp. 281–300. Copyright 1975, with permission from Elsevier [http://www.sciencedirect.com/science/journal/00928674]. Photo courtesy of Pierre Chambon, College of France.

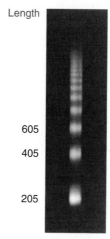

FIGURE 10.3 Micrococcal nuclease digests chromatin in nuclei into a multimeric series of DNA bands that can be separated by gel electrophoresis. Photo courtesy of Markus Noll, Universität Zürich.

When interphase nuclei are suspended in a solution of low ionic strength, they swell and rupture to release fibers of chromatin. FIGURE 10.1 shows a lysed nucleus in which fibers are streaming out. In some regions, the fibers consist of tightly packed material, but in regions that have become stretched they can be seen to consist of discrete particles. These are the **nucleosomes**. In especially extended regions, individual nucleosomes are visibly connected by a fine thread, which is a free duplex of DNA. A continuous duplex thread of DNA runs through the series of particles.

Individual nucleosomes can be obtained by treating chromatin with the endonuclease **micrococcal nuclease (MNase)**, which cuts the DNA thread at the junction between nucleosomes, a region known as **linker DNA**. Ongoing digestion with MNase releases groups of particles, and eventually single nucleosomes. Individual nucleosomes can be seen in FIGURE 10.2 as compact particles measuring ~10 nm in diameter.

When chromatin is digested with MNase, the DNA is cleaved into integral multiples of a unit length. Fractionation by gel electrophoresis reveals the "ladder" presented in FIGURE 10.3. Such ladders extend for ~10 steps, and the unit length, determined by the increments between successive steps, is ~200 bp.

FIGURE 10.4 shows that the ladder is generated by groups of nucleosomes. When nucleosomes are fractionated on a sucrose gradient, they give a series of discrete peaks that correspond to monomers, dimers, trimers, and so on. When the DNA is extracted from the individual fractions and electrophoresed, each fraction yields a band of DNA whose size corresponds with a step on the micrococcal nuclease ladder. The monomeric nucleosome contains DNA of the unit length, the nucleosome dimer contains DNA of twice the unit length, and so on. More than 95% of the DNA of chromatin can be recovered in the form of the 200-bp ladder, indicating that almost all DNA must be organized in nucleosomes.

The length of DNA present in the nucleosome can vary from the "typical" value of

200 bp. The chromatin of any particular cell type has a characteristic average value (±5 bp). The average most often is between 180 and 200, but there are extremes as low as 154 bp (in a fungus) or as high as 260 bp (in sea urchin sperm). The average value may be different in individual tissues of the adult organism, and there can be differences between different parts of the genome in a single cell type. Variations from the genome average often include tandemly repeated sequences, such as clusters of 5S RNA genes.

A common structure underlies the varying amount of DNA that is contained in nucleo-somes of different sources. The association of DNA with the histone octamer forms a core particle containing 146 bp of DNA, irrespective of the total length of DNA in the nucleosome. The variation in total length of DNA per nucleosome is superimposed on this basic core structure.

The core particle is defined by the effects of MNase on the nucleosome monomer. The initial reaction of the enzyme is to cut between nucleosomes, but if it is allowed to continue after monomers have been generated, it proceeds to digest some of the DNA of the individual nucleosome, as shown in **FIGURE 10.5**. Initial cleavage results in nucleosome monomers with (in this example) ~200 bp of DNA. After the first step, some monomers are found in which the length of DNA has been "trimmed" to ~165 bp. Finally, this is reduced to the length of the DNA of the core particle, 146 bp.

As a result of this type of analysis, nucleosomal DNA is functionally divided into two regions:

- *Core DNA* has an invariant length of 146 bp, the minimum length of DNA needed to form a stable monomeric nucleosome, and is relatively resistant to digestion by nucleases.
- Linker DNA comprises the rest of the repeating unit. Its length varies from as little as 8 bp to as much as 114 bp per nucleosome.

Core particles have properties similar to those of the nucleosomes themselves, although they are smaller. Their shape and size are similar to those of nucleosomes; this suggests that the essential geometry of the particle is established by the interactions between DNA and the protein octamer in the core particle. Core particles are readily obtained as a homogeneous population, and as a result they are often used for structural studies in preference to nucleosome preparations.

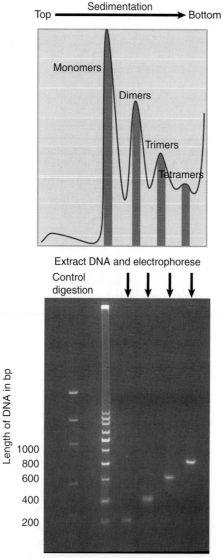

FIGURE 10.4 Each multimer of nucleosomes contains the appropriate number of unit lengths of DNA. In the photo, artificial bands simulate a DNA ladder that would be produced by MNase digestion. The image was constructed using PCR fragments with sizes corresponding to actual band sizes. Photo courtesy of Jan Kieleczawa, Wyeth Research.

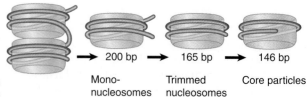

FIGURE 10.5 Micrococcal nuclease initially cleaves between nucleosomes. Mononucleosomes typically have ~200 bp DNA. End-trimming reduces the length of DNA first to ~165 bp, and then generates core particles with 146 bp.

10.3 The Nucleosome Is the Subunit of All Chromatin

Key concepts

- A nucleosome contains ~200 bp of DNA and two copies of each core histone (H2A, H2B, H3, and H4).
- DNA is wrapped around the outside surface of the protein octamer.
- The histone octamer has a structure of an $H3_2$-$H4_2$ tetramer associated with two H2A-H2B dimers.
- Each histone is extensively interdigitated with its partner.
- All core histones have the structural motif of the histone fold. N- and C-terminal histone tails extend out of the nucleosome.
- H1 is associated with linker DNA and may lie at the point where DNA enters or exits the nucleosome.

The 10 nm particles seen in Figure 10.2 represent the fundamental building block of all chromatin, the nucleosome. The nucleosome contains ~200 bp of DNA associated with a **histone octamer** that consists of two copies each of histones H2A, H2B, H3, and H4. These are known as the **core histones**. Their association is illustrated diagrammatically in **FIGURE 10.6**.

The histones are small, basic proteins (rich in arginine and lysine residues), resulting in a high affinity for DNA. Histones H3 and H4 are among the most conserved proteins known, and the core histones are responsible for DNA packaging in all eukaryotes. H2A and H2B are also conserved among eukaryotes, but show appreciable species-specific variation in sequence, particularly in the histone tails. The core regions of the histones are even conserved

in Archaea and appear to play a similar role in compaction of archeal DNA.

The shape of the nucleosome corresponds to a flat disk or cylinder of diameter 11 nm and height 6 nm. The length of the DNA is roughly twice the ~34 nm circumference of the particle. The DNA follows a symmetrical path around the octamer. **FIGURE 10.7** shows the DNA path diagrammatically as a helical coil that makes ~1⅔ turns around the cylindrical octamer. Note that the DNA "enters" and "exits" on one side of the nucleosome.

Viewing a cross-section through the nucleosome, in **FIGURE 10.8** we see that the two circumferences made by the DNA lie close to one another. The height of the cylinder is 6 nm, of which 4 nm is occupied by the two turns of DNA (each of diameter 2 nm). The pattern of the two turns has a possible functional consequence. One turn around the nucleosome takes ~80 bp of DNA, so two points separated by 80 bp in the free double helix may actually be close on the nucleosome surface, as illustrated in **FIGURE 10.9**.

The core histones tend to form two types of subcomplexes. H3 and H4 form a very stable tetramer in solution ($H3_2$–$H4_2$). H2A and H2B most typically form a dimer (H2A–H2B). A space-filling model of the structure of the

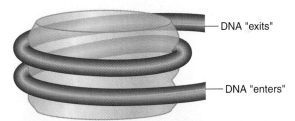

FIGURE 10.7 The nucleosome is a cylinder with DNA organized into ~1⅔ turns around the surface.

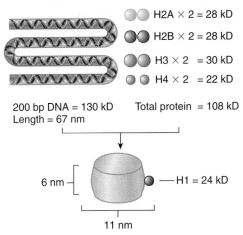

200 bp DNA = 130 kD   Total protein = 108 kD
Length = 67 nm

H2A × 2 = 28 kD
H2B × 2 = 28 kD
H3 × 2 = 30 kD
H4 × 2 = 22 kD

6 nm — H1 = 24 kD

11 nm

FIGURE 10.6 The nucleosome consists of approximately equal masses of DNA and histones (including H1). The predicted mass of the nucleosome is 262 kD.

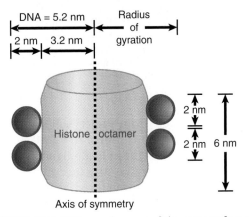

DNA = 5.2 nm   Radius of gyration
2 nm   3.2 nm
Histone octamer
2 nm
2 nm   6 nm
Axis of symmetry

FIGURE 10.8 DNA occupies most of the outer surface of the nucleosome.

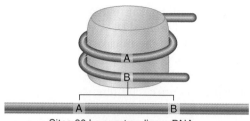

Sites 80 bp apart on linear DNA
are close together on nucleosome

FIGURE 10.9 Sequences on the DNA that lie on different turns around the nucleosome may be close together.

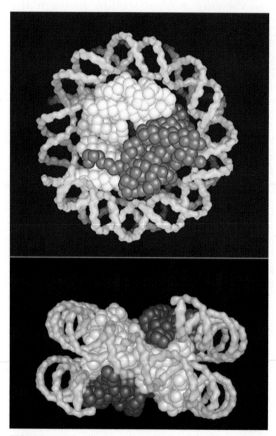

FIGURE 10.10 The crystal structure of the histone core octamer is represented in a space-filling model with the H3$_2$-H4$_2$ tetramer shown in white and the H2A-H2B dimers shown in blue. Only one of the H2A-H2B dimers is visible in the top view, because the other is hidden underneath. The path of the DNA is modeled in green. Photos courtesy of E. N. Moudrianakis, Johns Hopkins University.

histone octamer (from the crystal structure at 3.1 Å resolution) is shown in **FIGURE 10.10**. Tracing the paths of the individual polypeptide backbones in the crystal structure shows that the histones are not organized as individual globular proteins, but that each is interdigitated with its partner: H3 with H4, and H2A with H2B. Thus this figure emphasizes the H3$_2$-

H4$_2$ tetramer (white) and the H2A-H2B dimer (blue) substructure of the nucleosome, but does not show individual histones.

In the top view, it can be seen that the H3$_2$-H4$_2$ tetramer accounts for the diameter of the octamer. It forms the shape of a horseshoe. The H3$_2$-H4$_2$ tetramer alone can organize DNA *in vitro* into particles that display some of the properties of the core particle. The H2A-H2B pairs fit in as two dimers, but only one can be seen in this view. In the side view, the responsibilities of the H3$_2$-H4$_2$ tetramer and of the separate H2A-H2B dimers can be distinguished. The protein forms a sort of spool, with a superhelical path that could correspond to the binding site for DNA, which would be wound in ~1⅔ turns in a nucleosome. The model displays twofold symmetry about an axis that would run perpendicular through the side view.

All four core histones show a similar type of structure in which three α-helices are connected by two loops. This highly conserved structure is called the **histone fold** and is shown in **FIGURE 10.11**. These regions interact to form crescent-shaped heterodimers; each heterodimer binds 2.5 turns of the DNA double helix. Binding is mostly to the phosphodiester backbones though ionic interactions with the many basic amino acids in the histones (consistent with the need to package any DNA irrespective of sequence). A high-resolution view of the nucleosome (based on the crystal structure at 2.8 Å) is shown in **FIGURE 10.12**. The H3$_2$-H4$_2$ tetramer is formed by interactions between the two H3 subunits, as can be seen at the top of the nucleosome (in green) in the left panel of the figure. The association of the two H2A-H2B dimers on opposite faces of the nucleosome is visible in the right panel (in turquoise and yellow).

Each of the core histones has a histone fold domain that contributes to the central protein mass of the nucleosome, sometimes referred to as the *globular core*. Each histone also has a flexible N-terminal tail (H2A and H2B have C-terminal tails as well), which contains sites for covalent modification that are important in chromatin function. The tails, which account for about one quarter of the protein mass, are too flexible to be visualized by X-ray crystallography. Therefore, their positions in the nucleosome are not well defined, and they are generally depicted schematically, as seen in **FIGURE 10.13**. However, the points at which the tails exit the nucleosome core are known, and the tails of both H3 and H2B can be seen to pass

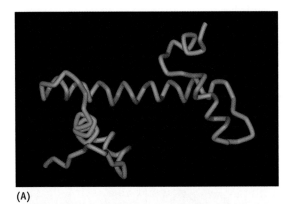

(A)

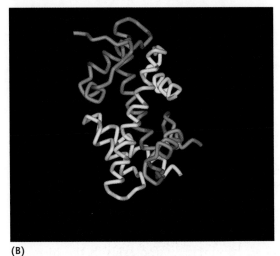

(B)

FIGURE 10.11 The histone fold (A) consists of two short α-helices flanking a longer α-helix. Histone pairs (H3 + H4 and H2A + H2B) interact to form histone dimers (B). Structures from Protein Data Bank 1HIO. G. Arents, et al., *Proc. Natl. Acad. Sci. USA* 88 (1991): 10145–10152.

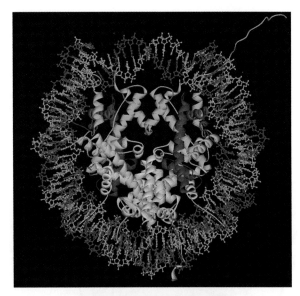

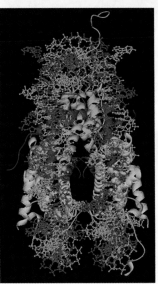

FIGURE 10.12 The crystal structure of the histone core octamer is represented in a ribbon model, including the 146-bp DNA phosphodiester backbones (orange and blue) and eight histone protein main chains (green: H3; purple: H4; turquoise: H2A; yellow: H2B). Structures from Protein Data Bank 1AOI. K. Luger, et al., *Nature* 389 (1997): 251–260.

between the turns of the DNA superhelix and extend out of the nucleosome, as shown in **FIGURE 10.14**. The tails of H4 and H2A extend from both faces of the nucleosome. When histone tails are crosslinked to DNA by UV irradiation, more products are obtained with nucleosomes compared to core particles, which could mean that the tails contact the linker DNA. The tail of H4 appears to contact an H2A-H2B dimer in an adjacent nucleosome, which may contribute to the formation of higher-order structures (see *Section 10.7, The Path of Nucleosomes in the Chromatin Fiber*).

The **linker histones** also play an important role in the formation of higher-order chromatin structures. The linker histone family, typified by histone H1, comprises a set of closely related proteins that show appreciable variation both between tissues and between species. The role of H1 is different from that of the core histones. It is present in half the amount of a core histone and can be extracted more readily from chromatin. H1 can be removed without affecting the structure of the nucleosome, consistent with a location external to the particle. Nucleosomes that contain linker histones are sometimes referred to as **chromatosomes**.

The interaction of histone H1 with the nucleosome is poorly understood. H1 is retained on nucleosome monomers that have at least 165 bp of DNA, but does not

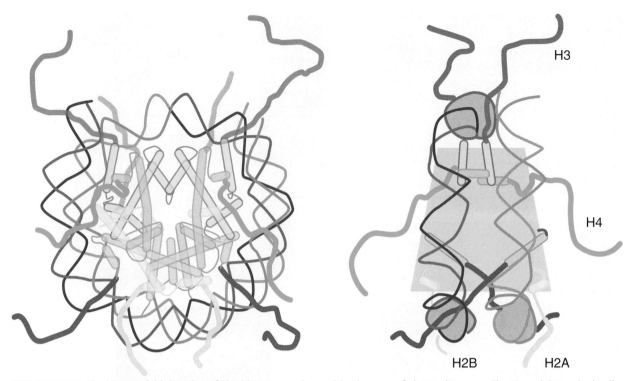

FIGURE 10.13 The histone fold domains of the histones are located in the core of the nucleosome. The N- and C-terminal tails, which carry many sites for modification, are flexible and their positions cannot be determined by crystallography.

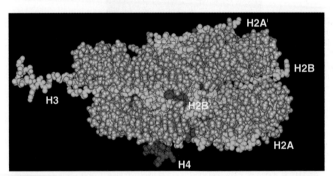

FIGURE 10.14 The histone tails are disordered and exit from both faces of the nucleosome and between turns of the DNA. Note this figure shows only the first few amino acids of the tails, as the complete tails were not present in the crystal structure. Structure from Protein Data Bank 1AOI. K. Luger, et al., *Nature* 389 (1997): 251–260.

FIGURE 10.15 Possible model for the interaction of histone H1 with the nucleosome. H1 may interact with both the central gyre of the DNA at the dyad axis, as well as with the linker DNA at either the entry or exit.

bind to the 146-bp core particle. This suggests that H1 could be located in the region of the linker DNA immediately adjacent to the core DNA. While the precise positioning of linker histones remains somewhat controversial, recent models suggest H1 may interact with either the entry or exit DNA in addition to the central turn of DNA on the nucleosome, as shown in **FIGURE 10.15**. In this position, H1 has the potential to influence the angle of DNA entry or exit, which may contribute to the formation of higher order structures (see *Section 10.7, The Path of Nucleosomes in the Chromatin Fiber*).

10.4 Nucleosomes Are Covalently Modified

Key concepts

- Histones are modified by methylation, acetylation, phosphorylation, and other modifications.
- Combinations of specific histone modifications define the function of local regions of chromatin; this is known as the histone code.
- The bromodomain is found in a variety of proteins that interact with chromatin; it is used to recognize acetylated sites on histones.
- Several protein motifs recognize methyl lysines, such as chromodomains, PHD domains, and Tudor domains.

All of the histones are subject to numerous covalent modifications, most of which occur in the histone tails. All of the histones can be modified at numerous sites by methylation, acetylation, or phosphorylation, as shown in **FIGURE 10.16**. While these modifications are relatively small, other, more dramatic modifications occur as well, such as mono-ubiquitylation, sumoylation, and ADP-ribosylation. Many of the functions of these modifications are yet to be characterized.

Lysines in the histone tails are the most common targets of modification. Acetylation, methylation, ubiquitylation and sumoylation all occur on the free epsilon (ε) amino group of lysine. As seen in **FIGURE 10.17**, acetylation neutralizes the positive charge that resides on the NH_3 form of the ε-amino group. In contrast, lysine methylation retains the positive charge, and lysine can be mono-, di-, or trimethylated. Arginine can be mono- or dimethylated. Phosphorylation occurs on the hydroxyl group of serine and threonine. This introduces a negative charge in the form of the phosphate group.

These modifications are transient. They can change the charge of the protein molecule, and as a result they are potentially able to change the functional properties of the octamers. For example, extensive lysine acetylation reduces the overall positive charge of the tails, leading to release of the tails from interactions with DNA on their own or other nucleosomes. Modification of histones is associated with structural changes that occur in chromatin at replication and transcription, and specific modifications also facilitate DNA repair. Modifications at *specific* positions on *specific* histones define different functional states of chromatin. Newly synthesized core histones carry specific patterns of acetylation that are removed after the histones are assembled into chromatin, as shown in **FIGURE 10.18**. Other modifications are dynamically added and removed to regulate transcription, replication, repair, and chromosome condensation. These other modifications are usually added and removed from nucleosomes that are incorporated into chromatin, as depicted for acetylation in **FIGURE 10.19**.

The specificity of the modifications is controlled by the fact that many of the modifying enzymes have individual target sites in specific histones. **FIGURE 10.20** summarizes the effects of some of the modifications that occur on histones H3 and H4. Many modified sites are subject to only a single type of modification *in vivo*, but others can be subject to alternative modification states (such as lysine 9 of histone H3, which is acetylated or methylated under different conditions). In some cases, modification

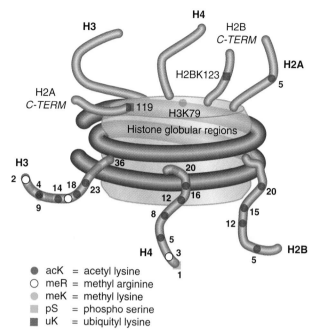

FIGURE 10.16 The histone tails can be acetylated, methylated, phosphorylated, and ubiquitylated at numerous sites. Not all possible modifications are shown. Adapted from *The Scientist* 17 (2003): p. 27.

of one site may activate or inhibit modification of another site. The idea that combinations of signals may be used to define chromatin function has sometimes been called the **histone code**. This hypothesis proposes that the *collective impact* of multiple modifications at particular sites defines the function of a chromatin domain. These modifications are not restricted to a single histone; the code is derived from all the modifications within a nucleosome or even nearby nucleosomes.

The changes in charge caused by some histone modifications can directly alter the structure of chromatin, but a major function of histone modification lies in the *creation of binding sites* for the attachment of nonhistone proteins that change the properties of chromatin. In recent years, a number of protein domains have been identified that bind to specifically modified histone tails.

The **bromodomain** is found in a variety of proteins that interact with chromatin. Bromodomains recognize acetylated lysine, and different bromodomain-containing proteins recognize different acetylated targets. The bromodomain itself recognizes only a very short sequence of four amino acids, including the acetylated lysine, so specificity for target recognition must depend on interactions involving other regions. The structure of a bromodomain bound to its acetylated lysine target is shown

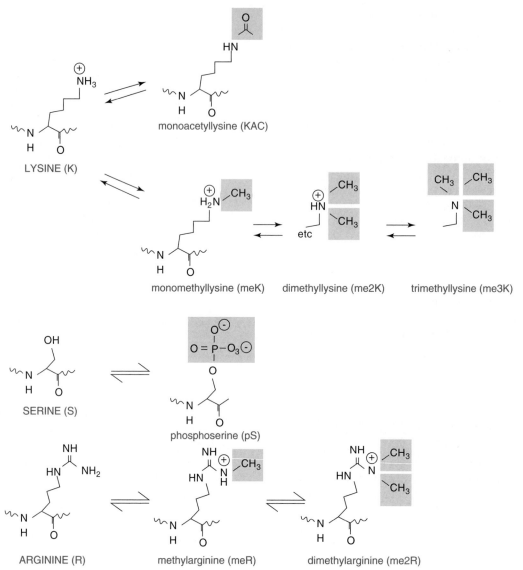

FIGURE 10.17 The positive charge on lysine is neutralized upon acetylation, while methylated lysine and arginine retain their positive charges. Lysine can be mono-, di- or triacetylated, while arginine can be mono- or diacetylated. Serine or threonine phosphorylation results in a negative charge.

in **FIGURE 10.21**. The bromodomain is found in a range of proteins that interact with chromatin, including components of the transcription apparatus and some of the enzymes that actually acetylate histones (discussed in *Section 28.9, Histone Acetylation Is Associated with Transcription Activation*).

Methylated lysines (and arginines) are recognized by a number of different domains, which not only can recognize specific modified sites but also can distinguish between mono-, di- or trimethylated lysines. The **chromodomain** is a common protein motif of 60 amino acids present in a number of chromatin-associated proteins. A number of other methyl lysine binding domains have been identified, as shown in Figure 10.22, such as the **PHD** (plant home-odomain) and **Tudor** domains; the number of different motifs designed to recognize particular methylated sites emphasizes the importance and complexity of histone modifications.

The idea that *combinations* of modifications are critical, as proposed in the histone code hypothesis, has been reinforced by recent discoveries of proteins or complexes that can recognize multiple sites of modification. For example, some proteins have tandem bromo-domains or chromodomains, with particular

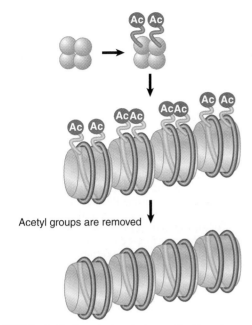

Acetyl groups are removed

FIGURE 10.18 Acetylation during replication occurs on specific sites on histones before they are incorporated into nucleosomes.

Inactive gene

Active gene

FIGURE 10.19 Acetylation associated with gene activation occurs by directly modifying specific sites on histones that are already incorporated into nucleosomes.

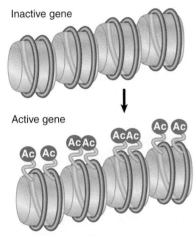

Histone	Site	Modification	Function
H3	K-4	Methylation	Transcription activation
H3	K-9	Methylation	Chromatin condensation
	K-9	Methylation	Required for DNA methylation
	K-9	Acetylation	Transcription activation
H3	S-10	Phosphorylation	Transcription activation
H3	K-14	Acetylation	Prevents methylation at Lys-9
H3	K-79	Methylation	Telomeric silencing
H4	R-3	Methylation	
H4	K-5	Acetylation	Nucleosome assembly
H4	K-12	Acetylation	Nucleosome assembly
H4	K-16	Acetylation	Nucleosome assembly
	K-16	Acetylation	Fly X activation

FIGURE 10.20 Most modified sites in histones have a single, specific type of modification, but some sites can have more than one type of modification. Individual functions can be associated with some of the modifications.

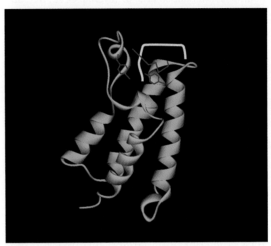

FIGURE 10.21 Bromodomains are protein motifs that bind acetyllysines. The bromodomain fold consists of a cluster of four α-helices with an acetyllysine binding pocket at one end. Figure shows the bromodomain of yeast Gcn5 bound to an H4K16ac peptide. Structure from Protein Data Bank 1E6I. D. J. Owen, et al., *EMBO J.* 19 (2000): 6141–6149.

10.5 Histone Variants Produce Alternative Nucleosomes

Key concepts

- All core histones except H4 are members of families of related variants.
- Histone variants can be closely related or highly divergent from canonical histones.
- Different variants serve different functions in the cell.

While all nucleosomes share a related core structure, some nucleosomes exhibit subtle or dramatic differences resulting from the incorporation of **histone variants**. Histone variants comprise a large group of histones that are

spacing, which could promote binding to histones that are acetylated or methylated at two specific sites. There are also cases in which modification at one site can prevent a protein from recognizing its target modification at another site. It is clear that the effects of a single modification may not always be predictable, and the context of other modifications must be accounted for in order to assign a function to a region of chromatin.

(A)

(B)

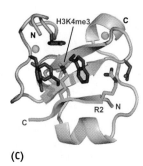

(C)

FIGURE 10.22 Numerous protein motifs recognize methylated lysines. (A) The chromodomain of HP1 binds trimethylated K9 of histone H3. Structure from Protein Data Bank 1KNE. S. A. Jacobs and S. Khorasanizadeh, *Science* 295 (2002): 2080–2083. (B) The Tudor domain of JMJD2A binds trimethylated K4 of histone H3. Both chromodomains and Tudor domains are members of the "royal superfamily," which bind their targets via a partial β-barrel structure. Structure from Protein Data Bank 2GFA. Y. Huang, et al., *Science* 12 (2006): 748–751. (C) The PHD finger of BPTF also binds trimethylated K4 of histone H3, using a structure related to DNA-binding zinc finger domains. Photo courtesy of Sean D. Taverna, Johns Hopkins University School of Medicine, and Haitao Li, Memorial Sloan-Kettering Cancer Center. Additional information at S. D. Taverna, et al., *Nat. Struct. Mol. Biol.* 14 (2007): 1025-1040.

related to the histones we have already discussed, but have differences in sequence from the "canonical" histones. These sequence differences can be small (as few as four amino acid differences) or extensive (such as alternative tail sequences).

Variants have been identified for all core histones except histone H4. The best-characterized histone variants are summarized in **FIGURE 10.23**. Most variants have significant differences between them, particularly in the N- and C-terminal tails. At one extreme, macroH2A is nearly three times larger than conventional H2A, and contains a large C-terminal tail that is not related to any other histone. At the other end of the spectrum, canonical H3 (also known as H3.1) differs from the H3.3 variant at only four amino acid positions, three in the histone core and one in the N-terminal tail.

Histone variants have been implicated in a number of different functions, and their incorporation changes the nature of the chromatin containing the variant. We have already discussed one type of histone variant, the centromeric H3 (or CenH3) histone, known as Cse4 in yeast. CenH3 histones are incorporated into specialized nucleosomes present at centromeres in all eukaryotes (see *Sections 9.13, Regional Centromeres Contain a Centromeric Histone H3 Variant and Repetitive DNA,* and *9.15, The* S. cerevisiae *Centromere Binds a Protein Complex*). In yeast, it has been shown that these centromeric nucleosomes consist of Cse4, H4, and a nonhistone protein Scm3, which replaces H2A/H2B dimers. In *Drosophila,* the centromeric chromatin appears to consist of "hemisomes" containing one copy each of CenH3, H4, H2A, and H2B. It is not known whether any centromeric chromatin in higher eukaryotes contains an Scm3-like protein at a subset of centromeric nucleosomes.

The other major H3 variant is histone H3.3. In multicellular eukaryotes this variant is a minority component of the total H3 in the cell, but in yeast, the major H3 is actually of the H3.3 type. H3.3 is expressed throughout the cell cycle, in contrast to most histones that are expressed during S phase, when new chromatin assembly is required during DNA replication. As a result, H3.3 is available for assembly at any time in the cell cycle, and is incorporated at sites of active transcription, where nucleosomes become disrupted. As a result of this, H3.3 is often referred to as a "replacement" histone, in contrast to the "replicative" histone H3.1 (see *Section 10.8, Replication of Chromatin Requires Assembly of Nucleosomes*).

The H2A variants are the largest and most diverse family of core histone variants, and have been implicated in a variety of distinct functions. One of the best studied is the variant H2AX. H2AX is normally present in only 10%–15% of the nucleosomes in multicellular eukaryotes, though again (like H3.3) this subtype is the major H2A present in yeast. This variant has a C-terminal tail that is distinct from the canonical H2A, which is characterized by a SQEL/Y motif at the end. This motif is the target of phosphorylation by ATM/ATR kinases, activated by DNA damage, and this histone variant is involved in DNA repair, particularly repair of double-strand breaks (see Chapter 16, *Repair Systems Handle Damage to DNA*). H2AX phosphorylated at the SQEL/Y motif is referred to as "γ-H2AX," and is required to stabilize binding of various repair factors at DNA breaks and to maintain checkpoint arrest. γ-H2AX appears

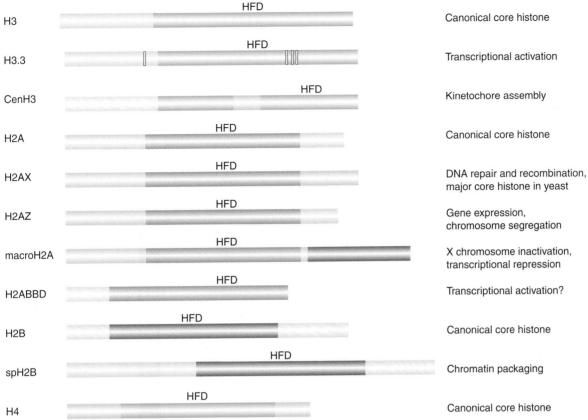

FIGURE 10.23 The major core histones contain a conserved histone-fold domain. In the histone H3.3 variant, the residues that differ from the major histone H3 (also known as H3.1) are highlighted in yellow. The centromeric histone CenH3 has a unique N terminus, which does not resemble other core histones. Most H2A variants contain alternative C-termini, except H2ABbd, which contains a distinct N terminus. The sperm-specific SpH2B has a long N-terminus. Proposed functions of the variants are listed. Adapted from K. Sarma and D. Reinberg, *Nat. Rev. Mol. Cell Biol.* 6 (2005): 139–149.

within moments at broken DNA ends, as can be seen in **FIGURE 10.24**, which shows foci of γ-H2AX forming along the path of double-strand breaks induced by a laser.

Other H2A variants have different roles. The H2AZ variant, which has ~60% sequence identity with canonical H2A, has been shown to be important in several processes, such as gene activation, heterochromatin–euchromatin boundary formation, and cell-cycle progression. The vertebrate-specific macroH2A is named for its extremely long C-terminal tail, which contains a leucine-zipper dimerization motif that may mediate chromatin compaction by facilitating internucleosome interactions. Mammalian macroH2A is enriched in the inactive X chromosome in females, which is assembled into a silent, heterochromatic state. In contrast, the mammalian H2ABbd variant is *excluded* from the inactive X, and forms a less stable nucleosome than canonical H2A; perhaps this histone is designed to be more easily displaced in transcriptionally active regions of euchromatin.

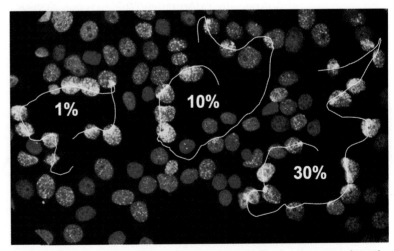

FIGURE 10.24 γ-H2AX is detected by an antibody (yellow) and appears along the path traced by a laser that produces double-strand breaks. The percentages refer to the relative laser energy used in each transit. Nuclei are red. © Rogakou et al., 1999. Originally published in **The Journal of Cell Biology**, 146: 905–915. Photo courtesy of William M. Bonner, National Cancer Institute, NIH.

Still other variants are expressed in limited tissues, such as spH2B, present in sperm and required for chromatin compaction. The presence and distribution of histone variants shows that individual chromatin regions, entire chromosomes, or even specific tissues can have unique "flavors" of chromatin specialized for different functions. In addition, the histone variants, like the canonical histones, are subject to numerous covalent modifications, adding levels of complexity to the roles chromatin plays in nuclear processes.

10.6 DNA Structure Varies on the Nucleosomal Surface

Key concepts

- DNA is wrapped 1.67 times around the histone octamer.
- DNA on the nucleosome shows regions of smooth curvature and regions of abrupt kinks.
- The structure of the DNA is altered so that it has an increased number of base pairs/turn in the middle, but a decreased number at the ends.
- ~0.6 negative turns of DNA are absorbed by the change in bp/turn from 10.5 in solution to an average of 10.2 on the nucleosomal surface, which explains the linking-number paradox.

So far we have focused on the protein components of the nucleosome. The DNA wrapped around these proteins is in an unusual conformation. The exposure of DNA on the surface of the nucleosome explains why it is accessible to cleavage by certain nucleases. The reaction with nucleases that attack single strands has been especially informative. The enzymes DNase I and DNase II make single-strand nicks in DNA; they cleave a bond in one strand, but the other strand remains intact. No effect is visible in double-stranded DNA, but when this DNA is

denatured, short fragments are released instead of full-length single strands. If the DNA has been labeled at its ends, the end fragments can be identified by detection of the label, as summarized in **FIGURE 10.25**. When DNA is free in solution, it is nicked (relatively) at random. The DNA on nucleosomes can also be nicked by the enzymes, but only at regular intervals. When the points of cutting are determined by using end-labeled DNA and the DNA is denatured and electrophoresed, a ladder of the sort displayed in **FIGURE 10.26** is obtained.

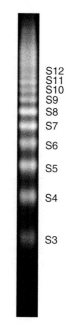

FIGURE 10.26 Sites for nicking lie at regular intervals along core DNA, as seen in a DNase I digest of nuclei. Photo courtesy of Leonard C. Lutter, Molecular Biology Research Program, Henry Ford Hospital.

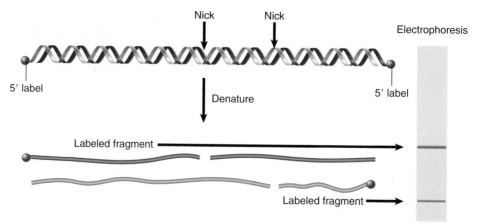

FIGURE 10.25 Nicks in double-stranded DNA are revealed by fragments when the DNA is denatured to give single strands. If the DNA is labeled at (say) the 5′ ends, only the 5′ fragments are visible by autoradiography. The size of the fragment identifies the distance of the nick from the labeled end.

The interval between successive steps on the ladder is 10-11 bases. The ladder extends for the full distance of core DNA. The cleavage sites are numbered as S1 through S13 (where S1 is ~10 bases from the labeled 5′ end, S2 is ~20 bases from it, and so on). The enzymes DNase I and DNase II generate essentially the same ladder, and the same pattern is obtained by cleaving with a hydroxyl radical, which argues that the pattern reflects the structure of the DNA itself rather than any sequence preference. The sensitivity of nucleosomal DNA to nucleases is analogous to a footprinting experiment. Thus we can assign the lack of reaction at particular target sites to the structure of the nucleosome, in which certain positions on DNA are rendered inaccessible.

There are two strands of DNA in the core particle, so in an end-labeling experiment both of the 5′ (or 3′) ends are labeled, one on each strand. Thus the cutting pattern includes fragments derived from both strands. This is visible in Figure 10.25, where each labeled fragment is derived from a different strand. The corollary is that, in an experiment, each labeled band may actually represent two fragments that are generated by cutting the same distance from either of the labeled ends.

How, then, should we interpret discrete preferences at particular sites? One view is that the path of DNA on the particle is symmetrical (about a horizontal axis through the nucleosome, as drawn in Figure 10.7). If, for example, no 80-base fragment is generated by DNase I, this must mean that the position at 80 bases from the 5′ end of either strand is not susceptible to the enzyme.

When DNA is immobilized on a flat surface, sites are cut with a regular separation. **FIGURE 10.27** shows that this reflects the recurrence of the exposed site with the helical periodicity of B-form DNA. The cutting periodicity (the spacing between cleavage points) coincides with—indeed, is a reflection of—the structural periodicity (the number of base pairs per turn of the double helix). Thus the distance between the sites corresponds to the number of base pairs per turn. Measurements of this type yield the average value for double-helical B-type DNA of 10.5 bp/turn.

A similar analysis of DNA on the surface of the nucleosome reveals striking variation in the structural periodicity at different points. At the ends of the DNA, the average distance between pairs of DNase I digestion sites is about 10.0 bases each, significantly less than the usual 10.5 bp/turn. In the center of the particle, the separation between cleavage sites averages 10.7 bases. This variation in cutting periodicity along the core DNA means that there is variation in the structural periodicity of core DNA. The DNA has more bp/turn than its solution value in the middle, but has fewer bp/turn at the ends. The average periodicity over the entire nucleosome is only 10.17 bp/turn, which is significantly less than the 10.5 bp/turn of DNA in solution.

The crystal structure of the core particle (Figure 10.12) shows that DNA is wound into a *solenoidal* (spring-shaped) supercoil, with 1.67 turns wound around the histone octamer. The pitch of the superhelix varies and has a discontinuity in the middle. Regions of high curvature are arranged symmetrically, and are the sites least sensitive to DNase I.

The high-resolution structure of the nucleosome core shows in detail how the structure of DNA is distorted. Most of the supercoiling occurs in the central 129 bp, which are coiled into 1.59 left-handed superhelical turns with a diameter of 80 Å (only four times the diameter of the DNA duplex itself). The terminal sequences on either end make only a very small contribution to the overall curvature.

The central 129 bp are in the form of B-DNA, but with a substantial curvature that is needed to form the superhelix. The major groove is smoothly bent, but the minor groove has abrupt kinks, as shown in **FIGURE 10.28**. These conformational changes may explain why the central part of nucleosomal DNA is not usually a target for binding by regulatory proteins, which typically bind to the terminal parts of the core DNA or to the linker sequences.

Some insights into the structure of nucleosomal DNA emerge when we compare predictions for supercoiling in the path that DNA follows with actual measurements of supercoiling of nucleosomal DNA. Circular "minichromosomes" that are fully assembled into nucleosomes can be isolated from eukaryotic cells. The degree of supercoiling on the individual nucleosomes of the minichromosome can be measured as illustrated in **FIGURE 10.29**. First,

Sites exposed to DNase I

FIGURE 10.27 The most exposed positions on DNA recur with a periodicity that reflects the structure of the double helix. (For clarity, sites are shown for only one strand.)

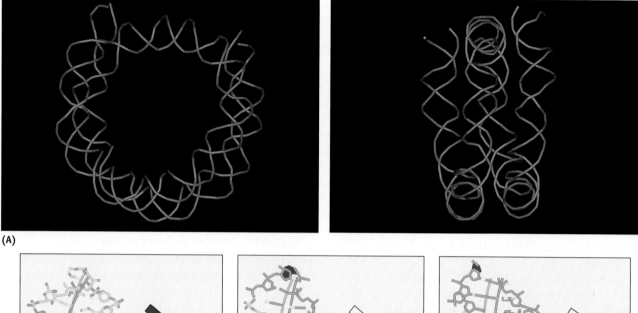

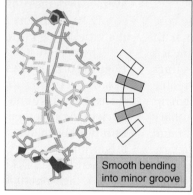

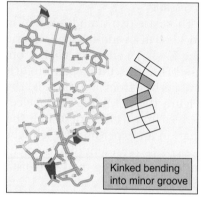

(A)

(B)

Smooth bending into major groove

Smooth bending into minor groove

Kinked bending into minor groove

FIGURE 10.28 DNA structure in nucleosomal DNA. (A) The trace of the DNA backbone in the nucleosome is shown in the absence of protein for clarity. Structures from Protein Data Bank: 1P34. U. M. Muthurajan, et al., *EMBO J.* 23 (2004): 260–271. (B) Regions of curvature in nucleosomal DNA. Actual structures (left) and schematic representations (right) show uniformity of curvature along the major groove (blue) and both smooth and kinked bending into the minor groove (orange). Also indicated are the DNA axes for the experimental (pink) and ideal (gray) superhelices. Adapted from T. J. Richmond, and C. A. Davey, *Nature* 423 (2003): 145–150.

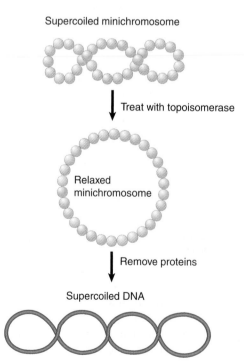

Supercoiled minichromosome

Treat with topoisomerase

Relaxed minichromosome

Remove proteins

Supercoiled DNA

the free supercoils of the minichromosome itself are relaxed, so that the nucleosomes form a circular string with a superhelical density of 0. Next, the histone octamers are extracted. This releases the DNA to follow a free path. Every supercoil that was present but constrained in the minichromosome will appear in the deproteinized DNA as −1 turn. Now the total number of supercoils in the DNA is measured.

The observed value is close to the number of nucleosomes. Thus the DNA follows a path on the nucleosomal surface that generates ~1 negative supercoiled turn when the restraining protein is removed. The path that DNA follows on the nucleosome, though, corresponds to −1.67 superhelical turns. This discrepancy is sometimes called the *linking number paradox*.

FIGURE 10.29 The supercoils of the SV40 minichromosome can be relaxed to generate a circular structure, whose loss of histones then generates supercoils in the free DNA.

The discrepancy is explained by the difference between the 10.17 average bp/turn of nucleosomal DNA and the 10.5 bp/turn of free DNA. In a nucleosome of 200 bp, there are 200/10.17 = 19.67 turns. When DNA is released from the nucleosome, it now has 200/10.5 = 19.0 turns. The path of the less tightly wound DNA on the nucleosome absorbs –0.67 turns, which explains the discrepancy between the physical path of –1.67 and the measurement of –1.0 superhelical turns. In effect, some of the torsional strain in nucleosomal DNA goes into increasing the number of bp/turn; only the rest is left to be measured as a supercoil.

10.7 The Path of Nucleosomes in the Chromatin Fiber

Key concepts

- 10 nm chromatin fibers consist of a string of nucleosomes.
- 30 nm fibers have six nucleosomes/turn, which are organized into a two-start helix.
- Histone H1, histone tails, and increased ionic strength all promote the formation of the 30 nm fiber.

When chromatin is examined in the electron microscope, two types of fibers are seen: the 10 nm fiber and 30 nm fiber. They are described by the approximate diameter of the thread (that of the 30 nm fiber actually varies from ~25–30 nm).

The **10 nm fiber** is essentially a continuous string of nucleosomes and represents the least compacted level of chromatin structure. In fact, in a stretched-out 10 nm fiber, linker DNA and nucleosomes can be easily distinguished and the fiber resembles a string of beads, as seen in the example of **FIGURE 10.30**. The 10 nm fiber structure is obtained under conditions of low ionic strength and does not require the presence of histone H1. This means that it is a function strictly of the nucleosomes themselves. A depiction of the continuous series of nucleosomes in this fiber is shown in **FIGURE 10.31**.

When chromatin is visualized in conditions of greater ionic strength, the **30 nm fiber** is obtained. An example is given in **FIGURE 10.32**. The fiber can be seen to have an underlying coiled structure. It has ~6 nucleosomes for every turn, which corresponds to a packing ratio of 40 (that is, each μm along the axis of the fiber contains 40 μm of DNA). The formation of this fiber requires the histone tails, which are involved in internucleosomal contacts, and

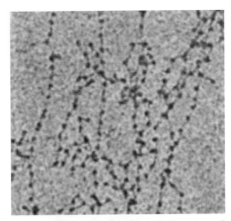

FIGURE 10.30 The 10 nm fiber in partially unwound state can be seen to consist of a string of nucleosomes. Photo courtesy of Barbara Hamkalo, University of California, Irvine.

FIGURE 10.31 The 10 nm fiber is a continuous string of nucleosomes.

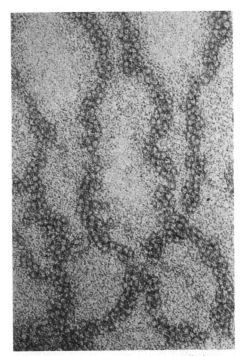

FIGURE 10.32 The 30 nm fiber has a coiled structure. Photo courtesy of Barbara Hamkalo, University of California, Irvine.

is facilitated by the presence of a linker histone such as H1. This fiber is thought to be the basic constituent of both interphase chromatin and mitotic chromosomes, though it has been difficult to observe this directly *in vivo*.

The most likely arrangement for packing nucleosomes into the fiber is a solenoid, in which the nucleosomes turn in a helical array that is coiled around a central cavity (which is likely occupied by linker DNA). The two main forms of a solenoid are a single-start, which forms from a single linear array, and a two-start, which in effect consists of a double row of nucleosomes. **FIGURE 10.33** shows a two-start model suggested by recent crosslinking data identifying a double stack of nucleosomes in the 30 nm fiber. This model is also supported by the crystal structure of a tetranucleosome complex.

Levels of folding beyond the 30 nm fiber are very poorly understood, but it is obvious that the 40-fold compaction provided by the 30 nm fiber is still a long way from the levels of compaction required for interphase or mitotic packaging of chromosomes. Chromatin fibers with diameters of 60–300 nm (called "chromonema fibers") have been observed by both light and electron microscopy. Such fibers are presumed to consist of folded 30 nm fibers and would represent a major level of compaction (a 30 nm fiber running just across the width of a 100 nm fiber would contain >10 kb of DNA), but the actual substructures of these large fibers remain unknown. **FIGURE 10.34** shows a hypothetical depiction of higher order folding.

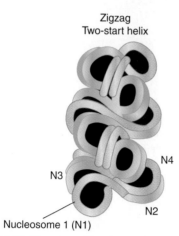

FIGURE 10.33 The 30 nm fiber is a two start helix consisting of two rows of nucleosomes coiled into a solenoid. Reprinted from *Cell*, vol. 128, D. J. Tremethick, Higher-order structure of chromatin . . . , pp. 651–654. Copyright 2007, with permission from Elsevier [http://www.sciencedirect.com/science/journal/00928674].

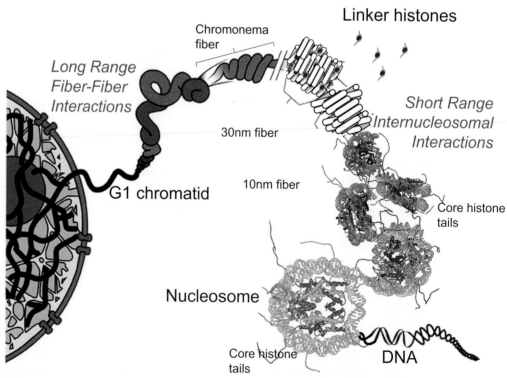

FIGURE 10.34 Levels of chromatin packaging. 10 nm fibers are folded into 30 nm fibers as a result of short range internucleosomal interactions, stabilized by linker histones. 30 nm fibers are further folded into large chromonema fibers, which are ultimately organized through long-range fiber-fiber interactions and other interactions to form interphase chromatids or metaphase chromosomes. Modified courtesy of Karolin Luger and Jeffrey C. Hansen, Colorado State University.

10.8 Replication of Chromatin Requires Assembly of Nucleosomes

Key concepts

- Histone octamers are not conserved during replication, but H2A-H2B dimers and $H3_2$-$H4_2$ tetramers are conserved.
- There are different pathways for the assembly of nucleosomes during replication and independently of replication.
- Accessory proteins are required to assist the assembly of nucleosomes.
- CAF-1 and ASF1 are histone assembly proteins that are linked to the replication machinery.
- A different assembly protein, HIRA, and the histone H3.3 variant are used for replication-independent assembly.

Replication separates the strands of DNA and therefore must inevitably disrupt the structure of the nucleosome. However, this disruption is confined to the immediate vicinity of the replication fork. Once DNA has been replicated, nucleosomes are quickly generated on both the duplicates. The transience of the replication event is a major difficulty in analyzing the structure of a particular region while it is being replicated.

Replication of chromatin does not involve any protracted period during which the DNA is free of histones. Once DNA has been replicated, nucleosomes are quickly generated on both the duplicates. This point is illustrated by the electron micrograph of **FIGURE 10.35**, which shows a recently replicated stretch of DNA that is already packaged into nucleosomes on both daughter duplex segments.

Both biochemical analysis and visualization of the replication fork indicate that the disrup-

tion of nucleosome structure is limited to a short region immediately around the fork. Progress of the fork disrupts nucleosomes, but they form very rapidly on the daughter duplexes as the fork moves forward. In fact, the assembly of nucleosomes is directly linked to the replisome that is replicating DNA.

How do histones associate with DNA to generate nucleosomes? Do the histones preform a protein octamer around which the DNA is subsequently wrapped? Or does the histone octamer assemble on DNA from free histones? Either of these pathways can be used *in vitro* to assemble nucleosomes, depending on the conditions that are employed. In one pathway, a preformed octamer binds to DNA. In the other pathway, a tetramer of $H3_2$-$H4_2$ binds first, and then two H2A-H2B dimers are added. This latter stepwise assembly is the pathway that is used in replication, shown in **FIGURE 10.36**.

Accessory proteins are involved in assisting histones to associate with DNA. Accessory proteins can act as "molecular chaperones" that bind to the histones in order to release either individual histones or complexes ($H3_2$-$H4_2$ or H2A-H2B) to the DNA in a controlled manner. This could be necessary because the histones, as basic proteins, have a general high affinity for DNA. Such interactions allow histones to form nucleosomes without becoming trapped in other kinetic intermediates (that is, other complexes resulting from indiscreet binding of histones to DNA).

Numerous histone chaperones have been identified. Chromatin assembly factor (CAF)-1 and Anti-silencing function 1 (ASF1) are two chaperones that function at the replication fork. CAF-1 is a conserved three-subunit complex that is directly recruited to the replication fork by proliferating cell nuclear antigen (PCNA), the processivity factor for DNA polymerase. ASF1 interacts with the replicative helicase that unwinds the replication fork. Furthermore, CAF-1 and ASF1 interact with each other. These interactions provide the link between replication and nucleosome assembly, ensuring that nucleosomes are assembled as soon as DNA has been replicated.

CAF-1 acts stoichiometrically, and functions by binding to newly synthesized H3 and H4. New nucleosomes form by assembling first the $H3_2$-$H4_2$ tetramer, and then adding the H2A-H2B dimers. ASF1 appears to play an important role in transfer of parental nucleosomes from ahead of the replication fork to the newly synthesized region behind the fork,

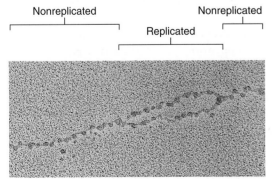

FIGURE 10.35 Replicated DNA is immediately incorporated into nucleosomes. Photo courtesy of Steven L. McKnight, UT Southwestern Medical Center at Dallas.

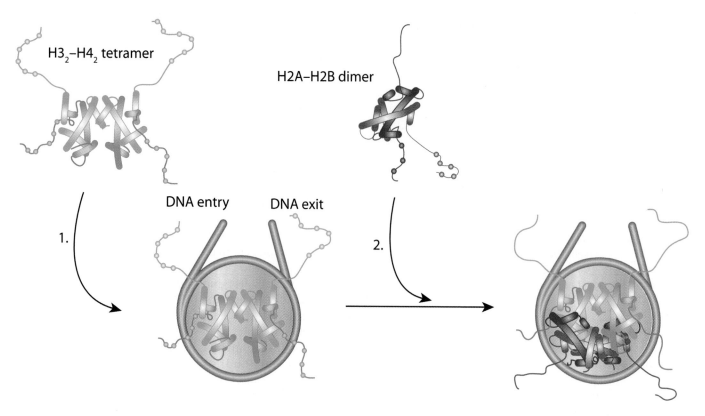

FIGURE 10.36 During nucleosome assembly *in vivo*, H3-H4 tetramers form and bind DNA first, then two H2A-H2B dimers are added to form the complete nucleosome.

although ASF1 can bind and assemble newly synthesized histones as well.

The pattern of disassembly and reassembly has been difficult to characterize in detail, but a working model is illustrated in **FIGURE 10.37**. The replication fork displaces histone octamers, which then dissociate into $H3_2$-$H4_2$ tetramers and H2A-H2B dimers. These "old" tetramers and dimers enter a pool that also includes "new" tetramers and dimers, which are assembled from newly synthesized histones. Nucleosomes assemble ~600 bp behind the replication fork. Assembly is initiated when $H3_2$-$H4_2$ tetramers bind to each of the daughter duplexes, assisted by CAF-1 or ASF1. Two H2A-H2B dimers then bind to each $H3_2$-$H4_2$ tetramer to complete the histone octamer. The assembly of tetramers and dimers is random with respect to "old" and "new" subunits. It appears that nucleosomes are disrupted and reassembled in a similar way during transcription, though different histone chaperones are involved in this process (see *Section 10.10, Nucleosomes Are Displaced and Reassembled During Transcription*).

During S phase (the period of DNA replication) in a eukaryotic cell, the duplication of chromatin requires synthesis of sufficient histone proteins to package an entire genome—

basically the same quantity of histones must be synthesized that are already contained in nucleosomes. The synthesis of histone mRNAs is controlled as part of the cell cycle, and increases enormously in S phase. The pathway for assembling chromatin from this equal mix of old and new histones during S phase is called the *replication-coupled (RC) pathway*.

Another pathway, called the *replication-independent (RI) pathway*, exists for assembling nucleosomes during other phases of cell cycle, when DNA is not being synthesized. This may become necessary as the result of damage to DNA or because nucleosomes are displaced during transcription. The assembly process must necessarily have some differences from the replication-coupled pathway, because it cannot be linked to the replication apparatus. The replication-independent pathway uses the histone H3.3 variant, which was introduced in *Section 10.5, Histone Variants Produce Alternative Nucleosomes*.

The histone H3.3 variant differs from the highly conserved H3 histone at four amino acid positions (see Figure 10.21). H3.3 slowly replaces H3 in differentiating cells that do not have replication cycles. This happens as the result of assembly of new histone octamers to

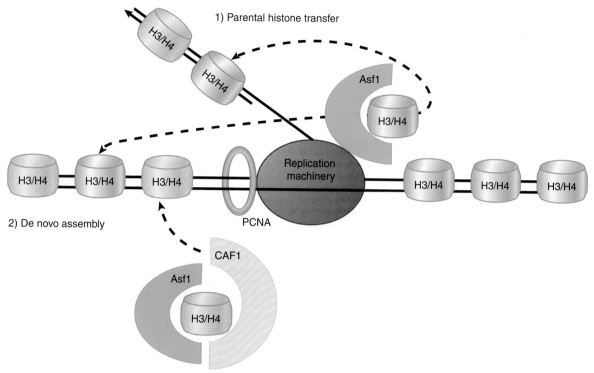

FIGURE 10.37 Replication fork passage displaces histone octamers from DNA. They disassemble into H3-H4 tetramers and H2A-H2B dimers. H3-H4 tetramers (blue) are directly transferred behind the replication forks. Newly synthesized histones (orange) are assembled into H3-H4 tetramers and H2A-H2B dimers. The old and new tetramers and dimers are assembled with the aid of histone chaperones into new nucleosomes immediately behind the replication fork. H2A-H2B dimers are omitted from the figure for simplicity; chaperones responsible for dimer assembly have not been identified. Adapted from W. Rocha and A. Verreault, *FEBS Lett.* 582 (2008): 1938–1949.

replace those that have been displaced from DNA for whatever reason. The mechanism that is used to ensure the use of H3.3 in the replication-independent pathway is different in two cases that have been investigated.

In the protozoan *Tetrahymena*, histone usage is determined exclusively by availability. Histone H3 is synthesized only during the cell cycle; the variant replacement histone is synthesized only in nonreplicating cells. In *Drosophila*, however, there is an active pathway that ensures the usage of H3.3 by the replication-independent pathway. New nucleosomes containing H3.3 assemble at sites of transcription, presumably replacing nucleosomes that were displaced by RNA polymerase. The assembly process discriminates between H3 and H3.3 on the basis of their sequences, specifically excluding H3 from being utilized. By contrast, replication-coupled assembly uses both types of H3 (although H3.3 is available at much lower levels than H3, and therefore enters only a small proportion of nucleosomes).

CAF-1 is not involved in replication-independent assembly. (There also are organisms such as yeast and *Arabidopsis* for which its gene is not essential, implying that alternative assembly processes may be used in replication-

coupled assembly.) Instead, replication-independent assembly uses a factor called HIRA. Depletion of HIRA from *in vitro* systems for nucleosome assembly inhibits the formation of nucleosomes on nonreplicated DNA, but not on replicating DNA, which indicates that the pathways do indeed use different assembly mechanisms. Like CAF-1 and ASF1, HIRA functions as a chaperone to assist the incorporation of histones into nucleosomes. This pathway appears to be generally responsible for replication-independent assembly; for example, HIRA is required for the decondensation of the sperm nucleus, when protamines are replaced by histones, in order to generate chromatin that is competent to be replicated following fertilization.

As described earlier, assembly of nucleosomes containing an alternative to H3 also occurs at centromeres (see *Section 9.15, The S. cerevisiae Centromere Binds a Protein Complex*). Centromeric DNA replicates early during the replication phase of the cell cycle. The incorporation of H3 at the centromeres is inhibited, and instead a protein called CENP-A is incorporated in higher eukaryotic cells (in *Drosophila* it is called Cid, and in yeast it is called Cse4). This occurs by the replication-independent assembly pathway, apparently because the replication-coupled pathway is inhibited for a brief period while centromeric DNA replicates.

10.9 Do Nucleosomes Lie at Specific Positions?

Key concepts

- Nucleosomes may form at specific positions as the result of either the local structure of DNA or proteins that interact with specific sequences.
- A common cause of nucleosome positioning is when proteins binding to DNA establish a boundary.
- Positioning may affect which regions of DNA are in the linker and which face of DNA is exposed on the nucleosome surface.
- DNA sequence determinants (exclusion or preferential binding) may be responsible for half of the *in vivo* nucleosome positions.

Does a particular DNA sequence always lie in a certain position *in vivo* with regard to the topography of the nucleosome? Or are nucleosomes arranged randomly on DNA, so that a particular sequence may occur at any location—for example, in the core region in one copy of the genome and in the linker region in another?

To investigate this question, it is necessary to use a defined sequence of DNA; more precisely, we need to determine the position relative to the nucleosome of a defined point in the DNA. **FIGURE 10.38** illustrates the principle of a procedure used to achieve this.

Suppose that the DNA sequence is organized into nucleosomes in only one particular configuration, so that each site on the DNA always is located at a particular position on the nucleosome. This type of organization is called **nucleosome positioning** (or sometimes nucleosome phasing). In a series of positioned nucleosomes, the linker regions of DNA comprise unique sites.

Consider the consequences for just a single nucleosome. Cleavage with MNase generates a monomeric fragment that constitutes a specific sequence. If the DNA is isolated and cleaved with a restriction enzyme that has only one target site in this fragment, it should be cut at a unique point. This produces two fragments, each of unique size.

The products of the MNase/restriction enzyme double digest are separated by gel electrophoresis. A probe representing the sequence on one side of the restriction site is used to iden-

Positioning places target sequence (red) at unique position

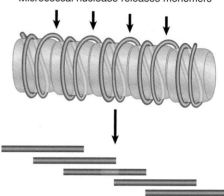

Micrococcal nuclease releases monomers

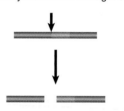

Restriction enzyme cleaves at target sequence

Fragment has restriction cut at one end, micrococcal cut at other end; electrophoresis gives unique band

FIGURE 10.38 Nucleosome positioning places restriction sites at unique positions relative to the linker sites cleaved by micrococcal nuclease.

tify the corresponding fragment in the double digest. This technique is called **indirect end labeling** (because it is not possible to label the end of the nucleosomal DNA fragment itself, so it must be detected indirectly with a probe).

Reversing the argument, the identification of a single sharp band demonstrates that the position of the restriction site is uniquely defined with respect to the end of the nucleosomal DNA (as defined by the MNase cut). Thus the nucleosome has a unique sequence of DNA.

What happens if the nucleosomes do not lie at a single position? Now the linkers consist of different DNA sequences in each copy of the genome. Thus the restriction site lies at

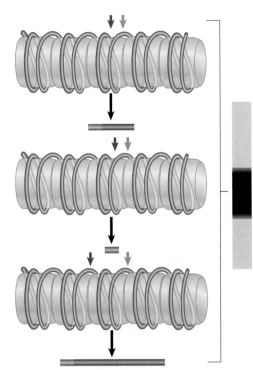

FIGURE 10.39 In the absence of nucleosome positioning, a restriction site can lie at any possible location in different copies of the genome. Fragments of all possible sizes are produced when a restriction enzyme cuts at a target site (red) and micrococcal nuclease cuts at the junctions between nucleosomes (green).

a different position each time; in fact, it lies at all possible locations relative to the ends of the monomeric nucleosomal DNA. **FIGURE 10.39** shows that the double cleavage then generates a broad smear, ranging from the smallest detectable fragment (~20 bases) to the length of the monomeric DNA.

In discussing these experiments, we have treated MNase as an enzyme that cleaves DNA at the exposed linker regions without any sort of sequence specificity. MNase does have some sequence specificity, though, which is biased toward selection of A-T-rich sequences. Thus we cannot assume that the existence of a specific band in the indirect end-labeling technique represents the distance from a restriction cut to the linker region. It could instead represent the distance from the restriction cut to a preferred micrococcal nuclease cleavage site.

This possibility is controlled by treating the naked DNA in exactly the same way as the chromatin. If there are preferred sites for MNase in the particular region, specific bands are found. This pattern of bands can then be compared with the pattern generated from chromatin.

A difference between the control DNA band pattern and the chromatin pattern provides evidence for nucleosome positioning. Some of the bands present in the control DNA digest may disappear from the nucleosome digest, indicating that preferentially cleaved positions are unavailable. New bands may appear in the nucleosome digest when new sites are rendered preferentially accessible by the nucleosomal organization.

Nucleosome positioning might be accomplished in either of two ways:

- Intrinsic mechanisms: *Nucleosomes are deposited specifically at particular DNA sequences, or are excluded by specific sequences.* This modifies our view of the nucleosome as a subunit able to form between any sequence of DNA and a histone octamer.
- Extrinsic mechanisms: *The first nucleosome in a region is preferentially assembled at a particular site due to action of other protein(s).* A preferential starting point for nucleosome positioning can result either from the exclusion of a nucleosome from a particular region (due to competition with another protein binding that region), or by specific deposition of a nucleosome at a particular site. The excluded region or the positioned nucleosome provides a boundary that restricts the positions available to the adjacent nucleosome. A series of nucleosomes may then be assembled sequentially, with a defined repeat length.

It is now clear that the deposition of histone octamers on DNA is not random with regard to sequence. The pattern is intrinsic in some cases, in which it is determined by structural features in DNA. It is extrinsic in other cases, resulting from the interactions of other proteins with the DNA and/or histones.

Certain structural features of DNA affect placement of histone octamers. DNA has intrinsic tendencies to bend in one direction rather than another. For example, AT dinucleotides bend easily, and thus A-T-rich sequences are easier to wrap tightly in a nucleosome. A-T-rich regions locate so that the minor groove faces in toward the octamer, whereas G-C-rich regions are arranged so that the minor groove points out. Long runs of dA-dT (>8 bp), in contrast, stiffen the DNA and avoid positioning in the central tight superhelical turn of the core. It is not yet possible to sum all of the relevant structural effects and thus entirely predict the location of a particular DNA sequence with regard

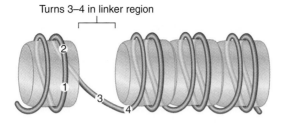

Turns 3–4 in linker region

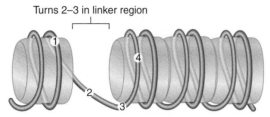

Turns 2–3 in linker region

FIGURE 10.40 Translational positioning describes the linear position of DNA relative to the histone octamer. Displacement of the DNA by 10 bp changes the sequences that are in the more exposed linker regions, but does not necessarily alter which face of DNA is protected by the histone surface and which is exposed to the exterior.

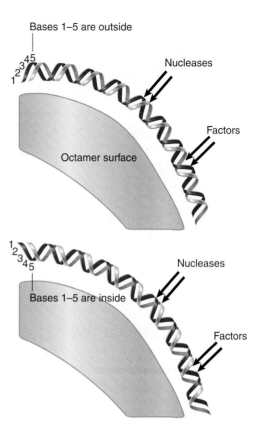

FIGURE 10.41 Rotational positioning describes the exposure of DNA on the surface of the nucleosome. Any movement that differs from the helical repeat (~10.2 bp/turn) displaces DNA with reference to the histone surface. Nucleotides on the inside are more protected against nucleases than nucleotides on the outside.

to the nucleosome. Sequences that cause DNA to take up more extreme structures may have effects such as the exclusion of nucleosomes, and thus could cause boundary effects.

Positioning of nucleosomes near boundaries is common. If there is some variability in the construction of nucleosomes—for example, if the length of the linker can vary by, say, 10 bp—the specificity of positioning would decline proceeding away from the first, defined nucleosome at the boundary. In this case, we might expect the positioning to be maintained rigorously only relatively near the boundary.

The location of DNA on nucleosomes can be described in two ways. **FIGURE 10.40** shows that **translational positioning** describes the position of DNA with regard to the boundaries of the nucleosome. In particular, it determines which sequences are found in the linker regions. Shifting the DNA by 10 bp brings the next turn into a linker region. Thus translational positioning determines which regions are more accessible (at least as judged by sensitivity to MNase).

DNA lies on the outside of the histone octamer. As a result, one face of any particular sequence is obscured by the histones, whereas the other face is exposed on the surface of the nucleosome. Depending upon its positioning with regard to the nucleosome, a site in DNA that must be recognized by a regulatory protein

could be inaccessible or available. The exact position of the histone octamer with respect to DNA sequence can therefore be important. **FIGURE 10.41** shows the effect of **rotational positioning** of the double helix with regard to the octamer surface. If the DNA is moved by a partial number of turns (imagine the DNA as rotating relative to the protein surface), there is a change in the exposure of sequence to the outside.

Both translational and rotational positioning can be important in controlling access to DNA. The best-characterized cases of positioning involve the specific placement of nucleosomes at promoters. Translational positioning and/or the exclusion of nucleosomes from a particular sequence may be necessary to allow a transcription complex to form. Some regulatory factors can bind to DNA only if a nucleosome is excluded to make the DNA freely accessible, and this creates a boundary for translational positioning. In other cases, regulatory factors can bind to DNA on the surface of the nucleosome, but rotational positioning is important to ensure that the face of DNA with the appropriate contact points is exposed.

We discuss the connection between nucleosomal organization and transcription in Chapter 28, *Eukaryotic Transcription Regulation*, but note for now that promoters (and some other structures) often have short regions that exclude nucleosomes. These regions typically form a boundary next to which nucleosome positions are restricted. A survey of an extensive region in the *Saccharomyces cerevisiae* genome (mapping 2278 nucleosomes over 482 kb of DNA) showed that in fact 60% of the nucleosomes have specific positions as the result of boundary effects, most often from promoters. Nucleosome positioning is a complex output of both intrinsic and extrinsic positioning mechanisms; thus it has been difficult to predict nucleosome positioning based on sequence alone, though there have been some successes. Large-scale sequencing studies of isolated nucleosomal DNA have revealed intriguing sequence patterns found in positioned nucleosomes *in vivo*, and some researchers have estimated that 50% or more of *in vivo* nucleosome positioning is the result of intrinsic sequence determinants encoded in the genomic DNA.

10.10 Nucleosomes Are Displaced and Reassembled During Transcription

Key concepts

- Most transcribed genes retain a nucleosomal structure, though the organization of the chromatin changes during transcription.
- Some heavily transcribed genes appear to be exceptional cases that are devoid of nucleosomes.
- RNA polymerase displaces histone octamers during transcription *in vitro*, but octamers reassociate with DNA as soon as the polymerase has passed.
- Nucleosomes are reorganized when transcription passes through a gene.
- Additional factors are required both for RNA polymerase to displace octamers during transcription and for the histones to reassemble into nucleosomes after transcription.

Heavily transcribed chromatin adopts structures that are visibly too extended to still be contained in nucleosomes. In the intensively transcribed genes coding for rRNA shown in **FIGURE 10.42**, the extreme packing of RNA polymerases makes it hard to see the DNA. We can

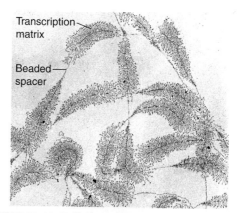

FIGURE 10.42 Individual rDNA transcription units alternate with nontranscribed DNA segments. Reproduced from O. L. Miller and B. R. Beatty, *Science* 164 (1969): 955–957. Photo courtesy of Oscar Miller.

not directly measure the lengths of the rRNA transcripts because the RNA is compacted by proteins, but we know (from the sequence of the rRNA) how long the transcript must be. The length of the transcribed DNA segment, which is measured by the length of the axis of the "Christmas tree," is ~85% of the length of the rRNA. This means that the DNA is almost completely extended.

On the other hand, transcriptionally active complexes of SV40 minichromosomes can be extracted from infected cells. They contain the usual complement of histones and display a beaded structure. Chains of RNA can be seen to extend from the minichromosome, as in the example of **FIGURE 10.43**. This argues that transcription can proceed while the SV40 DNA is organized into nucleosomes. Of course, the SV40 minichromosome is transcribed less intensively than the rRNA genes.

Transcription involves the unwinding of DNA, thus it seems obvious that some "elbow room" must be needed for the process. In thinking about transcription, we must bear in mind the relative sizes of RNA polymerase and the nucleosome. Eukaryotic RNA polymerases are large multisubunit proteins, typically >500 kD. Compare this with the ~260 kD of the nucleosome. **FIGURE 10.44** illustrates the relative sizes of RNA polymerase and the nucleosome. Consider the two turns that DNA makes around the nucleosome. Would RNA polymerase have sufficient access to DNA if the nucleic acid were confined to this path? During transcription, as RNA polymerase moves along the template, it binds tightly to a region of ~50 bp, including a locally unwound

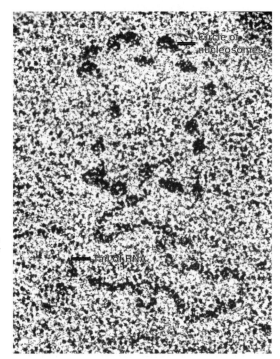

FIGURE 10.43 An SV40 minichromosome is transcribed while maintaining a nucleosomal structure. Reprinted from *J. Mol. Bio.*, vol. 131, P. Gariglio, et al., The template of the isolated native . . . , pp. 75–105. Copyright 1979, with permission from Elsevier [http://www.sciencedirect.com/science/journal/00222836]. Photo courtesy of Pierre Chambon, College of France.

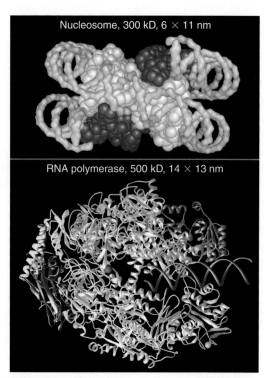

FIGURE 10.44 RNA polymerase is comparable in size to the nucleosome and might encounter difficulties in following the DNA around the histone octamer. Top photo courtesy of E. N. Moudrianakis, Johns Hopkins University. Bottom photo courtesy of Roger Kornberg, Stanford University School of Medicine.

segment of ~12 bp. The need to unwind DNA makes it seem unlikely that the segment engaged by RNA polymerase could remain on the surface of the histone octamer.

It therefore seems inevitable that transcription must involve a structural change. Thus the first question to ask about the structure of active genes is whether DNA being transcribed remains organized in nucleosomes. Experiments to test whether an RNA polymerase can transcribe directly through a nucleosome suggest that the histone octamer is displaced by the act of transcription. FIGURE 10.45 shows what happens when the phage T7 RNA polymerase transcribes a short piece of DNA containing a single octamer core *in vitro*. The core remains associated with the DNA after the polymerase passes, but is found in a different location. The core is most likely to rebind to the same DNA molecule from which it was displaced. Crosslinking the histones within the octamer does not create an obstacle to transcription, suggesting that (at least *in vitro*) transcription does not require dissociation of the octamer into its component histones.

Thus a small RNA polymerase can displace a single nucleosome, which reforms behind it, during transcription. Of course, the situation is more complex in a eukaryotic nucleus. Eukaryotic RNA polymerases are much larger, and the impediment to progress is a string of connected nucleosomes (which can also be folded into higher order structures). Overcoming these obstacles requires additional factors that act on chromatin (discussed in *Section 20.8, Initiation Is Followed by Promoter Clearance and Elongation*, and in detail in Chapter 28, *Eukaryotic Transcription Regulation*).

The organization of nucleosomes may be dramatically changed by transcription. FIGURE 10.46 a summary of what happens to the yeast HIS3 gene when it is induced by histidine starvation. Nucleosome positioning is examined by using MNase to cleave linker regions and a primer extension method to precisely map the boundaries of each nucleosome. Initially the gene displays a single dominant pattern of nucleosomes that are organized from the promoter and throughout the coding region. When the gene is activated, the nucleosomes become highly mobilized and adopt a num-

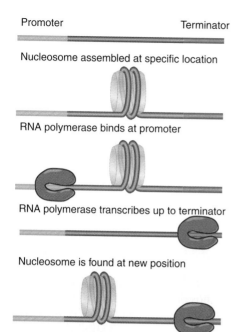

Promoter   Terminator

Nucleosome assembled at specific location

RNA polymerase binds at promoter

RNA polymerase transcribes up to terminator

Nucleosome is found at new position

FIGURE 10.45 An experiment to test the effect of transcription on nucleosomes shows that the histone octamer is displaced from DNA and rebinds at a new position.

Basal *HIS3* chromatin is static:
Dominant array with little remodelling activity.

D1   D2   D3   D4   D5

Activated *HIS3* chromatin is dynamic:
Nucleosomes are in flux as they are continually mobilised into different arrays by the competing activities of SWI/SNF, Isw1 and other remodelling complexes.

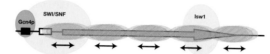

FIGURE 10.46 Basal/unactivated *HIS3* chromatin is static and shows a single dominant nucleosomal array with little remodeling activity (top panel). Activated HIS3 chromatin is dynamic: Nucleosomes are in flux as they are continually mobilized into different arrays by the competing activities of chromatin remodelling complexes. Reproduced from *Mol. Cell Biol.*, 2006, vol. 26, pp. 8252–8266, DOI and reproduced with permission from the American Society of Microbiology. Photo courtesy of David J. Clark, National Institutes of Health.

ber of alternative positions. The nucleosomes remain present at the same overall density but are no longer organized in phase. The action of chromatin remodelers is required to alter the nucleosomal positioning; this is discussed in detail in Section 28.7, Chromatin Remodeling Is an Active Process. When repression is reestablished, positioning reappears.

The unifying model is to suppose that RNA polymerase displaces histone octamers (either as a whole, or as dimers and tetramers) as it progresses. If the DNA behind the polymerase is available, the nucleosome is reassembled there. If the DNA is not available—for example, because another polymerase continues immediately behind the first—then the octamer may be permanently displaced, and the DNA may persist in an extended form.

The displacement and reassembly of nucleosomes does not occur solely as a result of the passage of RNA polymerase, but is facilitated by factors that help regulate this process. These include factors known as *ATP-dependent chromatin remodelers*, which use the energy of ATP hydrolysis to move or displace nucleosomes. These remodelers have been studied extensively, particularly in the context of transcription initiation, and will be discussed further in Chapter 28, *Eukaryotic Transcription Regulation*.

Other factors that are critical during transcription elongation, when nucleosomes are being rapidly displaced and reassembled, have been identified. The first of these to be characterized is a heterodimeric factor called FACT (*f*acilitates *c*hromatin *t*ranscription), which behaves like a transcription elongation factor. FACT is not part of RNA polymerase, but associates with it specifically during the elongation phase of transcription. FACT consists of two subunits that are well conserved in all eukaryotes, and it is associated with the chromatin of active genes.

When FACT is added to isolated nucleosomes, it causes them to lose H2A-H2B dimers. During transcription *in vitro*, it converts nucleosomes to "hexasomes" that have lost H2A-H2B dimers. This suggests that FACT is part of a mechanism for displacing octamers during transcription. FACT may also be involved in the reassembly of nucleosomes after transcription, because it assists formation of nucleosomes from core histones, thus acting like a histone chaperone. There is evidence *in vivo* that H2A-H2B dimers are displaced more readily during transcription than H3-H4 tetramers, suggesting that tetramers and dimers may be reassembled sequentially after transcription as they are after passage of a replication fork (see *Section 10.8, Replication of Chromatin Requires Assembly of Nucleosomes*).

This suggests the model shown in **FIGURE 10.47**, in which FACT (or a similar factor) detaches H2A-H2B from a nucleosome in front of RNA polymerase and then helps to add it to a nucleosome that is reassembling behind the

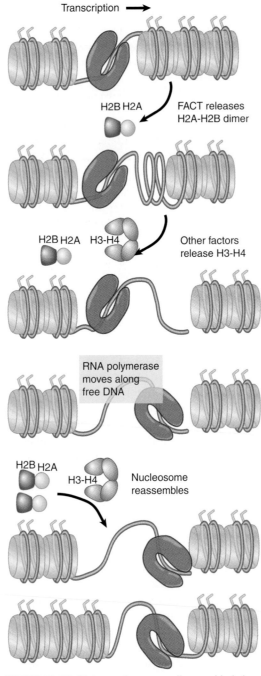

Transcription ➡

H2B H2A FACT releases H2A-H2B dimer

H2B H2A H3-H4 Other factors release H3-H4

RNA polymerase moves along free DNA

H2B H2A H3-H4 Nucleosome reassembles

FIGURE 10.47 Histone octamers are disassembled ahead of transcription to remove nucleosomes. They reform following transcription. Release of H2A-H2B dimers probably initiates the disassembly process.

enzyme. Other factors are likely to be required to complete the process. FACT's role may be more complex than this, as FACT has also been implicated in transcription initiation as well as in replication elongation. Another intriguing model that has been proposed is that FACT stabilizes a "reorganized" nucleosome, in which the dimers and tetramer remain locally tethered via FACT but are not stably organized into a canonical nucleosome. The model presumes the H2A-H2B dimers are less stable in this reorganized state, and thus more easily displaced. In this state, the nucleosomal DNA is highly accessible, and the reorganized nucleosome can either revert to the stable canonical organization or be displaced as needed for transcription.

Several other factors have been identified that play key roles in either nucleosome displacement or reassembly during transcription. These include the Spt6 protein, a factor involved in "resetting" chromatin structure after transcription. Spt6, like FACT, colocalizes with actively transcribed regions and can act as a histone chaperone to promote nucleosome assembly. Although CAF-1 is known to be involved only in replication-dependent histone deposition, one of CAF-1's partners in replication may in fact play a role in transcription as well. The CAF-1-associated protein Rtt106 is an H3-H4 chaperone that has recently been shown to play a role in H3 deposition during transcription.

10.11 DNase Sensitivity Detects Changes in Chromatin Structure

Key concepts

- Hypersensitive sites are found at the promoters of expressed genes, as well as other important sites such as origins of replication and centromeres.
- Hypersensitive sites are generated by the binding of factors that exclude histone octamers.
- A domain containing a transcribed gene is defined by increased sensitivity to degradation by DNase I.

There are numerous changes to chromatin that occur in active or potentially active regions. These include distinctive structural changes that occur at specific sites associated with initiation of transcription or with certain structural features in DNA. These changes were first detected by the effects of digestion with very low concentrations of the enzyme DNase I.

When chromatin is digested with DNase I, the first effect is the introduction of breaks in the duplex at specific, **hypersensitive sites**. Susceptibility to DNase I reflects the availability of DNA in chromatin, thus these sites represent chromatin regions in which the DNA is particularly exposed because it is not organized in the usual nucleosomal structure. A typical hypersensitive site is 100× more sensitive to enzyme attack than bulk chromatin. These sites

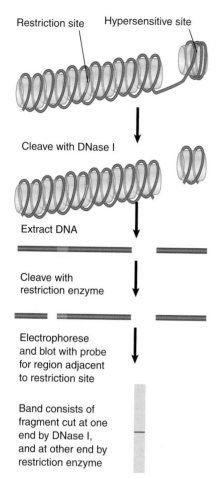

Restriction site   Hypersensitive site

Cleave with DNase I

Extract DNA

Cleave with
restriction enzyme

Electrophorese
and blot with probe
for region adjacent
to restriction site

Band consists of
fragment cut at one
end by DNase I,
and at other end by
restriction enzyme

FIGURE 10.48 Indirect end-labeling identifies the distance of a DNase hypersensitive site from a restriction cleavage site. The existence of a particular cutting site for DNase I generates a discrete fragment, whose size indicates the distance of the DNase I hypersensitive site from the restriction site.

are also hypersensitive to other nucleases and to chemical agents.

Hypersensitive sites are created by the local structure of chromatin, which may be tissue specific. Their locations can be determined by the technique of indirect end labeling that we introduced earlier in the context of nucleosome positioning. This application of the technique is recapitulated in **FIGURE 10.48**. In this case, cleavage at the hypersensitive site by DNase I is used to generate one end of the fragment. Its distance is measured from the other end, which is generated by cleavage with a restriction enzyme.

Many hypersensitive sites are related to gene expression. Every active gene has a site, or sometimes more than one site, in the region of the promoter. Most hypersensitive sites are found only in chromatin of cells in which the associated gene is either being expressed or is poised for expression; they do not occur when the gene is inactive. The 5' hypersensitive

site(s) appear before transcription begins and occur in DNA sequences that are required for gene expression.

What is the structure of a hypersensitive site? Its preferential accessibility to nucleases indicates that it is not protected by histone octamers, but this does not necessarily imply that it is free of protein. A region of free DNA might be vulnerable to damage, and in any case, how would it be able to exclude nucleosomes? In fact, hypersensitive sites typically result from the binding of specific regulatory proteins that exclude nucleosomes. It is very common to find pairs of hypersensitive sites that flank a nuclease-resistant core; the binding of nucleosome-excluding proteins is probably the basis for the existence of the protected region within the hypersensitive sites.

The proteins that generate hypersensitive sites are likely to be regulatory factors of various types, because hypersensitive sites are found associated with promoters and other elements that regulate transcription, origins of replication, centromeres, and sites with other structural significance. In some cases, they are associated with more extensive organization of chromatin structure. A hypersensitive site may provide a boundary for a series of positioned nucleosomes. Hypersensitive sites associated with transcription may be generated by transcription factors when they bind to the promoter as part of the process that makes it accessible to RNA polymerase.

In addition to detecting hypersensitive sites, DNase I digestion can also be used to assess the relative accessibility of a genomic region. A region of the genome that contains an active gene may have an altered overall structure, in addition to specific hypersensitive sites. The change in structure precedes, and is different from, the disruption of nucleosome structure that may be caused by the actual passage of RNA polymerase. DNase I sensitivity defines a *chromosomal **domain***, which is a region of altered structure including at least one active transcription unit, and sometimes extending farther. (Note that use of the term "domain" does not imply any necessary connection with the structural domains identified by the loops of chromatin or chromosomes.)

When chromatin is digested with DNase I, it is eventually degraded into acid-soluble material (very small fragments of DNA). The progress of the overall reaction can be followed in terms of the proportion of DNA that is rendered acid soluble. When only 10% of the total DNA

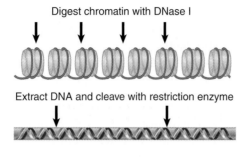

Digest chromatin with DNase I

Extract DNA and cleave with restriction enzyme

Electrophorese fragments and denature DNA;
probe for expressed and nonexpressed genes

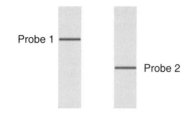

Probe 1

Probe 2

Compare intensities of bands in preparations
in which chromatin was digested with increasing
concentrations of DNase

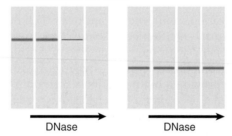

DNase

DNase

Probe 1 DNA is
preferentially digested

Probe 2 DNA is not
preferentially digested

FIGURE 10.49 Sensitivity to DNase I can be measured by determining the rate of disappearance of the material hybridizing with a particular probe.

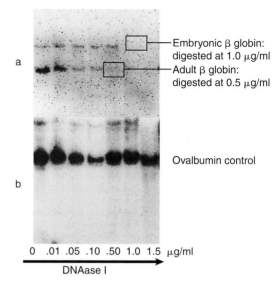

a

Embryonic β globin:
digested at 1.0 μg/ml

Adult β globin:
digested at 0.5 μg/ml

Ovalbumin control

b

0   .01 .05 .10 .50 1.0 1.5   μg/ml

DNAase I

FIGURE 10.50 In adult erythroid cells, the adult β-globin gene is highly sensitive to DNase I digestion; the embryonic β-globin gene (now known as ε-globin) is partially sensitive (probably due to spreading effects), but ovalbumin is not sensitive. Photo courtesy of Harold Weintraub, Fred Hutchinson Cancer Research Center. Used with permission of Mark T. Groudine.

has become acid soluble, more than 50% of the DNA of an active gene has been lost. This suggests that active genes are preferentially degraded.

The fate of individual genes can be followed by quantitating the amount of DNA that survives to react with a specific probe. The protocol is outlined in **FIGURE 10.49**. The principle is that the loss of a particular band indicates that the corresponding region of DNA has been degraded by the enzyme.

FIGURE 10.50 shows what happens to β-globin genes and an ovalbumin gene in chromatin extracted from chicken red blood cells (in which globin genes are expressed and the ovalbumin gene is inactive). The restriction fragments representing the β-globin genes are rapidly lost, whereas those representing the ovalbumin gene show little degradation. The

ovalbumin gene in fact is digested at the same rate as the bulk of DNA.

Thus the bulk of chromatin is relatively resistant to DNase I and contains nonexpressed genes (as well as other sequences). A gene becomes relatively susceptible to nuclease digestion specifically in the tissue(s) in which it is expressed or is poised to be expressed.

Is preferential susceptibility a characteristic only of highly expressed genes, such as globin, or of all active genes? Experiments using probes representing the entire cellular mRNA population suggest that all active genes, whether coding for abundant or for rare mRNAs, are preferentially susceptible to DNase I, though with some variations in the degree of susceptibility. The rarely expressed genes are likely to have very few RNA polymerase molecules actually engaged in transcription at any moment; this implies that the sensitivity to DNase I does not result from the act of transcription, but instead is a feature of genes that are able to be transcribed.

What is the extent of the preferentially sensitive region? This can be determined by using a series of probes representing the flanking regions as well as the transcription unit itself. The sensitive region always extends over the entire transcribed region; an additional region of several kb on either side may show an intermediate level of sensitivity (probably as the result of spreading effects).

The critical concept implicit in the description of the domain is that a region of high sensitivity to DNase I extends over a considerable distance. Often we think of regulation as residing in events that occur at a discrete site in DNA—for example, in the ability to initiate transcription at the promoter. Even if this is true, such regulation must determine, or must be accompanied by, a more wide-ranging change in structure.

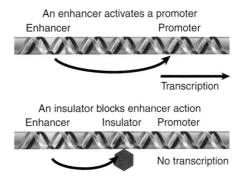

FIGURE 10.51 An enhancer activates a promoter in its vicinity, but can be blocked from doing so by an insulator located between them.

10.12 Insulators Define Transcriptionally Independent Domains

Key concepts

- Insulators are able to block passage of any activating or inactivating effects from enhancers, silencers, and other control elements.
- Insulators can provide barriers against the spread of heterochromatin.
- Insulators are specialized chromatin structures that typically contain hypersensitive sites.
- In most cases, two insulators can protect the region between them from all external effects.
- Different insulators are bound by different factors, and may use alternative mechanisms for enhancer blocking and/or heterochromatin barrier formation.

Different regions of the chromosome have different functions that are typically marked by specific chromatin structures or modification states. We will see later that many of the elements that control gene transcription can act from very large distances (see Chapter 20, *Eukaryotic Transcription*), and that highly compacted heterochromatin (introduced in Chapter 9, *Chromosomes*) can also spread over large distances (see *Section 29.2, Heterochromatin Propagates from a Nucleation Event*). The existence of these long-range interactions suggests that chromosomes must also contain functional elements that serve to partition chromosomes into domains that can be regulated independently of one another. **Insulators** are a class of elements that appear to fulfill this function. Insulators (also known as "barrier" or "boundary" elements) prevent the passage of activating or inactivating effects. They have either or both of two key properties:

- When an insulator is located between an enhancer and a promoter, it prevents the enhancer from activating the promoter. This enhancer-blocking effect

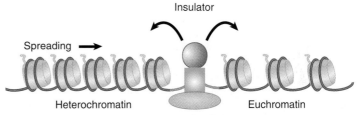

FIGURE 10.52 Heterochromatin may spread from a center and then blocks any promoters that it covers. An insulator may be a barrier to propagation of heterochromatin that allows the promoter to remain active.

is shown in **FIGURE 10.51**. This activity may explain how the action of an enhancer is limited to a particular promoter despite the ability of enhancers to activate promoters from long distances away (and the ability of enhancers to indiscriminately activate any promoter in the vicinity).

- When an insulator is located between an active gene and heterochromatin, it provides a barrier that protects the gene against the inactivating effect that spreads from the heterochromatin. This barrier effect is shown in **FIGURE 10.52**.

Some insulators possess both these properties, but others have only one, or the blocking and barrier functions can be separated. Although both actions are likely to be mediated by changing chromatin structure, they may involve different effects. In either case, however, the insulator defines a limit for long-range effects. By restricting enhancers so they can act only on specific promoters, and preventing the inadvertent spreading of heterochromatin into

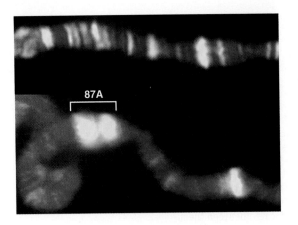

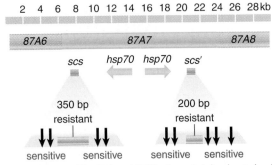

FIGURE 10.53 The 87A and 87C loci, containing heat shock genes, expand upon heat shock in *Drosophila* polytene chromosomes. Specialized chromatin structures that include hypersensitive sites mark the ends of the 8787 domain and insulate genes between them from the effects of surrounding sequences. Photo courtesy of Victor G. Corces, Emory University.

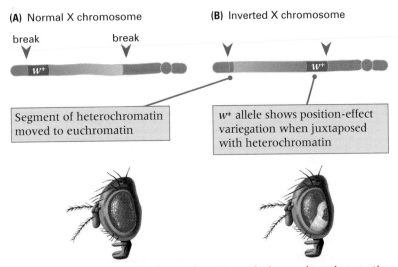

FIGURE 10.54 Position effects are often observed when an inversion or other chromosome rearrangement repositions a gene normally in euchromatin to a new location in or near heterochromatin. In this example, an inversion in the X chromosome of *Drosophila melanogaster* repositions the wild-type allele of the *white* gene near heterochromatin. Differences in expression due to position effects on the w+ allele are observed as mottled red and white eyes.

active regions, insulators function as elements for increasing the precision of gene regulation.

Insulators were first discovered during the analysis of a region of the *Drosophila melanogaster* genome shown in **FIGURE 10.53**. Two genes for the protein Hsp (heat-shock protein) 70 lie within an 18-kb region that constitutes band 87A7. Researchers had noted that when subjected to heat shock, a puff forms at 87A7 in polytene chromosomes, and there is a distinct boundary between the decondensed and condensed regions of the chromosomes. Special structures, called *scs* and *scs'* (specialized chromatin structures), are found at the ends of the band. Each element consists of a region that is highly resistant to degradation by DNase I flanked on either side by hypersensitive sites that are spaced at about 100 bp. The cleavage pattern at these sites is altered when the genes are turned on by heat shock.

The *scs* elements insulate the *hsp70* genes from the effects of surrounding regions (and presumably also protect the surrounding regions from the effects of heat shock activation at the *hsp70* loci). In the first assay for insulator function, *scs* elements were tested for their ability to protect a reporter gene from "position effects." In this experiment, *scs* elements were placed in constructs flanking the *white* gene, the gene responsible for producing red pigment in the *Drosophila* eye, and these constructs were randomly integrated into the fly genome. If the *white* gene is integrated without *scs* elements, its expression is subject to position effects; i.e., the chromatin context in which the gene is inserted strongly influences whether the gene is transcribed. This can be detected as a variegated color phenotype in the fly eye, as shown in **FIGURE 10.54**. However, when *scs* elements are placed on either side of the *white* gene, the gene can function anywhere it is placed in the genome—even in sites where it would normally be repressed by context (such as in heterochromatic regions), resulting in uniformly red eyes.

The *scs* and *scs'* elements, like many other insulators, do not themselves play positive or negative roles in controlling gene expression, but just restrict effects from passing from one region to the next. Unexpectedly, the *scs* elements themselves are not responsible for controlling the precise boundary between the condensed and decondensed regions at the heat shock puff, but instead serve to prevent regulatory cross-talk between the *hsp70* genes and the many other genes in the region.

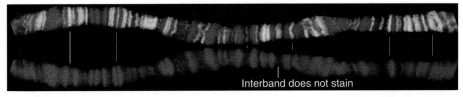

Red = DNA = bands　　　Green = BEAF in interband　　　Yellow = BEAF + DNA

FIGURE 10.55 A protein that binds to the insulator *scs'* is localized at interbands in *Drosophila* polytene chromosomes. Red staining identifies the DNA (the bands) on both the upper and lower samples; green staining identifies BEAF32 (often at interbands) on the upper sample. Yellow shows coincidence of the two labels (meaning that BEAF32 is in a band). Reprinted from *Cell*, vol. 81, K. Zhao, C. M. Hart, and U. K. Laemmli, Visualization of chromosomal domains . . . , pp. 879–889. Copyright 1995, with permission from Elsevier [http://www.sciencedirect.com/science/journal/00928674]. Photo courtesy of Ulrich K. Laemmli, University of Geneva, Switzerland.

The *scs* and *scs'* elements have different structures, and each appears to have a different basis for its insulator activity. The key sequence in the scs element is a stretch of 24 bp that binds the product of the *zw5* (*zeste white 5*) gene. The insulator property of *scs'* resides in a series of CGATA repeats. The repeats bind a pair of related proteins (encoded by the same gene) called BEAF-32. BEAF-32 is localized to ~50% of the interbands on polytene chromosomes, as seen in **FIGURE 10.55**. This suggests that there are many insulators in the genome (though BEAF-32 may bind noninsulators as well), and that BEAF-32 is a common part of the insulating apparatus.

Another insulator that has been extensively characterized in *Drosophila* is found in the transposon *gypsy*. Some experiments that initially defined the behavior of this insulator were based on a series of *gypsy* insertions into the *yellow* (*y*) locus. Different insertions cause loss of *y* gene function in some tissues, but not in others. The reason is that the *y* locus is regulated by four enhancers, as shown in **FIGURE 10.56**. Wherever *gypsy* is inserted, it blocks expression of all enhancers that it separates from the promoter, but not those that lie on the other side. The sequence responsible for this effect is an insulator that lies at one end of the transposon. The insulator works irrespective of its orientation of insertion.

The function of the *gypsy* insulator depends on several proteins, including Su(Hw) (*Suppressor of Hairy wing*), CP190, mod(mdg4), and dTopors. Mutations in the *su(Hw)* gene completely abolish insulation; *su(Hw)* encodes a protein that binds twelve 26-bp reiterated sites in the insulator and is necessary for its action. Su(Hw) has a zinc finger DNA-motif; mapping to polytene chromosomes shows that Su(Hw) is bound to hundreds of sites that include both *gypsy*

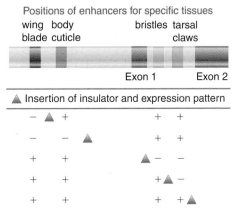

FIGURE 10.56 The insulator of the *gypsy* transposon blocks the action of an enhancer when it is placed between the enhancer and the promoter.

insertions and non-*gypsy* sites. Manipulations show that the strength of the insulator is determined by the number of copies of the binding sequence. CP190 is a centrosomal protein that assists Su(Hw) in binding site recognition.

mod(mdg4) and dTopors have a specific role in the creation of "insulator bodies," which appear to be clusters of Su(Hw)-bound insulators that can be observed in normal diploid nuclei. Despite the presence of >500 Su(Hw) binding sites in the *Drosophila* genome, visualization of Su(Hw) or mod(mdg4) shows that they are colocalized at ~25 discrete sites around the nuclear periphery. This suggests the model of **FIGURE 10.57**, in which Su(Hw) proteins bound at different sites on DNA are brought together by binding to mod(mdg4). The Su(Hw)/mod(mdg4) complex is localized at the nuclear periphery. The DNA bound to it is organized into loops. An average complex might have ~20 such loops. Enhancer–promoter actions can occur only within a loop, and cannot propagate between them. This model is supported by "insulator bypass" experiments,

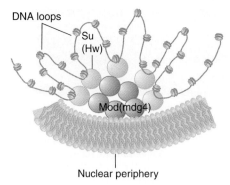

FIGURE 10.57 Su(Hw)/mod(mdg4) complexes are found in clusters at the nuclear periphery. They may organize DNA into loops that limit enhancer–promoter interactions.

in which placing a *pair* of insulators between an enhancer and promoter actually eliminates insulator activity—somehow the two insulators cancel each other out. This could be explained by the formation of a minidomain between the duplicated insulator (perhaps too small to create an anchored loop), which would essentially result in what should have been two adjacent loops fused into one. Not all insulators can be bypassed in this way, though; this and other evidence suggests that there are multiple mechanisms for insulator function.

The complexity of insulators and their roles is indicated by the behavior of another *Drosophila* insulator: the *Fab-7* element found in the *bithorax* locus (BX-C). This locus contains a series of *cis*-acting regulatory elements that control the activities of three homeotic genes (*Ubx*, *abd-A*, and *Abd-B*), which are differentially expressed along the anterior–posterior axis of the *Drosophila* embryo. The locus also contains at least three insulators that are not interchangeable; *Fab-7* is the best studied of these. The relevant part of the locus is shown in FIGURE 10.58. The regulatory elements *iab-6* and *iab-7* control expression of the adjacent gene *Abd-B* in successive regions of the embryo (segments A6 and A7). A deletion of *Fab-7* causes A6 to develop like A7, resulting in two "A7-like" segments (this is known as a homeotic transformation). This is a dominant effect, which suggests that *iab-7* has taken over control from *iab-6*. We can interpret this in molecular terms by supposing that *Fab-7* provides a boundary that prevents *iab-7* from acting when *iab-6* is usually active. In fact, in the absence of *Fab-7* it appears that *iab-6* and *iab-7* fuse into a single regulatory domain, which shows different behavior depending on the position along

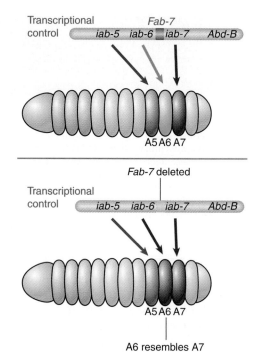

FIGURE 10.58 *Fab-7* is a boundary element that is necessary for the independence of regulatory elements *iab-6* and *iab-7*.

the AP axis. The insulator activity of *Fab-7* is also developmentally regulated, with a protein called Elba (*e*arly *b*oundary *a*ctivity) responsible for *Fab-7*'s blocking function early in development, but not later in development or in the adult. *Fab-7* is also associated with the *Drosophila* homolog of the CTCF protein, a mammalian insulator-binding protein that shows regulated binding to its targets (see *Section 29.9, Oppositely Imprinted Genes Can Be Controlled by a Single Center*). Finally, both *Fab-7* and a nearby insulator (*Fab-8*) are known to lie near "anti-insulator elements" (also called promoter-targeting sequences or PTS elements), which may allow an enhancer to overcome the blocking effects of an insulator.

The diversity of insulator behaviors and of the factors responsible for insulator function makes it impossible to propose a single model to explain the behavior of all insulators. Instead, it is clear that the term "insulator" refers to a variety of elements that use a number of distinct mechanisms to achieve similar (but not identical) functions. Notably, the mechanisms used to block enhancers may be very different from those used to block the spread of heterochromatin.

10.13 An LCR May Control a Domain

Key concepts

- LCRs are located at the 5′ end of a chromosomal domain and typically consist of multiple DNase hypersensitive sites.
- LCRs regulate gene clusters.
- LCRs usually regulate loci that show complex developmental or cell-type specific patterns of gene expression.
- LCRs control the transcription of target genes in the locus by direct interactions, forming looped structures.

Every gene is controlled by its proximal promoter, and most genes also respond to enhancers (containing similar regulatory elements located farther away); see Chapter 20, *Eukaryotic Transcription*. These local controls are not sufficient for all genes, though. In some cases, a gene lies within a domain of several genes, all of which are influenced by specialized regulatory elements that act on the whole domain. The existence of these elements was identified by the inability of a region of DNA including a gene and all its known regulatory elements to be properly expressed when introduced into an animal as a transgene.

The best-characterized example of a regulated gene cluster is provided by the mammalian β-globin genes. Recall from Figures 8.20 and 8.21 that the α- and β-globin genes in mammals each exist as clusters of related genes that are expressed at different times and different tissues during embryonic and adult development. These genes are associated with a large number of regulatory elements, which have been analyzed in detail. In the case of the adult human β-globin gene, regulatory sequences are located both 5′ and 3′ to the gene. The regulatory sequences include both positive and negative elements in the promoter region, as well as additional positive elements within and downstream of the gene.

All of these control regions are not, however, sufficient for proper expression of the human β-globin gene in a transgenic mouse within an order of magnitude of wild-type levels. Some further regulatory sequence is required. Regions that provide the additional regulatory function are identified by DNase I hypersensitive sites that are found at the ends of the β-globin cluster. The map of **FIGURE 10.59** shows that the 20 kb upstream of the ε gene

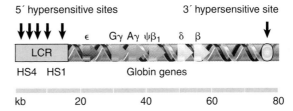

FIGURE 10.59 The β-globin locus is marked by hypersensitive sites at either end. The group of sites at the 5′ side constitutes the LCR and is essential for the function of all genes in the cluster.

contains a group of five hypersensitive sites, and that there is a single site 30 kb downstream of the β gene.

The 5′ regulatory sites are the primary regulators, and the region containing the cluster of hypersensitive sites is called the **locus control region (LCR)**. The role of the LCR is complex, but in some ways it behaves as a "super enhancer" that poises the entire locus for transcription. The precise function of the 3′ hypersensitive site in the mammalian locus is not clear, but it is known to physically interact with the LCR. A 3′ hypersensitive site in the chicken β-globin locus acts as an insulator, as does a fifth 5′ site upstream of the mammalian LCR. The LCR is absolutely required for expression of *each* of the globin genes in the locus. Each gene is then further regulated by its own specific controls. Some of these controls are autonomous: expression of the ε and γ genes appears intrinsic to those loci in conjunction with the LCR. Other controls appear to rely upon position in the cluster, which provides a suggestion that gene order in a cluster is important for regulation.

The entire region containing the globin genes, and extending well beyond them, constitutes a chromosomal domain. It shows increased sensitivity to digestion by DNase I (see Figure 10.49). Deletion of the 5′ LCR restores normal resistance to DNase over the whole region. In addition to increases in the general accessibility of the locus, the LCR is also apparently required to directly activate the individual promoters. The exact nature of the sequential interactions between the LCR and the individual promoters has not yet been fully defined, but it has recently become clear that the LCR contacts individual promoters directly, forming loops when these promoters are active. The domain controlled by the LCR also shows distinctive patterns of histone modifications (see *Section 28.9, Histone Acetylation Is Associated with Transcription Activation*) that are dependent on LCR function.

This model appears to apply to other gene clusters as well. The α-globin locus has a similar organization of genes that are expressed at different times, with a group of hypersensitive sites at one end of the cluster and increased sensitivity to DNase I throughout the region. So far, though, only a small number of other cases are known in which an LCR controls a group of genes.

One of these cases involves an LCR that controls genes on more than one chromosome. The T_H2 LCR coordinately regulates the T helper type 2 cytokine locus, a group of genes encoding a number of interleukins (important signaling molecules in the immune system). These genes are spread out over 120 kb on chromosome 11, and the T_H2 LCR controls these genes by interacting with their promoters. It also interacts with the promoter of the *IFN*γ gene on chromosome 10. The two types of interaction are alternatives that comprise two different cell fates; that is, in one group of cells the LCR causes expression of the genes on chromosome 11, whereas in the other group it causes the gene on chromosome 10 to be expressed.

The idea that looping interactions are important for chromosome structure and function was introduced in Chapter 9, *Chromosomes*, and was also discussed above in models for insulator function (see *Section 10.12, Insulators Define Transcriptionally Independent Domains*). New methods have been developed to begin to dissect the physical interactions between chromosomal loci *in vivo*, leading to new understanding of how these interactions result in regulatory functions. Direct interactions between the β-globin and T_H2 LCRs and their target loci have been mapped using a method known as chromosome conformation capture (3C). While there are now many variations of this procedure, the basic method is outlined in the top panel of **FIGURE 10.60**. Interacting regions of chromatin *in vivo* are captured using formaldehyde treatment to crosslink to fix the DNA and proteins that are in close contact. Next, the chromatin is digested with a restriction enzyme, then ligated under dilute conditions

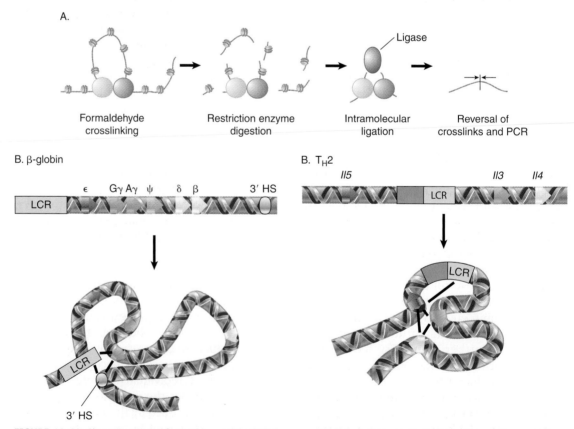

FIGURE 10.60 Chromosome conformation capture (3C) is one method to detect physical interactions between regions of chromatin *in vivo*. Looping interactions controlled by the β-globin and T_H2 LCRs have been mapped by 3C and some of the known contacts are shown. Adapted from A. Miele and J. Dekker, *Mol. Biosyst.* 4 (2008): 1046–1057.

to favor intra-molecular ligation. This results in preferential ligation of DNA fragments that are held in close proximity as a result of cross-linking. Finally, the proteins are removed by reversing the crosslinking and the new ligated junctions are detected by PCR.

As shown in the lower part of the figure, 3C and similar methods have allowed researchers to begin to unravel the complex and dynamic interactions that occur at loci regulated by LCRs. The β-globin LCR sequentially interacts with each globin gene at the developmental stage in which that gene is active; the figure shows the interactions that occur between the LCR, 3' HS, and the γ-globin genes in the fetal stage. Interestingly, the T_H2 LCR appears to interact with all three of its target genes (*Il3, -4,* and *-5*) simultaneously. These interactions occur in all T-cells whether or not these genes are expressed, but the precise organization of loops alters upon activation of the interleukin genes. This reorganization, which depends on the protein SATB1 (*special AT-rich binding protein*), suggests that the T_H2 LCR brings all the genes together in a poised state in T cells, awaiting the trigger of specific transcription factors to activate the genes rapidly when needed.

10.14 Summary

All eukaryotic chromatin consists of nucleosomes. A nucleosome contains a characteristic length of DNA, usually ~200 bp, which is wrapped around an octamer containing two copies each of histones H2A, H2B, H3, and H4. A single H1 (or other linker histone) may associate with a nucleosome. Virtually all genomic DNA is organized into nucleosomes. Treatment with micrococcal nuclease shows that the DNA packaged into each nucleosome can be divided operationally into two regions. The linker region is digested rapidly by the nuclease; the core region of 146 bp is resistant to digestion. Histones H3 and H4 are the most highly conserved, and an $H3_2$-$H4_2$ tetramer accounts for the diameter of the particle. Histones H2A and H2B are organized as two H2A-H2B dimers. Octamers are assembled by the successive addition of two H2A-H2B dimers to the $H3_2$-$H4_2$ tetramer. A large number of histone variants exist that can also be incorporated into nucleosomes; different variants perform different functions in chromatin and some are cell-type specific.

The path of DNA around the histone octamer creates −1.67 supercoils. The DNA "enters" and "exits" the nucleosome on the same side, and the entry or exit angle could be altered by histone H1. Removal of the core histones releases −1.0 supercoils. The difference can be largely explained by a change in the helical pitch of DNA, from an average of 10.2 bp/turn in nucleosomal form to 10.5 bp/turn when free in solution. There is variation in the structure of DNA from a periodicity of 10.0 bp/turn at the nucleosome ends to 10.7 bp/turn in the center. There are kinks in the path of DNA on the nucleosome.

Nucleosomes are organized into a fiber of 30 nm diameter that has six nucleosomes per turn and a packing ratio of 40. Removal of H1 or reduced ionic strength allows this fiber to unfold into a 10 nm fiber that consists of a linear string of nucleosomes. The 30 nm fiber probably consists of the 10 nm fiber wound into a two-start solenoid. The 30 nm fiber is the basic constituent of both euchromatin and heterochromatin; nonhistone proteins are responsible for further organization of the fiber into chromatin or chromosome ultrastructure.

There are two pathways for nucleosome assembly. In the replication-coupled pathway, the PCNA processivity subunit of the replisome recruits CAF-1, which is a nucleosome assembly factor or histone chaperone. CAF-1 assists the deposition of $H3_2$-$H4_2$ tetramers onto the daughter duplexes resulting from replication. The tetramers may be produced either by disruption of existing nucleosomes by the replication fork or as the result of assembly from newly synthesized histones. CAF-1 assembles newly synthesized tetramers, while the ASF1 chaperone also assists with deposition of $H3_2$-$H4_2$ tetramers that have been displaced by the replication fork. Similar sources provide the H2A-H2B dimers that then assemble with the $H3_2$-$H4_2$ tetramer to complete the nucleosome. The $H3_2$-$H4_2$ tetramer and the H2A-H2B dimers assemble at random, so the new nucleosomes may include both preexisting and newly synthesized histones. Nucleosome placement is not random throughout the genome, but is controlled by a combination of intrinsic (DNA sequence-dependent) and extrinsic (dependent on *trans*-factors) mechanisms that result in specific patterns of nucleosome deposition.

RNA polymerase displaces histone octamers during transcription. Nucleosomes reform on DNA after the polymerase has passed, unless transcription is very intensive (such as in rDNA) when they may be displaced completely. The replication-independent pathway for nucleosome assembly is responsible for replacing histone octamers that have been displaced by transcription. It uses the histone variant H3.3 instead of H3. A similar pathway, with another alternative to H3, is used for assembling nucleosomes at centromeric DNA sequences following replication.

Two types of changes in sensitivity to nucleases are associated with gene activity. Chromatin capable of being transcribed has a generally increased sensitivity to DNase I, reflecting a change in structure over an extensive region that can be defined as a domain containing active or potentially active genes. Hypersensitive sites in DNA occur at discrete locations, and are identified by greatly increased sensitivity to DNase I. A hypersensitive site consists of a sequence of typically >200 bp from which nucleosomes are excluded by the presence of other proteins. A hypersensitive site forms a boundary that may cause adjacent nucleosomes to be restricted in position. Nucleosome positioning may be important in controlling access of regulatory proteins to DNA.

An insulator blocks the transmission of activating or inactivating effects in chromatin. An insulator that is located between an enhancer and a promoter prevents the enhancer from activating the promoter. Two insulators define the region between them as a regulatory domain; regulatory interactions within the domain are limited to it, and the domain is insulated from outside effects. Most insulators block regulatory effects from passing in either direction, but some are directional. Insulators usually can block both activating effects (enhancer–promoter interactions) and inactivating effects (mediated by spread of heterochromatin), but some are limited to one or the other. Insulators are thought to act via changing higher-order chromatin structure, but the details are not certain.

LCRs function at a distance and may be required for any and all genes in a domain to be expressed. When a domain has an LCR, its function is essential for all genes in the domain, but LCRs do not seem to be common. LCRs contain enhancer-like hypersensitive site(s) that are needed for the full activity of promoter(s) within the domain, and to create a general domain of DNase sensitivity. LCRs also act by creating loops between LCR sequences and the promoters of active genes within the domain.

Hypersensitive sites occur at several types of regulators. Those that regulate transcription include promoters, enhancers, and LCRs. Other sites include insulators, origins of replication, and centromeres. A promoter or enhancer acts on a single gene, whereas an LCR contains a group of hypersensitive sites and may regulate a domain containing several genes.

References

10.3 The Nucleosome Is the Subunit of All Chromatin

Reviews

Izzo, A., Kamieniarz, K., and Schneider, R. (2008). The histone H1 family: specific members, specific functions? *Biol. Chem.* 389, 333–343.

Kornberg, R. D. (1977). Structure of chromatin. *Annu. Rev. Biochem.* 46, 931–954.

McGhee, J. D., and Felsenfeld, G. (1980). Nucleosome structure. *Annu. Rev. Biochem.* 49, 1115–1156.

Research

Angelov, D., Vitolo, J. M., Mutskov, V., Dimitrov, S., and Hayes, J. J. (2001). Preferential interaction of the core histone tail domains with linker DNA. *Proc. Natl. Acad. Sci. USA* 98, 6599–6604.

Arents, G., Burlingame, R. W., Wang, B.-C., Love, W. E., and Moudrianakis, E. N. (1991). The nucleosomal core histone octamer at 31 Å resolution: a tripartite protein assembly and a left-handed superhelix. *Proc. Natl. Acad. Sci. USA* 88, 10148–10152.

Finch, J. T. et al. (1977). Structure of nucleosome core particles of chromatin. *Nature* 269, 29–36.

Kornberg, R. D. (1974). Chromatin structure: a repeating unit of histones and DNA. *Science* 184, 868–871.

Luger, K. et al. (1997). Crystal structure of the nucleosome core particle at 28 Å resolution. *Nature* 389, 251–260.

Richmond, T. J., Finch, J. T., Rushton, B., Rhodes, D., and Klug, A. (1984). Structure of the nucleosome core particle at 7 Å resolution. *Nature* 311, 532–537.

Shen, X. et al. (1995). Linker histones are not essential and affect chromatin condensation *in vitro*. Cell 82, 47–56.

10.6 DNA Structure Varies on the Nucleosomal Surface

Reviews

Travers, A. A. and Klug, A. (1987). The bending of DNA in nucleosomes and its wider implications. *Philos. Trans. R. Soc. Lond. B Biol. Sci.* 317, 537–561.

Wang, J. (1982). The path of DNA in the nucleosome. *Cell* 29, 724–726.

Research

Richmond, T. J. and Davey, C. A. (2003). The structure of DNA in the nucleosome core. *Nature* 423, 145–150.

10.7 The Path of Nucleosomes in the Chromatin Fiber

Review

Tremethick, D. J. (2007). Higher-order structures of chromatin: the elusive 30 nm fiber. *Cell* 128, 651–654.

Research

Dorigo, B., Schalch, T., Kulangara, A., Duda, S., Schroeder, R. R., and Richmond, T. J. (2004). Nucleosome arrays reveal the two-start organization of the chromatin fiber. *Science* 306, 1571–1573.

Schalch, T., Duda, S., Sargent, D. F., and Richmond, T. J. (2005). X-ray structure of a tetranucleosome and its implications for the chromatin fibre. *Nature* 436, 138–141.

10.8 Replication of Chromatin Requires Assembly of Nucleosomes

Reviews

Corpet, A. and Almouzni, G. (2008). Making copies of chromatin: the challenge of nucleosomal organization and epigenetic information. *Trends Cell Biol.* 19, 29–41.

Eitoku, M., Sato, L., Senda, T., and Horikoshi M. (2008). Histone chaperones: 30 years from isolation to elucidation of the mechanisms of nucleosome assembly and disassembly. *Cell Mol. Life Sci.* 65, 414–444.

Osley, M. A. (1991). The regulation of histone synthesis in the cell cycle. *Annu. Rev. Biochem.* 60, 827–861.

Verreault, A. (2000). *De novo* nucleosome assembly: new pieces in an old puzzle. *Genes Dev.* 14, 1430–1438.

Research

Ahmad, K. and Henikoff, S. (2001). Centromeres are specialized replication domains in heterochromatin. *J. Cell Biol.* 153, 101–110.

Ahmad, K. and Henikoff, S. (2002). The histone variant H3.3 marks active chromatin by replication-independent nucleosome assembly. *Mol. Cell* 9, 1191–1200.

Gruss, C., Wu, J., Koller, T., and Sogo, J. M. (1993). Disruption of the nucleosomes at the replication fork. *EMBO J.* 12, 4533–4545.

Loppin, B., Bonnefoy, E., Anselme, C., Laurencon, A., Karr, T. L., and Couble, P. (2005). The histone H3.3 chaperone HIRA is essential for chromatin assembly in the male pronucleus. *Nature* 437, 1386–1390.

Ray-Gallet, D., Quivy, J. P., Scamps, C., Martini, E. M., Lipinski, M., and Almouzni, G. (2002). HIRA is critical for a nucleosome assembly pathway independent of DNA synthesis. *Mol. Cell* 9, 1091–1100.

Shibahara, K., and Stillman, B. (1999). Replication-dependent marking of DNA by PCNA facilitates CAF-1-coupled inheritance of chromatin. *Cell* 96, 575–585.

Smith, S. and Stillman, B. (1989). Purification and characterization of CAF-I, a human cell factor required for chromatin assembly during DNA replication *in vitro. Cell* 58, 15–25.

Smith, S. and Stillman, B. (1991). Stepwise assembly of chromatin during DNA replication in vitro. *EMBO J.* 10, 971–980.

Tagami, H., Ray-Gallet, D., Almouzni, G., and Nakatani, Y. (2004). Histone H3.1 and H3.3 complexes mediate nucleosome assembly pathways dependent or independent of DNA synthesis. *Cell* 116, 51–61.

Yu, L. and Gorovsky, M. A. (1997). Constitutive expression, not a particular primary sequence, is the important feature of the H3 replacement variant hv2 in *Tetrahymena thermophila. Mol. Cell. Biol.* 17, 6303–6310.

10.9 Do Nucleosomes Lie in Specific Positions?

Research

Chung, H.-R. and Vingron, M. (2009). Sequence-dependent nucleosome positioning, *J. Mol. Biol.* 386, 1411–1422.

Field, Y., Kaplan, N., Fondufe-Mittendorf, Y., Moore, I. K., Sharon, E., Lubling, Y., Widom, J., and Segal, E. (2008). Distinct modes of regulation by chromatin encoded through nucleosome positioning signals. *PLoS Comput. Biol.* 4(11): e1000216.

Peckham, H. E., Thurman, R. E., Fu, Y., Stamatoyannopoulos, J. A., Noble, W. S., Struhl, K. and Weng, Z. (2007). Nucleosome positioning signals in genomic DNA. *Genome Res.* 17, 1170–1177.

Segal, E., Fondufe-Mittendorf, Y., Chen, L., Thastrom, A., Field, Y., Moore, I. K., Wang, J. Z., and Widom, J. (2006). A genomic code for nucleosome positioning. *Nature* 442, 772–778.

Yuan, G. C., Liu, Y. J., Dion, M. F., Slack, M. D., Wu, L. F., Altschuler, S. J., Rando, O. J. (2005) Genome-scale identification of nucleosome positions in *S. cerevisiae. Science* 309, 626–630.

10.10 Nucleosomes Are Displaced and Reassembled During Transcription

Reviews

Formosa, T. (2008). FACT and the reorganized nucleosome. *Mol. BioSyst.* 4, 1085–1093.

Kornberg, R. D. and Lorch, Y. (1992). Chromatin structure and transcription. *Annu. Rev. Cell Biol.* 8, 563–587.

Kulaeva, O. I., and Studitsky, V. M., 2007. Transcription through chromatin by RNA polymerase II: histone displacement and exchange. *Mutat. Res.* 618, 116–129.

Thiriet, C., and Hayes, J. J., 2006. Histone dynamics during transcription: exchange of H2A/H2B dimers and H3/H4 tetramers during pol II elongation. *Results Probl. Cell Differ.* 41, 77–90.

Workman, J. L., 2006. Nucleosome displacement during transcription. Genes Dev. 20, 2507–2512.

Research

Belotserkovskaya, R., Oh, S., Bondarenko, V. A., Orphanides, G., Studitsky, V. M., and Reinberg, D. (2003). FACT facilitates transcription-dependent nucleosome alteration. *Science* 301, 1090–1093.

Bortvin, A. and Winston, F. (1996). Evidence that Spt6p controls chromatin structure by a direct interaction with histones. *Science* 272, 1473–1476.

Cavalli, G. and Thoma, F. (1993). Chromatin transitions during activation and repression of galactose-regulated genes in yeast. *EMBO J.* 12, 4603–4613.

Imbeault, D., Gamar, L., Rufiange, A., Paquet, E., and Nourani, A. (2008). The Rtt106 histone chaperone is functionally linked to transcription elongation and is involved in the regulation of spurious transcription from cryptic promoters in yeast. *J. Biol. Chem.* 283, 27350–27354.

Saunders, A., Werner, J., Andrulis, E. D., Nakayama, T., Hirose, S., Reinberg, D., and Lis, J. T. (2003). Tracking FACT and the RNA polymerase II elongation complex through chromatin *in vivo*. *Science* 301, 1094–1096.

Studitsky, V. M., Clark, D. J., and Felsenfeld, G. (1994). A histone octamer can step around a transcribing polymerase without leaving the template. *Cell* 76, 371–382.

10.11 DNase Sensitivity Detects Changes in Chromatin Structure

Reviews

Gross, D. S. and Garrard, W. T. (1988). Nuclease hypersensitive sites in chromatin. *Annu. Rev. Biochem.* 57, 159–197.

Krebs, J. E. and Peterson, C. L. (2000). Understanding "active" chromatin: a historical perspective of chromatin remodeling. *Crit Rev. Eukaryot. Gene Expr.* 10, 1–12.

Research

Groudine, M. and Weintraub, H. (1982). Propagation of globin DNAase I-hypersensitive sites in absence of factors required for induction: a possible mechanism for determination. *Cell* 30, 131–139.

Stalder, J. et al. (1980). Tissue-specific DNA cleavage in the globin chromatin domain introduced by DNAase I. *Cell* 20, 451–460.

10.12 Insulators Define Transcriptionally Independent Domains

Reviews

Bushey, A. M., Dorman, E. R. and Corces, V. G. (2008). Chromatin insulators: regulatory mechanisms and epigenetic inheritance. *Mol. Cell* 32, 1–9.

Gaszner, M. and Felsenfeld, G. (2006). Insulators: exploiting transcriptional and epigenetic mechanisms. *Nat. Rev. Genet.* 7, 703–713.

Maeda, R. K. and Karch, F. (2007). Making connections: boundaries and insulators in *Drosophila*. *Curr. Opin. Genet. Dev.* 17, 394–399.

Valenzuela, L. and Kamakaka, R. T. (2006). Chromatin insulators. *Annu. Rev. Genet.* 40, 107–138.

West, A. G., Gaszner, M., and Felsenfeld, G. (2002). Insulators: many functions, many mechanisms. *Genes Dev.* 16, 271–288.

Research

Aoki, T., Schweinsberg, S., Manasson, J., and Schedl, P. (2008). A stage-specific factor confers *Fab-7* boundary activity during early embryogenesis in *Drosophila*. *Mol. Cell Biol.* 28, 1047–1060.

Chung, J. H., Whiteley, M., and Felsenfeld, G. (1993). A 5' element of the chicken β-globin domain serves as an insulator in human erythroid cells and protects against position effect in *Drosophila*. *Cell* 74, 505–514.

Cuvier, O., Hart, C. M., and Laemmli, U. K. (1998). Identification of a class of chromatin boundary elements. *Mol. Cell Biol.* 18, 7478–7486.

Gaszner, M., Vazquez, J., and Schedl, P. (1999). The Zw5 protein, a component of the *scs* chromatin domain boundary, is able to block enhancer–promoter interaction. *Genes Dev.* 13, 2098–2107.

Gerasimova, T. I., Byrd, K., and Corces, V. G. (2000). A chromatin insulator determines the nuclear localization of DNA. *Mol. Cell* 6, 1025–1035.

Hagstrom, K., Muller, M., and Schedl, P. (1996). *Fab-7* functions as a chromatin domain boundary to ensure proper segment specification by the *Drosophila bithorax* complex. *Genes Dev.* 10, 3202–3215.

Harrison, D. A., Gdula, D. A., Cyne, R. S., and Corces, V. G. (1993). A leucine zipper domain of the suppressor of hairy-wing protein mediates its repressive effect on enhancer function. *Genes Dev.* 7, 1966–1978.

Kellum, R. and Schedl, P. (1991). A position-effect assay for boundaries of higher order chromosomal domains. *Cell* 64, 941–950.

Kuhn, E. J., Hart, C. M., and Geyer, P. K. (2004). Studies of the role of the *Drosophila scs* and *scs'* insulators in defining boundaries of a chromosome puff. *Mol. Cell Biol.* 24, 1470–1480.

Mihaly, J. et al. (1997). *In situ* dissection of the *Fab-7* region of the *bithorax* complex into a chromatin domain boundary and a polycomb-response element. *Development* 124, 1809–1820.

Pikaart, M. J., Recillas-Targa, F., and Felsenfeld, G. (1998). Loss of transcriptional activity of a transgene is accompanied by DNA methylation and histone deacetylation and is prevented by insulators. *Genes Dev.* 12, 2852–2862.

Roseman, R. R., Pirrotta, V., and Geyer, P. K. (1993). The su(Hw) protein insulates expression of the *D. melanogaster white* gene from chromosomal position-effects. *EMBO J.* 12, 435–442.

Zhao, K., Hart, C. M., and Laemmli, U. K. (1995). Visualization of chromosomal domains with boundary element-associated factor BEAF-32. *Cell* 81, 879–889.

Zhou, J. and Levine, M. (1999). A novel *cis*-regulatory element, the PTS, mediates an anti-insulator activity in the *Drosophila* embryo. *Cell* 99, 567–575.

10.13 An LCR May Control a Domain

Reviews

Bulger, M. and Groudine, M. (1999). Looping versus linking: toward a model for long-distance gene activation. *Genes Dev.* 13, 2465–2477.

Grosveld, F., Antoniou, M., Berry, M., De Boer, E., Dillon, N., Ellis, J., Fraser, P., Hanscombe, O., Hurst, J., and Imam, A. (1993). The regulation of human globin gene switching. *Philos. Trans. R. Soc. Lond. B Biol. Sci.* 339, 183–191.

Miele, A. and Dekker, J. (2008). Long-range chromosomal interactions and gene regulation. *Mol. BioSyst.* 4, 1046–1057.

Research

Cai, S., Lee, C. C., and Kohwi-Shigematsu, T. (2006). SATB1 packages densely looped, transcriptionally active chromatin for coordinated expression of cytokine genes. *Nat. Genet.* 38, 1278–1288.

Gribnau, J., de Boer, E., Trimborn, T., Wijgerde, M., Milot, E., Grosveld, F., and Fraser, P. (1998). Chromatin interaction mechanism of transcriptional control *in vitro*. *EMBO J.* 17, 6020–6027.

Spilianakis, C. G., Lalioti, M. D., Town, T., Lee, G. R., and Flavell, R. A. (2005). Inter-chromosomal associations between alternatively expressed loci. *Nature* 435, 637–645.

van Assendelft, G. B., Hanscombe, O., Grosveld, F., and Greaves, D. R. (1989). The β-globin dominant control region activates homologous and heterologous promoters in a tissue-specific manner. *Cell* 56, 969–977.

2 DNA Replication and Recombination

Image courtesy of Teresa Larsen, The Foundation for Scientific Literacy.

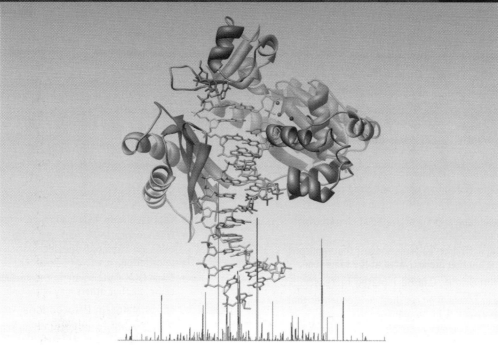

Reproduced from H. Zang, et al., *J. Biol. Chem.* 280 (2005): 29750–29764. © 2005, The American Society for Biochemistry and Molecular Biology. Photo courtesy of F. Peter Guengerich, Dept. of Biochemistry, Vanderbilt University School of Medicine.

The Replicon

Edited by Stephen D. Bell

CHAPTER OUTLINE

11.1 Introduction

Whether a cell has only one chromosome (as in most prokaryotes) or has many chromosomes (as in eukaryotes), the entire genome must be replicated precisely once for every cell division. How is the act of replication linked to the cell cycle?

Two general principles are used to compare the state of replication with the condition of the cell cycle:

- *Initiation of DNA replication commits the cell (prokaryotic or eukaryotic) to a further division.* From this standpoint, the number of descendants that a cell generates is determined by a series of decisions on whether or not to initiate DNA replication. Replication is controlled at the stage of initiation. *Once replication has started, it continues until the entire genome has been duplicated.*

- If replication proceeds, the consequent division cannot be permitted to occur until the replication event has been completed. Indeed, the completion of replication may provide a trigger for cell division. The duplicate genomes

are then segregated one to each daughter cell. The unit of segregation is the chromosome.

The unit of DNA in which an individual act of replication occurs is called the **replicon**. Each replicon "fires" once, and only once, in each cell cycle. The replicon is defined by its possession of the control elements needed for replication. It has an **origin** at which replication is initiated. It may also have a **terminus** at which replication stops. Any sequence attached to an origin—or, more precisely, not separated from an origin by a terminus—is replicated as part of that replicon. The origin is a *cis*-acting site, able to affect only that molecule of DNA on which it resides.

(The original formulation of the replicon [in bacteria] viewed it as a unit possessing both the origin *and* the gene coding for the regulator protein. Now, however, "replicon" is usually applied to eukaryotic chromosomes to describe a unit of replication that contains an origin; *trans*-acting regulator protein(s) may be encoded elsewhere.)

Bacteria and archaea may contain additional genetic information in the form of plasmids. *A plasmid is an autonomous circular DNA that constitutes a separate replicon.* Plasmid replicons may show **single-copy replication control** which means that they replicate once every time the bacterial chromosome replicates, or they may be under **multicopy replication control**, when they are present in a greater number of copies than the bacterial chromosome. Each phage or virus DNA also constitutes a replicon, and thus is able to initiate many times during an infectious cycle. Perhaps a better way to view the prokaryotic replicon, therefore, is to reverse the definition: Any DNA molecule that contains an origin can be replicated autonomously in the cell.

A major difference in the organization of bacterial, archaeal, and eukaryotic genomes is seen in their replication. A genome in a bacterial cell has a single replication origin and thus constitutes a single replicon; therefore the units of replication and segregation coincide. Initiation at a single origin sponsors replication of the entire genome, once for every cell division. Each haploid bacterium has a single chromosome, so this type of replication control is called **single copy**. The other prokaryotic domain of life, the archaea, is more complex. While some archaeal species have chromosomes with a bacterial-like situation of a single replication origin, other species initiate replication from multiple sites on a single chromosome. For example, the single circular chromosomes of *Sulfolobus* species have three origins and thus are composed of three replicons. This complexity is further heightened in eukaryotes. Each eukaryotic chromosome (usually a very long linear molecule of DNA) contains a large number of replicons spaced unevenly throughout the chromosomes. The presence of multiple origins per chromosome adds another dimension to the problem of control: All of the replicons on a chromosome must be fired during one cell cycle. They are not necessarily, however, active simultaneously. Each replicon must be activated over a fairly protracted period, *and each must be activated no more than once in each cell cycle.*

Some signal must distinguish replicated from nonreplicated replicons to ensure that replicons do not fire a second time. Many replicons are activated independently, so another signal must exist to indicate when the entire process of replicating all replicons has been completed.

We have begun to collect information about the construction of individual replicons, but we still have little information about the relationship between replicons. We do not know whether the pattern of replication is the same in every cell cycle. Are all origins always used, or are some origins sometimes silent? Do origins always fire in the same order? If there are different classes of origins, what distinguishes them?

In contrast with nuclear chromosomes, which have a single-copy type of control, the DNA of mitochondria and chloroplasts may be regulated more like plasmids that exist in multiple copies per bacterium. There are multiple copies of each organelle DNA per cell, and the control of organelle DNA replication must be related to the cell cycle.

In all these systems, the key question is to define the DNA sequences that function as origins and to determine how they are recognized by the appropriate proteins of the replication apparatus. We start by considering the basic construction of replicons and the various forms that they take in bacteria and eukaryotic cells. In Chapter 12, *Extrachromosomal Replicons*, we consider autonomously replicating units in bacteria. In Chapter 13, *Bacterial Replication Is Connected to the Cell Cycle*, we turn to the question of how replication of the genome is coordinated with bacterial division and what is responsible for segregating the genomes to daughter bacteria. In Chapter 14, *DNA Replication*, we examine the biochemistry of DNA synthesis.

11.2 Replicons Can Be Linear or Circular

Key concepts

- A replicated region appears as a bubble within nonreplicated DNA.
- A replication fork is initiated at the origin and then moves sequentially along DNA.
- Replication is unidirectional when a single replication fork is created at an origin.
- Replication is bidirectional when an origin creates two replication forks that move in opposite directions.

Replication starts at an origin by separating or melting the two strands of the DNA duplex. **FIGURE 11.1** shows that each of the parental strands then acts as a template to synthesize a complementary daughter strand. This model of replication, in which a parental duplex gives rise to two daughter duplexes, each containing one original parental strand and one

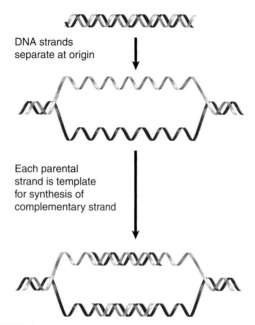

DNA strands separate at origin

Each parental strand is template for synthesis of complementary strand

FIGURE 11.1 An origin is a sequence of DNA at which replication is initiated by separating the parental strands and initiating synthesis of new DNA strands. Each new strand is complementary to the parental strand that acts as the template for its synthesis.

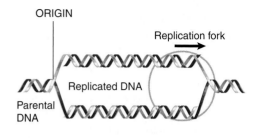

ORIGIN

Replication fork

Replicated DNA

Parental DNA

BIDIRECTIONAL REPLICATION

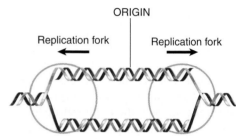

ORIGIN

Replication fork   Replication fork

FIGURE 11.3 Replicons may be unidirectional or bidirectional, depending on whether one or two replication forks are formed at the origin.

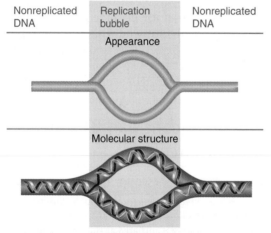

| Nonreplicated DNA | Replication bubble | Nonreplicated DNA |

Appearance

Molecular structure

FIGURE 11.2 Replicated DNA is seen as a replication bubble flanked by nonreplicated DNA.

new strand, is called **semiconservative replication**.

A molecule of DNA engaged in replication has two types of regions. **FIGURE 11.2** shows that when replicating DNA is viewed by electron microscopy, the replicated region appears as a **replication bubble** within the nonreplicated DNA. The nonreplicated region consists of the parental duplex; this opens into the replicated region where the two daughter duplexes have formed.

The point at which replication occurs is called the **replication fork** (also known as the **growing point**). *A replication fork moves sequentially along the DNA from its starting point at the origin.* The origin may be used to start either **unidirectional replication** or **bidirectional replication**. The type of event is determined by whether one or two replication forks set out from the origin. In unidirectional replication, one replication fork leaves the origin and proceeds along the DNA. In bidirectional replication, two replication forks are formed; they proceed away from the origin in opposite directions.

The appearance of a replication bubble does not distinguish between unidirectional and bidirectional replication. As depicted in **FIGURE 11.3**, the bubble can represent either of two structures. If generated by unidirectional replication, the bubble represents one fixed origin and one moving replication fork. If generated by bidirectional replication, the bubble represents a pair of replication forks. In either case, the progress of replication expands the bubble until ultimately it encompasses the whole replicon.

When a replicon is circular, the presence of a bubble forms the θ structure shown in **FIGURE 11.4**. The successive stages of replication of the circular DNA of polyoma virus are visualized by electron microscopy in **FIGURE 11.5**.

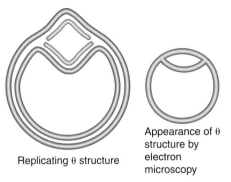

Replicating θ structure

Appearance of θ structure by electron microscopy

FIGURE 11.4 A replication bubble forms a θ structure in circular DNA.

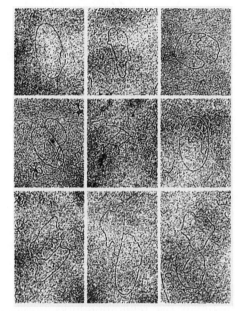

FIGURE 11.5 The replication bubble becomes larger as the replication forks proceed along the replicon. Note that the "bubble" becomes larger than the nonreplicated segment. The two sides of the bubble can be defined because they are both the same length. Photo courtesy of Bernhard Hirt, Swiss Institute for Experimental Cancer Research (ISREC).

11.3 Origins Can Be Mapped by Autoradiography and Electrophoresis

Key concepts

- Replication fork movement can be detected by labeling of newly synthesized DNA with pulses of either radioactive or fluorescent nucleotides.
- Replication forks create Y-shaped structures that change the electrophoretic migration of DNA molecules.

Whether a replicating bubble has one or two replication forks can be determined in two ways. The choice of method depends on whether the DNA is a defined molecule or an unidentified region of a cellular genome.

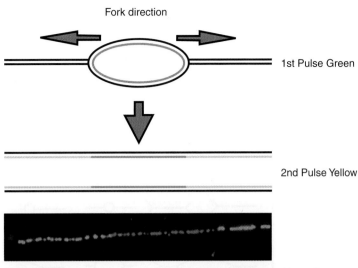

Fork direction

1st Pulse Green

2nd Pulse Yellow

FIGURE 11.6 Initial replication is detected by incorporation of a green label. A subsequent addition of a second label leads to a yellow signal. Photo reproduced from K. Marheineke, et al., Visualization of bidirectional initiation of chromosomal DNA replication in a human cell free system, *Nucleic Acids Res.,* vol. 33 (21), pp. 6931–6941, by permission of Oxford University Press. Photo courtesy of Kathrin Marheineke and Torsten Krude.

With a defined linear molecule, we can use electron microscopy to measure the distance of each end of the bubble from the end of the DNA, and then compare the positions of the ends of the bubbles in molecules that have bubbles of different sizes. If replication is unidirectional, only one of the ends will move; the other is the fixed origin. If replication is bidirectional, both will move; the origin is the point midway between them.

With undefined regions of large genomes, two successive pulses of labeled nucleotides can detect DNA replication. Traditionally this was performed with radioactive DNA precursors; however, recent advances in fluorescence labeling methods have made this latter approach the system of choice. FIGURE 11.6 shows the pattern of bidirectional replication generated by initial labeling of DNA, resulting in green DNA; the addition of a second fluorescent label generates yellow DNA. Another method for mapping origins with greater resolution takes advantage of the effects that changes in shape have upon electrophoretic migration of DNA. FIGURE 11.7 illustrates the two-dimensional mapping technique, in which restriction fragments of replicating DNA are electrophoresed in a first dimension that separates by mass and a second dimension where movement is determined more by shape. Different types of replicating molecules follow characteristic paths, measured by their deviation from the line that would be followed by a linear molecule of DNA that doubled in size.

A simple Y-structure, in which one fork moves along a linear fragment, follows a continuous path. An inflection point occurs when all three branches are the same length, and the structure therefore deviates most extensively from linear DNA. Analogous considerations determine the paths of double Y-structures or bubbles. An asymmetric bubble follows a discontinuous path, with a break at the point at which the bubble is converted to a Y-structure as one fork runs off the end.

Taken together, these techniques for characterizing replicating DNA show that origins are used most often to initiate bidirectional replication. From this level of resolution, we must now proceed to the molecular level to identify the *cis*-acting sequences that comprise the origin and the *trans*-acting factors that recognize it.

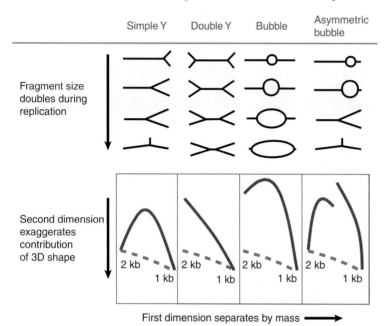

FIGURE 11.7 The position of the origin and the number of replicating forks determine the shape of a replicating restriction fragment, which can be followed by its electrophoretic path (solid line). The dashed line shows the path for a linear DNA.

11.4 The Bacterial Genome Is (Usually) a Single Circular Replicon

Key concepts
- Bacterial replicons are usually circles that replicate bidirectionally from a single origin.
- The origin of *E. coli*, *oriC*, is 245 bp in length.
- The two replication forks usually meet halfway around the circle, but there are ter sites that cause termination if the replication forks go too far.

Prokaryote replicons are usually circular, so that the DNA forms a closed circle with no free ends. Circular structures include the bacterial chromosome itself, all plasmids, and many bacteriophages, and are also common in chloroplasts and mitochondrial DNAs. **FIGURE 11.8** summarizes the stages of replicating a circular chromosome. After replication has initiated at the origin, two replication forks proceed in

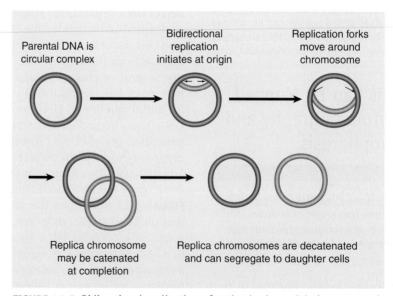

FIGURE 11.8 Bidirectional replication of a circular bacterial chromosome is initiated at a single origin. The replication forks move around the chromosome. If the replicated chromosomes are catenated, they must be disentangled before they can segregate to daughter cells.

opposite directions. The circular chromosome is sometimes described as a θ (theta) structure at this stage, because of its appearance. An important consequence of circularity is that the completion of the process can generate two chromosomes that are linked because one passes through the other (they are said to be catenated), and specific enzyme systems may be required to separate them (see *Section 13.7, Chromosomal Segregation May Require Site-Specific Recombination*).

The genome of *E. coli* is replicated bidirectionally from a single unique site called the origin, identified as the genetic locus *oriC*. Two replication forks initiate at *oriC* and move around the genome at approximately the same speed to a special termination region. One interesting question is what ensures that the DNA is replicated right across the region where the two forks meet.

Sequences that are involved with termination are called *ter* sites. A *ter* site contains a short, ~23-bp sequence. The termination sequences are unidirectional; that is, they function in only one orientation. The *ter* site is recognized by a unidirectional contrahelicase (called Tus in *E. coli* and RTP in *B. subtilis*) that recognizes the consensus sequence and prevents the replication fork from proceeding. The *E. coli* enzyme acts by antagonizing the replication helicase in a directional manner by direct contact between the DnaB helicase and Tus. Deletion of the *ter* sites does not, however, prevent normal replication cycles from occurring, although it does affect segregation of the daughter chromosomes.

Termination in *E. coli* and *B. subtilis* has the interesting features shown in **FIGURE 11.9**. The two replication forks meet and halt in a region approximately halfway around the chromosome from the origin. In *E. coli*, two clusters of five *ter* sites each, including *terE, D, A* on one side and *terC* and *B* on the other, are located ~100 kb on either side of this termination region. Each set of *ter* sites is specific for one direction of fork movement; that is, each set of *ter* sites allows a replication fork into the termination region, but does not allow it out the other side. For example, replication fork 1 can pass through *terC* and *terB* into the region, but it cannot continue past *terE, D,* and *A*. This arrangement creates a "replication fork trap." If, for some reason, one fork is delayed, so that the forks fail to meet in the middle, the faster fork will be trapped at the distal *ter* sites to wait for the slower fork.

What happens when a replication fork encounters a protein bound to DNA? We assume that repressors, for example, are displaced and then rebind. A particularly interesting question is what happens when a replication fork encounters an RNA polymerase engaged in transcription. A replication fork moves 10× faster than RNA polymerase. If they are proceeding in the same direction, either the replication fork must displace the RNA polymerase or it must slow down as it waits for the RNA polymerase to reach its terminator. It appears that a DNA polymerase moving in the same direction as an RNA polymerase can "bypass" it without disrupting transcription, but we do not understand how this happens.

A conflict arises when the replication fork meets an RNA polymerase traveling in the opposite direction (i.e., toward it). Can it displace the polymerase? Or do both replication and transcription come to a halt? An indication that these encounters cannot be easily resolved is provided by the gene organization on the *E. coli* chromosome. Almost all active transcription units are oriented so that they are expressed in the same direction as the replication fork that passes them. The exceptions all comprise small transcription units that are infrequently expressed. The difficulty of generating inversions containing highly expressed genes suggests that head-on encounters between a replication fork and a series of transcribing RNA polymerases may be lethal.

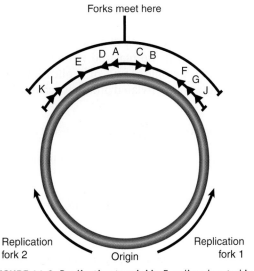

FIGURE 11.9 Replication termini in *E. coli* are located in a region between two sets of *ter* sites.

11.5 Methylation of the Bacterial Origin Regulates Initiation

Key concepts

- *oriC* contains binding sites for DnaA—dnaA-boxes.
- *oriC* also contains eleven GATC/CTAG repeats that are methylated on adenine on both strands.
- Replication generates hemimethylated DNA, which cannot initiate replication.
- There is a 13-minute delay before the GATC/CTAG repeats are remethylated.

The bacterial DnaA protein is the replication initiator; it binds sequence specifically to multiple sites (*dnaA* boxes) in *oriC*, the replication origin. DnaA is an ATP-binding protein and its binding to DNA is affected depending on whether ATP, ADP, or no nucleotide is bound. One mechanism by which the activity of the replication origin is controlled is DNA methylation. The *E. coli oriC* contains eleven copies of the sequence $\frac{\text{GATC}}{\text{CTAG}}$, which is a target for methylation at the N^6 position of adenine by the Dam methylase enzyme. These sites are also found throughout the genome. Note, though, that several of these methylation sites overlap *dnaA* boxes. This is illustrated in **FIGURE 11.10**.

Before replication, the palindromic target site is methylated on the adenines of each strand. Replication inserts the normal (non-modified) bases into the daughter strands. This generates **hemimethylated DNA**, in which one strand is methylated and one strand is unmethylated. Thus the replication event converts Dam target sites from fully methylated to hemimethylated condition.

What is the consequence for replication? The ability of a plasmid relying upon *oriC* to replicate in *dam⁻ E. coli* depends on its state of methylation. If the plasmid is methylated it undergoes a single round of replication, and then the hemimethylated products accumulate, as described in **FIGURE 11.11**. The hemimethylated plasmids then accumulate, rather than being replaced by unmethylated plasmids, suggesting that a hemimethylated origin cannot be used to initiate a replication cycle.

This suggests two explanations: Initiation may require full methylation of the Dam target sites in the origin, or it may be inhibited by hemimethylation of these sites. The latter seems to be the case, because an origin of nonmethylated DNA can function effectively.

Thus hemimethylated origins cannot initiate again until the Dam methylase has converted them into fully methylated origins. The GATC sites at the origin remain hemimethylated for ~13 minutes after replication. This long period is unusual because at typical GATC sites elsewhere in the genome, remethylation begins immediately (<1.5 minutes) following replication. One other region behaves like *oriC*: The promoter of the *dnaA* gene also shows a delay before remethylation begins. While it is hemimethylated the *dnaA* promoter is repressed, which causes a reduction in the level of DnaA protein. Thus the origin itself is inert, and production of the crucial initiator protein is repressed during this period.

DNA methylation in bacteria serves a second function as well: It allows the DNA mismatch recognition machinery to distinguish the old template strand from the new strand. If the DNA polymerase has made an error, such as creating an A–C base pair, the repair system will use the methylated strand as a template to replace the base on the nonmethylated strand. Without that methylation, the enzyme would have no way to determine which is the new strand.

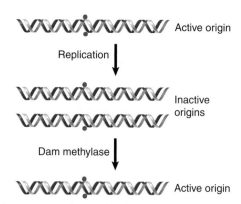

FIGURE 11.11 Only fully methylated origins can initiate replication; hemimethylated daughter origins cannot be used again until they have been restored to the fully methylated state.

High affinity DnaA binding sites   GATC Dam methylation site

DnaA-ATP binding sites   Site of initial DNA unwinding

Architecture of *E. coli oriC*

FIGURE 11.10 The *E. coli* origin of replication, *oriC*, contains multiple binding sites for the DnaA initiator protein. In a number of cases these sites overlap Dam methylation sites.

11.6 Origins May Be Sequestered After Replication

Key concepts

- SeqA binds to hemimethylated DNA and is required for delaying rereplication.
- SeqA may interact with DnaA.
- As the origins are hemimethylated they bind to the cell membrane and may be unavailable to methylases.
- The nature of the connection between the origin and the membrane is still unclear.

What is responsible for the delay in remethylation at *oriC* and *dnaA*? The most likely explanation is that these regions are sequestered in a form in which they are inaccessible to the Dam methylase.

A circuit responsible for controlling reuse of origins is identified by mutations in the gene *seqA*. The mutants reduce the delay in remethylation at both *oriC* and *dnaA*. As a result, they initiate DNA replication too soon, thereby accumulating an excessive number of origins. This suggests that *seqA* is part of a negative regulatory circuit that prevents origins from being remethylated. SeqA binds to hemimethylated DNA more strongly than to fully methylated DNA. It may initiate binding when the DNA becomes hemimethylated, at which point its continued presence prevents formation of an open complex at the origin. SeqA does not have specificity for the *oriC* sequence, and it seems likely that this is conferred by DnaA. This would explain the genetic interactions between *seqA* and *dnaA*.

The full scope of the system used to control reinitiation is not clear, but several mechanisms may be involved: physical sequestration of the origin, delay in remethylation, inhibition of DnaA binding, and repression of *dnaA* transcription. It is not immediately obvious which of these events cause the others, and whether their effects on initiation are direct or indirect. Indeed, we still have to come to grips with the central issue of which feature has the basic responsibility for timing. The period of sequestration appears to increase with the length of the cell cycle, which suggests that it directly reflects the clock that controls reinitiation. One aspect of the control may lie in the observation that hemimethylation of *oriC* is required for its association with cell membranes *in vitro*. This may reflect a physical repositioning to a region of the cell that is not permissive for replication initiation.

As the only member of the replication apparatus uniquely required at the origin, DnaA has attracted much attention. DnaA is a target for several regulatory systems. It may be that no one of these systems alone is adequate to control frequency of initiation, but that when combined they achieve the desired result. Some mutations in *dnaA* render replication asynchronous, which suggests that DnaA could be the "titrator" or "clock" that measures the number of origins relative to cell mass. Overproduction of DnaA yields conflicting results, which vary from no effect to causing initiation to take place at reduced mass.

It has been difficult to identify the protein component(s) that mediate membrane attachment. A hint that this is a function of DnaA is provided by its response to phospholipids. Phospholipids promote the exchange of ATP with ADP bound to DnaA. We do not know what role this plays in controlling the activity of DnaA (which requires ATP), but the reaction implies that DnaA is likely to interact with the membrane. This would imply that more than one event is involved in associating with the membrane. Perhaps a hemimethylated origin is bound by the membrane-associated inhibitor, but when the origin becomes fully methylated, the inhibitor is displaced by DnaA associated with the membrane.

If DnaA is the initiator that triggers a replication cycle, the key event will be its accumulation at the origin to a critical level. There are no cyclic variations in the overall concentration or expression of DnaA, which suggests that local events must be responsible. To be active in initiating replication, DnaA must be in the ATP-bound form. Thus hydrolysis of ATP to ADP by DnaA has the potential to regulate its activity. While DnaA has a weak intrinsic activity that converts the ATP to ADP, this is enhanced by a factor termed Hda. In a conceptually elegant feedback loop, Hda is recruited to a replication origin via the β subunit of the DNA polymerase. Thus, only once the origin has been activated and the full replication machinery assembled is Hda recruited, whence it acts to switch off DnaA, preventing a second round of replication.

Another factor that controls availability of DnaA at the origin is the competition for binding it to other sites on DNA. In particular, a locus called *dat* has a large concentration of DnaA-binding sites. It binds about 8× more DnaA than the origin. Deletion of *dat* causes initiation to occur more frequently. This significantly reduces the amount of DnaA available to the origin, but we do not yet understand exactly what role this may play in controlling the timing of initiation.

11.7 Archaeal Chromosomes Can Contain Multiple Replicons

Key concepts

- Some Archaea have multiple replication origins.
- These origins are bound by homologs of eukaryotic replication initiation factors.
- Origin binding introduces underwinding of DNA.

Although archaea, like most bacteria, have small circular chromosomes, some archaea possess multiple replication origins. Sequence motifs within these origins are recognized and bound specifically by archaeal homologs of the eukaryotic replication initiation factors Orc1 and Cdc6. These proteins bind to several sites in the origin and, in doing so, deform the DNA, as shown in **FIGURE 11.12**. In the archaeal species *Sulfolobus* all three origins are activated within a few minutes of each other, however, the mechanism of coordinate control of the origins is not yet known.

11.8 Each Eukaryotic Chromosome Contains Many Replicons

Key concepts

- A chromosome is divided into many replicons.
- The progression into S phase is tightly controlled.
- Eukaryotic replicons are 40 to 100 kb in length.
- Individual replicons are activated at characteristic times during S phase.
- Regional activation patterns suggest that replicons near one another are activated at the same time.

In eukaryotic cells, the replication of DNA is confined to the second part of the cell cycle, called **S phase**, which follows the G1 phase (see **FIGURE 11.13**). The eukaryote cell cycle is composed of alternating rounds of growth followed by DNA replication and cell division. After the cell divides into two daughter cells, each has to grow back to approximately the size of the original mother cell before cell division can occur again. The G1 phase of the cell cycle is primarily concerned with growth (although G1 is an abbreviation for *first gap* because the early cytologists could not see any activity). In G1, everything except DNA begins to be doubled: RNA, protein, lipids, and carbohydrate. The progression from G_1 into S is very tightly

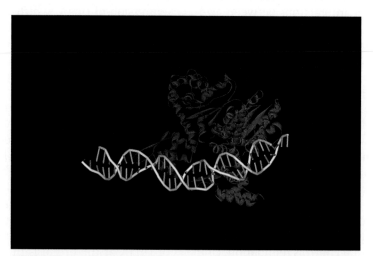

FIGURE 11.12 The crystal structure of a heterodimer of archaeal Orc1/Cdc6 replication initiator proteins bound to an origin of replication. The two proteins in blue and purple underwind and bend the DNA upon binding. Structure from Protein Data Bank 2QBY. E. L. Dueber, et al., *Science* 317 (2007): 1210–1213.

regulated and is controlled by a **checkpoint**. In order for a cell to be allowed to progress into S phase, there must be a certain minimum amount of growth, which is biochemically measured. In addition, there must not be any damage to the DNA. Damaged DNA or too little growth prevents the cell from progressing into S phase. When S phase is completed, G2 phase commences. There is no control point and no sharp demarcation.

Replication of the large amount of DNA contained in a eukaryotic chromosome is accomplished by dividing it into many individual replicons as seen in **FIGURE 11.14**. Only some of these replicons are engaged in replication at any point in S phase. Presumably each replicon is activated at a specific time during S phase, although the evidence on this issue is not decisive.

The start of S phase is signaled by the activation of the first replicons. Over the next few hours, initiation events occur at other replicons in an ordered manner. Chromosomal replicons usually display bidirectional replication.

Individual replicons in eukaryotic genomes are relatively small, typically ~40 kb in yeast or fly and ~100 kb in animal cells. They can, however, vary more than tenfold in length within a genome. The rate of replication is ~2000 bp/min, which is much slower than the 50,000 bp/min of bacterial replication fork movement, presumably because the chromosome is assembled into chromatin, not naked DNA.

From the speed of replication, it is evident that a mammalian genome could be replicated in ~1 hour if all replicons functioned simulta-

neously. S phase actually lasts for >6 hours in a typical somatic cell, though, which implies that no more than 15% of the replicons are likely to be active at any given moment. There are some exceptional cases, such as the early embryonic divisions of *Drosophila* embryos, where the duration of S phase is compressed by the simultaneous functioning of a large number of replicons.

How are origins selected for initiation at different times during S phase? In *Saccharomyces cerevisiae*, the default appears to be for replicons to replicate early, but *cis*-acting sequences can cause origins linked to them to replicate at a later time. In other organisms, there is a general hierarchy to the order of replication. Replicons near active genes are replicated earliest and replicons in heterochromatin replicate last.

Available evidence suggests that most chromosomal replicons do not have a termination region like that of bacteria, at which the replication forks cease movement and (presumably) dissociate from the DNA. It seems more likely that a replication fork continues from its origin until it meets a fork proceeding toward it from the adjacent replicon. We have already mentioned the potential topological problem of joining the newly synthesized DNA at the junction of the replication forks.

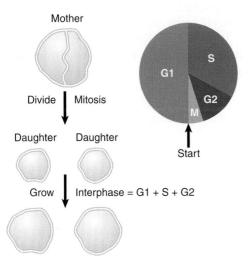

FIGURE 11.13 A growing cell alternates between cell division of a mother cell into two daughter cells and growth back to the original size.

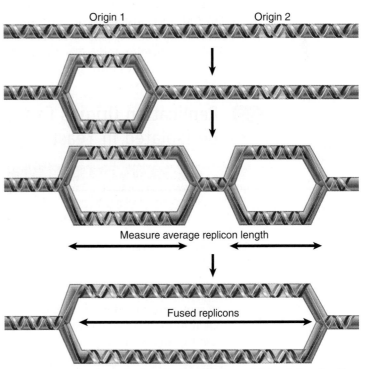

FIGURE 11.14 A eukaryotic chromosome contains multiple origins of replication that ultimately merge during replication.

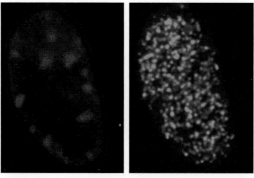

FIGURE 11.15 Replication forks are organized into foci in the nucleus. Cells were labeled with BrdU. The leftmost panel was stained with propidium iodide to identify bulk DNA. The right panel was stained using an antibody to BrdU to identify replicating DNA. Photos courtesy of Anthony D. Mills and Ron Laskey, Hutchinson/MRC Research Center, University of Cambridge.

The propensity of replicons located in the same vicinity to be active at the same time could be explained by "regional" controls, in which groups of replicons are initiated more or less coordinately, as opposed to a mechanism in which individual replicons are activated one by one in dispersed areas of the genome. Two structural features suggest the possibility of large-scale organization. Quite large regions of the chromosome can be characterized as "early replicating" or "late replicating," implying that there is little interspersion of replicons that fire at early or late times. Visualization of replicating forks by labeling with DNA precursors identifies 100 to 300 "foci" instead of uniform staining; each focus shown in FIGURE 11.15 probably contains >300 replication forks. The foci could represent fixed structures through which replicating DNA must move.

11.9 Replication Origins Can Be Isolated in Yeast

Key concepts

- Origins in *S. cerevisiae* are short A-T sequences that have an essential 11-bp sequence.
- The ORC is a complex of six proteins that binds to an ARS.
- Related ORC complexes are found in multicellular eukaryotes.

Any segment of DNA that has an origin should be able to replicate, so although plasmids are rare in eukaryotes, it may be possible to construct them by suitable manipulation *in vitro*. This has been accomplished in yeast, although not in multicellular eukaryotes.

S. cerevisiae mutants can be "transformed" to the wild-type phenotype by addition of DNA

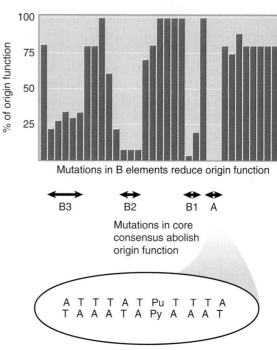

FIGURE 11.16 An *ARS* extends for ~50 bp and includes a consensus sequence (A) and additional elements (B1–B3).

that carries a wild-type copy of the gene. The discovery of yeast origins resulted from the observation that some yeast DNA fragments (when circularized) are able to transform defective cells very efficiently. These fragments can survive in the cell in the unintegrated (autonomous) state; that is, as self-replicating plasmids.

A high-frequency transforming fragment possesses a sequence that confers the ability to replicate efficiently in yeast. This segment is called an **ARS** (for autonomously replicating sequence). *ARS* elements are derived from origins of replication.

Where *ARS* elements have been systematically mapped over extended chromosomal regions, it seems that only some of them are actually used to initiate replication. The others are silent, or possibly used only occasionally. If it is true that some origins have varying probabilities of being used, it follows that there can be no fixed termini between replicons. In this case, a given region of a chromosome could be replicated from different origins in different cell cycles.

An *ARS* element consists of an A-T-rich region that contains discrete sites in which mutations affect origin function. Base composition rather than sequence may be important in the rest of the region. **FIGURE 11.16** shows a systematic mutational analysis along the length of an origin. Origin function is abolished completely by mutations in a 14-bp "core" region, called the **A domain**, which contains an 11-bp consensus sequence consisting of A-T base pairs. This consensus sequence (sometimes called the ACS, for ARS consensus sequence) is the only homology between known ARS elements.

Mutations in three adjacent elements, numbered B1 to B3, reduce origin function. An origin can function effectively with any two of the B elements, so long as a functional A element is present. (Imperfect copies of the core consensus, typically conforming at 9/11 positions, are found close to, or overlapping with, each B element, but they do not appear to be necessary for origin function.)

The **ORC** (origin recognition complex) is a complex of six proteins with a mass of ~400 kD. ORC binds to the A and B1 elements on the A-T-rich strand and is associated with ARS elements throughout the cell cycle. This means that initiation depends on changes in its condition rather than *de novo* association with an origin (see *Section 11.11, Licensing Factor Consists of MCM Proteins*). By counting the number of sites to which ORC binds, we can estimate that there are about 400 origins of replication in the yeast genome. This means that the average length of a replicon is ~35,000 bp. Counterparts to ORC are found in cells of multicellular eukaryotes.

ORC was first found in *S. cerevisiae* (where it is called scORC), but similar complexes have now been characterized in *Schizosaccharomyces pombe* (spORC), *Drosophila* (DmORC), and *Xenopus* (XlORC). All of the ORC complexes bind to DNA. Although none of the binding sites have been characterized in the same detail as in *S. cerevisiae*, in several cases they are at locations associated with the initiation of replication. It seems clear that ORC is an initiation complex whose binding identifies an origin of replication. Details of the interaction, however, are clear only in *S. cerevisiae*; it is possible that additional components are required to recognize the origin in the other cases.

ARS elements satisfy the classic definition of an origin as a *cis*-acting sequence that causes DNA replication to initiate. Are similar elements to be found in multicellular eukaryotes? The conservation of the ORC suggests that origins are likely to take the same sort of form in other eukaryotes, but in spite of this, there is little conservation of sequence among putative origins in different organisms.

Difficulties in finding consensus origin sequences suggest the possibility that origins may be more complex (or determined by features other than discrete *cis*-acting sequences).

There are suggestions that some animal cell replicons may have complex patterns of initiation: In some cases, many small replication bubbles are found in one region, posing the question of whether there are alternative or multiple starts to replication, and whether there is a small discrete origin.

Reconciliation between this phenomenon and the use of ORCs is suggested by the discovery that environmental effects can influence the use of origins. At one location where multiple bubbles are found, there is a primary origin that is used predominantly when the nucleotide supply is high. When the nucleotide supply is limiting, though, many secondary origins are also used, giving rise to a pattern of multiple bubbles. One possible molecular explanation is that ORCs dissociate from the primary origin and initiate elsewhere in the vicinity if the supply of nucleotides is insufficient for the initiation reaction to occur quickly. At all events, it now seems likely that we will be able in due course to characterize discrete sequences that function as origins of replication in multicellular eukaryotes.

11.10 Licensing Factor Controls Eukaryotic Rereplication

Key concepts

- Licensing factor is necessary for initiation of replication at each origin.
- Licensing factor is present in the nucleus prior to replication, but is removed, inactivated, or destroyed by replication.
- Initiation of another replication cycle becomes possible only after licensing factor reenters the nucleus after mitosis.

A eukaryotic genome is divided into multiple replicons, and the origin in each replicon is activated once and only once in a single division cycle. This could be achieved by the provision of some rate-limiting component that functions only once at an origin or by the presence of a repressor that prevents rereplication at origins that have been used. The critical questions about the nature of this regulatory system are how the system determines whether any particular origin has been replicated and what protein components are involved.

Insights into the nature of the protein components have been provided by using a system in which a substrate DNA undergoes only one cycle of replication. *Xenopus* eggs have all the components needed to replicate

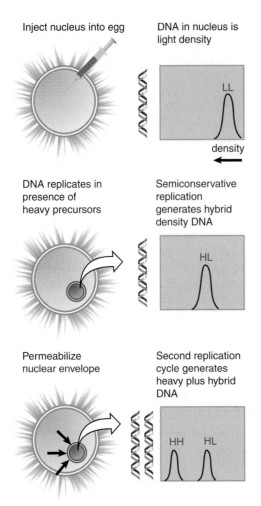

Inject nucleus into egg

DNA in nucleus is light density

LL

density

DNA replicates in presence of heavy precursors

Semiconservative replication generates hybrid density DNA

HL

Permeabilize nuclear envelope

Second replication cycle generates heavy plus hybrid DNA

HH   HL

FIGURE 11.17 A nucleus injected into a *Xenopus* egg can replicate only once unless the nuclear membrane is permeabilized to allow subsequent replication cycles.

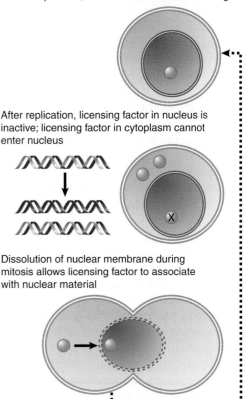

Prior to replication, nucleus contains active licensing factor

After replication, licensing factor in nucleus is inactive; licensing factor in cytoplasm cannot enter nucleus

X

Dissolution of nuclear membrane during mitosis allows licensing factor to associate with nuclear material

Cell division generates daughter nuclei competent to support replication

FIGURE 11.18 Licensing factor in the nucleus is inactivated after replication. A new supply of licensing factor can enter only when the nuclear membrane breaks down at mitosis.

DNA—in the first few hours after fertilization they undertake eleven division cycles without new gene expression—and they can replicate the DNA in a nucleus that is injected into the egg. **FIGURE 11.17** summarizes the features of this system.

When a sperm or interphase nucleus is injected into the egg, its DNA is replicated only once. (This can be followed by use of a density label, just like the original experiment that characterized semiconservative replication; see Figure 1.15.) If protein synthesis is blocked in the egg, the membrane around the injected material remains intact and the DNA cannot replicate again. In the presence of protein synthesis, however, the nuclear membrane breaks down just as it would for a normal cell division, and in this case subsequent replication cycles can occur. The same result can be achieved by using agents that permeabilize the nuclear membrane. This suggests that the nucleus contains a protein(s) needed for replication that is

used up in some way by a replication cycle, so even though more of the protein is present in the egg cytoplasm, it can only enter the nucleus if the nuclear membrane breaks down. The system can in principle be taken further by developing an *in vitro* extract that supports nuclear replication, thus allowing the components of the extract to be isolated and the relevant factors identified.

FIGURE 11.18 explains the control of reinitiation by proposing that this protein is a **licensing factor**. It is present in the nucleus prior to replication. One round of replication either inactivates or destroys the factor, and another round cannot occur until further factor is provided. Factor in the cytoplasm can gain access to the nuclear material only at the subsequent mitosis when the nuclear envelope breaks down. This regulatory system achieves two purposes. By removing a necessary component after replication, it prevents more than one cycle of replication from occurring. It also provides a feedback loop that makes the initiation of replication dependent on passing through the cell cycle.

11.11 Licensing Factor Consists of MCM Proteins

Key concepts

- The ORC is a protein complex that is associated with yeast origins throughout the cell cycle.
- Cdc6 protein is an unstable protein that is synthesized only in G1.
- Cdc6 binds to ORC and allows MCM proteins to bind.
- Cdt1 facilitates MCM loading on origins.
- When replication is initiated, Cdc6, Cdt1, and MCM proteins are displaced. The degradation of Cdc6 prevents reinitiation.
- Some MCM proteins are in the nucleus throughout the cell cycle, but others may enter only after mitosis.

The key event in controlling replication is the behavior of the ORC complex at the origin. Recall that ORC is a 400-kD complex that binds to the *S. cerevisiae ARS* sequence (see *Section 11.9, Replication Origins Can Be Isolated in Yeast*). The origin (*ARS*) consists of the A consensus sequence and three B elements (see Figure 11.16). The ORC complex of six proteins (all of which are coded by essential genes) binds to the A and adjacent B1 element. ATP is required for the binding, but is not hydrolyzed until some later stage. The transcription factor ABF1 binds to the B3 element; this assists initiation, but it is the events that occur at the A and B1 elements that actually cause initiation. Most origins are localized in regions between genes, which suggests that it may be important for the local chromatin structure to be in a nontranscribed condition.

The striking feature is that ORC remains bound at the origin through the entire cell cycle. However, changes occur in the pattern of protection of DNA as a result of binding of other proteins to the ORC-origin complex. **FIGURE 11.19** summarizes the cycle of events at the origin.

At the end of the cell cycle, ORC is bound to A–B1 elements of the origin and generates a pattern of protection *in vivo* that is similar to that found when it binds to free DNA *in vitro*. Basically, the region across A–B1 is protected against degradation by DNase, but there is a site that is hypersensitive to the enzyme in the center of B1.

There is a change during G1, seen most strikingly by the loss of the hypersensitive site. This results from the binding of Cdc6 protein to the ORC. In yeast, Cdc6 is a highly unstable protein, with a half-life of <5 minutes. It is syn-

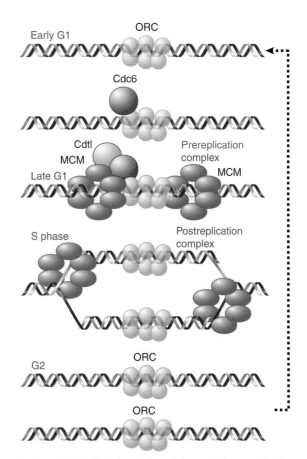

FIGURE 11.19 Proteins at the origin control susceptibility to initiation.

thesized during G1 and typically binds to the ORC between the exit from mitosis and late G1. Its rapid degradation means that no protein is available later in the cycle. In mammalian cells Cdc6 is controlled differently; it is phosphorylated during S phase, and as a result it is exported from the nucleus. This feature makes Cdc6 the key licensing factor. Cdc6 also provides the connection between ORC and a complex of proteins that is involved in initiation. Cdc6 has an ATPase activity that is required for it to support initiation.

The licensing factor and the system that controls its availability in yeast are identified by two different types of mutations:

- The licensing factor is identified by mutations in *MCM2,3,5*, which prevent initiation of replication.
- Mutations that have the opposite effect, and allow the accumulation of excess quantities of DNA, are found in genes that code for components of the ubiquitination system that is responsible for programmed degradation of specific

proteins. This suggests that licensing factor may be destroyed after the start of the replication cycle.

In yeast, free MCM2,3,5 enter the nucleus only during mitosis. Homologs are found in animal cells, where MCM3 is bound to chromosomal material before replication, but is released after replication. The animal cell MCM2,3,5 complex remains in the nucleus throughout the cell cycle, suggesting that it may be only one component of the licensing factor. Another component, able to enter only at mitosis, may be necessary for MCM2,3,5 to associate with chromosomal material.

The presence of Cdc6 at the yeast origin allows Cdt1 and MCM proteins to bind to the complex. Their presence is necessary for initiation. The origin therefore enters S phase in the condition of a **prereplication complex**, which contains ORC, Cdc6, Cdt1, and MCM proteins. When initiation occurs, Cdc6, Cdt1, and MCM are displaced, returning the origin to the state of the **postreplication complex**, which contains only ORC. Cdc6 is rapidly degraded during S phase, and as a result it is not available to support reloading of MCM proteins. Thus the origin cannot be used for a second cycle of initiation during S phase. In mammalian cells, Cdt1 is targeted for degradation by the action of a protein complex that is recruited to the origin of replication by PCNA, the eukaryotic counterpart of the bacterial β clamp.

If Cdc6 is made available to bind to the origin during G2 (by ectopic expression), MCM proteins do not bind until the following G1, which suggests that there is a secondary mechanism to ensure that they associate with origins only at the right time. This could be another part of licensing control. At least in *S. cerevisiae*, this control does not seem to be exercised at the level of nuclear entry, but this could be a difference between yeasts and animal cells. The MCM2-7 proteins form a six-member ring-shaped complex around DNA. Some of the ORC proteins have similarities to replication proteins that load DNA polymerase on to DNA. It is possible that ORC uses hydrolysis of ATP to load the MCM ring on to DNA. In *Xenopus* extracts, replication can be initiated if ORC is removed after it has loaded Cdc6 and MCM proteins. This shows that the major role of ORC is to identify the origin to the Cdc6 and MCM proteins that control initiation and licensing.

The MCM proteins are required for elongation as well as for initiation, and they continue to function at the replication fork. Biochemical studies have revealed that this complex of proteins possesses a helicase activity that unwinds DNA.

11.12 D Loops Maintain Mitochondrial Origins

Key concepts

- Mitochondria use different origin sequences to initiate replication of each DNA strand.
- Replication of the H strand is initiated in a D loop.
- Replication of the L strand is initiated when its origin is exposed by the movement of the first replication fork.

The origins of replicons in both prokaryotic and eukaryotic chromosomes are static structures: They comprise sequences of DNA that are recognized in duplex form and used to initiate replication at the appropriate time. Initiation requires separating the DNA strands and commencing bidirectional DNA synthesis. A different type of arrangement is found in mitochondria.

Replication starts at a specific origin in the circular duplex DNA. Initially, though, only one of the two parental strands (the H strand in mammalian mitochondrial DNA) is used as a template for synthesis of a new strand. Synthesis proceeds for only a short distance, displacing the original partner (L) strand, which remains single-stranded, as illustrated in FIGURE 11.20. The condition of this region gives rise to its name as the *displacement loop*, or **D loop**.

DNA polymerases cannot initiate synthesis, but require a priming 3′ end (see *Section 14.9, Priming Is Required to Start DNA Synthesis*). Replication at the H-strand origin is initiated when RNA polymerase transcribes a primer. The 3′ ends are generated in the primer by an endonuclease that cleaves the DNA–RNA hybrid at several discrete sites. The endonuclease is specific for the triple structure of DNA–RNA hybrid plus the displaced DNA single strand. The 3′ end is then extended into DNA by the DNA polymerase.

A single D loop is found as an opening of 500 to 600 bases in mammalian mitochondria. The short strand that maintains the D loop is unstable and turns over; it is frequently degraded and resynthesized to maintain the opening of the duplex at this site. Some mitochondrial DNAs possess several D loops, reflecting the use of multiple origins. The same mechanism is employed in chloroplast DNA, where (in complex plants) there are two D loops.

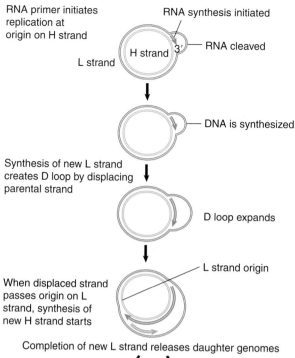

RNA primer initiates replication at origin on H strand

RNA synthesis initiated

RNA cleaved

H strand 3'

L strand

DNA is synthesized

Synthesis of new L strand creates D loop by displacing parental strand

D loop expands

L strand origin

When displaced strand passes origin on L strand, synthesis of new H strand starts

Completion of new L strand releases daughter genomes

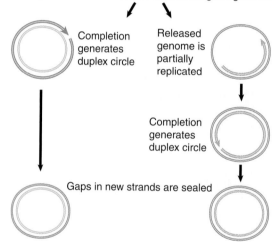

Completion generates duplex circle

Released genome is partially replicated

Completion generates duplex circle

Gaps in new strands are sealed

FIGURE 11.20 The D loop maintains an opening in mammalian mitochondrial DNA, which has separate origins for the replication of each strand.

To replicate mammalian mitochondrial DNA, the short strand in the D loop is extended. The displaced region of the original L strand becomes longer, expanding the D loop. This expansion continues until it reaches a point about two-thirds of the way around the circle. Replication of this region exposes an origin in the displaced L strand. Synthesis of an H strand initiates at this site, which is used by a special primase that synthesizes a short RNA. The RNA is then extended by DNA polymerase, proceeding around the displaced single-stranded L template in the opposite direction from L-strand synthesis.

As a result of the lag in its start, H-strand synthesis has proceeded only a third of the way around the circle when L-strand synthesis finishes. This releases one completed duplex circle and one gapped circle, the latter of which remains partially single-stranded until synthesis of the H strand is completed. Finally, the new strands are sealed to become covalently intact.

The existence of D loops exposes a general principle: *An origin can be a sequence of DNA that serves to initiate DNA synthesis using one strand as a template.* The opening of the duplex does not necessarily lead to the initiation of replication on the other strand. In the case of mitochondrial DNA replication, the origins for replicating the complementary strands lie at different locations. Origins that sponsor replication of only one strand are also found in the rolling circle mode of replication (see *Section 12.4, Rolling Circles Produce Multimers of a Replicon*).

11.13 Summary

Replicons in bacterial or eukaryotic chromosomes have a single unifying feature: Replication is initiated at an origin once, and only once, in each cell cycle. The origin is located within the replicon, and replication typically is bidirectional, with replication forks proceeding away from the origin in both directions. Replication is not usually terminated at specific sequences, but continues until DNA polymerase meets another DNA polymerase halfway around a circular replicon, or at the junction between two linear replicons.

An origin consists of a discrete sequence at which replication of DNA is initiated. Origins of replication tend to be rich in A-T base pairs. A bacterial chromosome contains a single origin, which is responsible for initiating replication once every cell cycle. The *oriC* in *E. coli* is a sequence of 245 bp. Any DNA molecule with this sequence can replicate in *E. coli*. Replication of the circular bacterial chromosome produces a θ structure, in which the replicated DNA starts out as a small replicating eye. Replication proceeds until the eye occupies the whole chromosome. The bacterial origin contains sequences that are methylated on both strands of DNA. Replication produces hemimethylated DNA, which cannot function as an origin. There is a delay before the hemimethylated origins are remethylated to convert them to a functional state, and this is responsible for preventing improper reinitiation.

A eukaryotic chromosome is divided into many individual replicons. Replication occurs during a discrete part of the cell cycle called S phase. Not all replicons are active simultaneously, though, so the process may take several hours. Eukaryotic replication is at least an order of magnitude slower than bacterial replication. Origins sponsor bidirectional replication and are probably used in a fixed order during S phase. Each replicon is activated only once in each cycle. Origins of replication were isolated as *ARS* sequences in yeast by virtue of their ability to support replication of any sequence attached to them. The core of an *ARS* is an 11-bp A-T rich sequence that is bound by the ORC protein complex, which remains bound throughout the cell cycle. Utilization of the origin is controlled by the MCM licensing factors that associate with the ORC.

After cell division, nuclei of eukaryotic cells have a licensing factor that is needed to initiate replication. In yeast, its destruction after initiation of replication prevents further replication cycles from occurring. Licensing factor cannot be imported into the nucleus from the cytoplasm, and can be replaced only when the nuclear membrane breaks down during mitosis.

The origin in yeast is recognized by the ORC proteins, which in yeast remain bound throughout the cell cycle. The protein Cdc6 is available only at S phase. In yeast it is synthesized during S phase and rapidly degraded. In animal cells it is synthesized continuously, but is exported from the nucleus during S phase. The presence of Cdc6 allows the MCM proteins to bind to the origin. The MCM proteins are required for initiation. The action of Cdc6 and the MCM proteins provides the licensing function.

References

11.1   Introduction

Research

Jacob, F., Brenner, S., and Cuzin, F. (1963). On the regulation of DNA replication in bacteria. Cold Spring Harbor Symp. *Quant. Biol.* 28, 329–348.

11.2   Replicons Can Be Linear or Circular

Review

Brewer, B. J. (1988). When polymerases collide: replication and transcriptional organization of the *E. coli* chromosome. *Cell* 53, 679–686.

Research

Cairns, J. (1963). The bacterial chromosome and its manner of replication as seen by autoradiography. *J. Mol. Biol.* 6, 208–213.

Iismaa, T. P. and Wake, R. G. (1987). The normal replication terminus of the *B. subtilis* chromosome, *terC*, is dispensable for vegetative growth and sporulation. *J. Mol. Biol.* 195, 299–310.

Liu, B., Wong, M. L., and Alberts, B. (1994). A transcribing RNA polymerase molecule survives DNA replication without aborting its growing RNA chain. *Proc. Natl. Acad. Sci. USA* 91, 10660–10664.

Steck, T. R. and Drlica, K. (1984). Bacterial chromosome segregation: evidence for DNA gyrase involvement in decatenation. *Cell* 36, 1081–1088.

Zyskind, J. W. and Smith, D. W. (1980). Nucleotide sequence of the *S. typhimurium* origin of DNA replication. *Proc. Natl. Acad. Sci. USA* 77, 2460–2464.

11.3   Origins Can Be Mapped by Autoradiography and Electrophoresis

Research

Huberman, J. and Riggs, A. D. (1968). On the mechanism of DNA replication in mammalian chromosomes. *J. Mol. Biol.* 32, 327–341.

11.4   The Bacterial Genome Is (Usually) a Single Circular Replicon

Research

Bastia, D., Zzaman, S., Krings, G., Saxena, M., Peng, X., and Greenberg, M. M. (2008). Replication termination mechanism as revealed by Tus-mediated polar arrest of a sliding helicase. *Proc. Natl. Acad. Sci. USA* 105, 12831–12836.

11.7   Archaeal Chromosomes Can Contain Multiple Replicons

Review

Barry, E. R. and Bell, S. D. (2006) DNA replication in the Archaea. *Micro. Mol. Biol. Rev.* 70, 876–887.

Research

Cunningham Dueber, E. L. Corn, J. E., Bell S. D., and Berger J. M. (2007). Replication origin recognition and deformation by a heterodimeric archaeal Orc1 complex. *Science* 317, 1210–1213.

11.8   Each Eukaryotic Chromosome Contains Many Replicons

Review

Fangman, W. L. and Brewer, B. J. (1991). Activation of replication origins within yeast chromosomes. *Annu. Rev. Cell Biol.* 7, 375–402.

Research

Blumenthal, A. B., Kriegstein, H. J., and Hogness, D. S. (1974). The units of DNA replication in *D. melanogaster* chromosomes. Cold Spring Harbor Symp. *Quant. Biol.* 38, 205–223.

11.9 Replication Origins Can Be Isolated in Yeast

Reviews

Bell, S. P. and Dutta, A. (2002). DNA replication in eukaryotic cells. *Annu. Rev. Biochem.* 71, 333–374.

DePamphlis, M. L. (1993). Eukaryotic DNA replication: anatomy of an origin. *Annu. Rev. Biochem.* 62, 29–63.

Gilbert, D. M. (2001). Making sense of eukaryotic DNA replication origins. *Science* 294, 96–100.

Kelly, T. J. and Brown, G. W. (2000). Regulation of chromosome replication. *Annu. Rev. Biochem.* 69, 829–880.

Research

Anglana, M., Apiou, F., Bensimon, A., and Debatisse, M. (2003). Dynamics of DNA replication in mammalian somatic cells: nucleotide pool modulates origin choice and interorigin spacing. *Cell* 114, 385–394.

Chesnokov, I., Remus, D., and Botchan, M. (2001). Functional analysis of mutant and wild-type *Drosophila* origin recognition complex. *Proc. Natl. Acad. Sci. USA* 98, 11997–12002.

Marahrens, Y. and Stillman, B. (1992). A yeast chromosomal origin of DNA replication defined by multiple functional elements. *Science* 255, 817–823.

Wyrick, J. J., Aparicio, J. G., Chen, T., Barnett, J. D., Jennings, E. G., Young, R. A., Bell, S. P., and Aparicio, O. M. (2001). Genome-wide distribution of ORC and MCM proteins in *S. cerevisiae*: high-resolution mapping of replication origins. *Science* 294, 2357–2360.

11.12 D Loops Maintain Mitochondrial Origins

Reviews

Clayton, D. (1982). Replication of animal mitochondrial DNA. *Cell* 28, 693–705.

Falkenberg, M., Larsson, N.-G., and Gustafsson, C. M. (2007) DNA replication and transcription in mammalian mitochondria. *Annu. Rev. Biochem.* 76, 679–700.

Shadel, G. S. and Clayton, D. A. (1997). Mitochondrial DNA maintenance in vertebrates. *Annu. Rev. Biochem.* 66, 409–435.

© Deco Images II/Almay Images

Extrachromosomal Replicons

Edited by Søren Johannes Sørensen and Lars Hestbjerg Hansen

CHAPTER OUTLINE

12.1 Introduction

A bacterium may be a host for independently replicating genetic units in addition to its chromosome. These extrachromosomal genomes fall into two general types: plasmids and bacteriophages (phages). Some plasmids, and all phages, have the ability to transfer from a donor bacterium to a recipient by an infective process. An important distinction between them is that plasmids exist only as free DNA genomes, whereas bacteriophages are viruses that package a nucleic acid genome into a protein coat and are released from the bacterium at the end of an infective cycle.

Plasmids are self-replicating circular molecules of DNA that are maintained in the cell in a stable and characteristic number of copies; that is, the average number remains constant from generation to generation. Low-copy number plasmids are maintained at a constant quantity relative to the bacterial host chromosome, often between one to ten per bacterium, depending on the plasmid. As with the host chromosome, they rely on a specific apparatus to be segregated equally at each bacterial division. Multicopy plasmids exist in many copies per unit bacterium and may be segregated to daughter bacteria stochastically (meaning that there are enough copies to ensure that each daughter cell always gains some by a random distribution).

Plasmids and phages are defined by their ability to reside in a bacterium as independent genetic units. Certain plasmids, and some phages, can also exist as sequences within the bacterial genome, though. In this case, the same sequence that constitutes the independent plasmid or phage genome is found within the chromosome, and is inherited like any other bacterial gene. Phages that are found as part of the bacterial chromosome are said to show **lysogeny**; plasmids that also have the ability to integrate into the chromosome are called **episomes**. Related processes are used by phages and episomes to insert into and excise from the bacterial chromosome.

A parallel between lysogenic phages and plasmids and episomes is that they maintain a selfish possession of their bacterium and often make it impossible for another element of the same type to become established. This effect is called **immunity**, although the molecular basis for plasmid immunity is different from lysogenic immunity, and is a consequence of the replication control system.

Several types of genetic units can be propagated in bacteria as independent genomes. Lytic phages may have genomes of any type of nucleic acid; they transfer between cells by release of infective particles. Lysogenic phages have double-stranded DNA genomes, as do plasmids and episomes. Some plasmids transfer between cells by a conjugative process (with direct contact between donor and recipient cells). A feature of the transfer process in both cases is that on occasion some bacterial host genes are transferred with the phage or plasmid DNA, so these events play a role in allowing exchange of genetic information between bacteria. When plasmids have the ability either to exist as extrachromosomal elements or to integrate into a bacterial genome, they are called episomes. All episomes are plasmids, but not all plasmids are episomes.

The key feature in determining the behavior of each type of unit is how its origin is used. An origin in a bacterial or eukaryotic chromosome is used to initiate a single replication event that extends across the replicon. Replicons, however, can also be used to sponsor other forms of replication. The most common alternative is used by the small, independently replicating units of viruses. The objective of a viral replication cycle is to produce many copies of the viral genome before the host cell is lysed to release them. Some viruses replicate in the same way as a host genome, with an initiation event leading to production of duplicate copies, each of which then replicates again, and so on. Others use a mode of replication in which many copies are produced as a tandem array following a single initiation event. A similar type of event is triggered by episomes when an integrated plasmid DNA ceases to be inert and initiates a replication cycle.

Many prokaryotic replicons are circular, and this indeed is a necessary feature for replication modes that produce multiple tandem copies. Some extrachromosomal replicons are linear, though, and in such cases we have to account for the ability to replicate the end of the replicon. (Of course, eukaryotic chromosomes are linear, so the same problem applies to the replicons at each end. These replicons, however, have a special system for resolving the problem.)

12.2 The Ends of Linear DNA Are a Problem for Replication

Key concept

- Special arrangements must be made to replicate the DNA strand with a 5′ end.

None of the replicons that we have considered so far have a linear end: either they are circular (as in the *E. coli* or mitochondrial genomes), or they are part of longer segregation units (as in eukaryotic chromosomes). Linear replicons do occur, though—in some cases as single extrachromosomal units, and of course at the ends of eukaryotic chromosomes.

The ability of all known nucleic acid polymerases, DNA or RNA, to proceed only in the 5′–3′ direction poses a problem for synthesizing DNA at the end of a linear replicon. Consider the two parental strands depicted in FIGURE 12.1. The lower strand presents no problem: It can act as template to synthesize a daughter strand that runs right up to the end, where presumably the polymerase falls off. To synthesize a complement at the end of the upper strand, however, synthesis must start right at the very last base, or else this strand would become shorter in successive cycles of replication.

We do not know whether initiation right at the end of a linear DNA is feasible. We usually think of a polymerase as binding at a site *surrounding* the position at which a base is to be incorporated. Thus a special mechanism must be employed for replication at the ends of linear replicons. Several types of solution may be

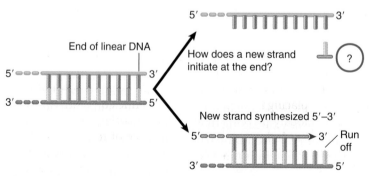

FIGURE 12.1 Replication could run off the 3′ end of a newly synthesized linear strand, but could it initiate at a 5′ end?

imagined to accommodate the need to copy a terminus:

- The problem may be circumvented by converting a linear replicon into a circular or multimeric molecule. Phages such as T4 or lambda use such mechanisms (see *Section 12.4, Rolling Circles Produce Multimers of a Replicon*).
- The DNA may form an unusual structure—for example, by creating a hairpin at the terminus, so that there is no free end. Formation of a crosslink is involved in replication of the linear mitochondrial DNA of *Paramecium*.
- Instead of being precisely determined, the end may be variable. Eukaryotic chromosomes may adopt this solution, in which the number of copies of a short repeating unit at the end of the DNA changes (see *Section 9.18, Telomeres Are Synthesized by a Ribonucleoprotein Enzyme*). A mechanism to add or remove units makes it unnecessary to replicate right up to the very end.
- A protein may intervene to make initiation possible at the actual terminus. Several linear viral nucleic acids have proteins that are *covalently linked to the 5′ terminal base*. The best characterized examples are adenovirus DNA, phage φ29 DNA, and poliovirus RNA.

12.3 Terminal Proteins Enable Initiation at the Ends of Viral DNAs

Key concept

- A terminal protein binds to the 5′ end of DNA and provides a cytidine nucleotide with a 3′–OH end that primes replication.

An example of initiation at a linear end is provided by adenovirus and φ29 DNAs, which actually replicate from both ends using the mechanism of **strand displacement** illustrated in FIGURE 12.2. The same events can occur independently at either end. Synthesis of a new strand starts at one end, displacing the homologous strand that was previously paired in the duplex. When the replication fork reaches the other end of the molecule, the displaced strand is released as a free single strand. It is then replicated independently; this requires the formation of a duplex origin by base pairing between

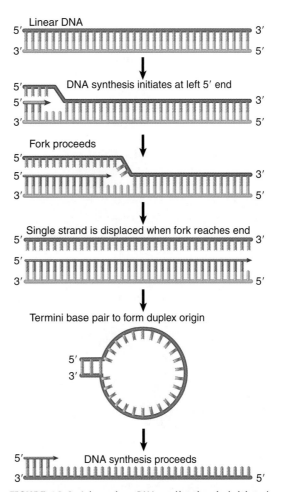

FIGURE 12.2 Adenovirus DNA replication is initiated separately at the two ends of the molecule and proceeds by strand displacement.

some short complementary sequences at the ends of the molecule.

In several viruses that use such mechanisms, a protein is found covalently attached to each 5′ end. In the case of adenovirus, a **terminal protein** is linked to the mature viral DNA via a phosphodiester bond to serine, as indicated in FIGURE 12.3.

How does the attachment of the protein overcome the initiation problem? The terminal protein has a dual role: It carries a cytidine nucleotide that provides the primer, and it is associated with DNA polymerase. In fact, linkage of terminal protein to a nucleotide is undertaken by DNA polymerase in the presence of adenovirus DNA. This suggests the model illustrated in FIGURE 12.4. The complex of polymerase and terminal protein, bearing the

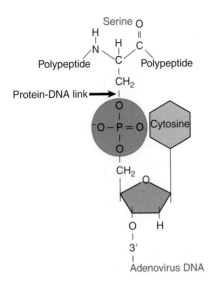

FIGURE 12.3 The 5' terminal phosphate at each end of adenovirus DNA is covalently linked to serine in the 55 kD Ad-binding protein.

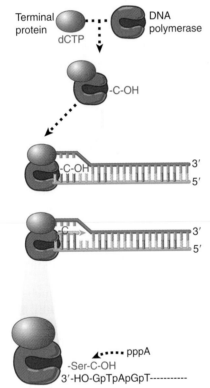

FIGURE 12.4 Adenovirus terminal protein binds to the 5' end of DNA and provides a C-OH end to prime synthesis of a new DNA strand.

priming C nucleotide, binds to the end of the adenovirus DNA. The free 3'–OH end of the C nucleotide is used to prime the elongation reaction by the DNA polymerase. This generates a new strand whose 5' end is covalently linked to the initiating C nucleotide. (The reaction actually involves displacement of protein from

DNA rather than binding *de novo*. The 5' end of adenovirus DNA is bound to the terminal protein that was used in the previous replication cycle. The old terminal protein is displaced by the new terminal protein for each new replication cycle.)

Terminal protein binds to the region located between 9 and 18 bp from the end of the DNA. The adjacent region, between positions 17 and 48, is essential for the binding of a host protein, nuclear factor I, which is also required for the initiation reaction. The initiation complex may therefore form between positions 9 and 48, a fixed distance from the actual end of the DNA.

12.4 Rolling Circles Produce Multimers of a Replicon

Key concept

- A rolling circle generates single-stranded multimers of the original sequence.

The structures generated by replication depend on the relationship between the template and the replication fork. The critical features are whether the template is circular or linear, and whether the replication fork is engaged in synthesizing both strands of DNA or only one.

Replication of only one strand is used to generate copies of some circular molecules. A nick opens one strand, and then the free 3'–OH end generated by the nick is extended by the DNA polymerase. The newly synthesized strand displaces the original parental strand. The ensuing events are depicted in FIGURE 12.5.

This type of structure is called a **rolling circle**, because the growing point can be envisaged as rolling around the circular template strand. It could in principle continue to do so indefinitely. As it moves, the replication fork extends the outer strand and displaces the previous partner. An example is shown in the electron micrograph of FIGURE 12.6.

The newly synthesized material is covalently linked to the original material, and as a result the displaced strand has the original unit genome at its 5' end. The original unit is followed by any number of unit genomes, synthesized by continuing revolutions of the template. Each revolution displaces the material synthesized in the previous cycle.

The rolling circle is put to several uses *in vivo*. Some pathways that are used to replicate DNA are depicted in FIGURE 12.7.

Cleavage of a unit length tail generates a copy of the original circular replicon in lin-

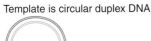

Template is circular duplex DNA

Initiation occurs on one strand

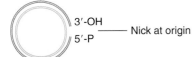

3'-OH

5'-P — Nick at origin

Elongation of growing strand displaces
old strand

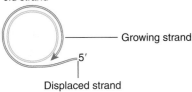

Growing strand

5'

Displaced strand

After one revolution displaced strand
reaches unit length

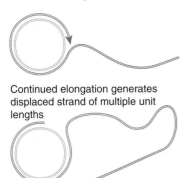

Continued elongation generates
displaced strand of multiple unit
lengths

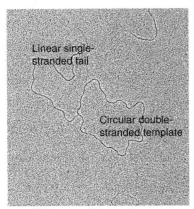

FIGURE 12.5 The rolling circle generates a multimeric single-stranded tail.

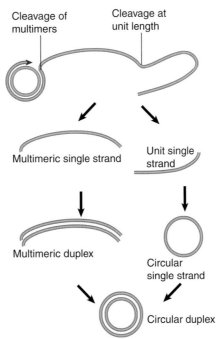

Cleavage of multimers

Cleavage at unit length

Multimeric single strand

Unit single strand

Multimeric duplex

Circular single strand

Circular duplex

FIGURE 12.7 The fate of the displaced tail determines the types of products generated by rolling circles. Cleavage at unit length generates monomers, which can be converted to duplex and circular forms. Cleavage of multimers generates a series of tandemly repeated copies of the original unit. Note that the conversion to double-stranded form could occur earlier, before the tail is cleaved from the rolling circle.

Linear single-stranded tail

Circular double-stranded template

FIGURE 12.6 A rolling circle appears as a circular molecule with a linear tail by electron microscopy. Photo courtesy of Ross B. Inman, Institute of Molecular Virology, Bock Laboratory and Department of Biochemistry, University of Wisconsin, Madison, Wisconsin, USA.

ear form. The linear form may be maintained as a single strand, or may be converted into a duplex by synthesis of the complementary strand (which is identical in sequence to the template strand of the original rolling circle).

The rolling circle provides a means for amplifying the original (unit) replicon. This mechanism is used to generate amplified ribosomal DNA (rDNA) in the *Xenopus* oocyte. The genes for ribosomal RNA (rRNA) are organized as a large number of contiguous repeats in the genome. A single repeating unit from the genome is converted into a rolling circle. The displaced tail, which contains many units, is converted into duplex DNA; later it is cleaved from the circle so that the two ends can be joined together to generate a large circle of amplified rDNA. The amplified material therefore consists of a large number of identical repeating units.

12.5 Rolling Circles Are Used to Replicate Phage Genomes

Key concept

- The φX A protein is a *cis*-acting relaxase that generates single-stranded circles from the tail produced by rolling circle replication.

Replication by rolling circles is common among bacteriophages. Unit genomes can be cleaved from the displaced tail, generating monomers that can be packaged into phage particles or used for further replication cycles. A more detailed view of a phage replication cycle that is centered on the rolling circle is given in **FIGURE 12.8**.

Phage φX174 consists of a single-stranded circular DNA known as the plus (+) strand. A

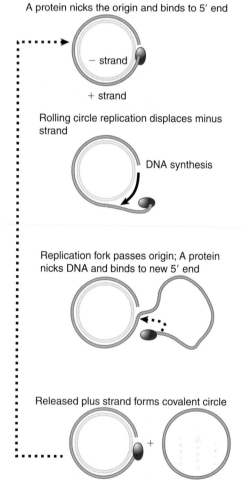

A protein nicks the origin and binds to 5′ end

− strand

+ strand

Rolling circle replication displaces minus strand

DNA synthesis

Replication fork passes origin; A protein nicks DNA and binds to new 5′ end

Released plus strand forms covalent circle

FIGURE 12.8 φX174 RF DNA is a template for synthesizing single-stranded viral circles. The A protein remains attached to the same genome through indefinite revolutions, each time nicking the origin on the viral (+) strand and transferring to the new 5′ end. At the same time, the released viral strand is circularized.

complementary strand, called the minus (−) strand, is synthesized. This action generates the duplex circle shown at the top of the figure, which is then replicated by a rolling circle mechanism.

The duplex circle is converted to a covalently closed form, which becomes supercoiled. A protein coded by the phage genome, the A protein, nicks the (+) strand of the duplex DNA at a specific site that defines the origin for replication. After nicking the origin, the A protein remains connected to the 5′ end that it generates, while the 3′ end is extended by DNA polymerase.

The structure of the DNA plays an important role in this reaction, for the DNA can be nicked *only when it is negatively supercoiled* (i.e., wound about its axis in space in the opposite sense from the handedness of the double helix; see *Section 1.5, Supercoiling Affects the Structure of DNA*). The A protein is able to bind to a single-stranded decamer fragment of DNA that surrounds the site of the nick. This suggests that the supercoiling is needed to assist the formation of a single-stranded region that provides the A protein with its binding site. (An enzymatic activity in which a protein cleaves duplex DNA and binds to a released 5′ end is sometimes called a **relaxase**.) The nick generates a 3′–OH end and a 5′–phosphate end (covalently attached to the A protein), both of which have roles to play in φX174 replication.

Using the rolling circle, the 3′–OH end of the nick is extended into a new chain. The chain is elongated around the circular (−) strand template until it reaches the starting point and displaces the origin. Now the A protein functions again. It remains connected with the rolling circle as well as to the 5′ end of the displaced tail, and is therefore in the vicinity as the growing point returns past the origin. Thus the same A protein is available again to recognize the origin and nick it, now attaching to the end generated by the new nick. The cycle can be repeated indefinitely.

Following this nicking event, the displaced single (+) strand is freed as a circle. The A protein is involved in the circularization. In fact, the joining of the 3′ and 5′ ends of the (+) strand product is accomplished by the A protein as part of the reaction by which it is released at the end of one cycle of replication, and starts another cycle.

The A protein has an unusual property that may be connected with these activities. It is *cis*-acting *in vivo*. (This behavior is not reproduced *in vitro*, as can be seen from its activity on any DNA template in a cell-free system.) *The implication is that* in vivo *the A protein synthesized by a particular genome can attach only to the DNA of that genome*. We do not know how this is accomplished. Its activity *in vitro*, however, shows how it remains associated with the same parental (–) strand template. The A protein has two active sites; this may allow it to cleave the "new" origin while still retaining the "old" origin; it then ligates the displaced strand into a circle.

The displaced (+) strand may follow either of two fates after circularization. During the replication phase of viral infection, it may be used as a template to synthesize the complementary (–) strand. The duplex circle may then be used as a rolling circle to generate more progeny. During phage morphogenesis, the displaced (+) strand is packaged into the phage virion.

12.6 The F Plasmid Is Transferred by Conjugation between Bacteria

Key concepts

- The free F plasmid is a replicon that is maintained at the level of one plasmid per bacterial chromosome.
- An F plasmid can integrate into the bacterial chromosome, in which case its own replication system is suppressed.
- The F plasmid codes for a DNA translocation complex and specific pili that form on the surface of the bacterium.
- An F-pilus enables an F-positive bacterium to contact an F-negative bacterium and to initiate conjugation.

Another example of a connection between replication and the propagation of a genetic unit is provided by bacterial **conjugation**, in which a plasmid genome or host chromosome is transferred from one bacterium to another.

Conjugation is mediated by the **F plasmid**, which is the classic example of an episome—an element that may exist as a free circular plasmid, or that may become integrated into the bacterial chromosome as a linear sequence (like a lysogenic bacteriophage). The F plasmid is a large circular DNA ~100 kb in length.

The F plasmid can integrate at several sites in the *E. coli* chromosome, often by a recombination event involving certain sequences (called IS sequences; see *Section 17.4, Transposons*

Cause Rearrangement of DNA) that are present on both the host chromosome and F plasmid. In its free (plasmid) form, the F plasmid utilizes its own replication origin (*oriV*) and control system, and is maintained at a level of one copy per bacterial chromosome. When it is integrated into the bacterial chromosome, this system is suppressed, and F DNA is replicated as a part of the chromosome.

The presence of the F plasmid, whether free or integrated, has important consequences for the host bacterium. Bacteria that are F-positive are able to conjugate (or mate) with bacteria that are F-negative. Conjugation involves direct, physical contact between donor (F-positive) and recipient (F-negative) bacteria; contact is followed by transfer of the F plasmid. If the F plasmid exists as a free plasmid in the donor bacterium, it is transferred as a plasmid, and the infective process converts the F-negative recipient into an F-positive state. If the F plasmid is present in an integrated form in the donor, the transfer process may also cause some or all of the bacterial chromosome to be transferred. Many plasmids have conjugation systems that operate in a generally similar manner, but the F plasmid was the first to be discovered and remains the paradigm for this type of genetic transfer.

A large (~33 kb) region of the F plasmid called the **transfer region** is required for conjugation. It contains ~40 genes that are required for the transmission of DNA; their organization is summarized in **FIGURE 12.9**. The genes are arranged in loci named *tra* and *trb*. Most of them are expressed coordinately as part of a single 32-kb transcription unit (the *traY-I* unit). *traM* and *traJ* are expressed separately. *traJ* is a regulator that turns on both *traM* and *traY-I*. On the opposite strand, *finP* is a regulator that codes for a small antisense RNA that turns off *traJ*. Its activity requires expression of another gene, *finO*. Only four of the *tra* and *trb* genes, *traD*,

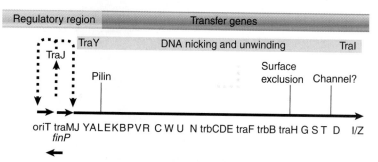

FIGURE 12.9 The *tra* region of the F plasmid contains the genes needed for bacterial conjugation.

traI, *traM*, and *traY*, in the major transcription unit are concerned directly with the transfer of DNA; most of these genes encode proteins that form a large membrane-spanning protein complex called a type 4 secretion system (T4SS). These systems are common in bacteria, where they have been shown to be involved in the transport of various proteins and DNA across the bacterial cell envelope and are responsible for maintaining contacts between mating bacteria.

F-positive bacteria possess surface appendages called **pili** (singular **pilus**) that are coded by the F plasmid. The gene *traA* codes for the single subunit protein, **pilin**, that is polymerized into the pilus extending from the inner to the outer membrane at the T4SS. At least 12 *tra* genes are required for the modification and assembly of pilin into the pilus and the stabilization of the T4SS. The F-pili are hairlike structures, 2 to 3 μm long, that protrude from the bacterial surface. A typical F-positive cell has two to three pili. The pilin subunits are polymerized into a hollow cylinder, ~8 nm in diameter, with a 2-nm axial hole.

Mating is initiated when the tip of the F-pilus contacts the surface of the recipient cell. **FIGURE 12.10** shows an example of *E. coli* cells beginning to mate. A donor cell does not contact other cells carrying the F plasmid, because the genes *traS* and *traT* code for "surface exclusion" proteins that make the cell a poor recipient in such contacts. This effectively restricts donor cells to mating with F-negative cells. (The presence of F-pili has secondary consequences; they provide the sites to which RNA phages and some single-stranded DNA phages attach, so F-positive bacteria are susceptible to infection by these phages, whereas F-negative bacteria are resistant.)

The initial contact between donor and recipient cells is easily broken, but other *tra* genes act to stabilize the association; this brings the mating cells closer together. The F-pili are essential for initiating pairing, but retract or disassemble as part of the process by which the mating cells are brought into close contact. It is proposed that the T4SS provides the channel through which DNA is transferred. TraD is a so-called coupling protein encoded by F plasmids that is necessary for recruitment of plasmid DNA to the T4SS, and it may associate with the T4SS to be involved in the actual plasmid transfer.

12.7 Conjugation Transfers Single-Stranded DNA

Key concepts

- Transfer of an F plasmid is initiated when rolling circle replication begins at *oriT*.
- The formation of a relaxosome initiates transfer into the recipient bacterium.
- The transferred DNA is converted into double-stranded form in the recipient bacterium.
- When an F plasmid is free, conjugation "infects" the recipient bacterium with a copy of the F plasmid.
- When an F plasmid is integrated, conjugation causes transfer of the bacterial chromosome until the process is interrupted by (random) breakage of the contact between donor and recipient bacteria.

Transfer of the F plasmid is initiated at a site called *oriT*, the origin of transfer, which is located at one end of the transfer region. The transfer process may be initiated when TraM recognizes that a mating pair has formed. TraY then binds near *oriT* and causes TraI to bind to form the **relaxosome** in conjunction with host-encoded DNA binding proteins called Integration Host Factor (IHF). TraI is a relaxase, like φX174 A protein. TraI nicks *oriT* at a unique site (called *nic*), and then forms a covalent link to the 5' end that has been generated. TraI also catalyzes the unwinding of ~200 bp of DNA and remains attached to the DNA 5' end throughout the conjugation process (this is a helicase activity). The TraI-bound DNA is then transferred to the T4SS by the coupling protein TraD, where it is exported to the recipient cell. **FIGURE 12.11** shows that the relaxase-bound 5' end leads the way into the recipient bacterium. The transferred single strand is circularized and a

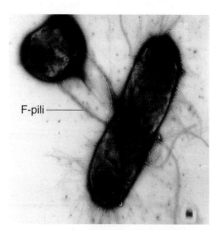

FIGURE 12.10 Mating bacteria are initially connected when donor F-pili contact the recipient bacterium. Photo courtesy of Emeritus Professor Ron Skurray, School of Biological Sciences, University of Sydney.

F-pili

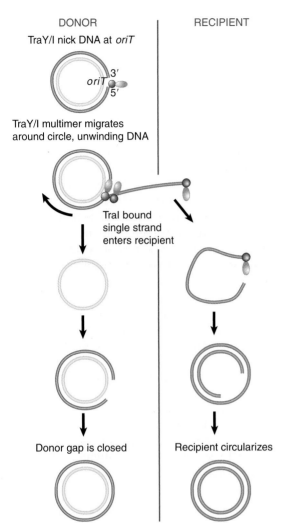

DONOR | **RECIPIENT**

TraY/I nick DNA at *oriT*

oriT 3'
5'

TraY/I multimer migrates
around circle, unwinding DNA

TraI bound
single strand
enters recipient

Donor gap is closed | Recipient circularizes

FIGURE 12.11 Transfer of DNA occurs when the F plasmid is nicked at *oriT* and a single strand is led by the 5' end bound to TraI into the recipient. Only one unit length is transferred. Complementary strands are synthesized to the single strand remaining in the donor and to the strand transferred into the recipient.

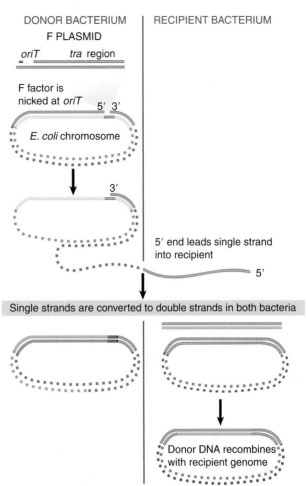

DONOR BACTERIUM | **RECIPIENT BACTERIUM**

F PLASMID

oriT   *tra* region

F factor is
nicked at *oriT*
5' 3'

E. coli chromosome

3'

5' end leads single strand
into recipient

5'

Single strands are converted to double strands in both bacteria

Donor DNA recombines
with recipient genome

FIGURE 12.12 Transfer of chromosomal DNA occurs when an integrated F plasmid is nicked at *oriT*. Transfer of DNA starts with a short sequence of F DNA and continues until prevented by loss of contact between the bacteria.

complement strand is synthesized in the recipient bacterium, which as a result is converted to the F-positive state.

A complementary strand must be synthesized in the donor bacterium to replace the strand that has been transferred. If this happens concomitantly with the transfer process, the state of the F plasmid will resemble the rolling circle of Figure 12.5. DNA synthesis could occur instantly, using the freed 3' end as a starting point. Conjugating DNA usually appears like a rolling circle, but replication as such is not necessary to provide the driving energy, and single-strand transfer is independent of DNA synthesis. Only a single unit length of the F plasmid is transferred to the recipient bacterium. This implies that some feature (perhaps TraI) ter-

minates the process after one revolution, after which the covalent integrity of the F plasmid is restored. TraI may also be involved in recircularization of the transferred DNA to which a complementary strand is then synthesized.

When an integrated F plasmid initiates conjugation, the orientation of transfer is directed away from the transfer region and into the bacterial chromosome. **FIGURE 12.12** shows that, following a short leading sequence of F DNA, bacterial DNA is transferred. The process continues until it is interrupted by the breaking of contacts between the mating bacteria. It takes ~100 minutes to transfer the entire bacterial chromosome, and under standard conditions contact is often broken before the completion of transfer.

Donor DNA that enters a recipient bacterium is converted to double-stranded form and may recombine with the recipient chromosome. (Note that two recombination events are required to insert the donor DNA.) Thus conjugation affords a means to exchange genetic material between bacteria, a contrast to their usual asexual growth (hence the original name Fertility factor or F factor). A strain of *E. coli* with an integrated F plasmid supports such recombination at relatively high frequencies (compared to strains that lack integrated F plasmids); such strains are described as **Hfr** (for high frequency recombination). Each position of integration for the F plasmid gives rise to a different Hfr strain, with a characteristic pattern of transferring bacterial markers to a recipient chromosome.

Contact between conjugating bacteria is usually broken before transfer of DNA is complete. As a result, the probability that a region of the bacterial chromosome will be transferred depends upon its distance from *oriT*. Bacterial genes located close to the site of F integration (in the direction of transfer) enter recipient bacteria first, and are therefore found at greater frequencies than those that are located farther away and enter later. This gives rise to a gradient of transfer frequencies around the chromosome, declining from the position of F integration. Marker positions on the donor chromosome can be assayed in terms of the time at which transfer occurs; this gave rise to the standard description of the *E. coli* chromosome as a map divided into 100 minutes. The map refers to transfer times from a particular Hfr strain; the starting point for the gradient of transfer is different for each Hfr strain because it is determined by the site where the F plasmid has integrated into the bacterial genome.

12.8 The Bacterial Ti Plasmid Causes Crown Gall Disease in Plants

Key concepts

- Infection with the bacterium *A. tumefaciens* can transform plant cells into tumors.
- The infectious agent is a plasmid carried by the bacterium.
- The plasmid also carries genes for synthesizing and metabolizing opines (arginine derivatives) that are used by the bacterium.

Most events in which DNA is rearranged or amplified occur within a genome, but the interaction between bacteria and certain plants involves the transfer of DNA from the bacterial genome to the plant genome. **Crown gall dis-**

ease, shown in **FIGURE 12.13**, can be induced in most dicotyledonous plants by the soil bacterium *Agrobacterium tumefaciens*. The bacterium is a parasite that effects a genetic change in the eukaryotic host cell, with consequences for both parasite and host: It improves conditions for survival of the parasite and causes the plant cell to grow as a tumor.

Agrobacteria are required to induce tumor formation, but the tumor cells do not require the continued presence of bacteria. As with animal tumors, the plant cells have been transformed into a state in which new mechanisms govern growth and differentiation. Transformation is caused by the expression within the plant cell of genetic information transferred from the bacterium.

The tumor-inducing principle of *Agrobacterium* resides in the **Ti plasmid**, which is perpetuated as an independent replicon within the bacterium. The plasmid carries genes involved in various bacterial and plant cell activities, including those required to generate the transformed state, and a set of genes concerned with synthesis or utilization of **opines** (novel derivatives of arginine).

Ti plasmids (and thus the *Agrobacteria* in which they reside) can be divided into four groups, according to the types of opine that are made:

- **Nopaline plasmids** carry genes for synthesizing nopaline in tumors and for utilizing it in bacteria. Nopaline

FIGURE 12.13 An *Agrobacterium* carrying a Ti plasmid of the nopaline type induces a teratoma, in which differentiated structures develop. Photo courtesy of the estate of Jeff Schell. Used with permission of the Max Planck Institute for Plant Breeding Research, Cologne.

tumors can differentiate into shoots with abnormal structures. They have been called **teratomas** by analogy with certain mammalian tumors that retain the ability to differentiate into early embryonic structures.

- **Octopine plasmids** are similar to nopaline plasmids, but the relevant opine is different. Octopine tumors are usually undifferentiated, however, and do not form teratoma shoots.
- **Agropine plasmids** carry genes for agropine metabolism; the tumors do not differentiate, and they develop poorly and die early.
- **Ri plasmids** can induce hairy root disease on some plants and crown gall on others. They have agropine-type genes, and may have segments derived from both nopaline and octopine plasmids.

The types of genes carried by a Ti plasmid are summarized in **FIGURE 12.14**. Genes utilized in the bacterium code for plasmid replication and incompatibility, transfer between bacteria, sensitivity to phages, and synthesis of other compounds, some of which are toxic to other soil bacteria. Genes used in the plant cell code for transfer of DNA into the plant, induction of the transformed state, and shoot and root induction.

The specificity of the opine genes depends on the type of plasmid. Genes needed for opine synthesis are linked to genes whose products catabolize the same opine; thus each strain of *Agrobacterium* causes crown gall tumor cells to synthesize opines that are useful for survival of the parasite. The opines can be used as the sole carbon and/or nitrogen source for the inducing *Agrobacterium* strain. The principle is that the transformed plant cell synthesizes those opines that the bacterium can use.

Locus	Function	Ti Plasmid
vir	DNA transfer into plant	all
shi	shoot induction	all
roi	root induction	all
nos	nopaline synthesis	nopaline
noc	nopaline catabolism	nopaline
ocs	octopine synthesis	octopine
occ	octopine catabolism	octopine
tra	bacterial transfer genes	all
Inc	incompatibility genes	all
oriV	origin for replication	all

FIGURE 12.14 Ti plasmids carry genes involved in both plant and bacterial functions.

12.9 T-DNA Carries Genes Required for Infection

Key concepts

- Part of the DNA of the Ti plasmid is transferred to the plant cell nucleus.
- The *vir* genes of the Ti plasmid are located outside the transferred region and are required for the transfer process.
- The *vir* genes are induced by phenolic compounds released by plants in response to wounding.
- The membrane protein VirA is autophosphorylated on histidine when it binds an inducer.
- VirA activates VirG by transferring the phosphate group to it.
- The VirA-VirG is one of several bacterial two-component systems that use a phosphohistidine relay.

The interaction between *Agrobacterium* and a plant cell is illustrated in **FIGURE 12.15**. The bacterium does not enter the plant cell, but rather transfers part of the Ti plasmid to the plant nucleus. The transferred part of the Ti genome is called **T-DNA**. It becomes integrated into the plant genome, where it expresses the functions

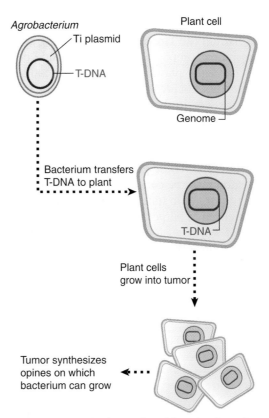

FIGURE 12.15 T-DNA is transferred from *Agrobacterium* carrying a Ti plasmid into a plant cell, where it becomes integrated into the nuclear genome and expresses functions that transform the host cell.

needed to synthesize opines and to transform the plant cell.

Transformation of plant cells requires three types of function carried in the *Agrobacterium*:

- Three loci on the *Agrobacterium* chromosome, *chvA*, *chvB*, and *pscA*, are required for the initial stage of binding the bacterium to the plant cell. They are responsible for synthesizing a polysaccharide on the bacterial cell surface.
- The *vir* region carried by the Ti plasmid outside the T-DNA region is required to release and initiate transfer of the T-DNA.
- The T-DNA is required to transform the plant cell.

The organization of the major two types of Ti plasmid is illustrated in **FIGURE 12.16**. About 30% of the ~200 kb Ti genome is common to nopaline and octopine plasmids. The common regions include genes involved in all stages of the interaction between *Agrobacterium* and a plant host, but considerable rearrangement of the sequences has occurred between the plasmids.

The T-region occupies ~23 kb. Some 9 kb is the same in the two types of plasmid. The Ti plasmids carry genes for opine synthesis (*Nos* or *Ocs*) within the T-region; corresponding genes for opine catabolism (*Noc* or *Occ*) reside elsewhere on the plasmid. The plasmids code for similar, but not identical, morphogenetic functions, as seen in the induction of characteristic types of tumors.

Functions affecting oncogenicity—the ability to form tumors—are not confined to the T-region. Those genes located outside the T-region must be concerned with establishing the tumorigenic state, but their products are not needed to perpetuate it. They may be concerned with transfer of T-DNA into the plant nucleus or perhaps with subsidiary functions

such as the balance of plant hormones in the infected tissue. Some of the mutations are host specific, preventing tumor formation by some plant species but not by others.

The virulence genes code for the functions required for the transfer of the T-DNA to the plant cell (whereas the proteins needed for conjugal transfer of the entire Ti plasmid to recipient bacteria are encoded by the *tra* region). Six loci (*virA*, *-B*, *-C*, *-D*, *-E*, and *-G*) reside in a 40-kb region outside the T-DNA. Each locus is transcribed as an individual unit; some contain more than one open reading frame. Some of the most important components and their role in the transformation process are illustrated in **FIGURE 12.17**.

We may divide the transforming process into (at least) two stages:

- *Agrobacterium* contacts a plant cell and the *vir* genes are induced.
- *vir* gene products cause T-DNA to be transferred to the plant cell nucleus, where it is integrated into the genome.

The *vir* genes fall into two groups that correspond to these stages. Genes *virA* and *virG* are regulators that respond to a change in the plant by inducing the other genes. Thus mutants in *virA* and *virG* are avirulent and cannot express the remaining *vir* genes. Genes *virB*, *-C*, *-D*, and *-E* code for proteins involved in the transfer of DNA. Mutants in *virB* and *virD* are avirulent in all plants, but the effects of mutations in *virC* and *virE* vary with the type of host plant.

virA and *virG* are expressed constitutively (at a rather low level). The signal to which they respond is provided by phenolic compounds generated by plants as a response to wounding. **FIGURE 12.18** presents an example. *Nicotiana tabacum* (tobacco) generates the molecules acetosyringone and α-hydroxyacetosyringone. Exposure to these compounds activates *virA*, which acts on *virG*, which in turn induces the expression *de novo* of *virB*, *-C*, *-D*, and *-E*. This reaction explains why *Agrobacterium* infection succeeds only on wounded plants.

VirA and VirG are an example of a classic type of bacterial system in which stimulation of a sensor protein causes autophosphorylation and transfer of the phosphate to the second protein. The relationship is illustrated in **FIGURE 12.19**.

VirA forms a homodimer that is located in the inner membrane; it may respond to the presence of the phenolic compounds in the periplasmic space. Exposure to these compounds causes VirA to become autophosphorylated on histi-

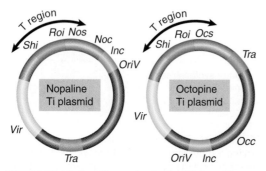

FIGURE 12.16 Nopaline and octopine Ti plasmids carry a variety of genes, including T-regions that have overlapping functions.

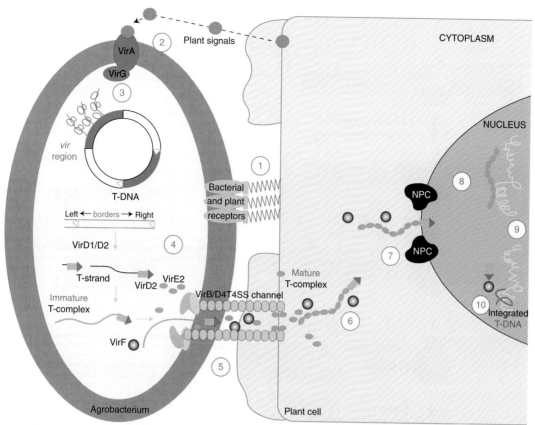

FIGURE 12.17 A model for the *Agrobacterium*-mediated genetic transformation. The transformation process comprises 10 major steps and begins with recognition and attachment of the *Agrobacterium* to the host cell (1) and the sensing of specific plant signals by the *Agrobacterium* VirA/VirG two component signal-transduction system (2). Following activation of the *vir* gene region (3), a mobile copy of the T-DNA is generated by the VirD1/VirD2 protein complex (4) and delivered as a VirD2-DNA complex (immature T-complex), together with several other Vir proteins, into the host cell cytoplasm (5). Following the association of VirE2 with the T-strand, the mature T-complex forms, travels through the host-cell cytoplasm (6) and is actively imported into the host-cell nucleus (7). Once inside the nucleus, the T-DNA is recruited to the point of integration (8), stripped of its escorting proteins (9), and integrated into the host genome (10). Reprinted from *Curr. Opin. Biotechnol.*, vol. 17, T. Tzfira and V. Citovsky, Agrobacterium-mediated genetic transformation of plants . . . , pp. 147–154. Copyright 2006, with permission from Elsevier [http://www.sciencedirect.com/science/journal/09581669].

dine. The phosphate group is then transferred to an Asp residue in VirG. The phosphorylated VirG binds to promoters of the *virB, -C, -D,* and *-E* genes to activate transcription. When *virG* is activated, its transcription is induced from a new start point—a different one from the one used for constitutive expression—with the result that the amount of VirG protein is increased.

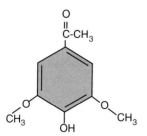

FIGURE 12.18 Acetosyringone (4-acetyl-2,6-dimethoxy-phenol) is produced by *N. tabacum* upon wounding and induces transfer of T-DNA from *Agrobacterium*.

12.10 Transfer of T-DNA Resembles Bacterial Conjugation

Key concepts

- T-DNA is generated when a nick at the right boundary creates a primer for synthesis of a new DNA strand.
- The preexisting single strand that is displaced by the new synthesis is transferred to the plant cell nucleus.
- Transfer is terminated when DNA synthesis reaches a nick at the left boundary.
- The T-DNA is transferred as a complex of single-stranded DNA with the VirE2 single strand-binding protein.
- The single-stranded T-DNA is converted into double-stranded DNA and integrated into the plant genome.
- The mechanism of integration is not known. T-DNA can be used to transfer genes into a plant nucleus.

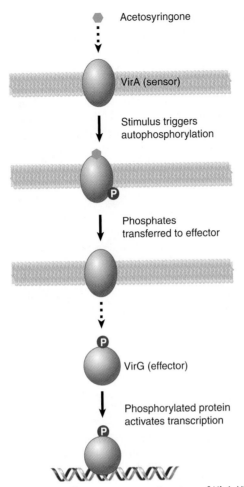

FIGURE 12.19 The two-component system of VirA-VirG responds to phenolic signals by activating transcription of target genes.

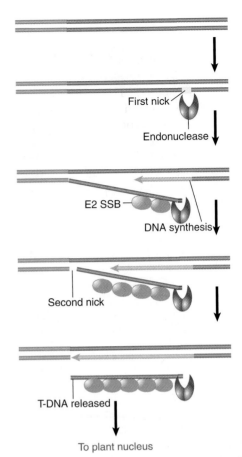

FIGURE 12.21 T-DNA is generated by displacement when DNA synthesis starts at a nick made at the right repeat. The reaction is terminated by a nick at the left repeat.

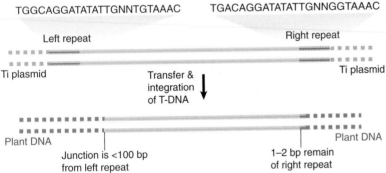

FIGURE 12.20 T-DNA has almost identical repeats of 25 bp at each end in the Ti plasmid. The right repeat is necessary for transfer and integration to a plant genome. T-DNA that is integrated in a plant genome has a precise junction that retains 1 to 2 bp of the right repeat, but the left junction varies and may be up to 100 bp short of the left repeat.

The transfer process actually selects the T-region for entry into the plant. FIGURE 12.20 shows that the T-DNA of a nopaline plasmid is demarcated from the flanking regions in the Ti plasmid by repeats of 25 bp, which differ at only two positions between the left and right ends. When

T-DNA is integrated into a plant genome, it has a well-defined right junction, which retains 1 to 2 bp of the right repeat. The left junction is variable; the boundary of T-DNA in the plant genome may be located at the 25-bp repeat or at one of a series of sites extending over ~100 bp within the T-DNA. At times multiple tandem copies of T-DNA are integrated at a single site.

The *virD* locus has four open reading frames. Two of the proteins coded at *virD*, VirD1 and VirD2, provide an endonuclease that initiates the transfer process by nicking T-DNA at a specific site. A model for transfer is illustrated in FIGURE 12.21. A nick is made at the right 25 bp repeat. It provides a priming end for synthesis of a DNA single strand. Synthesis of the new strand displaces the old strand, which is used in the transfer process. Transfer is terminated when DNA synthesis reaches a nick at the left repeat. This model explains why the right repeat is essential, and it accounts for the polarity of the process. If the left repeat fails to be nicked, transfer could continue farther along the Ti plasmid.

The transfer process involves production of a single molecule of single-stranded DNA in the infecting bacterium. It is transferred in the form of a DNA-protein complex, sometimes called the T-complex. The DNA is covered by the VirE2 single-strand binding protein, which has a nuclear localization signal and is responsible for transporting T-DNA into the plant cell nucleus. A single molecule of the D2 subunit of the endonuclease remains bound at the 5' end. The *virB* operon codes for eleven products that are involved in the transfer reaction.

Outside T-DNA, but immediately adjacent to the right border, is another short sequence, called *overdrive*, which greatly stimulates the transfer process. Overdrive functions like an enhancer: It must lie on the same molecule of DNA, but enhances the efficiency of transfer even when located several thousand base pairs away from the border. VirC1, and possibly VirC2, may act at the overdrive sequence.

Octopine plasmids have a more complex pattern of integrated T-DNA than nopaline plasmids. The pattern of T-strands is also more complex, and several discrete species can be found, corresponding to elements of T-DNA. This suggests that octopine T-DNA has several sequences that provide targets for nicking and/ or termination of DNA synthesis.

This model for transfer of T-DNA closely resembles the events involved in bacterial conjugation, when the *E. coli* chromosome is transferred from one cell to another in single-stranded form. The genes of the *virB* operon are homologous to the *tra* genes of certain bacterial plasmids (including the *tra* operons on Ti-plasmids) that are involved in conjugation (see *Section 12.7, Conjugation Transfers Single-Stranded DNA*). Together with VirD4 (a coupling protein), the gene products of the *virB* genes form a T4SS.

The T strand, along with several other Vir proteins, is then exported into the plant cell by the T4SS, a step that requires interaction of the bacterial T-pilus with at least one host-specific protein. The T-strand molecule is coated with numerous VirE2 molecules when entering the plant-cell cytoplasm. These molecules confer to the T-DNA the structure and protection needed for its travel to the plant-cell nucleus (see Figure 12.17).

We do not know how the transferred DNA is integrated into the plant genome. At some stage, the newly generated single strand must be converted into duplex DNA. Circles of T-DNA that are found in infected plant cells appear to be generated by recombination between the left and right 25-bp repeats, but we do not know

if they are intermediates. The actual event is likely to involve a nonhomologous recombination, because there is no homology between the T-DNA and the sites of integration.

What is the structure of the target site? Sequences flanking the integrated T-DNA tend to be rich in A-T base pairs (a feature displayed in target sites for some transposable elements). The sequence rearrangements that occur at the ends of the integrated T-DNA make it difficult to analyze the structure. We do not know whether the integration process generates new sequences in the target DNA comparable to the target repeats created in transposition.

T-DNA is expressed at its site of integration. The region contains several transcription units, each of which probably contains a gene expressed from an individual promoter. Their functions are concerned with the state of the plant cell, maintaining its tumorigenic properties, controlling shoot and root formation, and suppressing differentiation into other tissues. None of these genes is needed for T-DNA transfer.

The Ti plasmid presents an interesting organization of functions. Outside the T-region, it carries genes needed to initiate oncogenesis; at least some are concerned with the transfer of T-DNA, and we would like to know whether others function in the plant cell to affect its behavior at this stage. Also outside the T-region are the genes that enable the *Agrobacterium* to catabolize the opine that the transformed plant cell will produce. Within the T-region are the genes that control the transformed state of the plant, as well as the genes that cause it to synthesize the opines that will benefit the *Agrobacterium* that originally provided the T-DNA.

As a practical matter, the ability of *Agrobacterium* to transfer T-DNA to the plant genome makes it possible to introduce new genes into plants. The transfer/integration and oncogenic functions are separate; thus it is possible to engineer new Ti plasmids in which the oncogenic functions have been replaced by other genes whose effect on the plant we wish to test. The existence of a natural system for delivering genes to the plant genome has greatly facilitated genetic engineering of plants.

12.11 Summary

The rolling circle is an alternative form of replication for circular DNA molecules in which an origin is nicked to provide a priming end. One strand of DNA is synthesized from this end; this displaces the original partner strand, which is extruded as a tail. Multiple genomes can be produced by continuing revolutions of the circle.

Rolling circles are used to replicate some phages. The A protein that nicks the φX174 origin has the unusual property of *cis*-action. It acts only on the DNA from which it was synthesized. It remains attached to the displaced strand until an entire strand has been synthesized, and then nicks the origin again; this releases the displaced strand and starts another cycle of replication.

Rolling circles also characterize bacterial conjugation, which occurs when an F plasmid is transferred from a donor to a recipient cell following the initiation of contact between the cells by means of the F-pili. A free F plasmid infects new cells by this means; an integrated F plasmid creates an Hfr strain that may transfer chromosomal DNA. In conjugation, replication is used to synthesize complements to the single strand remaining in the donor and to the single strand transferred to the recipient, but does not provide the motive power.

Agrobacteria induce tumor formation in wounded plant cells. The wounded cells secrete phenolic compounds that activate *vir* genes carried by the Ti plasmid of the bacterium. The *vir* gene products cause a single strand of DNA from the T-DNA region of the plasmid to be transferred to the plant cell nucleus. Transfer is initiated at one boundary of T-DNA, but ends at variable sites. The single strand is converted into a double strand and integrated into the plant genome. Genes within the T-DNA transform the plant cell and cause it to produce particular opines (derivatives of arginine). Genes in the Ti plasmid allow *Agrobacteria* to metabolize the opines produced by the transformed plant cell. T-DNA has been used to develop vectors for transferring genes into plant cells.

References

12.4 Rolling Circles Produce Multimers of a Replicon

Research

Gilbert, W. and Dressler, D. (1968). DNA replication: the rolling circle model. *Cold Spring Harbor Symp. Quant. Biol.* 33, 473–484.

12.6 The F Plasmid Is Transferred by Conjugation between Bacteria

Research

Ihler, G. and Rupp, W. D. (1969). Strand-specific transfer of donor DNA during con-jugation in *E. coli. Proc. Natl. Acad. Sci. USA* 63, 138–143.

Lu, J., Wong, J. J., Edwards, R. A., Manchak, J., Frost, L. S., and Glover, J. N. (2008). Structural basis of specific TraD-TraM recognition during F plasmid-mediated bacterial conjugation. *Mol. Microbiol.* 70, 89–99.

12.7 Conjugation Transfers Single-Stranded DNA

Reviews

Frost, L. S., Ippen-Ihler, K., and Skurray, R. A. (1994). Analysis of the sequence and gene products of the transfer region of the F sex factor. *Microbiol. Rev.* 58, 162–210.

Ippen-Ihler, K. A. and Minkley, E. G. (1986). The conjugation system of F, the fertility factor of *E. coli. Annu. Rev. Genet.* 20, 593–624.

Lanka, E. and Wilkins, B. M. (1995). DNA processing reactions in bacterial conjugation. *Annu. Rev. Biochem.* 64, 141–169.

Willetts, N. and Skurray, R. (1987). Structure and function of the F factor and mechanism of conjugation. In Neidhardt, F. C., ed. *Escherichia coli and* Salmonella typhimurium. Washington, DC: American Society for Microbiology, pp. 1110–1133.

12.10 Transfer of T-DNA Resembles Bacterial Conjugation

Reviews

Gelvin, S. B. (2006). Agrobacterium virulence gene induction. *Methods Mol. Biol.* 343, 77–84.

Lacroix, B., Li, J., Tzfira, T., and Citovsky, V. (2006). Will you let me use your nucleus? How Agrobacterium gets its T-DNA expressed in the host plant cell. *Can. J. Physiol. Pharmacol.* 84, 333–345.

Research

Anand, A., Krichevsky, A., Schornack, S., Lahaye, T., Tzfira, T., Tang, Y., Citovsky, V., and Mysore, K. S. (2008). Arabidopsis VIRE2 INTERACTING PROTEIN2 is required for Agrobacterium T-DNA integration in plants. *Plant Cell.* 19, 695–708.

Lacroix, B., Loyter, A., and Citovsky, V. (2008). Association of the Agrobacterium T-DNA-protein complex with plant nucleosomes. *Proc. Natl. Acad. Sci. USA* 105, 15429–34.

Ulker, B., Li, Y., Rosso, M. G., Logemann, E., Somssich, I. E., and Weisshaar, B. (2008). T-DNA-mediated transfer of Agrobacterium tumefaciens chromosomal DNA into plants. *Nat. Biotechnol.* 26, 1015–1017.

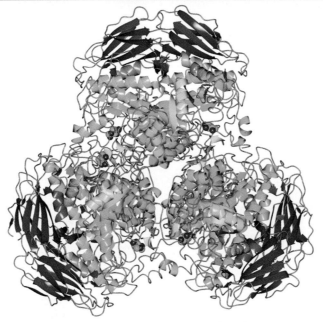

Structure from Protein Data Bank 2C8N. E. J. Taylor, et al., *Biochem. J.* 395 (2006): 31–37. Photo courtesy of Liz Potterton, York Structural Biology Laboratory, The University of York, and created using CCP4 Molecular Graphics [www.ysbl.york.ac.uk/~ccp4mg/].

Bacterial Replication Is Connected to the Cell Cycle

Edited by Barbara Funnell

CHAPTER OUTLINE

- SlmA/Noc proteins prevent septation from occurring in the space occupied by the bacterial chromosome.

- The Xer site-specific recombination system acts on a target sequence near the chromosome terminus to re-create monomers if a generalized recombination event has converted the bacterial chromosome to a dimer.

- Replicon origins are attached to the inner bacterial membrane.
- Chromosomes make abrupt movements from the mid-center to the one-quarter and three-quarter positions.

- Single-copy plasmids exist at one plasmid copy per bacterial chromosome origin.
- Multicopy plasmids exist at >1 plasmid copy per bacterial chromosome origin.
- Partition systems ensure that duplicated plasmids are segregated to different daughter cells produced by a division.

- Plasmids in a single compatibility group have origins that are regulated by a common control system.

- Replication of ColE1 requires transcription to pass through the origin, where the transcript is cleaved by RNase H to generate a primer end.
- The regulator RNA I is a short antisense RNA that pairs with the transcript and prevents the cleavage that generates the priming end.
- The Rom protein enhances pairing between RNA I and the transcript.

- mtDNA replication and segregation to daughter mitochondria is stochastic.
- Mitochondrial segregation to daughter cells is also stochastic.

13.1 Introduction

A major difference between prokaryotes and eukaryotes is the way in which replication is controlled and linked to the cell cycle.

In eukaryotes, the following are true:

- chromosomes reside in the nucleus,
- each chromosome consists of many replicons,
- replication requires coordination of these replicons to reproduce DNA during a discrete period of the cell cycle,
- the decision about whether to replicate is determined by a complex pathway that regulates the cell cycle, and
- the duplicated chromosomes are segregated to daughter cells during mitosis by means of a special apparatus.

FIGURE 13.1 shows that in bacteria, replication is triggered at a single origin when the cell mass increases past a threshold level, and the segregation of the daughter chromosomes is accomplished by ensuring that they find themselves on opposite sides of the septum that grows to divide the bacterium into two.

How does the cell know when to initiate the replication cycle? The initiation event

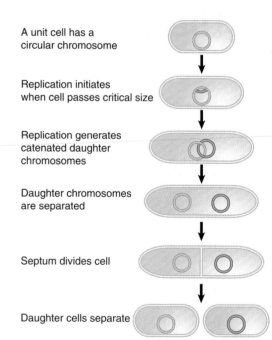

A unit cell has a circular chromosome

Replication initiates when cell passes critical size

Replication generates catenated daughter chromosomes

Daughter chromosomes are separated

Septum divides cell

Daughter cells separate

FIGURE 13.1 Replication initiates at the bacterial origin when a cell passes a critical threshold of size. Completion of replication produces daughter chromosomes that may be linked by recombination or that may be catenated. They are separated and moved to opposite sides of the septum before the bacterium is divided into two.

occurs once in each cell cycle and at the same time in every cell cycle. How is this timing set? An initiator protein could be synthesized continuously throughout the cell cycle; accumulation of a critical amount would trigger initiation. This is consistent with the fact that protein synthesis is needed for the initiation event. Another possibility is that an inhibitor protein might be synthesized or activated at a fixed point, and diluted below an effective level by the increase in cell volume. Current models suggest that variations of both possibilities operate to turn initiation on and then off precisely in each cell cycle. Synthesis of active DnaA protein, the bacterial initiator protein, reaches a threshold that turns on initiation, and the activity of inhibitors turns subsequent initiations off for the rest of the cell cycle (see Chapter 14, *DNA Replication*).

Bacterial chromosomes are specifically compacted and arranged inside the cell, and this organization is important for proper segregation, or partition, of daughter chromosomes at cell division. Some of the events in partitioning the daughter chromosomes are consequences of the circularity of the bacterial chromosome. Circular chromosomes are said to be catenated when one passes through another, connecting them. **Topoisomerases** are required to separate them. An alternative type of structure is formed when a recombination event occurs: A single recombination between two monomers converts them into a single dimer. This is resolved by a specialized recombination system that recreates the independent monomers.

13.2 Replication Is Connected to the Cell Cycle

Key concepts

- The doubling time of *E. coli* can vary over a 10x range, depending on growth conditions.
- It requires 40 minutes to replicate the bacterial chromosome (at normal temperature).
- Completion of a replication cycle triggers a bacterial division 20 minutes later.
- If the doubling time is >60 minutes, a replication cycle is initiated before the division resulting from the previous replication cycle.
- Fast rates of growth therefore produce multiforked chromosomes.

Bacteria have two links between replication and cell growth:

- The frequency of initiation of cycles of replication is adjusted to fit the rate at which the cell is growing.
- The completion of a replication cycle is connected with division of the cell.

The rate of bacterial growth is assessed by the **doubling time**, the period required for the number of cells to double. The shorter the doubling time, the faster the bacteria are growing. *E. coli* growth rates can range from doubling times as fast as 18 minutes to slower than 180 minutes. The bacterial chromosome is a single replicon; thus the frequency of replication cycles is controlled by the number of initiation events at the single origin. The replication cycle can be defined in terms of two constants:

- C is the fixed time of ~40 minutes required to replicate the entire bacterial chromosome. Its duration corresponds to a rate of replication fork movement of ~50,000 bp/minute. (The rate of DNA synthesis is more or less invariant at a constant temperature; it proceeds at the same speed unless and until the supply of precursors becomes limiting.)
- D is the fixed time of ~20 minutes that elapses between the completion of a round of replication and the cell division with which it is connected. This period may represent the time required to assemble the components needed for division.

(The constants C and D can be viewed as representing the maximum speed with which the bacterium is capable of completing these processes. They apply for all growth rates between doubling times of 18 and 60 minutes, but both constant phases become longer when the cell cycle occupies >60 minutes.)

A cycle of chromosome replication must be initiated at a fixed time of C. $D = 60$ minutes before a cell division. For bacteria dividing more frequently than every 60 minutes, a cycle of replication must be initiated before the end of the preceding division cycle. You might say that a cell is "born already pregnant" with the next generation.

Consider the example of cells dividing every 35 minutes. The cycle of replication connected with a division must have been initiated 25 minutes before the preceding division.

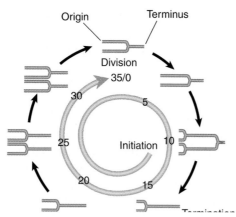

FIGURE 13.2 The fixed interval of 60 minutes between initiation of replication and cell division produces multiforked chromosomes in rapidly growing cells. Note that only the replication forks moving in one direction are shown; the chromosome actually is replicated symmetrically by two sets of forks moving in opposite directions on circular chromosomes.

This situation is illustrated in **FIGURE 13.2**, which shows the chromosomal complement of a bacterial cell at 5-minute intervals throughout the cycle.

At division (35/0 minutes), the cell receives a partially replicated chromosome. The replication fork continues to advance. At 10 minutes, when this "old" replication fork has not yet reached the terminus, initiation occurs at both origins on the partially replicated chromosome. The start of these "new" replication forks creates a **multiforked chromosome**.

At 15 minutes—that is, at 20 minutes before the next division—the old replication fork reaches the terminus. Its arrival allows the two daughter chromosomes to separate; each of them has already been partially replicated by the new replication forks (which now are the only replication forks). These forks continue to advance.

At the point of division, the two partially replicated chromosomes segregate. This recreates the point at which we started. The single replication fork becomes "old," it terminates at 15 minutes, and 20 minutes later, there is a division. We see that the initiation event occurs $1^{25}/_{35}$ cell cycles before the division event with which it is associated.

The general principle of the link between initiation and the cell cycle is that as cells grow more rapidly (the cycle is shorter), the initiation event occurs an increasing number of cycles before the related division. There are correspondingly more chromosomes in the individual bacterium. This relationship can be viewed as the cell's response to its inability to reduce the periods of *C* and *D* to keep pace with the shorter cycle.

13.3 The Septum Divides a Bacterium into Progeny That Each Contain a Chromosome

Key concepts

- Bacterial chromosomes are specifically arranged and positioned inside cells.
- Septum formation is initiated mid-cell, 50% of the distance from the septum to each end of the bacterium.
- The septum consists of the same peptidoglycans that comprise the bacterial envelope.
- The rod shape of *E. coli* is dependent on MreB, PBP2, and RodA.
- FtsZ is necessary to recruit the enzymes needed to form the septum.

Chromosome segregation in bacteria is especially interesting because the DNA itself is involved in the mechanism for partition. (This contrasts with eukaryotic cells, in which segregation is achieved by the complex apparatus of mitosis.) The bacterial apparatus is quite accurate; however, **anucleate cells**, which lack a **nucleoid**, form <0.03% of a bacterial population.

E. coli cells are shaped as cylindrical rods that end in two curved **poles**. The bacterial chromosome is compacted into a dense protein–DNA structure called the *nucleoid*, which takes up most of the space inside the cell (see *Section 9.3, The Bacterial Genome Is a Nucleoid*). It is not a disorganized mass of DNA; instead, specific DNA regions are localized to specific regions in the cell, and this positioning depends on the cell cycle. The arrangement is summarized in **FIGURE 13.3**. In newborn cells, the origin and terminus regions of the chromosome are at mid-cell. Following initiation, the new origins move toward the poles, or the ¼ and ¾ positions, and the terminus remains at mid-cell. Following cell division, the origins and termini reorient to mid-cell.

The division of a bacterium into two daughter cells is accomplished by the formation of a **septum**, a structure that forms in the center of the cell as an invagination from the surrounding envelope. The septum forms an impenetrable barrier between the two parts of the cell

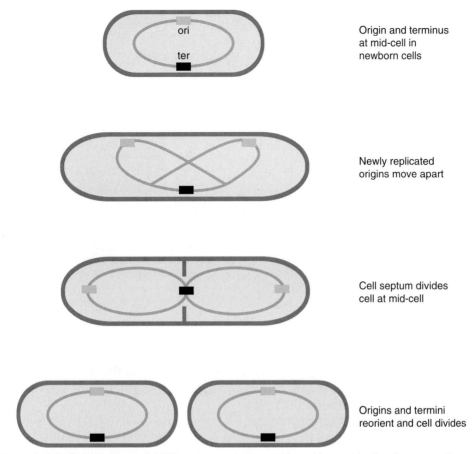

Origin and terminus
at mid-cell in
newborn cells

Newly replicated
origins move apart

Cell septum divides
cell at mid-cell

Origins and termini
reorient and cell divides

FIGURE 13.3 Attachment of bacterial DNA to the membrane could provide a mechanism for segregation.

and provides the site at which the two daughter cells eventually separate entirely. The septum then becomes the new pole of each daughter cell. Two related questions address the role of the septum in division: "What determines the location at which it forms?" and "What ensures that the daughter chromosomes lie on opposite sides of it?"

The septum consists of the same components as the cell envelope. There is a rigid layer of peptidoglycan in the periplasm, between the inner and outer membranes. The peptidoglycan is made by polymerization of tri- or pentapeptide-disaccharide units in a reaction involving connections between both types of subunit (transpeptidation and transglycosylation). The rodlike shape of the bacterium is maintained by several proteins, MreB, PBP2, and RodA. Mutations in any one of their genes and/or depletion of one of these proteins cause the bacterium to lose its extended shape and become round.

MreB is a bacterial cytoskeletal element. The structure of MreB protein resembles that of the eukaryotic protein actin, which polymerizes to form cytoskeletal filaments in eukary-

otic cells. Indeed, MreB polymerizes to form filaments that traverse a helical path along the inner membrane following the long axis of the cell. This network forms a scaffold that recruits the biosynthetic machinery for peptidoglycan synthesis, including PBP2, during elongation of the cells. RodA is a member of the SEDS family (SEDS stands for shape, elongation, division, and sporulation) present in all bacteria that have a peptidoglycan cell wall. Each SEDS protein functions together with a specific transpeptidase, which catalyzes the formation of the crosslinks in the peptidoglycan. PBP2 (penicillin-binding protein 2) is the transpeptidase that interacts with RodA. This demonstrates the important principle that shape and rigidity can be determined by the simple extension of a polymeric structure.

Another enzyme is responsible for generating the peptidoglycan in the septum (see *Section 13.5, FtsZ Is Necessary for Septum Formation*). The septum initially forms as a double layer of peptidoglycan, and the protein EnvA is required to split the covalent links between the layers so that the daughter cells may separate.

13.4 Mutations in Division or Segregation Affect Cell Shape

Key concepts

- *fts* mutants form long filaments because the septum that divides the daughter bacteria fails to form.
- Minicells form in mutants that produce too many septa; they are small and lack DNA.
- Anucleate cells of normal size are generated by partition mutants, in which the duplicate chromosomes fail to separate.

A difficulty in isolating mutants that affect cell division is that mutations in the critical functions may be lethal and/or pleiotropic. Most mutations in the division apparatus have been identified as conditional mutants (whose division is affected under nonpermissive conditions; typically they are temperature sensitive). Mutations that affect cell division or chromosome segregation cause striking phenotypic changes. **FIGURE 13.4** and **FIGURE 13.5** illustrate the opposite consequences of failure in the division process and failure in segregation:

- Long filaments form when septum formation is inhibited, but chromosome replication is unaffected. The bacteria continue to grow—and even continue to segregate their daughter chromosomes—but septa do not form. Thus the cell consists of a very long filamentous structure, with the nucleoids (bacterial chromosomes) regularly distributed along the length of the cell. This phenotype is displayed by *fts* mutants

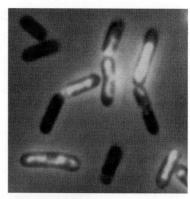

FIGURE 13.5 *E. coli* generate anucleate cells when chromosome segregation fails. Cells with chromosomes stain blue; daughter cells lacking chromosomes have no blue stain. This field shows cells of the *mukB* mutant; both normal and abnormal divisions can be seen. Photo courtesy of Sota Hiraga, Kyoto University.

(named for temperature-sensitive filamentation), which identify a defect or multiple defects that lie in the division process itself.

- **Minicells** form when septum formation occurs too frequently or in the wrong place, with the result that one of the new daughter cells lacks a chromosome. The minicell has a rather small size and lacks DNA, but otherwise appears morphologically normal. Anucleate cells form when segregation is aberrant; like minicells, they lack a chromosome, but because septum formation is normal, their size is unaltered. This phenotype is caused by *par* (partition) mutants (named because they are defective in chromosome segregation).

13.5 FtsZ Is Necessary for Septum Formation

Key concepts

- The product of *ftsZ* is required for septum formation at preexisting sites.
- FtsZ is a GTPase that forms a ring on the inside of the bacterial envelope. It is connected to other cytoskeletal components.

The gene *ftsZ* plays a central role in division. Mutations in *ftsZ* block septum formation and generate filaments. Overexpression induces minicells by causing an increased number of septation events per unit cell mass. FtsZ recruits a battery of cell division proteins that are responsible for synthesis of the new septum.

FtsZ functions at an early stage of septum formation. Early in the division cycle, FtsZ is

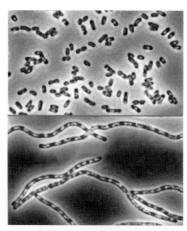

FIGURE 13.4 *Top panel:* Wild type cells. *Bottom panel:* Failure of cell division under nonpermissive temperatures generates multinucleated filaments. Photos courtesy of Sota Hiraga, Kyoto University.

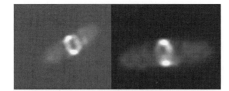

FIGURE 13.6 Immunofluorescence with an antibody against FtsZ shows that it is localized at the mid-cell. Photo courtesy of William Margolin, University of Texas Medical School at Houston.

localized throughout the cytoplasm but prior to cell division, FtsZ becomes localized in a ring around the circumference at the mid-cell position. The structure is called the **Z-ring**, which is shown in FIGURE 13.6. The formation of the Z-ring is the rate-limiting step in septum formation, and its assembly defines the position of the septum. In a typical division cycle, it forms in the center of the cell one to five minutes after division, remains for 15 minutes, and then quickly constricts to pinch the cell into two.

The structure of FtsZ resembles tubulin, suggesting that assembly of the ring could resemble the formation of microtubules in eukaryotic cells. FtsZ has GTPase activity, and GTP cleavage is used to support the oligomerization of FtsZ monomers into the ring structure. The Z-ring is a dynamic structure, in which there is continuous exchange of subunits with a cytoplasmic pool.

Two other proteins needed for division, ZipA and FtsA, interact directly and independently with FtsZ. ZipA is an integral membrane protein that is located in the inner bacterial membrane. It provides the means for linking FtsZ to the membrane. FtsA is a cytosolic protein, but is often found associated with the membrane. The Z-ring can form in the absence of either ZipA or FtsA, but it cannot form if both are absent. Both are needed for subsequent steps. This suggests that they have overlapping roles in stabilizing the Z-ring and perhaps in linking it to the membrane.

The products of several other *fts* genes join the Z-ring in a defined order after FtsA has been incorporated. They are all transmembrane proteins. The final structure is sometimes called the **septal ring**. It consists of a multiprotein complex that is presumed to have the ability to constrict the membrane. One of the last components to be incorporated into the septal ring is FtsW, which is a protein belonging to the SEDS family. *ftsW* is expressed as part of an operon with *ftsI*, which codes for a transpeptidase (also called PBP3 for penicillin-binding

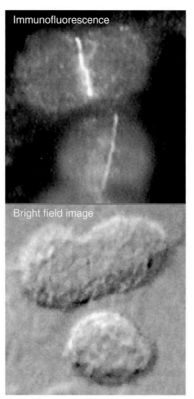

FIGURE 13.7 Immunofluorescence with antibodies against the Arabidopsis proteins FtsZ1 and FtsZ2 show that they are localized at the midpoint of the chloroplast (top panel). The bright field image (lower panel) shows the outline of the chloroplast more clearly. Photos courtesy of Katherine Osteryoung, Michigan State University.

protein 3), a membrane-bound protein that has its catalytic site in the periplasm. FtsW is responsible for incorporating FtsI into the septal ring. This suggests a model for septum formation in which the transpeptidase activity then causes the peptidoglycan to grow inward, thus pushing the inner membrane and pulling the outer membrane.

FtsZ is the major cytoskeletal component of septation. It is common in bacteria, and also is found in chloroplasts. FIGURE 13.7 shows the localization of the plant chloroplast homologs to a ring at the midpoint of the chloroplast. Chloroplasts also have other genes related to the bacterial division genes. Consistent with the common evolutionary origins of bacteria and chloroplasts, the apparatus for division generally seems to have been conserved.

Mitochondria, which also share an evolutionary origin with bacteria, usually do not have FtsZ. Instead, they use a variant of the protein dynamin, which is involved in pinching off vesicles from membranes of eukaryotic cytoplasm. This functions from the outside of the organelle, squeezing the membrane to generate a constriction.

The common feature, then, in the division of bacteria, chloroplasts, and mitochondria is the use of a cytoskeletal protein that forms a ring around the organelle and either pulls or pushes the membrane to form a constriction.

13.6 *min* and *noc/slm* Genes Regulate the Location of the Septum

Key concepts

- The location of the septum is controlled by *minC*, *-D*, and *-E* and by *noc/slmA*.
- The number and location of septa is determined by the ratio of MinE/MinC,D.
- Dynamic movement of the Min proteins in the cell sets up a pattern in which inhibition of Z-ring assembly is highest at the poles and lowest at mid-cell.
- SlmA/Noc proteins prevent septation from occurring in the space occupied by the bacterial chromosome.

Clues to the localization of the septum were first provided by minicell mutants. The original minicell mutation lies in the locus *minB*; deletion of *minB* generates minicells by allowing septation to occur at the poles instead of at mid-cell. As a result, the cell possesses the ability to initiate septum formation at mid-cell or at the poles, and the role of the wild-type *minB* locus is to suppress septation at the poles. The *minB* locus consists of three genes, *minC*, *-D*, and *-E*. The products of *minC* and *minD* form a division inhibitor. MinD is required to activate MinC, which prevents FtsZ from polymerizing into the Z-ring.

Expression of MinCD in the absence of MinE, or overexpression even in the presence of MinE, causes a generalized inhibition of division. The resulting cells grow as long filaments without septa (similar to those shown in Figure 13.4). Expression of MinE at levels comparable to MinCD confines the inhibition to the polar regions, thus restoring normal growth. The determinant of septation at the proper (mid-cell) site is, therefore, the ratio of MinCD to MinE.

The localization activities of the Min system are due to a remarkable dynamic behavior of MinD and MinE, which is illustrated in FIGURE 13.8. MinD, an ATPase, oscillates from one end of the cell to the other on a rapid time scale. MinD binds to and accumulates at the bacterial

Oscillation of Min proteins in *E. coli*

MinD binds to membrane at one pole
MinC binds to MinD
A MinE ring forms at the edge of MinD

MinE ring promotes dissociation of MinD from the membrane

MinD diffuses and binds to opposite pole
MinE ring dissociates and reforms at edge of new MinD zone

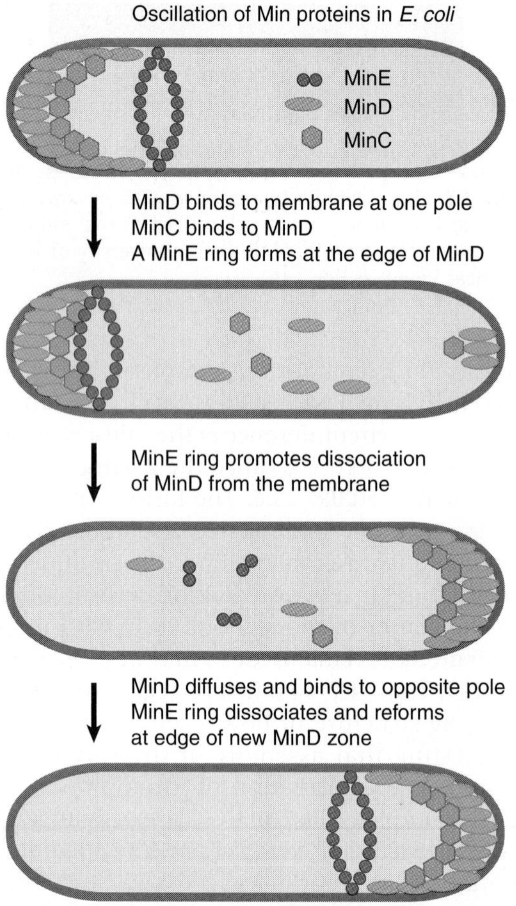

FIGURE 13.8 MinC/D is a division inhibitor whose action is confined to the polar sites by MinE.

membrane at one pole of the cell, is released, and then rebinds to the opposite pole. The periodicity of this process takes about 30 seconds, so that multiple oscillations occur within one bacterial cell generation. MinC, which cannot move on its own, oscillates as a passenger protein bound to MinD. MinE forms a ring around the cell at the edge of the zone of MinD. The MinE ring moves toward MinD at the poles and is necessary for the release of MinD from the membrane. The MinE ring then disassembles and reforms at the edge of the MinD zone that forms at the opposite pole. MinD and MinE are each required for the dynamics of the other. The consequence of this dynamic behavior is that the concentration of the MinC inhibitor is lowest at mid-cell and highest at the poles, which directs FtsZ assembly at mid-cell and inhibits its assembly at the poles.

Another process, called nucleoid occlusion, prevents Z-ring formation over the bacterial chromosome and thus prevents the septum from bisecting an individual chromosome at cell division. A protein called SlmA, which is an inhibitor of FtsZ, is necessary for nucleoid occlusion in *E. coli*. SlmA is a general DNA-binding protein, so SlmA bound to the bacterial chromosome acts on FtsZ to prevent septum formation in this region of the cell. In *Bacillus subtilis*, a different DNA-binding protein called Noc possesses a similar nucleoid occlusion function. The bacterial nucleoid takes up a large volume of the cell, and as a result this process restricts Z-ring assembly to the limited nucleoid-free spaces at the poles and mid-cell. The combination of nucleoid occlusion and the Min system promotes the Z-rings to form, and thus cell division to occur, at mid-cell.

13.7 Chromosomal Segregation May Require Site-Specific Recombination

Key concept

• The Xer site-specific recombination system acts on a target sequence near the chromosome terminus to recreate monomers if a generalized recombination event has converted the bacterial chromosome to a dimer.

After replication has created duplicate copies of a bacterial chromosome or plasmid, the copies can recombine. **FIGURE 13.9** demonstrates the consequences. A single intermolecular recombination event between two circles generates a dimeric circle; further recombination can generate higher multimeric forms. Such an event reduces the number of physically segregating units. In the extreme case of a single-copy plasmid that has just replicated, formation of a dimer by recombination means that the cell

only has one unit to segregate, and the plasmid therefore must inevitably be lost from one daughter cell. To counteract this effect, plasmids often have **site-specific recombination** systems that act upon particular sequences to sponsor an intramolecular recombination that restores the monomeric condition. For example, plasmid P1 encodes the Cre protein-*lox* site recombination system for this purpose. Scientists have further exploited the Cre-*lox* system extensively for genetic engineering in many different organisms (see *Section 3.12, Gene Knockouts and Transgenics*).

The same types of event can occur with the bacterial chromosome; **FIGURE 13.10** shows how they affect its segregation. If no recombination occurs, there is no problem, and the separate daughter chromosomes can segregate to the daughter cells. A dimer will be produced, however, if homologous recombination occurs between the daughter chromosomes produced by a replication cycle. If there has been such a recombination event, the daughter

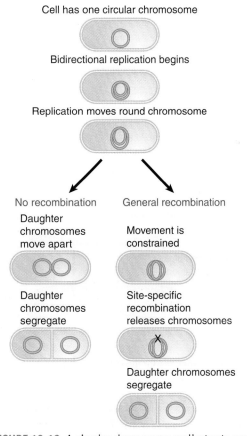

FIGURE 13.10 A circular chromosome replicates to produce two monomeric daughters that segregate to daughter cells. A generalized recombination event, however, generates a single dimeric molecule. This can be resolved into two monomers by a site-specific recombination.

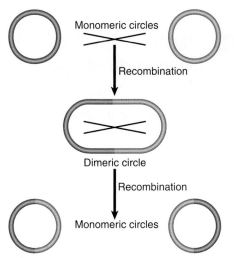

FIGURE 13.9 Intermolecular recombination merges monomers into dimers, and intramolecular recombination releases individual units from oligomers.

chromosomes cannot separate. In this case, a second recombination is required to achieve resolution in the same way as a plasmid dimer.

Most bacteria with circular chromosomes possess the Xer site-specific recombination system. In *E. coli*, this consists of two recombinases, XerC and XerD, which act on a 28-bp target site called *dif* that is located in the terminus region of the chromosome. The use of the Xer system is related to cell division in an interesting way. The relevant events are summarized in **FIGURE 13.11**. XerC can bind to a pair of *dif* sequences and form a Holliday junction between them. The complex may form soon after the replication fork passes over the *dif* sequence, which explains how the two copies of the target sequence can find one another consistently. Resolution of the junction to give recombinants, however, occurs only in the presence of FtsK, a protein located in the septum that is required for chromosome segregation and cell division. In addition, the *dif* target sequence must be located in a region of ~30 kb; if it is moved outside of this region, it cannot support the reaction. Remember that the termi-

nus region of the chromosome is located near the septum prior to cell division (see *Section 13.3, The Septum Divides a Bacterium into Progeny That Each Contain a Chromosome*).

The bacterium, however, should have site-specific recombination at *dif* only when there has already been a general recombination event to generate a dimer. (Otherwise the site-specific recombination would create the dimer!) How does the system know whether the daughter chromosomes exist as independent monomers or have been recombined into a dimer? One answer may be that segregation of chromosomes starts soon after replication. If there has been no recombination, the two chromosomes move apart from one another. The ability to move apart from one another, however, will be constrained if a dimer has been formed. This forces the terminus region to remain in the vicinity of the septum, where sites are exposed to the Xer system.

Bacteria that have the Xer system always have an FtsK homolog, and vice versa, which suggests that the system has evolved so that resolution is connected to the septum. FtsK is a large transmembrane protein. Its N-terminal domain is associated with the membrane and causes it to be localized to the septum. Its C-terminal domain has two functions. One is to cause Xer to resolve a dimer into two monomers. It also has an ATPase activity, which it can use to translocate along DNA *in vitro*. This could be used to pump DNA through the septum, in the same way that SpoIIIE transports DNA from the mother compartment into the prespore during sporulation. (See *Section 13.8, Partition Involves Separation of the Chromosomes*.)

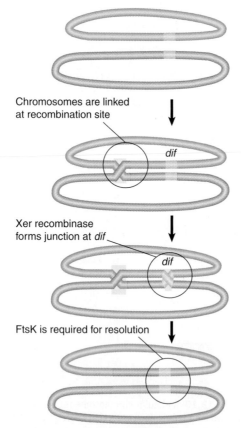

Chromosomes are linked
at recombination site

dif

Xer recombinase
forms junction at *dif*

dif

FtsK is required for resolution

FIGURE 13.11 A recombination event creates two linked chromosomes. Xer creates a Holliday junction at the *dif* site, but can resolve it only in the presence of FtsK.

13.8 Partition Involves Separation of the Chromosomes

Key concepts

- Replicon origins are attached to the inner bacterial membrane.
- Chromosomes make abrupt movements from the mid-center to the one-quarter and three-quarter positions.

Partition is the process by which the two daughter chromosomes find themselves on either side of the position at which the septum forms. Two types of event are required for proper partition:

- The two daughter chromosomes must be released from one another so that they can segregate following termination. This requires disentangling of DNA regions that are coiled around each other in the vicinity of the terminus. Most mutations affecting partition map in genes coding for topoisomerases—enzymes with the ability to pass DNA strands through one another. The mutations prevent the daughter chromosomes from segregating, with the result that the DNA is located in a single large mass at mid-cell. Septum formation then releases an anucleate cell and a cell containing both daughter chromosomes. This tells us that the bacterium must be able to disentangle its chromosomes topologically in order to be able to segregate them into different daughter cells.

- Mutations that affect the partition process itself are rare. We expect to find two classes: (1) *cis*-acting mutations should occur in DNA sequences that are the targets for the partition process; and (2) *trans*-acting mutations should occur in genes that code for the protein(s) that cause segregation, which could include proteins that bind to DNA or activities that control the locations in the cell. Both types of mutation have been found in the systems responsible for partition plasmids, but only *trans*-acting functions have been found in the bacterial chromosome. In addition, mutations in plasmid site-specific recombination systems increase plasmid loss (because the dividing cell has only one dimer to partition instead of two monomers), and therefore have a phenotype that is similar to partition mutants.

The original models for chromosome segregation suggested that the cell envelope grows by insertion of material between membrane-attachment sites of the two chromosomes, thus pushing them apart. In fact, the cell wall and membrane grow heterogeneously over the whole cell surface. Furthermore, replicated chromosomes are capable of abrupt movements to their final positions at one quarter and three quarters of the cell length. If protein synthesis is inhibited before the termination of replication, the chromosomes fail to segregate and thus remain close to the mid-cell position. When protein synthesis is allowed to resume, though, the chromosomes move to the quarter positions in the absence of any further envelope elongation. This suggests that an active process—one that requires protein synthesis—may move the chromosomes to specific locations.

Segregation is interrupted by mutations of the *muk* class, which give rise to anucleate progeny at a much increased frequency: both daughter chromosomes remain on the same side of the septum instead of segregating. Mutations in the *muk* genes are not lethal, and they may identify components of the apparatus that segregates the chromosomes. The gene *mukB* codes for a large (180 kD) protein, which has the same general type of organization as the two groups of structural maintenance of chromosomes (SMC) proteins that are involved in condensing and in holding together eukaryotic chromosomes. SMC-like proteins have also been found in other bacteria and mutations in their genes also increase the frequency of anucleate cells.

The insight into the role of MukB was the discovery that some mutations in *mukB* can be suppressed by mutations in *topA*, the gene that codes for topoisomerase I. MukB forms a complex with two other proteins, MukE and MukF, and the MukBEF complex is considered to be a condensin analogous to eukaryotic condensins. It uses a supercoiling mechanism to condense the chromosome. A defect in this function is the cause of failure to segregate properly. The defect can be compensated for by preventing topoisomerases from relaxing negative supercoils; the resulting increase in supercoil density helps to restore the proper state of condensation and thus allows segregation.

We still do not understand how genomes are positioned in the cell, but the process may be connected with condensation. FIGURE 13.12 shows a current model. The parental genome is centrally positioned. It must be decondensed in

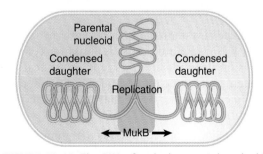

FIGURE 13.12 The DNA of a single parental nucleoid becomes decondensed during replication. MukB is an essential component of the apparatus that recondenses the daughter nucleoids.

order to pass through the replication apparatus. The daughter chromosomes emerge from replication, are disentangled by topoisomerases, and then passed in an uncondensed state to Muk-BEF, which causes them to form condensed masses at the positions that will become the centers of the daughter cells.

A physical link either directly or indirectly through chromosome-bound proteins exists between bacterial DNA and the membrane. Bacterial DNA can be found in membrane fractions, which tend to be enriched in genetic markers near the origin, the replication fork, and the terminus. The proteins present in these membrane fractions may be affected by mutations that interfere with the initiation of replication. The growth site could be a structure on the membrane to which the origin must be attached for initiation.

During sporulation in *B. subtilis*, one daughter chromosome must be segregated into the small forespore compartment. This is an unusual process that involves transfer of the chromosome across the nascent septum. One of the sporulation genes, *spoIIIE*, is required for this process. The SpoIIIE protein is located at the septum and has a translocation function that pumps DNA through to the forespore compartment. In addition, a protein called RacA tethers the replication origin to the pole of the new spore.

13.9 Single-Copy Plasmids Have a Partitioning System

Key concepts

- Single-copy plasmids exist at one plasmid copy per bacterial chromosome origin.
- Multicopy plasmids exist at >1 plasmid copy per bacterial chromosome origin.
- Partition systems ensure that duplicated plasmids are segregated to different daughter cells produced by a division.

The type of system that a plasmid uses to ensure that it is distributed to both daughter cells at division depends upon its type of replication system. Each type of plasmid is maintained in its bacterial host at a characteristic **copy number**:

- Single-copy control systems resemble that of the bacterial chromosome and result in one replication per cell division. A single-copy plasmid effectively maintains parity with the bacterial chromosome.

- Multicopy control systems allow multiple initiation events per cell cycle, with the result that there are several copies of the plasmid per bacterium. Multicopy plasmids exist in a characteristic number (typically 10 to 20) per bacterial chromosome.

Copy number is primarily a consequence of the type of replication control mechanism. The system responsible for initiating replication determines how many origins can be present in the bacterium. Each plasmid consists of a single replicon, and as a result the number of origins is the same as the number of plasmid molecules.

Single-copy plasmids have a system for replication control whose consequences are similar to those of the system for replication governing the bacterial chromosome. A single origin can be replicated once, and then the daughter origins are segregated to the different daughter cells.

Multicopy plasmids have a replication system that allows a pool of origins to exist. If the number is great enough (in practice, >10 per bacterium), an active segregation system becomes unnecessary, because even a statistical distribution of plasmids to daughter cells will result in the loss of plasmids at frequencies of $<10^{-6}$.

Plasmids are maintained in bacterial populations with very low rates of loss ($<10^{-7}$ per cell division is typical, even for a single-copy plasmid). The systems that control plasmid segregation can be identified by mutations that increase the frequency of loss, but that do not act upon replication itself. Several types of mechanism are used to ensure the survival of a plasmid in a bacterial population. It is common for a plasmid to carry several systems, often of different types, all acting independently to ensure its survival. Some of these systems act indirectly, whereas others are concerned directly with regulating the partition event. In terms of evolution, however, all serve the same purpose: to help ensure perpetuation of the plasmid to the maximum number of progeny bacteria.

Single-copy plasmids require partition systems to ensure that the duplicate copies find themselves on opposite sides of the septum at cell division, and are therefore segregated to a different daughter cell. In fact, functions involved in partition were first identified in plasmids. The components of a common system are summarized in **FIGURE 13.13**. Typically there are two *trans*-acting loci (usually called *parA* and *parB*) and a *cis*-acting element (usu-

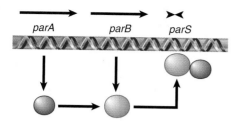

FIGURE 13.13 A common segregation system consists of genes *parA* and *parB* and the target site *parS*.

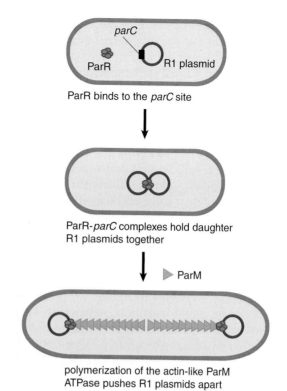

ParR binds to the *parC* site

ParR-*parC* complexes hold daughter R1 plasmids together

▷ ParM

polymerization of the actin-like ParM ATPase pushes R1 plasmids apart

FIGURE 13.14 The partition of plasmid R1 involves polymerization of the ParM ATPase between plasmids.

ally called *parS*) located next to the two genes. ParA is a partition ATPase. It binds to ParB, which binds to the *parS* site on DNA. Deletions of any of the three loci prevent proper partition of the plasmid. Systems of this type have been characterized for the plasmids F, P1, and R1. Partition systems generally fall into two major classes that depend on properties of the system's ATPase. In one group, such as the system in plasmid R1, the ATPase resembles actin and acts via polymerization (see below). The other group, which includes plasmids P1 and F, have a different type of ATPase (based on protein sequence homologies), and the mode of action of this type of ParA is unknown.

parS plays a role for the plasmid that is equivalent to the centromere in a eukaryotic cell. Binding of the ParB protein to it creates a structure that segregates the plasmid copies to opposite daughter cells. In some plasmids, such as P1, a bacterial protein, IHF, also binds at this site to form part of the structure. The complex of ParB (and IHF in some cases) with *parS* is called the *partition complex*. Formation of this initial complex enables further molecules of ParB to bind cooperatively, forming a very large protein–DNA complex. These complexes may hold daughter plasmids together in pairs until ready to interact with ParA. The activity of ParA is then necessary to position the plasmids in the cell so that at least one copy is on each side of the dividing cell septum.

The partition ATPase of plasmid R1, called ParM in this system, acts as a cytoskeletal element. The structure of ParM resembles eukaryotic actin and bacterial MreB protein (see *Section 13.3, The Septum Divides a Bacterium into Progeny That Each Contain a Chromosome*), and polymerizes into filamentous structures in the presence of ATP. In the R1 system, the partition site is called *parC* and the ParB-like protein is called ParR. Binding of ParM to the ParR/*parC* partition complexes stimulates the polymerization of ParM between complexes on daughter plasmids, effectively pushing the plasmids apart and

to opposite ends of the dividing cell (illustrated in **FIGURE 13.14**).

In the other, non-actin class of partition ATPases, it is not known how these ParA proteins work to position plasmids. There are no sequence nor structural similarities with ParM. It is possible that ParA proteins of plasmids such as P1 and F also act via polymerization. These ParA proteins do share some sequence similarities with the MinD ATPase that helps position the septum (see *Section 13.6, min and noc/slm Genes Regulate the Location of the Septum*). Intriguingly, some ParAs have been shown to oscillate inside the cell. The role of this oscillation is still a mystery, but these properties suggest that dynamic behavior of the ParA proteins is necessary for the partition reaction.

Proteins related to ParA and ParB are found in several bacteria. In *B. subtilis*, they are called SojJ and Spo0J, respectively. Mutations in these loci prevent sporulation because of a failure to segregate one daughter chromosome into the forespore. Mutations in the *spo0J* gene cause a 100-fold increase in the frequency of anucleate cells in vegetatively growing cells, suggesting that wild-type Spo0J contributes to chromosome segregation in normal cell cycles as well as during sporulation. Spo0J binds to a *parS* sequence that is present in multiple copies that are dispersed over ~20% of the chromosome in the vicinity of the origin. It is possible that Spo0J binds both old and newly synthesized origins, maintaining a status equivalent to chromosome pairing until the chromosomes are segregated to the opposite poles. In *Caulobacter crescentus*, ParA and ParB localize to the poles of the bacterium and ParB binds sequences close to the origin, thus localizing the origin to the pole. These results suggest that a specific apparatus is responsible for localizing the origin to the pole. The next stage of the analysis will be to identify the cellular components with which this apparatus interacts.

The importance to the plasmid of ensuring that all daughter cells gain replica plasmids is emphasized by the existence of multiple, independent systems in individual plasmids that ensure proper partition. **Addiction systems**, which operate on the basis that "we hang together or we hang separately," ensure that a bacterium carrying a plasmid can survive only as long as it retains the plasmid. There are several ways to ensure that a cell dies if it is "cured" of a plasmid, all of which share the principle illustrated in **FIGURE 13.15** that the plasmid produces both a poison and an antidote. The poison is a killer substance that is relatively stable,

whereas the antidote consists of a substance that blocks killer action but is relatively short lived. When the plasmid is lost the antidote decays, and then the killer substance causes the death of the cell. Thus bacteria that lose the plasmid inevitably die, and the population is condemned to retain the plasmid indefinitely. These systems take various forms. One specified by the F plasmid consists of killer and blocking proteins. The plasmid R1 has a killer that is the mRNA for a toxic protein; the antidote is a small antisense RNA that prevents expression of the mRNA.

13.10 Plasmid Incompatibility Is Determined by the Replicon

Key concept
• Plasmids in a single compatibility group have origins that are regulated by a common control system.

The phenomenon of plasmid incompatibility is related to the regulation of plasmid copy number and segregation. A **compatibility group** is defined as a set of plasmids whose members are unable to coexist in the same bacterial cell. The reason for their incompatibility is that they cannot be distinguished from one another at some stage that is essential for plasmid maintenance. DNA replication and segregation are stages at which this may apply.

The negative control model for plasmid incompatibility follows the idea that copy number control is achieved by synthesizing a repressor that measures the concentration of origins. (Formally, this is the same as the titration model for regulating replication of the bacterial chromosome.)

The introduction of a new origin in the form of a second plasmid of the same compatibility group mimics the result of replication of the resident plasmid; two origins now are present. Thus any further replication is prevented until after the two plasmids have been segregated to different cells to create the correct prereplication copy number, as illustrated in **FIGURE 13.16**.

A similar effect would be produced if the system for segregating the products to daughter cells could not distinguish between two plasmids. For example, if two plasmids have the same *cis*-acting partition sites, competition between them would ensure that they would be segregated to different cells, and therefore could not survive in the same line.

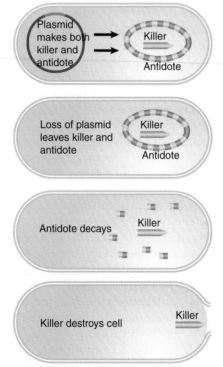

FIGURE 13.15 Plasmids may ensure that bacteria cannot live without them by synthesizing a long-lived killer and a short-lived antidote.

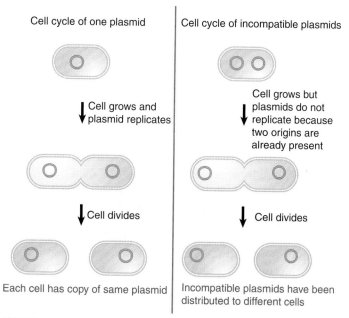

FIGURE 13.16 Two plasmids are incompatible (they belong to the same compatibility group) if their origins cannot be distinguished at the stage of initiation. The same model could apply to segregation.

The presence of a member of one compatibility group does not directly affect the survival of a plasmid belonging to a different group. Only one replicon of a given compatibility group (of a single-copy plasmid) can be maintained in the bacterium, but it does not interact with replicons of other compatibility groups.

13.11 The ColE1 Compatibility System Is Controlled by an RNA Regulator

Key concepts

- Replication of ColE1 requires transcription to pass through the origin, where the transcript is cleaved by RNase H to generate a primer end.
- The regulator RNA I is a short antisense RNA that pairs with the transcript and prevents the cleavage that generates the priming end.
- The Rom protein enhances pairing between RNA I and the transcript.

The best characterized copy number and incompatibility system is that of the plasmid ColE1, a multicopy plasmid that is maintained at a steady level of ~20 copies per *E. coli* cell. The system for maintaining the copy number depends on the mechanism for initiating replication at the ColE1 origin, as illustrated in **FIGURE 13.17**.

Replication starts with the transcription of an RNA that initiates 555 bp upstream of the origin. Transcription continues through

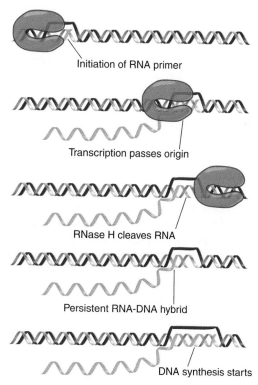

FIGURE 13.17 Replication of ColE1 DNA is initiated by cleaving the primer RNA to generate a 3′–OH end. The primer forms a persistent hybrid in the origin region.

the origin. The enzyme RNase H (whose name reflects its specificity for a substrate of RNA hybridized with DNA) cleaves the transcript at the origin. This generates a 3′–OH end that is used as the "primer" at which DNA synthesis is initiated (the use of primers is discussed in more detail in *Section 14.9, Priming Is Required to Start DNA Synthesis*). The primer RNA forms a persistent hybrid with the DNA. Pairing between the RNA and DNA occurs just upstream of the origin (around position −20) and also farther upstream (around position −265).

Two regulatory systems exert their effects on the RNA primer. One involves synthesis of an RNA complementary to the primer; the other involves a protein encoded by a nearby locus.

The regulatory species RNA I is a molecule of ~108 bases and is coded by the opposite strand from that specifying primer RNA. The relationship between the primer RNA and RNA I is illustrated in **FIGURE 13.18**. The RNA I molecule is initiated within the primer region and terminates close to the site where the primer RNA initiates. Thus RNA I is complementary to the 5′–terminal region of the primer RNA. Base pairing between the two RNAs controls the availability of the primer RNA to initiate a cycle of replication.

An RNA molecule such as RNA I that functions by virtue of its complementarity with another RNA coded in the same region is called a **countertranscript**. This type of mechanism, of course, is another example of the use of antisense RNA (see *Section 30.3, Noncoding RNAs Can Be Used to Regulate Gene Expression*).

Mutations that reduce or eliminate incompatibility between plasmids can be obtained by selecting plasmids of the same group for their ability to coexist. Incompatibility mutations in ColE1 map in the region of overlap between RNA I and primer RNA. This region is represented in two different RNAs, so either or both might be involved in the effect.

When RNA I is added to a system for replicating ColE1 DNA *in vitro*, it inhibits the formation of active primer RNA. The presence of RNA I, however, does not inhibit the initiation or elongation of primer RNA synthesis. This suggests that RNA I prevents RNase H from generating the 3′ end of the primer RNA. The basis for this effect lies in base pairing between RNA I and primer RNA.

Both RNA molecules have the same potential secondary structure in this region, with three duplex hairpins terminating in single-stranded loops. Mutations reducing incompatibility are located in these loops, which suggests that the initial step in base pairing between RNA I and primer RNA is contact between the unpaired loops.

How does pairing with RNA I prevent cleavage to form primer RNA? A model is illustrated in **FIGURE 13.19**. In the absence of RNA I, the primer RNA forms its own secondary structure (involving loops and stems). When RNA I is present, though, the two molecules pair and become completely double-stranded for the entire length of RNA I. The new secondary structure prevents the formation of the primer, probably by affecting the ability of the RNA to form the persistent hybrid.

The model resembles the mechanism involved in attenuation of transcription, in which the alternative pairings of an RNA sequence permit or prevent formation of the secondary structure needed for termination by RNA polymerase (see *Section 26.13, The trp Operon Is also Controlled by Attenuation*). The

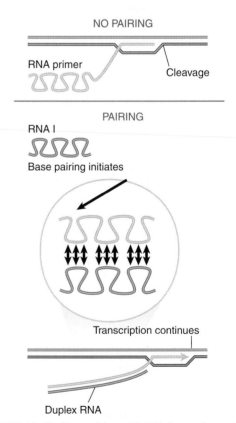

FIGURE 13.19 Base pairing with RNA I may change the secondary structure of the primer RNA sequence and thus prevent cleavage from generating a 3′–OH end.

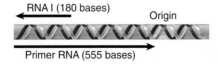

FIGURE 13.18 The sequence of RNA I is complementary to the 5′ region of primer RNA.

action of RNA I is exercised by its ability to affect distant regions of the primer precursor.

Formally, the model is equivalent to postulating a control circuit involving two RNA species. A large RNA primer precursor is a positive regulator and is needed to initiate replication. The small RNA I is a negative regulator that is able to inhibit the action of the positive regulator.

In its ability to act on any plasmid present in the cell, RNA I provides a repressor that prevents newly introduced DNA from functioning. This is analogous to the role of the lambda lysogenic repressor (see *Section 27.9, The Lambda Repressor and Its Operators Define the Immunity Region*). Instead of a repressor protein that binds the new DNA, an RNA binds the newly synthesized precursor to the RNA primer.

Binding between RNA I and primer RNA can be influenced by the Rom protein, which is coded by a gene located downstream of the origin. Rom enhances binding between RNA I and primer RNA transcripts of >200 bases. The result is to inhibit formation of the primer.

How do mutations in the RNAs affect incompatibility? **FIGURE 13.20** shows the situation when a cell contains two types of RNA I/primer RNA sequence. The RNA I and primer RNA made from each type of genome can interact, but RNA I from one genome does not interact with primer RNA from the other genome. This situation would arise when a mutation in the region that is common to RNA I and primer RNA occurred at a location involved in the base pairing between them. Each RNA I would continue to pair with the primer RNA coded by the same plasmid, but might be unable to pair with the primer RNA coded by the other plasmid. This would cause the original and the mutant plasmids to behave as members of different compatibility groups.

13.12 How Do Mitochondria Replicate and Segregate?

Key concepts

- mtDNA replication and segregation to daughter mitochondria is stochastic.
- Mitochondrial segregation to daughter cells is also stochastic.

Mitochondria must be duplicated during the cell cycle and segregated to the daughter cells. We understand some of the mechanics of this process, but not its regulation.

At each stage in the duplication of mitochondria—DNA replication, DNA segregation to duplicated mitochondria, and organelle segregation to daughter cells—the process appears to be stochastic, governed by a random distribution of each copy. The theory of distribution in this case is analogous to that of multicopy bacterial plasmids, with the same conclusion that ~10 copies are required to ensure that each daughter gains at least one copy (see *Section 13.9, Single-Copy Plasmids Have a Partitioning System*). When there are mtDNAs with allelic variations in the same cell, called **heteroplasmy** (either because of inheritance from different parents or because of mutation), the stochastic distribution may generate cells that have only one of the alleles.

Replication of mtDNA may be stochastic because there is no control over which particular copies are replicated, so that in any cycle some mtDNA molecules may replicate more times than others. The total number of copies of the genome may be controlled by titrating mass in a way similar to bacteria (see *Section 13.2, Replication Is Connected to the Cell Cycle*).

A mitochondrion divides by developing a ring around the organelle that constricts to pinch it into two halves. The mechanism is similar in principle to that involved in bacterial division. The apparatus that is used in plant cell mitochondria is similar to that used in bacteria and uses a homolog of the bacterial protein FtsZ (see *Section 13.5, FtsZ Is Necessary for Septum Formation*). The molecular apparatus is different in animal cell mitochondria and uses the protein dynamin, which is involved in formation of

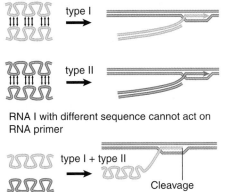

RNA I acts on any RNA primer coded by its own genome

type I

type II

RNA I with different sequence cannot act on RNA primer

type I + type II

Cleavage

FIGURE 13.20 Mutations in the region coding for RNA I and the primer precursor need not affect their ability to pair; but they may prevent pairing with the complementary RNA coded by a different plasmid.

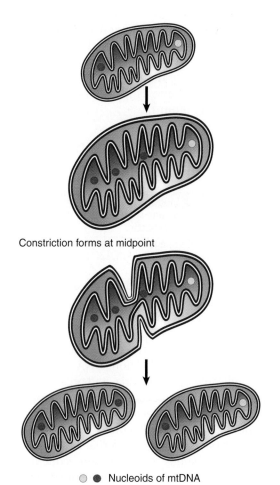

Constriction forms at midpoint

○ ● Nucleoids of mtDNA

FIGURE 13.21 Mitochondrial DNA replicates by increasing the number of genomes in proportion to mitochondrial mass but without ensuring that each genome replicates the same number of times. This can lead to changes in the representation of alleles in the daughter mitochondria.

membranous vesicles. An individual organelle may have more than one copy of its genome.

We do not know whether there is a partition mechanism for segregating mtDNA molecules within the mitochondrion, or whether they are simply inherited by daughter mitochondria according to which half of the mitochondrion they happen to lie in. **FIGURE 13.21** shows that the combination of replication and segregation mechanisms can result in a stochastic assignment of DNA to each of the copies; that is, so that the distribution of mitochondrial genomes to daughter mitochondria does not depend on their parental origins.

The assignment of mitochondria to daughter cells at mitosis also appears to be random. Indeed, it was the observation of somatic variation in plants that first suggested the existence of genes that could be lost from one of the daughter cells because they were not inherited according to Mendel's laws (see Figure 5.12).

In some situations a mitochondrion has both paternal and maternal alleles. This has two requirements: that both parents provide alleles to the zygote (which of course is not the case when there is maternal inheritance; see *Section 5.8, Some Organelles Have DNA*); and that the parental alleles are found in the same mitochondrion. For this to happen, parental mitochondria must have fused.

The size of the individual mitochondrion may not be precisely defined. Indeed, there is a continuing question as to whether an individual mitochondrion represents a unique and discrete copy of the organelle or whether it is in a dynamic flux in which it can fuse with other mitochondria. We know that mitochondria can fuse in yeast, because recombination between mtDNAs can occur after two haploid yeast strains have mated to produce a diploid strain. This implies that the two mtDNAs must have been exposed to one another in the same mitochondrial compartment. Attempts have been made to test for the occurrence of similar events in animal cells by looking for complementation between alleles after two cells have been fused, but the results are not clear.

13.13 Summary

A fixed time of 40 minutes is required to replicate the *E. coli* chromosome and a further 20 minutes is required before the cell can divide. When cells divide more rapidly than every 60 minutes, a replication cycle is initiated before the end of the preceding division cycle. This generates multiforked chromosomes. The initiation event occurs once and at a specific time in each cell cycle. Initiation timing depends on accumulating the active initiator protein DnaA and on inhibitors that turn off newly synthesized origins until the next cell cycle.

E. coli grows as a rod-shaped cell that divides into daughter cells by formation of a septum that forms at mid-cell. The shape is maintained by an envelope of peptidoglycan that surrounds the cell. The rod shape is dependent on the MreB actin-like protein that forms a scaffold for recruiting the enzymes necessary for peptidoglycan synthesis. The septum is dependent on FtsZ, which is a tubulin-like protein that can polymerize into a filamentous structure called a Z-ring. FtsZ recruits the enzymes necessary to make the septum. Absence of septum formation generates multinucleated filaments; an excess of septum formation generates anucleate minicells.

Many transmembrane proteins interact to form the septum. ZipA is located in the inner bacterial membrane and binds to FtsZ. Several other *fts* products, most of which are transmembrane proteins, join the Z-ring in an ordered process that generates a septal ring. The last proteins to bind are the SEDS protein FtsW and the transpeptidase FtsI (PBP3), which together function to produce the peptidoglycans of the septum. Chloroplasts use a related division mechanism that has an FtsZ-like protein, but mitochondria use a different process in which the membrane is constricted by a dynamin-like protein.

Plasmids and bacteria have site-specific recombination systems that regenerate pairs of monomers by resolving dimers created by general recombination. The Xer system acts on a target sequence located in the terminus region of the chromosome. The system is active only in the presence of the FtsK protein of the septum, which may ensure that it acts only when a dimer needs to be resolved.

Chromosome segregation involves several processes, including separation of catenated products by topoisomerases, site-specific recombination, and the action of MukB/SMC proteins in chromosome condensation following DNA replication.

Plasmids have a variety of systems that ensure or assist their stable inheritance in bacterial cells, and an individual plasmid may carry systems of several types. Plasmid localization is promoted by ParA and ParB partition proteins that act on a plasmid site, called *parS*. Plasmid addiction systems kill bacterial cells that fail to inherit a plasmid copy. The copy number of a plasmid describes whether it is present at the same level as the bacterial chromosome (one per unit cell) or in greater numbers. Plasmid incompatibility can be a consequence of the mechanisms involved in either replication or partition (for single-copy plasmids). Two plasmids that share the same control system for replication are incompatible because the number of replication events ensures that there is only one plasmid for each bacterial genome.

References

13.2 Replication Is Connected to the Cell Cycle

Review

Haeusser, D. P. and Levin, P. A. (2008). The great divide: coordinating cell cycle events during bacteria growth and division. *Curr. Opin. Microbiol.* 11, 94–99.

Research

Donachie, W. D. and Begg, K. J. (1970). Growth of the bacterial cell. *Nature* 227, 1220–1224.

Lobner-Olesen, A., Skarstad, K., Hansen, F. G., von-Meyenburg, K., and Boye, E. (1989). The DnaA protein determines the initiation mass of *Escherichia coli* K-12. *Cell* 57, 881–889.

13.3 The Septum Divides a Bacterium into Progeny That Each Contain a Chromosome

Reviews

Osborn, M. J. and Rothfield, L. (2007). Cell shape determination in *Escherichia coli*. *Curr. Opin. Microbiol.* 10, 606–610.

Reyes-Larnothe, R., Wang, X. D., and Sherratt, D. (2008). *Escherichia coli* and its chromosome. *Trends Microbiol.* 16, 238–245.

Thanbichler, M. and Shapiro, L. (2008). Getting organized—how bacterial cells move proteins and DNA. *Nature Rev. Microbiol.* 6, 28–40.

Research

Jones, L. J. F., CarballidoLopez, R., and Errington, J. (2001). Control of cell shape in bacteria: Helical, actin-like filaments in *Bacillus subtilis*. *Cell*, 104, 913–922.

Spratt, B. G. (1975). Distinct penicillin binding proteins involved in the division, elongation, and shape of *E. coli* K12. *Proc. Natl. Acad. Sci. USA* 72, 2999–3003.

13.4 Mutations in Division or Segregation Affect Cell Shape

Research

Adler, H. I. et al. (1967). Miniature *E. coli* cells deficient in DNA. *Proc. Natl. Acad. Sci. USA* 57, 321–326.

13.5 FtsZ Is Necessary for Septum Formation

Reviews

Errington, J., Daniel, R. A., and Scheffers, D. J. (2003). Cytokinesis in bacteria. *Microbiol. Mol. Biol. Rev.* 67, 52–65.

Weiss, D. S. (2004). Bacterial cell division and the septal ring. *Mol. Microbiol.* 54, 588–597.

Research

Bi, E. F. and Lutkenhaus, J. (1991). FtsZ ring structure associated with division in *Escherichia coli*. *Nature* 354, 161–164.

Mercer, K. L. and Weiss, D. S. (2002). The *E. coli* cell division protein FtsW is required to recruit its cognate transpeptidase, FtsI (PBP3), to the division site. *J. Bacteriol.* 184, 904–912.

Pichoff, S. and Lutkenhaus, J. (2002). Unique and overlapping roles for ZipA and FtsA in septal ring assembly in *Escherichia coli*. *EMBO J.* 21, 685–693.

13.6 *min* and *noc/slm* Genes Regulate the Location of the Septum

Review

Lutkenhaus, J. (2007). Assembly dynamics of the bacterial MinCDE system and spatial regulation of the Z Ring. *Annu. Rev. Biochem.* 76, 539–562.

Research

Bernhardt, T. G. and de Boer, P. A. J. (2005). SlmA, a nucleoid-associated, FtsZ binding protein required for blocking septal ring assembly over chromosomes in *E. coli. Mol. Cell* 18, 555–564.

Fu, X. L., Shih, Y. L., Zhang, Y., and Rothfield, L. I. (2001). The MinE ring required for proper placement of the division site is a mobile structure that changes its cellular location during the *Escherichia coli* division cycle. *Proc. Natl. Acad. Sci. USA* 98, 980–985.

Pichoff, S. and Lutkenhaus, J. (2001). *Escherichia coli* division inhibitor MinCD blocks septation by preventing Z-ring formation. *J. Bacteriol.* 183, 6630–6635.

Raskin, D. M. and de Boer, P. A. J. (1999). Rapid pole-to-pole oscillation of a protein required for directing division to the middle of *Escherichia coli. Proc. Natl. Acad. Sci. USA* 96, 4971–4976.

13.7 Chromosomal Segregation May Require Site-Specific Recombination

Research

Aussel, L., Barre, F. X., Aroyo, M., Stasiak, A., Stasiak, A. Z., and Sherratt, D. (2002). FtsK is a DNA motor protein that activates chromosome dimer resolution by switching the catalytic state of the XerC and XerD recombinases. *Cell* 108, 195–205.

Barre, F. X., Aroyo, M., Colloms, S. D., Helfrich, A., Cornet, F., and Sherratt, D. J. (2000). FtsK functions in the processing of a Holliday junction intermediate during bacterial chromosome segregation. *Genes Dev.* 14, 2976–2988.

Blakely, G., May, G., McCulloch, R., Arciszewska, L. K., Burke, M., Lovett, S. T., and Sherratt, D. J. (1993). Two related recombinases are required for site-specific recombination at dif and cer in *E. coli* K12. *Cell* 75, 351–361.

13.8 Partition Involves Separation of the Chromosomes

Reviews

Draper, G. C. and Gober, J. W. (2002). Bacterial chromosome segregation. *Annu. Rev. Microbiol.* 56, 567–597.

Ghosh, S. K., Hajra, S., Paek, A., and Jayaram, M. (2006). Mechanisms for chromosomal and plasmid segregation. *Annu. Rev. Biochem.* 75, 211–241.

Research

Ben-Yehuda, S., Rudner, D. Z., and Losick, R. (2003). RacA, a bacterial protein that anchors chromosomes to the cell poles. *Science* 299, 532–536.

Case, R. B., Chang, Y. P., Smith, S. B., Gore, J., Cozzarelli, N. R., and Bustamante, C. (2004). The bacterial condensin MukBEF compacts DNA into a repetitive, stable structure. *Science* 305, 222–227.

Jacob, F., Ryter, A., and Cuzin, F. (1966). On the association between DNA and the membrane in bacteria. *Proc. Roy. Soc. Lond. B Biol. Sci.* 164, 267–348.

Sawitzke, J. A. and Austin, S. (2000). Suppression of chromosome segregation defects of *E. coli muk* mutants by mutations in topoisomerase I. *Proc. Natl. Acad. Sci. USA* 97, 1671–1676.

13.9 Single-Copy Plasmids Have a Partitioning System

Reviews

Ebersbach, G. and Gerdes, K. (2005). Plasmid segregation mechanisms. *Annu. Rev. Genet.* 39, 453–479.

Hayes, F. and Barilla, D. (2006) The bacterial segrosome: a dynamic nucleoprotein machine for DNA trafficking and segregation. *Nat. Rev. Microbiol.* 4, 133–143.

Research

Ireton, K., Gunther, N. W., and Grossman, A. D. (1994). *spo0J* is required for normal chromosome segregation as well as the initiation of sporulation in *Bacillus subtilis. J. Bacteriol.* 176, 5320–5329.

Moller-Jensen, J., Borch, J., Dam, M., Jensen, R. B., Roepstorff, P., and Gerdes, K. (2003). Bacterial mitosis: ParM of plasmid R1 moves plasmid DNA by an actin-like insertional polymerization mechanism. *Mol. Cell* 12, 1477–1487.

Surtees, J. A. and Funnell, B. E. (2001). The DNA binding domains of P1 ParB and the architecture of the P1 plasmid partition complex. *J. Biol. Chem.* 276, 12385–12394.

13.10 Plasmid Incompatibility Is Determined by the Replicon

Reviews

Nordstrom, K. and Austin, S. J. (1989). Mechanisms that contribute to the stable segregation of plasmids. *Annu. Rev. Genet.* 23, 37–69.

Scott, J. R. (1984). Regulation of plasmid replication. *Microbiol. Rev.* 48, 1–23.

13.11 The ColE1 Compatibility System Is Controlled by an RNA Regulator

Research

Masukata, H. and Tomizawa, J. (1990). A mechanism of formation of a persistent hybrid between elongating RNA and template DNA. *Cell* 62, 331–338.

Tomizawa, J.-I. and Itoh, T. (1981). Plasmid ColE1 incompatibility determined by interaction of RNA with primer transcript. *Proc. Natl. Acad. Sci. USA* 78, 6096–6100.

13.12 How Do Mitochondria Replicate and Segregate?

Review

Birky, C. W. (2001). The inheritance of genes in mitochondria and chloroplasts: laws, mechanisms, and models. *Annu. Rev. Genet.* 35, 125–148.

14

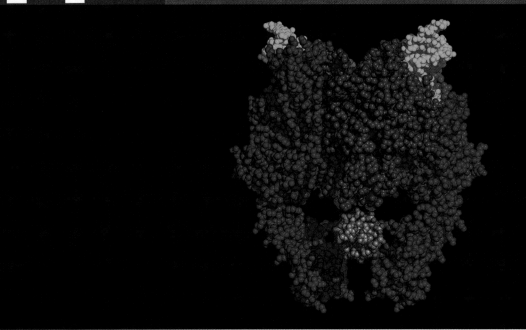

Photo courtesy of James Berger, University of California, Berkeley.

DNA Replication

Edited by Peter Burgers

14.1 Introduction

Replication of duplex DNA is a complicated endeavor involving multiple enzyme complexes. Different activities are involved in the stages of initiation, elongation, and termination. Before initiation can occur, however, the supercoiled chromosome must be relaxed (see *Section 1.5, Supercoiling Affects the Structure of DNA*). This occurs in segments beginning with the replication origin region. This alteration to the structure of the chromosome is accomplished by the enzyme **topoisomerase**. Replication cannot occur on supercoiled DNA, only the relaxed form. **FIGURE 14.1** shows an overview of the first stages of the process.

- *Initiation* involves recognition of an origin by a complex of proteins. Before DNA synthesis begins, the parental strands must be separated and (transiently) stabilized in the single-stranded state, creating a replication bubble. After this stage, synthesis of daughter strands can be initiated at the replication fork.

- *Elongation* is undertaken by another complex of proteins. The **replisome**

Proteins bind to origin and separate DNA strands

DNA polymerase and other proteins assemble into replisome

Replisome synthesizes daughter strands

FIGURE 14.1 Replication initiates when a protein complex binds to the origin and melts the DNA there. Then the components of the replisome, including DNA polymerase, assemble. The replisome moves along DNA, synthesizing both new strands.

exists only as a protein complex associated with the particular structure that DNA takes at the replication fork. It does not exist as an independent unit (for example, analogous to the ribosome), but assembles *de novo* at the origin for each replication cycle. As the replisome moves along DNA, the parental strands unwind and daughter strands are synthesized.

- At the end of the replicon, *joining* and/or *termination* reactions are necessary. Following termination, the duplicate chromosomes must be separated from one another, which requires manipulation of higher-order DNA structure.

Inability to replicate DNA is fatal for a growing cell. Mutants for replication must therefore be obtained as **conditional lethals**. These are able to accomplish replication under *permissive* conditions (typically provided by the normal temperature of incubation), but they are defective under *nonpermissive*, or *restrictive*, conditions (provided by the higher temperature of 42°C). A comprehensive series of such temperature-sensitive mutants in *E. coli* identifies a set of loci called the *dna* genes. The **dna mutants** distinguish two stages of replication by their behavior when the temperature is raised:

- The members of the major class of **quick-stop mutants** cease replication immediately upon a temperature increase. They are defective in the components of the replication apparatus, typically in the enzymes needed for elongation (but also include defects in the supply of essential precursors).
- The members of the smaller class of **slow-stop mutants** complete the current round of replication, but cannot start another. They are defective in the events involved in initiating a cycle of replication at the origin.

An important assay used to identify the components of the replication apparatus is called ***in vitro* complementation**. An *in vitro* system for replication is prepared from a *dna* mutant and is operated under conditions in which the mutant gene product is inactive. Extracts from wild-type cells are tested for their ability to restore activity. The protein encoded by the *dna* locus can be purified by identifying the active component in the extract.

Each component of the bacterial replication apparatus is now available for study *in vitro* as a biochemically pure product, and is implicated *in vivo* by mutations in its gene. Analogous eukaryotic chromosomal replication systems remain to be developed. Even so, studies of individual replisome components show a high structural and functional similarity with the bacterial replisome.

14.2 Initiation: Creating the Replication Forks at the Origin *oriC*

Key concepts

- Initiation at *oriC* requires the sequential assembly of a large protein complex on the membrane.
- *oriC* must be fully methylated.
- DnaA-ATP binds to short repeated sequences and forms an oligomeric complex that melts DNA.
- Six DnaC monomers bind each hexamer of DnaB, and this complex binds to the origin.
- A hexamer of DnaB forms the replication fork. Gyrase and SSB are also required.
- A short region of A-T-rich DNA is melted.
- DnaG is bound to the helicase complex and creates the replication forks.

Initiation of replication of duplex DNA in *E. coli* at the origin of replication, *oriC*, requires several successive activities. Some events that are required for initiation occur uniquely at the origin; others recur with the initiation of each Okazaki fragment during the elongation phase (see *Section 14.7, The Two New DNA Strands Have Different Modes of Synthesis*):

- Protein synthesis is required to synthesize the origin recognition protein, DnaA. This is the *E. coli* **licensing factor** that must be made anew for each round of replication. Drugs that block protein synthesis block a new round of replication, but not continuation of replication.
- There is a requirement for transcription activation. This is not synthesis of the mRNA for DnaA, but rather either one of two genes that flank *oriC* must be transcribed. This transcription near the origin aids DnaA in twisting open the origin.
- There must be membrane/cell wall synthesis. Drugs (like penicillin) that inhibit cell wall synthesis block initiation of replication.

Initiation of replication at *oriC* starts with formation of a complex that ultimately requires six proteins: DnaA, DnaB, DnaC, HU, gyrase, and SSB. Of the six proteins, DnaA draws our attention as the one uniquely involved in the initiation process. DnaB, an ATP hydrolysis-dependent 5′ to 3′ **helicase**, provides the "engine" of initiation after the origin has been opened (and the DNA is single-stranded) by its ability to further unwind the DNA. These events will only happen if the DNA at the origin is fully methylated on both strands.

DnaA is an ATP binding protein. The first stage in initiation is binding of the DnaA-ATP protein complex to the fully methylated *oriC* sequence. This takes place in association with the inner membrane. DnaA is in the active form only when bound to ATP. DnaA has intrinsic ATPase activity that hydrolyzes ATP to ADP and thus inactivates itself when the initiation stage ends. This ATPase activity is stimulated by membrane phospholipids and single-stranded DNA. Single-stranded DNA forms once the origin is open. This mechanism is used to prevent reinitiation of replication. The origin of replication region remains attached to the membrane for about one third of the cell cycle as part of the mechanism to prevent reinitiation. While sequestered in the membrane, the newly synthesized strand of *oriC* cannot be methylated and so *oriC* remains hemimethylated until DnaA is degraded.

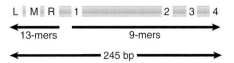

FIGURE 14.2 The minimal origin is defined by the distance between the outside members of the 13-mer and 9-mer repeats.

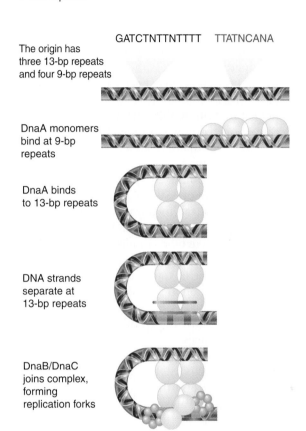

FIGURE 14.3 Prepriming involves formation of a complex by sequential association of proteins, which leads to the separation of DNA strands.

Opening *oriC* involves action at two types of sequence in the origin: 9 bp and 13 bp repeats. Together the 9 bp and 13 bp repeats define the limits of the 245 bp minimal origin, as indicated in **FIGURE 14.2**. An origin is activated by the sequence of events summarized in **FIGURE 14.3**, in which binding of DnaA-ATP is succeeded by association with the other proteins.

The four 9 bp consensus sequences on the right side of *oriC* provide the initial binding sites for DnaA-ATP. It binds cooperatively to form a central core around which *oriC* DNA is wrapped. DnaA then acts at three A-T-rich 13 bp tandem repeats located on the left side of *oriC*. In its active form, DnaA-ATP twists open the DNA strands at each of these sites to form an open bubble complex. All three 13 bp repeats must be opened for the reaction to proceed to the next stage. Transcription of either gene flanking *oriC* provides additional torsional stress to help snap apart the double-stranded DNA.

Altogether, two to four monomers of DnaA bind at the origin, and they recruit two "prepriming" complexes of the DnaB helicase bound to DnaC, so that there is one DnaB–DnaC complex for each of the two (bidirectional) replication forks. The only function of DnaC is that of a chaperone to repress the helicase activity of DnaB until it is needed. Each DnaB–DnaC complex consists of six DnaC monomers bound to a hexamer of DnaB. Note that the DnaB helicase cannot open double-stranded DNA; it can only unwind DNA that has already been opened, in this case by DnaA.

The prepriming complex generates a protein aggregate of 480 kD, which corresponds to a sphere of radius 6 nm. The formation of a complex at *oriC* is detectable in the form of the large protein blob visualized in Figure 14.3. When replication begins, a replication bubble becomes visible next to the blob. The region of strand separation in the open complex is large enough for both DnaB hexamers to bind, which initiates the two replication forks. As DnaB binds, it displaces DnaA from the 13 bp repeats and extends the length of the open region using its helicase activity. It then uses its helicase activity to extend the region of unwinding. Each DnaB activates a DnaG primase—in one case to initiate the leading strand, and in the other to initiate the first Okazaki fragment of the lagging strand.

Some additional proteins are required to support the unwinding reaction. **Gyrase**, a type II topoisomerase, provides a swivel that allows one DNA strand to rotate around the other. Without this reaction, unwinding would generate torsional strain (overwinding) in the DNA that would resist unwinding by the helicase. The protein **SSB** (**single-strand binding protein**) stabilizes the single-stranded DNA as it is formed and modulates the helicase activity. The length of duplex DNA that usually is unwound to initiate replication is probably <60 bp. The protein HU is a general DNA-binding protein in *E. coli*. Its presence is not absolutely required to initiate replication *in vitro*, but it stimulates the reaction. HU has the capacity to bend DNA, and is involved in building the structure that leads to formation of the open complex.

Input of energy in the form of ATP is required at several stages for the prepriming reaction, and it is required for unwinding DNA.

The helicase action of DnaB depends on ATP hydrolysis, and the swivel action of gyrase requires ATP hydrolysis. ATP also is needed for the action of primase and to load the β subunit of Pol III in order to initiate DNA synthesis.

Once the prepriming complex is loaded onto the replication forks, the next step is the recruitment of the **primase**, DnaG, which is then loaded onto the DnaB hexamer. This entails release of DnaC, which allows the DnaB helicase to become active. DnaC hydrolyzes ATP in order to release DnaB. This step marks the transition from initiation to elongation.

14.3 DNA Polymerases Are the Enzymes That Make DNA

Key concepts

- DNA is synthesized in both semiconservative replication and repair reactions.
- A bacterium or eukaryotic cell has several different DNA polymerase enzymes.
- One bacterial DNA polymerase undertakes semiconservative replication; the others are involved in repair reactions.

There are two basic types of DNA synthesis.

FIGURE 14.4 shows the result of **semiconservative replication**. The two strands of the parental duplex are separated, and each serves as a template for synthesis of a new strand. The parental duplex is replaced with two daughter duplexes, each of which has one parental strand and one newly synthesized strand.

FIGURE 14.5 shows the consequences of a **DNA repair** reaction. One strand of DNA has been damaged. It is excised and new material is synthesized to replace it. An enzyme that can synthesize a new DNA strand on a template strand is called a **DNA polymerase** (or more

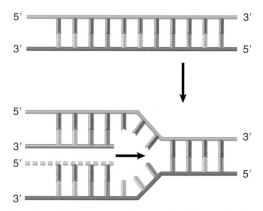

FIGURE 14.4 Semiconservative replication synthesizes two new strands of DNA.

properly, DNA-dependent DNA polymerase). Both prokaryotic and eukaryotic cells contain multiple DNA polymerase activities. Only a few of these enzymes actually undertake replication; those that do sometimes are called **DNA replicases**. The remaining enzymes are involved in repair synthesis or participate in subsidiary roles in replication.

All prokaryotic and eukaryotic DNA polymerases share the same fundamental type of synthetic activity, synthesis from 5′ to 3′ from a template that is 3′ to 5′. This means adding nucleotides one at a time to a 3′–OH end, as illustrated diagrammatically in **FIGURE 14.6**. The choice of the nucleotide to add to the chain is dictated by base pairing with the template strand.

Some DNA polymerases, such as the repair polymerases, function as independent enzymes, whereas others (notably the replication polymerases) are incorporated into large protein assemblies called **holoenzymes**. The DNA-synthesizing subunit is only one of several functions of the holoenzyme, which typically contains other activities concerned with fidelity.

FIGURE 14.7 summarizes the DNA polymerases that have been characterized in *E. coli*. DNA polymerase III, a multisubunit protein, is the replication polymerase responsible for *de novo* synthesis of new strands of DNA. DNA polymerase I (coded by *polA*) is involved in the repair of damaged DNA and, in a subsidiary role, in semiconservative replication. DNA polymerase II is required to restart a replication fork when its progress is blocked by damage in DNA. DNA polymerases IV and V are involved in allowing replication to bypass certain types of damage and are called **error-prone polymerases**.

When extracts of *E. coli* are assayed for their ability to synthesize DNA, the predominant enzyme activity is DNA polymerase I. Its activity is so great that it makes it impossible to detect the activities of the enzymes actually responsible for DNA replication! To develop *in vitro* systems in which replication can be followed, extracts are therefore prepared from *polA* mutant cells.

Several classes of eukaryotic DNA polymerases have been identified. DNA polymerases δ and ε are required for nuclear replication; DNA polymerase α is concerned with "priming" (initiating) replication. Other DNA polymerases are involved in repairing damaged nuclear DNA, or in translesion replication

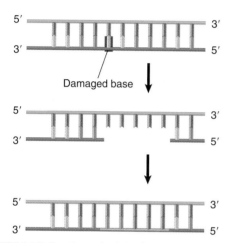

FIGURE 14.5 Repair synthesis replaces a short stretch of one strand of DNA containing a damaged base.

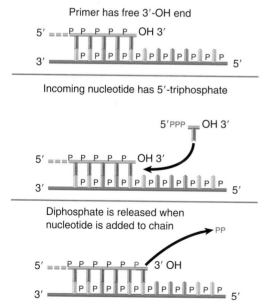

Primer has free 3′-OH end

Incoming nucleotide has 5′-triphosphate

Diphosphate is released when nucleotide is added to chain

FIGURE 14.6 DNA is synthesized by adding nucleotides to the 3′–OH end of the growing chain, so that the new chain grows in the 5′ → 3′ direction. The precursor for DNA synthesis is a nucleoside triphosphate, which loses the terminal two phosphate groups in the reaction.

Enzyme	Gene	Function
I	*polA*	major repair enzyme
II	*polB*	replication restart
III	*polC*	replicase
IV	*dinB*	translesion replication
V	*umuD′$_2$C*	translesion replication

FIGURE 14.7 Only one DNA polymerase is the replication enzyme. The others participate in repair of damaged DNA, restarting stalled replication forks, or bypassing damage in DNA.

of damaged DNA when repair of damage is impossible. Mitochondrial DNA replication is carried out by DNA polymerase γ (see *Section 14.14, Separate Eukaryotic DNA Polymerases Undertake Initiation and Elongation*).

14.4 DNA Polymerases Have Various Nuclease Activities

Key concept

- DNA polymerase I has a unique 5'–3' exonuclease activity that can be combined with DNA synthesis to perform nick translation.

Replicases often have nuclease activities as well as the ability to synthesize DNA. A 3'–5' exonuclease activity is typically used to excise bases that have been added to DNA incorrectly. This provides a "proofreading" error-control system (see *Section 14.5, DNA Polymerases Control the Fidelity of Replication*).

The first DNA-synthesizing enzyme to be characterized was DNA polymerase I, which is a single polypeptide of 103 kD. The chain can be cleaved into two parts by proteolytic treatment. The larger cleavage product (68 kD) is called the Klenow fragment. It is used in synthetic reactions *in vitro*. It contains the polymerase and the proofreading 3'–5' exonuclease activities. The active sites are ~30 Å apart in the protein, which indicates that there is spatial separation between adding a base and removing one.

The small fragment (35 kD) possesses a 5'–3' exonucleolytic activity, which excises small groups of nucleotides, up to ~10 bases at a time. This activity is coordinated with the synthetic/proofreading activity. It provides DNA polymerase I with a unique ability to start replication *in vitro* at a nick in DNA. (No other DNA polymerase has this ability.) At a point where a phosphodiester bond has been broken in a double-stranded DNA, the enzyme extends the 3'–OH end. As the new segment of DNA is synthesized, it displaces the existing homologous strand in the duplex. The displaced strand is degraded by the 5'–3' exonucleolytic activity of the enzyme.

This process of **nick translation** is illustrated in **FIGURE 14.8**. The displaced strand is degraded by the 5'–3' exonuclease activity of the enzyme. The properties of the DNA are unaltered, except that a segment of one strand has been replaced with newly synthesized material, and the position of the nick has

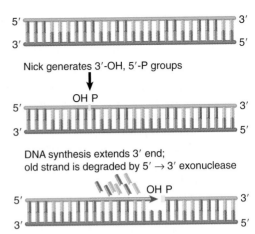

FIGURE 14.8 Nick translation replaces part of a preexisting strand of duplex DNA with newly synthesized material.

been moved along the duplex. This is of great practical use; nick translation has been a major technique for introducing radioactively labeled nucleotides into DNA *in vitro*.

The coupled 5'–3' synthetic/3'–5' exonucleolytic action is used most extensively for filling in short single-stranded regions in double-stranded DNA. These regions arise during lagging strand DNA replication (see *Section 14.6, DNA Polymerases Have a Common Structure*), and during DNA repair (see Figure 14.5).

14.5 DNA Polymerases Control the Fidelity of Replication

Key concepts

- High-fidelity DNA polymerases involved in replication have a precisely constrained active site that favors binding of Watson–Crick base pairs.
- DNA polymerases often have a 3'–5' exonuclease activity that is used to excise incorrectly paired bases.
- The fidelity of replication is improved by proofreading by a factor of ~100.

The fidelity of replication poses the same sort of problem we have encountered already in considering (for example) the accuracy of translation. It relies on the specificity of base pairing. Yet when we consider the energetics involved in base pairing, we would expect errors to occur with a frequency of ~10^{-2} per base pair replicated. The actual rate in bacteria seems to be ~10^{-8} to 10^{-10}. This corresponds to ~1 error per genome per 1000 bacterial replication cycles, or ~10^{-6} per gene per generation.

We can divide the errors that DNA polymerase makes during replication into two classes:

- *Substitutions* occur when the wrong (improperly paired) nucleotide is incorporated. The error level is determined by the efficiency of **proofreading**, in which the enzyme scrutinizes the newly formed base pair and removes the nucleotide if it is mispaired.
- *Frameshifts* occur when an extra nucleotide is inserted or omitted. Fidelity with regard to frameshifts is affected by the **processivity** of the enzyme: the tendency to remain on a single template rather than to dissociate and reassociate. This is particularly important for the replication of a homopolymeric stretch—for example, a long sequence of $dT_n:dA_n$, in which "replication slippage" can change the length of the homopolymeric run. As a general rule, increased processivity reduces the likelihood of such events. In multimeric DNA polymerases, processivity is usually increased by a particular subunit that is not needed for catalytic activity *per se*.

Bacterial replication enzymes have multiple error reduction systems. As discussed in Chapter 1 (*Genes Are DNA*), the geometry of an A-T base pair is very similar to that of a G-C base pair. This geometry is used by high-fidelity DNA polymerases as a fidelity mechanism. Only an incoming dNTP that base pairs properly with the template nucleotide fits in the active site, whereas mispairs such as A-C or A-A have the wrong geometry to fit into the active site. On the other hand, low-fidelity DNA polymerases, such as *E. coli* DNA polymerase IV used for damage bypass replication, have a more open active site that accommodates damaged nucleotides, but also mispairs. Thus either the expression or activity of these error-prone DNA polymerases is tightly regulated so that they are only active after DNA damage.

All of the bacterial enzymes possess a 3'–5' exonucleolytic activity that proceeds in the reverse direction from DNA synthesis. This provides a proofreading function illustrated diagrammatically in **FIGURE 14.9**. In the chain elongation step, a precursor nucleotide enters the position at the end of the growing chain. A bond is formed. The enzyme moves one base pair farther, and then is ready for the next precursor nucleotide to enter. If a mistake has been made, the DNA is structurally warped by the incorpo-

Enzyme adds base to growing strand

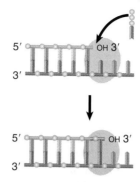

Enzyme moves on if new base is correct

Base is hydrolyzed and expelled if incorrect

FIGURE 14.9 DNA polymerases scrutinize the base pair at the end of the growing chain and excise the nucleotide added in the case of a misfit.

ration of the incorrect base that will cause the polymerase to pause or slow down. This will allow the enzyme to back up and remove the incorrect base (see *Section 14.6, DNA Polymerases Have a Common Structure*). In some regions errors occur more frequently than in others; that is, **mutation hotspots** occur in the DNA. This is caused by the underlying sequence context; that is, some sequences cause the polymerase to move faster or slower, which affects the ability to catch an error.

As noted in *Section 14.3, DNA Polymerases Are the Enzymes That Make DNA*, replication enzymes typically are found as multisubunit holoenzyme complexes, whereas repair DNA polymerases are typically found as single subunit enzymes. An advantage to a holoenzyme system is the availability of a specialized subunit responsible for error correction. In *E. coli* DNA polymerase III, this activity, a 3' to 5' exonuclease, resides in a separate subunit, the ε subunit. This subunit gives the replication enzyme a greater fidelity than the repair enzymes.

Different DNA polymerases handle the relationship between the polymerizing and proofreading activities in different ways. In some cases, the activities are part of the same protein subunit, but in others they are contained in different subunits. Each DNA polymerase has a characteristic error rate that is reduced by its proofreading activity. Proofreading typically decreases the error rate in replication from ~10^{-5} to ~10^{-7} per base pair replicated. Systems that recognize errors and correct them following replication then eliminate some of the errors, bringing the overall rate to <10^{-9} per base pair replicated (see *Section 16.7, Controlling the Direction of Mismatch Repair*).

The replicase activity of DNA polymerase III was originally discovered by a conditional lethal mutation in the *dnaE* locus, which codes for the 130 kD α subunit that possesses the DNA synthetic activity. The 3′–5′ exonucleolytic proofreading activity is found in another subunit, ε, coded by the *dnaQ* gene. The basic role of the ε subunit in controlling the fidelity of replication *in vivo* is demonstrated by the effect of mutations in *dnaQ*: The frequency with which mutations occur in the bacterial strain is increased by >10^3-fold.

14.6 DNA Polymerases Have a Common Structure

Key concepts

- Many DNA polymerases have a large cleft composed of three domains that resemble a hand.
- DNA lies across the "palm" in a groove created by the "fingers" and "thumb."

The first DNA polymerase for which the structure was determined was the Klenow fragment of the *E. coli* DNA polymerase I. From that data, **FIGURE 14.10** shows the common structural features that all DNA polymerases share. The enzyme structure can be divided into several independent domains, which are described by analogy with a human right hand. DNA binds in a large cleft composed of three domains. The "palm" domain has important conserved sequence motifs that provide the catalytic active site. The "fingers" are involved in positioning the template correctly at the active site. The "thumb" binds the DNA as it exits the enzyme, and is important in processivity. The most important conserved regions of each of these three domains converge to form a continuous surface at the catalytic site. The exonuclease activity resides in an independent domain with its own catalytic site. The N-terminal domain extends into the nuclease domain. DNA polymerases fall into five families based on sequence homologies; the palm is well conserved among

them, but the thumb and fingers provide analogous secondary structure elements from different sequences.

The catalytic reaction in a DNA polymerase occurs at an active site in which a nucleotide triphosphate pairs with an (unpaired) single strand of DNA. The DNA lies across the palm in a groove that is created by the thumb and fingers. **FIGURE 14.11** shows the crystal structure of the T7 enzyme complexed with DNA (in the form of a primer annealed to a template strand) and an incoming nucleotide that is about to be

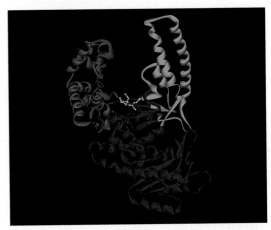

FIGURE 14.10 The structure of the Klenow fragment from *E. coli* DNA polymerase I. It has a right hand with fingers (blue), a palm (red), and a thumb (green). The Klenow fragment also includes an exonuclease domain. Structure from Protein Data Bank 1KFD. L. S. Beese, J. M. Friedman, and T. A. Steitz, *Biochemistry* 32 (1993): 14095–14101.

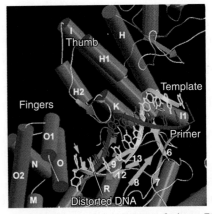

FIGURE 14.11 The crystal structure of phage T7 DNA polymerase shows that the template strand takes a sharp turn that exposes it to the incoming nucleotide. Photo courtesy of Charles Richardson and Thomas Ellenberger, Washington University School of Medicine.

added to the primer. The DNA is in the classic B-form duplex up to the last two base pairs at the 3' end of the primer, which are in the more open A-form. A sharp turn in the DNA exposes the template base to the incoming nucleotide. The 3' end of the primer (to which bases are added) is anchored by the fingers and palm. The DNA is held in position by contacts that are made principally with the phosphodiester backbone (thus enabling the polymerase to function with DNA of any sequence).

In structures of DNA polymerases of this family complexed only with DNA (that is, lacking the incoming nucleotide), the orientation of the fingers and thumb relative to the palm is more open, with the O helix (O, O1, O2; see Figure 14.11) rotated away from the palm. This suggests that an inward rotation of the O helix occurs to grasp the incoming nucleotide and create the active catalytic site. When a nucleotide binds, the fingers domain rotates 60° toward the palm, with the tops of the fingers moving by 30 Å. The thumb domain also rotates toward the palm by 8°. These changes are cyclical: they are reversed when the nucleotide is incorporated into the DNA chain, which then translocates through the enzyme to recreate an empty site.

The exonuclease activity is responsible for removing mispaired bases. The catalytic site of the exonuclease domain is distant from the active site of the catalytic domain, though. The enzyme alternates between polymerizing and editing modes, as determined by a competition between the two active sites for the 3' primer end of the DNA. Amino acids in the active site contact the incoming base in such a way that the enzyme structure is affected by the structure of a mismatched base. When a mismatched base pair occupies the catalytic site, the fingers cannot rotate toward the palm to bind the incoming nucleotide. This leaves the 3' end free to bind to the active site in the exonuclease domain, which is accomplished by a rotation of the DNA in the enzyme structure.

14.7 The Two New DNA Strands Have Different Modes of Synthesis

Key concept

- The DNA polymerase advances continuously when it synthesizes the leading strand (5'–3'), but synthesizes the lagging strand by making short fragments that are subsequently joined together.

The antiparallel structure of the two strands of duplex DNA poses a problem for replication. As the replication fork advances, daughter strands must be synthesized on both of the exposed parental single strands. The fork template strand moves in the direction from 5'–3' on one strand and in the direction from 3'–5' on the other strand. Yet DNA is synthesized only from a 5' end toward a 3' end (by adding a new nucleotide to the growing 3' end) on a template that is 3' to 5'. The problem is solved by synthesizing the new strand on the 5' to 3' template in a series of short fragments, each actually synthesized in the "backward" direction; that is, with the customary 5'–3' polarity.

Consider the region immediately behind the replication fork, as illustrated in **FIGURE 14.12**. We describe events in terms of the different properties of each of the newly synthesized strands:

- On the **leading strand** (sometimes called the *forward strand*) DNA synthesis can proceed continuously in the 5' to 3' direction as the parental duplex is unwound.
- On the **lagging strand** a stretch of single-stranded parental DNA must be exposed, and then a segment is synthesized in the reverse direction (relative to fork movement). A series of these fragments are synthesized, each 5'–3'; they then are joined together to create an intact lagging strand.

Discontinuous replication can be followed by the fate of a very brief label of radioactivity. The label enters newly synthesized DNA in the form of short fragments of ~1000 to 2000 bases in length. These **Okazaki fragments** are found in replicating DNA in both prokaryotes and eukaryotes. After longer periods of incubation, the label enters larger segments of DNA. The transition results from covalent linkages between Okazaki fragments.

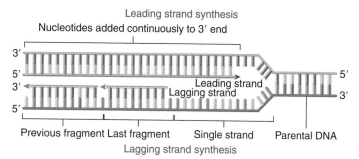

FIGURE 14.12 The leading strand is synthesized continuously, whereas the lagging strand is synthesized discontinuously.

The lagging strand *must* be synthesized in the form of Okazaki fragments. For a long time it was unclear whether the leading strand is synthesized in the same way or is synthesized continuously. All newly synthesized DNA is found as short fragments in *E. coli*. Superficially, this suggests that both strands are synthesized discontinuously. It turns out, however, that not all of the fragment population represents *bona fide* Okazaki fragments; some are pseudofragments that have been generated by breakage in a DNA strand that actually was synthesized as a continuous chain. The source of this breakage is the incorporation of some uracil into DNA in place of thymine. When the uracil is removed by a repair system, the leading strand has breaks until a thymine is inserted.

Thus the lagging strand is synthesized discontinuously and the leading strand is synthesized continuously. This is called **semidiscontinuous replication**.

14.8 Replication Requires a Helicase and a Single-Strand Binding Protein

Key concepts

- Replication requires a helicase to separate the strands of DNA using energy provided by hydrolysis of ATP.
- A single-stranded binding protein is required to maintain the separated strands.

As the replication fork advances, it unwinds the duplex DNA. One of the template strands is rapidly converted to duplex DNA as the leading daughter strand is synthesized. The other remains single stranded until a sufficient length has been exposed to initiate synthesis of an Okazaki fragment complementary to the lagging strand in the backward direction. The generation and maintenance of single-stranded DNA is therefore a crucial aspect of replication. Two types of function are needed to convert double-stranded DNA to the single-stranded state:

- A *helicase* is an enzyme that separates (or melts) the strands of DNA, usually using the hydrolysis of ATP to provide the necessary energy.
- A *single-strand binding protein (SSB)* binds to the single-stranded DNA, protecting it and preventing it from reforming the duplex state. The SSB binds typically in a cooperative manner in which the binding of additional monomers to the

existing complex is enhanced. The *E. coli* SSB is a tetramer; eukaryotic SSB (also known as RPA) is a trimer.

Helicases separate the strands of a duplex nucleic acid in a variety of situations, ranging from strand separation at the growing point of a replication fork to catalyzing migration of Holliday (recombination) junctions along DNA. There are twelve different helicases in *E. coli*. A helicase is generally multimeric. A common form of helicase is a hexamer. This typically translocates along DNA by using its multimeric structure to provide multiple DNA-binding sites.

FIGURE 14.13 shows a generalized schematic model for the action of a hexameric helicase. It is likely to have one conformation that binds to duplex DNA and another that binds to single-stranded DNA. Alternation between them drives the motor that melts the duplex and requires ATP hydrolysis—typically 1 ATP is hydrolyzed for each base pair that is unwound. A helicase usually initiates unwinding at a single-stranded region adjacent to a duplex. It may function with a particular polarity, preferring single-stranded DNA with a 3' end (3'–5' helicase) or with a 5' end (5'–3' helicase). A 5'–3' helicase is shown in Figure 14.13. Hexameric helicases typically encircle the DNA, which allows them to unwind DNA processively for many kilobases. This property makes them ideally suited as replicative DNA helicases.

Unwinding of double-stranded DNA by a helicase generates two single strands that are bound by SSB. *E. coli* SSB is a tetramer of 74 kD that binds single-stranded DNA cooperatively. The significance of the cooperative mode of binding is that the binding of one protein molecule makes it much easier for another to bind.

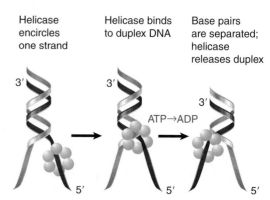

Helicase encircles one strand | Helicase binds to duplex DNA | Base pairs are separated; helicase releases duplex

ATP→ADP

FIGURE 14.13 A hexameric helicase moves along one strand of DNA. It probably changes conformation when it binds to the duplex, uses ATP hydrolysis to separate the strands, and then returns to the conformation it has when bound only to a single strand.

Thus once the binding reaction has started on a particular DNA molecule, it is rapidly extended until all of the single-stranded DNA is covered with the SSB protein. Note that this protein is not a DNA-unwinding protein; its function is to stabilize DNA that is already in the single-stranded condition.

Under normal circumstances *in vivo*, the unwinding, coating, and replication reactions proceed in tandem. The SSB binds to DNA as the replication fork advances, keeping the two parental strands separate so that they are in the appropriate condition to act as templates. SSB is needed in stoichiometric amounts at the replication fork. It is required for more than one stage of replication; *ssb* mutants have a quick-stop phenotype, and are defective in repair and recombination as well as in replication.

14.9 Priming Is Required to Start DNA Synthesis

Key concepts

- All DNA polymerases require a 3′–OH priming end to initiate DNA synthesis.
- The priming end can be provided by an RNA primer, a nick in DNA, or a priming protein.
- For DNA replication, a special RNA polymerase called a primase synthesizes an RNA chain that provides the priming end.
- *E. coli* has two types of priming reaction, which occur at the bacterial origin (*oriC*) and the φX174 origin.
- Priming of replication on double-stranded DNA always requires a replicase, SSB, and primase.
- DnaB is the helicase that unwinds DNA for replication in *E. coli*.

A common feature of all DNA polymerases is that they cannot initiate synthesis of a chain of DNA *de novo*, but can only elongate a chain. **FIGURE 14.14** shows the features required for initiation. Synthesis of the new strand can only start from a preexisting 3′–OH end, and the template strand must be converted to a single-stranded condition.

The 3′–OH end is called a **primer**. The primer can take various forms. Types of priming reaction are summarized in **FIGURE 14.15**.

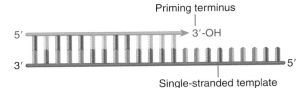

FIGURE 14.14 A DNA polymerase requires a 3′–OH end to initiate replication.

- A sequence of RNA is synthesized on the template, so that the free 3′–OH end of the RNA chain is extended by the DNA polymerase. This is commonly used in replication of cellular DNA and by some viruses.
- A preformed RNA (often a tRNA) pairs with the template, allowing its 3′–OH end to be used to prime DNA synthesis. This mechanism is used by retroviruses to prime reverse transcription of RNA (see Figure 17.28 in *Section 17.13, Viral RNA Is Generated by Reverse Transcription*).
- A primer terminus is generated within duplex DNA. The most common

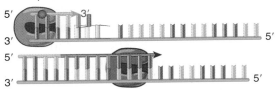

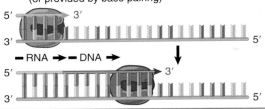

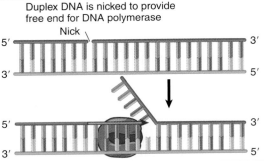

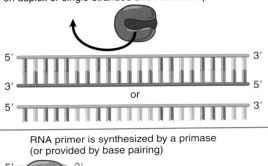

FIGURE 14.15 There are several methods for providing the free 3′–OH end that DNA polymerases require to initiate DNA synthesis.

mechanism is the introduction of a nick, as used to initiate rolling circle replication. In this case, the preexisting strand is displaced by new synthesis.

- A protein primes the reaction directly by presenting a nucleotide to the DNA polymerase. This reaction is used by certain viruses (see Figure 12.5 in *Section 12.3, Terminal Proteins Enable Initiation at the Ends of Viral DNAs*).

Priming activity is required to provide 3′–OH ends to start off the DNA chains on both the leading and lagging strands. The leading strand requires only one such initiation event, which occurs at the origin. There must be a series of initiation events on the lagging strand, though, because each Okazaki fragment requires its own start *de novo*. Each Okazaki fragment starts with a primer sequence of RNA ~10 bases long that provides the 3′–OH end for extension by DNA polymerase.

A primase is required to catalyze the actual priming reaction. This is provided by a special RNA polymerase activity, the product of the *dnaG* gene. The enzyme is a single polypeptide of 60 kD (much smaller than RNA polymerase). The primase is an RNA polymerase that is used only under specific circumstances; that is, to synthesize short stretches of RNA that are used as primers for DNA synthesis. DnaG primase associates transiently with the replication complex, and typically synthesizes a ~10-base primer. Primers start with the sequence pppAG positioned opposite the sequence 3′–GTC-5′ in the template.

There are two types of priming reaction in *E. coli*:

- The *oriC* system, named for the bacterial origin, basically involves the association of the DnaG primase with the protein complex at the replication fork.
- The φX system, named originally for phage φX174, requires an initiation complex consisting of additional components, called the primosome. This system is used when damage causes the replication fork to collapse and it must be restarted (see *Section 14.16, Lesion Bypass Requires Polymerase Replacement*).

At times replicons are referred to as being of the φX or *oriC* type. The types of activities involved in the initiation reaction are summarized in **FIGURE 14.16**. Although other replicons in *E. coli* may have alternatives for some of these particular proteins, the same general types of activity are required in every case. A helicase

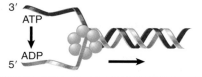

FIGURE 14.16 Initiation requires several enzymatic activities, including helicases, single-strand binding proteins, and synthesis of the primer.

is required to generate single strands, a single-strand binding protein is required to maintain the single-stranded state, and the primase synthesizes the RNA primer.

DnaB is the central component in both φX and *oriC* replicons. It provides the 5′–3′ helicase activity that unwinds DNA. Energy for the reaction is provided by cleavage of ATP. Basically, DnaB is the active component required to advance the replication fork. In *oriC* replicons, DnaB is initially loaded at the origin as part of a large complex (see *Section 14.2, Initiation: Creating the Replication Forks at the Origin oriC*). It forms the growing point at which the DNA strands are separated as the replication fork advances. It is part of the DNA polymerase complex and interacts with the DnaG primase to initiate synthesis of each Okazaki fragment on the lagging strand.

14.10 Coordinating Synthesis of the Lagging and Leading Strands

Key concepts

- Different enzyme units are required to synthesize the leading and lagging strands.
- In *E. coli*, both these units contain the same catalytic subunit (DnaE).
- In other organisms, different catalytic subunits may be required for each strand.

Each new DNA strand, leading and lagging, is synthesized by an individual catalytic unit. **FIGURE 14.17** shows that the behavior of these two units is different because the new DNA strands are growing in opposite directions. One enzyme unit is moving in the same direction as the unwinding point of the replication fork and synthesizing the leading strand continuously. The other unit is moving "backward" relative to the DNA, along the exposed single strand. Only short segments of template are exposed at any one time. When synthesis of one Okazaki fragment is completed, synthesis of the next Okazaki fragment is required to start at a new location approximately in the vicinity of the growing point for the leading strand. This requires that DNA polymerase III on the lagging strand disengage from the template, move to a new location, and be reconnected to the template at a primer to start a new Okazaki fragment.

The term "enzyme unit" avoids the issue of whether the DNA polymerase that synthesizes the leading strand is the same type of enzyme as the DNA polymerase that synthesizes the lagging strand. In the case we know best, *E. coli*, there is only a single DNA polymerase catalytic subunit used in replication, the DnaE polypeptide. Some bacteria and eukaryotes have multiple replication DNA polymerases (see *Section 14.14, Separate Eukaryotic DNA Polymerases Undertake Initiation and Elongation*). The active replicase is a dimer (see *Section 14.11, DNA Polymerase Holoenzyme Consists of Subcomplexes*), and each half of the dimer contains DnaE as the catalytic subunit. DnaE is supported by other proteins (which differ between the leading and lagging strands).

The use of a single type of catalytic subunit, however, may be atypical. In the bacterium *Bacillus subtilis*, there are two different catalytic subunits. PolC is the homolog to *E. coli*'s DnaE, and is responsible for synthesizing the leading strand. A related protein, DnaE$_{BS}$, is the catalytic subunit that synthesizes the lagging strand. Eukaryotic DNA polymerases have the same general structure, with different enzyme units synthesizing the leading and lagging strands. (see *Section 14.14, Separate Eukaryotic DNA Polymerases Undertake Initiation and Elongation*).

A major problem of the semidiscontinuous mode of replication follows from the use of different enzyme units to synthesize each new DNA strand: How is synthesis of the lagging strand coordinated with synthesis of the leading strand? As the replisome moves along DNA,

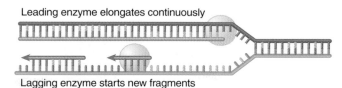

Leading enzyme elongates continuously

Lagging enzyme starts new fragments

FIGURE 14.17 A replication complex contains separate catalytic units for synthesizing the leading and lagging strands.

unwinding the parental strands, one enzyme unit elongates the leading strand. Periodically the primosome activity initiates an Okazaki fragment on the lagging strand, and the other enzyme unit must then move in the reverse direction to synthesize DNA. We will see in next sections how leading and lagging strand replication is coordinated by interactions between the leading and lagging strand enzyme units.

14.11 DNA Polymerase Holoenzyme Consists of Subcomplexes

Key concepts

- The *E. coli* replicase DNA polymerase III is a 900-kD complex with a dimeric structure.
- Each monomeric unit has a catalytic core, a dimerization subunit, and a processivity component.
- A clamp loader places the processivity subunits on DNA, where they form a circular clamp around the nucleic acid.
- One catalytic core is associated with each template strand.

We can now relate the subunit structure of *E. coli* DNA polymerase III to the activities required for DNA synthesis and propose a model for its action. The replisome consists of two DNA polymerase III holoenzyme complexes and associated proteins necessary for dimerization and function. The holoenzyme is a complex of 900 kD that contains ten proteins organized into four types of subcomplex:

- There are at least two copies of the catalytic core. Each catalytic core contains the α subunit (the DNA polymerase activity), the ε subunit (the 3′–5′ proofreading exonuclease), and the θ subunit (which stimulates the exonuclease).
- There are two copies of the dimerizing subunit, τ, which link the two catalytic cores together.

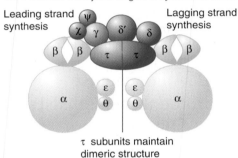

Clamp loader cleaves ATP to load clamp on DNA

Clamp loader

ATP → ADP + P

Clamp

Core enzyme joins

Core enzyme

tau + second core joins to give a symmetric dimer

Leading strand synthesis

Lagging strand synthesis

τ subunits maintain dimeric structure

FIGURE 14.18 DNA polymerase III holoenzyme assembles in stages, generating an enzyme complex that synthesizes the DNA of both new strands.

- There are two copies of the **clamp**, which is responsible for holding catalytic cores on to their template strands. Each clamp consists of a homodimer of β subunits, the β ring, which binds around the DNA and ensures processivity.
- The γ complex is a group of five proteins that comprise the **clamp loader**; the clamp loader places the clamp on DNA.

A model for the assembly of DNA polymerase III is shown in **FIGURE 14.18**. The holoenzyme assembles on DNA in three stages:

- First the clamp loader uses hydrolysis of ATP to bind β subunits to a template-primer complex.
- Binding to DNA changes the conformation of the site on β that binds to the clamp loader, and as a result it now has a high affinity for the core polymerase. This enables core polymerase to bind, and this is the means by which the core polymerase is brought to DNA.

- A τ dimer binds to the core polymerase, and provides a dimerization function that binds a second core polymerase (associated with another β clamp). The replisome is an asymmetric dimer because it has only one clamp loader. The clamp loader is responsible for adding a pair of β dimers to each parental strand of DNA.

Each of the core complexes of the holoenzyme synthesizes one of the new strands of DNA. The clamp loader is also needed for unloading the β complex from DNA; as a result, the two cores have different abilities to dissociate from DNA. This corresponds to the need to synthesize a continuous leading strand (where polymerase remains associated with the template) and a discontinuous lagging strand (where polymerase repetitively dissociates and reassociates). The clamp loader is associated with the core polymerase that synthesizes the lagging strand, and plays a key role in the ability to synthesize individual Okazaki fragments.

14.12 The Clamp Controls Association of Core Enzyme with DNA

Key concepts

- The core on the leading strand is processive because its clamp keeps it on the DNA.
- The clamp associated with the core on the lagging strand dissociates at the end of each Okazaki fragment and reassembles for the next fragment.
- The helicase DnaB is responsible for interacting with the primase DnaG to initiate each Okazaki fragment.

The β-ring dimer makes the holoenzyme highly *processive*. β is strongly bound to DNA, but can slide along a duplex molecule. The crystal structure of β shows that it forms a ring-shaped dimer. The model in **FIGURE 14.19** shows the β ring in relationship to a DNA double helix. The ring has an external diameter of 80 Å and an internal cavity of 35 Å, almost twice the diameter of the DNA double helix (20 Å). The space between the protein ring and the DNA is filled by water. Each of the β subunits has three globular domains with similar organization (although their sequences are different). As a result, the dimer has sixfold symmetry that is reflected in twelve α-helices that line the inside of the ring.

The β-ring dimer surrounds the duplex, providing the "sliding clamp" that allows the holoenzyme to slide along DNA. The structure

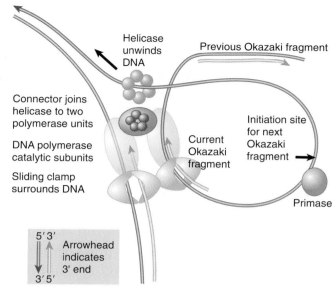

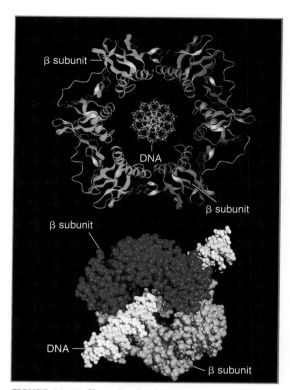

FIGURE 14.19 The subunit of DNA polymerase III holoenzyme consists of a head-to-tail dimer (the two subunits are shown in red and orange) that forms a ring completely surrounding a DNA duplex (shown in the center). Reprinted from *Cell*, vol. 69, X. P. Kong, et al., Three-dimensional structure of the β . . . , pp. 425–437. Copyright 1992, with permission from Elsevier [http://www.sciencedirect.com/science/journal/00928674]. Photo courtesy of John Kuriyan, University of California, Berkeley.

FIGURE 14.20 The helicase creating the replication fork is connected to two DNA polymerase catalytic subunits, each of which is held onto DNA by a sliding clamp. The polymerase that synthesizes the leading strand moves continuously. The polymerase that synthesizes the lagging strand dissociates at the end of an Okazaki fragment and then reassociates with a primer in the single-stranded template loop to synthesize the next fragment.

explains the high processivity—the enzyme can transiently dissociate, but cannot fall off and diffuse away. The α-helices on the inside have some positive charges that may interact with the DNA via the intermediate water molecules. The protein clamp does not directly contact the DNA, and as a result it may be able to "ice skate" along the DNA, making and breaking contacts via the water molecules.

How does the clamp get on to the DNA? The clamp is a circle of subunits surrounding DNA; thus its assembly or removal requires the use of an energy-dependent process by the clamp loader. The γ clamp loader is a pentameric circular structure that binds an open form of the β ring preparatory to loading it onto DNA. In effect, the ring is opened at one of the interfaces between the two β subunits by the δ subunit of the clamp loader. The clamp loader binds on top of a closed circular clamp, with its ATPase site juxtaposed to the clamp, and uses hydrolysis of ATP to provide the energy to open the ring of the clamp and insert DNA into the central cavity.

The relationship between the β clamp and the γ clamp loader is a paradigm for similar systems used by DNA polymerases ranging from bacteriophages to animal cells. The clamp is a heteromer (or possibly a dimer or trimer) that forms a ring around DNA with a set of twelve α-helices forming sixfold symmetry for the structure as a whole. The clamp loader has some subunits that hydrolyze ATP to provide energy for the reaction.

The basic principle that is established by the dimeric polymerase model is that, while one polymerase subunit synthesizes the leading strand continuously, the other cyclically initiates and terminates the Okazaki fragments of the lagging strand within a large, single-stranded loop formed by its template strand. **FIGURE 14.20** draws a generic model for the operation of such a replicase. The replication fork is created by a helicase—which typically forms a hexameric ring—that translocates in the 5′–3′ direction on the template for the lagging strand. The helicase is connected to two DNA polymerase catalytic subunits, each of which is associated with a sliding clamp.

We can describe this model for DNA polymerase III in terms of the individual components of the enzyme complex, as illustrated in **FIGURE 14.21**. A catalytic core is associated with each template strand of DNA. The holoenzyme moves continuously along the template for the

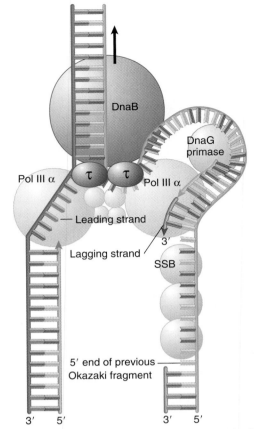

FIGURE 14.21 Each catalytic core of Pol III synthesizes a daughter strand. DnaB is responsible for forward movement at the replication fork.

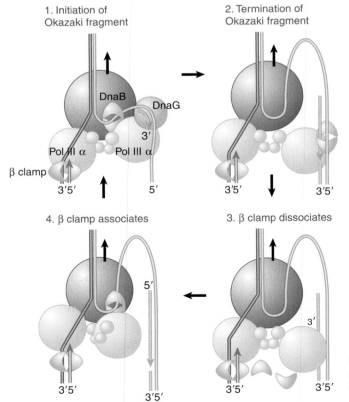

FIGURE 14.22 Core polymerase and the clamp dissociate at completion of Okazaki fragment synthesis and reassociate at the beginning.

leading strand; the template for the lagging strand is "pulled through," thus creating a loop in the DNA. DnaB creates the unwinding point, and translocates along the DNA in the "forward" direction.

DnaB contacts the τ subunit(s) of the clamp loader. This establishes a direct connection between the helicase–primase complex and the catalytic cores. This link has two effects. One is to increase the speed of DNA synthesis by increasing the rate of movement by DNA polymerase core by tenfold. The second is to prevent the leading strand polymerase from falling off; that is, to increase its processivity.

Synthesis of the leading strand creates a loop of single-stranded DNA that provides the template for lagging strand synthesis, and this loop becomes larger as the unwinding point advances. After initiation of an Okazaki fragment, the lagging strand core complex pulls the single-stranded template through the β clamp while synthesizing the new strand. The single-stranded template must extend for the length of at least one Okazaki fragment before the lagging polymerase completes one fragment and is ready to begin the next.

What happens when the Okazaki fragment is completed? All of the components of the replication apparatus function processively (that is, they remain associated with the DNA), except for the primase and the β clamp. **FIGURE 14.22** shows that the β clamp must be cracked open by the γ clamp loader when the synthesis of each fragment is completed, releasing the loop. We can think of the clamp loader here as a molecular wrench that is modulated by ATP. The clamp loader causes the β clamp to alter its conformation to an unstable configuration, which then springs open. A new β clamp is then recruited by the clamp loader to initiate the next Okazaki fragment. The lagging strand polymerase transfers from one β clamp to the next in each cycle, without dissociating from the replicating complex.

What is responsible for recognizing the sites for initiating synthesis of Okazaki fragments? In *oriC* replicons, the connection between priming and the replication fork is provided by the dual properties of DnaB: It is the helicase that propels the replication fork, and it interacts with the DnaG primase at an appropriate site. Following primer synthesis, the primase

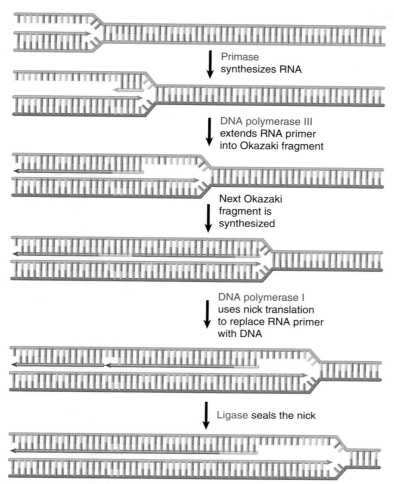

FIGURE 14.23 Synthesis of Okazaki fragments require priming, extension, removal of RNA primer, gap filling, and nick ligation.

is released. The length of the priming RNA is limited to eight to fourteen bases. Apparently DNA polymerase III is responsible for displacing the primase.

14.13 Okazaki Fragments Are Linked by Ligase

Key concepts

- Each Okazaki fragment starts with a primer and stops before the next fragment.
- DNA polymerase I removes the primer and replaces it with DNA.
- DNA ligase makes the bond that connects the 3′ end of one Okazaki fragment to the 5′ beginning of the next fragment.

We can now expand our view of the actions involved in joining Okazaki fragments, as illustrated in FIGURE 14.23. The complete order of events is uncertain, but it must involve synthesis of RNA primer, its extension with DNA, removal of the RNA primer, its replacement by a stretch of DNA, and the covalent linking of adjacent Okazaki fragments.

Synthesis of an Okazaki fragment terminates just before the start of the RNA primer of the preceding fragment. When the primer is removed, there will be a gap. The gap is filled by DNA polymerase I; polA mutants fail to join their Okazaki fragments properly. The 5′–3′ exonuclease activity removes the RNA primer while simultaneously replacing it with a DNA sequence extended from the 3′–OH end of the next Okazaki fragment. This is equivalent to nick translation, except that the new DNA replaces a stretch of RNA rather than a segment of DNA.

In mammalian systems (where the DNA polymerase does not have a 5′–3′ exonuclease activity), Okazaki fragments are connected by a two-step process. Synthesis of an Okazaki fragment displaces the RNA primer of the preceding fragment in the form of a "flap." FIGURE 14.24

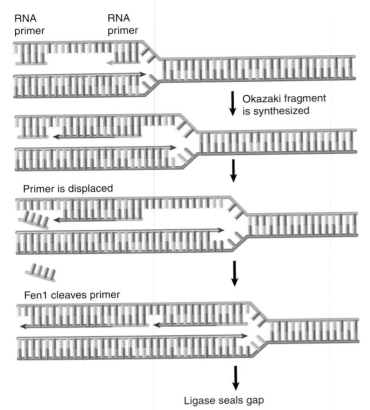

RNA primer   RNA primer

↓ Okazaki fragment is synthesized

Primer is displaced

Fen1 cleaves primer

↓ Ligase seals gap

FIGURE 14.24 FEN1 is an exo-/endonuclease that recognizes the structure created when one strand of DNA is displaced from a duplex as a "flap." In replication it cleaves at the base of the flap to remove the RNA primer.

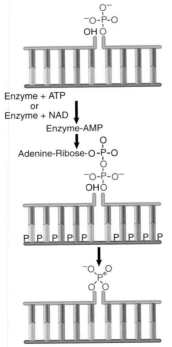

Enzyme + ATP
or
Enzyme + NAD

Enzyme-AMP

Adenine-Ribose-O-P-O

P P P P   P P P P

FIGURE 14.25 DNA ligase seals nicks between adjacent nucleotides by employing an enzyme-AMP intermediate.

shows that the base of the flap is cleaved by the enzyme FEN1 (*flap en*donuclease 1). In this reaction, FEN1 functions as an endonuclease, but it also has a 5′–3′ exonuclease activity. In DNA repair reactions, FEN1 may cleave next to a displaced nucleotide and then use its exonuclease activity to remove adjacent material.

Failure to remove a flap rapidly can have important consequences in regions of repeated sequences. Direct repeats can be displaced and misaligned with the template; palindromic sequences can form hairpins. These structures may change the number of repeats (see Figure 7.21). The general importance of FEN1 is that it prevents flaps of DNA from generating structures that may cause deletions or duplications in the genome.

Once the RNA has been removed and replaced, the adjacent Okazaki fragments must be linked together. The 3′–OH end of one fragment is adjacent to the 5′–phosphate end of the previous fragment. The enzyme **DNA ligase** makes a bond by using a complex with AMP. **FIGURE 14.25** shows that the AMP of the enzyme complex becomes attached to the 5′-phosphate of the nick and then a phosphodiester bond is formed with the 3′–OH terminus of the nick, releasing the enzyme and the AMP. Ligases are present in both prokaryotes and eukaryotes.

The *E. coli* and T4 ligases share the property of sealing nicks that have 3′–OH and 5′–phosphate termini, as illustrated in Figure 14.25. Both enzymes undertake a two-step reaction that involves an enzyme–AMP complex. (The *E. coli* and T4 enzymes use different cofactors. The *E. coli* enzyme uses NAD [*n*icotinamide *a*denine *d*inucleotide] as a cofactor, whereas the T4 enzyme uses ATP.) The AMP of the enzyme complex becomes attached to the 5′–phosphate of the nick, and then a phosphodiester bond is formed with the 3′–OH terminus of the nick, releasing the enzyme and the AMP.

14.14 Separate Eukaryotic DNA Polymerases Undertake Initiation and Elongation

Key concepts

• A replication fork has one complex of DNA polymerase α/primase, one complex of DNA polymerase δ, and one complex of DNA polymerase ε.

• The DNA polymerase α/primase complex initiates the synthesis of both DNA strands.

• DNA polymerase ε elongates the leading strand and a second DNA polymerase δ elongates the lagging strand.

DNA polymerase	Function	Structure
	High fidelity replicases	
α	Nuclear replication	350 kD tetramer
δ	Lagging strand	250 kD tetramer
ε	Leading strand	350 kD tetramer
γ	Mitochondrial replication	200 kD dimer
	High fidelity repair	
β	Base excision repair	39 kD monomer
	Low fidelity repair	
ζ	Base damage bypass	heteromer
η	Thymine dimer bypass	monomer
ι	Required in meiosis	monomer
κ	Deletion and base substitution	monomer

FIGURE 14.26 Eukaryotic cells have many DNA polymerases. The replication enzymes operate with high fidelity. Except for the β enzyme, the repair enzymes all have low fidelity. Replication enzymes have large structures, with separate subunits for different activities. Repair enzymes have much simpler structures.

Function	*E. coli*	Eukaryote	Phage T4
Helicase	DnaB	MCM complex	41
Loading helicase/primase	DnaC	cdc6	59
Single strand maintenance	SSB	RPA	32
Priming	DnaG	Polα/primase	61
Sliding clamp	β	PCNA	45
Clamp loading (ATPase)	γδ complex	RFC	44/62
Catalysis	*Pol III core*	Polδ + Pol ε	43
Holoenzyme dimerization	τ	?	43
RNA removal	*Pol I*	FEN1	43
Ligation	*Ligase*	Ligase 1	T4 ligase

FIGURE 14.27 Similar functions are required at all replication forks.

subunits has the responsibility for catalysis, and the others are concerned with ancillary functions, such as priming or processivity. These enzymes all replicate DNA with high fidelity, as does the slightly less complex mitochondrial enzyme. The repair polymerases have much simpler structures, which often consist of a single monomeric subunit (although it may function in the context of a complex of other repair enzymes). Of the enzymes involved in repair, DNA polymerase β has an intermediate fidelity; all of the others have much greater error rates and are called *error-prone polymerases*. All mitochondrial DNA replication and recombination is undertaken by DNA polymerase γ.

Each of the three nuclear DNA replication polymerases has a different function, as summarized in **FIGURE 14.27**.

- DNA polymerase α/primase initiates the synthesis of new strands.
- DNA polymerase ε then elongates the leading strand.
- DNA polymerase δ then elongates the lagging strand.

DNA polymerase α is unusual because it has the ability to initiate a new strand. It is used to initiate both the leading and lagging strands. The enzyme exists as a complex consisting of a 180-kD catalytic (DNA polymerase) subunit, which is associated with three other subunits: the B subunit that appears necessary for assembly, and two small subunits that provide the primase (RNA polymerase) activity. Reflecting its dual capacity to prime and extend chains, this complex is often called pol α/primase.

As shown in **FIGURE 14.28**, the pol α/primase enzyme binds to the initiation complex at the origin and synthesizes a short strand consisting of ~10 bases of RNA followed by 20 to 30 bases of DNA (sometimes called iDNA). It is then replaced by an enzyme that will extend

Eukaryotic replication is similar in most aspects to bacterial replication. It is semiconservative, bidirectional, and semidiscontinuous. As a result of the greater amount of DNA in a eukaryote, the genome has multiple replicons. Replication takes place during S phase of the cell cycle. Replicons in euchromatin initiate before replicons in heterochromatin; replicons near active genes initiate before replicons near inactive genes. Origins of replication in eukaryotes are not well defined, except for those in yeast (called ARS, _a_utonomously _r_eplicating _s_equences in *S. cerevisiae*). The number of replicons used in any one cycle is tightly controlled. During embryonic development more are activated than in slower growing adult cells.

Eukaryotes have a much larger number of DNA polymerases. They can be broadly divided into those required for replication and repair polymerases involved in repairing damaged DNA. Nuclear DNA replication requires DNA polymerases α, β, and ε. All the other nuclear DNA polymerases are concerned with synthesizing stretches of new DNA to replace damaged material or using damaged DNA as a template (see *Section 14.16, The Primosome Is Needed to Restart Replication*, for the error-prone DNA polymerases). **FIGURE 14.26** shows that most of the nuclear replicases are large heterotetrameric enzymes. In each case, one of the

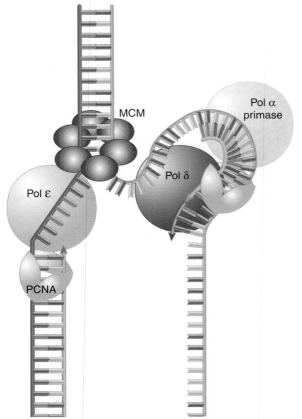

FIGURE 14.28 Three different DNA polymerases make up the eukaryotic replication fork. Pol α-primase is responsible for primer synthesis on the lagging strand. The MCM helicase (the eukaryotic homolog of DnaB) unwinds the dsDNA, while PCNA (homolog of β) endows the complex with processivity.

the chain. On the leading strand, this is DNA polymerase ε; on the lagging strand this is DNA polymerase δ. This event is called the *polymerase switch*. It involves interactions among several components of the initiation complex.

DNA polymerase ε is a highly processive enzyme that continuously synthesizes the leading strand. Its processivity results from its interaction with two other proteins, RF-C and PCNA (PCNA is called proliferating cell nuclear antigen for historical reasons).

The roles of RF-C and PCNA are analogous to the *E. coli* γ clamp loader and β processivity unit (see *Section 14.12, The Clamp Controls Association of Core Enzyme with DNA*). RF-C is a clamp loader that catalyzes the loading of PCNA on to DNA. It binds to the 3′ end of the DNA and uses ATP hydrolysis to open the ring of PCNA so that it can encircle the DNA. The processivity of DNA polymerase δ is maintained by PCNA, which tethers DNA polymerase δ to the template. The crystal structure of PCNA closely

resembles the *E. coli* β subunit: a trimer forms a ring that surrounds the DNA. The sequence and subunit organization are different from the dimeric β clamp; however, the function is likely to be similar.

DNA polymerase δ elongates the lagging strand. Like DNA polymerase ε on the leading strand, DNA polymerase δ forms a processive complex with the PCNA clamp. The exonuclease FEN1 removes the RNA primers of Okazaki fragments. The complex of DNA polymerase δ and FEN1 carries out the same type of nick translation that *E. coli* DNA polymerase I carries out during Okazaki fragment maturation (see Figure 14.24). The enzyme DNA ligase I is specifically required to seal the nicks between the completed Okazaki fragments. Currently, it is not known what factor takes on the function of the *E. coli* τ dimer that dimerizes the polymerase complexes in order to ensure coordinated DNA replication.

14.15 Phage T4 Provides Its Own Replication Apparatus

Key concept

- Phage T4 provides its own replication apparatus, which consists of DNA polymerase, the gene *32* SSB, a helicase, a primase, and accessory proteins that increase speed and processivity.

When phage T4 takes over an *E. coli* cell, it provides several functions of its own that either replace or augment the host functions. The phage places little reliance on expression of host functions. The degradation of host DNA is important in releasing nucleotides that are reused in the synthesis of phage DNA. (The phage DNA differs in base composition from cellular DNA in using hydroxymethylcytosine instead of the customary cytosine.)

The phage-coded functions concerned with DNA synthesis in the infected cell can be identified by mutations that impede the production of mature phages. Essential phage functions are identified by conditional lethal mutations, which fall into three phenotypic classes:

- Those in which there is no DNA synthesis at all identify genes whose products either are components of the replication apparatus or are involved in the provision of precursors (especially the hydroxymethylcytosine).

- Those in which the onset of DNA synthesis is delayed are concerned with the initiation of replication.
- Those in which DNA synthesis starts but then is arrested include regulatory functions, the DNA ligase, and some of the enzymes concerned with host DNA degradation.

There are also nonessential genes concerned with replication, including those involved in glucosylating the hydroxymethylcytosine in the DNA.

Synthesis of T4 DNA is catalyzed by a multienzyme aggregate assembled from the products of a small group of essential genes.

The gene 32 protein (gp32) is a highly cooperative single-strand binding protein, which is needed in stoichiometric amounts. It was the first example of its type to be characterized. The geometry of the T4 replication fork may specifically require the phage-coded protein, because the *E. coli* SSB cannot substitute. The gp32 forms a complex with the T4 DNA polymerase; this interaction could be important in constructing the replication fork.

The T4 system uses an RNA priming event that is similar to that of its host. With single-stranded T4 DNA as template, the gene *41* and *61* products act together to synthesize short primers. Their behavior is analogous to that of DnaB and DnaG in *E. coli*. The gene *41* protein is the counterpart to DnaB. It is a hexameric helicase that uses hydrolysis of GTP to provide the energy to unwind DNA. The p41/p61 complex moves processively in the 5'–3' direction in lagging strand synthesis, periodically initiating Okazaki fragments. Another protein, the product of gene *59*, loads the p41/p61 complex onto DNA; it is required to displace the p32 protein in order to allow the helicase to assemble on DNA.

The gene *61* protein is needed in much smaller amounts than most of the T4 replication proteins. There are as few as ten copies of gp61 per cell. (This impeded its characterization. It is required in such small amounts that originally it was missed as a necessary component, because enough was present as a contaminant of the gp32 preparation!) Gene *61* protein has the primase activity, which is analogous to DnaG of *E. coli*. The primase recognizes the template sequence 3'–TTG-5' and synthesizes pentaribonucleotide primers that have the general sequence pppApCpNpNpNp. If the complete replication apparatus is present, these primers are extended into DNA chains.

The gene *43* DNA polymerase has the usual 5'–3' synthetic activity, which is associated with a 3'–5' exonuclease proofreading activity. It catalyzes DNA synthesis and removes the primers. When T4 DNA polymerase uses a single-stranded DNA as template, its rate of progress is uneven. The enzyme moves rapidly through single-stranded regions, but proceeds much more slowly through regions that have a base-paired intrastrand secondary structure. The accessory proteins assist the DNA polymerase in passing these roadblocks and maintaining its speed.

The remaining three proteins are referred to as "polymerase accessory proteins." They increase the affinity of the DNA polymerase for the DNA, as well as increase its processivity and speed. The gene *45* product is a trimer that acts as a sliding clamp. The structure of the trimer is similar to that of the eukaryotic PCNA trimer or the *E. coli* β dimer, in that it forms a circle around DNA that holds the DNA polymerase subunit more tightly on the template.

The products of genes *44* and *62* form a tight complex that has ATPase activity. They are the equivalent of the γδ clamp loader complex, and their role is to load p45 onto DNA. Four molecules of ATP are hydrolyzed in loading the p45 clamp and the p43 DNA polymerase onto DNA.

The overall structure of the replisome is similar to that of *E. coli*. It consists of two coupled holoenzyme complexes, one synthesizing the leading strand and the other synthesizing the lagging strand. In this case, the dimerization involves a direct interaction between the p43 DNA polymerase subunits, and p32 plays a role in coordinating the actions of the two DNA polymerase units.

Thus far we have dealt with DNA replication solely in terms of the progression of the replication fork. The need for other functions is shown by the DNA-delay and DNA-arrest mutants. Three of the four genes of the DNA-delay mutants are *39*, *52*, and *60*, which code for the three subunits of T4 topoisomerase II, an activity needed for removing supercoils in the template. The essential role of this enzyme suggests that T4 DNA does not remain in a linear form, but rather becomes topologically constrained during some stage of replication. The topoisomerase could be needed to allow rotation of DNA ahead of the replication fork.

Comparison of the T4 apparatus with the *E. coli* apparatus suggests that DNA replication poses a set of problems that are solved in analogous ways in different systems. We may now compare the enzymatic and structural activities found at the replication fork in *E. coli*, T4, and eukaryotic cells. Figure 14.27 summarizes the functions and assigns them to individual proteins. We can interpret the known properties of replication complex proteins in terms of similar functions that involve the unwinding, priming, catalytic, and sealing reactions. The components of each system interact in restricted ways, as shown by the fact that phage T4 requires its own helicase, primase, clamp, and so on, and by the fact that bacterial proteins cannot substitute for their phage counterparts.

14.16 Lesion Bypass Requires Polymerase Replacement

Key concepts

- A replication fork stalls when it arrives at damaged DNA.
- The replication complex must be replaced by a specialized DNA polymerase for lesion bypass.
- After the damage has been repaired, the primosome is required to reinitiate replication by reinserting the replication complex.

Damage to chromosomes that is not repaired before replication can be catastrophic and lethal. When the replication complex encounters damaged and modified bases such that it cannot place a complementary base opposite it,

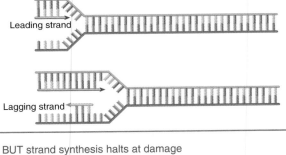

Replication fork advances on normal DNA

Leading strand

Lagging strand

BUT strand synthesis halts at damage

OR double strand break occurs at nick

FIGURE 14.29 The replication fork stalls and may collapse when it reaches a damaged base or a nick in DNA. Arrowheads indicate 3′ ends.

the polymerase stops and the replication fork collapses. A cell has two options to avoid death: recombination (see Chapter 15, *Homologous and Site-Specific Recombination*) or **lesion bypass**. Both bacteria and eukaryotes have multiple error-prone DNA polymerases that have the ability to synthesize past a lesion on the template (see Chapter 16, *Repair Systems*). These enzymes have this ability because they are not constrained to follow standard base pairing rules. Note that this DNA synthesis is not to repair the lesion, but simply to bypass it, to continue replication. That will allow the cell to return to the lesion to repair it.

FIGURE 14.29 compares an advancing replication fork with what happens when there is damage to a base in the DNA or a nick in one strand. In either case, DNA synthesis is halted, and the replication fork either is stalled or is disrupted and collapses. Replication-fork stalling appears to be quite common; estimates for the frequency in *E. coli* suggest that 18% to 50% of bacteria encounter a problem during a replication cycle. *E. coli* has two error-prone DNA polymerases that can replicate through a lesion, DNA polymerases IV and V (see *Section 16.6, Error-Prone Repair*), plus the repair DNA polymerase II, that are used for translesion synthesis. Eukaryotes have five error-prone DNA polymerases with different specificities.

There are two consequences when lesion bypass occurs. First, when the replication complex stalls at a lesion, the polymerase on the strand with the lesion must be removed from the template and replaced by an error-prone polymerase. Second, when the damage has been bypassed, the repair polymerase must be removed and the replication complex reinserted. When used for lesion bypass during replication, these error-prone DNA polymerases replace the replisome and are connected to the β clamp temporarily to allow the lesion bypass polymerase to insert nucleotides opposite the lesion. DNA polymerase III then replaces the error-prone polymerase. The consequences may be different, depending on whether the lesion has occurred on the lagging or leading strand. The replication polymerase on the lagging strand may be more easily replaced.

One model for how a stalled replication enzyme can be replaced, the "tool belt model," is shown in Figure 14.27. This model proposes that DNA pol IV can displace the stalled DNA pol III while it is still bound to the β clamp. This will allow DNA pol IV to access the 3′–OH of the primer and extend it.

Alternatively, the situation can be rescued by a recombination event that excises and replaces the damage or provides a new duplex to replace the region containing the double-strand break. The principle of the repair event is to use the built-in redundancy of information between the two DNA strands. FIGURE 14.30 shows the key events in such a repair event. Basically, information from the undamaged DNA daughter duplex is used to repair the damaged sequence. This creates a typical recombination junction that is resolved by the same systems that perform homologous recombination. In fact, one view is that the major importance of these systems for the cell is in repairing damaged DNA at stalled replication forks.

After the damage has been repaired, the replication fork must be restarted. FIGURE 14.31 shows that this may be accomplished by assembly of the **primosome**, which in effect reloads DnaB so that helicase action can continue. Early work on replication made extensive use of phage φX174, and led to the discovery of a complex system for priming. A primosome assembles at a unique phage site on its single-stranded DNA called the assembly site (*pas*). The *pas* is the equivalent of an origin for synthesis of the complementary strand of φX174. The

primosome consists of six proteins: PriA, PriB, PriC, DnaT, DnaB, and DnaC. Two alternative assembly pathways exist, one beginning with PriA and the other with PriC. This may reflect the many types of DNA damage that can occur.

On φX174 DNA, the primosome forms initially at the pas; primers are subsequently initiated at a variety of sites. PriA translocates along the DNA, displacing SSB, to reach additional sites at which priming occurs. As in the *E. coli oriC* replicon, DnaB plays a key role in unwinding and priming in φX174 replicons. The role of PriA is to load DnaB, which in turn recruits DnaG primase to prime DNA synthesis for the conversion of single-stranded viral DNA to the double-stranded DNA form.

It has always been puzzling that when replicating in *E. coli*, φX174 origins should use a complex structure that is not required to replicate the bacterial chromosome. Why does the bacterium provide this complex? The answer is provided by the fate of the stalled replication fork. The mechanism used at *oriC* is specific for origin DNA sequence and cannot be used to restart replication following lesion bypass because each lesion occurs in a different sequence. A separate mechanism employing structural rather than sequence recognition is used.

The proteins encoded by the *E. coli pri* genes form the core of the primosome. φX174 has

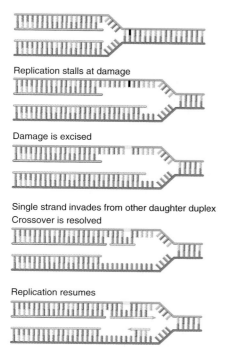

Replication stalls at damage

Damage is excised

Single strand invades from other daughter duplex
Crossover is resolved

Replication resumes

FIGURE 14.30 When replication halts at damaged DNA, the damaged sequence is excised and the complementary (newly synthesized) strand of the other daughter duplex crosses over to repair the gap. Replication can now resume, and the gaps are filled in.

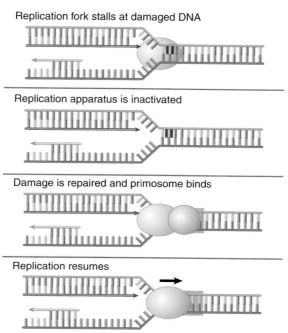

Replication fork stalls at damaged DNA

Replication apparatus is inactivated

Damage is repaired and primosome binds

Replication resumes

FIGURE 14.31 The primosome is required to restart a stalled replication fork after the DNA has been repaired.

simply co-opted the primosome for its own replication. The PriA DNA helicase binds first to the single strand region in cooperation with SSB. The key event in localizing the primosome is the ability of PriA to displace SSB from single-stranded DNA. PriA then recruits PriB and DnaT, which is then able to recruit the DnaB/C complex as described above (see *Section 14.2, Initiation: Creating the Replication Forks at the Origin oriC*). The alternate replisome loading system only requires PriC.

Replication fork reactivation is a common (and therefore important) reaction. It may be required in most chromosomal replication cycles. It is impeded by mutations in either the retrieval systems that replace the damaged DNA or in the components of the primosome.

14.17 Summary

The common mode of origin activation involves an initial limited melting of the double helix, followed by more general unwinding to create single strands. Several proteins act sequentially at the *E. coli* origin. Replication is initiated at *oriC* in *E. coli* when DnaA binds to a series of 9 bp repeats. This is followed by binding to a series of 13 bp repeats, where it uses hydrolysis of ATP to generate the energy to separate the DNA strands. The prepriming complex of DnaC–DnaB displaces DnaA. DnaC is released in a reaction that depends on ATP hydrolysis; DnaB is joined by the replicase enzyme, and replication is initiated by two forks that set out in opposite directions.

The availability of DnaA at the origin is an important component of the system that determines when replication cycles should initiate. Following initiation of replication, DnaA hydrolyzes its ATP under the stimulus of the β sliding clamp, thereby generating an inactive form of the protein.

Several sites that are methylated by the Dam methylase are present in the *E. coli* origin, including those of the 13-mer binding sites for DnaA. The origin remains hemimethylated and is in a sequestered state for ~10 minutes following initiation of a replication cycle. During this period it is associated with the membrane and reinitiation of replication is repressed.

DNA synthesis occurs by semidiscontinuous replication, in which the leading strand of DNA growing 5′–3′ is extended continuously, but the lagging strand that grows overall in the opposite 3′–5′ direction is made as short Okazaki fragments, each synthesized 5′–3′. The leading strand and each Okazaki fragment of the lagging strand initiate with an RNA primer that is extended by DNA polymerase. Bacteria and eukaryotes each possess more than one DNA polymerase activity. DNA polymerase III synthesizes both lagging and leading strands in *E. coli*. Many proteins are required for DNA polymerase III action and several constitute part of the replisome within which it functions.

The replisome contains an asymmetric dimer of DNA polymerase III; each new DNA strand is synthesized by a different core complex containing a catalytic (α) subunit. Processivity of the core complex is maintained by the β clamp, which forms a ring around DNA. The clamp is loaded on to DNA by the clamp loader complex. Clamp/clamp loader pairs with similar structural features are widely found in both prokaryotic and eukaryotic replication systems.

The looping model for the replication fork proposes that, as one half of the dimer advances to synthesize the leading strand, the other half of the dimer pulls DNA through as a single loop that provides the template for the lagging strand. The transition from completion of one Okazaki fragment to the start of the next requires the lagging strand catalytic subunit to dissociate from DNA and then reattach to a β clamp at the priming site for the next Okazaki fragment.

DnaB provides the helicase activity at a replication fork; this depends on ATP cleavage. DnaB may function by itself in oriC replicons to provide primosome activity by interacting periodically with DnaG, which provides the primase that synthesizes RNA.

Phage T4 codes for a replication apparatus consisting of seven proteins: DNA polymerase, helicase, single-strand binding protein, priming activities, and accessory proteins. Similar functions are required in other replication systems, including in eukaryotes. Different enzymes initiate and elongate the new strands of DNA. DNA polymerase α/primase primes the leading strands at origins and primes Okazaki fragments. DNA polymerase δ synthesizes Okazaki fragments and DNA polymerase ε synthesizes the leading strand.

The φX priming event also requires PriA, and DnaB, DnaC, and DnaT. The importance of the primosome for the bacterial cell is that it is used to restart replication at forks that stall when they encounter damaged DNA.

References

14.1   Introduction

Research

Hirota, Y., Ryter, A., and Jacob, F. (1968). Thermosensitive mutants of *E. coli* affected in the processes of DNA synthesis and cellular division. Cold Spring Harbor Symp. *Quant. Biol.* 33, 677–693.

14.2   Initiation: Creating the Replication Forks at the Origin *oriC*

Reviews

Johnson, A. and O'Donnell, M. (2005). Cellular DNA replicases: components and dynamics at the replication fork. *Annu. Rev. Biochem.* 74, 283–315.

Kaguni, J. M. (2006). DnaA: controlling the initiation of bacterial DNA replication and more. *Annu. Rev. Microbiol.* 60, 351–375.

Research

Bramhill, D. and Kornberg, A. (1988). Duplex opening by dnaA protein at novel sequences in initiation of replication at the origin of the *E. coli* chromosome. *Cell* 52, 743–755.

Erzberger, J. P., Mott, M. L., and Berger, J. M. (2006). Structural basis for ATP-dependent DnaA assembly and replication-origin remodeling. *Nat. Struct. Mol. Biol.* 13, 676–683.

Fuller, R. S., Funnell, B. E., and Kornberg, A. (1984). The dnaA protein complex with the *E. coli* chromosomal replication origin (*oriC*) and other DNA sites. *Cell* 38, 889–900.

Funnell, B. E. and Baker, T. A. (1987). *In vitro* assembly of a prepriming complex at the origin of the *E. coli* chromosome. *J. Biol. Chem.* 262, 10327–10334.

Hiasa, H., and Marians, K. J. (1999). Initiation of bidirectional replication at the chromosomal origin is directed by the interaction between helicase and primase. *J. Biol. Chem.* 274, 27244–27248.

Sekimizu, K., Bramhill, D., and Kornberg, A. (1987). ATP activates dnaA protein in initiating replication of plasmids bearing the origin of the *E. coli* chromosome. *Cell* 50, 259–265.

Wahle, E., Lasken, R. S., and Kornberg, A. (1989). The dnaB-dnaC replication protein complex of *Escherichia coli*. II. Role of the complex in mobilizing dnaB functions. *J. Biol. Chem.* 264, 2469–2475.

14.4   DNA Polymerases Have Various Nuclease Activities

Reviews

Hubscher, U., Maga, G., and Spadari, S. (2002). Eukaryotic DNA polymerases. *Annu. Rev. Biochem.* 71, 133–163.

Johnson, K. A. (1993). Conformational coupling in DNA polymerase fidelity. *Annu. Rev. Biochem.* 62, 685–713.

Joyce, C. M. and Steitz, T. A. (1994). Function and structure relationships in DNA polymerases. *Annu. Rev. Biochem.* 63, 777–822.

Research

Shamoo, Y. and Steitz, T. A. (1999). Building a replisome from interacting pieces: sliding clamp complexed to a peptide from DNA polymerase and a polymerase editing complex. Cell 99, 155–166.

14.8   Replication Requires a Helicase and Single-Strand Binding Protein

Review

Singleton, M. R., Dillingham, M. S., and Wigley, D. B. (2007). Structure and mechanism of helicases and nucleic acid translocases. *Annu. Rev. Biochem.* 76, 23–50.

14.10   Coordinating Synthesis of the Lagging and Leading Strands

Research

Dervyn, E., Suski, C., Daniel, R., Bruand, C., Chapuis, J., Errington, J., Janniere, L., and Ehrlich, S. D. (2001). Two essential DNA polymerases at the bacterial replication fork. *Science* 294, 1716–1719.

14.11   DNA Polymerase Holoenzyme Consists of Subcomplexes

Reference

Johnson, A., and O'Donnell, M. (2005). Cellular DNA replicases: components and dynamics at the replication fork. *Annu. Rev. Biochem.* 74, 283–315.

Research

Studwell-Vaughan, P. S. and O'Donnell, M. (1991). Constitution of the twin polymerase of DNA polymerase III holoenzyme. *J. Biol. Chem.* 266, 19833–19841.

Stukenberg, P. T., Studwell-Vaughan, P. S., and O'Donnell, M. (1991). Mechanism of the sliding beta-clamp of DNA polymerase III holoenzyme. *J. Biol. Chem.* 266, 11328–11334.

14.12   The Clamp Controls Association of Core Enzyme with DNA

Reviews

Benkovic, S. J., Valentine, A. M., and Salinas, F. (2001). Replisome-mediated DNA replication. *Annu. Rev. Biochem.* 70, 181–208.

Davey, M. J., Jeruzalmi, D., Kuriyan, J., and O'Donnell, M. (2002). Motors and switches: AAA+ machines within the replisome. *Nat. Rev. Mol. Cell Biol.* 3, 826–835.

Research

Bowman, G. D., O'Donnell, M., and Kuriyan, J. (2004). Structural analysis of a eukaryotic sliding DNA clamp-clamp loader complex. *Nature* 429, 724–730.

Jeruzalmi, D., O'Donnell, M., and Kuriyan, J. (2001). Crystal structure of the processivity clamp loader gamma (γ) complex of *E. coli* DNA polymerase III. *Cell* 106, 429–441.

Stukenberg, P. T., Turner, J., and O'Donnell, M. (1994). An explanation for lagging strand replication: polymerase hopping among DNA sliding clamps. *Cell* 78, 877–887.

14.13 Okazaki Fragments Are Linked by Ligase

Review

Liu, Y., Kao, H. I., and Bambara, R. A. (2004). Flap endonuclease 1: a central component of DNA metabolism. *Annu. Rev. Biochem.* 73, 589–615.

Research

Garg, P., Stith, C. M., Sabouri, N., Johansson, E., and Burgers, P. M. (2004). Idling by DNA polymerase δ maintains a ligatable nick during lagging-strand DNA replication. *Genes Dev.* 18, 2764–2773.

14.14 Separate Eukaryotic DNA Polymerases Undertake Initiation and Elongation

Reviews

Goodman, M. F. (2002). Error-prone repair DNA polymerases in prokaryotes and eukaryotes. *Annu. Rev. Biochem.* 71, 17–50.

Hubscher, U., Maga, G., and Spadari, S. (2002). Eukaryotic DNA polymerases. *Annu. Rev. Biochem.* 71, 133–163.

Kaguni, L. S. (2004). DNA polymerase gamma, the mitochondrial replicase. *Annu. Rev. Biochem.* 73, 293–320.

Kunkel, T. A., and Burgers, P. M. (2008). Dividing the workload at a eukaryotic replication fork. *Trends Cell Biol.* 18, 521–527.

Research

Bowman, G. D., O'Donnell, M., and Kuriyan, J. (2004). Structural analysis of a eukaryotic sliding DNA clamp-clamp loader complex. *Nature* 429, 724–730.

Karthikeyan, R., Vonarx, E. J., Straffon, A. F., Simon, M., Faye, G., and Kunz, B. A. (2000). Evidence from mutational specificity studies that yeast DNA polymerases delta and epsilon replicate different DNA strands at an intracellular replication fork. *J. Mol. Biol.* 299, 405–419.

McElhinny, S. A., Gordenin, D. A., Stith, C. M., Burgers, P. M., and Kunkel, T. A. (2008). Division of labor at the eukaryotic replication fork. *Mol. Cell* 30, 137–144.

Pursell, Z. F., Isoz, I., Lundström, E.-B., Johansson, E., and Kunkel, T. A. (2007). Yeast DNA polymerase ε participates in leading-strand DNA replication. *Science* 317, 127–130.

Shiomi, Y., Usukura, J., Masamura, Y., Takeyasu, K., Nakayama, Y., Obuse, C., Yoshikawa, H., and Tsurimoto, T. (2000). ATP-dependent structural change of the eukaryotic clamp-loader protein, replication factor C. *Proc. Natl. Acad. Sci. USA* 97, 14127–14132.

Waga, S., Masuda, T., Takisawa, H., and Sugino, A. (2001). DNA polymerase epsilon is required for coordinated and efficient chromosomal DNA replication in Xenopus egg extracts. *Proc. Natl. Acad. Sci. USA* 98, 4978–4983.

Zuo, S., Bermudez, V., Zhang, G., Kelman, Z., and Hurwitz, J. (2000). Structure and activity associated with multiple forms of *S. pombe* DNA polymerase delta. *J. Biol. Chem.* 275, 5153–5162.

14.15 Phage T4 Provides Its Own Replication Apparatus

Research

Ishmael, F. T., Alley, S. C., and Benkovic, S. J. (2002). Assembly of the bacteriophage T4 helicase: architecture and stoichiometry of the gp41-gp59 complex. *J. Biol. Chem.* 277, 20555–20562.

Salinas, F., and Benkovic, S. J. (2000). Characterization of bacteriophage T4-coordinated leading- and lagging-strand synthesis on a minicircle substrate. *Proc. Natl. Acad. Sci. USA* 97, 7196–7201.

Schrock, R. D. and Alberts, B. (1996). Processivity of the gene 41 DNA helicase at the bacteriophage T4 DNA replication fork. *J. Biol. Chem.* 271, 16678–16682.

Yang, J., Nelson, S. W., and Benkovic, S. J. (2006). The control mechanism for lagging strand polymerase recycling during bacteriophage T4 DNA replication. *Mol. Cell* 21, 153–164.

14.16 Lesion Bypass Requires Polymerase Replacement

Reviews

Cox, M. M. (2001). Recombinational DNA repair of damaged replication forks in *E. coli*: questions. *Annu. Rev. Genet.* 35, 53–82.

Heller, R. C. and Marians, K. J. (2006). Replisome assembly and the direct restart of stalled replication forks. *Nat. Rev. Mol. Cell Biol.* 7, 932–943.

Kuzminov, A. (1995). Collapse and repair of replication forks in *E. coli. Mol. Microbiol.* 16, 373–384.

McGlynn, P. and Lloyd, R. G. (2002). Recombinational repair and restart of damaged replication forks. *Nat. Rev. Mol. Cell Biol.* 3, 859–870.

Prakash, S., Johnson, R. E., and Prakash, L. (2005). Eukaryotic translesion synthesis DNA polymerases: specificity of structure and function. *Annu. Rev. Biochem.* 74, 317–353.

Research

Furukohri, A., Goodman, M. F., and Maki, H. (2008). A dynamic polymerase exchange with *E. coli* DNA polymerase IV replacing DNA polymerase III on the sliding clamp. *J. Biol. Chem.* 283, 11260–11269.

Lecointe, F. et al. (2007). Anticipating chromosomal replication fork arrest: SSB targets repair DNA helicases to active forks. *EMBO J.* 26, 4239–4251.

Loper, M., Boonsombat, R., Sandler, S. J., and Keck, J. L. (2007). A hand-off mechanism for primosome assembly in replication restart. *Mol. Cell* 26, 781–793.

Seigneur, M., Bidnenko, V., Ehrlich, S. D., and Michel, B. (1998). RuvAB acts at arrested replication forks. *Cell* 95, 419–430.

15

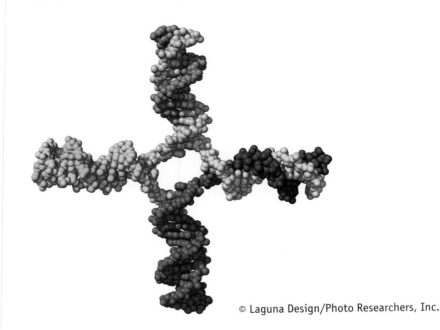

© Laguna Design/Photo Researchers, Inc.

Homologous and Site-Specific Recombination

Edited by Hannah L. Klein and Samantha Hoot

CHAPTER OUTLINE

15.1 Introduction
- Homologous recombination is essential in meiosis for generating diversity and for chromosome segregation, and in mitosis to repair DNA damage and stalled replication forks.
- Site-specific recombination involves specific DNA sequences.
- Recombination systems have been adapted for experimental use.

15.2 Homologous Recombination Occurs between Synapsed Chromosomes in Meiosis
- Chromosomes must synapse (pair) in order for chiasmata to form where crossing-over occurs.
- The stages of meiosis can be correlated with the molecular events at the DNA level.

15.3 Double-Strand Breaks Initiate Recombination
- The double-strand break repair (DSBR) model of recombination is initiated by making a double-strand break in one (recipient) DNA duplex and is relevant for meiotic and mitotic homologous recombination.
- Exonuclease action generates 3′–single-stranded ends that invade the other (donor) duplex.

- When a single strand from one duplex displaces its counterpart in the other duplex, it creates a branched structure called a D-loop.
- Strand exchange generates a stretch of heteroduplex DNA consisting of one strand from each parent.
- New DNA synthesis replaces the material that has been degraded.
- Capture of the second DSB end by annealing generates a recombinant joint molecule in which the two DNA duplexes are connected by heteroduplex DNA and two Holliday junctions.
- The joint molecule is resolved into two separate duplex molecules by nicking two of the connecting strands.
- Whether recombinants are formed depends on whether the strands involved in the original exchange or the other pair of strands are nicked during resolution.

15.4 Gene Conversion Accounts for Interallelic Recombination
- Heteroduplex DNA that is created by recombination can have mismatched sequences where the recombining alleles are not identical.
- Repair systems may remove mismatches by changing one of the strands so its sequence is complementary to the other.

- Mismatch repair of heteroduplex DNA generates nonreciprocal recombinant products called gene conversions.

15.5 The Synthesis-Dependent Strand-Annealing Model

- The synthesis-dependent strand-annealing model (SDSA) is relevant for mitotic recombination as it produces gene conversions from double-strand breaks without associated crossovers.

15.6 Nonhomologous End-Joining Can Repair Double-Strand Breaks

- Repair of double-strand breaks when homologous sequence is not available occurs through a nonhomologous end-joining (NHEJ) reaction.
- Immune receptor V(D)J recombination occurs through a specialized NHEJ pathway.

15.7 The Single-Strand Annealing Mechanism Functions at Some Double-Strand Breaks

- Single-strand annealing (SSA) occurs at double-strand breaks between direct repeats.
- Resection of double-strand break ends results in 3' single-stranded tails.
- Complementarity between the repeats allows for annealing of the single strands.
- The sequence between the direct repeats is deleted after SSA is completed.

15.8 Break-Induced Replication Can Repair Double-Strand Breaks

- Break-induced replication (BIR) is initiated by a one-ended double-strand break.
- BIR at repeated sequences can result in translocations.

15.9 Recombining Meiotic Chromosomes Are Connected by the Synaptonemal Complex

- During the early part of meiosis, homologous chromosomes are paired in the synaptonemal complex.
- The mass of chromatin of each homolog is separated from the other by a proteinaceous complex.

15.10 The Synaptonemal Complex Forms after Double-Strand Breaks

- Double-strand breaks that initiate recombination occur before the synaptonemal complex forms.
- If recombination is blocked, the synaptonemal complex cannot form.
- Meiotic recombination involves two phases: one that results in gene conversion without crossover, and one that results in crossover products.

15.11 Pairing and Synaptonemal Complex Formation Are Independent

- Mutations can occur in either chromosome pairing or synaptonemal complex formation without affecting the other process.

15.12 The Bacterial RecBCD System Is Stimulated by *chi* Sequences

- The RecBCD complex has nuclease and helicase activities.

- RecBCD binds to DNA downstream of a *chi* sequence, unwinds the duplex, and degrades one strand from 3'–5' as it moves to the *chi* site.
- The *chi* site triggers loss of the RecD subunit and nuclease activity.

15.13 Strand-Transfer Proteins Catalyze Single-Strand Assimilation

- RecA forms filaments with single-stranded or duplex DNA and catalyzes the ability of a single-stranded DNA with a free 3' end to displace its counterpart in a DNA duplex.

15.14 Holliday Junctions Must Be Resolved

- The bacterial Ruv complex acts on recombinant junctions.
- RuvA recognizes the structure of the junction and RuvB is a helicase that catalyzes branch migration.
- RuvC cleaves junctions to generate recombination intermediates.
- Resolution in eukaryotes is less well understood, but a number of meiotic and mitotic proteins are implicated.

15.15 Eukaryotic Genes Involved in Homologous Recombination

- The MRX complex, Exo1, and Sgs1/Dna2 in yeast and the MRN complex and BLM in mammalian cells resect double-strand breaks.
- The Rad51 recombinase binds to single-stranded DNA with the aid of mediator proteins, which overcome the inhibitory effects of RPA.
- Strand invasion is dependent on Rad54 and Rdh54 in yeast and Rad54 and Rad54B in mammalian cells.
- Yeast Sgs1, Mus81/Mms4 and human BLM, MUS81/EME1 are implicated in resolution of Holliday junctions.

15.16 Specialized Recombination Involves Specific Sites

- Specialized recombination involves reaction between specific sites that are not necessarily homologous.
- Phage lambda integrates into the bacterial chromosome by recombination between a site on the phage and the *att* site on the *E. coli* chromosome.
- The phage is excised from the chromosome by recombination between the sites at the end of the linear prophage.
- Phage lambda *int* codes for an integrase that catalyzes the integration reaction.

15.17 Site-Specific Recombination Involves Breakage and Reunion

- Cleavages staggered by 7 bp are made in both *attB* and *attP* and the ends are joined crosswise.

15.18 Site-Specific Recombination Resembles Topoisomerase Activity

- Integrases are related to topoisomerases, and the recombination reaction resembles topoisomerase action except that nicked strands from *different* duplexes are sealed together.

- The reaction conserves energy by using a catalytic tyrosine in the enzyme to break a phosphodiester bond and link to the broken 3' end.
- Two enzyme units bind to each recombination site and the two dimers synapse to form a complex in which the transfer reactions occur.

15.19 Lambda Recombination Occurs in an Intasome
- Lambda integration takes place in a large complex that also includes the host protein IHF.
- The excision reaction requires Int and Xis and recognizes the ends of the prophage DNA as substrates.

15.20 Yeast Can Switch Silent and Active Loci for Mating Type
- The yeast mating type locus *MAT* has either the *MAT***a** or *MAT*α genotype.
- Yeast with the dominant allele *HO* switch their mating type at a frequency ~10^{-6}.
- The allele at *MAT* is called the active cassette.
- There are also two silent cassettes, *HML*α and *HMR***a**.
- Switching occurs if *MAT***a** is replaced by *HMR*α or *MAT*α is replaced by *HMR***a**.

15.21 Unidirectional Gene Conversion Is Initiated by the Recipient *MAT* Locus
- Mating type switching is initiated by a double-strand break made at the *MAT* locus by the HO endonuclease.
- The recombination event is a synthesis-dependent strand-annealing reaction.

15.22 Antigenic Variation in Trypanosomes Uses Homologous Recombination
- Variant surface glycoprotein (VSG) switching in *Trypanosoma brucei* evades host immunity.
- VSG switching requires recombination events to move VSG genes to specific expression sites.

15.23 Recombination Pathways Adapted for Experimental Systems
- Mitotic homologous recombination allows for targeted transformation.
- The Cre/*lox* and FLP/*FRT* systems allow for targeted recombination and gene knockout construction.
- The FLP/*FRT* system has been adapted to construct recyclable selectable markers for gene deletion.

15.24 Summary

15.1 Introduction

Key concepts

- Homologous recombination is essential in meiosis for generating diversity and for chromosome segregation, and in mitosis to repair DNA damage and stalled replication forks.
- Site-specific recombination involves specific DNA sequences.
- Recombination systems have been adapted for experimental use.

Homologous recombination is an essential cellular process required for generating genetic diversity, ensuring proper chromosome segregation, and repairing certain types of DNA damage. Evolution could not happen without genetic recombination. If it were not possible to exchange material between (homologous) chromosomes, the content of each individual chromosome would be irretrievably fixed in its particular alleles. When mutations occurred, it would not be possible to separate favorable and unfavorable changes. The length of the target for mutation damage would effectively be increased from the gene to the chromosome. Ultimately a chromosome would accumulate so many deleterious mutations that it would fail to function.

By shuffling the genes, recombination allows favorable and unfavorable mutations to be separated and tested as individual units in new assortments. It provides a means of escape and spreading for favorable alleles, and a means to eliminate an unfavorable allele without bringing down all the other genes with which this allele is associated. This is the basis for natural selection.

In addition to its role in genetic diversity, homologous recombination is also required in mitosis for repair of lesions at replication forks and for restarting replication that has stalled at these lesions. The importance of mitotic recombination events is highlighted by examples of human diseases that result from defects in recombination repair of DNA damage where altered activity of homologous recombination proteins is seen in some types of cancers. Homologous recombination is also essential for a process known as antigenic switching, which allows disease-causing parasites known as trypanosomes to evade the human immune system.

Recombination occurs between precisely corresponding sequences, so that not a single base pair is added to or lost from the recombinant chromosomes. Three types of recombination share the feature that the process involves physical exchange of material between duplex DNAs:

Recombination involving reaction between homologous sequences of DNA is called *generalized* or **homologous recombination**. In eukaryotes, it occurs at meiosis, usually both in males (during spermatogenesis) and females (during oogenesis). We recall that it happens at the "four strand" stage of meiosis and involves only two nonsister strands of the four strands (see *Section 2.7, Recombination Occurs by Physical Exchange of DNA*).

Another type of event sponsors recombination between specific pairs of sequences. This was first characterized in prokaryotes where *specialized recombination*, also known as **site-specific recombination**, is responsible for the integration of phage genomes into the bacterial chromosome. The recombination event involves specific sequences of the phage DNA and bacterial DNA, which include a short stretch of homology. The enzymes involved in this event act only on the particular pair of target sequences in an intermolecular reaction. Some related intramolecular reactions are responsible during bacterial division for regenerating two monomeric circular chromosomes when a dimer has been generated by generalized recombination. This latter class also includes recombination events that invert specific regions of the bacterial chromosome.

In special circumstances, gene rearrangement is used to control expression. Rearrangement may create new genes, which are needed for expression in particular circumstances, as in the case of the immunoglobulins. This example of **somatic recombination** will be discussed in Chapter 18, *Recombination in the Immune System*. Recombination events also may be responsible for switching expression from one preexisting gene to another, as in the example of yeast mating type, where the sequence at an active locus can be replaced by a sequence from a silent locus. Rearrangements are also required to control expression of surface antigens in the parasites known as trypanosomes, in which silent

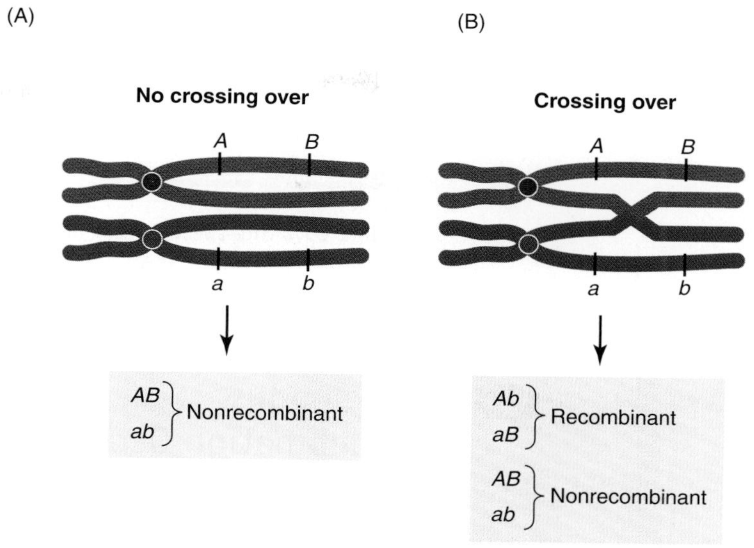

alleles of surface antigen genes are duplicated into active expression sites. Some of these types of rearrangement share mechanistic similarities with transposition; in fact, they can be viewed as specially directed cases of transposition.

Let's consider the nature and consequences of the generalized and specialized recombination reactions.

FIGURE 15.1 makes the point that generalized recombination occurs between two homologous DNA duplexes and can occur at any point along their length. The cross-over is the point at which each becomes joined to the other. There is no change in the overall organization of DNA; the products have the same structure as the parents, and both parents and products are homologous.

Specialized recombination occurs only between specific sites. The results depend on the locations of the two recombining sites. **FIGURE 15.2** shows that an intermolecular recombination between a circular DNA and a linear DNA inserts the circular DNA into the linear DNA. Specialized recombination is often used to make changes such as this in the organization of DNA. The change in organization is a consequence of the locations of the recombining sites. We have a large amount of information about the enzymes that undertake specialized recombination, which are related to the **topoisomerases** that act to change the supercoiling of DNA in space (see *Section 1.5, Supercoiling Affects the Structure of DNA*).

(A) No crossing over

(B) Crossing over

AB
ab } Nonrecombinant

Ab
aB } Recombinant

AB
ab } Nonrecombinant

FIGURE 15.1 No crossing over between the A and B genes gives rise to only nonrecombinant gametes. Crossing over between the A and B genes gives rise to the recombinant gametes Ab and aB and the nonrecombinant gametes AB and ab.

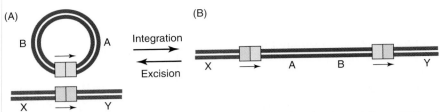

FIGURE 15.2 Site-specific recombination occurs between the circular and linear DNAs at the boxed region (A). Integration results in an insertion of the A and B sequences between the X and Y sequences (B). The reaction is promoted by integrase enzymes. Reversal of the reaction results in a precise excision of the A and B sequences. Adapted from B. Alberts, et al. *Molecular Biology of the Cell, Fourth edition.* Garland Science, 2002.

15.2 Homologous Recombination Occurs between Synapsed Chromosomes in Meiosis

Progress through meiosis

Key concepts

- Chromosomes must synapse (pair) in order for chiasmata to form where crossing-over occurs.
- The stages of meiosis can be correlated with the molecular events at the DNA level.

Homologous recombination is a reaction between two duplexes of DNA. Its critical feature is that the enzymes responsible can use any pair of homologous sequences as substrates (although some types of sequences may be favored over others). The frequency of recombination is not constant throughout the genome, but is influenced by both global and local effects. The overall frequency may be different in oocytes and in sperm; recombination occurs twice as frequently in female as in male humans. Within the genome, its frequency depends upon chromosome structure; for example, crossing–over is suppressed in the vicinity of the condensed and inactive regions of heterochromatin.

Recombination occurs during the protracted prophase of meiosis. **FIGURE 15.3** shows the visible progress of chromosomes through the five stages of meiotic prophase. Studies in yeast have shown that all of the molecular events of homologous recombination are finished by late pachytene.

The beginning of meiosis is marked by the point at which individual chromosomes become visible. Each of these chromosomes has replicated previously and consists of two

Leptotene
Condensed chromosomes become visible, often attached to nuclear envelope

Zygotene
Chromosomes begin pairing in limited region or regions

Pachytene
Synaptonemal complex extends along entire length of paired chromosomes

Diplotene
Chromosomes separate, but are held together by chiasmata

Diakinesis
Chromosomes condense, detach from envelope; chiasmata remain. All four chromatids become visible

FIGURE 15.3 Recombination occurs during the first meiotic prophase. The stages of prophase are defined by the appearance of the chromosomes, each of which consists of two replicas (sister chromatids), although the duplicated state becomes visible only at the end.

sister chromatids, each of which contains a duplex DNA. The homologous chromosomes approach one another and begin to pair in one or more regions, forming **bivalents**. Pairing extends until the entire length of each chromosome is apposed with its homolog. The process is called **synapsis** or **chromosome pairing**. When the process is completed, the chromosomes are laterally associated in the form of a **synaptonemal complex**, which has a characteristic structure in each species, although there is wide variation in the details between species.

Recombination between chromosomes involves a physical exchange of parts (achieved through a double-strand break on one chromatid to initiate recombination), formation of a **joint molecule** between the chromatids, and resolution to break the joint and form intact chromatids that have new genetic information. When the chromosomes begin to separate, they can be seen to be held together at discrete sites called **chiasmata**. The number and distribution of chiasmata parallel the features of genetic crossing–over. Traditional analysis holds that a chiasma represents the crossingover event. The chiasmata remain visible when the chromosomes condense and all four chromatids become evident.

What is the molecular basis for these events? Each sister chromatid contains a single DNA duplex, so each bivalent contains four duplex molecules of DNA. Recombination requires a mechanism that allows the duplex DNA of one sister chromatid to interact with the duplex DNA of a sister chromatid from the other chromosome. It must be possible for this reaction to occur between any pair of corresponding sequences in the two molecules in a highly specific manner that allows material to be exchanged with precision at the level of the individual base pair.

We know of only one mechanism for nucleic acids to recognize one another on the basis of sequence: complementarity between single strands. If (at least) one strand displaces the corresponding strand in the other duplex, the two duplex molecules will be specifically connected at corresponding sequences. If the strand exchange is extended, there can be more extensive connection between the duplexes.

15.3 Double-Strand Breaks Initiate Recombination

Key concepts

- The double-strand break repair (DSBR) model of recombination is initiated by making a double-strand break in one (recipient) DNA duplex and is relevant for meiotic and mitotic homologous recombination.
- Exonuclease action generates 3'–single-stranded ends that invade the other (donor) duplex.
- When a single strand from one duplex displaces its counterpart in the other duplex, it creates a branched structure called a D-loop.
- Strand exchange generates a stretch of heteroduplex DNA consisting of one strand from each parent.
- New DNA synthesis replaces the material that has been degraded.
- Capture of the second DSB end by annealing generates a recombinant joint molecule in which the two DNA duplexes are connected by heteroduplex DNA and two Holliday junctions.
- The joint molecule is resolved into two separate duplex molecules by nicking two of the connecting strands.
- Whether recombinants are formed depends on whether the strands involved in the original exchange or the other pair of strands are nicked during resolution.

Genetic exchange is initiated by a **double-strand break (DSB)**. The double-strand break repair (DSBR) model is illustrated in **FIGURE 15.4**. Recombination is initiated by an endonuclease that cleaves one of the partner DNA duplexes, the "recipient." In meiosis this is performed by the Spo11 protein, which is related to DNA topoisomerases (**FIGURE 15.5**). DNA topoisomerases are enzymes that catalyze changes in the topology of DNA by transiently breaking one or both strands of DNA, passing the unbroken strand(s) through the gap, and then resealing the gap. The ends that are generated by the break are never free, but instead are manipulated exclusively within the confines of the enzyme—in fact, they are covalently linked to the enzyme. Spo11 undergoes a similar covalent attachment when it forms DSBs during meiosis.

In mitotic cells DSBs form spontaneously as a result of DNA damage or through the action of specific processes that are programmed to form

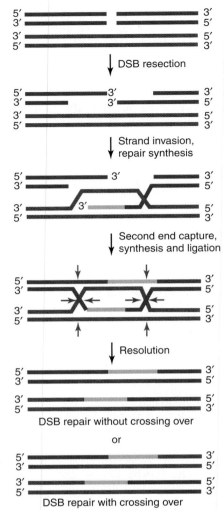

FIGURE 15.4 Double-strand break repair model of homologous recombination. Recombination is initiated by a double-strand break. Following nuclease degradation of the ends, called DNA resection, single-strand tails with 3′-OH ends are formed. Strand invasion by one end into homologous sequences forms a D-loop. Extension of the 3′-OH end by DNA synthesis enlarges the D-loop. Once the displaced loop can pair with the other side of the break, the second double-strand break end is captured. DNA synthesis to complete the break repair, followed by ligation results in the formation of two Holliday junctions. Resolution at the blue arrowheads results in a noncrossover product. Resolution of one Holliday junction at the blue arrowheads and the other Holliday junction at the red arrowheads results in a crossover product.

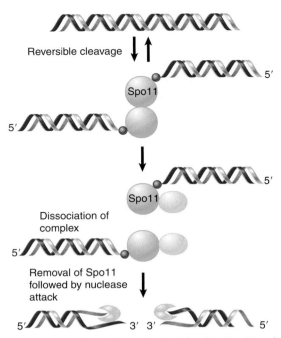

FIGURE 15.5 Spo11 is covalently joined to the 5′ ends of double-strand breaks.

called **single-strand invasion**. The formation of **heteroduplex DNA** generates a **D-loop (displacement loop)**, in which one strand of the donor duplex is displaced. The point at which an individual strand of DNA crosses from one duplex to the other is called the **recombinant joint**. An important feature of a recombinant joint is its ability to move along the duplex. Such mobility is called **branch migration**. The D-loop is extended by repair DNA synthesis, using the free 3′ end as a primer to generate double-stranded DNA. **FIGURE 15.6** illustrates the migration of a single strand in a duplex. The branching point can migrate in either direction as one strand is displaced by the other.

Branch migration is important for both theoretical and practical reasons. As a matter of principle, it confers a dynamic property on recombining structures. As a practical feature, its existence means that the point of branching cannot be established by examining a molecule *in vitro* (because the branch may have migrated since the molecule was isolated).

Branch migration could allow the point of crossover in the recombination intermediate to move in either direction. The rate of branch migration is uncertain, but as seen *in vitro* is probably inadequate to support the formation of extensive regions of heteroduplex DNA in natural conditions. Any extensive branch migration *in vivo* must therefore be catalyzed by a recombination enzyme.

breaks such as V(D)J recombination or mating-type switching in yeast. The DSB is enlarged to a gap by exonuclease action. The exonuclease(s), which can work in concert with a DNA helicase, nibble away one strand on either side of the break, generating 3′ single-stranded termini; this process is known as **5′-end resection**. One of the free 3′ ends then invades a homologous region in the other ("donor") duplex. This is

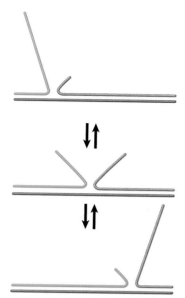

FIGURE 15.6 Branch migration can occur in either direction when an unpaired single strand displaces a paired strand.

Eventually the D-loop becomes large enough to correspond to the entire length of the gap on the recipient chromatid. When the extruded single strand reaches the far side of the gap, the complementary single-stranded sequences anneal, and the second DSB end can be captured. The second strand of the gap is filled in by repair synthesis and there is heteroduplex DNA on either side of the gap. The gap itself is flanked by crossed strands or recombinant joints called **Holliday junctions**. Overall, the gap has been repaired by two individual rounds of single-strand DNA synthesis. The joints must be resolved by cutting.

If both joints are resolved in the same way, the original noncrossover molecules will be released, each with a region of altered genetic information that is a footprint of the exchange event. If the two joints are resolved in opposite ways, a genetic cross-over is produced.

The involvement of DSBs at first seems surprising. Once a break has been made right across a DNA molecule, there is no going back. In the DSBR model, the initial cleavage is immediately followed by loss of information. Any error in retrieving the information could be fatal. On the other hand, the very ability to retrieve lost information by resynthesizing it from another duplex provides a major safety net for the cell.

The joint molecule formed by strand exchange must be resolved into two separate duplex molecules. **Resolution** requires a further pair of nicks. We can most easily visualize

the outcome by viewing the joint molecule in one plane as a Holliday junction. This is illustrated in the bottom half of Figure 15.4, which represents the resolution reaction. The outcome of the reaction depends on which pair of strands is nicked.

If the nicks are made in the pair of strands that were not originally nicked (the pair that did not initiate the strand exchange), all four of the original strands have been nicked. This releases *crossover recombinant* DNA molecules. The duplex of one DNA parent is covalently linked to the duplex of the other DNA parent via a stretch of heteroduplex DNA.

If the same two strands involved in the original nicking are nicked again, the other two strands remain intact. The nicking releases the original parental duplexes, which remain intact with the exception that each has a residuum of the event in the form of a length of heteroduplex DNA. These are noncrossover products that nonetheless contain sequence from the donor DNA duplex and as such are considered recombinant.

What is the minimum length of the region required to establish the connection between the recombining duplexes? Experiments in which short homologous sequences carried by plasmids or phages are introduced into bacteria suggest that the rate of recombination is substantially reduced if the homologous region is <75 bp. This distance is appreciably longer than the ~10 bp required for association between complementary single-stranded regions, which suggests that recombination imposes demands beyond annealing of complements as such.

15.4 Gene Conversion Accounts for Interallelic Recombination

Key concepts

- Heteroduplex DNA that is created by recombination can have mismatched sequences where the recombining alleles are not identical.
- Repair systems may remove mismatches by changing one of the strands so its sequence is complementary to the other.
- Mismatch repair of heteroduplex DNA generates nonreciprocal recombinant products called gene conversions.

The involvement of heteroduplex DNA explains the characteristics of recombination between alleles; indeed, allelic recombination provided the impetus for the development of a recombination model that invoked heterodu-plex DNA as an intermediate. When recombi-nation between alleles was discovered, the natural assumption was that it takes place by the same mechanism of reciprocal recombina-tion that applies to more distant loci. That is to say, both events are initiated in the same man-ner: A DSB repair event can occur within a locus to generate a reciprocal pair of recombi-nant chromosomes. In the close quarters of a single gene, however, formation and repair of heteroduplex DNA itself is responsible for the gene conversion event.

Individual recombination events can be studied in the ascomycetes fungi, because the products of a single meiosis are held together in a large cell called the *ascus* (or less commonly, the *tetrad*). Even better is that in some fungi, the four haploid nuclei produced by meiosis are arranged in a linear order. (Actually, a mitosis occurs after the production of these four nuclei, giving a linear series of eight haploid nuclei.) **FIGURE 15.7** shows that each of these nuclei effectively represents the genetic character of one of the eight strands of the four chromo-somes produced by the meiosis.

Meiosis in a heterozygous diploid should generate four copies of each allele in these fungi. This is seen in the majority of spores. There are some spores, though, with abnormal ratios. They are explained by the formation and correction of heteroduplex DNA in the region in which the alleles differ. The figure illustrates a recombination event in which a length of

hybrid DNA occurs on one of the four meiotic chromosomes, a possible outcome of recombi-nation initiated by a DSB.

Suppose that two alleles differ by a single point mutation. When a strand exchange occurs to generate heteroduplex DNA, the two strands of the heteroduplex will be mispaired at the site of mutation. Thus each strand of DNA carries different genetic information. If no change is made in the sequence, the strands separate at the ensuing replication, each giving rise to a duplex that perpetuates its information. This event is called *postmeiotic segregation*, because it reflects the separation of DNA strands after meiosis. Its importance is that it demonstrates directly the existence of heteroduplex DNA in recombining alleles.

Another effect is seen when examining recombination between alleles: The proportions of the alleles differ from the initial 4:4 ratio. This effect is called **gene conversion**. It describes a nonreciprocal transfer of information from one chromatid to another.

Gene conversion results from exchange of strands between DNA molecules, and the change in sequence may have either of two causes at the molecular level, known as **gap repair** or **mismatch repair**:

- *Gap repair*: As indicated by the DSBR model in Figure 15.4, one DNA duplex may act as a donor of genetic infor-mation that directly replaces the cor-responding sequences in the recipient duplex by a process of gap generation, strand exchange, and gap filling.

- *Mismatch repair*: As part of the exchange process, heteroduplex DNA is gener-ated when a single strand from one duplex pairs with its complement in the other duplex. Repair systems recognize mispaired bases in heteroduplex DNA, and then may excise and replace one of the strands to restore complementarity (see *Section 16.7, Controlling the Direction of Mismatch Repair*). Such an event con-verts the strand of DNA representing one allele into the sequence of the other allele.

Gene conversion does not depend on crossing-over, but is correlated with it. A large proportion of the aberrant asci show genetic recombination between two markers on either side of a site of interallelic gene conversion. This is exactly what would be predicted if the aber-rant ratios result from initiation of the recombi-nation process as shown in Figure 15.4, but with

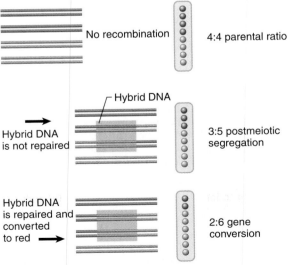

FIGURE 15.7 Spore formation in the ascomycetes allows determination of the genetic constitution of each of the DNA strands involved in meiosis.

No recombination   4:4 parental ratio

Hybrid DNA

Hybrid DNA is not repaired   3:5 postmeiotic segregation

Hybrid DNA is repaired and converted to red   2:6 gene conversion an approximately equal probability of resolving the structure with or without recombination. The implication is that fungal chromosomes initiate crossing-over about twice as often as would be expected from the measured frequency of recombination between distant genes.

Various biases are seen when recombination is examined at the molecular level. Either direction of gene conversion may be equally likely, or allele-specific effects may create a preference for one direction. Gradients of recombination may fall away from hotspots. We now know that recombination **hotspots** represent sites at which double-strand breaks are initiated, and that the gradient is correlated with the extent to which the gap at the hotspot is enlarged and converted to long single-stranded ends (see *Section 15.10, The Synaptonemal Complex Forms after Double-Strand Breaks*).

Some information about the extent of gene conversion is provided by the sequences of members of gene clusters. Usually, the products of a recombination event will separate and become unavailable for analysis at the level of DNA sequence. When a chromosome carries two (nonallelic) genes that are related, though, they may recombine by an "unequal crossing-over" event (see *Section 7.2, Unequal Crossing-Over Rearranges Gene Clusters*). All we need to note for now is that a heteroduplex may be formed between the two nonallelic genes. Gene conversion effectively converts one of the nonallelic genes to the sequence of the other.

The presence of more than one gene copy on the same chromosome provides a footprint to trace these events. For example, if heteroduplex formation and gene conversion occurred over part of one gene, this part may have a sequence identical with, or very closely related to, the other gene, whereas the remaining part shows more divergence. Available sequences suggest that gene conversion events may extend for considerable distances, up to a few thousand bases.

15.5 The Synthesis-Dependent Strand-Annealing Model

Key concept

- The synthesis-dependent strand-annealing model (SDSA) is relevant for mitotic recombination as it produces gene conversions from double-strand breaks without associated crossovers.

The double-strand break repair model accounts for meiotic homologous recombination that gives crossover products, but cannot explain all homologous recombination as mitotic gene conversions are typically not accompanied by crossing-over. The *synthesis-dependent strand-annealing* (SDSA) model serves as a better model for what occurs during mitotic homologous recombination in which DSB repair events and gene conversion are not associated with crossing-over. Studies of the double-strand break that occurs during mating-type switching events in yeast (discussed later in this chapter) lead to the development of SDSA as a model for mitotic recombination.

The synthesis-dependent strand-annealing pathway, shown in **FIGURE 15.8**, is initiated in a mechanism similar to the DSBR model in that DSBs are processed by 5′-end resection. Following strand invasion and DNA synthesis, the second end is not captured as it is in the double-strand break repair model. In the

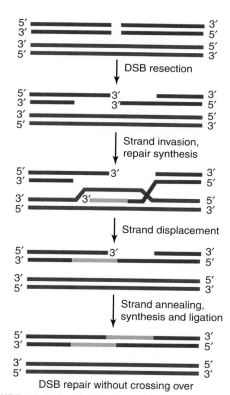

DSB repair without crossing over

FIGURE 15.8 Synthesis-dependent strand-annealing model of homologous recombination. Recombination is initiated by a double-strand break and is followed by end processing to form single-strand tails with 3′-OH ends. Strand invasion and DNA synthesis repairs on a strand of the break. Instead of second send capture as depicted in Figure 11.21, the strand in the D-loop is displaced. The single strand can anneal with the single strand of the other end. Repair synthesis then completes the double-strand break repair process. No Holliday junction is formed and the product is always noncrossover.

SDSA pathway, the invading strand, which contains newly synthesized DNA identical in sequence to the strand it displaced, is itself displaced. Following displacement, the invading strand reanneals with the other end of the double-strand break. This is followed by synthesis and ligation to repair the double-strand break. In this model, the break is repaired using the homologous sequence as a template but does not involve crossing-over. This feature of the synthesis-dependent strand-annealing model makes it suitable for mitotic gene conversions for which there is no associated crossing-over.

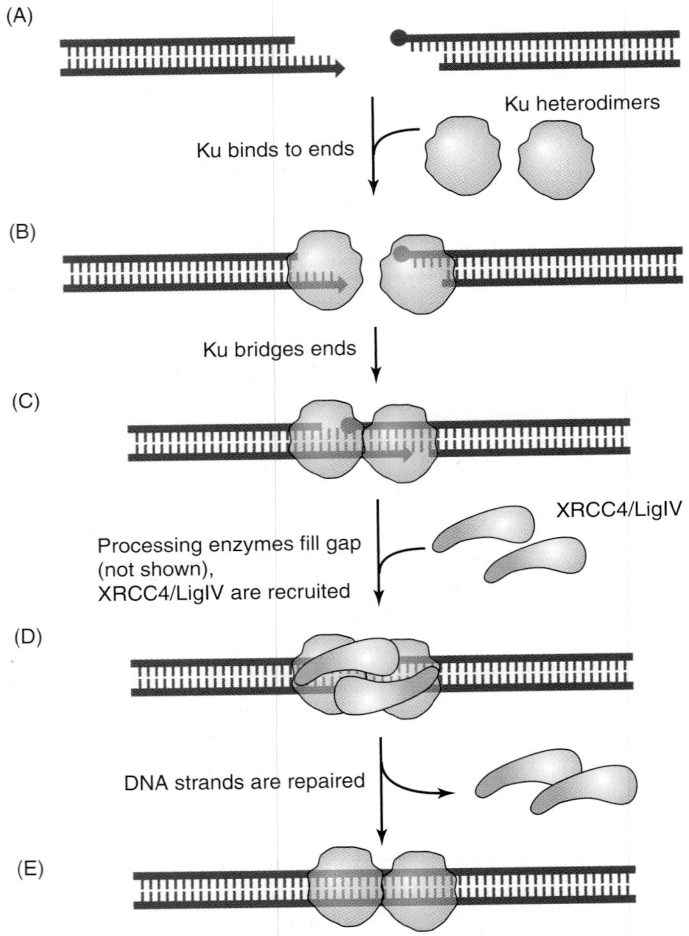

The synthesis-dependent strand-annealing pathway is also responsible for recombination without crossover in the first phase of meiosis (discussed in *Section 15.10, The Synaptonemal Complex Forms after Double-Strand Breaks*).

15.6 Nonhomologous End-Joining Can Repair Double-Strand Breaks

Key concepts

- Repair of double-strand breaks when homologous sequence is not available occurs through a nonhomologous end-joining (NHEJ) reaction.
- Immune receptor V(D)J recombination occurs through a specialized NHEJ pathway.

In circumstances where no or limited homology is present at double-strand breaks, a process known as **nonhomologous end-joining** (**NHEJ**) is used for repair (see also Chapter 16, *Repair Systems*). NHEJ typically functions when homologous sequences are not readily available, as in unreplicated G1 cells. NHEJ has an important role in immune system function, as it is essential to a process known as V(D)J recombination, in which rearrangements at immune receptor loci contribute to the vast diversity in immunoglobulin genes (discussed in Chapter 18, *Recombination in the Immune System*). NHEJ is initiated at double-strand breaks by a protein complex known as Ku that binds to each of the DNA ends, as seen in **FIGURE 15.9**. The Ku complex is a heterodimer of two subunits known as Ku70 and Ku80. After the DNA ends are bound by the Ku complex, the MRN complex (or MRX complex in yeast) assists in bringing the broken DNA ends together by acting as a bridge between the two molecules. The MRN complex consists of Mre11, Rad50, and Nbs1 (Xrs2 in yeast). In order to join the two DNA ends, a ligation reaction is carried out by DNA ligase IV (LigIV), which functions specifically in NHEJ. Also involved in this process is a protein called XRCC4 (Lif1 in yeast) that associates with LigIV to allow for the DNA ends to be ligated. Frequently during the NHEJ process, mutations are generated through nucleotide deletion and insertion that occurs during the processing steps prior to ligation.

FIGURE 15.9 Nonhomologous end-joining. The black dot on one of the two DSB ends signifies a nonligatable end (A). The double-strand break ends are bound by the Ku heterodimer (B). The Ku:DNA complexes are juxtaposed (C) to bridge the ends and the gap is filled in by processing enzymes and Pol λ or Pol μ. The ends are ligated by the specialized DNA ligase LigIV with its partner XRCC4 (D) to repair the double-strand break (E). Adapted from J. M. Jones, M. Gellert, and W. Yang, *Structure* 9 (2001): 881–884.

15.7 The Single-Strand Annealing Mechanism Functions at Some Double-Strand Breaks

Key concepts

- Single-strand annealing (SSA) occurs at double-strand breaks between direct repeats.
- Resection of double-strand break ends results in 3′ single-stranded tails.
- Complementarity between the repeats allows for annealing of the single strands.
- The sequence between the direct repeats is deleted after SSA is completed.

There are some homologous recombination events to repair double-strand breaks that are not dependent on strand invasion, D-loop formation, or the proteins that promote these processes. In order to account for these recombination events, which typically take place between direct repeats (repeat sequences that are oriented in the same direction) a model has been devised in which homology between single-strand overhangs is used to direct recombination, shown in **FIGURE 15.10**. When a double-strand break occurs between two direct repeats, the ends are resected to give single strands. When resection proceeds to the repeat sequences such that the 3′ single-strand tails are homologous, the single strands can anneal. Processing and ligation of the 3′ ends then seals the double-strand break. As shown in Figure 15.10, this resection, followed by annealing, eliminates the sequence between the two direct repeats and leaves only one copy of the repeated sequence. There are types of human disease that arise from loss of sequence between direct repeats, presumably through a single-strand annealing (SSA) mechanism. These diseases include insulin-dependent diabetes, Fabry disease, and α-thalassemia.

15.8 Break-Induced Replication Can Repair Double-Strand Breaks

Key concepts

- Break-induced replication (BIR) is initiated by a one-ended double-strand break.
- BIR at repeated sequences can result in translocations.

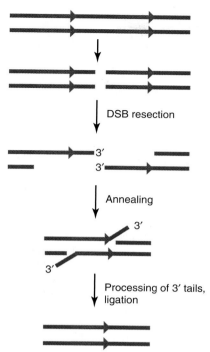

FIGURE 15.10 Single-strand annealing model of homologous recombination. A double-strand break occurs between direct repeats, depicted as red arrows. Following end processing to form single-strand tails with 3′-OH ends, the single strands anneal by homology at the red arrows. The single-strand tails are removed by endonucleases that recognize branch structures. The end product is double-strand break repair with a deletion of the sequences between the repeats and loss of one repeat sequence.

We saw in the previous section that double-strand breaks between direct repeats can induce the single-strand annealing mechanism. There are other types of repeat sequences at which double-strand breaks induce a repair mechanism known as *break-induced replication* (BIR). During DNA replication, certain sequences termed *fragile sites* are particularly susceptible to double-strand break formation. They often contain repeat sequences related to those found in transposable elements (discussed in Chapter 17, *Transposable Elements and Retroviruses*) and are located throughout the genome. Fragile sites are prone to breakage during DNA replication, creating a double-strand break at the site of replication. BIR can initiate repair from these DSBs by using the homologous sequence from a repeat on a nonhomologous chromosome, creating a nonreciprocal translocation, as seen in **FIGURE 15.11**.

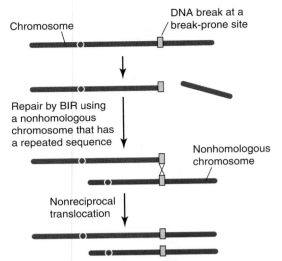

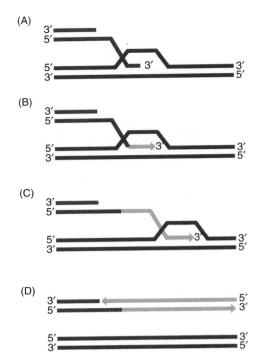

FIGURE 15.11 Break-induced replication can result in nonreciprocal translocations. A DNA break on the red chromosome results on loss of the chromosome end and a break with only one end. The end is repaired by recombination, using a homologous sequence found on a different chromosome, here the blue chromosome. Since there is only one end at the broken chromosome, repair occurs by copying the blue chromosome sequence to the end. This results in a translocation of some of the blue chromosome sequence to the red chromosome.

FIGURE 15.12 Possible mechanisms of break-induced replication. Strand invasion into homologous sequences by a single-strand tail with a 3'-OH end forms a D-loop. In (A), synthesis results in a single-strand region that is later converted into duplex DNA. In (B), a single replication fork is formed that moves in one direction to the end of the template sequence. Resolution of the Holliday junction results in newly synthesized DNA on both molecules. In (C), the Holliday junction branch migrates to result in newly synthesized DNA only on the broken strand, as in (A). Adapted from M. J. McEachern and J. E. Haber, *Annu. Rev. Biochem.* 75 (2006): 111–135.

The mechanism of BIR involves resection of the double-strand break end to leave a 3'–OH single-strand overhang, which can then undergo strand invasion at a homologous sequence, shown in **FIGURE 15.12**. The invading strand causes the formation of a D-loop that can be thought of as a replication bubble. The invading strand is then extended using the donor DNA as template for replication. When the invading strand is displaced, it can then act as a single-stranded template on which synthesis can be primed to create double-stranded DNA. The template strand is used until replication reaches the end of the chromosome; as a result, gene conversions from BIR events can be hundreds of kilobases long. Additionally, chromosome translocations can occur from this process if the homology used during strand invasion is a result of repeat sequences present at various sites in the genome. Template switching that occurs during break-induced replication can result in some of the complex chromosomal rearrangements that are seen in tumor cells.

15.9 Recombining Meiotic Chromosomes Are Connected by the Synaptonemal Complex

Key concepts

- During the early part of meiosis, homologous chromosomes are paired in the synaptonemal complex.
- The mass of chromatin of each homolog is separated from the other by a proteinaceous complex.

A basic paradox in recombination is that the parental chromosomes never seem to be in close enough contact for recombination of DNA to occur. The chromosomes enter meiosis in the form of replicated (sister chromatid) pairs, which are visible as a mass of chromatin.

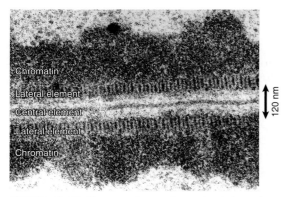

FIGURE 15.13 The synaptonemal complex brings chromosomes into juxtaposition. Reproduced from D. von Wettstein. *Proc. Natl. Acad. Sci. USA* 68 (1971): 851–855. Photo courtesy of Diter von Wettstein, Washington State University.

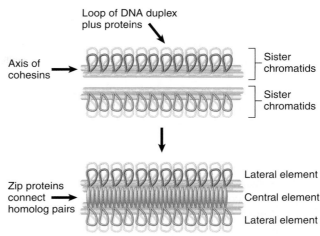

FIGURE 15.14 Each pair of sister chromatids has an axis made of cohesins. Loops of chromatin project from the axis. The synaptonemal complex is formed by linking together the axes via Zip proteins.

They pair to form the synaptonemal complex, and it has been assumed for many years that this represents some stage involved with recombination—possibly a necessary preliminary to exchange of DNA. A more recent view is that the synaptonemal complex is a consequence rather than a cause of recombination, but we have yet to define how the structure of the synaptonemal complex relates to molecular contacts between DNA molecules.

Synapsis begins when each chromosome (sister chromatid pair) condenses around a proteinaceous structure called the **axial element**. The axial elements of corresponding chromosomes then become aligned, and the synaptonemal complex forms as a tripartite structure, in which the axial elements, now called **lateral elements**, are separated from each other by a **central element**. FIGURE 15.13 shows an example.

Each chromosome at this stage appears as a mass of chromatin bounded by a lateral element. The two lateral elements are separated from each other by a fine, but dense, central element. The triplet of parallel dense strands lies in a single plane that curves and twists along its axis. The distance between the homologous chromosomes is considerable in molecular terms, at more than 200 nm (the diameter of DNA is 2 nm). Thus a major problem in understanding the role of the complex is that, although it aligns homologous chromosomes, it is far from bringing homologous DNA molecules into contact.

The only visible link between the two sides of the synaptonemal complex is provided by spherical or cylindrical structures observed in fungi and insects. They lie across the complex and are called **nodes** or **recombination nodules**; they occur with the same frequency and distribution as the chiasmata. Their name reflects the possibility that they may prove to be the sites of recombination.

From mutations that affect synaptonemal complex formation, we can relate the types of proteins that are involved to its structure. FIGURE 15.14 presents a molecular view of the synaptonemal complex. Its distinctive structural features are due to two groups of proteins:

- The cohesins form a single linear axis for each pair of sister chromatids from which loops of chromatin extend. This is equivalent to the lateral element of Figure 15.13. (The cohesins belong to a general group of proteins involved in connecting sister chromatids so that they segregate properly at mitosis or meiosis.)
- The lateral elements are connected by transverse filaments that are equivalent to the central element of Figure 15.13. These are formed from Zip proteins.

Mutations in proteins that are needed for lateral elements to form are found in the genes coding for cohesins. The cohesins that are used in meiosis include Smc3 (which is also used in mitosis) and Rec8 (which is specific to meiosis and is related to the mitotic cohesin Scc1). The cohesins appear to bind to specific sites along the chromosomes in both mitosis and meiosis. They are likely to play a structural role in chromosome segregation. At meiosis, the formation of the lateral elements may be necessary for the later stages of recombination, because although these mutations do not prevent the formation of double-strand breaks, they do block formation of recombinants.

The *zip1* mutation allows lateral elements to form and to become aligned, but they do not become closely synapsed. The N-terminal domain of the Zip1 protein is localized in the central element, but the C-terminal domain is localized in the lateral elements. Two other proteins, Zip2 and Zip3, are also localized with Zip1. The group of Zip proteins form transverse filaments that connect the lateral elements of the sister chromatid pairs.

15.10 The Synaptonemal Complex Forms after Double-Strand Breaks

Key concepts

- Double-strand breaks that initiate recombination occur before the synaptonemal complex forms.
- If recombination is blocked, the synaptonemal complex cannot form.
- Meiotic recombination involves two phases: one that results in gene conversion without crossover, and one that results in crossover products.

There is good evidence in yeast that double-strand breaks initiate recombination in both homologous and site-specific recombination. Double-strand breaks were initially implicated in the change of mating type, which involves the replacement of one sequence by another (see *Section 15.21, Unidirectional Gene Conversion Is Initiated by the Recipient* MAT *Locus*). Double-strand breaks also occur early in meiosis at sites that provide hotspots for recombination. Their locations are not sequence specific. They tend to occur in promoter regions and in general to coincide with more accessible regions of chromatin. The frequency of recombination declines in a gradient on one or both sides of the hotspot. The hotspot identifies the site at which recombination is initiated, and the gradient reflects the probability that the recombination events will spread from it.

We may now interpret the role of double-strand breaks in molecular terms. The blunt ends created by the double-strand break are rapidly converted on both sides into long 3' single-stranded ends, as shown in the model of Figure 15.4. A yeast mutation (*rad50*) that blocks the conversion of the blunt end into the single-stranded protrusion is defective in recombination. This suggests that double-strand breaks are necessary for recombination. The gradient is determined by the declining probability that a single-stranded region will be generated as distance increases from the site of the DSB.

In *rad50* mutants, the 5' ends of the double-strand breaks are connected to the protein Spo11, which, as discussed previously, is homologous to the catalytic subunits of a family of type II topoisomerases. This suggests that Spo11 may be a topoisomerase-like enzyme that generates the double-strand breaks. Recall the model for this reaction shown in Figure 15.5, which suggests that Spo11 interacts reversibly with DNA; the break is converted into a permanent structure by an interaction with another protein that dissociates the Spo11 complex. Removal of Spo11 is then followed by nuclease action. At least nine other proteins are required to process the double-strand breaks. One group of proteins is required to convert the double-strand breaks into protruding 3'–OH single-stranded ends. Another group then enables the single-stranded ends to invade homologous duplex DNA.

The correlation between recombination and synaptonemal complex formation is well established, and recent work has shown that all mutations that abolish chromosome pairing in *Drosophila* or in yeast also prevent recombination. The system for generating the double-strand breaks that initiate recombination is generally conserved. Spo11 homologs have been identified in several higher eukaryotes, and a mutation in the *Drosophila* gene blocks all meiotic recombination.

There are few systems in which it is possible to compare molecular and cytological events at recombination, but recently there has been progress in analyzing meiosis in *Saccharomyces cerevisiae*. The relative timing of events is summarized in **FIGURE 15.15**.

Double-strand breaks appear and then disappear over a 60-minute period. The first joint molecules, which are putative recombination intermediates, appear soon after the DSBs disappear. The sequence of events suggests that double-strand breaks, individual pairing reactions, and formation of recombinant structures occur in succession at the same chromosomal site.

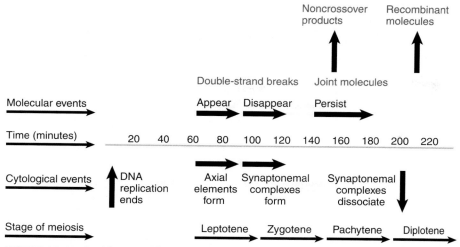

FIGURE 15.15 Double-strand breaks appear when axial elements form and disappear during the extension of synaptonemal complexes. Joint molecules appear and persist until DNA recombinants are detected at the end of pachytene.

Double-strand breaks appear during the period when axial elements form. They disappear during the conversion of the paired chromosomes into synaptonemal complexes. This relative timing of events suggests that formation of the synaptonemal complex results from the initiation of recombination via the introduction of double-strand breaks and their conversion into later intermediates of recombination. This idea is supported by the observation that the *rad50* mutant cannot convert axial elements into synaptonemal complexes. This refutes the traditional view of meiosis that the synaptonemal complex represents the need for chromosome pairing to precede the molecular events of recombination.·

It has been difficult to determine whether recombination occurs at the stage of synapsis, because recombination is assessed by the appearance of recombinants after the completion of meiosis. By assessing the appearance of recombinants in yeast directly in terms of the production of DNA molecules containing diagnostic restriction sites, though, it has been possible to show that recombinants appear at the end of pachytene. This clearly places the completion of the recombination event after the formation of synaptonemal complexes.

Thus the synaptonemal complex forms after the double-strand breaks that initiate recombination, and it persists until the formation of recombinant molecules. It does not appear to be necessary for recombination as such, because some mutants that lack a normal synaptonemal complex can generate recombi-

nants. Mutations that abolish recombination, however, also fail to develop a synaptonemal complex. This suggests that the synaptonemal complex forms as a consequence of recombination, following chromosome pairing, and is required for later stages of meiosis.

The double-strand break repair model proposes that resolution of Holliday junctions gives rise to either noncrossover products (with a residual stretch of hybrid DNA) or to crossovers (recombinants), depending on which strands are involved in resolution (see Figure 15.4). Recent measurements of the times of production of noncrossover and crossover molecules, however, suggest that this may not be true. Crossovers do not appear until well after the first appearance of joint molecules, whereas noncrossovers appear almost simultaneously with the joint molecules (see Figure 15.15). The appearance of these two types of products correspond to what is considered two independent phases of meiotic recombination. In the first phase, double-strand breaks are repaired through a synthesis-dependent strand-annealing reaction, whereas in the second phase, the double-strand break repair pathway is predominant and results largely in crossover products. The molecular outcomes of these phases are illustrated in **FIGURE 15.16**. If both types of product were produced by the same resolution process, however, we would expect them to appear at the same time. The discrepancy in timing suggests that crossovers are produced as previously thought—by resolution of joint molecules—but that there may be some other route for the production of noncrossovers.

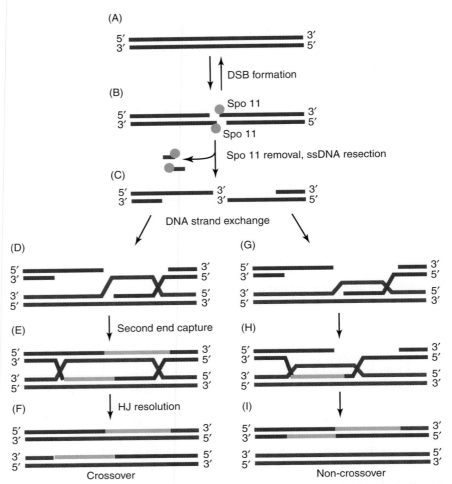

FIGURE 15.16 Model of meiotic homologous recombination. A DNA duplex (A) is cleaved by Spo11 to form a double-strand break with Spo11 covalently attached to the ends (B). After Spo11 is removed the ends are resected by the MRX/N complex to give single-strand tails with 3'-OH ends, which are complex with Rad51 and Dmc1. Strand exchange occurs by strand invasion (D and G). Second end capture results in a double Holliday junction, which is resolved to form crossover products (E and F). Most of the double-strand breaks do not engage in a second end capture mechanism and instead engage in a synthesis-dependent strand-annealing mechanism (H and I), which results in noncrossover products. Adapted from M. J. Neale and S. Keeney, *Nature* 442 (2006): 153–158.

15.11 Pairing and Synaptonemal Complex Formation Are Independent

Key concept

- Mutations can occur in either chromosome pairing or synaptonemal complex formation without affecting the other process.

We can distinguish the processes of pairing and synaptonemal complex formation by the effects of two mutations, each of which blocks one of the processes without affecting the other.

The *zip2* mutation allows chromosomes to pair, but they do not form synaptonemal complexes. Thus recognition between homologs is independent of recombination or synaptonemal complex formation.

The specificity of association between homologous chromosomes is controlled by the gene *HOP2* in *S. cerevisiae*. In *hop2* mutants, normal amounts of synaptonemal complex form at meiosis, but the individual complexes contain nonhomologous chromosomes. This suggests that the formation of synaptonemal complexes as such is independent of homology (and therefore cannot be based on any extensive comparison of DNA sequences). The usual role of Hop2 is to prevent nonhomologous chromosomes from interacting.

Double-strand breaks form in the mispaired chromosomes in the synaptonemal complexes of *hop2* mutants, but they are not repaired. This suggests that, if formation of the synaptonemal complex requires double-strand breaks, it does not require any extensive reaction of these breaks with homologous DNA.

It is not clear what usually happens during pachytene, before DNA recombinants are observed. It may be that this period is occupied by the subsequent steps of recombination, which involve the extension of strand exchange, DNA synthesis, and resolution.

At the next stage of meiosis (diplotene), the chromosomes shed the synaptonemal complex; the chiasmata then become visible as points at which the chromosomes are connected. This has been presumed to indicate the occurrence of a genetic exchange, but the molecular nature of a chiasma is unknown. It is possible that it represents the residuum of a completed exchange, or that it represents a connection between homologous chromosomes where a genetic exchange has not yet been resolved. Later in meiosis, the chiasmata move toward the ends of the chromosomes. This flexibility suggests that they represent some remnant of the recombination event rather than providing the actual intermediate.

Recombination events occur at discrete points on meiotic chromosomes, but we cannot as yet correlate their occurrences with the discrete structures that have been observed; that is, recombination nodules and chiasmata. Insights into the molecular basis for the formation of discontinuous structures, however, are provided by the identification of proteins involved in yeast recombination that can be localized to discrete sites. These include Msh4 (which is homologous to bacterial proteins involved in mismatch repair) and Dmc1 and Rad51 (which are homologs of the *E. coli* RecA protein). The exact roles of these proteins in recombination remain to be established.

Recombination events are subject to a general control. Only a minority of interactions actually mature as crossovers, but these are distributed in such a way that, in general, each pair of homologs acquires only one to two crossovers, yet the probability of zero crossovers for a homologous pair is very low (<0.1%). This process is probably the result of a single crossover control, because the nonrandomness of crossovers is generally disrupted in certain mutants. Furthermore, the occurrence of recombination is necessary for progress through meiosis, and a "checkpoint" system exists to block meiosis if recombination has not

occurred. (The block is lifted when recombination has been successfully completed; this system provides a safeguard to ensure that cells do not try to segregate their chromosomes until recombination has occurred.)

15.12 The Bacterial RecBCD System Is Stimulated by *chi* Sequences

Key concepts

- The RecBCD complex has nuclease and helicase activities.
- RecBCD binds to DNA downstream of a *chi* sequence, unwinds the duplex, and degrades one strand from 3'–5' as it moves to the *chi* site.
- The *chi* site triggers loss of the RecD subunit and nuclease activity.

The nature of the events involved in exchange of sequences between DNA molecules was first described in bacterial systems. Here the recognition reaction is part and parcel of the recombination mechanism and involves restricted regions of DNA molecules rather than intact chromosomes. The general order of molecular events is similar, though: A single strand from a broken molecule interacts with a partner duplex, the region of pairing is extended, and an endonuclease resolves the partner duplexes. Enzymes involved in each stage are known, although they probably represent only some of the components required for recombination.

Bacterial enzymes implicated in recombination have been identified by the occurrence of *rec⁻* mutations in their genes. The phenotype of *rec⁻* mutants is the inability to undertake generalized recombination. Some ten to twenty loci have been identified.

Bacteria do not usually exchange large amounts of duplex DNA, but there may be various routes to initiate recombination in prokaryotes. In some cases, DNA may be available with free single-stranded 3' ends: DNA may be provided in single-stranded form (as in conjugation; see *Section 12.7, Conjugation Transfers Single-Stranded DNA*), single-stranded gaps may be generated by irradiation damage, or single-stranded tails may be generated by phage genomes undergoing replication by a rolling circle. In circumstances involving two duplex molecules (as in recombination at meiosis in eukaryotes), however, single-stranded regions and 3' ends must be generated.

One mechanism for generating suitable ends has been discovered as a result of the existence of certain hotspots that stimulate recombination. These hotspots, which were discovered in phage lambda in the form of mutants called *chi*, have single base-pair changes that create sequences that stimulate recombination. These sites lead us to the role of other proteins involved in recombination.

These sites share a constant nonsymmetrical sequence of 8 bp:

5' GCTGGTGG 3'

3' CGACCACC 5'

The *chi* sequence occurs naturally in *E. coli* DNA about once every 5 to 10 kb. Its absence from wild-type lambda DNA, and also from other genetic elements, shows that it is not essential for recombination.

A *chi* sequence stimulates recombination in its general vicinity, within about a distance of up to 10 kb from the site. A *chi* site can be activated by a double-strand break made several kb away on one particular side (to the right of the sequence shown above). This dependence on orientation suggests that the recombination apparatus must associate with DNA at a broken end, and then can move along the duplex only in one direction.

chi sites are targets for the action of an enzyme coded by the genes *recBCD*. This complex possesses several activities: It is a potent nuclease that degrades DNA (originally identified as the activity exonuclease V); it has helicase activities that can unwind duplex DNA in the presence of a single-strand binding protein (SSB); and it has ATPase activity. Its role in recombination may be to provide a single-stranded region with a free 3' end.

FIGURE 15.17 shows how these reactions are coordinated on a substrate DNA that has a *chi* site. RecBCD binds to DNA at a double-stranded end. Two of its subunits have helicase activities: RecD functions with 5'–3' polarity, and RecB functions with 3'–5' polarity. Translocation along DNA and unwinding the double helix is initially driven by the RecD subunit. As RecBCD advances, it degrades the released single strand with the 3' end. When it reaches the *chi* site, it recognizes the top strand of the *chi* site in single-stranded form. This causes the enzyme to pause. It then cleaves the top strand of the DNA at a position between four and six bases to the right of *chi*. Recognition of the *chi* site causes the RecD subunit to dissociate or become inactivated, at which point the enzyme loses its nuclease activity. It continues, however,

RecBCD binds 5' a double-strand break 3'

RecBCD unwinds and degrades DNA
RecD subunit translocates 5'–3'

RecBCD cleaves single strand at *chi*

RecD dissociates

RecBC continues as helicase
RecB subunit translocates 3'–5'

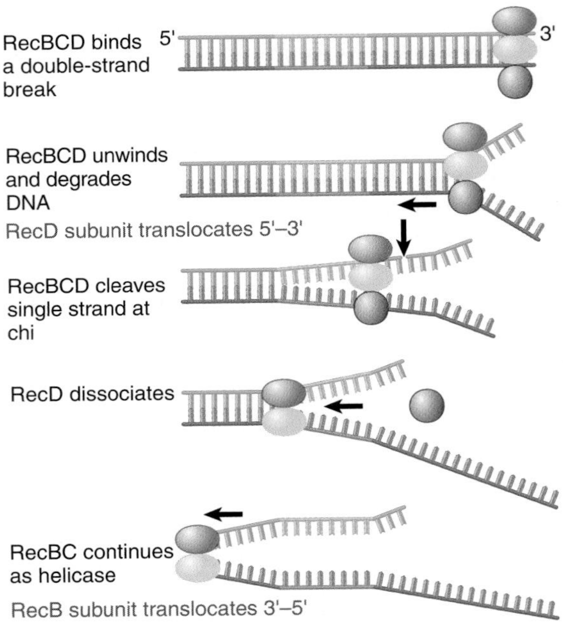

FIGURE 15.17 RecBCD nuclease approaches a *chi* sequence from one side, degrading DNA as it proceeds; at the *chi* site, it makes an endonucleolytic cut, loses RecD, and retains only the helicase activity.

to function as a helicase—now using only the RecB subunit to drive translocation—at about half the previous speed. The overall result of this interaction is to generate single-stranded DNA with a 3' end at the *chi* sequence. This is a substrate for recombination.

15.13 Strand-Transfer Proteins Catalyze Single-Strand Assimilation

Key concept

- RecA forms filaments with single-stranded or duplex DNA and catalyzes the ability of a single-stranded DNA with a free 3' end to displace its counterpart in a DNA duplex.

The *E. coli* protein RecA was the first example to be discovered of a DNA strand-transfer protein. It is the paradigm for a group that includes several other bacterial and archaeal proteins, as well as eukaryotic Rad51 and the meiotic protein Dmc1 (both discussed in detail in *Section 15.15, Eukaryotic Genes Involved in Homologous Recombination*). Analysis of yeast *rad51* mutants shows that this class of protein plays a central role in recombination. They accumulate double-strand breaks and fail to form normal synaptonemal complexes. This reinforces the idea that exchange of strands between DNA

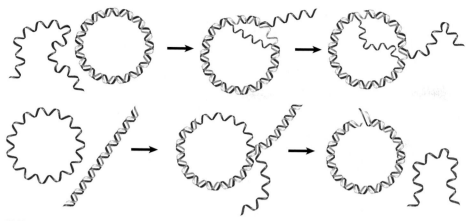

FIGURE 15.18 RecA promotes the assimilation of invading single strands into duplex DNA so long as one of the reacting strands has a free end.

duplexes is involved in formation of the synaptonemal complex, and raises the possibility that chromosome synapsis is related to the bacterial strand assimilation reaction.

RecA in bacteria has two quite different types of activity: It can stimulate protease activity in the SOS response (see *Section 16.13, RecA Triggers the SOS System*), and can promote base pairing between a single strand of DNA and its complement in a duplex molecule. Both activities are activated by single-stranded DNA in the presence of ATP.

The DNA-handling activity of RecA enables a single strand to displace its homolog in a duplex in a reaction that is called single-strand assimilation (or single-strand invasion). The displacement reaction can occur between DNA molecules in several configurations and has three general conditions:

- One of the DNA molecules must have a single-stranded region.
- One of the molecules must have a free 3′ end.
- The single-stranded region and the 3′ end must be located within a region that is complementary between the molecules.

The reaction is illustrated in **FIGURE 15.18**. When a linear single strand invades a duplex, it displaces the original partner to its complement. The reaction can be followed most easily by making either the donor or recipient a circular molecule. The reaction proceeds 5′–3′ along the strand whose partner is being displaced and replaced; that is, the reaction involves an exchange in which (at least) one of the exchanging strands has a free 3′ end.

Single-strand assimilation is potentially related to the initiation of recombination. All models call for an intermediate in which one or both single strands cross over from one duplex to the other (see Figure 15.4). RecA could catalyze this stage of the reaction. In the bacterial context, RecA acts on substrates generated by RecBCD. RecBCD-mediated unwinding and cleavage can be used to generate ends that initiate the formation of heteroduplex joints. RecA can take the single strand with the 3′ end that is released when RecBCD cuts at *chi*, and can use it to react with a homologous duplex sequence, thus creating a joint molecule.

All of the bacterial and archaeal proteins in the RecA family can aggregate into long filaments with single-stranded or duplex DNA. There are six RecA monomers per turn of the filament, which has a helical structure with a deep groove that contains the DNA. The stoichiometry of binding is three nucleotides (or base pairs) per RecA monomer. The DNA is held in a form that is extended 1.5 times relative to duplex B DNA, making a turn every 18.6 nucleotides (or base pairs). When duplex DNA is bound, it contacts RecA via its minor groove, leaving the major groove accessible for possible reaction with a second DNA molecule.

The interaction between two DNA molecules occurs within these filaments. When a single strand is assimilated into a duplex, the first step is for RecA to bind the single strand into a **presynaptic filament**. The duplex is then incorporated, probably forming some sort of triple-stranded structure. In this system, synapsis precedes physical exchange of material, because the pairing reaction can take place even in the absence of free ends, when strand exchange is impossible. A free 3′ end is required for strand exchange. The reaction occurs within the filament, and RecA remains bound to the strand that was originally single, so that at the end of the reaction RecA is bound to the duplex molecule.

All of the proteins in this family can promote the basic process of strand exchange without a requirement for energy input. RecA, however, augments this activity by using ATP hydrolysis. Large amounts of ATP are hydrolyzed during the reaction. The ATP may act through an allosteric effect on RecA conformation. When bound to ATP, the DNA-binding site of RecA has a high affinity for DNA; this is needed to bind DNA and for the pairing reaction. Hydrolysis of ATP converts the binding site to low affinity, which is needed to release the heteroduplex DNA.

We can divide the reaction that RecA catalyzes between single-stranded and duplex DNA into three phases:

- a slow presynaptic phase in which RecA polymerizes on single-stranded DNA;
- a fast pairing reaction between the single-stranded DNA and its complement in the duplex to produce a heteroduplex joint, and
- a slow displacement of one strand from the duplex to produce a long region of heteroduplex DNA.

The presence of SSB stimulates the reaction by ensuring that the substrate lacks secondary structure. It is not clear yet how SSB and RecA both can act on the same stretch of DNA. Like SSB, RecA is required in stoichiometric amounts, which suggests that its action in strand assimilation involves binding cooperatively to DNA to form a structure related to the filament.

When a single-stranded molecule reacts with a duplex DNA, the duplex molecule becomes unwound in the region of the recombinant joint. The initial region of heteroduplex DNA may not even lie in the conventional double helical form, but could consist of the two strands associated side by side. A region of this type is called a *paranemic joint*, as compared with the classical intertwined *plectonemic* relationship of strands in a double helix, depicted in **FIGURE 15.19**. A paranemic joint is unstable; further progress of the reaction requires its conversion to the double-helical form. This reaction is equivalent to removing negative supercoils and may require an enzyme that solves the unwinding/rewinding problem by making transient breaks that allow the strands to rotate about each other.

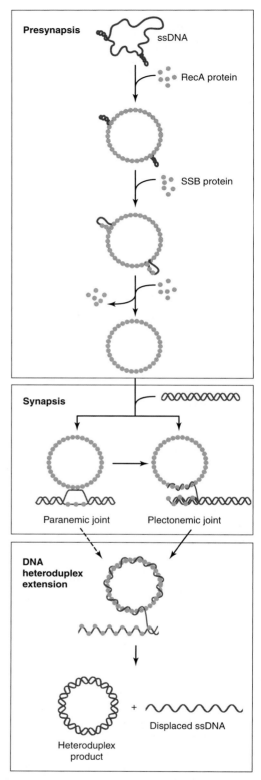

FIGURE 15.19 Formation of paranemic and plectonemic joints. Once homology is found side by side pairing is formed, called paranemic pairing, which then transitions to plectonemic pairing where the paired DNA strands are in a double helix configuration. Note that these pairing stages involve strand invasion and D-loop formation. Adapted from P. R. Bianco and S. C. Kowalczykowski. *Encyclopedia of Life Sciences*. John Wiley & Sons, Ltd., 2005.

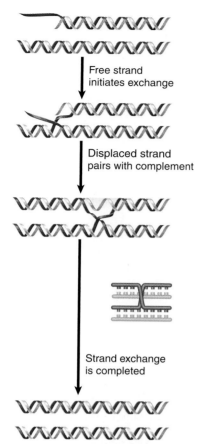

FIGURE 15.20 RecA-mediated strand exchange between partially duplex and entirely duplex DNA generates a joint molecule with the same structure as a recombination intermediate.

All of the reactions we have discussed so far represent only a part of the potential recombination event: the invasion of one duplex by a single strand. Two duplex molecules can interact with each other under the sponsorship of RecA, provided that one of them has a single-stranded region of at least fifty bases. The single-stranded region can take the form of a tail on a linear molecule or of a gap in a circular molecule.

The reaction between a partially duplex molecule and an entirely duplex molecule leads to the exchange of strands. An example is illustrated in **FIGURE 15.20**. Assimilation starts at one end of the linear molecule, where the invading single strand displaces its homolog in the duplex in the customary way. When the reaction reaches the region that is duplex in both molecules, though, the invading strand unpairs from its partner, which then pairs with the other displaced strand.

At this stage, the molecule has a structure indistinguishable from the recombinant joint in Figure 15.4. The reaction sponsored *in vitro* by RecA can generate Holliday junctions, which suggests that the enzyme can mediate reciprocal strand transfer. We know less about the geometry of four-strand intermediates bound by RecA, but presumably two duplex molecules can lie side by side in a way consistent with the requirements of the exchange reaction.

The biochemical reactions characterized *in vitro* leave open many possibilities for the functions of strand-transfer proteins *in vivo*. Their involvement is triggered by the availability of a single-stranded 3' end. In bacteria, this is most likely generated when RecBCD processes a double-strand break to generate a single-stranded end. One of the main circumstances in which this is invoked may be when a replication fork stalls at a site of DNA damage (see *Section 16.9, Recombination Is an Important Mechanism to Recover from Replication Errors*). The introduction of DNA during conjugation, when RecA is required for recombination with the host chromosome, is more closely related to conventional recombination. In yeast, DSBs may be generated by DNA damage or as part of the normal process of recombination. In either case, processing of the break to generate a 3' single-stranded end is followed by loading the single strand into a filament with Rad51, followed by a search for matching duplex sequences. This can be used in both repair and recombination reactions.

15.14 Holliday Junctions Must Be Resolved

Key concepts

- The bacterial Ruv complex acts on recombinant junctions.
- RuvA recognizes the structure of the junction and RuvB is a helicase that catalyzes branch migration.
- RuvC cleaves junctions to generate recombination intermediates.
- Resolution in eukaryotes is less well understood, but a number of meiotic and mitotic proteins are implicated.

One of the most critical steps in recombination is the resolution of the Holliday junction, which determines whether there is a reciprocal recombination or a reversal of the structure that leaves only a short stretch of hybrid

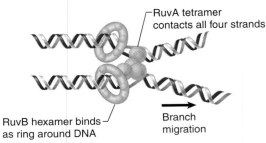

FIGURE 15.21 RuvAB is an asymmetric complex that promotes branch migration of a Holliday junction.

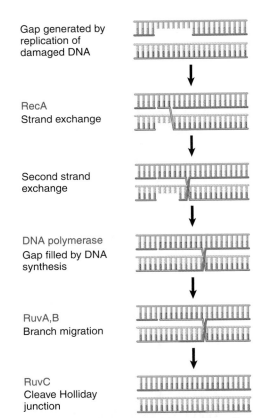

FIGURE 15.22 Bacterial enzymes can catalyze all stages of recombination in the repair pathway following the production of suitable substrate DNA molecules.

DNA (see Figure 15.4). Branch migration from the exchange site (see Figure 15.6) determines the length of the region of hybrid DNA (with or without recombination). The proteins involved in stabilizing and resolving Holliday junctions have been identified as the products of the *ruv* genes in *E. coli*. RuvA and RuvB increase the formation of heteroduplex structures. RuvA recognizes the structure of the Holliday junction. RuvA binds to all four strands of DNA at the crossover point and forms two tetramers that sandwich the DNA. RuvB is a hexameric helicase with an ATPase activity that provides the motor for branch migration. Hexameric rings of RuvB bind around each duplex of DNA upstream of the crossover point. A diagram of the complex is shown in FIGURE 15.21.

The RuvAB complex can cause the branch to migrate as fast as 10 to 20 bp/sec. A similar activity is provided by another helicase, RecG. RuvAB displaces RecA from DNA during its action. The RuvAB and RecG activities both can act on Holliday junctions, but if both are mutant, *E. coli* is completely defective in recombination activity.

The third gene, *ruvC*, codes for an endonuclease that specifically recognizes Holliday junctions. It can cleave the junctions *in vitro* to resolve recombination intermediates. A common tetranucleotide sequence provides a hotspot for RuvC to resolve the Holliday junction. The tetranucleotide (ATTG) is asymmetric, and thus may direct resolution with regard to which pair of strands is nicked. This determines whether the outcome is **patch recombinant** formation (no overall recombination) or **splice recombinant** formation (recombination between flanking markers). Crystal structures of RuvC and other junction-resolving enzymes show that there is little structural similarity among the group, in spite of their common function.

We may now account for the stages of recombination in *E. coli* in terms of individual proteins. FIGURE 15.22 shows the events that are involved in using recombination to repair a gap in one duplex by retrieving material from the other duplex. The major caveat in applying these conclusions to recombination in eukaryotes is that bacterial recombination generally involves interaction between a fragment of DNA and a whole chromosome. It occurs as a repair reaction that is stimulated by damage to DNA, but this is not entirely equivalent to recombination between genomes at meiosis. Nonetheless, similar molecular activities are involved in manipulating DNA.

All of this suggests that recombination uses a "resolvasome" complex that includes enzymes catalyzing branch migration as well as junction-resolving activity. It is possible that mammalian cells contain a similar complex.

Although resolution in eukaryotic cells is less well understood, a number of proteins have been implicated in mitotic and meiotic resolution. *S. cerevisiae* strains that contain *mus81* mutations are defective in recombination.

Mus81 is a component of an endonuclease that resolves Holliday junctions into duplex structures. The resolvase is important both in meiosis and for restarting stalled replication forks (see *Section 16.9, Recombination Is an Important Mechanism to Recover from Replication Errors*). Other proteins known to be involved in the resolution process are described in the broader context of eukaryotic homologous recombination factors in the following section.

15.15 Eukaryotic Genes Involved in Homologous Recombination

Key concepts

- The MRX complex, Exo1, and Sgs1/Dna2 in yeast and the MRN complex and BLM in mammalian cells resect double-strand breaks.
- The Rad51 recombinase binds to single-stranded DNA with the aid of mediator proteins, which overcome the inhibitory effects of RPA.
- Strand invasion is dependent on Rad54 and Rdh54 in yeast and Rad54 and Rad54B in mammalian cells.
- Yeast Sgs1, Mus81/Mms4 and human BLM, MUS81/EME1 are implicated in resolution of Holliday junctions.

Previously we have briefly mentioned some of the proteins involved in homologous recombination in eukaryotes. In this section, the proteins involved in homologous recombination will be discussed in more detail, focusing on the double-strand break repair and synthesis-dependent strand-annealing models. Additionally, the steps in the single-strand annealing and break-induced replication mechanisms that overlap with those of double-strand break repair and synthesis-dependent strand annealing proceed by the same enzymatic processes. Many of the eukaryotic homologous recombination genes are called *RAD* genes, because they were first isolated in screens for mutants with increased sensitivity to X-ray irradiation. X-rays make DSBs in DNA; thus it is not surprising that *rad* mutants sensitive to X-rays also are defective in mitotic and meiotic recombination. The double-strand break repair model shown in Figure 15.4 indicates at which step the proteins described below act.

1. End processing/presynapsis. In mitotic cells, double-strand breaks are produced by exogenous sources such as irradiation or chemical treatment, and from endogenous sources such as topoisomerases and nicks on the template strand. During replication nicks are converted to double-strand breaks. The ends of these breaks are processed by exonucleolytic degradation to have single-strand tails with 3′–OH ends. In meiosis, double-strand breaks are induced by Spo11-dependent cleavage. The first step in end processing entails binding of the broken end by the MRN or MRX complex, in association with the endonuclease Sae2 (CtIP in mammalian cells).

Mre11 works as part of a complex with two other factors, called Rad50 and Xrs2 in yeast and Rad50 and Nbs1 in humans. Xrs2 and Nbs1 have no similarity to each other. Rad50, mentioned previously in *Section 15.10*, is thought to help hold double-strand break ends together via dimers connected at the tips by a hook structure that becomes active in the presence of zinc ion, as shown in **FIGURE 15.23**. Rad50 and Mre11 are related to the bacterial proteins SbcC and SbcD, which have double-stranded DNA exonuclease and single-stranded endonuclease activities. Xrs2 and Nbs1 have DNA binding activity. Nbs1 is so named because a mutant allele was first discovered in individuals with *Nijmegen breakage syndrome*, a rare DNA damage syndrome that is associated with defective DNA damage checkpoint signaling and lymphoid tumors. Rare mutations that produce MRE11 with low activity have been found in humans, with the syndrome called ATLD (ataxia-telangiectasia-like disorder). Patients with this syndrome have not been reported to be cancer prone, but they have developmental problems and show defects in DNA damage checkpoint signaling. Mutations in *MRE11*, *RAD50*, or *XRS2* render cells sensitive to ionizing radiation and diploids have a poor meiotic outcome. Null mutations of *MRE11*, *RAD50*, or *NBS1* in mice are lethal.

After MRN/MRX and CtIP/Sae2 have prepared the double-strand break ends and removed any attached proteins or adduct that would inhibit end resection, the ends are resected by nucleases that act in concert with DNA helicases that unwind the duplex to expose single-strand DNA ends. Recent studies have identified the Exo1 and Dna2 exonucleases and the Sgs1 (in yeast) and BLM (in mammalian cells) helicases as critical factors for end processing.

After the double-strand breaks have been processed to have 3′–OH single-strand tails, the single-strand DNA is bound first by the single-strand DNA-binding protein RPA to remove any secondary structure. Next, with the aid of *mediator proteins* that help Rad51 displace

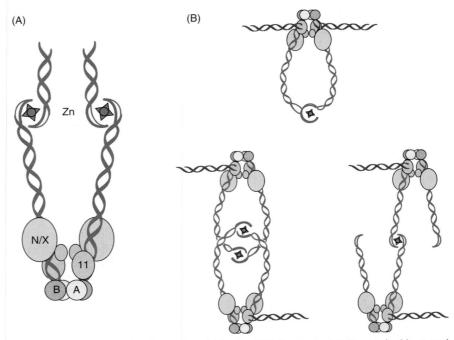

(A)   (B)

Zn

N/X

11

B A

FIGURE 15.23 Structure of Rad50 and model for the MRX/N complex binding to double-strand breaks. Rad50 has a coiled-coil domain similar to SMC proteins. The globular end contains two ATP binding and hydrolysis regions (A and B) and forms a complex with Mre11 and Nbs1 (N) or Xrs2 (X). The other end of the coil binds zinc cation and forms a dimer with another MRN/X molecule. The globular end binds to chromatin. The complex binds to double-strand breaks and can bring then together in a reaction involving two ends and one MRN/X complex (top right figure) or through an interaction between two MRN/X dimers as depicted in the bottom right figure. Adapted from M. Lichten, *Nat. Struct. Mol. Biol.* 12 (2005): 392–393.

RPA and bind the single-strand DNA, Rad51 forms a nucleofilament. Rad51 is related to RecA with 30% identity and forms a right-handed helical nucleofilament in an ATP-dependent process, with six Rad51 molecules and 18 nucleotides of single-strand DNA per helical turn. This binding stretches the DNA by approximately 1.5-fold, compared to B-form DNA. Rad51 is required for all homologous recombination processes, with the exception of the single-strand annealing and nonhomologous end-joining mechanisms. *RAD51* is not an essential gene in yeast, but null mutants are reduced in mitotic recombination and are sensitive to ionizing radiation. Double-strand breaks form but become degraded. In mice, *RAD51* is essential, and mice that are homozygous for mutant *rad51* do not survive past early stages of embryogenesis. This is thought to reflect the fact that, in vertebrates, at least one double-strand break occurs spontaneously during every replication cycle as a result of unrepaired template strand nicks.

In vitro, the mediators help in the removal of RPA and in the assembly of Rad51 on the single-stranded DNA, and promote *in vitro*

strand exchange reactions. In yeast, the mediators are Rad52 and Rad55/Rad57. Rad55 and Rad57, which form a stable heterodimer, have some homology to Rad51, but have no strand exchange activity *in vitro*.

In human cells, the mediators are also related to *RAD51*, with 20%–30% sequence identity, and are called *RAD51B*, *RAD51C*, *RAD51D*, *XRCC2*, and *XRCC3*, or the "*RAD51* paralogs." (Recall that paralogs are genes that have arisen by duplication within an organism and therefore are related by sequence, but have often evolved to have different functions.) The human mediator proteins form three complexes: one composed of RAD51B and RAD51C, a second composed of RAD51D and XRCC2, and the third composed of RAD51C and XRCC3. The paralogous genes have been deleted in chicken cell lines and knocked down in mammalian cells. Although the cell lines are viable, they are subject to numerous chromosome breaks and rearrangements and have reduced viability compared to normal cell lines. Mice in which the paralogous genes have been deleted are not viable and undergo early embryonic death.

The human BRCA2 protein, which is mutated in familial breast and ovarian cancers, and in the DNA damage syndrome Fanconi anemia, has mediator activity *in vitro*. As BRCA2 interacts physically with RAD51 protein and can bind to single-stranded DNA, this is not an unexpected activity for BRCA2. Indeed, genetic studies in mouse cells have shown that BRCA2 is required for homologous recombination. The related Brh2 protein of the pathogenic fungus *Ustilago maydis* binds in a complex to Rad51 protein and recruits it to single-strand DNA coated with RPA, to initiate Rad51 nucleofilament formation.

Yeast mutants deleted for *RAD55* or *RAD57* show temperature-dependent ionizing radiation sensitivity, and are reduced in homologous recombination. Neither mutant undergoes successful meiosis.

Rad52 is not essential for recombination *in vivo* in mammalian cells and does not appear to have a mediator role in these cells. It is, however, the most critical homologous recombination protein in yeast as *rad52* null mutants are extremely sensitive to ionizing radiation and are defective in all types of homologous recombination assayed. *RAD52*-deficient cells never complete meiosis.

2. Synapsis. Once the Rad51 filament has formed on single-strand DNA in the double-strand break-repair and synthesis-dependent strand annealing processes, a search for homology with another DNA molecule begins, and once found, strand invasion to form a D loop occurs. Strand invasion requires the Rad54 protein and the related Rdh54/Tid1 protein in yeast, and RAD54B in mammalian cells. Rad54 and Rdh54 are members of the SWI/SNF chromatin remodeling superfamily (see *Section 28.7, Chromatin Remodeling Is an Active Process*). They possess a double-strand DNA-dependent ATPase activity, can promote chromatin remodeling, and can translocate on double-stranded DNA, inducing superhelical stress in double-stranded DNA. Although Rad54, Rdh54, and RAD54B are not DNA helicases, the translocase activity causes local opening of double-strands, which may serve to stimulate D-loop formation. In yeast, *RAD54* is required for efficient mitotic recombination and for double-strand break repair as *RAD54*-deficient cells are sensitive to ionizing radiation and other DNA damaging compounds. *RDH54*-deficient cells have a modest defect in recombination and are slightly DNA damage sensitive. This sensitivity is enhanced when both *RAD54* and *RDH54* are

deleted. In meiotic cells, *rad54* mutants can complete meiosis, but have reduced spore viability. *rdh54* mutants are more deficient in meiosis, and have a stronger effect on spore viability. The double mutant does not complete meiosis. In chicken cells and mouse cells, *RAD54* and *RAD54B* deletion mutants are viable, in contrast to other homologous recombination gene deletion mutants. The cells show increased sensitivity to ionizing radiation and other clastogens (agents that cause chromosomal breaks) and have reduced rates of recombination.

3. DNA heteroduplex extension and branch migration. The proteins involved in this step are not as well defined as those required in the early steps of homologous recombination, yet the homologous recombination pathways of double-strand break repair and synthesis-dependent strand annealing both have D-loop extension as an important part of the process. D-loop formation results in Rad51 filament being formed on double-stranded DNA. Rad54 protein has the ability to remove Rad51 from double-stranded DNA. This step might be important for DNA polymerase extension from the 3' terminus. DNA polymerase delta is thought to be the polymerase for repair synthesis in double-strand break-mediated recombination; however, some recent studies have also implicated DNA polymerase η/Rad30 as being able to extend from the strand invasion intermediate terminus.

4. Resolution. The search for eukaryotic resolvase proteins has been a long process. Mutants of the DNA helicases Sgs1 of yeast and BLM in humans result in higher crossover rates, and as a result these helicases have been proposed to prevent crossover formation by Holliday junction resolution as noncrossovers. This is proposed to occur by branch migration of the double Holliday junctions to convergence, through the DNA helicase action, as seen in FIGURE 15.24 and FIGURE 15.25. The end structure is suggested to be a hemicatenane, where DNA strands are looped around each other. This structure is then resolved by the action of an associated DNA topoisomerase: Top3 in the case of Sgs1 and hTOPOIIIα in the case of BLM. *In vitro*, BLM and hTOPOIIIα can dissolve double Holliday junctions into a noncrossover molecule.

While the helicase-topoisomerase complex can resolve Holliday junctions as noncrossover in mitotic cells, the meiotic Holliday junction resolvase that can result in crossovers has not been fully identified. Additional endonuclease

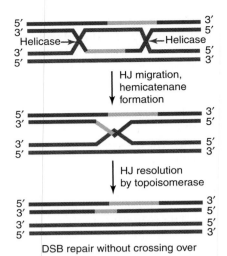

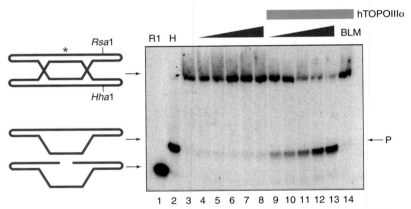

FIGURE 15.24 Double Holliday junction dissolution by the action of a DNA helicase and topoisomerase. The two Holliday junctions are pushed toward each other by branch migration using the DNA helicase activity. The resulting structure is a hemicatenane where single strands from two different DNA helices are wound around each other. This is cut by a DNA topoisomerase, unwinding and releasing the two DNA molecules and forming non-crossover products.

FIGURE 15.25 Holliday junction dissolution by BLM helicase and HTOPO IIIα. Artificial Holliday junctions are formed by annealing two sequences with hairpin ends. The blue and red ends contain different restriction endonuclease sites for identification. Digestion of the substrate with Rsa1 (R1) or Hha1 (H) results in a nicked or closed red molecule, which is radioactively labeled for detection. Adding increasing amounts of BLM helicase alone does not release the red molecule, but adding hTOPO IIIα with BLM helicase results in release of the intact red molecule, as shown in lanes 9–13. Photo reprinted by permission from Macmillan Publishers Ltd: *Nature*, L. Wu and I.D. Hickson, vol. 426, pp. 870–874, copyright 2003.

activities contained in the Mus81/Mms4 complex in yeast and MUS81/EME1 complex in mammalian cells can cleave nicked Holliday junction-like structures and branched DNA structures *in vitro*. The relationship of this activity to meiotic crossover formation, however, is not fully defined. Recently, eukaryotic resolvase homologs were identified in humans and *S. cerevisiae*. The proteins, GEN1 in humans

and Yen1 in yeast, are capable of resolving Holliday structures *in vitro*, although an *in vivo* role in homologous recombination has yet to be demonstrated.

15.16 Specialized Recombination Involves Specific Sites

Key concepts

- Specialized recombination involves reaction between specific sites that are not necessarily homologous.
- Phage lambda integrates into the bacterial chromosome by recombination between a site on the phage and the *att* site on the *E. coli* chromosome.
- The phage is excised from the chromosome by recombination between the sites at the end of the linear prophage.
- Phage lambda *int* codes for an integrase that catalyzes the integration reaction.

Specialized recombination involves a reaction between two specific sites. The lengths of target sites are short, and are typically in a range of 14 to 50 bp. In some cases the two sites have the same sequence, but in other cases they are nonhomologous. The reaction is used to insert a free phage DNA into the bacterial chromosome or to excise an integrated phage DNA from the chromosome, and in this case the two recombining sequences are different from one another. It is also used before division to regenerate monomeric circular chromosomes from a dimer that has been created by a generalized recombination event (see *Section 13.7, Chromosomal Segregation May Require Site-Specific Recombination*). In this case the recombining sequences are identical.

The enzymes that catalyze site-specific recombination are generally called **recombinases**, and more than one hundred of them are now known. Those involved in phage integration or related to these enzymes are also known as the integrase family. Prominent members of the integrase family are the prototype Int from phage lambda, Cre from phage P1, and the yeast FLP enzyme (which catalyzes a chromosomal inversion).

The classic model for site-specific recombination is illustrated by phage lambda. The conversion of lambda DNA between its different life forms involves two types of event. The pattern of gene expression is regulated as described in Chapter 27, *Phage Strategies*. The physical condition of the DNA is different in the lysogenic and lytic states:

- In the lytic lifestyle, lambda DNA exists as an independent, circular molecule in the infected bacterium.
- In the lysogenic state, the phage DNA is an integral part of the bacterial chromosome (called **prophage**).

Transition between these states involves site-specific recombination:

- To enter the lysogenic condition, free lambda DNA must be inserted into the host DNA. This is called **integration**.
- To be released from lysogeny into the lytic cycle, prophage DNA must be released from the chromosome. This is called **excision**.

Integration and excision occur by recombination at specific loci on the bacterial and phage DNAs called attachment (**att**) **sites**. The attachment site on the bacterial chromosome is called att^λ in bacterial genetics. The locus is defined by mutations that prevent integration of lambda; it is occupied by prophage λ in lysogenic strains. When the att^λ site is deleted from the *E. coli* chromosome, an infecting lambda phage can establish lysogeny by integrating elsewhere, although the efficiency of the reaction is <0.1% of the frequency of integration at att^λ. This inefficient integration occurs at *secondary attachment sites,* which resemble the authentic *att* sequences.

For describing the integration/excision reactions, the bacterial attachment site (att^λ) is called *attB*, consisting of the sequence components *BOB'*. The attachment site on the phage, *attP*, consists of the components *POP'*. **FIGURE 15.26** outlines the recombination reaction between these sites. The sequence O is common to *attB* and *attP*. It is called the **core sequence**, and the recombination event occurs within it. The flanking regions *B, B'* and *P, P'* are referred to as the *arms*; each is distinct in sequence. The phage DNA is circular, so the recombination event inserts it into the bacterial chromosome as a linear sequence. The prophage is bounded by two new *att* sites (the products of the recombination) called *attL* and *attR*.

An important consequence of the constitution of the *att* sites is that the integration and excision reactions do not involve the same pair of reacting sequences. Integration requires recognition between *attP* and *attB*, whereas excision requires recognition between *attL* and *attR*. The directional character of site-specific

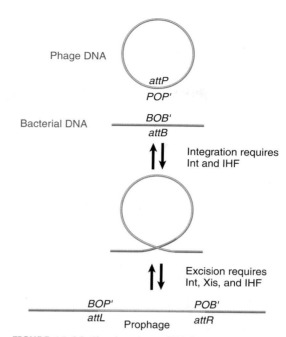

FIGURE 15.26 Circular phage DNA is converted to an integrated prophage by a reciprocal recombination between *attP* and *attB*; the prophage is excised by reciprocal recombination between *attL* and *attR*.

recombination is controlled by the identity of the recombining sites.

The recombination event is reversible, but different conditions prevail for each direction of the reaction. This is an important feature in the life of the phage, because it offers a means to ensure that an integration event is not immediately reversed by an excision, and vice versa.

The difference in the pairs of sites reacting at integration and excision is reflected by a difference in the proteins that mediate the two reactions:

- Integration ($attB \times attP$) requires the product of the phage gene *int*, which codes for an integrase enzyme, and a bacterial protein called integration host factor (IHF).
- Excision ($attL \times attR$) requires the product of phage gene *xis*, in addition to Int and IHF.

Thus Int and IHF are required for both reactions. Xis plays an important role in controlling the direction; it is required for excision, but inhibits integration.

A similar system, but with somewhat simpler requirements for both sequence and protein components, is found in the bacteriophage P1. The Cre recombinase coded by the phage catalyzes a recombination between two target sequences. Unlike phage lambda, for which the recombining sequences are different, in phage P1 they are identical. Each consists of a 34 bp-long sequence called *loxP*. The Cre recombinase is sufficient for the reaction; no accessory proteins are required. As a result of its simplicity and its efficiency, what is now known as the Cre/*lox* system has been adapted for use in eukaryotic cells, where it has become one of the standard techniques for undertaking site-specific recombination.

15.17 Site-Specific Recombination Involves Breakage and Reunion

Key concept

- Cleavages staggered by 7 bp are made in both *attB* and *attP* and the ends are joined crosswise.

The *att* sites have distinct sequence requirements, and *attP* is much larger than *attB*. The function of *attP* requires a stretch of 240 bp, whereas the function of *attB* can be exercised by the 23 bp fragment extending from –11 to +11, in which there are only 4 bp on either side of the core. The disparity in their sizes suggests that *attP* and *attB* play different roles in the recombination, with *attP* providing additional information necessary to distinguish it from *attB*.

Does the reaction proceed by a concerted mechanism in which the strands in *attP* and *attB* are cut simultaneously and exchanged? Or are the strands exchanged one pair at a time, with the first exchange generating a Holliday junction and the second cycle of nicking and ligation occurring to release the structure? The alternatives are depicted in **FIGURE 15.27**.

The recombination reaction has been halted at intermediate stages by the use of "suicide substrates," in which the core sequence is nicked. The presence of the nick interferes with the recombination process. This makes it possible to identify molecules in which recombination has commenced but has not been completed. The structures of these intermediates suggest that exchanges of single strands take place sequentially.

The model illustrated in **FIGURE 15.28** shows that if *attP* and *attB* sites each suffer the same staggered cleavage, complementary single-stranded ends could be available for crosswise hybridization. The distance between the lambda crossover points is 7 bp, and the reac-

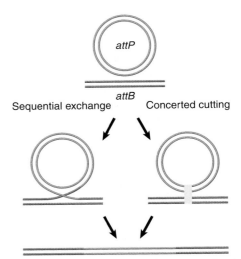

FIGURE 15.27 Does recombination between *attP* and *attB* proceed by sequential exchange or concerted cutting?

tion generates 3′–phosphate and 5′–OH ends. The reaction is shown for simplicity as generating overlapping single-stranded ends that anneal, but actually occurs by a process akin to the recombination event of Figure 15.4. The corresponding strands on each duplex are cut at the same position, the free 3′ ends exchange between duplexes, the branch migrates for a distance of 7 bp along the region of homology, and then the structure is resolved by cutting the other pair of corresponding strands.

15.18 Site-Specific Recombination Resembles Topoisomerase Activity

Key concepts

- Integrases are related to topoisomerases, and the recombination reaction resembles topoisomerase action except that nicked strands from *different* duplexes are sealed together.
- The reaction conserves energy by using a catalytic tyrosine in the enzyme to break a phosphodiester bond and link to the broken 3′ end.
- Two enzyme units bind to each recombination site and the two dimers synapse to form a complex in which the transfer reactions occur.

Integrases use a mechanism similar to that of type I topoisomerases, in which a break is made in one DNA strand at a time. The difference is that a recombinase reconnects the ends crosswise, whereas a topoisomerase makes a break, manipulates the ends, and then rejoins the original ends. The basic principle of the system

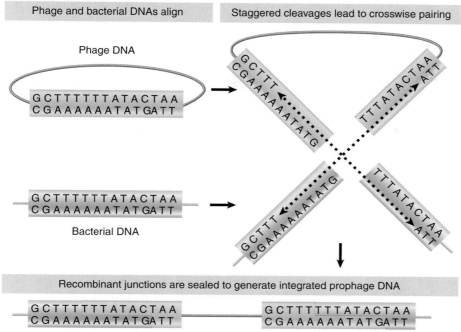

Phage and bacterial DNAs align

Phage DNA

```
GCTTTTTTATACTAA
CGAAAAAATATGATT
```

Bacterial DNA

```
GCTTTTTTATACTAA
CGAAAAAATATGATT
```

Staggered cleavages lead to crosswise pairing

Recombinant junctions are sealed to generate integrated prophage DNA

```
GCTTTTTTATACTAA          GCTTTTTTATACTAA
CGAAAAAATATGATT          CGAAAAAATATGATT
```

FIGURE 15.28 Staggered cleavages in the common core sequence of *attP* and *attB* allow crosswise reunion to generate reciprocal recombinant junctions.

is that four molecules of the recombinase are required, one to cut each of the four strands of the two duplexes that are recombining.

FIGURE 15.29 shows the nature of the reaction catalyzed by an integrase. The enzyme is a monomeric protein that has an active site capable of cutting and ligating DNA. The reaction involves an attack by a tyrosine on a phosphodiester bond. The 3′ end of the DNA chain is linked through a phosphodiester bond to a tyrosine in the enzyme. This releases a free 5′–hydroxyl end.

Two enzyme units are bound to each of the recombination sites. At each site, only one of the units attacks the DNA. The symmetry of the system ensures that complementary strands are broken in each recombination site. The free 5′–OH end in each site attacks the 3′–phosphotyrosine link in the other site. This generates a Holliday junction.

The structure is resolved when the other two enzyme units (which had not been involved

FIGURE 15.29 Integrases catalyze recombination by a mechanism similar to that of topoisomerases. Staggered cuts are made in DNA and the 3′-phosphate end is covalently linked to a tyrosine in the enzyme. The free hydroxyl group of each strand then attacks the P-Tyr link of the other strand. The first exchange shown in the figure generates a Holliday structure. The structure is resolved by repeating the process with the other pair of strands.

1. Two enzyme subunits bind to each duplex DNA

2. Each duplex is cleaved on one strand to generate a P-Tyr bond and an -OH end

3. Each hydroxyl attacks the Tyr-phosphate link in the other duplex

4. The reactions are repeated by the other subunits to join the other strands

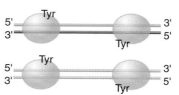

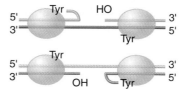

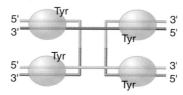

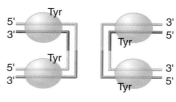

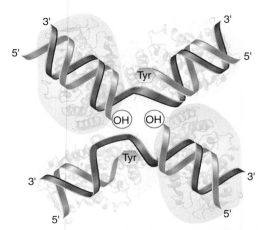

FIGURE 15.30 A synapsed loxA recombination complex has a tetramer of Cre recombinases, with one enzyme monomer bound to each half site. Two of the four active sites are in use, acting on complementary strands of the two DNA sites.

in the first cycle of breakage and reunion) act on the other pair of complementary strands.

The successive interactions accomplish a conservative strand exchange, in which there are no deletions or additions of nucleotides at the exchange site, and there is no need for input of energy. The transient 3'–phosphotyrosine link between protein and DNA conserves the energy of the cleaved phosphodiester bond.

FIGURE 15.30 shows the reaction intermediate, based on the crystal structure. (Trapping the intermediate was made possible by using a suicide substrate like that described for *att* recombination, which consists of a synthetic DNA duplex with a missing phosphodiester bond, so that the attack by the enzyme does not generate a free 5'–OH end.) The structure of the Cre-*lox* complex shows two Cre molecules, each of which is bound to a 15-bp length of DNA. The DNA is bent by ~100° at the center of symmetry. Two of these complexes assemble in an antiparallel way to form a tetrameric protein structure bound to two synapsed DNA molecules. Strand exchange takes place in a central cavity of the protein structure that contains the central six bases of the crossover region.

The tyrosine that is responsible for cleaving DNA in any particular half site is provided by the enzyme subunit that is bound to that half site. This is called *cis* cleavage. This is true also for the Int integrase and XerD recombinase. The FLP recombinase cleaves in *trans*, however, which involves a mechanism in which the enzyme subunit that provides the tyrosine

is not the subunit bound to that half site, but rather is one of the other subunits.

15.19 Lambda Recombination Occurs in an Intasome

Key concepts

• Lambda integration takes place in a large complex that also includes the host protein IHF.
• The excision reaction requires Int and Xis and recognizes the ends of the prophage DNA as substrates.

Unlike the Cre/*lox* recombination system, which requires only the enzyme and the two recombining sites, phage lambda recombination occurs in a large structure and has different components for each direction of the reaction (integration versus excision).

The host protein IHF is required for both integration and excision. IHF is a 20-kD protein of two different subunits, which are encoded by the genes *himA* and *himD*. IHF is not an essential protein in *E. coli*, and is not required for homologous bacterial recombination. It is one of several proteins with the ability to wrap DNA on a surface. Mutations in the *him* genes prevent lambda site-specific recombination and can be suppressed by mutations in λ*int*, which suggests that IHF and Int interact. Site-specific recombination can be performed *in vitro* by Int and IHF.

The *in vitro* reaction requires supercoiling in *attP*, but not in *attB*. When the reaction is performed *in vitro* between two supercoiled DNA molecules, almost all of the supercoiling is retained by the products. Thus there cannot be any free intermediates in which strand rotation could occur. This was one of the early hints that the reaction proceeds through a Holliday junction. We now know that the reaction proceeds by the mechanism typical of this class of enzymes, which is related to the topoisomerase I mechanism (see *Section 15.18, Site-Specific Recombination Resembles Topoisomerase Activity*).

Int has two different modes of binding. The C-terminal domain behaves like the Cre recombinase. It binds to inverted sites at the core sequence, positioning itself to make the cleavage and ligation reactions on each strand at the positions illustrated in **FIGURE 15.31**. The N-terminal domain binds to sites in the arms of *attP* that have a different consensus sequence. This binding is responsible for the aggregation

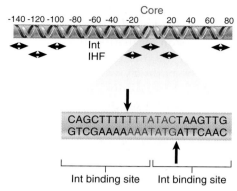

FIGURE 15.31 Int and IHF bind to different sites in *attP*. The Int recognition sequences in the core region include the sites of cutting.

```
CAGCTTTTTTTATACTAAGTTG
GTCGAAAAAAATATGATTCAAC
```

Int binding site · Int binding site

of subunits into the intasome. The two domains probably bind DNA simultaneously, thus bringing the arms of *attP* close to the core.

IHF binds to sequences of ~20 bp in *attP*. The IHF binding sites are approximately adjacent to sites where Int binds. Xis binds to two sites located close to one another in *attP*, so that the protected region extends over 30 to 40 bp. Together, Int, Xis, and IHF cover virtually all of *attP*. The binding of Xis changes the organization of the DNA so that it becomes inert as a substrate for the integration reaction.

When Int and IHF bind to *attP*, they generate a complex in which all the binding sites are pulled together on the surface of a protein. Supercoiling of *attP* is needed for the formation of this *intasome*. The only binding sites in *attB* are the two Int sites in the core. Int does not bind directly to *attB* in the form of free DNA, though. The intasome is the intermediate that "captures" *attB*, as indicated schematically in **FIGURE 15.32**.

According to this model, the initial recognition between *attP* and *attB* does not depend directly on DNA homology, but instead is determined by the ability of Int proteins to recognize both *att* sequences. The two *att* sites then are brought together in an orientation predetermined by the structure of the intasome. Sequence homology becomes important at this stage, when it is required for the strand-exchange reaction.

The asymmetry of the integration and excision reactions is shown by the fact that Int can form a similar complex with *attR* only if Xis is added. This complex can pair with a condensed complex that Int forms at *attL*. IHF is

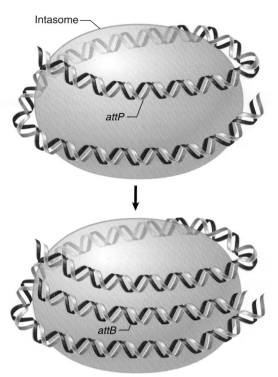

FIGURE 15.32 Multiple copies of Int protein may organize *attP* into an intasome, which initiates site-specific recombination by recognizing *attB* on free DNA.

not needed for this reaction. A significant difference between lambda integration/excision and the recombination reactions catalyzed by Cre or Flp is that Int-catalyzed reactions bind the regulatory sequences in the arms of the target sites, bending the DNA and allowing interactions between arm and core sites that drive each reaction to its conclusion. This is why each lambda reaction is irreversible, whereas recombination catalyzed by Cre or Flp is reversible. Crystal structures of λ-Int tetramers show that, like other recombinases, the tetramer has two active and two inactive subunits that switch roles during recombination. Allosteric interactions triggered by arm-binding control structural transitions in the tetramer that drive the reaction.

Much of the complexity of site-specific recombination may be caused by the need to regulate the reaction so that integration occurs preferentially when the virus is entering the lysogenic state, whereas excision is preferred when the prophage is entering the lytic cycle. By controlling the amounts of Int and Xis, the appropriate reaction will occur.

15.20 Yeast Can Switch Silent and Active Loci for Mating Type

Key concepts

- The yeast mating type locus *MAT* has either the *MAT***a** or *MAT*α genotype.
- Yeast with the dominant allele *HO* switch their mating type at a frequency ~10^{-6}.
- The allele at *MAT* is called the active cassette.
- There are also two silent cassettes, *HML*α and *HMR***a**.
- Switching occurs if *MAT***a** is replaced by *HMR*α or *MAT*α is replaced by *HMR***a**.

The yeast *S. cerevisiae* can propagate in either the haploid or diploid condition. Conversion between these states takes place by mating (fusion of haploid cells to give a diploid) and by sporulation (meiosis of diploids to give haploid spores). The ability to engage in these activities is determined by the mating type of the strain, which can be either **a** or α. Haploid cells of type **a** can mate only with haploid cells of type α to generate diploid cells of type **a**/α. The diploid cells can sporulate to regenerate haploid spores of either type.

Mating behavior is determined by the genetic information present at the *MAT* locus. Cells that carry the *MAT***a** allele at this locus are type **a**; likewise, cells that carry the *MAT*α allele are type α. Recognition between cells of opposite mating type is accomplished by the secretion of pheromones: α cells secrete the small polypeptide α-factor; **a** cells secrete **a**-factor. A cell of one mating type carries a surface receptor for the pheromone of the opposite type. When an **a** cell and an α cell encounter one another, their pheromones act on their receptors to arrest the cells in the G1 phase of the cell cycle, and various morphological changes occur. In a successful mating, the cell cycle arrest is followed by cell and nuclear fusion to produce an **a**/αdiploid cell.

Mating is a symmetrical process that is initiated by the interaction of pheromone secreted by one cell type with the receptor carried by the other cell type. The only genes that are uniquely required for the response pathway in a particular mating type are those coding for the receptors. Either the **a** factor–receptor interaction or the α factor–receptor interaction switches on the same response pathway. Mutations that eliminate steps in the common pathway have the same effects in both cell types. The pathway consists of a signal transduction cascade that

leads to the synthesis of products that make the necessary changes in cell morphology and gene expression for mating to occur.

Much of the information about the yeast mating-type pathway was deduced from the properties of mutations that eliminate the ability of **a** and/or α cells to mate. The genes identified by such mutations are called *STE* (for sterile). Mutations in the genes for the pheromones or receptors are specific for individual mating types, whereas mutations in the other *STE* genes eliminate mating in both **a** and α cells. This situation is explained by the fact that the events that follow the interaction of factor with receptor are identical for both types.

Some yeast strains have the remarkable ability to switch their mating types. These strains carry a dominant allele *HO* and change their mating type frequently—as often as once every generation. Strains with the recessive allele *ho* have a stable mating type, which is subject to change with a frequency ~10^{-6}.

The presence of *HO* causes the genotype of a yeast population to change. Irrespective of the initial mating type, within a very few generations there are large numbers of cells of both mating types, leading to the formation of *MAT***a**/*MAT*α diploids that take over the population. The production of stable diploids from a haploid population can be viewed as the raison d'être for switching.

The existence of switching suggests that all cells contain the potential information needed to be either *MAT***a** or *MAT*α but express only one type. Where does the information to change mating types come from? Two additional loci are needed for switching. *HML*α is needed for switching to give a *MAT*α type; *HMR***a** is needed for switching to give a *MAT***a** type. These loci lie on the same chromosome that carries *MAT*. *HML* is far to the left and *HMR* is far to the right.

The **mating type cassette** model is illustrated in **FIGURE 15.33**. It proposes that *MAT* has an *active cassette* of either type α or type **a**. *HML* and *HMR* have *silent cassettes*. In general, *HML* carries an α cassette, whereas *HMR* carries an **a** cassette. All cassettes carry information that codes for mating type, but only the active cassette at *MAT* is expressed. Mating-type switching occurs when the active cassette is replaced by information from a silent cassette. The newly installed cassette is then expressed.

Switching is nonreciprocal; the copy at *HML* or *HMR* replaces the allele at *MAT*. We know this because a mutation at *MAT* is lost permanently when it is replaced by switching—it

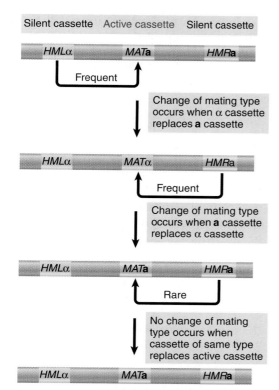

FIGURE 15.33 Changes of mating type occur when silent cassettes replace active cassettes of opposite genotype; recombination occurs between cassettes of the same type, the mating type remains unaltered.

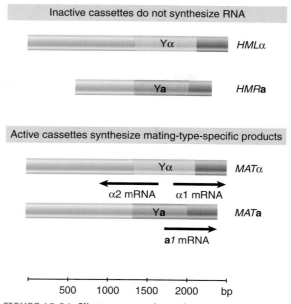

FIGURE 15.34 Silent cassettes have the same sequences as the corresponding active cassettes, except for the absence of the extreme flanking sequences in *HMR*a. Only the Y region changes between **a** and α types.

does not exchange with the copy that replaces it. This is in effect a directed gene conversion event. The directionality is established by the double-strand break initiation event, which occurs in the active *MAT* gene and not in the silent cassettes.

If the silent copy present at *HML* or *HMR* is mutated, switching introduces a mutant allele into the *MAT* locus. The mutant copy at *HML* or *HMR* remains there through an indefinite number of switches.

Mating-type switching is a directed event, in which there is only one recipient (*MAT*), but two potential donors (*HML* and *HMR*). Switching usually involves replacement of *MAT*a by the copy at *HML*α or replacement of *MAT*α by the copy at *HMR*a. In 80%–90% of switches, the *MAT* allele is replaced by one of opposite type. This is determined by the phenotype of the cell. Cells of **a** phenotype preferentially choose *HML* as donor; cells of α phenotype preferentially choose *HMR*.

Several groups of genes are involved in establishing and switching mating type. In addition to the genes that directly determine mating type, they include genes needed to repress

the silent cassettes, to switch mating type, or to execute the functions involved in mating, and most importantly, the homologous recombination factors described in *Section 15.15*.

By comparing the sequences of the two silent cassettes (*HML*α and *HMR*a) with the sequences of the two types of active cassette (*MAT*a and *MAT*α), we can delineate the sequences that determine mating type. The organization of the mating type loci is summarized in FIGURE 15.34. Each cassette contains common sequences that flank a central region that differs in the **a** and α types of cassette (called Ya or Yα). On either side of this region, the flanking sequences are virtually identical, although they are shorter at *HMR*. The active cassette at *MAT* is transcribed from a promoter within the Y region.

15.21 Unidirectional Gene Conversion Is Initiated by the Recipient *MAT* Locus

Key concepts

- Mating type switching is initiated by a double-strand break made at the *MAT* locus by the HO endonuclease.
- The recombination event is a synthesis-dependent strand-annealing reaction.

A switch in mating type is accomplished by a gene conversion in which the recipient site (*MAT*) acquires the sequence of the donor type (*HML* or *HMR*). Sites needed for the recombination have been identified by mutations at *MAT* that prevent switching. The unidirectional nature of the process is indicated by lack of mutations in *HML* or *HMR*.

The mutations identify a site at the right boundary of *Y* at *MAT* that is crucial for the switching event. The nature of the boundary is shown by analyzing the locations of these point mutations relative to the site of switching (this is done by examining the results of rare switches that occur in spite of the mutation). Some mutations lie within the region that is replaced (and thus disappear from *MAT* after a switch), whereas others lie just outside the replaced region (and therefore continue to impede switching). Thus sequences both within and outside the replaced region are needed for the switching event.

Switching is initiated by a double-strand break close to the *Y-Z* boundary that coincides with a site that is sensitive to attack by DNase. (This is a common feature of chromosomal sites that are involved in initiating transcription or recombination.) It is recognized by an endonuclease coded by the *HO* locus. The HO endonuclease makes a staggered double-strand break just to the right of the *Y* boundary. Cleavage generates the single-stranded ends of four bases drawn in **FIGURE 15.35**. The nuclease does not attack mutant *MAT* loci that cannot switch. Deletion analysis shows that most or all of the sequence of 24 bp surrounding the *Y* junction is required for cleavage *in vitro*. The recognition site is relatively large

for a nuclease, and it occurs only at the three mating-type cassettes.

Only the *MAT* locus, and not the *HML* or *HMR* loci, is a target for the endonuclease. It seems plausible that the same mechanisms that keep the silent cassettes from being transcribed also keep them inaccessible to the HO endonuclease. This inaccessibility ensures that switching is unidirectional.

The reaction triggered by the cleavage is illustrated schematically in **FIGURE 15.36** in terms of the general reaction between donor and recipient regions. The recombination occurs through a synthesis-dependent strand-annealing mechanism, as described earlier in *Section 15.5*. As expected, the stages following the initial cut require the enzymes involved in general recombination. Mutations in some of these genes prevent switching. In fact, studies of switching at the *MAT* locus were important in the development of the SDSA model.

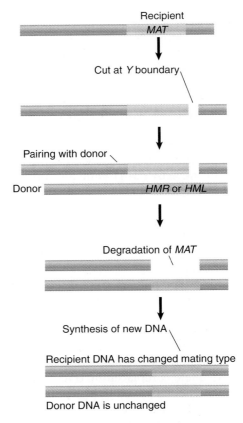

FIGURE 15.36 Cassette substitution is initiated by a double-strand break in the recipient (*MAT*) locus, and may involve pairing on either side of the *Y* region with the donor (*HMR* or *HML*) locus.

Y region

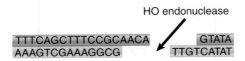

TTTCAGCTTTCCGCAACAGTATA
AAAGTCGAAAGGCGTTGTCATAT

HO endonuclease

TTTCAGCTTTCCGCAACA   GTATA
AAAGTCGAAAGGCG   TTGTCATAT

FIGURE 15.35 HO endonuclease cleaves *MAT* just to the right of the *Y* region, which generates sticky ends with a 4-base overhang.

15.22 Antigenic Variation in Trypanosomes Uses Homologous Recombination

Key concepts

- Variant surface glycoprotein (VSG) switching in *Trypanosoma brucei* evades host immunity.
- VSG switching requires recombination events to move VSG genes to specific expression sites.

The single-celled parasites known as trypanosomes cause two major types of human disease: African sleeping sickness and Chagas disease. These organisms are able to evade the host immune response through a process known as *antigenic variation*, in which expression of the major surface antigen is altered in a cyclical pattern in response to immune pressure. The variant surface glycoprotein (VSG) of trypanosomes is the major target of the immune system, but once antibodies are present to a given VSG, trypanosomes are able to switch expression to one of the many hundreds of VSG genes in their genomes. The VSG genes are organized into multiple subtelomeric tandem arrays and are also located in telomeric arrays on minichromosomes. Although all the genes in these arrays are silenced, they are either intact genes or pseudogenes. The switch is controlled by a recombination event in which a silent VSG gene is moved to a transcriptionally active, subtelomeric site known as an expression site (ES). This is illustrated in **FIGURE 15.37**. There are 20 subtelomeric expression sites, but only one of these is actively transcribed at one time. The transcriptionally active ES is thought to be a hot spot for recombination due to the open chromatin in this region. In fact, VSG recombination occurs at a higher frequency than would be expected for random events, leading to a VSG switch rate ranging from 10^{-2} to 10^{-3} switch events per cell per generation. Segmental gene conversion events using different VSGs can create chimeric VSG genes at the active expression site that contain sequences from multiple donor VSG genes.

DNA rearrangement through gene conversion, telomere exchange, and other unidentified processes are responsible for replacing an inactive VSG allele for the one in the active ES. The gene conversion event results in a duplication of the inactive VSG gene at the active

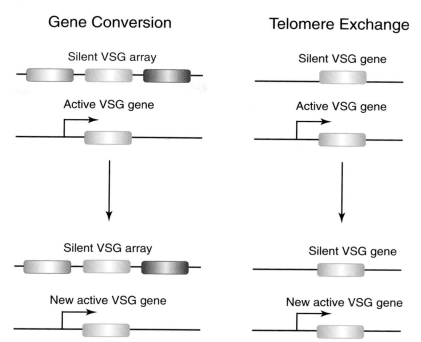

FIGURE 15.37 Switching mechanisms in trypanosome antigenic variation. Most of the VSG genes are arranged in arrays in subtelomeric locations, and consist of silent complete genes and pseudogenes. Gene conversion of the active VSG gene using information from one of the silent genes in the arrays results in a change in the sequence information in the active gene and a change in the surface antigen of the trypanosome. A second mode of variation comes from telomere exchange, to switch an inactive telomeric VSG gene from minichromosomes to the site of the active VSG gene. Both mechanisms use homologous recombination factors but the precise mechanism of exchange is not known. Reprinted from *Trends Genet.*, vol. 22, J. E. Taylor and G. Rudenko, Switching trypanosome coats . . . , pp. 614–620. Copyright 2006, with permission from Elsevier [http://www.sciencedirect.com/science/journal/01689525].

ES locus, allowing for expression of the previously inactive VSG. Despite the specificity of the genomic loci involved in the VSG switching event itself, the process has been shown to depend on general recombination factors.

Initial work has shown that the NHEJ pathway is not required for VSG switching, as mutants in Ku70/80 are not impaired for antigenic variation. In contrast, trypanosome mutants that do not express Rad51 are greatly impaired in VSG switching, indicating that homologous recombination is essential for this process. Further work has demonstrated a role for the trypanosome homologue of BRCA2 in VSG switching. It is unclear whether enzymes specific to VSG switch recombination are involved in this process as well.

Despite the fact that gene conversion is required for VSG switching, defects in mismatch repair pathway genes in trypanosomes do not affect antigenic variation.

15.23 Recombination Pathways Adapted for Experimental Systems

Key concepts

- Mitotic homologous recombination allows for targeted transformation.
- The Cre/*lox* and FLP/*FRT* systems allow for targeted recombination and gene knockout construction.
- The FLP/*FRT* system has been adapted to construct recyclable selectable markers for gene deletion.

Site-specific recombination not only has important biological roles discussed above, but has also been exploited to create targeted recombination events in experimental systems. Two classic examples of site-specific recombination have been adapted for experimental use: the Cre-*lox* and FLP-*FRT* systems.

The Cre/*lox* system, which is derived from bacteriophage P1, functions in a similar manner. The Cre enzyme recognizes and cleaves *lox* sites. One of the most common uses of the Cre/*lox* system is in gene targeting in mouse, as shown in **FIGURE 15.38**. Cre/*lox* can be used to conditionally turn off or turn on a gene in mouse. A construct is designed that is flanked by *lox* sites, with the *Cre* gene under control of an inducible promoter that can be turned on by temperature or hormones. Expression of *Cre* results in production of the Cre protein; the Cre protein then recognizes and cleaves the *lox* sites and promotes rejoining of the cut *lox* sites to leave behind a single *lox* site with the material between the *lox* sites having been excised.

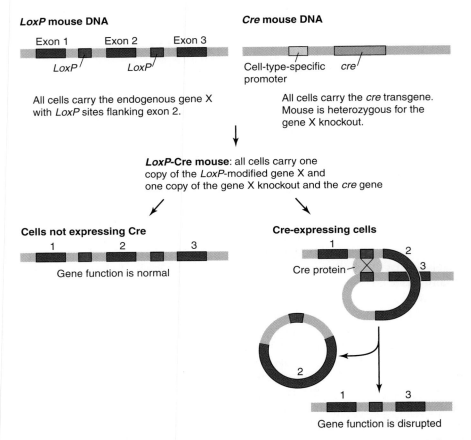

FIGURE 15.38 Using Cre/*lox* to make cell-type specific gene knockouts in mouse. *LoxP* sites are inserted into the chromosome to flank exon 2 of the gene X. The second copy of the X gene has been knocked out. The mouse formed with this construct is called the *LoxP* mouse. Another mouse, called the Cre mouse, has the *cre* gene inserted into the genome. Adjacent to the *cre* gene is a promoter that directs expression of the *cre* gene only in certain cell types. This mouse also carries a knockout of one copy of gene X. When the two mice are crossed, progeny that carry the *LoxP* construct, the gene X knockout, and the *cre* gene are produced. When Cre protein is expressed in cells that activate the promoter, it catalyzes site-specific recombination between the *LoxP* sites, and exon 2 of gene X is deleted. This inactivates the one functional copy of gene X in those cells expressing Cre. Adapted from H. Lodish, et al. *Molecular Cell Biology, Fifth edition.* W. H. Freeman & Company, 2003.

The Cre/*lox* system can be used to conditionally remove an exon from a mouse gene, resulting in a gene knockout (see *Section 3.12, Gene Knockouts and Transgenics*), or it can fuse the gene of interest to a promoter and thereby control expression of the gene of interest. Expression of a gene in tissues where it is not normally expressed, or at a time when the gene is not normally expressed, is called *ectopic expression*. Ectopic expression studies can reveal information about gene redundancy, specificity, and cell autonomy.

Another system that has been adapted for experimental use is derived from the yeast *S. cerevisiae*. The yeast two-micron plasmid is an autonomously replicating episome that is present in high copy numbers. The plasmid, which has no apparent benefit to the cell, is amplified through a site-specific recombination reaction that is carried out by a specialized recom-

binase known as FLP (flip). FLP recognizes inverted repeat sequences known as *FRT* (*FLP recombinase target*) sites. During replication, FLP-mediated recombination promotes rolling-circle replication that results in amplification of the two-micron plasmid. The FLP-*FRT* system is used in *Drosophila* to induce site-specific mitotic recombination events that can be used to create homozygous mutations or to make conditional knockouts, as shown in **FIGURE 15.39**.

To use the FLP/*FRT* system in *Drosophila*, the *FLP* gene expression is regulated. When FLP is expressed, it cuts the *FRT* sites, which have been inserted on a chromosome where there is a gene of interest centromere distal to the *FRT* site. The cutting of the *FRT* site, which is not 100% efficient, induces a double-strand break at the *FRT* site. The double-strand breaks are repaired by homologous recombination, and some of them will result in crossing-over.

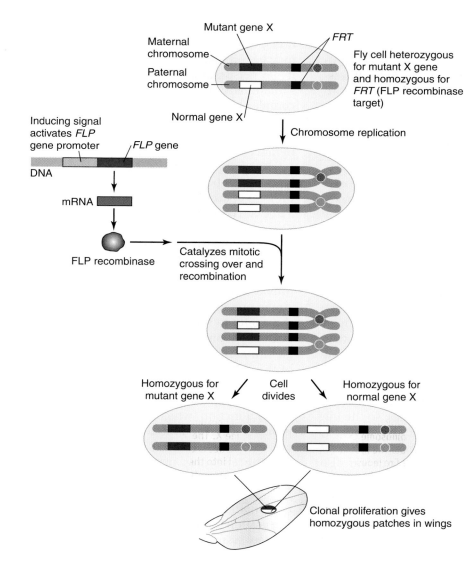

FIGURE 15.39 Using FLP/*FRT* to make homozygous recessive cells by homologous recombination. A fly is heterozygous for a mutant gene and homozygous insertion of the *FRT* site on the same chromosome. Induction of the FLP gene allows the FLP recombinase protein to be made. FLP recognizes the *FRT* site and makes a double-strand break, which promotes homologous recombination. Some of the recombination events occur by the double-strand break repair mechanism and result in crossing over. Following chromosome segregation, one daughter cell receives two mutant copies of the gene and the other daughter cell receives two normal copies of the gene. In the example shown, a patch of mutant cells is formed on the wing of a *Drosophila*. This technique allows assessment of a recessive mutant phenotype at a late stage in development. Adapted from B. Alberts, et al. *Molecular Biology of the Cell, Fourth Edition*. Garland Science, 2002.

Depending how the chromosomes then segregate, some cells will now be homozygous for the mutant gene. In genetic studies, the chromosome is often marked by a gene that affects a pigment, to give a visual readout for the recombination. The mitotic recombination uncovers the recessive pigmentation mutation and the mutant gene of interest, making them homozygous recessive. One use of this system is to see the effects of a recessive mutation that is lethal: when homozygous recessive in the zygote, the mutation will be lethal. If it is carried in the heterozygous state, though, the organism will be viable. Then the gene is rendered homozygous in clones of cells by induction of FLP, either by temperature or a tissue-specific transcription regulation, enabling the investigator to ask about effects of loss of the gene in specific cells at a specific time during development.

In recent years, FLP-*FRT* has been further adapted to construct recyclable selectable marker cassettes. In these systems, a selectable marker is placed between two flanking *FRT* sites. Also contained within the cassette is the *FLP* gene under the control of a regulatable promoter. Targeted integration of the FLP-*FRT* cassette is used to replace a locus of interest with the *FLP* marker cassette. Following integration, induced expression of the FLP recombinase catalyzes recombination between the flanking *FRT* sites, resulting in excision of the selectable marker cassette. This recyclable marker strategy is advantageous in diploid organisms because it allows for sequential rounds of targeted integration to make homozygous deletions of a gene of interest.

15.24 Summary

Recombination is initiated by a double-strand break in DNA. The break is enlarged to a gap with a single-stranded end; the free single-stranded end then forms a heteroduplex with the allelic sequence. Correction events may occur at sites that are mismatched within the heteroduplex DNA. The DNA in which the break occurs actually incorporates the sequence of the chromosome that it invades, so the initiating DNA is called the recipient. Gap repair, using the donor genetic information to repair the gap in the recipient DNA molecule, can also result in a gene conversion event. Hotspots for recombination are sites where double-strand breaks are initiated. A gradient of gene conversion is determined by the likelihood that a sequence near the free end will be converted to a single strand; this decreases with distance from the break. After gap repair, if the invading strain disengages from the recombination intermediate and anneals with the other end of the break, only gene conversion occurs. This is called the SDSA model. If instead the second end of the break is captured into the recombination intermediate, two Holliday junctions are formed. Resolution of the Holliday junctions can give crossover products if resolved in the appropriate direction. Recombination initiated by a DSB and processed to yield a double Holliday junction intermediate is called double-strand break repair (DSBR).

Meiotic recombination is initiated in yeast by Spo11, a topoisomerase-like enzyme that creates double-strand breaks and becomes linked to the free 5' ends of DNA. The DSB is then processed by generating single-stranded DNA that can anneal with its complement in the other chromosome. Yeast mutations that block synaptonemal complex formation show that recombination is required for its formation. Formation of the synaptonemal complex may be initiated by double-strand breaks, and it may persist until recombination is completed. Mutations in components of the synaptonemal complex block its formation but do not prevent chromosome pairing, so homolog recognition is independent of recombination and synaptonemal complex formation.

The full set of reactions required for recombination can be undertaken by the Rec and Ruv proteins of *E. coli*. A single-stranded region with a free end is generated by the RecBCD nuclease. The enzyme binds to DNA on one side of a *chi* sequence and then moves to the *chi* sequence, unwinding DNA as it progresses. A single-strand break is made at the *chi* sequence. *chi* sequences provide hotspots for recombination. The single-strand provides a substrate for RecA, which has the ability to synapse homologous DNA molecules by sponsoring a reaction in which a single strand from one molecule invades a duplex of the other molecule. Heteroduplex DNA is formed by displacing one of the original strands of the duplex. These actions create a recombination junction, which is resolved by the Ruv proteins. RuvA and RuvB act at a heteroduplex, and RuvC cleaves Holliday junctions.

The enzymes involved in site-specific recombination have actions related to those of topoisomerases. Among this general class of recombinases, those concerned with phage integration form the subclass of integrases. The Cre/*lox* system uses two molecules of Cre to bind to each *lox* site, so that the recombining complex is a tetramer. This is one of the standard systems for inserting DNA into a foreign genome. Phage lambda integration requires the phage Int protein and host IHF protein and involves a precise breakage and reunion in the absence of any synthesis of DNA. The reaction involves wrapping of the *attP* sequence of phage DNA into the nucleoprotein structure of the intasome, which contains several copies of Int and IHF; the host *attB* sequence is then bound and recombination occurs. Reaction in the reverse direction requires the phage protein Xis. Some integrases function by *cis*-cleavage, where the tyrosine that reacts with DNA in a half site is provided by the enzyme subunit bound to that half site; others function by *trans*-cleavage, for which a different protein subunit provides the tyrosine.

The yeast *S. cerevisiae* can propagate in either the haploid or diploid condition. Conversion between these states takes place by mating (fusion of haploid cells to give a diploid) and by sporulation (meiosis of diploids to give haploid spores). The ability to engage in these activities is determined by the mating type of the strain. The mating type is determined by the sequence of the *MAT* locus, and can be changed by a recombination event that substitutes a different sequence at this locus. The recombination event is initiated by a DSB—such as a homologous recombination event—but then the subsequent events ensure a unidirectional replacement of the sequence at the *MAT* locus.

Replacement is regulated so that *MATa* is usually replaced by the sequence from *HMLα*, whereas *MATα* is usually replaced by the sequence from *HMRa*. The endonuclease *HO* triggers the reaction by recognizing a unique target site at *MAT*. *HO* is regulated at the level of transcription by a system that ensures its expression in mother cells but not daughter cells, with the consequence that both progeny have the same (new) mating type.

Homologous recombination is also essential for the process of antigenic variation in the trypanosomes. Recombination is required to switch inactive VSG genes into active VSG expression sites. The molecular mechanisms behind this phenomenon are not completely understood, but it is clear that it does not involve NHEJ or mismatch repair enzymes. Rad51 is essential for this process, indicating the importance of homologous recombination.

Recombination pathways have been exploited as experimental tools for generation of gene knockouts and other recombination mediated events. Two major examples of these experimental tools include the Cre/*lox* and FLP/*FRT* systems. These tools both rely on site-specific recombination to create targeted recombination events in experimental systems.

References

15.3 Double-Strand Breaks Initiate Recombination

Reviews

Lichten, M. and Goldman, A. S. (1995). Meiotic recombination hotspots. *Annu. Rev. Genet.* 29, 423–444.

Szostak, J. W., Orr-Weaver, T. L., Rothstein, R. J., and Stahl, F. W. (1983). The double-strand-break repair model for recombination. *Cell* 33, 25–35.

Research

Hunter, N. and Kleckner, N. (2001). The single-end invasion: an asymmetric intermediate at the double-strand break to double-Holliday junction transition of meiotic recombination. *Cell* 106, 59–70.

15.5 The Synthesis-Dependent Strand-Annealing Model

Review

Paques, F. and Haber, J. E. (1999). Multiple pathways of recombination induced by double-strand breaks in *Saccharomyces cerevisiae*. *Microbiol. Mol. Biol. Rev.* 63, 349–404.

Research

Ferguson, D. O. and Holloman, W. K. (1996). Recombinational repair of gaps in DNA is asymmetric in *Ustilago maydis* and can be explained by a migrating D-loop model. *Proc. Natl. Acad. Sci. USA* 93, 5419–5424.

Keeney, S. and Neale, M. J. (2006). Initiation of meiotic recombination by formation of DNA double-strand breaks: mechanism and regulation. *Biochem. Soc. Trans.* 34, 523–525.

Nassif, N., Penney, J., Pal, S., Engels, W. R., and Gloor, G. B. (1994). Efficient copying of nonhomologous sequences from ectopic sites via P-element-induced gap repair. *Mol. Cell Biol.* 14, 1613–1625.

15.6 Nonhomologous End-Joining Can Repair Double-Strand Breaks

Reviews

Daley, J. M., Palmbos, P. L., Wu, D., and Wilson, T. E. (2005). Nonhomologous end joining in yeast. *Annu. Rev. Genet.* 39, 431–451.

Lieber, M. R., Ma, Y., Pannicke, U., and Schwarz, K. (2003). Mechanism and regulation of human nonhomologous DNA end-joining. *Nat. Rev. Mol. Cell Biol.* 4, 712–720.

Lieber, M. R., Ma, Y., Pannicke, U., and Schwarz, K. (2004). The mechanism of vertebrate non-homologous DNA end joining and its role in V(D)J recombination. *DNA Repair (Amst)* 3, 817–826.

15.7 The Single-Strand Annealing Mechanism Functions at Some Double-Strand Breaks

Research

Ivanov, E. L., Sugawara, N., Fishman-Lobell, J., and Haber, J. E. (1996). Genetic requirements for the single-strand annealing pathway of double-strand break repair in *Saccharomyces cerevisiae*. *Genetics* 142, 693–704.

15.8 Break-Induced Replication Can Repair Double-Strand Breaks

Reviews

Kraus, E., Leung, W. Y., and Haber, J. E. (2001). Break-induced replication: a review and an example in budding yeast. *Proc. Natl. Acad. Sci. USA* 98, 8255–8262.

Llorente, B., Smith, C. E., and Symington, L. S. (2008). Break-induced replication: what is it and what is it for? *Cell Cycle* 7, 859–864.

15.9 Recombining Meiotic Chromosomes Are Connected by the Synaptonemal Complex

Reviews

Roeder, G. S. (1997). Meiotic chromosomes: it takes two to tango. *Genes Dev.* 11, 2600–2621.

Zickler, D. and Kleckner, N. (1999). Meiotic chromosomes: integrating structure and function. *Annu. Rev. Genet.* 33, 603–754.

Research

Blat, Y. and Kleckner, N. (1999). Cohesins bind to preferential sites along yeast chromosome III, with differential regulation along arms versus the central region. *Cell* 98, 249–259.

Dong, H. and Roeder, G. S. (2000). Organization of the yeast Zip1 protein within the central region of the synaptonemal complex. *J. Cell Biol.* 148, 417–426.

Klein, F. et al. (1999). A central role for cohesins in sister chromatid cohesion, formation of axial elements, and recombination during yeast meiosis. *Cell* 98, 91–103.

Sym, M., Engebrecht, J. A., and Roeder, G. S. (1993). ZIP1 is a synaptonemal complex protein required for meiotic chromosome synapsis. *Cell* 72, 365–378.

15.10 The Synaptonemal Complex Forms after Double-Strand Breaks

Reviews

McKim, K. S., Jang, J. K., and Manheim, E. A. (2002). Meiotic recombination and chromosome segregation in *Drosophila* females. *Annu. Rev. Genet.* 36, 205–232.

Petes, T. D. (2001). Meiotic recombination hot spots and cold spots. *Nat. Rev. Genet.* 2, 360–369.

Research

Allers, T. and Lichten, M. (2001). Differential timing and control of noncrossover and crossover recombination during meiosis. *Cell* 106, 47–57.

Weiner, B. M. and Kleckner, N. (1994). Chromosome pairing via multiple interstitial interactions before and during meiosis in yeast. *Cell* 77, 977–991.

15.12 The Bacterial RecBCD System Is Stimulated by *chi* Sequences

Research

Dillingham, M. S., Spies, M., and Kowalczykowski, S. C. (2003). RecBCD enzyme is a bipolar DNA helicase. *Nature* 423, 893–897.

Spies, M., Bianco, P. R., Dillingham, M. S., Handa, N., Baskin, R. J., and Kowalczykowski, S. C. (2003). A molecular throttle: the recombination hotspot chi controls DNA translocation by the RecBCD helicase. *Cell* 114, 647–654.

Taylor, A. F. and Smith, G. R. (2003). RecBCD enzyme is a DNA helicase with fast and slow motors of opposite polarity. *Nature* 423, 889–893.

15.13 Strand-Transfer Proteins Catalyze Single-Strand Assimilation

Reviews

Kowalczykowski, S. C., Dixon, D. A., Eggleston, A. K., Lauder, S. D., and Rehrauer, W. M. (1994). Biochemistry of homologous recombination in *Escherichia coli*. *Microbiol. Rev.* 58, 401–465.

Kowalczykowski, S. C. and Eggleston, A. K. (1994). Homologous pairing and DNA strand-exchange proteins. *Annu. Rev. Biochem.* 63, 991–1043.

Lusetti, S. L. and Cox, M. M. (2002). The bacterial RecA protein and the recombinational DNA repair of stalled replication forks. *Annu. Rev. Biochem.* 71, 71–100.

15.14 Holliday Junctions Must Be Resolved

Reviews

Lilley, D. M. and White, M. F. (2001). The junction-resolving enzymes. *Nat. Rev. Mol. Cell Biol.* 2, 433–443.

West, S. C. (1997). Processing of recombination intermediates by the RuvABC proteins. *Annu. Rev. Genet.* 31, 213–244.

Research

Boddy, M. N., Gaillard, P. H., McDonald, W. H., Shanahan, P., Yates, J. R., and Russell, P. (2001). Mus81-Eme1 are essential components of a Holliday junction resolvase. *Cell* 107, 537–548.

Chen, X. B., Melchionna, R., Denis, C. M., Gaillard, P. H., Blasina, A., Van de Weyer, I., Boddy, M. N., Russell, P., Vialard, J., and McGowan, C. H. (2001). Human Mus81-associated endonuclease cleaves Holliday junctions *in vitro*. *Mol. Cell* 8, 1117–1127.

Constantinou, A., Davies, A. A., and West, S. C. (2001). Branch migration and Holliday junction resolution catalyzed by activities from mammalian cells. *Cell* 104, 259–268.

Kaliraman, V., Mullen, J. R., Fricke, W. M., Bastin-Shanower, S. A., and Brill, S. J. (2001). Functional overlap between Sgs1-Top3 and the Mms4-Mus81 endonuclease. *Genes Dev.* 15, 2730–2740.

15.15 Eukaryotic Genes Involved in Homologous Recombination

Reviews

Krogh, B. O. and Symington, L. S. (2004). Recombination proteins in yeast. *Annu. Rev. Genet.* 38 233-271.

San Filippo, J., Sung, P., and Klein, H. (2008). Mechanism of eukaryotic homologous recombination. *Annu. Rev. Biochem.* 77, 229-257.

Sung, P. and Klein, H. (2006). Mechanism of homologous recombination: mediators and helicases take on regulatory functions. *Nat. Rev. Mol. Cell Biol.* 7, 739-750.

Research

Gravel, S., Chapman, J. R., Magill, C., and Jackson, S. P. (2008). DNA helicases Sgs1 and BLM promote DNA double-strand break resection. *Genes Dev.* 22, 2767–2772.

Hollingsworth, N. M. and Brill, S. J. (2004). The Mus81 solution to resolution: generating meiotic crossovers without Holliday junctions. *Genes Dev.* 18, 117–125.

Ip, S. C., Rass, U., Blanco, M. G., Flynn, H. R., Skehel, J. M., and West, S. C. (2008). Identification of Holliday junction resolvases from humans and yeast. *Nature* 456, 357–361.

Mimitou, E. P. and Symington, L. S. (2008). Sae2, Exo1 and Sgs1 collaborate in DNA double-strand break processing. *Nature* 455, 770–774.

Zhu, Z., Chung, W. H., Shim, E.Y., Lee, S. E., and Ira, G. (2008). Sgs1 helicase and two nucleases Dna2 and Exo1 resect DNA double-strand break ends. *Cell* 134, 981–994.

15.16 Specialized Recombination Involves Specific Sites

Review

Craig, N. L. (1988). The mechanism of conservative site-specific recombination. *Annu. Rev. Genet.* 22, 77–105.

Research

Metzger, D., Clifford, J., Chiba, H., and Chambon, P. (1995). Conditional site-specific recombination in mammalian cells using a ligand-dependent chimeric Cre recombinase. *Proc. Natl. Acad. Sci. USA* 92, 6991–6995.

Nunes-Duby, S. E., Kwon, H. J., Tirumalai, R. S., Ellenberger, T., and Landy, A. (1998). Similarities and differences among 105 members of the Int family of site-specific recombinases. *Nucleic Acids Res.* 26, 391–406.

15.18 Site-Specific Recombination Resembles Topoisomerase Activity

Research

Guo, F., Gopaul, D. N., and van Duyne, G. D. (1997). Structure of Cre recombinase complexed with DNA in a site-specific recombination synapse. *Nature* 389, 40–46.

15.19 Lambda Recombination Occurs in an Intasome

Research

Biswas, T., Aihara, H., Radman-Livaja, M., Filman, D., Landy, A., and Ellenberger, T. (2005). A structural basis for allosteric control of DNA recombination by lambda integrase. *Nature* 435, 1059–1066.

Wojciak, J. M., Sarkar, D., Landy, A., and Clubb, R. T. (2002). Arm-site binding by lambda integrase: solution structure and functional characterization of its amino-terminal domain. *Proc. Natl. Acad. Sci. USA* 99, 3434–3439.

15.22 Antigenic Variation in Trypanosomes Uses Homologous Recombination

Review

Taylor, J. E. and Rudenko, G. (2006). Switching trypanosome coats: what's in the wardrobe? *Trends Genet.* 22, 614–620.

Research

Machado-Silva, A., Teixeira, S. M., Franco, G. R., Macedo, A. M., Pena, S. D., McCulloch, R., and Machado, C. R. (2008). Mismatch repair in *Trypanosoma brucei:* heterologous expression of MSH2 from *Trypanosoma cruzi* provides new insights into the response to oxidative damage. *Gene* 411, 19–26.

Proudfoot, C. and McCulloch, R. (2005). Distinct roles for two RAD51-related genes in *Trypanosoma brucei* antigenic variation. *Nucleic Acids Res.* 33, 6906–6919.

15.23 Recombination Pathways Adapted for Experimental Systems

Research

Egli, D., Hafen, E., and Schaffner, W. (2004). An efficient method to generate chromosomal rearrangements by targeted DNA double-strand breaks in *Drosophila melanogaster.* *Genome Res.* 14, 1382–1393.

Le, Y. and Sauer, B. (2001). Conditional gene knockout using Cre recombinase. *Mol. Biotechnol.* 17, 269-275.

Reuss, O., Vik, A., Kolter, R., and Morschhauser, J. (2004). The SAT1 flipper, an optimized tool for gene disruption in *Candida albicans. Gene* 341, 119–127.

Photo courtesy of Oscar Llorca, Centro de Investigaciones Biológicas (CIB)-CSIC.
Additional information at A. Rivera-Calzada, et al., *Structure* 13 (2005): 243–255.

Repair Systems

CHAPTER OUTLINE

16.1 Introduction

16.2 Repair Systems Correct Damage to DNA

- Repair systems recognize DNA sequences that do not conform to standard base pairs.
- Excision systems remove one strand of DNA at the site of damage and then replace it.
- Recombination-repair systems use recombination to replace the double-stranded region that has been damaged.
- All these systems are prone to introducing errors during the repair process.
- Photoreactivation is a nonmutagenic repair system that acts specifically on pyrimidine dimers.

16.3 Excision Repair Systems in *E. coli*

- The Uvr system makes incisions ~12 bases apart on both sides of damaged DNA, removes the DNA between them, and resynthesizes new DNA.
- Transcribed genes are preferentially repaired when DNA damage occurs.

16.4 Eukaryotic Nucleotide Excision Repair Pathways

- Xeroderma pigmentosum (XP) is a human disease caused by mutations in any one of several nucleotide excision repair genes.
- Numerous proteins, including XP products and the transcription factor TF$_{II}$H, are involved in eukaryotic nucleotide excision repair.

- Global genome repair recognizes damage anywhere in the genome.
- Transcriptionally active genes are preferentially repaired via transcription-coupled repair.
- Global genome repair and transcription-coupled repair differ in their mechanisms of damage recognition (XPC vs. RNA polymerase II).
- TF$_{II}$H provides the link to a complex of repair enzymes.
- Mutations in the XPD component of TF$_{II}$H cause three types of human diseases.

16.5 Base Excision Repair Systems Require Glycosylases

- Base excision repair is triggered by directly removing a damaged base from DNA.
- Base removal triggers the removal and replacement of a stretch of polynucleotides.
- The nature of the base removal reaction determines which of two pathways for excision repair is activated.
- The polδ/ϵ pathway replaces a long polynucleotide stretch; the polβ pathway replaces a short stretch.
- Uracil and alkylated bases are recognized by glycosylases and removed directly from DNA.
- Glycosylases and photolyase act by flipping the base out of the double helix, where, depending on the reaction, it is either removed or modified and returned to the helix.

16.1 Introduction

Any event that introduces a deviation from the usual double-helical structure of DNA is a threat to the genetic constitution of the cell. Injury to DNA is minimized by systems that recognize and correct the damage. The repair systems are as complex as the replication apparatus itself, which indicates their importance for the survival of the cell. When a repair system reverses a change to DNA, there is no consequence. A mutation may result, though, when it fails to do so. The measured rate of mutation reflects a balance between the number of damaging events occurring in DNA and the number that have been corrected (or miscorrected).

Repair systems often can recognize a range of distortions in DNA as signals for action, and a cell is likely to have several systems able to deal with DNA damage. The importance of DNA repair in eukaryotes is indicated by the identification of >130 repair genes in the human genome. We can divide the repair systems into several general types, as summarized in **FIGURE 16.1**.

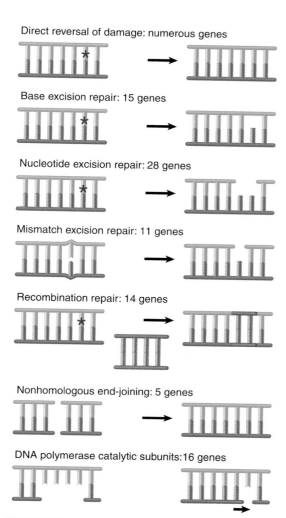

Direct reversal of damage: numerous genes

Base excision repair: 15 genes

Nucleotide excision repair: 28 genes

Mismatch excision repair: 11 genes

Recombination repair: 14 genes

Nonhomologous end-joining: 5 genes

DNA polymerase catalytic subunits: 16 genes

FIGURE 16.1 Repair genes can be classified into pathways that use different mechanisms to reverse or bypass damage to DNA.

- Some enzymes directly reverse specific sorts of damage to DNA.
- There are pathways for base excision repair, nucleotide excision repair, and mismatch repair, all of which function by removing and replacing material.
- There are systems that function by using recombination to retrieve an undamaged copy that is then used to replace a damaged duplex sequence.
- The nonhomologous end-joining pathway rejoins broken double-stranded ends.
- Several different DNA polymerases can resynthesize stretches of replacement DNA.

Direct repair is rare and involves the reversal or simple removal of the damage. One good example is **photoreactivation** of pyrimidine dimers, in which inappropriate covalent bonds

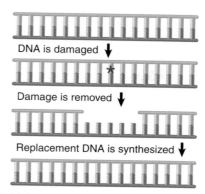

DNA is damaged ↓

Damage is removed ↓

Replacement DNA is synthesized ↓

FIGURE 16.2 Excision repair directly replaces damaged DNA and then resynthesizes a replacement stretch for the damaged strand.

between adjacent bases are reversed by a light-dependent enzyme. This system is widespread in nature, occurring in all but placental mammals, and appears to be especially important in plants. In *E. coli* it depends on the product of a single gene (*phr*) that codes for an enzyme called photolyase.

Mismatches between the strands of DNA are one of the major targets for repair systems. **Mismatch repair (MMR)** is accomplished by scrutinizing DNA for apposed bases that do not pair properly. Mismatches that arise during replication are corrected by distinguishing between the "new" and "old" strands and preferentially correcting the sequence of the newly synthesized strand. Other systems deal with mismatches generated by base conversions, such as the result of deamination. The importance of these systems is emphasized by the fact that cancer is caused in human populations by mutation of genes related to those involved in mismatch repair in yeast.

Mismatches are usually corrected by **excision repair**, which is initiated by a recognition enzyme that sees an actual damaged base or a change in the spatial path of DNA. There are two types of excision repair systems:

- **Base excision repair (BER)** systems directly remove the damaged base and replace it in DNA. A good example is DNA uracil glycosylase, which removes uracils that are mispaired with guanines (see *Section 16.5, Base Excision Repair Systems Require Glycosylases*).
- **Nucleotide excision repair (NER)** systems excise a sequence that includes the damaged base(s); a new stretch of DNA is then synthesized to replace the excised material. **FIGURE 16.2** summarizes the main events in the operation

of such a system. Such systems are common. Some recognize general damage to DNA; others act upon specific types of base damage. There are usually multiple excision repair systems in a single cell type.

Recombination-repair systems handle situations in which damage remains in a daughter molecule and replication has been forced to bypass the site, which typically creates a gap in the daughter strand. A retrieval system uses recombination to obtain another copy of the sequence from an undamaged source; the copy is then used to repair the gap.

A major feature in recombination and repair is the need to handle double-strand breaks (DSBs). DSBs initiate crossovers in homologous recombination. They can also be created by problems in replication, when they may trigger the use of recombination-repair systems. When DSBs are created by environmental damage (for example, by radiation damage), or are the result of the shortening of telomeres, they can cause mutations. In addition to recombination repair, DSBs can also be repaired by joining together nonhomologous DNA ends.

Mutations that affect the ability of *E. coli* cells to engage in DNA repair fall into groups that correspond to several repair pathways (not necessarily all independent). The major known pathways are the *uvr* excision repair system, the methyl-directed mismatch repair system, and the *recB* and *recF* recombination and recombination-repair pathways. The enzyme activities associated with these systems are endonucleases and exonucleases (important in removing damaged DNA); resolvases (endonucleases that act specifically on recombinant junctions); helicases to unwind DNA; and DNA polymerases to synthesize new DNA. Some of these enzyme activities are unique to particular repair pathways, whereas others participate in multiple pathways.

The replication apparatus devotes a lot of attention to quality control. DNA polymerases use proofreading to check the daughter strand sequence and to remove errors. Some of the repair systems are less accurate when they synthesize DNA to replace damaged material. For this reason, these systems have been known historically as *error-prone* systems.

16.2 Repair Systems Correct Damage to DNA

Key concepts

- Repair systems recognize DNA sequences that do not conform to standard base pairs.
- Excision systems remove one strand of DNA at the site of damage and then replace it.
- Recombination-repair systems use recombination to replace the double-stranded region that has been damaged.
- All these systems are prone to introducing errors during the repair process.
- Photoreactivation is a nonmutagenic repair system that acts specifically on pyrimidine dimers.

The types of damage that trigger repair systems can be divided into two general classes:

- *Single-base changes* affect the sequence of DNA but do not grossly distort its overall structure. They do not affect transcription or replication, when the strands of the DNA duplex are separated. Thus these changes exert their damaging effects on future generations through the consequences of the change in DNA sequence. The reason for this type of effect is the conversion of one base into another that is not properly paired with the partner base. Single-base changes may happen as the result of mutation of a base *in situ* or by replication errors. **FIGURE 16.3** shows that deamination of cytosine to uracil (spontaneously or by chemical mutagen) creates a mismatched U-G pair. **FIGURE 16.4** shows that a replication error might insert adenine instead of cytosine to create an A-G pair. Similar consequences could result from covalent addition of a small group to a base that modifies its ability to base pair. These changes may result in very minor structural distortion (as in the case of a U-G pair) or quite significant change (as in the case of an A-G pair), but the common feature is that the mismatch persists only until the next replication. Thus only limited time is available to repair the damage before it is made permanent by replication.

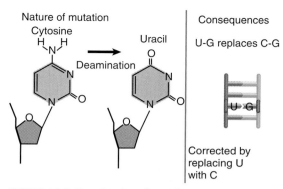

FIGURE 16.3 Deamination of cytosine creates a U-G base pair. Uracil is preferentially removed from the mismatched pair.

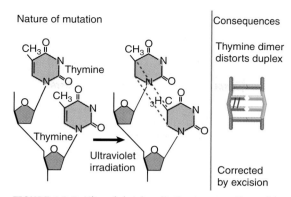

FIGURE 16.5 Ultraviolet irradiation causes dimer formation between adjacent thymines. The dimer blocks replication and transcription.

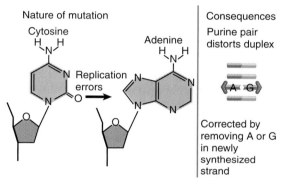

FIGURE 16.4 A replication error creates a mismatched pair that may be corrected by replacing one base; if uncorrected, a mutation is fixed in one daughter duplex.

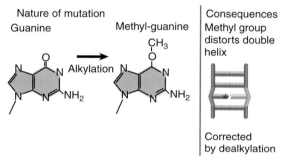

FIGURE 16.6 Methylation of a base distorts the double helix and causes mispairing at replication. Star indicates the methyl group.

- Structural distortions may provide a physical impediment to replication or transcription. Introduction of covalent links between bases on one strand of DNA or between bases on opposite strands inhibits replication and transcription. **FIGURE 16.5** shows the example of ultraviolet (UV) irradiation, which introduces covalent bonds between two adjacent thymine bases and results in an intrastrand **pyrimidine dimer**. **FIGURE 16.6** shows that similar consequences can result from the addition of a bulky adduct to a base that distorts the structure of the double helix. A single-strand nick or the removal of a base, as shown in **FIGURE 16.7**, prevents a strand from serving as a proper template for synthesis of RNA or DNA. The common feature in all these changes is that the

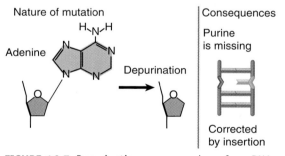

FIGURE 16.7 Depurination removes a base from DNA, blocking replication and transcription.

damaged adduct remains in the DNA and continues to cause structural problems and/or induce mutations until it is removed.

When a repair system is eliminated, cells become exceedingly sensitive to agents that cause DNA damage, particularly the type of damage recognized by the missing system.

16.3 Excision Repair Systems in *E. coli*

Key concepts

- The Uvr system makes incisions ~12 bases apart on both sides of damaged DNA, removes the DNA between them, and resynthesizes new DNA.
- Transcribed genes are preferentially repaired when DNA damage occurs.

Systems vary in their specificity, but share the same general features. Each system removes mispaired or damaged bases from DNA and then synthesizes a new stretch of DNA to replace them. The general pathway for excision repair is illustrated in **FIGURE 16.8**.

In the **incision** step, the damaged structure is recognized by an endonuclease that cleaves the DNA strand on both sides of the damage.

In the **excision** step, a 5′–3′ exonuclease removes a stretch of the damaged strand. Alternatively, a helicase can displace the damaged strand, which is subsequently degraded.

In the *synthesis* step, the resulting single-stranded region serves as a template for a DNA polymerase to synthesize a replacement for the excised sequence. (Synthesis of the new strand can be associated with removal of the old strand, in one coordinated action.) Finally, DNA ligase covalently links the 3′ end of the new DNA strand to the original DNA.

The *E. coli uvr* system of excision repair includes three genes (*uvrA, -B*, and -*C*), which code for the components of a repair endonuclease. It functions in the stages indicated in **FIGURE 16.9**. First, a UvrAB dimer recognizes pyrimidine dimers and other bulky lesions. Next, UvrA dissociates (this requires adenosine triphosphate [ATP]), and UvrC joins UvrB. The UvrBC complex makes an incision on each side—one that is seven nucleotides from the 5′ side of the damaged site and another that is three to four nucleotides away from the 3′ side. This also requires ATP. UvrD is a helicase that helps to unwind the DNA to allow release of the single strand between the two cuts. The enzyme that excises the damaged strand is DNA polymerase I. The enzyme involved in the repair synthesis also is likely to be DNA polymerase I (although DNA polymerases II and III can substitute for it).

UvrABC repair accounts for virtually all of the excision repair events in *E. coli*. In almost all (99%) of cases, the average length of replaced DNA is ~12 nucleotides. (For this reason, the process is sometimes described as *short-patch repair*.) The remaining 1% of cases involve the replacement of stretches of DNA mostly ~1500 nucleotides long, but extending as much as >9000

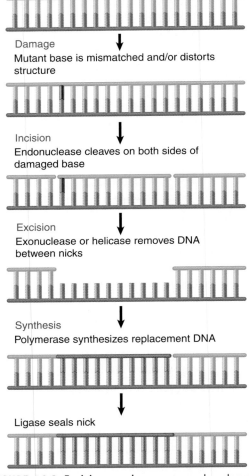

Damage
Mutant base is mismatched and/or distorts structure

Incision
Endonuclease cleaves on both sides of damaged base

Excision
Exonuclease or helicase removes DNA between nicks

Synthesis
Polymerase synthesizes replacement DNA

Ligase seals nick

FIGURE 16.8 Excision repair removes and replaces a stretch of DNA that includes the damaged base(s).

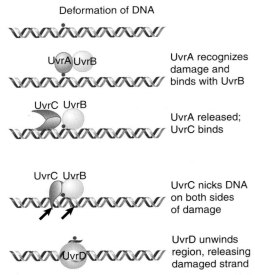

Deformation of DNA

UvrA recognizes damage and binds with UvrB

UvrA released; UvrC binds

UvrC nicks DNA on both sides of damage

UvrD unwinds region, releasing damaged strand

FIGURE 16.9 The Uvr system operates in stages in which UvrAB recognizes damage, UvrBC nicks the DNA, and UvrD unwinds the marked region.

nucleotides (sometimes called *long-patch repair*). We do not know why some events trigger the long-patch rather than the short-patch mode.

The Uvr complex can also be directed to sites of damage by other proteins. Damage to DNA can result in stalled transcription, in which case a protein called Mfd displaces the RNA polymerase and recruits the Uvr complex. **FIGURE 16.10** shows a model for the link between transcription and repair. When RNA polymerase encounters DNA damage in the template strand, it stalls because it cannot use the damaged sequences as a template to direct complementary base pairing. This explains the specificity of the effect for the template strand (damage in the nontemplate strand does not impede progress of the RNA polymerase).

The Mfd protein has two roles. First, it displaces the ternary complex of RNA polymerase from DNA. Second, it causes the UvrABC enzyme to bind to the damaged DNA, direct-ing excision repair to the damaged strand. After the DNA has been repaired, the next RNA polymerase to traverse the gene is able to produce a normal transcript.

16.4 Eukaryotic Nucleotide Excision Repair Pathways

Key concepts

- Xeroderma pigmentosum (XP) is a human disease caused by mutations in any one of several nucleotide excision repair genes.
- Numerous proteins, including XP products and the transcription factor TF$_{II}$H, are involved in eukaryotic nucleotide excision repair.
- Global genome repair recognizes damage anywhere in the genome.
- Transcriptionally active genes are preferentially repaired via transcription-coupled repair.
- Global genome repair and transcription-coupled repair differ in their mechanisms of damage recognition (XPC vs. RNA polymerase II).
- TF$_{II}$H provides the link to a complex of repair enzymes.
- Mutations in the XPD component of TF$_{II}$H cause three types of human diseases.

The general principle of excision repair in eukaryotic cells is similar to that of bacteria. Bulky lesions, such as those created by UV damage, crosslinking agents, and numerous chemical carcinogens, are also recognized and repaired by a nucleotide excision repair system. The critical role of mammalian nucleotide excision repair is seen in certain human hereditary disorders. The best investigated of these is **xeroderma pigmentosum (XP)**, a recessive disease resulting in hypersensitivity to sunlight, and in particular, ultraviolet light. The deficiency results in skin disorders and cancer predisposition.

The disease is caused by a deficiency in nucleotide excision repair. XP patients cannot excise pyrimidine dimers and other bulky adducts. Mutations occur in one of eight genes called *XPA* to *XPG*, all of which encode proteins involved in various stages of nucleotide excision repair. There are actually two major pathways of nucleotide excision repair in eukaryotes, illustrated in **FIGURE 16.11**.

The major difference between the two pathways is how the damage is initially recognized. In *global genome repair* (GG-NER), the XPC protein detects the damage and initiates the repair pathway. XPC can recognize damage anywhere in the genome. In mammals, XPC is a component of a lesion-sensing complex that also

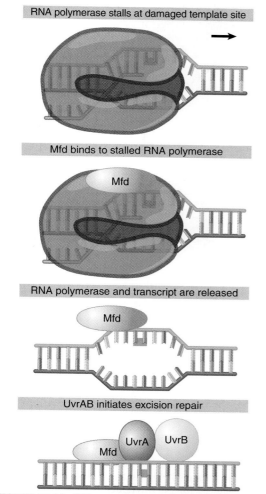

FIGURE 16.10 Mfd recognizes a stalled RNA polymerase and directs DNA repair to the damaged template strand. Photo courtesy of Barbara Hamkalo, University of California, Irvine.

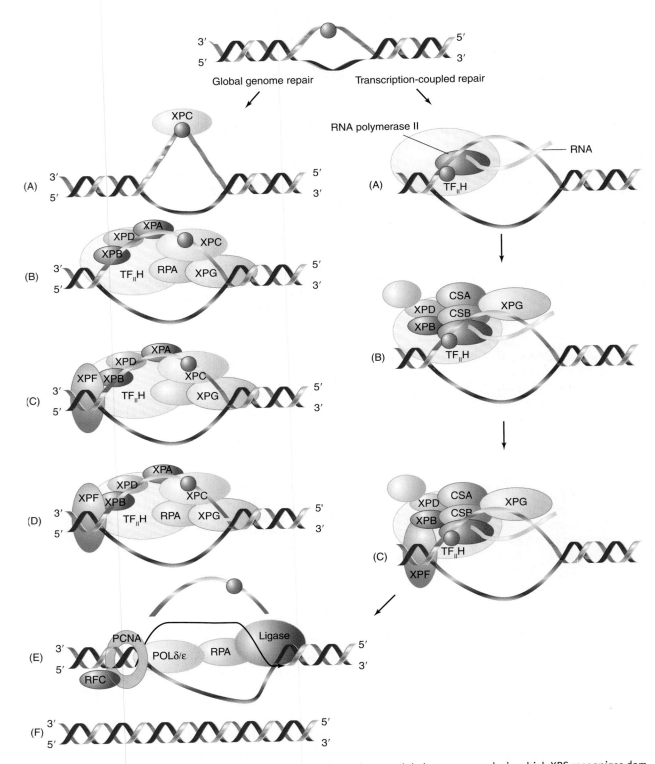

FIGURE 16.11 Nucleotide excision repair occurs via two major pathways: global genome repair, in which XPC recognizes damage anywhere in the genome, and transcription-coupled repair, in which the transcribed strand of active genes is preferentially repaired, and the damage is recognized by an elongating RNA polymerase. Adapted from E. C. Friedberg, et al., *Nature Rev. Cancer* 1 (2001): 22–23.

includes the proteins HR23B and centrin2. XPC also detects distortions that are not repaired by NER (such as small unwound regions of DNA), suggesting other proteins are required to verify the damage bound by XPC. Although XPC recognizes many types of lesions, some types of damage, such as UV-induced cyclobutane pyrimidine dimers (CPDs), are not well recognized by XPC. In this case, the DNA damage binding (DDB) complex assists in recruiting XPC to this type of damage.

On the other hand, *transcription-coupled repair* (TC-NER), as the name suggests, is responsible for repairing lesions that occur in the transcribed strand of active genes. In this case, the damage is recognized by RNA polymerase II itself, which stalls when it encounters a bulky lesion. Interestingly, the repair function may require modification or degradation of RNA polymerase. The large subunit of RNA polymerase is degraded when the enzyme stalls at sites of UV damage.

The two pathways eventually merge and use a common set of proteins to effect the repair itself. The strands of DNA are unwound for ~20 bp around the damaged site. This action is performed by the helicase activity of the transcription factor $TF_{II}H$, itself a large complex, which includes the products of two XP genes, *XPB* and *XPD*. XPB and XPD are both helicases; the XPB helicase is required for promoter melting during transcription, while the XPD helicase performs the unwinding function in NER (though the ATPase activity of XPB is also required during this stage). $TF_{II}H$ is already present in a stalled transcription complex; as a result, repair of transcribed strands is extremely efficient compared to repair of nontranscribed regions.

In the next step, cleavages are made on either side of the lesion by endonucleases encoded by the *XPG* and *XPF* genes. XPG is related to an endonuclease called FEN1 that cleaves DNA during the base excision repair pathway (see *Section 16.5, Base Excision Repair Systems Require Glycosylases*) XPF is found as part of a two-protein incision complex with ERCC1, which may assist XPF in binding DNA at the site of incision. Typically, about 25–30 nucleotides are excised during NER.

Finally, the single-stranded stretch including the damaged bases can then be replaced by new synthesis, and the final remaining nick is ligated by a complex of ligase III and XRCC1.

$TF_{II}H$, particularly the XPB and XPD subunits, plays numerous and complex roles in NER and transcription. The degradation of the large subunit of RNA polymerase II is deficient in cells from patients with Cockayne syndrome, a repair disorder characterized by neurological impairment and growth deficiency, which may also show photosensitivity similar to that of XP, but without the cancer predisposition. Cockayne syndrome can be caused by mutations in either of two genes (*CSA* and *CSB*), both of whose products appear to be part of or bound to $TF_{II}H$, and can also be caused by specific mutations in *XPB* or *XPD*.

Another disease that can be caused by mutations in *XPD* is trichothiodystrophy, which has little in common with XP or Cockayne (it is marked by brittle hair and may also include mental retardation). All of this marks XPD as a pleiotropic protein, in which different mutations can affect different functions. In fact, XPD is required for the stability of the $TF_{II}H$ complex during transcription, but its helicase activity is not needed during transcription. Mutations that prevent XPD from stabilizing the complex cause trichothiodystrophy. The helicase activity is required for the repair function. Mutations that affect the helicase activity cause the repair deficiency that results in XP or Cockayne syndrome.

In cases where replication encounters a thymine dimer that has not been removed, replication requires DNA polymerase η activity in order to proceed past the dimer. This polymerase is encoded by *XPV*. This bypass mechanism allows cell division to proceed even in the presence of unrepaired damage, but this is generally a last resort as cells prefer to put a hold on cell division until all damage is repaired.

16.5 Base Excision Repair Systems Require Glycosylases

Key concepts

- Base excision repair is triggered by directly removing a damaged base from DNA.
- Base removal triggers the removal and replacement of a stretch of polynucleotides.
- The nature of the base removal reaction determines which of two pathways for excision repair is activated.
- The polδ/ε pathway replaces a long polynucleotide stretch; the polβ pathway replaces a short stretch.
- Uracil and alkylated bases are recognized by glycosylases and removed directly from DNA.
- Glycosylases and photolyase act by flipping the base out of the double helix, where, depending on the reaction, it is either removed or modified and returned to the helix.

Base excision repair is similar to the nucleotide excision repair pathways described in the previous section. The process usually starts in a different way, however, with the removal of an *individual* damaged base. This serves as the trigger to activate the enzymes that excise and replace a stretch of DNA, including the damaged site.

Enzymes that remove bases from DNA are called **glycosylases** and **lyases**. FIGURE 16.12 shows that a glycosylase cleaves the bond between the damaged or mismatched base and the deoxyribose. FIGURE 16.13 shows that some glycosylases are also lyases that can take the reaction a stage further by using an amino (NH_2) group to attack the deoxyribose ring. This is usually followed by a reaction that introduces a nick into the polynucleotide chain.

FIGURE 16.14 shows that the exact form of the pathway depends on whether the damaged base is removed by a glycosylase or lyase.

Glycosylase action is followed by the endonuclease APE1, which cleaves the polynucleotide chain on the 5′ side. This in turn attracts a replication complex including the DNA polymerase δ/ε and ancillary components, which performs a short synthesis reaction extending for two to ten nucleotides. The displaced material is removed by the endonuclease FEN1. The enzyme ligase-1 seals the chain. This is called the *long-patch* pathway. (Note these names refer to mammalian enzymes, but the descriptions are generally applicable for all eukaryotes.)

When the initial removal involves lyase action, the endonuclease APE1 instead recruits DNA polymerase β to replace a single nucleotide. The nick is then sealed by the ligase

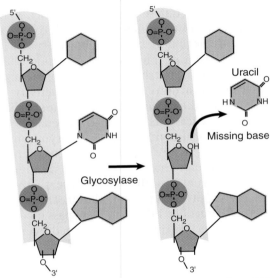

FIGURE 16.12 A glycosylase removes a base from DNA by cleaving the bond to the deoxyribose.

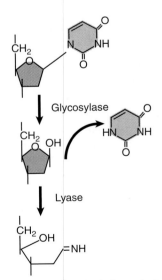

FIGURE 16.13 A glycosylase hydrolyzes the bond between base and deoxyribose (using H_2O), but a lyase takes the reaction further by opening the sugar ring (using NH_2).

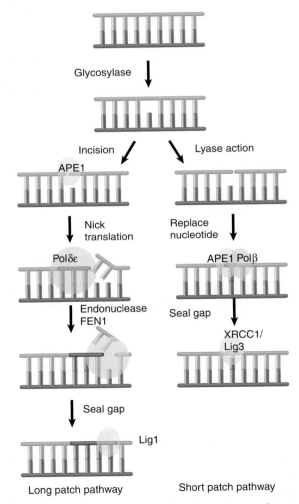

FIGURE 16.14 Base removal by glycosylase or lyase action triggers mammalian excision-repair pathways.

XRCC1/ligase-3. This is called the *short-patch* pathway.

Several enzymes that remove or modify individual bases in DNA use a remarkable reaction in which a base is "flipped" out of the double helix. This type of interaction was first demonstrated for methyltransferases—enzymes that add a methyl group to cytosine in DNA. This base-flipping mechanism places the base directly into the active site of the enzyme, where it can be modified and returned to its normal position in the helix, or, in the case of DNA damage, it can be immediately excised. Alkylated bases (typically in which a methyl group has been added to a base) are removed by this mechanism. A human enzyme, alkyladenine DNA glycosylase (AAG), recognizes and removes a variety of alkylated substrates, including 3-methyladenine, 7-methylguanine, and hypoxanthine. **FIGURE 16.15** shows the structure of AAG bound to a methylated adenine, in which the adenine is flipped out and bound in the glycosylase's active site.

By contrast with this mechanism, 1-methyladenine is corrected by an enzyme that uses an oxygenating mechanism (encoded in *E. coli* by the gene *alkB*, which has homologs in numerous eukaryotes, including three human genes). The methyl group is oxidized to a CH_2OH group, and then the release of the HCHO moiety (formaldehyde) restores the structure of adenine. A very interesting discovery is that the bacterial enzyme, and one of the human enzymes, can also repair the same damaged base in RNA. In the case of the human enzyme, the main target may be ribosomal RNA. This is the first known repair event with RNA as a target.

One of the most common reactions in which a base is directly removed from DNA is catalyzed by uracil-DNA glycosylase. Uracil typically only occurs in DNA because of a (spontaneous) deamination of cytosine. It is recognized by the glycosylase and removed. The reaction is similar to that shown in Figure 16.15: The uracil is flipped out of the helix and into the active site in the glycosylase. It appears that most or all glycosylases and lyases (in both prokaryotes and eukaryotes) work in a similar way.

Another enzyme that uses base flipping is the photolyase that reverses the bonds between pyrimidine dimers (see Figure 16.5). The pyrimidine dimer is flipped into a cavity in the enzyme. Close to this cavity is an active site that contains an electron donor, which provides the electrons to break the bonds. Energy for the reaction is provided by light in the visible wavelength. While most prokaryotic and eukaryotic species possess photolyase, placental mammals (but not marsupials) have lost this activity.

The common feature of these enzymes is the flipping of the target base into the enzyme structure. Recent work has shown that Rad4, the yeast XPC homolog (the protein that recognizes UV damage and other lesions during nucleotide excision repair), uses an interesting variation on this theme. Rad4 flips out the two adenine bases that are complementary to the linked thymines in a pyrimidine dimer, rather than flipping out the damaged pyrimidine dimer itself. In fact, it is believed that the ease with which these unpaired adenines are flipped out is actually the mechanism by which Rad4 detects the damage. Thus in this case, the target for the subsequent repair is not directly recognized by Rad4 at all, and instead the protein uses flipping as an indirect mechanism to detect the loss of a normal base-paired DNA double helix.

When a base is removed from DNA, the reaction is followed by excision of the phosphodiester backbone by an endonuclease, DNA synthesis by a DNA polymerase to fill the gap, and ligation by a ligase to restore the integrity of the polynucleotide chain, as described for the nucleotide excision repair pathways in the previous section.

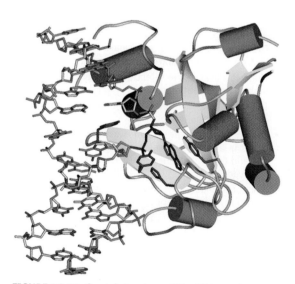

FIGURE 16.15 Crystal structure of the DNA repair enzyme alkyladenine DNA glycosylase (AAG) bound to a damaged base (3-methyladenine). The base (black) is flipped out of the DNA double helix (blue) and into AAG's active site (orange and green). Reproduced from A. Y. Lau, et al., *Proc. Natl. Acad. Sci. USA* 97 (2000): 13573–13578. Photo courtesy of Tom Ellenberger, Washington University School of Medicine.

16.6 Error-Prone Repair

Key concepts

- Damaged DNA that has not been repaired causes DNA polymerase III to stall during replication.
- DNA polymerase V (coded by *umuCD*) or DNA polymerase IV (coded by *dinB*) can synthesize a complement to the damaged strand.
- The DNA synthesized by repair DNA polymerases often has errors in its sequence.

The existence of repair systems that engage in DNA synthesis raises the question of whether their quality control is comparable with that of DNA replication. As far as we know, most systems, including *uvr*-controlled excision repair, do not differ significantly from DNA replication in the frequency of mistakes. **Error-prone synthesis** of DNA, however, occurs in *E. coli* under certain circumstances.

The error-prone pathway, also known as *translesion synthesis*, was first observed when it was found that the repair of damaged λ phage DNA is accompanied by the induction of mutations if the phage is introduced into cells that had previously been irradiated with UV. This suggests that the UV irradiation of the host has activated functions that generate mutations when repairing λ DNA. The mutagenic response also operates on the bacterial host DNA.

What is the actual error-prone activity? It is a specialized DNA polymerase that inserts random (usually incorrect) bases when it passes any site at which it cannot insert complementary base pairs in the daughter strand. Mutations in the genes *umuD* and *umuC* abolish UV-induced mutagenesis. This implies that the UmuC and UmuD proteins cause mutations to occur after UV irradiation. The genes constitute the *umuDC* operon, whose expression is induced by DNA damage. Their products form a complex UmuD′$_2$C, which consists of two subunits of a truncated UmuD protein and one subunit of UmuC. UmuD is cleaved by RecA, which is activated by DNA damage.

The UmuD′$_2$C complex has DNA polymerase activity. It is called DNA polymerase V, and is responsible for synthesizing new DNA to replace sequences that have been damaged by UV. This is the only enzyme in *E. coli* that can bypass the classic pyrimidine dimers produced by UV (or other bulky adducts). The polymerase activity is error prone. Mutations in either *umuC* or *umuD* inactivate the enzyme, which makes high doses of UV irradiation lethal.

How does an alternative DNA polymerase get access to the DNA? When the replicase (DNA polymerase III) encounters a block, such as a thymidine dimer, it stalls. It is then displaced from the replication fork and replaced by DNA polymerase V. In fact, DNA polymerase V uses some of the same ancillary proteins as DNA polymerase III. The same situation is true for DNA polymerase IV, the product of *dinB*, which is another enzyme that acts on damaged DNA.

DNA polymerases IV and V are part of a larger family of *translesion polymerases*, which includes eukaryotic DNA polymerases and whose members are specialized for repairing damaged DNA. In addition to the *dinB* and *umuCD* genes that code for DNA polymerases IV and V in *E. coli*, this family also includes the *RAD30* gene coding for DNA polymerase η of *S. cerevisiae*, and the *XPV* gene described previously that encodes the human homolog. A difference between the bacterial and eukaryotic enzymes is that the latter are not error prone at thymine dimers: They accurately introduce an A-A pair opposite a T-T dimer. When they replicate through other sites of damage, however, they are more prone to introduce errors.

16.7 Controlling the Direction of Mismatch Repair

Key concepts

- The *mut* genes code for a mismatch repair system that deals with mismatched base pairs.
- There is a bias in the selection of which strand to replace at mismatches.
- The strand lacking methylation at a hemimethylated $^{GATC}_{CTAG}$ is usually replaced.
- The mismatch repair system is used to remove errors in a newly synthesized strand of DNA. At G-T and C-T mismatches, the T is preferentially removed.
- Eukaryotic MutS/L systems repair mismatches and insertion/deletion loops.

Genes whose products are involved in controlling the fidelity of DNA synthesis during either replication or repair may be identified by mutations that have a **mutator** phenotype. A mutator mutant has an increased frequency of spontaneous mutation. If identified originally by the mutator phenotype, a gene is described as *mut*; often, though, a *mut* gene is later found to be equivalent with a known replication or repair activity.

Many *mut* genes turn out to be components of *mismatch-repair systems*. Failure to remove a damaged or mispaired base before replication allows it to induce a mutation. Functions in this group include the Dam methylase that identifies the target for repair, and enzymes that participate directly or indirectly in the removal of particular types of damage (MutH, -S, -L, and -Y).

When a structural distortion is removed from DNA, the wild-type sequence is restored. In most cases, the distortion is due to the creation of a base that is not naturally found in DNA, and that is therefore recognized and removed by the repair system.

A problem arises if the target for repair is a mispaired partnership of (normal) bases created when one was mutated or misinserted during replication. The repair system has no intrinsic means of knowing which is the wild-type base and which is the mutant. All it sees are two improperly paired bases, either of which can provide the target for excision repair.

If the mutated base is excised, the wild-type sequence is restored. If it happens to be the original (wild-type) base that is excised, though, the new (mutant) sequence becomes fixed. Often, however, the direction of excision repair is not random, but instead is biased in a way that is likely to lead to restoration of the wild-type sequence.

Some precautions are taken to direct repair in the right direction. For example, for cases such as the deamination of 5-methylcytosine to thymine, there is a special system to restore the proper sequence (see also *Section 1.14, Mutations Are Concentrated at Hotspots*). The deamination generates a G-T pair, and the system that acts on such pairs, has a bias to correct them to G-C pairs (rather than to A-T pairs). The system that undertakes this reaction includes the MutL and MutS products that remove T from both G-T and C-T mismatches.

The mutT, M, Y system handles the consequences of oxidative damage. A major type of chemical damage is caused by oxidation of G to 8-oxo-G. **FIGURE 16.16** shows that the system operates at three levels. MutT hydrolyzes the damaged precursor (8-oxo-dGTP), which prevents it from being incorporated into DNA. When guanine is oxidized in DNA its partner is cytosine, and MutM preferentially removes the 8-oxo-G from 8-oxo-G-C pairs. Oxidized guanine mispairs with A, and so when 8-oxo-G survives and is replicated, it generates an 8-oxo-G-A pair. MutY removes A from these pairs. MutM and MutY are glycosylases that directly remove a base from DNA. This

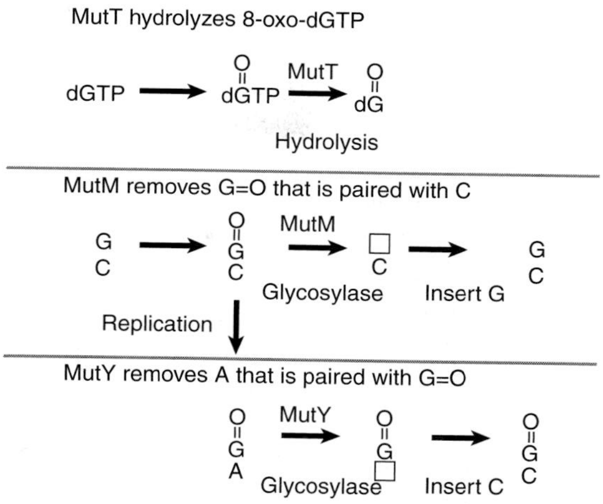

FIGURE 16.16 Preferential removal of bases in pairs that have oxidized guanine is designed to minimize mutations.

creates an apurinic site that is recognized by an endonuclease whose action triggers the involvement of the excision repair system.

When mismatch errors occur during replication in *E. coli*, it is possible to distinguish the original strand of DNA. Immediately after replication of methylated DNA, only the original parental strand carries methyl groups. In the period during which the newly synthesized strand awaits the introduction of methyl groups, the two strands can be distinguished. This provides the basis for a system to correct replication errors. The *dam* gene codes for a methylase whose target is the adenine in the sequence $^{GATC}_{CTAG}$. The hemimethylated state is used to distinguish replicated origins from nonreplicated origins. The same target sites are used by a replication-related mismatch repair system.

FIGURE 16.17 shows that DNA containing mismatched base partners is repaired preferentially by excising the strand that lacks the methylation. The excision is quite extensive; mismatches can be repaired preferentially for >1 kb around a GATC site. The result is that the newly synthesized strand is corrected to the sequence of the parental strand.

E. coli dam− mutants show an increased rate of spontaneous mutation. This repair system therefore helps reduce the number of mutations caused by errors in replication. It consists of several proteins, coded by *mut* genes. MutS binds to the mismatch and is joined by MutL. MutS can use two DNA-binding sites, as illustrated in **FIGURE 16.18**. The first specifically recognizes mismatches. The second is not specific for sequence or structure, and is used to translocate along DNA until a GATC sequence

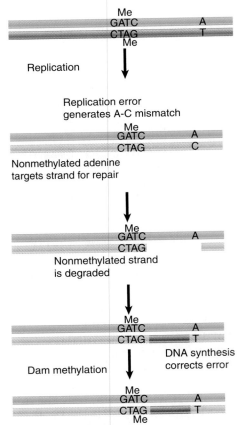

FIGURE 16.17 GATC sequences are targets for the Dam methylase after replication. During the period before this methylation occurs, the nonmethylated strand is the target for repair of mismatched bases.

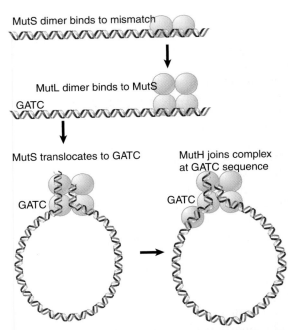

FIGURE 16.18 MutS recognizes a mismatch and translocates to a GATC site. MutH cleaves the unmethylated strand at the GATC. Endonucleases degrade the strand from the GATC to the mismatch site.

is encountered. Hydrolysis of ATP is used to drive the translocation. MutS is bound to both the mismatch site and to DNA as it translocates, and as a result it creates a loop in the DNA.

Recognition of the GATC sequence causes the MutH endonuclease to bind to MutSL. The endonuclease then cleaves the unmethylated strand. This strand is then excised from the GATC site to the mismatch site. The excision can occur in either the 5′–3′ direction (using RecJ or exonuclease VII) or in the 3′–5′ direction (using exonuclease I), and is assisted by the helicase UvrD. A new DNA strand is then synthesized by DNA polymerase III.

Eukaryotic cells have systems homologous to the *E. coli mut* system. Msh2 ("MutS homolog 2") provides a scaffold for the apparatus that recognizes mismatches. Msh3 and Msh6 provide specificity factors. In addition to repairing single-base mismatches, they are responsible for repairing mismatches that arise as the result of replication slippage. The Msh2-Msh3 hMutβ complex binds mismatched insertion/deletion loops, while the Msh2-Msh6 hMutα complex binds to single-base mismatches. Other proteins, including MutL homologs, are required for the repair process itself. Surprisingly, even though higher eukaryotes possess DNA methylation, eukaryotic mismatch repair systems do not use DNA methylation to select the daughter strand for repair. It is not known how eukaryotes recognize the daughter strand during mismatch repair, but MutSL homologs interact directly with the replication machinery.

The eukaryotic MutS/L system is particularly important for repairing errors caused by *replication slippage*. In a region such as a microsatellite, where a very short sequence is repeated several times, realignment between the newly synthesized daughter strand and its template can lead to a stuttering in which the DNA polymerase slips backward and synthesizes extra repeating units. These units in the daughter strand are extruded as a single-stranded loop from the double helix, which is repaired by homologs of the MutS/L system, as shown in FIGURE 16.19.

The importance of the MutS/L system for mismatch repair is indicated by the high rate at which it is found to be defective in human cancers. Loss of this system leads to an increased mutation rate, and mutations in MutS/L components can lead to hereditary nonpolyposis colorectal cancer (HNPCC). A characteristic feature of HNPCC is *microsatellite instability*, in which the lengths (num-

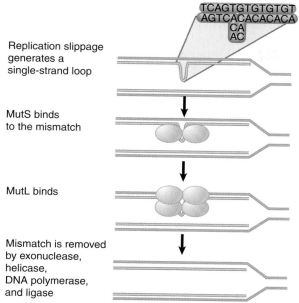

FIGURE 16.19 The MutS/MutL system initiates repair of mismatches produced by replication slippage.

Replication slippage generates a single-strand loop

MutS binds to the mismatch

MutL binds

Mismatch is removed by exonuclease, helicase, DNA polymerase, and ligase

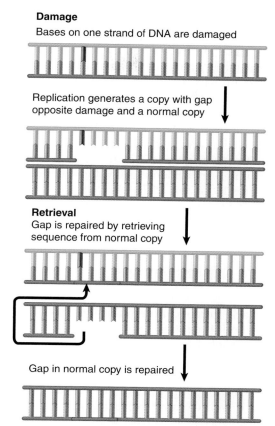

Damage
Bases on one strand of DNA are damaged

Replication generates a copy with gap opposite damage and a normal copy

Retrieval
Gap is repaired by retrieving sequence from normal copy

Gap in normal copy is repaired

FIGURE 16.20 An *E. coli* retrieval system uses a normal strand of DNA to replace the gap left in a newly synthesized strand opposite a site of unrepaired damage.

bers of repeats) of microsatellite sequences change rapidly in the tumor cells due to the loss of the mismatch repair system to correct replication slippage in these sequences. This instability can be used diagnostically to identify HNPCC.

16.8 Recombination-Repair Systems in *E. coli*

Key concepts

- The *rec* genes of *E. coli* code for the principal recombination-repair system.
- The recombination-repair system functions when replication leaves a gap in a newly synthesized strand that is opposite a damaged sequence.
- The single strand of another duplex is used to replace the gap.
- The damaged sequence is then removed and resynthesized.

Recombination-repair systems use activities that overlap with those involved in genetic recombination. They are also sometimes called "post-replication repair" because they function after replication. Such systems are effective in dealing with the defects produced in daughter duplexes by replication of a template that contains damaged bases. An example is illustrated in **FIGURE 16.20**.

Consider a structural distortion, such as a pyrimidine dimer, on one strand of a double helix. When the DNA is replicated, the dimer prevents the damaged site from acting as a template. Replication is forced to skip past it.

DNA polymerase probably proceeds up to or close to the pyrimidine dimer. The polymerase then ceases synthesis of the corresponding daughter strand. Replication restarts some distance farther along. This replication may be performed by translesion polymerases, which can replace the main DNA polymerase at such sites of unrepaired damage (see *Section 16.6, Error-Prone Repair*). A substantial gap is left in the newly synthesized strand.

The resulting daughter duplexes are different in nature. One has the parental strand containing the damaged adduct, which faces a newly synthesized strand with a lengthy gap. The other duplicate has the undamaged parental strand, which has been copied into a normal complementary strand. The retrieval system takes advantage of the normal daughter.

The gap opposite the damaged site in the first duplex is filled by utilizing the homologous single strand of DNA from the normal duplex. Following this **single-strand exchange**, the recipient duplex has a parental (damaged) strand facing a wild-type strand. The donor duplex has a normal parental strand facing a gap; the gap can be filled by repair synthesis in the usual way, generating a normal duplex. Thus the damage is confined to the original distortion (although the same recombination-repair events must be repeated after every replication cycle unless and until the damage is removed by an excision repair system).

The principal pathway for recombination-repair in *E. coli* is identified by the *rec* genes (see Figures 15.17 and 15.18). In *E. coli* deficient in excision repair, mutation of the *recA* gene essentially abolishes all the remaining repair and recovery facilities. Attempts to replicate DNA in *uvr⁻ recA⁻* cells produce fragments of DNA whose size corresponds with the expected distance between thymine dimers. This result implies that the dimers provide a lethal obstacle to replication in the absence of RecA function. It explains why the double mutant cannot tolerate >1 to 2 dimers in its genome (compared with the ability of a wild-type bacterium to handle as many as 50).

One *rec* pathway involves the *recBC* genes and is well characterized; the other involves *recF* and is not so well defined. They fulfill different functions *in vivo*. The RecBC pathway is involved in restarting stalled replication forks (see *Section 16.9, Recombination Is an Important Mechanism to Recover from Replication Errors*). The RecF pathway is involved in repairing the gaps in a daughter strand that are left after replicating past a pyrimidine dimer.

The RecBC and RecF pathways both function prior to the action of RecA (although in different ways). They lead to the association of RecA with a single-stranded DNA. The ability of RecA to exchange single strands allows it to perform the retrieval step in Figure 16.20. Nuclease and polymerase activities then complete the repair action.

The RecF pathway contains a group of three genes: *recF*, *recO*, and *recR*. The proteins form two types of complex, RecOR and RecOF. They promote the formation of RecA filaments on single-stranded DNA. One of their functions is to make it possible for the filaments to assemble in spite of the presence of single strand binding protein (SSB), which is inhibitory to RecA assembly.

The designations of repair and recombination genes are based on the phenotypes of the mutants, but sometimes a mutation isolated in one set of conditions and named as a *uvr* gene turns out to have been isolated in another set of conditions as a *rec* gene. This illustrates the point that the *uvr* and *rec* pathways are not independent, because *uvr* mutants show reduced efficiency in recombination-repair. We must expect to find a network of nuclease, polymerase, and other activities, which constitute repair systems that are partially overlapping (or in which an enzyme usually used to provide some function can be substituted by another from a different pathway).

16.9 Recombination Is an Important Mechanism to Recover from Replication Errors

Key concepts

- A replication fork may stall when it encounters a damaged site or a nick in DNA.
- A stalled fork may reverse by pairing between the two newly synthesized strands.
- A stalled fork may restart after repairing the damage and use a helicase to move the fork forward.
- The structure of the stalled fork is the same as a Holliday junction and may be converted to a duplex and DSB by resolvases.

In many cases, rather than skipping a DNA lesion, DNA polymerase instead stops replicating when it encounters DNA damage. **FIGURE 16.21** shows one possible outcome when a replication fork stalls. The fork stops moving forward when it encounters the damage. The replication apparatus disassembles, at least partially. This allows branch migration to occur, when the fork effectively moves backward, and the new daughter strands pair to form a duplex structure. After the damage has been repaired, a helicase rolls the fork forward to restore its structure. Then the replication apparatus can reassemble, and replication is restarted (see *Section 14.16, The Primosome Is Needed to Restart Replication*).

The pathway for handling a stalled replication fork requires repair enzymes, and restarting stalled replication forks is thought to be a major role of the recombination-repair systems. In *E. coli*, the RecA and RecBC systems have an important role in this reaction (in fact, this may be their major function in the bacterium). One possible pathway is for RecA to stabilize single-stranded DNA by binding to it at the stalled replication fork and possibly acting as the sensor that detects the stalling event. RecBC is involved in excision repair of the damage. After the damage has been repaired, replication can resume.

Another pathway may use recombination-repair—possibly the strand-exchange reactions

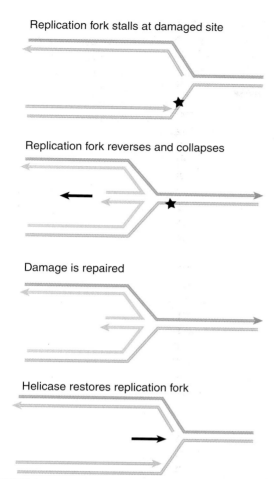

Replication fork stalls at damaged site

Replication fork reverses and collapses

Damage is repaired

Helicase restores replication fork

FIGURE 16.21 A replication fork stalls when it reaches a damaged site in DNA. Reversing the fork allows the two daughter strands to pair. After the damage has been repaired, the fork is restored by forward-branch migration catalyzed by a helicase. Arrowheads indicate 3′ ends.

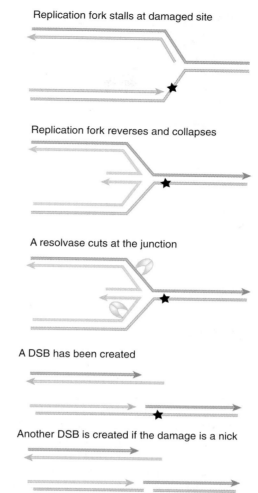

Replication fork stalls at damaged site

Replication fork reverses and collapses

A resolvase cuts at the junction

A DSB has been created

Another DSB is created if the damage is a nick

FIGURE 16.22 The structure of a stalled replication fork resembles a Holliday junction and can be resolved in the same way by resolvases. The results depend on whether the site of damage contains a nick. Result 1 shows that a double-strand break is generated by cutting a pair of strands at the junction. Result 2 shows a second DSB is generated at the site of damage if it contains a nick. Arrowheads indicate 3′ ends.

of RecA. FIGURE 16.22 shows that the structure of the stalled fork is essentially the same as a Holliday junction created by recombination between two duplex DNAs (see *Section.15.3, Double-Strand Breaks Initiate Recombination*). This makes it a target for resolvases. A double-strand break is generated if a resolvase cleaves either pair of complementary strands. In addition, if the damage is in fact a nick, another double-strand break is created at this site.

Stalled replication forks can be rescued by recombination-repair. We don't know the exact sequence of events, but one possible scenario is outlined in FIGURE 16.23. The principle is that a recombination event occurs on either side of the damaged site, allowing an undamaged single strand to pair with the damaged strand. This allows the replication fork to be reconstructed so that replication can continue, effectively bypassing the damaged site.

16.10 Recombination-Repair of Double-Strand Breaks in Eukaryotes

Key concepts

- The yeast *RAD* mutations, identified by radiation-sensitive phenotypes, are in genes that code for repair systems.
- The *RAD52* group of genes is required for recombination repair.
- The MRX (yeast) or MRN (mammals) complex is required to form a single-stranded region at each DNA end.
- The RecA homolog Rad51 forms a nucleoprotein filament on the single-stranded regions, assisted by Rad52 and Rad55/57.
- Rad54 and Rdh54/Rad54B are involved in homology search and strand invasion.

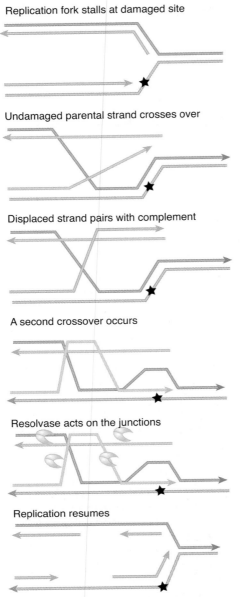

Replication fork stalls at damaged site

Undamaged parental strand crosses over

Displaced strand pairs with complement

A second crossover occurs

Resolvase acts on the junctions

Replication resumes

FIGURE 16.23 When a replication fork stalls, recombination-repair can place an undamaged strand opposite the damaged site. This allows replication to continue.

When a replication fork encounters a lesion in a single stand, it can result in the formation of a double-strand break (DSB). DSBs are one of the most severe types of DNA damage that can occur, particularly in eukaryotes. If a DSB on a linear chromosome is not repaired, the portion of the chromosome lacking a centromere will not be segregated at the next cell division. In addition to their occurrence during replication, DSBs can be generated in a number of other ways, including ionizing radiation, oxygen radicals generated by cellular metabolism, or action of endonucleases. The preferred mechanism for repairing DSBs is to use recombination-repair, as this ensures that no critical genetic information is lost due to sequence loss at the breakpoint.

Several of the genes required for recombination repair in eukaryotes have already been discussed in the context of homologous recombination (see *Section 15.15, Eukaryotic Genes Involved in Homologous Recombination*). Many eukaryotic repair genes are named *RAD* genes; they were initially characterized genetically in yeast by virtue of their sensitivity to *radia*tion. There are three general groups of repair genes in the yeast *S. cerevisiae*, identified by the *RAD3* group (involved in excision repair), the *RAD6* group (required for postreplication repair), and the *RAD52* group (concerned with recombination-like mechanisms). Homologs of these genes are present in higher eukaryotes as well.

The *RAD52* group plays essential roles in homologous recombination, and includes a large number of genes such as *RAD50, RAD51, RAD54, RAD55, RAD57*, and *RAD59*. These Rad proteins are all required at different stages of repair of a double-strand break. As occurs during meiotic recombination, the Mre11/Rad50/Xbs1 (MRX) complex (MRN in mammals) binds to the free DNA ends, and may tether the ends together, as shown in **FIGURE 16.24**. In concert with exonucleases and helicases, the MRX complex is required to resect the ends of the double-strand break to generate single-stranded tails with 3'–OH overhangs. This single-stranded DNA serves to activate a DNA damage checkpoint, stopping cell division until the damage can be repaired. The RecA homolog Rad51 binds to the single-stranded DNA to form a nucleoprotein filament, which is used for strand invasion of a homologous sequence. Rad52 and the Rad55/57 complex are required to form a stable Rad51 filament, and Rad54 and its homolog Rdh54 (Rad54B in mammals) assist in the search for homologous donor DNA and subsequent strand invasion. Rad54 and Rdh54 are members of the SWI2/SNF2 superfamily of chromatin remodeling enzymes (see *Section 28.7, Chromatin Remodeling Is an Active Process*), and may be necessary for reconfiguring chromatin structure at both the damage site and at the donor DNA. Following repair synthesis, the resulting structure (which resembles a Holliday junction) is resolved (see Figure 15.4 in *Section 15.3, Double-Strand Breaks Initiate Recombination*, for an illustration of these events).

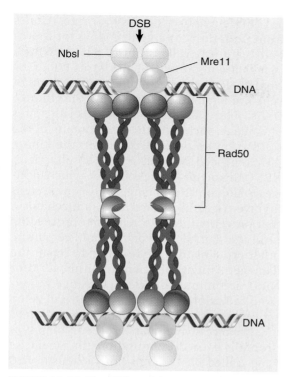

FIGURE 16.24 The MRN complex, required for 5′ end resection, also serves as a DNA bridge to prevent broken ends from separating. The "head" region of Rad50, bound to Mre11, binds DNA, while the extensive coiled coil region of Rad50 ends with a "zinc hook" that mediates interaction with another MRN complex. The precise position of Nbs1 within the complex is unknown but it interacts directly with Mre11.

16.11 Nonhomologous End-Joining Also Repairs Double-Strand Breaks

Key concepts

- The NHEJ pathway can ligate blunt ends of duplex DNA.
- Mutations in double-strand break repair pathways cause human diseases.

Repair of DSBs by homologous recombination ensures no genetic information is lost from a broken DNA end. In many cases, though, a sister chromatid or homologous chromosome is not easily available to use as a template for repair. In addition, some DSBs are specifically repaired using error-prone mechanisms as an intermediate in the recombination of immunoglobulin genes (see *Section 18.12 RAG1/RAG2 Catalyze Breakage and Religation of V(D)J Gene Segments*). In these cases, the mechanism used to repair these breaks is called **nonhomologous end-joining (NHEJ)**, and consists of ligating the blunt ends together.

The steps involved in NHEJ are summarized in **FIGURE 16.25**. The same enzyme com-

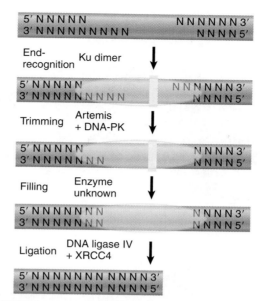

FIGURE 16.25 Nonhomologous end-joining requires recognition of the broken ends, trimming of overhanging ends and/or filling, followed by ligation.

plex undertakes the process in both NHEJ and immune recombination. The first stage is recognition of the broken ends by a heterodimer consisting of the proteins Ku70 and Ku80. They form a scaffold that holds the ends together and allows other enzymes to act on them. A key component is the DNA-dependent protein kinase (DNA-PK$_{cs}$), which is activated by DNA to phosphorylate protein targets. One of these targets is the protein Artemis, which in its activated form has both exonuclease and endonuclease activities, and can both trim overhanging ends and cleave the hairpins generated by recombination of immunoglobulin genes. The DNA polymerase activity that fills in any remaining single-stranded protrusions is not known. The actual joining of the double-stranded ends is performed by DNA ligase IV, which functions in conjunction with the protein XRCC4. Mutations in any of these components may render eukaryotic cells more sensitive to radiation. Some of the genes for these proteins are mutated in patients who have diseases due to deficiencies in DNA repair.

The Ku heterodimer is the sensor that detects DNA damage by binding to the broken ends. Ku can bring broken ends together by binding two DNA molecules. The crystal structure in **FIGURE 16.26** shows why it binds only to ends: The bulk of the protein extends for about two turns along one face of DNA (visible in the lower panel), but a narrow bridge between the subunits, located in the center of the structure, completely encircles DNA. This means that the heterodimer needs to slip onto a free end.

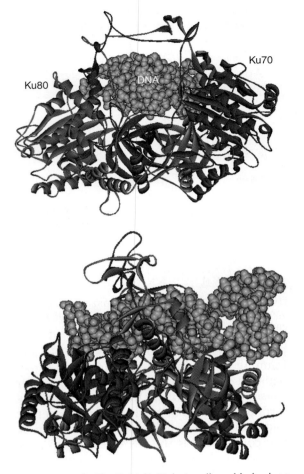

FIGURE 16.26 The Ku70-Ku80 heterodimer binds along two turns of the DNA double helix and surrounds the helix at the center of the binding site. Structures from Protein Data Bank 1JEY. J. R. Walker, R. A. Corpina, and J. Goldberg, *Nature* 412 (2001): 607–614.

All of the repair pathways we have discussed are conserved in mammals, yeast, and bacteria. Deficiency in DNA repair causes several human diseases. The inability to repair double-strand breaks in DNA is particularly severe and leads to chromosomal instability. The instability is revealed by chromosomal aberrations, which are associated with an increased rate of mutation, which in turn leads to an increased susceptibility to cancer in patients with the disease. The basic cause can be mutation in pathways that control DNA repair or in the genes that code for enzymes of the repair complexes. The phenotypes can be very similar, as in the case of *ataxia telangiectasia* (AT), which is caused by failure of a cell cycle checkpoint pathway, and *Nijmegen breakage syndrome* (NBS), which is caused by a mutation of a repair enzyme.

Nijmegen breakage syndrome results from mutations in a gene coding for a protein (variously called Nibrin, p95, or NBS1) that is a component of the Mre11/Rad50/Nbs1 (MRN) repair complex. When human cells are irradiated with agents that induce DSBs, many factors accumulate at the sites of damage, including the components of the MRN complex. After irradiation, the kinase ATM (encoded by the *AT* gene) phosphorylates NBS1; this activates the complex, which localizes to sites of DNA damage. Subsequent steps involve triggering a *checkpoint* (a mechanism that prevents the cell cycle from proceeding until the damage is repaired) and recruiting other proteins that are required to repair the damage. Patients deficient in either ATM or NBS1 are immunodeficient, sensitive to ionizing radiation, and predisposed to develop cancer, especially lymphoid cancers.

The recessive human disorder Bloom syndrome is caused by mutations in a helicase gene (called *BLM*) that is homologous to *recQ* of *E. coli*. The mutation results in an increased frequency of chromosomal breaks and sister chromatid exchanges. BLM associates with other repair proteins as part of a large complex. One of the proteins with which it interacts is hMLH1, a mismatch-repair protein that is the human homolog of bacterial mutL. The yeast homologs of these two proteins, Sgs1 and MLH1, also associate, identifying these genes as parts of a well-conserved repair pathway, and illustrating that there is cross-talk between different repair pathways.

16.12 DNA Repair in Eukaryotes Occurs in the Context of Chromatin

Key concepts

- Both histone modification and chromatin remodeling are essential for repair of DNA damage in chromatin.
- H2A phosphorylation (γ-H2AX) is a conserved double-strand break-dependent modification that recruits chromatin modifying activities and facilitates assembly of repair factors.
- Different patterns of histone modifications may distinguish stages of repair or different pathways of repair.
- Remodelers and chaperones are required to reset chromatin structure after completion of repair.

DNA repair in eukaryotic cells involves an additional layer of complexity: the nucleoso-

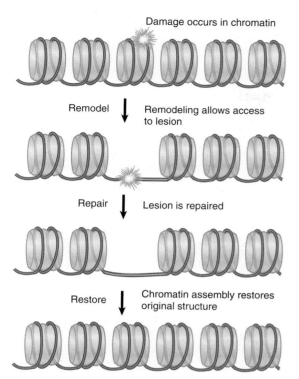

Damage occurs in chromatin

Remodel — Remodeling allows access to lesion

Repair — Lesion is repaired

Restore — Chromatin assembly restores original structure

FIGURE 16.27 DNA damage in chromatin requires chromatin remodeling and histone modification for efficient repair, and after repair the original chromatin structure must be restored.

mal packaging of the DNA substrate. Chromatin presents an obstacle to DNA repair, as it does to replication and transcription, as nucleosomes must be displaced in order for processes such as strand unwinding, excision, or resection to occur. Chromatin in the vicinity of DNA damage must therefore be modified and remodeled before or during repair, and then the original chromatin state must be restored after repair is completed, as shown in FIGURE 16.27.

Access to DNA in chromatin is controlled by a combination of covalent histone modifications, which change the structure of chromatin and create alternative binding sites for chromatin-binding proteins (discussed in *Section 10.4, Nucleosomes Are Covalently Modified*), and ATP-dependent chromatin remodeling (discussed in *Section 28.7, Chromatin Remodeling Is an Active Process*), in which remodeling complexes use the energy of ATP to slide or displace nucleosomes. Both histone modification and chromatin remodeling have been implicated in all of the eukaryotic repair pathways discussed in this chapter; for example, both the global-genome and transcription-coupled pathways of nucleotide excision repair depend

on specific chromatin remodeling enzymes, and repair of UV-damaged DNA is facilitated by histone acetylation. The best understanding of the roles of chromatin modification, however, is in the repair of DNA double-strand breaks.

Much of our understanding of the role of chromatin modification in double-strand break repair comes from studies in yeast utilizing a system derived from the yeast mating-type switching apparatus, which was introduced in *Section 15.20, Yeast Can Switch Silent and Active Loci for Mating Type*, and *Section 15.21, Unidirectional Transposition Is Initiated by the Recipient MAT Locus*. In this experimental system, yeast strains contain a galactose-inducible HO endonuclease, which generates a unique double-strand break at the active mating type locus (*MAT*) when cells are grown in galactose. These breaks are repaired using the same recombination-repair factors described in *Section 16.10, Recombination-Repair of Double-Strand Breaks in Eukaryotes*, using homologous sequences present at the silent mating type loci *HML* or *HMR*. In the absence of homologous donor sequences (or, for haploid yeast, a sister chromatid during S/G2), cells utilize the second major pathway of DSB repair, nonhomologous end-joining (NHEJ), to directly ligate broken chromosome ends.

Using this system (and other methods for inducing double-strand breaks in mammalian systems as well), researchers have identified numerous histone modifications and chromatin-remodeling events that take place during repair. The best characterized of these is the phosphorylation of the histone H2AX variant (see *Section 10.5, Histone Variants Produce Alternative Nucleosomes*). The major H2A in yeast is actually of the H2AX type, which is distinguished by an SQEL/Y motif at the end of the C-terminal tail. (This variant makes up only 5%–15% of the total H2A in mammalian cells.) The serine in the SQEL/Y sequence is the substrate for phosphorylation by the Mec1/Tel1 kinases in yeast, homologs of the mammalian ATM/ATR kinases (ATM is the checkpoint kinase affected in AT patients, discussed in the previous section). H2AX phosphorylated at this site (serine 129 in yeast, 139 in mammals) is referred to as **γ-H2AX**.

γ-H2AX is a universal marker for double-strand breaks, whether they occur as a result of damage, or during their normal

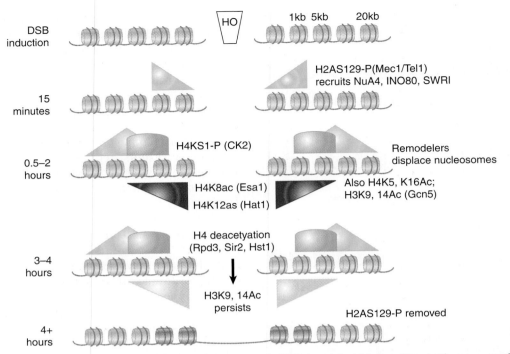

FIGURE 16.28 Summary of known histone modifications at an HO-induced double-strand break. The arrow on the left indicates approximate timing of events. Repair rates for homologous recombination and nonhomologous end-joining differ in this experimental system, so the precise timing of different modification events relative to one another is not always directly comparable between pathways. The relative distances from the breakpoint are indicated in the upper right (not to scale). Shaded triangles and arcs show distributions and relative levels of the indicated modifications.

appearance during mating type switching in yeast, or during meiotic recombination in numerous species. H2AX phosphorylation is one of the earliest events to occur at a double-strand break, appearing close to the breakpoint within minutes of damage, and spreading to include ~50 kb of chromatin in yeast, and megabases of chromatin in mammals. γ-H2AX is detectable throughout the repair process, and is linked to checkpoint recovery after repair. H2AX phosphorylation stabilizes the association of repair factors at the breakpoint, and also serves to recruit chromatin remodeling enzymes and a histone acetyltransferase to facilitate subsequent stages of repair.

In addition to γ-H2AX, numerous other histone modification events occur at double-strand breaks, at defined times during the repair process. These are summarized in **FIGURE 16.28**. They include transient phosphorylation of H4S1 by casein kinase 2, a modification more important for NHEJ than DSBR; and complex, asynchronous waves of acetylation of both histones H3 and H4, controlled by at least three different acetyltransferases and three different deacety-

lases. It is not fully understood how each modification promotes different steps in the repair process, but it is important to note that the patterns of modification differ between homologous recombination and end-joining pathways, suggesting that these modifications may recruit factors specific for the different repair mechanisms.

A number of chromatin-remodeling enzymes also act at double-strand breaks. All chromatin remodeling enzymes are members of the SWI2/SNF2 superfamily of enzymes, but there are numerous subfamilies within this group (see Figure 28.18). At least three different subfamilies are implicated in double-strand break repair: the SWI/SNF and RSC complexes of the SNF2 subfamily, the INO80 and SWR1 complexes of the INO80 group, and Rad54 and Rdh54 of the Rad54 subfamily. As discussed in *Section 16.10, Recombination-Repair of Double-Strand Breaks in Eukaryotes*, the Rad54 and Rdh54 enzymes play roles during the search for homologous donors and strand invasions stages of repair, but other chromatin remodelers appear important during every stage, including initial damage recognition, strand resection, and in the resetting of chromatin as repair is completed. This final stage also requires the activities of the histone chaperones Asf1 and CAF-1 (introduced in *Section 10.8, Replication of Chromatin Requires Assembly of Nucleosomes*), which are needed to restore chromatin structure on the newly repaired region and allow recovery from the DNA damage checkpoint.

16.13 RecA Triggers the SOS System

Key concepts

- Damage to DNA causes RecA to trigger the SOS response, which consists of genes coding for many repair enzymes.
- RecA activates the autocleavage activity of LexA.
- LexA represses the SOS system; its autocleavage activates those genes.

When cells respond to DNA damage, the actual repair of the lesion is only one part of the overall response. Eukaryotic cells also engage in two other key types of activities when damage is detected: activation of checkpoints to arrest the cell cycle until the damage is repaired (see *Section 16.11*), and induction of a suite of transcriptional changes that facilitate the damage response (such as production of repair enzymes).

Bacteria also engage in a more global response to damage than just the repair event, known as the *SOS response*. This response depends on the recombination protein RecA, discussed in *Sections 16.8 and 16.9*. RecA's role in recombination-repair is only one of its activities. This extraordinary protein also has another, quite distinct function: It can be activated by many treatments that damage DNA or inhibit replication in *E. coli*. This causes it to trigger the SOS response, a complex series of phenotypic changes that involves the expression of many genes whose products include repair functions. These dual activities of the RecA protein make it difficult to know whether a deficiency in repair in recA mutant cells is due to loss of the DNA strand-exchange function of RecA or to some other function whose induction depends on the protease activity.

The inducing damage can take the form of ultraviolet irradiation (the most studied case) or can be caused by crosslinking or alkylating agents. Inhibition of replication by any of several means—including deprivation of thymine, addition of drugs, or mutations in several of the *dna* genes—has the same effect.

The response takes the form of increased capacity to repair damaged DNA, which is achieved by inducing synthesis of the components of both the long-patch excision repair system and the Rec recombination-repair pathways. In addition, cell division is inhibited. Lysogenic prophages may be induced.

The initial event in the response is the activation of RecA by the damaging treatment. We do not know very much about the relationship between the damaging event and the sudden change in RecA activity. A variety of damaging events can induce the SOS response; thus current work focuses on the idea that RecA is activated by some common intermediate in DNA metabolism.

The inducing signal could consist of a small molecule released from DNA, or it might be some structure formed in the DNA itself. *In vitro*, the activation of RecA requires the presence of single-stranded DNA and ATP. Thus the activating signal could be the presence of a single-stranded region at a site of damage. Whatever form the signal takes, its interaction with RecA is rapid: The SOS response occurs within a few minutes of the damaging treatment.

Activation of RecA causes proteolytic cleavage of the product of the *lexA* gene. LexA is a small (22 kD) protein that is relatively stable in untreated cells, where it functions as a repressor at many operons. The cleavage reaction is unusual: LexA has a latent protease activity that is activated by RecA. When RecA is activated, it causes LexA to undertake an autocatalytic cleavage; this inactivates the LexA repressor function, and coordinately induces all the operons to which it was bound. The pathway is illustrated in FIGURE 16.29.

The target genes for LexA repression include many with repair functions. Some of these SOS genes are active only in treated cells; others are active in untreated cells, but the level of expression is increased by cleavage of LexA. In the case of *uvrB*, which is a component of the excision repair system, the gene has two promoters; one functions independently of LexA, the other is subject to its control. Thus after cleavage of LexA, the gene can be expressed from the second promoter as well as from the first.

LexA represses its target genes by binding to a 20-bp stretch of DNA called an SOS box, which includes a consensus sequence with eight absolutely conserved positions. As is common with other operators, the SOS boxes overlap with the respective promoters. At the *lexA* locus—the subject of autogenous repression—there are two adjacent SOS boxes.

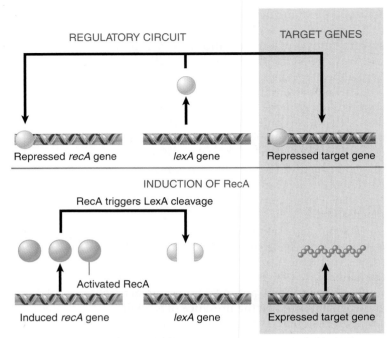

TARGET GENES

Repressed *recA* gene

lexA gene

Repressed target gene

INDUCTION OF RecA

RecA triggers LexA cleavage

Activated RecA

Induced *recA* gene

lexA gene

Expressed target gene

FIGURE 16.29 The LexA protein represses many genes, including repair genes, *recA* and *lexA*. Activation of RecA leads to proteolytic cleavage of LexA and induces all of these genes.

RecA and LexA are mutual targets in the SOS circuit: RecA triggers cleavage of LexA, which represses *recA* and itself. The SOS response therefore causes amplification of both the RecA protein and the LexA repressor. The results are not so contradictory as might at first appear.

The increase in expression of RecA protein is necessary (presumably) for its direct role in the recombination-repair pathways. On induction, the level of RecA is increased from its basal level of ~1200 molecules/cell by up to 50×. The high level in induced cells means there is sufficient RecA to ensure that all the LexA protein is cleaved. This should prevent LexA from reestablishing repression of the target genes.

The main importance of this circuit for the cell, however, lies in the cell's ability to return rapidly to normalcy. When the inducing signal is removed, the RecA protein loses the ability to destabilize LexA. At this moment, the *lexA* gene is being expressed at a high level; in the absence of activated RecA, the LexA protein rapidly accumulates in the uncleaved form and turns off the SOS genes. This explains why the SOS response is freely reversible.

RecA also triggers cleavage of other cellular targets, sometimes with more direct consequences. The UmuD protein is cleaved when RecA is activated; the cleavage event activates UmuD and the error-prone repair system. The current model for the reaction is that the UmuD$_2$UmuC complex binds to a RecA filament near a site of damage, RecA activates the complex by cleaving UmuD to generate UmuD', and the complex then synthesizes a stretch of DNA to replace the damaged material.

Activation of RecA also causes cleavage of some other repressor proteins, including those of several prophages. Among these is the lambda repressor (with which the protease activity was discovered). This explains why lambda is induced by ultraviolet irradiation: the lysogenic repressor is cleaved, releasing the phage to enter the lytic cycle.

This reaction is not a cellular SOS response, but instead represents recognition by the prophage that the cell is in trouble. Survival is then best assured by entering the lytic cycle to generate progeny phages. In this sense, prophage induction is piggybacking onto the cellular system by responding to the same indicator (activation of RecA).

The two activities of RecA are relatively independent. The recA441 mutation allows the SOS response to occur without inducing treatment, probably because RecA remains spontaneously in the activated state. Other mutations abolish the ability to be activated. Neither type of mutation affects the ability of RecA to handle DNA. The reverse type of mutation, inactivating the recombination function but leaving intact the ability to induce the SOS response, would be useful in disentangling the direct and indirect effects of RecA in the repair pathways.

16.14 Summary

All cells contain systems that maintain the integrity of their DNA sequences in the face of damage or errors of replication and that distinguish the DNA from sequences of a foreign source.

Repair systems can recognize mispaired, altered, or missing bases in DNA, as well as other structural distortions of the double helix. Excision repair systems cleave DNA near a site of damage, remove one strand, and synthesize a new sequence to replace the excised material. The *Uvr* system provides the main excision-repair pathway in *E. coli*. The *mut* and *dam* systems are involved in correcting mismatches generated by incorporation of incorrect bases during replication and function by preferentially removing the base on the strand of DNA that is not methylated at a *dam* target sequence.

Eukaryotic homologs of the *E. coli* MutSL system are involved in repairing mismatches that result from replication slippage; mutations in this pathway are common in certain types of cancer.

Repair systems can be connected with transcription in both prokaryotes and eukaryotes. Eukaryotes have two major nucleotide excision repair pathways: one that repairs damage anywhere in the genome, and one that specializes in the repair to transcribed strands of DNA. Both pathways depend on subunits of the transcription factor TF$_{II}$H. Human diseases are caused by mutations in genes coding for nucleotide excision repair activities, including the TF$_{II}$H subunits. They have homologs in the conserved *RAD* genes of yeast.

Recombination-repair systems retrieve information from a DNA duplex and use it to repair a sequence that has been damaged on both strands. The prokaryotic *RecBC* and *RecF* pathways both act prior to RecA, whose strand-transfer function is involved in all bacterial recombination. A major use of recombination-repair may be to recover from the situation created when a replication fork stalls. Genes in the *RAD52* group are involved in homologous recombination in eukaryotes.

Nonhomologous end-joining (NHEJ) is a general mechanism for repairing broken ends in eukaryotic DNA when homologous recombination is not possible. The Ku heterodimer brings the broken ends together so they can be ligated. Several human diseases are caused by mutations in enzymes of both the homologous recombination and nonhomologous end-joining pathways.

All repair occurs in the context of chromatin. Histone modifications and chromatin remodeling enzymes are required to facilitate repair, and histone chaperones are also needed to reset chromatin structure after repair is completed.

RecA has the ability to induce the SOS response. RecA is activated by damaged DNA in an unknown manner. It triggers cleavage of the LexA repressor protein, thus releasing repression of many loci, and inducing synthesis of the enzymes of both excision repair and recombination-repair pathways. Genes under LexA control possess an operator SOS box. RecA also directly activates some repair activities. Cleavage of repressors of lysogenic phages may induce the phages to enter the lytic cycle.

References

16.2 Repair Systems Correct Damage to DNA

Reviews

Sancar, A., Lindsey-Boltz, L. A., Unsal-Kaçmaz, K., and Linn, S. (2004). Molecular mechanisms of mammalian DNA repair and the DNA damage checkpoints. *Annu. Rev. Biochem.* 73, 39–85.

Wood, R. D., Mitchell, M., Sgouros, J., and Lindahl, T. (2001). Human DNA repair genes. *Science* 291, 1284–1289.

16.3 Excision Repair Systems in *E. coli*

Review

Goosen, N. and Moolenaar, G.F. (2008). Repair of UV damage in bacteria. *DNA Repair* 7, 353–379.

16.4 Eukaryotic Nucleotide Excision Repair Pathways

Reviews

Barnes, D. E. and Lindahl, T. (2004). Repair and genetic consequences of endogenous DNA base damage in mammalian cells. *Annu. Rev. Genet.* 38, 445–476.

Bergoglio, V. and Magnaldo, T. (2006). Nucleotide excision repair and related human diseases. *Genome Dynamics* 1, 35–52.

McCullough, A. K., Dodson, M. L., and Lloyd, R. S. (1999). Initiation of base excision repair: glycosylase mechanisms and structures. *Annu. Rev. Biochem.* 68, 255–285.

Nouspikel, T (2009). Nucleotide excision repair: variations on versatility. *Cell Mol. Life Sci. PMID* 66, 994–1009.

Sancar, A., Lindsey-Boltz, L. A., Unsal-Kaçmaz, K., and Linn, S. (2004). Molecular mechanisms of mammalian DNA repair and the DNA damage checkpoints. *Annu. Rev. Biochem.* 73, 39–85.

Research

Klungland, A. and Lindahl, T. (1997). Second pathway for completion of human DNA base excision-repair: reconstitution with purified proteins and requirement for DNase IV (FEN1). *EMBO J.* 16, 3341–3348.

Matsumoto, Y. and Kim, K. (1995). Excision of deoxyribose phosphate residues by DNA polymerase beta during DNA repair. *Science* 269, 699–702.

Reardon, J. T. and Sancar, A. (2003). Recognition and repair of the cyclobutane thymine dimer, a major cause of skin cancers, by the human excision nuclease. *Genes Dev.* 17, 2539–2551.

16.5 Base Excision Repair Systems Require Glycosylases

Review

Baute, J. and Depicker, A. (2008). Base excision repair and its role in maintaining genome stability. *Crit. Rev. Biochem. Mol. Biol.* 43, 239–276.

Research

Aas, P. A., Otterlei, M., Falnes, P. A., Vagbe, C. B., Skorpen, F., Akbari, M., Sundheim, O., Bjoras, M., Slupphaug, G., Seeberg, E., and Krokan, H. E.(2003). Human and bacterial oxidative demethylases repair alkylation damage in both RNA and DNA. *Nature* 421, 859–863.

Falnes, P. A., Johansen, R. F., and Seeberg, E. (2002). AlkB-mediated oxidative demethylation reverses DNA damage in *E. coli*. *Nature* 419, 178–182.

Klimasauskas, S., Kumar, S., Roberts, R. J., and Cheng, X. (1994). HhaI methyltransferase flips its target base out of the DNA helix. *Cell* 76, 357–369.

Lau, A. Y., Glassner, B. J., Samson, L. D., and Ellenberger, T. (2000). Molecular basis for discriminating between normal and damaged bases by the human alkyladenine glycosylase, AAG. *Proc. Natl. Acad. Sci. USA* 97, 13573–13578.

Lau, A. Y., Scherer, O. D., Samson, L., Verdine, G. L., and Ellenberger, T. (1998). Crystal structure of a human alkylbase-DNA repair enzyme complexed to DNA: mechanisms for nucleotide flipping and base excision. *Cell* 95, 249–258.

Mol, D. D. et al. (1995). Crystal structure and mutational analysis of human uracil-DNA glycosylase: structural basis for specificity and catalysis. *Cell* 80, 869–878.

Park, H. W., Kim, S. T., Sancar, A., and Deisenhofer, J. (1995). Crystal structure of DNA photolyase from *E. coli*. *Science* 268, 1866–1872.

Savva, R. et al. (1995). The structural basis of specific base-excision repair by uracil-DNA glycosylase. *Nature* 373, 487–493.

Trewick, S. C., Henshaw, T. F., Hausinger, R. P., Lindahl, T., and Sedgwick, B. (2002). Oxidative demethylation by *E. coli* AlkB directly reverts DNA base damage. *Nature* 419, 174–178.

Vassylyev, D. G. et al. (1995). Atomic model of a pyrimidine dimer excision repair enzyme complexed with a DNA substrate: structural basis for damaged DNA recognition. *Cell* 83, 773–782.

16.6 Error-Prone Repair

Review

Green, C. M. and Lehmann, A. R. (2005). Translesion synthesis and error-prone polymerases. *Adv. Exp. Med. Biol.* 570, 199–223.

Rattray, A. J. and Strathern, J. N. (2003). Error-prone DNA polymerases: when making a mistake is the only way to get ahead. *Annu. Rev. Genet.* 37, 31–66.

Research

Friedberg, E. C., Feaver, W. J., and Gerlach, V. L. (2000). The many faces of DNA polymerases: strategies for mutagenesis and for mutational avoidance. *Proc. Natl. Acad. Sci. USA* 97, 5681–5683.

Goldsmith, M., Sarov-Blat, L., and Livneh, Z. (2000). Plasmid-encoded MucB protein is a DNA polymerase (pol RI) specialized for lesion bypass in the presence of MucA, RecA, and SSB. *Proc. Natl. Acad. Sci. USA* 97, 11227–11231.

Johnson, R. E., Prakash, S., and Prakash, L. (1999). Efficient bypass of a thymine-thymine dimer by yeast DNA polymerase, Pol eta. *Science* 283, 1001–1004.

Maor-Shoshani, A., Reuven, N. B., Tomer, G., and Livneh, Z. (2000). Highly mutagenic replication by DNA polymerase V (UmuC) provides a mechanistic basis for SOS untargeted mutagenesis. *Proc. Natl. Acad. Sci. USA* 97, 565–570.

Wagner, J., Gruz, P., Kim, S. R., Yamada, M., Matsui, K., Fuchs, R. P., and Nohmi, T. (1999). The dinB gene encodes a novel *E. coli* DNA polymerase, DNA pol IV, involved in mutagenesis. *Mol. Cell* 4, 281–286.

16.7 Controlling the Direction of Mismatch Repair

Reviews

Hsieh, P. and Yamane, K. (2008). DNA mismatch repair: molecular mechanism, cancer, and ageing. *Mech. Ageing Dev.* 129, 391–407.

Kunkel, T. A. and Erie, D. A. (2005). DNA mismatch repair. *Annu. Rev. Biochem.* 74, 681–710.

Research

Strand, M., Prolla, T. A., Liskay, R. M., and Petes, T. D. (1993). Destabilization of tracts of simple repetitive DNA in yeast by mutations affecting DNA mismatch repair. *Nature* 365, 274–276.

16.8   Recombination-Repair Systems in *E. coli*

Review

West, S. C. (1997). Processing of recombination intermediates by the RuvABC proteins. *Annu. Rev. Genet.* 31, 213–244.

Research

Bork, J. M. and Inman, R. B. (2001). The RecOR proteins modulate RecA protein function at 5′ ends of single-stranded *DNA. EMBO J.* 20, 7313–7322.

16.9   Recombination Is an Important Mechanism to Recover from Replication Errors

Reviews

Cox, M. M., Goodman, M. F., Kreuzer, K. N., Sherratt, D. J., Sandler, S. J., and Marians, K. J. (2000). The importance of repairing stalled replication forks. *Nature* 404, 37–41.

McGlynn, P. and Lloyd, R. G. (2002). Recombinational repair and restart of damaged replication forks. *Nat. Rev. Mol. Cell Biol.* 3, 859–870.

Michel, B., Viguera, E., Grompone, G., Seigneur, M., and Bidnenko, V. (2001). Rescue of arrested replication forks by homologous recombination. *Proc. Natl. Acad. Sci. USA* 98, 8181–8188.

Research

Courcelle, J. and Hanawalt, P. C. (2003). RecA-dependent recovery of arrested DNA replication forks. *Annu. Rev. Genet.* 37, 611–646.

Kuzminov, A. (2001). Single-strand interruptions in replicating chromosomes cause double-strand breaks. *Proc. Natl. Acad. Sci. USA* 98, 8241–8246.

Rangarajan, S., Woodgate, R., and Goodman, M. F. (1999). A phenotype for enigmatic DNA polymerase II: a pivotal role for pol II in replication restart in UV-irradiated *Escherichia coli. Proc. Natl. Acad. Sci. USA* 96, 9224–9229.

16.10   Recombination-Repair of Double-Strand Breaks in Eukaryotes

Reviews

Krogh, B. O. and Symington, L. S. (2004). Recombination proteins in yeast. *Annu. Rev. Genet.* 38, 233–271.

Li X. and Heyer W.D. (2008). Homologous recombination in DNA repair and DNA damage tolerance. *Cell Res.* 18, 99–113.

Pardo, B., Gómez-González, B., and Aguilera, A. (2009). DNA double-strand break repair: how to fix a broken relationship. *Cell Mol. Life Sci.*

66, 1039–1056.

Prakash, S. and Prakash, L. (2002). Translesion DNA synthesis in eukaryotes: a one- or two-polymerase affair. *Genes Dev.* 16, 1872–1883.

Sancar, A., Lindsey-Boltz, L. A., Unsal-Kaçmaz, K., and Linn, S. (2004). Molecular mechanisms of mammalian DNA repair and the DNA damage checkpoints. *Annu. Rev. Biochem.* 73, 39–85.

Research

Wolner, B., van Komen, S., Sung, P., and Peterson, C. L. (2003). Recruitment of the recombinational repair machinery to a DNA double-strand break in yeast. *Mol. Cell* 12, 221–232.

16.11   Nonhomologous End-Joining Also Repairs Double-Strand Breaks

Reviews

D'Amours, D. and Jackson, S. P. (2002). The Mre11 complex: at the crossroads of DNA repair and checkpoint signalling. *Nat. Rev. Mol. Cell Biol.* 3, 317–327.

Pardo, B., Gómez-González, B., and Aguilera, A. (2009). DNA double-strand break repair: how to fix a broken relationship. *Cell Mol. Life Sci.* 66, 1039–1056.

Weterings E. and Chen, D. J. (2008). The endless tale of nonhomologous end-joining. *Cell Research* 18:114-124.

Research

Carney, J. P., Maser, R. S., Olivares, H., Davis, E. M., Le Beau, M., Yates, J. R., Hays, L., Morgan, W. F., and Petrini, J. H. (1998). The hMre11/hRad50 protein complex and Nijmegen breakage syndrome: linkage of double-strand break repair to the cellular DNA damage response. *Cell* 93, 477–486.

Cary, R. B., Peterson, S. R., Wang, J., Bear, D. G., Bradbury, E. M., and Chen, D. J. (1997). DNA looping by Ku and the DNA-dependent protein kinase. *Proc. Natl. Acad. Sci. USA* 94, 4267–4272.

Ellis, N. A., Groden, J., Ye, T. Z., Straughen, J., Lennon, D. J., Ciocci, S., Proytcheva, M., and German, J. (1995). The Bloom's syndrome gene product is homologous to RecQ helicases. *Cell* 83, 655–666.

Ma, Y., Pannicke, U., Schwarz, K., and Lieber, M. R. (2002). Hairpin opening and overhang processing by an Artemis/DNA-Dependent protein kinase complex in nonhomologous end joining and V(D)J recombination. *Cell* 108, 781–794.

Ramsden, D. A. and Gellert, M. (1998). Ku protein stimulates DNA end joining by mammalian DNA ligases: a direct role for Ku in repair of DNA double-strand breaks. *EMBO J.* 17, 609–614.

Varon, R. et al. (1998). Nibrin, a novel DNA double-strand break repair protein, is mutated in Nijmegen breakage syndrome. *Cell* 93, 467–476.

Walker, J. R., Corpina, R. A., and Goldberg, J. (2001). Structure of the Ku heterodimer bound to DNA and its implications for double-strand break repair. *Nature* 412, 607–614.

16.12 DNA Repair in Eukaryotes Occurs in the Context of Chromatin

Reviews

Ataian, Y. and Krebs, J. E. (2006). Five repair pathways in one context: chromatin modification during DNA repair. *Biochem. Cell Biol.* 84, 490–504.

Bao, Y. and Shen, X. (2007). Chromatin remodeling in DNA double-strand break repair. *Curr. Opin. Genet. Dev.* 17, 126–131.

Downs J. A., Nussenzweig M. C., Nussenzweig A. (2007). Chromatin dynamics and the preservation of genetic information. *Nature* 447, 951–958.

Humpal, S. E., Robinson, D. A., and Krebs, J. E. (2009). Marks to stop the clock: histone modifications and checkpoint regulation in the DNA damage response. *Biochem. Cell Biol.* 87, 243–253.

Kim, J. A. and Haber, J. E. (2009). Chromatin assembly factors Asf1 and CAF-1 have overlapping roles in deactivating the DNA damage checkpoint when DNA repair is complete. *Proc. Natl. Acad. Sci.* 106, 1151–1156.

Krebs, J. E. (2007). Moving marks: dynamic histone modifications in yeast. *Mol. Biosyst.* 3, 590–597.

Osley, M. A., Tsukuda, T., and Nickoloff, J. A. (2007). ATP-dependent chromatin remodeling factors and DNA damage repair. *Mutat. Res.* 618, 65–80.

Research

Chen, C. C., Carson, J. J., Feser, J., Tamburini, B., Zabaronick, S., Linger, J., and Tyler, J. K. (2008). Acetylated lysine 56 on histone H3 drives chromatin assembly after repair and signals for the completion of repair. *Cell* 134, 231–243.

Cheung W. L., Turner F. B., Krishnamoorthy T., Wolner B., Ahn S. H., Foley M., Dorsey J. A.,

Peterson C. L., Berger S. L., and Allis C. D. (2005). Phosphorylation of histone H4 serine 1 during DNA damage requires casein kinase II in *S. cerevisiae. Curr. Biol.* 15, 656–660.

Downs J. A., Allard S., Jobin-Robitaille O., Javaheri A., Auger A., Bouchard N., Kron S. J., Jackson S. P., and Cote J. (2004). Binding of chromatin-modifying activities to phosphorylated histone H2A at DNA damage sites. *Mol. Cell* 16, 979–990.

Downs, J. A., Lowndes, N. F., and Jackson, S. P. (2000). A role for *Saccharomyces cerevisiae* histone H2A in DNA repair. *Nature* 408, 1001–1004.

Moore, J. D., Yazgan, O., Ataian, Y., and Krebs, J. E. (2007). Diverse roles for histone H2A modifications in DNA damage response pathways in yeast. *Genetics* 176, 15–25.

Morrison, A. J., Highland, J., Krogan, N. J., Arbel-Eden, A., Greenblatt, J. F., Haber, J. E., and Shen, X. (2004). INO80 and gamma-H2AX interaction links ATP-dependent chromatin remodeling to DNA damage repair. *Cell* 119, 767–775.

Papamichos-Chronakis, M., Krebs, J. E., and Peterson, C. L. (2006). Interplay between Ino80 and Swr1 chromatin remodeling enzymes regulates cell cycle checkpoint adaptation in response to DNA damage. *Genes Dev.* 20, 2437–2449.

Rogakou E. P., Boon C., Redon C., and Bonner W. M. (1999). Megabase chromatin domains involved in DNA double-strand breaks *in vivo. J. Cell Biol.* 146, 905–916.

Tamburini, B. A. and Tyler, J. K. (2005). Localized histone acetylation and deacetylation triggered by the homologous recombination pathways of double-strand DNA repair. *Mol. Cell Biol.* 25, 4903–4913.

Tsukuda, T., Fleming, A. B., Nickoloff, J. A., and Osley, M. A., (2005). Chromatin remodeling at a DNA double strand break site in *Saccharomyces cerevisiae. Nature* 438, 379–383.

van Attikum, H., Fritsch, O., Hohn, B., and Gasser, S. M. (2004). Recruitment of the INO80 complex by H2A phosphorylation links ATP-dependent chromatin remodeling with DNA double-strand break repair. *Cell* 119, 777–788.

16.13 RecA Triggers the SOS System

Research

Tang, M. et al. (1999). UmuD'$_2$C is an error-prone DNA polymerase, *E. coli* pol V. *Proc. Natl. Acad. Sci. USA* 96, 8919–8924.

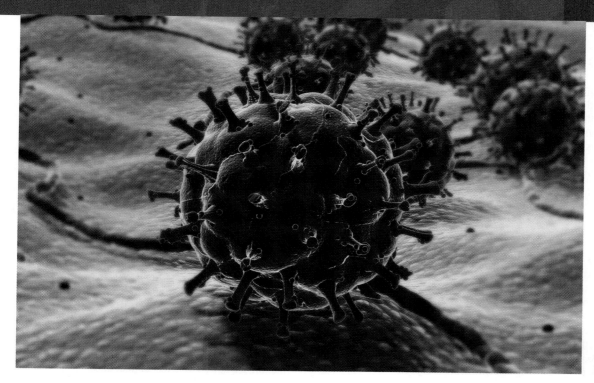

© MedicalRF/Photo
Researchers, Inc.

Transposable Elements
and Retroviruses

Edited by Damon Lisch

CHAPTER OUTLINE

17.6 Nonreplicative Transposition Proceeds by Breakage and Reunion

- Nonreplicative transposition results if a crossover structure is nicked on the unbroken pair of donor strands and the target strands on either side of the transposon are ligated.
- Two pathways for nonreplicative transposition differ according to whether the first pair of transposon strands are joined to the target before the second pair are cut (Tn5), or whether all four strands are cut before joining to the target (Tn10).

17.7 Maize Transposons Can Cause Breakage and Rearrangements

- Transposition in maize was discovered because of the effects of the chromosome breaks generated by transposition of "controlling elements."
- The break generates one chromosome that has a centromere, a broken end, and one acentric fragment.
- The acentric fragment is lost during mitosis; this can be detected by the disappearance of dominant alleles in a heterozygote.
- Fusion between the broken ends of the chromosome generates dicentric chromosomes, which undergo further cycles of breakage and fusion.
- The fusion-breakage-bridge cycle is responsible for the occurrence of somatic variegation.

17.8 Transposons Form Families in Maize

- Each family of transposons in maize has both autonomous and nonautonomous members.
- Autonomous transposons code for proteins that enable them to transpose.
- Nonautonomous transposons cannot catalyze transposition, but they can transpose when an autonomous element provides the necessary proteins.
- Autonomous transposons have changes of phase, when their properties alter in association with changes in the state of methylation.

17.9 The Role of Transposable Elements in Hybrid Dysgenesis

- P elements are transposons that are carried in P strains of *Drosophila melanogaster*, but not in M strains.
- When a P male is crossed with an M female, transposition is activated.
- The insertion of P elements at new sites in these crosses inactivates many genes and makes the cross infertile.

17.10 P Elements Are Activated in the Germline

- P elements are activated in the germline of P male × M female crosses because a tissue-specific splicing event removes one intron, which generates the coding sequence for the transposase.
- The P element also produces a repressor of transposition, which is inherited maternally in the cytoplasm.

- The presence of the repressor explains why M male × P female crosses remain fertile.

17.11 The Retrovirus Life Cycle Involves Transposition-Like Events

- A retrovirus has two copies of its genome of single-stranded RNA.
- An integrated provirus is a double-stranded DNA sequence.
- A retrovirus generates a provirus by reverse transcription of the retroviral genome.

17.12 Retroviral Genes Code for Polyproteins

- A typical retrovirus has three genes: *gag*, *pol*, and *env*.
- Gag and Pol proteins are translated from a full-length transcript of the genome.
- Translation of Pol requires a frameshift by the ribosome.
- Env is translated from a separate mRNA that is generated by splicing.
- Each of the three protein products is processed by proteases to give multiple proteins.

17.13 Viral DNA Is Generated by Reverse Transcription

- A short sequence (R) is repeated at each end of the viral RNA, so the 5' and 3' ends are R-U5 and U3-R, respectively.
- Reverse transcriptase starts synthesis when a tRNA primer binds to a site 100 to 200 bases from the 5' end.
- When the enzyme reaches the end, the 5'-terminal bases of RNA are degraded, exposing the 3' end of the DNA product.
- The exposed 3' end of the DNA product base pairs with the 3' terminus of another RNA genome.
- Synthesis continues, generating a product in which the 5' and 3' regions are repeated, giving each end the structure U3-R-U5.
- Similar strand-switching events occur when reverse transcriptase uses the DNA product to generate a complementary strand.
- Strand switching is an example of the copy choice mechanism of recombination.

17.14 Viral DNA Integrates into the Chromosome

- The organization of proviral DNA in a chromosome is the same as a transposon, with the provirus flanked by short direct repeats of a sequence at the target site.
- Linear DNA is inserted directly into the host chromosome by the retroviral integrase enzyme.
- Two base pairs of DNA are lost from each end of the retroviral sequence during the integration reaction.

17.15 Retroviruses May Transduce Cellular Sequences

- Transforming retroviruses are generated by a recombination event in which a cellular RNA sequence replaces part of the retroviral RNA.

17.1 Introduction

A major cause of variation in nearly all genomes is provided by **transposable elements** or **transposons**: these are discrete sequences in the genome that are mobile—they are able to transport themselves to other locations within the genome. The mark of a transposon is that it does not utilize an independent form of the element (such as phage or plasmid DNA), but moves directly from one site in the genome to another. Unlike most other processes involved in genome restructuring, transposition does not rely on any relationship between the sequences at the donor and recipient sites. Transposons are restricted to moving themselves, and sometimes additional sequences, to new sites elsewhere within the same genome; they are, therefore, an internal counterpart to the vectors that can transport sequences from one genome to another. They can be a major source of mutations in the genome, as shown in **FIGURE 17.1**, and have had a significant impact on the overall size of many genomes, including our own.

Transposons fall into two general classes: those that are able directly to manipulate DNA so as to propagate themselves within the genome (Class II elements, or DNA-type elements), and those whose source of mobility is the ability to make DNA copies of their RNA transcripts, which are integrated at new sites in the genome (Class I elements, or retroelements).

Transposons that mobilize via DNA are widespread in both prokaryotes and eukary-

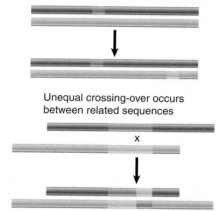

Transposon generates new copy at random site

Unequal crossing-over occurs between related sequences

x

FIGURE 17.1 A major cause of sequence change within a genome is the movement of a transposon to a new site. This may have direct consequences on gene expression. Further, unequal crossing over between related sequences causes rearrangements. Copies of transposons can provide targets for such events.

otes. Each transposon carries gene(s) that code for the enzyme activities required for its own transposition, although it may also require ancillary functions of the genome in which it resides (such as DNA polymerase or DNA gyrase).

Transposition that involves an obligatory intermediate of RNA is primarily confined to eukaryotes. Transposons that employ an RNA intermediate all use some form of reverse transcriptase to translate RNA into DNA. Some of these elements are closely related to retroviral proviruses in their general organization and mechanism of transposition. As a class, these elements are called LTR retrotransposons, or simply **retrotransposons**. Members of a second class of elements that also uses reverse transcriptase but lack LTRs, and that employ a distinct mode of transposition, are referred to as non-LTR retrotransposons, or simply **retroposons**. [The nomenclature of transposable elements is somewhat confusing in the literature, but this system of distinguishing elements by the presence or absence of the LTR reflects the modern understanding of both the evolution and the transposition mechanisms of these elements.]

Like any other reproductive cycle, the cycle of a **retrovirus** or retrotransposon is continuous; it is arbitrary to consider the point at which we interrupt it a "beginning." Our perspectives of these elements are biased, though, by the forms in which we usually observe them. The interlinked cycles of retroviruses and retrotransposons are depicted in **FIGURE 17.2**. Retroviruses were first observed as infectious virus particles that were capable of transmission between cells, and so the intracellular cycle (involving duplex DNA) is thought of as the means of reproducing the RNA virus. Retrotransposons were discov-

ered as components of the genome, and the RNA forms have been mostly characterized for their functions as mRNAs and transposition intermediates. Thus we think of retrotransposons as genomic (duplex DNA) sequences and retroviruses as RNA/protein complexes, but this obscures the close relationship between these elements. Indeed, recent phylogenetic evidence suggests that retroviruses as a class are simply retrotransposons that have acquired envelope proteins, the inverse of the previously assumed relationship.

A genome may contain both functional and nonfunctional (defective) elements of either class of element. In most cases the majority of elements in a eukaryotic genome are defective, and have lost the ability to transpose independently, although they may still be recognized as substrates for transposition by the enzymes produced by functional transposons. A eukaryotic genome contains a large number and variety of transposons. The relatively small fly genome has 1572 identified transposons belonging to 96 distinct families. Larger genomes, such as those of maize and humans, can harbor hundreds of thousands of transposons. Roughly half of the genetic material of each of these species is composed of transposons.

Transposable elements of all kinds can promote rearrangements of the genome directly or indirectly:

- The transposition event itself may cause deletions or inversions or lead to the movement of a host sequence to a new location.
- Transposons serve as substrates for cellular recombination systems by functioning as "portable regions of homology"; two copies of a transposon at different locations (even on different chromosomes) may provide sites for reciprocal recombination. Such exchanges result in deletions, insertions, inversions, or translocations.

The intermittent activities of a transposon seem to provide a somewhat nebulous target for natural selection. This concern has prompted suggestions that most transposable elements confer neither advantage nor disadvantage on the phenotype, but could constitute "selfish DNA"—DNA concerned only with its own propagation. Indeed, in considering transposi-

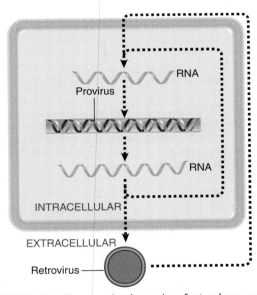

FIGURE 17.2 The reproductive cycles of retroviruses and retrotransposons alternate reverse transcription from RNA to DNA with transcription from DNA to RNA. Only retroviruses can generate infectious particles. Retrotransposons are confined to an intracellular cycle.

tion as an event that is distinct from other cellular recombination systems, we tacitly accept the view that the transposon is an independent entity that resides in the genome.

Such a relationship of the transposon to the genome would resemble that of a parasite with its host. Presumably the propagation of an element by transposition is balanced by the harm done if a transposition event inactivates a necessary gene, or if the number of transposons becomes a burden on cellular systems. Yet we must remember that any transposition event conferring a selective advantage—for example, a genetic rearrangement—will lead to preferential survival of the genome carrying the active transposon.

17.2 Insertion Sequences Are Simple Transposition Modules

Key concepts

- An insertion sequence is a transposon that codes for the enzyme(s) needed for transposition flanked by short inverted terminal repeats.
- The target site at which an insertion sequence is inserted is duplicated during the insertion process to form two repeats in direct orientation at the ends of the transposon.
- The length of the direct repeat is 5 to 9 bp and is characteristic for any particular insertion sequence.

Transposable elements were first identified at the molecular level in the form of spontaneous insertions in bacterial operons. Such an insertion prevents transcription and/or translation of the gene in which it is inserted. Many different types of transposable elements have now been characterized in both prokaryotes and eukaryotes (in which they are far more abundant), but the basic principles and biochemistry of elements first described in bacteria apply to DNA-type elements in many species.

The simplest bacterial transposons are called **insertion sequences** (reflecting the way in which they were detected). Each type is given the prefix **IS**, followed by a number that identifies the type. (The original classes were numbered IS1 to IS4; later classes have

numbers reflecting the history of their isolation, but not corresponding to the more than 700 elements so far identified!)

The IS elements are normal constituents of bacterial chromosomes and plasmids. A standard strain of *E. coli* is likely to contain several (<10) copies of any one of the more common IS elements. To describe an insertion into a particular site, a double colon is used; thus λ::IS1 describes an IS1 element inserted into phage lambda. Most IS elements insert at a variety of sites within host DNA. Some, though, show varying degrees of preference for particular hotspots.

The IS elements are autonomous units, each of which codes only for the proteins needed to sponsor its own transposition. Each IS element is different in sequence, but there are some common features in organization. The structure of a generic transposon before and after insertion at a target site is illustrated in **FIGURE 17.3**, which also summarizes the details of some common IS elements.

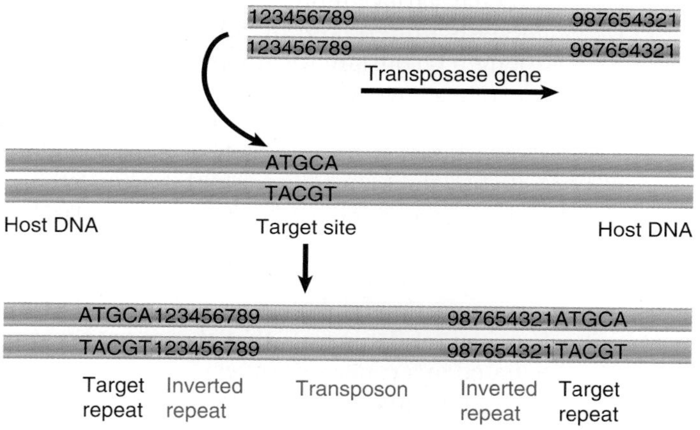

Transposon	Target repeat (bp)	Inverted repeat (bp)	Overall length (bp)	Target selection
IS1	9	23	768	random
IS2	5	41	1327	hotspots
IS4	11–13	18	1428	$AAAN_{20}TTT$
IS5	4	16	1195	hotspots
IS10R	9	22	1329	NGCTNAGCN
IS50R	9	9	1531	hotspots
IS903	9	18	1057	random

FIGURE 17.3 IS elements have inverted terminal repeats and generate direct repeats of flanking DNA at the target site. In this example, the target is a 5 bp sequence. The ends of the transposon consist of inverted repeats of 9 bp, where the numbers 1 through 9 indicate a sequence of base pairs.

An IS element ends in short **inverted terminal repeats**; usually the two copies of the repeat are closely related rather than identical. As illustrated in the figure, the presence of the inverted terminal repeats means that the same sequence is encountered proceeding toward the element from the flanking DNA on either side of it.

When an IS element transposes, a sequence of host DNA at the site of insertion is duplicated. The nature of the duplication is revealed by comparing the sequence of the target site before and after an insertion has occurred. Figure 17.3 shows that at the site of insertion, the IS DNA is always flanked by very short **direct repeats**. (In this context, "direct" indicates that two copies of a sequence are repeated in the same orientation, not that the repeats are adjacent.) In the original gene (prior to insertion), however, the target site has the sequence of only one of these repeats. In the figure, the target site consists of the sequence $_{\mathrm{TACGT}}^{\mathrm{ATGCA}}$. After transposition, one copy of this sequence is present on either side of the transposon. The sequence of the direct repeat varies among individual transposition events undertaken by a transposon, but the length is constant for any particular IS element (a reflection of the mechanism of transposition).

An IS element therefore displays a characteristic structure in which its ends are identified by the inverted terminal repeats, whereas the adjacent ends of the flanking host DNA are identified by the short direct repeats. When observed in a sequence of DNA, this type of organization is taken to be diagnostic of a transposon, and suggests that the sequence originated in a transposition event.

The inverted repeats define the ends of a transposon. Recognition of the ends is common to transposition events sponsored by all types of DNA-type transposon. *cis*-acting mutations that prevent transposition are located in the ends, which are recognized by a protein(s) responsible for transposition. The protein is called a **transposase**.

Many of the IS elements contain a single long coding region, which starts just inside the inverted repeat at one end and terminates just before or within the inverted repeat at the other end. This region codes for the transposase. Some elements have a more complex organization. IS1, for instance, has two separate reading frames; the transposase is produced by making a frameshift during translation to allow both reading frames to be used.

The frequency of transposition varies among different elements. Under most circumstances the overall rate of transposition is $\sim 10^{-3}$ to 10^{-4} per element per generation. Insertions in individual targets occur at a level comparable with the spontaneous mutation rate, usually $\sim 10^{-5}$ to 10^{-7} per generation. Reversion (by precise excision of the IS element) is usually infrequent, with a range of rates of 10^{-6} to 10^{-10} per generation, which is $\sim 10^{3}$ times less frequent than insertion.

17.3 Transposition Occurs by Both Replicative and Nonreplicative Mechanisms

Key concepts

- Most transposons use a common mechanism in which staggered nicks are made in target DNA, the transposon is joined to the protruding ends, and the gaps are filled.
- The order of events and exact nature of the connections between transposon and target DNA determine whether transposition is replicative or nonreplicative.

The insertion of a transposon into a new site is illustrated in **FIGURE 17.4**. It consists of making staggered breaks in the target DNA, joining the transposon to the protruding single-stranded

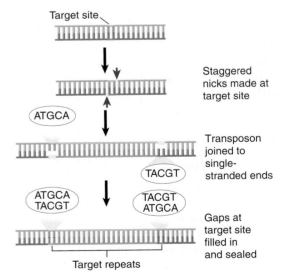

FIGURE 17.4 The direct repeats of target DNA flanking a transposon are generated by the introduction of staggered cuts whose protruding ends are linked to the transposon.

ends, and filling in the gaps. The generation and filling of the staggered ends explain the occurrence of the direct repeats of target DNA at the site of insertion. The stagger between the cuts on the two strands determines the length of the direct repeats; thus the target repeat characteristic of each transposon reflects the geometry of the enzyme involved in cutting target DNA.

The use of staggered ends is common to most means of transposition, but we can distinguish two major types of mechanism by which a transposon moves:

- In **replicative transposition**, the element is duplicated during the reaction, so that the transposing entity is a copy of the original element. **FIGURE 17.5** summarizes the results of such a transposition. The transposon is copied as part of its movement. One copy remains at the original site, whereas the other inserts at the new site. Thus transposition is accompanied by an increase in the number of copies of the transposon. Replicative transposition involves two types of enzymatic activity: a transposase that acts on the ends of the original transposon, and a **resolvase** that acts on the duplicated copies. While one group of transposons move only by replicative transposition (see *Section 17.5, Replicative Transposition Proceeds through a Cointegrate*), true replicative transposition is relatively rare among transposons in general.

- In **nonreplicative transposition**, the transposing element moves as a physical entity directly from one site to another and is conserved. The insertion sequences and **composite transposons** (**Tn**) Tn10 and Tn5 (as well as many eukaryotic transposons) use the

mechanism shown in **FIGURE 17.6**, which involves the release of the transposon from the flanking donor DNA during transfer. This type of mechanism, often referred to as "cut-and-paste," requires only a transposase. Another mechanism utilizes the connection of donor and target DNA sequences and shares some steps with replicative transposition. Both mechanisms of nonreplicative transposition cause the element to be inserted at the target site and lost from the donor site. What happens to the donor molecule after a nonreplicative transposition? Its survival requires that host repair systems recognize the double-strand break and repair it.

Some bacterial transposons use only one type of pathway for transposition, whereas others may be able to use multiple pathways. The elements IS1 and IS903 use both nonreplicative and replicative pathways, and the ability of phage Mu to turn to either type of pathway from a common intermediate has been well characterized.

The same basic types of reaction are involved in all classes of transposition events. The ends of the transposon are disconnected from the donor DNA by cleavage reactions that generate 3'–OH ends. The exposed ends are then joined to the target DNA by transfer reactions, involving transesterification in which the 3'–OH end directly attacks the target DNA. These reactions take place within a nucleoprotein complex that contains the necessary enzymes and both ends of the transposon. Transposons differ as to whether the target DNA is recognized before or after the cleavage of the transposon itself, and whether one or both strands at the ends of the transposon are cleaved prior to integration.

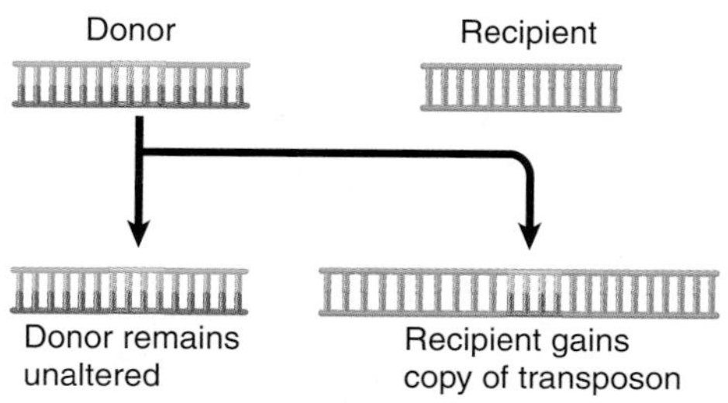

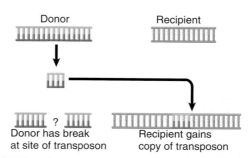

FIGURE 17.5 Replicative transposition creates a copy of the transposon, which inserts at a recipient site. The donor site remains unchanged, so both donor and recipient have a copy of the transposon.

FIGURE 17.6 Nonreplicative transposition allows a transposon to move as a physical entity from a donor to a recipient site. This leaves a break at the donor site, which is lethal unless it can be repaired.

The choice of target site is in effect made by the transposase, sometimes in conjunction with accessory proteins. In some cases, the target is chosen virtually at random. In others, there is specificity for a consensus sequence or for some other feature in the target. The feature can take the form of a structure in DNA, such as bent DNA, or a protein–DNA complex. In the latter case, the nature of the target complex can cause the transposon to insert at specific promoters (such as Ty1 or Ty3, which select pol III promoters in yeast), inactive regions of the chromosome, or replicating DNA.

17.4 Transposons Cause Rearrangement of DNA

Key concepts
- Homologous recombination between multiple copies of a transposon causes rearrangement of host DNA.
- Homologous recombination between the repeats of a transposon may lead to precise or imprecise excision.

In addition to the "simple" intermolecular transposition that results in insertion at a new site, transposons promote other types of DNA rearrangements. Some of these events are consequences of the relationship between the multiple copies of the transposon. Others represent alternative outcomes of the transposition mechanism, and they leave clues about the nature of the underlying events.

Rearrangements of host DNA may result when a transposon inserts a copy at a second site near its original location. Host systems may undertake reciprocal recombination between the two copies of the transposon; the consequences are determined by whether the repeats are the same or in inverted orientation.

FIGURE 17.7 illustrates the general rule that recombination between any pair of direct repeats will delete the material between them. The intervening region is excised as a circle of DNA (which is lost from the cell); the chromosome retains a single copy of the direct repeat. A recombination between the directly repeated IS1 modules of the composite transposon Tn9 would replace the transposon with a single IS1 module.

Deletion of sequences adjacent to a transposon could therefore result from a two-stage process; transposition generates a direct repeat of a transposon, and recombination occurs between the repeats. The majority of deletions that arise in the vicinity of transposons, however, probably result from a variation in the pathway followed in the transposition event itself.

FIGURE 17.8 depicts the consequences of a reciprocal recombination between a pair of inverted repeats. The region between the repeats becomes inverted; the repeats themselves remain available to sponsor further inversions. A composite transposon whose modules are inverted is a stable component of the genome, although the direction of the central region with regard to the modules could be inverted by recombination.

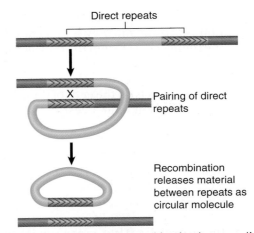

FIGURE 17.7 Reciprocal recombination between direct repeats excises the material between them; each product of recombination has one copy of the direct repeat.

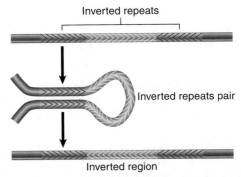

FIGURE 17.8 Reciprocal recombination between inverted repeats inverts the region between them.

Excision in this case is not supported by transposons themselves, but occurs when bacterial enzymes recognize homologous regions in the transposons. This is important because the loss of a transposon may restore function at the site of insertion. **Precise excision** requires removal of the transposon, plus one copy of the duplicated sequence. This is rare; it occurs at a frequency of ~10^{-6} for Tn5 and ~10^{-9} for Tn10. It probably involves a recombination between the duplicated target sites.

Imprecise excision leaves a remnant of the transposon. The remnant may be sufficient to prevent reactivation of the target gene, but it may be insufficient to cause polar effects in adjacent genes so that a change of phenotype occurs. Imprecise excision occurs at a frequency of ~10^{-6} for Tn10. It involves recombination between sequences of 24 bp in the IS10 modules; these sequences are inverted repeats, but since the IS10 modules themselves are inverted, they form direct repeats in Tn10.

The greater frequency of imprecise excision compared with precise excision probably reflects the increase in the length of the direct repeats (24 bp as opposed to 9 bp). Neither type of excision relies on transposon-coded functions, but the mechanism is not known. Excision is RecA-independent and could occur by some cellular mechanism that generates spontaneous deletions between closely spaced repeated sequences.

Both precise and imprecise excisions can also arise as a consequence of transposition of "cut-and-paste" elements in eukaryotes. In this case, the outcome depends on the nature of the repair of the double-stranded DNA break introduced by excision of the element. This break can be repaired using the homologous chromosome or the sister chromatid, resulting in a transfer of DNA from those templates. Repair using a chromosome that lacks the transposon insertion can result in precise restoration of sequences surrounding the original insertion. Repair using the sister chromatid results in restoration of the transposon insertion. Incomplete repair can result in deletions, either of sequences flanking the insertion or of portions of the transposon. Alternatively, the break can be repaired using nonhomologous end joining, which results in the addition or deletion of short stretches of DNA.

17.5 Replicative Transposition Proceeds Through a Cointegrate

Key concepts

- Replication of a strand transfer complex generates a cointegrate, which is a fusion of the donor and target replicons.
- The cointegrate has two copies of the transposon, which lie between the original replicons.
- Recombination between the transposon copies regenerates the original replicons, but the recipient has gained a copy of the transposon.
- The recombination reaction is catalyzed by a resolvase coded by the transposon.

The basic structures involved in replicative transposition are illustrated in **FIGURE 17.9**: The 3' ends of the strand transfer complex are used as primers for replication. This generates a structure called a **cointegrate**, which represents a fusion of the two original molecules. The cointegrate has two copies of the transposon, one at each junction between the original replicons, oriented as direct repeats. The crossover is formed by the transposase. Its conversion into the cointegrate requires host replication functions.

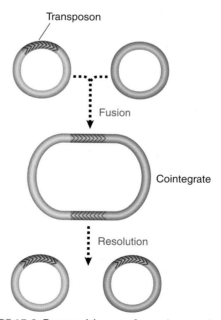

FIGURE 17.9 Transposition may fuse a donor and recipient replicon into a cointegrate. Resolution releases two replicons, each containing a copy of the transposon.

A homologous recombination between the two copies of the transposon releases two individual replicons, each of which has a copy of the transposon. One of the replicons is the original donor replicon. The other is a target replicon that has gained a transposon flanked by short direct repeats of the host target sequence. The recombination reaction is called **resolution**; the enzyme activity responsible is called the **resolvase**.

The reactions involved in generating a cointegrate have been defined in detail for phage Mu and are illustrated in **FIGURE 17.10**. The process starts with the formation of the strand transfer complex (sometimes called a crossover complex). The donor and target strands are ligated

so that each end of the transposon sequence is joined to one of the protruding single strands generated at the target site. The strand transfer complex generates a crossover-shaped structure held together at the duplex transposon. The fate of the crossover structure determines the mode of transposition.

The principle of replicative transposition is that replication through the transposon duplicates it, which creates copies at both the target and donor sites. The product is a cointegrate.

The crossover structure contains a single-stranded region at each of the staggered ends. These regions are pseudoreplication forks that provide a template for DNA synthesis. (Use of the ends as primers for replication implies that the strand breakage must occur with a polarity that generates a 3'–OH terminus at this point.)

If replication continues from both the pseudoreplication forks, it will proceed through the transposon, separating its strands and terminating at its ends. Replication is accomplished by host-coded functions. At this juncture, the structure has become a cointegrate, possessing direct repeats of the transposon at the junctions between the replicons (as can be seen by tracing the path around the cointegrate).

Transposon   Target

Nicking
Single-strand cuts generate staggered ends in both transposon and target

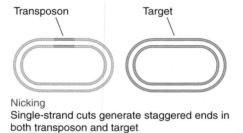

Crossover structure (strand transfer complex): Nicked ends of transposon are joined to nicked ends of target

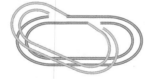

Replication from free 3' ends generates cointegrate: Single molecule has two copies of transposon

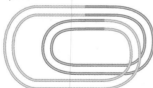

Cointegrate drawn as continuous path shows that transposons are at junctions between replicons

FIGURE 17.10 Mu transposition generates a crossover structure, which is converted by replication into a cointegrate.

17.6 Nonreplicative Transposition Proceeds by Breakage and Reunion

Key concepts

- Nonreplicative transposition results if a crossover structure is nicked on the unbroken pair of donor strands and the target strands on either side of the transposon are ligated.
- Two pathways for nonreplicative transposition differ according to whether the first pair of transposon strands are joined to the target before the second pair are cut (Tn5), or whether all four strands are cut before joining to the target (Tn10).

The crossover structure can also be used in nonreplicative transposition. The principle of nonreplicative transposition by this mechanism is that a breakage and reunion reaction allows the target to be reconstructed with the insertion of the transposon; the donor remains broken. No cointegrate is formed.

FIGURE 17.11 shows the cleavage events that generate nonreplicative transposition of phage

Mu. Once the unbroken donor strands have been nicked, the target strands on either side of the transposon can be ligated. The single-stranded regions generated by the staggered cuts must be filled in by repair synthesis. The product of this reaction is a target replicon in which the transposon has been inserted between repeats of the sequence created by the original single-strand nicks. The donor replicon has a double-strand break across the site where the transposon was originally located.

Nonreplicative transposition can also occur by an alternative pathway in which nicks are made in target DNA, but a double-strand break is made on either side of the transposon, releasing it entirely from flanking donor sequences (as envisaged in Figure 17.6). This "cut and paste" pathway is used by Tn10 and by many eukaryotic transposons and is illustrated in FIGURE 17.12.

A simple experiment to prove that Tn10 transposes nonreplicatively made use of an artificially constructed heteroduplex of Tn10 that contained single-base mismatches. If transposition involves replication, the transposon at the new site will contain information from only one of the parent Tn10 strands. If, however, transposition takes place by physical movement of the existing transposon, the mismatches will be conserved at the new site. This proved to be the case.

The basic difference in Figure 17.11 from the model of Figure 17.12 is that both strands of Tn10 are cleaved before any connection is made to the target site. The first step in the reaction is recognition of the transposon ends by the transposase, forming a proteinaceous structure within which the reaction occurs. At each end of the transposon, the strands are cleaved in a specific order: the transferred strand (the one to be connected to the target site) is cleaved first, followed by the other strand. (This is the same order as in the Mu transposition of Figure 17.10 and Figure 17.11.)

Tn5 also transposes by nonreplicative transposition. FIGURE 17.13 shows the interesting cleavage reaction that separates the transposon from the flanking sequences: First one

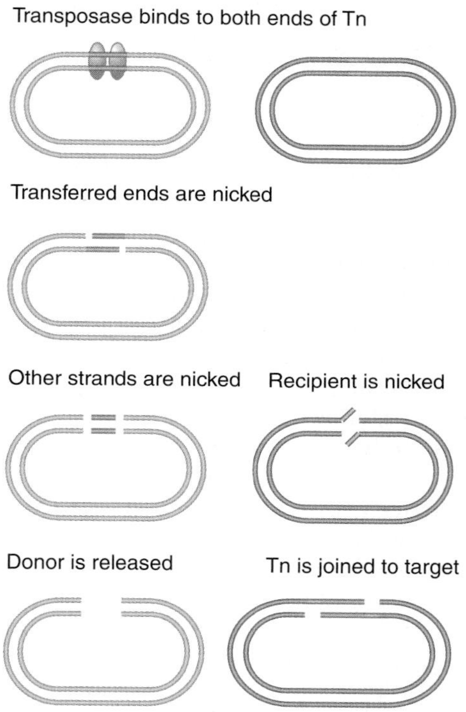

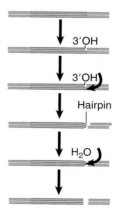

Transposase binds to both ends of Tn

Transferred ends are nicked

Other strands are nicked   Recipient is nicked

Donor is released   Tn is joined to target

FIGURE 17.12 Both strands of Tn10 are cleaved sequentially, and then the transposon is joined to the nicked target site.

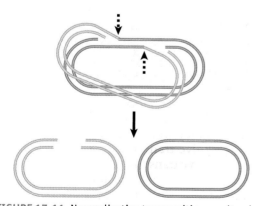

FIGURE 17.11 Nonreplicative transposition results when a crossover structure is released by nicking. This inserts the transposon into the target DNA, flanked by the direct repeats of the target, and the donor is left with a double-strand break.

3'OH
3'OH
Hairpin
H_2O

FIGURE 17.13 Cleavage of Tn5 from flanking DNA involves nicking, interstrand reaction, and hairpin cleavage.

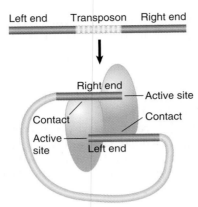

FIGURE 17.14 Each subunit of the Tn5 transposase has one end of the transposon located in its active site and also makes contact at a different site with the other end of the transposon.

Left end   Transposon   Right end

Right end — Active site
Contact
Active site — Contact
Left end

Maize Transposons Can Cause Breakage and Rearrangements

Key concepts

- Transposition in maize was discovered because of the effects of the chromosome breaks generated by transposition of "controlling elements."
- The break generates one chromosome that has a centromere, a broken end, and one acentric fragment.
- The acentric fragment is lost during mitosis; this can be detected by the disappearance of dominant alleles in a heterozygote.
- Fusion between the broken ends of the chromosome generates dicentric chromosomes, which undergo further cycles of breakage and fusion.
- The fusion-breakage-bridge cycle is responsible for the occurrence of somatic variation.

DNA strand is nicked. The 3'–OH end that is released then attacks the other strand of DNA. This releases the flanking sequence and joins the two strands of the transposon in a hairpin. An activated water molecule then attacks the hairpin to generate free ends for each strand of the transposon.

In the next step, the cleaved donor DNA is released, and the transposon is joined to the nicked ends at the target site. The transposon and the target site remain constrained in the proteinaceous structure created by the transposase (and other proteins). The double-strand cleavage at each end of the transposon precludes any replicative-type transposition and forces the reaction to proceed by nonreplicative transposition, thus giving the same outcome as in Figure 17.12, but with the individual cleavage and joining steps occurring in a different order.

The Tn5 and Tn10 transposases both function as dimers. Each subunit in the dimer has an active site that successively catalyzes the double-strand breakage of the two strands at one end of the transposon, and then catalyzes staggered cleavage of the target site. **FIGURE 17.14** illustrates the structure of the Tn5 transposase bound to the cleaved transposon. Each end of the transposon is located in the active site of one subunit. One end of the subunit also contacts the other end of the transposon. This controls the geometry of the transposition reaction. Each of the active sites will cleave one strand of the target DNA. It is the geometry of the complex that determines the distance between these sites on the two target strands (nine base pairs in the case of Tn5).

One of the most visible consequences of the existence and mobility of transposons occurs during plant development, when somatic variation occurs. This is due to changes in the location or behavior of **controlling elements** (the name that transposons were given in maize before their molecular nature was discovered).

Two features of maize have helped to follow transposition events. Transposons in eukaryotes often insert near genes that have visible but nonlethal effects on the phenotype. Maize displays clonal development, which means that the occurrence and timing of a transposition event can be visualized as depicted diagrammatically in **FIGURE 17.15**. The nature of the event does not matter: It may be a point mutation, insertion, excision, or chromosome break. What is important is that it occurs in a heterozygote to alter the expression of one allele. The descendants of a cell that has suffered the event then display a new phenotype, whereas the descendants of cells not affected by the event continue to display the original phenotype.

Mitotic descendants of a given cell remain in the same location and give rise to a **sector** of tissue. A change in phenotype during somatic development is called **variegation**; it is revealed by a sector of the new phenotype residing within the tissue of the original phenotype. The size of the sector depends on the number of divisions in the lineage giving rise to it, so the size of the area of the new phenotype is determined by the timing of the change in genotype: the earlier its occurrence in the cell

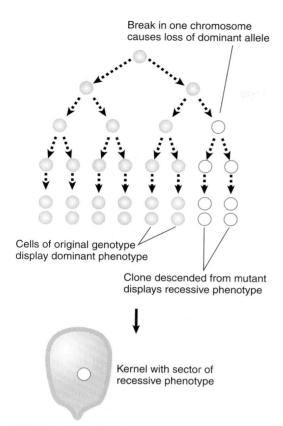

FIGURE 17.15 Clonal analysis identifies a group of cells descended from a single ancestor in which a transposition-mediated event altered the phenotype. Timing of the event during development is indicated by the number of cells; tissue specificity of the event may be indicated by the location of the cells.

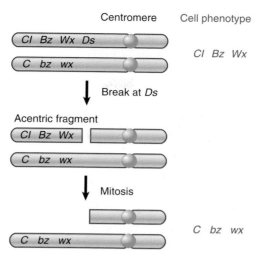

FIGURE 17.16 A break at a controlling element causes loss of an acentric fragment; if the fragment carries the dominant markers of a heterozygote, its loss changes the phenotype. The effects of the dominant markers, *CI*, *Bz*, and *Wx*, can be visualized by the color of the cells or by appropriate staining.

lineage, the greater the number of descendants and thus the size of patch in the mature tissue. This is seen most vividly in the variation in kernel color, when patches of one color appear within another color.

Insertion of a transposon may affect the activity of adjacent genes. Deletions, duplications, inversions, and translocations all occur at the sites where transposons are present. Chromosome breakage is a common consequence of the presence of some elements. In maize and other plants, the activities of the controlling elements are often regulated during development. The elements transpose and promote genetic rearrangements at characteristic times and frequencies during plant development.

Some forms of the maize *Ds* (for disassociator) element are particularly prone to chromosome breakage, which is why its activity was so easily detected by Barbara McClintock. The consequences are illustrated in FIGURE 17.16. Consider a heterozygote in which *Ds* lies on

one homolog between the centromere and a series of dominant markers. The other homolog lacks *Ds* and has recessive markers (*C, bz,* and *wx*). Breakage at *Ds* generates an **acentric fragment** carrying the dominant markers. As a result of its lack of a centromere, this fragment is lost at mitosis. Thus the descendant cells have only the recessive markers carried by the intact chromosome. This gives the type of situation whose results are depicted in Figure 17.15.

FIGURE 17.17 shows that breakage at *Ds* leads to the formation of two unusual chromosomes. These are generated by joining the broken ends of the products of replication. One is a U-shaped acentric fragment consisting of the joined sister chromatids for the region distal to *Ds* (on the left as drawn in the figure). The other is a U-shaped **dicentric chromosome** comprising the sister chromatids proximal to *Ds* (on its right in the figure). The latter structure leads to the classic **breakage-fusion-bridge cycle** illustrated in the figure.

Follow the fate of the dicentric chromosome when it attempts to segregate on the mitotic spindle. Each of its two centromeres pulls toward an opposite pole. The tension breaks the chromosome at a random site between the centromeres. In the example of the figure, breakage occurs between loci *A* and *B*, with the result that one daughter chromosome has a duplication of *A*, whereas the other has a deletion. If *A* is a dominant marker, the cells

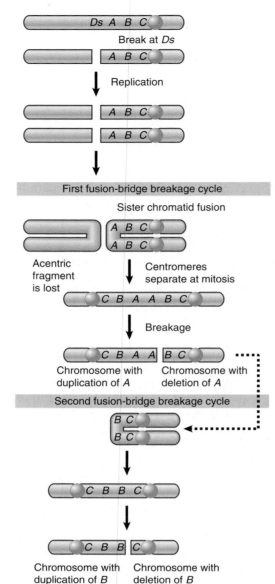

Ds A B C

Break at *Ds*

A B C

↓ Replication

A B C

A B C

First fusion-bridge breakage cycle

Sister chromatid fusion

A B C
A B C

Acentric fragment is lost

Centromeres separate at mitosis

C B A A B C

↓ Breakage

C B A A | **B C**

Chromosome with duplication of *A* | Chromosome with deletion of *A*

Second fusion-bridge breakage cycle

B C
B C

↓

C B B C

↓

C B B | **C**

Chromosome with duplication of *B* | Chromosome with deletion of *B*

FIGURE 17.17 *Ds* provides a site to initiate the chromatid breakage-fusion-bridge cycle. The products can be followed by clonal analysis.

with the duplication will retain the A phenotype, but cells with the deletion will display the recessive *a* phenotype.

The breakage-fusion-bridge cycle continues through further cell generations, allowing genetic changes to continue in the descendants. For example, consider the deletion chromosome that has lost *A*. In the next cycle, a break occurs between *B* and *C*, so that the descendants are divided into those with a duplication of *B* and those with a deletion. Successive losses of dominant markers are revealed by subsectors within sectors.

17.8 Transposons Form Families in Maize

Key concepts

- Each family of transposons in maize has both autonomous and nonautonomous members.
- Autonomous transposons code for proteins that enable them to transpose.
- Nonautonomous transposons cannot catalyze transposition, but they can transpose when an autonomous element provides the necessary proteins.
- Autonomous transposons have changes of phase, when their properties alter in association with changes in the state of methylation.

The maize genome, like that of most eukaryotes, contains many families of transposons. The numbers, types, and locations of the elements are characteristic for each individual maize strain. They may occupy a significant part of the genome; in fact, the overall size of the maize genome has roughly doubled in size in the last six million years due to transposon activity. At present, though, only a limited number of transposons are known to be active in maize. Most of these elements are DNA-type elements that are present in relatively low numbers in the maize genome. As a result of their effects on maize gene expression, they were called "controlling elements" by McClintock. They have been studied extensively over the past several decades.

The members of each family are divided into two classes:

- **Autonomous transposons** have the ability to excise and transpose. As a result of the continuing activity of an autonomous transposon, its insertion at any locus creates an unstable or "mutable" allele. Loss of the autonomous transposon itself, or of its ability to transpose, converts a mutable allele to a stable allele.

- **Nonautonomous transposons** are stable; they do not transpose or suffer other spontaneous changes in condition. They become unstable only when an autonomous member of the same family is present elsewhere in the genome. When complemented in *trans* by an autonomous element, a nonautonomous element displays the usual range of activities associated with autonomous elements, including the ability to transpose to new sites. Non-

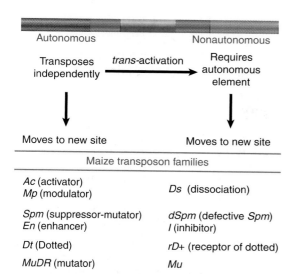

FIGURE 17.18 Each controlling element family has both autonomous and nonautonomous members. Autonomous elements are capable of transposition. Nonautonomous elements are deficient in transposition. Pairs of autonomous and nonautonomous elements can be classified in >4 families.

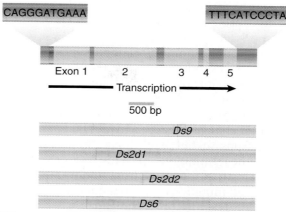

FIGURE 17.19 The *Ac* element has five exons that code for a transposase; *Ds* elements have internal deletions.

autonomous transposons are derived from autonomous transposons by loss of *trans*-acting functions needed for transposition.

Families of transposons are defined by the interactions between autonomous and nonautonomous elements. A family consists of a single type of autonomous element accompanied by many varieties of nonautonomous elements. A nonautonomous element is placed in a family by its ability to be activated in *trans* by the autonomous elements. The major families of active transposons in maize are summarized in FIGURE 17.18.

Characterized at the molecular level, the classical maize transposons share the usual form of organization—inverted repeats at the ends and short direct repeats in the adjacent target DNA—but otherwise vary in size and coding capacity. All families of transposons share the same type of relationship between the autonomous and nonautonomous elements. The autonomous elements have open reading frames between the terminal repeats, whereas the nonautonomous elements do not code for functional proteins. Sometimes the internal sequences are related to those of autonomous elements; at other times they are composed of fragments of genes that have been captured between transposon-inverted repeats.

The Mutator transposon is the most active and mutagenic of all maize transposons. The autonomous element *MuDR* codes for the genes *mudrA* (which codes for the MURA transposase) and *mudrB* (which codes for an accessory protein required for integration). The ends of the elements are marked by 200 bp inverted repeats. Nonautonomous Mutator elements—basically any unit that has the inverted repeats, but which may not have any internal sequence relationship to MuDR—are also mobilized by MURA and MURB. Mutator elements in maize are the founding members of the MULE (*Mu-like Element*) superfamily of transposons, which are present in bacteria, fungi, plants, and animals.

In maize lines with active transposons, there are typically several members of a given transposon family. By analyzing autonomous and nonautonomous elements of the *Ac/Ds* family, we have molecular information about many individual examples of these elements. FIGURE 17.19 summarizes their structures.

Most of the length of the autonomous *Ac* **(activator) element** is occupied by a single gene consisting of five exons. The product is the transposase. The element itself ends in inverted repeats of 11 bp, and a target sequence of 8 bp is duplicated at the site of insertion.

Ds **elements** vary in both length and sequence, but are related to *Ac*. They end in the same 11 bp inverted repeats. They are shorter than *Ac*, and the length of deletion varies. At one extreme, the element *Ds9* has a deletion of only 194 bp. In a more extensive deletion, the *Ds6* element retains a length of only 2 kb, representing 1 kb from each end of *Ac*. A complex double *Ds* element has one *Ds6* sequence inserted in reverse orientation into another. Double *Ds* elements are particularly prone to cause chromosomal breaks such as those observed by McClintock.

Nonautonomous elements lack internal sequences, but possess the terminal inverted repeats (and possibly other sequence features). Some nonautonomous elements are derived from autonomous elements by deletions (or other changes) that inactivate the *trans*-acting transposase, but leave intact the sites (including the termini) on which the transposase acts. Their structures range from minor (but inactivating) mutations of *Ac* to sequences that have major deletions or rearrangements.

At another extreme, the *Ds1* family members comprise short sequences whose only relationship to *Ac* lies in the possession of terminal inverted repeats. Elements of this class need not be directly derived from *Ac*, but could be derived by any event that generates the inverted repeats. Their existence suggests that the transposase recognizes only the terminal inverted repeats, or possibly the terminal repeats in conjunction with some short internal sequence.

Ds1 elements are just one example of a widespread form of DNA-type elements called MITEs (*miniature inverted repeat transposable elements*). These are very short derivatives of autonomous elements found in many eukaryotes that can be present in tens or hundreds of thousands of copies in a given genome. They range from 300 to 500 bp, and generate 2–3 bp target site duplications. Unlike many other classes of transposons in plants, MITEs are often found in or near genes.

Transposition of *Ac/Ds* occurs by a nonreplicative "cut and paste" mechanism that involves double-stranded breaks followed by integration of the released element. The mechanism of transposition is similar to that described for Tn5 and Tn10 (17.6). It is accompanied by its disappearance from the donor location. Clonal analysis suggests that transposition of *Ac/Ds* almost always occurs soon after the donor element has been replicated. These features resemble transposition of the bacterial element Tn10. The cause is the same: transposition does not occur when the DNA of the transposon is methylated on both strands (the typical state before replication), and is activated when the DNA is hemimethylated (the typical state immediately after replication). The recipient site is frequently on the same chromosome as the donor site, and often is quite close to it. Note that if transposition is from a replicated region of a chromosome into an unreplicated region, the transposition event will result in a net increase in the copy number of the element; one chromatid will carry a single copy of the transposon, and the second chromatid will carry two copies. This ensures that elements such as *Ac* can increase their copy number, even though transposition is not duplicative.

Replication generates two copies of a potential *Ac/Ds* donor, but usually only one copy actually transposes. What happens to the donor site? The rearrangements that are found at sites from which controlling elements have been lost can be explained in terms of the consequences of a chromosome break. Based on the sequence of the donor site following excision, the majority of the breaks caused by *Ac* excision appear to be repaired using nonhomologous end-joining, which usually creates sequence alterations, or transposon footprints, at the excision sites. If the resulting transposon footprint restores functionality to the gene in which the *Ac* element had been inserted, the result is a reversion event. Otherwise, the result is a stable, nonfunctional gene. In contrast, the mode of Mu element transposition appears to vary depending on the tissue type. Late during somatic development, transposition is similar to that observed for *Ac*. In germinal tissues, though, the vast majority of transposition events are effectively replicative, perhaps due to gap repair using the sister chromatid as a template.

Autonomous and nonautonomous elements are subject to a variety of changes in their condition. Some of these changes are genetic; others are epigenetic. The major change is (of course) the conversion of an autonomous element into a nonautonomous element, but further changes may occur in the nonautonomous element. *cis*-acting defects may render a nonautonomous element impervious to autonomous elements. Thus a nonautonomous element may become permanently stable because it can no longer be activated to transpose.

Autonomous elements are subject to "changes of phase," which are heritable but often unstable alterations in their properties. These may take the form of a reversible inactivation in which the element cycles between an active and inactive condition during plant development, or they may result in stably inactive elements.

Phase changes in both the *Ac* and Mu types of autonomous element are associated with changes in the methylation of DNA. The inactive form of all elements are methylated at cytosine residues. In most cases, it is not known what triggers this loss of activity, but in the case of *MuDR* epigenetic silencing can be triggered by a derivative of *MuDR* that is duplicated and inverted relative to itself. This rearrangement results in the production of a hairpin RNA, in which two parts of the transcript are perfect complements to each other. The resulting double-stranded RNA is processed by cellular factors into small RNAs that in turn trigger methylation and transcriptional gene silencing of the *MuDR* element (see Chapter 30, *Regulatory RNA*).

The effect of methylation is common generally among transposons in plants and other organisms that methylate their DNA. The best demonstration of the effect of methylation on activity comes from observations made with the *Arabidopsis* mutant *ddm1*, which causes a genome-wide loss of methylation. Among the targets that lose methyl groups is a family of transposons related to *MuDR*. Direct analysis of genome sequences shows that the demethylation and associated modification of histone tails (see *Section 28.10, Methylation of Histones and DNA Is Connected*) allow transposition events to occur. Methylation is probably the major mechanism that is used to prevent transposons from damaging the genome by transposing too frequently. Transposons appear to be targeted for methylation because they are far more likely to produce double-stranded or otherwise aberrant transcripts that can be used to guide sequence-specific DNA methylation using small RNA produced from those transcripts. Once methylation of a transposon has been established, it can be heritably maintained over many generations. In both plants and animals that methylate their DNA, the vast majority of transposons are epigenetically silenced in this way.

There may be self-regulating controls of transposition, analogous to the immunity effects displayed by bacterial transposons. An increase in the number of *Ac* elements in the genome decreases the frequency of transposition. The *Ac* element may code for a repressor of transposition; the activity could be carried by the same protein that provides transposase function. Additionally, derivatives of some transposons, such as those of P elements in *Drosophila*, encode truncated proteins that can repress the activity of autonomous elements in somatic tissue (see *Section 17.10, P Elements Are Activated in the Germline*).

17.9 The Role of Transposable Elements in Hybrid Dysgenesis

Key concepts

- P elements are transposons that are carried in P strains of *Drosophila melanogaster*, but not in M strains.
- When a P male is crossed with an M female, transposition is activated.
- The insertion of P elements at new sites in these crosses inactivates many genes and makes the cross infertile.

Certain strains of *D. melanogaster* encounter difficulties in interbreeding. When flies from two of these strains are crossed, the progeny display "dysgenic traits"—a series of defects including mutations, chromosomal aberrations, distorted segregation at meiosis, and reduced fertility. The appearance of these correlated defects is called **hybrid dysgenesis**.

Two systems responsible for hybrid dysgenesis have been identified in *D. melanogaster*. In the first, flies are divided into the types I (inducer) and R (reactive). Reduced fertility is seen in crosses of I males with R females, but not in the reverse direction. In the second system, flies are divided into the two types P (paternal contributing) and M (maternal contributing). **FIGURE 17.20** illustrates the asymmetry of the system; a cross between a P male and an M female causes dysgenesis, but the reverse cross does not.

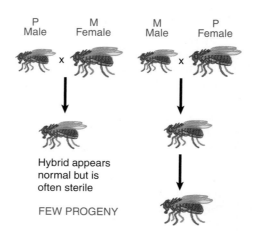

FIGURE 17.20 Hybrid dysgenesis is asymmetrical; it is induced by P male × M female crosses, but not by M male × P female crosses.

Dysgenesis is principally a phenomenon of the germ cells. In crosses involving the P-M system, the F1 hybrid flies have normal somatic tissues. Their gonads, however, do not develop normally and the hybrids are often sterile, particularly at higher temperatures. The morphological defect in gamete development dates from the stage at which rapid cell divisions commence in the germline.

Any one of the chromosomes of a P male can induce dysgenesis in a cross with an M female. The construction of recombinant chromosomes shows that several regions within each P chromosome are able to cause dysgenesis. This suggests that a P male has sequences at many different chromosomal locations that can induce dysgenesis. The locations differ between individual P strains. The P-specific sequences are absent from chromosomes of M flies.

The nature of the P-specific sequences was first identified by mapping the DNA of *w* mutants found among the dysgenic hybrids. All the mutations result from the insertion of DNA into the *white* (*w*) locus. (The insertion inactivates the gene, which is required for red eye color, causing the white-eye phenotype for which the locus is named.) The inserted sequence is called the **P element**.

The P element insertions form a classic transposable system. Individual elements vary in length but are homologous in sequence. All P elements possess inverted terminal repeats of 31 bp, and generate direct repeats of target DNA of 8 bp upon transposition. The longest P elements are ~2.9 kb long and have four open reading frames. The shorter elements arise, apparently rather frequently, by internal deletions of a full-length P factor. At least some of the shorter P elements have lost the capacity to produce the transposase, but may be activated in *trans* by the enzyme coded by a complete P element.

A P strain carries 30 to 50 copies of the P element, about a third of them full length. The elements are absent from M strains. In a P strain the elements are carried as inert components of the genome, but they become activated to transpose when a P male is crossed with an M female.

Chromosomes from P-M hybrid dysgenic flies have P elements inserted at many new sites. The insertions inactivate the genes in which they are located and often cause chromosomal breaks. The result of the transpositions is therefore to dramatically alter the genome.

17.10 P Elements Are Activated in the Germline

Key concepts

- P elements are activated in the germline of P male × M female crosses because a tissue-specific splicing event removes one intron, which generates the coding sequence for the transposase.
- The P element also produces a repressor of transposition, which is inherited maternally in the cytoplasm.
- The presence of the repressor explains why M male × P female crosses remain fertile.

Activation of P elements is tissue-specific: It occurs only in the germline. P elements are transcribed, though, in both germline and somatic tissues. Tissue-specificity is conferred by a change in the splicing pattern.

FIGURE 17.21 depicts the organization of the element and its transcripts. The primary transcript extends for 2.5 or 3.0 kb, the difference probably reflecting merely the leakiness of the termination site. Two protein products can be produced:

- In somatic tissues, only the first two introns are excised, creating a coding region of *ORF0-ORF1-ORF2*. Translation of this RNA yields a protein of 66 kD.

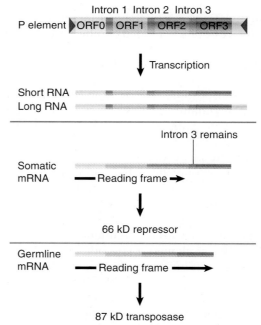

FIGURE 17.21 The P element has four exons. The first three are spliced together in somatic expression; all four are spliced together in germline expression.

This protein is a repressor of transposon activity.

- In germline tissues, an additional splicing event occurs to remove intron 3. This connects all four open reading frames into an mRNA that is translated to generate a protein of 87 kD. This protein is the transposase.

Two types of experiment have demonstrated that splicing of the third intron is needed for transposition. First, if the splicing junctions are mutated *in vitro* and the P element is reintroduced into flies, its transposition activity is abolished. Second, if the third intron is deleted, so that *ORF3* is constitutively included in the mRNA in all tissues, transposition occurs in somatic tissues as well as the germline. Thus whenever *ORF3* is spliced to the preceding reading frame, the P element becomes active. This is the crucial regulatory event, and usually it occurs only in the germline.

What is responsible for the tissue-specific splicing? Somatic cells contain a protein that binds to sequences in exon 3 to prevent splicing of the last intron (see *Section 21.12, Alternative Splicing Is a Rule, Rather Than an Exception, in Multicellular Eukaryotes*). The absence of this protein in germline cells allows splicing to generate the mRNA that codes for the transposase.

Transposition of a P element requires ~150 bp of terminal DNA. The transposase binds to 10 bp sequences that are adjacent to the 31 bp inverted repeats. Transposition occurs by a non-replicative "cut and paste" mechanism resembling that of Tn10. (It contributes to hybrid dysgenesis in two ways: Insertion of the transposed element at a new site may cause mutations, and the break that is left at the donor site—see Figure 17.6—can have a deleterious effect.)

It is interesting that, in a significant proportion of cases, the break in donor DNA is repaired by using the sequence of the homologous chromosome. If the homolog has a P element the presence of a P element at the donor site may be restored (so the event resembles the result of a replicative transposition). If the homolog lacks a P element, repair may generate a sequence lacking the P element, thus apparently providing a precise excision (an unusual event in other transposable systems).

The dependence of hybrid dysgenesis on the origin of the female in a cross shows that the cytoplasm is important, as are the P factors themselves. The contribution of the cytoplasm is described as the **cytotype**; a line of flies containing P elements has P cytotype, whereas a line of flies lacking P elements has M cytotype. Hybrid dysgenesis occurs only when chromosomes containing P factors find themselves in M cytotype, that is, when the male parent has P elements and the female parent does not.

Cytotype shows an inheritable cytoplasmic effect; when a cross occurs through P cytotype (the female parent has P elements), hybrid dysgenesis is suppressed for several generations of crosses with M female parents. Thus something in P cytotype, which can be diluted out over some generations, suppresses hybrid dysgenesis.

The effect of cytotype has been a particularly puzzling phenomenon. All explanations assume that a repressor molecule is deposited into the egg cell cytoplasm, as is illustrated in **FIGURE 17.22**. The repressor is provided as a maternal factor in the egg. In a P line, there must be sufficient repressor to prevent transposition from occurring, even though the P elements are present. In any cross involving a P female, its presence prevents either synthesis or activity of the transposase. When the female parent is M type, though, there is no

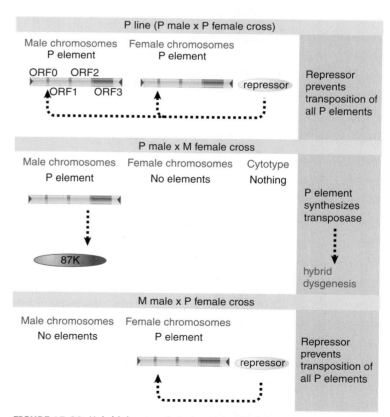

FIGURE 17.22 Hybrid dysgenesis is determined by the interactions between P elements in the genome and repressors in the cytotype.

repressor in the egg, and the introduction of a P element from the male parent results in activity of transposase in the germline. The ability of P cytotype to exert an effect through more than one generation suggests that there must be enough repressor protein in the egg, and it must be stable enough, to be passed on through the adult to be present in the eggs of the next generation.

For many years, the best candidate for the repressor was the 66 kD protein. There are, however, strains of flies that lack P elements capable of producing a 66 kD repressor protein but that do exhibit P cytotype. More recent evidence has implicated small RNAs in P element repression; genes important in processing small RNAs derived from P elements transcripts (and those of several other transposons as well) are also required for efficient transposon silencing. This observation has lead to a model in which P cytotype is conditioned by P elements at particular positions that produce transcripts that are processed into a specific class of small RNAs, called piRNAs (see *Section 30.5, microRNAs are Widespread Regulators in Eukaryotes*). In this case, it is the presence of these small RNAs in the cytoplasm that are responsible for P element cytotype repression. Like the small RNAs involved in RNA interference (see Chapter 30, *Regulatory RNA*), piRNAs are hypothesized to direct the degradation of P element transcript. An appealing feature of this model is that it suggests that P element cytotype repression is a particular example of a widespread mechanism by which transposon activity is repressed in plants, fungi and animals.

Remarkably, P elements have only been detectable in the *D. melanogaster* genome for a few decades. They came from a second species of *Drosophila*, *D. willisoni*, through a horizontal transfer of P element sequence. Subsequent to that transfer, P elements rapidly spread throughout the worldwide population of *D. melanogaster*. Analysis of P elements in a variety of *Drosophila* species reveal that horizontal transfer of this transposon has occurred repeatedly throughout its history. This propensity to move between species has been documented among a number of transposons, leading to the suggestion that an important component to the transposon life-cycle is the ability to regularly invade "naive" genomes that lack sequences (such as those that produce piRNAs) that can repress transposon activity.

17.11 The Retrovirus Life Cycle Involves Transposition-Like Events

Key concepts

- A retrovirus has two copies of its genome of single-stranded RNA.
- An integrated provirus is a double-stranded DNA sequence.
- A retrovirus generates a provirus by reverse transcription of the retroviral genome.

Retroviruses have genomes of single-stranded RNA that are replicated through a double-stranded DNA intermediate. The life cycle of the virus involves an obligatory stage in which the double-stranded DNA is inserted into the host genome by a transposition-like event that generates short direct repeats of target DNA. This similarity is not surprising, given evidence that new retroviruses have arisen repeatedly over evolutionary time as a consequence of the capture by retrotransposons of genes encoding envelope proteins, which makes infection possible.

The significance of this integration reaction extends beyond the perpetuation of the virus. Some of its consequences are that:

- a retroviral sequence that is integrated in the germline remains in the cellular genome as an endogenous **provirus**. Like a lysogenic bacteriophage, a provirus behaves as part of the genetic material of the organism.
- cellular sequences occasionally recombine with the retroviral sequence and then are transposed with it; these sequences may be inserted in to the genome as duplex sequences in new locations.
- cellular sequences that are transposed by a retrovirus may change the properties of a cell that becomes infected with the virus.

The particulars of the retroviral life cycle are expanded in **FIGURE 17.23**. The crucial steps are that the viral RNA is converted into DNA, the DNA becomes integrated into the host genome, and then the DNA provirus is transcribed into RNA. The enzyme responsible for generating the initial DNA copy of the RNA is **reverse transcriptase**. The enzyme converts the RNA into a linear duplex of DNA in the cytoplasm of the infected cell. The DNA also is converted into circular forms, but these do not appear to be involved in reproduction.

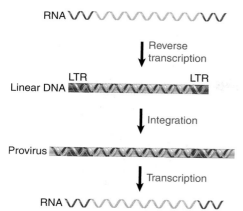

FIGURE 17.23 The retroviral life cycle proceeds by reverse transcribing the RNA genome into duplex DNA, which is inserted into the host genome, in order to be transcribed into RNA.

The linear DNA makes its way to the nucleus. One or more DNA copies become integrated into the host genome. A single enzyme called **integrase** is responsible for integration. Retroviral integrases are related by sequence, structure, and function to the transposases encoded by transposons. The provirus is transcribed by the host machinery to produce viral RNAs, which serve both as mRNAs and as genomes for packaging into virions. Integration is a normal part of the life cycle and is necessary for transcription.

Two copies of the RNA genome are packaged into each virion, making the individual virus particle effectively diploid. When a cell is simultaneously infected by two different but related viruses, it is possible to generate heterozygous virus particles carrying one genome of each type. The diploidy may be important in allowing the virus to acquire cellular sequences. The enzymes reverse transcriptase and integrase are carried with the genome in the viral particle.

17.12 Retroviral Genes Code for Polyproteins

Key concepts

- A typical retrovirus has three genes: *gag*, *pol*, and *env*.
- Gag and Pol proteins are translated from a full-length transcript of the genome.
- Translation of Pol requires a frameshift by the ribosome.
- Env is translated from a separate mRNA that is generated by splicing.
- Each of the three protein products is processed by proteases to give multiple proteins.

A typical retroviral sequence contains three or four "genes." (In this context, the term *genes* is used to identify coding regions, each of which actually gives rise to multiple proteins by processing reactions.) A typical retrovirus genome with three genes is organized in the sequence *gag-pol-env*, as indicated in **FIGURE 17.24**.

Retroviral mRNA has a conventional structure; it is capped at the 5′ end and polyadenylated at the 3′ end. It is represented in two mRNAs. The full-length mRNA is translated to give the Gag and Pol polyproteins. The Gag product is translated by reading from the initiation codon to the first termination codon. This termination codon must be bypassed to express Pol.

Different mechanisms are used in different viruses to proceed beyond the *gag* termination codon, depending on the relationship between the *gag* and *pol* reading frames. When *gag* and *pol* follow continuously, suppression by a glutamyl-tRNA that recognizes the termination codon allows a single protein to be generated. When *gag* and *pol* are in different reading frames, a ribosomal frameshift occurs to generate a single protein. Usually the readthrough is ~5% efficient, so Gag protein outnumbers Gag-Pol protein about 20-fold.

The Env polyprotein is expressed by another means: Splicing generates a shorter *subgenomic* mRNA that is translated into the Env product.

The *gag* gene gives rise to the protein components of the nucleoprotein core of the virion. The *pol* gene codes for functions concerned with nucleic acid synthesis and recombination. The *env* gene codes for components of the envelope of the particle, which also sequesters components from the cellular cytoplasmic membrane.

Both the Gag or Gag-Pol and the Env products are polyproteins that are cleaved by a protease to release the individual proteins that are found in mature virions. The protease activity is coded by the virus in various forms: It may be part of Gag or Pol, and at times it takes the form of an additional independent reading frame.

The production of a retroviral particle involves packaging the RNA into a core, surrounding it with capsid proteins, and pinching off a segment of membrane from the host cell. The release of infective particles by such means is shown in **FIGURE 17.25**. The process is reversed during infection: A virus infects a new host cell by fusing with the plasma membrane and then releasing the contents of the virion.

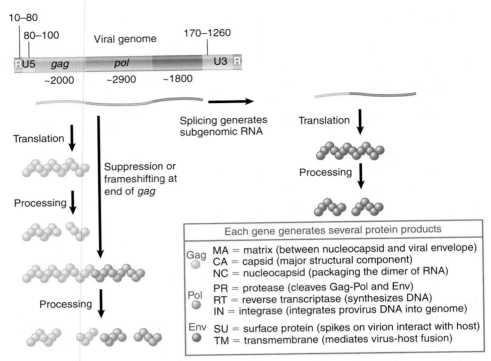

FIGURE 17.24 The genes of the retrovirus are expressed as polyproteins that are processed into individual products.

Each gene generates several protein products

Gag
MA = matrix (between nucleocapsid and viral envelope)
CA = capsid (major structural component)
NC = nucleocapsid (packaging the dimer of RNA)

Pol
PR = protease (cleaves Gag-Pol and Env)
RT = reverse transcriptase (synthesizes DNA)
IN = integrase (integrates provirus DNA into genome)

Env
SU = surface protein (spikes on virion interact with host)
TM = transmembrane (mediates virus-host fusion)

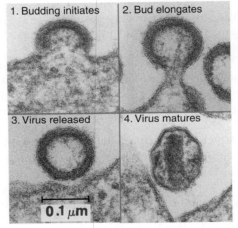

1. Budding initiates
2. Bud elongates
3. Virus released
4. Virus matures

0.1 μm

FIGURE 17.25 Retroviruses (HIV) bud from the plasma membrane of an infected cell. Photos courtesy of Matthew A. Gonda, Ph.D., Chief Executive Officer, International Medical Innovations, Inc.

17.13 Viral DNA Is Generated by Reverse Transcription

Key concepts

- A short sequence (R) is repeated at each end of the viral RNA, so the 5′ and 3′ ends are R-U5 and U3-R, respectively.
- Reverse transcriptase starts synthesis when a tRNA primer binds to a site 100 to 200 bases from the 5′ end.
- When the enzyme reaches the end, the 5′-terminal bases of RNA are degraded, exposing the 3′ end of the DNA product.
- The exposed 3′ end of the DNA product base pairs with the 3′ terminus of another RNA genome.
- Synthesis continues, generating a product in which the 5′ and 3′ regions are repeated, giving each end the structure U3-R-U5.
- Similar strand-switching events occur when reverse transcriptase uses the DNA product to generate a complementary strand.
- Strand switching is an example of the copy choice mechanism of recombination.

Retroviruses are called **plus strand viruses**, because the viral RNA itself codes for the protein products. As its name implies, reverse transcriptase is responsible for converting the genome (plus strand RNA) into a complementary DNA strand, which is called the **minus strand DNA**. Reverse transcriptase also catalyzes subsequent stages in the production of duplex DNA. It has a DNA polymerase activity, which enables it to synthesize a duplex DNA from the single-stranded reverse transcript of the RNA. The second DNA strand in this duplex is called the **plus strand DNA**. As a necessary adjunct to this activity, the enzyme has an RNase H activity, which can degrade the RNA part of the RNA–DNA hybrid. All retroviral reverse transcriptases share considerable similarities of amino acid sequence, and homologous sequences can be recognized in all other retroelements.

The structures of the DNA forms of the virus are compared with the RNA in **FIGURE 17.26**. The viral RNA has direct repeats at its ends. These **R segments** vary in different strains of virus from 10 to 80 nucleotides. The sequence at the 5′ end of the virus is R-**U5**, and the sequence at the 3′ end is **U3**-R. The R segments are used during the conversion from the RNA to the DNA form to generate the more extensive direct repeats that are found in linear DNA, as shown in **FIGURE 17.27** and **FIGURE 17.28**. The shortening of 2 bp at each end in the integrated form is a consequence of the mechanism of integration (see Figure 17.30).

Like other DNA polymerases, reverse transcriptase requires a primer. For retroviruses, the native primer is tRNA. An uncharged host tRNA is present in the virion. A sequence of 18 bases at the 3′ end of the tRNA is base paired to a site 100 to 200 bases from the 5′ end of one of the viral RNA molecules. The tRNA may also be base paired to another site near the 5′ end of the other viral RNA, thus assisting in dimer formation between the viral RNAs.

Here is a dilemma: Reverse transcriptase starts to synthesize DNA at a site only 100 to 200 bases downstream from the 5′ end. How can DNA be generated to represent the intact RNA genome? (This is an extreme variant of the general problem in replicating the ends of any linear nucleic acid; see *Section 12.2, The Ends of Linear DNA Are a Problem for Replication.*)

Synthesis *in vitro* proceeds to the end, generating a short DNA sequence called *strong-stop minus DNA.* This molecule is not found *in vivo* because synthesis continues by the reaction illustrated in Figure 17.27. Reverse transcriptase switches templates, carrying the nascent DNA with it to the new template. This is the first of two jumps between templates.

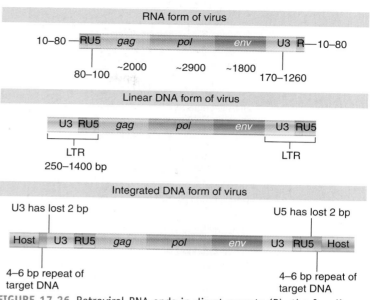

FIGURE 17.26 Retroviral RNA ends in direct repeats (R), the free linear DNA ends in LTRs, and the provirus ends in LTRs that are shortened by two bases each.

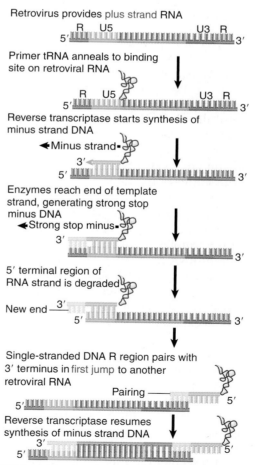

FIGURE 17.27 Minus strand DNA is generated by switching templates during reverse transcription.

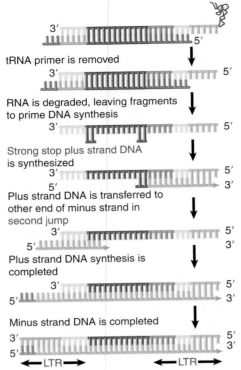

FIGURE 17.28 Synthesis of plus strand DNA requires a second jump.

Within the figure:

3′ — tRNA primer is removed — 5′

RNA is degraded, leaving fragments to prime DNA synthesis

Strong stop plus strand DNA is synthesized

Plus strand DNA is transferred to other end of minus strand in second jump

Plus strand DNA synthesis is completed

Minus strand DNA is completed

←LTR→   ←LTR→

In this reaction, the R region at the 5′ terminus of the RNA template is degraded by the RNase H activity of reverse transcriptase. Its removal allows the R region at a 3′ end to base pair with the newly synthesized DNA. Reverse transcription then continues through the U3 region into the body of the RNA.

The source of the R region that pairs with the strong-stop minus DNA can be either the 3′ end of the same RNA molecule (intramolecular pairing) or the 3′ end of a different RNA molecule (intermolecular pairing). The switch to a different RNA template is used in the figure because there is evidence that the sequence of the tRNA primer is not inherited in a retroposon life cycle. (If intramolecular pairing occurred, we would expect the sequence to be inherited, because it would provide the only source for the primer binding sequence in the next cycle. Intermolecular pairing allows another retroviral RNA to provide this sequence.)

The result of the switch and extension is to add a U3 segment to the 5′ end. The stretch of sequence U3-R-U5 is called the **long terminal repeat (LTR)** because a similar series of events adds a U5 segment to the 3′ end, giving it the same structure of U5-R-U3. Its length varies from 250 to 1400 bp (see Figure 17.26).

We now need to generate the plus strand of DNA and to generate the LTR at the other end. The reaction is shown in Figure 17.28. Reverse transcriptase primes synthesis of plus strand DNA from a fragment of RNA that is left after degrading the original RNA molecule. A *strong-stop plus strand DNA* is generated when the enzyme reaches the end of the template. This DNA is then transferred to the other end of a minus strand, where it is probably released by a displacement reaction when a second round of DNA synthesis occurs from a primer fragment farther upstream (to its left in the figure). It uses the R region to pair with the 3′ end of a minus strand DNA. This double-stranded DNA then requires completion of both strands to generate a duplex LTR at each end.

Each retroviral particle carries two RNA genomes. This makes it possible for recombination to occur during a viral life cycle. In principle this could occur during minus strand synthesis and/or during plus strand synthesis:

- The intermolecular pairing shown in Figure 17.27 allows a recombination to occur between sequences of the two successive RNA templates when minus strand DNA is synthesized. Retroviral recombination is mostly due to strand transfer at this stage, when the nascent DNA strand is transferred from one RNA template to another during reverse transcription.

- Plus strand DNA may be synthesized discontinuously, in a reaction that involves several internal initiations. Strand transfer during this reaction can also occur, but is less common.

The common feature of both events is that recombination results from a change in the template during the act of DNA synthesis. This is a general example of a mechanism for recombination called **copy choice**. For many years this was regarded as a possible mechanism for general recombination. It is unlikely to be employed by cellular systems, but is a common basis for recombination during infection by RNA viruses, including those that replicate exclusively through RNA forms, such as poliovirus.

Strand switching occurs with a certain frequency during each cycle of reverse transcription, that is, in addition to the transfer reaction that is forced at the end of the template strand. The principle is illustrated in **FIGURE 17.29**, although we do not know much about the mechanism. Reverse transcription

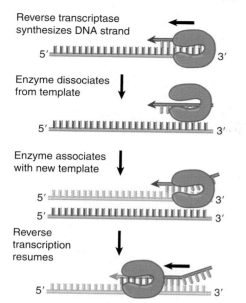

Reverse transcriptase
synthesizes DNA strand

Enzyme dissociates
from template

Enzyme associates
with new template

Reverse
transcription
resumes

FIGURE 17.29 Copy choice recombination occurs when reverse transcriptase releases its template and resumes DNA synthesis using a new template. Transfer between template strands probably occurs directly, but is shown here in separate steps to illustrate the process.

in vivo occurs in a ribonucleoprotein complex, in which the RNA template strand is bound to virion components, including the major protein of the capsid. In the case of HIV, addition of this protein (NCp7) to an *in vitro* system causes recombination to occur. The effect is probably indirect: NCp7 affects the structure of the RNA template, which in turn affects the likelihood that reverse transcriptase will switch from one template strand to another.

17.14 ## Viral DNA Integrates into the Chromosome

Key concepts

- The organization of proviral DNA in a chromosome is the same as a transposon, with the provirus flanked by short direct repeats of a sequence at the target site.
- Linear DNA is inserted directly into the host chromosome by the retroviral integrase enzyme.
- Two base pairs of DNA are lost from each end of the retroviral sequence during the integration reaction.

The organization of the integrated provirus resembles that of the linear DNA. The LTRs at each end of the provirus are identical. The 3' end of U5 consists of a short inverted repeat relative to the 5' end of U3, so the LTR itself

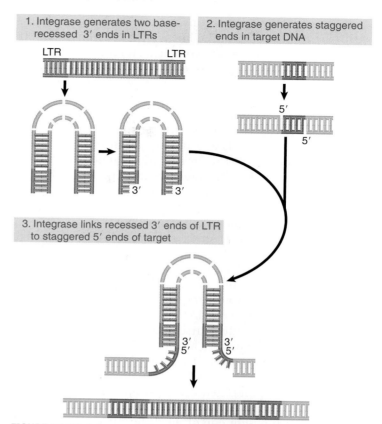

1. Integrase generates two base-recessed 3' ends in LTRs

2. Integrase generates staggered ends in target DNA

3. Integrase links recessed 3' ends of LTR to staggered 5' ends of target

FIGURE 17.30 Integrase is the only viral protein required for the integration reaction, in which each LTR loses 2 bp and is inserted between 4 bp repeats of target DNA.

ends in short inverted repeats. The integrated proviral DNA is like a transposon: The proviral sequence ends in inverted repeats and is flanked by short direct repeats of target DNA.

The provirus is generated by directly inserting a linear DNA into a target site. In addition to linear DNA, there are circular forms of the viral sequences. One has two adjacent LTR sequences generated by joining the linear ends. The other has only one LTR—presumably generated by a recombination event and actually comprising the majority of circles. For a long time it appeared that the circle might be an integration intermediate (by analogy with the integration of lambda DNA). We now know, though, that the linear form is used for integration.

Integration of linear DNA is catalyzed by a single viral product, the integrase. Integrase acts on both the retroviral linear DNA and the target DNA. The reaction is illustrated in **FIGURE 17.30**.

The ends of the viral DNA are important, just as they are for transposons. The most conserved feature is the presence of the dinucleotide sequence CA close to the end of each LTR. This CA dinucleotide is conserved among all retroviruses, viral retrotransposons, and many

DNA transposons as well. The integrase brings the ends of the linear DNA together in a ribonucleoprotein complex, and then converts the blunt ends into recessed ends by removing the bases beyond the conserved CA. In general, this involves a loss of two bases.

Target sites are chosen at random with respect to sequence. The integrase makes staggered cuts at a target site. In the example of Figure 17.30, the cuts are separated by 4 bp. The length of the target repeat depends on the particular virus; it may be 4, 5, or 6 bp. Presumedly it is determined by the geometry of the reaction of integrase with target DNA.

The 5′ ends generated by the cleavage of target DNA are covalently joined to the 3′ recessed ends of the viral DNA. At this point, both termini of the viral DNA are joined by one strand to the target DNA. The single-stranded region is repaired by enzymes of the host cell, and in the course of this reaction the protruding two bases at each 5′ end of the viral DNA are removed. The result is that the integrated viral DNA has lost 2 bp at each LTR; this corresponds to the loss of 2 bp from the left end of the 5′ terminal U3 and to the loss of 2 bp from the right end of the 3′ terminal U5. There is a characteristic short direct repeat of target DNA at each end of the integrated retroviral genome.

The viral DNA integrates into the host genome at randomly selected sites. A successfully infected cell gains one to ten copies of the provirus. (An infectious virus enters the cytoplasm, of course, but the DNA form becomes integrated into the genome in the nucleus. Some retroviruses can replicate only in proliferating cells, because entry into the nucleus requires the cell to pass through mitosis, when the viral genome gains access to the nuclear material. Others, such as HIV, can be actively transported into the nucleus even in the absence of cell division.)

The U3 region of each LTR carries a promoter. The promoter in the left LTR is responsible for initiating transcription of the provirus. Recall that the generation of proviral DNA is required to place the U3 sequence at the left LTR; thus we see that the promoter is in fact generated by the conversion of the RNA into duplex DNA.

Sometimes (probably rather rarely), the promoter in the right LTR sponsors transcription of the host sequences that are adjacent to the site of integration. The LTR also carries an enhancer (a sequence that activates promoters in the vicinity) that can act on cellular as well as viral sequences. Integration of a retrovirus can be responsible for converting a host cell into a tumorigenic state when certain types of genes are activated in this way.

We have dealt thus far with retroviruses in terms of the infective cycle, in which integration is necessary for the production of further copies of the RNA. When a viral DNA integrates in a germline cell, though, it becomes an inherited "endogenous provirus" of the organism. Endogenous viruses usually are not expressed, but sometimes they are activated by external events, such as infection with another virus.

17.15 Retroviruses May Transduce Cellular Sequences

Key concept

- Transforming retroviruses are generated by a recombination event in which a cellular RNA sequence replaces part of the retroviral RNA.

An interesting light on the viral life cycle is cast by the occurrence of **transducing viruses**, which are variants that have acquired cellular sequences in the form illustrated in **FIGURE 17.31**. Part of the viral sequence has been replaced by the *v-onc* gene. Protein synthesis generates a Gag-v-Onc protein instead of the usual Gag, Pol, and Env proteins. The resulting virus is **replication-defective**; it cannot sustain an infective cycle by itself. It can, however, be perpetuated in the company of a **helper virus** that provides the missing viral functions.

Onc is an abbreviation for *oncogenesis*, the ability to *transform* cultured cells so that the usual regulation of growth is released to allow unrestricted division. Both viral and cellular *onc*

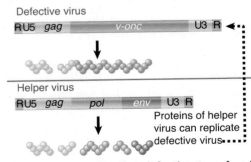

FIGURE 17.31 Replication-defective transforming viruses have a cellular sequence substituted for part of the viral sequence. The defective virus may replicate with the assistance of a helper virus that carries the wild-type functions.

genes may be responsible for creating tumorigenic cells.

A *v-onc* gene confers upon a virus the ability to transform a certain type of host cell. Loci with homologous sequences found in the host genome are called *c-onc* genes. How are the *onc* genes acquired by the retroviruses? A revealing feature is the discrepancy in the structures of *c-onc* and *v-onc* genes. The *c-onc* genes usually are interrupted by introns, whereas the *v-onc* genes are uninterrupted. This suggests that the *v-onc* genes originate from spliced RNA copies of the *c-onc* genes.

A model for the formation of transforming viruses is illustrated in **FIGURE 17.32**. A retrovirus has integrated near a *c-onc* gene. A deletion occurs to fuse the provirus to the *c-onc* gene; transcription then generates a joint RNA, which contains viral sequences at one end and cellular *onc* sequences at the other end. Splicing removes the introns in the cellular parts of the RNA. The RNA has the appropriate signals for packaging into the virion, which will be present if the cell also contains another, intact copy of the provirus. At this point, some of the diploid virus particles may contain one fused RNA and one viral RNA.

A recombination between these sequences could generate the transforming genome, in which the viral repeats are present at both ends. (Recombination occurs by various means at a high frequency during the retroviral infective cycle. We do not know anything about its demands for homology in the substrates, but we assume that the nonhomologous reaction between a viral genome and the cellular part of the fused RNA proceeds by the same mechanisms responsible for viral recombination.)

The common features of the entire retroviral class suggest that it may be derived from a single ancestor. This is supported by phylogenetic analysis of reverse transcriptases from a wide variety of retroelements, including both retrotransposons and retroviruses. The fact that this class of element has features common to both DNA-type transposons (integrase/transposase) and non-LTR retroposons (reverse transcriptase) has led to the suggestion that LTR retrotransposons arose as a consequence of a fusion between these two, more ancient element classes. Other functions, such as Env proteins and transforming genes, would have been incorporated later. (There is no reason to suppose that the mechanism involved in acquisition of *env* and *onc* genes; viruses carry-

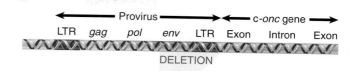

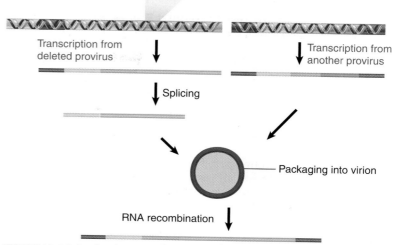

FIGURE 17.32 Replication-defective viruses may be generated through integration and deletion of a viral genome to generate a fused viral-cellular transcript that is packaged with a normal RNA genome. Nonhomologous recombination is necessary to generate the replication-defective transforming genome.

ing these genes may have a selective advantage, though).

17.16 Yeast *Ty* Elements Resemble Retroviruses

Key concepts

- *Ty* transposons have a similar organization to endogenous retroviruses.
- *Ty* transposons are retrotransposons (with a reverse transcriptase activity) that transpose via an RNA intermediate.

Ty elements comprise a family of dispersed repetitive DNA sequences that are found at different sites in different strains of yeast. ***Ty*** is an abbreviation for "transposon yeast." Five types of *Ty* elements in yeast (*Ty1–Ty5*) have been identified. All are LTR retrotransposons, with characteristic LTRs and *gag* and *pol* genes with homology to those encoded by retroviruses. These elements are representative of two of the major classes of retrotransposons in eukaryotes, the *Ty1/copia* class (*Ty1, Ty2, Ty4,* and *Ty5*) and the *Ty3/gypsy* class. Each class is phylogenetically distinct, and each contains a characteristic order of open reading frames.

In the yeast *Saccharomyces cerevisiae, Ty1* is the most abundant and the most well-characterized

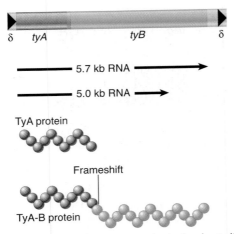

FIGURE 17.33 *Ty* elements terminate in short direct repeats and are transcribed into two overlapping RNAs. They have two reading frames, with sequences related to the retroviral *gag* and *pol* genes.

retroelement. A *Ty1* transposition event creates a characteristic footprint: 5 bp of target DNA are repeated on either side of the inserted *Ty1* element. Under most circumstances the frequency of *Ty1* transposition is lower than that of most bacterial transposons, ~10^{-7}–10^{-8}, but it can be increased by a variety of factors that stress the organism, such as mutagens and nutrient depletion.

The general organization of *Ty1* elements is illustrated in **FIGURE 17.33**. Each element is 5.9 kb long; the last 334 bp at each end constitute LTRs, called δ for historical reasons but referred to here simply as LTRs. Individual *Ty1* elements have many changes from the prototype of their class, including base pair substitutions, insertions, and deletions. There are ~30 copies of the *Ty1* and 13 copies of the closely related *Ty2* in a typical yeast genome. In addition, there are ~180 independent solo *Ty1/Ty2* LTRs.

The LTR sequences also show considerable heterogeneity, although the two repeats of an individual *Ty1* element are often identical or at least very closely related. The LTR sequences associated with *Ty1* elements show greater conservation of sequence than the solo LTRs. This is because transposition of *Ty1* elements, like replication of retroviruses, involves duplication of the LTRs (see below). Thus, recently inserted elements carry identical LTRs, but solo LTRs diverge over time due to random mutations.

The *Ty1* element is transcribed into two poly(A)⁺ RNA species, which constitute as much as 8% of the total mRNA of a haploid yeast cell. Both species initiate within a promoter in the LTR at the left end. One terminates after 5 kb; the other terminates after 5.7 kb, within the LTR sequence at the right end.

The sequence of the *Ty1* element has two open reading frames. These frames are expressed in the same direction, but are read in different phases and overlap by 13 amino acids. *TyA* is related to retroviral *gag* genes and encodes a capsid protein. *TyB* contains regions that have homologies with reverse transcriptase, protease, and integrase sequences of retroviruses.

The organization and functions of *TyA* and *TyB* are analogous to the behavior of the retroviral *gag* and *pol* functions. The reading frames *TyA* and *TyB* are expressed in two forms. The TyA protein represents the *TyA* reading frame and terminates at its end. The *TyB* reading frame, however, is expressed only as part of a joint protein, in which the *TyA* region is fused to the *TyB* region by a specific frameshift event that allows the termination codon to be bypassed. (This is analogous to *gag-pol* translation in retroviruses.)

Recombination between *Ty1* elements seems to occur in bursts; when one event is detected, there is an increased probability of finding others. Gene conversion occurs between *Ty1* elements at different locations, with the result that one element is "replaced" by the sequence of the other.

Ty elements can be deleted via homologous recombination between the directly repeated LTR sequences. The large number of solo LTR elements may be footprints of such events. A deletion of this nature may be associated with reversion of a mutation caused by the insertion of *Ty*; the level of reversion may depend on the exact LTR sequences left behind and the nature of the insertion site.

A paradox is that both LTRs have the same sequence, yet a promoter is active in the LTR at one end and a terminator is active in the LTR at the other end. (A similar feature is found in other transposable elements, including the retroviruses.)

Ty elements are classic retrotransposons, in that they transpose through an RNA intermediate. An ingenious protocol used to detect this event is illustrated in **FIGURE 17.34**. An intron was inserted into an element to generate a unique *Ty* sequence. This sequence was placed under the control of a *GAL* promoter on a plasmid and introduced into yeast cells. Transposition results in the appearance of multiple copies of the transposon in the yeast genome, but the copies all lack the intron.

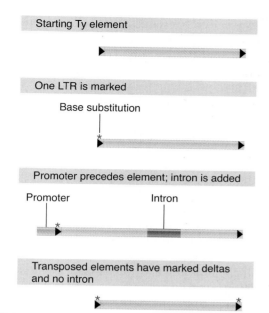

Starting Ty element

One LTR is marked

Base substitution

Promoter precedes element; intron is added

Promoter   Intron

Transposed elements have marked deltas and no intron

FIGURE 17.34 A unique *Ty* element, engineered to contain an intron, transposes to give copies that lack the intron. The copies possess identical terminal repeats, which are generated from one of the termini of the original Ty element.

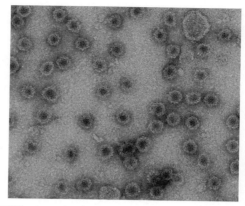

FIGURE 17.35 *Ty* elements generate virus-like particles. Reprinted from *J. Mol. Biol.*, vol. 292, H. A. AL-Khayat, et al., Yeast Ty retrotransposons . . . , pp. 65–73. Copyright 1999, with permission from Elsevier [http://www.sciencedirect.com/science/journal/00222836]. Photo courtesy of Dr. Hind A. AL-Khayat, Imperial College London, United Kingdom.

We know of only one way to remove introns: RNA splicing. This suggests that transposition occurs by the same mechanism as with retroviruses. The *Ty* element is transcribed into an RNA that is recognized by the splicing apparatus. The spliced RNA is recognized by a reverse transcriptase and regenerates a duplex DNA copy, which is then integrated back into the genome using the integrase protein.

The analogy with retroviruses extends further. The original *Ty1* element has a difference in sequence between its two LTRs. The transposed elements possess identical LTR sequences, however, which are derived from the 5′ delta of the original element. Just as shown for retroviruses in Figures 17.26 and 17.27, the complete LTR is regenerated by adding a U5 to the 3′ end and a U3 to the 5′ end.

Transposition is controlled by genes within the *Ty1* element. The *GAL* promoter used to control transcription of the marked *Ty1* element is inducible: It is turned on by the addition of galactose. Induction of the promoter has two effects. It is necessary to activate transposition of the marked element, and its activation also increases the frequency of transposition of the other *Ty1* elements on the yeast chromosome. This implies that the products of the *Ty1* element can act in *trans* on other elements (actually on their RNAs).

The *Ty* element does not give rise to infectious particles, but virus-like particles (VLPs) accumulate within the cells in which transposition has been induced. The particles, which can be seen in **FIGURE 17.35**, contain full-length RNA, double-stranded DNA, reverse transcriptase activity, and a TyB product with integrase activity. The TyA product is cleaved like a *gag* precursor to produce the mature core proteins of the VLP.

Not all of the *Ty1* elements in any yeast genome are active: Some have lost the ability to transpose (and are analogous to inert endogenous proviruses). These "dead" elements retain LTRs, though, and as a result they provide targets for transposition in response to the proteins synthesized by an active element.

17.17 Many Kinds of Transposable Elements Reside in *Drosophila melanogaster*

Key concept
- *copia* is a retrotransposon that is abundant in *D. melanogaster*.

The presence of transposable elements in *D. melanogaster* was first inferred from observations analogous to those that identified the first insertion sequences in *E. coli*. Unstable mutations are found that revert to wild type by deletion, or that generate deletions of the flanking

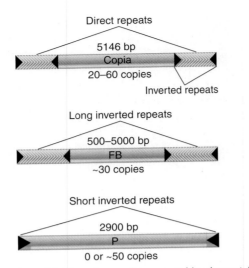

FIGURE 17.36 Three types of transposable element in *D. melanogaster* have different structures.

material with an endpoint at the original site of mutation. They are caused by several types of transposable sequence, which are illustrated in **FIGURE 17.36**. Some of these sequences include the *copia* retroposon, the FB family, and the P elements discussed previously in *Section 17.10, P Elements Are Activated in the Germline.*

A well-characterized family of retrotransposons in *D. melanogaster* is *copia*. Its name reflects the presence of a large (copious) number of closely related sequences that code for abundant mRNAs. The *copia* family is part of a widespread family of LTR retrotransposons (the *Ty1/copia* group) that include *Ty1* elements in yeast and BARE1 elements in barley, where they are present in tens of thousands of copies.

In *D. melanogaster*, the number of copies of the *copia* element depends on the strain of fly; usually it is 20 to 60. The members of the family are widely dispersed. The locations of *copia* elements show a different (although overlapping) spectrum in each strain of *D. melanogaster*.

These differences have developed over evolutionary periods. Comparisons of strains that have diverged recently (over the past 40 years or so) as the result of their propagation in the laboratory reveal few changes. We cannot estimate the rate of change, but the nature of the underlying events is indicated by the result of growing cells in culture. The number of *copia* elements per genome then increases substantially, by as much as threefold. The additional elements represent insertions of *copia* sequences at new sites. Adaptation to culture in some unknown way transiently increases the rate of

transposition to a range of 10^{-3} to 10^{-4} events per generation. Similar experiments in plants suggest that transposon activation may be a universal response to cell or tissue culture.

The *copia* element is ~5000 bp long, with identical direct terminal repeats of 276 bp. Each of the direct repeats itself ends in related inverted repeats. A direct repeat of 5 bp of target DNA is generated at the site of insertion. The divergence between individual members of the *copia* family is slight, at <5%; variants often contain small deletions. All of these features are common to the other *copia*-like families, although their individual members display greater divergence.

The identity of the two direct repeats of each *copia* element implies either that they interact to permit correction events, or that both are generated from one of the direct repeats of a progenitor element during transposition. As in the similar case of *Ty* elements, this is suggestive of a relationship with retroviruses.

The *copia* elements in the genome are always intact; individual copies of the terminal repeats have not been detected (although we would expect them to be generated if recombination deleted the intervening material). At times *copia* elements are found in the form of free circular DNA; like retroviral DNA circles, the longer form has two terminal repeats and the shorter form has only one. As with *Ty1* elements, particles containing *copia* RNA have also been noted.

The *copia* sequence contains a single long-reading frame of 4227 bp. There are homologies between parts of the *copia* open reading frame and the *gag* and *pol* sequences of retroviruses. A notable absence from the homologies is any relationship with retroviral *env* sequences required for the envelope of the virus, which means that *copia*, like *Ty1*, is unlikely to be able to generate virus-like particles.

Transcripts of *copia* are found as abundant poly(A)⁺ mRNAs, representing both full- and part-length transcripts. The mRNAs have a common 5′ terminus, which results from initiation in the middle of one of the terminal repeats. Several proteins are produced, probably involving events such as splicing of RNA and cleavage of polyproteins.

We lack direct evidence for *copia*'s mode of transposition. Given its similarity to many other retroviruses and retrotransposons, though, it almost certainly transposes in a manner typical of those elements.

17.18 Retroelements Fall into Three Classes

Key concepts

- LTR retrotransposons mobilize via an RNA that is similar to retroviral RNA, but does not form an infectious particle.
- Although retroelements that lack LTRs, or retroposons, also transpose via reverse transcriptase, they employ a distinct method of integration and are phylogenetically distinct from both retroviruses and LTR retrotransposons.
- Other elements can be found that were generated by an RNA-mediated transposition event, but they do not themselves code for enzymes that can catalyze transposition.
- Retroelements constitute almost half of the human genome.

Retroelements are defined by their use of mechanisms for transposition that involve reverse transcription of RNA into DNA. Three classes of retroelements are distinguished in FIGURE 17.37: LTR retrotransposons, non-LTR retroposons, and the nonautonomous SINEs.

LTR retrotransposons, or simply retrotransposons, have LTRs and code for reverse transcriptase and integrase activities. They reproduce in the same manner as retroviruses but differ from them in not passing through an independent infectious form. They are best characterized in the *Ty*, *copia*, and *Tos17* elements of yeast, flies, and rice, respectively.

The non-LTR retrotransposons, or retroposons, also have reverse transcriptase activity but constitute a phylogenetically distinct family of elements that employ a distinct transposition mechanism. Unlike retrotransposons and retroviruses, retroposons lack LTRs and use a different mechanism from retroviruses to prime the reverse transcription reaction. They are derived from RNA polymerase II transcripts. A minority of the elements in a given genome are fully functional and can transpose autonomously; others have mutations, and thus can only transpose as the result of the action of a *trans*-acting autonomous element. The most common elements of this class in the human genome are the **LINEs**, or *long-interspersed nuclear elements*.

In addition to LTR retrotransposons and non-LTR retroposons, many genomes contain large numbers of sequences whose external and internal features suggest that they originated in RNA sequences. In these cases, though, we can only speculate on how a DNA copy was generated. We assume that they were targets for a transposition event by an enzyme system coded elsewhere—that is, they are always nonautonomous—and that they originated in cellular transcripts. They do not code for proteins that have transposition functions. The most prominent components of this family are called **short-interspersed nuclear elements (SINEs)**. These elements are derived from RNA polymerase III transcripts, usually 7SL RNAs, 5S rRNAs, and tRNAs. Many of these elements also include portions of a cognate LINE, leading to the hypothesis that SINEs can use the enzymatic machinery of LINEs for replication.

	LTR retrotransposons	non-LTR retroposons	SINES
Common types	Ty (*S. cerevisiae*) copia (*D.melanogaster*)	L1 (human) B1, B2 ID, B4 (mouse)	SINES (mammals) Pseudogenes of pol III transcripts
Termini	Long terminal repeats	No repeats	No repeats
Target repeats	4–6 bp	7–21 bp	7–21 bp
Enzyme activities	Reverse transcriptase and/or integrase	Reverse transcriptase /endonuclease	None (or none coding for transposon products)
Organization	May contain introns (removed in subgenomic mRNA)	One or two uninterrupted ORFs	No introns

FIGURE 17.37 Retroelements can be divided into LTR retrotransposons, non-LTR retroposons, and the nonautonomous SINEs.

FIGURE 17.38 shows the organization and sequence relationships of elements that code for reverse transcriptase. Like retroviruses, the LTR-retrotransposons can be classified into groups according to the number of independent reading frames for *gag*, *pol*, and *int*, and the order of the genes. In spite of these superficial differences of organization, the common features are the presence LTRs as well as reverse transcriptase and integrase activities. In contrast, non-LTR retroposons such as the mammalian LINE elements lack LTRs. They have two reading frames; one codes for a nucleic acid-binding protein and the other codes for reverse transcriptase and endonuclease activity.

LTR-containing elements can vary from integrated retroviruses to retrotransposons that do not have the capacity to generate infectious particles. Yeast and fly genomes have the *Ty* and *copia* elements that cannot generate infectious particles. Mammalian genomes have some endogenous retroviruses that, when active, can generate infectious particles. The mouse genome has several active endogenous retroviruses, which are able to generate particles that propagate horizontal infections. By contrast, almost all endogenous retroviruses lost their activity some 50 million years ago in the human lineage, and the genome now has mostly inactive remnants of the endogenous retroviruses.

LINEs and SINEs comprise a major part of the animal genome. They were defined originally by the existence of a large number of relatively short sequences that are related to one another (comprising the moderately repetitive DNA described in *Section 5.5, Eukaryotic Genomes Contain Both Nonrepetitive and Repetitive DNA Sequences*). They are described as interspersed sequences or interspersed repeats because of their common occurrence and widespread distribution. In many higher eukaryotic genomes, particularly metazoans, LINEs and SINEs can make up half of the total DNA. In contrast, in plant genomes, LTR retrotransposons tend to predominate.

FIGURE 17.39 summarizes the distribution of the different types of transposons that constitute almost half of the human genome. Except for the SINES, which never encode functional proteins, the other types of elements all consist of both functional elements and elements that have suffered deletions that eliminated parts of the reading frames that code for the protein(s) needed for transposition. The relative proportions of these types of transposons are generally similar in the mouse genome.

The most common LINE in mammalian genomes is called L1. The typical member is ~6500 bp long and terminates in an A-rich tract. The two open reading frames of a full-length element are called ORF1 and ORF2. The number of full-length elements is usually small (~50), and the remainder of the copies are truncated. Transcripts can be found. As implied by its presence in repetitive DNA, the LINE family shows sequence variation among individual members. The members of the family within a species, however, are relatively homogeneous compared to the variation shown between species. L1 is the only member of the LINE family that has been active in either the mouse or human lineages. It seems to have remained highly active in the mouse, but has declined in the human lineage.

Only one SINE has been active in the human lineage: the common **Alu element**.

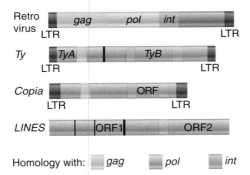

FIGURE 17.38 Retrotransposons are closely related to retroviruses have a similar organization, but non-LTR retroposons such as LINEs share only the reverse transcriptase activity and lack LTRs.

Element	Organization				Length (Kb)	Human genome	
						Number	Fraction
Retrovirus/LTR retrotransposon	LTR	*gag*	*pol* (*env*)	LTR	1–11	450,000	8%
LINES (autonomous), e.g., L1		*ORF1*	(*pol*)	(A)n	6–8	850,000	17%
SINES (nonautonomous), e.g., Alu				(A)n	<0.3	1,500,000	15%
DNA transposon			Transposase		2–3	300,000	3%

FIGURE 17.39 Four types of transposable elements constitute almost half of the human genome.

The mouse genome has a counterpart to this element (B1), and also other SINES (B2, ID, B4) that have been active. Human Alu and mouse B1 SINEs are probably derived from the 7SL RNA (see *Section 17.19, The Alu Family Has Many Widely Dispersed Members*). The other mouse SINEs appear to have originated from reverse transcripts of tRNAs. The transposition of the SINES probably results from their recognition as substrates by an active L1 element.

17.19 The Alu Family Has Many Widely Dispersed Members

Key concept

• A major part of repetitive DNA in mammalian genomes consists of repeats of a single family organized like transposons and derived from RNA polymerase III transcripts.

The most prominent SINE comprises members of a single family. Its short length and high degree of repetition make it comparable to simple sequence (satellite) DNA, except that the individual members of the family are dispersed around the genome instead of being confined to tandem clusters. Again, there is significant similarity between the members within a species compared with variation between species.

In the human genome, a large part of the moderately repetitive DNA exists as sequences of ~300 bp that are interspersed with nonrepetitive DNA. At least half of the renatured duplex material is cleaved by the restriction enzyme AluI at a single site located 170 bp along the sequence. The cleaved sequences all are members of a single family known as the Alu family, after the means of its identification. There are about one million members in the human genome (equivalent to one member per 3 kb of DNA). The individual Alu sequences are widely dispersed. A related sequence family is present in the mouse (where the ~350,000 members are called the B1 family), in the Chinese hamster (where it is called the Alu-equivalent family), and in other mammals.

The individual members of the Alu family are related rather than identical. The human family seems to have originated by means of a 130 bp tandem duplication, with an unrelated sequence of 31 bp inserted in the right half of the dimer. The two repeats are sometimes called the "left half" and the "right half" of the Alu sequence. The individual members of the Alu family have an average identity with the consensus sequence of 87%. The mouse B1 repeating unit is 130 bp long and corresponds to a monomer of the human unit. It has 70%–80% homology with the human sequence.

The Alu sequence is related to 7SL RNA, a component of the signal recognition particle involved in protein targeting to the endoplasmic reticulum, and Alu elements are likely derived from 7SL RNA transcripts. The 7SL RNA corresponds to the left half of an Alu sequence with an insertion in the middle. Thus the ninety 5′ terminal bases of 7SL RNA are homologous to the left end of Alu, the central 160 bases of 7SL RNA have no homology to Alu, and the 39 terminal bases of 7SL RNA are homologous to the right end of Alu. Like 7SL RNA genes, active Alu elements contain a functional internal RNA polymerase III promoter and are actively transcribed by this enzyme.

The members of the Alu family resemble transposons in being flanked by short direct repeats. They display, however, the curious feature that the lengths of the repeats are different for individual members of the family.

A variety of properties have been found for the Alu family, and its ubiquity has prompted many suggestions for its function. It is not yet possible, though, to discern its true role, if any (it may simply be a particularly successful selfish DNA). At least some members of the family can be transcribed into independent RNAs. In the Chinese hamster, some (though not all) members of the Alu-equivalent family appear to be transcribed *in vivo*. Transcription units of this sort are found in the vicinity of other transcription units.

Members of the Alu family may be included within structural gene transcription units, as seen by their presence in long nuclear RNA. The presence of multiple copies of the Alu sequence in a single nuclear molecule can generate secondary structure. In fact, the presence of Alu family members in the form of inverted repeats is responsible for most of the secondary structure found in mammalian nuclear RNA.

17.20 LINEs Use an Endonuclease to Generate a Priming End

Key concept

• LINES do not have LTRs and require the retroposon to code for an endonuclease that generates a nick to prime reverse transcription.

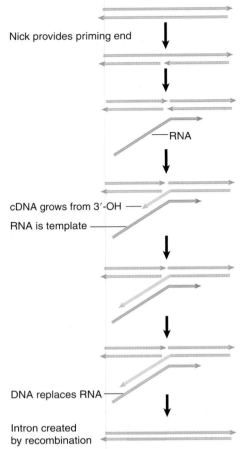

Nick provides priming end

RNA

cDNA grows from 3'-OH

RNA is template

DNA replaces RNA

Intron created
by recombination

FIGURE 17.40 Retrotransposition of non-LTR retroposons occurs by nicking the target to provide a primer for cDNA synthesis on an RNA template. The arrowheads indicate 3' ends.

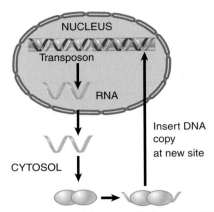

FIGURE 17.41 A LINE is transcribed into an RNA that is translated into proteins that assemble into a complex with the RNA. The complex translocates to the nucleus, where it inserts a DNA copy into the genome.

LINE elements, like all retroposons, do not terminate in the LTRs that are typical of retroviral elements. This poses the question: How is reverse transcription primed? It does not involve the typical reaction in which a tRNA primer pairs with the LTR (see Figure 17.27). The open reading frames in these elements lack many of the retroviral functions, such as protease or integrase domains, but typically have reverse transcriptase-like sequences and code for an endonuclease activity. In the human LINE L1, *ORF1* is a DNA-binding protein and *ORF2* has both reverse transcriptase and endonuclease activities; both products are required for transposition.

FIGURE 17.40 shows how these activities support transposition. A nick is made in the DNA target site by an endonuclease activity coded by the retroposon. The RNA product of the element associates with the protein bound at the nick. The nick provides a 3'–OH end that primes synthesis of cDNA on the RNA template.

A second cleavage event is required to open the other strand of DNA, and the RNA/DNA hybrid is linked to the other end of the gap either at this stage or after it has been converted into a DNA duplex. A similar mechanism is used by some mobile introns (see Figure 23.11).

One of the reasons why LINE elements are so effective lies with their method of propagation. When a LINE mRNA is translated, the protein products show a *cis*-preference for binding to the mRNA from which they were translated. **FIGURE 17.41** shows that the ribonucleoprotein complex then moves to the nucleus, where the proteins insert a DNA copy into the genome. Reverse transcription often does not proceed fully to the end, resulting in a truncated and inactive element. There is, however, the potential for insertion of an active copy, because the proteins are acting in *cis* on a transcript of the original active element.

By contrast, the proteins produced by the DNA transposons must be imported into the nucleus after being synthesized in the cytoplasm, but they have no means of distinguishing full-length transposons from inactive deleted transposons. **FIGURE 17.42** shows that instead of distinguishing these two types of transposons, the proteins will indiscriminately recognize any element by virtue of the repeats that mark the ends. This greatly reduces their chance of acting on a full-length element as opposed to one that has been deleted, resulting in an inability to replicate the autonomous elements efficiently. This can potentially lead to extinction of the entire family of elements.

Are transposition events of retroelements currently occurring in these genomes, or are we seeing only the footprints of ancient systems?

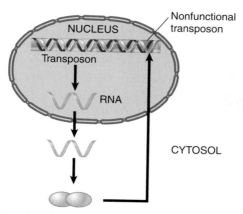

FIGURE 17.42 A transposon is transcribed into an RNA that is translated into proteins that move independently to the nucleus, where they act on any pair of inverted repeats with the same sequence as the original transposon.

This varies with the species. There are only a few currently active transposons in the human genome, but by contrast several active transposons are known in the mouse genome. This explains the fact that spontaneous mutations caused by LINE insertions occur at a rate of ~3% in mouse, but only 0.1% in humans. There appear to be ~80 to 100 active LINE elements in the human genome. Some human diseases can be pinpointed as the result of transposition of L1 into genes, and others result from unequal crossing-over events involving repeated copies of L1. A model system in which LINE transposition occurs in tissue culture cells suggests that a transposition event can introduce several types of collateral damage as well as inserting into a new site; the damage includes chromosomal rearrangements and deletions. Such events may be viewed as agents of genetic change. Neither DNA transposons nor retroviral-like retrotransposons seem to have been active in the human genome for 40 to 50 million years, but several active examples of both are found in the mouse.

Note that for transpositions to survive, they must occur in the germline. Similar events occur in somatic cells, but do not survive beyond one generation.

17.21 Summary

Prokaryotic and eukaryotic cells contain a variety of transposons that mobilize by moving or copying DNA sequences. The transposon can be identified only as an entity within the genome; its mobility does not involve an independent form. The transposon could be selfish DNA, concerned only with perpetuating itself within the resident genome; if it conveys any selective advantage upon the genome, this must be indirect. All transposons have systems to limit the extent of transposition, because unbridled transposition is presumably damaging, but the molecular mechanisms are different in each case.

The archetypal transposon has inverted repeats at its termini and generates direct repeats of a short sequence at the site of insertion. The simplest types are the bacterial insertion sequences (IS), which consist essentially of the inverted terminal repeats flanking a coding frame(s) whose product(s) provide transposition activity.

The generation of target repeats flanking a transposon reflects a common feature of transposition. The target site is cleaved at points that are staggered on each DNA strand by a fixed distance (often five or nine base pairs). The transposon is in effect inserted between protruding single-stranded ends generated by the staggered cuts. Target repeats are generated by filling in the single-stranded regions.

IS elements, composite transposons, P elements, and the "controlling elements" in maize mobilize by nonreplicative transposition, in which the element moves directly from a donor site to a recipient site. A single transposase enzyme undertakes the reaction. It occurs by a "cut and paste" mechanism in which the transposon is separated from flanking DNA. Cleavage of the transposon ends, nicking of the target site, and connection of the transposon ends to the staggered nicks all occur in a nucleoprotein complex containing the transposase. Loss of the transposon from the donor creates a double-strand break whose fate can vary depending on the host repair mechanisms and the timing of excision. In the case of Tn10, transposition becomes possible immediately after DNA replication, when sites recognized by the *dam* methylation system are transiently hemimethylated. This imposes a demand for the existence of two copies of the donor site, which may enhance the cell's chances for survival.

Phage Mu can undergo either replicative or nonreplicative transposition. In replicative transposition, after the transposon at the donor site becomes connected to the target site, replication generates a cointegrate molecule that has two copies of the transposon. A resolution reaction that involves recombination between two particular sites then frees the two copies of the transposon, so that one remains at the donor site and one appears at the target site. Two enzymes coded by the transposon are required: Transposase recognizes the ends of the transposon and connects them to the target site, and resolvase provides a site-specific recombination function. Mu can also can use its cointegrate intermediate to transpose by a non-replicative mechanism. The difference between this reaction and the nonreplicative transposition of IS elements is that the cleavage events occur in a different order.

The best characterized transposons in plants are the controlling elements of maize, which fall into several families. Each family contains a single type of autonomous element that is analogous to bacterial transposons in its ability to mobilize. A family also contains many different nonautonomous elements that are derived by mutations of the autonomous element. The nonautonomous elements lack the ability to transpose, but display transposition activity and other abilities of the autonomous element when an autonomous element is present to provide the necessary *trans*-acting functions.

Transposition of maize elements is non-replicative, and probably requires only the enzymes coded by the element. Transposition occurs preferentially after replication of the element. It is likely that there are mechanisms to limit the frequency of transposition. Advantageous rearrangements of the maize genome may have been connected with the presence of the elements.

P elements in *D. melanogaster* are responsible for hybrid dysgenesis. A cross between a male carrying P elements and a female lacking them generates hybrids that are sterile. A P element has four open reading frames, which are separated by introns. Splicing of the first three ORFs generates a 66 kD repressor and occurs in somatic cells. Splicing of all four ORFs to generate the 87 kD transposase occurs only in the germline by a tissue-specific splicing event. P elements mobilize when exposed to cytoplasm lacking the repressor. The burst of transposition events inactivates the genome by random insertions. Only a complete P element can generate transposase, but defective elements can be mobilized in *trans* by the enzyme.

Reverse transcription is the unifying mechanism for reproduction of retroviruses and perpetuation of retroelements. The cycle of each type of element is in principle similar, although retroviruses are usually regarded from the per-spective of the free viral (RNA) form, whereas retrotransposons are regarded from the stance of the genomic (duplex DNA) form.

Retroviruses have genomes of single-stranded RNA that are replicated through a double-stranded DNA intermediate. An individual retrovirus contains two copies of its genome. The genome contains the *gag*, *pol*, and *env* genes that are translated into polyproteins, each of which is cleaved into smaller functional proteins. The Gag and Env components are concerned with packing RNA and generating the virion; the Pol components are concerned with nucleic acid synthesis.

Reverse transcriptase is the major component of Pol, and is responsible for synthesizing a DNA (minus strand) copy of the viral (plus strand) RNA. The DNA product is longer than the RNA template; by switching template strands, reverse transcriptase copies the 3′ sequence of the RNA to the 5′ end of the DNA, and copies the 5′ sequence of the RNA to the 3′ end of the DNA. This generates the characteristic LTRs (long terminal repeats) of the DNA. A similar switch of templates occurs when the plus strand of DNA is synthesized using the minus strand as a template. Linear duplex DNA is inserted into a host genome by the integrase enzyme. Transcription of the integrated DNA from a promoter in the left LTR generates further copies of the RNA sequence.

Switches in template during nucleic acid synthesis allow recombination to occur by copy choice. During an infective cycle, a retrovirus may exchange part of its usual sequence for a cellular sequence; the resulting virus is usually replication-defective, but can be perpetuated in the course of a joint infection with a helper virus. Many of the defective viruses have gained an RNA version (*v-onc*) of a cellular gene (*c-onc*). The *onc* sequence may be any one of a number of genes whose expression in *v-onc* form causes the cell to be transformed into a tumorigenic phenotype.

The integration event generates direct target repeats (like transposons that mobilize via DNA). An inserted provirus therefore has direct terminal repeats of the LTRs, flanked by short repeats of target DNA. Mammalian and avian genomes have endogenous (inactive) proviruses with such structures. Other elements with this organization have been found in plants, animals, and fungi. *Ty* elements of yeast and *copia* elements of flies have coding sequences with homology to reverse transcriptase and mobilize via an RNA form. They may generate particles resembling viruses, but do not have infectious capability. The LINE sequences of mammalian genomes are further removed from the retroviruses, but retain enough similarities to suggest a common origin. They use a different type of priming event to initiate reverse transcription, in which an endonuclease activity associated with the reverse transcriptase makes a nick that provides a 3'–OH end for priming synthesis on an RNA template. The frequency of LINE transposition is increased because its protein products are *cis*-acting; they associate with the mRNA from which they were translated to form a ribonucleoprotein complex that is transported into the nucleus.

The members of another class of retroelements have the hallmarks of transposition via RNA, but have no coding sequences (or at least none resembling retroviral functions). They may have originated as passengers in a retroviral-like transposition event, in which an RNA was a target for a reverse transcriptase. A particularly prominent family that appears to have originated from a processing event are represented by SINEs; it includes the human Alu family. Some snRNAs, including 7SL snRNA (a component of the SRP), are related to this family.

References

17.1   Introduction

Reviews

Craig, N. L., Craigie, R., Gellert, M., and Lambowitz, A., eds. (2002). *Mobile DNA II*. Washington, DC: American Society for Microbiology Press.

Deininger, P. L. and Roy-Engel, A. M. (2002). Mobile elements in animal and plant genomes. In Craig, N.L, Craigie, R., Gellert, M., and Lambowitz, A., eds. *Mobile DNA II*. Washington, DC: American Society for Microbiology Press, pp. 1074–1092.

Feschotte C. and Pritham E. J. (2007). DNA transposons and the evolution of eukaryotic genomes. *Ann. Rev. Genet.* 41, 331–368.

17.2   Insertion Sequences Are Simple Transposition Modules

Reviews

Chandler, M. and Mahillon, J. (2002). Bacterial insertion sequences revisited. In Craig, N.L, Craigie, R., Gellert, M., and Lambowitz, A., eds. *Mobile DNA II*. Washington, DC: American Society for Microbiology Press, pp. 305–366.

Craig, N. L. (1997). Target site selection in transposition. *Annu. Rev. Biochem.* 66, 437–474.

Research

Grindley, N. D. (1978). IS1 insertion generates duplication of a 9 bp sequence at its target site. *Cell* 13, 419–426.

17.3   Transposition Occurs by Both Replicative and Nonreplicative Mechanisms

Reviews

Craig, N. L. (1997). Target site selection in transposition. *Annu. Rev. Biochem.* 66, 437–474.

Grindley, N. D. and Reed, R. R. (1985). Transpositional recombination in prokaryotes. *Annu. Rev. Biochem.* 54, 863–896.

Haren, L., Ton-Hoang, B., and Chandler, M. (1999). Integrating DNA: transposases and retroviral integrases. *Annu. Rev. Microbiol.* 53, 245–281.

17.6   Nonreplicative Transposition Proceeds by Breakage and Reunion

Reviews

Reznikoff, W. S. (2008). Transposon Tn5. *Annu. Rev. Genet.* 42, 269–286.

Research

Bender, J. and Kleckner, N. (1986). Genetic evidence that Tn10 transposes by a nonreplicative mechanism. *Cell* 45, 801–815.

Bolland, S. and Kleckner, N. (1996). The three chemical steps of Tn10/IS10 transposition involve repeated utilization of a single active site. *Cell* 84, 223–233.

Davies, D. R., Goryshin, I. Y., Reznikoff, W. S., and Rayment, I. (2000). Three-dimensional structure of the Tn5 synaptic complex transposition intermediate. *Science* 289, 77–85.

Haniford, D. B., Benjamin, H. W., and Kleckner, N. (1991). Kinetic and structural analysis of a cleaved donor intermediate and a strand transfer intermediate in Tn10 transposition. *Cell* 64, 171–179.

Kennedy, A. K., Guhathakurta, A., Kleckner, N., and Haniford, D. B. (1998). Tn10 transposition via a DNA hairpin intermediate. *Cell* 95, 125–134.

17.7   Maize Transposons Can Cause Breakage and Rearrangements

Review

Jones, R. N. (2005). McClintock's controlling elements: the full story. *Cytogenet. Genome Res.* 109, 90–103.

Research

Huang, J. T. and Dooner, H. K. (2008). Macrotransposition and other complex chromosomal restructuring in maize by closely linked transposons in direct orientation. *Plant Cell* 8, 2019–2032.

17.8 Transposons Form Families in Maize

Reviews

Feschotte, C, Jiang, N., and Wessler, S. R. (2002). Plant transposable elements: where genetics meets genomics. *Nat. Rev. Genet.* 3, 329–341.

Gierl, A., Saedler, H., and Peterson, P. A. (1989). Maize transposable elements. *Annu. Rev. Genet.* 23, 71–85.

Kunz, R. and Weil, C. F. (2002). The hAT and CACTA superfamilies of plant transposons. In Craig, N. L, Craigie, R., Gellert, M., and Lambowitz, A., eds. *Mobile DNA II.* Washington, DC: American Society for Microbiology Press, pp. 400–600.

Research

Benito, M. I. and Walbot, V. (1997). Characterization of the maize Mutator transposable element MURA transposase as a DNA-binding protein. *Mol. Cell Biol.* 17, 5165–5175.

Jiang, N., Bao, Z., Zhang, X., Hirochika, H., Eddy, S. R., McCouch, S. R., and Wessler, S. R. (2004). An active DNA transposon family in rice. *Nature* 421, 163–167.

Ros, F. and Kunze, R. (2001). Regulation of activator/dissociation transposition by replication and DNA methylation. *Genetics* 157, 1723–1733.

Singer, T., Yordan, C., and Martienssen, R. A. (2001). Robertson's *Mutator* transposons in *A. thaliana* are regulated by the chromatin-remodeling gene decrease in DNA Methylation (DDM1). *Genes Dev.* 15, 591–602.

Slotkin, K. R., Freeling, M., and Lisch, D. (2005). Heritable silencing of a transposon family is initiated by a naturally occurring inverted repeat derivative. *Nature Genetics* 137, 641–644.

Zhou, L., Mitra, R., Atkinson, P. W., Hickman, A. B., Dyda, F., and Craig, N. L. (2004). Transposition of hAT elements links transposable elements and V(D)J recombination. *Nature* 432, 960–961.

17.9 The Role of Transposable Elements in Hybrid Dysgenesis

Reviews

Engels, W. R. (1983). The P family of transposable elements in *Drosophila. Annu. Rev. Genet.* 17, 315–344.

Rio, D. C. (2002). P transposable elements in *Drosophila melanogaster.* In Craig, N. L, Craigie, R., Gellert, M., and Lambowitz, A., eds. *Mobile DNA II.* Washington, DC: American Society for Microbiology Press, pp. 484–518.

Research

Daniels, S. B., Peterson, K. R., Strausbaugh, L. D., Kidwell, M. G., and Chovnick, A. (1990). Evidence for horizontal transmission of the P transposable element between *Drosophila* species. *Genetics* 124, 339–355.

Engels, W. R., Johnson-Schlitz, D. M., Eggleston, W. B., and Sved, J. (1990). High-frequency P element loss in *Drosophila* is homolog dependent. *Cell* 62, 515–525.

17.10 P Elements Are Activated in the Germline

Research

Brennecke J., Malone C. D., Aravin, A. A., Sachidanandam, R., Stark, A., and Hannon, G. J. (2008). An epigenetic role for maternally inherited piRNAs in transposon silencing. *Science* 322, 1387–1392.

Laski, F. A., Rio, D. C., and Rubin, G. M. (1986). Tissue specificity of *Drosophila* P element transposition is regulated at the level of mRNA splicing. *Cell* 44, 7–19.

17.11 The Retrovirus Life Cycle Involves Transposition-Like Events

Review

Varmus, H. E. and Brown, P. O. (1989). Retroviruses. In Howe, M. M. and Berg, D. E., eds., *Mobile DNA.* Washington, DC: American Society for Microbiology, pp. 53–108.

Research

Baltimore, D. (1970). RNA-dependent DNA polymerase in virions of RNA tumor viruses. *Nature* 226, 1209–1211.

Temin, H. M. and Mizutani, S. (1970). RNA-dependent DNA polymerase in virions of Rous sarcoma virus. *Nature* 226, 1211–1213.

17.13 Viral DNA Is Generated by Reverse Transcription

Reviews

Katz, R. A. and Skalka, A. M. (1994). The retroviral enzymes. *Annu. Rev. Biochem.* 63, 133–173.

Lai, M. M. C. (1992). RNA recombination in animal and plant viruses. *Microbiol. Rev.* 56, 61–79.

Negroni, M. and Buc, H. (2001). Mechanisms of retroviral recombination. *Annu. Rev. Genet.* 35, 275–302.

Research

Hu, W. S. and Temin, H. M. (1990). Retroviral recombination and reverse transcription. *Science* 250, 1227–1233.

Negroni, M. and Buc, H. (2000). Copy-choice recombination by reverse transcriptases: reshuffling of genetic markers mediated by RNA chaperones. *Proc. Natl. Acad. Sci. USA* 97, 6385–6390.

17.14 Viral DNA Integrates into the Chromosome

Reviews

Craigie, R. (2002). Retroviral integration. In Craig, N. L, Craigie, R., Gellert, M., and Lambowitz, A., eds. *Mobile DNA II.*

Washington, DC: American Society for Microbiology Press, pp. 613–630.

Craigie, R., Fujiwara, T., and Bushman, F. (1990). The IN protein of Moloney murine leukemia virus processes the viral DNA ends and accomplishes their integration *in vitro*. *Cell* 62, 829–837.

17.16 Yeast *Ty* Elements Resemble Retroviruses

Research

Boeke, J. D. et al. (1985). Ty elements transpose through an RNA intermediate. *Cell* 40, 491–500.

Lauermann, V. and Boeke, J. D. (1994). The primer tRNA sequence is not inherited during *Ty1* retrotransposition. *Proc. Natl. Acad. Sci. USA* 91, 9847–9851.

17.17 Many Kinds of Transposable Elements Reside in *D. melanogaster*

Research

Mount, S. M. and Rubin, G. M. (1985). Complete nucleotide sequence of the *Drosophila* transposable element *copia*: homology between *copia* and retroviral proteins. *Mol. Cell Biol.* 5, 1630–1638.

17.18 Retroelements Fall into Three Classes

Reviews

Deininger, P. L. (1989). SINEs: short interspersed repeated DNA elements in higher eukaryotes. In Howe, M. M. and Berg, D. E., eds. *Mobile DNA*. Washington, DC: American Society for Microbiology, pp. 619–636.

Moran, J. and Gilbert, N. (2002). Mammalian LINE-1 retrotransposons and related elements. In Craig, N. L, Craigie, R., Gellert, M. and Lambowitz, A., eds. *Mobile DNA II*. Washington, DC: American Society for Microbiology Press, pp. 836-869.

Research

Chinwalla, A. T. et al. (2002). Initial sequencing and comparative analysis of the mouse genome. *Nature* 420, 520–562.

Dewannieux, M., Esnault, C., and Heidmann, T. (2003). LINE-mediated retrotransposition of marked Alu sequences. *Nature Genet.* 35, 41–48.

Loeb, D. D. et al. (1986). The sequence of a large L1Md element reveals a tandemly repeated 5' end and several features found in retrotransposons. *Mol. Cell Biol.* 6, 168–182.

Sachidanandam, R. et al. (2001). A map of human genome sequence variation containing 1.42 million single nucleotide polymorphisms. The International SNP Map Working Group. *Nature* 409, 928–933.

17.20 LINEs Use an Endonuclease to Generate a Priming End

Review

Ostertag, E. M. and Kazazian, H. H. (2001). Biology of mammalian L1 retrotransposons. *Annu. Rev. Genet.* 35, 501–538.

Research

Feng, Q., Moran, J. V., Kazazian, H. H., and Boeke, J. D. (1996). Human L1 retrotransposon encodes a conserved endonuclease required for retrotransposition. *Cell* 87, 905–916.

Gilbert, N., Lutz-Prigge, S., and Moran, J. V. (2002). Genomic deletions created upon LINE-1 retrotransposition. *Cell* 110, 315–325.

Luan, D. D. et al. (1993). Reverse transcription of R2Bm RNA is primed by a nick at the chromosomal target site: a mechanism for non-LTR retrotransposition. *Cell* 72, 595–605.

Moran, J. V., Holmes, S. E., Naas, T. P., DeBerardinis, R. J., Boeke, J. D., and Kazazian, H. H. (1996). High frequency retrotransposition in cultured mammalian cells. *Cell* 87, 917–927.

Symer, D. E., Connelly, C., Szak, S. T., Caputo, E. M., Cost, G. J., Parmigiani, G., and Boeke, J. D. (2002). Human l1 retrotransposition is associated with genetic instability *in vitro*. *Cell* 110, 327–338.

18

© Laguna Design/Photo Researchers, Inc.

Somatic Recombination and Hypermutation in the Immune System

Edited by Paolo Casali

CHAPTER OUTLINE

- There are two families of L chains (Igλ and Igκ) and a single family of IgH chains.
- Each chain has an N-terminal variable (V) region and a C-terminal constant (C) region.
- The V region recognizes antigen and the C region mediates the effector response.
- V and C regions are separately encoded by V gene segments and C gene segments.
- A gene coding for an intact Ig chain is generated by somatic recombination of V(D)J genes (variable, diversity, and joining genes in the H chain; variable and joining genes in the L chain) giving raise to V domains, to be expressed together with a given C gene (C domain).

18.6 L Chains Are Assembled by a Single Recombination Event

- A λ chain is assembled through a single recombination event involving a Vλ gene segment and a JλCλ gene segment.
- The Vλ gene segment has a leader exon, intron, and V-coding region.
- The JλCλ gene segment has a short Jλ-coding exon, an intron, and a Cλ-coding region.
- A κ chain is assembled by a single recombination event involving a Vκ gene segment and one of five Jκ segments preceding the Cκ gene.

18.7 H Chains Are Assembled by Two Sequential Recombination Events

- The units for H chain recombination are a V_H gene, a D segment, and a J_HC_H gene segment.
- The first recombination joins D to J_HC_H.
- The second recombination joins V_H to $D\text{-}J_HC_H$ to yield $V_H\text{-}D\text{-}J_HC_H$.
- The C_H segment consists of four exons.

18.8 Recombination Generates Extensive Diversity

- The human IgH locus can generate in excess of 10^8 $V_H\text{-}D\text{-}J_H$ sequences.
- Recombined $V_H\text{-}D\text{-}J_HC_H$ can be paired with in excess of 10^6 recombined Vκ-JκCκ or Vλ-JλCλ chains.

18.9 Immune Recombination Uses Two Types of Consensus Sequence

- The consensus sequence used for recombination is a heptamer separated by either 12 or 23 base pairs from a nonamer.
- Recombination occurs between two consensus sequences that have different spacers.

18.10 V(D)J DNA Recombination Occurs by Deletion or Inversion

- Recombination occurs by double-strand DNA breaks (DSBs) at the heptamers of two RSSs.
- The signal ends of the linear DNA excised between two DSBs are joined to generate a DNA circle.
- The coding ends are covalently ligated to join V_L to J_LC_L (L chain), or D to J_HC_H and V_H to $D\text{-}J_HC_H$ (H chain).

- If the recombining genes lie in an inverted instead of a direct orientation, the intervening DNA is inverted, but retained, instead of being excised as a circle.

18.11 Allelic Exclusion Is Triggered by Productive Rearrangements

- V(D)J gene rearrangement is productive if it leads to expression of a protein.
- A productive V(D)J gene rearrangement prevents any further rearrangement of the same kind from occurring, whereas a nonproductive rearrangement does not.
- Allelic exclusion applies separately to L chains (only one κ or λ may be productively rearranged) and to H chains (one H chain is productively rearranged).

18.12 RAG1/RAG2 Catalyze Breakage and Religation of V(D)J Gene Segments

- The RAG proteins are necessary and sufficient for the cleavage reaction.
- RAG1 recognizes the nonamer consensus sequences for recombination. RAG2 binds to RAG1 and cleaves DNA at the heptamer. The reaction resembles the topoisomerase-like resolution reaction that occurs in transposition.
- The reaction proceeds through a hairpin intermediate at the coding end; opening of the hairpin is responsible for insertion of extra bases (P nucleotides) in the recombined gene.
- Terminal deoxynucleotidyl transferase (TdT) inserts additional unencoded N nucleotides at the V(D)J junctions.
- The DSBs at the coding joints are repaired by the same mechanism that has generated the whole V(D)J sequence.

18.13 Early IgH Chain Expression Is Modulated by RNA Processing

- All B lymphocytes newly emerging from the bone marrow express the membrane-bound monomeric form of IgM (μm). A change in RNA splicing causes μm to be replaced by the secreted form (μs) after a mature B cell is activated and begins differentiation to antibody-producing cells in the periphery.

18.14 Class Switching Is Effected by DNA Recombination (Class Switching DNA Recombination CSR)

- Igs comprise five classes according to the type of C_H chain.
- Class switching is effected by a recombination between S regions that deletes the DNA between the upstream C_H region gene cluster and the (new) downstream C_H region gene cluster.

18.15 CSR Involves Elements of the NHEJ Pathway

- CSR requires activation of intervening promoters (I_H promoters) that lie upstream of the S regions involved in the recombination event.
- $I_H\text{-}C_H$ transcription through the S region is required.
- S regions contain highly repetitive motifs with 5'-AGCT-3' as a major component.

18.1 The Immune System: Innate and Adaptive Immunity

In general, differential control of gene expression, rather than changes in DNA sequence, explains the different phenotypes of given somatic cells. The immune system is a most important exception to the axiom of genetics that the genetic constitution created in the zygote by the combination of sperm and egg is inherited by all somatic cells of the organism. In developing immune cells (B and T lymphocytes), the genome changes through extensive somatic DNA recombination to create functional genes. Other cases of somatic recombination are represented by the substitution of one sequence for another to generate new surface antigens in trypanosomes or to change the mating type of yeast. In mature B cells, additional DNA recombination and somatic hypermutation of recombined DNA segments occur to further diversify the function of these effector lymphocytes.

The immune system of vertebrates mounts a protective response that distinguishes in general, foreign (nonself) soluble molecules or molecules on microorganisms from components (molecules or cells) of the organism itself (self). Nonself and self-components capable of inducing a specific immune response are referred to as **antigens**. In general, an antigen is a protein or protein-attached moiety that has entered the bloodstream—for example, the coat protein of an infecting virus or bacterium. Exposure to an antigen triggers the unfolding of an immune response aimed at *specifically recognizing the antigen*, thereby destroying the infecting virus or bacterium expressing it.

Immune reactions are effected by white blood cells—B and T lymphocytes, macrophages, and dendritic cells. Lymphocytes are named after the organ in which they differentiate or mature. In mammals, **B cells** mature in the bone marrow, whereas **T cells** mature in the thymus. (The "B" in "B cells" originally stemmed from *bursa* of *Hieronimus Fabricius*, after the Paduan anatomist who recognized in the sixteenth century this lymphoid organ in birds as equivalent of mammalian bone marrow.) *Each class of lymphocytes uses the rearrangement of DNA as a mechanism for producing the proteins that enable it to participate in the immune response.*

Responses to antigens on viruses and bacteria, such as an antibody response to *Streptococcus* (*Pneumococcus*) *pneumoniae* or a killer T lymphocyte-mediated response to influenza virus-infected cells, are highly specific and are the expression of **adaptive (acquired) immunity**. The adaptive immune response is characterized by a latency period—in general a few days—which is required for the clonal selection and expansion of the B cells and/or T cells specific for the antigen. The antigen, which is driving the response, can be on a bacterium, a virus, or other microorganism. Clonal selection of B cells or T cells relies on binding of antigen to **B cell receptors (BCR)** and **T cell receptors (TCR)**, both of which possess a high affinity for that antigen. The structural basis for this selection process is provided by the generation of a very large number of BCRs/TCRs, so as to create a high probability of recognizing any foreign molecule. BCRs/TCRs that recognize the body's own proteins are screened out early in the process. Activation of the BCR on B cells triggers the pathways of the **humoral response**; activation of the TCR on T cells triggers the pathways of the **cell-mediated response**. The organism retains a **memory** of the specific B and/or T cell response. Such memory enables the organism to respond more rapidly once exposed again to the same pathogen. Immunological memory provides protective immunity against the same antigen that drove the original response. The principles of adaptive immunity are similar, albeit somewhat different in details, throughout the vertebrates.

In contrast to adaptive immunity, **innate immunity** provides an immediate (without latency) first line of defense against invading microorganisms. The innate response depends on receptors encoded in the germline to recognize shared structural patterns, as occurs on microbial pathogens. The innate response is nonspecific for any given pathogen and cannot generate memory. It is triggered in different ways and to different degrees, as determined by the nature of the foreign microbial antigen inducing it. Through differential modulation of the innate response, the nature of the antigen also directs the character of the adaptive response eventually mounted to the same antigen.

18.2 The Innate Response Utilizes Conserved Recognition Molecules and Signaling Pathways

Key concepts

- Innate immunity is triggered by pattern recognition receptors (PRRs) that recognize highly conserved microbe-associated molecular patterns (MAMPs) found in bacteria, viruses, and other infectious agents.
- Toll-like receptors (TLRs) are important PRRs that directly activate innate immune responses.
- TLR signaling pathways are highly conserved from invertebrates to vertebrates and an analogous pathway is found in plants.
- TLRs are expressed in dendritic cells (DCs), macrophages, and neutrophils.
- TLRs are expressed in B lymphocytes and some T lymphocytes.

Innate immunity provides a first line of defense against microbial pathogens. It is present in virtually all multicellular organisms, albeit in different forms and relying on different effector mechanisms. The innate response depends on the recognition of certain predefined patterns in pathogens. These patterns are motifs that are conserved in microorganisms, but they are not found in multicellular eukaryotes, thus allowing the immune system to quickly and with high probability distinguish dangerous nonself patterns from self-patterns. Furthermore, because these molecular patterns are synthesized by several sequential microbial enzyme reactions, the genes controlling them mutate much more slowly compared to protein antigens. These conserved microbial components are now known as **MAMPs**, replacing the original term 'pathogen-associated molecular patterns' **(PAMPs)**, to reflect the fact that non-pathogenic bacteria, such commensal bacteria residing in the gut, also display conserved MAMPs. Each conserved microbial motif or pattern is typically recognized by a receptor dedicated to the purpose of triggering the innate response upon an infection. For example, Gram-negative bacterial lipopolysaccharide (LPS) is a well-known MAMP, and other MAMPs include bacterial flagellin, lipoteichoic acid from Gram-positive bacteria, peptidoglycans, and nucleic acid variants normally associated with viruses, such as single- or double-stranded RNA (ssRNA or dsRNA) or certain unmethylated CpG DNA. Upon sensing their ligands, these receptors rapidly activate innate immune responses by identifying non-self molecules, protecting the host from infection.

Receptors that trigger the innate response are known as **PRRs (pattern recognition receptors)**. These are found on innate immune cells such as neutrophils, macrophages, and dendritic cells (DCs) and cause the pathogen to be phagocytosed and killed. Some PRRs are also expressed in cells important for adaptive immune responses such as B lymphocytes and some T lymphocyte subsets. The response is rapid, because the set of receptors is already present on the cells and does not have to be rearranged and amplified by selection, unlike BCRs and TCRs. Innate response pathways are widely conserved and are found in organisms ranging from flies to humans. In general, the innate response somewhat contains the first wave of invasion by pathogens, but cannot deal effectively with later stages of the infection, which require the potency and specificity of the adaptive response. There is some overlap and crosstalk between the innate and adaptive responses in that cells activated by the innate response subsequently participate in the adaptive response, and some PRRs function directly in some lymphocyte subsets. A number of important MAMPs and their corresponding PRRs are summarized in **FIGURE 18.1**.

A key insight into the nature of innate immunity was the discovery of the role of **TLRs (Toll-like receptors)** in this response. In *Drosophila*, the receptor Toll, which is related to the mammalian IL-1 receptor, triggers the pathway that specifies dorsal–ventral development. This pathway entails the activation of the transcription factor dorsal, a member of the Rel family, which is related to the mammalian factor NF-κB. The pathway of innate immunity in vertebrates is parallel to the Toll receptor pathway, with similar components. In fact, one of the first indications of the nature of innate immunity in flies was the discovery of the transcription factor Dif (dorsal-related immunity factor), which is activated by one of the pathways.

Flies have no system of adaptive immunity, but are resistant to microbial infections. This is because their innate immune systems trigger synthesis of potent antimicrobial peptides. More than twenty distinct peptides have been

Microorganism	MAMP	Location	PRR
Bacteria	Triacyl lipopeptides (Pam$_3$CSK$_4$)	Cell wall	TLR1/2
Bacteria	Muramyl dipeptide	Cell wall	NOD2
Flagellated bacteria	Flagellin	Flagellum	TLR5
Gram^{+ve} bacteria	Peptidoglycan	Cell wall	TLR2/6
Gram^{-ve} bacteria	Lipoteichoic acid	Cell wall	TLR2/6
Gram^{-ve} bacteria	Lipopolysaccharide	Cell wall	TLR4
Bacteria and viruses	ssRNA	Inside cell/caspid	TLR7/8; NALP3
RNA viruses	dsRNA		TLR3/RIG-1 helicase
Fungi	β-glycans	Cell wall	Dectin-1
Mycoplasma	Triacyl lipopeptides (Pam$_3$CSK$_4$)	Cell wall	TLR2/6
DNA-containing microorganisms	Unmethylated CpG DNA	Inside cell/caspid	TLR9

FIGURE 18.1 Innate immunity; a summary of MAMPs and corresponding PRRs.

identified in *Drosophila*, where they are synthesized in the fat body (the organ analogous to the vertebrate liver). Two of the peptides are antifungal and five act primarily on bacteria. The general mode of action is to kill the target organism by permeabilizing its membrane. All of these peptides are encoded by genes whose promoters respond to transcription factors of the Rel family (FIGURE 18.2).

Two innate response pathways function in *Drosophila*: one responds principally to fungi, whereas the other responds principally to Gram-negative bacteria. Gram-positive bacteria may be able to trigger both pathways, each consisting of multiple steps (FIGURE 18.3). Fungi and Gram-positive bacteria activate a proteolytic cascade that generates an insect cytokine called Spatzle, which binds to and activates the *Drosophila*'s Toll receptor. This is the NF-κB-like pathway. The Toll receptor activates the transcription factor Dif (a relative of NF-κB), leading ultimately to activation of the antifungal peptide drosomycin. Gram-negative bacteria trigger a pathway via a different receptor that activates the transcription factor Relish, leading to production of the bactericidal peptide

attacin. This pathway is called the Imd pathway after one of its components, a protein that has a "death domain" related to those found in the pathways for apoptosis.

The key receptors sensing the bacteria are called **peptoglycan recognition proteins (PGRPs)** because of their high affinities for bacterial peptidoglycans. There are two types of these proteins. First, PGRP-SAs are short extracellular proteins. They function by activating the proteases that produce Spatzle, which triggers the Toll receptor pathway. Second, PGRP-LCs are transmembrane proteins with an extracellular PGRP domain; they are the receptors that activate the Imd pathway.

The innate immune response is highly conserved. Mice with a mutation in the *TLR4* gene do not respond to LPS and are resistant to septic shock, as induced by LPS. About ten human homologs of the TLRs can activate several immune response genes, demonstrating that the pathway of innate immunity also functions in humans. The pathway downstream of the TLRs is generally similar in all cases, typically leading to their activation by homo- or heterodimerization and/or

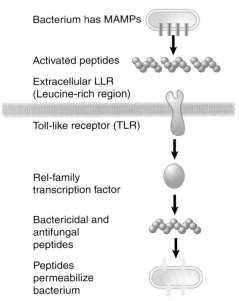

FIGURE 18.2 Innate immunity is triggered by MAMPs. In flies, MAMPs cause the production of peptides that activate Toll-like receptors. The receptors lead to a pathway that activates a transcription factor for the Rel family. Target genes for this factor include bactericidal and antifungal peptides. The peptides act by permeabilizing the membrane of the pathogenic organism.

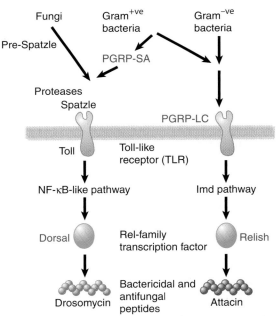

FIGURE 18.3 One of *Drosophila*'s innate immunity pathways is closely related to the mammalian pathway for activating NF-κB; the other has components related to those of apoptosis pathways.

conformational changes, ultimately resulting in the induction of transcription factors such as NF-κB, AP-1, IRFs, or cell-specific proteins such as AID in B lymphocytes. Once a TLR is activated, it interacts with one or more of five known **Toll/interleukin 1/resistance (TIR) domain**-containing adapters: MyD88, TRIF/TICAM-1, TRAM/TICAM-2, MAL/TIRAP, and SARM, by means of homotypic TIR-TIR interactions, relaying the signal through one or more of these adapters. The pathway upstream of the TLRs is different in mammals and flies, because the pathogen ligands directly activate mammalian TLRs, whereas they activate cytokines like Spatzle in insects. For example, LPS binds to TLR4, and unmethylated CpG DNA from bacteria or viruses is first internalized and then binds to TLR9 present in intracellular endosomes. The downstream TLR pathways are similar in insects and mammals, though they are more expanded and versatile in mammals. In mammals, TLR4 signals through both MyD88 and TRIF, whereas TLR9 signals through MyD88 only, triggering the appropriate response depending on cell type and conditions. While TLRs were thought to directly activate only innate immune cells, such as macrophages and DCs, it is now known that they are also highly expressed in and directly activate B lymphocytes, which are critical components of adaptive immunity. The exact roles of TLR and BCR signaling in B cell antibody responses have only recently received some attention, and their relative roles and crosstalk should be further elucidated in the near future.

Plants have extensive defense mechanisms, with pathways analogous to the innate response in animals. The same principle applies: MAMPs are the motifs that identify the infecting agent as a pathogen. The proteins that respond to the pathogens are coded by a class of genes called the disease-resistance genes. Many of these genes encode receptors that share a property with the TLR class of animal receptors: The extracellular domain has a motif called the *leucine-rich region* (LRR). The response mechanism is different from that of animal cells, and is directed to activate a mitogen-activated protein kinase (MAPK) cascade. Many different pathogens activate the same cascade, which suggests that a variety of pathogen-receptor interactions converge at or before the activation of the first MAPK.

18.3 Adaptive Immunity

Key concepts

- Helper T (T$_h$) cells produce signals required by B cells to enable them to differentiate into antibody-producing cells.
- Cytotoxic T cells (CTLs) or killer T cells are responsible for the cell-mediate d response in which fragments of foreign antigens are displayed on the surface of a cell. These fragments are recognized by the TCR expressed on the surface of T cells.
- In TCR recognition, the antigen must be presented in conjunction with a major histocompatibility complex (MHC) molecule.

The specific (adaptive) immune response is defined according to whether it is effected mainly by B cells (antibodies) or T cells. Most naturally occurring antigens, such as those on bacteria and viruses, elicit both specific antibodies and specific effector T cells.

The *antibody response* depends on B lymphocytes, the cells that secrete **antibodies** or **immunoglobulins (Igs)**. Specific recognition and binding of an antigen by the BCR expressed on the surface of B cells is the first step in B cell activation, proliferation, and differentiation to production of large amounts of antibodies specific for the same antigen. *The structure and specificity of the antibody produced by a given B cell are identical to those of the BCR borne on the same B cell.* Binding of antigen by a BCR, and later on by the corresponding antibody, requires the recognition of a small region or structure on the antigen. Antibodies recognize naturally occurring proteins, carbohydrate, or phospholipid antigens, such as structural components of bacteria and viruses (**FIGURE 18.4**) as well as bacterial toxins. Binding of antigen by antibody gives rise to an antigen–antibody complex. This complex then recruits other components of the immune system to mediate biological effector functions.

The antibody response depends on several cellular and soluble elements. B cells need signals provided by T cells to enable them to differentiate to antibody-producing cells. These T cells are called **helper T (T$_h$) cells**, because they help the activation, proliferation, and differentiation of B cells. The antigen–antibody complex triggers the activation of soluble mediators and phagocytic cells (macrophages) that can eventually lead to the disruption of the antibody-bound bacterium or virus. The major soluble mediator pathway is provided

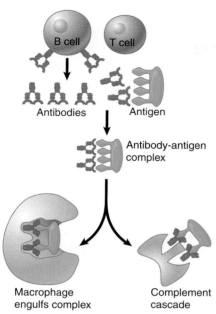

Secretion of antibodies by B cell requires helper T cells

Antibodies

Antigen

Antibody-antigen complex

Macrophage engulfs complex

Complement cascade

FIGURE 18.4 Free antibodies bind to antigens to form antigen–antibody complexes that are removed from the bloodstream by macrophages or that are attacked directly by the activated complement cascade.

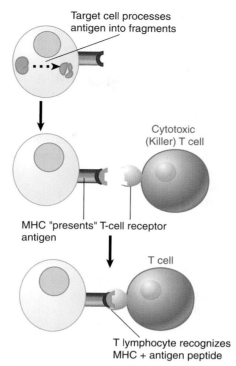

Target cell processes antigen into fragments

Cytotoxic (Killer) T cell

MHC "presents" T-cell receptor antigen

T cell

T lymphocyte recognizes MHC + antigen peptide

FIGURE 18.5 In cell-mediated immunity, cytotoxic T cells use the T cell receptor (TCR) to recognize a peptide fragment of the antigen that is presented on the surface of the target cell by the MHC molecule.

by **complement**, a multiprotein/enzymatic cascade whose name reflects its ability to "complement" the action of the antibody itself. Complement consists of a set of ~20 proteins that function through a cascade of proteolytic actions. Complement is an important element in innate immunity, but it also integrates innate effector functions with adaptive responses. If the target antigen is part of a cell—for example, an infecting bacterium—the action of complement culminates in the lysis of the bacterium. The activation of complement also releases pro-inflammatory soluble mediators and chemotactic mediators; that is, molecules that can attract phagocytic cells, such macrophages and granulocytes, which scavenge the target cells or their products. Alternatively, the antigen–antibody complex may be taken up directly by macrophages (scavenger cells) and destroyed.

The **cell-mediated response** is effected by a class of T lymphocytes called **cytotoxic T cells (CTLs)** or killer T cells (**FIGURE 18.5**). A cell-mediated response is typically elicited by an intracellular parasite, such as a virus that infects the body's own cells. As a result of the viral infection, fragments of foreign (viral) antigens are displayed on the surface of the cell.

These fragments are recognized by the TCR expressed on the surface of T cells. Unlike the BCR, which is dimeric in structure with two identical antigen-binding sites, the TCR has only one antigen-binding site.

A crucial feature of TCR recognition is that *the antigen must be presented in conjunction with a cellular protein that is a member of the* **major histocompatibility complex (MHC)**. The MHC protein possesses a groove on its surface that binds a peptide fragment derived from the foreign antigen. The TCR recognizes the combination of a peptide fragment and MHC protein. Each individual has a characteristic set of MHC proteins. These are important in graft reactions; transplantation of tissue from one individual to another can be rejected because of the difference in MHC proteins between the donor and the recipient, an issue of major medical importance. The requirement that T lymphocytes recognize (foreign) antigen in the context of (self) MHC protein ensures that the cell-mediated response acts only on host cells that have been infected with a foreign antigen. MHC proteins fall into the general clusters of class I and class II.

The immune response has evolved to cope with invading microorganisms by specifically targeting them and eventually leading to their neutralization. Specific target recognition is the prerogative of BCR/Igs and TCRs. A crucial aspect of this function lies in the ability to distinguish "self" from "nonself." Components of the self must *never* be attacked. Foreign targets must be *destroyed*. The property of failing to attack foreign or self-components is referred to as *tolerance*. Loss of self-tolerance results in an attack of the self (autoimmunity) and, eventually, **autoimmune disease**.

An active process of learning of the "self" prevents the emergence or persistence of a lymphocyte repertoire capable of responding to self-components—in most cases proteins—but also glycoproteins and phospholipids. Tolerance arises early in lymphocyte development when B cells and T cells that recognize self-antigens with high affinity are purged by *clonal deletion*, a process that is also referred to as negative selection. In addition to negative selection, a process of positive selection of T cells expressing certain TCRs also occurs. Positive selection is critical for the survival and differentiation of T cells. Positive selection of BCRs that bind with moderate affinity to a surrounding antigen may also be required for the differentiation of B cells. A corollary of tolerance is that it can be difficult to obtain antibodies against proteins that are closely related to those of the organism itself. As a result, it may be difficult to use (for example) mice or rabbits to obtain antibodies against human proteins that have been highly conserved in mammalian evolution. This obstacle is in most cases overcome by the use of immunopotentiators or **adjuvants**, particularly the complete form of Freund's adjuvant, an emulsion in mineral oil of inactivated and dried *Mycobacterium tuberculosis* extract, which allows for the induction of a strong immune response to otherwise weak antigens.

Each of the three groups of structures required for the immune response—BCR, TCR, and MHC—is highly diverse. In a large number of individuals, many variants of each protein exist. A large family of genes codes each protein; in both BCRs and TCRs, the germline-encoded diversity in the population is increased by DNA rearrangements, which occur in both B and T lymphocytes.

BCRs/Igs and TCRs are direct counterparts expressed by B and T lymphocytes, respectively. They are related in structure and their genes are related in organization. The sources of variability are similar. MHC proteins also share some common structural features with antibodies, as do other lymphocyte-specific proteins. The immune system relies on a series of related gene families, indeed a **superfamily** of genes, which may have evolved from some common ancestor encoding a primitive defense element.

18.4 Clonal Selection Amplifies Lymphocytes That Respond to Given Antigens

Key concepts

- Each B cell expresses a single BCR/Ig and each T cell expresses a single TCR.
- A broad repertoire of BCRs/Igs and TCRs exists at any time in an organism.
- Antigen binding to a BCR or TCR triggers the clonal proliferation of that B or T cell.

After an organism has been exposed to an antigen, such as one on an infectious agent, it becomes *immune* to infection by the same agent. Before exposure to a particular antigen, the organism lacks adequate capacity to deal with any toxic effects mediated by or associated with that agent. This ability is acquired through the induction of a specific immune response. After the infection has been defeated, the organism retains the ability to respond rapidly in the event of a reinfection by the same microorganism. This is brought about by **clonal selection** (**FIGURE 18.6**). The repertoire of both B and T lymphocytes comprises a large variety of BCRs or TCRs. *Any individual B lymphocyte expresses one BCR/Ig, which is capable of recognizing specifically only a single antigen; likewise, any individual T lymphocyte expresses only one particular TCR.* In the lymphocyte repertoire, unstimulated B cells and T cells are morphologically indistinguishable. Upon exposure to antigen, though, a B cell whose BCR is able to bind the antigen, or a T cell whose TCR can recognize it, is activated and induced to divide by signaling from the surface of the cell through the BCR/TCR and associated signaling molecules. The induced cell then differentiates into an antibody-producing cell or effector T lymphocyte through morphological changes that include an increase in cell size. This is especially pronounced in B cells.

The initial expansion of a specific B or T cell upon first exposure to antigen underlies the primary immune response. Large numbers of B or T lymphocytes with specificity for the target antigen are produced. Each population represents a clone of the original responding cell. Selected B cells secrete large quantities of antibodies and they may even come to dominate the antibody response. After a successful primary immune response has been mounted, the organism retains the selected B and T cell clones expressing the corresponding BCRs and TCRs for antigen. These *memory cells* respond promptly and vigorously with clonal expansion upon encounter with the same antigen that induced their differentiation, leading to a secondary (or *memory* or *anamnestic*) immune response. Thus, both memory B and T cells are critical elements in the specific resistance to infections after first exposure to a microbial pathogen or vaccine.

The repertoire of B lymphocytes in a mammal comprises at least ~10^{12} specificities (i.e., clones). The T cell repertoire is less expansive. Some B and T cell clones are poorly represented; that is, they consist of a few cells each, as the corresponding antigen had never been encountered before. Others consist of as many as to 10^6 cells, because clonal selection has expanded the specific pool from a progenitor lymphocyte in response to a specific antigen. Naturally occurring antigens are in general relatively large molecules and efficient *immunogens*; that is, inducers of an effective immune response. Small molecules may identify antigenic determinants and can be recognized by antibodies, although owing to their small size, they are not effective in inducing an immune response. They do, however, induce a response when conjugated with a larger carrier molecule, usually a protein such as ovalbumin (OVA), keyhole limpet hemocyanin (KLH), or chicken gamma globulin (CGG). A small molecule that is not immunogenic per se, but can elicit a specific response upon conjugation with a protein carrier, is defined as a **hapten**.

Only a small part of the surface of a macromolecular antigen is actually recognized by any one antibody. The binding site consists of only five or six amino acids. Any given protein may have more than one such binding site, in which case it induces antibodies with specificities for different sites. The site or region inducing a response is called an **antigenic determinant** or **epitope**. In an antigen containing several

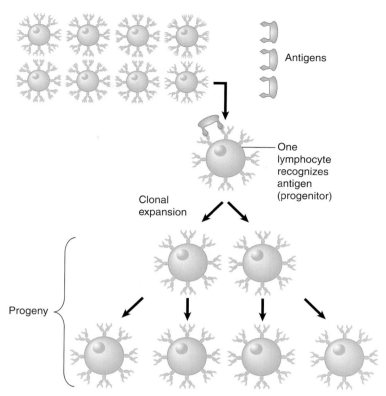

FIGURE 18.6 The B cell and T cell repertoires include BCRs and TCRs with a variety of specificities. Reaction with an antigen leads to clonal expansion of the lymphocyte with the BCR or TCR that can recognize the antigen.

epitopes, some epitopes may be more effective than others in inducing a specific immune response. In fact, they may be so effective that they dominate the response, in that they are the targets of all specific elicited antibodies and/or effector T cells.

The dynamic distribution of mature B and T lymphocytes maximizes their chances to encounter their target antigens. Lymphocytes are peripatetic cells. They develop from immature stem cells in the adult bone marrow. They migrate to the peripheral lymphoid tissues—such as spleen, lymph nodes, Peyer's patches, and tonsils—either directly via the bloodstream (B cells) or through the thymus (T cells). Lymphocytes recirculate between blood and lymph; the process of dispersion ensures that an antigen will be exposed to lymphocytes of all possible specificities. When a lymphocyte encounters an antigen that binds its BCR or TCR, clonal expansion ensues and, under appropriate conditions, a specific immune response is elicited.

18.5 Ig Genes Are Assembled from Discrete DNA Segments in B Lymphocytes

Key concepts

- An Ig is a tetramer of two identical L chains and two identical H chains.
- There are two families of L chains (Igλ and Igκ) and a single family of IgH chains.
- Each chain has an N-terminal variable (V) region and a C-terminal constant (C) region.
- The V region recognizes antigen and the C region mediates the effector response.
- V and C regions are separately encoded by V gene segments and C gene segments.
- A gene coding for an intact Ig chain is generated by somatic recombination of V(D)J genes (variable, diversity, and joining genes in the H chain; variable and joining genes in the L chain) giving raise to V domains, to be expressed together with a given C gene (C domain).

Sophisticated evolutionary mechanisms have evolved to guarantee that the organism is prepared to produce specific antibodies for a broad variety of naturally occurring and man-made components that it has never encountered before. Each antibody is a tetramer consisting of two identical immunoglobulin light **chains (L)** and two identical immunoglobulin **heavy (H) chains** (FIGURE 18.7). In humans, there are two types of L chain (λ and κ) and nine types of H chain. The class is determined by the H chain **constant region (C region)**, which mediates the effector functions. Different

classes of Igs have different effector functions. L chains and H chains share the same general type of organization in which each protein chain consists of two principal domains: the N-terminal **variable region (V region)** and the C-terminal C region. These were defined originally by comparing the amino acid sequences of different Ig chains secreted by monoclonal B cell tumors (plasmacytomas). As the names suggest, the V regions show considerable changes in sequence from one protein to the next, whereas the C regions show substantial homology.

Corresponding regions of the L chains and H chains associate to generate distinct domains in the Ig protein. The V domain is generated by association between a recombined H chain V_H-D-J_H segment and a recombined L chain Vλ-Jλ or Vκ-Jκ segment. *The V domain is responsible for recognizing the antigen.* Production of V domains of different specificities creates the ability to respond to diverse antigens. The total number of V region genes for either L- or H-chain proteins is measured in hundreds. *Thus, the protein displays the maximum versatility in the region responsible for binding the antigen.* The number of C regions is vastly smaller than the number of V regions. The C regions in the subunits of the Ig tetramer associate to generate several individual C domains. The first domain results from association of the single C region of the L chain (C_L) with the C_{H1} part of the H chain C region (C_H). The two copies of this domain complete the arms of the Y-shaped antibody molecule. Association between the C regions of the H chains generates the remaining C domains, which vary in number depending on the type of H chain.

There are many genes coding for V regions, but only a few genes coding for C regions. In this context, *"gene" means a sequence of DNA coding for a discrete part of the final Ig polypeptide* (H or L chain). Thus, **V genes** code for variable regions and **C genes** code for constant regions, although *neither type of gene is expressed as an independent unit.* To construct a unit that can be expressed in the form of a whole L or H chain, a V gene must be joined physically to a C gene. In this system, two "genes" code for one polypeptide.

The sequences coding for L chains and H chains are assembled in the same way: *any one of several V gene segments may be joined to any one of a few C gene segments.* This **somatic recombination** occurs *in the B lymphocyte in which the BCR/antibody is expressed.* The large number of

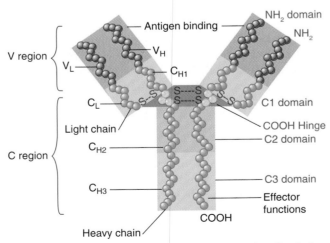

FIGURE 18.7 An antibody (immunoglobulin, or Ig) molecule is a heterodimer consisting of two identical heavy chains and two identical light chains. Schematized here is an IgG1, which comprises an N-terminal variable (V) region and a C-terminal constant (C) region.

available V gene segments is responsible for a major part of the diversity of Igs. Not all diversity is encoded in the genome, though; more is generated by changes that occur during the process of constructing a functional gene.

Essentially the same mechanisms underlie the generation of functional genes encoding the protein chains of the TCR. Two types of receptor are found on T cells—one consisting of two types of chain called α and β, and the other consisting of γ and δ chains. Like the genes coding for Igs, the genes coding for the individual chains in TCRs consist of separate parts, including recombining V(D)J gene segments and C region genes (see *Section 18.20, The TCR Is Related to the BCR*).

The organism does not possess the functional genes for producing a particular BCR/antibody or TCR. It possesses a large number of V gene segments and a smaller number of C gene segments. The subsequent assembly of a productive gene from these parts allows the BCR/TCR to be expressed on both B and T cells so that it is available to react with the antigen. V(D)J DNA rearrangement occurs *before the exposure to antigen*. Productive V(D)J rearrangements are expressed by B cells and T cells as surface BCRs and TCRs, which provide the structural substrate for *selection* of the those clones capable of binding the antigen. The entire process occurs in somatic cells and does not affect the germline; thus, the progeny of the organism does not inherit the specific response to an antigen. The crucial fact about the synthesis of Igs, therefore, is that *the arrangement of V gene segments and C gene segments is different in the cells expressing BCR/Ig or TCR from all other somatic cells or germ cells*.

There are two families of Ig L chains, κ and λ, and one family comprising all the types of H chain. Each family resides on a different chromosome and consists of its own set of both V and C gene segments. This is referred to as the *germline organization*, and is found in the germline and in somatic cells of all lineages other than the immune system. In a cell expressing an antibody, though, each of its chains—one L type (either κ or λ) and one H type—is encoded by a single intact DNA sequence. The recombination event that brings a V gene segment in proximity to, and to be expressed with, a C gene segment creates a productive gene consisting of exons that correspond precisely with the functional domains of the protein. After transcription of the whole DNA sequence, the intronic sequences are removed by RNA splicing.

V(D)J recombination occurs in developing B lymphocytes. A B lymphocyte in general has only one productive rearrangement of L chain gene segments (either κ or λ) and one of H chain gene segments. Likewise, a T lymphocyte productively rearranges an α gene and a β gene, or a δ gene and a γ gene. The BCR/Ig and TCR expressed by any one cell is determined by the particular configuration of V gene segments and C gene segments that has been joined.

The principles by which functional genes are assembled are the same in each family, but there are differences in the details of the organization of both the V and C gene segments, and correspondingly of the recombination reaction between them. In addition to these segments, other short DNA sequences (D segments and J segments) are included in the functional somatic loci.

If any L chain can pair with any H chain, ~ 10^6 different L chains and ~ 10^8 different H chains can pair to generate ~ 10^{14} antibodies. Indeed, a mammal has the ability to generate 10^{14} or more different antibody specificities.

18.6 L Chains Are Assembled by a Single Recombination Event

Key concepts

- A λ chain is assembled through a single recombination event involving a Vλ gene segment and a JλCλ gene segment.
- The Vλ gene segment has a leader exon, intron, and V-coding region.
- The JλCλ gene segment has a short Jλ-coding exon, an intron, and a Cλ-coding region.
- A κ chain is assembled by a single recombination event involving a Vκ gene segment and one of five Jκ segments preceding the Cκ gene.

A λ chain is assembled from two DNA segments (**FIGURE 18.8**). The Vλ gene segment consists of the leader exon (L) separated by a single intron from the V segment. The JλCλ gene segment consists of the Jλ segment separated by a single intron from the Cλ exon.

J is an abbreviation for joining, because the **J segment** identifies the region to which the Vλ segment becomes connected. Thus, the joining reaction does not directly involve Vλ and Cλ gene segments, but occurs via the Jλ segment (Vλ-JλCλ joining). The Jλ segment is short and codes for the last few amino acids of the variable region, as defined by amino acid sequence. In the whole gene generated by recombination,

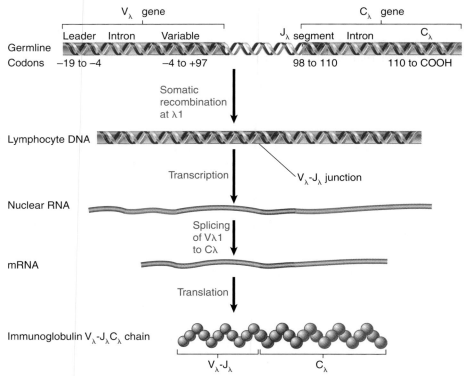

V_λ gene

C_λ gene

Leader Intron Variable

J_λ segment Intron C_λ

Germline

Codons −19 to −4 −4 to +97 98 to 110 110 to COOH

Somatic
recombination
at λ1

Lymphocyte DNA

Transcription

V_λ-J_λ junction

Nuclear RNA

Splicing
of Vλ1
to Cλ

mRNA

Translation

Immunoglobulin V_λ-$J_\lambda C_\lambda$ chain

V_λ-J_λ

C_λ

FIGURE 18.8 The C_λ gene segment is preceded by a Jλ segment, so that V_λ-J_λ recombination generates a productive V_λ-$J_\lambda C_\lambda$.

the Vλ-Jλ segment constitutes a single exon coding for the entire variable region.

A κ chain is also assembled from two DNA segments (**FIGURE 18.9**). There are, however, differences in the organization of the Cκ locus as compared to the Cλ locus. A group of five Jκ segments is spread over a region of 500 to 700 bp, separated by an intron of 2 to 3 Kb from the Cκ exon. In the mouse, the central Jκ segment is nonfunctional (ψJ3). A Vκ segment (which contains a leader exon like Vλ) may be joined to any one of the Jκ segments. Whichever Jκ segment is used, it becomes the terminal part of the intact variable exon. Any Jκ segment upstream of the recombining Jκ segment is lost (Jκ1 has been lost in the figure); any Jκ segment downstream of the recombining Jκ segment is treated as part of the intron between the V and C exons (Jκ3 is included in the intron that is spliced out in the figure).

All functional J segments possess a sequence signal at their 5′ boundary that makes it possible to recombine with the V segment; they also possess a signal at the 3′ boundary that can be used for splicing to the C exon. Whichever J segment is recognized in DNA V-J joining, it will use its splicing signal in RNA processing.

18.7 H Chains Are Assembled by Two Sequential Recombination Events

Key concepts

- The units for H chain recombination are a V_H gene, a D segment, and a $J_H C_H$ gene segment.
- The first recombination joins D to $J_H C_H$.
- The second recombination joins V_H to D-$J_H C_H$ to yield V_H-D-$J_H C_H$.
- The C_H segment consists of four exons.

The assembly of a complete H chain involves an additional segment. The **D segment** (for 'diversity') was discovered by the presence in the protein of an extra two to thirteen amino acids between the sequences coded by the V_H and J_H segments. An array of D segments lies on the chromosome between the cluster of V_H segments and that of J_H segments.

V_H-D-J_H joining takes place in two stages (**FIGURE 18.10**). First, one of the D segments recombines with a J_H segment; second, a V_H segment recombines with the already recombined D-J_H segment. The resulting V_H-D-J_H DNA sequence is then expressed with the nearest

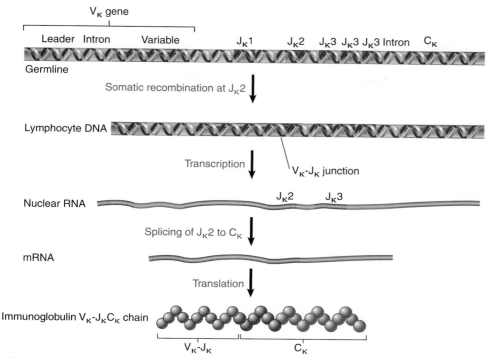

FIGURE 18.9 The C_κ gene segment is preceded by multiple Jκ segments in the germ line. Vκ-Jκ joining may recognize any one of the J segments, which is then spliced to the C gene segment during RNA processing.

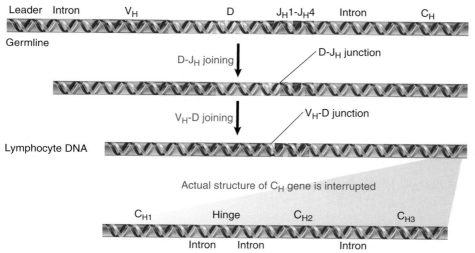

FIGURE 18.10 Heavy genes are assembled by sequential recombination events. First a D_H segment is recombined with a J_H segment, and then a V_H gene segment is recombined with the D_H segment.

downstream C_H gene, which consists of a cluster of four exons. (The use of different C_H genes is discussed in *Section 18.14, Class Switching Is Effected by DNA Recombination.*) The D segments are organized in a tandem array. The human locus comprises ~30 D segments, followed by a cluster of six J_H gene segments. Mechanisms yet to be identified ensure that the *same* D segment is involved in the D-J_H recombination and related V_H-D-J_H recombination.

The structure of recombined V(D)J segments is similar in organization in the H chain and λ and κ chain loci. The leader exon codes for the signal sequence, which is involved in membrane attachment, and the second exon codes for the major part of the variable region itself, which is about 100 codons long. The remainder of the variable region is provided by the D segment (in H chain locus only) and by a J segment (in all three loci).

The structure of the C region is different in different H and L chains. In both κ and λ chains, the C region is coded by a single exon, which becomes the third exon of the recombined Vκ-JκCκ or Vλ-JλCλ gene. In H chains, the C_H region is coded by multiple and discrete exons, coding for four regions: C_H1, $C_H$hinge, C_H2, and C_H3 (IgG) or C_H1, C_H2, C_H3, and C_H4 (IgM). Each C_H exon is ~100 codons long, with the hinge exon being shorter; the introns are ~300 bp each.

18.8 Recombination Generates Extensive Diversity

Key concepts

- The human IgH locus can generate in excess of 10^8 V_H-D-J_H sequences.
- Recombined V_H-D-$J_H C_H$ can be paired with in excess of 10^6 recombined Vκ-JκCκ or Vλ-JλCλ chains.

A census of the available V, D, J, and C gene segments provides a measure of the diversity that can be accommodated by the variety of the coding regions carried in the germline. In both the IgH and L chain loci, many V gene segments are linked to a much smaller number of C gene segments.

The human λ locus (chromosome 22) has four Cλ genes, each preceded by its own Jλ segment (**FIGURE 18.11**). The mouse λ locus (chromosome 16) is much less diverse. The main difference is that in a mouse there are only two V_λ gene segments, each of which is linked to two JλCλ regions. One of the C_λ gene segments is inactive. This configuration suggests that the mouse suffered in its evolutionary history a severe deletion of most of its germline V_λ gene segments.

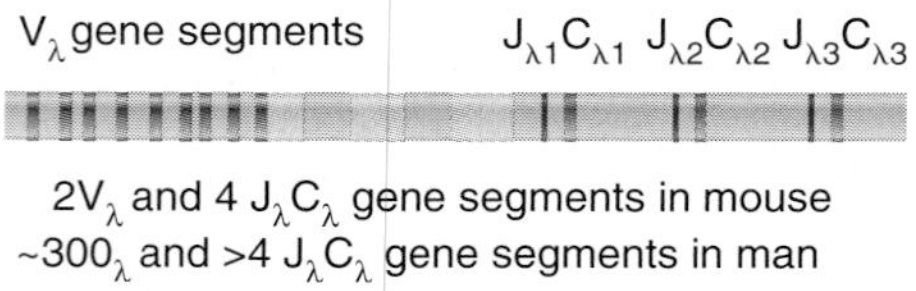

Both the human κ locus (chromosome 2) and the mouse κ locus (chromosome 6) have only one Cκ gene segment, preceded by six Jκ gene segments (one of them inactive) (**FIGURE 18.12**). The Vκ gene segments occupy a large cluster on the chromosome, upstream of the Cκ region. The human cluster has two regions. Just upstream of the C_κ gene segment, a region of 600 Kb contains the J_κ segments and 40 V_κ gene segments. A gap of 800 Kb separates this region from another cluster of 36 Vκ gene gene segments.

The V_H, V_λ, and V_κ gene segments are segregated into families. A family comprises members that share more than 80% amino acid identity. In humans, the V_H locus comprises six V_H families (V1 through V6). V3 and V4 are the largest families, each with more than 10 functional members; V6 is the smallest family, consisting of one functional member only. In mice, the Vκ locus comprises about 18 V_κ families, which vary in size from 2 to 100 members. Like other families of related genes, related V gene segments form subclusters, which were generated by duplication and divergence of individual ancestral members. Many of the V segments are inactive pseudogenes. Although nonfunctional, some of these may function as donor of partial V sequences in secondary pseudorearrangements.

A given lymphocyte expresses *either* a κ *or* a λ chain to be paired with a V_H-D-$J_H C_H$ chain. In humans, ~60% of B cells express κ chains and ~40% express λ. In mice, 95% of B cells express a κ chain, presumably because of the reduced number of λ gene segments available.

The single IgH chain locus on human chromosome 14 consists of multiple discrete segments (**FIGURE 18.13**). The furthest 3' member of the V_H cluster is separated by only 20 Kb from the first D segment. The D segments are spread over ~50 Kb, followed by the cluster of six J_H segments. Over the next 220 Kb lie all the C_H genes. In addition to the nine functional C_H genes, there are two pseudogenes. The human IgH locus organization suggests that a Cγ gene was duplicated to give the subcluster of Cγ-Cγ-Cε-Cα, after which the entire subcluster was then tandem duplicated. In the mouse IgH locus (chromosome 12), there are more V_H gene segments, fewer D and J_H segments, and eight (instead of nine) C_H genes.

The human IgH locus alone can produce more than 10^4 V_H-D-J_H sequences by combining 51 V genes, 30 D segments, and 6 J_H

V_λ gene segments $J_{\lambda1}C_{\lambda1}$ $J_{\lambda2}C_{\lambda2}$ $J_{\lambda3}C_{\lambda3}$

2V_λ and 4 $J_\lambda C_\lambda$ gene segments in mouse
~300$_\lambda$ and >4 $J_\lambda C_\lambda$ gene segments in man

FIGURE 18.11 The lambda family consists of V_λ gene segments and a small number of J_λ-C_λ gene segments.

36 V_κ 40 V_κ $J_{\kappa1}$-$J_{\kappa5}$ C_κ

FIGURE 18.12 The human and mouse Igκ families consist of V_κ gene segments and five functional J_κ segments linked to a single C_κ gene segment. V_κ genes include nonfunctional pseudogenes.

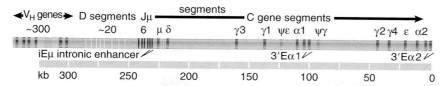

FIGURE 18.13 A single gene cluster in humans contains all the information for the IgH chain. Depicted is a schematic map of the human IgH chain locus.

segments. This degree of diversity is further compounded by the imprecision in the V_H-D-J_H joinings, the insertion of unencoded nucleotide (N) additions and use, particularly in humans, of multiple D-D segments. By combining any one of more 50 V gene segments with any one of 5 Jκ segments, the human κ locus has the potential to produce 300 different Vκ–Jκ segments. These, however, are conservative estimates as more diversity is introduced by insertion of untemplated N nucleotides, albeit at lower frequency than in V_H-D-J_H. Further diversification in individual genes during or after V_H-D-J_H, Vκ–Jκ, and Vλ–Jλ recombination occurs by somatic changes (see *Section 18.16, SHM Generates Additional Diversity in Mice and Humans*).

18.9 Immune Recombination Uses Two Types of Consensus Sequence

Key concepts

- The consensus sequence used for recombination is a heptamer separated by either 12 or 23 base pairs from a nonamer.
- Recombination occurs between two consensus sequences that have different spacers.

The recombination of Igκ, Igλ, and IgH chain genes involves the same mechanism, although the number and nature of recombining elements are different. The same consensus sequences are found at the boundaries of all germline segments that participate in the joining reactions. Each consensus sequence consists of a *heptamer* (7 bp sequence) separated by either 12 or 23 bp from a *nonamer* (9 bp sequence). These sequences are referred to as **recombination signal sequences (RSS)** **(FIGURE 18.14)**. In the κ locus, each Vκ gene segment is followed by an RSS sequence with a 12-bp spacer. Each Jκ segment is preceded by an RSS with a 23-bp spacer. The Vκ and J_κ RSS are inverted in orientation. In the λ locus, each Vλ gene segment is followed by an RSS with a

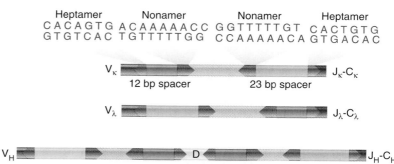

FIGURE 18.14 RSS sequences are present in inverted orientation at each pair of recombining sites. One member of each pair has a 12-bp spacer between its components; the other has a 23-bp spacer.

23-bp spacer; each Jλ gene segment is preceded by an RSS with a 12-bp spacer. The rule that governs the joining reaction is that *an RSS with one type of spacer can be joined only to an RSS with the other type of spacer*. This is referred to as the "12/23 rule."

The RSSs at the V and J segments can lie in either order; thus the different spacers do not impart any directional information, but instead serve to prevent one V or J gene segment from recombining with another of the same. This concept is borne out by the structure of the components of the IgH locus segments. Each V_H gene segment is followed by an RSS with a 23-bp spacer. The D segments are flanked on either side by RSSs with 12-bp spacers. The J_H segments are preceded by RSSs with 23-bp spacers. Thus, a V_H segment must recombine with a D segment, and a D segment must recombine with J_H segment. A V_H gene segment cannot be joined directly to a J_H segment, because both possess the same type of RSS. The spacer between the components of the RSS corresponds to close to one (12 bp) or two turns (23 bp) of the double helix. This may reflect geometric constraints in the recombination reaction. The recombination protein(s) may approach the DNA from one side, in the same way that RNA polymerase and repressors approach recognition elements, such as promoters and operators.

18.10 V(D)J DNA Recombination Occurs by Deletion or Inversion

Key concepts

- Recombination occurs by double-strand DNA breaks (DSBs) at the heptamers of two RSSs.
- The signal ends of the linear DNA excised between two DSBs are joined to generate a DNA circle.
- The coding ends are covalently ligated to join V_L to $J_L C_L$ (L chain), or D to $J_H C_H$ and V_H to $D-J_H C_H$ (H chain).
- If the recombining genes lie in an inverted instead of a direct orientation, the intervening DNA is inverted, but retained, instead of being excised as a circle.

Recombination of the components of Ig genes is accomplished by a physical rearrangement of different DNA segments, which involves DNA breakage and ligation. In the H chain locus, there are two recombination events: first $D-J_H$, then V_H-D-J_H. DNA breakage and ligation occur as separate reactions. A double-strand DNA break (DSB) is made at the heptamers that lie at the ends of the coding units. This releases the DNA between the V and the JC gene segments; the cleaved termini of this fragment are called **signal ends**. The cleaved termini of the V and J-C loci are called **coding ends**. The two coding ends are covalently linked to form a coding V-C joint.

Most V_L and $J_L C_L$ gene segments are organized in the same orientation. As a result, the cleavage at each RSS releases the intervening DNA as a linear fragment, which, when religated at the signal ends, gives rise to a circle (**FIGURE 18.15**). Deletion to release an excised DNA circle is the predominant mode of recombination at the Ig and TCR loci.

In some cases, the Vλ gene segment in the germline configuration is inverted in orientation on the chromosome relative to the JλCλ DNA. In such a case, breakage and ligation inverts the intervening DNA instead of deleting it. The outcomes of deletion versus inversion in terms of the coding sequence are the same. Recombination with an inverted V gene segment, however, makes it *necessary* for the signal ends to be joined or a DSB in the locus is generated (Figure 18.15). Recombination by inversion occurs also in some cases in the κ locus, the IgH locus, and the TCR locus.

18.11 Allelic Exclusion Is Triggered by Productive Rearrangements

Key concepts

- V(D)J gene rearrangement is productive if it leads to expression of a protein.
- A productive V(D)J gene rearrangement prevents any further rearrangement of the same kind from occurring, whereas a nonproductive rearrangement does not.
- Allelic exclusion applies separately to L chains (only one κ or λ may be productively rearranged) and to H chains (one H chain is productively rearranged).

Each B cell expresses a single κ or λ chain and a single type (isotype) of IgH chain, because only a single productive rearrangement of each type occurs in a given lymphocyte in order to express only one L and one H chain. Each event involves the genes of only *one* of the homologous chromosomes. Thus, *the alleles on the other chromosome are not expressed in the same cell*. This phenomenon is called **allelic exclusion**.

The occurrence of allelic exclusion complicates the analysis of somatic recombination. A probe reacting with a region that has rearranged on one homolog will also detect the allelic sequences on the other homolog. This means that one has to analyze V(D)J configuration on the two chromosomes in order to understand the V(D)J rearrangement history of a given B cell.

Two different configurations of Ig locus can exist in B cells:

- A DNA probe specific for the expressed V gene may reveal one rearranged copy and one germline copy, indicating that joining has occurred on one chromosome, whereas the other chromosome has remained unaltered.
- A DNA probe specific for the expressed V gene reveals two different rearranged patterns, indicating that both chromosomes underwent independent V(D)J rearrangement events involving the same gene.

In general, in those cases in which both chromosomes in a B cell lost the germline configuration, only one of them underwent through a **productive rearrangement** to express a functional IgH or L chain. The other suffered

(a) RAG binding and nicking

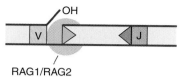

RAG1/RAG2

(b) Synapsis

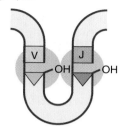

(c) Hairpin formation and cleavage

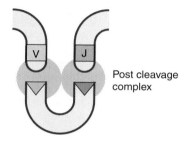

Post cleavage complex

(d) Hairpin opening and joining

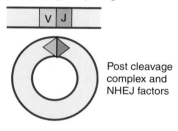

Post cleavage complex and NHEJ factors

FIGURE 18.15 Breakage and recombination at RSSs generate VJC sequences. A generic V-J rearrangement is shown for simplicity. In most cases, the V and J segments undergoing recombination are arranged in the same transcriptional orientation and rearrangement occurs by deletion of the intervening DNA, as shown. Less commonly, V and J segments undergoing recombination are arranged in opposite transcriptional directions and rearrangement occurs by inversion (not shown). Adapted from D. B. Roth, *Nat. Rev. Immunol.* 3 (2003): 656–666.

a **nonproductive rearrangement**. This can occur in different ways, but in each case the gene sequence cannot be expressed as an Ig chain. The rearrangement may be incomplete (e.g., because D-J_H joining has occurred but V_H-D-J_H joining has not followed), or it may be aberrant, with the process completed but failing to generate a gene that encodes a functional protein.

The coexistence of productive and nonproductive rearrangements suggests the existence of a feedback mechanism controlling the recombination process (**FIGURE 18.16**). Let's assume that a B cell starts with two IgH chain loci in the (unrearranged) germline configuration Ig^0. Either locus may recombine V_H, D, and $J_H C_H$ to generate a productive gene (IgH$^+$) or a nonproductive gene (IgH$^-$) rearrangement. If the first rearrangement is productive, the expression of a functional IgH chain provides an inhibitory signal to the B cell to prevent rearrangement of the other IgH allele. As a result, the configuration of this B cell with respect to the IgH locus will be IgH$^+$/Ig^0. If the first rearrangement is nonproductive, it will result in a configuration Ig$^-$/Ig^0. The lack of an expressed IgH chain will not provide an inhibitory (negative) feedback for rearrangement of the remaining germline allele. If this undergoes a productive rearrangement, the B cell will have the configuration Ig$^+$/Ig$^-$. Two successive nonproductive rearrangements will result in an Ig$^-$/Ig$^-$

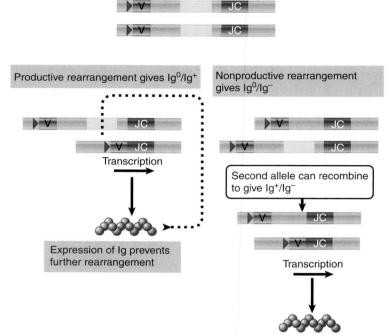

Germline genes (Ig⁰/Ig⁰)

Productive rearrangement gives Ig⁰/Ig⁺

Transcription

Expression of Ig prevents further rearrangement

Nonproductive rearrangement gives Ig⁰/Ig⁻

Second allele can recombine to give Ig⁺/Ig⁻

Transcription

FIGURE 18.16 A successful rearrangement to produce an active light (depicted) or heavy chain suppresses further rearrangements of the same type, resulting in allelic exclusion.

configuration. In some cases, a B cell in an Ig⁻/Ig⁻ configuration can attempt an atypical rearrangement utilizing cryptic RSSs embedded in the coding DNA of a V gene. Indeed, certain Ig locus DNA configurations found in B cells can only be explained as having been generated by successive rearrangements.

Thus, allelic exclusion is caused by the suppression of further rearrangements as soon as a productive IgH or L chain rearrangement is achieved. Allelic exclusion *in vivo* is exemplified by the creation of transgenic mice in which a rearranged V_H-D-$J_H C_H$, $V\kappa$–$J\kappa C\kappa$, or $V\lambda$–$J\lambda C\lambda$ DNA has been inserted into the Ig locus. Expression of the transgene in B cells suppresses the corresponding rearrangement of endogenous V(D)J genes. Allelic exclusion is independent for the IgH and L chain loci. IgH chain genes usually rearrange first. Allelic exclusion for L chains applies equally to both families (cells may express *either* productive κ or λ L chains). In most cases, a B cell rearranges its κ locus first. It then tries to rearrange the λ locus only if both κ rearrangement attempts are unsuccessful.

The same consensus sequences and the same V(D)J recombinase are involved in the recombination reactions at IgH, Igκ, and Igλ loci, and yet the three loci rearrange in a sequential order. It is unclear why the IgH

rearrangement precedes IgL rearrangement and why Igκ precedes Igλ. The DNA in the different loci may become accessible to the enzyme(s) effecting the rearrangement at different times, possibly reflecting each locus transcription status. Transcription starts before rearrangement, although some Ig locus mRNA, such as I_H-C_H (germline I_H-C_H transcripts), has no coding function. Transcription events may change the structure of chromatin, making the consensus sequences for recombination available to the enzyme effecting the rearrangement.

18.12 RAG1/RAG2 Catalyze Breakage and Religation of V(D)J Gene Segments

Key concepts

- The RAG proteins are necessary and sufficient for the cleavage reaction.
- RAG1 recognizes the nonamer consensus sequences for recombination. RAG2 binds to RAG1 and cleaves DNA at the heptamer. The reaction resembles the topoisomerase-like resolution reaction that occurs in transposition.
- The reaction proceeds through a hairpin intermediate at the coding end; opening of the hairpin is responsible for insertion of extra bases (P nucleotides) in the recombined gene.
- Terminal deoxynucleotidyl transferase (TdT) inserts additional unencoded N nucleotides at the V(D)J junctions.
- The DSBs at the coding joints are repaired by the same mechanism that has generated the whole V(D)J sequence.

The proteins RAG1 and RAG2 are necessary and sufficient for DNA cleavage in V(D)J recombination. They are encoded by two genes, separated by <10 Kb on the chromosome: *RAG1* and *RAG2*. *RAG1/RAG2* gene transfection into fibroblasts causes a suitable DNA substrate to undergo the V(D)J recombination. Mice that lack *RAG1* or *RAG2* are unable to recombine their BCR/Ig and TCR, and as a result abort B lymphocyte and T lymphocyte development. RAG1/RAG2 proteins together undertake the catalytic reactions of cleaving and rejoining DNA, and also provide a structural framework within which the whole recombination reaction occurs.

RAG1 recognizes the RSS (heptamer/nonamer signals with the appropriate 12/23 spacing) and recruits RAG2 to the complex. The nonamer provides the site for initial recognition, and the heptamer directs the site of cleavage. The complex nicks one strand at each

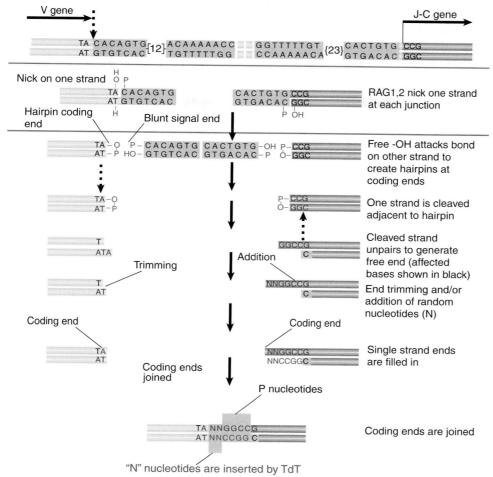

FIGURE 18.17 Processing of coding ends introduces variability at V_κ-J_κ, V_λ-J_λ, or V_H-D-J_H junctions. Depicted is a V_κ-J_κ junction.

junction (**FIGURE 18.17**). The nick has 3′–OH and 5′–P ends. The free 3′–OH end then attacks the phosphate bond at the corresponding position *in the other strand of the duplex*. This creates a hairpin at the coding end, in which the 3′ end of one strand is covalently linked to the 5′ end of the other strand, and leaves a blunt double-strand break at the signal end.

This second cleavage is a transesterification reaction in which bond energies are conserved. It resembles the topoisomerase-like reactions catalyzed by the resolvase proteins of bacterial transposons (see *Section 17.3, Transposition Occurs by Both Replicative and Nonreplicative Mechanisms*). The parallel with these reactions is further supported by a homology between RAG1 and bacterial invertase proteins, which invert specific segments of DNA by similar recombination reactions. In fact, the RAG proteins can insert a donor DNA whose free ends consist of the appropriate signal sequences (heptamer-12/23 spacer-nonamer) into an unrelated target DNA in an *in vitro* transposition reaction, suggesting that somatic recombination of immune genes evolved from an ancestral transposon.

The hairpins at the coding ends provide the substrate for the next stage of the reaction. The Ku70:Ku80 heterodimer binds to the DNA ends and a protein called Artemis opens the hairpins. The joining reaction that works on the coding end uses the same pathway of nonhomologous end-joining (NHEJ) that repairs DSBs in all cells (see *Section 16.11, Nonhomologous End-Joining Also Repairs Double-Strand Breaks*). If a single-strand break is introduced into one strand close to the hairpin, an unpairing reaction at the end generates a single-stranded protrusion. Synthesis of a complement to the exposed single strand then converts the coding end to an extended duplex. This reaction explains the introduction of **P nucleotides** at coding ends. P nucleotides are a few extra base pairs related to, but reversed in orientation from, the original coding end.

In addition to P nucleotides, some extra bases called **N nucleotides** can also be inserted between the coding ends in an untemplated and random fashion. Their insertion occurs via the activity of the enzyme **terminal deoxynucleotidyl transferase (TdT)** which, like RAG1/RAG2, is expressed at those stages of B and T lymphocyte development when V(D)J recombination occurs at a free 3' coding end generated during the joining process.

The initial stages of the reaction were identified by isolating intermediates from lymphocytes of mice with a **severe combined immunodeficiency (SCID)** mutation, which results in a much-reduced level of BCR/Ig and TCR V(D)J gene recombination. *SCID* mice accumulate DSBs at Ig V gene segment coding ends and cannot complete the V(D)J joining reaction. This *SCID* mutation displays a defective DNA-dependent protein kinase (DNA-PK). This kinase is recruited to DNA by the Ku70:Ku80 heterodimer, which binds to the broken DNA ends. DNA-PK$_{cs}$ (DNA-PK catalytic subunit) phosphorylates and thereby activates Artemis, which in turn nicks the hairpin ends; Artemis also possesses exonuclease and endonuclease activities that function in the NHEJ pathway. The actual ligation is undertaken by DNA ligase IV and also requires XRCC4. Mutations in Ku proteins, XRCC4, or DNA ligase IV are found in patients with congenital diseases involving deficiencies in DNA repair that result in increased sensitivity to radiation. The free (signal) 5'-phosphorylated blunt ends at the heptamer sequences of the intervening DNA (which is looped out by the V(D)J recombinations) also bind Ku70:Ku80. Without further modification, a complex of DNA ligase IV:XRCC4 joins the two signal ends to form the signal joint.

Thus, changes in DNA sequence during V(D)J recombination are a consequence of the enzymatic mechanisms involved in breaking and rejoining the DNA. In IgH chain V_H-D-J_H recombination, base pairs are lost and/or N nucleotides inserted at the V_H-D or D-J_H junctions. Deletions also occur in Vκ-Jκ and Vλ-Jλ joining, but N insertions at these joints are less frequent than in V_H-D or D-J_H junctions. The changes in sequence affect the amino acid coded at V_H-D-J_H junctions or at V_L-J_L junctions.

The above mechanisms will ensure that most coding joints will display a different sequence from that predicted as a result of direct joining of the coding ends of the V, D, and J segments involved in each recombination. Variations in the sequence of V_L-J_L junctions make it possible for different amino acid residues to be encoded here, generating diverse structures at the site that contacts antigen. The amino acid at position 96 is created by Vκ-Jκ and Vλ-Jλ recombination. It forms part of the antigen-binding site and also is involved in making contacts between the L chains and the H chains. Thus, maximum diversity is generated at the site that contacts the target antigen.

Changes in the number of base pairs at coding joints affect the reading frame. V_L-J_L recombination appears to be random with regard to reading frame, so that only one-third of the joined sequences retain the proper frame of reading through the junctions. If a Vκ-Jκ or Vλ-Jλ recombination occurs so that the J_L segment is out of frame, translation is terminated prematurely by a nonsense codon in the incorrect frame. This may be the price a B cell pays for being able to generate maximal diversity of the expressed Vκ-Jκ and Vλ-Jλ sequences. Even greater diversity is generated by recombinations that involve the V_H, D, and J_H gene segments of the Ig H chain, mainly due to random and variable "chopping off" of D and J_H DNA as well as random and variable N nucleotide insertions. Nonproductive recombinations are generated by a joining that places V_H out of frame with the rearranged D-J_H gene segment.

Germline (unrearranged) V gene segments about to undergo recombination are transcribed, albeit at a moderate level. Once V(D)J gene segments are productively recombined, the resulting sequence is transcribed at a higher rate. The sequence upstream of a V gene segment is not altered by the joining reaction, though, and as a result *the promoter is conserved in unrearranged, nonproductively rearranged, and productively rearranged V genes*. The V promoter lies upstream of every V gene segment but is only moderately active when in germline configuration. Its activation is significantly enhanced by its downstream relocation closer to the C region after V(D)J rearrangement, suggesting that the V promoter activation depends on downstream *cis*-elements. Indeed, an enhancer located within or downstream of the V, D, and J gene clusters significantly enhances the activation of V promoter. This enhancer is referred to as intronic enhancer (iEμ in the H chain and iEκ in the κ chain). It is tissue-specific, being active only in B cells (**FIGURE 18.18**).

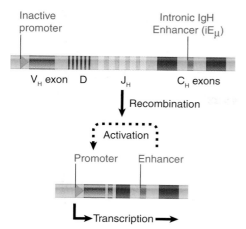

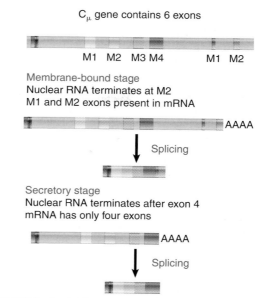

FIGURE 18.18 A V gene promoter is inactive until recombination brings it into the proximity (and therefore under the influence) of the iEμ enhancer that lies downstream of the Sμ region and upstream of the Cμ exon cluster. The enhancer is active only in B lymphocytes.

FIGURE 18.19 The 3' end of each C_H (Cμ, Cγ, Cα, or Cδ) gene cluster controls the use of splicing junctions so that alternative forms (membrane or secretory) of the heavy gene are expressed.

18.13 Early IgH Chain Expression Is Modulated by RNA Processing

Key concept

- All B lymphocytes newly emerging from the bone marrow express the membrane-bound monomeric form of IgM (μm). A change in RNA splicing causes mm to be replaced by the secreted form (μs) after a mature B cell is activated and begins differentiation to antibody-producing cells in the periphery.

As a stem cell differentiates to a Pro-B cell and subsequently a Pre-B cell, a surrogate L chain (λ-Vpre-B) is expressed and paired with the Cμ chain to give rise to a monomeric IgM molecule ($L_2\mu_2$). This form of IgM is expressed on the surface of the B cell and contains the $C\mu_m$ version of the constant region ('m' indicates that IgM is located in the membrane); it is referred to as Pre-BCR. The membrane location may be related to the need to initiate cell proliferation in response to the initial recognition of a surrounding antigen by the Pre-BCR. After a few divisions, the Pre-B cell rearranges the Ig L chain locus to express a full-fledged BCR consisting of two identical V_H-D-J_HCμ chains as paired with two identical Vκ-JκCκ or Vλ-JλCλ chains.

When, after encounter with antigen, the B lymphocyte differentiates further into an antibody-producing cell and plasma cell, the $C\mu_s$ (secreted) version of the C_H region is expressed. The IgM actually is secreted as a pentamer IgM_5J, in which J is a joining polypeptide (with no connection to the J region gene), which forms disulfide linkages with μ chains. Secretion of the IgM in pentameric form characterizes the early stage of an antibody response.

The $C\mu_m$ and $C\mu_s$ versions of the Igμ chain differ only at the C-terminal end. The μ_m chain ends with a hydrophobic sequence that probably secures it in the membrane. This sequence is replaced by a shorter hydrophilic sequence in μ_s, which allows the Igμ chain to pass through the membrane. The change of C-terminus is accomplished by an alternative splicing event, which is controlled by the 3' end of the nuclear RNA (FIGURE 18.19).

At the membrane-bound stage, the RNA terminates after exon M2, and the C_H region is generated by splicing together six exons. The first four exons code for the four domains of the C_H region. The last two exons, M1 and M2, code for the 41-residue hydrophobic C_H-terminal region and its nontranslated-tail. The 5' splice junction within exon 4 is connected to the 3' splice junction at the beginning of M1. At the secreted stage, the nuclear RNA terminates after exon 4. The 5' splice junction within this exon that had been linked to M1 in the membrane form is ignored. This allows the exon to extend for an additional 20 codons. A similar transition from membrane to secreted forms occurs for the other C_H regions ($C\gamma$, $C\alpha$, and $C\varepsilon$), suggesting that the mechanism for expression of membrane and secreted forms of all Ig classes is the same, as further indicated by the conservation of exon structures.

18.14 Class Switching Is Effected by DNA Recombination

Key concepts

- Ig's comprise five classes according to the type of C_H chain.
- Class switching is effected by a recombination between S regions that deletes the DNA between the upstream C_H region gene cluster and the downstream C_H region gene cluster.

The *class* of Ig is defined by the type of C_H region. There are five C_H classes: IgM, IgD, IgG, IgA, and IgE (FIGURE 18.20). IgM is the first Ig to be produced by a differentiating B cell and activates complement efficiently. IgD is subsequently expressed when the mature B-cell exits the bone marrow. IgG comprises four subclasses (IgG1, IgG2, IgG3, and IgG4 in humans and IgG1, IgG2a, IgG2b, and IgG3 in mice), and is the most abundant Ig in the circulation. Unlike IgM, which is confined to circulation, IgG passes into the extravascular spaces. IgA is abundant on mucosal surfaces and on secretions in the respiratory tract and the intestine. IgE is associated with the allergic response and with defense against parasites.

B lymphocytes start their "productive" life expressing IgM and IgD on their surfaces. A B lymphocyte expresses only a single class of Ig at any one time, but after encountering antigen, a B cell undergoes activation, proliferation, and differentiation from an IgM- to an IgG-, IgA-, or IgE-producing cell. This process occurs in peripheral lymphoid organs, such as the lymph nodes and spleen, and is referred to as **class switching**. Class switching is induced in a T-dependent fashion through engagement of surface B cell CD40 by CD154 expressed on the surface of T_h cells and exposure to T cell-derived cytokines, such as IL-4 (IgG and IgE) and TGF-β (IgA), or in a T-independent fashion through, for instance, engagement of TLRs on B cells by conserved molecules on bacteria or viruses (MAMPs), such as bacterial lipolysaccharides, CpG, or viral double-strand RNA.

Class switching is effected by *class switch DNA recombination (CSR)* and involves only C_H genes; the V_H-D-J_H segment originally expressed as part of an IgM continues to be expressed in a new context (IgG, IgA, or IgE). A given recombined V_H-D-J_H segment can be expressed sequentially in combination with more than one C_H gene region. The same $V\kappa$–$J\kappa C\kappa$ or $V\lambda$–$J\lambda C\lambda$ chain continues to be expressed throughout the lineage of the cell. CSR, therefore, allows the type of biological effector response (mediated by the C_H region) to change while maintaining the same specificity of antigen recognition (mediated by the combination of V_H-D-J_H and $V\kappa$–$J\kappa$ or V_H-D-J_H and $V\lambda$–$J\lambda$ regions).

CSR involves a mechanism different from that effecting V(D)J recombination and is active later in B cell development, at the stage of B cell differentiation in peripheral lymphoid organs. B cells that underwent CSR show deletions of the DNA encompassing C_μ and all the other C_H gene segments preceding the expressed C_H gene. CSR entails a recombination that brings a (new) downstream C_H gene segment into juxtaposition with the expressed V_H-D-J_H unit. The sequences of switched V_H-D-$J_H C_H$ units show that the sites of switching (i.e., DSBs) lie upstream of each C_H gene. The switching sites segregate within specialized DNA sequences, the **S regions**. The S regions lie within the introns

Type	IgM	IgD	IgG	IgA	IgE
C_H chain	μ	δ	γ	α	ε
Structure	$(\mu_2 L_2)_5 J$	$\delta_2 L_2$	$\gamma_2 L_2$	$(\alpha_2 L_2)_2 J$	$\varepsilon_2 L_2$
Proportion in circulating blood	5%	1%	80%	14%	<1%
Effector function	Activates complement	Development of tolerance (?)	Activates complement	Found in secretions	Allergic response

FIGURE 18.20 Immunoglobulin type and functions are determined by the H chain. J is a joining protein in IgM, unrelated to J (joining) gene segments. IgM exists mainly as a pentamer (i.e., 5 IgM $\mu_2 L_2$ tetramers) and IgA as a dimer. IgD, IgG, and IgE exist as single $H_2 L_2$ tetramers.

that precede the C_H coding regions—all C_H gene regions have S regions upstream of the coding sequences. As a result, CSR does not alter the translational IgH reading frame. In a first CSR event, such as from $C\mu$ to $C\gamma1$, expression of $C\mu$ is succeeded by expression of $C\gamma1$. The $C\gamma1$ gene segment is brought into its new functional location by recombination between $S\mu$ and $S\gamma1$. The $S\mu$ site lies between V_H-D-J_H and the $C\mu$ gene segment. The $S\gamma1$ site lies upstream of the $C\gamma1$ gene. The DNA sequence between the two S region DSBs is excised as circular DNA that is transiently transcribed. This deletional event imposes a restriction on the IgH locus: *Once a CSR event has occurred, a B cell cannot express any C_H gene segment that used to lie between the first C_H and the new C_H gene segment.* For instance, human B cells expressing $C\gamma1$ cannot give rise to cells expressing $C\gamma3$, which has been deleted. They can, however, undergo CSR to any C_H gene segment *downstream* of the expressed $C\gamma1$ gene, such as $C\alpha$. This is accomplished by recombination between $S\mu S\gamma1$ (generated by the original CSR event) and $S\alpha1$ to give rise to a new $S\mu S\alpha_1$ DNA junction (**FIGURE 18.21**). Multiple sequential CSR events can occur, but they are not an obligatory means to proceed to later C_H gene segments. IgM can switch directly to *any* other Ig class.

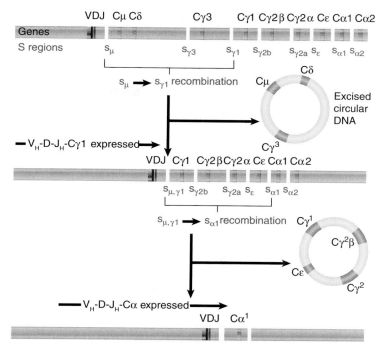

FIGURE 18.21 Class switching of C_H genes occurs by recombination between switch (S) regions and deletion of the intervening DNA between the recombining S sites as S circles. Circles are transiently transcribed in the switching cell. Sequential recombinations can occur. The mouse IgH locus is depicted.

18.15 CSR Involves Elements of the NHEJ Pathway

Key concepts

- CSR requires activation of Intervening promoters (I_H promoters) that lie upstream of the S regions involved in the recombination event.
- I_H-C_H transcription through the S region is required.
- S regions contain highly repetitive motifs with 5'-AGCT-3' as a major component.
- DSBs target mainly 5'-AGCT-3' within S regions; the DSBs' free ends are then religated through an NHEJ-like reaction.

CSR initiates with transcription from the I_H promoters of the C_H regions that will be involved in the DNA recombination event. An I_H promoter lies immediately upstream of each S region. Such I_H promoters are activated upon binding of transcription factors induced by CD40-signaling, TLR-signaling, and occupancy of receptors by cytokines, such as IL-4 or TGF-β. The I_H promoters that lie upstream of each of the S regions that will be involved in the CSR event are activated to induce "germline" I_H-C_H transcripts, which are then spliced at the I_H

region to join with the corresponding C_H region (**FIGURE 18.22**).

S regions vary in length, as defined by the limits of the sites involved in recombination, from 1 to 10 Kb. They contain clusters of repeating units that vary from 20 to 80 nucleotides in length, with the major component being 5'-AGCT-3'. Most S regions are located ~2 Kb upstream of their respective C_H gene clusters. The CSR process continues with the occurrence of DSBs in S regions followed by rejoining of the cleaved ends. The DSBs do not occur at obligatory sites within S regions, as different B cells expressing the same Ig class prove to have broken the upstream and downstream S regions at different points and recombined them, resulting in different S-S sequences.

Ku70:Ku80 and DNA-PKcs, which are required for the joining phase of V(D)J recombination and for the general NHEJ, are also required for CSR, indicating that the CSR joining reaction uses the NHEJ pathway. CSR can occur, though, albeit at a lower efficiency, in the absence of XRCC4 or DNA Ligase IV, suggesting that an alternative pathway can be used in the ligation of S region DSB ends.

The key insight into the mechanism of CSR has been the discovery of the requirement for

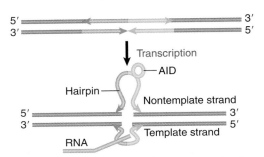

FIGURE 18.23 When transcription separates the strands of DNA, one strand forms a single-strand loop if 5'-AGCT-3' motifs in the same strand are juxtaposed.

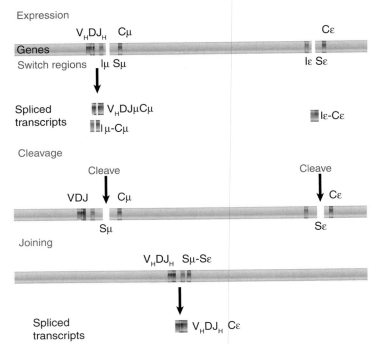

FIGURE 18.22 Class switching occurs through sequential and discrete stages. The I_H promoters initiate transcription of sterile transcripts. The S regions are cleaved and recombination occurs at the cleaved regions. Depicted is class switch DNA recombination from Sμ to Sε.

the enzyme **activation-induced (cytidine) deaminase (AID)**. In the absence of AID, CSR aborts before the DNA nicking or breaking stage. SHM is also blocked, revealing an important connection between these two processes that is central to the maturation of the antibody response and the generation of high-affinity antibodies (see *Section 18.17, SHM Is Mediated by AID, Ung, Elements of the Mismatch DNA Repair (MMR) Machinery, and Translesion DNA Synthesis (TLS) Polymerases*).

AID is expressed late in the natural history of a B lymphocyte, after the B cell encounters antigen and differentiates in germinal centers of peripheral lymphoid organs, restricting the processes of CSR and SHM to this stage. AID possesses structural similarities to the members of a class of enzymes that act on RNA to change a cytidine to a uridine (see *Section 23.10, RNA Editing Occurs at Individual Bases*). AID, however, deaminates cytidines in DNA.

Another enzyme is required for both CSR and SHM: Ung. Ung, a uracil DNA glycosylase, deglycosylates the uracil generated by the AID-mediated deamination of cytidine to give

rise to an abasic site. Mice that are deficient in Ung have a tenfold reduction in CSR, suggesting that the sequential intervention of AID and Ung creates abasic sites that are critical for the generation of DSBs. Different events follow in the CSR and SHM processes.

AID more efficiently deaminates cytidine in single-strand DNA, such as in DNA that is being transcribed and, therefore, exists as a functionally single-strand DNA. Functionally single-strand DNA exists in germline I_H-C_H transcription, in which the S region nontemplate strand of DNA is displaced when the bottom strand is used as a template for RNA synthesis (**FIGURE 18.23**). Although this has been proposed as an operational model for DNA deamination by AID, it would not explain how AID deaminates both DNA strands, which it does. The abasic site emerging after sequential AID-mediated deamination of cytidine and Ung-mediated deglycosylation of uridine is attacked by an apyridinic/apurinic endonuclease (APE), which creates a nick in the DNA strand. Generation of nicks in a nearby location on opposite DNA strands would give rise to DSBs in S regions. The DSB free ends in an upstream and downstream S region are joined by NHEJ, a repair system that acts on DSBs (see *Section 16.11, Nonhomologous End-Joining Also Repairs Double-Strand Breaks*). How the CSR machinery specifically targets S regions, and what determines the targeting of the upstream and downstream S regions recruited into the recombination process, is not yet understood. Recent data suggest a role of 5'-AGCT-3' repeats in targeting AID to S regions. 5'-AGCT-3' repeats account for more than 40% of all residues of Sμ and constitute the primary targets of DSBs.

18.16 Somatic Hypermutation (SHM) Generates Additional Diversity in Mice and Humans

Key concepts

- SHM introduces mutations in the antigen-binding V(D)J sequence.
- The mutations occur mostly as substitutions of individual bases.
- In the IgH chain locus, SHM depends on the iEμ and 3'Eα that enhance V_H-D-$J_H C_H$ transcription.
- In the Igκ chain locus, SHM depends on iEκ and 3'Eκ that enhance Vκ-JκCκ transcription. The λ locus transcription depends on the weaker λ2-4 and λ3-1 enhancers.

Comparison between the sequences of rearranged and expressed Ig V(D)J genes in B cells, which underwent proliferation and differentiation in the periphery after encountering antigen, and the corresponding germline V, D, and J gene segment templates, often reveals that expressed V(D)J sequences are changed at several locations. Some of these changes result from sequence changes at the V-J or V-D-J junctions that occur during the recombination process. Other changes are superimposed on these and accumulate within the coding sequences of the germline V(D)J templates, as a result of different mechanisms in different species. In mice and humans, the mechanism is **somatic hypermutation (SHM)**. In chickens, rabbits, and pigs, a different mechanism—**gene conversion**—is at work

SHM inserts mostly point mutations in the expressed V(D)J sequence. The process is referred to as hypermutation because it introduces mutations at a rate that is at least 10^6-fold higher (10^{-3} change/base/cell division) than that of the spontaneous mutation rate in the genome at large (10^{-9} change/base/cell division). In contrast, in chickens, rabbits, and pigs, gene conversion substitutes a rearranged and expressed V gene segment with a sequence from a different V gene (see *Section 18.18, Avian Igs Are Assembled from Pseudogenes*).

An oligonucleotide probe synthesized according to the sequence of an expressed unmutated V gene segment can be used to identify the possible corresponding template segment(s) in the germline. Any expressed V gene whose sequence is different from any germline V gene in the same organism must have been generated by somatic changes. Until a few years ago, not every potential germline V gene segment template had actually been identified. This was not a limitation, however, in the mouse λ chain system, as this is a relatively simple locus. A census of several myelomas producing λ1 chains showed that the same germline gene segment encoded many expressed V genes. Others, however, expressed new sequences that must have been generated by mutation of the germline gene segment. The current availability of mouse and human genomic DNA maps, including the Ig locus, has made it possible to readily identify germline Ig V gene templates.

To analyze the frequency and nature of **somatic mutations** accumulating during an ongoing immune response, one can examine a large number of cells in which the same V gene segment is expressed. A potentially productive approach is to characterize the Ig V(D)J sequences of a cohort of B cells, all of which respond to a given antigen or better antigenic determinant. Haptens are used for this purpose. Unlike a large protein, for which different parts induce different antibodies, haptens are small molecules whose discrete structure induces a consistently restricted antibody response. A hapten is not immunogenic *per se*, in that it does not induce an immune response if injected as such. It does, however, induce an immune response after conjugation with a "carrier" protein to form an antigen. A hapten-carrier conjugate is then used to immunize mice of a single strain. After induction of a strong antibody response, B lymphocytes (in general from the spleen) are obtained and fused with non-Ig expressing myeloma fusion partner (immortal tumor) cells to generate a monoclonal "hybridoma" that secretes indefinitely the antibody expressed by the primary B cell used for the fusion. In one example, 10 out of 19 different B cell lines producing monoclonal antibodies directed against the hapten phosphorylcholine utilized the same V_H sequence. This sequence was that of the V_H gene segment T15, one of four related V_H genes. The other nine expressed gene segments, which differed from each other and from all four germline members of the family. They were more closely related to the T15

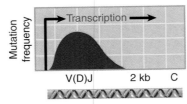

FIGURE 18.24 Somatic mutation occurs in the region surrounding the V segment and extends over the recombined V(D)J segment.

germline sequence than to any of the others, and their flanking sequences were the same as those around T15. This suggested that they arose from the T15 member through SHM.

The sequence changes (mutations) were concentrated in the V_H-D-J_H DNA, which encodes the IgH chain antigen-binding site, but tapered off throughout a region downstream of the V_H gene promoter for ~1.5 Kb (**FIGURE 18.24**). The mutations consisted in all cases of substitutions of individual nucleotide pairs. Most sequences bore ~3 to 15 substitutions, corresponding to <10 amino acid changes in the protein. Only some mutations were replacement mutations, as they affected the amino acid sequence; others were silent mutations, as they lie in the third-base coding position as well as in nontranslated regions. The large proportion of silent mutations suggests that SHM targets randomly the expressed V(D)J DNA sequence and extends beyond it. There is a tendency for some mutations to recur on multiple occasions in the same residue(s). These constitute mutational "hotspots," as a result of some intrinsic preference in the SHM machinery. The best-characterized hotspot is 5'-RGYW-3', where R is a purine (A or G), G is G, Y is a pyrimidine (C or T), and W is A or T. Interestingly, the 5'-AGCT-3' iteration of 5'-RGYW-3' is not only a preferential target of DSBs in S regions, but a major target of SHM. Like CSR, which requires germline I_H-C_H transcription of the targets S_X-C_X sequences, SHM requires transcription of the target V_H-D-J_H, $V\kappa$-$J\kappa$ and $V\lambda$-$J\lambda$ sequences. This is emphasized by the requirement for the enhancer that activates transcription at each Ig locus.

Upon exposure to antigen, B cells expressing a BCR with highest intrinsic affinity to that antigen are selected, activated, and proliferate. SHM occurs during B clonal proliferation. It randomly inserts one point mutation in the V(D)J sequence of approximately half of the progeny cells; as a result, B cells expressing mutated antibodies become a high fraction of the clone within a few divisions. Random replacement

mutations have unpredictable effects on protein function; some decrease the affinity of the BCR for the antigen driving the response, whereas others increase the intrinsic affinity for that antigen. The B cell clone(s) expressing a BCR with the highest affinity for antigen is positively selected and acquires a growth advantage over all other clones; the other clones are gradually counterselected for survival and proliferation. Further positive selection of the clone(s) that accumulated mutations conferring the highest affinity for antigen will result in narrowing clonal restriction and accumulation of high affinity clones.

18.17 SHM Is Mediated by AID, Ung, Elements of the Mismatch DNA Repair (MMR) Machinery, and Translesion DNA Synthesis (TLS) Polymerases

Key concepts

- SHM uses some of the same critical elements of CSR.
- Like CSR, SHM requires AID.
- Ung intervention influences the pattern of somatic mutations.
- Elements of the MMR pathway and TLS polymerases are involved in SHM and CSR.

The deamination or removal of a cytosine base leads to insertion of somatic mutation(s) in different ways (**FIGURE 18.25**). When AID deaminates cytosine, it gives rise to a uracil. This is not germane to DNA and can be handled by the cell in different ways. The uracil can be "replicated over"; it will pair with adenine during replication. The emerging mutation is an obligatory C→T transition and a G→A transition on the complementary strand. The net result is the replacement of the original C-G pair with a T-A pair in half of the progeny cells. Ung can be blocked by introducing into cells the gene encoding a protein that inhibits cytosine deglycosylation by the enzyme. The gene is a component of the bacteriophage PSB-2, whose genome is unusual in containing uracil, so that the enzyme needs to be blocked during a phage infection. When the gene is introduced into a lymphocyte cell line, there is a dramatic change in the pattern of mutations, with almost all mutations comprising the predicted transition

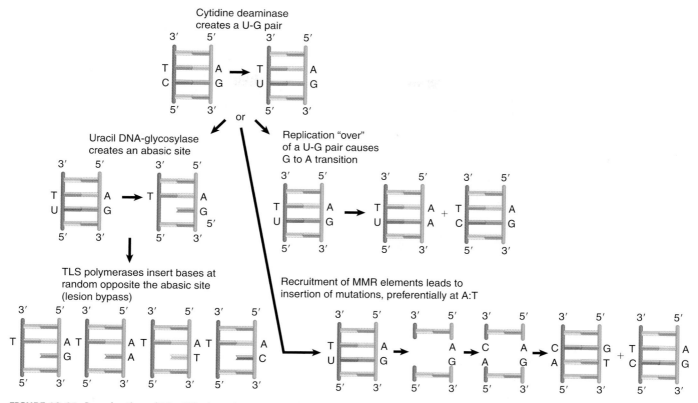

FIGURE 18.25 Deamination of C by AID gives rise to a U:G mispair. U can be replicated over, resulting in C:G→A:T transitions in 50% of progeny B cells. When the action of cytidine deaminase (top) is followed by that of uracil DNA-glycosylase, an abasic site is created. Replication past this site should insert all four bases at random into the daughter strand (center). If the uracil is not removed from the DNA, its replication gives rise to a C:G→T:A transition. Alternatively, the U:G mispair is recognized by the MMR machinery, which excises DNA containing the mismatch and then fills in the resulting gap using an error-prone DNA polymerase. This will lead to insertion of further mismatches (mutations).

from C-G to A-T. Alternatively, the uracil can be removed from DNA by Ung to give rise to an abasic site. Indeed, the key event in generating a random spectrum of mutations is therefore to create the abasic site. This can be replicated over by an error-prone TLS polymerase, such as polymerase ζ, polymerase η, or polymerase θ, each of which can insert all three possible mismatches (mutations) across the abasic site (see *Section 16.6, Error-Prone Repair*). In another mechanism, the U:G mispair recruits the MMR machinery, starting with Msh2/Msh6, to excise the stretch of DNA containing the damage, thereby creating a gap that needs to be filled in by resynthesis of the missing DNA strand (see *Section 16.7, Controlling the Direction of Mismatch Repair*). This resynthesis is carried out by an error-prone TLS polymerase that will introduce mutations. What restricts the activity of the SHM machinery to target V(D)J regions is still unknown.

The main difference between CSR and SHM is at the end of the process. DSBs are introduced and are obligatory in CSR, whereas individual point mutations are inserted by SHM. We do not know where the two processes diverge. One possibility is that DSBs are introduced at abasic sites in CSR, but the sites are erratically repaired in SHM. Another possibility is that DSBs are introduced in both cases, but are repaired in an error-prone manner in SHM.

18.18 Avian Igs Are Assembled from Pseudogenes

Key concept

- An Ig gene in chickens is generated by copying a sequence from one of 25 pseudogenes into the recombined (acceptor) V gene.

The chicken Ig locus is the paradigm for the Ig somatic diversification mechanism utilized by rabbits, cows, and pigs, which rely upon using the diversity that is encoded in the genome. A similar mechanism is used by both the single L chain locus (of the λ type) and the H chain loci. The chicken λ locus comprises only one functional V gene segment, one Jλ segment, and

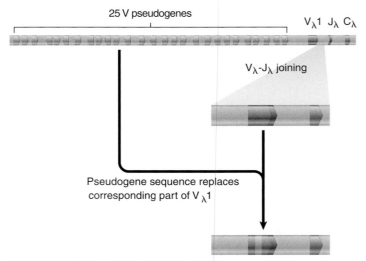

FIGURE 18.26 The chicken λ light chain locus has 25 V pseudogenes upstream of the single functional V_λ-J_λ-C region. Sequences derived from the pseudogenes, however, are found in active rearranged VJC genes.

one Cλ gene segment (**FIGURE 18.26**). Upstream of the functional Vλ1 gene segment lie 25 Vλ pseudogenes, organized in either orientation. They are classified as pseudogenes because either the coding segment is deleted at one or both ends, or proper RSSs are missing, or both. This is confirmed by the fact that only the Vλ1 gene segment recombines with the JλCλ gene segment.

Nevertheless, sequences of rearranged Vλ-JλCλ gene segments show considerable diversity. A rearranged gene has one or more positions at which a cluster of changes occurred in its sequence. A sequence identical to the new sequence can almost always be found in one of the pseudogenes, which remain unchanged. The exceptional sequences that are not found in a pseudogene always result from changes at the junction between the original sequence and the altered sequence. In general, sequences from the pseudogenes, between 10 and 120 bp in length, are integrated into the active $V\lambda_1$ region by gene conversion. The unmodified $V\lambda_1$ sequence is not expressed, even at early times during the immune response. A successful conversion event probably occurs every ten to twenty cell divisions to every rearranged $V\lambda_1$ sequence. At the end of the immune maturation period, a rearranged $V\lambda_1$ sequence has four to six converted segments spanning its entire length, which are derived from different donor pseudogenes. If all pseudogenes can participate in this gene conversion process, more than 2.5×10^8 possible combinations are allowed.

The enzymatic basis for copying pseudogene sequences into the recombined Ig V gene depends on enzymes involved in recombination, and is related to the mechanism of human and mouse SHM. Some of the genes involved in recombination are required for the gene conversion process (see *Section 15.15, Eukaryotic Genes Involved in Homologous Recombination*). For example, gene conversion is prevented by deletion of *RAD54*. Deletion of other recombination genes (such as *XRCC2*, *XRCC3*, and *RAD51B*) has another, interesting effect: somatic mutations occur in the V gene of the expressed locus. The frequency of the somatic mutations is ~tenfold greater than the rate of gene conversion.

Thus, the absence of SHM in chickens is not due to a deficiency in the enzymatic systems that are responsible for SHM in humans and mice. The most likely explanation for a connection between (lack of) recombination and SHM is that unrepaired DSBs in the locus trigger the induction of mutations. The reason why SHM occurs in mice and humans but not in chickens may, therefore, lie with the nature of the repair system that operates on DSBs in the Ig locus. It would be more efficient in chickens, so that DSBs in the Ig locus are repaired through gene conversion before mutations can be induced.

18.19 B Cell Memory Allows for the Mounting of a Prompt and Strong Secondary Response

Key concepts

- Most B cells that mount a primary response to an antigen do not survive beyond the end of the primary response.
- Toward the end of the primary response, memory B cells are generated that are highly specific for the antigen driving the response. These B cells are in a resting state.
- Reexposure to the same antigen triggers a secondary response through rapid activation and clonal expansion of memory B cells.

CSR and SHM are the two central processes that underlie the antigen-driven differentiation of mature B cells in high affinity, class-switched, antibody-producing cells and memory B cells. B cells derive from a self-renewing population of stem cells in the bone marrow. The process that eventually gives rise to mature B cells depends upon Ig V(D)J gene rearrangement, which requires *RAG1/RAG2* genes, the Ku70:Ku86 heterodimer, DNA-PK, Artemis, and DNA ligase IV in association with XRCC4, and can involve TdT. If gene V(D)J rearrange-

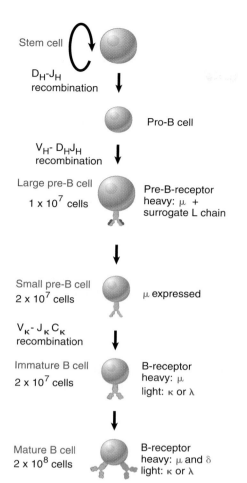

FIGURE 18.27 B cell development proceeds through sequential stages of H chain and L chain V(D)J gene rearrangement.

Stem cell

D_H-J_H recombination

Pro-B cell

V_H- $D_H J_H$ recombination

Large pre-B cell
1×10^7 cells

Pre-B-receptor heavy: μ + surrogate L chain

Small pre-B cell
2×10^7 cells

μ expressed

V_κ- $J_\kappa C_\kappa$ recombination

Immature B cell
2×10^7 cells

B-receptor heavy: μ
light: κ or λ

Mature B cell
2×10^8 cells

B-receptor heavy: μ and δ
light: κ or λ

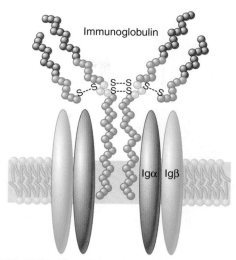

FIGURE 18.28 The BCR consists of an immunoglobulin tetramer ($H_2 L_2$) linked to two copies of the signal-transducing heterodimer (IG$\alpha\beta$).

ment is blocked, B cell development is aborted. The BCRs expressed by B cells display specificities that are determined by the particular V(D)J gene recombinations and any additional N nucleotides incorporated during the V(D)J joining process.

B cell development in the bone marrow entails a first step in which an IgH D segment is recombined with a J_H segment (FIGURE 18.27). Cells at this stage (recombined D-J_H) are referred to as Pro-B cells. D-J_H recombination is followed by V_H-D-J_H recombination, which generates an IgH μ chain. Several recombination events involving a succession of nonproductive and productive rearrangements may occur, as discussed previously (see Figure 18.15). These B cells express the productively recombined IgH chain (V_H-D-$J_H C\mu$) paired with a *surrogate* L chain (a protein resembling a λ chain) to form the Pre-BCR, and are referred to as Pre-B cells. The Pre-BCR is very similar in function and structure to a BCR, although once engaged it

signals in a different way. The expression of the Pre-BCR drives the Pre-B cell through five or six divisions (large Pre-B cells), after which the Pre-B cell stops dividing to revert back to a small size, thereby signaling the rearrangement of a V gene segment with a J gene segment in the κ or λ locus. A κ or λ L chain is then expressed as paired with the rearranged V_H-D-$J_H C\mu$ as a BCR on the surface of the immature B cell. As the cells transition into mature B cells, the expression of an IgH δ chain is added to that of the IgH μ chain.

Thus, a complete Ig molecule functions both as a BCR and as a secreted antibody (FIGURE 18.28). The intracytoplasmic tails of the two IgH chains are associated with transmembrane proteins called Igα and Igβ. These proteins provide the structures that trigger the intracellular signaling pathways in response to BCR engagement by antigen.

The activation of the mature B cell through BCR cross-linking by antigen is also influenced by interactions with other surface receptors, such as the engagement of CD40 by CD40 ligand (CD154) expressed on T_h cells. Exposure of the mature B cell to antigen results in a differentiation process that will eventually give rise to a "mature" antibody response. A primary immune response is initiated following clonal expansion of the B cells responding to the antigen. This generates plasma cells, which produce mostly unmutated IgM with a low intrinsic affinity but high avidity for the antigen. CSR and SHM take place toward the late stages of the primary response to generate B cells capable of producing more specific

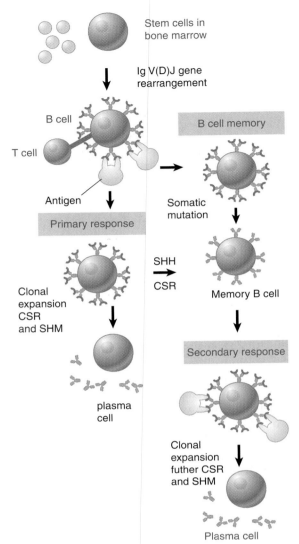

antigen, they can mount a secondary response very rapidly, by vigorous clonal expansion. Further somatic mutations are accumulated and more CSR events occur during the secondary response, eventually giving rise to switched and hypermutated B cells with a very high affinity for antigens. Most of these cells will terminally differentiate into plasma cells producing large amounts of antibodies; the remaining B cells will differentiate to memory B cells. Memory B cells will be "frozen" with respect to their V(D)J somatic mutations and IgH chain class, and will be ready to give rise to a very vigorous, high affinity memory or anamnestic antibody response when they reencounter antigen.

Virtually all B cells recruited in an antigen-specific antibody response to undergo CSR and SHM (**FIGURE 18.29**) are "conventional" B cells or B-2 cells. In addition to these cells, a separate set of B cells exists, referred to as B-1 cells. B-1 cells also undergo the V(D)J gene rearrangement and apparently are selected for expression of a particular repertoire of antibody specificities. They may be involved in natural immunity; that is, they may possess the intrinsic ability to respond in a T-independent fashion to certain naturally occurring antigens, particularly bacterial components such as polysaccharides.

18.20 The TCR Is Related to the BCR

Key concepts

- T cells use a mechanism of V(D)J recombination similar to that of B cells to express either of two types of TCR.
- TCRαβ is found on >95% and TCRγδ on <5% of T lymphocytes in the adult.

Both B and T cells use similar evolutionary conserved mechanisms to express significant diversity in BCR and TCR variable regions. T cells express on their surface TCRαβ or TCRγδ, each of which is expressed at different times during T cell development (**FIGURE 18.30**). In adult mice, TCRγδ is expressed by <5% of T lymphocytes and TCRαβ by >95% of T lymphocytes. TCRγδ is synthesized at an early stage of T cell development. TCRγδ is the only TCR expressed during the first 15 days of gestation, but is virtually lost by birth, at day 20. TCRαβ is synthesized later in T cell development than TCRγδ, being first expressed at day 15 to 17 of gestation. At birth, TCRαβ is the predominant TCR. TCRαβ is synthesized by a separate lin-

FIGURE 18.29 B cell differentiation is responsible for acquired immunity. Initial exposure of mature B cells to antigen results in a primary response and generation of memory cells. Subsequent exposure to antigen induces a secondary response through activation of the memory cells.

IgG, IgA, and/or IgE antibodies. Most of these B cells revert back to small resting lymphocytes (memory B cells).

A secondary immune response is elicited upon reexposure to the antigen that has induced the primary response as a result of the generation of memory B cells. These B cells comprise a minor proportion of the B cells generated at the end of the primary response. They express mutated V(D)J gene segments coding for a BCR that displays increased affinity for antigen and that possibly underwent CSR to mainly IgG. They are in a resting state, but are rapidly activated when they reencounter the same antigen. Having been preselected by and for the same

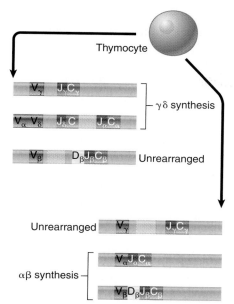

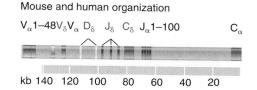

Mouse and human organization

$V_\alpha 1-48 V_\delta V_\alpha$ D_δ J_δ C_δ $J_\alpha 1-100$ C_α

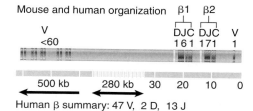

kb 140 120 100 80 60 40 20

Human α summary: 42 V   61 J

FIGURE 18.31 The human TCRα locus contains interspersed α and δ segments. A Vδ segment is located within the Vα cluster. The D-J-Cδ segments lie between the V gene segments and the J-Cα segments. The mouse locus is similar, but includes more Vδ segments.

Mouse and human organization   β1   β2

V
<60   DJC DJC V
                   1 6 1 1 7 1 1

500 kb   280 kb   30   20   10   0

Human β summary: 47 V, 2 D, 13 J

FIGURE 18.32 The TCRβ locus contains many V gene segments spread over ~500 kb that lie ~280 kb upstream of the two D-J-C clusters.

FIGURE 18.30 The TCRγδ receptor is synthesized early in T cell development. TCRαβ is synthesized later and is responsible for cell-mediated immunity, in which target antigen and host MHC are recognized together.

eage of cells from those expressing TCRγδ and involves independent rearrangement events.

Like the BCR, the TCR must recognize a foreign antigen of virtually any possible structure. The TCR resembles the BCR in structure. The V sequences have the same general internal organization in both the TCR and the BCR. The TCR C region is related to the Ig C regions, but has a single C domain followed by transmembrane and cytoplasmic portions. The exon–intron structure reflects the protein function. The organization and configuration of the TCR genes are highly similar to those of the BCR/Ig genes. *Each TCR locus (α, β, γ, and δ) is organized in a fashion similar to that of the Ig locus, with separate gene segments that are brought together by a recombination reaction specific to the lymphocyte.* The components are similar to those found in the three Ig loci: IgH, κ, and λ.

The TCRα locus resembles the Igκ locus, with Vα gene segments separated from a cluster of Jα segments that precedes a single Cα gene segment (**FIGURE 18.31**). The organization of the TCRα locus is similar in both humans and mice, with some differences only in the number of Vα gene segments and Jα segments. In addition to the α segments, this locus also contains embedded δ segments. The organization of the TCRβ locus resembles that of the IgH locus, although the large cluster of Vβ gene segments lie upstream of two clusters, each containing

a D segment, several Jβ segments, and a Cβ gene segment (**FIGURE 18.32**). Again, the only differences between humans and mice are in the numbers of Vβ and Jβ genes.

Diversity in the TCR is generated by the same mechanisms as in the BCR/Ig. Germline encoded (intrinsic) diversity results from the combination of a variety of V, D, and J segments; some additional diversity results from the introduction of new sequences at the junctions between these components, in the form of P and/or N nucleotides. The recombination of TCR gene segments occurs through mechanisms highly similar to those of the BCR in B cells. Appropriate nonamer-spacer-heptamer RSSs direct it. These RSSs are identical to those used by the Ig genes and are handled by the same enzymes. As in the BCR/Ig loci, most rearrangements in the TCR loci occur by deletion. Rearrangements of TCR gene segments, like those of TCR/Ig genes, may be productive or nonproductive. Like the Ig locus in B cells, the transcription factors that control and mediate the rearrangement of the TCR locus in T cells have just begun to be appreciated.

The organization of the TCRγ locus resembles that of the Igλ locus, with V gene segments separated from a series of JC segments (**FIGURE 18.33**). The TCRγ locus displays relatively little diversity, with ~8 functional V segments. The organization is different in humans and

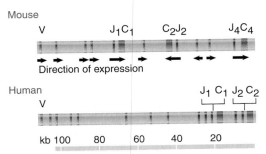

FIGURE 18.33 The TCRγ locus contains a small number of functional V gene segments (and also some pseudogenes not shown) that lie upstream of the J-C loci.

mice. The mouse TCRγ locus has three functional JγCγ segments. The human TCRγ locus has multiple Jγ segments for each Cγ gene segment.

The cluster of genes encoding the TCRδ chain lies entirely embedded in the TCRα locus, between the Vα and Cα genes (Figure 18.31). The Vδ gene segments are interspersed within the Vα gene segments. Overall, the number of TCR Vγ and Vδ gene segments is much lower than that of the Vα and Vβ gene segments. Nevertheless, great diversity is generated at the TCRδ locus, as D-D rearrangements occur frequently, each of them entailing N nucleotide additions. The embedding of the TCRδ cluster of Dδ and Jδ genes and the Cδ gene in the TCRα locus implies that expression of TCRαβ and TCRγδ is mutually exclusive at any one allele, because all the Dδ, Jδ, and Cδ gene segments are lost once a Vα-Jα rearrangement occurs.

D-D rearrangements also occur at the TCRαβ locus, resulting from D-D joinings. The TCRβ locus shows allelic exclusion in much the same way as the Ig locus; rearrangement is suppressed once a productive allele has been generated. The TCRα locus may be different; several cases of continued rearrangements suggest the possibility that substitution of Vα sequences may continue after a productive allele has been generated. Unlike the IgH, Igκ, and Igλ loci, neither TCRαβ nor TCRγδ undergoes SHM.

18.21 The TCR Functions in Conjunction with the MHC

Key concept

- The TCR recognizes a short peptide set in the groove of an MHC molecule on the surface of an antigen-presenting cell (APC).

T cells with TCRαβ comprise subtypes that have a variety of functions related to interactions with other cells of the immune system. CTLs possess the ability to lyse a target cell. Th cells help the activation/generation of CTLs or help the differentiation of B cells into antibody-producing cells.

The BCR/antibody and the TCR differ in their modalities of interaction with their ligands. A BCR/antibody recognizes a small area (epitope) within the antigen, which can be composed of a linear sequence (six to eight amino acids) identifying a linear determinant or a cluster of amino acids brought together by the three-dimensional structure of the antigen (conformation determinant). A TCR binds a peptide derived from the antigen upon processing by an **antigen-presenting cell (APC)**. (The peptide is generated when the proteasome degrades the antigen protein within the APC.) The peptide fragment is "presented" to the T cell by the APC in the context of an MHC protein, in a groove on the surface of the MHC. Thus, the T cell simultaneously recognizes the peptide and an MHC protein carried by the APC (**FIGURE 18.34**). Both Th cells and CTLs recognize the antigen in this fashion, but with different requirements; i.e., they recognize peptides of different sizes and as presented in conjunction with different types of MHC proteins (see *Section 18.22, The Major Histocompatibility Locus Comprises a Cohort of Genes Involved in Immune Recognition*). Th cells recognize peptide antigens, ≥13 amino acids long, presented by MHC class II proteins, whereas CTLs recognize peptide antigens no more than eight to ten amino acids long, presented by MHC class I proteins. The TCRαβ provides the structural correlate for the helper Th cell function and for the CTL function. In both cases, TCRαβ recognizes both the antigenic peptide and the self-MHC protein. A given TCR has specificity for a particular MHC, as well as for the associated antigen peptide. The basis for this dual recognition capacity the most interesting structural features of the TCRαβ.

Recombination to generate functional TCR chains is linked to the development of the T lymphocyte (**FIGURE 18.35**). The first stage is rearrangement to form an active TCRβ chain. This binds a nonrearranging surrogate TCRα chain called pre-TCRα. At this stage, the lymphocyte has not yet expressed on the surface either CD4 or CD8. The pre-TCR heterodimer then associates with the CD3 signaling complex (see the next paragraph). Signaling

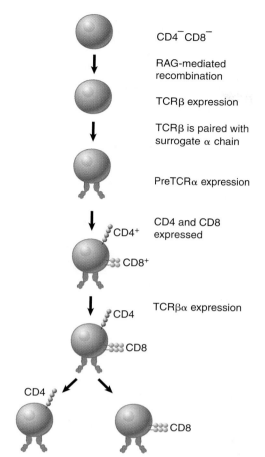

FIGURE 18.34 T cell development proceeds through sequential stages.

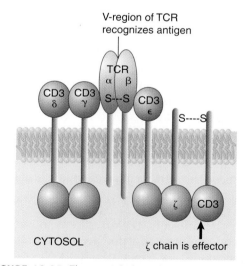

FIGURE 18.35 The two chains of the T cell receptor associate with the polypeptides of the CD3 complex. The variable regions of the TCR are exposed on the cell surface. The cytoplasmic domains of the ζ chains of CD3 provide the effector function.

from the complex triggers several rounds of cell division, during which TCRα chains are rearranged, and the CD4 and CD8 genes are turned on, so that the lymphocyte transitions from CD4⁻CD8⁻ or double negative (DN) thymocyte to CD4⁺CD8⁺ or double positive (DP) thymocyte. TCRα chain rearrangement continues in the DP thymocytes. The maturation process continues through both positive selection (for mature TCR complexes able to bind a self ligand with moderate affinity) and negative selection (against complexes that interact with self-ligands at high affinity). Both positive and negative selection involves interaction with MHC proteins. DP thymocytes either die within three to four days or become mature lymphocytes as the result of the selection process. The surface TCRαβ heterodimer becomes cross-linked on the surface during positive selection, which rescues the thymocyte from apoptosis (nonnecrotic cell death). If thymocytes survive the subsequent negative selection, they give rise to the separate T lymphocyte subsets CD4⁺CD8⁻ and CD4⁻CD8⁺ cells.

The TCR is associated with the CD3 complex of proteins, which are involved in transmitting a signal from the surface of the cell to nucleus, when the TCR is activated by binding of antigen (Figure 18.35). The interaction of the TCR variable regions with antigen causes the ζ chain of the CD3 complex to signal T cell activation, in a fashion comparable to the BCR Igα and Igβ complex signaling B cell activation.

Considerable diversity is required in both recognition of the foreign antigen, which requires the ability to respond to novel structures, and recognition of the MHC protein, which is restricted to one of the many different MHC proteins encoded in the genome. T_h cells and CTLs rely upon different classes of MHC proteins; however, they use the same pool of α and β gene segments to assemble their TCRs. Even allowing for the introduction of additional variation during the TCR recombination process, the number of different TCRs generated is relatively limited, but nevertheless sufficient to satisfy the diversity demands imposed by the variety of TCR ligands. This is made possible by the relatively low binding affinity requirements by the TCR-peptide/MHC interaction, which allows for one TCR to interact with multiple different ligands displaying some similarities.

18.22 The Major Histocompatibility Locus Comprises a Cohort of Genes Involved in Immune Recognition

Key concepts

- The MHC locus codes for class I and class II molecules, as well as for other proteins of the immune system.
- Class I proteins are the transplantation antigens distinguishing "self" from "nonself."
- Class II proteins are involved in interactions of T cells with APCs.
- MHC class I molecules are heterodimers consisting of a variant α chain and the invariant β₂ microglobulin.
- MHC class II molecules are heterodimers consisting of an α chain and a β chain.

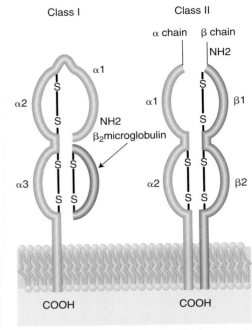

FIGURE 18.36 Class I and class II MHC have a related structure. Class I antigens consist of a single polypeptide (α) with three external domains (α1, α2, and α3) that interacts with β2 microglobulin (β2 M). Class II antigens consist of two polypeptides (α and β), each with two domains (α1 and α2, and β1 and β2) with a similar overall structure.

The MHC molecules have evolved to maximize the efficacy and flexibility of their function: to bind peptides derived from microbial pathogens and present them to T cells. In response to a strong evolutionary pressure to eliminate many kinds of microorganisms, these MHC proteins have evolved as encoded by polygenic (several sets of MHC genes in all individuals) and polymorphic (multiple variants of gene within the population at large) cohorts of genes, the MHC genes. MHC proteins are dimers inserted in the plasma membrane, with a major part of the protein protruding on the extracellular side. The structure of MHC class I and class II molecules are related, although they are made up of different components (**FIGURE 18.36**).

MHC class I molecules consist of a heterodimer of the class I chain (α) itself and the β2-microglobulin protein. The class I chain is a 45 kD transmembrane component that has three external domains (each ~90 amino acids long, one of which interacts with β2 microglobulin), a transmembrane domain (~ 40 residues), and a short cytoplasmic domain (~30 residues). MHC class II molecules consist of two chains, α and β, whose combination generates an overall structure in which there are two extracellular domains. There are three class Iα-chains in humans: HLA-A, HLA-B, and HLA-C. The β2 microglobulin is a secreted protein of 12 kD. It is needed for the class I chain to be transported to the cell surface. Mice lacking the β2 microglobulin gene express no MHC class I antigens on the cell surface. There are three pairs of class IIα- and β-chain genes in the human: *HLA-DR*, *HLA-DP*, and *HLA-DQ*.

The MHC locus occupies a small region of a single chromosome in mice (*H2* locus, chromosome 17) and in humans (human leukocyte antigen or *HLA* locus, chromosome 6). These regions contain multiple genes encoding proteins of the immune system. In general, such genes are highly polymorphic; that is, they are different from one another in individual genomes. Also located in this region are genes coding for proteins found on lymphocytes and macrophages that have a related structure and are important in the function of cells of the immune system.

The genes of the *MHC* locus are grouped in three clusters according to the structures and immunological properties of the respective products. The *MHC* region was originally defined by genetics in the mouse, where the classical *H2* region occupies 0.3 map units. Together with the adjacent region where mutations affecting immune function are also found, this corresponds to a region of ~2000 Kb of DNA. The *MHC* region, which has been completely sequenced in several mammals, as well as in some birds and fish, is generally quite conserved. The genomic regions where the class I and class II genes are located mark the original boundaries of the locus, from telomere to cen-

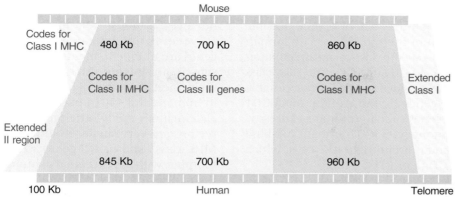

FIGURE 18.37 The MHC region extends for >2 Mb. MHC proteins of classes I and II are encoded by two separate regions. The class III region is defined as the segment between them. The extended regions describe segments that are syntenic on either end of the cluster. The major difference between mouse and human is the presence of *H2* class I genes in the extended region on the left. The murine locus is located on chromosome 17, and the human locus is located on chromosome 6.

tromere (**FIGURE 18.37**: right to left). The genes in the class III region, which separates class I from class II genes, encode many proteins with a variety of functions. Defining the ends of the locus varies with the species, and the region beyond the class I genes on the telomeric side is called the extended class I region. Likewise, the region beyond the class II gene cluster on the centromeric side is referred to as the extended class II region. The major difference between mice and humans is that the extended class II region contains some class I (*H2-K*) genes in mice.

The organization of class I genes is based on the structure of their products (**FIGURE 18.38**). The first exon encodes a signal sequence, cleaved from the protein during membrane passage. The next three exons encode each of the external domains. The fifth exon encodes the transmembrane domain. The last three rather small exons together encode the cytoplasmic domain. The only difference in the genes for human transplantation antigens is that their cytoplasmic domain is encoded by only two exons. The exon encoding the third external domain of the class I genes is highly conserved relative to the other exons. The conserved domain probably represents the region that interacts with β2-microglobulin, which explains the need for constancy of structure. This domain also exhibits homologies with the constant region domains of Igs. Most of the sequence variation between class I alleles occurs in the first and second external domains, sometimes taking the form of a cluster of base substitutions in a small region.

The gene for β2-microglobulin is located on a separate chromosome. It has four exons,

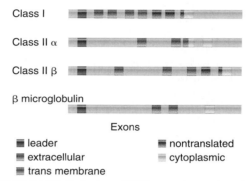

FIGURE 18.38 Each class of MHC genes has a characteristic organization, in which exons represent individual protein domains

the first encoding a signal sequence, the second encoding the bulk of the protein (from amino acids 3 to 95), the third encoding the last four amino acids and some of the nontranslated UTR, and the last encoding the rest of the UTR. The length of β2-microglobulin is similar to that of an Ig V gene; there are certain similarities in amino acid constitution, and there are some (limited) homologies of nucleotide sequence between β2-microglobulin and Ig constant domains or type I gene third external domains.

MHC class I genes encode **transplantation antigens**. They are present on every cell of the mammal. As their name suggests, these proteins are responsible for the rejection of foreign tissue, which is recognized as such by virtue of its particular array of transplantation antigens. In the immune system, their presence on target cells is required for cell-mediated responses. The types of class I proteins are defined serologically by their antigenic properties. The murine class I genes code for the H2-K and H2-D/L proteins. Each mouse strain has one of several possible alleles for each of these proteins. The human class I genes encode the classical transplantation antigens, HLA-A, -B, and -C. Some HLA class I-like genes lie outside the MHC locus. Notable among these genes are those of the small CD1 family. CD1 genes code for proteins expressed on DCs and monocytes. CD1 proteins can bind glycolipids and present them to T cells, which are neither CD4 nor CD8.

MHC class II genes encode the MHC class II proteins. These are expressed on the surfaces of both B and T lymphocytes, as well as on macrophages and dendritic cells. MHC class II molecules are critically involved in antigen presentation and communications between cells that are necessary to induce a specific immune response. In particular, they are required for T_h cell function. The murine class II genes were originally identified as immune response (Ir) genes; that is, genes whose expression made possible for an immune response to a given antigen to be triggered—hence the I-A and I-E terminology. The human class II region (also called *HLA-D*) is arranged into *DR*, *DP*, and *DQ* subregions. This region also includes several genes that are related to the initiation of antigen-specific response, namely, antigen presentation. These genes include those encoding TAP and LMP, as well as those encoding the DM and DO molecules. Class II gene expression is induced by IFN-γ through CIITA, the MHC class II transcriptional activator.

MHC class III genes occupy a "transitional" region between the class I and class II regions. The class III region includes genes coding for **complement** components, including C2, C4, and factor B. The role of complement factors is to interact with antibody–antigen complexes and mediate activation of the complement cascade, eventually lysing cells, bacteria, or viruses. Other genes lying in this transitional region include those encoding tumor necrosis factor-α (TNF-α), lymphotoxin-α (LTA), and lymphotoxin-β (LTB).

There are several hundred genes in the *MHC* regions of mammals, but it is possible for MHC functions to be provided by far fewer genes, as in the case of chickens, where the *MHC* region is 92 Kb and comprises only nine genes. In comparison to other gene families, there are differences in the exact numbers of genes devoted to each function. The *MHC* locus shows extensive variation between individuals, and a number of genes may be differ-

ent in different individuals. As a general rule, however, a mouse genome has fewer active *H2* genes than a human genome. The class II genes are unique to mammals (except for one subgroup); birds and fish have different genes in their place. There are ~8 functional class I genes in humans and ~30 in mice. The class I region also includes many other genes. The class III regions are very similar in humans and mice. MHC class I and class II genes are highly polymorphic, with the exception of human *DRα* and the mouse homologue *Eα*, and likely arose as a result of extensive gene duplications. Further divergence arose through mutations and gene conversion.

18.23 Summary

Virtually all the genes discussed in this chapter likely descended from a common ancestor gene that encoded a primitive protein domain. Such a gene would have encoded a protein that mediated nonspecific defense against a variety of microbial pathogens. It is possibly the ancestor of the conserved genes coding for the more than twenty antifungal, antibacterial, and antiviral peptides found in *Drosophila*. Further duplication and evolution of these genes likely gave rise to the diverse repertoire of Ig V(D)J, and C genes in the Ig and TCR loci, as well as the genes in the MHC locus.

The immune system has evolved to respond to an enormous variety of microbial pathogens, such as bacteria, viruses, and other infectious agents. This is accomplished by triggering a virtually immediate response that recognizes common structures or MAMPS shared by many pathogens using PRRs. The diversity of these receptors is limited and encoded in the germline. The receptors involved are typically members of the Toll-like class of receptors, and the related signaling pathways resemble the pathway triggered by Toll receptors during embryonic development. The pathway culminates in activation of transcription factors that cause genes to be expressed, and whose products inactivate the infective agent, typically by permeabilizing its membrane.

The innate immune response is triggered in different ways and to different degrees, depending on the nature of the foreign microbial antigen inducing it. It contains (to some degree) the invading microorganism during the early stages of infection, but fails in general to limit the spreading of the infection in later stages or to eradicate the invading microbial pathogen.

The innate immune response is nonspecific and does not generate immunological memory. Nevertheless, through differential modulation of the innate response, the nature of the antigen determines the nature of the adaptive response eventually mounted to the same antigen.

The adaptive immune response relies on BCR/Ig and TCR, which are molecules that play analogous recognition functions on B cells and T cells, respectively. The BCR/Ig or TCR components are generated by rearrangement of DNA in a single lymphocyte. Many different rearrangements occur early in the development of the immune system, thereby creating a large repertoire of cells of different specificities. Exposure to an antigen recognized by the BCR or TCR leads to clonal expansion to give rise to many progeny cells that have the same specificity as the original (parental) cell. The very large number of BCRs/TCRs available in the primary B and T cell repertoire, so as to create a high probability of recognizing any foreign molecule, provides the structural basis for this selection process.

Each Ig protein is a tetramer containing two identical H chains and two identical L chains. A TCR is a dimer containing two different chains. Each polypeptide chain is expressed from a gene created by recombining one of many V gene segments with D segments and J segments, as linked to one of a few C segments. Ig L chains (either Igκ or Igλ) have the general structure V-JC, IgH chains have the structure V-D-JC, TCR α and γ chains resemble IgL chains, and TCRδ and TCRβ resemble IgH chains.

The V(D)J gene segments and their organization are different for each type of chain, but the principle and mechanism of recombination appear to be the same. The same nonamer-spacer-heptamer RSSs are involved in each recombination; the reaction always involves the joining of an RSS with 23 bp spacing to an RSS with 12 bp spacing. The cleavage reaction is catalyzed by the RAG1/RAG2 proteins, and the joining reaction is catalyzed by the same elements of the general NHEJ pathway that repairs DSBs. The mechanism of action of the RAG proteins is related to the action of site-specific recombination catalyzed by resolvases. Recombining different V(D)J segments generates considerable diversity; however, additional variations are introduced in the form of truncations and/or additions of N nucleotides at the junctions between V(D)J DNA segments during the recombination process. A productive rearrangement inhibits the occurrence of further rearrangements (allelic exclusion). Allelic exclusion ensures that a given lymphocyte synthesizes only a single BCR/Ig or TCR.

Mature B cells express surface IgM and IgD BCR. After encounter of antigen and activation, these B cells start secreting the corresponding IgM antibodies using a mechanism of differential or alternative splicing. This underlies the expression of a membrane-bound version of a BCR and its corresponding secreted version (antibody). BCRs/TCRs that recognize the body's own proteins are screened out early in the process. B and T cell clones are expanded and further selected in response to antigen during the primary immune response. Activation of the BCR on B cells triggers the pathways of the humoral response; activation of the TCR on T cells triggers the pathways of the cell-mediated response. The primary immune (adaptive) response is characterized by a latency period—in general a few days—required for the clonal selection and proliferation of the B cells and/or T cells specific for the antigen, be it on a bacterium or a virus or other microorganism, driving the response. Clonal selection of B or T cells relies on binding of antigen to BCR and TCR on selected B and T cells (clones). These clones are significantly expanded in size and undergo SHM and CSR in the late stages of the primary response. Reexposure to the same antigen induces a secondary response, which has virtually no latency period and is much bigger in magnitude and more specific than the primary response.

SHM and CSR continue to occur in the secondary response, upon reexposure to the same antigen. SHM inserts point-mutation changes in Ig V(D)J gene sequences. It requires the actions of the AID cytidine deaminase and the Ung glycosylase. Mutations induced by AID lead in most cases to removal of uracil by Ung, and the bypassing of abasic sites by TLS polymerases and/or recruitment of elements of the MMR machinery. The use of the V region is fixed by the first productive rearrangement, but B cells undergo CSR, thereby switching use of C_H genes from the initial Cμ chain to one of the C_H chains lying farther downstream. This process involves a different type of recombination in which the DNA intervening between the V_H-D-J_H region and the new C_H gene is deleted and rejoined as a switch circle. More than one CSR event can occur in a B cell. CSR requires the same AID cytidine deaminase and Ung that are required for SHM. It also uses elements of the NHEJ pathway of DNA repair. Differential or alternative splicing also underlies the expression of membrane and secreted forms of all switched isotypes: IgG, IgA, and IgE.

SHM and CSR occur in peripheral lymphoid organs and are critical in the maturation of the antibody response and the generation of immunological B cell but not T cell memory. Immunological memory provides protective immunity against the same antigen that drove the original response. Thus, the organism retains a memory of the specific B and/or T cell response. The principles of adaptive immunity are similar, albeit somewhat different in details, throughout the vertebrates. Such memory enables the organism to respond more rapidly and vigorously once exposed again to the same pathogen, and provides the cellular and molecular basis for the use of vaccines.

References

18.2 The Innate Response Utilizes Conserved Recognition Molecules and Signaling Pathways

Reviews

Coutinho, A. and Poltorack, A. (2003). Innate immunity: from lymphocyte mitogens to Toll-like receptors and back. *Curr. Opin. Immunol.* 15, 599–602.

Hoffmann, J. A., Kafatos, F. C., Janeway, C. A. and Ezekowitz, R. A. (1999). Phylogenetic perspectives in innate immunity. *Science* 284, 1313–1318.

Hoffmann, J. A. (2003). The immune response of Drosophila. *Nature* 426, 33–38.

Janeway, C. A., Jr. and Medzhitov, R. (2002). Innate immune recognition. *Annu. Rev. Immunol.* 20, 197–216.

Lanzavecchia, A. and Sallusto, F. (2007). Toll-like receptors and innate immunity in B-cell activation and antibody responses. *Curr. Opin. Immunol.* 19, 268–274.

Lee, M. S. and Kim, Y. J. (2007). Signaling pathways downstream of pattern-recognition receptors and their cross talk. *Annu. Rev. Biochem.* 76, 447–480.

Medzhitov, R. and Janeway, C. A., Jr. (1997). Innate immunity: the virtues of a nonclonal system of recognition. *Cell* 91, 295–298.

O'Neill, L. A. and Bowie, A. G. (2007). The family of five: TIR-domain-containing adaptors in Toll-like receptor signalling. *Nat. Rev. Immunol.* 7, 353–364.

Palm, N. W. and Medzhitov, R. (2009). Pattern recognition receptors and control of adaptive immunity. *Immunol. Rev.* 227, 221–233.

Rast, J. P., Smith, L. C., Loza-Coll, M., Hibino, T., and Litman, G. W. (2006). Genomic insights into the immune system of the sea urchin. *Science* 314, 952–956.

Research

Baeuerle, P. A. and Baltimore, D. (1988). I kappa B: a specific inhibitor of the NF-kappa B transcription factor. *Science* 242, 540–546.

Carty, M., Goodbody, R., Schroder, M., Stack, J., Moynagh, P. N., and Bowie, A.G. (2006). The human adaptor SARM negatively regulates adaptor protein TRIF-dependent Toll-like receptor signaling. *Nat. Immunol.* 7, 1074–1081.

Fitzgerald, K. A., Palsson-McDermott, E. M., Bowie, A. G., Jefferies, C. A., Mansell, A. S., Brady, G., Brint, E., Dunne, A., Gray, P., Harte, M. T., McMurray, D., Smith, D. E., Sims, J. E., Bird, T. A., and O'Neill, L. A. (2001). Mal (MyD88-adapter-like) is required for Toll-like receptor-4 signal transduction. *Nature* 413, 78–83.

Fitzgerald, K. A., Rowe, D. C., Barnes, B. J., Caffrey, D. R., Visintin, A., Latz, E., Monks, B., Pitha, P. M., and Golenbock, D. T. (2003). LPS-TLR4 signaling to IRF-3/7 and NF-kappaB involves the toll adapters TRAM and TRIF. *J. Exp. Med.* 198, 1043–1055.

Ghosh, S., Gifford, A. M., Riviere, L. R., Tempst, P., Nolan, G. P., and Baltimore, D. (1990). Cloning of the p50 DNA binding subunit of NF-kappa B: homology to rel and dorsal. *Cell* 62, 1019–1029.

Grosshans, J., Bergmann, A., Haffter, P., and Nusslein-Volhard, C. (1994). Activation of the kinase Pelle by Tube in the dorsoventral signal transduction pathway of Drosophila embryo. *Nature* 372, 563–566.

Jiang, Z., Georgel, P., Li, C., Choe, J., Crozat, K., Rutschmann, S., Du, X., Bigby, T., Mudd, S., Sovath, S., Wilson, I. A., Olson, A., and Beutler, B. (2006). Details of Toll-like receptor:adapter interaction revealed by germ-line mutagenesis. *Proc. Natl. Acad. Sci. USA* 103, 10961–10966.

Kagan, J. C., Su, T., Horng, T., Chow, A., Akira, S., and Medzhitov, R. (2008). TRAM couples endocytosis of Toll-like receptor 4 to the induction of interferon-beta. *Nat. Immunol.* 9, 361–368.

Lemaitre, B., Nicolas, E., Michaut, L., Reichhart, J. M., and Hoffmann, J. A. (1996). The dorsoventral regulatory gene cassette spatzle/Toll/cactus controls the potent antifungal response in Drosophila adults. *Cell* 86, 973–983.

Medzhitov, R., Preston-Hurlburt, P., and Janeway, C. A., Jr. (1997). A human homologue of the Drosophila Toll protein signals activation of adaptive immunity. *Nature* 388, 394–397.

Oshiumi, H., Matsumoto, M., Funami, K., Akazawa, T., and Seya, T. (2003). TICAM-1, an adaptor molecule that participates in Toll-like receptor 3-mediated interferon-beta induction. *Nat. Immunol.* 4, 161–167.

Poltorak, A., He, X., Smirnova, I., Liu, M. Y., Van Huffel, C., Du, X., Birdwell, D., Alejos, E., Silva, M., Galanos, C., Freudenberg, M.,

Ricciardi-Castagnoli, P., Layton, B., and Beutler, B. (1998). Defective LPS signaling in C3H/HeJ and C57BL/10ScCr mice: mutations in Tlr4 gene. *Science* 282, 2085–2088.

Rock, F. L., Hardiman, G., Timans, J. C., Kastelein, R. A., and Bazan, J. F. (1998). A family of human receptors structurally related to Drosophila Toll. *Proc. Natl. Acad. Sci. USA* 95, 588–593.

Rogozin, I. B., Iyer, L. M., Liang, L., Glazko, G. V., Liston, V. G., Pavlov, Y. I., Aravind, L., and Pancer, Z. (2007). Evolution and diversification of lamprey antigen receptors: evidence for involvement of an AID-APOBEC family cytosine deaminase. *Nat. Immunol.* 8, 647–656.

Sen, R. and Baltimore, D. (1986). Inducibility of kappa immunoglobulin enhancer-binding protein Nf-kappa B by a posttranslational mechanism. *Cell* 47, 921–928.

Wesche, H., Henzel, W. J., Shillinglaw, W., Li, S., and Cao, Z. (1997). MyD88: an adapter that recruits IRAK to the IL-1 receptor complex. *Immunity* 7, 837–847.

18.5 Ig Genes Are Assembled from Discrete DNA Segments in B Lymphocytes

Reviews

Cobb, R. M., Oestreich, K. J., Osipovich, O. A., and Oltz, E.M. (2006). Accessibility control of V(D)J recombination. *Adv. Immunol.* 91, 45–109.

Jung, D., Giallourakis, C., Mostoslavsky, R., and Alt, F. W. (2006). Mechanism and control of V(D)J recombination at the immunoglobulin heavy chain locus. *Annu. Rev. Immunol.* 24, 541–570.

Kuo, T. C. and Schlissel, M. S. (2009). Mechanisms controlling expression of the RAG locus during lymphocyte development. *Curr. Opin. Immunol.* 21, 173–178.

Schatz, D. G. (2004). V(D)J recombination. *Immunol. Rev.* 200, 5–11.

Research

Hozumi, N. and Tonegawa, S. (1976). Evidence for somatic rearrangement of immunoglobulin genes coding for variable and constant regions. *Proc. Natl. Acad. Sci. USA* 73, 3628–3632.

Schatz, D. G., Oettinger, M.A., and Baltimore, D. (1989). The V(D)J recombination activating gene, RAG-1. *Cell* 59, 1035–1048.

18.6 L Chains Are Assembled by a Single Recombination Event

Reviews

Langerak, A. W. and van Dongen, J. J. (2006). Recombination in the human Igκ locus. *Crit. Rev. Immunol.* 26, 23–42.

Schlissel, M. S. (2004). Regulation of activation and recombination of the murine Igκ locus. *Immunol. Rev.* 200, 215–223.

Research

Johnson, K., Hashimshony, T., Sawai, C. M., Pongubala, J. M., Skok, J. A., Aifantis, I., and Singh, H. (2008). Regulation of immunoglobulin light-chain recombination by the transcription factor IRF-4 and the attenuation of interleukin-7 signaling. *Immunity* 28, 335–345.

Lewis, S., Gifford, A., and Baltimore, D. (1985). DNA elements are asymmetrically joined during the site-specific recombination of kappa immunoglobulin genes. *Science* 228, 677–685.

Max, E. E., Seidman, J. G., and Leder, P. (1979). Sequences of five potential recombination sites encoded close to an immunoglobulin kappa constant region gene. *Proc. Natl. Acad. Sci. USA* 76, 3450–3454.

18.8 Recombination Generates Extensive Diversity

Reviews

Fugmann, S. D., Lee, A. I., Shockett, P. E., Villey, I. J., and Schatz, D. G. (2000). The RAG proteins and V(D)J recombination: complexes, ends, and transposition. *Annu. Rev. Immunol.* 18, 495–527.

Schatz, D. G. and Spanopoulou, E. (2005). Biochemistry of V(D)J recombination. *Curr. Top. Microbiol. Immunol.* 290, 49–85.

Research

Curry, J. D., Geier, J. K., and Schlissel, M.S. (2005). Single-strand recombination signal sequence nicks *in vivo*: evidence for a capture model of synapsis. *Nat. Immunol.* 6, 1272–1279.

Du, H., Ishii, H., Pazin, M. J., and Sen, R. (2008). Activation of 12/23-RSS-dependent RAG cleavage by hSWI/SNF complex in the absence of transcription. *Mol. Cell* 31, 641–649.

Melek, M. and Gellert, M. (2000). RAG1/2-mediated resolution of transposition intermediates: two pathways and possible consequences. *Cell* 101, 625–633.

Qiu, J. X., Kale, S. B., Yarnell Schultz, H., and Roth, D. B. (2001). Separation-of-function mutants reveal critical roles for RAG2 in both the cleavage and joining steps of V(D)J recombination. *Mol. Cell* 7, 77–87.

18.11 Allelic Exclusion Is Triggered by Productive Rearrangements

Reviews

Cedar, H. and Bergman, Y. (2008). Choreography of Ig allelic exclusion. *Curr. Opin. Immunol.* 20, 308–317.

Corcoran, A. E. (2005). Immunoglobulin locus silencing and allelic exclusion. *Semin. Immunol.* 17, 141–154.

Perlot, T. and Alt, F. W. (2008). Cis-regulatory elements and epigenetic changes control genomic rearrangements of the IgH locus. *Adv. Immunol.* 99, 1–32.

Research

Hewitt, S. L., Farmer, D., Marszalek, K., Cadera, E., Liang, H. E., Xu, Y., Schlissel, M. S., and Skok, J. A. (2008). Association between the Igκ and Igh immunoglobulin loci mediated by the 3' Igκ enhancer induces 'decontraction' of the IgH locus in pre-B cells. *Nat. Immunol.* 9, 396–404.

Liang, H. E., Hsu, L. Y., Cado, D., and Schlissel, M. S. (2004). Variegated transcriptional activation of the immunoglobulin kappa locus in pre-B cells contributes to the allelic exclusion of light-chain expression. *Cell* 118, 19–29.

18.12 RAG1/RAG2 Catalyze Breakage and Religation of V(D)J Gene Segments

Reviews

Bergeron, S., Anderson, D. K., and Swanson, P. C. (2006). RAG and HMGB1 proteins: purification and biochemical analysis of recombination signal complexes. *Methods Enzymol.* 408, 511–528.

Jung, D., Giallourakis, C., Mostoslavsky, R., and Alt, F. W. (2006). Mechanism and control of V(D)J recombination at the immunoglobulin heavy chain locus. *Annu. Rev. Immunol.* 24, 541–570.

Schatz, D. G. and Spanopoulou, E. (2005). Biochemistry of V(D)J recombination. *Curr. Top. Microbiol. Immunol.* 290, 49–85.

Research

Deriano, L., Stracker, T. H., Baker, A., Petrini, J. H., and Roth, D. B. (2009). Roles for NBS1 in alternative nonhomologous end-joining of V(D)J recombination intermediates. *Mol. Cell* 34, 13–25.

Difilippantonio, S., Gapud, E., Wong, N., Huang, C. Y., Mahowald, G., Chen, H. T., Kruhlak, M. J., Callen, E., Livak, F., Nussenzweig, M. C., Sleckman, B. P., and Nussenzweig, A. (2008). 53BP1 facilitates long-range DNA end-joining during V(D)J recombination. *Nature* 456, 529–533.

Lu, C. P., Sandoval, H., Brandt, V. L., Rice, P. A., and Roth, D. B. (2006). Amino acid residues in Rag1 crucial for DNA hairpin formation. *Nat. Struct. Mol. Biol.* 13, 1010–1015.

Ma, Y., Pannicke, U., Schwarz, K., and Lieber, M.R. (2002). Hairpin opening and overhang processing by an Artemis/DNA-dependent protein kinase complex in nonhomologous end joining and V(D)J recombination. *Cell* 108, 781–794.

Tsai, C. L., Drejer, A. H., and Schatz, D. G. (2002). Evidence of a critical architectural function for the RAG proteins in end processing, protection, and joining in V(D)J recombination. *Genes Dev.* 16, 1934–1949.

Yarnell Schultz, H., Landree, M. A., Qiu, J. X., Kale, S. B., and Roth, D. B. (2001). Joining-deficient RAG1 mutants block V(D)J recombination *in vivo* and hairpin opening in vitro. *Mol. Cell* 7, 65–75.

18.14 Class Switching Is Effected by DNA Recombination

Reviews

Honjo, T., Kinoshita, K., and Muramatsu, M. (2002). Molecular mechanism of class switch recombination: linkage with somatic hypermutation. *Annu. Rev. Immunol.* 20, 165–196.

Li, Z., Woo, C. J., Iglesias-Ussel, M. D., Ronai, D., and Scharff, M. D. (2004). The generation of antibody diversity through somatic hypermutation and class switch recombination. *Genes Dev.* 18, 1–11.

Stavnezer, J., Guikema, J. E., and Schrader, C. E. (2008). Mechanism and regulation of class switch recombination. *Annu. Rev. Immunol.* 26, 261–292.

Wang, C. L. and Wabl, M. (2004). DNA acrobats of the Ig class switch. *J. Immunol.* 172, 5815–5821.

Xu, Z., Pone, E. J., Al-Qahtani, A., Park, S. R., Zan, H., and Casali, P. (2007). Regulation of aicda expression and AID activity: relevance to somatic hypermutation and class switch DNA recombination. *Crit. Rev. Immunol.* 27, 367–397.

Yang, S. Y. and Schatz, D. G. (2007). Targeting of AID-mediated sequence diversification by cis-acting determinants. *Adv. Immunol.* 94, 109–125.

Research

Basu, U., Chaudhuri, J., Alpert, C., Dutt, S., Ranganath, S., Li, G., Schrum, J. P., Manis, J. P., and Alt, F. W. (2005). The AID antibody diversification enzyme is regulated by protein kinase A phosphorylation. *Nature* 438, 508–511.

Geisberger, R., Rada, C., and Neuberger, M.S. (2009). The stability of AID and its function in class-switching are critically sensitive to the identity of its nuclear-export sequence. *Proc. Natl. Acad. Sci. USA* 106, 6736–6741.

Iwasato, T., Shimizu, A., Honjo, T., and Yamagishi, H. (1990). Circular DNA is excised by immunoglobulin class switch recombination. *Cell* 62, 143–149.

Kinoshita, K., Tashiro, J., Tomita, S., Lee, C. G., and Honjo, T. (1998). Target specificity of immunoglobulin class switch recombination is not determined by nucleotide sequences of S regions. *Immunity* 9, 849–858.

Kinoshita, K., Harigai, M., Fagarasan, S., Muramatsu, M., and Honjo, T. (2001). A hallmark of active class switch recombination: transcripts directed by I promoters on looped-out circular DNAs. *Proc. Natl. Acad. Sci. USA* 98, 12620–12623.

Matsuoka, M., Yoshida, K., Maeda, T., Usuda, S., and Sakano, H. (1990). Switch circular DNA formed in cytokine-treated mouse splenocytes: evidence for intramolecular DNA deletion in immunoglobulin class switching. *Cell* 62, 135–142.

Muramatsu, M., Sankaranand, V. S., Anant, S., Sugai, M., Kinoshita, K., Davidson, N. O., and Honjo, T. (1999). Specific expression of activation-induced cytidine deaminase (AID), a novel member of the RNA-editing deaminase family in germinal center B cells. *J. Biol. Chem.* 274, 18470–18476.

Muramatsu, M., Kinoshita, K., Fagarasan, S., Yamada, S., Shinkai, Y., and Honjo, T. (2000). Class switch recombination and hypermutation require activation-induced cytidine deaminase (AID), a potential RNA editing enzyme. *Cell* 102, 553–563.

Muto, T., Muramatsu, M., Taniwaki, M., Kinoshita, K., and Honjo, T. (2000). Isolation, tissue distribution, and chromosomal localization of the human activation-induced cytidine deaminase (AID) gene. *Genomics* 68, 85–88.

Nagaoka, H., Muramatsu, M., Yamamura, N., Kinoshita, K., and Honjo, T. (2002). Activation-induced deaminase (AID)-directed hypermutation in the immunoglobulin Smu region: implication of AID involvement in a common step of class switch recombination and somatic hypermutation. *J. Exp. Med.* 195, 529–534.

Okazaki, I. M., Kinoshita, K., Muramatsu, M., Yoshikawa, K., and Honjo, T. (2002). The AID enzyme induces class switch recombination in fibroblasts. *Nature* 416, 340–345.

Park, S. R., Zan, H., Pal, Z., Zhang, J., Al-Qahtani, A., Pone, E. J., Xu, Z., Mai, T., and Casali, P. (2009). HoxC4 binds to the promoter of the cytidine deaminase AID gene to induce AID expression, class-switch DNA recombination and somatic hypermutation. *Nat. Immunol.* 10, 540–550.

Petersen-Mahrt, S. K., Harris, R. S., and Neuberger, M. S. (2002). AID mutates *E. coli* suggesting a DNA deamination mechanism for antibody diversification. *Nature* 418, 99–103.

Rada, C., Williams, G. T., Nilsen, H., Barnes, D. E., Lindahl, T., and Neuberger, M. S. (2002). Immunoglobulin isotype switching is inhibited and somatic hypermutation perturbed in UNG-deficient mice. *Curr. Biol.* 12, 1748–1755.

Revy, P., Muto, T., Levy, Y., Geissmann, F., Plebani, A., Sanal, O., Catalan, N., Forveille, M., Dufourcq-Labelouse, R., Gennery, A., Tezcan, I., Ersoy, F., Kayserili, H., Ugazio, A. G., Brousse, N., Muramatsu, M., Notarangelo, L. D., Kinoshita, K., Honjo, T., Fischer, A., and Durandy, A. (2000). Activation-induced cytidine deaminase (AID) deficiency causes the autosomal recessive form of the Hyper-IgM syndrome (HIGM2). *Cell* 102, 565–575.

Zarrin, A. A., Alt, F. W., Chaudhuri, J., Stokes, N., Kaushal, D., Du Pasquier, L., and Tian, M. (2004). An evolutionarily conserved target motif for immunoglobulin class-switch recombination. *Nat. Immunol.* 5, 1275–1281.

18.16 Somatic Hypermutation (SHM) Generates Additional Diversity in Mice and Humans

Reviews

Honjo, T., Kinoshita, K. and Muramatsu, M. (2002). Molecular mechanism of class switch recombination: linkage with somatic hypermutation. *Annu. Rev. Immunol.* 20, 165–196.

Kinoshita, K. and Honjo, T. (2001). Linking class-switch recombination with somatic hypermutation. *Nat. Rev. Mol. Cell. Biol.* 2, 493–503.

Neuberger, M. S. (2008). Antibody diversification by somatic mutation: from Burnet onwards. *Immunol. Cell Biol.* 86, 124–132.

Tarlinton, D. M. (2008). Evolution in miniature: selection, survival and distribution of antigen reactive cells in the germinal centre. *Immunol. Cell. Biol.* 86, 133–138.

Teng, G. and Papavasiliou, F.N. (2007). Immunoglobulin somatic hypermutation. *Annu. Rev. Genet.* 41, 107–120.

Research

Di Noia, J. and Neuberger, M. S. (2002). Altering the pathway of immunoglobulin hypermutation by inhibiting uracil-DNA glycosylase. *Nature* 419, 43–48.

Muramatsu, M., Kinoshita, K., Fagarasan, S., Yamada, S., Shinkai, Y., and Honjo, T. (2000). Class switch recombination and hypermutation require activation-induced cytidine deaminase (AID), a potential RNA editing enzyme. *Cell* 102, 553–563.

Peters, A. and Storb, U. (1996). Somatic hypermutation of immunoglobulin genes is linked to transcription initiation. *Immunity* 4, 57–65.

18.17 SHM Is Mediated by AID, Ung, Elements of the Mismatch DNA Repair (MMR) Machinery, and Translesion DNA Synthesis (TLS) Polymerases

Reviews

Casali, P., Pal, Z., Xu, Z., and Zan, H. (2006). DNA repair in antibody somatic hypermutation. *Trends Immunol.* 27, 313–321.

Di Noia, J. M. and Neuberger, M. S. (2007). Molecular mechanisms of antibody somatic hypermutation. *Annu. Rev. Biochem.* 76, 1–22.

Honjo, T., Kinoshita, K., and Muramatsu, M. (2002). Molecular mechanism of class switch recombination: linkage with somatic hypermutation. *Annu. Rev. Immunol.* 20, 165–196.

Jiricny, J. (2006). The multifaceted mismatch-repair system. *Nat. Rev. Mol. Cell. Biol.* 7, 335–346.

Kinoshita, K. and Honjo, T. (2001). Linking class-switch recombination with somatic hypermutation. *Nat. Rev. Mol. Cell Biol.* 2, 493–503.

Odegard, V. H. and Schatz, D. G. (2006). Targeting of somatic hypermutation. *Nat. Rev. Immunol.* 6, 573–583.

Peled, J. U., Kuang, F. L., Iglesias-Ussel, M. D., Roa, S., Kalis, S. L., Goodman, M. F., and Scharff, M. D. (2008). The biochemistry of somatic hypermutation. *Annu. Rev. Immunol.* 26, 481–511.

Reynaud, C. A., Delbos, F., Faili, A., Gueranger, Q., Aoufouchi, S., and Weill, J. C. (2009). Competitive repair pathways in immunoglobulin gene hypermutation. *Philos. Trans. R. Soc. Lond. B Biol. Sci.* 364, 613–619.

Weill, J. C. and Reynaud, C. A. (2008). DNA polymerases in adaptive immunity. *Nat. Rev. Immunol.* 8, 302–312.

Xu, Z., Fulop, Z., Zhong, Y., Evinger, A. J., 3rd, Zan, H., and Casali, P. (2005). DNA lesions and repair in immunoglobulin class switch recombination and somatic hypermutation. *Ann. N. Y. Acad. Sci.* 1050, 146–162.

Xu, Z., Zan, H., Pal, Z., and Casali, P. (2007). DNA replication to aid somatic hypermutation. *Adv. Exp. Med. Biol.* 596, 111–127.

Research

Aoufouchi, S., Faili, A., Zober, C., D'Orlando, O., Weller, S., Weill, J. C., and Reynaud, C. A. (2008). Proteasomal degradation restricts the nuclear lifespan of AID. *J. Exp. Med.* 205, 1357–1368.

Di Noia, J. and Neuberger, M. S. (2002). Altering the pathway of immunoglobulin hypermutation by inhibiting uracil-DNA glycosylase. *Nature* 419, 43–48.

Muramatsu, M., Kinoshita, K., Fagarasan, S., Yamada, S., Shinkai, Y., and Honjo, T. (2000). Class switch recombination and hypermutation require activation-induced cytidine deaminase (AID), a potential RNA editing enzyme. *Cell* 102, 553–563.

Peters, A. and Storb, U. (1996). Somatic hypermutation of immunoglobulin genes is linked to transcription initiation. *Immunity* 4, 57–65.

Rada, C., Di Noia, J. M., and Neuberger, M. S. (2004). Mismatch recognition and uracil excision provide complementary paths to both Ig switching and the A/T-focused phase of somatic mutation. *Mol. Cell* 16, 163–171.

Zan, H., Komori, A., Li, Z., Cerutti, A., Schaffer, A., Flajnik, M. F., Diaz, M., and Casali, P. (2001). The translesion DNA polymerase zeta plays a major role in Ig and bcl-6 somatic hypermutation. *Immunity* 14: 643–653.

Zan, H., Wu, X., Komori, A., Holloman, W. K., and Casali, P. (2003). AID-dependent generation of resected double-strand DNA breaks and recruitment of Rad52/Rad51 in somatic hypermutation. *Immunity* 18, 727–738.

Zan, H., Shima, N., Xu, Z., Al-Qahtani, A., Evinger Iii, A. J., Zhong, Y., Schimenti, J. C., and Casali, P. (2005). The translesion DNA polymerase theta plays a dominant role in immunoglobu-lin gene somatic hypermutation. *EMBO J.* 24, 3757–3769.

18.18 Avian Igs Are Assembled from Pseudogenes

Reviews

Arakawa, H. and Buerstedde, J. M. (2004). Immunoglobulin gene conversion: insights from bursal B cells and the DT40 cell line. *Dev. Dyn.* 229, 458–464.

Ratcliffe, M. J. (2006). Antibodies, immunoglobulin genes and the bursa of Fabricius in chicken B cell development. *Dev. Comp. Immunol.* 30, 101–118.

Research

Reynaud, C. A., Anquez, V., Grimal, H., and Weill, J. C. (1987). A hyperconversion mechanism generates the chicken light chain preimmune repertoire. *Cell* 48, 379–388.

Sale, J. E., Calandrini, D. M., Takata, M., Takeda, S., and Neuberger, M. S. (2001). Ablation of XRCC2/3 transforms immunoglobulin V gene conversion into somatic hypermutation. *Nature* 412, 921–926.

18.19 B Cell Memory Allows for the Mounting of a Prompt and Strong Secondary Response

Reviews

Chappell, C. P. and Jacob, J. (2007). Germinal-center-derived B-cell memory. *Adv. Exp. Med. Biol.* 590, 139–148.

Crotty, S. and Ahmed, R. (2004). Immunological memory in humans. *Semin. Immunol.* 16, 197–203.

Gourley, T. S., Wherry, E. J., Masopust, D., and Ahmed, R. (2004). Generation and maintenance of immunological memory. *Semin. Immunol.* 16, 323–333.

Lanzavecchia, A., Bernasconi, N., Traggiai, E., Ruprecht, C. R., Corti, D., and Sallusto, F. (2006). Understanding and making use of human memory B cells. *Immunol. Rev.* 211, 303–309.

Pulendran, B. and Ahmed, R. (2006). Translating innate immunity into immunological memory: implications for vaccine development. *Cell* 124, 849–863.

Rajewsky, K. (1996). Clonal selection and learning in the antibody system. *Nature* 381, 751–758.

Research

Bernasconi, N. L., Traggiai, E., and Lanzavecchia, A. (2002). Maintenance of serological memory by polyclonal activation of human memory B cells. *Science* 298, 2199–2202.

18.20 The TCR Is Related to the BCR

Reviews

Cobb, R. M., Oestreich, K. J., Osipovich, O. A., and Oltz, E. M. (2006). Accessibility control of V(D)J recombination. *Adv. Immunol.* 91, 45–109.

Davis, M. M. (1990). T cell receptor gene diversity and selection. *Annu. Rev. Biochem.* 59, 475–496.

Goldrath, A. W. and Bevan, M. J. (1999). Selecting and maintaining a diverse T-cell repertoire. *Nature* 402, 255–262.

Surh, C. D. and Sprent, J. (2008). Homeostasis of naive and memory T cells. *Immunity* 29, 848–862.

Taghon, T. and Rothenberg, E. V. (2008). Molecular mechanisms that control mouse and human TCR-alphabeta and TCR-gammadelta T cell development. *Semin. Immunopathol.* 30, 383–398.

Research

Abarrategui, I. and Krangel, M.S. (2006). Regulation of T cell receptor-alpha gene recombination by transcription. *Nat. Immunol.* 7, 1109–1115.

Goldrath, A. W. and Bevan, M. J. (1999). Low-affinity ligands for the TCR drive proliferation of mature CD8+ T cells in lymphopenic hosts. *Immunity* 11, 183–190.

Jackson, A. M. and Krangel, M. S. (2006). Turning T-cell receptor beta recombination on and off: more questions than answers. *Immunol. Rev.* 209, 129–141.

Oestreich, K. J., Cobb, R. M., Pierce, S., Chen, J., Ferrier, P., and Oltz, E. M. (2006). Regulation of TCRbeta gene assembly by a promoter/enhancer holocomplex. *Immunity* 24, 381–391.

Wucherpfennig, K. W. (2005). The structural interactions between T cell receptors and MHC-peptide complexes place physical limits on self-nonself discrimination. *Curr. Top. Microbiol. Immunol.* 296, 19–37.

18.21 The TCR Functions in Conjunction with the MHC

Reviews

Collins, E. J. and Riddle, D. S. (2008). TCR-MHC docking orientation: natural selection, or thymic selection? *Immunol. Res.* 41, 267–294.

Garcia, K. C., Adams, J. J., Feng, D., and Ely, L. K. (2009). The molecular basis of TCR germline bias for MHC is surprisingly simple. *Nat. Immunol.* 10, 143–147.

Godfrey, D. I., Rossjohn, J., and McCluskey, J. (2008). The fidelity, occasional promiscuity, and versatility of T cell receptor recognition. *Immunity* 28, 304–314.

Mazza, C. and Malissen, B. (2007). What guides MHC-restricted TCR recognition? *Semin. Immunol.* 19, 225–235.

Peterson, P., Org, T., and Rebane, A. (2008). Transcriptional regulation by AIRE: molecular mechanisms of central tolerance. *Nat. Rev. Immunol.* 8, 948–957.

Rudolph, M. G., Stanfield, R. L., and Wilson, I. A. (2006). How TCRs bind MHCs, peptides, and coreceptors. *Annu. Rev. Immunol.* 24: 419–466.

Wucherpfennig, K. W. (2005). The structural interactions between T cell receptors and MHC-peptide complexes place physical limits on self-nonself discrimination. *Curr. Top. Microbiol. Immunol.* 296, 19–37.

Research

Borg, N. A., Ely, L. K., Beddoe, T., Macdonald, W. A., Reid, H. H., Clements, C. S., Purcell, A. W., Kjer-Nielsen, L., Miles, J. J., Burrows, S. R., McCluskey, J., and Rossjohn, J. (2005). The CDR3 regions of an immunodominant T cell receptor dictate the 'energetic landscape' of peptide-MHC recognition. *Nat. Immunol.* 6, 171–180.

Feng, D., Bond, C. J., Ely, L. K., Maynard, J., and Garcia, K. C. (2007). Structural evidence for a germline-encoded T cell receptor-major histocompatibility complex interaction 'codon.' *Nat. Immunol.* 8, 975–983.

Gras, S., Burrows, S. R., Kjer-Nielsen, L., Clements, C. S., Liu, Y. C., Sullivan, L. C., Bell, M. J., Brooks, A. G., Purcell, A. W., McCluskey, J., and Rossjohn, J. (2009). The shaping of T cell receptor recognition by self-tolerance. *Immunity* 30, 193–203.

Kosmrlj, A., Jha, A. K., Huseby, E. S., Kardar, M., and Chakraborty, A. K. (2008). How the thymus designs antigen-specific and self-tolerant T cell receptor sequences. *Proc. Natl. Acad. Sci. USA* 105, 16671–16676.

18.22 The Major Histocompatibility Locus Comprises a Cohort of Genes Involved in Immune Recognition

Reviews

Blackwell, J. M., Jamieson, S. E., and Burgner, D. (2009). HLA and infectious diseases. *Clin. Microbiol. Rev.* 22, 370–385.

Charron, D. (2005). Immunogenetics today: HLA, MHC and much more. *Curr. Opin. Immunol.* 17, 493–497.

Kumanovics, A., Takada, T., and Lindahl, K. F. (2003). Genomic organization of the mammalian MHC. *Annu. Rev. Immunol.* 21, 629–657.

Piertney, S. B. and Oliver, M. K. (2006). The evolutionary ecology of the major histocompatibility complex. *Heredity* 96, 7–21.

Shiina, T., Hosomichi, K., Inoko, H., and Kulski, J. K. (2009). The HLA genomic loci map: expression, interaction, diversity and disease. *J. Hum. Genet.* 54: 15–39.

Steinmetz, M. and Hood, L. (1983). Genes of the major histocompatibility complex in mouse and man. *Science* 222, 727–733.

Trowsdale, J. (2005). HLA genomics in the third millennium. *Curr. Opin. Immunol.* 17, 498–504.

Research

de Bakker, P. I., McVean, G., Sabeti, P. C., Miretti, M. M., Green, T., Marchini, J., Ke, X.,

Monsuur, A. J., Whittaker, P., Delgado, M., Morrison, J., Richardson, A., Walsh, E.C., Gao, X., Galver, L., Hart, J., Hafler, D. A., Pericak-Vance, M., Todd, J. A., Daly, M. J., Trowsdale, J., Wijmenga, C., Vyse, T. J., Beck, S., Murray, S.S ., Carrington, M., Gregory, S., Deloukas, P., and Rioux, J. D. (2006). A high-resolution HLA and SNP haplotype map for disease association studies in the extended human MHC. *Nat. Genet.* 38, 1166–1172.

Gregersen, J. W., Kranc, K. R., Ke, X., Svendsen, P., Madsen, L. S., Thomsen, A. R., Cardon, L. R., Bell, J. I., and Fugger, L. (2006). Functional epis-tasis on a common MHC haplotype associated with multiple sclerosis. *Nature* 443, 574–577.

Guo, Z., Hood, L., Malkki, M., and Petersdorf, E.W. (2006). Long-range multilocus haplotype phasing of the MHC. *Proc. Natl. Acad. Sci. USA* 103, 6964–6969.

Nejentsev, S., Howson, J. M., Walker, N. M., Szeszko, J., Field, S. F., Stevens, H. E., Reynolds, P., Hardy, M., King, E., Masters, J., Hulme, J., Maier, L. M., Smyth, D., Bailey, R., Cooper, J. D., Ribas, G., Campbell, R. D., Clayton, D. G., and Todd, J. A. (2007). Localization of type 1 diabetes susceptibility to the MHC class I genes HLA-B and HLA-A. *Nature* 450, 887–892.

Transcription and Posttranscriptional Mechanisms

PART 3

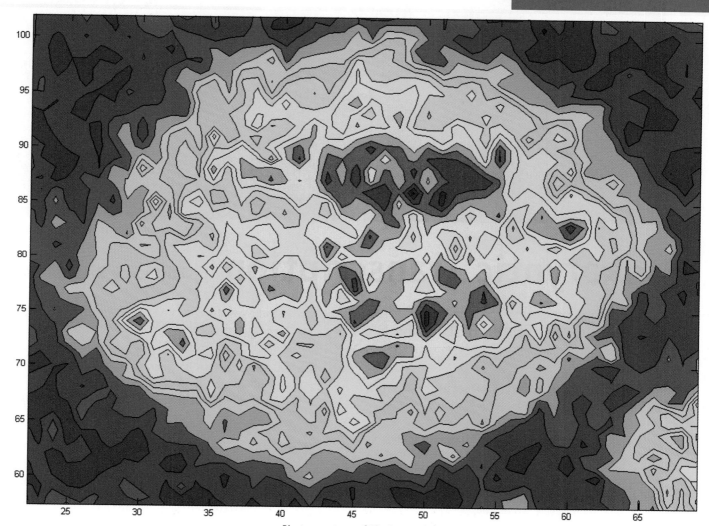

Photo courtesy of Jie Ren and Jun Ma, University of Cincinnati College of Medicine and Cincinnati Children's Hospital.

19

© Phantatomix/Photo Researchers, Inc.

Prokaryotic Transcription **Edited by Richard Gourse**

CHAPTER OUTLINE

19.1 Introduction
- Transcription is 5′ to 3′ on a template that is 3′ to 5′.

19.2 Transcription Occurs by Base Pairing in a "Bubble" of Unpaired DNA
- RNA polymerase separates the two strands of DNA in a transient "bubble" and uses one strand as a template to direct synthesis of a complementary sequence of RNA.
- The bubble is 12 to 14 bp, and the RNA–DNA hybrid within the bubble is 8 to 9 bp.

19.3 The Transcription Reaction Has Three Stages
- RNA polymerase binds to a promoter site on DNA to form a closed complex.
- RNA polymerase initiates transcription after opening the DNA duplex to form a transcription bubble.
- During elongation the transcription bubble moves along DNA and the RNA chain is extended in the 5′→3′ direction by adding nucleotides to the 3′ end.
- Transcription stops and the DNA duplex reforms when RNA polymerase dissociates at a terminator site.

19.4 Bacterial RNA Polymerase Consists of Multiple Subunits
- Bacterial RNA core polymerases are ~400 kD multisubunit complexes with the general structure $\alpha_2\beta\beta'\omega$.
- Catalysis derives from the β and β' subunits.

19.5 RNA Polymerase Holoenzyme Consists of the Core Enzyme and Sigma Factor
- Bacterial RNA polymerase can be divided into the $\alpha_2\beta\beta'\omega$ core enzyme that catalyzes transcription and the σ subunit that is required only for initiation.
- Sigma factor changes the DNA-binding properties of RNA polymerase so that its affinity for general DNA is reduced and its affinity for promoters is increased.

19.6 How Does RNA Polymerase Find Promoter Sequences?
- The rate at which RNA polymerase binds to promoters can be too fast to be accounted for by simple diffusion.
- RNA polymerase binds to random sites on DNA and exchanges them with other sequences until a promoter is found.

19.7 The Holoenzyme Goes through Transitions in the Process of Recognizing and Escaping from Promoters

- When RNA polymerase binds to a promoter, it separates the DNA strands to form a transcription bubble and incorporates nucleotides into RNA.
- There may be a cycle of abortive initiations before the enzyme moves to the next phase.
- Sigma factor is usually released from RNA polymerase when the nascent RNA chain reaches ~10 bases in length.

19.8 Sigma Factor Controls Binding to DNA by Recognizing Specific Sequences in Promoters

- A promoter is defined by the presence of short consensus sequences at specific locations.
- The promoter consensus sequences usually consist of a purine at the startpoint, a hexamer with a sequence close to TATAAT centered at ~ −10, and another hexamer with a sequence similar to TTGACA centered at ~ −35.
- Individual promoters usually differ from the consensus at one or more positions.
- Promoter efficiency can be affected by additional elements as well.

19.9 Promoter Efficiencies Can Be Increased or Decreased by Mutation

- Down mutations to decrease promoter efficiency usually decrease conformance to the consensus sequences, whereas up mutations have the opposite effect.
- Mutations in the −35 sequence can affect initial binding of RNA polymerase.
- Mutations in the −10 sequence can affect binding or the melting reaction that converts a closed to an open complex.

19.10 Multiple Regions in RNA Polymerase Directly Contact Promoter DNA

- The structure of σ^{70} changes when it associates with core enzyme, allowing its DNA-binding regions to interact with the promoter.
- Multiple regions in σ^{70} interact with the promoter.
- The α subunit also contributes to promoter recognition.

19.11 Footprinting Is a High Resolution Method for Characterizing RNA Polymerase–Promoter and DNA–Protein Interactions in General

- The consensus sequences at −35 and −10 provide most of the contact points for RNA polymerase in the promoter.
- The points of contact lie primarily on one face of the DNA.

19.12 Interactions between Sigma Factor and Core RNA Polymerase Change During Promoter Escape

- A domain in sigma occupies the RNA exit channel and must be displaced to accommodate RNA synthesis.

- Abortive initiations usually occur before the enzyme forms a true elongation complex.
- Sigma factor is usually released from RNA polymerase by the time the nascent RNA chain reaches ~10 nt in length.

19.13 A Model for Enzyme Movement Is Suggested by the Crystal Structure

- DNA moves through a channel in RNA polymerase and makes a sharp turn at the active site.
- Changes in the conformations of certain flexible modules within the enzyme control the entry of nucleotides to the active site.

19.14 A Stalled RNA Polymerase Can Restart

- An arrested RNA polymerase can restart transcription by cleaving the RNA transcript to generate a new 3′ end.

19.15 Bacterial RNA Polymerase Terminates at Discrete Sites

- There are two classes of terminators: Those recognized solely by RNA polymerase itself without the requirement for any cellular factors are usually referred to as "intrinsic terminators." Others require a cellular protein called rho and are referred to as "rho-dependent terminators."
- Intrinsic termination requires recognition of a terminator sequence in DNA that codes for a hairpin structure in the RNA product.
- The signals for termination lie mostly within *sequences already transcribed* by RNA polymerase, and thus termination relies on scrutiny of the template and/or the RNA product that the polymerase is transcribing.

19.16 How Does Rho Factor Work?

- Rho factor is a protein that binds to nascent RNA and tracks along the RNA to interact with RNA polymerase and release it from the elongation complex.

19.17 Supercoiling Is an Important Feature of Transcription

- Negative supercoiling increases the efficiency of some promoters by assisting the melting reaction.
- Transcription generates positive supercoils ahead of the enzyme and negative supercoils behind it, and these must be removed by gyrase and topoisomerase.

19.18 Phage T7 RNA Polymerase Is a Useful Model System

- The T7 family of RNA polymerases are single polypeptides with the ability to recognize phage promoters and carry out many of the activities of the multisubunit RNA polymerases.
- Crystal structures of T7 family RNA polymerases with DNA identify the DNA-binding region, the active site, and suggest models for promoter escape.

19.19 Competition for Sigma Factors Can Regulate Initiation

- *E. coli* has seven sigma factors, each of which causes RNA polymerase to initiate at a set of promoters defined by specific −35 and −10 sequences.

- The activities of the different sigma factors are regulated by different mechanisms.

19.20 Sigma Factors May Be Organized into Cascades

- A cascade of sigma factors is created when one sigma factor is required to transcribe the gene coding for the next sigma factor.
- The early genes of phage SPO1 are transcribed by host RNA polymerase.
- One of the early genes codes for a sigma factor that causes RNA polymerase to transcribe the middle genes.
- Two of the middle genes code for subunits of a sigma factor that causes RNA polymerase to transcribe the late genes.

19.21 Sporulation Is Controlled by Sigma Factors

- Sporulation divides a bacterium into a mother cell that is lysed and a spore that is released.
- Each compartment advances to the next stage of development by synthesizing a new sigma factor that displaces the previous sigma factor.
- Communication between the two compartments coordinates the timing of sigma factor substitutions.

19.22 Antitermination Can Be a Regulatory Event

- An antitermination complex allows RNA polymerase to read through terminators.
- Phage lambda uses antitermination systems for regulation of both its early and late transcripts, but the two systems work by completely different mechanisms.
- Binding of factors to the nascent RNA links the antitermination proteins to the terminator site through an RNA loop.
- Antitermination of transcription also occurs in rRNA operons.

19.23 The Cycle of Bacterial Messenger RNA

- Transcription and translation occur simultaneously in bacteria, coupled transcription/translation, as ribosomes begin translating an mRNA before its synthesis has been completed.
- Bacterial mRNA is unstable and has a half-life of only a few minutes.
- A bacterial mRNA may be polycistronic in having several coding regions that represent different genes.

19.24 Summary

19.1 Introduction

Key concept

- Transcription is 5′ to 3′ on a template that is 3′ to 5′.

Transcription produces an RNA chain *identical in sequence* with one strand of the DNA, sometimes called the **coding strand**. This strand is made 5′→3′ and is *complementary* to (i.e., it base-pairs with) the **template**, which is 3′→5′. The RNA-like strand therefore is called the **non-template strand**, and the one that serves as the template for synthesis of the RNA is called the *template strand*, as seen in **FIGURE 19.1**.

RNA synthesis is catalyzed by the enzyme **RNA polymerase**. Transcription starts when RNA polymerase binds to a special region, called the **promoter**, at the start of the gene. The promoter includes the first base pair that is transcribed into RNA (the **startpoint**), as well as surrounding bases. From this position, RNA polymerase moves along the template, synthesizing RNA until it reaches a **terminator** sequence, where the transcript ends. Thus, a **transcription unit** extends from the promoter to the terminator. The critical feature of the transcription unit, depicted in **FIGURE 19.2**,

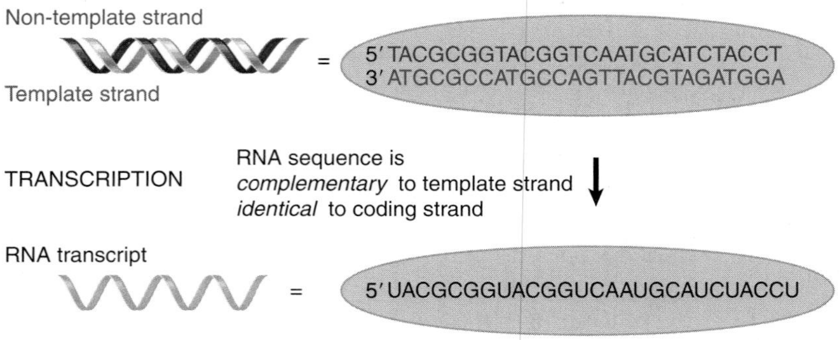

Non-template strand

Template strand

= 5′ TACGCGGTACGGTCAATGCATCTACCT
3′ ATGCGCCATGCCAGTTACGTAGATGGA

TRANSCRIPTION

RNA sequence is *complementary* to template strand *identical* to coding strand

RNA transcript

= 5′ UACGCGGUACGGUCAAUGCAUCUACCU

FIGURE 19.1 The function of RNA polymerase is to copy one strand of duplex DNA into RNA.

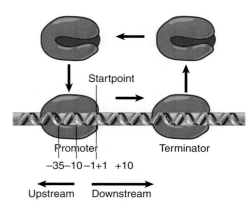

FIGURE 19.2 A transcription unit is a sequence of DNA transcribed into a single RNA, starting at the promoter and ending at the terminator.

Startpoint

Promoter Terminator

−35−10−1+1 +10

Upstream Downstream

is that it constitutes a stretch of DNA used as a template for the production of a *single* RNA molecule. A transcription unit may encode more than one gene.

Sequences prior to the startpoint are described as **upstream** of it; those after the startpoint (within the transcribed sequence) are **downstream** of it. Sequences are usually written so that transcription proceeds from left (upstream) to right (downstream). This corresponds to writing the mRNA in the usual 5'→3' direction.

The DNA sequence often is written to show only the nontemplate strand, which (as mentioned above) has the same sequence as the RNA. Base positions are numbered in both directions away from the startpoint, which is called +1; numbers increase as they go downstream. The base before the startpoint is numbered −1, and the negative numbers increase going upstream. (There is no base assigned the number 0.)

The initial transcription product, containing the original 5' end, is called the **primary transcript**. rRNA and tRNA primary transcripts go through a maturation process in which sequences at the ends are cleaved off ("processed") by endonucleases. The mature products from rRNA and tRNA operons are stable, approaching the generation time of the bacterium. In contrast, mRNA primary transcripts are subject to almost immediate attack by endonucleases and exonucleases. Thus, bacterial mRNA lifetimes average only one to three minutes. In eukaryotes, rRNA and tRNA transcripts are processed and the resulting products are stable, as in bacteria, but mRNAs are much more stable than in bacteria. Modification and decay of mRNAs will be discussed in *Section 19.23, The Cycle of Bacterial Messenger RNA.*

Transcription is the first stage in gene expression and is the step at which it is regulated most often. Regulatory factors often determine whether a particular gene is transcribed by RNA polymerase, and subsequent stages in transcription and other steps in gene expression are also regulated frequently.

Two important questions in transcription are:

- How does RNA polymerase find promoters on DNA? This is a particular example of a more general question: How do proteins distinguish their specific binding sites in DNA from other sequences?

- How do regulatory proteins interact with RNA polymerase (and with one another) to activate or to inhibit specific steps during initiation, elongation, or termination of transcription?

In this chapter, we describe the interactions of bacterial RNA polymerase with DNA from its initial contact with the promoter, through the act of transcription, to its release from the DNA when the transcript has been completed. In Chapter 20, *Eukaryotic Transcription*, we consider the analogous reactions between eukaryotic RNA polymerases and their templates. Chapter 26, *The Operon*, describes various means by which regulatory proteins and factors can assist or prevent bacterial RNA polymerase from transcribing a particular gene. In Chapter 27, *Phage Strategies*, we consider how individual regulatory interactions can be connected into more complex networks. Chapter 30, *Regulatory RNA*, discusses additional means of regulation, including the use of small RNAs, and considers how these interactions can be connected into larger regulatory networks.

19.2 Transcription Occurs by Base Pairing in a "Bubble" of Unpaired DNA

Key concepts

- RNA polymerase separates the two strands of DNA in a transient "bubble" and uses one strand as a template to direct synthesis of a complementary sequence of RNA.

- The bubble is 12 to 14 bp, and the RNA–DNA hybrid within the bubble is 8 to 9 bp.

Transcription utilizes complementary base pairing, in common with the other polymerization reactions: replication and translation. **FIGURE 19.3** illustrates the general principle of transcription. RNA synthesis takes place within a "transcription bubble," in which DNA is transiently separated into its single strands and the template strand is used to direct synthesis of the RNA strand.

The RNA chain is synthesized from the 5' end toward the 3' end by adding new nucleotides to the 3' end of the growing chain. The 3'–OH group of the last nucleotide added to the chain reacts with an incoming nucleoside 5' triphosphate. The incoming nucleotide loses its terminal two phosphate groups (γ and β); its α group is used in the phosphodiester bond linking it to the

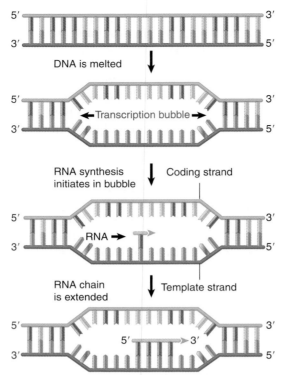

FIGURE 19.3 DNA strands separate to form a transcription bubble. RNA is synthesized by complementary base pairing with one of the DNA strands.

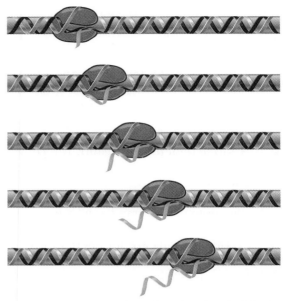

FIGURE 19.4 Transcription takes place in a bubble, in which RNA is synthesized by base pairing with one strand of DNA in the transiently unwound region. As the bubble progresses, the DNA duplex reforms behind it, displacing the RNA in the form of a single polynucleotide chain.

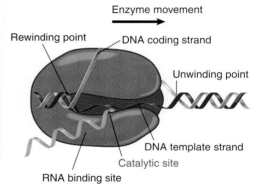

FIGURE 19.5 During transcription, the bubble is maintained within bacterial RNA polymerase, which unwinds and rewinds DNA and synthesizes RNA.

chain. The overall reaction rate for the bacterial RNA polymerase can be as fast as ~40–50 nucleotides/second at 37°C for most transcripts; this is about the same as the rate of translation (15 amino acids/sec), but much slower than the rate of DNA replication (~800 bp/sec).

RNA polymerase creates the transcription bubble when it binds to a promoter. **FIGURE 19.4** illustrates the RNA polymerase moving along the DNA, with the bubble moving with it and the RNA chain growing in length. The process of base pairing and base addition within the bubble is catalyzed and scrutinized by the RNA polymerase itself.

The structure of the bubble within the transcription complex is shown in the expanded view of **FIGURE 19.5**. As RNA polymerase moves along the DNA template, it unwinds the duplex at the front of the bubble (the unwinding point), and rewinds the DNA at the back (the rewinding point). The length of the transcription bubble is ~12 to 14 bp, but the length of the RNA–DNA hybrid within the bubble is only 8 to 9 bp. As the enzyme moves along the template, the DNA duplex reforms, and the RNA is displaced as a free polynucleotide chain. The last 14 ribonucleotides in the growing RNA are complexed with the DNA and/or the enzyme at any given moment.

19.3 The Transcription Reaction Has Three Stages

Key concepts

- RNA polymerase binds to a promoter site on DNA to form a closed complex.
- RNA polymerase initiates transcription after opening the DNA duplex to form a transcription bubble.
- During elongation the transcription bubble moves along DNA and the RNA chain is extended in the 5′→3′ direction by adding nucleotides to the 3′ end.
- Transcription stops and the DNA duplex reforms when RNA polymerase dissociates at a terminator site.

INITIATION
Template recognition: RNA polymerase binds to duplex DNA

DNA is unwound at promoter

Very short chains
are synthesized and released

ELONGATION:
Polymerase synthesizes RNA

TERMINATION:
RNA polymerase and RNA are released

FIGURE 19.6 Transcription has three stages: The enzyme binds to the promoter and melts DNA and remains stationary during initiation; moves along the template during elongation; and dissociates at termination.

The transcription reaction can be divided into the three stages illustrated in **FIGURE 19.6**: **initiation**, in which the promoter is recognized, a bubble is created, and RNA synthesis begins; **elongation**, in which the bubble moves along the DNA; and **termination**, in which the RNA transcript is released and the bubble closes.

Initiation itself can be divided into multiple steps. *Template recognition begins with the binding of RNA polymerase to the double-stranded DNA at a DNA sequence called the promoter.* The enzyme first forms a **closed complex** in which the DNA remains double-stranded. Next the enzyme locally unwinds the section of promoter DNA that includes the transcription start site to form the **open complex**. Separation of the DNA double strands makes the template strand available for base pairing with incoming ribonucleotides and synthesis of the first

nucleotide bonds in RNA. The initiation phase can be protracted by the occurrence of abortive events, in which the enzyme makes short transcripts, typically shorter than ~10 nucleotides (nt), while still bound at the promoter. The enzyme often makes successive rounds of abortive transcripts by releasing them and starting RNA synthesis again. The initiation phase ends when the enzyme finally succeeds in extending the chain and clearing the promoter.

Elongation involves processive movement of the enzyme by disruption of base pairing in double-stranded DNA, exposing the template strand for nucleotide addition, and translocation of the transcription bubble downstream. As the enzyme moves, the template strand of the transiently unwound region is paired with the nascent RNA at the point of growth. Nucleotides are added covalently to the 3' end of the growing RNA chain, forming an RNA–DNA hybrid within the unwound region. Behind the unwound region, the DNA template strand pairs with its original partner to reform the double helix, and the growing strand of RNA emerges from the enzyme.

The traditional view of elongation as a monotonic process, in which the enzyme moves forward along the DNA at a steady pace corresponding to nucleotide addition, has been revised in recent years. RNA polymerase pauses or even arrests at certain sequences. Displacement of the 3' end of the RNA from the active site can cause the polymerase to "backtrack" and remove a few nucleotides from the growing RNA chain before restarting.

Termination involves recognition of sequences that signal the enzyme to halt further nucleotide addition to the RNA chain. The transcription bubble collapses as the RNA–DNA hybrid is disrupted and the DNA reforms a duplex, phosphodiester bond formation ceases, and the transcription complex dissociates into its component parts: RNA polymerase, DNA, and RNA transcript. *The sequence of DNA that directs the end of transcription is called the terminator.*

19.4 Bacterial RNA Polymerase Consists of Multiple Subunits

Key concepts

- Bacterial RNA core polymerases are ~400 kD multisubunit complexes with the general structure $\alpha_2\beta\beta'\omega$.
- Catalysis derives from the β and β' subunits.

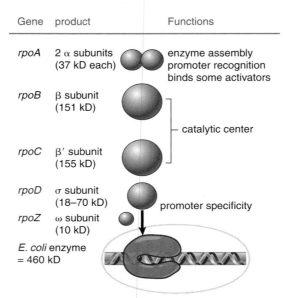

Gene	product	Functions
rpoA	2 α subunits (37 kD each)	enzyme assembly promoter recognition binds some activators
rpoB	β subunit (151 kD)	catalytic center
rpoC	β′ subunit (155 kD)	catalytic center
rpoD	σ subunit (18–70 kD)	promoter specificity
rpoZ	ω subunit (10 kD)	

E. coli enzyme = 460 kD

FIGURE 19.7 Eubacterial RNA polymerases have five types of subunits: α, β, β′ and ω have rather constant sizes in different bacterial species, but σ varies more widely.

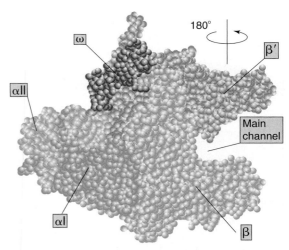

FIGURE 19.8 The upstream face of the core RNA polymerase, illustrating the 'crabclaw' shape of the enzyme. β (cyan) and β′ subunit (pink) of RNA polymerase have a channel for the DNA template. α I is shown in green and α II in yellow, ω is red. Adapted from K. M. Geszvain and R. Landick (ed. N. P. Higgins). *The Bacterial Chromosome.* American Society for Microbiology, 2004.

The best genetically and biochemically characterized RNA polymerases are from bacteria, especially *Escherichia coli*. The only bacterial RNA polymerases for which high resolution crystal structures have been solved, however, are from two thermophilic bacterial species, *Thermus aquaticus* and *Thermus thermophilus*. Nevertheless, in all bacteria *a single type of RNA polymerase is responsible for the synthesis of rRNA, mRNA, and tRNA*, unlike the situation in eukaryotes where rRNAs, mRNAs, and tRNAs typically are transcribed by different RNA polymerases, Pol I, II, and III. About 13,000 RNA polymerase molecules are present in an *E. coli* cell, although the precise number varies with the growth conditions. Although not all the RNA polymerases are actually engaged in transcription at any one time, almost all are bound either specifically or nonspecifically to DNA.

The *complete enzyme* or **holoenzyme** in *E. coli* has a molecular weight of ~460 kD. The holoenzyme ($\alpha_2\beta\beta'\omega\sigma$) can be separated into two components, the core enzyme ($\alpha_2\beta\beta'\omega$) and the sigma factor (the σ polypeptide), which is concerned specifically with promoter recognition. Its subunit composition is summarized in **FIGURE 19.7**. The β and β′ subunits together account for RNA catalysis and make up most of the enzyme by mass. Their amino acid sequences and their three-dimensional structures are conserved with those of the largest subunits of the RNA polymerases from all three domains of life: bacteria, archaea, and eukaryotes (see *Section 20.2, Eukaryotic RNA*

Polymerases Consist of Many Subunits), indicating that the basic features of transcription are shared among the multisubunit RNA polymerases of all organisms. β and β′ together form the enzyme's active center, the main channel through which the DNA passes during the transcription cycle, the secondary channel through which the substrate ribonucleotides enter the enzyme on their path to the active site, and the exit channel through which the nascent RNA leaves the enzyme. Consistent with the role of these subunits in all these functions, mutations in *rpoB* and *rpoC*, the genes coding for β and β′, affect all stages of transcription.

The dimer formed by the two α subunits serves as a scaffold for assembly of the core enzyme. The **C-terminal domain (CTD)** of the α subunits also contacts promoter DNA directly and thereby contributes to promoter recognition (see below). Furthermore, the α and σ subunits are the major surfaces on RNA polymerase for interactions of the enzyme with factors that regulate transcription initiation. The ω subunit also plays a role in enzyme assembly and may also play a role certain regulatory functions.

The σ subunit is primarily responsible for promoter recognition. The crystal structure of the bacterial core enzyme shows that it has a crab clawlike shape, with one claw formed primarily by the β subunit and the other primarily by the β′ subunit, as seen in **FIGURE 19.8**. The main channel for DNA lies at the interface of the β and β′ subunits, which stabilize the separated

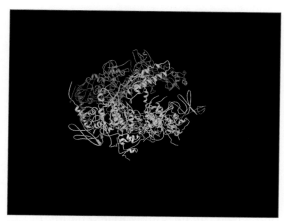

FIGURE 19.9 The structure of RNA polymerase core enzyme for the bacterium *Thermus aquaticus*, with the β subunit in blue and the β′ subunit in green. Structure from Protein Data Bank 1HQM. L. Minakhin, et al., *Proc. Natl. Acad. Sci. USA* 98 (2001): 892–897.

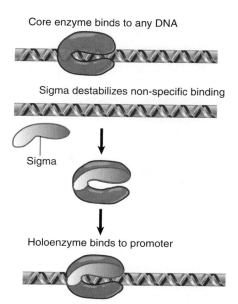

FIGURE 19.10 Core enzyme binds indiscriminately to any DNA. Sigma factor reduces the affinity for sequence-independent binding and confers specificity for promoters.

single strands in the transcription bubble, as seen in **FIGURE 19.9**.

The catalytic site is at the base of the cleft formed by the β and β′ "jaws." One of the two catalytic Mg^{2+} ions needed for the mechanism of catalysis is tightly bound to the enzyme in the active site (see *Section 19.18, Phage T7 RNA Polymerase Is a Useful Model System*). The other Mg^{2+} arrives at the active site in complex with the incoming nucleotide triphosphate (NTP). As indicated earlier, the eukaryotic core enzyme has the same basic structure as the bacterial enzyme (Figure 20.2), although it contains some additional subunits and sequence features not found in the bacterial enzyme. The major differences between the bacterial and eukaryotic enzymes are almost exclusively at the periphery of the enzyme, far from the active site.

19.5 RNA Polymerase Holoenzyme Consists of the Core Enzyme and Sigma Factor

Key concepts

- Bacterial RNA polymerase can be divided into the $\alpha_2\beta\beta'\omega$ core enzyme that catalyzes transcription and the σ subunit that is required only for initiation.
- Sigma factor changes the DNA-binding properties of RNA polymerase so that its affinity for general DNA is reduced and its affinity for promoters is increased.

The core enzyme has general affinity for DNA, primarily because of electrostatic interactions between the protein, which is basic, and the DNA, which is acidic. When bound to DNA in this fashion, the DNA remains in duplex form. *Core enzyme has the ability to synthesize RNA on a DNA template, but it cannot recognize promoters.*

The form of the enzyme responsible for initiating transcription from promoters is called the holoenzyme ($\alpha_2\beta\beta'\omega\sigma$) (**FIGURE 19.10**). It differs from the core enzyme by containing a σ factor. *Sigma factor not only ensures that bacterial RNA polymerase initiates transcription from specific sites, but it also reduces binding to nonspecific sequences.* The association constant for binding of core to DNA is reduced by a factor of ~10^4, and the half-life of the complex is <1 second, whereas holoenzyme binds to promoters much more tightly, with an association constant ~1000 times higher on average and a half-life that can be as long as several hours. Thus, σ factor substantially destabilizes promoter-nonspecific binding.

There is wide variation in the rate at which the holoenzyme binds to different promoter sequences, and thus this is an important parameter in determining *"promoter strength,"* the efficiency of an individual promoter in initiating transcription. The frequency of initiation varies from ~1/sec for rRNA genes under optimal conditions to < 1/30 min for some other promoters. Sigma factor is usually released when the RNA chain reaches less than ~10 nt in length, leaving the core enzyme responsible for elongation.

19.6 How Does RNA Polymerase Find Promoter Sequences?

Key concepts

- The rate at which RNA polymerase binds to promoters can be too fast to be accounted for by simple diffusion.
- RNA polymerase binds to random sites on DNA and exchanges them with other sequences until a promoter is found.

RNA polymerase must find promoters within the context of the genome. How are promoters distinguished from the 4×10^6 bp that comprise the rest of the *E. coli* genome? **FIGURE 19.11** illustrates simple models for how RNA polymerase might find promoter sequences from among all the sequences it can access. RNA polymerase holoenzyme locates the chromosome by random diffusion and binds sequence-nonspecifically to the negatively charged DNA. In this mode, holoenzyme dissociates very rapidly. Diffusion sets an upper limit for the rate constant for associating with a 75 bp target of $<10^8$ M^{-1} sec^{-1}. The actual forward rate constant for some promoters *in vitro*, however, appears to be $\geq 10^8$ M^{-1} sec^{-1}, at or above the diffusion limit. Making and breaking a series of complexes until (by chance) RNA polymerase encounters a promoter and progresses to an open complex capable of making RNA would be a relatively slow process. Thus, the time required for random cycles of successive association and dissociation at loose binding sites is too great to account for the way RNA polymerase finds its promoter. RNA polymerase must therefore use some other means to seek its binding sites.

Figure 19.11 shows that the process is likely to be speeded up because the initial target for RNA polymerase is the whole genome, not just a specific promoter sequence. By increasing the target size, the rate constant for diffusion to DNA is correspondingly increased and is no longer limiting. How does the enzyme move from a random binding site on DNA to a promoter? There is considerable evidence that at least three different processes contribute to the rate of promoter search by RNA polymerase. First, the enzyme may move in a one-dimensional random walk along the DNA ("sliding"). Second, given the intricately folded nature of the chromosome in the bacterial nucleoid, having bound to one sequence on the chromosome, the enzyme is now closer to other sites, reducing the time needed for dissociation and rebinding to another site ("intersegment transfer" or "hopping"). Third, while bound nonspecifically to one site, the enzyme may exchange DNA sites until a promoter is found ("direct transfer").

19.7 The Holoenzyme Goes through Transitions in the Process of Recognizing and Escaping from Promoters

Key concepts

- When RNA polymerase binds to a promoter, it separates the DNA strands to form a transcription bubble and incorporates nucleotides into RNA.
- There may be a cycle of abortive initiations before the enzyme moves to the next phase.
- Sigma factor is usually released from RNA polymerase when the nascent RNA chain reaches ~10 bases in length.

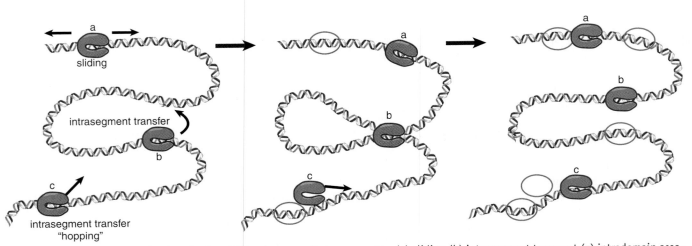

FIGURE 19.11 Proposed mechanisms for how RNA polymerase finds a promoter. (a) sliding (b) intersegment transport (c) intradomain association and dissociation or hopping. Adapted from C. Bustamante, et al., *J. Biol. Chem.* 274 (1999): 166665–166668.

We can now describe the stages of transcription in terms of the interactions between different forms of RNA polymerase and the DNA template. The initiation reaction can be described by the parameters that are summarized in **FIGURE 19.12**:

- The holoenzyme-promoter reaction starts by forming a *closed binary complex* as seen in Figure 19.12(a). "Closed" means that the DNA remains duplex. The formation of the closed binary complex is reversible; thus it is usually described by an equilibrium constant (K_B). There is a wide range in values of the equilibrium constant for forming the closed complex.

- The closed complex is converted into an *open complex* by "melting" a short region of DNA within the sequence bound by the enzyme as seen in Figure 19.12(b). For most promoters, conversion from the closed to the open complex is irreversible, and this reaction can be described by the forward rate constant (k_f). Some promoters (e.g., rRNA promoters), though, do not form stable open complexes, and this is a key to their regulation. Sigma factor plays an essential role in the melting reaction (see *Section 19.19, Competition for Sigma Factors Can Regulate Initiation*). The transitions that occur from initiation to elongation are also accompanied by major changes in the structure and composition of the complex.

Changes in the shape of RNA polymerase accompany the kinetic transitions described earlier, as well as the transition to the elongation complex (illustrated in **FIGURE 19.13**). In the closed complex, RNA polymerase holoenzyme covers about 55 bp of DNA, extending from ~−55 to ~+1. The double-stranded DNA binds primarily along one face of the holoenzyme, contacting the C-terminal domains of the α subunits as well as regions 2 and 4 of the σ subunit (see below). During the transition to the open complex, the conformation of both the RNA polymerase and the DNA change. The most dramatic changes in the structure of the complex are depicted in Figure 19.12: (a) an ~90 degree bend in the DNA, which allows the template strand to approach the active site of the enzyme; (b) strand opening of the promoter DNA between ~−11 and +3 with respect to the transcription start site; and (c) closing of the jaws of the enzyme to encircle the section of the promoter downstream section of the tran-

scription start site. Thus, promoter contacts in the open complex extend from ~−55 to ~+20.

The next step is to incorporate the first two nucleotides and formation of a phosphodiester bond between them. This generates a **ternary complex** containing RNA as well as DNA and the enzyme. At most promoters, an RNA chain forms that is several bases long without movement of the enzyme down the template.

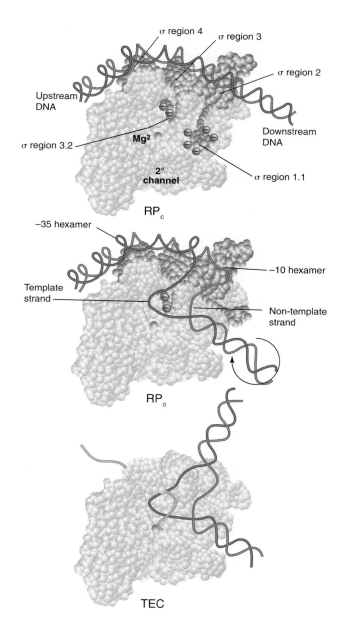

FIGURE 19.12 RNA polymerase passes through several steps prior to elongation. A closed binary complex is converted to an open form and then into a ternary complex. Adapted from S. P. Haugen, W. Ross, and R. L. Gourse, *Nat. Rev. Microbiol.* 6 (2008): 507-519.

Initiation complex contains sigma and covers ~75 bp

−50 −40 −30 −20 −10 1 +10 +20 +30

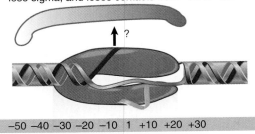

Initial elongation complex forms at 10 bases, may lose sigma, and loses contacts from −35 to −55

−50 −40 −30 −20 −10 1 +10 +20 +30

General elongation complex forms at 15–20 bases and covers 30–40 bp

−50 −40 −30 −20 −10 1 +10 +20 +30

FIGURE 19.13 RNA polymerase initially contacts the region from −55 to +20. When sigma dissociates, the core enzyme contracts to −30; when the enzyme moves a few base pairs, it becomes more compactly organized into the general elongation complex.

After each base is added, there is a certain probability that the enzyme will release the RNA chain, resulting in **abortive initiation** products. After release of the abortive product, the enzyme again begins synthesizing RNA at position +1. Repeated cycles of abortive initiation generate oligonucleotides that usually are only a few bases long, but can be almost 20 nt in length, before the enzyme actually succeeds in escaping from the promoter.

Interactions with RNA polymerase ultimately dissolve during the process of promoter escape. By the time the RNA chain has been extended to 15 to 20 nt, the enzyme generally has gone through all the transitions that typify an elongation complex. The two most obvious of these transitions are the release of the σ factor, shown in Figure 19.12(c), and the formation of a complex covering only ~35 bp of DNA, rather than the ~70 bp characteristic of promoter complexes. Although σ release usually occurs during the process of promoter escape, this is not obligatory for the transition

to elongation. In some cases σ has been identified in elongation complexes, but its association with the enzyme may reflect rebinding to the core enzyme during the elongation phase.

Sigma Factor Controls Binding to DNA by Recognizing Specific Sequences in Promoters

Key concepts

- A promoter is defined by the presence of short consensus sequences at specific locations.
- The promoter consensus sequences usually consist of a purine at the startpoint, a hexamer with a sequence close to TATAAT centered at ~ −10, and another hexamer with a sequence similar to TTGACA centered at ~−35.
- Individual promoters usually differ from the consensus at one or more positions.
- Promoter efficiency can be affected by additional elements as well.

As a sequence of DNA whose function is to be *recognized by proteins*, a promoter differs from sequences whose role is to be transcribed. The information for promoter function is provided directly by the DNA sequence: its structure is the signal. This is a classic example of a *cis*-acting site, as defined previously in Figure 2.16 and Figure 2.17. By contrast, expressed regions gain their meaning only after the information is transferred into the form of some other nucleic acid or protein.

One way to design a promoter would be for a particular sequence of DNA to be recognized by RNA polymerase. Every promoter would consist of, or at least include, this sequence. In the bacterial genome, the minimum length that could provide an adequate signal is 12 bp. (Any shorter sequence is likely to occur—just by chance—a sufficient number of additional times to provide false signals. The minimum length required for unique recognition increases with the size of genome, a problem in eukaryotic genomes.) The 12 bp sequence need not be contiguous. If a specific number of base pairs separates two constant shorter sequences, their combined length could be less than 12 bp, because the *distance* of separation itself provides a part of the signal (even if the intermediate *sequence* is itself irrelevant). In fact, RNA polymerase recognizes promoter DNA sequences in large part from "direct readout" of specific bases in the DNA by specific amino acids in the holoenzyme. The dramatic differences in the strengths of different bacterial promoters derives in large part from variation in how well the different promoter sequences are able to be read out by the amino acid sequences present in the sigma and alpha subunits.

Attempts to identify the features in DNA that are necessary for RNA polymerase binding started by comparing the sequences of different promoters. Any essential nucleotide sequence should be present in all the promoters. Such a sequence is said to be **conserved**. A conserved sequence need not necessarily be conserved at every single position, though; some variation is permitted. How do we analyze a sequence of DNA to determine whether it is sufficiently conserved to constitute a recognizable signal?

Putative DNA recognition sites can be defined in terms of an idealized sequence that represents the base most often present at each position. A **consensus sequence** is defined by aligning all known examples so as to maximize their homology. For a sequence to be accepted as a consensus, each particular base must be reasonably predominant at its position, and most of the actual examples must be related to the consensus by only one or two substitutions.

A striking feature in the sequence of promoters in *E. coli* is the *lack of extensive conservation of sequence* over the entire 75 bp associated with RNA polymerase. Some short stretches within the promoter are conserved, however, and they are critical for its function. *Conservation of only very short consensus sequences is a typical feature of regulatory sites (such as promoters) in both prokaryotic and eukaryotic genomes.*

There are several elements in bacterial promoters that contribute to their recognition by RNA polymerase holoenzyme. Two six bp elements, referred to as the **–10 element** and **–35 element** (as well as the length of the "spacer" sequence between them) are usually the most important of these recognition sequences. The promoter sequence at and directly adjacent to the transcription start point, the sequences on either side of the –10 element (referred to as the "extended –10 element" on the upstream side and the "discriminator" on the downstream side), and the 10–20 bp directly upstream of the –35 element (referred to as the "UP element," however, also interact sequence-specifically with RNA polymerase and contribute to promoter efficiency.

- A 6 bp region is recognizable centered approximately 10 bp upstream of the

startpoint in most promoters (the actual distance from the start site varies slightly from promoter to promoter). This hexameric sequence is usually called the –10 element, the *Pribnow Box*, or the **TATA box**. Its consensus, *TATAAT*, can be summarized in the form

$$T_{80} \ A_{95} \ T_{45} \ A_{60} \ A_{50} \ T_{96}$$

where the subscript denotes the percent occurrence of the most frequently found base, which varies from 45% to 96%. (A position at which there is no discernible preference for any base would be indicated by N.) The frequency of occurrence corresponds to the importance of these base pairs in binding RNA polymerase. Thus, the initial highly conserved TA and the final, almost completely conserved, T in the –10 sequence are often crucial for promoter recognition. We now know that the –10 element makes sequence-specific contacts to the sigma factor regions 2.3 and 2.4 (see below). This region of the promoter is double-stranded in the closed complex and single-stranded in the open complex, though, so interactions between the –10 element and RNA polymerase are complex and change at different stages in the process of transcription initiation.

- The conserved hexamer centered at ~35 bp upstream of the startpoint is called the **–35 element**. The consensus is TTGACA; in more detailed form, the conservation can be written

$$T_{82} \ T_{84} \ G_{78} \ A_{65} \ C_{54} \ A_{45}.$$

Bases in this element interact directly with region 4.2 of the sigma factor (see below) similarly in both the closed and open complex.

- The distance separating the –35 and –10 sites is between 16 and 18 bp in ~90% of promoters; in the exceptions, it is as little as 15 bp or as great as 20 bp. *Although the actual sequence in most of the intervening region is relatively unimportant, the distance is critical because, given the helical nature of the DNA, it determines not only the appropriate separation of the two interacting regions in RNA polymerase but also the geometrical orientation of the two sites with respect to one another.*

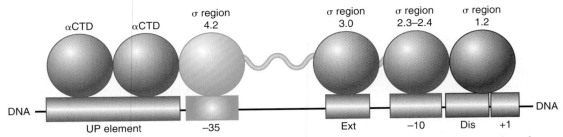

FIGURE 19.14 DNA elements and RNA polymerase modules that contribute to promoter recognition by sigma factor. Adapted from S. P. Haugen, W. Ross, and R. L. Gourse, *Nat. Rev. Microbiol.* 6 (2008): 507–519.

- The *startpoint* is usually (>90% of the time) a purine. It is common for the startpoint to be the central base in the sequence CAT, but the conservation of this triplet is not great enough to regard it as an obligatory signal.
- Certain base pairs in the region between the startpoint and the −10 element are contacted by the sigma factor region 1.2 (see below). For example, a sequence-specific interaction between a guanine residue on the nontemplate strand two positions downstream of the −10 element are especially important in determining the stability of the open complex. Thus, differences in promoter sequence at positions that are not highly conserved can contribute to the variation in the strengths of different promoters.
- Bases in the extended −10 element are contacted by region 3.0 of the sigma factor (see below). The sequence TGN at the upstream end of the −10 element results in interactions that are especially essential for transcription initiation when the promoter lacks a −35 element sequence that closely matches the consensus. This illustrates the modularity of promoter sequences: a weak match to the consensus in one module can be compensated for by a strong match to the consensus in another.
- The ~20 bp region upstream of the −35 element interacts with the CTDs of the two α subunits. Effects of these interactions on promoter activity can be quite substantial, increasing transcription well over an order of magnitude for highly expressed promoters like those in rRNA genes. When these sequences closely match the consensus, this region is referred to as the **UP element**.

The structure of a promoter, showing the permitted range of variation from this optimum, is illustrated in **FIGURE 19.14**.

19.9 Promoter Efficiencies Can Be Increased or Decreased by Mutation

Key concepts

- Down mutations to decrease promoter efficiency usually decrease conformance to the consensus sequences, whereas up mutations have the opposite effect.
- Mutations in the −35 sequence can affect initial binding of RNA polymerase.
- Mutations in the −10 sequence can affect binding or the melting reaction that converts a closed to an open complex.

Effects of mutations can provide information about promoter function. Mutations in promoters affect the level of expression of the gene(s) they control without altering the gene products themselves. Most are identified as bacterial mutants that have lost, or have very much reduced, transcription of the adjacent genes. They are known as **down mutations**. Mutants are also found with **up mutations** in which there is increased transcription from the promoter.

It is important to remember that "up" and "down" mutations are defined relative to the *usual* efficiency with which a particular promoter functions. This varies widely. Thus a change that is recognized as a down mutation in one promoter might never have been isolated in another (which in its wild-type state could be even less efficient than the mutant form of the first promoter). Information gained from studies *in vivo* simply identifies the overall direction of the change caused by mutation.

Mutations that increase the similarity of the −10 or −35 elements to the consensus sequences or bring the distance between them closer to 17 bp usually increase promoter activity. Likewise, mutations that decrease the resemblance of either site to the consensus or make the distance between them more different from 17 bp result in decreased promoter activity. Down mutations tend to be concentrated in the most highly conserved promoter positions, confirming the particular importance of these bases as determinants of promoter efficiency. There are, however, occasional exceptions to these rules.

For example, a promoter with consensus sequences in all the modules described above is illustrated in Figure 19.14. There are, however, no such natural promoters in the *E. coli* genome, and artificial promoters with "perfect" matches to the consensus at all these positions are actually weaker than promoters with at least one mismatch in the −10 or −35 consensus hexamers. This is because they bind to RNA polymerase so tightly that this actually impedes promoter escape.

To determine the absolute effects of promoter mutations, the affinity of RNA polymerase for wild-type and mutant promoters have been measured *in vitro*. Variation in the rate at which RNA polymerase binds to different promoters *in vitro* correlates well with the frequencies of transcription when their genes are expressed *in vivo*. Taking this analysis further, the stage at which a mutation influences the efficiency of a promoter can be determined. Does it change the affinity of the promoter for binding RNA polymerase? Does it leave the enzyme able to bind but unable to initiate? Is the influence of an ancillary factor altered?

By measuring the kinetic constants for formation of a closed complex and its conversion to an open complex, we can dissect the two stages of the initiation reaction:

- Down mutations in the −35 sequence usually reduce the rate of closed complex formation, but they do not inhibit the conversion to an open complex.
- Down mutations in the −10 sequence can reduce either the initial formation of a closed complex, its conversion to the open form, or affect both.

The consensus sequence of the −10 site consists exclusively of A-T base pairs, a configuration that assists the initial melting of DNA into single strands. The lower energy needed to disrupt A-T pairs compared with G-C pairs means that a stretch of A-T pairs demands the minimum amount of energy for strand separation. The sequences immediately around and downstream from the startpoint also influence the initiation event. Furthermore, the initial transcribed region (from ~+1 to ~+20) influences the rate at which RNA polymerase clears the promoter and therefore has an effect upon promoter strength. Thus the overall strength of a promoter cannot always be predicted from its consensus sequences, even when taking into consideration the other RNA polymerase recognition elements in addition to the −10 and −35 elements.

It is important to emphasize that although similarity to consensus is a useful tool for identifying promoters by DNA sequence alone, and "typical" promoters contain easily recognized −35 and −10 sequences, many promoters lack recognizable −10 and/or −35 elements. In many of these cases, the promoter cannot be recognized by RNA polymerase alone and requires an ancillary protein (an "activator"; see Chapter 26, *The Operon*), which overcomes the deficiency in intrinsic interaction between RNA polymerase and the promoter. It is also important to emphasize that "optimal activity" does not mean "maximal activity." Many promoters have evolved with sequences far from consensus precisely because it is not optimal for the cell to make too much of the product their RNA transcript encodes.

19.10 Multiple Regions in RNA Polymerase Directly Contact Promoter DNA

Key concepts

- The structure of σ^{70} changes when it associates with core enzyme, allowing its DNA-binding regions to interact with the promoter.
- Multiple regions in σ^{70} interact with the promoter.
- The α subunit also contributes to promoter recognition.

As mentioned briefly in *Section 19.8*, several domains in the sigma factor subunit and the CTD in the alpha subunit contact promoter DNA. The identification of a series of different consensus sequences recognized by holoenzymes containing different sigma factors (as seen in **FIGURE 19.15**) implies that the sigma factor subunit must itself contact DNA. This suggests further that the different sigma factors must bind similarly to core enzyme so that the DNA recognition surfaces on the different sigma factors would be positioned similarly to make critical contacts with the promoter sequences in the vicinity of the –35 and –10 sequences.

Subunit/gene	Size (# aa)	Approx. # of promoters	Promoter sequence recognized
Sigma 70 (*rpoD*)	613	1000	TTGACA-16 to 18 bp-TATAAT
Sigma 54 (*rpoN*)	477	5	CTGGNA-6 bp-TTGCA
Sigma S (*rpoS*)	330	100	TTGACA-16 to 18 bp-TATAAT
Sigma 32 (*rpoH*)	284	30	CCCTTGAA-13 to 15 bp-CCCGATNT
Sigma F (*rpoF*)	239	40	CTAAA-15 bp-GCCGATAA
Sigma E (*rpoE*)	202	20	GAA-16 bp-YCTGA
Sigma FecI (*fecI*)	173	1–2	?

FIGURE 19.15 *E. coli* sigma factors recognize promoters with different consensus sequences.

Further evidence that sigma factor contacts the promoter directly at both the –35 and –10 consensus sequences was provided by substitutions in the sigma factor that suppressed mutations in the consensus sequences. When a mutation at a particular position in the promoter prevents recognition by RNA polymerase, and a compensating mutation in sigma factor allows the polymerase to use the mutant promoter, the most likely explanation is that the relevant base pair in DNA is contacted by the amino acid that has been substituted.

Comparisons of the sequences of several bacterial sigma factors suggested conserved regions in *E. coli* σ⁷⁰ (**FIGURE 19.16**) that interact directly with promoters, and these inferences were substantiated by the identification of a crystal structure of RNA polymerase holoenzyme in complex with a promoter fragment. The bacteria *Thermus aquaticus* and *Thermus thermophilus* illustrate how the DNA-binding regions of the sigma factor fold into independent domains in the protein regions 1.2, 2.3–2.4, 3.0, and 4.1–4.2.

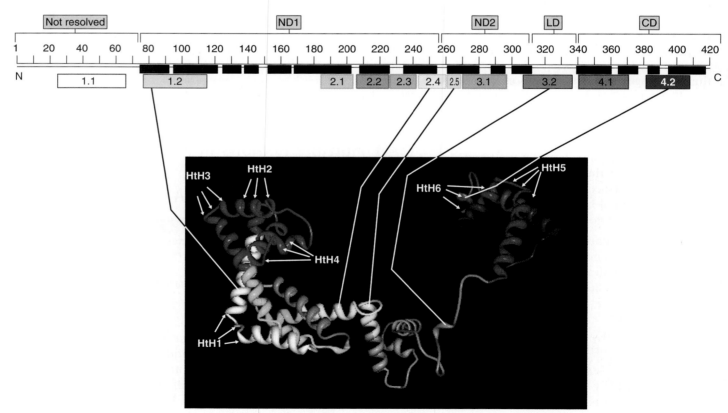

FIGURE 19.16 The structure of sigma factor in the context of the holoenzyme: –10 and –35 interactions. Sigma factor is extended and its domains are connected by flexible linkers. Illustration adapted from D. G. Vassylyev, et al., *Nature* 417 (2002): 712–719. Structure from Protein Data Bank 1IW7. D. G. Vassylyev, et al., *Nature* 417 (2002): 712-719.

Figure 19.16 illustrates the sections of sigma factor that play direct roles in promoter recognition. This figure shows the structure of the major sigma factor as it exists in the context of the holoenzyme. Two short parts of region 2 and one part of region 4 (2.3, 2.4, and 4.2) contact bases in the –10 and –35 elements, respectively; sigma factor region 1.2 contacts the promoter region just downstream from the –10 element, and region 3.0 contacts the promoter region just upstream from the –10 element. Each of these regions forms short stretches of α-helix in the protein. A crystal structure of the holoenzyme in complex with a promoter fragment, in conjunction with experiments with promoters in which the DNA strands were built to contain mismatches ("heteroduplexes") showed that σ^{70} makes contacts bases principally on the nontemplate strand of the –10 element, the extended –10 element, and the discriminator region, and it continues to hold these contacts after the DNA has been unwound in this region. This confirms that sigma factor is important in the melting reaction.

The use of α-helical motifs in proteins to recognize duplex DNA sequences is common (see *Section 28.6, There Are Many Types of DNA-Binding Domains*). Amino acids separated by three to four positions lie on the same face of an α-helix and are therefore in a position to contact adjacent base pairs. **FIGURE 19.17** shows that amino acids lying along one face of the 2.4 region α-helix contact the bases at positions –12 to –10 of the –10 promoter sequence.

Region 2.3 resembles proteins that bind single-stranded nucleic acids and is involved in the melting reaction. Regions 2.1 and 2.2 (which comprise the most highly conserved part of sigma factor) are involved in the interaction with the core enzyme. It is assumed that all sigma factors bind the same regions of the core polymerase, which ensures that the sigma factors compete for limiting core RNA polymerase.

Although sigma factor has domains that recognize specific bases in promoter DNA, the N-terminal region of free sigma factor (region 1.1), acting as an autoinhibitory domain, masks the DNA-binding region; only once the conformation of the sigma factor has been altered by its association with the core enzyme can it bind specifically to promoter sequences (**FIGURE 19.18**). The inability of free sigma factor to recognize promoter sequences is important: if sigma factor could bind to promoters as a free subunit, it might block holoenzyme from initiating transcription. Figure 19.18 schematizes the conformational change in sigma factor at open complex formation.

When sigma factor binds to the core polymerase, the N-terminal domain swings ~20 Å away from the DNA-binding domains, and the DNA-binding domains separate from one another by ~15 Å, presumably to acquire a more elongated conformation appropriate for contacting DNA. Mutations in either the –10 or –35 sequences prevent an N-terminal-deleted σ^{70} from binding to DNA, which suggests that σ^{70} contacts both sequences simultaneously. This fits with the information from the crystal structure of the holoenzyme (Figure 19.16), in which it is clear that the sigma factor has a rather elongated structure, extending over the ~68 Å of two turns of DNA.

Although sigma factor region 1.1 is not resolved in the crystal structure, biophysical

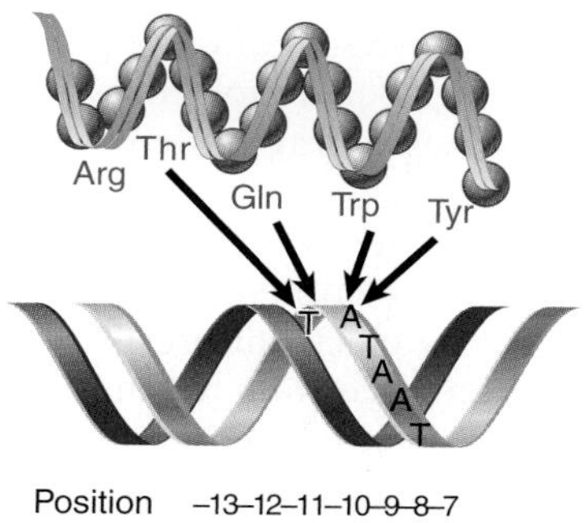

FIGURE 19.17 Amino acids in the 2.4 α-helix of σ^{70} contact specific bases in the coding strand of the –10 promoter sequence.

Protein

DNA

Position   –13–12–11–10–9–8–7

Thr
Arg
Gln   Trp   Tyr

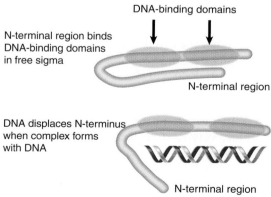

FIGURE 19.18 The N-terminus of sigma blocks the DNA-binding regions from binding to DNA. When an open complex forms, the N-terminus swings 20 Å away, and the two DNA-binding regions separate by 15 Å.

DNA-binding domains

N-terminal region binds DNA-binding domains in free sigma

N-terminal region

DNA displaces N-terminus when complex forms with DNA

N-terminal region

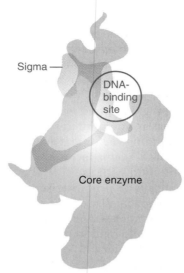

FIGURE 19.19 Sigma factor has an elongated structure that extends along the surface of the core subunits when the holoenzyme is formed.

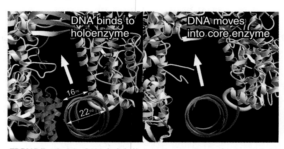

FIGURE 19.20 DNA initially contacts sigma factor (pink) and core enzyme (gray). It moves deeper into the core enzyme to make contacts at the –10 sequence. When sigma is released, the width of the passage containing DNA increases. Reprinted by permission from Macmillan Publishers Ltd: *Nature*, D. G. Vassylyev, et al., vol. 417, pp. 712–719, copyright 2002. Photo courtesy of Shigeyuki Yokoyama, The University of Tokyo.

measurements of its position in the holoenzyme versus the open complex suggest that in the free holoenzyme, the N-terminal domain (region 1.1) is located in the main DNA channel of the enzyme, essentially mimicking the location that the promoter will occupy when a transcription complex is formed (**FIGURE 19.19**). When the holoenzyme forms an open complex on DNA, the N-terminal sigma factor domain is displaced from the main channel. Its position with respect to the rest of the protein is therefore very flexible; it changes when sigma factor binds to core enzyme and again when the holoenzyme binds to DNA. The DNA helix has to move some 16 Å from its initial position in order to enter the main DNA channel, and then it has to move again to allow DNA to enter the channel during open complex formation. **FIGURE 19.20** illustrates this movement, looking in cross-section down the helical axis of the DNA.

Although it was first thought that sigma factor is the only subunit of RNA polymerase that contributes to the promoter region, the *C-terminal domains* of the two alpha subunits also can play a major role in contacting promoter DNA by binding to UP elements (see *Section 19.8*). Because the αCTDs are tethered flexibly to the rest of RNA polymerase (see Figure 19.14), the enzyme can reach regions quite far upstream while still bound to the –10 and –35 elements. The αCTDs thereby provide mobile domains for contacting transcription factors bound at different distances upstream from the transcription start site in different promoters (see Chapter 26, *The Operon*).

19.11 Footprinting Is a High Resolution Method for Characterizing RNA Polymerase–Promoter and DNA–Protein Interactions in General

Key concepts

- The consensus sequences at –35 and –10 provide most of the contact points for RNA polymerase in the promoter.
- The points of contact lie primarily on one face of the DNA.

The ability of RNA polymerase (or indeed any protein) to recognize DNA can be characterized by **footprinting**. A sequence of DNA bound to the protein is *partially* digested with an endonuclease to attack individual phosphodiester bonds within the nucleic acid. Under appropriate conditions, any particular phosphodiester bond is broken in some, but not in all, DNA molecules. The positions that are cleaved can be identified by using DNA labeled on one strand at one end only. The principle is the same as that involved in DNA sequencing: partial cleavage of an end-labeled molecule at a susceptible site creates a fragment of unique length.

FIGURE 19.21 shows that following the nuclease treatment the broken DNA fragments can be separated by electrophoresis on a gel that separates them according to length. Each fragment that retains a labeled end produces a radioactive band. The position of the band corresponds to the number of bases in the fragment. The shortest fragments move the fastest, so distance from the labeled end is counted up from the bottom of the gel.

In free DNA, virtually *every* susceptible bond position is broken in one or another molecule. Figure 19.20 illustrates that when the DNA is complexed with a protein, the positions covered by the DNA-binding protein are protected from cleavage. Thus when two reactions are run in parallel—a control DNA in which no protein was present and an experimental mixture containing molecules of DNA bound to the protein—a characteristic pattern emerges. When a bound protein blocks access of the nuclease to DNA, the bonds in the bound sequence fail to be broken in the *experimental mixture and that part of the gel remains unrepresented by labeled DNA fragments.*

In the control, virtually every bond is broken generating a ladder of bands, with one band representing each base. There are thirty-one bands in the figure. In the protected fragment, bonds cannot be broken in the region bound by the protein, so bands representing fragments of the corresponding sizes are not generated. The absence of bands 9–18 in the figure identifies a protein-binding site covering the region located 9–18 bases from the labeled end of the DNA. By comparing the control and experimental lanes with a sequencing reaction that is run in parallel, it becomes possible to "read off" the corresponding sequence directly, thus identifying the nucleotide sequence of the binding site.

As described previously (see Figure 19.13), RNA polymerase binds to the promoter region from −55 to +20. The points at which RNA polymerase actually contacts the promoter can be identified by modifying the footprinting technique to treat RNA polymerase-promoter complexes with reagents that modify particular bases. We can then perform the experiment in two ways:

- The DNA can be modified before it is bound to RNA polymerase. In this case, if the modification prevents RNA polymerase from binding, we have identified a base position where contact is essential.
- The RNA polymerase–DNA complex can be modified. We then can compare the pattern of protected bands with that of free DNA and of the unmodified complex. Some bands disappear, thus identifying sites at which the enzyme has protected the promoter against modification. Other bands increase in intensity, thus identifying sites at which the DNA must be held in a conforma-

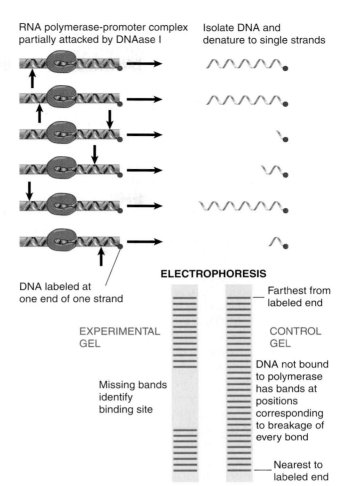

FIGURE 19.21 Footprinting identifies DNA-binding sites for proteins by their protection against nicking.

tion in which it is more exposed to the cleaving agent.

These changes in sensitivity revealed the geometry of the complex, as summarized in **FIGURE 19.22** for a typical promoter. The regions at −35 and −10 contain most of the contact points for the enzyme. Within these regions, the same sets of positions tend both to prevent binding if previously modified, and to show increased or decreased susceptibility to modification after binding. The points of contact do not coincide completely with sites of mutation; however, they occur in the same limited region.

It is noteworthy that the same *positions* in different promoters provide many of the contact points, even though a different base is present. This indicates that there is a common mechanism for RNA polymerase binding, although the reaction does not depend on the presence of particular bases at some of the points of contact. This model explains why some of the points of contact are not sites of mutation. In addition, not every mutation lies in a point of contact;

the mutations may influence the neighborhood without actually being touched by the enzyme.

It is especially significant that the experiments using premodification identify sites in the same region that is protected by the enzyme against subsequent modification. These two experiments measure different things. Premodification identifies all those sites that the enzyme must recognize in order to bind to DNA. Protection experiments recognize all those sites that actually make contact in the binary complex. The protected sites include all the recognition sites and also some additional positions, which suggests that the enzyme first recognizes a set of bases necessary for it to "touch down" and then extends its points of contact to additional bases.

The region of DNA that is unwound in the binary complex can be identified directly by chemical changes in its availability. When the strands of DNA are separated, the unpaired bases become susceptible to reagents that cannot reach them in the double helix. Such experiments implicate positions between −9 and +3 in the initial melting reaction. The region unwound during initiation therefore includes the right end of the −10 sequence and extends just past the startpoint.

Viewed in three dimensions, the points of contact upstream of the −10 sequence all lie on one face of DNA. This can be seen in the lower drawing in Figure 19.22, in which the contact points are marked on a double helix viewed from one side. Most lie on the nontemplate strand. These bases are probably recognized in the initial formation of a closed binary complex. This would make it possible for RNA polymerase to approach DNA from one side and recognize that face of the DNA.

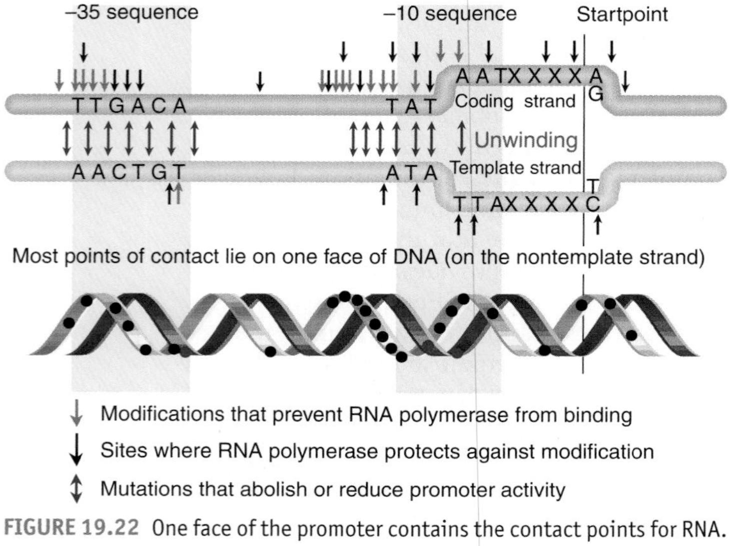

As DNA unwinding commences, further sites that originally lay on the other face of DNA can be recognized and bound.

19.12 Interactions between Sigma Factor and Core RNA Polymerase Change During Promoter Escape

Key concepts

- A domain in sigma occupies the RNA exit channel and must be displaced to accommodate RNA synthesis.
- Abortive initiations usually occur before the enzyme forms a true elongation complex.
- Sigma factor is usually released from RNA polymerase by the time the nascent RNA chain reaches ~10 nt in length.

RNA polymerase encounters a dilemma in reconciling its needs for initiation with those for elongation. First, the RNA exit channel is actually occupied by part of the sigma factor, the linker connecting domains 3 and 4. Therefore, promoter escape must involve rearrangement of the sigma factor, displacing it from the RNA exit channel so that RNA synthesis can proceed. Second, initiation requires tight binding *only* to particular sequences (promoters), whereas elongation requires association with *all* sequences that the enzyme encounters during transcription. FIGURE 19.23 illustrates how the dilemma is solved by the reversible association of sigma factor with core enzyme. As mentioned earlier, the enzyme usually undergoes cycles of abortive initiation in the process of escaping from the promoter. The enzyme does not move down the template while it undergoes these abortive cycles. Rather, it pulls the first few nucleotides of downstream DNA into itself, extruding these single-strands onto the surface of the enzyme in a process called "*DNA scrunching.*" By a mechanism that is not completely understood, the enzyme then escapes from this abortive cycling mode and enters the elongation phase (discussed shortly).

Although the release of sigma factor from the complex is not essential for promoter escape, dissociation of sigma factor from core usually occurs concurrently with or soon after promoter escape. Sigma factor is in excess of core RNA polymerase, so release of sigma from holoenzyme is not simply to make it available for use in additional copies of holoenzyme. In fact, sigma factors compete for limiting copies of core RNA polymerase as a means of changing

FIGURE 19.22 One face of the promoter contains the contact points for RNA.

↓ Modifications that prevent RNA polymerase from binding

↓ Sites where RNA polymerase protects against modification

↕ Mutations that abolish or reduce promoter activity

the transcription profile (see *Section 19.19, Competition for Sigma Factors Can Regulate Initiation*).

The core enzyme in the ternary complex (which comprises DNA, nascent RNA, and RNA polymerase) is essentially "locked in" until elongation has been completed. As will be described shortly, this processivity results in part from the way the enzyme encircles the DNA and in part from the increase in the affinity of the enzyme for the complex afforded by interactions with the nascent RNA.

The drug rifampicin (a member of the rifamycin antibiotic family) blocks transcription by bacterial RNA polymerase. It is the major antibiotic used against tuberculosis. The crystal structure of RNA polymerase bound to rifampicin explains its action: it binds in a pocket of the β subunit, >12 Å away from the active site, but in a position where it blocks the path of the elongating RNA. By preventing the RNA chain from extending beyond two to three nucleotides, it blocks transcription.

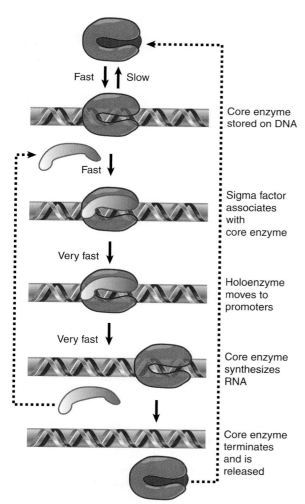

FIGURE 19.23 Sigma factor and core enzyme recycle at different points in transcription.

Fast ↓ ↑ Slow

Core enzyme stored on DNA

Fast ↓

Sigma factor associates with core enzyme

Very fast ↓

Holoenzyme moves to promoters

Very fast ↓

Core enzyme synthesizes RNA

Core enzyme terminates and is released

Key concepts

- DNA moves through a channel in RNA polymerase and makes a sharp turn at the active site.
- Changes in the conformations of certain flexible modules within the enzyme control the entry of nucleotides to the active site.

As a result of the crystal structures of the bacterial and yeast enzymes in complex with NTPs and/or with DNA, we now have considerable information about the structure and function of RNA polymerase during elongation. Bacterial RNA polymerase has overall dimensions of ~90 × 95 × 160 Å, and the archaeal and eukaryotic RNA polymerases are only slightly larger, primarily from additional stretches of amino acids and/or extra subunits situated on the periphery of the enzyme. Nevertheless, the core enzymes share not only a common structure, in which there is a "channel" ~25 Å wide that accommodates the DNA, but a common mechanism for nucleotide addition.

A model of this channel in bacterial RNA polymerase is illustrated in **FIGURE 19.24**. The groove holds ~17 bp of DNA. In conjunction with the ~13 nt of DNA accommodated by the enzyme's active site region, this accounts for the ~30–35 nt long protected region observed

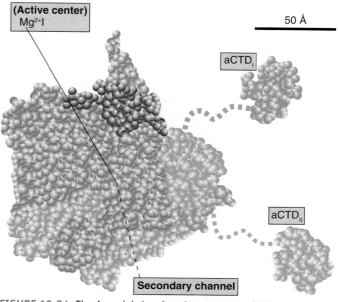

(Active center) Mg²⁺|

50 Å

aCTD$_I$

aCTD$_{II}$

Secondary channel

FIGURE 19.24 The A model showing the structure of RNA polymerase through the main channel. Subunits are colorcodes as follows: β', pink, β, cyan, aI, green, aII, yellow, ω, red. Adapted from K. M. Geszvain and R. Landick (ed. N. P. Higgins). *The Bacterial Chromosome*. American Society for Microbiology, 2004.

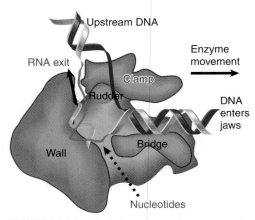

FIGURE 19.25 DNA is forced to make a turn at the active site by a wall of protein. Nucleotides may enter the active site through a pore in the protein.

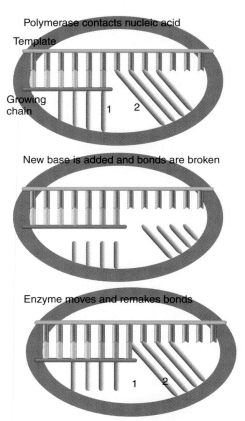

FIGURE 19.26 Movement of a nucleic acid polymerase requires breaking and remaking bonds to the nucleotides at fixed positions relative to the enzyme structure. The nucleotides in these positions change each time the enzyme moves a base along the template

in footprints of the elongation complex. The groove is lined with positive charges, enabling it to interact with the negatively charged phosphate groups of DNA. The catalytic site is formed by a cleft between the two large subunits which grasp DNA downstream in its "jaws" as it enters the RNA polymerase. RNA polymerase surrounds the DNA, and a catalytic Mg^{2+} ion is found at the active site. The DNA is held in position by the downstream clamp, another name for one of the jaws. FIGURE 19.25 illustrates the 90-degree turn that the DNA takes at the entrance to the active site because of an adjacent wall of protein. The length of the RNA hybrid is limited by another protein obstruction, called the lid. Nucleotides are thought to enter the active site from below, via the secondary channel (called the "pore" in yeast RNA polymerase). The transcription bubble includes 8 to 9 bp of DNA–RNA hybrid. The lid separates the DNA and RNA bases at one end of the hybrid (see Figure 19.25), and the DNA base on the template strand at the other end of the hybrid is flipped out to allow pairing with the incoming NTP.

Once DNA has been melted, the trajectory of the individual strands within the enzyme is no longer constrained by the rigidity of the double helix, allowing DNA to make its 90-degree turn at the active site. Furthermore, there is a large conformational change in the enzyme itself involving the "clamp," which makes up one of the jaws of the enzyme that holds the downstream DNA in place.

One of the dilemmas of any nucleic acid polymerase is that the enzyme must make tight contacts with the nucleic acid substrate and product, but must break these contacts

and remake them with each cycle of nucleotide addition. Consider the situation illustrated in FIGURE 19.26. A polymerase makes a series of specific contacts with the bases at particular positions. For example, contact "1" is made with the base at the end of the growing chain and contact "2" is made with the base in the template strand that is complementary to the next base to be added. Note, however, that the bases that occupy these locations in the nucleic acid chains change every time a nucleotide is added!

The top and bottom panels of the figure show the same situation: a base is about to be added to the growing chain. The difference is that the growing chain has been extended by one base in the bottom panel. The geometry of both complexes is exactly the same, but contacts "1" and "2" in the bottom panel are made to bases in the nucleic acid chains that are located one position farther along the chain. The middle panel shows that this must mean that, after the base is added, and before the enzyme moves relative to the nucleic acid, the contacts made to specific positions must be broken so that they can be remade to bases that occupy those positions after the movement.

There are RNA polymerase crystal structures that provide considerable insight into how the enzyme retains contact with its substrate while breaking and remaking bonds in the process of the nucleotide addition cycle and undergoing translocation. A flexible module called the trigger loop appears to be unfolded before nucleotide addition, but becomes folded once the NTP enters the active site. Once bond formation and translocation of the enzyme to the next position are complete, the trigger loop unfolds again, ready for the next cycle. Thus, a structural change in the trigger loop coordinates the sequence of events in catalysis.

19.14 A Stalled RNA Polymerase Can Restart

Key concept

- An arrested RNA polymerase can restart transcription by cleaving the RNA transcript to generate a new 3′ end.

RNA polymerase must be able to handle situations when transcription is blocked. This can happen, for example, when DNA is damaged. A model system for such situations is provided by arresting elongation *in vitro* by omitting one of the necessary precursor nucleotides, allowing fraying of the end of the RNA. Any event that causes misalignment of the 3′ terminus of the RNA with the active site results in the same problem, though: something is needed to reposition the 3′–OH of the nascent RNA with the active site so that it can undergo attack from the next NTP and phosphodiester bond formation. Realignment is accomplished by cleavage of the RNA to place the terminus in the right location for addition of further bases.

Although the cleavage activity is intrinsic to RNA polymerase itself, it is stimulated greatly by accessory factors that are ubiquitous in the three biological kingdoms. There are two such factors in *E. coli*, GreA and GreB, and eukaryotic RNA polymerase II uses TFIIS for the same purpose. TFIIS displays little similarity in sequence or structure to the Gre factors, but it binds to the same part of the enzyme, the RNA polymerase secondary channel (pore).

The Gre factors/TFIIS enable the polymerase to cleave a few ribonucleotides from the 3′ terminus of the RNA product, thereby allowing the catalytic site of RNA polymerase to be realigned with the 3′–OH. Each of the factors inserts a narrow protein domain (in TFIIS this is a zinc ribbon, in the bacterial enzyme it is a coiled-coil) deep into RNA polymerase, approaching very close to the catalytic center. Two acidic amino acids at the tip of the factor approach the primary catalytic magnesium ion in the active site, allowing a second magnesium ion to enter and convert the catalytic site to turn into a ribonuclease.

In summary, the elongating RNA polymerase has the ability to unwind and rewind DNA, to keep hold of the separated strands of DNA as well as the RNA product, to catalyze the addition of ribonucleotides to the growing RNA chain, to monitor the progress of this reaction, and—with the assistance of an accessory factor or two—to fix problems that occur by cleaving off a few nt of the RNA product and restarting RNA synthesis.

19.15 Bacterial RNA Polymerase Terminates at Discrete Sites

Key concepts

- There are two classes of terminators: Those recognized solely by RNA polymerase itself without the requirement for any cellular factors are usually referred to as "intrinsic terminators." Others require a cellular protein called rho and are referred to as "rho-dependent terminators."
- Intrinsic termination requires recognition of a terminator sequence in DNA that codes for a hairpin structure in the RNA product.
- The signals for termination lie mostly within *sequences already transcribed* by RNA polymerase, and thus termination relies on scrutiny of the template and/or the RNA product that the polymerase is transcribing.

Once RNA polymerase has started transcription, the enzyme moves along the template, synthesizing RNA, until it meets a *terminator* sequence. At this point, the enzyme stops adding nucleotides to the growing RNA chain, releases the completed product, and dissociates from the DNA template. Termination requires that all hydrogen bonds holding the RNA–DNA hybrid together must be broken, after which the DNA duplex reforms.

It is sometimes difficult to define the termination site for an RNA that has been synthesized in the living cell, because the 3′ end of the molecule can be degraded by a 3′ exonuclease or cleaved by an endonuclease, leaving no history

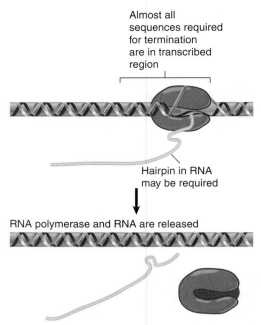

FIGURE 19.27 The DNA sequences required for termination are located upstream of the terminator sequence. Formation of a hairpin in the RNA may be necessary.

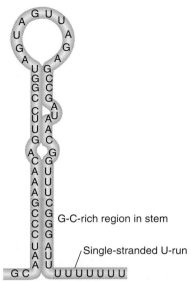

FIGURE 19.28 Intrinsic terminators include palindromic regions that form hairpins varying in length from 7 to 20 bp. The stem-loop structure includes a G-C-rich region and is followed by a run of U residues.

of the actual site at which RNA polymerase terminated in the remaining transcript; this is especially common in eukaryotes. Therefore, termination sites are often best characterized *in vitro*. The ability of the enzyme to terminate *in vitro*, however, is strongly influenced by parameters such as the ionic strength and temperature at which the reaction is performed; as a result, termination at a particular position *in vitro* does not prove that this is the same site where it occurs in cells. If the same 3′ end is detected *in vivo* and with purified components *in vitro*, though, this is generally recognized as good evidence for the authentic site of termination.

FIGURE 19.27 and FIGURE 19.28 summarizes the two major features found in **intrinsic terminators**. First, intrinsic terminators—i.e., those that do not require auxiliary **rho factor** (ρ), as described shortly—require a G+C-rich **hairpin** to form in the secondary structure of the RNA being transcribed. *Thus, termination depends on the RNA product and is not determined simply by scrutiny of the DNA sequence during transcription.* The second feature is a series of up to 7 U residues (T residues in the DNA) following the hairpin stem but preceding the actual position of termination. There are ~1100 sequences in the *E. coli* genome that fit these criteria, suggesting more than half of the cell's transcripts are terminated at intrinsic terminators. **Rho-dependent terminators** are defined by the

need for addition of rho factor *in vitro*, and mutations show that the factor is involved in termination *in vivo*.

Terminators vary widely in their efficiencies. **Readthrough** transcripts refer to the fraction of transcripts that are not stopped by the terminator. (Readthrough is the same term used in *Section 25.14, Suppressors May Compete with Wild-Type Reading of the Code*, to describe a ribosome's suppression of termination codons.) Furthermore, the termination event can be *prevented* by specific ancillary factors that interact with RNA and/or RNA polymerase, a situation referred to as **antitermination**. Thus, as in the case of initiation or elongation, termination can be regulated as a mechanism for controlling gene expression, as described in Chapter 27, *Phage Strategies*.

There are other parallels between initiation and termination. Both require breaking of hydrogen bonds (initial melting of DNA at initiation and RNA–DNA dissociation at termination), and both can utilize additional proteins (sigma factors, activators, repressors, and rho factor) that interact with the core enzyme. Whereas initiation relies solely upon the interaction between RNA polymerase and duplex DNA, however, the termination event also involves recognition of signals in the transcript by RNA polymerase.

Point mutations that reduce termination efficiency usually occur within the stem region of the hairpin or in the U-rich sequence, supporting the importance of these sequences in the mechanism of termination. The RNA–DNA hybrid makes a large contribution to the forces holding the elongation complex together. Thus breaking the hybrid would destabilize the elongation complex, leading to termination. Interactions of the hairpin with the RNA polymerase or forces exerted by formation of the hairpin as the RNA emerges from the RNA exit channel can transiently misalign the 3' end of the RNA with the active center in the enzyme. This misalignment, combined with the unusually weak RNA–DNA hybrid formed from the rU-dA RNA–DNA base pairs resulting from the stretch of U residues, destabilize the elongation complex.

Termination efficiency *in vitro* can vary widely, though; for example, from 2% to 90%. The efficiency of termination depends not only on the sequences in the hairpin and the number and positions of U residues downstream of the hairpin, but also on sequences both further upstream and downstream of the site of termination. Instead of terminating, the enzyme may simply pause before resuming elongation. These pause sites can serve regulatory purposes on their own (see *Section 26.13, The trp Operon Is Also Controlled by Attenuation*). Whether RNA polymerase arrests and releases the RNA chain or whether it merely pauses before resuming transcription (i.e., the duration of the pause and the efficiency of escape from the pause) is determined by a complex set of kinetic and thermodynamic considerations resulting from the characteristics of the hairpin and the U-rich stretch in the RNA, and the upstream and downstream sequences in the DNA. For example, pausing can occur at sites that resemble terminators, but where the separation between the hairpin and the U-run is longer than optimal for termination.

Considerably less is known about the signals and ancillary factors involved in termination of eukaryotic RNA polymerases. Each class of polymerase appears to use a distinct mechanism (see Chapter 21, *RNA Splicing and Processing*).

19.16 How Does Rho Factor Work?

Key concept

- Rho factor is a protein that binds to nascent RNA and tracks along the RNA to interact with RNA polymerase and release it from the elongation complex.

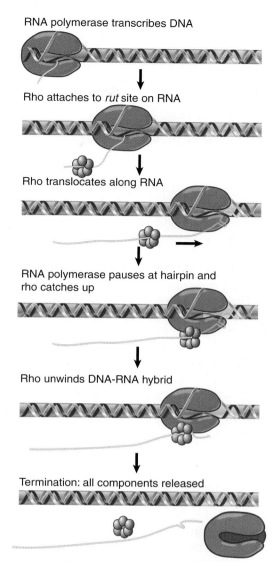

RNA polymerase transcribes DNA

Rho attaches to *rut* site on RNA

Rho translocates along RNA

RNA polymerase pauses at hairpin and rho catches up

Rho unwinds DNA-RNA hybrid

Termination: all components released

FIGURE 19.29 Rho factor binds to RNA at a rut site and translocates along RNA until it reaches the RNA–DNA hybrid in RNA polymerase, where it releases the RNA from the DNA.

Rho factor is an essential protein in *E. coli* that causes transcription termination. The Rho concentration may be as high as ~10% the concentration of RNA polymerase. Rho-independent termination accounts for almost half of *E. coli* terminators.

FIGURE 19.29 illustrates a model for rho function. First it binds to a sequence within the transcript upstream of the site of termination. This sequence is called a ***rut*** site (an acronym for *rho utilization*). The rho factor then tracks along the RNA until it catches up to RNA polymerase. When the RNA polymerase reaches the termination site, rho causes RNA polymerase to release the RNA. Pausing by the polymerase at the site of termination allows time for rho

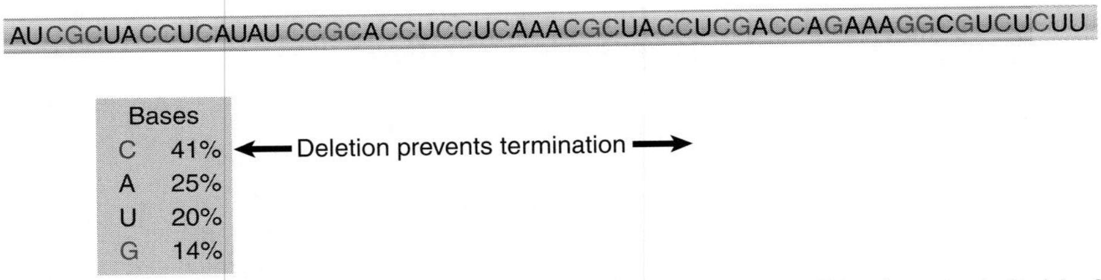

AUCGCUACCUCAUAU CCGCACCUCCUCAAACGCUACCUCGACCAGAAAGGCGUCUCUU

Bases	
C	41%
A	25%
U	20%
G	14%

← Deletion prevents termination →

FIGURE 19.30 A *rut* site has a sequence rich in C and poor in G preceding the actual site(s) of termination. The sequence corresponds to the 3′ end of the RNA.

factor to translocate to the hybrid stretch and is an important feature of termination.

We see an important general principle here. When we know the site on DNA at which some protein exercises its effect, we cannot assume that this coincides with the DNA sequence that it initially recognizes. They can be separate, and there need not be a fixed relationship between them. In fact, *rut* sites in different transcription units are found at varying distances preceding the sites of termination. A similar distinction is made by antitermination factors (see *Section 19.22, Antitermination Can Be a Regulatory Event*).

What actually constitutes a *rut* site is somewhat unclear. The common feature of *rut* sites is that the sequence is rich in C residues and poor in G residues and has no secondary structure. An example is given in **FIGURE 19.30**. C is by far the most common base (41%) and G is the least common base (14%). *rut* sites vary in length. As a general rule, the efficiency of a *rut* site increases with the length of the C-rich/G-poor region.

Rho is a member of the family of hexameric ATP-dependent helicases. Each subunit has an RNA-binding domain and an ATP hydrolysis domain. The hexamer functions by passing nucleic acid through the hole in the middle of the assembly formed from the RNA-binding domains of the subunits (**FIGURE 19.31**). The structure of rho gives some hints about how it might function. It winds RNA from the 3′ end around the exterior of the N-terminal domains, and pushes the 5′ end of the bound region into the interior, where it is bound by a secondary RNA-binding domain in the C-terminal domains. The initial form of rho is a gapped ring, but binding of the RNA converts it to a closed ring.

After binding to the *rut* site, rho uses its helicase activity, driven by ATP hydrolysis, to translocate along RNA until it reaches the RNA polymerase. It then may utilize its helicase activity to unwind the duplex structure

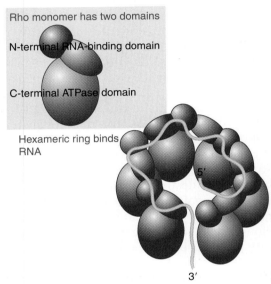

Rho monomer has two domains

N-terminal RNA-binding domain

C-terminal ATPase domain

Hexameric ring binds RNA

5′

3′

FIGURE 19.31 Rho has an N-terminal RNA-binding domain and a C-terminal ATPase domain. A hexamer in the form of a gapped ring binds RNA along the exterior of the N-terminal domains. The 5′ end of the RNA is bound by a secondary binding site in the interior of the hexamer.

and/or interact with RNA polymerase to help release RNA.

Rho needs to translocate along RNA from the *rut* site to the actual point of termination. This requires the factor to move faster than RNA polymerase. The enzyme pauses when it reaches a terminator, and termination occurs if rho catches it there. Pausing is therefore important in rho-dependent termination, just as in intrinsic termination, because it gives time for the other necessary events to occur.

The coupling between transcription and translation has important consequences for rho action. Rho must first have access to RNA upstream of the transcription complex and then moves along the RNA to catch up with RNA polymerase. As a result, its activity is impeded when ribosomes are translating an mRNA. This model explains a phenomenon that puzzled early bacterial geneticists. In some cases, a non-

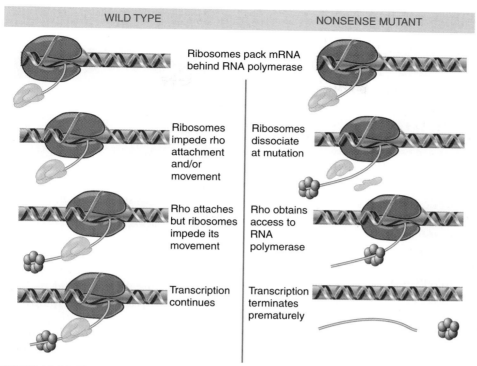

WILD TYPE		NONSENSE MUTANT

Ribosomes pack mRNA behind RNA polymerase

Ribosomes impede rho attachment and/or movement

Ribosomes dissociate at mutation

Rho attaches but ribosomes impede its movement

Rho obtains access to RNA polymerase

Transcription continues

Transcription terminates prematurely

FIGURE 19.32 The action of rho factor may create a link between transcription and translation when a rho-dependent terminator lies soon after a nonsense mutation.

sense mutation in one gene of a transcription unit was found to prevent the expression of subsequent genes in the unit even though both genes had their own ribosome binding sites, an effect called **polarity**.

Rho-dependent termination sites *within* a transcription unit are usually masked by translating ribosomes (**FIGURE 19.32**), and therefore rho cannot act on downstream RNA polymerases. Nonsense mutations release ribosomes within the RNA of a multigene operon, though, enabling rho to terminate transcription prematurely and prevent expression of distal genes in the transcription unit even though their open reading frames contained wild-type sequences.

Why are stable RNAs (rRNAs and tRNAs) not subject to polarity? tRNAs are short and form extensive secondary structures that probably prevent rho binding. Parts of rRNAs also have extensive structure but rRNAs are much longer than tRNAs, leaving ample opportunity for rho action. Cells have evolved another mechanism for preventing premature termination of rRNA transcripts, though: There are proteins that bind to so-called *nut sites* in the leader regions of both the 16S and 23S rRNA

transcripts, forming **antitermination complexes** that inhibit the action of Rho.

rho mutations show wide variations in their influence on termination. The basic nature of the effect is a failure to terminate. The magnitude of the failure, however, as seen in the percent of readthrough *in vivo*, depends on the particular target locus. Similarly, the need for rho factor *in vitro* is variable. Some (rho-dependent) terminators require relatively high concentrations of rho, whereas others function just as well at lower levels. This suggests that different terminators require different levels of rho factor for termination and therefore respond differently to the residual levels of rho factor in the mutants (*rho* mutants are usually leaky).

Some *rho* mutations can be suppressed by mutations in other genes. This approach provides an excellent way to identify proteins that interact with rho. The β subunit of RNA polymerase is implicated by two types of mutation. First, mutations in the *rpoB* gene can reduce termination at a rho-dependent site. Second, mutations in *rpoB* can restore the ability to terminate transcription at rho-dependent sites in *rho* mutant bacteria. We do not, however, know what function the interaction plays.

19.17 Supercoiling Is an Important Feature of Transcription

Key concepts

- Negative supercoiling increases the efficiency of some promoters by assisting the melting reaction.
- Transcription generates positive supercoils ahead of the enzyme and negative supercoils behind it, and these must be removed by gyrase and topoisomerase.

Both prokaryotic and eukaryotic RNA polymerases usually seem to initiate transcription more efficiently *in vitro* when the template is supercoiled, and in some cases, promoter efficiency is aided tremendously by negative supercoiling. Why are different promoters influenced more by the extent of supercoiling than others? The most likely possibility is that the dependence of a promoter on supercoiling is determined by the free energy needed to melt the DNA in the initiation complex. The free energy of melting in turn is dependent on the DNA sequence of the promoter. The more G+C-rich the promoter sequence corresponding to the position of the transcription bubble, the more dependent the promoter would be on supercoiling to help melt the DNA.

However, whether a particular promoter's activity is facilitated by supercoiling is much more complicated. The dependence of different promoters on the degree of supercoiling is also affected by DNA sequences outside of the bubble, because supercoiling changes the geometry of the complex, affecting the angles and distances between bases in space. Therefore, differences in the degree of supercoiling can alter interactions between bases in the promoter and amino acids in RNA polymerase. Furthermore, because different parts of the chromosome exhibit different degrees of supercoiling, the effect of supercoiling on a promoter's activity can be influenced by the location of the promoter on the chromosome.

As RNA polymerase continually unwinds and rewinds the DNA as it moves down the template (illustrated in Figure 19.4), either the entire transcription complex must rotate around the DNA, or the DNA itself must rotate about its helical axis. It is thought that the latter situation is closer to reality: the DNA threads through the enzyme like a screw through a bolt.

One consequence of the rotation of DNA is illustrated in **FIGURE 19.33**. In the *twin domain* model for transcription, as RNA poly-

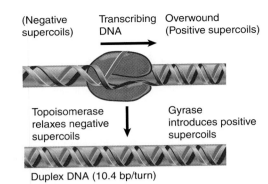

FIGURE 19.33 Transcription generates more tightly wound (positively supercoiled) DNA ahead of RNA polymerase, while the DNA behind becomes less tightly wound (negatively supercoiled).

merase moves with respect to the double helix, it generates positive supercoils (more tightly wound DNA) ahead of it and leaves negative supercoils (partially unwound DNA) behind it. For each helical turn traversed by RNA polymerase, +1 turn is generated ahead and −1 turn behind. Transcription therefore not only is affected by the local structure of DNA, but it also affects the actual structure of the DNA. The enzymes DNA gyrase, which introduces negative supercoils into DNA, and DNA topoisomerase I, which removes negative supercoils in DNA, are required to prevent topological stresses from building up in the course of transcription and replication. Blocking the activities of gyrase and topoisomerase therefore result in major changes in DNA supercoiling which in turn affects transcription and replication. This was discussed earlier in the context of replication (see *Section 14.2, Initiation: Creating the Replication Forks at the Origin oriC*).

19.18 Phage T7 RNA Polymerase Is a Useful Model System

Key concepts

- The T7 family of RNA polymerases are single polypeptides with the ability to recognize phage promoters and carry out many of the activities of the multisubunit RNA polymerases.
- Crystal structures of T7 family RNA polymerases with DNA identify the DNA-binding region, the active site, and suggest models for promoter escape.

Certain bacteriophages (e.g., T3, T7, N4) make their own RNA polymerases, consisting of single polypeptide chains. These RNA polymerases recognize just a few promoters on the phage DNA, but they carry out many of the activities of the multisubunit RNA polymerases.

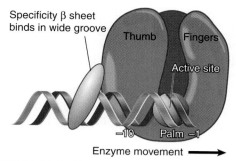

Specificity β sheet binds in wide groove

Thumb   Fingers

Active site

−10   Palm −1

Enzyme movement ⟶

FIGURE 19.34 T7 RNA polymerase has a specificity loop that binds positions −7 to −11 of the promoter while positions −1 to −4 enter the active site.

Thus, they provide model systems for the study of specific transcription functions.

For example, the T7 RNA polymerase is a single polypeptide chain of <100 kD. It synthesizes RNA at a rate of ~300 nucleotides / second at 37°C, a rate that is much faster than that from the multisubunit RNA polymerase of its bacterial host and faster than the ribosomes that translate its mRNAs. Thus, T7-directed transcription would be subject to transcriptional polarity if it were not for the fact that transcription by T7 RNA polymerase occurs only later in infection, when Rho expression is limited.

The T7 RNA polymerase is homologous to DNA and RNA polymerases in that the catalytic cores of all three enzymes have similar structures. DNA lies in a "palm" surrounded by "fingers" and a "thumb" (see Figure 14.7), and the enzyme uses an identical catalytic mechanism. We now have several crystal structures of the T7 and N4 RNA polymerases.

T7 RNA polymerase recognizes its target sequence in DNA by binding to bases in the major groove, as shown in **FIGURE 19.34**, using a *specificity loop* formed by a β ribbon. This feature is unique to the single-subunit RNA polymerases (it is not found in DNA polymerases). Like the multisubunit RNA polymerases, the promoter consists of specific bases in DNA upstream of the transcription start site, although T7 promoters consists of fewer bases than promoters typically recognized by multisubunit RNA polymerases.

The transition from the promoter complex to the elongation complex is accomplished by two major conformational changes in the enzyme. First, as with the multisubunit RNA polymerases, the template is "scrunched" in the active site, and the enzyme remains bound to the promoter as the polymerase undergoes *abortive synthesis*, producing short transcripts from 2 to 12 nt in length. The promoter-binding domain would present an obstacle to abortive

product formation if it were not for the fact that it is moved out of the way by a rotation of approximately 45 degrees, allowing the polymerase to maintain promoter contacts while synthesis of the initial RNA transcript. This is analogous to the displacement of the sigma factor domain 3-domain 4 linker from the RNA exit channel during the initial stages of RNA synthesis in the multisubunit bacterial RNA polymerase. The RNA emerges to the surface of the enzyme when twelve to fourteen nucleotides have been synthesized. An even larger conformational change occurs next, in which a subdomain called region H moves more than 70 Å from its location in the initiation complex. This massive structural reorganization of the N-terminal domain upon formation of the elongation complex creates a tunnel through which the RNA transcript can exit, as well as a binding site for the single-stranded nontemplate DNA of the transcription bubble.

19.19 Competition for Sigma Factors Can Regulate Initiation

Key concepts

- *E. coli* has seven sigma factors, each of which causes RNA polymerase to initiate at a set of promoters defined by specific −35 and −10 sequences.
- The activities of the different sigma factors are regulated by different mechanisms.

In the next few sections, we provide a few examples of regulation of initiation, elongation, and termination. Other examples will be presented in Chapter 26, *The Operon* and Chapter 27, *Phage Strategies*.

The division of labor between a core enzyme responsible for chain elongation and a sigma factor responsible for promoter selection raised the question of whether there would be more than one type of sigma factor, each specific for a different set of promoters. **FIGURE 19.35** shows the principle of a system in which a substitution of the sigma factor changes the choice of promoter.

E. coli often uses alternative sigma factors to respond to changes in environmental or nutritional conditions; they are listed in **FIGURE 19.36** (sigma factors are named by the molecular weight of the product or by the function of the genes they transcribe). The most abundant sigma factor, responsible for transcription of most genes under normal conditions, is σ^{70}

(called sigma A in most bacterial species) and is encoded by the *rpoD* gene. The alternative sigma factor σ^S (σ^{38}) is used for making many stress-related products; σ^H (σ^{32}) and σ^E (σ^{24}) are required for making products needed for responding to conditions that unfold proteins in the cytoplasm and periplasm, respectively; σ^N (σ^{54}) makes products needed primarily for nitrogen assimilation; σ^{FecI} (σ^{19}) makes a few products needed for iron transport; and σ^F (σ^{28}) expresses products needed for synthesis of flagella.

The unfolded protein response is one of the most conserved regulatory responses in all of biology. Originally discovered as a response to an increase in temperature (and therefore called the **heat shock response**), a similar set of proteins is synthesized in all three biological kingdoms that protect cells against environmental stress. Many of these heat shock proteins are *chaperones*, which reduce the levels of unfolded proteins by refolding them or degrading them. In *E. coli*, the induction of heat shock proteins occurs at the transcription level. The gene *rpoH* is a regulator needed to switch on the heat shock response. Its product is σ^{32} is an alternative sigma factor that recognizes the promoters of the heat shock genes.

The heat shock response (mostly chaperones and proteases) is feedback regulated. The key to the control of σ^{32} is that the availability of these cytoplasmic proteases and chaperones is dependent on whether or not they are titrated away by unfolded proteins. Thus, when unfolded protein levels go down (either because the heat shock proteins refold or degrade them, or because the temperature is lowered), they no longer titrate away the proteases that degrade σ^{32}, and σ^{32} levels return to normal. Because σ^{70} and σ^{32} compete for available core enzyme, transcription from heat shock gene promoters returns to basal levels as σ^{24} and σ^{32} levels go back to normal. Thus, the set of gene products made during heat shock depends on the balance between σ^{70} and σ^{32}. Consistent with the importance of sigma competition, the concentration of σ^{70} is greater than that of core RNA polymerase under σ^{32} noninducing conditions.

σ^{32} is not the only sigma factor that controls the unfolded protein response. σ^E is induced by accumulation of unfolded proteins in the periplasmic space and outer membrane (rather than in the cytoplasm). As with σ^{32}, proteolysis is the key to induction of transcription $E\sigma^E$-dependent promoters. The intricate circuit responsible for regulation of σ^E activity is summarized in **FIGURE 19.37**. σ^E binds to a protein (RseA) that is located in the inner membrane. RseA is an example of an **anti-sigma factor**. When bound to σ^E, RseA prevents σ^E from binding to core RNA polymerase and activating $E\sigma^E$ promoters. These promoters transcribe products needed for refolding denatured periplasmic proteins or degrading them. Thus, the periplasmic heat shock response is a transient, feedback response controlled by the concentrations of its own gene products. The σ^E regulon respond to the levels of unfolded and denatured periplasmic proteins rather than unfolded and denatured cytoplasmic proteins.

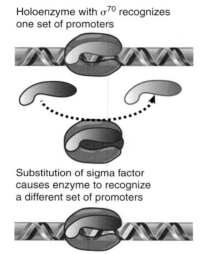

Holoenzyme with σ^{70} recognizes one set of promoters

Substitution of sigma factor causes enzyme to recognize a different set of promoters

FIGURE 19.35 The sigma factor associated with core enzyme determines the set of promoters at which transcription is initiated.

Gene	Factor	Use
rpoD	σ^{70}	most required functions
rpoS	σ^S	stationary phase/some stress responses
rpoH	σ^{32}	heat shock
rpoE	σ^E	periplasmic/extracellular proteins
rpoN	σ^{54}	nitrogen assimilation
rpoF	σ^F	flagellar synthesis/chemotaxis
fecI	σ^{fecI}	iron metabolism/transport

FIGURE 19.36 In addition to σ^{70}, *E. coli* has several sigma factors that are induced by particular environmental conditions. (A number in the name of a factor indicates its mass.)

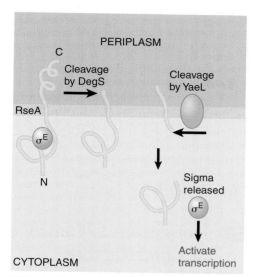

FIGURE 19.37 RseA is synthesized as a protein in the inner membrane. Its cytoplasmic domain binds the σ^E factor. RseA is cleaved sequentially in the periplasmic space and then in the cytoplasm. The cytoplasmic cleavage releases σ^E.

How does RseA know when to release σ^E? The mechanism involves regulated, sequential proteolysis of RseA. The accumulation of unfolded proteins activates a protease (DegS) in the periplasmic space, which cleaves off the C-terminal end of the RseA protein. This cleavage activates another protease, RseP, this time on the cytoplasmic face of the inner membrane. RseP cleaves the N-terminal region of RseA, ultimately releasing σ^E. σ^E can then bind core RNA polymerase and activate transcription. Thus, accumulation of unfolded proteins at the periphery of the bacterium activates the set of genes controlled by the sigma factor.

19.20 Sigma Factors May Be Organized into Cascades

Key concepts

- A cascade of sigma factors is created when one sigma factor is required to transcribe the gene coding for the next sigma factor.
- The early genes of phage SPO1 are transcribed by host RNA polymerase.
- One of the early genes codes for a sigma factor that causes RNA polymerase to transcribe the middle genes.
- Two of the middle genes code for subunits of a sigma factor that causes RNA polymerase to transcribe the late genes.

As in *E. coli*, sigma factors are used extensively to control initiation of transcription in the bacterium *Bacillus subtilis*. The *B. subtilis* genome encodes at least 18 different sigma factors, compared to the seven found in *E. coli*. Larger numbers of sigma factors than in *E. coli* is not unusual. In fact, the *Streptomyces coelicolor* genome encodes >60!

In *B. subtilis*, some of the sigma factors are present in vegetative cells, whereas others are produced only in the special circumstances of phage infection or during the change from vegetative growth to sporulation. The major RNA polymerase engaged in normal vegetative growth contains the same subunits and has the same overall structure as that of *E. coli*, $\alpha_2\beta\beta'\omega\sigma$, but in addition it has another subunit called δ. Its major sigma factor (σ^A) recognizes promoters with the same consensus sequences used by the *E. coli* enzyme under direction from σ^{70}. Alternative RNA polymerases containing different sigma factors are found in much smaller amounts and recognize promoters with different consensus sequences in −35 and −10 regions.

Transitions from expression of one set of genes to another set are a feature of bacteriophage infection. This is the case in *B. subtilis* infection by the phage SPO1, as it is in *E. coli* infection by phages like T7, N4, or λ. In all but the very simplest cases, the development of the phage involves shifts in the pattern of transcription during the infective cycle. These shifts may be accomplished by the synthesis of a phage-encoded RNA polymerase or by the efforts of phage-encoded ancillary factors that control the bacterial RNA polymerase. During infection of *B. subtilis* by phage SPO1, the different stages of infection are controlled via the production of new sigma factors.

The infective cycle of SPO1 has three stages of gene expression. Immediately on infection, the **early genes** of the phage are transcribed. After four to five minutes, the early genes cease transcription and the **middle genes** are transcribed. At eight to twelve minutes, middle gene transcription is replaced by transcription of **late genes**.

The early genes are transcribed by the holoenzyme of the host bacterium. They are essentially indistinguishable from host genes whose promoters have the intrinsic ability to be recognized by the RNA polymerase $\alpha_2\beta\beta'\omega\sigma^A$.

Expression of phage genes is required for the transitions to middle and late gene transcription. Three regulatory genes, *28*, *33*, and

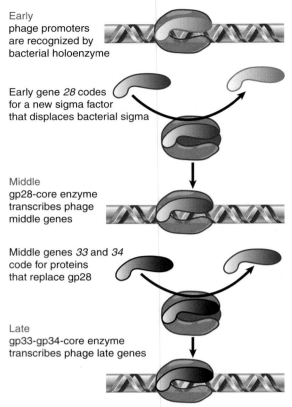

Early
phage promoters are recognized by bacterial holoenzyme

Early gene *28* codes for a new sigma factor that displaces bacterial sigma

Middle
gp28-core enzyme transcribes phage middle genes

Middle genes *33* and *34* code for proteins that replace gp28

Late
gp33-gp34-core enzyme transcribes phage late genes

FIGURE 19.38 Transcription of phage SP01 genes is controlled by two successive substitutions of the sigma factor that change the initiation specificity.

34, control the course of transcription. Their functions are summarized in **FIGURE 19.38**. The pattern of regulation resembles a **cascade**, in which the host enzyme transcribes an early gene whose product is needed to transcribe the middle genes. After this transcription, two of the middle genes code for products that are needed to transcribe the late genes.

Mutants in the early gene *28* cannot transcribe the middle genes. The product of gene *28* (called gp28) is a protein of 26 kD that replaces the host sigma factor on the core enzyme. *This substitution is the sole event required to make the transition from early to middle gene expression.* It creates a holoenzyme that can no longer transcribe the host genes but instead specifically transcribes the middle genes. We do not know how gp28 displaces σ^{43} or what happens to the host sigma polypeptide.

Two of the middle genes are involved in the next transition. Mutations in either gene 33 or 34 prevent transcription of the late genes. The products of these genes form a dimer that

replaces gp28 on the core polymerase. Again, we do not know how gp33 and gp34 exclude gp28 (or any residual host σ^A), *but once they have bound to the core enzyme, it is able to initiate transcription only at the promoters for late genes.*

The successive replacements of sigma factor have dual consequences. Each time the subunit is changed, the RNA polymerase becomes able to recognize a new class of genes *and* it no longer recognizes the previous class. These switches therefore constitute global changes in the activity of RNA polymerase.

19.21 Sporulation Is Controlled by Sigma Factors

Key concepts

- Sporulation divides a bacterium into a mother cell that is lysed and a spore that is released.
- Each compartment advances to the next stage of development by synthesizing a new sigma factor that displaces the previous sigma factor.
- Communication between the two compartments coordinates the timing of sigma factor substitutions.

A good example of the use of switching of holoenzymes to control changes in gene expression is provided by **sporulation**, an alternative lifestyle that occurs in many bacterial species. When logarithmic growth ceases because nutrients in the medium become depleted, the **vegetative phase** in growth of these bacteria ends. This triggers sporulation, a developmental stage in which the cell is resistant to many kinds of environmental and nutritional stresses (illustrated in **FIGURE 19.39**). During spore formation in *B subtilis*, one of the daughter genomes that result from DNA replication is segregated at one end of the cell, attached to the cell pole. A septum forms, generating two independent compartments: the mother cell and the forespore. The growing septum traps part of one chromosome in the forespore, and then a translocase (SpoIIIE) pumps the rest of the chromosome into the forespore. Eventually the forespore, with its engulfed chromosome, is surrounded by a tough coat, and this spore is stable almost indefinitely.

Sporulation takes approximately eight hours. It can be viewed as a primitive sort of differentiation, in which a parent cell (the vegetative bacterium) gives rise to two different daughter cells with distinct fates: the mother

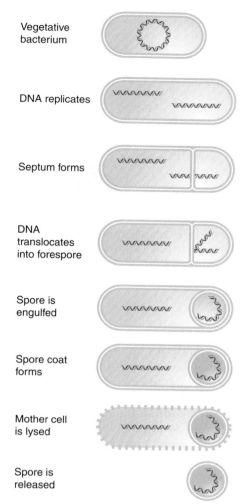

FIGURE 19.39 Sporulation involves the differentiation of a vegetative bacterium into a mother cell that is lysed and a spore that is released.

cell is eventually lysed, and the spore that is released has an entirely different structure from the original bacterium.

Sporulation involves a drastic change in the biosynthetic activities of the bacterium, in which many genes are involved. Changes in gene expression resulting ultimately in the formation of the spore result primarily from changes in transcription initiation. Some of the genes that function in the vegetative phase are turned off during sporulation, but most continue to be expressed. Many genes specific for sporulation are expressed only during this period, though. At the end of sporulation, ~40% of the bacterial mRNA is sporulation specific.

New forms of RNA polymerase become active in sporulating cells; they contain the same core enzyme as vegetative cells, but have different proteins in place of the vegetative sigma factor, σ^A. The changes in transcriptional specificity are summarized in **FIGURE 19.40**. The principle is that in each compartment the existing sigma factor is successively displaced by a new sigma factor that causes transcription of a different set of genes. Communication between the compartments occurs in order to coordinate the timing of the changes in the forespore and mother cell.

The sporulation cascade is initiated when environmental conditions trigger a **phosphorelay**, in which a phosphate group is passed along a series of proteins until it reaches a transcriptional regulator called SpoOA. Many gene products are involved in this process, whose complexity reflects the utilization of checkpoints—times when the bacterium confirms that it wishes to continue on the pathway to differentiation. This is not a regulatory course that should be undertaken unnecessarily, as the ultimate decision is irreversible.

Activation of SpoOA by phosphorylation marks the beginning of sporulation. In its phosphorylated form, SpoOA activates transcription of two operons, each of which is transcribed by a different form of the host RNA polymerase. Host enzyme utilizing the general sigma factor σ^A transcribes the gene coding for σ^F, and host enzyme under the direction of another sigma factor, σ^H, transcribes the gene coding for a precursor to the sigma factor σ^E. The precursor sigma factor is referred to as pro-σ^E. Both σ^F and pro-σ^E are produced before septum formation, but become active later.

Transcription directed by σ^F is inhibited because an antisigma factor (SpoIIAB) binds to it, preventing it from forming a holoenzyme. In the forespore, however, an anti-antisigma factor (SpoIIAA) inhibits the inhibitor. Inactivation of the anti-antisigma is controlled by a series of phosphorylation / dephosphorylation events, in which dephosphorylation by a phosphatase called SpoIIE is the first step. SpoIIE is an integral membrane protein that accumulates at the cell pole, with the result that its phosphatase domain becomes more concentrated in the forespore. In summary, dephosphorylation activates SpoIIAA, which in turn displaces SpoIIAB from σ^F. Release of σ^F activates it.

Activation of sigma F marks the start of cell-specific gene expression. Under the direction of σ^F, RNA polymerase transcribes the first set of

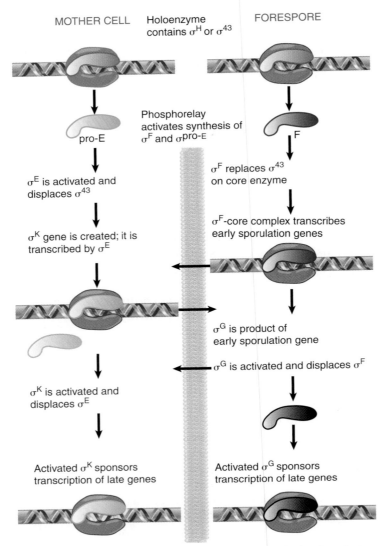

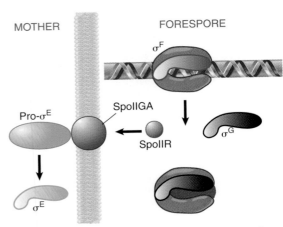

FIGURE 19.41 σ^F triggers synthesis of the next sigma factor in the forespore (σ^G) and turns on SpoIIR, which causes SpoIIGA to cleave pro-σ^E.

FIGURE 19.40 Sporulation involves successive changes in the sigma factors that control the initiation specificity of RNA polymerase. The cascades in the mother cell (left) and the forespore (right) are related by signals passed across the septum (indicated by horizontal arrows).

sporulation genes. Not all transcription in the forespore comes from Eσ^F. σ^A is not destroyed during sporulation, and, therefore, the vegetative holoenzyme, Eσ^A, remains in sporulating cells.

The cascade continues as products derived from promoters recognized by Eσ^F are made in the forespore (see **FIGURE 19.41**). For example, Eσ^F makes a transcript coding for $\sigma^{G,}$ which in turn forms the holoenzyme that transcribes the late sporulation genes. Eσ^F also recognizes a promoter controlling expression of a product responsible for communicating with the mother

cell compartment, SpoIIR, which is secreted from the forespore into the membrane separating the two compartments. In the membrane, SpoIIR activates the membrane-bound protein SpoIIGA which cleaves inactive precursor pro-σ^E into active σ^E in the mother cell. (σ^E produced in the forespore is degraded.)

The cascade continues when σ^E in the mother cell is replaced by σ^K. (The production of σ^K is quite complex, because its gene is created by a site-specific recombination event!) Like σ^E, σ^K is also synthesized as an inactive precursor, pro-σ^K. Thus, σ^K has to be activated by cleavage of its precursor form before it can replace σ^E and transcribe late genes in the mother cell. The timing of these events in the two compartments is coordinated by still other signals. In summary, the activity of σ^E in the mother cell is necessary for activation of σ^G in the forespore, and the activity of σ^G is required to generate a signal that is transmitted across the septum to activate σ^K.

Sporulation is thus controlled by a cascade in which sigma factors in each compartment are successively activated by sigmas F, E, G, and K, each directing the synthesis of a particular set of genes. The cascade can be represented by a crisscross pattern of signals crossing the septum, connecting gene expression in one compartment with that in the other, as illustrated in **FIGURE 19.42**. As new sigma factors become active, old sigma factors are displaced, turning sets of different genes on and off in the two compartments.

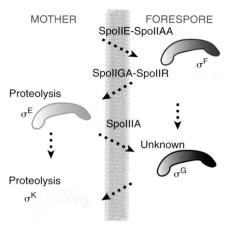

FIGURE 19.42 The crisscross regulation of sporulation coordinates timing of events in the mother cell and forespore.

Antitermination Can Be a Regulatory Event

Key concepts

- An antitermination complex allows RNA polymerase to read through terminators.
- Phage lambda uses antitermination systems for regulation of both its early and late transcripts, but the two systems work by completely different mechanisms.
- Binding of factors to the nascent RNA links the antitermination proteins to the terminator site through an RNA loop.
- Antitermination of transcription also occurs in rRNA operons.

Antitermination is used as a mechanism for control of transcription in both phage and bacterial operons. As shown in **FIGURE 19.43**, antitermination refers to modification of the enzyme, which allows it to read past a terminator into genes that lie downstream. In the example shown in the figure, the default pathway is for RNA polymerase to terminate at the end of region 1, but antitermination results in continued transcription through region 2.

Antitermination systems are common in lambdoid bacteriophages (phages similar to phage lambda and described in Chapter 27, *Phage Strategies*). Unlike the *E. coli* T7-like phages and the *B. subtilis* SPO1 phages discussed above, lambda does not encode either its own dedicated RNA polymerase or even its own dedicated sigma factors. Rather, it uses the host multisubunit RNA polymerase for all of its transcription. Shortly after phage infection,

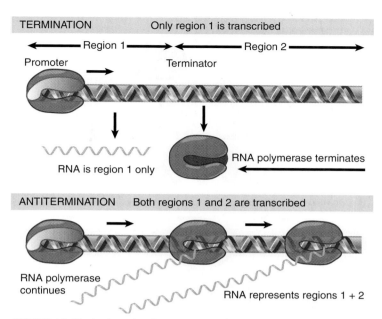

FIGURE 19.43 Antitermination can control transcription by determining whether RNA polymerase terminates or reads through a particular terminator into the following region.

transcription begins at two early promoters, P_R and P_L. There are, however, terminators in each of these operons that follow the transcription start site before most of the genes that encode most early functions, and termination of transcription at these positions aborts the infection. If RNA polymerase reads through the terminators and transcribes the early genes responsible for replication of the phage genome, though, lambda development proceeds.

The first termination decision is controlled by an antitermination protein named "N," which is the first protein produced by expression from P_L. N forms a complex with host proteins called Nus factors (*N u*tilization *s*ubstances) to modify RNA polymerase in such a way that it no longer responds to the terminators. The antitermination complex actually forms on the nascent RNA at a sequence called ***nut*** (*N u*tilization *s*ite). *nut* sites consist primarily of RNA sequences called *boxA* and *boxB* where the host factors NusA, NusB, NusE (ribosomal protein S10), and NusG assemble. The antitermination proteins remain bound to these RNA sites as *a persistent antitermination complex* as RNA polymerase synthesizes the two transcripts to the right and the left. Thus, the nascent RNA physically connects the antitermination proteins bound to the *nut* site with the RNA polymerase as it approaches

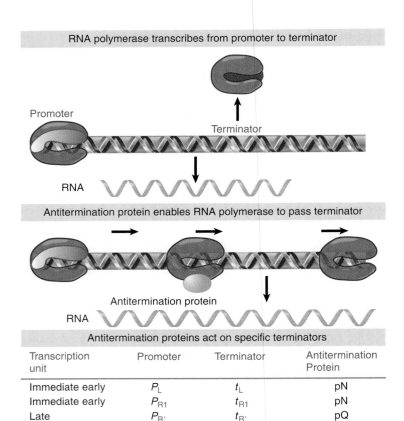

RNA polymerase transcribes from promoter to terminator

Promoter

Terminator

RNA

Antitermination protein enables RNA polymerase to pass terminator

Antitermination protein

RNA

Antitermination proteins act on specific terminators

Transcription unit	Promoter	Terminator	Antitermination Protein
Immediate early	P_L	t_L	pN
Immediate early	P_{R1}	t_{R1}	pN
Late	$P_{R'}$	$t_{R'}$	pQ

FIGURE 19.44 An antitermination protein can act on RNA polymerase to enable it to read through a specific terminator.

terminators. Although the actual mechanism by which the antitermination complex prevents termination is still not understood, tethering of the antitermination proteins to RNA polymerase through the nascent RNA explains its ability to antiterminate at successive terminators spaced hundreds or even thousands of bases downstream. The last protein produced by the N-antiterminated transcript from the other early promoter, P_R, is named "Q." Like N, Q is an antitermination protein. Q antiterminates transcription from the late promoter P_R, which produces a transcript coding for the phage's head and tail proteins. Thus, lambda gene expression occurs in two stages, each of which is controlled by antitermination (see *Section 27.7, The Lytic Cycle Depends on Antitermination by pN* and **FIGURE 19.44**). Q enables RNA polymerase to read through terminators in the late transcription unit, but it does so by a completely different mechanism than N. Unlike N, Q binds DNA (at the *qut*, *Q ut*ilization, site), but like N it travels with RNA polymerase and somehow interferes with the action of terminators throughout the late operon. It appears that

the action of Q involves acceleration of RNA polymerase through pause sites. (We discuss the overall regulation of lambda development in Chapter 27, *Phage Strategies*.)

rRNA operons might be expected to exhibit polarity, because they are long but are not translated. Each of the rRNA operons of *E. coli*, however, contains *boxA* and *boxB*-like sequences that assemble antitermination complexes on the transcripts consisting of at least some of the same Nus factors as those utilized by phage lambda. These complexes do not contain an N- or Q-like factor, which are encoded only by phage genomes, but they are sufficient to prevent premature termination at the hairpin sequences and weak Rho-dependent terminators that occur fortuitously within the rRNA structural genes. Antitermination is needed for efficient rRNA production all the time, not just when lambda infects cells. Thus bacterial evolution did not select for the Nus factors to facilitate lambda gene expression. Rather, these factors undoubtedly evolved to prevent polarity in rRNA operons. The leader regions of the *rrn* operons contain *boxA* sequences that assemble the Nus factors as the *boxA* sequences in RNA emerge from the RNA exit channel. As with antitermination in lambda, this process somehow changes the properties of RNA polymerase in such a way that it can now read through terminators, although the mechanism remains unclear.

19.23 The Cycle of Bacterial Messenger RNA

Key concepts

- Transcription and translation occur simultaneously in bacteria, coupled transcription/translation, as ribosomes begin translating an mRNA before its synthesis has been completed.
- Bacterial mRNA is unstable and has a half-life of only a few minutes.
- A bacterial mRNA may be polycistronic in having several coding regions that represent different genes.

Messenger RNA has the same function in all cells, but there are important differences in the details of the synthesis and structure of prokaryotic and eukaryotic mRNA.

A major difference in the production of mRNA depends on the locations where transcription and translation occur:

- In bacteria, mRNA is transcribed and translated in the single cellular compartment; the two processes are so closely linked that they occur simultaneously. Ribosomes attach to bacterial mRNA even before its transcription has been completed, so the *polysome* is likely still to be attached to DNA. Bacterial mRNA usually is unstable, and is therefore translated into polypeptides for only a few minutes. This process is called **coupled transcription/translation**.
- In a eukaryotic cell, synthesis and maturation of mRNA occur exclusively in the nucleus. Only after these events are completed is the mRNA exported to the cytoplasm, where it is translated by ribosomes. A typical eukaryotic mRNA is relatively stable and continues to be translated for several hours, although there is a great deal of variation in the stability of specific mRNAs.

FIGURE 19.45 shows that transcription and translation are intimately related in bacteria. Transcription begins when the enzyme RNA polymerase binds to DNA and then moves along, making a copy of one strand. Very soon after transcription begins, ribosomes attach to the 5' end of the mRNA and start translation, even before the rest of the message has been synthesized. Multiple ribosomes move along the mRNA while it is being synthesized. The 3' end of the mRNA is generated when transcription terminates. Ribosomes continue to translate the mRNA while it survives, but it is degraded in the overall 5'→3' direction quite rapidly. The mRNA is synthesized, translated by the ribosomes, and degraded, all in rapid succession. An individual molecule of mRNA survives for only a matter of minutes at most.

Bacterial transcription and translation take place at similar rates. At 37°C, transcription of mRNA occurs at ~40 to 50 nucleotides/second. This is very close to the rate of protein synthesis, which is roughly 15 amino acids/second. It therefore takes ~1 minute to transcribe and translate an mRNA of 2500 bp, corresponding to a 90 kD polypeptide. When expression of a new gene is initiated, its mRNA typically will appear in the cell within ~1.5 minutes. The corresponding polypeptide will appear within perhaps another 0.5 minute.

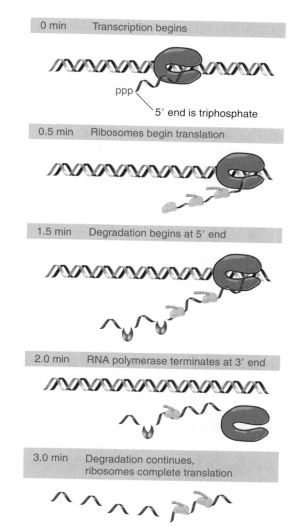

FIGURE 19.45 Overview: mRNA is transcribed, translated, and degraded simultaneously in bacteria.

Bacterial translation is very efficient, and most mRNAs are translated by a large number of tightly packed ribosomes. In one example (*trp* mRNA), about 15 initiations of transcription occur every minute, and each of the 15 mRNAs probably is translated by ~30 ribosomes in the interval between its transcription and degradation.

The instability of most bacterial mRNAs is striking. Degradation of mRNA closely follows its translation and likely begins within one minute of the start of transcription. The 5' end of the mRNA starts to decay before the 3' end has been synthesized or translated. Degradation seems to follow the last ribosome of the convoy along the mRNA. Degradation proceeds more slowly, though—probably at about half the speed of transcription or translation.

The stability of mRNA has a major influence on the amount of polypeptide that is produced. It is usually expressed in terms of the half-life. The mRNA representing any particular gene has a characteristic half-life, but the average is ~2 minutes in bacteria.

This series of events is only possible, of course, because transcription, translation, and degradation all occur in the same direction. The dynamics of gene expression have been "caught in the act" in the electron micrograph of **FIGURE 19.46**. In these (unknown) transcription units, several mRNAs are under synthesis simultaneously, and each carries many ribosomes engaged in translation. (This corresponds to the stage shown in the second panel in Figure 19.45.) An RNA whose synthesis has not yet been completed is often called a **nascent RNA**.

Bacterial mRNAs vary greatly in the number of proteins for which they code. Some mRNAs carry only a single ORF; they are **monocistronic**. Others (the majority) carry sequences coding for several polypeptides; they are **polycistronic**. In these cases, a single mRNA is transcribed from a group of adjacent cistrons. (Such a cluster of cistrons constitutes an operon that is controlled as a single genetic unit; see Chapter 26, *The Operon*.)

All mRNAs contain three regions. The coding region (the open reading frame) or ORF consists of a series of codons representing the amino acid sequence of the polypeptide, starting (usually) with AUG and ending with one of the three termination codons. The mRNA is always longer than the coding region, though, as extra regions are present at both ends. An additional sequence at the 5′ end, upstream of the coding region, is described as the leader or **5′ UTR** (untranslated region). An additional sequence downstream from the termination signal, forming the 3′ end, is called the trailer or **3′ UTR**. Although they do not encode a polypeptide, these sequences may contain important regulatory instructions, especially in eukaryote mRNA.

A polycistronic mRNA also contains **intercistronic regions**, as illustrated in **FIGURE 19.47**. They vary greatly in size. They may be as long as 30 nucleotides in bacterial mRNAs (and even longer in phage RNAs), or they may be very short, with as few as one or two nucleotides separating the termination codon for one polypeptide from the initiation codon for the next. In an extreme case, two genes actually overlap, so that the last base of one coding region is also the first base of the next coding region.

The number of ribosomes engaged in translating a particular cistron depends on the efficiency of its initiation site in the 5′ UTR. The initiation site for the first cistron becomes available as soon as the 5′ end of the mRNA is synthesized. How are subsequent cistrons translated? Are the several coding regions in a polycistronic mRNA translated independently or is their expression connected? Is the mechanism of initiation the same for all cistrons, or is it different for the first cistron and the internal cistrons?

Translation of a bacterial mRNA proceeds sequentially through its cistrons. At the time when ribosomes attach to the first coding region, the subsequent coding regions have not yet even been transcribed. By the time the second ribosome site is available, translation is well under way through the first cistron. Typically ribosomes terminate translation at the end of the first cistron (and dissociate into subunits), and a new ribosome assembles independently at the start of the next coding region. (We discuss the processes of initiation and termination in Chapter 24, *Translation*.)

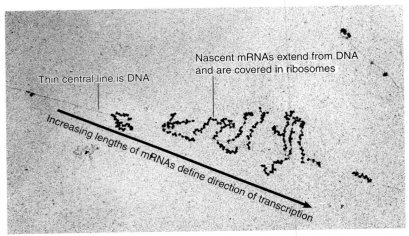

FIGURE 19.46 Transcription units can be visualized in bacteria. Photo courtesy of Oscar Miller.

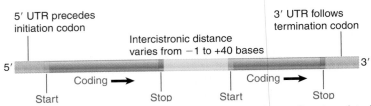

FIGURE 19.47 Bacterial mRNA includes untranslated as well as translated regions. Each coding region has its own initiation and termination signals. A typical mRNA may have several coding regions.

19.24 Summary

A transcription unit comprises the DNA between a promoter, where transcription initiates, and a terminator, where it ends. One strand of the DNA in this region serves as a template for synthesis of a complementary strand of RNA. The RNA–DNA hybrid region is short and transient, as the transcription "bubble" moves along DNA. The RNA polymerase holoenzyme that synthesizes bacterial RNA can be separated into two components. Core enzyme is a multimer containing the subunits $\alpha_2\beta\beta'\omega$ that is sufficient for elongating the RNA chain. Sigma factor (σ) is a single subunit that is required only at the stage of initiation for recognizing the promoter.

Core enzyme has a general affinity for DNA. The addition of sigma factor reduces the affinity of the enzyme for nonspecific binding to DNA, and increases its affinity for promoters. The rate at which RNA polymerase finds its promoters can be too rapid to be accounted for by random encounters with DNA by simple diffusion; transcription factors that recruit RNA polymerase to the DNA and direct exchange of the enzyme between one DNA sequence and another are likely to play a role in the promoter search.

Many bacterial promoters can be identified from the sequences of two 6 bp sequences centered at –35 and –10 relative to the startpoint, although other accessory promoter elements upstream from the –35 element (the UP element) and surrounding the –10 element (the extended –10 and discriminator regions) also contribute to promoter recognition. The distance separating the consensus sequences is almost always 16 to 18 bp. The enzyme can cover as much as ~75 bp of DNA. The initial "closed" binary complex is converted to an "open" binary complex by sequential melting of a sequence of ~14 bp that begins in the –10 region and extends to about 3 bp downstream from the startpoint. The A-T-rich base pair composition of the –10 sequence contributes to the melting reaction.

The binary complex is converted to a ternary complex by the incorporation of ribonucleotide precursors. There are usually multiple cycles of abortive initiation, during which RNA polymerase synthesizes and releases very short RNA chains without escaping from the promoter. At the end of this stage, sigma is usually released, and the resulting core enzyme covers only ~35 bp of DNA rather than the twice that amount observed in the initiation complex. The core enzyme then moves down the template, unwinding the DNA as it synthesizes the RNA transcript.

The core enzyme can be directed to recognize promoters with different consensus sequences by alternative sigma factors. In *E. coli*, these sigma factors are activated by adverse conditions, such as heat shock or nitrogen starvation. The geometry of the RNA polymerase-promoter complex is relatively similar for all holoenzymes. All sigma factors except σ^{54} recognize consensus elements located about 35 and 10 upstream from the transcription start site, making direct contacts with bases in these elements. The σ^{70} factor of *E. coli* has an N-terminal autoinhibitory domain that prevents the DNA-binding regions from recognizing DNA. The autoinhibitory region is displaced by DNA when the holoenzyme forms an open complex.

The "strength" of a promoter describes the frequency at which RNA polymerase initiates transcription; it is related to the closeness with which its promoter elements –35, –10, and other accessory elements conform to the ideal consensus sequences. Negative supercoiling increases the strength of certain promoters. Transcription generates positive supercoils ahead of RNA polymerase and leaves negative supercoils behind the enzyme.

B. subtilis contains a single major sigma factor with the same specificity as the major *E. coli* sigma factor, but it also contains a variety of minor sigma factors, some of which are activated sequentially during the process of sporulation; sporulation is regulated by a sigma factor cascade in which sigma factor replacements occur in the forespore and mother cell. Cascades involving sequential utilization of different RNA polymerases can also regulate transcription during bacteriophage infection and development.

Bacterial RNA polymerase terminates transcription at two types of sites. Intrinsic terminators contain a G-C-rich hairpin followed by a U-rich region. They are recognized *in vitro* by core enzyme alone. Rho-dependent terminators require rho factor both *in vitro* and *in vivo*; rho binds to *rut* sites that are rich in C and poor in G residues that precede the actual site of termination. Rho is a hexameric ATP-dependent helicase that translocates along the RNA until it reaches the RNA polymerase, where it dissociates the RNA polymerase from DNA. In both types of termination, pausing by RNA polymerase likely contributes to the termination event.

Antitermination is used by lambdoid phages to regulate progression from one stage of gene expression to the next. Multiprotein complexes containing the lambda phage N protein or Q protein, as well as Nus factors can associate with RNA polymerase through RNA and perhaps DNA loops, respectively, and prevent transcription termination. The N-containing antitermination complex allows RNA polymerase to read through terminators located at the ends of the immediate early genes, whereas Q-containing antitermination complexes are required later in phage infection.

References

19.2 Transcription Occurs by Base Pairing in a "Bubble" of Unpaired DNA

Review

Losick, R. and Chamberlin, M. (1976). RNA polymerase. *Cold Spring Harbor Monograph Series* 6.

Research

Revyakin, A., Liu, C., Ebright, R. H., and Strick T. R. (2006). Abortive initiation and productive initiation by RNA polymerase involve DNA scrunching. *Science*, 314, 1139–1143.

19.3 The Transcription Reaction Has Three Stages

Research

Rice, G. A., Kane, C. M., and Chamberlin, M. (1991). Footprinting analysis of mammalian RNA polymerase II along its transcript: an alternative view of transcription elongation. *Proc. Natl. Acad. Sci. USA* 88, 4245–4281.

Wang, D. et al. (1995). Discontinuous movements of DNA and RNA in RNA polymerase accompany formation of a paused transcription complex. *Cell* 81, 341–350.

19.4 Bacterial RNA Polymerase Consists of Multiple Subunits

Review

Helmann, J. D. and Chamberlin, M. (1988). Structure and function of bacterial sigma factors. *Annu. Rev. Biochem.* 57, 839–872.

Shilatifard, A., Conway, R. C., and Conway, J. W. (2003). The RNA polymerase II elongation complex. *Annu. Rev. Biochem.* 72, 693–715.

Research

Campbell, E. A., Korzheva, N., Mustaev, A., Murakami, K., Nair, S., Goldfarb, A., and Darst, S. A. (2001). Structural mechanism for rifampicin inhibition of bacterial RNA polymerase. *Cell* 104, 901–912.

Geszvain, K. and Landick, R. (2005). The structure of bacterial RNA polymerase. In *The Bacterial Chromosome* Higgins, N. P. (ed.). Washington, DC: American Society for Microbiology Press, pp. 283–296.

Korzheva, N., Mustaev, A., Kozlov, M., Malhotra, A., Nikiforov, V., Goldfarb, A., and Darst, S. A. (2000). A structural model of transcription elongation. *Science* 289, 619–625.

Vassylyev, D. G., Vassylyeva, M. N., Perederina, A., Tahirov, T. H., and Artsimovitch, I. (2007). Structural basis for transcription elongation by bacterial RNA polymerase. *Nature* 448, 157–162.

Zhang, G., Campbell, E. A., Zhang, E. A., Minakhin, L., Richter, C., Severinov, K., and Darst, S. A. (1999). Crystal structure of *Thermus aquaticus* core RNA polymerase at 3.3 Å resolution. *Cell* 98, 811–824.

19.5 RNA Polymerase Holoenzyme Consists of the Core Enzyme and Sigma Factor

Research

Travers, A. A. and Burgess, R. R. (1969). Cyclic reuse of the RNA polymerase sigma factor. *Nature* 222, 537–540.

19.6 How Does RNA Polymerase Find Promoter Sequences?

Review

Bustamante, C., Guthold, M., Zhu, X., and Yang G. (1999). Facilitated target location on DNA by individual *Escherichia coli* RNA polymerase molecules observed with the scanning force microscope operating in liquid. *J. Bio. Chem.* 274, 16665–16669.

19.7 The Holoenzyme Goes through Transitions in the Process of Recognizing and Escaping from Promoters

Research

Bar-Nahum, G. and Nudler, E. (2001). Isolation and characterization of sigma(70)-retaining transcription elongation complexes from *E. coli. Cell* 106, 443–451.

Kapanidis, A. N., Margeat, E., Ho, S. O., Kortkhonjia, E., Weiss, S., and Ebright, R. H. (2006). Initial transcription by RNA polymerase proceeds through a DNA-scrunching mechanism. *Science* 314, 1144–1147.

Krummel, B. and Chamberlin, M. J. (1989). RNA chain initiation by *E. coli* RNA polymerase. Structural transitions of the enzyme in early ternary complexes. *Biochemistry* 28, 7829–7842.

Mukhopadhyay, J., Kapanidis, A. N., Mekler, V., Kortkhonjia, E., Ebright, Y. W., and Ebright, R. H. (2001). Translocation of sigma(70) with RNA polymerase during transcription.

Fluorescence resonance energy transfer assay for movement relative to DNA. *Cell* 106, 453–463.

Wang, Q., Tullius, T. D., and Levin, J. R. (2007). Effects of discontinuities in the DNA template on abortive initiation and promoter escape by *E. coli* RNA polymerase. *J. Biol. Chem* 282, 26917–26927.

19.8 Sigma Factor Controls Binding to DNA by Recognizing Specific Sequences in Promoters

Reviews

Haugen, S. P., Ross, W., and Gourse R. L. (2008). Advances in bacterial promoter recognition and its control by factors that do not bind DNA. *Nature Rev. Micro.* 6, 507–520.

McClure, W. R. (1985). Mechanism and control of transcription initiation in prokaryotes. *Annu. Rev. Biochem.* 54, 171–204.

Research

Bar-Nahum, G. and Nudler, E. (2001). Isolation and characterization of sigma(70)-retaining transcription elongation complexes from *E. coli. Cell* 106, 443–451.

Haugen, S. P., Ross.,W., Manrique, M., and Gourse, R. L. (2008). Fine structure of the promoter–σ region 1.2 interaction. *Proc. Natl. Acad. Sci.* 105, 3292–3297.

Mukhopadhyay, J., Kapanidis, A. N., Mekler, V., Kortkhonjia, E., Ebright, Y. W., and Ebright, R. H. (2001). Translocation of sigma(70) with RNA polymerase during transcription. Fluorescence resonance energy transfer assay for movement relative to DNA. *Cell* 106, 453–463.

Ross, W., Gosink, K. K., Salomon, J., Igarashi, K., Zou, C., Ishihama, A., Severinov, K., and Gourse, R. L. (1993). A third recognition element in bacterial promoters: DNA binding by the alpha subunit of RNA polymerase. *Science* 262, 1407–1413.

19.9 Promoter Efficiencies Can Be Increased or Decreased by Mutation

Review

McClure, W. R. (1985). Mechanism and control of transcription initiation in prokaryotes. *Annu. Rev. Biochem.* 54, 171–204.

19.10 Multiple Regions in RNA polymerase Directly Contact Promoter DNA

Research

Campbell, E. A., Muzzin, O., Chlenov, M., Sun, J. L., Olson, C. A., Weinman, O., Trester-Zedlitz, M. L., and Darst, S. A. (2002). Structure of the bacterial RNA polymerase promoter specificity sigma subunit. *Mol. Cell* 9, 527–539.

Dombrowski, A. J. et al. (1992). Polypeptides containing highly conserved regions of transcription initiation factor σ70 exhibit specificity of binding to promoter DNA. *Cell* 70, 501–512.

Mekler, V., Kortkhonjia, E., Mukhopadhyay, J., Knight, J., Revyakin, A., Kapanidis, A. N., Niu, W., Ebright, Y. W., Levy, R., and Ebright, R. H. (2002). Structural organization of bacterial RNA polymerase holoenzyme and the RNA polymerase-promoter open complex. *Cell* 108, 599–614.

Vassylyev, D. G., Sekine, S., Laptenko, O., Lee, J., Vassylyeva, M. N., Borukhov, S., and Yokoyama S. (2002). Crystal structure of a bacterial RNA polymerase holoenzyme at 2.6 Å resolution. *Nature* 417, 712–719.

19.11 Footprinting Is a High Resolution Method For Characterizing RNA Polymerase-Promoter and DNA-Protein Interactions in General

Review

Siebenlist, U., Simpson, R. B., and Gilbert, W. (1980). *E. coli* RNA polymerase interacts homologously with two different promoters. *Cell* 20, 269–281.

19.13 A Model for Enzyme Movement Is Suggested by the Crystal Structure

Reviews

Herbert, K. M., Greenleaf, W. J., and Block, S. M. (2008). Single-molecule studies of RNA polymerase: motoring along. *Annu. Rev. Biochem.* 77, 149–176.

Shilatifard, A., Conaway, R. C., and Conaway, J. W. (2003). The RNA polymerase II elongation complex. *Annu. Rev. Biochem.* 72, 693–715.

Research

Cramer, P., Bushnell, D. A., Fu, J., Gnatt, A. L., Maier-Davis, B., Thompson, N. E., Burgess, R. R., Edwards, A. M., David, P. R., and Kornberg, R. D. (2000). Architecture of RNA polymerase II and implications for the transcription mechanism. *Science* 288, 640–649.

Cramer, P., Bushnell, P., and Kornberg, R. D. (2001). Structural basis of transcription: RNA polymerase II at 2.8 Å resolution. *Science* 292, 1863–1876.

Gnatt, A. L., Cramer, P., Fu, J., Bushnell, D. A., and Kornberg, R. D. (2001). Structural basis of transcription: an RNA polymerase II elongation complex at 3.3 Å resolution. *Science* 292, 1876–1882.

19.14 A Stalled RNA Polymerase Can Restart

Research

Kettenberger, H., Armache, K. J., and Cramer, P. (2003). Architecture of the RNA polymerase II-TFIIS complex and implications for mRNA cleavage. *Cell* 114, 347–357.

Opalka, N., Chlenov, M., Chacon, P., Rice, W. J., Wriggers, W., and Darst, S. A. (2003). Structure and function of the transcription elongation factor GreB bound to bacterial RNA polymerase. *Cell* 114, 335–345.

19.15 Bacterial RNA Polymerase Terminates at Discrete Sites

Reviews
Adhya, S. and Gottesman, M. (1978). Control of transcription termination. *Annu. Rev. Biochem.* 47, 967–996.

Friedman, D. I., Imperiale, M. J., and Adhya, S. L. (1987). RNA 3′ end formation in the control of gene expression. *Annu. Rev. Genet.* 21, 453–488.

Greenblat, J. F. (2008). Transcription termination: pulling out all the stops. *Cell* 132, 917–919.

Platt, T. (1986). Transcription termination and the regulation of gene expression. *Annu. Rev. Biochem.* 55, 339–372.

von Hippel, P. H. (1998). An integrated model of the transcription complex in elongation, termination, and editing. *Science* 281, 660–665.

Research
Lee, D. N., Phung, L., Stewart, J., and Landick, R. (1990). Transcription pausing by *E. coli* RNA polymerase is modulated by downstream DNA sequences. *J. Biol. Chem.* 265, 15145–15153.

Lesnik, E. A., Sampath, R., Levene, H. B., Henderson, T. J., McNeil, J. A., and Ecker, D. J. (2001). Prediction of rho-independent transcriptional terminators in *E. coli. Nucleic Acids Res.* 29, 3583–3594.

Reynolds, R., Bermadez-Cruz, R. M., and Chamberlin, M. J. (1992). Parameters affecting transcription termination by *E. coli* RNA polymerase. I. Analysis of 13 rho-independent terminators. *J. Mol. Biol.* 224, 31–51.

19.16 How Does Rho Factor Work?

Reviews
Das, A. (1993). Control of transcription termination by RNA-binding proteins. *Annu. Rev. Biochem.* 62, 893–930.

Richardson, J. P. (1996). Structural organization of transcription termination factor Rho. *J. Biol. Chem.* 271, 1251–1254.

von Hippel, P. H. (1998). An integrated model of the transcription complex in elongation, termination, and editing. *Science* 281, 660–665.

Research
Brennan, C. A., Dombroski, A. J., and Platt, T. (1987). Transcription termination factor rho is an RNA-DNA helicase. *Cell* 48, 945–952.

Geiselmann, J., Wang, Y., Seifried, S. E., and von Hippel, P. H. (1993). A physical model for the translocation and helicase activities of *E. coli*

transcription termination protein Rho. *Proc. Natl. Acad. Sci. USA* 90, 7754–7758.

Roberts, J. W. (1969). Termination factor for RNA synthesis. *Nature* 224, 1168–1174.

Skordalakes, E. and Berger, J. M. (2003). Structure of the Rho transcription terminator: mechanism of mRNA recognition and helicase loading. *Cell* 114, 135–146.

19.17 Supercoiling Is an Important Feature of Transcription

Research
Wu, H.-Y. et al. (1988). Transcription generates positively and negatively supercoiled domains in the template. *Cell* 53, 433–440.

19.18 Phage T7 RNA Polymerase Is a Useful Model System

Research
Cheetham, G. M., Jeruzalmi, D., and Steitz, T. A. (1999). Structural basis for initiation of transcription from an RNA polymerase-promoter complex. *Nature* 399, 80–83.

Cheetham, G. M. T. and Steitz, T. A. (1999). Structure of a transcribing T7 RNA polymerase initiation complex. *Science* 286, 2305–2309.

Temiakov, D., Mentesana, D., Temiakov, D., Ma, K., Mustaev, A., Borukhov, S., and McAllister, W. T. (2000). The specificity loop of T7 RNA polymerase interacts first with the promoter and then with the elongating transcript, suggesting a mechanism for promoter clearance. *Proc. Natl. Acad. Sci. USA* 97, 14109–14114.

19.19 Competition for Sigma Factors Can Regulate Initiation

Review
Hengge-Aronis, R. (2002). Signal transduction and regulatory mechanisms involved in control of the sigma(S) (RpoS) subunit of RNA polymerase. *Microbiol. Mol. Biol. Rev.* 66, 373–393.

Research
Alba, B. M., Onufryk, C., Lu, C. Z., and Gross, C. A. (2002). DegS and YaeL participate sequentially in the cleavage of RseA to activate the sigma(E)-dependent extracytoplasmic stress response. *Genes Dev.* 16, 2156–2168.

Grossman, A. D., Erickson, J. W., and Gross, C. A. (1984). The htpR gene product of *E. coli* is a sigma factor for heat-shock promoters. *Cell* 38, 383–390.

Kanehara, K., Ito, K., and Akiyama, Y. (2002). YaeL (EcfE) activates the sigma(E) pathway of stress response through a site-2 cleavage of anti-sigma(E), RseA. *Genes Dev.* 16, 2147–2155.

Sakai, J., Duncan, E. A., Rawson, R. B., Hua, X., Brown, M. S., and Goldstein, J. L. (1996). Sterol-regulated release of SREBP-2 from cell membranes requires two sequential cleavages, one within a transmembrane segment. *Cell* 85, 1037–1046.

19.21 Sporulation Is Controlled by Sigma Factors

Reviews

Errington, J. (1993). *B. subtilis* sporulation: regulation of gene expression and control of morphogenesis. *Microbiol. Rev.* 57, 1–33.

Haldenwang, W. G. (1995). The sigma factors of *B. subtilis*. *Microbiol. Rev.* 59, 1–30.

Losick, R. and Stragier, P. (1992). Crisscross regulation of cell-type specific gene expression during development in *B. subtilis*. *Nature* 355, 601–604.

Losick, R. et al. (1986). Genetics of endospore formation in *B. subtilis*. *Annu. Rev. Genet.* 20, 625–669.

Stragier, P. and Losick, R. (1996). Molecular genetics of sporulation in *B. subtilis*. *Annu. Rev. Genet.* 30, 297–341.

Research

Haldenwang, W. G., Lang, N., and Losick, R. (1981). A sporulation-induced sigma-like regulatory protein from *B. subtilis*. *Cell* 23, 615–624.

Haldenwang, W. G. and Losick, R. (1980). A novel RNA polymerase sigma factor from *B. subtilis*. *Proc. Natl. Acad. Sci. USA* 77, 7000–7004.

19.22 Antitermination Can Be a Regulatory Event

Review

Greenblatt, J., Nodwell, J. R., and Mason, S. W. (1993). Transcriptional antitermination. *Nature* 364, 401–406.

Research

Legault, P., Li, J., Mogridge, J., Kay, L. E., and Greenblatt, J. (1998). NMR structure of the bacteriophage lambda N peptide/boxB RNA complex: recognition of a GNRA fold by an arginine-rich motif. *Cell* 93, 289–299.

Mah, T. F., Kuznedelov, K., Mushegian, A., Severinov, K., and Greenblatt, J. (2000). The alpha subunit of *E. coli* RNA polymerase activates RNA binding by NusA. *Genes Dev.* 14, 2664–2675.

Mogridge, J., Mah, J., and Greenblatt, J. (1995). A protein-RNA interaction network facilitates the template-independent cooperative assembly on RNA polymerase of a stable antitermination complex containing the lambda N protein. *Genes Dev.* 9, 2831–2845.

Olson, E. R., Flamm, E. L., and Friedman, D. I. (1982). Analysis of nutR: a region of phage lambda required for antitermination of transcription. *Cell* 31, 61–70.

19.23 The Cycle of Bacterial Messenger RNA

Research

Brenner, S., Jacob, F. and Meselson, M. (1961). An unstable intermediate carrying information from genes to ribosomes for protein synthesis. *Nature* 190, 576–581.

© Carol & Mike Werner/Visuals Unlimited

Eukaryotic Transcription

CHAPTER OUTLINE

20.1 Introduction
- Chromatin must be opened before RNA polymerase can bind the promoter.

20.2 Eukaryotic RNA Polymerases Consist of Many Subunits
- RNA polymerase I synthesizes rRNA in the nucleolus.
- RNA polymerase II synthesizes mRNA in the nucleoplasm.
- RNA polymerase III synthesizes small RNAs in the nucleoplasm.
- All eukaryotic RNA polymerases have ~12 subunits and are complexes of ~500 kD.
- Some subunits are common to all three RNA polymerases.
- The largest subunit in RNA polymerase II has a CTD (carboxy-terminal domain) consisting of multiple repeats of a heptamer.

20.3 RNA Polymerase I Has a Bipartite Promoter
- The RNA polymerase I promoter consists of a core promoter and an upstream promoter element (UPE).

- The factor UBF1 wraps DNA around a protein structure to bring the core and UPE into proximity.
- SL1 includes the factor TBP that is involved in initiation by all three RNA polymerases.
- RNA polymerase I binds to the UBF1-SL1 complex at the core promoter.

20.4 RNA Polymerase III Uses Both Downstream and Upstream Promoters
- RNA polymerase III has two types of promoters.
- Internal promoters have short consensus sequences located within the transcription unit and cause initiation to occur at a fixed distance upstream.
- Upstream promoters contain three short consensus sequences upstream of the startpoint that are bound by transcription factors.
- $TF_{III}A$ and $TF_{III}C$ bind to the consensus sequences and enable $TF_{III}B$ to bind at the startpoint.
- $TF_{III}B$ has TBP as one subunit and enables RNA polymerase to bind.

20.1 Introduction

Key concept

- Chromatin must be opened before RNA polymerase can bind the promoter.

Initiation of transcription on a chromatin template that is already opened requires the enzyme RNA polymerase to bind at the promoter and transcription factors to bind to enhancers. *In vitro* transcription on a DNA template requires a different subset of transcription factors than are needed to transcribe a chromatin template (we will examine how chromatin is opened in Chapter 28, *Eukaryotic Transcription Regulation*). Any protein that is needed for the initiation of transcription, but that is not itself part of RNA polymerase, is defined as a transcription factor. Many transcription factors act by recognizing *cis*-acting sites on DNA. Binding to DNA, however, is not the only means of action for a transcription factor. A factor may recognize another factor, may recognize RNA polymerase, or may be incorporated into an initiation complex only in the presence of several other proteins. The ultimate test for membership in the transcription apparatus is functional: a protein must be needed for transcription to occur at a specific promoter or set of promoters.

A significant difference between the transcription of eukaryotic and prokaryotic RNAs is that in bacteria, transcription takes place on a DNA template, whereas in eukaryotes, transcription takes place on a chromatin template. Chromatin changes everything and must be taken into account at every step. The chromatin must be in an open structure and even in an open structure, nucleosome octamers must be removed from the promoter before RNA polymerase can bind. This can sometimes require transcription from a silent or cryptic promoter, either on the same strand or on the antisense strand.

A second major difference is that the bacterial RNA polymerase, with its sigma factor subunit, can read the DNA sequence to find and bind to its promoter. A eukaryotic RNA polymerase cannot read the DNA. Initiation at eukaryotic promoters therefore involves a large number of factors that must prebind to a variety of cis-acting elements before the RNA polymerase can bind. These factors are called **basal transcription factors**. The RNA polymerase then binds to this basal transcription factor/DNA complex. This binding region is defined as the **core promoter,** the region containing all the binding sites necessary for RNA polymerase to bind and function. RNA polymerase itself binds around the **startpoint** of transcription, but does not directly contact the extended upstream region of the promoter. By contrast, the bacterial promoters discussed in Chapter 19, *Prokaryotic Transcription*, are largely defined in terms of the binding site for RNA polymerase in the immediate vicinity of the startpoint.

While bacteria have a single RNA polymerase that transcribes all three major classes of genes, transcription in eukaryotic cells is divided into three classes. Each class is transcribed by a different RNA polymerase:

- RNA polymerase I transcribes 18S/28S rRNA.
- RNA polymerase II transcribes mRNA and a few small RNAs.
- RNA polymerase III transcribes tRNA, 5S ribosomal RNA, and other small RNAs.

This is the picture that we have of the major classes of genes. As we will see in Chapter 30, *Regulatory RNA*, recent discoveries by whole genome tiling arrays have uncovered a new world of antisense transcripts, intergenic transcripts, and heterochromatin transcripts. We do not yet know anything about the promoters for these classes or their regulation, but we do know that many (possibly most) of these transcripts are produced by RNA polymerase II.

Basal transcription factors are needed for initiation, but most are not required subsequently. For the three eukaryotic RNA polymerases, the transcription factors, rather than the RNA polymerases themselves, are responsible for recognizing the promoter DNA sequence. For all eukaryotic RNA polymerases, the basal transcription factors create a structure at the promoter to provide the target that is recognized by the RNA polymerase. For RNA polymerases I and III, these factors are relatively simple, but for RNA polymerase II they form a sizeable group. The basal factors join with RNA polymerase II to form a complex surrounding the startpoint, and they determine the site of initiation. The basal factors together with RNA polymerase constitute the basal transcription apparatus.

The promoters for RNA polymerases I and II are (mostly) upstream of the startpoint, but a large number of promoters for RNA polymerase III lie downstream (within the transcription unit) of the startpoint. Each promoter contains characteristic sets of short conserved sequences that are recognized by the appropriate class of basal transcription factors. RNA polymerases I and III each recognize a relatively restricted set of promoters, and rely upon a small number of accessory factors.

Promoters utilized by RNA polymerase II show much more variation in sequence, and have a modular organization. All RNA polymerase II promoters have sequence elements close to the startpoint that are bound by the basal apparatus and the polymerase to establish the site of initiation. Other sequences farther upstream (or downstream), called **enhancer** sequences, determine whether the promoter is expressed, and if expressed, whether this occurs in all cell types or is cell type-specific. An enhancer is another type of site involved in transcription and is identified by sequences that stimulate initiation, but that are located a variable distance from the core promoter. Enhancer elements are often targets for tissue-specific or temporal regulation. Some enhanc-

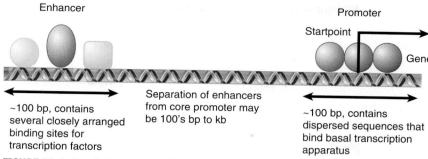

Enhancer

Promoter

Startpoint

Gene

~100 bp, contains several closely arranged binding sites for transcription factors

Separation of enhancers from core promoter may be 100's bp to kb

~100 bp, contains dispersed sequences that bind basal transcription apparatus

FIGURE 20.1 A typical gene transcribed by RNA polymerase II has a promoter that extends upstream from the site where transcription is initiated. The promoter contains several short (~10 bp) sequence elements that bind transcription factors, dispersed over ~100 bp. An enhancer containing a more closely packed array of elements that also bind transcription factors may be located several hundred bp to several kb distant. (DNA may be coiled or otherwise rearranged so that transcription factors at the promoter and at the enhancer interact to form a large protein complex.)

ers bind transcription factors that function by short-range interactions and are located near the promoter, whereas others can be located thousands of base pairs away. **FIGURE 20.1** illustrates the general properties of promoters and enhancers. A regulatory site that binds more negative regulators than positive regulators to control transcription is called a **silencer**.

Promoters that are constitutively expressed and needed in all cells (their genes are sometimes called **housekeeping genes**) have upstream sequence elements that are recognized by ubiquitous activators. No one element/factor combination is an essential component of the promoter, which suggests that initiation by RNA polymerase II may be regulated in many different ways. Promoters that are expressed only in certain times or places have sequence elements that require activators that are available only at those times or places.

The components of an enhancer or silencer resemble those of the promoter, in that they consist of a variety of modular elements that can bind positive regulators or negative regulators in a closely packed array. Enhancers do not need to be near the promoter. They can be upstream, inside a gene, or beyond the end of a gene, and their orientation relative to the gene does not matter. Proteins bound at enhancer elements interact with proteins bound at promoter elements, very often through intermediates called **coactivators**.

Eukaryotic transcription is most often under positive regulation: A transcription factor is provided under tissue-specific control to activate a promoter or set of promoters that contain a common target sequence. This is a

multistep process that first involves opening the chromatin and then binding the basal transcription factors, and then binding the polymerase. Regulation by specific repression of a target promoter is less common.

A eukaryotic transcription unit generally contains a single gene, and termination occurs beyond the end of the coding region. Termination lacks the regulatory importance that applies in prokaryotic systems. RNA polymerases I and III terminate at discrete sequences in defined reactions, but the mode of termination by RNA polymerase II is not clear. The significant event in generating the 3' end of an mRNA, however, is not the termination event itself, but instead results from a cleavage reaction in the primary transcript (see Chapter 21, *RNA Splicing and Processing*).

20.2 Eukaryotic RNA Polymerases Consist of Many Subunits

Key concepts

- RNA polymerase I synthesizes rRNA in the nucleolus.
- RNA polymerase II synthesizes mRNA in the nucleoplasm.
- RNA polymerase III synthesizes small RNAs in the nucleoplasm.
- All eukaryotic RNA polymerases have ~12 subunits and are complexes of ~500 kD.
- Some subunits are common to all three RNA polymerases.
- The largest subunit in RNA polymerase II has a CTD (carboxy-terminal domain) consisting of multiple repeats of a heptamer.

The three eukaryotic RNA polymerases have different locations in the nucleus that correspond with the different genes that they transcribe.

The most prominent activity is the enzyme RNA polymerase I, which resides in the nucleolus and is responsible for transcribing the genes coding for the 18S and 28S rRNA. It accounts for most cellular RNA synthesis (in terms of quantity).

The other major enzyme is RNA polymerase II, which is located in the nucleoplasm (the part of the nucleus excluding the nucleolus). It represents most of the remaining cellular activity and is responsible for synthesizing most of the **heterogeneous nuclear RNA (hnRNA)**, the precursor for most mRNA and a lot more. The classical definition was that hnRNA includes everything but rRNA and tRNA in the nucleus (again, classically, mRNA is only found in the cytoplasm). With modern molecular tools, we can now look a little closer at hnRNA and find many low-abundance RNAs that are very important, plus a lot that we are just now starting to understand. The mRNA is the least abundant of the three major RNAs, accounting for just 2%–5% of the cytoplasmic RNA.

RNA polymerase III is a minor enzyme in terms of activity, but it produces a collection of stable, essential RNAs. This nucleoplasmic enzyme synthesizes the 5S rRNA, tRNAs, and other small RNAs that constitute over a quarter of the cytoplasmic RNAs.

All eukaryotic RNA polymerases are large proteins, functioning as complexes of ~500 kD. They typically have ~12 subunits. The purified enzyme can undertake template-dependent transcription of RNA, but is not able to initiate selectively at promoters. The general constitution of a eukaryotic RNA polymerase II enzyme as typified in *Saccharomyces cerevisiae* is illustrated in **FIGURE 20.2**. The two largest subunits are homologous to the β and β′ subunits of bacterial RNA polymerase. Three of the remaining subunits are common to all the RNA polymerases; that is, they are also components of RNA polymerases I and III. Note that there is no subunit related to the bacterial sigma factor. Its function is contained in the basal transcription factors.

The largest subunit in RNA polymerase II has a **carboxy-terminal domain (CTD)**, which consists of multiple repeats of a consensus sequence of seven amino acids. The

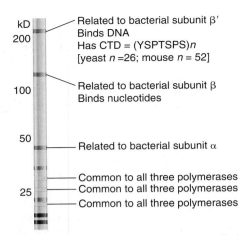

FIGURE 20.2 Some subunits are common to all classes of eukaryotic RNA polymerases and some are related to bacterial RNA polymerase. This drawing is a simulation of purified yeast RNA polymerase II run on an SDS gel to separate the subunits by size.

sequence is unique to RNA polymerase II. There are ~26 repeats in yeast and ~50 in mammals. The number of repeats is important because deletions that remove (typically) more than half of the repeats are lethal. The CTD can be highly phosphorylated on serine or threonine residues. The CTD is involved in regulating the initiation reaction (see *Section 20.8, Initiation Is Followed by Promoter Clearance and Elongation*), transcription elongation, and all aspects of mRNA processing, even export of mRNA to the cytoplasm.

The RNA polymerases of mitochondria and chloroplasts are smaller, and they resemble bacterial RNA polymerase rather than any of the nuclear enzymes (because they evolved from eubacteria). Of course, the organelle genomes are much smaller, thus the resident polymerase needs to transcribe relatively few genes, and the control of transcription is likely to be very much simpler. These enzymes are more similar to bacteriophage enzymes that do not need to respond to a more complex environment.

A major practical distinction between the eukaryote enzymes is drawn from their response to the bicyclic octapeptide α-amanitin (the toxic compound in *Amanita* mushroom species). In essentially all eukaryotic cells, the activity of RNA polymerase II is rapidly inhibited by low concentrations of α-amanitin. RNA polymerase I is not inhibited. The response of RNA polymerase III is less well conserved; in animal cells it is inhibited by high levels, but in yeast and insects it is not inhibited.

20.3 RNA Polymerase I Has a Bipartite Promoter

Key concepts

- The RNA polymerase I promoter consists of a core promoter and an upstream promoter element (UPE).
- The factor UBF1 wraps DNA around a protein structure to bring the core and UPE into proximity.
- SL1 includes the factor TBP that is involved in initiation by all three RNA polymerases.
- RNA polymerase I binds to the UBF1-SL1 complex at the core promoter.

RNA polymerase I transcribes only the genes for ribosomal RNA, from a single type of promoter. The precursor transcript includes the sequences of both large 28S and small 18S rRNAs, which are later processed by cleavages and modifications. There are many copies of the transcription unit. They alternate with **nontranscribed spacers** and are organized in a cluster, as discussed in *Section 7.3, Genes for rRNA Form Tandem Repeats Including an Invariant Transcription Unit*. The organization of the promoter, and the events involved in initiation, are illustrated in **FIGURE 20.3**. RNA polymerase I exists as a holoenzyme that contains additional factors required for initiation, and is recruited by its transcription factors directly as a giant complex to the promoter.

The promoter consists of two separate regions. The core promoter surrounds the startpoint, extending from −45 to +20, and is sufficient for transcription to initiate. It is generally G-C-rich (unusual for a promoter), except for the only conserved sequence element, a short A-T-rich sequence around the startpoint. The core promoter's efficiency, however, is very much increased by the upstream promoter element (UPE, sometimes also called the upstream control element, or UCE). The UPE is another G-C-rich sequence related to the core promoter sequence, and extends from −180 to −107. This type of organization is common to pol I promoters in many species, although the actual sequences vary widely.

RNA polymerase I requires two ancillary transcription factors. The factor that binds to the core promoter is SL1 (or TIF-1B and Rib1 in different species), which consists of four protein subunits. One of the components of SL1 is the **TATA-binding protein (TBP)**, a factor that also is required for initiation by RNA

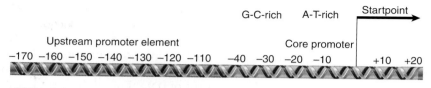

UBF binds to upstream promoter element

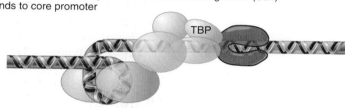

RNA polymerase I holoenzyme includes core binding factor (SL1) that binds to core promoter

FIGURE 20.3 Transcription units for RNA polymerase I have a core promoter separated by ~70 bp from the upstream promoter element. UBF binding to the UPE increases the ability of core-binding factor to bind to the core promoter. Core-binding factor (SL1) positions RNA polymerase I at the startpoint.

polymerases II and III (see *Section 20.6, TBP Is a Universal Factor*). TBP does not bind directly to G-C-rich DNA, and DNA binding is the responsibility of the other components of SL1. It is likely that TBP interacts with RNA polymerase, probably with a common subunit or a feature that has been conserved among polymerases. SL1 enables RNA polymerase I to initiate from the promoter at a low basal frequency.

SL1 has primary responsibility for ensuring that the RNA polymerase is properly localized at the startpoint. We will see shortly that a comparable function is provided for RNA polymerases II and III by a factor that consists of TBP and other proteins. Thus a common feature in initiation by all three polymerases is a reliance on a "positioning factor" that consists of TBP associated with proteins that are specific for each type of promoter. The exact mode of action is different for each of the TBP-dependent positioning factors; at the promoter for RNA polymerase I it does not bind DNA, whereas at the promoter for RNA polymerase II it is the principal means for locating the factor on DNA.

For high-frequency initiation, the factor transcription factor UBF is required. This is a single polypeptide that binds to a G-C-rich element in the UPE. UBF has two functions: it stimulates promoter release by the RNA polymerase and, as described below, it stimulates SL1. One indication of how UBF interacts with SL1 is given by the importance of the spacing between UBF and the core promoter. This can be changed by distances involving integral numbers of turns of DNA, but not by distances that introduce half turns. UBF binds to the minor groove of DNA and wraps the DNA in a loop of almost 360° turn on the protein surface, with the result that the core promoter and UPE come into close proximity, enabling UBF to stimulate binding of SL1 to the promoter.

Figure 20.3 shows initiation as a series of sequential interactions. RNA polymerase I, however, exists as a holoenzyme that contains most or all of the factors required for initiation, and is probably recruited directly to the promoter.

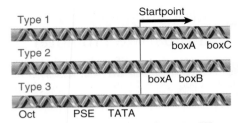

FIGURE 20.4 Promoters for RNA polymerase III may consist of bipartite sequences downstream of the startpoint, with *boxA* separated from either *boxC* or *boxB*, or they may consist of separated sequences upstream of the startpoint (Oct, PSE, TATA).

20.4 RNA Polymerase III Uses Both Downstream and Upstream Promoters

Key concepts

- RNA polymerase III has two types of promoters.
- Internal promoters have short consensus sequences located within the transcription unit and cause initiation to occur at a fixed distance upstream.
- Upstream promoters contain three short consensus sequences upstream of the startpoint that are bound by transcription factors.
- TF$_{III}$A and TF$_{III}$C bind to the consensus sequences and enable TF$_{III}$B to bind at the startpoint.
- TF$_{III}$B has TBP as one subunit and enables RNA polymerase to bind.

Recognition of promoters by RNA polymerase III strikingly illustrates the relative roles of transcription factors and the polymerase enzyme. The promoters fall into two general classes that are recognized in different ways by different groups of factors. The promoters for 5S and tRNA genes are *internal*; they lie downstream of the startpoint. The promoters for snRNA (small nuclear RNA) genes lie upstream of the startpoint in the more conventional manner of other promoters. In both cases, the individual elements that are necessary for promoter function consist exclusively of sequences recognized by transcription factors, which in turn direct the binding of RNA polymerase.

The structures of three types of promoter for RNA polymerase III are summarized in **FIGURE 20.4**. There are two types of internal promoter. Each contains a bipartite structure,

in which two short sequence elements are separated by a variable sequence. The 5S ribosomal gene Type 1 promoter consists of a *boxA* sequence separated by an intermediate element (IE) from a *boxC* sequence; the entire *boxA-IE-boxC* region is often referred to as the 5S internal control region (ICR). In yeast, only the *boxC* element is required for transcription. The tRNA type 2 promoter consists of a *boxA* sequence separated from a *boxB* sequence. A common group of type 3 promoters coding for other small RNAs have three sequence elements that are all located upstream of the startpoint; these same elements are also present in a number of RNA polymerase II promoters.

The detailed interactions are different at the two types of internal promoter, but the principle is the same. TF$_{III}$C binds downstream of the startpoint, either independently (tRNA type 2 promoters) or in conjunction with TF$_{III}$A (5S type 1 promoters). The presence of TF$_{III}$C enables the positioning factor TF$_{III}$B to bind at the startpoint. RNA polymerase III is then recruited.

FIGURE 20.5 summarizes the stages of reaction at type 2 internal promoters used for tRNA genes. The distance between *boxA* and *boxB* can vary since many tRNA genes contain a small intron. TF$_{III}$C binds to both *boxA* and *boxB*. This enables TF$_{III}$B to bind at the startpoint. At this point RNA polymerase III can bind.

The difference at type 1 internal promoters (for 5S genes) is that TF$_{III}$A must bind at *boxA* to enable TF$_{III}$C to bind at *boxC*. TF$_{III}$A is a 5S sequence-specific binding factor that binds to the promoter and to the 5S RNA as a chaperone and gene regulator. **FIGURE 20.6** shows that once TF$_{III}$C has bound, events follow the same course as at type 2 promoters, with TF$_{III}$B (which contains the ubiquitous TBP) binding at the startpoint, and RNA polymerase III joining the complex. Type 1 promoters are found only in the genes for 5S rRNA.

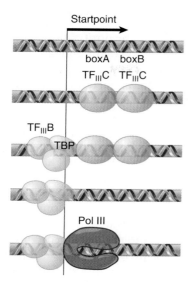

FIGURE 20.5 Internal type 2 pol III promoters use binding of TF$_{III}$C to *boxA* and *boxB* sequences to recruit the positioning factor TF$_{III}$B, which recruits RNA polymerase III.

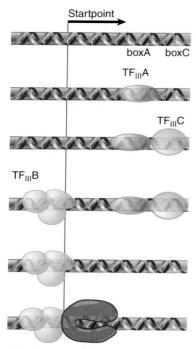

FIGURE 20.6 Internal type 1 pol III promoters use the assembly factors TF$_{III}$A and TF$_{III}$C, at *boxA* and *boxC*, to recruit the positioning factor TF$_{III}$B, which recruits RNA polymerase III.

TF$_{III}$A and TF$_{III}$C are **assembly factors**, whose sole role is to assist the binding of the positioning factor TF$_{III}$B at the correct location. Once TF$_{III}$B has bound, TF$_{III}$A and TF$_{III}$C can be removed from the promoter without affecting the initiation reaction. *TF$_{III}$B remains bound in the vicinity of the startpoint, and its presence is suf-*

ficient to allow RNA polymerase III to identify and bind at the startpoint. Thus TF$_{III}$B is the only true initiation factor required by RNA polymerase III. This sequence of events explains how the promoter boxes downstream can cause RNA polymerase to bind at the startpoint, farther upstream. Although the ability to transcribe these genes is conferred by the internal promoter, changes in the region immediately upstream of the startpoint can alter the efficiency of transcription.

TF$_{III}$C is a large protein complex (>500kD), which is comparable in size to RNA polymerase itself and contains six subunits. TF$_{III}$A is a member of an interesting class of proteins containing a nucleic acid-binding motif called a zinc finger (see *Section 28.6, There Are Many Types of DNA-Binding Domains*). The positioning factor TF$_{III}$B consists of three subunits. It includes the same protein factor TBP that is present in the core-binding factor SL1 used for pol I promoters, and (as we will see in *Section 20.6, TBP Is a Universal Factor*) in the corresponding transcription factor TF$_{II}$D used by RNA polymerase II. It also contains Brf, which is related to the transcription factor TF$_{II}$B that is used by RNA polymerase II. The third subunit is called B''; it is dispensable if the DNA duplex is partially melted, which suggests that its function is to initiate the transcription bubble. The role of B'' may be comparable to the role played by sigma factor in bacterial RNA polymerase (see *Section 19.19, Competition for Sigma Factors Can Regulate Initiation*).

The upstream region has a conventional role in the third class of polymerase III promoters. In the example shown in Figure 20.4, there are three upstream elements. These elements are also found in promoters for snRNA genes that are transcribed by RNA polymerase II. (Genes for some snRNAs are transcribed by RNA polymerase II, whereas others are transcribed by RNA polymerase III.) The upstream elements function in a similar manner in promoters for both RNA polymerases II and III.

Initiation at an upstream promoter for RNA polymerase III can occur on a short region that immediately precedes the startpoint and contains only the TATA element. Efficiency of transcription, however, is much increased by the presence of the enhancer PSE (proximal sequence element) and OCT (named because it has an 8-base-pair binding sequence) elements. The factors that bind at these elements interact cooperatively. The PSE element may be essential at promoters used by RNA polymerase II, whereas it is stimulatory in promoters used by

RNA polymerase III; its name stands for proximal sequence element.

The TATA element confers specificity for the type of polymerase (II or III) that is recognized by an snRNA promoter. It is bound by a factor that includes TBP, which actually recognizes the sequence in DNA. TBP is associated with other proteins, which are specific for the type of promoter. The function of TBP and its associated proteins is to position the RNA polymerase correctly at the startpoint. We discuss this in more detail for RNA polymerase II (see *Section 20.6, TBP Is a Universal Factor*).

The factors work in the same way for both types of promoters for RNA polymerase III. *The factors bind at the promoter before RNA polymerase itself can bind.* They form a **preinitiation complex** that directs binding of the RNA polymerase. RNA polymerase III does not itself recognize the promoter sequence, but binds adjacent to factors that are themselves bound just upstream of the startpoint. For the type I and type II internal promoters, the assembly factors ensure that $TF_{III}B$ (which includes TBP) is bound just upstream of the startpoint, thereby providing the positioning information. For the upstream promoters, $TF_{III}B$ binds directly to the region including the TATA box. This means that irrespective of the location of the promoter sequences, factor(s) are bound close to the startpoint in order to direct binding of RNA polymerase III. In all cases, the chromatin must be modified and in an open configuration.

20.5 The Startpoint for RNA Polymerase II

Key concepts

- RNA polymerase II requires general transcription factors (called $TF_{II}X$) to initiate transcription.
- RNA polymerase II promoters frequently have a short conserved sequence Py_2CAPy_5 (the initiator Inr) at the startpoint.
- The TATA box is a common component of RNA polymerase II promoters and consists of an A-T-rich octamer located ~25 bp upstream of the startpoint.
- The DPE is a common component of RNA polymerase II promoters that do not contain a TATA box.
- A core promoter for RNA polymerase II includes the Inr and, commonly, either a TATA box or a DPE. It may also contain other minor elements.

The basic organization of the apparatus for transcribing protein-coding genes was revealed by the discovery that purified RNA polymerase II can catalyze synthesis of mRNA, but cannot initiate transcription unless an additional extract is added. The purification of this extract led to the definition of the general transcription factors, or *basal transcription factors*—a group of proteins that are needed for initiation by RNA polymerase II at all promoters. RNA polymerase II in conjunction with these factors constitutes the basal transcription apparatus that is needed to transcribe any promoter. The general factors are described as $TF_{II}X$, where "X" is a letter that identifies the individual factor. The subunits of RNA polymerase II and the general transcription factors are conserved among eukaryotes.

Our starting point for considering promoter organization is to define the core promoter as the shortest sequence at which RNA polymerase II can initiate transcription. A core promoter can in principle be expressed in any cell (though in practice a core promoter alone results in little or no transcription in the chromatin context *in vivo*). It is the minimum sequence that enables the general transcription factors to assemble at the startpoint. These factors are involved in the mechanics of binding to DNA and enable RNA polymerase II to initiate transcription. A core promoter functions at only a low efficiency. Other proteins, called *activators*, another class of transcription factors, are required for a proper level of function (see *Section 20.9, Enhancers Contain Bidirectional Elements That Assist Initiation*). The activators are not described systematically, but have casual names reflecting their histories of identification.

We might expect any sequence components involved in the binding of RNA polymerase and general transcription factors to be conserved at most or all promoters, as is the case for pol I and pol III promoters. As with bacterial promoters, when promoters for RNA polymerase II are compared, homologies in the regions near the startpoint are restricted to rather short sequences. These elements correspond with the sequences implicated in promoter function by mutation. **FIGURE 20.7** shows the construction of a typical pol II core promoter with three of the most common pol II promoter elements. The eukaryotic pol II promoter is far more structurally diverse than the bacterial promoter, though. In addition to the three major elements, there are a number of minor elements that can also serve to define the promoter.

At the startpoint, there is no extensive homology of sequence, but there is a tendency for the first base of mRNA to be A, flanked on either side by pyrimidines. (This description is also valid for the CAT start sequence of

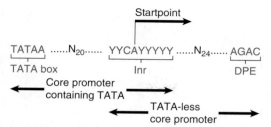

FIGURE 20.7 A minimal pol II promoter may have a TATA box ~25 bp upstream of the Inr. The TATA box has the consensus sequence of TATAA. The Inr has pyrimidines (Y) surrounding the CA at the startpoint. The DPE is downstream of the startpoint. The sequence shows the coding strand.

bacterial promoters.) This region is called the **initiator (Inr)**, and may be described in the general form Py_2CAPy_5, where Py stands for any pyrimidine. The Inr is contained between positions −3 and +5.

Many promoters have a sequence called the **TATA box**, usually located ~25 bp upstream of the startpoint in higher eukaryotes. It constitutes the only upstream promoter element that has a relatively fixed location with respect to the startpoint. The consensus sequence of this core element is TATAA, usually followed by three more A-T base pairs (see *Section 19.8, Sigma Factor Controls Binding to DNA by Recognizing Specific Sequences in Promoters*, for a discussion of consensus sequence). The TATA box tends to be surrounded by G-C-rich sequences, which could be a factor in its function. It is almost identical with the −10 TATA box sequence found in bacterial promoters; in fact, it could pass for one except for the difference in its location at −25 instead of −10. (The exception is in yeast, where the TATA box is more typically found at −90.) Single-base substitutions in the TATA box may act as up or down mutations, depending on how close the original sequence matches the consensus sequence and how different the mutant sequence is. Typically, substitutions that introduce a G-C base pair are the most severe.

Promoters that do not contain a TATA element are called **TATA-less promoters**. Surveys of promoter sequences suggest that 50% or more of promoters may be TATA-less. When a promoter does not contain a TATA box, it often contains another element, the **DPE (downstream promoter element)**, which is located at +28 to +32.

Typical core promoters consist either of a TATA box plus Inr, or of an Inr plus DPE, although other combinations with minor elements exist as well.

20.6 TBP Is a Universal Factor

Key concepts

- TBP is a component of the positioning factor that is required for each type of RNA polymerase to bind its promoter.
- The factor for RNA polymerase II is $TF_{II}D$, which consists of TBP and ~14 TAFs, with a total mass ~800 kD.
- TBP binds to the TATA box in the minor groove of DNA.
- TBP forms a saddle around the DNA and bends it by ~80°.

Before transcription initiation can begin, the chromatin has to be modified and remodeled to the open configuration, and any nucleosome octamer positioned over the promoter has to be moved or removed at all classes of eukaryotic promoters (we will examine this aspect of transcription control more closely in Chapter 28, *Eukaryotic Transcription Regulation*). At that point it is possible for a positioning factor to bind to the promoter. Each class of RNA polymerase is assisted by a positioning factor that contains TBP associated with other components. The name TBP stands for "TATA binding protein"; it was initially so-named because it was a protein that bound to the TATA box in RNA polymerase II genes. It was subsequently discovered to also be part of the positioning factors SL1 for RNA polymerase I (see *Section 20.3, RNA Polymerase I Has a Bipartite Promoter*) and $TF_{III}B$ RNA polymerase III (see *Section 20.4, RNA Polymerase III Uses Both Downstream and Upstream Promoters*). For these latter two RNA polymerases, TBP does not recognize the TATA box sequence (except in type 3 pol III promoters); thus the name is misleading. In addition, many RNA polymerase II promoters lack TATA boxes, but still require the presence of TBP.

For RNA polymerase II, the positioning factor is $TF_{II}D$, which consists of TBP associated with up to 14 other subunits called **TAFs** (for TBP-associated factors). Some TAFs are stoichiometric with TBP; others are present in lesser amounts, which means that there are multiple $TF_{II}D$ variants. $TF_{II}Ds$ containing different TAFs could recognize promoters with different combinations of conserved elements described above in *Section 20.5, The Startpoint for RNA Polymerase II*. Some TAFs are tissue-specific. The total mass of $TF_{II}D$ typically is ~800 kD. The TAFs in $TF_{II}D$ were originally named in the form $TAF_{II}00$, for example, where the number "00" gives the molecular mass of the subunit. Recently, the RNA polymerase II TAFs have been renamed TAF1, TAF2, and so forth;

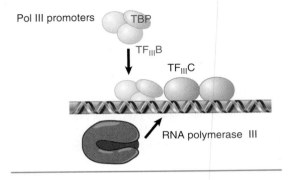

Pol III promoters

TBP

TF$_{III}$B

TF$_{III}$C

RNA polymerase III

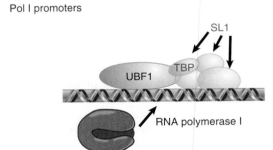

Pol I promoters

SL1

UBF1

TBP

RNA polymerase I

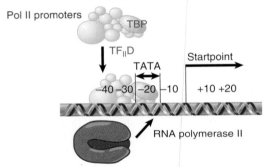

Pol II promoters

TBP

TF$_{II}$D

Startpoint

TATA

−40 −30 −20 −10   +10 +20

RNA polymerase II

FIGURE 20.8 RNA polymerases are positioned at all promoters by a factor that contains TBP.

FIGURE 20.9 A view in cross-section shows that TBP surrounds DNA from the side of the narrow groove. TBP consists of two related (40% identical) conserved domains, which are shown in light and dark blue. The N-terminal region varies extensively and is shown in green. The two strands of the DNA double helix are in light and dark gray. Photo courtesy of Stephen K. Burley.

in this nomenclature TAF1 is the largest TAF, TAF2 is the next largest, and homologous TAFs in different species thus have the same names.

FIGURE 20.8 shows that the positioning factor recognizes the promoter in a different way in each case. At promoters for RNA polymerase III, TF$_{III}$B binds adjacent to TF$_{III}$C. At promoters for RNA polymerase I, SL1 binds in conjunction with UBF. TF$_{II}$D is solely responsible for recognizing promoters for RNA polymerase II. At a promoter that has a TATA element, TBP binds specifically to the TATA box, but at TATA-less promoters, the TAFs have the role of recognizing other promoter elements, including the Inr and DPE. Whatever its means of entry into the initiation complex, it has the common purpose of interaction with the RNA polymerase.

TBP has the unusual property of binding to DNA in the minor groove. (The vast majority of DNA-binding proteins bind in the major

groove.) The crystal structure of TBP suggests a detailed model for its binding to DNA. **FIGURE 20.9** shows that it surrounds one face of DNA, forming a "saddle" around a stretch of the minor groove, which is bent to fit into this saddle. In effect, the inner surface of TBP binds to DNA, and the larger outer surface is available to extend contacts to other proteins. The DNA-binding site consists of a C-terminal domain that is conserved between species, and the variable N-terminal tail is exposed to interact with other proteins. It is a measure of the conservation of mechanism in transcriptional initiation that the DNA-binding sequence of TBP is 80% conserved between yeast and humans.

Binding of TBP may be inconsistent with the presence of nucleosome octamers. Nucleosomes form preferentially by placing A-T-rich sequences with the minor grooves facing inward (see *Section 10.9, Do Nucleosomes Lie at Specific Positions?*); as a result, they could prevent binding of TBP. This may explain why the presence of a nucleosome at the promoter prevents initiation of transcription.

TBP binds to the minor groove and bends the DNA by ~80°, as illustrated in **FIGURE 20.10**. The TATA box bends toward the major groove, widening the minor groove. The distortion is restricted to the 8 bp of the TATA box; at each end of the sequence, the minor groove has its usual width of ~5 Å, but at the center of the sequence the minor groove is >9 Å. This is a deformation of the structure, but does not actually separate the strands of DNA because base pairing is maintained. The extent of the bend

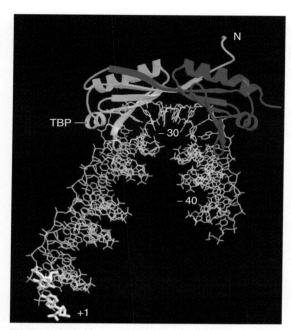

FIGURE 20.10 The cocrystal structure of TBP with DNA from −40 to the startpoint shows a bend at the TATA box that widens the narrow groove where TBP binds. Photo courtesy of Stephen K. Burley.

can vary with the exact sequence of the TATA box, and is correlated with the efficiency of the promoter.

This structure has several functional implications. By changing the spatial organization of DNA on either side of the TATA box, it allows the transcription factors and RNA polymerase to form a closer association than would be possible on linear DNA. The bending at the TATA box corresponds energetically to unwinding of about one-third of a turn of DNA, and is compensated by a positive writhe.

The presence of TBP in the minor groove, combined with other proteins binding in the major groove, creates a high density of protein–DNA contacts in this region. Binding of purified TBP to DNA *in vitro* protects ~1 turn of the double helix at the TATA box, typically extending from −37 to −25. Binding of the TF$_{II}$D complex in the initiation reaction, however, regularly protects the region from −45 to −10.

Within TF$_{II}$D as a free protein complex, the factor TAF1 binds to TBP, where it occupies the concave DNA-binding surface. In fact, the structure of the binding site, which lies in the N-terminal domain of TAF1, mimics the surface of the minor groove in DNA. This molecular mimicry allows TAF1 to control the ability of TBP to bind to DNA; the N-terminal domain of TAF1 must be displaced from the DNA-binding surface of TBP in order for TF$_{II}$D to bind to DNA.

Strikingly, a number of TAFs resemble histones: nine of 14 TAFs contain a histone fold domain, though in most cases the TAFs lack the residues of this domain that are responsible for DNA binding. Four TAFs do have some intrinsic DNA binding ability: TAF4b, TAF12, TAF9, and TAF6 are (distant) homologs of histones H2A, H2B, H3, and H4, respectively. (The histones form the basic complex that binds DNA in eukaryotic chromatin; see *Section 10.3, The Nucleosome Is the Subunit of All Chromatin.*) TAF4b/TAF12 and TAF9/TAF6 form heterodimers using the histone fold motif; together they may form the basis for a structure resembling a histone octamer. Such a structure may be responsible for non-sequence-specific interactions of TF$_{II}$D with DNA. Histone folds are also used in pairwise interactions between other TAF$_{II}$s.

Some of the TAF$_{II}$s may be found in other complexes as well as in TF$_{II}$D. In particular, the histone-like TAF$_{II}$s also are found in protein complexes that modify the structure of chromatin prior to transcription (see *Section 28.9, Histone Acetylation Is Associated with Transcription Activation*).

20.7 The Basal Apparatus Assembles at the Promoter

Key concepts

- The upstream elements and the factors that bind to them increase the frequency of initiation.
- Binding of TF$_{II}$D to the TATA box or Inr is the first step in initiation.
- Other transcription factors bind to the complex in a defined order, extending the length of the protected region on DNA.
- When RNA polymerase II binds to the complex, it initiates transcription.

In a cell, gene promoters can be found in three basic types of chromatin with respect to activity. The first is an inactive gene in closed chromatin. The second is a potentially active gene in open chromatin, called a poised gene. This class may assemble the basal apparatus, but cannot proceed to transcribe the gene without a second signal to start transcription. Heat shock genes are poised so that they can be activated immediately upon a rise in temperature. The third class (which we will examine shortly) is a gene being turned on in open chromatin.

What has been largely unexplored until recently is the involvement of noncoding RNA (ncRNA) transcripts in gene activation. Numerous recent examples have been described in which transcription of ncRNAs regulates transcription of nearby or overlapping protein-coding genes. A recent example describes the involvement of ncRNAs that initiate upstream from the *S. pombe fbp1* gene promoter, in the stepwise removal of two nucleosome octamers from that promoter region. The production of these functional ncRNAs (also referred to as cryptic unstable transcripts, or CUTs) may be much more common than originally believed. A significant number of active promoters have transcripts generated upstream of the promoters (known as promoter upstream transcripts, or PROMPTs). PROMPTs are transcribed in both sense and antisense orientations relative to the downstream promoter, and may play a regulatory role in transcription. The many roles of ncRNAs in transcriptional regulation will be discussed further in *Section 30.3, Noncoding RNAs Can Be Used to Regulate Gene Expression*.

The initiation process requires the basal transcription factors to act in a defined order to build a complex that will be joined by RNA polymerase. The series of events is summarized in **FIGURE 20.11**. Once a polymerase is bound, its activity then is controlled by enhancer-binding transcription factors.

A promoter for RNA polymerase II often consists of two types of region. The core promoter contains the startpoint itself, typically identified by the Inr, and often includes either the TATA box or DPE close by; additional less common elements may be found as well. The efficiency and specificity with which a promoter is recognized, however, depend upon short sequences farther upstream, which are recognized by a different group of transcription factors, sometimes called *activators*. In general, the target sequences are ~100 bp upstream of the startpoint, but sometimes they are more distant. Binding of activators at these sites may influence the formation of the initiation com-

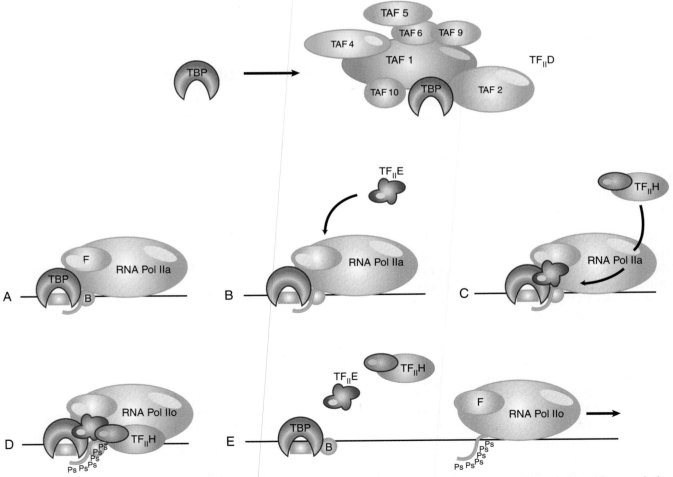

FIGURE 20.11 An initiation complex assembles at promoters for RNA polymerase II by an ordered sequence of association with transcription factors. TF$_{II}$D consists of TBP plus its associated TAFs as shown in the top panel; TBP alone, rather than TF$_{II}$D, is shown in the remaining panels for simplicity. Adapted from M. E. Maxon, J. A. Goodrich, and R. Tijan, *Genes Dev.* 8 (1994): 515–524.

plex at (probably) any one of several stages. Promoters are organized on a principle of "mix and match." A variety of elements can contribute to promoter function, but none is essential for all promoters.

The first step in activating a TATA box-containing promoter in open chromatin is initiated when TF$_{II}$D binds the TATA box. This may be enhanced by upstream elements acting through a coactivator. (TF$_{II}$D also recognizes the Inr sequence at the startpoint, the DPE, and possibly other promoter elements.) When TF$_{II}$A joins the complex, TF$_{II}$D becomes able to bind to a region extending farther upstream. TF$_{II}$A may activate TBP by relieving the repression that is caused by the TAF1.

TF$_{II}$B binds downstream of the TATA box, adjacent to TBP, thus extending contacts along one face of the DNA from −10 to +10. The crystal structure shown in **FIGURE 20.12** extends this model. It makes contacts in the minor groove downstream of the TATA box, and contacts the major groove upstream of the TATA box in a region called the BRE. In archaeans, the homolog of TF$_{II}$B actually makes sequence-specific contacts with the promoter in the BRE region.

This step is believed to be the major determinant in the establishment of promoter polarity, which way the RNA polymerase faces, and which strand is the template strand. TF$_{II}$B may provide the surface that is in turn recognized by RNA polymerase, so that it is responsible for the directionality of the polymerase binding.

The crystal structure of TF$_{II}$B with RNA polymerase shows that three domains of the factor interact with the enzyme. As illustrated schematically in **FIGURE 20.13**, an N-terminal zinc ribbon from TF$_{II}$B contacts the enzyme near the site where RNA exits; it is possible that this interferes with the exit of RNA and influences the switch from abortive initiation to promoter escape. An elongated "finger" of TF$_{II}$B is inserted into the polymerase active center. The C-terminal domain interacts with the RNA polymerase and with TF$_{II}$D to orient the DNA. It also determines the path of the DNA where it contacts the factors TF$_{II}$E, TF$_{II}$F, and TF$_{II}$H, which may align them in the basal factor complex.

The factor TF$_{II}$F is a heterotetramer consisting of two types of subunit. The larger subunit (RAP74) has an ATP-dependent DNA helicase activity that could be involved in melting the DNA at initiation. The smaller subunit (RAP38) has some homology to the regions of bacterial sigma factor that contact the core polymerase; it binds tightly to RNA polymerase II. TF$_{II}$F may bring RNA polymerase II to the assembling transcription complex and provide the means by which it binds. The complex of TBP and TAFs may interact with the CTD tail of RNA polymerase, and interaction with TF$_{II}$B may also be important when TF$_{II}$F/polymerase joins the complex.

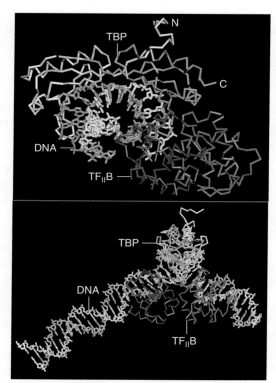

FIGURE 20.12 Two views of the ternary complex of TF$_{II}$B-TBP-DNA show that TF$_{II}$B binds along the bent face of DNA. The two strands of DNA are green and yellow, TBP is blue and TF$_{II}$B is red and purple. Photo courtesy of Stephen K. Burley.

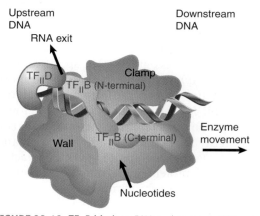

FIGURE 20.13 TF$_{II}$B binds to DNA and contacts RNA polymerase near the RNA exit site and at the active center, and orients it on DNA. Compare with Figure 20.12, which shows the polymerase structure engaged in transcription.

Polymerase binding extends the sites that are protected downstream to +15 on the template strand and +20 on the nontemplate strand. The enzyme extends the full length of the complex because additional protection is seen at the upstream boundary.

What happens at TATA-less promoters? The same general transcription factors, including TF$_{II}$D, are needed. The Inr provides the positioning element; TF$_{II}$D binds to it via an ability of one or more of the TAFs to recognize the Inr directly. Other TAFs in TF$_{II}$D also recognize the DPE element downstream from the startpoint. The function of TBP at these promoters is more like that at promoters for RNA polymerase I and at internal promoters for RNA polymerase III.

When a TATA box is present, it determines the location of the startpoint. Its deletion causes the site of initiation to become erratic, although any overall reduction in transcription is relatively small. Indeed, some TATA-less promoters lack unique startpoints, so initiation occurs within a cluster of startpoints. The TATA box aligns the RNA polymerase via the interaction with TF$_{II}$D and other factors so that it initiates at the proper site. Binding of TBP to TATA is the predominant feature in recognition of the promoter, but two large TAFs (TAF1 and TAF2) also contact DNA in the vicinity of the startpoint and influence the efficiency of the reaction.

Assembly of the RNA polymerase II initiation complex provides an interesting contrast with prokaryotic transcription. Bacterial RNA polymerase is essentially a coherent aggregate with intrinsic ability to recognize and bind the promoter DNA; the sigma factor, needed for initiation but not for elongation, becomes part of the enzyme before DNA is bound, although it may be later released. RNA polymerase II can bind to the promoter, but only after separate transcription factors have bound. The transcription factors play a role analogous to that of bacterial sigma factor—to allow the basic polymerase to recognize DNA specifically at promoter sequences—but have evolved more independence. Indeed, the factors are primarily responsible for the specificity of promoter recognition. Only some of the factors participate in protein–DNA contacts (and only TBP and certain TAFs make sequence-specific contacts); thus protein–protein interactions are important in the assembly of the complex.

Although assembly can take place just at the core promoter *in vitro*, this reaction is not sufficient for transcription *in vivo*, where interactions with activators that recognize the more upstream elements are required. The activators interact with the basal apparatus at various stages during its assembly (see *Section 28.5, Activators Interact with the Basal Apparatus*).

20.8 Initiation Is Followed by Promoter Clearance and Elongation

Key concepts

- TF$_{II}$E and TF$_{II}$H are required to melt DNA to allow polymerase movement.
- Phosphorylation of the CTD is required for promoter clearance and elongation to begin.
- Further phosphorylation of the CTD is required at some promoters to end abortive initiation.
- The histone octamers must be temporarily modified during the transit of the RNA polymerase.
- The CTD coordinates processing of RNA with transcription.
- Transcribed genes are preferentially repaired when DNA damage occurs.
- TF$_{II}$H provides the link to a complex of repair enzymes.

Some final steps are needed to release the RNA polymerase from the promoter once the first nucleotide bonds have been formed. This step is called *promoter clearance* and is the key regulated step in determining if a poised gene or an active gene will be transcribed. This step is controlled by enhancers. (Remember, the key step in bacterial transcription is conversion of the closed complex to the open complex; see *Section 19.3, The Transcription Reaction Has Three Stages*.) Most of the general transcription factors are required solely to bind RNA polymerase to the promoter, but some act at a later stage.

The transcription factors that bind enhancers usually do not directly contact elements at the promoter to control it, but rather bind to a coactivator that binds to the promoter elements. The coactivator Mediator is one of the most common coactivators. This is a very large multisubunit protein complex, conserved from yeast to humans, that integrates signals from many enhancer-bound transcription factors. Both poised and active genes require the interaction of the transcription factors bound to enhancers with the promoter.

The last factors to join the initiation complex are TF$_{II}$E and TF$_{II}$H. They act at the later stages of initiation. Binding of TF$_{II}$E causes the boundary of the region protected downstream to be extended by another turn of the double helix, to +30. TF$_{II}$H is the only general transcription factor that has multiple independent enzymatic activities. Its several activities include an ATPase, helicases of both polarities, and a kinase activity that can phosphorylate the CTD tail of RNA polymerase II (on serine 5 of the heptapeptide repeat). TF$_{II}$H is an exceptional factor that may also play a role in elongation. Its interaction with DNA downstream of the startpoint is required for RNA polymerase to escape from the promoter. TF$_{II}$H is also involved in repair of damage to DNA (see *Section 16.4, Eukaryotic Nucleotide Excision Repair Pathways*).

On a linear template, ATP hydrolysis, TF$_{II}$E, and the helicase activity of TF$_{II}$H (provided by the XPB and XPD subunits) are required for polymerase movement. This requirement is bypassed with a supercoiled template. This suggests that TF$_{II}$E and TF$_{II}$H are required to melt DNA to allow polymerase movement to begin. The helicase activity of the XPB subunit of TF$_{II}$H is responsible for the actual melting of DNA.

RNA polymerase II stutters at some genes when it starts transcription. (The result is not dissimilar to the abortive initiation of bacterial RNA polymerase discussed in *Section 19.12, Interactions between Sigma Factor and Core RNA Polymerase Change During Promoter Escape*, although the mechanism is different.) At many genes, RNA polymerase II terminates after a short distance. The short RNA product is degraded rapidly. To extend elongation into the transcription unit, a kinase complex called P-TEFb is required. P-TEFb contains the CDK9 kinase, which is a member of the kinase family that controls the cell cycle. P-TEFb acts on the CTD to phosphorylate it further (on serine 2 of the heptapeptide repeat). We do not yet understand why this effect is required at some promoters but not others or how it is regulated.

Phosphorylation of the CTD tail is needed to release RNA polymerase II from the promoter and transcription factors so that it can make the transition to the elongating form, as shown in **FIGURE 20.14**. The phosphorylation pattern on the CTD is dynamic during the elongation process, controlled and catalyzed by multiple protein kinases and phosphatases. Most of the basal transcription factors are released from the promoter at this stage. Mediator specifically interacts with polymerase with an unphosphorylated CTD, and phosphorylation appears to serve to disrupt this interaction.

The CTD is involved, directly or indirectly, in processing mRNA while it is being synthesized and after it has been released by RNA polymerase II. Each site of phosphorylation on

the CTD serves as a recognition or anchor point for other proteins to dock with the polymerase. The capping enzyme (guanylyl transferase), which adds the G residue to the 5' end of newly synthesized mRNA, binds to CTD phosphorylated at serine 5, the first phosphorylation event catalyzed by TF$_{II}$H. This may be important in enabling it to modify (and thus protect) the 5' end as soon as it is synthesized. Subsequently, serine 2 phosphorylation by P-TEFb leads to recruitment of a set of proteins called SCAFs to the CTD, and they in turn bind to splicing factors. This may be a means of coordinating transcription and splicing. Finally, some components of the cleavage/polyadenylation apparatus used during transcription termination also bind to the CTD phosphorylated at serine 2. Oddly enough, they do so at the time of initiation, so that RNA polymerase is ready for the 3' end processing reactions as soon as it sets out. Export from the nucleus through the nuclear pore is also controlled by the CTD and may be coordinated with 3' end processing. All of this suggests that the CTD may be a general focus for connecting other processes with transcription. In the cases of capping and splicing, the CTD functions indirectly to promote formation of the protein complexes that undertake the reactions. In the case of 3' end generation, it may participate directly in the reaction.

The key event in determining whether (and when, in the case of a poised polymerase) a gene will be expressed is *promoter clearance*, release from the promoter. Once that has occurred and initiation factors are released, there is a transition to the elongation phase. The transcription complex now consists of the RNA polymerase II, the basal factors TF$_{II}$E and TF$_{II}$H, elongation factors like TF$_{II}$S to prevent inappropriate pausing, and all of the enzymes and factors bound to the CTD. This complex now has to transcribe a chromatin template, through nucleosomes. The whole gene may be in open chromatin, especially if it is not too large, or only the area around the promoter. Some genes, like the Muscular Dystrophy gene (*DMD*), can be megabases in size and require many hours to transcribe. The histone octamers must be transiently modified—in some cases temporarily disassembled—and then reassembled on the template (see Chapter 10, *Chromatin*, and Chapter 28, *Eukaryotic Transcription Regulation*). The octamer itself is different, having the canonical H3 replaced by the variant H3.3.

There is a model in which the first polymerase to leave the promoter acts as a pathfinder

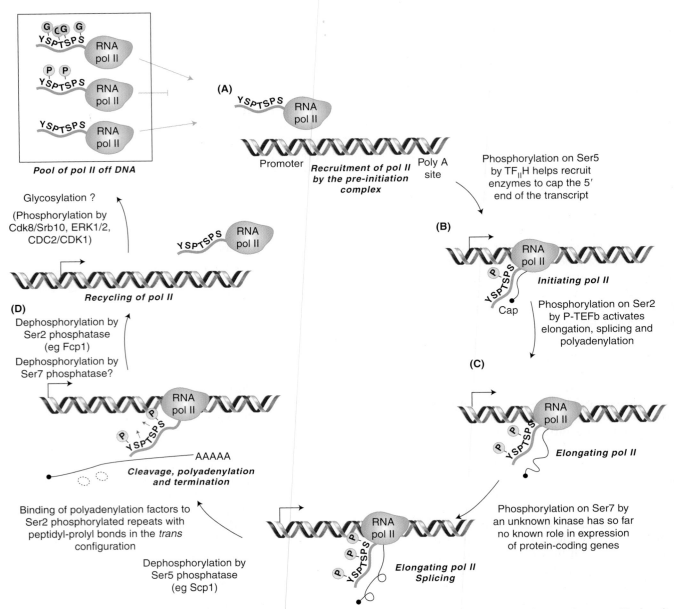

FIGURE 20.14 Modification of the RNA polymerase II CTD heptapeptide during transcription. The CTD of RNA polymerase II when it enters the preinitiation complex is unphosphorylated. Phosphorylation of Ser residues serves as binding sites for both mRNA processing enzymes and kinases that catalyze further phosphorylation as described in the figure. Reprinted from *Trends Genet.*, vol. 24, P. P. Gardner and J. Vinther, Mutation of miRNA target sequences . . . , pp. 262–265. Copyright 2008, with permission from Elsevier [http://www.sciencedirect.com/science/journal/01689525].

polymerase. Its major function is to ensure that the entire gene is in open chromatin. It carries with it enzyme complexes to modify the histones and remodel the chromatin. Histone H2B is dynamically monoubiquitinated in actively transcribed chromatin. This is required in order for the second step, methylation of histone H3, which is, in turn, required for the recruitment of chromatin remodelers (see *Section 10.4, Nucleosomes Are Covalently Modified*, and *Section 28.11, Promoter Activation Involves Multiple Changes to Chromatin*).

The most recent model has each polymerase using a chromatin remodeling complex together with a histone chaperone to remove an H2A/H2B dimer, leaving a hexamer (in place of the octamer), which is easier to temporarily displace. These modifications are also necessary to reassemble the nucleosome octamer on the DNA in the wake of the RNA polymerase (see *Section 10.10, Nucleosomes Are Displaced and Reassembled During Transcription*).

As discussed above in *Section 20.7, The Basal Apparatus Assembles at the Promoter*, there can be considerable heterogeneity in the DNA sequence elements that comprise the core promoter that can lead to promoter specificity. One of these elements is known as the *pause button*, a GC-rich sequence typically located downstream from the start of initiation. This element has been found in a surprising number of *Drosophila* developmental genes.

The general process of initiation is similar to that catalyzed by bacterial RNA polymerase. Binding of RNA polymerase generates a closed complex, which is converted at a later stage to an open complex in which the DNA strands have been separated. In the bacterial reaction, formation of the open complex completes the necessary structural change to DNA; a difference in the eukaryotic reaction is that further unwinding of the template is needed after this stage.

In both bacteria and eukaryotes, there is a direct link from RNA polymerase to the activation of repair. The basic phenomenon was first observed because transcribed genes are preferentially repaired. It was then discovered that it is only the template strand of DNA that is the target—the nontemplate strand is repaired at the same rate as bulk DNA. When RNA polymerase encounters DNA damage in the template strand, it stalls because it cannot use the damaged sequences as a template to direct complementary base pairing. This explains the specificity of the effect for the template strand (damage in the nontemplate strand does not impede progress of the RNA polymerase). The general transcription factor TF$_{II}$H is involved. TF$_{II}$H is found in alternative forms, which consist of a core associated with other subunits.

TF$_{II}$H has a common function in both initiating transcription and repairing damage. The same helicase subunits (XPB and XPD) create the initial transcription bubble and melt DNA at a damaged site. Subunits with the name XP are coded for by genes in which mutations cause the disease *xeroderma pigmentosum*, which causes a predisposition to cancer. The role of TF$_{II}$H subunits in DNA repair is discussed in detail in *Section 16.4, Eukaryotic Nucleotide Excision Repair Pathways*.

The repair function may require modification or degradation of a stalled RNA polymerase. The large subunit of RNA polymerase is degraded when the enzyme stalls at sites of UV damage. We do not yet understand the connection between the transcription/repair apparatus as such and the degradation of RNA polymerase. It is possible that removal of the polymerase is necessary once it has become stalled.

20.9 Enhancers Contain Bidirectional Elements That Assist Initiation

Key concepts

- An enhancer activates the promoter nearest to itself, and can be any distance either upstream or downstream of the promoter.
- A UAS (upstream activating sequence) in yeast behaves like an enhancer, but works only upstream of the promoter.
- Enhancers form complexes of activators that interact directly or indirectly with the promoter.

We have largely considered the promoter as an isolated region responsible for binding RNA polymerase. Eukaryotic promoters do not necessarily function alone, though. In most cases, the activity of a promoter is enormously increased by the presence of an enhancer located at a variable distance from the core promoter. Some enhancers function through long-range interactions of tens of kilobases; others function through short-range interactions and may lie quite close to the core promoter.

One of the first common elements to be described near the promoter was the sequence at −75 now called the CAAT box, named for its consensus sequence. It is often located close to −80, but it can function at distances that vary considerably from the startpoint. It functions in either orientation. Susceptibility to mutations suggests that the CAAT box plays a strong role in determining the efficiency of the promoter, but does not influence its specificity. A second common upstream element is the GC box at −90, which contains the sequence GGGCGG. Often, multiple copies are present in the promoter, and they occur in either orientation. The GC box, too, is a relatively common element near the promoter.

The concept that the enhancer is distinct from the promoter reflects two characteristics. The position of the enhancer relative to the promoter need not be fixed, but can vary substantially. **FIGURE 20.15** shows that it can be upstream, downstream, or within a gene (typically in introns). In addition, it can function in either orientation (that is, it can be inverted) relative to the promoter. Manipulations of DNA show that an enhancer can stimulate any promoter placed in its vicinity, even tens of kilobases away in either direction.

Like the promoter, an enhancer (or its alter ego, a silencer) is a modular element constructed of short DNA sequence elements that bind various types of transcription factors.

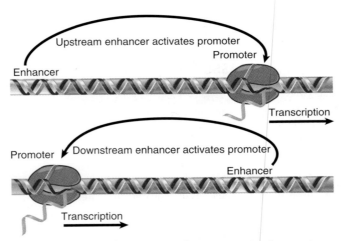

FIGURE 20.15 An enhancer can activate a promoter from upstream or downstream locations, and its sequence can be inverted relative to the promoter.

Enhancers can be simple or complex depending on the number of binding elements and the type of transcription factors they bind.

One way to divide up the world of enhancer-binding transcription factors is to consider positive and negative factors. Transcription factors can be positive and stimulate transcription (as **activators**) or can be negative and repress transcription (as **repressors**). At any given time in a cell, as determined by its developmental history, that cell will contain a mixture of transcription factors that can bind to an enhancer. If more activators bind than repressors, the element will be an enhancer. If more repressors bind than activators, the element will be a silencer.

Another way to examine the transcription factors that bind enhancers is by function. The first class we will consider is called *true activators*; that is, they function by both binding specific DNA sites and making contact with the basal machinery at the promoter, either directly by themselves, or, more commonly, through coactivators like Mediator. This class functions equally well on a DNA template or a chromatin template. There are two additional classes of activators that have a completely different mechanism of activation. One includes activators that function by recruiting chromatin modification enzymes and chromatin remodeling complexes. There are many activators that actually function both as true activators and by recruiting chromatin modifiers. The third class includes architectural transcription factors. Their sole function is to change the structure of the DNA, typically to bend it. This can then result in bringing together two transcription factors separated by a short distance to synergize. In the next section, we will examine more closely how the different classes of activators and repressors work together in an enhancer, and in Chapter 28, *Eukaryotic Transcription Regulation*, we will examine transcription regulation in more detail.

Elements analogous to enhancers, called **upstream activating sequences (UAS)**, are found in yeast. They can function in either orientation at variable distances upstream of the promoter, but cannot function when located downstream. They have a regulatory role: the UAS is bound by the regulatory protein(s) that activates the genes downstream.

Reconstruction experiments in which the enhancer sequence is removed from the DNA and then is inserted elsewhere show that normal transcription can be sustained as long as it is present anywhere on the DNA molecule (as long as no insulators are present in the intervening DNA; see *Section 10.12, Insulators Define Transcriptionally Independent Domains*). If a β-globin gene is placed on a DNA molecule that contains an enhancer, its transcription is increased *in vivo* more than 200-fold, even when the enhancer is several kb upstream or downstream of the startpoint, in either orientation. We have yet to discover at what distance the enhancer fails to work.

20.10 Enhancers Work by Increasing the Concentration of Activators Near the Promoter

Key concepts

- Enhancers usually work only in *cis* configuration with a target promoter.
- The principle is that an enhancer works in any situation in which it is constrained to be in the same proximity as the promoter.

Enhancers function by binding combinations of transcription factors, either positive or negative, that control the promoter and, by extension, gene expression. The promoter is the site where, in open chromatin, basal transcription factors prebind so that RNA polymerase can find the promoter. How can an enhancer stimulate initiation at a promoter that can be located any distance away on either side of it?

Enhancer function involves interaction with the basal apparatus at the core promoter

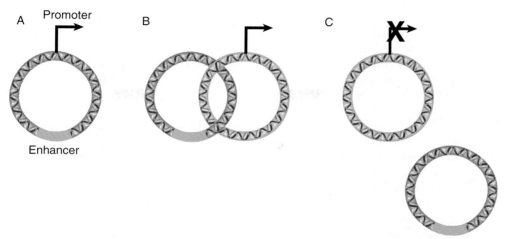

FIGURE 20.16 An enhancer may function by bringing proteins into the vicinity of the promoter. An enhancer and promoter on separate circular DNAs do not interact as in (C), but can interact when the two molecules are catenated as in (B).

element. Enhancers are modular, like promoters. Some elements are found in both long range enhancers and enhancers near promoters. Some individual elements found near promoters share with distal enhancers the ability to function at variable distance and in either orientation. Thus the distinction between long-range enhancers and short-range enhancers is blurred.

The essential role of the enhancer may be to increase the concentration of activator in the vicinity of the promoter (vicinity in this sense being a relative term) in *cis*. Numerous experiments have demonstrated that the level of gene expression (that is, the rate of transcription) is proportional to the net number of activator binding sites. The more activators bound at an enhancer site, the higher the level of expression.

The *Xenopus laevis* ribosomal RNA enhancer is able to stimulate transcription from its RNA polymerase I promoter. This stimulation is relatively independent of location and is able to function when removed from the chromosome and placed with its promoter on a circular plasmid. There is, however, no stimulation when the enhancer and promoter are on separated plasmids. Yet, when the enhancer is placed on a plasmid that is catenated (interlocked) with a second plasmid that contains the promoter, initiation is almost as effective as when the enhancer and promoter are on the same circular molecule as shown in **FIGURE 20.16** (even though, in this case, the enhancer is acting on its promoter in *trans*). Again, this suggests that the critical feature is localization of the protein bound at the enhancer, which increases the enhancer's chance of contacting a protein bound at the promoter.

If proteins bound at an enhancer several kb distant from a promoter interact directly with proteins bound in the vicinity of the startpoint, the organization of DNA must be flexible enough to allow the enhancer and promoter to be closely located. This requires the intervening DNA to be extruded as a large "loop." Such loops have been directly observed in the case of bacterial enhancers.

What limits the activity of an enhancer? Typically it works upon the nearest promoter. There are situations in which an enhancer is located between two promoters, but activates only one of them on the basis of specific protein–protein contacts between the complexes bound at the two elements. The action of an enhancer may be limited by an insulator—an element in DNA that prevents the enhancer from acting on promoters beyond the insulator (see *Section 10.12, Insulators Define Transcriptionally Independent Domains*).

20.11 Gene Expression Is Associated with Demethylation

Key concept

- Demethylation at the 5′ end of the gene is necessary for transcription.

Methylation of DNA is one of several regulatory events that influence the activity of a promoter. Methylation at the promoter may prevent transcription, and those methyl groups must be removed in order to activate a promoter. This effect is well characterized at promoters for both RNA polymerase I and RNA polymerase II. In effect, methylation is a reversible regulatory event. It is triggered by modifications to histones that include deacetylation and protein methylation (see *Section 28.10, Methylation of Histones and DNA Is Connected*).

Methylation also occurs as an epigenetic event. In this case, modification may occur specifically in sperm or oocyte, with the result that there may be a difference between two alleles in the next generation. This can result in differences in the expression of the paternal and maternal alleles (see *Section 29.8, DNA Methylation Is Responsible for Imprinting*).

In this chapter we are concerned with the means by which methylation influences transcription, which is the same whether the methyl groups were added or removed as a local regulatory event or as an epigenetic event.

Methylation at promoters for RNA polymerase II occurs at CG doublets (also referred to as "CpG" doublets). The distribution of methyl groups can be examined by taking advantage of restriction enzymes that cleave target sites containing the CG doublet. Two types of restriction activity are compared in **FIGURE 20.17**. These isoschizomers are enzymes that cleave the same target sequence in DNA, but have a different response to its state of methylation.

The enzyme HpaII cleaves the sequence CCGG (writing the sequence of only one strand of DNA). If the second C is methylated, though, the enzyme can no longer recognize the site.

The enzyme MspI, however, cleaves the same target site irrespective of the state of methylation at this C. Thus MspI can be used to identify all the CCGG sequences, and HpaII can be used to determine whether they are methylated.

With a substrate of unmethylated DNA, the two enzymes would generate the same restriction bands. In methylated DNA, however, the modified positions are not cleaved by HpaII. For every such position, one larger HpaII fragment replaces two MspI fragments. **FIGURE 20.18** gives an example.

Many genes show a pattern in which the state of methylation is constant at most sites but varies at others. Some of the sites are methylated in all tissues examined; some sites are unmethylated in all tissues. A minority of sites are methylated in tissues in which the gene is not expressed, but are not methylated in tissues in which the gene is active. Thus an active gene may be described as undermethylated.

Experiments with the drug 5-azacytidine produce indirect evidence that demethylation can result in gene expression. The drug is incorporated into DNA in place of cytidine and cannot be methylated, because the 5' position is blocked. This leads to the appearance of demethylated sites in DNA as the consequence of replication (see Figure 11.10).

The phenotypic effects of 5-azacytidine include the induction of changes in the state of cellular differentiation. For example, muscle cells are induced to develop from nonmuscle cell precursors. The drug also activates genes on a silent X chromosome, which is consistent with the idea that the state of methylation is connected with chromosomal inactivity.

As well as examining the state of methylation of resident genes, we can compare the results of introducing methylated or nonmethylated DNA into new host cells. Such experi-

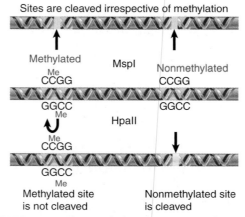

Sites are cleaved irrespective of methylation

FIGURE 20.17 The restriction enzyme MspI cleaves all CCGG sequences whether or not they are methylated at the second C, but HpaII cleaves only unmethylated CCGG tetramers.

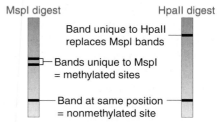

FIGURE 20.18 The results of MspI and HpaII cleavage are compared by gel electrophoresis of the fragments.

ments show a clear correlation: The methylated gene is inactive, but the unmethylated gene is active.

What is the extent of the undermethylated region? In the chicken α-globin gene cluster in adult erythroid cells, the undermethylation is confined to sites that extend from ~500 bp upstream of the first of the two adult α genes to ~500 bp downstream of the second. Sites of undermethylation are present in the entire region, including the spacer between the genes. The region of undermethylation coincides with the region of maximum sensitivity to DNase I (see *Section 10.11, DNase Sensitivity Detects Changes in Chromatin Structure*). This argues that undermethylation is a feature of a domain that contains a transcribed gene or genes. As with many changes in chromatin, it seems likely that the absence of methyl groups is associated with the ability to be transcribed rather than with the act of transcription itself.

Our problem in interpreting the general association between undermethylation and gene activation is that only a minority (sometimes a small minority) of the methylated sites are involved. It is likely that the state of methylation is critical at specific sites or in a restricted region. It is also possible that a reduction in the level of methylation (or even the complete removal of methyl groups from some stretch of DNA) is part of some structural change needed to permit transcription to proceed.

In particular, demethylation at the promoter may be necessary to make it available for the initiation of transcription. In the γ-globin gene, for example, the presence of methyl groups in the region around the startpoint, between –200 and +90, suppresses transcription. Removal of the three methyl groups located upstream of the startpoint, or of the three methyl groups located downstream, does not relieve the suppression. Removal of all methyl groups, though, allows the promoter to function. Transcription may therefore require a methyl-free region at the promoter (see *Section 20.12, CpG Islands Are Regulatory Targets*). There are exceptions to this general relationship.

Some genes can be expressed even when they are extensively methylated. Any connection between methylation and expression thus is not universal in an organism, but the general rule is that methylation prevents gene expression and demethylation is required for expression.

20.12 CpG Islands Are Regulatory Targets

Key concepts

- CpG islands surround the promoters of constitutively expressed genes where they are unmethylated.
- CpG islands also are found at the promoters of some tissue-regulated genes.
- There are ~29,000 CpG islands in the human genome.
- Methylation of a CpG island prevents activation of a promoter within it.
- Repression is caused by proteins that bind to methylated CpG doublets.

The presence of **CpG islands** in the 5′ regions of some genes is connected with the effect of methylation on gene expression. These islands are detected by the presence of an increased density of the dinucleotide sequence, CpG (CpG = 5′-CG-3′).

The CpG doublet occurs in vertebrate DNA at only ~20% of the frequency that would be expected from the proportion of G-C base pairs. (This may be because when CpG doublets are methylated on C, spontaneous deamination of methyl-C converts it to T, which if incorrectly repaired introduces a mutation that removes the doublet.) In certain regions, however, the density of CpG doublets reaches the predicted value; in fact, it is increased by 10× relative to the rest of the genome. The CpG doublets in these regions are generally unmethylated.

These CpG-rich islands have an average G-C content of ~60%, compared with the 20% average in bulk DNA. They take the form of stretches of DNA typically 1 to 2 kb long. There are ~45,000 such islands in the human genome. Some of the islands are present in repeated Alu elements and may just be the consequence of their high G-C-content. The human genome sequence confirms that, excluding these, there are ~29,000 islands. There are fewer in the mouse genome, ~15,500. About 10,000 of the predicted islands in both species appear to reside in a context of sequences that are conserved between the species, suggesting that these may be the islands with regulatory significance. The structure of chromatin in these regions has changes associated with gene expression when the CpG islands are unmethylated (see *Section 28.11, Promoter Activation Involves Multiple Changes to Chromatin*): there is a reduced content

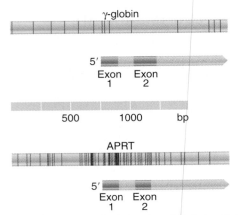

FIGURE 20.19 The typical density of CpG doublets in mammalian DNA is ~1/100 bp, as seen for a γ-globin gene. In a CpG-rich island, the density is increased to >10 doublets/100 bp. The island in the APRT gene starts ~100 bp upstream of the promoter and extends ~400 bp into the gene. Each vertical line represents a CpG doublet.

of histone H1 (which probably means that the structure is less compact), the other histones are extensively acetylated (a feature that tends to be associated with gene expression), and there are DNase hypersensitive sites, or sites nearly devoid of histone octamers (as would be expected of active promoters). The presence of methylated CpG sites precludes the presence of the histone variant H2A.Z in nucleosomes.

In several cases, CpG-rich islands begin just upstream of a promoter and extend downstream into the transcribed region before petering out. **FIGURE 20.19** compares the density of CpG doublets in a "general" region of the genome with a CpG island identified from the DNA sequence. The CpG island surrounds the 5′ region of the *APRT* gene, which is constitutively expressed.

All of the housekeeping genes that are constitutively expressed have CpG islands; this accounts for about half of the islands. The remaining islands occur at the promoters of tissue-regulated genes; approximately 50% of these genes have islands. In these cases, the islands are unmethylated irrespective of the state of expression of the gene, so that CpG island methylation is not correlated with transcriptional state for tissue-specific genes. The presence of unmethylated CpG-rich islands may be necessary, but is not sufficient, for transcription. Thus the presence of unmethylated CpG islands may be taken as an indication that a gene is potentially active rather than inevitably transcribed. Many islands that are unmethylated in the animal become methylated in cell lines in tissue culture (or in some cancers); this could be connected with the inability of these

lines to express all of the functions typical of the tissue from which they were derived. There is one clear example in which there is a strong correlation between promoter methylation and gene expression: promoter CpG islands become methylated in the mammalian inactive X chromosome (see *Section 29.5, X Chromosomes Undergo Global Changes*).

Methylation of a CpG island can affect transcription. One of two mechanisms can be involved:

- Methylation of a binding site for some factor may prevent it from binding. This happens in a case of binding to a regulatory site other than the promoter (see *Section 29.9, Oppositely Imprinted Genes Can Be Controlled by a Single Center*).
- Methylation may cause specific repressors to bind to the DNA.

Repression is caused by either of two types of protein that bind to methylated CpG sequences. The protein MeCP1 requires the presence of several methyl groups to bind to DNA, whereas MeCP2 and a family of related proteins can bind to a single methylated CpG base pair. This explains why a methylation-free zone is required for initiation of transcription. Binding of proteins of either type prevents transcription *in vitro* by a nuclear extract.

MeCP2, which directly represses transcription by interacting with complexes at the promoter, also interacts with the Sin3 repressor complex, which contains histone deacetylase activities (see Figure 28.24). This observation provides a direct connection between two types of repressive modifications: methylation of DNA and deacetylation of histones.

While promoters that contain CpG islands (~60% CpG density) or show no CpG enrichment (~20% CpG density) exhibit a generally poor correlation between promoter methylation and transcription, there is a third class of promoters that appears to be consistently regulated by CpG methylation. Approximately 12% of human genes contain so-called "weak" CpG islands, in which the density of CpGs is ~30%, intermediate between the other two classes of promoters. These genes show a strong inverse relationship between promoter CpG methylation and RNA polymerase II occupancy.

The absence of methyl groups is associated with gene expression (or at least the potential for expression). There are, however, some difficulties in supposing that the state of methylation provides a general means for controlling gene expression. In the case of *Drosophila melanogaster* (and other Dipteran insects), there is very little methylation of DNA (although one methyltransferase, Dnmt2, has been identified; its importance is unclear), and there is no methylation of DNA in the nematode *Caenorhabditis elegans* or in yeast. The other differences between inactive and active chromatin appear to be the same as in species that display methylation. Thus in these organisms, any role that methylation has in vertebrates is replaced by some other mechanism.

Three changes that occur in active genes are:

- A hypersensitive site(s) is established near the promoter.
- The chromatin of a domain, including the transcribed region, becomes more sensitive to DNase I.
- The DNA of the same region is undermethylated.

All of these changes are necessary for transcription.

20.13 Summary

Of the three eukaryotic RNA polymerases, RNA polymerase I transcribes rDNA and accounts for the majority of activity, RNA polymerase II transcribes structural genes for mRNA and has the greatest diversity of products, and RNA polymerase III transcribes small RNAs. The enzymes have similar structures, with two large subunits and many smaller subunits; there are some common subunits among the enzymes.

None of the three RNA polymerases recognize their promoters directly. A unifying principle is that transcription factors have primary responsibility for recognizing the characteristic sequence elements of any particular promoter, and they serve in turn to bind the RNA polymerase and to position it correctly at the startpoint. At each type of promoter, histone octamers must be removed or moved. The initiation complex is then assembled by a series of reactions in which individual factors join (or leave) the complex. The factor TBP is required for initiation by all three RNA polymerases. In each case it provides one subunit of a transcription factor that binds in the vicinity of the startpoint.

An RNA polymerase II promoter consists of a number of short-sequence elements in the region upstream of the startpoint. Each element is bound by one or more transcription factors. The basal apparatus, which consists of the TF_{II} factors, assembles at the startpoint and enables

RNA polymerase to bind. The TATA box (if there is one) near the startpoint, and the initiator region immediately at the startpoint, are responsible for selection of the exact startpoint at promoters for RNA polymerase II. TBP binds directly to the TATA box when there is one; in TATA-less promoters it is located near the startpoint by binding to the Inr or to the DPE downstream. After binding of $TF_{II}D$, the other general transcription factors for RNA polymerase II assemble the basal transcription apparatus at the promoter. Other elements in the promoter, located upstream of the TATA box, bind activators that interact with the basal apparatus. The activators and basal factors are released when RNA polymerase begins elongation.

The CTD of RNA polymerase II is phosphorylated during the initiation reaction. It provides a point of contact for proteins that modify the RNA transcript, including the 5' capping enzyme, splicing factors, the 3' processing complex, and mRNA export from the nucleus. As the RNA polymerase moves through the transcription unit, histone octamers must be modified to allow passage.

Promoters may be stimulated by enhancers, sequences that can act at great distances and in either orientation on either side of a gene. Enhancers also consist of sets of elements, although they are more compactly organized. Some elements are found both close to promoters and in distant enhancers. Enhancers probably function by assembling a protein complex that interacts with the proteins bound at the promoter, requiring that DNA between is "looped out."

CpG islands contain concentrations of CpG doublets and often surround the promoters of constitutively expressed genes, although they are also found at the promoters of regulated genes. The island including a promoter must be unmethylated for that promoter to be able to initiate transcription. A specific protein binds to the methylated CpG doublets and prevents initiation of transcription.

References

20.2 Eukaryotic RNA Polymerases Consist of Many Subunits

Reviews

Doi, R. H. and Wang, L.-F. (1986). Multiple prokaryotic RNA polymerase sigma factors. *Microbiol. Rev.* 50, 227–243.

Young, R. A. (1991). RNA polymerase II. *Annu. Rev. Biochem.* 60, 689–715.

20.3 RNA Polymerase I Has a Bipartite Promoter

Reviews

Grummt, I. (2003). Life on a planet of its own: regulation of RNA polymerase I transcription in the nucleolus. *Genes Dev.* 17, 1691–1702.

Mathews, D. A. and Olson, W. M. (2006). What is new in the nucleolus? *EMBO Rep.* 7, 870–873.

Paule, M. R. and White, R. J. (2000). Survey and summary: transcription by RNA polymerases I and III. *Nucleic Acids Res.* 28, 1283–1298.

Research

Bell, S. P., Learned, R. M., Jantzen, H. M., and Tjian, R. (1988). Functional cooperativity between transcription factors UBF1 and SL1 mediates human ribosomal RNA synthesis. *Science* 241, 1192–1197.

Kuhn, C.-D., Geiger, S. R., Baumli, S., Gartmann, M., Gerber, J., Jennebach, S., Mielke, T., Tschochner, H., Beckmann, R., and Cramer P. (2007). Functional architecture of RNA polymerase I. *Cell* 131, 1260–1273.

20.4 RNA Polymerase III Uses Both Downstream and Upstream Promoters

Reviews

Geiduschek, E. P. and Tocchini-Valentini, G. P. (1988). Transcription by RNA polymerase III. *Annu. Rev. Biochem.* 57, 873–914.

Schramm, L. and Hernandez, N. (2002). Recruitment of RNA polymerase III to its target promoters. *Genes Dev.* 16, 2593–2620.

Research

Bogenhagen, D. F., Sakonju, S., and Brown, D. D. (1980). A control region in the center of the 5S RNA gene directs specific initiation of transcription: II the 3' border of the region. *Cell* 19, 27–35.

Galli, G., Hofstetter, H., and Birnstiel, M. L. (1981). Two conserved sequence blocks within eukaryotic tRNA genes are major promoter elements. *Nature* 294, 626–631.

Kassavatis, G. A., Braun, B. R., Nguyen, L. H., and Geiduschek, E. P. (1990). S. cerevisiae TFIIIB is the transcription initiation factor proper of RNA polymerase III, while TFIIIA and TFIIIC are assembly factors. *Cell* 60, 235–245.

Kassavetis, G. A., Joazeiro, C. A., Pisano, M., Geiduschek, E. P., Colbert, T., Hahn, S., and Blanco, J. A. (1992). The role of the TATA-binding protein in the assembly and function of the multisubunit yeast RNA polymerase III transcription factor, TFIIIB. *Cell* 71, 1055–1064.

Kassavetis, G. A., Letts, G. A., and Geiduschek, E. P. (1999). A minimal RNA polymerase III transcription system. *EMBO J.* 18, 5042–5051.

Kunkel, G. R. and Pederson, T. (1988). Upstream elements required for efficient transcription of a human U6 RNA gene resemble those of U1 and U2 genes even though a different polymerase is used. *Genes Dev.* 2, 196–204.

Pieler, T., Hamm, J., and Roeder, R. G. (1987). The 5S gene internal control region is composed of three distinct sequence elements, organized as two functional domains with variable spacing. *Cell* 48, 91–100.

Sakonju, S., Bogenhagen, D. F., and Brown, D. D. (1980). A control region in the center of the 5S RNA gene directs specific initiation of transcription: I the 5' border of the region. *Cell* 19, 13–25.

20.5 The Startpoint for RNA Polymerase II

Reviews

Butler, J. E. and Kadonaga, J. T. (2002). The RNA polymerase II core promoter: a key component in the regulation of gene expression. *Genes Dev.* 16, 2583–2592.

Smale, S. T., Jain, A., Kaufmann, J., Emami, K. H., Lo, K., and Garraway, I. P. (1998). The initiator element: a paradigm for core promoter heterogeneity within metazoan protein-coding genes. *Cold Spring Harb Symp Quant Biol.* 63, 21–31.

Smale, S. T. and Kadonaga, J. T. (2003). The RNA polymerase II core promoter. *Annu. Rev. Biochem.* 72, 449–479.

Woychik, N. A. and Hampsey, M. (2002). The RNA polymerase II machinery: structure illuminates function. *Cell* 108, 453–463.

Research

Burke, T. W. and Kadonaga, J. T. (1996). Drosophila TFIID binds to a conserved downstream basal promoter element that is present in many TATA-box-deficient promoters. *Genes Dev.* 10, 711–724.

Singer, V. L., Wobbe, C. R., and Struhl, K. (1990). A wide variety of DNA sequences can functionally replace a yeast TATA element for transcriptional activation. *Genes Dev.* 4, 636–645.

Smale, S. T. and Baltimore, D. (1989). The "initiator" as a transcription control element. *Cell* 57, 103–113.

20.6 TBP Is a Universal Factor

Reviews

Berk, A. J. (2000). TBP-like factors come into focus. *Cell* 103, 5–8.

Burley, S. K. and Roeder, R. G. (1996). Biochemistry and structural biology of TFIID. *Annu. Rev. Biochem.* 65, 769–799.

Hernandez, N. (1993). TBP, a universal eukaryotic transcription factor? *Genes Dev.* 7, 1291–1308.

Lee, T. I. and Young, R. A. (1998). Regulation of gene expression by TBP-associated proteins. *Genes Dev.* 12, 1398–1408.

Orphanides, G., Lagrange, T., and Reinberg, D. (1996). The general transcription factors of RNA polymerase II. *Genes Dev.* 10, 2657–2683.

Research

Crowley, T. E., Hoey, T., Liu, J. K., Jan, Y. N., Jan, L. Y., and Tjian, R. (1993). A new factor related to TATA-binding protein has highly restricted expression patterns in Drosophila. *Nature* 361, 557–561.

Horikoshi, M. et al. (1988). Transcription factor ATD interacts with a TATA factor to facilitate establishment of a preinitiation complex. *Cell* 54, 1033–1042.

Kim, J. L., Nikolov, D. B., and Burley, S. K. (1993). Cocrystal structure of TBP recognizing the minor groove of a TATA element. *Nature* 365, 520–527.

Kim, Y. et al. (1993). Crystal structure of a yeast TBP/TATA box complex. *Nature* 365, 512–520.

Liu, D. et al. (1998). Solution structure of a TBP-TAFII230 complex: protein mimicry of the minor groove surface of the TATA box unwound by TBP. *Cell* 94, 573–583.

Martinez, E. et al. (1994). TATA-binding protein-associated factors in TFIID function through the initiator to direct basal transcription from a TATA-less class II promoter. *EMBO J.* 13, 3115–3126.

Nikolov, D. B. et al. (1992). Crystal structure of TFIID TATA-box binding protein. *Nature* 360, 40–46.

Ogryzko, V. V. et al. (1998). Histone-like TAFs within the PCAF histone acetylase complex. *Cell* 94, 35–44.

Sprouse R. O., Karpova, T. A., Mueller, F., Dasgupta, A., McNally, J. G., and Auble, D. T. (2008). Regulation of TATA-binding protein dynamics in living yeast cells. *Proc. Natl Acad. Sci. USA* 105, 13304–13308.

Verrijzer, C. P. et al. (1995). Binding of TAFs to core elements directs promoter selectivity by RNA polymerase II. *Cell* 81, 1115–1125.

Wu, J., Parkhurst, K. M., Powell, R. M., Brenowitz, M., and Parkhurst, L. J. (2001). DNA bends in TATA-binding protein-TATA complexes in solution are DNA sequence-dependent. *J. Biol. Chem.* 276, 14614–14622.

20.7 The Basal Apparatus Assembles at the Promoter

Reviews

Egloff, S. and Murphy, S. (2008). Cracking the RNA polymerase II CTD code. *Trends Genet.* 24, 280–288.

Muller F., Demeny, M. A., and Tora, L. (2007). New problems in RNA polymerase II transcription initiation: matching the diversity of core promoters with a variety of promoter recognition factors. *J. Biol. Chem.* 282, 14685–14689.

Nikolov, D. B. and Burley, S. K. (1997). RNA polymerase II transcription initiation: a structural view. *Proc. Natl. Acad. Sci. USA* 94, 15–22.

Zawel, L. and Reinberg, D. (1993). Initiation of transcription by RNA polymerase II: a multistep process. *Prog. Nucleic Acid Res. Mol. Biol.* 44, 67–108.

Research

Buratowski, S., Hahn, S., Guarente, L., and Sharp, P. A. (1989). Five intermediate complexes in transcription initiation by RNA polymerase II. *Cell* 56, 549–561.

Burke, T. W. and Kadonaga, J. T. (1996). Drosophila TFIID binds to a conserved downstream basal promoter element that is present in many TATA-box-deficient promoters. *Genes Dev.* 10, 711–724.

Bushnell, D. A., Westover, K. D., Davis, R. E., and Kornberg, R. D. (2004). Structural basis of transcription: an RNA polymerase II-TFIIB cocrystal at 4.5 Angstroms. *Science* 303, 983–988.

Carninci, P., et al. (2006) Genome-wide analysis of mammalian promoter architecture and evolution. *Nat. Gen.* 38, 626–635.

Littlefield, O., Korkhin, Y., and Sigler, P. B. (1999). The structural basis for the oriented assembly of a TBP/TFB/promoter complex. *Proc. Natl. Acad. Sci. USA* 96, 13668–13673.

Nikolov, D. B. et al. (1995). Crystal structure of a TFIIB-TBP-TATA-element ternary complex. *Nature* 377, 119–128.

20.8 Initiation Is Followed by Promoter Clearance and Elongation

Reviews

Ares, M. Jr. and Proudfoot, N. J. (2005). The Spanish connection: transcription and mRNA processing get even closer. *Cell* 120:163–166.

Calvo, O. and Manley, J. L. (2003). Strange bedfellows: polyadenylation factors at the promoter. *Genes Dev.* 17, 1321–1327.

Hartzog, G. A. and Quan, T. K. (2008). Just the FACTs: Histone H2B ubiquitylation and nucleosome dynamics. *Mol. Cell* 31, 2–4.

Hirose, Y. and Manley, J. L. (2000). RNA polymerase II and the integration of nuclear events. *Genes Dev.* 14, 1415–1429.

Lehmann, A. R. (2001). The xeroderma pigmentosum group D (XPD) gene: one gene, two functions, three diseases. *Genes Dev.* 15, 15–23.

Price, D. H. (2000). P-TEFb, a cyclin dependent kinase controlling elongation by RNA polymerase II. *Mol. Cell Biol.* 20, 2629–2634.

Proudfoot, N. J., Furger, A., and Dye, M. J. (2002). Integrating mRNA processing with transcription. *Cell* 108, 501–512.

Shilatifard, A., Conaway, R. C., and Conaway, J. W. (2003). The RNA polymerase II elongation complex. *Annu. Rev. Biochem.* 72, 693–715.

Woychik, N. A. and Hampsey, M. (2002). The RNA polymerase II machinery: structure illuminates function. *Cell* 108, 453–463.

Research

Douziech, M., Coin, F., Chipoulet, J. M., Arai, Y., Ohkuma, Y., Egly, J. M., and Coulombe, B. (2000). Mechanism of promoter melting by the xeroderma pigmentosum complementation group B helicase of transcription factor IIH revealed by protein-DNA photo-cross-linking. *Mol. Cell Biol.* 20, 8168–8177.

Fong, N. and Bentley, D. L. (2001). Capping, splicing, and 3′ processing are independently stimulated by RNA polymerase II: different functions for different segments of the CTD. *Genes Dev.* 15, 1783–1795.

Goodrich, J. A. and Tjian, R. (1994). Transcription factors IIE and IIH and ATP hydrolysis direct promoter clearance by RNA polymerase II. *Cell* 77, 145–156.

Hendrix, D. A., Hong, J.-W., Zeitlinger, J., Rokhsar, D. S., and Levine, M. S. (2008). Promoter elements associated with RNA polymerase II stalling in the Drosophila embryo. *Proc. Natl. Acad. Sci. USA* 105, 7762–7767.

Hirota, K., Miyosha, T., Kugou, K., Hoffman, C. S., Shibata, T., and Ohta, K. (2008). Stepwise chromatin remodeling by a cascade of transcription initiation of non coding RNAs. *Nature* 456, 130–135.

Holstege, F. C., van der Vliet, P. C., and Timmers, H. T. (1996). Opening of an RNA polymerase II promoter occurs in two distinct steps and requires the basal transcription factors IIE and IIH. *EMBO J.* 15, 1666–1677.

Kim, T. K., Ebright, R. H., and Reinberg, D. (2000). Mechanism of ATP-dependent promoter melting by transcription factor IIH. *Science* 288, 1418–1422.

Montanuy, I., Torremocha, R., Hernandez-Munain, C., and Suñé, C. (2008). Promoter influences transcription elongation: TATA-BOX element mediates the assembly of processive transcription complexes responsive to cyclin-dependent kinase 9. *J. Biol. Chem.* 283, 7368–7378.

Spangler, L., Wang, X., Conaway, J. W., Conaway, R. C., and Dvir, A. (2001). TFIIH action in transcription initiation and promoter escape requires distinct regions of downstream promoter DNA. *Proc. Natl. Acad. Sci. USA* 98, 5544–5549.

20.9 Enhancers Contain Bidirectional Elements That Assist Initiation

Reviews

Maniatis, T., Falvo, J. V., Kim, T. H., Kim, T. K., Lin, C. H., Parekh, B. S., and Wathelet, M. G. (1998). Structure and function of the interferon-beta enhanceosome. *Cold Spring Harbor Symp. Quant. Biol.* 63, 609–620.

Muller, M. M., Gerster, T., and Schaffner, W. (1988). Enhancer sequences and the regulation of gene transcription. *Eur. J. Biochem.* 176, 485–495.

Munshi, N., Yie, Y., Merika, M., Senger, K., Lomvardas, S., Agalioti, T., and Thanos, D. (1999). The IFN-beta enhancer: a paradigm for understanding activation and repression of inducible gene expression. *Cold Spring Harbor Symp. Quant. Biol.* 64, 149–159.

Research

Banerji, J., Rusconi, S., and Schaffner, W. (1981). Expression of β-globin gene is enhanced by remote SV40 DNA sequences. *Cell* 27, 299–308.

20.10 Enhancers Work by Increasing the Concentration of Activators Near the Promoter

Review

Blackwood, E. M. and Kadonaga, J. T. (1998). Going the distance: a current view of enhancer action. *Science* 281, 60–63.

Research

Mueller-Storm, H. P., Sogo, J. M., and Schaffner, W. (1989). An enhancer stimulates transcription in trans when attached to the promoter via a protein bridge. *Cell* 58, 767–777.

Zenke, M. et al. (1986). Multiple sequence motifs are involved in SV40 enhancer function. *EMBO J.* 5, 387–397.

20.12 CpG Islands Are Regulatory Targets

Review

Bird, A. (2002). DNA methylation patterns and epigenetic memory. *Genes Dev.* 16, 6–21.

Research

Antequera, F. and Bird, A. (1993). Number of CpG islands and genes in human and mouse. *Proc. Natl. Acad. Sci. USA* 90, 11995–11999.

Bird, A. et al. (1985). A fraction of the mouse genome that is derived from islands of non-methylated, Cp-G-rich DNA. *Cell* 40, 91–99.

Boyes, J. and Bird, A. (1991). DNA methylation inhibits transcription indirectly via a methyl-CpG binding protein. *Cell* 64, 1123–1134.

Zilberman, D., Coleman-Derr, D., Ballinger, T., and Henikoff, S. (2008). Histone H2A.Z and DNA methylation are mutually antagonistic chromatin marks. *Nature* 456, 125–130.

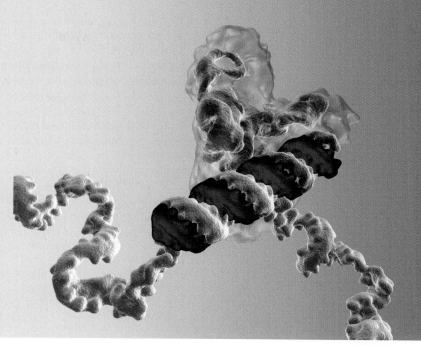

© Kenneth Eward/Photo Researchers, Inc.

RNA Splicing and Processing

**Edited by
Xiang-Dong Fu**

21.6 snRNAs Are Required for Splicing

- The five snRNPs involved in splicing are U1, U2, U5, U4, and U6.
- Together with some additional proteins, the snRNPs form the spliceosome.
- All the snRNPs except U6 contain a conserved sequence that binds the Sm proteins that are recognized by antibodies generated in autoimmune disease.

21.7 Commitment of Pre-mRNA to the Splicing Pathway

- U1 snRNP initiates splicing by binding to the 5' splice site by means of an RNA–RNA pairing reaction.
- The commitment complex contains U1 snRNP bound at the 5' splice site and the protein U2AF bound to a pyrimidine tract between the branch site and the 3' splice site.
- In multicellular eukaryotic cells, SR proteins play an essential role in initiating the formation of the commitment complex.
- Pairing splice sites can be accomplished by intron definition or exon definition.

21.8 The Spliceosome Assembly Pathway

- The commitment complex progresses to pre-spliceosome (the A complex) in the presence of ATP.
- Recruitment of U5 and U4/U6 snRNPs converts the pre-spliceosome to the mature spliceosome (the B1 complex).
- The B1 complex is next converted to the B2 complex in which U1 snRNP is released to allow U6 snRNA to interact with the 5' splice site.
- U4 dissociates from U6 snRNP to allow U6 snRNA to pair with U2 snRNA to form the catalytic center for splicing.
- Both transesterification reactions take place in the activated spliceosome (the C complex).
- The splicing reaction is reversible at all steps.

21.9 An Alternative Spliceosome Uses Different snRNPs to Process the Minor Class of Introns

- An alternative splicing pathway uses another set of snRNPs with only U5 snRNP in common with the major spliceosome.
- The target introns are defined by longer consensus sequences at the splice junctions, rather than strictly according to the GU-AG or AU-AC rules.
- Major and minor spliceosomes share critical protein factors, including SR proteins.

21.10 Pre-mRNA Splicing Likely Shares the Mechanism with Autocatalytic Group II Introns

- Group II introns excise themselves from RNA by an autocatalytic splicing event.
- The splice junctions and mechanism of splicing of group II introns are similar to splicing of nuclear introns.

- A group II intron folds into a secondary structure that generates a catalytic site resembling the structure of U6-U2-nuclear intron.

21.11 Splicing Is Temporally and Functionally Coupled with Multiple Steps in Gene Expression

- Splicing can occur during or after transcription.
- The transcription and splicing machineries are physically and functionally integrated.
- Splicing is connected to mRNA export and stability control.
- Splicing in the nucleus can influence mRNA translation in the cytoplasm.

21.12 Alternative Splicing Is a Rule, Rather Than an Exception, in Multicellular Eukaryotes

- Specific exons or exonic sequences may be excluded or included in the mRNA products by using alternative splicing sites.
- Alternative splicing contributes to structural and functional diversity of gene products.
- Sex determination in *Drosophila* involves a series of alternative splicing events in genes coding for successive products of a pathway.

21.13 Splicing Can Be Regulated by Exonic and Intronic Splicing Enhancers and Silencers

- Alternative splicing is often associated with weak splice sites.
- Sequences surrounding alternative exons are often more evolutionarily conserved than sequences flanking constitutive exons.
- Specific exonic and intronic sequences can enhance or suppress splice site selection.
- The effect of splicing enhancers and silencers is mediated by sequence-specific RNA binding proteins, many of which may be developmentally regulated and/or expressed in a tissue-specific manner.
- The rate of transcription can directly affect the outcome of alternative splicing.

21.14 *trans*-Splicing Reactions Use Small RNAs

- Splicing reactions usually occur only in *cis* between splice junctions on the same molecule of RNA.
- *trans*-splicing occurs in trypanosomes and worms where a short sequence (SL RNA) is spliced to the 5' ends of many precursor mRNAs.
- SL RNAs have a structure resembling the Sm-binding site of U snRNAs.

21.15 The 3' Ends of mRNAs Are Generated by Cleavage and Polyadenylation

- The sequence AAUAAA is a signal for cleavage to generate a 3' end of mRNA that is polyadenylated.
- The reaction requires a protein complex that contains a specificity factor, an endonuclease, and poly(A) polymerase.

- The specificity factor and endonuclease cleave RNA downstream of AAUAAA.
- The specificity factor and poly(A) polymerase add ~200 A residues processively to the 3' end.
- The poly(A) tail controls mRNA stability and influences translation.
- Cytoplasmic polyadenylation plays a role in *Xenopus* embryonic development.

21.16 The 3' mRNA End Processing Is Critical for Transcriptional Termination

- There are various ways to end transcription by different RNA polymerases.
- The mRNA 3' end formation signals termination of Pol II transcription.

21.17 The 3' End Formation of Histone mRNA Requires U7 snRNA

- The expression of histone mRNAs is replication dependent and is regulated during the cell cycle.
- Histone mRNAs are not polyadenylated; their 3' ends are generated by a cleavage reaction that depends on the structure of the mRNA.
- The cleavage reaction requires the SLBP to bind to a stem-loop structure and the U7 snRNA to pair with an adjacent single-stranded region.
- The cleavage reaction is catalyzed by a factor shared with the polyadenylation complex.

21.18 tRNA Splicing Involves Cutting and Rejoining in Separate Reactions

- RNA polymerase III terminates transcription in poly(U)$_4$ sequence embedded in a GC-rich sequence.
- tRNA splicing occurs by successive cleavage and ligation reactions.
- An endonuclease cleaves the tRNA precursors at both ends of the intron.
- Release of the intron generates two half-tRNAs with unusual ends that contain 5' hydroxyl and 2'–3' cyclic phosphate.

- The 5'–OH end is phosphorylated by a polynucleotide kinase, the cyclic phosphate group is opened by phosphodiesterase to generate a 2'–phosphate terminus and 3'–OH group, exon ends are joined by an RNA ligase, and the 2'–phosphate is removed by a phosphatase.

21.19 The Unfolded Protein Response Is Related to tRNA Splicing

- Ire1 is an inner nuclear membrane protein with its N-terminal domain in the ER lumen and its C-terminal domain in the nucleus; the C-terminal domain exhibits both kinase and endonuclease activities.
- Binding of an unfolded protein to the N-terminal domain activates the C-terminal endonuclease by autophosphorylation.
- The activated endonuclease cleaves *HAC1* (*Xbp1* in vertebrates) mRNA to release an intron and generate exons that are ligated by a tRNA ligase.
- Only spliced *HAC1* mRNA can be translated to a transcription factor that activates genes coding for chaperones that help to fold unfolded proteins.
- Activated Ire1 induces apoptosis when the cell is over stressed by unfolded proteins.

21.20 Production of rRNA Requires Cleavage Events and Involves Small RNAs

- RNA polymerase I terminates transcription at an 18-base terminator sequence.
- The large and small rRNAs are released by cleavage from a common precursor rRNA; the 5S rRNA is separately transcribed.
- The C/D group of snoRNAs is required for modifying the 2' position of ribose with a methyl group.
- The H/ACA group of snoRNAs is required for converting uridine to pseudouridine.
- In each case the snoRNA base pairs with a sequence of rRNA that contains the target base to generate a typical structure that is the substrate for modification.

21.21 Summary

21.1 Introduction

RNA is a central player in gene expression. It was first characterized as an intermediate in protein synthesis, but since then many other RNAs have been discovered that play structural or functional roles at various stages of gene expression. The involvement of RNA in many functions concerned with gene expression supports the general view that life may have evolved from an "RNA world" in which RNA was originally the active component in maintaining and expressing genetic information. Many of these functions were subsequently assisted or taken over by proteins, with a consequent increase in versatility and probably efficiency.

All RNAs transcribed from their prospective genes require further processing to become mature and functional. Interrupted genes are found in all classes of eukaryotic organisms. They represent a minor proportion of the genes of the very simplest eukaryotes, but the vast majority of genes in higher eukaryotic genomes. Genes vary widely according to the numbers and lengths of introns, but a typical mammalian gene has seven to eight exons spread out over ~16 kb. The exons are relatively short (~100 to 200 bp) and the introns are relatively long (>1 kb) (see *Section 4.7, Genes Show a Wide Distribution of Sizes*).

The discrepancy between the interrupted organization of the gene and the uninterrupted organization of its mRNA requires processing of the primary transcription product. The primary transcript has the same organization as the gene and is called the **pre-mRNA**. Removal of the introns from pre-mRNA leaves a typical messenger of ~2.2 kb. The process by which the introns are removed is called **RNA splicing**. Removal of introns is a major part of the production of RNA in all eukaryotes. Although interrupted genes are relatively rare in most unicellular/oligocellular eukaryotes (such as the yeast *Saccharomyces cerevisiae*), the overall proportion underestimates the importance of introns because most of the genes that are interrupted code for relatively abundant proteins. Splicing is therefore involved in the production of a greater proportion of total mRNA than would be apparent from analysis of the genome, perhaps as much as 50%.

One of the first clues about the nature of the discrepancy in size between nuclear genes and their products in higher eukaryotes was provided by the properties of nuclear RNA. Its average size is much larger than mRNA, it is very unstable, and it has a much greater sequence complexity. Taking its name from its broad size distribution, it is called **heterogeneous nuclear RNA (hnRNA)**.

The physical form of hnRNA is a ribonucleoprotein particle **(hnRNP)**, in which the hnRNA is bound by a set of abundant RNA-binding proteins. Some of the proteins may have a structural role in packaging the hnRNA; several are known to affect RNA processing or facilitate RNA export out of the nucleus.

Splicing occurs in the nucleus, together with the other modifications that are made to newly synthesized RNAs. The process of expressing an interrupted gene is reviewed in **FIGURE 21.1**. The transcript is capped at the 5′ end, has the introns removed, and is polyadenylated at the 3′ end. The RNA is then transported through nuclear pores to the cytoplasm, where it is available to be translated.

With regard to the various processing reactions that occur in the nucleus, we should like to know at what point splicing occurs *vis-à-vis* the other modifications of RNA. Does splicing occur at a particular location in the nucleus, and is it connected with other events—for example, transcription and/or nucleocytoplasmic transport? Does the lack of splicing make an important difference in the expression of uninterrupted genes?

With regard to the splicing reaction itself, one of the main questions is how its specificity is controlled. What ensures that the ends of each intron are recognized in pairs so that the correct sequence is removed from the RNA? Are introns excised from a precursor in a particular order? Is the maturation of RNA used to regulate gene expression by discriminating among the available precursors or by changing the pattern of splicing?

Besides RNA splicing to remove introns, many noncoding RNAs also require processing to mature, and they play roles in diverse aspects of gene expression.

21.2 The 5′ End of Eukaryotic mRNA Is Capped

Key concepts

- A 5′ cap is formed by adding a G to the terminal base of the transcript via a 5′–5′ link.
- The capping process takes place during the transcription, which may be important for transcription reinitiation.
- The 5′ cap of most mRNA is monomethylated, but some small noncoding RNAs are trimethylated.
- The cap structure is recognized by protein factors to influence mRNA stability, splicing, export, and translation.

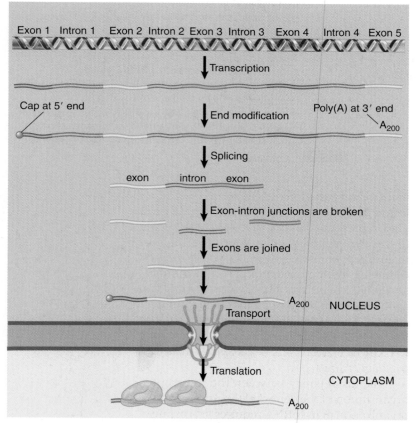

FIGURE 21.1 RNA is modified in the nucleus by additions to the 5′ and 3′ ends and by splicing to remove the introns. The splicing event requires breakage of the exon–intron junctions and joining of the ends of the exons. Mature mRNA is transported through nuclear pores to the cytoplasm, where it is translated.

Transcription starts with a nucleoside triphosphate (usually a purine, A or G). The first nucleotide retains its 5' triphosphate group and makes the usual phosphodiester bond from its 3' position to the 5' position of the next nucleotide. The initial sequence of the transcript can be represented as:

$$5'ppp^A/_GpNpNpNp \ldots$$

When the mature mRNA is treated *in vitro* with enzymes that should degrade it into individual nucleotides, however, the 5' end does not give rise to the expected nucleoside triphosphate. Instead it contains two nucleotides, which are connected by a 5'–5' triphosphate linkage and also bear a methyl group. The terminal base is always a guanine that is added to the original RNA molecule after transcription.

Addition of the 5' terminal G is catalyzed by a nuclear enzyme, guanylyl-transferase (GT). In mammals, GT has two enzymatic activities, one functioning as the triphosphatase to remove the two phosphates in GTP and the other as the quanylyl-transferase to fuse the guanine to the original 5' triphosphate terminus of the RNA. In yeast, these two activities are carried out by two separate enzymes. The new G residue added to the end of the RNA is in the reverse orientation from all the other nucleotides:

$$5'Gppp + 5' pppApNpNp \ldots \rightarrow$$
$$Gppp5'-5'ApNpNp \ldots + pp + p$$

This structure is called a **cap**. It is a substrate for several methylation events. **FIGURE 21.2** shows the full structure of a cap after all possible methyl groups have been added. The most important event is the addition of a single methyl group at the 7 position of the terminal guanine, which is carried out by guanine-7-methyltransferase (MT).

Although the capping process can be accomplished *in vitro* using purified enzymes, the reaction normally takes place during transcription. Shortly after transcription initiation, Pol II is paused ~30 nucleotides downstream from the initiation site, waiting for the recruitment of the capping enzymes to add the cap to the 5' end of nascent RNA. Without this protection, nascent RNA may be vulnerable to attack by 5'–3' exonucleases, and such trimming may induce the Pol II complex to fall off from the DNA template. Thus, the process of capping is important for Pol II to enter the productive mode of elongation to transcribe the rest of the gene. In this regard, the evolvement of the pausing mechanism for 5' capping represents a checkpoint for transcription reinitiation from the initial pausing site.

In a population of eukaryotic mRNAs, every molecule contains only one methyl group in the terminal guanine, generally referred to as monomethylated cap. In contrast, some other small noncoding RNAs, such as those involved in RNA splicing in the spliceosome (see *Section 21.6, snRNAs Are Required for Splicing*) are further methylated to contain three methyl groups in the terminal guanine. This structure is called a trimethylated cap. The enzymes for these additional methyl-transfers are present in the cytoplasm. This may ensure that only some specialized RNAs are further modified at their caps.

One of the major functions for the formation of a cap is to protect the mRNA from degradation. In fact, enzymatic decapping represents one of the major mechanisms in eukaryotic cells to regulate mRNA turnover (see *Section 21.11, Splicing Is Temporally and Functionally Coupled with Multiple Steps in Gene Expression*). In the nucleus, the cap is recognized and bound by the cap binding CBP20/80 heterodimer. This binding event stimulates splicing of the first intron, and *via* a direct interaction with the mRNA export machinery (TREX complex), facilitates mRNA export out of the nucleus. Once reaching the cytoplasm, a different set of proteins (eIF4F) binds the cap to initiate translation of the mRNA in the cytoplasm.

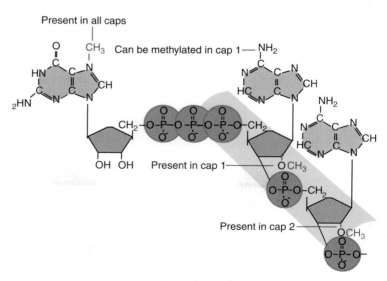

FIGURE 21.2 The cap blocks the 5' end of mRNA and is methylated at several positions.

21.3 Nuclear Splice Junctions Are Short Sequences

Key concepts

- Splice sites are the sequences immediately surrounding the exon–intron boundaries. They are named for their positions relative to the intron.
- The 5′ splice site at the 5′ (left) end of the intron includes the consensus sequence GU.
- The 3′ splice site at the 3′ (right) end of the intron includes the consensus sequence AG.
- The GU-AG rule (originally called the GT-AG rule in terms of DNA sequence) describes the requirement for these constant dinucleotides at the first two and last two positions of introns in pre-mRNAs.
- There exist minor introns relative to the major introns that follow the GU-AG rule.
- Minor introns follow a general AU-AC rule with a different set of consensus sequences at the exon–intron boundaries.

To focus on the molecular events involved in nuclear intron splicing, we must consider the nature of the *splice sites*, the two exon–intron boundaries that include the sites of breakage and reunion. By comparing the nucleotide sequence of mRNA with that of the structural gene, the junctions between exons and introns can be assigned.

There is no extensive homology or complementarity between the two ends of an intron. The junctions do, however, have well conserved, though rather short, consensus sequences. It is possible to assign a specific end to every intron by relying on the conservation of exon–intron junctions. They can all be aligned to conform to the consensus sequence given in the upper portion of **FIGURE 21.3**.

The height of each letter indicates the percent occurrence of the specified base at each consensus position. High conservation is found only immediately within the intron at the presumed junctions. This identifies the sequence of a generic intron as:

$$\text{GU} \ldots \ldots \text{AG}$$

The intron defined in this way starts with the dinucleotide GU and ends with the dinucleotide AG; as a result, the junctions are often described as conforming to the **GU-AG rule**.

Note that the two sites have different sequences and so they define the ends of the intron directionally. They are named proceeding from left to right along the intron as the 5′ splice site (sometimes called the left or donor site) and the 3′ splice site (also called the right or acceptor site). The consensus sequences are implicated as the sites recognized in splicing by point mutations that prevent splicing *in vivo* and *in vitro*.

In addition to the majority of introns that follow the GU-AG rule, a small fraction of introns are exceptions with a different set of consensus sequences at the exon–intron boundaries as shown in the lower portion of Figure 21.3. These introns were initially described as minor introns that follow the AU-AC role because of the conserved AU-AC dinucleotides at both ends of each intron as shown in the middle panel of Figure 21.3. The major and minor introns, however, are better described as U2-type and U12-type introns based on the distinct splicing machineries that process them (see *Section 21.9, An Alternative Spliceosome Uses Different snRNPs to Process the Minor Class of Introns*). As a result, some introns that appear to the follow the GU-AG rule are actually processed as the U12-type of introns as indicated in the lower panel of Figure 21.3.

21.4 Splice Junctions Are Read in Pairs

Key concepts

- Splicing depends only on recognition of pairs of splice junctions.
- All 5′ splice sites are functionally equivalent, and all 3′ splice sites are functionally equivalent.
- Additional conserved sequences at both 5′ and 3′ splice sites define functional splice sites among numerous other potential sites in the pre-mRNA.

A typical mammalian mRNA has many introns. The basic problem of pre-mRNA splicing results from the simplicity of the splice sites and is illustrated in **FIGURE 21.4**. What ensures that the correct pairs of sites are recognized and spliced together in the presence of numerous sequences that match the consensus of *bona fide* splice sites in the intron? The corresponding GU-AG pairs must be connected across great distances (some introns are >100 kb long). We can imagine two types of mechanism that might be responsible for pairing the appropriate 5′ and 3′ sites:

- It could be an *intrinsic property* of the RNA to connect the sites at the ends of a particular intron. This would require matching of specific sequences or struc-

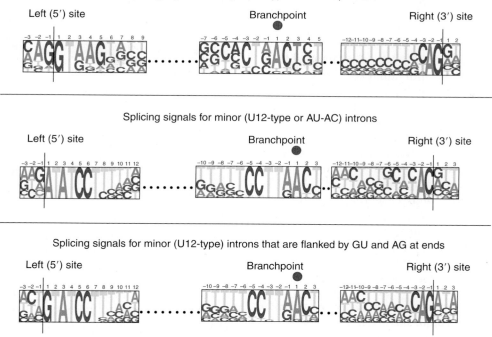

Splicing signals for major (US-type or GU-AG) introns

Left (5') site   Branchpoint   Right (3') site

Splicing signals for minor (U12-type or AU-AC) introns

Left (5') site   Branchpoint   Right (3') site

Splicing signals for minor (U12-type) introns that are flanked by GU and AG at ends

Left (5') site   Branchpoint   Right (3') site

FIGURE 21.3 The ends of nuclear introns are defined by the GU-AG rule. Minor introns are defined by different consensus sequences at the 5' splice site, branchpoint, and 3' splice site.

tures, which has been seen in certain insect genes, but this does not seem to be the case for most eukaryotic genes.

- It could be that all 5' sites may be functionally equivalent and all 3' sites may be similarly indistinguishable, but splicing could follow *rules* that ensure a 5' site is always connected to the 3' site that comes next in the RNA.

Neither the splice sites nor the surrounding regions have any sequence complementarity, which excludes models for complementary base pairing between intron ends. Experiments using hybrid RNA precursors show that any 5' splice site can in principle be connected to any 3' splice site. For example, when the first exon of the early SV40 transcription unit is linked to the third exon of mouse β globin, the hybrid intron can be excised to generate a perfect connection between the SV40 exon and the β-globin exon. Indeed, this interchangeability

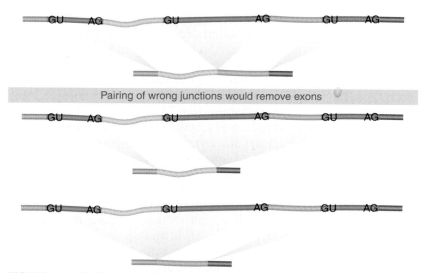

FIGURE 21.4 Splicing junctions are recognized only in the correct pairwise combinations.

is the basis for the exon-trapping technique described previously in Figure 5.8. Such experiments make two general points:

- *Splice sites are generic*: They do not have specificity for individual RNA precursors, and individual precursors do not convey specific information (such as secondary structure) that is needed for splicing. In some cases, however, specific RNA-binding proteins (e.g., hnRNP A1) have been shown to promote splice site pairing by binding to adjacent prospective splice sites.
- *The apparatus for splicing is not tissue specific*: An RNA can usually be properly spliced by any cell, whether or not it is usually synthesized in that cell. (We discuss exceptions in which there are tissue-specific alternative splicing patterns in *Section 21.12, Alternative Splicing Is a Rule, Rather Than an Exception, in Multicellular Eukaryotes*)

If all 5′ splice sites and all 3′ splice sites look similar to the splicing apparatus, what rules ensure that recognition of splice sites is restricted so that only the 5′ and 3′ sites of the same intron are spliced? Are introns removed in a specific order from a particular RNA?

Splicing is temporally coupled with transcription (e.g., many splicing events are already completed before the RNA polymerase reaches the end of the gene); as a result it is reasonable to assume that transcription provides a rough order of splicing in the 5′ to 3′ direction (something like a first-come, first-served mechanism). Secondly, a functional splice site is often surrounded by a series of sequence elements that can enhance or suppress the site (see *Section 21.13, Splicing Can Be Regulated by Exonic and Intronic Splicing Enhancers and Silencers*). Thus, sequences in both exons and introns can also function as regulatory elements for splice site selection.

We can imagine that, in order to be efficiently recognized by the splicing machinery, a functional splice site has to have the right sequence context, including specific consensus sequences and surrounding splicing enhancing elements that are dominant over splicing suppressing elements. These mechanisms together may ensure that splice signals are read in pairs in a relatively linear order.

21.5 Pre-mRNA Splicing Proceeds through a Lariat

Key concepts

- Splicing requires the 5′ and 3′ splice sites and a branch site just upstream of the 3′ splice site.
- The branch sequence is conserved in yeast but less well conserved in multicellular eukaryotes.
- A lariat is formed when the intron is cleaved at the 5′ splice site, and the 5′ end is joined to a 2′ position at an A at the branch site in the intron.
- The intron is released as a lariat when it is cleaved at the 3′ splice site, and the left and right exons are then ligated together.

The mechanism of splicing has been characterized *in vitro* using cell-free systems in which introns can be removed from RNA precursors. Nuclear extracts can splice purified RNA precursors; this shows that the action of splicing is not obligated to link to the process of transcription. Splicing can occur in RNAs that are neither capped nor polyadenylated even though these events normally occur in the cell in a coordinated manner, and the efficiency of splicing may be influenced by transcription and other processing events (see *Section 21.11, Splicing Is Temporally and Functionally Coupled with Multiple Steps in Gene Expression*).

The stages of splicing *in vitro* are illustrated in the pathway of **FIGURE 21.5**. We discuss the reaction in terms of the individual RNA species that can be identified, but remember that *in vivo* the species containing exons are not released as free molecules, but remain held together by the splicing apparatus.

FIGURE 21.6 shows that the first step of the splicing reaction is a nucleophilic attack by the 2′–OH on the 5′ splice site. The left exon takes the form of a linear molecule. The right intron–exon molecule forms a **lariat**, in which the 5′ terminus generated at the end of the intron simultaneously transesterificates to become linked by a 2′–5′ bond to a base within the intron. The target base is an A in a sequence that is called the **branch site**.

In the second step, the free 3′–OH of the exon that was released by the first reaction now attacks the bond at the 3′ splice site. Note that the number of phosphodiester bonds is conserved. There were originally two 5′–3′ bonds at the exon–intron splice sites; one has been replaced by the 5′–3′ bond between the exons,

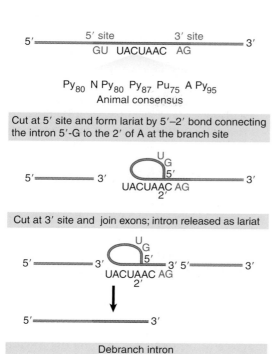

FIGURE 21.5 Splicing occurs in two stages. First the 5' exon is cleaved off, and then it is joined to the 3' exon.

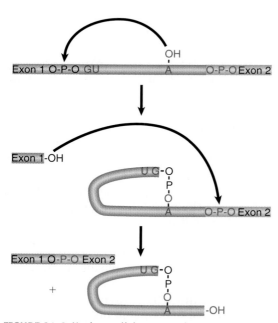

FIGURE 21.6 Nuclear splicing occurs by two transesterification reactions in which an OH group attacks a phosphodiester bond.

and the other has been replaced by the 2'–5' bond that forms the lariat. The lariat is then "debranched" to give a linear excised intron, which is rapidly degraded.

The sequences needed for splicing are the short consensus sequences at the 5' and 3' splice sites and at the branch site. Together with the knowledge that most of the sequence of an intron can be deleted without impeding splicing, this indicates that there is no demand for specific conformation in the intron (or exon).

The branch site plays an important role in identifying the 3' splice site. The branch site in yeast is highly conserved and has the consensus sequence UACUAAC. The branch site in multicellular eukaryotes is not well conserved, but has a preference for purines or pyrimidines at each position and retains the target A nucleotide (see Figure 21.5).

The branch site lies 18 to 40 nucleotides upstream of the 3' splice site. Mutations or deletions of the branch site in yeast prevent splicing. In multicellular eukaryotes, the relaxed constraints in its sequence result in the ability to use related sequences (called cryptic sites) when the authentic branch is deleted or mutated. Proximity to the 3' splice site appears to be important, because the cryptic site is always close to the authentic site. A cryptic site is used only when the branch site has been inactivated. When a cryptic branch sequence is used in this manner, splicing otherwise appears to be normal, and the exons give the same products as wild type. *The role of the branch site therefore is to identify the nearest 3' splice site as the target for connection to the 5' splice site.* This can be explained by the fact that an interaction occurs between protein complexes that bind to these two sites.

21.6 snRNAs Are Required for Splicing

Key concepts

- The five snRNPs involved in splicing are U1, U2, U5, U4, and U6.
- Together with some additional proteins, the snRNPs form the spliceosome.
- All the snRNPs except U6 contain a conserved sequence that binds the Sm proteins that are recognized by antibodies generated in autoimmune disease.

The 5′ and 3′ splice sites and the branch sequence are recognized by components of the splicing apparatus that assemble to form a large complex. This complex brings together the 5′ and 3′ splice sites before any reaction occurs, which explains why a deficiency in any one of the sites may prevent the reaction from initiating. The complex assembles sequentially on the pre-mRNA, and several intermediates can be identified by fractionating complexes of different sizes. Splicing occurs only after all the components have assembled.

The splicing apparatus contains both proteins and RNAs (in addition to the pre-mRNA). The RNAs take the form of small molecules that exist as ribonucleoprotein particles. Both the nucleus and cytoplasm of eukaryotic cells contain many discrete small RNA species. They range in size from 100 to 300 bases in higher eukaryotes and extend in length to ~1000 bases in yeast. They vary considerably in abundance, from 10^5 to 10^6 molecules per cell to concentrations too low to be detected directly.

Those restricted to the nucleus are called **small nuclear RNAs (snRNAs)**; those found in the cytoplasm are called **small cytoplasmic RNAs (scRNAs)**. In their natural state, they exist as ribonucleoprotein particles (snRNP and scRNP). Colloquially, they are sometimes known as **snurps** and **scyrps**, respectively. There is also a class of small RNAs found in the nucleolus, called **snoRNAs (small nucleolar RNAs)**, which are involved in processing ribosomal RNA (see *Section 21.20, Production of rRNA Requires Cleavage Events and Involves Small RNAs*).

The snRNPs involved in splicing, together with many additional proteins, form a large particulate complex called the **spliceosome**. Isolated from the *in vitro* splicing systems, it comprises a 50S to 60S ribonucleoprotein particle. The spliceosome may be formed in stages as the snRNPs join, proceeding through several "presplicing complexes." The spliceosome is a large body, greater in mass than the ribosome.

FIGURE 21.7 summarizes the components of the spliceosome. The five snRNAs account for more than a quarter of the mass; together with their 41 associated proteins, they account for almost half of the mass. Some 70 other proteins found in the spliceosome are described as **splicing factors**. They include proteins required for assembly of the spliceosome, proteins required for it to bind to the RNA substrate, and proteins involved in constructing an RNA-based center for **transesterification** reactions. In addition to these proteins, another ~30 proteins associ-

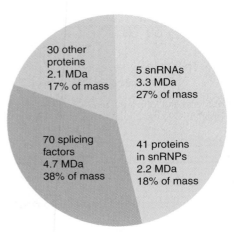

FIGURE 21.7 The spliceosome is ~12 MDa. Five snRNPs account for almost half of the mass. The remaining proteins include known splicing factors, as well as proteins that are involved in other stages of gene expression.

ated with the spliceosome have been implicated in acting at other stages of gene expression, which suggests splicing may be connected to other steps in gene expression (see *Section 21.11, Splicing Is Temporally and Functionally Coupled with Multiple Steps in Gene Expression*).

The spliceosome forms on the intact precursor RNA and passes through an intermediate state in which it contains the individual 5′ exon linear molecule and the right-lariat intron–exon. Little spliced product is found in the complex, which suggests that it is usually released immediately following the cleavage of the 3′ site and ligation of the exons.

We may think of the snRNP particles as being involved in building the structure of the spliceosome. Like the ribosome, the spliceosome depends on RNA–RNA interactions as well as protein–RNA and protein–protein interactions. Some of the reactions involving the snRNPs require their RNAs to base pair directly with sequences in the RNA being spliced; other reactions require recognition between snRNPs or between their proteins and other components of the spliceosome.

The importance of snRNA molecules can be tested directly in yeast by making mutations in their genes, or in *in vitro* splicing reactions by targeted degradation of individual snRNAs in the nuclear extract. Inactivation of five snRNAs, individually or in combination, prevents splicing. All of the snRNAs involved in splicing can be recognized in conserved forms in all eukaryotes, including plants. The corresponding RNAs in yeast are often rather larger, but conserved regions include features that are similar to the snRNAs of multicellular eukaryotes.

The snRNPs involved in splicing are U1, U2, U5, U4, and U6. They are named according to the snRNAs that are present. Each snRNP contains a single snRNA and several (<20) proteins. The U4 and U6 snRNPs are usually found together as a di-snRNP (U4/U6) particle. A common structural core for each snRNP consists of a group of eight proteins, all of which are recognized by an autoimmune antiserum called **anti-Sm**; conserved sequences in the proteins form the target for the antibodies. The other proteins in each snRNP are unique to it. The Sm proteins bind to the conserved sequence $PuAU_{3-6}Gpu$, which is present in all snRNAs except U6. The U6 snRNP instead contains a set of Sm-like (Lsm) proteins.

Some of the proteins in the snRNPs may be involved directly in splicing; others may be required in structural roles or just for assembly or interactions between the snRNP particles. About one third of the proteins involved in splicing are components of the snRNPs. Increasing evidence suggests a direct role of RNA in catalysis; most splicing factors may therefore provide structural or assembly roles in the spliceosome.

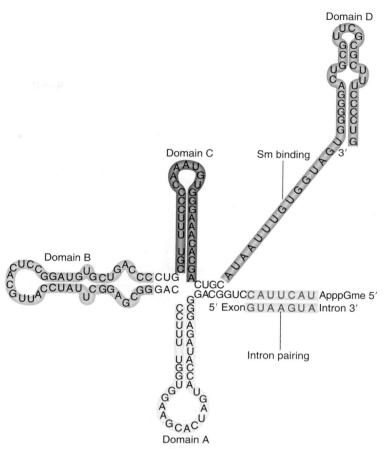

FIGURE 21.8 U1 snRNA has a base-paired structure that creates several domains. The 5' end remains single stranded and can base pair with the 5' splice site.

21.7 Commitment of Pre-mRNA to the Splicing Pathway

Key concepts

- U1 snRNP initiates splicing by binding to the 5' splice site by means of an RNA–RNA pairing reaction.
- The commitment complex contains U1 snRNP bound at the 5' splice site and the protein U2AF bound to a pyrimidine tract between the branch site and the 3' splice site.
- In multicellular eukaryotic cells, SR proteins play an essential role in initiating the formation of the commitment complex.
- Pairing splice sites can be accomplished by intron definition or exon definition.

Recognition of the consensus splicing signals involves both RNAs and proteins. Certain snRNAs have sequences that are complementary to the consensus sequences or to one another, and base pairing between snRNA and pre-mRNA, or between snRNAs, plays an important role in splicing.

The human U1 snRNP contains the core Sm proteins, three U1-specific proteins (U1-70k, U1A, and U1C), and U1 snRNA. The secondary structure of the U1 snRNA is drawn in

FIGURE 21.8. It contains several domains. The Sm-binding site is required for interaction with the common snRNP proteins. Domains identified by the individual stem-loop structures provide binding sites for proteins that are unique to U1 snRNP. Binding of U1 snRNP to the 5' splice site is the first step in splicing. U1 snRNA interacts with the 5' splice site by base pairing between its single-stranded 5' terminus and a stretch of four to six bases of the 5' splice site.

Mutations in the 5' splice site and U1 snRNA can be used to test directly whether pairing between them is necessary. The results of such an experiment are illustrated in FIGURE 21.9. The wild-type sequence of the splice site of the 12S adenovirus pre-mRNA pairs at five out of six positions with U1 snRNA. A mutant in the 12S RNA that cannot be spliced has two sequence changes; the GG residues at positions 5 to 6 in the intron are changed to AU. When a mutation is introduced into U1 snRNA that restores pairing at position 5, normal splicing is regained.

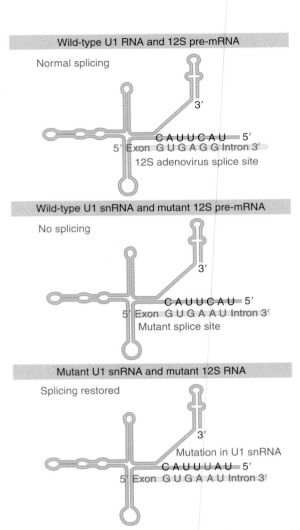

Wild-type U1 RNA and 12S pre-mRNA

Normal splicing

3'

C A U U C A U — 5'
5' Exon G U G A G G Intron 3'
12S adenovirus splice site

Wild-type U1 snRNA and mutant 12S pre-mRNA

No splicing

3'

C A U U C A U — 5'
5' Exon G U G A A U Intron 3'
Mutant splice site

Mutant U1 snRNA and mutant 12S RNA

Splicing restored

3'

Mutation in U1 snRNA
C A U U U A U — 5'
5' Exon G U G A A U Intron 3'

FIGURE 21.9 Mutations that abolish function of the 5' splice site can be suppressed by compensating mutations in U1 snRNA that restore base pairing.

Other cases, in which corresponding mutations are made in U1 snRNA to see whether they can suppress the mutation in the splice site, suggest the general rule: complementarity between U1 snRNA and the 5' splice site is necessary for splicing, but the efficiency of splicing is not determined solely by the number of base pairs that can form.

The U1 snRNA pairing reaction with the 5' splicing is stabilized by protein factors. Two such factors play a particular role: the branch-point binding protein (BBP, also known as SF1) interacts with the branchpoint sequence, and U2AF (a heterodimer consisting of U2AF65 and U2AF35 in multicellular eukaryotic cells or Mud2 in the yeast *Saccharomyces cerevisiae*) binds to the polypyrimidine tract between the branchpoint sequence and the invariant AG dinucleotide at the end of each intron. Each of these binding events is not very strong, but

together they bind in a cooperative fashion, resulting in the formation of a relatively stable complex called the *commitment complex*.

The commitment complex is also known as the **E complex** ("E" for "early") in mammalian cells, the formation of which does not require ATP (compared to all late ATP-dependent steps in the assembly of the spliceosome; see *Section 21.8, The Spliceosome Assembly Pathway*). Unlike in yeast, however, the consensus sequences at the splice sites in mammalian genes are only loosely conserved, and consequently additional protein factors are needed for the formation of E complex.

The factor or factors that play a central role in this and other spliceosome assembly processes are **SR proteins**, which constitute a family of splicing factors that contain one or two RNA recognition motifs at the N-terminus and a signature domain rich with multiple Arg/Ser dipeptide repeats (called the RS domain) at their C-terminus. Their RNA recognition motifs are responsible for sequence-specific binding to RNA, and the RS domain can bind to both RNA and other splicing factors via protein–protein interactions, thereby providing additional "glue" for various parts of the E complex.

As illustrated in **FIGURE 21.10**, SR proteins can bind to the 70kD component of U1 snRNP (the U1 70kD protein also contains an RS domain, but it is not considered a typical SR protein) to enhance or stabilize its base pairing with the 5' splice site. SR proteins can also bind to 3' splice site-bound U2AF (an RS domain is also present in both U2AF65 and U2AF35). These protein–protein interaction networks are thought to be critical for the formation of the E complex. SR proteins copurify with the Pol II complex and are able to kinetically commit RNA to the splicing pathway; thus they likely function as the splicing initiators in multicellular eukaryotic cells.

Typical SR proteins are neither encoded in the genome of *S. cerevisiae* nor needed for splicing by the organism where the splicing signals are nearly invariant, but they are absolutely essential for splicing in all multicellular eukaryotes where the splicing signals are highly divergent. The evolution of SR proteins in multicellular eukaryotes likely contributes to high efficacy and high fidelity splicing on loosely conserved splice sites. The recognition of functional splice sites during the formation of the E complex can take two routes, as illustrated in **FIGURE 21.11**. In *S. cerevisiae*, where nearly all intron-containing genes are interrupted by a single small intron

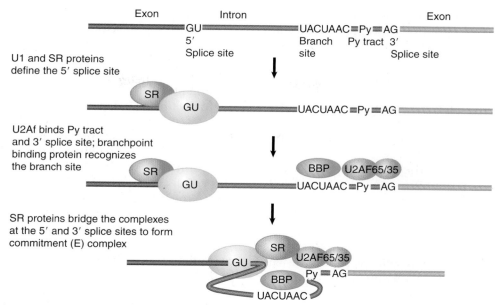

FIGURE 21.10 The commitment (E) complex forms by the successive addition of U1 snRNP to the 5′ splice site, U2AF to the pyrimidine tract/3′ splice site, and the bridging protein SF1/BBP.

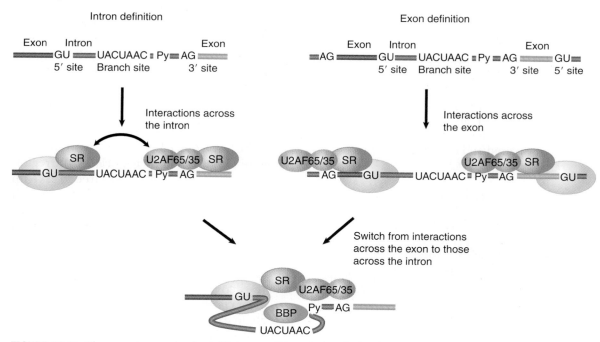

FIGURE 21.11 There are two routes for initial recognition of 5′ and 3′ splice sites by either intron definition or exon definition.

(~100 to 300 nucleotides in length), the 5′ and 3′ splice sites are simultaneously recognized by U1 snRNP, BBP, and Mud2, as discussed above. This process is referred to as **intron definition** as illustrated on the left in Figure 21.11. (Note that the intron definition mechanism applies to small introns in multicellular eukaryotic cells and thus the figure is drawn with the nomenclature for mammalian splicing factors involved in the process).

In comparison, introns are long and highly variable in length in multicellular eukaryotic genomes, and there are many sequences that resemble real splice sites in them. This makes the paired recognition of the 5′ and 3′ splice sites inefficient, if not impossible. The solution to this problem is the process of **exon definition**, which takes advantage of normally small exons (~100 to 300 nucleotides in length) in higher eukaryotic cells.

As shown on the right of Figure 21.11, during exon definition, the U2AF heterodimer binds to the 3′ splice site and U1 snRNP base pairs with the 5′ splice site downstream from the exon sequence. This process may be aided by SR proteins that bind to specific exon sequences between the 3′ and downstream 5′ splice sites. By an as-yet unknown mechanism, the complexes formed across the exon are then switched to the complexes that link the 3′ splice site to the upstream 5′ splice site and the downstream 5′ splice site to the next downstream 3′ splice sites across introns. This establishes the "permissive" configuration that allows later spliceosome assembly steps to occur.

Blockage of this transition is actually a means to regulate the selection of certain exons during regulated splicing (see *Section 21.13, Splicing Can Be Regulated by Exonic and Intronic Splicing Enhancers and Silencers*). Finally, the exon definition mechanism mediated by SR proteins also provides a mechanism to only allow adjacent 5′ and 3′ splice sites to be paired and linked by splicing.

21.8 The Spliceosome Assembly Pathway

Key concepts

- The commitment complex progresses to pre-spliceosome (the A complex) in the presence of ATP.
- Recruitment of U5 and U4/U6 snRNPs converts the pre-spliceosome to the mature spliceosome (the B1 complex).
- The B1 complex is next converted to the B2 complex in which U1 snRNP is released to allow U6 snRNA to interact with the 5′ splice site.
- U4 dissociates from U6 snRNP to allow U6 snRNA to pair with U2 snRNA to form the catalytic center for splicing.
- Both transesterification reactions take place in the activated spliceosome (the C complex).
- The splicing reaction is reversible at all steps.

Following formation of the E complex, the other snRNPs and factors involved in splicing associate with the complex in a defined order. **FIGURE 21.12** shows the components of the complexes that can be identified as the reaction proceeds.

In the first ATP-dependent step, U2 snRNP joins U1 snRNP on the pre-mRNA by binding to the branchpoint sequence, which also involves base pairing between the sequence in U2 snRNA and the branchpoint sequence. This results in the conversion of the E complex to the prespliceosome commonly known as the **A complex**, and this step requires ATP hydrolysis.

The B1 complex is formed when a trimer containing the U5 and U4/U6 snRNPs binds to the A complex. This complex is regarded as a spliceosome because it contains the components needed for the splicing reaction. It is converted to the B2 complex after U1 is released. The dissociation of U1 is necessary to allow other components to come into juxtaposition with the 5′ splice site, most notably U6 snRNA.

The catalytic reaction is triggered by the release of U4, which also takes place during the transition from the B1 to B2 complex. The role of U4 snRNA may be to sequester U6 snRNA until it is needed. **FIGURE 21.13** shows the changes that occur in the base pairing interactions between snRNAs during splicing. In the U6/U4 snRNP, a continuous length of 26 bases of U6 is paired with two separated regions of U4. When U4 dissociates, the region in U6 that is released becomes free to take up another structure. The first part of it pairs with U2; the second part forms an intramolecular hairpin. The interaction between U4 and U6 is mutually incompatible with the interaction between U2 and U6, so the release of U4 controls the ability of the spliceosome to proceed to the activated state.

For clarity, the figure shows the RNA substrate in extended form, but the 5′ splice site is actually close to the U6 sequence immediately on the 5′ side of the stretch bound to U2. This sequence in U6 snRNA pairs with sequences in the intron just downstream of the conserved GU at the 5′ splice site (mutations that enhance such pairing improve the efficiency of splicing).

Thus several pairing reactions between snRNAs and the substrate RNA occur in the course of splicing. They are summarized in

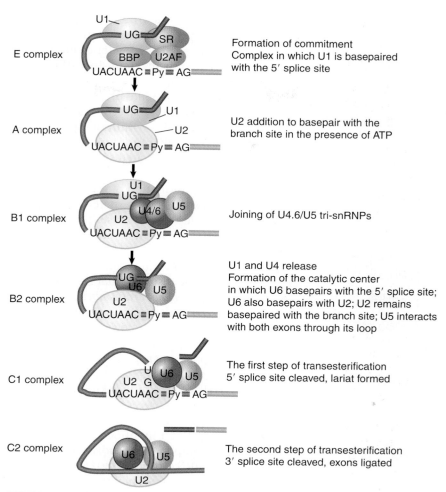

FIGURE 21.12 The splicing reaction proceeds through discrete stages in which spliceosome formation involves the interaction of components that recognize the consensus sequences.

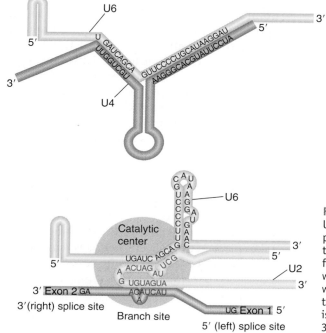

FIGURE 21.13 U6-U4 pairing is incompatible with U6-U2 pairing. When U6 joins the spliceosome it is paired with U4. Release of U4 allows a conformational change in U6; one part of the released sequence forms a hairpin (gray), and the other part (pink) pairs with U2. An adjacent region of U2 is already paired with the branch site, which brings U6 into juxtaposition with the branch. Note that the substrate RNA is reversed from the usual orientation and is shown 3′ to 5′.

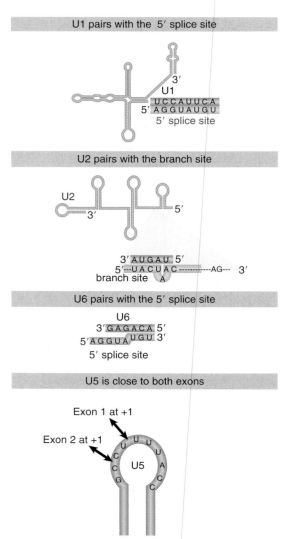

U1 pairs with the 5′ splice site

3′
U1
5′ UCCAUUCA
AGGUAUGU
5′ splice site

U2 pairs with the branch site

U2
3′   5′

3′ AUGAU 5′
5′---UACUAC----------AG--- 3′
branch site   A

U6 pairs with the 5′ splice site

U6
3′ GAGACA 5′
5′ AGGUA UGU 3′
5′ splice site

U5 is close to both exons

Exon 1 at +1

Exon 2 at +1

U U
C   U
U
C U5   A
C   C
G   C

FIGURE 21.14 Splicing utilizes a series of base pairing reactions between snRNAs and splice sites.

FIGURE 21.14. The snRNPs have sequences that pair with the substrate and with one another. They also have single-stranded regions in loops that are in close proximity to sequences in the substrate, and which play an important role, as judged by the ability of mutations in the loops to block splicing.

The base pairing between U2 and the branch point, and between U2 and U6, creates a structure that resembles the active center of group II self-splicing introns (see Figure 21.15 in *Section 21.9*). This suggests the possibility that the catalytic component could comprise an RNA structure generated by the U2–U6 interaction. U6 is paired with the 5′ splice site, and crosslinking experiments show that a loop

in U5 snRNA is immediately adjacent to the first base positions in both exons. Although the available evidence points to a RNA-based catalysis mechanism within the spliceosome, contribution(s) by proteins cannot be ruled out. One candidate protein is Prp8, a large scaffold protein that directly contacts both the 5′ and 3′ splice sites within the spliceosome.

Both transesterification reactions take place in the activated spliceosome (the C complex) after a series of RNA arrangements is completed. The formation of the lariat at the branch site is responsible for determining the use of the 3′ splice site, because the 3′ consensus sequence nearest to the 3′ side of the branch becomes the target for the second transesterification.

The important conclusion suggested by these results is that *the snRNA components of the splicing apparatus interact both among themselves and with the substrate RNA by means of base pairing interactions, and these interactions allow for changes in structure that may bring reacting groups into apposition and may even create catalytic centers.*

Although the spliceosome is likely a large RNA machine, like ribosomes, many protein factors are essential for the machine to run. Extensive mutational analyses undertaken in yeast identified both the RNA and protein components (known as PRP mutants for pre-mRNA processing). Several of the products of these genes have motifs that identify them as a family of ATP-dependent RNA helicases, which are crucial for a series of ATP-dependent RNA rearrangements in the spliceosome.

Prp5 is critical for U2 binding to the branch-point during the transition from the E to the A complex; Brr2 facilitates U1 and U4 release during the transition from the B1 to B2 complex; Prp2 is responsible for the activation of the spliceosome during the conversion of the B2 complex to the C complex; and Prp22 helps the release of the mature mRNA from the spliceosome. In addition, a number of RNA helicases are shown to play roles in recycling of snRNPs for the next round of spliceosome assembly.

These findings explain why ATP hydrolysis is required from various steps of the splicing reaction, although the actual transesterification reactions do not require ATP. Despite the fact that a sequential series of RNA arrangements takes place in the spliceosome, it is remarkable that the process seems to be reversible after both the first and second transesterification reactions.

21.9 An Alternative Spliceosome Uses Different snRNPs to Process the Minor Class of Introns

Key concepts

- An alternative splicing pathway uses another set of snRNPs with only U5 snRNP in common with the major spliceosome.
- The target introns are defined by longer consensus sequences at the splice junctions, rather than strictly according to the GU-AG or AU-AC rules.
- Major and minor spliceosomes share critical protein factors, including SR proteins.

GU-AG introns comprise the majority (>98% of splicing junctions in the human genome). Exceptions to this case are noncanonical splice AU-AC junctions and other variations. Initially, this minor class of introns was referred to as AU-AC introns compared to the major class of introns that follow the GU-AG rule during splicing. With the elucidation of the machinery for processing of both major and minor introns, it becomes clear that this nomenclature for the minor class of introns is not entirely accurate.

Guided by years of research on the major spliceosome, the machinery for processing the minor class of introns was quickly elucidated; it consists of U11 and U12 (related to U1 and U2, respectively), a common U5 shared with the major spliceosome, and the $U4_{atac}$ and $U6_{atac}$ snRNAs. The splicing reaction is essentially similar to that of the major class of introns, and the snRNAs play analogous roles: U11 basepairs with the 5' splice sites; U12 basepairs with the branchpoint sequence near the 3' splice site; and $U4_{atac}$ and $U6_{atac}$ provide analogous functions during the spliceosome assembly and activation of the spliceosome.

It turns out that the dependence on the type of spliceosome is also influenced by the sequences in other places in the intron, so that there are some GU-AG introns spliced by the U12-type spliceosome. A strong consensus sequence at the left end defines the U12-dependent type of intron: 5'^G_AUAUCCUUU. . . PyA^G_C 3'. In fact, most U12-dependent introns have the GU. . . AG termini. They have a highly conserved branch point, though (UCCUUPuAPy), which pairs with U12 (see Figure 21.3). This difference in branchpoint sequences is the primary distinction between the major and minor classes of introns. For this reason, the major class of introns is termed *U2-dependent introns* and the minor class is called *U12-dependent introns*, instead of AU-AC introns.

The two types of intron coexist in a variety of genomes, and in most cases are found in the same gene. U12-dependent introns tend to be flanked by U2-dependent introns. The phylogeny of these introns suggests that AU-AC U12-dependent introns may once have been more common, but tend to be converted to GU-AG termini, and to U2-dependence, in the course of evolution. The common evolution of the systems is emphasized by the fact that they use analogous sets of base pairing between the snRNAs and with the substrate pre-mRNA. In addition, all essential splicing factors (i.e., SR proteins) studied thus far are required for processing both U2-type and U12-type introns.

One noticeable difference between U2 and U12 types of intron is that U1 and U2 appear to independently recognize the 5' and 3' splice sites in the major class of introns during the formation of the E and A complexes, whereas U11 and U12 form a complex in the first place, which together contact the 5' and 3' splice sites to initiate the processing of the minor class of introns. This ensures that the splice sites in the minor class of introns are recognized simultaneously by the intron definition mechanism. It also avoids "confusing" the splicing machineries during the transition from exon definition to intron definition for processing the major and minor classes of introns that are present in the same gene.

21.10 Pre-mRNA Splicing Likely Shares the Mechanism with Group II Autocatalytic Introns

Key concepts

- Group II introns excise themselves from RNA by an autocatalytic splicing event.
- The splice junctions and mechanism of splicing of group II introns are similar to splicing of nuclear introns.
- A group II intron folds into a secondary structure that generates a catalytic site resembling the structure of U6-U2-nuclear intron.

Introns in protein-coding genes (in fact, in all genes except nuclear tRNA-coding genes) can be divided into three general classes. Nuclear pre-mRNA introns are identified only by the possession of the GU. . . AG dinucleotides at the 5' and 3' ends and the branch site/pyrimidine tract near the 3' end. They do not show any common features of secondary structure. In contrast, group I and group II introns found in organelles and in bacteria (group I introns are found also in the nucleus in unicellular/oligo-cellular eukaryotes) are classified according to their internal organization. Each can be folded into a typical type of secondary structure.

The group I and group II introns have the remarkable ability to excise themselves from an RNA. This is called **autosplicing** or **self-splicing**. Group I introns are more common than group II introns. There is little relationship between the two classes, but in each case the RNA can perform the splicing reaction *in vitro* by itself, without requiring enzymatic activities provided by proteins; however, proteins are almost certainly required *in vivo* to assist with folding (see Chapter 23, *Catalytic RNA*).

FIGURE 21.15 shows that three classes of introns are excised by two successive trans-esterifications (shown previously for nuclear introns in Figure 21.6). In the first reaction, the 5' exon–intron junction is attacked by a free hydroxyl group (provided by an internal 2'–OH position in nuclear and group II introns, or by a free guanine nucleotide in group I introns). In the second reaction, the free 3'–OH at the end of the released exon in turn attacks the 3' intron–exon junction.

There are parallels between group II introns and pre-mRNA splicing. Group II mitochondrial introns are excised by the same mechanism as nuclear pre-mRNAs via a lariat that is held together by a 2'–5' bond. When an isolated group II RNA is incubated *in vitro* in the absence of additional components, it is able to perform the splicing reaction. This means that the two transesterification reactions shown in Figure 21.15 can be performed by the group II intron RNA sequence itself. The number of phosphodiester bonds is conserved in the reaction, and as a result an external supply of energy is not required; this could have been an important feature in the evolution of splicing.

A group II intron forms into a secondary structure that contains several domains formed by base-paired stems and single-stranded loops. Domain 5 is separated by two bases from domain 6, which contains an A residue that

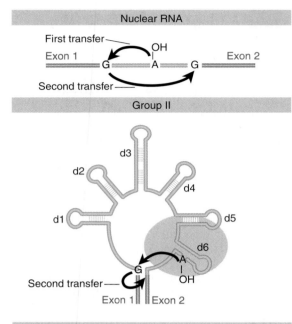

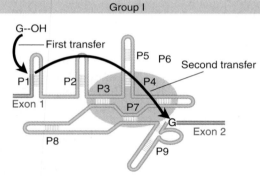

FIGURE 21.15 Three classes of splicing reactions proceed by two transesterifications. First, a free OH group attacks the exon 1–intron junction. Second, the OH created at the end of exon 1 attacks the intron–exon 2 junction.

donates the 2'–OH group for the first transesterification. This constitutes a catalytic domain in the RNA. **FIGURE 21.16** compares this secondary structure with the structure formed by the combination of U6 with U2 and of U2 with the branch site. The similarity suggests that U6 may have a catalytic role in pre-mRNA splicing.

The features of group II splicing suggest that splicing evolved from an autocatalytic reaction undertaken by an individual RNA molecule, in which it accomplished a controlled deletion of an internal sequence. It is likely that such a reaction requires the RNA to fold into a specific conformation, or series of conformations, and would occur exclusively in *cis* conformation.

The ability of group II introns to remove themselves by an autocatalytic splicing event stands in great contrast to the requirement of nuclear introns for a complex splicing appara-

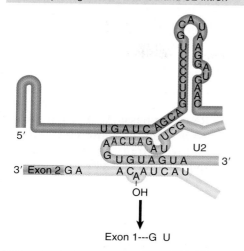

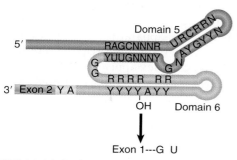

Group II splicing constructs an active center
from the base paired regions of domains 5 and 6

FIGURE 21.16 Nuclear splicing and group II splicing involve the formation of similar secondary structures. The sequences are more specific in nuclear splicing; group II splicing uses positions that may be occupied by either purine (R) or pyrimidine (Y).

tus. We may regard the snRNAs of the spliceosome as compensating for the lack of sequence information in the intron, and as providing the information required to form particular structures in RNA. The functions of the snRNAs may have evolved from the original autocatalytic system. These snRNAs act in *trans* upon the substrate pre-mRNA; we might imagine that the ability of U1 to pair with the 5′ splice site, or of U2 to pair with the branch sequence, replaced a similar reaction that required the relevant sequence to be carried by the intron. Thus the snRNAs may undergo reactions with the pre-mRNA substrate—and with one another—that have substituted for the series of conformational changes that occur in RNAs that splice by group II mechanisms. In effect, these changes have relieved the substrate pre-mRNA of the obligation to carry the sequences needed to

sponsor the reaction. As the splicing apparatus has become more complex (and as the number of potential substrates has increased), proteins have played a more important role.

21.11 Splicing Is Temporally and Functionally Coupled with Multiple Steps in Gene Expression

Key concepts

- Splicing can occur during or after transcription.
- The transcription and splicing machineries are physically and functionally integrated.
- Splicing is connected to mRNA export and stability control.
- Splicing in the nucleus can influence mRNA translation in the cytoplasm.

Pre-mRNA splicing has long been recognized to take place cotranscriptionally, although the two reactions can take place separately *in vitro* and have been studied as separate processes in gene expression. Major experimental evidence supporting cotranscriptional splicing came from the observations that many splicing events have been completed before the completion of transcription. In general, introns near the 5′ end of the gene are removed during transcription, but introns near the end of the gene can be processed either during or after transcription.

Besides temporal coupling between transcription and splicing, there are probably other reasons for these two key processes to be linked in a functional way. Indeed, the machineries for 5′ capping, intron removal, and even polyadenylation at the 3′ end (see *Section 21.16, The 3′ mRNA End Processing Is Critical for Transcriptional Termination*) show physical interactions with the core machinery for transcription. A common mechanism is to use the large C-terminal domain of the largest subunit of Pol II (known as CTD) as a loading pad for various RNA processing factors, although in most cases it is yet to be defined whether the tethering is direct or mediated by some common protein or even RNA factors (see *Section 20.8, Initiation Is Followed by Promoter Clearance and Elongation*).

Such physical integration would ensure efficient recognition of emerging splicing signals to pair adjacent functional splice sites during transcription, thus maintaining a rough order of splicing from the 5′ to 3′ direction. The recognition of the emerging splicing signals by the RNA processing factors and enzymes associated with the elongation Pol II complex would also allow these factors to compete effectively with other nonspecific RNA binding proteins, such as hnRNP proteins, that are abundantly present in the nucleus for RNA packaging.

If RNA splicing benefits from transcription, why not the other way around? In fact, increasing evidence has suggested so; as illustrated in **FIGURE 21.17**, the 5′ capping enzymes seem to help overcome initial transcriptional pausing near the promoter; splicing factors appear to play some roles in facilitating transcriptional elongation; and the 3′ end formation of mRNA is clearly instrumental to transcriptional termination (see *Section 21.16, The 3′ mRNA End Processing Is Critical for Transcriptional Termination*). Thus, transcription and RNA processing are highly coordinated in multicellular eukaryotic cells.

RNA processing is functionally linked not only to the upstream transcriptional events, but also to downstream steps, such as mRNA export and stability control. It has been known for a long time that intermediately processed RNA that still contains some introns cannot be exported efficiently, which may be due to the retention effect of the spliceosome in the nucleus. Splicing-facilitated mRNA export can be demonstrated by nuclear injection of intronless RNA derived from cDNA or pre-mRNA that will give rise to identical RNA upon splicing. The RNA that has gone through the splicing process is exported more efficiently than the RNA derived from the cDNA, indicating that the splicing process helps mRNA export.

As illustrated in **FIGURE 21.18**, a specific complex is deposited onto the exon–exon junction, which is called the **exon junction complex (EJC)**. This complex appears to directly recruit a number of RNA-binding proteins implicated in mRNA export. Apparently, these mechanisms may act in synergy to promote the export of mRNA coming out of transcription and the cotranscriptional RNA splicing apparatus. This process may start early in transcription. The cap binding CBP20/80 complex appears to directly bind to the mRNA export machinery (the TREX com-

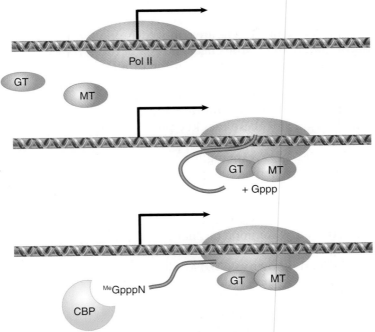

FIGURE 21.17 Coupling transcription with the 5′ capping reaction. Pol II transcription is initially paused near the transcription state. Both guanylyltransferase (GT) and 7-methyltransferase (MT) are recruited to the Pol II complex to catalyze 5′ capping and the cap is bound by the cap binding protein complex at the 5′ end of the nascent transcript. These reactions allow the paused Pol II to enter the mode of productive elongation.

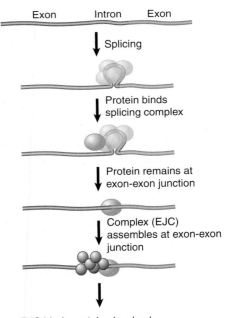

EJC binds proteins involved in RNA export, localization, decay

FIGURE 21.18 The EJC (exon junction complex) is deposited near the splice junction as a consequence of the splicing reaction.

plex) in a manner that depends on splicing to remove the first intron near the 5' end to facilitate mRNA export. A key factor in mediating mRNA export is REE (also named Aly, Yra1p in yeast), which is part of the EJC and can directly interact with the mRNA transporter TAP (Mex67p in yeast) as shown in **FIGURE 21.19**.

The EJC complex has an additional role in escorting mRNA out of the nucleus, which has a profound effect on mRNA stability in the cytoplasm. This is because an EJC that has retained some aberrant mRNAs can recruit other factors that promote decapping enzymes to remove the protective cap at the 5' end of the mRNA. As illustrated in **FIGURE 21.20**, the EJC is normally removed by the scanning ribosome during the first round of translation in the cytoplasm. If, however, for some reason a premature stop codon is introduced into a processed mRNA as a result of point mutation or alternative splicing (see *Section 21.12, Alternative Splicing Is a Rule Rather Than Exception in Multicellular Eukaryotes*), the ribosome will fall off before reaching the natural stop codon, which is typically located in the last exon. The

inability of the ribosome to strip off the EJC complex deposited after the premature stop codon will allow the recruitment of decapping enzymes to induce rapid degradation of the mRNA. This process is called **nonsense-mediated mRNA decay (NMD)**, which represents an mRNA surveillance mechanism that prevents translation of truncated proteins from the mRNA that carries a premature stop codon. NMD is discussed further in *Section 22.9, Quality Control of mRNA Translation Is Performed by Cytoplasmic Surveillance Systems*.

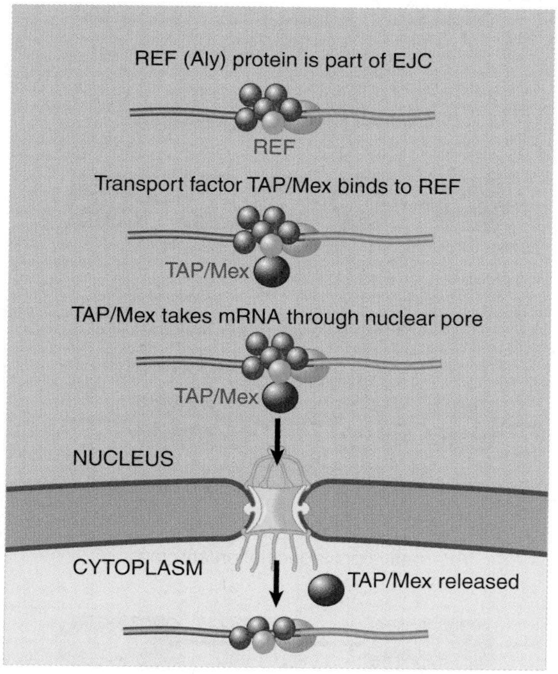

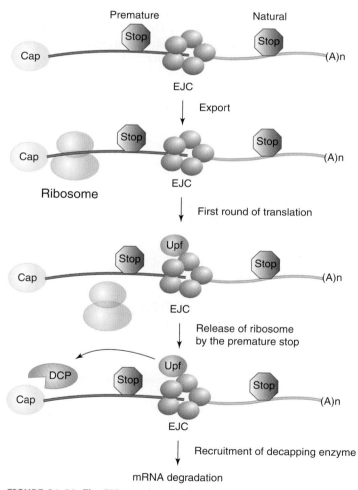

FIGURE 21.20 The EJC complex couples splicing with NMD. The EJC can also recruit Upr proteins if remains on the exported mRNA. After nuclear export, EJC should be tripped off by the scanning ribosome in the first round of translation. If an EJC remains on the mRNA because of a premature stop codon in the front, which releases the ribosome, the EJC will recruit additional proteins, such as Upf, which will then recruit the decapping enzyme (DCP). This will induce decapping at the 5' end and mRNA degradation from the 5' to 3' direction in the cytoplasm.

FIGURE 21.19 A REF protein binds to a splicing factor and remains with the spliced RNA product. REF binds to an export factor that binds to the nuclear pore.

21.12 Alternative Splicing Is a Rule, Rather Than an Exception, in Multicellular Eukaryotes

Key concepts
- Specific exons or exonic sequences may be excluded or included in the mRNA products by using alternative splicing sites.
- Alternative splicing contributes to structural and functional diversity of gene products.
- Sex determination in *Drosophila* involves a series of alternative splicing events in genes coding for successive products of a pathway.

When an interrupted gene is transcribed into an RNA that gives rise to a single type of spliced mRNA, there is no ambiguity in assignment of exons and introns. The RNAs of most genes, however, follow patterns of **alternative splicing**, which occurs when a single gene gives rise to more than one mRNA sequence. By large-scale cDNA cloning and sequencing, it has become apparent that more than 90% of the genes expressed in mammals are alternatively spliced. Thus alternative splicing is not just the result of mistakes made by the splicing machinery; it is part of the gene expression program that results in multiple gene products from a single gene locus.

There are various modes of alternative splicing, including intron retention, alternative 5′ splice site selection, alternative 3′ splice site selection, exon inclusion or skipping, and mutually exclusive selection of the alternative exons, as summarized in FIGURE 21.21. A single primary transcript may undergo more than one mode of alternative splicing. The mutually exclusive exons are normally regulated in a tissue-specific manner. Added to this complexity, in some cases, the ultimate pattern of expression is also dictated by the use of different transcription start points or the generation of alternative 3′ ends.

Alternative splicing can affect gene expression in the cell in at least in two ways. One way is to create structural diversity of gene products by including or omitting some coding sequences or by creating alternative reading frames for a

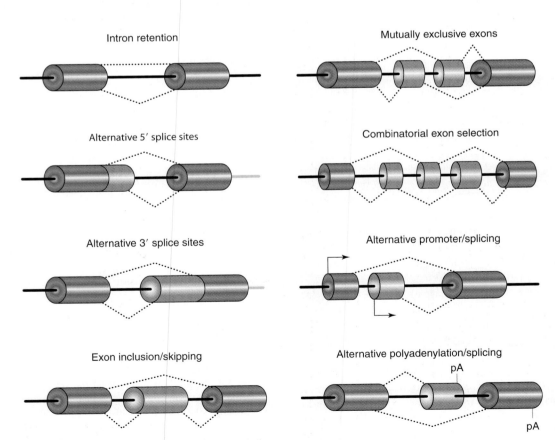

Intron retention

Alternative 5′ splice sites

Alternative 3′ splice sites

Exon inclusion/skipping

Mutually exclusive exons

Combinatorial exon selection

Alternative promoter/splicing

Alternative polyadenylation/splicing

pA

pA

FIGURE 21.21 Different modes of alternative splicing.

portion of the gene. This can often modify the functional property of encoded proteins. For example, the *CaMKIIδ* gene contains three alternatively spliced exons as shown in **FIGURE 21.22**. The gene is expressed in almost all cell types and tissues in mammals. When all three alternative exons are skipped, the mRNA encodes a cytoplasmic kinase that phosphorylates a large number of protein substrates. When exon 14 is included, the kinase is transported to the nucleus because exon 14 contains a nuclear localization signal. This allows the kinase to regulate transcription in the nucleus. When both exons 15 and 16 are included, which is normally detected in neurons, the kinase is targeted to the cell membrane where it can influence specific ion channel activities.

In other cases, the alternatively spliced products exhibit opposite functions. This applies to essentially all genes involved in the regulation of apoptosis; each gene expresses at least two isoforms, one functioning to promote apoptosis and the other protecting cells against apoptosis. It is thought that the isoform ratios of these apoptosis regulators may dictate whether the cell lives or dies.

Alternative splicing may also affect various properties of the mRNA by including or omitting certain regulatory RNA elements, which may significantly alter the half-life of the mRNA. In many cases, the main purpose of alternative splicing may be to cause a certain percentage of primary transcripts to carry a premature stop codon or codons so that those transcripts can be rapidly degraded. This may represent an alternative strategy to transcriptional regulation to control the abundance of specific mRNAs in the cell. This mechanism is used to achieve homeostatic expression for many splicing regulators in specific cell types or tissues. In such regulation, a specific positive splicing regulator may affect its own alternative splicing, resulting in the inclusion of an exon containing a premature stop codon. This siphons a fraction of its mRNA to degradation, thereby reducing the protein concentration. Thus, when the concentration of such positive splicing regulator fluctuates in the cell, its mRNA concentration will be shifted in the opposite direction.

Although many alternative splicing events have been characterized and the biological roles of the alternatively spliced products determined, the best-understood example is still the pathway of sex determination in *D. melanogaster*, which involves interactions between a series of genes in which alternative splicing events

distinguish males and females. The pathway takes the form illustrated in **FIGURE 21.23**, in which the ratio of X chromosomes to autosomes determines the expression of *sex lethal* (*sxl*), and changes in expression are passed sequentially through the other genes to *doublesex* (*dsx*), the last in the pathway.

The pathway starts with sex-specific splicing of *sxl*. Exon 3 of the *sxl* gene contains a termination codon that prevents synthesis of functional protein. This exon is included in the mRNA produced in males, but is skipped in females. As a result, only females produce Sxl protein. The protein has a concentration of basic amino acids that resembles other RNA-binding proteins.

The presence of Sxl protein changes the splicing of the *transformer* (*tra*) gene. Figure 21.23 shows that this involves splicing a constant 5′ site to alternative 3′ sites (note that this mode applies to both *sxl* and *tra* splicing, as illustrated). One splicing pattern occurs in both males and females, and results in an RNA that has an early termination codon. The presence of Sxl protein inhibits usage of the upstream 3′ splice site by binding to the polypyrimidine tract at its branch site. When this site is skipped, the next 3′ site is used. This generates a female-specific mRNA that codes for a protein.

Thus Sxl autoregulates its own splicing to ensure its expression in females, and *tra* produces a protein only in females; like Sxl, Tra protein is a splicing regulator. *tra2* has a similar function in females (but is also expressed in the males). The Tra and Tra2 proteins are SR splicing factors that act directly upon the target transcripts. Tra and Tra2 cooperate (in females) to affect the splicing of *dsx*. In the *dsx* gene, females splice the 5′ site of intron 3 to the 3′ site of that intron; as a result translation terminates at the end of exon 4. Males splice the 5′ site of

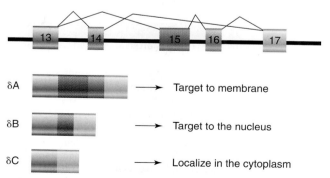

FIGURE 21.22 Alternative splicing of the *CaMKIIδ* gene: different alternative exons target the kinase to different cellular compartments.

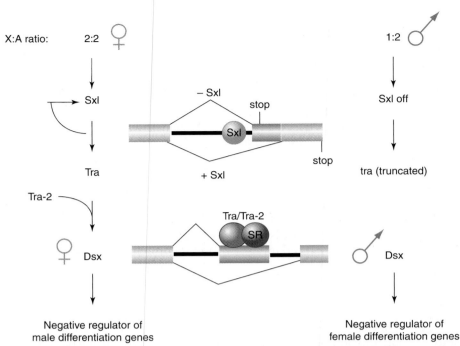

X:A ratio: 2:2 ♀   1:2 ♂

FIGURE 21.23 Sex determination in *D. melanogaster* involves a pathway in which different splicing events occur in females. Blocks at any stage of the pathway result in male development. Illustrated are *tra* pre-mRNA splicing controlled by the Sxl protein, which blocks the use of the alternative 3' splice site, and *dsx* pre-mRNA splicing regulated by both Tra and Tra2 proteins in conjunction with other SR proteins, which positively influence the inclusion of the alternative exon.

intron 3 directly to the 3' site of intron 4, thus omitting exon 4 from the mRNA and allowing translation to continue through exon 6. The result of the alternative splicing is that different Dsx proteins are produced in each sex: the male product blocks female sexual differentiation, whereas the female product represses expression of male-specific genes.

21.13 Splicing Can Be Regulated by Exonic and Intronic Splicing Enhancers and Silencers

Key concepts

- Alternative splicing is often associated with weak splice sites.
- Sequences surrounding alternative exons are often more evolutionarily conserved than sequences flanking constitutive exons.
- Specific exonic and intronic sequences can enhance or suppress splice site selection.
- The effect of splicing enhancers and silencers is mediated by sequence-specific RNA binding proteins, many of which may be developmentally regulated and/or expressed in a tissue-specific manner.
- The rate of transcription can directly affect the outcome of alternative splicing.

Alternative splicing is generally associated with weak splice sites, meaning that the splicing signals located at both ends of introns diverge from the consensus splicing signals. This allows these weak splicing signals to be modulated by various *trans*-acting factors generally known as *alternative splicing regulators*. Contrary to common assumptions, however, these weak splice sites are generally more conserved across mammalian genomes than are constitutive splice sites. This observation argues against the notion that alternative splicing might result from splicing mistakes by the splicing machinery and favors the possibility that many alternative splicing events might be evolutionarily conserved to preserve the regulation of gene expression at the level of RNA processing.

The regulation of alternative splicing is a complex process, involving a large number of RNA-binding *trans*-acting splicing regulators. As illustrated in **FIGURE 21.24**, these RNA-binding proteins may recognize RNA elements in both exons and introns near the alternative splice site and exert both positive and negative influence on the selection of the alternative splice site. Those that bind to exons to enhance the selection are positive splicing regulators and the corresponding *cis*-acting elements are referred to as *exonic*

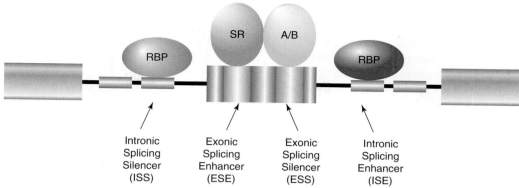

FIGURE 21.24 Exonic and intronic sequences can modulate the splice site selection by functioning as splicing enhancers or silencers. In general, SR proteins bind to exonic splicing enhancers and the hnRNP proteins (such as the A and B families of RNA binding proteins) bind to exonic silencers. Other RNA binding proteins (RBP) can function as splicing regulators by binding to intronic splicing enhancers or silencers.

splicing enhancers (or ESEs). SR proteins are among the best-characterized ESE-binding regulators. In contrast, some RNA binding proteins, such as hnRNP A and B, bind to exonic sequences to suppress splice site selection; the corresponding _cis_-acting elements are thus known as _exonic splicing silencers_ (ESSs). Similarly, many RNA-binding proteins affect splice site selection through intronic sequences. The corresponding positive and negative _cis_-acting elements in introns thus are called _intronic splicing enhancers_ (ISEs) or _intronic splicing silencers_ (ISSs).

Adding to this complexity are the positional effects of many splicing regulators. The best-known examples are the Nova and Fox families of RNA-binding splicing regulators, which can enhance or suppress splice site selection, depending on where they bind relative to the alternative exon. For example, as illustrated in FIGURE 21.25, binding of both Nova and Fox to intronic sequences upstream of the alternative exon generally results in the suppression of the exon, whereas their binding to intronic sequences downstream of the alternative splicing exon frequently enhances the selection of the exon. Both Nova and Fox are differentially expressed in different tissues, particularly in the brain. Thus, tissue-specific regulation of alternative splicing can be achieved by tissue-specific expression of _trans_-acting splicing regulators.

How a specific alternative splicing event is regulated by various positive and negative splicing regulators remains is not completely understood. In principle, these splicing regulators function to enhance or suppress the recognition of specific splicing signals by some of the core components of the splicing ma-

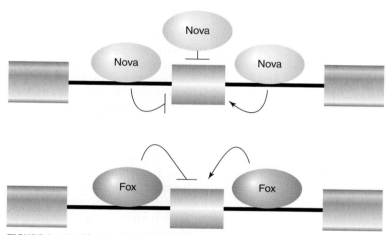

FIGURE 21.25 The Nova and Fox families of RNA binding proteins can promote or suppress splice site selection in a context dependent fashion. Binding of Nova to exons and flanking upstream introns inhibits the inclusion of the alternative exon while Nova binding to the downstream flanking intronic sequences promotes the inclusion of the alternative exon. Fox binding to the upstream intronic sequence inhibits the inclusion of the alternative exon whereas binding of Fox to the downstream intronic sequence promotes the inclusion of the alternative exon.

chinery. The best-understood cases are SR proteins and hnRNA A/B proteins for their positive and negative roles in enhancing or suppressing splice site recognition, respectively. Binding of SR proteins to ESEs promotes or stabilizes U1 binding to the 5' splice site and U2AF binding to the 3' splice site. Thus spliceosome assembly becomes more efficient in the presence of SR proteins. This role of SR proteins applies to both constitutive and alternative splicing, making SR proteins both essential splicing factors and alternative splicing regulators. In contrast, hnRNP A/B proteins seem to bind to RNA and compete with the binding by SR proteins and other core spliceosome components in the recognition of functional splicing signals.

SR proteins are able to commit a pre-mRNA to the splicing pathway, whereas hnRNP proteins antagonize this process. Given that hnRNP proteins are highly abundant in the nucleus, how do SR proteins effectively compete with hnRNPs in the nucleus to facilitate splicing? Apparently, this is accomplished by the cotranscriptional splicing mechanism inside the nucleus of the cell (see *Section 21.7, Commitment of Pre-mRNA to the Splicing Pathway*). It is thus conceivable that the transcription process can affect alternative splicing. This in fact has been shown to be the case. Alternative splicing appears to be affected by specific promoters used to drive gene expression as well as by the rate of transcription during the elongation phase.

Different promoters may attract different sets of transcription factors, which may in turn affect transcriptional elongation. Thus the same mechanism may underlie the influence of promoter usage and transcriptional elongation rate on alternative splicing. The current evidence suggests a kinetic model where a slow transcriptional elongation rate would afford a weak splice site emerging from the elongating Pol II complex sufficient time to pair with the upstream splice site before the appearance of the downstream competing splice site. This model stresses a functional consequence of the coupling between transcription and RNA splicing in the nucleus.

21.14 *trans*-Splicing Reactions Use Small RNAs

Key concepts

- Splicing reactions usually occur only in *cis* between splice junctions on the same molecule of RNA.
- *trans*-splicing occurs in trypanosomes and worms where a short sequence (SL RNA) is spliced to the 5′ ends of many precursor mRNAs.
- SL RNAs have a structure resembling the Sm-binding site of U snRNAs.

In both mechanistic and evolutionary terms, splicing has been viewed as an *intramolecular* reaction, essentially amounting to a controlled deletion of the intron sequences at the level of RNA. In genetic terms, splicing occurs only in *cis*. This means that *only sequences on the same molecule of RNA can be spliced together*.

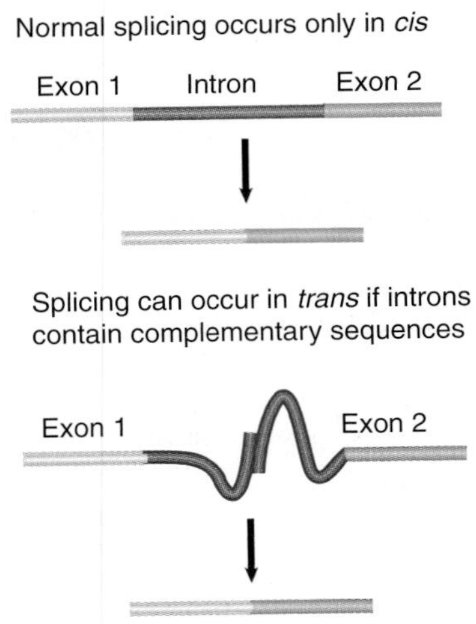

Normal splicing occurs only in *cis*

Splicing can occur in *trans* if introns contain complementary sequences

FIGURE 21.26 Splicing usually occurs only in *cis* between exons carried on the same physical RNA molecule, but *trans*-splicing can occur when special constructs are made that support base pairing between introns.

The upper part of **FIGURE 21.26** shows the normal situation. The introns can be removed from each RNA molecule, allowing the exons of that RNA molecule to be spliced together, but there is no *intermolecular* splicing of exons between different RNA molecules. Although we know that *trans*-splicing between pre-mRNA transcripts of the same gene does occur, it must be exceedingly rare, because if it were prevalent the exons of a gene would be able to complement one another genetically instead of belonging to a single complementation group.

Some manipulations can generate *trans*-splicing. In the example illustrated in the lower part of Figure 21.26, complementary sequences were introduced into the introns of two RNAs. Base pairing between the complements should create an H-shaped molecule. This molecule could be spliced in *cis*, to connect exons that are covalently connected by an intron, or it could be spliced in *trans*, to connect exons of the juxtaposed RNA molecules. Both reactions occur *in vitro*.

Another situation in which *trans*-splicing is possible *in vitro* occurs when substrate RNAs are provided in the form of one containing a 5′ splice site and the other containing a 3′ splice site together with appropriate downstream sequences (which may be either the next 5′ splice site or a splicing enhancer). In effect, this mimics splicing by exon definition (see the right side of Figure 21.11), and shows that *in vitro* it

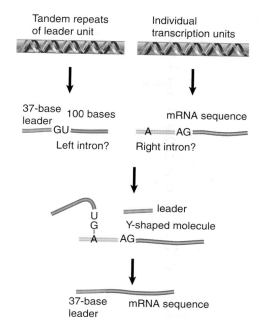

FIGURE 21.27 The SL RNA provides an exon that is connected to the first exon of an mRNA by *trans*-splicing. The reaction involves the same interactions as nuclear *cis*-splicing, but generates a Y-shaped RNA instead of a lariat.

is not necessary for the left and right splice sites to be on the same RNA molecule.

These results show that there is no *mechanistic* impediment to *trans*-splicing. They exclude models for splicing that require processive movement of a spliceosome along the RNA. It must be possible for a spliceosome to recognize the 5′ and 3′ splice sites of different RNAs when they are in close proximity.

Although *trans*-splicing is rare in multicellular eukaryotes, it occurs as the primary mechanism to process precursor RNA into mature, translatable mRNAs in some organisms, such as trypanosomes and nematodes. In trypanosomes, all genes are expressed as polycistronic transcripts like those in bacteria. The transcribed RNA, however, cannot be translated without a 37-nucleotide leader brought in by *trans*-splicing to convert a polycistronic RNA into individual monocistronic mRNAs for translation. The leader sequence is not encoded upstream of the individual transcription units, though. Instead it is transcribed into an independent RNA, carrying additional sequences at its 3′ end, from a repetitive unit located elsewhere in the genome. **FIGURE 21.27** shows that this RNA carries the leader sequence followed by a 5′ splice site sequence. The sequences

encoding the mRNAs carry a 3′ splice site just preceding the sequence found in the mature mRNA.

When the leader and the mRNA are connected by a *trans*-splicing reaction, the 3′ region of the leader RNA and the 5′ region of the mRNA in effect comprise the 5′ and 3′ halves of an intron. When splicing occurs, a 2′–5′ link forms by the usual reaction between the GU of the 5′ intron and the branch sequence near the AG of the 3′ intron. The two parts of the intron are covalently linked, but generate a Y-shaped molecule instead of a lariat.

The RNA that donates the 5′ exon for *trans*-splicing is called the **SL RNA (spliced leader RNA)**. The SL RNAs, which are 100 nucleotides in length, can fold into a common secondary structure that has three stem-loops and a single-stranded region that resembles the Sm-binding site. The SL RNAs therefore exist as snRNPs that count as members of the Sm snRNP class. During the *trans*-splicing reaction, SL RNA becomes part of the spliced product, as illustrated in the upper panel of **FIGURE 21.28**. Like other snRNPs involved in splicing (except U6), SL RNA carries a trimethylated cap, which is recognized by a variant cap binding factor eIF4E to facilitate translation.

In *C. elegans*, about 70% of genes are processed by the *trans*-splicing mechanism, which can be further divided into two classes. One class of gene produces monocistronic transcripts, which are processed by both *cis*- and *trans*-splicing. In these cases, while *cis*-splicing is used to remove internal intronic sequences, *trans*-splicing is employed to provide the 22-nucleotide leader sequence derived from the SL RNA for translation. The other class of gene is polycistronic. In these cases, *trans*-splicing is used to convert the polycistronic transcripts into monocistronic transcripts in addition to providing the SL leader sequence for their translation as illustrated in the bottom panel of Figure 21.28.

There are two types of SL RNA in *C. elegans*. SL1 RNA (the first to be discovered) is only used to remove the 5′ ends of pre-mRNAs transcribed from monocistronic genes. How does the SL RNA find the 3′ splice site to initiate *trans*-splicing, and in doing so, how does *trans*-splicing avoid competition or interference with *cis*-splicing? The ability to target a functional 3′ splice site is provided by the proteins as part of the SL snRNP. For example, purified SL snRNP from *Ascaris*, a parasitic nematode, contains two

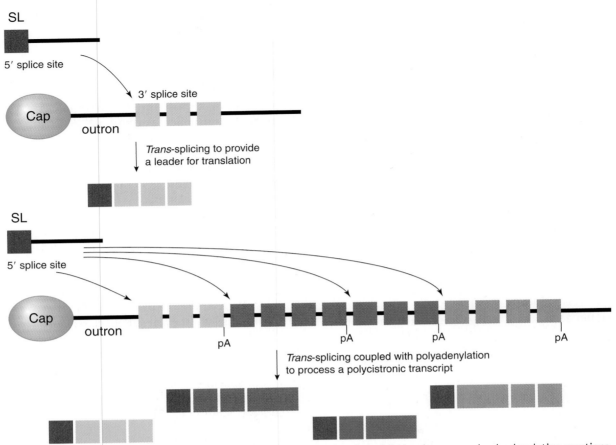

FIGURE 21.28 The SL RNA adds a leader to facilitate translation. Coupled with the cleavage and polyadenylation reactions, the addition of the SL RNA is also used to convert polycistronic transcripts to monocistronic units.

specific proteins, one of which (SL-30kD) can directly interact with the branchpoint binding protein at the 3′ splice site. The SL1 RNA is only *trans*-spliced to the first 5′ untranslated region, but does not interfere with downstream *cis*-splicing events. This is because only the 5′ untranslated region (called an *outron*; see Figure 21.28) contains a functional 3′ splice site, but it does not have the upstream 5′ splice site to pair with the downstream 3′ splice site.

The SL2 RNA is used in most cases to process polycistronic transcripts that are separated by a 100-nucleotide spacer sequence between the two adjacent gene units. In a small fraction of genes where the two adjacent gene units are linked without any spacer sequences, the SL1 RNA is used to break them up.

During processing of these polycistronic transcripts by either of the SL snRNAs, the *trans*-splicing reaction is tightly coupled with the cleavage and polyadenylation reactions at the end of each gene unit. Such coupling appears to be facilitated by direct protein–protein interactions between the SL2 snRNP and the cleavage stimulatory factor CstF that binds to the U-rich sequence downstream the AAUAAA signal (see *Section 21.15, The 3′ Ends of mRNAs Are Generated by Cleavage and Polyadenylation*). These mechanisms allow related genes to be coregulated at the level of transcription (because they are transcribed as polycistronic transcripts) and individually regulated after transcription (because individually gene units are separated as a result of RNA processing).

21.15 The 3' Ends of mRNAs Are Generated by Cleavage and Polyadenylation

Key concepts

- The sequence AAUAAA is a signal for cleavage to generate a 3' end of mRNA that is polyadenylated.
- The reaction requires a protein complex that contains a specificity factor, an endonuclease, and poly(A) polymerase.
- The specificity factor and endonuclease cleave RNA downstream of AAUAAA.
- The specificity factor and poly(A) polymerase add ~200 A residues processively to the 3' end.
- The poly(A) tail controls mRNA stability and influences translation.
- Cytoplasmic polyadenylation plays a role in *Xenopus* embryonic development.

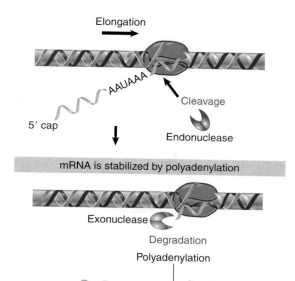

FIGURE 21.29 The sequence AAUAAA is necessary for cleavage to generate a 3' end for polyadenylation.

The 3' ends of Pol II transcribed mRNAs are generated by cleavage followed by polyadenylation. Addition of poly(A) to nuclear RNA can be prevented by the analog 3'–deoxyadenosine, which is also known as *cordycepin*. Although cordycepin does not stop the transcription of nuclear RNA, its addition prevents the appearance of mRNA in the cytoplasm. This shows that polyadenylation is *necessary* for the maturation of mRNA from nuclear RNA. The poly(A) tail is known to protect the mRNA from degradation by 3'–5' exonucleases. In yeast, the poly(A) tail is also suggested to play roles in facilitating nuclear export of matured mRNA and in cap stability.

Generation of the 3' end is illustrated in **FIGURE 21.29**. The RNA polymerase transcribes past the site corresponding to the 3' end, and sequences in the RNA are recognized as targets for an endonucleolytic cut followed by polyadenylation. RNA polymerase continues transcription after the cleavage, but the 5' end that is generated by the cleavage is unprotected, which signals transcriptional termination (see *Section 21.16, The 3' mRNA End Processing Is Critical for Transcriptional Termination*).

The site of cleavage/polyadenylation in most pre-mRNAs is flanked by two *cis*-acting signals: an upstream AAUAAA motif, which is usually located 11 to 30 nucleotides from the site, and a downstream U-rich or GU-rich element. The AAUAAA is needed for both cleav-

age and polyadenylation because deletion or mutation of the AAUAAA hexamer prevents generation of the polyadenylated 3' end.

The development of a system in which polyadenylation occurs *in vitro* opened the route to analyzing the reactions. The formation and functions of the complex that undertakes 3' processing are illustrated in **FIGURE 21.30**. Generation of the proper 3' terminal structure depends on the *cleavage and polyadenylation specific factor* (CPSF), which contains multiple subunits. One of the subunits binds directly to the AAUAAA motif and to the *cleavage stimulatory factor* (CstF), which is also a multicomponent complex. One of these components binds directly to a downstream GU-rich sequence. CPSF and CstF can enhance each other in recognizing the polyadenylation signals. The specific enzymes involved are an *endonuclease* (the 73kD subunit of CPSF) to cleave the RNA and a **poly(A) polymerase (PAP)** to synthesize the poly(A) tail.

The poly(A) polymerase has a nonspecific catalytic activity. When it is combined with the other components, the synthetic reaction becomes specific for RNA containing the sequence AAUAAA. The polyadenylation reaction passes through two stages. First, a rather short oligo(A) sequence (~10 residues) is added

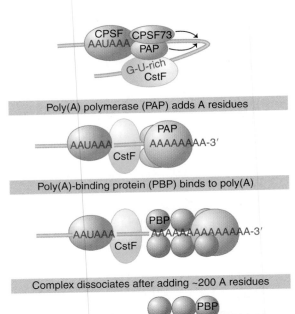

Poly(A) polymerase (PAP) adds A residues

Poly(A)-binding protein (PBP) binds to poly(A)

Complex dissociates after adding ~200 A residues

FIGURE 21.30 The 3′ processing complex consists of several activities. CPSF and CstF each consist of several subunits; the other components are monomeric. The total mass is >900 kD.

to the 3′ end. This reaction is absolutely dependent on the AAUAAA sequence, and poly(A) polymerase performs it under the direction of the specificity factor. In the second phase, the nuclear poly(A) binding protein (PABP II) binds the oligo(A) tail to allow extension of the poly(A) tail to the full ~200 residue length. The poly(A) polymerase by itself adds A residues individually to the 3′ position. Its intrinsic mode of action is distributive; it dissociates after each nucleotide has been added. In the presence of CPSF and PABP II, however, it functions processively to extend an individual poly(A) chain. After the polyadenylation reaction, PABP II binds stoichiometrically to the poly(A) stretch, which by some unknown mechanism limits the action of poly(A) polymerase to ~200 additions of A residues.

Upon export of mature mRNAs out of the nucleus, the poly(A) tail is bound by the cytoplasmic poly(A) binding protein (PABP I). PABP I not only protects the mRNA from degradation by the 3′ to 5′ exonucleases, but also binds to the translation initiation factor eIF4G to facilitate translation of the mRNA. Thus the mRNA in the cytoplasm forms a closed loop in which a protein complex contains both the 5′ and 3′ ends of the mRNA (see Figure 24.22 in *Section 24.9, Eukaryotes Use a Complex of Many Initia-*

tion Factors). Polyadenylation therefore affects both stability and initiation of translation in the cytoplasm.

During embryonic development of *Xenopus*, polyadenylation is carried out in the cytoplasm to provide a maternal control in early embryogenesis. Some stored maternal mRNAs may either be polyadenylated by the poly(A) polymerase in the cytoplasm to stimulate translation or deadenylated to terminate translation. A specific AU-rich *cis*-acting element (CPE) in the 3′ tail directs the meiotic maturation-specific polyadenylation in the cytoplasm to activate translation of some specific maternal mRNAs. To regulate mRNA degradation, there are at least two types of *cis*-acting sequences found in the 3′ tail that can trigger mRNA deadenylation: EDEN (embryonic deadenylation element) is a 17-nucleotide sequence and ARE elements are AU-rich, usually containing tandem repeats of AUUUA. A poly(A)-specific RNAase (PARN) is involved in mRNA degradation in the cytoplasm. Of course, mRNA deadenylation is always in competition with mRNA stabilization, which together determine the half-life of individual mRNAs in the cell (see Chapter 22, *mRNA Stability and Localization*).

21.16 The 3′ mRNA End Processing Is Critical for Transcriptional Termination

Key concepts

- There are various ways to end transcription by different RNA polymerases.
- The mRNA 3′ end formation signals termination of Pol II transcription.

Information about the termination reaction for eukaryotic RNA polymerases is less detailed than our knowledge of initiation. 3′ ends of RNAs can be generated in two ways. Some RNA polymerases terminate transcription at a defined (terminator) sequence in DNA, as shown in **FIGURE 21.31**. RNA polymerase III appears to use this strategy by having a discrete oligo(dT) sequence to signal the release of Pol III for transcription termination.

For RNA polymerase I, the sole product of transcription is a large precursor that contains the sequences of the major rRNA. Termination occurs at two discrete sites (T1 and

T2) downstream of the mature 3' end. These terminators are recognized by a specific DNA-binding Reb1p in yeast or TTF1 in mice. Pol I termination is also associated with a cleavage event mediated by the endonuclease Rnt1p, which cleaves the nascent RNA about 15 to 50 bases downstream from the 3' end of processed 28S rRNA (see *Section 21.20, Production of rRNA Requires Cleavage Events and Involves Small RNAs*). In this regard, Pol I termination is mechanistically related to Pol II termination in that both processes may involve an RNA cleavage event.

In contrast to Pol I and Pol III termination, RNA polymerase II usually does not show discrete termination, but continues to transcribe about 1.5 kb past the site corresponding to the 3' end. The cleavage event at the polyadenylation site provides a trigger for termination by RNA polymerase II, as shown in **FIGURE 21.32**.

Two models have been proposed for Pol II termination. The *allosteric model* suggests that RNA cleavage at the polyadenylation site may trigger some conformational changes in both the Pol II complex and local chromatin structure. This may be induced by factor exchanges during the polyadenylation reaction, resulting in Pol II pausing and then release from template DNA.

An alternative model known as the *torpedo model* proposes that a specific exonuclease binds to the 5' end of the RNA that is continuing to be transcribed after cleavage. It degrades the RNA faster than it is synthesized, so that it catches up with RNA polymerase. It then interacts with ancillary proteins that are bound to the carboxy-terminal domain of the polymerase; this interaction triggers the release of RNA polymerase from DNA, causing transcription to terminate. This model explains why the termination sites for RNA polymerase II are not well defined, but may occur at varying locations within a long region downstream of the site corresponding to the 3' end of the RNA. The major experimental evidence for the torpedo model is the role of the nuclear 5'–3' exonuclease Rat1 in yeast or Xrn2 in mammals. Deletion of the gene frequently causes readthrough transcription to the next gene. In some experimental systems, though, mutation of the AAUAAA signal to impair cleavage at the natural polyadenylation site does not necessarily trigger the release of the transcribing Pol II and cause transcriptional readthrough. This evidence, coupled with some local changes in chromatin structure, thus favors the allosteric model.

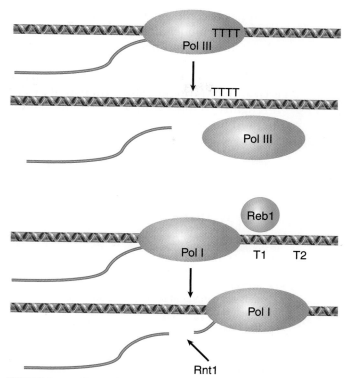

FIGURE 21.31 Transcription by Pol I and Pol III uses specific terminators to end transcription.

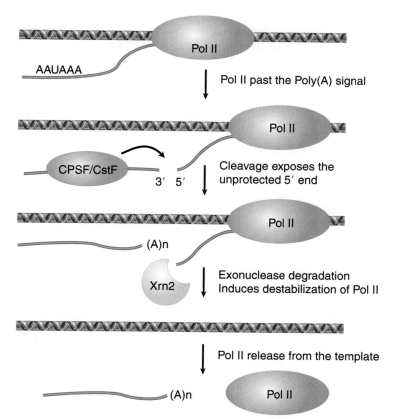

FIGURE 21.32 3' end formation of Pol II transcripts facilitates transcriptional termination.

It has become apparent that the allosteric and torpedo models are not necessarily mutually exclusive; both may reflect some critical aspects associated with Pol II transcriptional termination. By either or both mechanisms, it is clear that transcriptional termination by Pol II is tightly coupled with the 3′ end formation for most mRNAs in eukaryotic cells.

21.17 The 3′ End Formation of Histone mRNA Requires U7 snRNA

Key concepts

- The expression of histone mRNAs is replication dependent and is regulated during the cell cycle.
- Histone mRNAs are not polyadenylated; their 3′ ends are generated by a cleavage reaction that depends on the structure of the mRNA.
- The cleavage reaction requires the SLBP to bind to a stem-loop structure and the U7 snRNA to pair with an adjacent single-stranded region.
- The cleavage reaction is catalyzed by a factor shared with the polyadenylation complex.

Histone biogenesis is primarily controlled by the regulation of histone mRNA abundance during the cell cycle. At this G1/S transition, the abundance of histone mRNAs is increased >30-fold due to elevated transcription; this process is regulated by the Cyclin E/Cdk2 complex. The rise in histone mRNAs is followed by a rapid decay of histone mRNAs at the end of S phase.

Histone mRNAs are not polyadenylated (except in *S. cerevisiae*). The formation of their 3′ ends is therefore different from that of the coordinated cleavage/polyadenylation reaction. Formation of their 3′ ends depends upon a highly conserved stem-loop structure located 14 to 50 bases downstream from the termination codon and a <u>h</u>istone <u>d</u>ownstream <u>e</u>lement (HDE) located ~15 nucleotides downstream of the stem-loop. Cleavage occurs between the stem-loop and HDE, leaving five bases downstream of the stem-loop. Mutations that prevent formation of the duplex stem of the stem-loop prevent formation of the end of the RNA. Secondary mutations that restore duplex structure (though not necessarily the original sequence) restore 3′ end formation. This indicates that *formation of the secondary structure is more important than the exact sequence.*

The histone 3′ end formation reaction is shown in **FIGURE 21.33**. Two factors are required to specify the cleavage reaction: the stem-loop binding protein (SLBP) recognizes the stem-loop structure, and the 5′ end of U7 snRNA base pairs with a purine-rich sequence within HDE. U7 snRNP is a minor snRNP consisting of the 63-nucleotide U7 snRNA and a set of several proteins related to snRNPs involved in mRNA splicing (see *Section 21.6, snRNAs Are Required for Splicing*). Unique to U7 snRNP are two Sm-like proteins, LSM10 and LSM11, which replace Sm D1 and D2 in the splicing snRNPs. Prevention of base pairing between U7 snRNA and HDE impairs 3′ processing of the histone mRNAs,

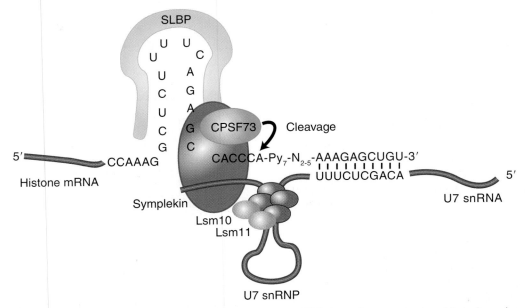

FIGURE 21.33 Generation of the 3′ end of histone H3 mRNA depends on a conserved hairpin and a sequence that base pairs with U7 snRNA.

and compensatory mutations in U7 snRNA that restore complementarity restore 3' processing. This indicates that U7 snRNA functions by base pairing with the histone mRNAs.

Cleavage to generate a 3' terminus occurs at a fixed distance from the site recognized by U7 snRNA, which suggests that the snRNA is involved in defining the cleavage site. The factor responsible for cleavage is a specific cleavage and *polyadenylation specificity factor* (CPSF73). Thus, this member of the metallo-β-lactamase family plays a key role in 3' end formation for both polyadenylated mRNAs and nonpolyadenylated histone mRNAs. Several other proteins have been identified to be important for histone 3' end formation, including CPSF100 and Symplekin, but their specific roles remain to be defined. These additional proteins may provide scaffold functions to stabilize the 3' end processing complex.

Interestingly, disruption of U7 base pairing with the target sequences in histone genes or siRNA-mediated depletion of other components involved in the histone 3' end formation all result in transcriptional readthrough and polyadenylation by using a poly(A) signal downstream from the DHE. Thus, similar to the role of mRNA cleavage/polyadenylation in Pol II transcriptional termination on most protein-coding genes, U7-mediated RNA cleavage during 3' end formation appears to be critical for transcriptional termination on histone genes.

21.18 tRNA Splicing Involves Cutting and Rejoining in Separate Reactions

Key concepts

- RNA polymerase III terminates transcription in poly(U)$_4$ sequence embedded in a GC-rich sequence.
- tRNA splicing occurs by successive cleavage and ligation reactions.
- An endonuclease cleaves the tRNA precursors at both ends of the intron.
- Release of the intron generates two half-tRNAs with unusual ends that contain 5' hydroxyl and 2'–3' cyclic phosphate.
- The 5'–OH end is phosphorylated by a polynucleotide kinase, the cyclic phosphate group is opened by phosphodiesterase to generate a 2'–phosphate terminus and 3'–OH group, exon ends are joined by an RNA ligase, and the 2'–phosphate is removed by a phosphatase.

Most splicing reactions depend on short consensus sequences and occur by transesterification reactions in which breaking and making of bonds is coordinated. The splicing of tRNA genes is achieved by a different mechanism that relies upon separate cleavage and ligation reactions.

Some 59 of the 272 nuclear tRNA genes in the yeast *S. cerevisiae* are interrupted. Each has a single intron that is located just one nucleotide beyond the 3' side of the anticodon. The introns vary in length from 14 to 60 bases. Those in related tRNA genes are related in sequence, but the introns in tRNA genes representing different amino acids are unrelated. *There is no consensus sequence that could be recognized by the splicing enzymes.* This is also true of interrupted nuclear tRNA genes of plants, amphibians, and mammals.

All the introns include a sequence that is complementary to the anticodon of the tRNA. This creates an alternative conformation for the anticodon arm in which the anticodon is base paired to form an extension of the usual arm. An example is shown in **FIGURE 21.34**. Only the anticodon arm is affected—the rest of the molecule retains its usual structure.

The exact sequence and size of the intron is not important. Most mutations in the intron do not prevent splicing. *Splicing of tRNA depends principally on recognition of a common secondary structure in tRNA rather than a common sequence of the intron.* Regions in various parts of the molecule are important, including the stretch between the acceptor arm and D arm, in the TψC arm, and especially the anticodon arm. This is reminiscent of the structural demands

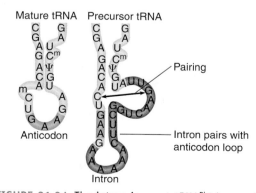

FIGURE 21.34 The intron in yeast tRNA^Phe base pairs with the anticodon to change the structure of the anticodon arm. Pairing between an excluded base in the stem and the intron loop in the precursor may be required for splicing.

placed on tRNA for translation (see Chapter 24, *Translation*).

The intron is not entirely irrelevant, however. Pairing between a base in the intron loop and an unpaired base in the stem is required for splicing. Mutations at other positions that influence this pairing (for example, to generate alternative patterns for pairing) influence splicing. The rules that govern availability of tRNA precursors for splicing resemble the rules that govern recognition by aminoacyl-tRNA synthetases (see *Section 25.9, tRNAs Are Selectively Paired with Amino Acids by Aminoacyl-tRNA Synthetases*).

In a temperature-sensitive mutant of yeast that fails to remove the introns, the interrupted precursors accumulate in the nucleus. The precursors can be used as substrates for a cell-free system extracted from wild-type cells. The splicing of the precursor can be followed by virtue of the resulting size reduction. This is seen by the change in position of the band on gel electrophoresis, as illustrated in **FIGURE 21.35**. The reduction in size can be accounted for by the appearance of a band representing the intron.

The cell-free extract can be fractionated by assaying the ability to splice the tRNA. The *in vitro* reaction requires ATP. Characterizing the reactions that occur with and without ATP shows that the *two separate stages of the reaction are catalyzed by different enzymes.*

- The first step does not require ATP. It involves phosphodiester bond cleavage by an atypical nuclease reaction. It is catalyzed by an endonuclease.

- The second step requires ATP and involves bond formation; it is a ligation reaction, and the responsible enzyme activity is described as an **RNA ligase**.

Splicing of pre-tRNA to remove introns is essential in all organisms, but different organisms use different mechanisms to accomplish pre-tRNA splicing. In bacteria, introns in pre-tRNAs are self-spliced as group I or group II autocatalytic introns. In Archaea and Eukarya, pre-tRNA splicing involves the action of three enzymes: (1) an endonuclease that recognizes and cleaves the precursor at both ends of the intron, a ligase that joins the tRNA exons, and 2' phosphotransferase that removes the 2'-phosphate on spliced tRNA.

The yeast endonuclease is a heterotetrameric protein consisting of two catalytic subunits, Sen34 and Sen2, and two structural subunits, Sen54 and Sen15. Its activities are illustrated in **FIGURE 21.36**. The related subunits Sen34 and Sen2 cleave the 3' and 5' splice sites, respectively. Subunit Sen54 may determine the sites of cleavage by "measuring" distance from a point in the tRNA structure. This point is in the elbow of the (mature) L-shaped structure. The role of subunit Sen15 is not known, but its gene is essential in yeast. The base pair that forms between the first base in the anticodon loop and the base preceding the 3' splice site is required for 3' splice site cleavage.

An interesting insight into the evolution of tRNA splicing is provided by the endonucleases of Archaea. These are homodimers or homotetramers, in which each subunit has an active site (although only two of the sites func-

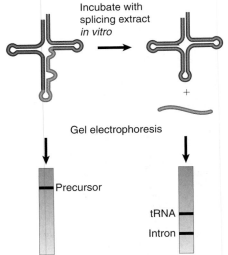

FIGURE 21.35 Splicing of yeast tRNA *in vitro* can be followed by assaying the RNA precursor and products by gel electrophoresis.

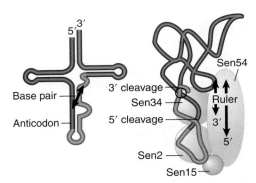

FIGURE 21.36 The 3' and 5' cleavages in *S. cerevisiae* pre-tRNA are catalyzed by different subunits of the endonuclease. Another subunit may determine location of the cleavage sites by measuring distance from the mature structure. The AI base pair is also important.

tion in the tetramer) that cleaves one of the splice sites. The subunit has sequences related to the sequences of the active sites in the Sen34 and Sen2 subunits of the yeast enzyme. The archaeal enzymes recognize their substrates in a different way, though. Instead of measuring distance from particular sequences, they recognize a structural feature called the bulge-helix-bulge. **FIGURE 21.37** shows that cleavage occurs

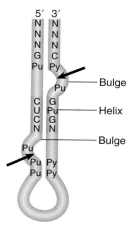

FIGURE 21.37 Archaeal tRNA splicing endonuclease cleaves each strand at a bulge in a bulge-helix-bulge motif.

in the two bulges. Thus the origin of splicing of tRNA precedes the separation of the Archaea and the eukaryotes. If it originated by insertion of the intron into tRNAs, this must have been a very ancient event.

The overall tRNA splicing reaction is summarized in **FIGURE 21.38**. The products of cleavage are a linear intron and two half-tRNA molecules. These intermediates have unique ends. Each 5′ terminus ends in a hydroxyl group; each 3′ terminus ends in a 2′,3′–cyclic phosphate group.

The two half-tRNAs base pair to form a tRNA-like structure. When ATP is added, the second reaction occurs, which is catalyzed by a single enzyme with multiple enzymatic activities.

1. *Cyclic phosphodiesterase activity.* Both of the unusual ends generated by the endonuclease must be altered prior to the ligation reaction. The cyclic phosphate group is first opened to generate a 2′–phosphate terminus.
2. *Kinase activity.* The product has a 2′–phosphate group and a 3′–OH group. The 5′–OH group generated by the endonuclease must be phosphorylated to give a 5′–phosphate. This generates

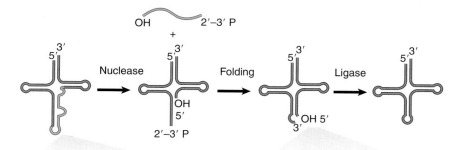

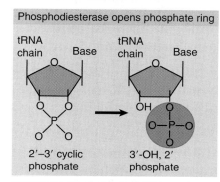

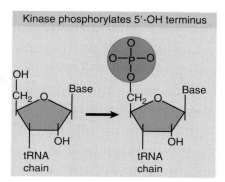

FIGURE 21.38 Splicing of tRNA requires separate nuclease and ligase activities. The exon–intron boundaries are cleaved by the nuclease to generate 2′ to 3′ cyclic phosphate and 5′ OH termini. The cyclic phosphate is opened to generate 3′-OH and 2′ phosphate groups. The 5′-OH is phosphorylated. After releasing the intron, the tRNA half molecules fold into a tRNA-like structure that now has a 3′-OH, 5′-P break. This is sealed by a ligase.

a site in which the 3'–OH is next to the 5'–phosphate.

3. *Ligase activity*. Covalent integrity of the polynucleotide chain is then restored by ligase activity. The spliced molecule is now uninterrupted, with a 5'–3' phosphate linkage at the site of splicing, but it also has a 2'-phosphate group marking the event on the spliced tRNA. In the last step, this surplus group is removed by a phosphatase, which transfers the 2'-phosphate to NDP to form ADP ribose 1',2'-cyclic phosphate.

The tRNA splicing pathway described above is slightly different from that of vertebrates. Before the action of the RNA ligases, a cyclase generates a cyclic 2',3' cyclic terminus from the initial 3'-phosphomonoester terminus via a 3' adenylalated intermediate. The RNA ligase is also different from that in yeast because it can join a 2',3'-cyclic phosphodiester and a 5'-OH to form a conventional 3',5'-phosphodiester bond, but these reactions leave no extra 2'-phosphate.

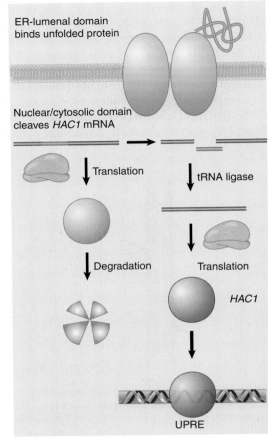

FIGURE 21.39 The unfolded protein response occurs by activating special splicing of *HAC1* mRNA to produce a transcription factor that recognizes the UPRE.

21.19 The Unfolded Protein Response Is Related to tRNA Splicing

Key concepts

- Ire1 is an inner nuclear membrane protein with its N-terminal domain in the ER lumen and its C-terminal domain in the nucleus; the C-terminal domain exhibits both kinase and endonuclease activities.
- Binding of an unfolded protein to the N-terminal domain activates the C-terminal endonuclease by autophosphorylation.
- The activated endonuclease cleaves *HAC1* (*Xbp1* in vertebrates) mRNA to release an intron and generate exons that are ligated by a tRNA ligase.
- Only spliced *HAC1* mRNA can be translated to a transcription factor that activates genes coding for chaperones that help to fold unfolded proteins.
- Activated Ire1 induces apoptosis when the cell is over stressed by unfolded proteins.

An unusual splicing system that is related to tRNA splicing is the <u>u</u>nfolded <u>p</u>rotein <u>r</u>esponse (UPR) pathway conserved from yeast to mammals. As summarized in **FIGURE 21.39**, the accumulation of unfolded proteins in the lumen of the endoplasmic reticulum (ER) triggers the UPR pathway. This leads to increased transcription of genes encoding chaperones that assist protein folding in the ER. A signal must therefore be transmitted from the lumen of the ER to the nucleus.

The sensor that activates the pathway is the inositol-requiring protein Ire1, which is localized in the ER and/or inner nuclear membrane. The N-terminal domain of Ire1 lies in the lumen of the ER where it detects the presence of unfolded proteins, presumably by binding to exposed motifs. The C-terminal half of Ire1 is located in either the cytoplasm or nucleus (because of the continuous membrane of the ER and the nucleus) and exhibits both Ser/Thr kinase activity and a specific endonuclease activity. Binding of unfolded proteins causes aggregation of Ire1 monomers on the ER membrane, leading to the activation of the C-terminal domain on the other side of the membrane by autophosphorylation.

The activated C-terminal endonuclease has, at present, only one (though important) substrate, which is the mRNA encoding the UPR-specific transcription factor Hac1 in yeast (Xbp1 in vertebrates). Under normal conditions, when the UPR pathway is not activated, *HAC1* mRNA contains a 252-nucleotide intron (Xbp1 contains a 26-nucleotide intron). The intron in *HAC1* prevents the mRNA from being translated into a functional protein in yeast whereas in mammalian cells the intron in *Xbp1* allows translation, but the protein is rapidly degraded by the proteosome. Unusual splicing components are involved in processing this intron. The activated Ire1 endonuclease acts directly on *HAC1* mRNA (*Xbp1* mRNA in vertebrates) to cleave the two splicing junctions, leaving 2′,3′-cyclic phosphate at the 3′ end of the 5′ exon and 5′-OH at the 5′ end of the 3′ exon. The two junctions are then ligated by the tRNA ligase that acts in the tRNA splicing pathway. Thus, the entire pathway for processing *HAC1* (*Xbp1*) pre-mRNA resembles the pre-tRNA pathway.

There are important differences between the two pathways, however. Ire1 and tRNA endonuclease share no sequence homology or subunit composition. The endonuclease activity of Ire1 is highly regulated in the ER and has only one substrate (*HAC1* pre-mRNA). In contrast, tRNA endonuclease has many substrates, all with common tRNA folding, with little preference on sequences surrounding the splice junctions.

By using such tRNA-like pathway to remove the intron in the *HAC1* (*Xbp1*) mRNA, the mature mRNA can be translated to produce a potent basic-leucine zipper (bZIP) transcription factor to bind to a common motif (UPRE) in the promoter of many downstream genes. The gene products protect the cell by increasing the expression of proteins to assist protein folding.

If the UPR system is overwhelmed by unfolded proteins, the activated kinase domain of Ire1 binds to the TRAF2 adaptor molecule in the cytoplasm to activate the apoptosis pathway and kill the cell. Thus, the cell uses an unusual tRNA processing strategy to respond to unfolded proteins. There is, however, no apparent relationship between the Ire1 endonuclease and the tRNA splicing endonuclease, so it is not obvious how this specialized system would have evolved.

21.20 Production of rRNA Requires Cleavage Events and Involves Small RNAs

Key concepts

- RNA polymerase I terminates transcription at an 18-base terminator sequence.
- The large and small rRNAs are released by cleavage from a common precursor rRNA; the 5S rRNA is separately transcribed.
- The C/D group of snoRNAs is required for modifying the 2′ position of ribose with a methyl group.
- The H/ACA group of snoRNAs is required for converting uridine to pseudouridine.
- In each case the snoRNA base pairs with a sequence of rRNA that contains the target base to generate a typical structure that is the substrate for modification.

The major rRNAs are synthesized as part of a single primary transcript that is processed to generate the mature products. The precursor contains the sequences of the 18S, 5.8S, and 28S rRNAs. (The nomenclature of different ribosomal RNAs is based on early sedimentation studies conducted on sucrose gradients in the 1970s.) In multicellular eukaryotes, the precursor is named for its sedimentation rate as *45S RNA*. In unicellular/oligocellular eukaryotes it is smaller (35S in yeast).

The mature rRNAs are released from the precursor by a combination of cleavage events and trimming reactions to remove both external transcribed spacers (ETS) and internal transcribed spacers (ITS). **FIGURE 21.40** shows the general pathway in yeast. There can be variations in the order of events, but basically similar reactions are involved in all eukaryotes. Most of the 5′ ends are generated directly by a cleavage event. Most of the 3′ ends are generated by cleavage followed by a 3′–5′ trimming reaction. These processes are specified by many *cis*-acting RNA motifs in both ETSs and ITSs and are acted upon by >150 processing factors.

Many ribonucleases have been implicated in processing rRNA, including some specific components of the exosome, which is an assembly of several exonucleases that also participates in mRNA degradation (see *Section 22.5, Most Eukaryotic mRNA Is Degraded via Two Deadenylation-Dependent Pathways*). Mutations in individual enzymes usually do not prevent processing, which suggests that their activities are redundant and that different combinations

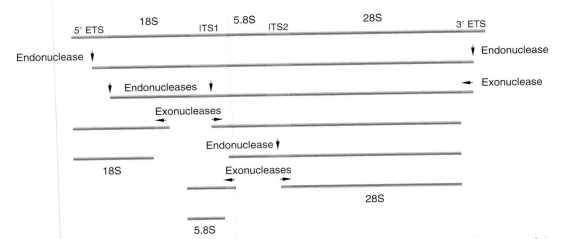

FIGURE 21.40 Mature eukaryotic rRNAs are generated by cleavage and trimming events from a primary transcript.

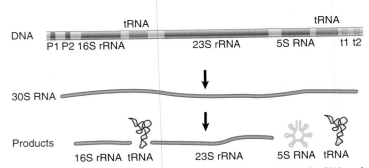

FIGURE 21.41 The *rrn* operons in *E. coli* contain genes for both rRNA and tRNA. The exact lengths of the transcripts depend on which promoters (P) and terminators (t) are used. Each RNA product must be released from the transcript by cuts on either side.

of cleavages can be used to generate the mature molecules.

There are always multiple copies of the transcription unit for the rRNAs. The copies are organized as tandem repeats (see *Section 7.3, Genes for rRNA Form Tandem Repeats Including an Invariant Transcription Unit*). The genes coding for rRNAs are transcribed by RNA polymerase I in the nucleolus. In contrast, 5S RNA is transcribed from separate genes by RNA polymerase III. In general, the 5S genes are clustered, but are separated from the genes for the major rRNAs.

There is a difference in the organization of the precursor in bacteria. The sequence corresponding to 5.8S rRNA forms the 5′ end of the large (23S) rRNA; that is, there is no processing between these sequences. FIGURE 21.41 shows that the precursor also contains the 5S rRNA and one or two tRNAs. In *E. coli*, the seven *rrn* operons are dispersed around the genome; four *rrn* loci contain one tRNA gene between the 16S and 23S rRNA sequences, and the other *rrn* loci contain two tRNA genes in this region. Addi-

tional tRNA genes may or may not be present between the 5S sequence and the 3′ end. Thus the processing reactions required to release the products depend on the content of the particular *rrn* locus.

In both prokaryotic and eukaryotic rRNA processing, both processing factors and ribosomal proteins (and possibly other proteins) bind to the precursor, so that the substrate for processing is not the free RNA, but rather a ribonucleoprotein complex. Like pre-mRNA processing, rRNA processing takes place cotranscriptionally. As a result, the processing factors are intertwined with ribosomal proteins in building the ribosomes, instead of first processing and then stepwise assembly on processed rRNAs.

Processing and modification of rRNA requires a class of small RNAs called snoRNAs (small nucleolar RNAs). There are hundreds of snoRNAs in *S. cerevisiae* and vertebrate genomes. Some of these snoRNAs are encoded by individual genes; others are expressed from polycistrons, and many are derived from introns of their host genes. These snoRNAs themselves undergo complex processing and maturation steps. Some snoRNAs are required for cleavage of the precursor to rRNA; one example is U3 snoRNA, which is required for the first cleavage event. The U3-containing complex corresponds to the "terminal knobs" at the 5′ end of nascent rRNA transcripts, which are visible under an electron microscope. We do not know what role the snoRNA plays in cleavage. It could be required to pair with specific rRNA sequences to form a secondary structure that is recognized by an endonuclease.

Two groups of snoRNAs are required for the modifications that are made to bases in the

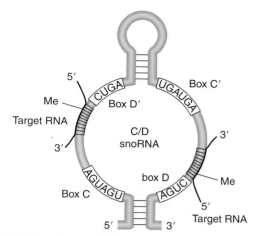

FIGURE 21.42 A snoRNA base pairs with a region of rRNA that is to be methylated.

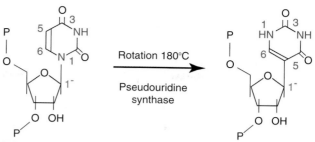

FIGURE 21.43 Uridine is converted to pseudouridine by replacing the N1-sugar bond with a C5-sugar bond and rotating the base relative to the sugar.

rRNA. The members of each group are identified by very short conserved sequences and common features of secondary structure.

The C/D group of snoRNAs is required for adding a methyl group to the 2' position of ribose. There are >100 2'–O–methyl groups at conserved locations in vertebrate rRNAs. This group takes its name from two short conserved sequences motifs called boxes C and D. Each snoRNA contains a sequence near the D box that is complementary to a region of the 18S or 28S rRNA that is methylated. Loss of a particular snoRNA prevents methylation in the rRNA region to which it is complementary.

FIGURE 21.42 shows that the snoRNA base pairs with the rRNA to create the duplex region that is recognized as a substrate for methylation. Methylation occurs within the region of complementarity at a position that is fixed five bases on the 5' side of the D box. It is likely that each methylation event is specified by a different snoRNA; ~40 snoRNAs have been implicated in this modification. Each C+D box snoRNA is associated with three proteins Nop1p (fibrillarin in vertebrates), Nop56p, and Nop58p. The methylase(s) have not been fully characterized, although the major snoRNP protein Nop1p/fibrillarin is structurally similar to methyltransferases.

Another group of snoRNAs is involved in base modification by converting uridine to pseudouridine. There are ~50 residues in yeast rRNAs and ~100 in vertebrate rRNAs that are modified by pseudouridination. The pseudouridination reaction is shown in FIGURE 21.43, in which the N1 bond from uridylic acid to ribose is broken, the base is rotated, and C5 is rejoined to the sugar.

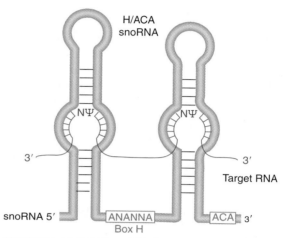

FIGURE 21.44 H/ACA snoRNAs have two short conserved sequences and two hairpin structures, each of which has regions in the stem that are complementary to rRNA. Pseudouridine is formed by converting an unpaired uridine within the complementary region of the rRNA.

Pseudouridine formation in rRNA requires the H/ACA group of snoRNAs. They are named for the presence of an ACA triplet three nucleotides from the 3' end and a partially conserved sequence (the H box) that lies between two stem-loop hairpin structures. Each of these snoRNAs has a sequence complementary to rRNA within the stem of each hairpin. FIGURE 21.44 shows the structure that would be produced by pairing with the rRNA. Within each pairing region, there are two unpaired bases, one of which is a uridine that is converted to pseudouridine.

The H/ACA snoRNAs are associated with four specific nucleolar proteins, Cbf5p (dyskerin in vertebrates), Nhp2p, Nop10p, and Gar1p. Importantly, Cbf5p/dyskerin is structurally similar to known pseudouridine synthases, and thus it likely provides the enzymatic activity in the snoRNA-guided pseudouridination reaction. Many snoRNAs are also used to guide base modifications in tRNAs as well as in snRNAs involved in pre-mRNA splicing, which are critical for their functions in prospective reactions. There is, however, a large number of snoRNAs that do not have apparent targets. These snoRNAs are called *orphan RNAs*. The existence of these orphan RNAs indicates that many biological processes may use RNA-guided mechanisms to functionally modify other expressed RNAs in a more diverse fashion than we currently understand.

21.21 Summary

Splicing accomplishes the removal of introns and the joining of exons into the mature sequence of RNA. There are at least four types of reaction, as distinguished by their requirements *in vitro* and the intermediates that they generate. The systems include eukaryotic nuclear introns, group I and group II introns, and tRNA introns. Each reaction involves a change of organization within an individual RNA molecule, and is therefore a *cis*-acting event.

Pre-mRNA splicing follows preferred but not obligatory pathways. Only very short consensus sequences are necessary; the rest of the intron appears largely irrelevant. Both exonic and intronic sequences can exert positive or negative influence on the selection of the nearby splice site, though. All 5' splice sites are probably equivalent, as are all 3' splice sites. The required sequences are given by the GU-AG rule, which describes the ends of the intron. The UACUAAC branch site of yeast, or a less well-conserved consensus in mammalian introns, is also required. The reaction with the 5' splice site involves formation of a lariat that joins the GU end of the intron via a 2'–5' linkage to the A at position 6 of the branch site. The 3'–OH end of the exon then attacks the 3' splice site, so that the exons are ligated and the intron is released as a lariat. Lariat formation is responsible for choice of the 3' splice site. Both reactions are transesterifications in which phosphodiester bonds are conserved. Several stages of the reaction require hydrolysis of ATP, probably to drive conformational changes in the RNA and/or protein components. Alternative splicing patterns are caused by protein factors that either stimulate use of a new site or that block use of the default site.

Pre-mRNA splicing requires formation of a spliceosome—a large particle that assembles the consensus sequences into a reactive conformation. The spliceosome forms by the process of intron definition, involving recognition of the 5' splice site, branch site, and 3' splice site. This applies to small introns, like those in yeast. If, however, introns are large, like those in vertebrates, recognition of the splice sites first follows the process of exon definition, involving the interactions across the exon between the 3' splice site and the downstream 5' splice site. This is then switched to paired interactions across the intron for later steps of spliceosome assembly. By either intron definition or exon definition, the initial process of splice site recognition commits the pre-mRNA substrate to the splicing pathway. The pre-mRNA complex contains U1 snRNP and a number of key protein splicing factors, including U2AF and the branch site binding factor. In multicellular eukaryotic cells, the formation of the commitment complex requires the participation of SR proteins.

The spliceosome contains the U1, U2, U4/U6, and U5 snRNPs, as well as a large number of splicing factors. The U1, U2, and U5 snRNPs each contain a single snRNA and several proteins; the U4/U6 snRNP contains two snRNAs and several proteins. Some proteins are common to all snRNP particles. U1 snRNA base pairs with the 5' splice site, U2 snRNA base pairs with the branch sequence, and U5 snRNP holds the 5' and 3' splice sites together *via* a looped sequence within the spliceosome. When U4 releases U6, the U6 snRNA base pairs with the 5' splice site and U2, which remains base paired with the branch sequence; this may create the catalytic center for splicing. An alternative set of snRNPs provides analogous functions for splicing the U12-dependent subclass of introns. The catalytic core resembles that in group II autocatalytic introns; as a result, it is likely that the spliceosome is a giant RNA machine (like the ribosome) in which key RNA elements are at the center of the reaction.

Splicing is usually intramolecular, but *trans*-splicing (intermolecular splicing) occurs in trypanosomes and nematodes. It involves a reaction between a small SL RNA and the pre-mRNA. In worms there are two types of SL RNA: one is used for splicing to the 5' end of an mRNA; the other is used for splicing to an internal site to break up the polycistronic precursor RNA. The introduction of the SL RNA to the processed mRNAs provides necessary signals for translation.

The termination capacity of RNA polymerase II is tightly linked to 3' end formation of the mRNA. The sequence AAUAAA, located 11 to 30 bases upstream of the cleavage site, provides the signal for both cleavage by an endonuclease and polyadenylation by the poly(A) polymerase. This is enhanced by the complex bound on the G/U-rich element downstream the cleavage site. Transcription is terminated when an exonuclease, which binds to the 5' end of the nascent RNA chain created by the cleavage, catches up to RNA polymerase.

All Pol II transcripts are polyadenylated with the exception of histone mRNAs, which neither contains an intron nor receives a poly(A) tail. The 3' end formation of histone mRNA depends on a stem-loop structure and base pairing of a downstream element with U7 snRNA to instrument a cleavage. The stem-loop structure may protect the end, as in bacteria.

tRNA splicing involves separate endonuclease and ligase reactions. The endonuclease recognizes the secondary (or tertiary) structure of the precursor and cleaves both ends of the intron. The two half-tRNAs released by loss of the intron can be ligated by the tRNA ligase in the presence of ATP. This tRNA maturation pathway is exploited by the unfolded protein response pathway in the ER.

rRNA processing takes place in the nucleolus where U3 snRNA initiates a series of actions of endonucleases and exonucleases to cut and trim extra sequences in the precursor rRNA to produce individual ribosomal RNAs. Hundreds to thousands of noncoding RNAs are expressed in eukaryotic cells. In the nucleolus, two groups of such noncoding RNAs, termed snoRNAs, are responsible for pairing with rRNAs at sites that are modified. Group C/D snoRNAs identify target sites for methylation, and group H/ACA snoRNAs specify sites where uridine is converted to pseudouridine.

References

21.1 Introduction

Review

Lewin, B. (1975). Units of transcription and translation: sequence components of hnRNA and mRNA. *Cell* 4, 77–93.

21.2 The 5' End of Eukaryotic mRNA Is Capped

Review

Bannerjee, A. K. (1980). 5' terminal cap structure in eukaryotic mRNAs. *Microbiol. Rev.* 44, 175–205.

Research

Mandal, S. S. et al. (2004). Functional interactions of RNA-capping enzyme with factors that positively and negatively regulated promoter escape by RNA polymerase II. *Proc. Natl. Acad. Sci. USA* 101, 7572–7577.

McCracken, S. et al. (1997). 5'-capping enzymes are targeted to pre-mRNA by binding to the phosphorylated carboxy-terminal domain of RNA polymerase II. *Genes Dev.* 11, 3306–3318.

21.3 Nuclear Splice Junctions Are Short Sequences

Reviews

Padgett, R. A. (1986). Splicing of messenger RNA precursors. *Annu. Rev. Biochem.* 55, 1119–1150.

Sharp, P. A. (1987). Splicing of mRNA precursors. *Science* 235, 766–771.

Sharp, P. A. and Burge, C. B. (1997). Classification of introns: U2-type or U12-type. *Cell* 91, 875–879.

Research

Graveley, B. R. (2005). Mutually exclusive splicing of the insect Dscam pre-mRNA directed by competing intronic RNA secondary structures. *Cell* 123, 65–73.

Krainer, A. R. et al. (1984). Normal and mutant human β-globin pre-mRNAs are accurately and efficiently spliced *in vitro*. *Cell* 36, 993–1005.

21.5 Pre-mRNA Splicing Proceeds through a Lariat

Review

Sharp, P. A. (1994). Split genes and RNA splicing. *Cell* 77, 805–815.

Research

Reed, R. and Maniatis, T. (1985). Intron sequences involved in lariat formation during pre-mRNA splicing. *Cell* 41, 95–105.

Ruskin, B. et al. (1984). Excision of an intact intron as a novel lariat structure during pre-mRNA splicing *in vitro*. *Cell* 38, 317–331.

21.6 snRNAs Are Required for Splicing

Reviews

Guthrie, C. (1991). Messenger RNA splicing in yeast: clues to why the spliceosome is a ribonucleoprotein. *Science* 253, 157–163.

Guthrie, C. and Patterson, B. (1988). Spliceosomal snRNAs. *Annu. Rev. Genet.* 22, 387–419.

Maniatis, T. and Reed, R. (1987). The role of small nuclear ribonucleoprotein particles in pre-mRNA splicing. *Nature* 325, 673–678.

Research

Black, D. L. et al. (1985). U2 as well as U1 small nuclear ribonucleoproteins are involved in premessenger RNA splicing. *Cell* 42, 737–750.

Black, D. L. and Steitz, J. A. (1986). Pre-mRNA splicing *in vitro* requires intact U4/U6 small nuclear ribonucleoprotein. *Cell* 46, 697–704.

Grabowski, P. J., Seiler, S. R., and Sharp, P. A. (1985). A multicomponent complex is involved in the splicing of messenger RNA precursors. *Cell* 42, 345–353.

Krainer, A. R. and Maniatis, T. (1985). Multiple components including the small nuclear ribonucleoproteins U1 and U2 are required for pre-mRNA splicing *in vitro*. *Cell* 42, 725–736.

21.7 Commitment of Pre-mRNA to the Splicing Pathway

Reviews

Berget. S. M. (1995). Exon recognition in vertebrate splicing. *J. Biol. Chem.* 270, 2411–2414.

Fu, X-D. (1995). The superfamily of arginine/serine-rich splicing factors. *RNA* 1, 663–680.

Reed, R. (1996). Initial splice-site recognition and pairing during pre-mRNA splicing. *Curr. Opin. Genet. Dev.* 6, 215–220.

Research

Abovich, N. and Rosbash, M. (1997). Cross-intron bridging interactions in the yeast commitment complex are conserved in mammals. *Cell* 89, 403–412.

Berglund, J. A. et al. (1997). The splicing factor BBP interacts specifically with the pre-mRNA branchpoint sequence UACUAAC. *Cell* 89, 781–787.

Fu, X.-D. (1993). Specific commitment of different pre-mRNA to splicing single SR proteins. *Nature* 365, 82–85.

Hoffman, B. E. and Grabowski, P. J. (1992). U1 snRNP targets an essential splicing factor, U2AF65, to the 3′ splice site by a network of interactions spanning the exon. *Genes Dev.* 6, 2554–2568.

Ibrahim, E. C. et al. (2005). Serine/arginine-rich protein-dependent suppression of exon skipping by exonic splicing enhancers. *Proc. Natl. Acad. Sci. USA* 102, 5002–5007.

Kohtz, J. D. et al. (1994). Protein-protein interactions and 5′ splice-site recognition in mammalian mRNA precursors. *Nature* 368, 119–124.

Robberson, B. L. and Berget, S. M. (1990). Exon definition may facilitate splice site selection in RNAs with multiple exons. *Mol. Cell Biol.* 10, 84–94.

Wu, J. Y. and Maniatis, T. (1993). Specific interactions between proteins implicated in splice site selection and regulated alternative splicing. *Cell* 75, 1061–1070.

21.8 The Spliceosome Assembly Pathway

Reviews

Burge, C. B., Tushl, T. H., and Sharp, P. A. (1999). Splicing of precursors to mRNAs by the spliceosome. In Gesteland, R. F., and Atkins, J. F. eds. *The RNA World, 2nd ed.*, Cold Spring Harbor Laboratory Press, Plainview, NY, pp. 525–560.

Research

Cheng, S. C. and Abelson, J. (1987). Spliceosome assembly in yeast. *Genes Dev.* 1, 1014–1027.

Konarska, M. M. and Sharp, P. A. (1987). Interactions between small nuclear ribonucleoprotein particles in formation of spliceosomes. *Cell* 49, 736–774.

Newman, A. and Norman, C. (1991). Mutations in yeast U5 snRNA alter the specificity of 5′ splice site cleavage. *Cell* 65, 115–123.

Tseng, C. K. and Cheng, S. C. (2008). Both catalytic steps of nuclear pre-mRNA splicing are reversible. *Science* 320, 1782–1784.

Zhuang, Y. and Weiner, A. M. (1986). A compensatory base change in U1 snRNA suppresses a 5′ splice site mutation. *Cell* 46, 827–835.

21.9 An Alternative Spliceosome Uses Different snRNPs to Process the Minor Class of Introns

Research

Burge, C. B., Padgett, R. A., and Sharp, P. A. (1998). Evolutionary fates and origins of U12-type introns. *Mol. Cell* 2, 773–785.

Dietrich, R. C., Incorvaia, R., and Padgett, R. A. (1997). Terminal intron dinucleotide sequences do not distinguish between U2- and U12-dependent introns. *Mol. Cell* 1, 151–160.

Hall, S. L. and Padgett, R. A. (1994). Conserved sequences in a class of rare eukaryotic introns with non-consensus splice sites. *J. Mol. Biol.* 239, 357–365.

Tarn, W.-Y. and Steitz, J. A. (1996). A novel spliceosome containing U11, U12, and U5 snRNPs excises a minor class AT-AC intron *in vitro*. *Cell* 84, 801–811.

Tarn, W.-Y. and Steitz, J. A. (1996). Highly diverged U4 and U6 small nuclear RNAs required for splicing rare AT-AC introns. *Science* 273, 1824–1832.

21.10 Pre-mRNA Splicing Likely Shares the Mechanisms with Group II Autocatalytic Introns

Reviews

Madhani, H. D. and Guthrie, C. (1994). Dynamic RNA-RNA interactions in the spliceosome. *Annu. Rev. Genet.* 28, 1–26.

Michel, F. and Ferat, J.-L. (1995). Structure and activities of group II introns. *Annu. Rev. Biochem.* 64, 435–461.

Research

Madhani, H. D. and Guthrie, C. (1992). A novel base-pairing interaction between U2 and U6 snRNAs suggests a mechanism for the catalytic activation of the spliceosome. *Cell* 71, 803–817.

21.11 Splicing Is Temporally and Functionally Coupled with Multiple Steps in Gene Expression

Reviews

Maniatis, T. and Reed, R. (2002). An extensive network of coupling among gene expression machines. *Nature* 416, 499–506.

Maquat, L. E. (2004). Nonsense-mediated mRNA decay: splicing, translation and mRNAdynamics. *Nature Rev. Mol. Cell Biol.* 5, 89–99.

Pandit, S., Wang, D., and Fu, X-D. (2008). Functional integration of transcriptional and RNA processing machineries. *Curr. Opin. Cell Biol.* 20, 260–265.

Proudfoot, N. J., Furger, A., and Dye, M. J. (2002). Integrating mRNA processing with transcription. *Cell* 108, 501–512.

Research

Cheng, H. et al. (2006). Human mRNA export Machinery recruited to the 5′ end of mRNA. *Cell* 127, 1389–1400.

Das, R. et al. (2007). SR proteins function in coupling RNAP II transcription to pre-mRNA splicing. *Mol. Cell* 26, 867–881.

Le Hir, H. et al. (2000). The spliceosome deposits multiple proteins 20–24 nucleotides upstream of mRNA exon-exon junctions. *EMBO J.* 19, 6860–6869.

Lin, S. et al. (2008). The splicing factor SC35 has an active role in transcriptional elongation. *Nature Struc. Mol. Biol.* 15, 819–826.

Luo, M. L. et al. (2001). Pre-mRNA splicing and mRNA export linked by direct interactions between UAP56 and Aly. *Nature* 413, 644–647.

Zhou, Z., et al. (2000). The protein Aly links pre-messenger-RNA splicing to nuclear export in metazoans. *Nature* 407, 401–405.

21.12 Alternative Splicing Is a Rule, Rather Than an Exception, in Multicellular Eukaryotes

Review

Black, D. (2003). Mechanisms of alternative pre-messenger RNA splicing. *Annu. Rev. Biochem.* 72, 291–336.

Research

Ge, H. and Manley, J. L. (1990). A protein, ASF, controls cell-specific alternative splicing of SV40 early pre-mRNA *in vitro*. *Cell* 62, 25–34.

Krainer, A. R. et al. (1990). The essential pre-mRNA splicing factor SF2 influences 5′ splice site selection by activating proximal sites. *Cell* 62, 35–42.

Lynch, K. W. and Maniatis, T. (1996). Assembly of specific SR protein complexes on distinct regulatory elements of the *Drosophila* doublesex splicing enhancer. *Genes Dev.* 10, 2089–2101.

Tian, M. and Maniatis, T. (1993). A splicing enhancer complex controls alternative splicing of doublesex pre-mRNA. *Cell* 74, 105–114.

Wang, E. T. et al. (2008). Alternative isoform regulation in human tissue transcriptomes. *Nature* 456, 470–476.

Xu, X.-D. et al. (2005). ASF/SF2-regulated CaMKIIdelta alternative splicing temporally reprograms excitation-contraction coupling in cardiac muscle. *Cell* 120, 59–72.

21.13 Splicing Can be Regulated by Exonic and Intornic Splicing Enhancers and Silencers

Review

Blencowe, B. J. (2006). Alternative splicing: New Insights from global analysis. *Cell* 126, 37–47.

Research

Cramer, P. et al. (1999). Coupling of transcription with alternative splicing: RNA Pol II promoters modulate SF2/ASF and 9G8 effects on an exonic splicing enhancer. *Mol. Cell* 4, 251–258.

de la Mata, M. et al. (2003). A slow RNA polymerase II affects alternative splicing *in vivo*. *Mol. Cell* 12, 525–532.

Fairbrother, W. G. et al. (2002). Predictive identification of exonic splicing enhancers in human genes. *Science* 297, 1007–1113.

Locatalosi, D. D. et al. (2008). HITS-CLIP yields genome-wide insights into brain alternative RNA processing. *Nature* 456, 464–470.

Sharma, S. et al. (2005). Polypyrimidine tract binding protein blocks the 5′ splice site-dependent assembly of U2AF and the prespliceosome E complex. *Mol. Cell* 19, 485–496.

Wang, Z. et al. (2004). Systematic identification and analysis of exonic splicing silencers. *Cell* 119, 831–845.

Yeo, G. et al. (2008). An RNA code for the Fox2 splicing regulator revealed by mapping RNA-protein interactions in stem cells. *Nature Struc. Mol. Biol.* 16, 130–137.

Zhang, X. H. and Chasin, L. A. (2004). Computational definition of sequence motifs governing constitute exon splicing. *Genes Dev.* 18, 1241–1250.

Zhu, J. et al. (2001). Exon identity established through differential antagonism between exonic splicing silencer-bound hnRNP A1 and enhancer-bound SR proteins. *Mol. Cell* 8, 1351–1361.

21.14 trans-Splicing Reactions Use Small RNAs

Review

Nilsen, T. (1993). *Trans*-splicing of nematode pre-mRNA. *Annu. Rev. Immunol.* 47, 413–440.

Research

Blumenthal, T. et al. (2002). A global analysis of *C. elegans* operons. *Nature* 417, 851–854.

Denker, J. A. et al. (2002). New components of the spliced leader RNP required for nematode *trans*-splicing. *Nature* 417, 667–670.

Fischer, S. E. J., Butler, M. D., Pan, Q., and Ruvkun, G. (2008). *trans*-Splicing in *C. elegans* generates the negative RNAi regulator ERI-6/7. *Nature* 455, 491–496.

Hannon, G. J. et al. (1990). *trans*-splicing of nematode pre-mRNA *in vitro*. *Cell* 61, 1247–1255.

Huang, X. Y. and Hirsh, D. (1989). A second *trans*-spliced RNA leader sequence in the nematode *C. elegans*. *Proc. Natl. Acad. Sci. USA* 86, 8640–8644.

Krause, M. and Hirsh, D. (1987). A *trans*-spliced leader sequence on actin mRNA in *C. elegans*. *Cell* 49, 753–761.

Murphy, W. J., Watkins, K. P., and Agabian, N. (1986). Identification of a novel Y branch structure as an intermediate in trypanosome mRNA processing: evidence for *trans*-splicing. *Cell* 47, 517–525.

Sutton, R. and Boothroyd, J. C. (1986). Evidence for *trans*-splicing in trypanosomes. *Cell* 47, 527–535.

21.15 The 3′ Ends of mRNAs Are Generated by Cleavage and Polyadenylation

Reviews

Colgan, D. F. and Manley, J. L. (1997). Mechanism and regulation of mRNA polyadenylation. *Genes Dev.* 11, 2755–2766.

Shatkin, A. J. and Manley, J. L. (2000). The ends of the affair: Capping and polyadenylation. *Nature Struct. Biol.* 7, 838–842.

Wahle, E. and Keller, W. (1992). The biochemistry of 3′-end cleavage and polyadenylation of messenger RNA precursors. *Annu. Rev. Biochem.* 61, 419–440.

Research

Conway, L. and Wickens, M. (1985). A sequence downstream of AAUAAA is required for formation of SV40 late mRNA 3′ termini in frog oocytes. *Proc. Natl. Acad. Sci. USA* 82, 3949–3953.

Fox, C. A., Sheets, M. D., and Wickens, M. P. (1989). Poly(A) addition during maturation of frog oocytes: distinct nuclear and cytoplasmic activities and regulation by the sequence UUUUUAU. *Genes Dev.* 3, 2151–2162.

Gil, A. and Proudfoot, N. (1987). Position-dependent sequence elements downstream of AAUAAA are required for efficient rabbit β-globin mRNA 3′ end formation. *Cell* 49, 399–406.

Karner, C. G. et al. (1998). The deadenylating nuclease (DAN) is involved in poly(A) tail removal during the meiotic maturation of *Xenopus* oocytes. *EMBO J.* 17, 5427–5437.

McGrew, L. L. et al. (1989). Poly(A) elongation during *Xenopus* oocyte maturation is required for translational recruitment and is mediated by a short sequence element. *Genes Dev.* 3, 803–815.

Takagaki, Y., Ryner, L. C., and Manley, J. L. (1988). Separation and characterization of a poly(A) polymerase and a cleavage/specificity factor required for pre-mRNA polyadenylation. *Cell* 52, 731–742.

21.16 The 3′ End Processing Is Critical for Transcriptional Termination

Review

Buratowski, S. (2005). Connection between mRNA 3′ end processing and transcription termination. *Curr. Opin. Cell Biol.* 17, 257–261.

Research

Dye, M. J. and Proudfoot, N. J. (1999). Terminal exon definition occurs cotranscriptionally and promotes termination of RNA polymerase II. *Mol. Cell* 3, 371–378.

Kim, M. et al. (2004). The yeast Rat1 exonuclease promotes transcription termination by RNA polymerase II. *Nature* 432, 517–522.

Luo, W., Johnson, A. W., and Bentley, D. L. (2006). The role of Rat1 in coupling mRNA 3′ end processing to transcription termination: implications for a unified allosteric-torpedo model. *Genes Dev.* 20, 954–965.

21.17 Cleavage of the 3′ End of Histone mRNA Requires the U7 snRNA

Review

Marzluff, W. F., Wagner, E. J., and Duronio, R. J. (2008). Metabolism and regulation of canonical histone mRNAs: life without a poly(A) tail. *Nature Rev. Genet.* 9, 843–854.

Research

Dominski, Z., Yang, X. C. and Marzluff, W. F. (2005). The polyadenylation factor CPSF73 is involved in histone pre-mRNA processing. *Cell* 123, 37–48.

Kolev, N. G. and Steitz, J. A. (2005). Symplekin and multiple other polyadenylation factors participate in 3′ end maturation of histone mRNAs. *Genes Dev.* 19, 283–2592.

Mowry, K. L. and Steitz, J. A. (1987). Identification of the human U7 snRNP as one of several factors involved in the 3' end maturation of histone premessenger RNAs. *Science* 238, 1682–1687.

Pillar, R. S. et al. (2003). Unique Sm core structure of U7 snRNPs: assembly by a specialized SMN complex and the role of a new component, Lsm 11, in histone RNA processing. *Genes Dev.* 17, 2321–2333.

Wang, Z. F. et al. (1996). The protein that binds the 3' end of histone mRNA: a novel RNA-binding protein required for histone pre-mRNA processing. *Genes Dev.* 10, 3028–3040.

21.18 tRNA Splicing Involves Cutting and Rejoining in Separate Reactions

Research

Diener, J. L. and Moore, P. B. (1998). Solution structure of a substrate for the archaeal pre-tRNA splicing endonucleases: the bulge-helix-bulge motif. *Mol. Cell* 1, 883–894.

Di Nicola Negri, E. et al. (1997). The eucaryal tRNA splicing endonuclease recognizes a tripartite set of RNA elements. *Cell* 89, 859–866.

Reyes, V. M. and Abelson, J. (1988). Substrate recognition and splice site determination in yeast tRNA splicing. *Cell* 55, 719–730.

Trotta, C. R. et al. (1997). The yeast tRNA splicing endonuclease: a tetrameric enzyme with two active site subunits homologous to the archaeal tRNA endonucleases. *Cell* 89, 849–858.

21.19 The Unfolded Protein Response Is Related to tRNA Splicing

Review

Lin, J. H., Walter, P., and Benedict Yen, T. S. (2008). Endoplasmic reticulum stress in disease pathogenesis. *Annu. Rev. Pathol. Mech. Dis.* 3, 399–425.

Research

Gonzalez, T. N. et al. (1999). Mechanism of non-spliceosomal mRNA splicing in the unfolded protein response pathway. *EMBO J.* 18, 3119–3132.

Sidrauski, C., Cox, J. S., and Walter, P. (1996). tRNA ligase is required for regulated mRNA splicing in the unfolded protein response. *Cell* 87, 405–413.

Sidrauski, C. and Walter, P. (1997). The transmembrane kinase Ire1p is a site-specific endonuclease that initiates mRNA splicing in the unfolded protein response. *Cell* 90, 1031–1039.

21.20 Production of rRNA Requires Cleavage Events and Involves Small RNAs

Review

Alessandro, F. and Tollervey, D. (2002). Making ribosomes. *Curr. Opin. Cell. Biol.* 14, 313–318.

Filipowicz, W. and Pogacic, V. (2002). Biogenesis of small nucleolar ribonucleoproteins. *Curr. Opin. Cell. Biol.* 14, 319–327.

Granneman, S. and Baserga, S. L. (2005). Crosstalk in gene expression: coupling and co-regulation of rDNA transcription, pre-ribosome assembly and pre-rRNA processing. *Curr. Opin. Cell Biol.* 17, 281–286.

Matera, A. G., Terns, R. M., and Terns, M. P. (2007). Non-coding RNAs: lessons from the small nuclear and small nucleolar RNAs. *Nature Rev. Mol. Cell Biol.* 8, 209–220.

Research

Balakin, A. G., Smith, L., and Fournier, M. J. (1996). The RNA world of the nucleolus: two major families of small RNAs defined by different box elements with related functions. *Cell* 86, 823–834.

Bousquet-Antonelli, C. et al. (1997). A small nucleolar RNP protein is required for pseudouridylation of eukaryotic ribosomal RNAs. *EMBO J.* 16, 4770–4776.

Ganot, P., Bortolin, M. L., and Kiss, T. (1997). Site-specific pseudouridine formation in preribosomal RNA is guided by small nucleolar RNAs. *Cell* 89, 799–809.

Ganot, P., Caizergues-Ferrer, M., and Kiss, T. (1997). The family of box ACA small nucleolar RNAs is defined by an evolutionarily conserved secondary structure and ubiquitous sequence elements essential for RNA accumulation. *Genes Dev.* 11, 941–956.

Kass, S. et al. (1990). The U3 small nucleolar ribonucleoprotein functions in the first step of preribosomal RNA processing. *Cell* 60, 897–908.

Kiss-Laszlo, Z., Henry, Y., and Kiss, T. (1998). Sequence and structural elements of methylation guide snoRNAs essential for site-specific ribose methylation of pre-rRNA. *EMBO J.* 17, 797–807.

Kiss-Laszlo, Z. et al. (1996). Site-specific ribose methylation of preribosomal RNA: a novel function for small nucleolar RNAs. *Cell* 85, 1077–1068.

Ni, J., Tien, A. L., and Fournier, M. J. (1997). Small nucleolar RNAs direct site-specific synthesis of pseudouridine in rRNA. *Cell* 89, 565–573.

22

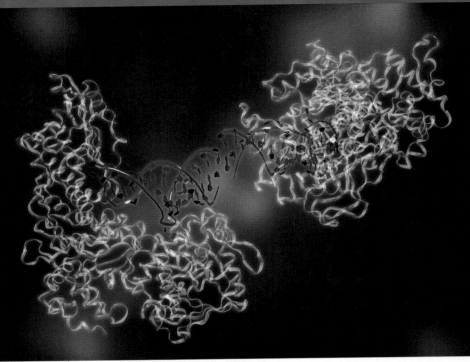

© Equinox Graphics/Photo Researchers, Inc.

mRNA Stability and Localization

Edited by Ellen Baker

22.1 Introduction

RNA is a central player in gene expression. It was first characterized as an intermediate in protein synthesis, but since then many other RNAs have been discovered that play structural or functional roles at other stages of gene expression. The involvement of RNA in many functions concerned with gene expression supports the general view that the entire process may have evolved in an "RNA world" in which RNA was originally the active component in maintaining and expressing genetic information. Many of these functions were subsequently assisted or taken over by proteins, with a consequent increase in versatility and probably efficiency. The focus in this chapter is messenger RNA (mRNA). The functions of other cellular RNAs are discussed in other chapters: snRNAs and snoRNAs in Chapter 21, *RNA Splicing and Processing*; tRNA and rRNA in Chapter 24, *Translation*; and miRNAs and siRNAs in Chapter 30, *Regulatory RNA*; the subset of RNAs

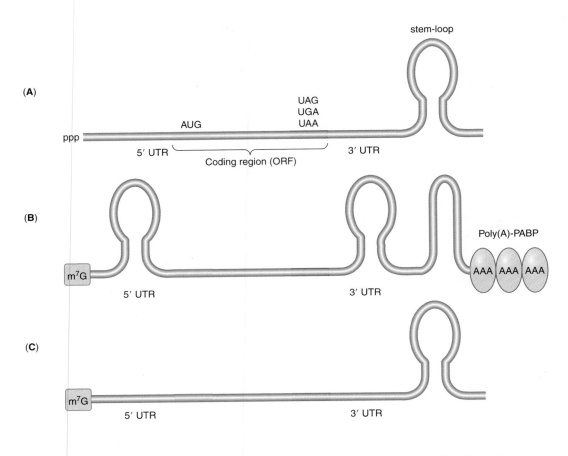

FIGURE 22.1 Features of prokaryotic and eukaryotic mRNAs. (a) A typical bacterial mRNA. This is a monocistronic mRNA, but bacterial mRNAs may also be polycistronic. Many bacterial mRNAs end in a terminal stem-loop. (b) All eukaryotic mRNAs begin with a cap (m^7G) and almost all end with a poly(A) tail. The poly(A) tail is coated with poly(A)-binding proteins (PABPs). Eukaryotic mRNAs may have one or more regions of secondary structure, typically in the 5′ and 3′ UTRs. (c) The major histone mRNAs in mammals have a 3′ terminal stem-loop in place of a poly(A) tail.

that have retained ancestral catalytic activity are discussed in Chapter 23, *Catalytic RNA*.

Messenger RNA plays the principal role in the expression of protein-coding genes. Each mRNA molecule carries the genetic code for synthesis of a specific polypeptide during the process of translation. An mRNA carries much more information as well: it may also carry information for how frequently it will be translated, how long it is likely to survive, and where in the cell it will be translated. This information is carried in the form of RNA *cis*-elements and associated proteins. Much of this information is located in parts of the mRNA sequence that are not directly involved in encoding protein.

FIGURE 22.1 shows some of the structural features typical of mRNAs in prokaryotes and eukaryotes. Bacterial mRNA termini are not modified after transcription, so they begin with the 5′ triphosphate nucleotide used in initiation

of transcription, and end with the final nucleotide added by RNA polymerase before termination. The 3′ end of many of *E. coli* mRNAs form a hairpin structure involved in intrinsic (rho-independent) transcription termination (see Chapter 19, *Prokaryotic Transcription*). Eukaryotic mRNAs are cotranscriptionally capped and polyadenylated (see Chapter 21, *RNA Splicing and Processing*). Most of the nonprotein-coding regulatory information is carried in the **5′** and **3′ untranslated regions (UTR)** of an mRNA, but some elements are present in the coding region. While all mRNAs are linear sequences of nucleotides, secondary and tertiary structures can be formed by intramolecular base-pairing. These structures can be simple, like the **stem-loop** structures illustrated in the figure, or more complex, involving branched structures or pairing of nucleotides from distant regions of the molecule. Investigation of the mechanisms by which

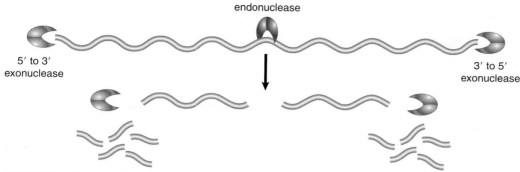

FIGURE 22.2 Types of ribonucleases. Exonucleases are unidirectional. They can digest RNA either from the 5' end or from the 3' end, liberating individual ribonucleotides. Endonucleases cleave RNA at internal phosphodiester linkages. An endonuclease usually targets specific sequences and/or secondary structures.

mRNA regulatory information is deciphered and acted upon by machinery responsible for mRNA degradation, translation, and localization is an important field in molecular biology today.

22.2 Messenger RNAs Are Unstable Molecules

Key concepts

- mRNA instability is due to the action of ribonucleases.
- Ribonucleases differ in their substrate preference and mode of attack.
- mRNAs exhibit a wide range of half-lives.
- Differential mRNA stability is an important contributor to mRNA abundance and therefore the spectrum of proteins made in a cell.

Messenger RNAs are relatively unstable molecules, unlike DNA and, to a lesser extent, rRNAs and tRNAs. While it is true that the phosphodiester bonds connecting ribonucleotides are somewhat weaker than those connecting deoxyribonucleotides due to the presence of the 2' hydroxyl group on the ribose sugar, this is not the primary reason for the instability of mRNA. Rather, cells contain a myriad of RNA degrading enzymes, called **ribonucleases**, some of which specifically target mRNA molecules.

Ribonucleases are enzymes that cleave the phosphodiester linkage connecting RNA ribonucleotides. They are diverse molecules because many different protein domains have evolved to have ribonuclease activity. The rare examples of known ribozymes (catalytic RNAs) include multiple ribonucleases, indicating the ancient origins of this important activity (see Chapter 23, *Catalytic RNA*). Ribonucleases, often

just called nucleases when the RNA nature of the substrate is obvious, have many roles in a cell, including participation in DNA replication, DNA repair, processing of new transcripts (including pre-mRNAs, tRNAs, rRNAs, snRNAs, and miRNAs) and the degradation of mRNA. Ribonucleases are either **endoribonucleases** or **exoribonucleases**, as depicted in **FIGURE 22.2** (and as discussed in *Section 3.2, Nucleases*). Endonucleases cleave an RNA molecule at an internal site, and may have a requirement or preference for a certain structure or sequence. Exonucleases remove nucleotides from an RNA terminus, and have a defined polarity of attack—either 5' to 3' or 3' to 5'. Some exonucleases are **processive**, remaining engaged with the substrate while sequentially removing nucleotides, while others are **distributive**, catalyzing the removal of only one or a few nucleotides before dissociating from the substrate.

Most mRNAs decay stochastically (like the decay of radioactive isotopes), and as a result mRNA stability is usually expressed as a **half-life ($t_{1/2}$)**. The term **mRNA decay** is often used interchangeably with mRNA degradation. mRNA-specific stability information is encoded in *cis*-sequences (see *Section 22.7, mRNA-Specific Half-Lives Are Controlled by Sequences or Structures within the mRNA*) and is therefore characteristic of each mRNA. Different mRNAs can exhibit remarkably different stabilities, varying by 100-fold or more. In *E. coli* the typical mRNA half-life is about three minutes, but half-lives of individual mRNAs may be as short as 20 seconds and as long as 90 minutes. In budding yeast, mRNA half-lives range from 3 to 100 minutes, whereas in metazoans, half-lives range from minutes to hours, and in rare cases, even days. Abnormal mRNAs can be targeted for very rapid

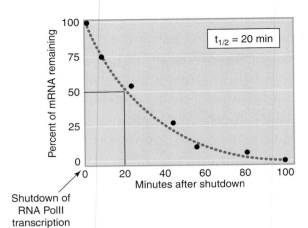

Shutdown of
RNA PolII
transcription

FIGURE 22.3 Method for determining mRNA half-lives. RNA polymerase II transcription is shut down, either by a drug or a temperature shift in strains with a temperature-sensitive mutation in a PolII gene. The levels of specific mRNAs are determined by northern blot or RT-PCR at various times following shut-down. RNA degradation, once initiated, is usually so rapid that intermediates in the process are not detectable. The half-life is the time required for the mRNA to fall to one half of its initial value.

destruction (see *Section 22.8, Newly Synthesized RNAs Are Checked for Defects via a Nuclear Surveillance System* and *Section 22.9, Quality Control of mRNA Translation Is Performed by Cytoplasmic Surveillance Systems*). Half-life values are generally determined by some version of the method illustrated in **FIGURE 22.3**.

The abundance of specific mRNAs in a cell is a consequence of their combined rates of synthesis (transcription and processing) and degradation. mRNA levels reach a **steady state** when these parameters remain constant. The spectrum of proteins synthesized by a cell is largely a reflection of the abundance of their mRNA templates (although differences in translational efficiency play a role). The importance of mRNA decay is highlighted by large-scale studies that have examined the relative contributions of decay rate and transcription rate to differential mRNA abundance. Decay rate predominates. The great advantage of unstable mRNAs is the ability to rapidly change the output of translation through changes in mRNA synthesis. Clearly this advantage is important enough to compensate for the seeming wastefulness of making and destroying mRNAs so quickly. Abnormal control of mRNA stability has been implicated in disease states, including cancer, chronic inflammatory responses, and coronary disease.

22.3 Eukaryotic mRNAs Exist in the Form of mRNPs from Their Birth to Their Death

Key concepts

- mRNA associates with a changing population of proteins during its nuclear maturation and cytoplasmic life.
- Some nuclear-acquired mRNP proteins have roles in the cytoplasm.
- A very large number of RNA-binding proteins exist, most of which remain uncharacterized.
- Different mRNAs are associated with distinct, but overlapping, sets of regulatory proteins, creating RNA regulons.

From the time pre-mRNAs are transcribed in the nucleus until their cytoplasmic destruction, eukaryotic mRNAs are associated with a changing repertoire of proteins. RNA–protein complexes are called **ribonucleoprotein particles (RNPs)**. Many of the pre-mRNA-binding proteins are involved in splicing and processing reactions (see Chapter 21, *RNA Splicing and Processing*), and others are involved in quality control (discussed in *Section 22.8, Newly Synthesized RNAs Are Checked for Defects via a Nuclear Surveillance System*). The nuclear maturation of an mRNA comprises multiple remodeling steps involving both the RNA sequence and its complement of proteins. The mature mRNA product is export-competent only when fully processed and associated with the correct protein complexes, including TREX (for *transcription export*), which mediates its association with the nuclear pore export receptor. Mature mRNAs retain multiple binding sites (*cis*-elements) for different regulatory proteins, most often within their 5' or 3' UTRs.

While many nuclear proteins are shed before or during mRNA export to the cytoplasm, others accompany the mRNA and have cytoplasmic roles. For example, once in the cytoplasm, the nuclear cap-binding complex participates in the new mRNA's first translation event, the so-called "pioneering round" of translation. This first translation initiation is critical for a new mRNA; if it is found to be a defective template it will be rapidly destroyed by a surveillance system (see *Section 22.9, Quality Control of mRNA Translation Is Performed by Cytoplasmic Surveillance Systems*). An mRNA that

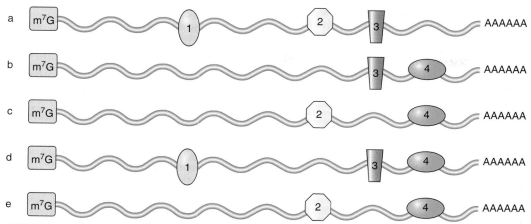

FIGURE 22.4 The concept of an RNA regulon. Eukaryotic mRNAs are bound by a variety of proteins that control its translation, localization and stability. The subset of mRNAs that have a binding protein in common are considered part of the same regulon. In the diagram, mRNAs a and d are part of regulon 1, mRNAs a, c and e are part of regulon 2, and so on.

passes its translation test will spend the rest of its existence associated with a variety of proteins that control its translation, its stability and sometimes its cellular location. The "nuclear history" of an mRNA is critical in determining its fate in the cytoplasm.

A large number of different **RNA-binding proteins (RBPs)** are known, and many more are predicted based on genome analysis. The *S. cerevisiae* genome encodes nearly 600 different proteins predicted to bind to RNA, about one-tenth of the total gene number for this organism. Based on similar proportions, the human genome would be expected to contain over 2000 such proteins. These estimates are based on the presence of characterized RNA-binding domains, and it is likely that additional RNA-binding domains remain to be found. The RNA targets and functions of the great majority of these RBPs are unknown, although it is considered likely that a large fraction of them interact with pre-mRNA or mRNA. This kind of analysis does not include the many proteins that do not bind RNA directly, but participate in RNA-binding complexes.

An important insight into why the number of different mRNA-binding proteins is so large has come from the finding that mRNAs are associated with distinct, but overlapping, sets of RBPs. Studies that have matched specific RBPs with their target mRNAs have revealed that those mRNAs encode proteins with shared features such as involvement in similar cellular processes or location. Thus, the repertoire

of bound proteins catalogues the mRNA. For example, hundreds of yeast mRNAs are bound by one or more of six related *Puf* proteins. Puf1 and Puf2 bind mostly mRNAs encoding membrane proteins, whereas Puf3 binds mostly mRNAs encoding mitochondrial proteins, and so on. A current model, illustrated in **FIGURE 22.4**, proposes that the coordinate control of posttranscriptional processes of mRNAs is mediated by the combinatorial action of multiple RBPs, much like the coordinate control of gene transcription is mediated by the right combinations of transcription factors (see Chapter 28, *Eukaryotic Transcription Regulation*). The set of mRNAs that share a particular type of RBP has been called an **RNA regulon**.

22.4 Prokaryotic mRNA Degradation Involves Multiple Enzymes

Key concepts

- Degradation of bacterial mRNAs is initiated by removal of a pyrophosphate from the 5′ terminus.
- Monophosphorylated mRNAs are degraded during translation in a two-step cycle involving endonucleolytic cleavages, followed by 3′ to 5′ digestion of the resulting fragments.
- 3′ polyadenylation can facilitate the degradation of mRNA fragments containing secondary structure.
- The main degradation enzymes work as a complex called the degradosome.

Our understanding of prokaryotic mRNA degradation comes mostly from studies of *E. coli*. So far, the general principles apply to the other bacterial species studied. In prokaryotes, mRNA degradation occurs during the process of translation. Prokaryotic ribosomes begin translation even before transcription is completed, attaching to the mRNA at an initiation site near the 5′ end and proceeding toward the 3′ end. Multiple ribosomes can initiate translation on the same mRNA sequentially, forming a **polyribosome** (or **polysome**): one mRNA with multiple ribosomes.

E. coli mRNAs are degraded by a combination of endonuclease and 3′→5′ exonuclease activities. The major mRNA degradation pathway in *E. coli* is a multistage process illustrated in **FIGURE 22.5**. The initiating step is removal of pyrophosphate from the 5′ terminus leaving a single phosphate. The monophosphorylated form stimulates the catalytic activity of an endonuclease (RNase E), which makes an initial cut near the 5′ end of the mRNA. This cleavage leaves a 3′-OH on the upstream fragment and a 5′ monophosphate on the downstream fragment. It functionally destroys a **monocistronic mRNA**, as ribosomes can no longer initiate translation. The upstream fragment is then degraded by a 3′→5′ exonuclease (PNPase = polynucleotide phosphorylase). This two-step ribonuclease cycle is repeated along the length of the mRNA in a 5′ to 3′ direction as more RNA gets exposed following passage of previously initiated ribosomes. This process proceeds very rapidly as the short fragments generated by RNase E can be detected only in mutant cells in which exonuclease activity is impaired.

PNPase, as well as the other known 3′→5′ exonucleases in *E. coli*, are unable to progress through double-stranded regions. Thus the stem-loop structure at the 3′ end of many bacterial mRNAs protects the mRNA from direct 3′ attack. Some internal fragments generated by RNase E cleavage also have regions of secondary structure that would impede exonuclease digestion. PNPase *is*, however, able to digest through double-stranded regions if there is a stretch of single-stranded RNA at least seven to ten nucleotides long located 3′ to the stem-loop. The single-stranded sequence seems to serve as a necessary staging platform for the enzyme. Rho-independent termination leaves a single-stranded region that is too short to serve as a platform. To solve this problem a bacterial **polymerase (PAP)** adds 10 to 40 nucleotide **poly(A)** tails to 3′ termini, making them susceptible to 3′→5′ degradation. RNA fragments terminating in particularly stable secondary structures may require repeated polyadenylation and exonuclease digestion steps. It is not known whether polyadenylation is ever the initiating step for degradation of mRNA or whether it is used only to help degrade fragments, including the 3′ terminal one. Some experiments indicate that RNase E cleavage of an mRNA may be required to activate the poly(A) polymerase. This would explain why intact mRNAs do not seem to be degraded from the 3′ end.

RNase E and PNPase, along with a helicase and another accessory enzyme, form a multiprotein complex called the **degradosome**.

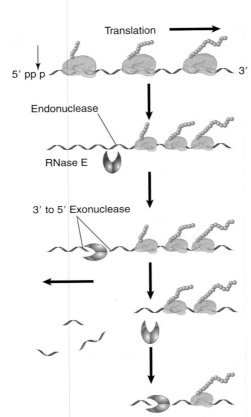

FIGURE 22.5 Degradation of bacterial mRNAs. Bacterial mRNA degradation is initiated by cleavage of the triphosphate 5′ terminus to yield a monophosphate. mRNAs are then degraded in a 2-step cycle: an endonucleolytic cleavage, followed by 3′ to 5′ exonuclease digestion of the released fragment. The endonucleolytic cleavages occur in a 5′ to 3′ direction on the mRNA, following the passage of the last ribosome.

Labels in figure: Translation; 5′ pp p; 3′; Endonuclease; RNase E; 3′ to 5′ Exonuclease

RNAase E plays dual roles in the complex. Its N-terminal domain provides the endonuclease activity, whereas its C-terminal domain provides a scaffold that holds together the other components. While RNase E and PNPase are the principal endo- and exonucleases active in mRNA degradation, others also exist, probably with more restricted roles. The role of other nucleases in mRNA degradation has been addressed by evaluating the phenotypes of mutants in each of the enzymes. For example, the inactivation of RNase E slows mRNA degradation without completely blocking it. Mutations that inactivate PNPase or either of the other two known 3′→5′ exonucleases have essentially no effect on overall mRNA stability. This reveals that any pair of the exonucleases can carry out apparently normal mRNA degradation. However, only two of the three exonucleases (PNPase and RNase R) can digest fragments with stable secondary structures. This was demonstrated in double mutant studies, in which both PNPase and RNase R are inactivated. In these mutants, mRNA fragments that contain secondary structures accumulated.

Many questions about mRNA degradation in *E. coli* remain to be answered. Half-lives for different mRNAs in *E. coli* can differ more than 100-fold. The basis for these extreme differences in stability is not fully understood, but appears to be largely due to two factors. Different mRNAs exhibit a range of susceptibilities to endonuclease cleavage, some protection being conferred by secondary structure in the 5′ end region. Some mRNAs are more efficiently translated than others, resulting in a denser packing of protective ribosomes. Whether or not there are additional pathways of mRNA degradation is not known. No 3′→5′ exonuclease has been found in *E. coli*, although one has been identified in *Bacillus subtilis*. It is likely that the different endonucleases and exonucleases have distinct roles. A genome-wide study using microarrays looked at the steady state levels of more than four thousand mRNAs in cells mutant for RNase E or PNPase or other degradosome components. Many mRNA levels increased in the mutants, as expected for a decrease in degradation. Others, however, remained at the same level or even decreased. The half-lives of specific mRNAs can be altered by different cellular physiological states such as starvation or other forms of stress, and mechanisms for these changes remain mostly unknown.

22.5 Most Eukaryotic mRNA Is Degraded via Two Deadenylation-Dependent Pathways

Key concepts

- The modifications at both ends of mRNA protect it against degradation by exonucleases.
- The two major mRNA decay pathways are initiated by deadenylation catalyzed by poly(A) nucleases.
- Deadenylation may be followed either by decapping and 5′ to 3′ exonuclease digestion, or by 3′ to 5′ exonuclease digestion.
- The decapping enzyme competes with the translation initiation complex for 5′ cap binding.
- The exosome, which catalyzes 3′ to 5′ mRNA digestion, is a large, evolutionarily conserved complex.
- Degradation may occur within discrete cytoplasmic particles called processing bodies (PBs).
- A variety of particles containing translationally repressed mRNAs exist in different cell types.

Eukaryotic mRNAs are protected from exonucleases by their modified ends (Figure 22.1). The 7-methyl guanosine cap protects against 5′ attack; the poly(A) tail, in association with bound proteins, protects against 3′ attack. Exceptions are the histone mRNAs in mammals, which terminate in a stem-loop structure rather than a poly(A) tail. A sequence-independent endonuclease attack—the initiating mechanism used by bacteria—is rare or absent in eukaryotes. mRNA decay has been characterized most extensively in budding yeast, although most findings apply to mammalian cells as well.

Degradation of the vast majority of mRNAs is deadenylation-dependent, i.e., degradation is initiated by breaching their protective poly(A) tail. The newly formed poly(A) tail (which is about 70–90 adenylate nucleotides in yeast and about 200 in mammals) is coated with **poly(A) binding proteins (PABP)**. The poly(A) tail is subject to gradual shortening upon entry into the cytoplasm, a process catalyzed by specific **poly(A) nucleases** (also called **deadenylases**). In both yeast and mammalian cells, the poly(A) tail is initially shortened by the PAN2/3 complex, followed by a more rapid digestion of the remaining 60 to 80 A tail by a second complex, CCR4-NOT, which contains the processive exonuclease Ccr4 and at least eight other subunits. Remarkably, similar

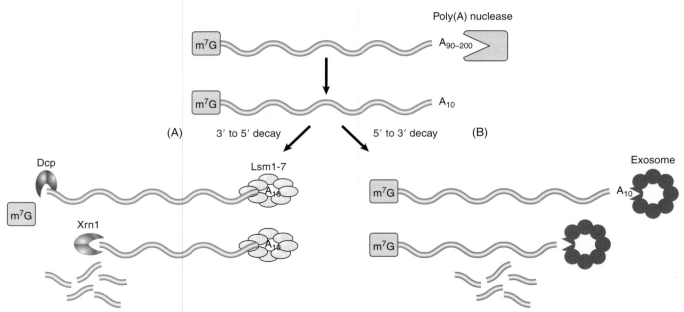

FIGURE 22.6 The major deadenylation-dependent decay pathways in eukaryotes. Two pathways are initiated by deadenylation. In both, poly(A) is shortened by a poly(A) nuclease until it reaches a length of about 10 As. Then an mRNA may be degraded by the 5′ to 3′ pathway or by the 3′ to 5′ pathway. The 5′ to 3′ pathway involves decapping by Dcp and digestion by the Xrn1 exonuclease. The 3′ to 5′ pathway involves digestion by the exosome complex.

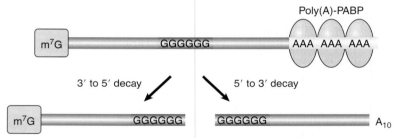

FIGURE 22.7 Use of a poly(G) sequence to determine direction of decay. A poly(G) sequence, engineered into an mRNA, will block the progression of exonucleases in yeast. The 5′ or 3′ mRNA fragment resistant to degradation accumulates in the cell and can be identified by northern blot.

CCR4-NOT complexes are involved in a variety of other processes in gene expression, including transcriptional activation. It is thought to be a global regulator of gene expression, integrating transcription and mRNA degradation. Other poly(A) nucleases exist in both yeast and mammalian cells, and the reason for this multiplicity is not yet clear.

Two different mRNA degradation pathways are initiated by poly(A) removal, illustrated in **FIGURE 22.6**. In the first pathway (Figure 22.6a), digestion of the poly(A) tail down to oligo(A) length (~10 to 12 A) triggers decapping at the 5 end of the mRNA. Decapping is catalyzed by a **decapping enzyme** complex consisting of two proteins in yeast (Dcp1 and Dcp2) and their

homologs plus additional proteins in mammals. Decapping yields a 5′ monophosphorylated RNA end (the substrate for the 5′ to 3′ processive exonuclease XRN1), which rapidly digests the mRNA. In fact this digestion is so fast that intermediates could not be identified until investigators discovered that a stretch of guanosine nucleotides (poly-G) could block Xrn1 progression in yeast. As illustrated in **FIGURE 22.7**, they engineered mRNAs to contain an internal poly-G tract and found that the oligoadenylated 3′ end of the mRNAs accumulated. This result showed (1) that 5′ to 3′ exonuclease digestion was the primary route of decay, and (2) that decapping preceded complete removal of the poly(A) tail.

The cap is normally resistant to decapping during active translation because it is bound by the **cytoplasmic cap-binding protein**, a component of the eukaryotic initiation factor 4F (eIF4F) complex required for translation (described in Chapter 24, *Translation*). Thus, the translation and decapping machineries compete for the cap. How does deadenylation at the 3′ end of the mRNA render the cap susceptible? Translation is known to involve a physical interaction between bound PABP at the 3′ end and the eIF4F complex at the 5′ end. Release of PABP by deadenylation is thought to destabilize the eIF4F-cap interaction, leaving the cap more frequently exposed.

The mechanism is not this simple, though, because additional proteins are known to be involved the decapping event. A complex of seven related proteins, Lsm1-7, binds to the oligo(A) tract after loss of PABP, and is required for decapping. Furthermore, a number of decapping enhancers have been discovered. The mechanisms by which these proteins stimulate decapping are not fully understood, although they appear to act either by recruiting/stimulating the decapping machinery or by inhibiting translation.

In the second pathway (Figure 22.6b), deadenylation to oligo(A) is followed by 3' to 5' exonuclease digestion of the body of the mRNA. This degradation step is catalyzed by the **exosome**, a ring-shaped complex consisting of a nine-subunit core with one or more additional proteins attached to its surface. A recent report showed that the exosome also has endonuclease activity, and the function of this activity in mRNA decay remains unknown. The exosome exists in similar form in Archaea, and is also analogous to the bacterial degradosome in that its core subunits are structurally related to PNPase. Thus, the exosome is an ancient piece of molecular machinery. The exosome also plays an important role in the nucleus, described in *Section 22.8, Newly Synthesized RNAs Are Checked for Defects via a Nuclear Surveillance System*.

The relative importance of each mechanism isn't known, although in yeast, the deadenylation-dependent decapping pathway seems to predominate. The pathways are at least partially redundant. Hundreds of yeast mRNAs were examined by microarray analysis in cells in which either the 5' to 3' or 3' to 5' pathway was inactivated. In either case, only a small percentage of transcripts increased in abundance relative to wild-type cells. This finding suggests that few yeast mRNAs have a requirement for one or the other pathway. It has been proposed that these deadenylation-dependent pathways represent the default degradation pathways for all polyadenylated mRNAs, though subsets of mRNAs can be targets for other specialized pathways, described in *Section 22.6, Other Degradation Pathways Target Specific mRNAs*. Even those mRNAs that are degraded by the default pathways, however, are degraded at different mRNA-specific rates.

New studies suggest that mRNA degradation occurs within discrete particles throughout the cytoplasm, called **processing bodies (PBs)**. These structures, which are large enough to be seen with a light microscope, are clusters of nontranslating mRNPs and a variety of proteins associated with translational repression and mRNA decay, including the decapping machinery and Xrn1 exonuclease. Poly(A)-binding proteins are not generally found in PBs, suggesting that deadenylation precedes localization into these structures. Processing bodies are dynamic, increasing and decreasing in size and number, and even disappearing, under different cellular and experimental conditions that affect translation and decay. For example, release of mRNAs from polysomes by a drug that inhibits translation initiation results in a large increase in PB number and size, as does slowing degradation by partial inactivation of decay components. PBs appear to be formed by assembly of translationally repressed mRNAs and PB protein components rather than being destinations to which targeted mRNAs migrate. Not all resident mRNAs are doomed for destruction, though; some can be released for translation, but which ones and why they are freed isn't yet clear. It is not known whether all mRNA degradation normally occurs in these bodies, or even what function(s) they serve. One obvious idea is that concentrating powerful destructive enzymes in isolated locations renders mRNA degradation more safe and efficient.

Other mRNA-containing particles related to PBs are present in specific cell types. Their similarities are based on the presence of most of the same proteins involved in translational repression and decay. **Maternal mRNA granules** are found in oocytes from a variety of organisms. These granules comprise collections of mRNAs that are held in a state of translational repression until they are activated during subsequent development. Repression is achieved by extensive deadenylation, and activation is achieved by polyadenylation. These granules may also carry mRNAs being transported to specific regions of this large cell (see *Section 22.10, Some Eukaryotic mRNAs Are Localized to Specific Regions of a Cell*). **Neuronal granules** have been identified in *Drosophila* neurons. Similar to the maternal mRNA granule, these granules function in the translational repression and transport of specific mRNAs. A fourth type of particle is called a **stress granule**. Stress granules are quite different in composition from the previous three types; however, they also contain translationally inactive mRNAs that aggregate in response to a general inhibition of translation initiation.

22.6 Other Degradation Pathways Target Specific mRNAs

Key concepts

- Four additional degradation pathways involve regulated degradation of specific mRNAs.
- Deadenylation-independent decapping proceeds in the presence of a long poly(A) tail.
- The degradation of the nonpolyadenylated histone mRNAs is initiated by 3' addition of a poly(U) tail.
- Degradation of some mRNAs may be initiated by sequence-specific or structure-specific endonucleolytic cleavage.
- An unknown number of mRNAs are target for degradation or translational repression by microRNAs.

Four other pathways for mRNA degradation have been described. **FIGURES 22.8** and **22.9** summarize these, along with the two major pathways. These pathways are specific for subsets of mRNAs, and typically involve regulated degradation events.

One pathway involves deadenylation-*in*dependent decapping, i.e., decapping proceeds in the presence of a still long poly(A) tail. Decapping is then followed by Xrn1 digestion. Bypassing the deadenylation step requires a mechanism to recruit the decapping machinery and inhibit

eIF4F binding without the help of the Lsm1-8 complex. One of the mRNAs degraded by this pathway is RPS28B mRNA, which encodes the ribosomal protein S28 and is an interesting autoregulation mechanism. A stem-loop in its 3' UTR is involved in recruiting a known decapping enhancer. The recruitment occurs only when the stem-loop is bound by S28 protein. Thus an excess of free S28 in the cell will cause the accelerated decay of its mRNA.

A second specialized pathway is used to degrade the cell-cycle regulated histone mRNAs in mammalian cells. These mRNAs are responsible for synthesis of the huge number of histone proteins needed during DNA replication. They accumulate only during S-phase and are rapidly degraded at its end. The nonpolyadenylated histone mRNAs terminate in a stem-loop structure similar to that of many bacterial mRNAs. Their mode of degradation has striking similarities to bacterial mRNA decay. A polymerase, structurally similar to the bacterial poly(A) polymerase, adds a short poly(U) tail instead of a poly(A) tail. This short tail serves as a platform for the Lsm1-7 complex and/or the exosome, activating the standard decay pathways. This mode of degradation provides an important evolutionary link between mRNA decay systems in prokaryotes and eukaryotes.

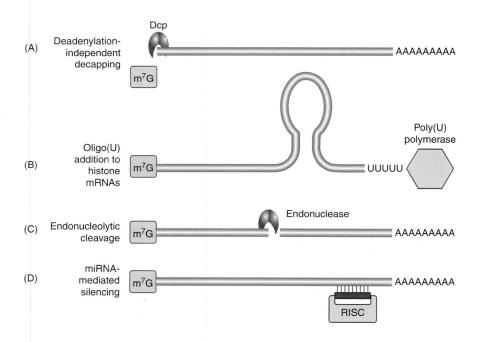

FIGURE 22.8 Other decay pathways in eukaryotic cells. The initiating event for each pathway is illustrated. (A) Some mRNAs may be decapped before deadenylation occurs. (B) Histone mRNAs receive a short poly(U) tail to become a decay substrate. (C) Degradation of some mRNAs can be initiated by a sequence-specific endonucleolytic cut. (D) Some mRNAs can be targeted for degradation or translational silencing by complementary guide miRNAs.

A third pathway is initiated by sequence- or structure-specific endonucleotic cleavage. The cleavage is followed by 5′ to 3′ and 3′ to 5′ digestion of the fragments, and a scavenging decapping enzyme, different from the Dcp complex, can remove the cap. Several endonucleases that cleave specific target sites in mRNAs have been identified. One interesting case is the targeted cleavage of yeast CLB2 (cyclin B2) mRNA, which occurs only at the end of mitosis. The endonuclease that catalyzes the cleavage, RNase MRP, is restricted to the nucleolus and mitochondria for most of the cell cycle where it is involved in RNA processing, but is transported to the cytoplasm in late mitosis.

The fourth pathway is the **microRNA (miRNA)** pathway, which leads directly to endonucleolytic cleavage of mRNA or to translational repression. In this case, an mRNA is targeted by the base-pairing of short (19 to 21 bp) complementary RNAs (guide miRNAs) in the context of a protein complex called RISC. The guide miRNAs are derived from transcribed miRNA genes, and are generated by cleavage from longer precursor RNAs. Thus, the destabilization of target mRNAs is controlled by regulated transcription of the miRNA genes. The details of this mechanism are described in Chapter 30, *Regulatory RNA*. The significance of this newly described pathway to total mRNA decay is not yet known, but could be substantial. At least one thousand miRNAs are predicted to function in humans.

An integrated model of mRNA degradation has been proposed. This model suggests that the deadenylation-dependent decay pathways represent the default systems for degrading all polyadenylated mRNAs. The rate of deadenylation and/or other steps in degradation by these pathways can be controlled by *cis*-acting elements in each mRNA and *trans*-acting factors present in the cell. Superimposed on the default system are the mRNA decay pathways described above for targeting specific mRNAs.

22.7 mRNA-Specific Half-Lives Are Controlled by Sequences or Structures within the mRNA

Key concepts

- Specific *cis*-elements in an mRNA affect its rate of degradation.
- Destabilizing elements (DEs) can accelerate mRNA decay, while stabilizing elements (SEs) can reduce it.
- AU-rich elements (AREs) are common destabilizing elements in mammals, and are bound by a variety of proteins.
- Some DE-binding proteins interact with components of the decay machinery and probably recruit them for degradation.
- Stabilizing elements occur on some highly stable mRNAs.
- mRNA degradation rates can be altered in response to a variety of signals.

Pathway	Initiating event	Secondary step(s)	Substrates
Deadenylation-dependent 5′ to 3′ digestion	Deadenylation to oligo(A)	Oligo (A) binding Lsm complex Decapping 5′ to 3′ exonuclease digestion by XRN1	Probably most polyadenylated mRNAs
Deadenylation-dependent 3′ to 5′ digestion	Deadenylation to oligo(A)	3′ to 5′ exonuclease digestion by exosome	Probably most polyadenylated mRNAs
Deadenylation-independent decapping	Decapping	5′ to 3′ exonuclease digestion	Few specific mRNAs
Endonucleolytic pathway	Endonuclease cleavage	5′ to 3′ and 3′ to 5′ exonuclease digestion	Few specific mRNAs
Histone mRNA pathway	Oligouridylation	Oligo(U) binding by Lsm complex Decapping and 5′ to 3′ exonuclease digestion by XRN1 3′ to 5′ digestion by exosome	Histone mRNAs in mammals
miRNA pathway	Base-pairing with miRNA in RISC	Endonucleolytic cleavage or translational repression	Many mRNAs (extent unknown)

FIGURE 22.9 Table summarizing key elements of mRNA decay pathways in eukaryotic cells.

What accounts for the large range of half-lives of different mRNAs in the same cell? Specific *cis*-elements within an mRNA are known to affect its stability. The most common location for such elements is within the 3' UTR, although they exist elsewhere. Whole genome studies have revealed many highly conserved 3' UTR motifs, but their roles remain mostly unknown. Some are target sites for miRNA base-pairing. Others are binding sites for RBPs, some of which have known functions in stability. Rates of deadenylation can vary widely for different mRNAs, and sequences that affect this rate have been described.

Destabilizing elements (DEs) have been the most widely studied. The criterion for defining a destabilizing sequence element is that its introduction into a more stable mRNA accelerates its degradation. Removal of an element from an mRNA does not necessarily stabilize it, indicating that an individual mRNA can have more than one destabilizing element. To complicate their identification further, the presence of a DE does not guarantee a short half-life under all conditions, because other sequence elements in the mRNA can modify its effectiveness.

The most well studied type of DE is the **AU-rich element (ARE)**, found in the 3' UTR of up to 8% of mammalian mRNAs. AREs are heterogeneous, and a number of subtypes have been characterized. One type consists of the pentamer sequence AUUUA present once or repeated multiple times in different sequence contexts. Another type does not contain AUUUA and is predominantly U-rich. A large number of ARE-binding proteins with specificity for certain ARE types and/or cell types have been identified. How do AREs work to stimulate rapid degradation? Many ARE-binding proteins have been found to interact with one or more components of the degradation machinery, including the exosome, deadenylases, and decapping enzyme, suggesting that they act by recruiting the degradation machinery. The exosome can bind some AREs directly. The AREs of a number of mRNAs have been shown to accelerate the deadenylation step of decay, although it is not likely that they all work this way. Another way they might act is by facilitating efficient engagement of the mRNA into processing bodies.

Many AU-rich DEs and other kinds of destabilizing elements have been identified in the mRNAs of budding yeast and other model organisms. For example, the previously mentioned Puf proteins of yeast bind to specific UG-rich elements and accelerate the degradation of target mRNAs. In this case, the destabilizing mechanism is accelerated deadenylation by recruitment of the CCR4-NOT deadenylase. A genomics analysis of yeast 3' UTRs has identified 53 sequence elements that correlate with the half-lives of mRNAs containing them, suggesting the number of different destabilizing elements may be large. **FIGURE 22.10** summarizes the known actions of destabilizing elements.

Stabilizing elements (SEs) have been identified in a few unusually stable mRNAs. Three mRNAs studied in mammalian cells have stabilizing pyrimidine-rich sequences in their 3' UTRs. Proteins that bind to this element in globin mRNA have been shown to interact with PABP, suggesting they might function to protect the poly(A) tail from degradation. In some cases, an mRNA can be stabilized by inhibition of its DE. For example, certain ARE-binding proteins act to prevent the ARE from destabilizing the mRNA, presumably by blocking the ARE binding site. An example of regulated mRNA stabilization occurs for the mammalian transferrin mRNA. It is stabilized when its 3' UTR **iron-response element (IRE)**, consisting of multiple stem-loop structures, is bound by a specific protein, as shown in **FIGURE 22.11**. The affinity of the IRE-binding protein for the IRE is altered by iron binding, exhibiting low affinity when its iron-binding site is full and high affinity when it is not. When the cellular iron concentration is low, more transferrin is needed

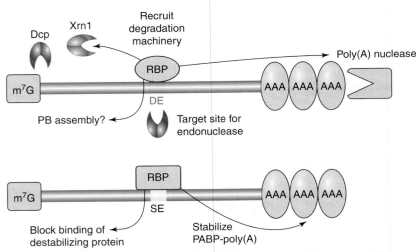

FIGURE 22.10 Mechanisms by which destabilizing elements (DEs) and stabilizing elements (SEs) function. Affects of DEs and SEs on mRNA stability are mediated primarily through the proteins that bind to them. One exception is a DE that acts as an endonuclease target site.

to import iron from the bloodstream, and under these conditions the transferrin mRNA is stabilized. The IRE-binding protein stabilizes the mRNA by inhibiting the function of destabilizing sequences in the vicinity. Interestingly, the same IRE-binding protein also binds an IRE in ferritin mRNA and regulates this mRNA in a very different way. Ferritin is an iron-binding protein that sequesters excess cellular iron. The IRE-binding protein binds IRE stem-loops in the 5' UTR of ferritin when iron is low, and blocks the interaction of the cap-binding complex with ferritin mRNA. Thus, translation of ferritin mRNA is prevented when cellular iron levels are low—the conditions under which transferrin mRNA is stabilized and translated.

Many *cis* element-binding proteins are subject to modifications that are likely to affect their function, including phosphorylations, methylations, conformational changes due to effector binding, and isomerizations. Such modifications may be responsible for changes in mRNA degradation rates induced by cellular signals. mRNA decay can be altered in response to a wide variety of environmental and internal stimuli, including cell cycle progression, cell differentiation, hormones, nutrient supply, and viral infection. Microarray studies have shown that almost 50% of changes in mRNA levels stimulated by cellular signals are due to mRNA stabilization or destabilization events, not to transcriptional changes. How these changes are effected remains largely unknown.

22.8 Newly Synthesized RNAs Are Checked for Defects via a Nuclear Surveillance System

Key concepts

- Aberrant nuclear RNAs are identified and destroyed by a surveillance system.
- The nuclear exosome functions both in the processing of normal substrate RNAs and in the destruction of aberrant RNAs.
- The yeast TRAMP complex recruits the exosome to aberrant RNAs and facilitates its 3' to 5' exonuclease activity.
- Substrates for TRAMP-exosome degradation include unspliced or aberrantly spliced pre-mRNAs and improperly terminated RNA Pol II transcripts lacking a poly(A) tail.
- The majority of RNA Pol II transcripts may be cryptic unstable transcripts (CUTs) that are rapidly destroyed in the nucleus.

All newly synthesized RNAs are subject to multiple processing steps after they are transcribed (see Chapter 21, *RNA Splicing and Processing*). At each step, errors may be made. While DNA errors are repaired by a variety of repair systems (see Chapter 16, *Repair Systems*), detectable errors in RNA are dealt with by destroying the defective RNA. **RNA surveillance systems** exist in both the nucleus and cytoplasm to handle different kinds of problems. Surveillance involves two kinds of activities: one to identify and tag the aberrant substrate RNA and another to destroy it.

The destroyer is the nuclear **exosome**. The nuclear exosome core is almost identical to the cytoplasmic exosome, though it interacts with different protein cofactors. It removes nucleotides from targeted RNAs by 3' to 5' exonuclease activity. The nuclear exosome has multiple functions involving RNA processing of some noncoding RNA transcripts (snRNA, snoRNA, and rRNA) and complete degradation of aberrant transcripts. The exosome is recruited to its processing substrates by protein complexes that recognize specific RNA sequences or RNA/RNP structures. For example, Nrd1-Nab3 is a sequence-specific protein dimer that recruits the exosome to normal sn/snoRNA processing substrates. This protein pair binds to GUA[A/G] and UCUU elements, respectively. The Nrd1-Nab3 cofactor is also involved in transcription termination of these nonpolyadenylated Pol II-transcribed RNAs, suggesting that the processing exosome may be recruited directly to the site of their synthesis.

Aberrantly processed, modified, or misfolded RNAs require other protein cofactors for identification and exosome recruitment. The major nuclear complex performing this

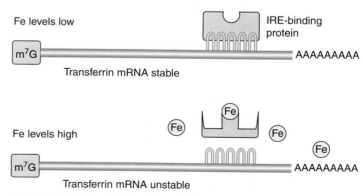

FIGURE 22.11 Regulation of transferring mRNA stability by iron levels. The IRE in the 3' UTR is the binding site for a protein that stabilizes the mRNA. The IRE-binding protein is sensitive to iron (Fe) levels in the cell, binding to the IRE only when iron is low.

function in yeast is called **TRAMP** (an acronym for the component proteins), and it exists in at least two forms differing in the type of poly(A) polymerase present. The TRAMP complex acts in several ways to effect degradation:

1. It interacts directly with the exosome, stimulating its exonuclease activity.
2. It includes a helicase, which is probably required to unwind secondary structure and/or move RNA-binding proteins from structured RNP substrates during degradation.
3. It adds a short 3′ **oligo(A) tail** to target substrates. The oligo(A) tail is thought to make the targeted RNP a better substrate for the degradation machinery in the same way that the oligo(A) tail functions in bacteria.

FIGURE 22.12 summarizes the roles of TRAMP and the exosome. It has become clear that RNA degradation in bacteria and Archaea and nuclear RNA degradation in eukaryotes are evolutionarily related processes. Their similarity suggests that the ancestral role of polyadenylation was to facilitate RNA degradation, and that poly(A) was later adapted in eukaryotes for the oddly reverse function of stabilizing mRNAs in the cytoplasm.

What are the substrates for TRAMP-exosome degradation? The TRAMP complex is remarkable in that it recognizes a wide variety of aberrant RNAs synthesized by all three transcribing polymerases. It is not known how this is accomplished given that the targeted RNAs share no recognizably common features. Some researchers favor a kinetic competition model, hypothesizing that RNAs that do not get processed and assembled into final RNP form *in a timely manner* will become substrates for exosome degradation. This mechanism avoids the need to posit specific recognition of innumerable possible defects.

What kinds of abnormalities condemn pre-mRNAs to nuclear destruction? Two kinds of substrate have been identified. One type is unspliced or aberrantly spliced pre-mRNAs. Components of the spliceosome retain such transcripts either until they are degraded by the exosome or until proper splicing is completed if possible. It is thought that the kinetic competition model probably applies here, too. A pre-mRNA that is not efficiently spliced and packaged is at increased risk of being accessed by the exosome degradation machinery. The basis for recognition of aberrantly spliced pre-mRNAs is not known. The second type of pre-mRNA substrate is one that has been improperly terminated, lacking a poly(A) tail. While polyadenylation is protective in true mRNAs, it may actually be destabilizing for **cryptic unstable transcripts (CUTs)**. These nonprotein-coding RNAs (also discussed in *Section 30.3, Noncoding RNAs Can Be Used to Regulate Gene Expression*) are transcribed by RNA Pol II and do not encode recognizable genes; however, they frequently overlap with (and may regulate) protein-coding genes. These transcripts are polyadenylated by a component of the TRAMP complex (Trf4). They are distinguished from other transcripts of unknown function by their extreme instability, normally being degraded by the TRAMP-exosome complex immediately after synthesis, possibly targeted by the Trf4-dependent polyadenylation. In fact, the existence of these transcripts was first convincingly demonstrated in yeast strains with impaired nuclear RNA degradation. More than three-quarters of RNA Pol II transcripts may be comprised of noncoding RNAs and be subject to rapid degradation by the exosome! Some CUTs appear to arise from spurious transcription initiation, and the short-lived RNA products themselves typically do not appear to have a function (i.e., these RNAs do not typically act in *trans*). There are, however, examples in which there is a role for the transcription process itself in regulating nearby or overlapping coding genes (one example is described in *Section 30.3, Noncoding RNAs Can Be Used to Regulate Gene Expression*).

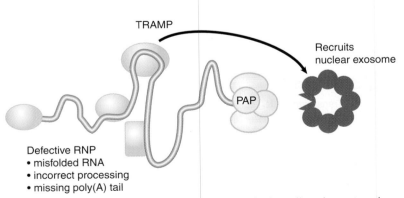

FIGURE 22.12 The role of TRAMP and the exosome in degrading aberrant nuclear RNAs. Defective RNPs are tagged by protein cofactors which then recruit the nuclear exosome. The cofactor in yeast cells is the complex TRAMP. The poly(A) polymerase (PAP, or Trf4) in TRAMP adds a short poly(A) tail to the 3′ end of the targeted RNA.

22.9 Quality Control of mRNA Translation Is Performed by Cytoplasmic Surveillance Systems

Key concepts

- Nonsense-mediated decay (NMD) targets mRNAs with premature stop codons.
- Targeting of NMD substrates requires a conserved set of UPF and SMG proteins.
- Recognition of a termination codon as premature involves unusual 3' UTR structure or length in many organisms and the presence of downstream exon junction complexes (EJC) in mammals.
- Nonstop decay (NSD) targets mRNAs lacking an in frame termination codon and requires a conserved set of SKI proteins.
- No-go decay (NGD) targets mRNAs with stalled ribosomes in their coding regions.

Some kinds of mRNA defects can be assessed only during translation. Surveillance systems have evolved to detect three types of mRNA defects that threaten translational fidelity and to target the defective mRNAs for rapid degradation. **FIGURE 22.13** shows the substrates for each of these three systems. All three systems involve abnormal translation termination events, so it is useful to review what happens during normal termination (see *Section 24.15, Termination Codons Are Recognized by Protein Factors,* for a more detailed description). When a translating ribosome reaches the termination (stop) codon, a pair of **release factors** (eRF1 and eRF2 in eukaryotes) enters the ribosomal A site, which is normally filled by incoming tRNAs during elongation. The release factor complex mediates the release of the completed polypeptide, followed by the mRNA, remaining tRNA, and ribosomal subunits.

Nonsense-mediated decay (NMD) targets mRNAs containing a premature termination codon (PTC). Its name comes from "nonsense mutation," which is only one way that mRNAs with a PTC can be generated. Genes without nonsense mutations can give rise to aberrant transcripts containing a PTC by (1) RNA polymerase error or (2) incomplete, incorrect, or alternative splicing. It has been estimated that almost half of alternatively spliced pre-mRNAs generate at least one form with PTC. About 30% of known disease-causing alleles probably encode an mRNA with a PTC. An mRNA with a PTC will produce C-terminal truncated polypeptides, which are considered to be particularly toxic

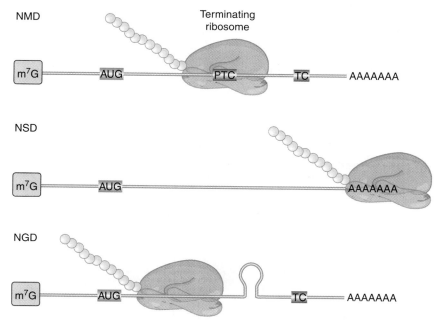

FIGURE 22.13 Substrates for cytoplasmic surveillance systems. Nonsense-mediated decay (NMD) degrades mRNAs with a premature termination codon (PTC) positioned ahead of its normal termination codon (TC). Nonstop decay (NSD) degrades mRNAs lacking an in-frame termination codon. No-go decay (NGD) degrades mRNAs having a ribosome stalled in the coding region.

to a cell due to their tendency to trap multiple binding partners in nonfunctional complexes. The NMD pathway has been found in all eukaryotes.

Targeting of PTC-containing mRNAs requires translation and a conserved set of protein factors. They include three **Upf proteins** (Upf1, Upf2, and Upf3), and four additional proteins (Smg1, 5, 6, and 7). Upf1 is the first NMD protein to act, binding to the terminating ribosome—specifically to its release factor complex. UPF attachment tags the mRNA for rapid decay. The specific roles of the NMD factors have not yet been defined, although phosphorylation of ribosome-bound Upf1 by Smg1 is critical. Their combined actions condemn the mRNA to the general decay machinery, and stimulate rapid deadenylation. The target mRNAs are degraded by both 5′ to 3′ and 3′ to 5′ pathways.

How are PTCs distinguished from the normal termination codon further downstream? The mechanism has been studied extensively both in yeast and in mammalian cells, where it is somewhat different; these are illustrated in **FIGURE 22.14**. The major signal that identifies a PTC in mammalian cells is the presence of a splice junction, marked by an **exon junction complex (EJC)** downstream of the premature termination codon. The majority of genes in higher eukaryotes do not have an intron interrupting the 3′ UTR, so authentic termination codons are not generally followed by a splice junction. During the **pioneer round of translation** for a normal mRNA, all EJCs occur within the coding region and are displaced by the transiting ribosome. During the pioneer round of translation for an NMD substrate, Upf2 and Upf3 proteins bind to the residual downstream EJC(s), targeting it for degradation.

Most *S. cerevisiae* genes are not interrupted by introns at all, so the mechanism for PTC detection must be different. In this case an abnormally long 3′ UTR is the warning sign. This was demonstrated by the finding that extension of the 3′ UTR of a normal mRNA could convert it into a substrate for NMD. A current model proposes that proper translation termination at a stop codon requires a signal from a nearby PABP. Although 3′ UTRs are highly variable in nucleotide length, the physical distance between the termination codon and the poly(A) tail is not strictly a function of length because secondary structures and interactions between bound RBPs can compress the distance. The requirement for PABP was demonstrated in multiple organisms by tethering a PABP close to the PTC, as illustrated in **FIGURE 22.15**. The mRNA was no longer targeted by NMD. PTC recognition also occurs independently of splicing in *Drosophila*, *C. elegans*, plants, and in some mammalian mRNAs, suggesting the length and structure of 3′ UTR is critical for the normal process of translation termination in all organisms.

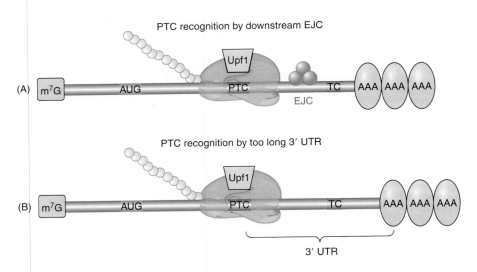

FIGURE 22.14 Two mechanisms by which a termination codon is recognized as premature. (A) In mammals, the presence of an Exon Junction Complex downstream of a termination codon targets the mRNA for NMD. (B) In probably all eukaryotes, an abnormally long 3′ UTR is recognized by the distance between the termination codon and the poly(A)-PABP complex. In either case, the Upf1 protein binds to the terminating ribosome to trigger decay.

Some normal mRNAs are targeted by NMD. These were identified by experiments in which Upf1 levels were reduced, resulting in a subset of transcripts that increased in abundance. The list of normal NMD substrates includes mRNAs with especially long 3' UTRs, mRNAs encoding selenoproteins (which use the termination codon UGA as a selenocysteine codon), and an unknown number of alternatively spliced mRNAs. Not all targeted mRNAs are predicted to be NMD substrates based on our current understanding. NMD may turn out to be an important rapid decay pathway for a variety of short-lived mRNAs.

Nonstop decay (NSD) targets mRNAs that lack an in-frame termination codon (middle panel in Figure 22.13). Failure to terminate results in a ribosome translating into the poly(A) tail and probably stalling at the 3' end. Nonstop decay substrates are generated mainly by premature transcription termination and polyadenylation in the nucleus. Such prematurely polyadenylated transcripts are surprisingly common. Analysis of random cDNA populations derived from yeast and human mRNAs suggests that 5%–10% of polyadenylation events may occur at upstream "cryptic" sites that resemble an authentic polyadenylation signal. Targeting nonstop substrates involves a set of factors called the **SKI proteins**. The ribosome is released from the mRNA by the action of Ski7. Ski7 has a GTPase domain similar to eEF3 and probably binds to the ribosome in the A-site to stimulate release. The subsequent recruitment of the other SKI proteins and the exosome results in 3' to 5' decay of the mRNA. Decay of nonstop substrates can also occur in the absence of Ski7, and proceeds by decapping and 5' to 3' digestion. Susceptibil-

ity to decapping could be due to the pioneer ribosome displacing PABPs as it traverses the poly(A) tail. Rapid decay of nonstop substrates results not only in prevention of toxic polypeptides, but also liberation of trapped ribosomes. Interestingly, *E. coli* uses a specialized noncoding RNA (tmRNA) that acts like both a tRNA and an mRNA to rescue ribosomes stalled on a nonstop mRNA. tmRNA directs the addition of a short peptide that targets the defective protein product for degradation, provides a stop codon to allow recycling of the ribosome, and targets degradation of the defective mRNA by RNAse R.

No-go decay (NGD) targets mRNAs with ribosomes stalled in the coding region codon (bottom panel of Figure 22.13). Transient or prolonged stalling can be caused by natural features of some mRNAs, including strong secondary structures and rarely used codons (whose cognate tRNAs are in low abundance). This newly discovered surveillance pathway has been studied only in yeast, and is the least understood of the three. Targeting of the mRNA involves recruitment of two proteins, Dom34 and Hbs1, which are homologous to eRF1 and eRF3, respectively. mRNA degradation is initiated by an endonucleoytic cut, and the 5' and 3' fragments are digested by the exosome and Xrn1. Dom34 might be the endonuclease, as one of its domains is nuclease-like. Why would a normal mRNA have hard-to-translate sequences that might condemn it to rapid degradation? Such sequences can be thought of another kind of destabilizing element. Evolutionary retention of impediments to efficient translation suggests that they serve an important function in controlling the half-life of these mRNAs.

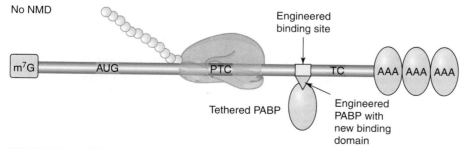

FIGURE 22.15 Effect of tethering a PABP near to a premature termination codon. A PABP gene was altered to express a phage RNA-binding domain. Its binding site was engineered into a test NMD substrate gene. The tethered PABP prevented the usual rapid degradation of this mRNA by NMD. This method has many applications in molecular biology.

22.10 Some Eukaryotic mRNAs Are Localized to Specific Regions of a Cell

Key concepts

- Localization of mRNAs serves diverse functions in single cells and developing embryos.
- Three mechanisms for the localization of mRNA have been documented.
- Localization requires *cis*-elements on the target mRNA and *trans*-factors to mediate the localization.
- The predominant active transport mechanism involves the directed movement of mRNPs along cytoskeletal tracks.

The cytoplasm is a crowded place occupied by a high concentration of proteins. It is not clear how freely polysomes can diffuse, and most mRNAs are probably translated in random locations that are determined by their point of entry into the cytoplasm and the distance that they may have moved away from it. Some mRNAs are translated only at specific sites, though— their translation is repressed until they reach their destinations. There are over 100 specific mRNAs whose regulated localization has been described, a number that certainly represents a small fraction of the total. mRNA localization serves a number of important functions in eukaryotic organisms of all types. Three key functions are illustrated in **FIGURE 22.16**, and are discussed as follows.

1. Localization of specific mRNAs in the oocytes of many animals serves to set up future patterns in the embryo (such as axis polarity) and to assign developmental fates to cells residing in different regions. These localized maternal mRNAs encode transcription factors or other proteins that regulate gene expression. In *Drosophila* oocytes, *bicoid* and *nanos* mRNAs are localized to the anterior and posterior poles, respectively, and their translation following fertilization results in gradients of their protein products. The gradients are used by cells in early development for the specification of their anterior–posterior position in the embryo. *Bicoid* encodes a transcription factor and *nanos* encodes a translational repressor. Some localized mRNAs encode determinants of cell fate. For example, *oskar* mRNA localizes in the posterior of the oocyte and initiates the process leading to development of primordial germ cells in the embryo.

2. mRNA localization also plays a role in asymmetric cell divisions; i.e., mitotic divisions that result in daughter cells that differ from one another. One way this is accomplished is by asymmetric segregation of cell fate determinants, which may be proteins and/or the mRNAs that encode them. In *Drosophila* embryos, *prospero* mRNA and its product (a transcription factor) are localized to a region of the peripheral cortex of the embryo. Later in development, oriented cell division of neuroblasts assures that only the outermost daughter cell receives *prospero*, committing it to a ganglion mother-cell fate. Asymmetric cell division is also used by budding yeast to generate a daughter cell of a different mating type than the mother cell, an event described later in this section.

3. mRNA localization in adult, differentiated cell types is a mechanism for the compartmentalization of the cell into specialized regions. Localization may be used to assure that components of multiprotein complexes are synthesized in proximity to one another and that proteins targeted to organelles or specialized areas of cells are synthesized conveniently nearby. mRNA localization is particularly important for highly polarized cells such as neurons. While

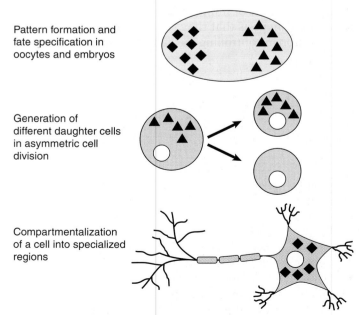

Pattern formation and fate specification in oocytes and embryos

Generation of different daughter cells in asymmetric cell division

Compartmentalization of a cell into specialized regions

FIGURE 22.16 Three main functions of mRNA localization.

most mRNAs are translated in the neuron cell body, many mRNAs are localized to its dendritic and axonal extensions. Among those is β-actin mRNA, whose product participates in dendrite and axon growth. β-actin mRNA localizes to sites of active movement in a wide variety of motile cell types. Interestingly, localization of mRNA at neuronal postsynaptic sites seems to be essential for modifications accompanying learning. In glial cells, the myelin basic protein (MBP) mRNA, which encodes a component of the myelin sheath, is localized to a specific myelin-synthesizing compartment. Plants localize mRNAs to the cortical region of cells and to regions of polar cell growth.

In some cases, mRNA localization involves transport from one cell to another. Maternal mRNPs in *Drosophila* are synthesized and assembled in surrounding nurse cells, and are transferred to the developing oocyte through cytoplasmic canals. Plants can export RNAs through plasmodesmata and transport them for long distances via the phloem vascular system. mRNAs are sometimes transported *en masse* in **mRNP granules**. The compositions of these granules are not yet well defined.

Three mechanisms for the localization of mRNA have been well documented:

- The mRNA is uniformly distributed but degraded at all sites except the site of translation.
- The mRNA is freely diffusible but becomes trapped at the site of translation.
- The mRNA is actively transported to a site where it is translated.

Active transport is the predominant mechanism for localization. Transport is achieved by translocation of motor proteins along cytoskeletal tracks. All three molecular motor types are exploited: dyneins and kinesins, which travel along microtubules in opposite directions, and myosins, which travel along actin fibers. This mode of localization requires at least four components: (1) *cis*-elements on the target mRNA, (2) *trans*-factors that directly or indirectly attach the mRNA to the correct motor protein, (3) *trans*-factors that repress translation, and (4) an anchoring system at the desired location.

Only a few *cis*-elements, sometimes called **zipcodes**, have been characterized. They are diverse, include examples of both sequence and structural RNA elements, and can occur anywhere in the mRNA, though most are in the 3′ UTR. Zipcodes have been difficult to identify, presumably because many consist of complex secondary and tertiary structures. A large number of *trans*-factors have been associated with localized mRNA transport and translational repression, some of which are highly conserved in different organisms. For example, the double-stranded RNA-binding protein *staufen* is involved in localizing mRNAs in the oocytes of *Drosophila* and *Xenopus*, as well as the nervous systems of *Drosophila*, mammals, and probably worms and zebrafish. This multitalented factor has multiple domains that can couple complexes to both actin- and microtubule-dependent transport pathways. Almost nothing is known about the fourth required component—anchoring mechanisms. Two examples of localization mechanisms are discussed below.

The localization of β-actin mRNA has been studied in cultured fibroblasts and neurons. The zipcode is a 54-nucleotide element in the 3′ UTR. Cotranscriptional binding of the zipcode element by the protein ZBP1 is required for localization, suggesting this mRNA is committed to localization before it is even processed and exported from the nucleus. Interestingly, β-actin mRNA localization is dependent on intact actin fibers in fibroblasts and intact microtubules in neurons.

Genetic analysis of Ash1 mRNA localization in yeast has provided the most complete picture of a localization mechanism to date, and is illustrated in **FIGURE 22.17**. During budding, the Ash1 mRNA is localized to the developing bud tip, resulting in Ash1 synthesis only in the newly formed daughter cell. Ash1 is a transcriptional repressor that disallows expression of the HO endonuclease, a protein required for mating-type switching (see *Section 15.20, Yeast Can Switch Silent and Active Loci for Mating Type*). The result is that mating-type switching occurs only in the mother cell. The Ash1 mRNA has four stem-loop localization elements in its coding region to which the protein She2 binds, probably in the nucleus. The protein She3 serves as an adaptor, binding both to She2 and to the myosin motor protein Myo4 (also called She1). A Puf protein, Puf6, binds to the mRNA, repressing its translation. The motor transports the Ash1 mRNP along the polarized actin fibers that lead from the mother cell to the developing bud. Additional proteins are required for proper localization and expression of the Ash1

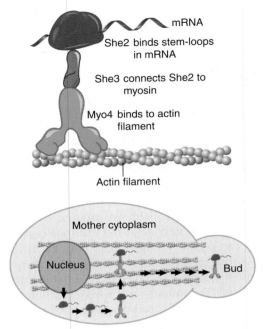

FIGURE 22.17 Localization of *Ash1* mRNA. Newly exported *Ash1* mRNA is attached to the myosin motor Myo4 via a complex with the She2 and She3 proteins. The motor transports the mRNA along actin filaments to the developing bud.

mRNA. More than 20 yeast mRNAs use the same localization pathway.

Localization mechanisms that do not involve active transport have been clearly demonstrated for only a few localized mRNAs in oocytes and early embryos. The mechanism of local entrapment of diffusible mRNAs requires the participation of previously localized anchors, which have not been identified. In *Drosophila* oocytes, diffusing *nanos* mRNA is trapped at the posterior "germ plasm," a specialized region of the cytoplasm underlying the cortex. In *Xenopus* oocytes, mRNAs localized to the vegetal pole are first trapped in a somewhat mysterious, membrane-laden structure called the mitochondrial cloud (MC), which later migrates to the vegetal pole carrying mRNAs with it. The mechanism of localized mRNA stabilization has been described for an mRNA that also localizes to the posterior pole of the *Drosophila* embryo. Early in development, the *hsp83* mRNA is uniformly distributed through the embryonic cytoplasm, but later it is degraded everywhere except at the pole. A protein called smaug is involved in destabilizing the majority of the *hsp83* mRNAs, most likely by recruiting the CCR4/NOT complex. How the pole-localized mRNAs escape is not known.

22.11 Summary

Cellular RNAs are relatively unstable molecules due to the presence of cellular ribonucleases. Ribonucleases differ in mode of attack and are specialized for different RNA substrates. These RNA-degrading enzymes have many roles in a cell, including the decay of messenger RNA. The fact that mRNAs are short-lived allows rapid adjustment of the spectrum of proteins synthesized by a cell by regulating gene transcription rates. Messenger RNAs of different sequence exhibit very different susceptibilities to nuclease action, with half-lives varying by 100-fold or more.

mRNA associates with a changing population of proteins during its nuclear maturation and cytoplasmic life. A very large number of RNA-binding proteins exist, most of which remain uncharacterized. Many proteins with nuclear roles are shed before or during mRNA export to the cytoplasm. Others accompany the mature mRNA and have cytoplasmic roles. mRNAs are associated with distinct, but overlapping, sets of RNA-binding proteins (RBPs) with roles in translation, stability, and localization. The group of mRNAs that share a particular type of RBP has been called an RNA regulon.

Degradation of bacterial mRNAs is initiated by removal of a pyrophosphate from the 5' terminus. This step triggers a cycle of endonucleolytic cleavages, followed by 3' to 5' exonucleolytic digestion of released fragments. The 3' stem-loop on many mRNAs protects them from 3' attack. The 3' to 5' exonuclease activity is facilitated by polyadenylation of 3' ends, forming a platform for the enzyme. The main proteins involved in mRNA degradation function as a complex called the degradosome.

Degradation of most eukaryotic mRNAs in yeast, and probably in mammals, requires deadenylation as the first step. Extensive shortening of the poly(A) tail allows one of two degradation pathways to proceed. The 5' to 3' decay pathway involves decapping and 5' to 3' exonuclease digestion. The 3' to 5' decay pathway is catalyzed by the exosome, a large exonuclease complex. Translation and decay by the 5' to 3' pathway are competing processes because the translation initiation complex and the decapping enzyme both bind to the cap. Particles called processing bodies (PBs) contain mRNAs and proteins involved in both decay and translational repression, and are thought to be the sites of mRNA degradation.

Four other pathways for mRNA degradation have been described that target specific mRNAs. Each uses the same degradation machinery as the deadenylation-dependent pathways but are initiated differently. They are initiated by (1) deadenylation-independent decapping, (2) addition of a 3' poly(U) tail, (3) sequence/structure-specific endonucleolytic cleavage, and (4) base-pairing of microRNAs.

Differences in the characteristic half-lives of mRNAs are due to specific *cis*-elements within an mRNA. Destabilizing elements and stabilizing elements have been described. They are most commonly located in the 3' UTR, and act by serving as binding sites for proteins or microRNAs. AU-rich elements (AREs) destabilize a large number of mRNAs in mammalian cells. Proteins that bind to destabilizing elements probably act primarily by recruiting some component(s) of the degradation machinery. mRNA stability can be regulated in response to cellular signals by modification of binding proteins.

There are quality-control surveillance systems operating in both the nucleus and cytoplasm that target defective RNAs for degradation. In the nucleus, the exosome has a role in both processing of certain normal RNAs and destruction of abnormal ones. Defective RNAs are identified by a variety of exosome cofactors that then recruit the exosome. The major cofactor in yeast cells is the TRAMP complex, which has homologs in other eukaryotic organisms. RNA Pol II transcripts that are substrates for nuclear degradation include those that are not spliced correctly or lack normal poly(A) tails. The majority of RNA Pol II transcripts may be cryptic unstable transcripts (CUTs).

A variety of mRNAs are targeted by cytoplasmic surveillance systems. All three systems involve abnormal translation termination events. Nonsense-mediated decay (NMD) targets mRNAs with premature termination codons. A conserved set of factors (the UPF and SMG proteins) are involved in identifying and committing an NMD substrate to the general decay machinery. A premature termination codon is recognized during the pioneer round of translation by a downstream exon junction complex or by an unusually distant 3' mRNA terminus. NMD also is involved in degrading certain normal unstable mRNAs. Nonstop decay (NSD) targets mRNAs lacking an in-frame termination codon and requires a conserved set of SKI proteins to force release of the trapped ribosome and recruit degradation machinery. No-go decay (NGD) targets mRNAs with stalled ribosomes in their coding regions, and causes ribosome release and degradation.

Some mRNAs are localized to specific regions of cells, and are not translated until their cellular destinations are reached. Localization requires *cis*-elements on the target mRNA and *trans*-factors to mediate the localization. Localization serves three main functions:

1. In oocytes, it serves to set up future patterns in the embryo and to assign developmental fates to cells residing in different regions.
2. In cells that divide asymmetrically, it is a mechanism to segregate protein factors to only one of the daughter cells.
3. In some cells, especially polarized cell types, it is a mechanism to establish subcellular compartments. Three mechanisms for localization are known:
   a. degradation of the mRNA at all sites other than the target site;
   b. selective anchoring of diffusing mRNA at the target site, and
   c. directed transport of the mRNA on cytoskeletal tracks.

The latter is the most common method and exploits actin- and microtubule-based molecular motors.

References

General

Houseley J., and Tollervey, D. (2009). The many pathways of RNA degradation. *Cell* 136, 763–776.

22.2 Messenger RNAs Are Unstable Molecules

Research

Dölken, L., Ruzsics Z., Rädle B., Friedel C. C., Zimmer R., Mages J., Hoffmann R., Dickinson P., Forster T., Ghaza P., and Koszinowski, U. H. (2008). High-resolution gene expression profiling for simultaneous kinetic parameter analysis of RNA synthesis and decay. *RNA* 14, 1959–72.

Foat, B. C., Houshmandi, S. S., Olivas, W. M., and Bussemaker, H. J. (2005). Profiling condition-specific, genome-wide regulation of mRNA stability in yeast. *Proc. Natl. Acad. Sci. (USA)* 102P, 17675–80.

22.3 Eukaryotic mRNAs Exist in the Form of mRNPs from Their Birth to Their Death

Research

Hogan, D. J. et al. (2008). Diverse RNA-binding proteins interact with functionally related sets of RNAs, suggesting an extensive regulatory system. *PLoS Biology* 6(10), e255.

Reviews

Keene, J. D. (2007). RNA regulons: coordination of post-transcriptional events. *Nature Reviews/ Genetics* 8, 533–543.

Moore, M. J. (2005). From birth to death: the complex lives of eukaryotic mRNAs. *Science* 309, 1514–1518.

22.4 Prokaryotic mRNA Degradation Involves Multiple Enzymes

Research

Bernstein, J. A., Khodursky, A. B., Lin P. H., Lin-Chao, S., and Cohen, S. N. (2002). Global analysis of mRNA decay and abundance in *Escherichia coli* at single-gene resolution using two-color fluorescent DNA microarrays. *Proc. Natl. Acad. Sci. (USA)* 99, 9697–9702.

Celesnik, H., Deana, A., and Belasco, J. G. (2007). Initiation of RNA decay in *Escherichia coli* by 5′ pyrophosphate removal. *Mol. Cell* 27, 79–90.

Mohanty, B. K. and Kushner, S. R. (2006). The majority of *Escherichia coli* mRNAs undergo post-transcriptional modification in exponentially growing cells. *Nucleic Acids Res.* 34(19), 5695–5704.

Reviews

Carpousis, A. J. (2007). The RNA degradosome of *Escherichia coli*: an mRNA-degrading machine assembled on RNase E. *Annu. Rev. Microbiol.* 61, 71–87.

Condon, C. (2007). Maturation and degradation of RNA in bacteria. *Current Opinion in Microbiology* 10, 271–278.

Deana, A. and Belasco, J. G. (2005). Lost in translation: the influence of ribosomes on bacterial mRNA decay. *Genes Dev.* 19, 2526–2533.

22.5 Most Eukaryotic mRNA Is Degraded via Two Deadenylation-Dependent Pathways

Reviews

Franks, T. M. and Lykke-Andersen, J. (2008). The control of mRNA decapping and P-body formation. *Mol. Cell* 32, 605–615.

Parker, R. and Song, H. (2004). The enzymes and control of eukaryotic mRNA turnover. *Nat. Struct. Mol. Biol.* 11, 121–127.

Parker, R. and Sheth, U. (2007). P Bodies and the control of mRNA translation and degradation. *Mol. Cell* 25, 635–646.

Research

Sheth, U. and Parker, R. (2003). Decapping and decay of messenger RNA occur in cytoplasmic processing bodies. *Science* 300, 805–808.

Zheng, D., Ezzeddine, N., Chen, C. Y., Zhu, W., He, X., and Shyu, A. B. (2008). Deadenylation is prerequisite for P-body formation and mRNA decay in mammalian cells. *J. Cell Biol.* 182, 89–101.

22.6 Other Degradation Pathways Target Specific mRNAs

Reviews

Filipowicz, W., Bhattacharyya, S. N., and Sonenberg, N. (2008). Mechanisms of post-transcriptional regulation by microRNAs: are the answers in sight? *Nature Rev. Genet.* 9, 102–114.

Garneau, N. L., Wilusz, J., and Wilusz, C. J. (2007). The highways and byways of mRNA decay. *Nature Rev. Mol. Cell Biol.* 8, 113–126.

Research

Mullen, T. E. and Marzluff, W. F. (2008). Degradation of histone mRNA requires oligouridylation followed by decapping and simultaneous degradation of the mRNA both 5′ to 3′ and 3′ to 5′. *Genes Dev.* 22, 50–65.

22.7 mRNA-Specific Half-Lives Are Controlled by Sequences or Structures within the mRNA

Reviews

Chen, C. Y. A. and Shyu, A. B. (1995). AU-rich elements: characterization and importance in mRNA degradation. *Trends Biochem. Sci.* 20, 465–470.

von Roretz, C. and Gallouzi, I. E. (2008). Decoding ARE-mediated decay: is microRNA part of the equation? *J. Cell Biol.* 181, 189–194.

22.8 Newly Synthesized RNAs Are Checked for Defects via a Nuclear Surveillance System

Research

Arigo, J. T., Eyler, D. E., Carroll, K. L., and Corden, J. L. (2006). Termination of cryptic unstable transcripts is directed by yeast RNA-binding proteins Nrd1 and Nab3. *Mol. Cell* 24, 735–746.

Davis, C. A. and Ares, M. (2006). Accumulation of unstable promoter-associated transcripts upon loss of the nuclear exosome subunit Rrp6p in *Saccharomyces cerevisiae*. *Proc. Natl. Acad. Sci. (USA)* 103, 3262–3267.

Kadaba, S., Wang, X., and Anserson, J. T. (2006). Nuclear RNA surveillance in *Saccharomyces cerevisiae*: Trf4p-dependent polyadenylation of nascent hypomethylated tRNA and an aberrant form of 5S RNA. *RNA* 12, 508–521.

Reviews

Houseley, J., LaCava, J., and Tollervey, D. (2006). RNA-quality control by the exosome. *Nature Rev. Mol. Cell Biol.* 7, 529–539.

Houseley, J. and Tollervey, D. (2008). The nuclear RNA surveillance machinery: the link between ncRNAs and genome structure in budding yeast? *Biochem. Biophys. Acta* 1779, 239–246.

Villa, T., Rougemaille, M., and Libri, D. (2008). Nuclear quality control of RNA polymerase II ribonucleoproteins in yeast: tilting the balance to shape the transcriptome. *Biochem. Biophys. Acta* 1779, 524–531.

22.9 Quality Control of mRNA Translation Is Performed by Cytoplasmic Surveillance Systems

Reviews

Isken, O. and Maquat, L. E. (2007). Quality control of eukaryotic mRNA: safeguarding cells from abnormal mRNA function. *Genes Dev.* 21, 1833–1856.

McGlincy, N. J. and Smith, C. W. J. (2008). Alternative splicing resulting in nonsense-mediated mRNA decay: what is the meaning of nonsense? *Trends Biochem. Sci.* 33, 385–393.

Shyu, A. B., Wilkinson, M. F., and van Hoof, A. (2008). Messenger RNA regulation: to translate or to degrade. *EMBO J.* 27, 471–481.

Stalder, L. and Mühlemann, O. (2008). The meaning of nonsense. *Trends Cell Biol.* 18(7), 315–321.

Research

Wilson, M. A., Meaux, S., and van Hoof, A. (2008). Diverse aberrancies target yeast mRNAs to cytoplasmic mRNA surveillance pathways. *Biochem. Biophys. Acta* 1779, 550–557.

22.10 Some Eukaryotic mRNA Are Localized to Specific Regions of a Cell

Research

Blower, M. D., Feric, E., Weis, K., and Heald, R. (2007). Genome-wide analysis demonstrates conserved localization of messenger RNAs to mitotic microtubules. *J. Cell Biol.* 179, 1365–1373.

Lecuyer, E., Yoshida, H., Parthasarathy, N., Alm, C., Babak, T., Cerovina, T., Hughes, T. R., Tomancak, P., and Krause, H. M. (2007). Global analysis of mRNA localization reveals a prominent role in organizing cellular architecture and function. *Cell* 131, 174–187.

Reviews

Bullock, S. L. (2007). Translocation of mRNAs by molecular motors: think complex? *Seminars Cell Devel. Biol.* 18, 194–201.

Du, T. G., Schmid, M., and Jansen, R. P. (2007). Why cells move messages: the biological functions of mRNA localization. *Semin. Cell Dev. Biol.* 18, 171–177.

Giorgi, C. and Moore, M. J. (2007). The nuclear nurture and cytoplasmic nature of localized mRNPs. *Semin. Cell Devel. Biol.* 18, 186–193.

Martin, K. C. and Ephrussi, A. (2009). mRNA localization: gene expression in the spatial dimension. *Cell* 136, 719–30.

23

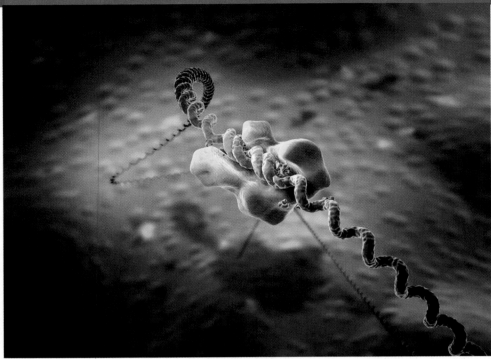

© Nucleus Medical Art, Inc./Phototake/Alamy Images

Catalytic RNA

Edited by Douglas J. Briant

CHAPTER OUTLINE

23.1 Introduction

The idea that only proteins could possess enzymatic activity was deeply rooted in biochemistry. The rationale for the identification of enzymes with proteins resided in the view that only proteins, with their varied three-dimensional structures and variety of side-chain groups, had the flexibility to create the active sites that catalyze biochemical reactions. Critical studies of systems involved in RNA processing, however, have shown this view to be an oversimplification.

The first examples of RNA-based catalysis were identified in the bacterial tRNA processing enzyme, ribonuclease P (RNase P), and self-splicing group I introns in RNA from *Tetrahymena thermophilus*. For their pioneering work on RNA catalysts, Sidney Altman and Thomas Cech were awarded with the 1989 Nobel Prize in Chemistry. Since the initial discovery of catalytic RNA, several other types of catalytic reactions mediated by RNA have been identified. Importantly, ribosomes, the RNA-protein complexes that manufacture peptides (see Chapter 24, *Translation*), have been identified as ribozymes, with RNA acting as the catalytic component and protein acting as a scaffold.

Ribozyme has become a general term used to describe an RNA with catalytic activity, and it is possible to characterize the enzymatic activity in the same way as a more conventional enzyme. Some RNA catalytic activities are directed against separate substrates (intermolecular), whereas others are intramolecular, which limits the catalytic action to a single cycle.

The enzyme RNase P is a ribonucleoprotein that contains a single RNA molecule bound to a protein. RNase P functions intermolecularly and is an example of a ribozyme that catalyzes multiple-turnover reactions. While originally identified in *E. coli*, RNase P is now known to be required for the viability of both prokaryotes and eukaryotes. The RNA possesses the ability to catalyze cleavage in a tRNA substrate, whereas the protein component plays an indirect role, probably to maintain the structure of the catalytic RNA.

The two classes of self-splicing introns, group I and group II, are good examples of ribozymes that function intramolecularly. Both group I and group II introns possess the ability to splice themselves out of their respective pre-mRNAs. While under normal conditions the self-splicing reaction is intramolecular and therefore single-turnover, group I introns can be engineered to generate RNA molecules that have several other catalytic activities related to the original activity.

The common theme of the reactions performed by catalytic RNA is that the RNA can perform an intramolecular or intermolecular reaction that involves cleavage or joining of phosphodiester bonds *in vitro*. Although the specificity of the reaction and the basic catalytic activity is provided by RNA, proteins associated with the RNA may be needed for the reaction to occur efficiently *in vivo*.

RNA splicing is not the only means by which changes can be introduced in the informational content of RNA. In the process of **RNA editing**, changes are introduced at individual bases, or bases are added at particular positions within an mRNA. The insertion of bases (most commonly uridine residues) occurs for several genes in the mitochondria of certain unicellular/oligocellular eukaryotes. Like splicing, RNA editing involves the breakage and reunion of bonds between nucleotides, but also requires a template for coding the information of the new sequence.

23.2 Group I Introns Undertake Self-Splicing by Transesterification

Key concepts

- The only factors required for autosplicing *in vitro* by group I introns are two metal ions and a guanosine nucleotide.
- Splicing occurs by two transesterifications, without requiring input of energy.
- The 3'–OH end of the guanosine cofactor attacks the 5' end of the intron in the first transesterification.
- The 3'–OH end generated at the end of the first exon attacks the junction between the intron and second exon in the second transesterification.
- The intron is released as a linear molecule that circularizes when its 3'–OH terminus attacks a bond at one of two internal positions.
- In *Tetrahymena*, an internal bond of the excised intron can also be attacked by other nucleotides in a *trans*-splicing region.

Group I introns are found in diverse locations, with more than two thousand identified to date. Unlike RNase P, group I introns are not essential for viability. Group I introns occur in the genes coding for rRNA in the nuclei of the unicellular/oligocellular eukaryotes *Tetrahymena thermophila* (a ciliate) and *Physarum polycephalum* (a slime mold). They are common in the genes of fungi and protists as well as occurring rarely in prokaryotes and animals. Group I introns have an intrinsic ability to splice themselves. This is called **self-splicing** or **autosplicing**. (This property also is found in the group II introns discussed in *Section 23.6, Group II Introns May Code for Multifunction Proteins*.)

Self-splicing was discovered as a property of the transcripts of the rRNA genes in *T. thermophila*. The genes for the two major rRNAs follow the usual organization, in which both are expressed as part of a common transcription unit. The product is a 35S precursor RNA with the sequence of the small (17S) rRNA in the 5' end, and the sequence of the larger (26S) rRNA toward the 3' end.

In some strains of *T. thermophila*, the sequence coding for 26S rRNA is interrupted by a single, short intron. When the 35S precursor RNA is incubated *in vitro*, splicing occurs as an autonomous reaction. The intron is excised from the precursor and accumulates as a linear fragment of 400 bases, which is subsequently converted to a circular RNA. These events are summarized in **FIGURE 23.1**.

The reaction requires two metal ions and a guanosine nucleotide cofactor. No other base

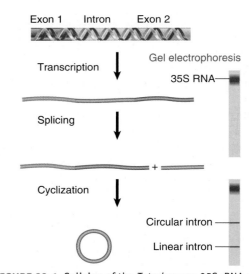

FIGURE 23.1 Splicing of the *Tetrahymena* 35S rRNA precursor can be followed by gel electrophoresis. The removal of the intron is revealed by the appearance of a rapidly moving small band. When the intron becomes circular, it electrophoreses more slowly, as seen by a higher band.

can be substituted for G, but a triphosphate is not needed: GTP, GDP, GMP, and guanosine itself all can be used, so there is no net energy requirement. The guanosine nucleotide must have a 3'–OH group.

The fate of the guanosine nucleotide can be followed by using a radioactive label. The radioactivity initially enters the excised linear intron fragment. The G residue becomes linked to the 5' end of the linear intron by a normal phosphodiester bond.

FIGURE 23.2 shows that three transfer reactions occur. In the first transfer, the guanosine nucleotide behaves as a cofactor providing a free 3'–OH group that attacks the 5' end of the intron. This reaction creates the G–intron link and generates a 3'–OH group at the end of the 5' exon. The second transfer involves a similar chemical reaction, in which the newly formed 3'–OH at the end of exon 1 attacks the second exon. The two transfers are connected; no free exons have been observed, so their ligation may occur as part of the same reaction that releases the intron. The intron is released as a linear molecule, but the third transfer reaction converts it to a circle.

Each stage of the self-splicing reaction occurs by a transesterification, in which one phosphate ester is converted directly into another without any intermediary hydrolysis. Bonds are exchanged directly and energy is conserved, so the reaction does not require input of energy from hydrolysis of ATP or GTP. Each consecutive transesterification reaction involves no net change of energy. In the cell, the concentration of GTP is high relative to that of RNA and therefore drives the reaction forward whereupon a change in secondary structure in the RNA prevents the reverse reaction. This allows the reaction to proceed to completion, instead of coming to equilibrium between spliced product and nonspliced precursors.

The ability to splice is intrinsic to the RNA and the system is able to proceed *in vitro* without addition of any protein components. The RNA forms a specific secondary/tertiary structure in which the relevant groups are brought into juxtaposition so that a guanosine nucleotide can be bound to a specific site and then the bond breakage and reunion reactions shown in Figure 23.2 can occur. Although a property of the RNA itself, the reaction is very slow *in vitro*. This is because group I intron splicing is assisted *in vivo* by proteins that serve to stabilize the RNA structure in a favorable conformation for splicing.

The ability to engage in these transfer reactions resides with the sequence of the intron, which continues to be reactive after its excision as a linear molecule. **FIGURE 23.3** summarizes catalytic activities of the excised intron from *Tetrahymena*, with residue numbers corresponding to that organism.

The intron can circularize when the 3' terminal G (ΩG) attacks an internal position near the 5' end. The internal bond is broken and the new 5' end is transferred to the 3'–OH end of the intron, circularizing the intron. The previous 5' end with the original exogenous guanosine nucleotide (exoG) is released as a linear fragment. The circularized intron can be linearized by specifically hydrolyzing the bond between ΩG and the internal residue that had closed the circle. This is called a *reverse cyclization*. Depending on the position of the primary cyclization, the linear molecule generated by hydrolysis remains reactive and can perform a secondary cyclization.

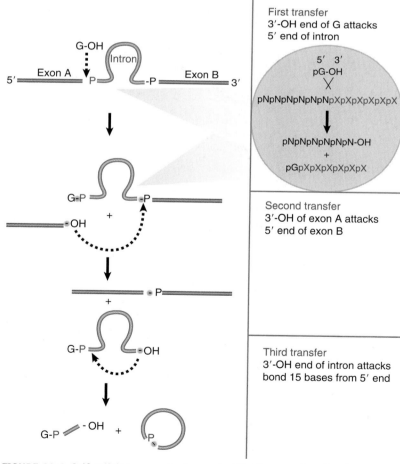

FIGURE 23.2 Self-splicing occurs by transesterification reactions in which bonds are exchanged directly. The bonds that have been generated at each stage are indicated by the shaded boxes.

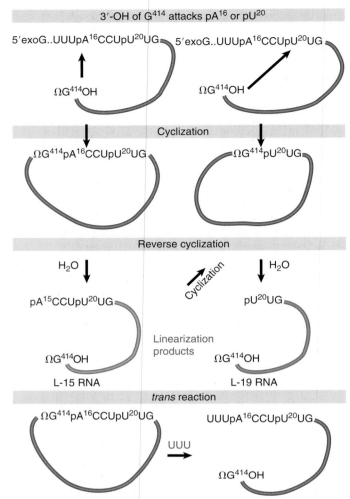

3′-OH of G414 attacks pA16 or pU20

5′exoG..UUUpA^{16}CCUpU^{20}UG   5′exoG..UUUpA^{16}CCUpU^{20}UG

ΩG^{414}OH   ΩG^{414}OH

Cyclization

ΩG^{414}pA^{16}CCUpU^{20}UG   ΩG^{414}pU^{20}UG

Reverse cyclization

H$_2$O   Cyclization   H$_2$O

pA^{15}CCUpU^{20}UG   pU^{20}UG

Linearization products

ΩG^{414}OH   ΩG^{414}OH

L-15 RNA   L-19 RNA

trans reaction

ΩG^{414}pA^{16}CCUpU^{20}UG   UUUpA^{16}CCUpU^{20}UG

UUU

ΩG^{414}OH

FIGURE 23.3 The excised intron can form circles by using either of two internal sites for reaction with the 5′ end, and can reopen the circles by reaction with water or oligonucleotides.

The final product of the spontaneous reactions following release of the *Tetrahymena* group I intron is the L-19 RNA, a linear molecule generated by reversing the shorter circular form. This molecule has an enzymatic activity that allows it to catalyze the extension of short oligonucleotides. The reactivity of the released intron extends beyond merely reversing the cyclization reaction. Addition of the oligonucleotide UUU reopens the primary circle by reacting with the ΩG–internal nucleotide bond. The UUU (which resembles the 3′ end of the 15-mer released by the primary cyclization) becomes the 5′ end of the linear molecule that is formed. This is an *intermolecular* reaction, and thus demonstrates the ability to connect together two different RNA molecules.

This series of reactions demonstrates vividly that the autocatalytic activity reflects a generalized ability of the RNA molecule to form an active center that can bind guanosine cofactors, recognize oligonucleotides, and bring together the reacting groups in a conformation that allows bonds to be broken and rejoined. Other group I introns have not been investigated in as much detail as the *Tetrahymena* intron, but their properties are generally similar.

The autosplicing reaction is an intrinsic property of RNA *in vitro*, but the role of proteins *in vivo* is not fully characterized. Some indications for the involvement of proteins are provided by mitochondrial systems, where splicing of group I introns requires the *trans*-acting products of other genes. One striking case is presented by the *cyt18* mutant of *Neurospora crassa*, which is defective in splicing several mitochondrial group I introns. The product of this gene turns out to be the mitochondrial tyrosyl-tRNA synthetase. This is explained by the fact that the intron can take up a tRNA-like tertiary structure that is stabilized by the synthetase, thereby promoting the catalytic reaction. This relationship between the synthetase and splicing is consistent with the idea that splicing originated as an RNA-mediated reaction, subsequently assisted by RNA-binding proteins that originally had other functions. The *in vitro* self-splicing ability may represent the basic biochemical interaction. The RNA structure creates the active site, but is able to function efficiently *in vivo* only when assisted by a protein complex.

23.3 Group I Introns Form a Characteristic Secondary Structure

Key concepts

- Group I introns form a secondary structure with nine duplex regions.
- The cores of regions P3, P4, P6, and P7 have catalytic activity.
- Regions P4 and P7 are both formed by pairing between conserved consensus sequences.
- A sequence adjacent to P7 base pairs with the sequence that contains the reactive G.

All group I introns can be organized into a characteristic secondary structure with nine helices (P1–P9). **FIGURE 23.4** shows a model for the secondary structure of the *Tetrahymena* intron. While structural analyses were able to elucidate the secondary structure of the group I intron, it was not until the recent determination of the crystal structure that the tertiary structure of the intron was revealed. Several crystal struc-

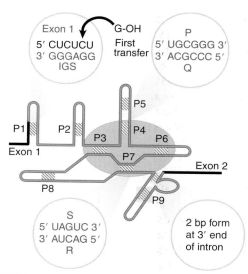

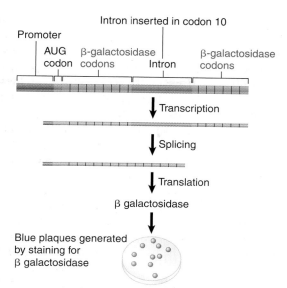

FIGURE 23.4 Group I introns have a common secondary structure that is formed by nine base-paired regions. The sequences of regions P4 and P7 are conserved, and identify the individual sequence elements P, Q, R, and S. P1 is created by pairing between the end of the left exon and the IGS of the intron; a region between P7 and P9 pairs with the 3′ end of the intron.

FIGURE 23.5 Placing the *Tetrahymena* intron within the β-galactosidase coding sequence creates an assay for self-splicing in *E. coli*. Synthesis of β-galactosidase can be tested by adding a compound that is turned blue by the enzyme. The sequence is carried by a bacteriophage, so the presence of blue plaques (containing infected bacteria) indicates successful splicing.

tures of group I introns have been solved and these confirm previous models of the secondary structure. Two of the base-paired regions are generated by pairing between conserved sequence elements that are common to group I introns. P4 is constructed from the sequences *P* and *Q*; P7 is formed from the sequences *R* and *S*. The other base-paired regions vary in sequence in individual introns. Mutational analysis identifies an intron "core" containing P3, P4, P6, and P7, which provides the minimal region that can undertake a catalytic reaction. The lengths of group I introns vary widely and the consensus sequences are located a considerable distance from the actual splice junctions.

Some of the pairing reactions are directly involved in bringing the splice junctions into a conformation that supports the enzymatic reaction. P1 includes the 3′ end of the 5′ exon. The sequence within the intron that pairs with the exon is called the internal guide sequence (IGS). The name IGS reflects the fact that originally the region immediately 3′ to the IGS sequence shown in Figure 23.4 was thought to pair with the 3′ splice junction, thus bringing the two junctions together. This interaction may occur but does not seem to be essential. A very short sequence—sometimes as short as two bases—between P7 and P9 base pairs with the sequence that immediately precedes the reactive G (ΩG, position 414 in *Tetrahymena*) at the 3′ end of the intron.

The importance of base pairing in creating the necessary core structure in the RNA is emphasized by the properties of *cis*-acting mutations that prevent splicing of group I introns. Such mutations have been isolated for the mitochondrial introns through mutants that cannot remove an intron *in vivo*, and they have been isolated for the *Tetrahymena* intron by transferring the splicing reaction into a bacterial environment. The construct shown in **FIGURE 23.5** allows the splicing reaction to be followed in *E. coli*. The self-splicing intron is placed at a location that interrupts the tenth codon of the β-galactosidase coding sequence. The protein can therefore be successfully translated from an RNA only after the intron has been removed. The synthesis of β-galactosidase by *E. coli* in this system indicates that splicing can occur in conditions quite distant from those prevailing in *Tetrahymena* or even *in vitro*. While the group I intron from *Tetrahymena* can autosplice from the β-galactosidase mRNA in *E. coli*, it is not clear whether or not the reaction is assisted by bacterial proteins. In this assay, mutations in the group I consensus sequences that disrupt their base pairing stop splicing and therefore prevent expression of β-galactosidase. The mutations can be reverted by making compensating changes that restore base pairing.

Mutations in the corresponding consensus sequences in mitochondrial group I introns have similar effects to those observed in *Tetrahymena*. A mutation in one consensus sequence may be reverted by a mutation in the complementary consensus sequence to restore pairing; for example, mutations in the R consensus can be compensated by mutations in the S consensus.

Together these results suggest that the group I splicing reaction depends on the formation of secondary structure between pairs of consensus sequences within the intron. The principle established by this work is that *sequences distant from the splice junctions themselves are required to form the active site that makes self-splicing possible.*

Ribozymes Have Various Catalytic Activities

Key concepts

- By changing the substrate binding-site of a group I intron, it is possible to introduce alternative sequences that interact with the reactive G.
- The reactions follow classical enzyme kinetics with a low catalytic rate.
- Reactions using 2'–OH bonds could have been the basis for evolving the original catalytic activities in RNA.

The catalytic activity of group I introns was discovered by virtue of their ability to autosplice, but they are able to undertake other catalytic reactions *in vitro*. All of these reactions are based on transesterifications. We analyze these reactions in terms of their relationship to the splicing reaction itself.

The catalytic activity of a group I intron is conferred by its ability to generate particular secondary and tertiary structures that create active sites that are equivalent to the active sites of conventional (proteinaceous) enzymes. **FIGURE 23.6** illustrates the splicing reaction in terms of these sites (this is the same series of reactions shown in Figure 23.2).

The substrate-binding site is formed from the P1 helix, in which the 3' end of the first intron base pairs with the IGS. A guanosine-binding site is formed by sequences in P7. This site may be occupied either by a free exogenous guanosine nucleotide (exoG) or by the ΩG residue (position 414 in *Tetrahymena*). In the first transfer reaction, the guanosine-binding site is occupied by free guanosine nucleotide. Following release of the intron it is occupied by ΩG. The second transfer releases the joined exons. The third transfer creates the circular intron.

Binding to the substrate involves a change of conformation. Before substrate binding, the

Catalytic RNA has a guanosine-binding site and substrate-binding site

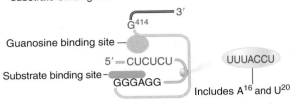

First transfer G-OH occupies G-binding site; 5' exon occupies substrate-binding site

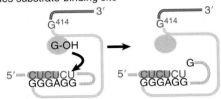

Second transfer G^{414} is in G-binding site; 5' exon is in substrate-binding site

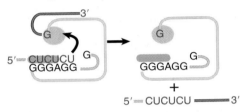

Third transfer G^{414} is in G-binding site; 5' end of intron is in substrate-binding site

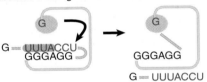

FIGURE 23.6 Excision of the group I intron in *Tetrahymena* rRNA occurs by successive reactions between the occupants of the guanosine-binding site and the substrate-binding site. The left exon is pink, and the right exon is purple.

5' end of the IGS is close to P2 and P8; after binding, when it forms the P1 helix, it is close to conserved bases that lie between P4 and P5. The reaction is visualized by contacts that are detected in the secondary structure in **FIGURE 23.7**. In the tertiary structure, the two sites alternatively contacted by P1 are 37 Å apart, which implies a substantial movement in the position of P1.

Some further enzymatic reactions that *Tetrahymena* group I introns can perform are characterized in **FIGURE 23.8**. The ribozyme can function as a sequence-specific endoribonuclease by utilizing the ability of the IGS to bind complementary sequences. In this example,

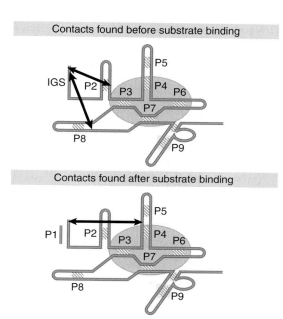

Contacts found before substrate binding

Contacts found after substrate binding

FIGURE 23.7 The position of the IGS in the tertiary structure changes when P1 is formed by substrate binding.

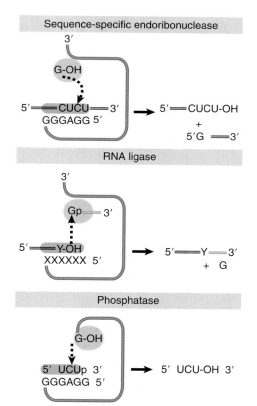

Sequence-specific endoribonuclease

RNA ligase

Phosphatase

FIGURE 23.8 Catalytic reactions of the ribozyme involve transesterifications between a group in the substrate-binding site and a group in the G-binding site.

it binds an external substrate containing the sequence CUCU, instead of binding the analogous sequence that is usually contained at the end of the 5′ exon. A guanosine-containing nucleotide is present in the G-binding site, and

Enzyme	Substrate	K_M (mM)	Turnover (/min)
19-base virusoid	24-base RNA	0.0006	0.5
L-19 Intron	CCCCCC	0.04	1.7
RNase P RNA	pre-tRNA	0.00003	0.4
RNase P complete	pre-tRNA	0.00003	29
RNase T1	GpA	0.05	5,700
β galactosidase	lactose	4.0	12,500

FIGURE 23.9 Reactions catalyzed by RNA have the same features as those catalyzed by proteins, although the rate is slower. The K_M gives the concentration of substrate required for half-maximum velocity; this is an inverse measure of the affinity of the enzyme for substrate. The turnover number gives the number of substrate molecules transformed in unit time by a single catalytic site.

attacks the CUCU sequence in precisely the same way that the exon is usually attacked in the first transfer reaction. This cleaves the target sequence into a 5′ molecule that resembles the 5′ exon and a 3′ molecule that bears a terminal G residue.

By mutating the IGS element, it is possible to change the specificity of the ribozyme so that it recognizes sequences complementary to the new sequence at the IGS region. This alteration of the IGS to change the specificity of the substrate-binding site enables other RNA targets to be processed by the ribozyme, which can also be used to perform RNA–ligase reactions. An RNA terminating in a 3′–OH is bound in the substrate site and an RNA terminating in a 5′–G residue is bound in the G-binding site. An attack by the hydroxyl on the phosphate bond connects the two RNA molecules, with the loss of the G residue.

The phosphatase reaction is not directly related to the splicing transfer reactions. An oligonucleotide sequence that is complementary to the IGS and terminates in a 3′–phosphate can be attacked by the ΩG. The phosphate is transferred to the ΩG and an oligonucleotide with a free 3′–OH end is then released. The phosphate can then be transferred either to an oligonucleotide terminating in 3′–OH (effectively reversing the reaction) or indeed to water releasing inorganic phosphate and completing an authentic phosphatase reaction.

The reactions catalyzed by RNA can be characterized in the same way as classical enzymatic reactions in terms of Michaelis–Menten kinetics. FIGURE 23.9 analyzes the reactions catalyzed by RNA. The K_M values for RNA-catalyzed reactions are low and therefore imply that the RNA can bind its substrate with high specificity. The turnover numbers for RNA catalyzed reactions, however, are low, which reflects a low catalytic rate. In effect, the RNA molecules behave in the same general manner as traditionally defined for enzymes, although they are relatively slow compared to protein catalysts (where a typical range of turnover numbers is 10^3 to 10^6 min^{-1}).

A powerful extension of the activities of ribozymes has been made with the discovery that they can be regulated by ligands (see *Section 30.3, Noncoding RNAs Can Be Used to Regulate Gene Expression*). These *cis*-acting regulatory RNA regions are called **riboswitches**. In almost all riboswitches, a conformational change determines the on or off state of the switch. One notable exception is the *glmS* gene in Gram-positive bacteria, which forms a self-cleaving ribozyme in the presence of glucosamine-6-phosphate (GlcN6P). **FIGURE 23.10** summarizes the regulation of the *glmS* riboswitch. (See *Section 30.2, A Riboswitch Can Alter Its Structure According to Its Environment*, for additional details.)

If an active center is a surface that exposes a series of active groups in a fixed relationship, it is possible to understand how RNA is capable of providing a catalytic center. In a protein, the active groups are provided by the side chains of the amino acids. The amino acid side chains have appreciable variety, including positive and negative ionic groups and hydrophobic groups. In RNA, the available moieties are more restricted, consisting primarily of the exposed groups of bases. Short regions of RNA are held in a particular secondary/tertiary conformation, providing an active surface and maintaining an environment in which bonds can be broken and formed. It seems inevitable that the interaction between the RNA catalyst and the RNA substrate will rely on base pairing to create the active environment. Divalent cations (usually Mg^{2+}) play an important role in structure, typically being present at the active site where they coordinate the positions of the various groups. Divalent metal cations also play a direct role in the endonucleolytic activity of virusoid ribozymes (see *Section 23.9, Viroids Have Catalytic Activity*).

The evolutionary implications of these discoveries are intriguing. The split personality of the genetic apparatus—in which RNA is present in all components, but proteins undertake catalytic reactions—has always been puzzling. It seems unlikely that the very first replicating systems could have contained both nucleic acid and protein. Suppose, though, that the first systems contained only a self-replicating nucleic acid with primitive catalytic activities—just those needed to make and break phosphodiester bonds. If it is also assumed that the involvement

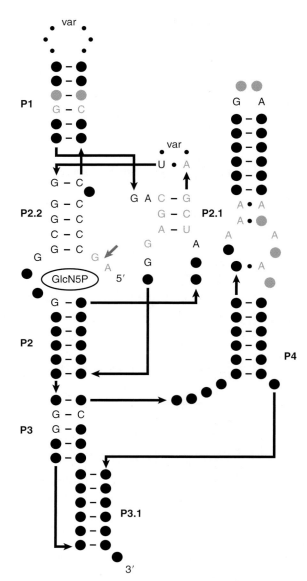

FIGURE 23.10 A ribozyme is contained within the 5' untranslated region of the mRNA coding for the enzyme that produces glucosamine-6-phosphate (GlcN6P). When GlcN6P binds to the ribozyme, it cleaves off the 5' end of the mRNA, thereby inactivating it and preventing further production of the enzyme. Regions important for maintaining the active tertiary structure are shown in blue, while the cleavage site is indicated by a red arrow. Reprinted from *Curr. Opin. Struct. Biol.*, vol. 17, T. E. Edwards, D. J. Klein, and A. R. Ferré-D' Amaré, Riboswitches: small-molecule recognition . . . , pp. 273–279. Copyright 2007, with permission from Elsevier [http://www.sciencedirect.com/science/journal/0959440X].

of 2'–OH bonds in current splicing reactions is derived from these primitive catalytic activities, it can be argued that the original nucleic acid was RNA, because DNA lacks the 2'–OH group and therefore could not undertake such reactions. Proteins could have been added for their ability to stabilize the RNA structure. The greater versatility of proteins then could have allowed them to take over catalytic reactions, leading eventually to the complex and sophisticated apparatus of modern gene expression.

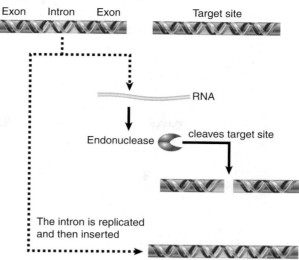

FIGURE 23.11 An intron codes for an endonuclease that makes a double-strand break in DNA. The sequence of the intron is duplicated and then inserted at the break.

23.5 Some Group I Introns Code for Endonucleases That Sponsor Mobility

Key concepts

- Mobile introns are able to insert themselves into new sites.
- Mobile group I introns code for an endonuclease that makes a double-strand break at a target site.
- The intron transposes into the site of the double-strand break by a DNA-mediated replicative mechanism.

Certain introns of both the group I and group II classes contain open reading frames that are translated into proteins. Expression of the proteins allows the intron (either in its original RNA form or as a DNA copy of the RNA) to be *mobile*: It is able to insert itself into a new genomic site. Introns of both groups I and II are widespread, being found in both prokaryotes and eukaryotes. Group I introns migrate by DNA-mediated mechanisms, whereas group II introns migrate by RNA-mediated mechanisms.

Intron mobility was first detected by crosses in which the alleles for the relevant gene differ with regard to their possession of the intron. Polymorphisms for the presence or absence of introns are common in fungal mitochondria. This is consistent with the view that these introns originated by insertion into the gene. Some light on the process that could be involved is cast by an analysis of recombination in crosses involving the large rRNA gene of the yeast mitochondrion.

The large rRNA gene of the yeast mitochondrion has a group I intron that contains a coding sequence. The intron is present in some strains of yeast (called ω^+) but absent in others (ω^-). Progeny of genetic crosses between ω^+ and ω^- do not result in the expected genotypic ratio; the progeny are usually ω^+. If we think of the ω^+ strain as a donor and the ω^- strain as a recipient, we form the view that in $\omega^+ \times \omega^-$ crosses a new copy of the intron is generated in the ω^- genome. As a result, the progeny are all ω^+. Mutations can occur in either parent to abolish the non-Mendelian genotypic assortment. Certain mutants show normal segregation, with equal numbers of ω^+ and ω^- progeny. When mapped, mutations in the ω^- strain occur close to the site where the intron would be inserted.

Mutations in the ω^+ strain lie in the reading frame of the intron and prevent production of the protein. This suggests the model of **FIGURE 23.11**, in which the protein coded by the intron in an ω^+ strain recognizes the site where the intron should be inserted in an ω^- strain and causes it to be preferentially inherited.

Some group I introns encode endonucleases that make them mobile. There are at least five families of homing endonuclease genes (HEGs). Two common families of HEGs are the LAGLIDADG and His-Cys Box endonucleases. These HEG-containing group I introns, however, constitute a small portion of the overall number of nuclear group I introns. While approximately 1200 nuclear group I introns have been identified, less than 30 of these contain HEGs.

The ω intron contains an HEG, the product of which is an endonuclease known as I-SceI. *I-SceI recognizes the ω^- gene as a target for a double-strand break.* I-SceI recognizes an 18-bp target sequence that contains the site where the intron is inserted. The target sequence is cleaved on each strand of DNA two bases to the 3' side of the insertion site. Thus the cleavage sites are 4 bp apart and generate overhanging single strands. This type of cleavage is related to the cleavage characteristic of transposons when they migrate to new sites (see *Chapter 17, Transposable Elements and Retroviruses*). The double-strand break probably initiates a gene conversion process in which the sequence of the ω^+ gene is copied to replace the sequence of the ω^- gene. The reaction involves transposition by a duplicative mechanism and occurs solely at the level of DNA. Insertion of the intron interrupts the sequence recognized by the endonuclease, thus ensuring stability.

Similar introns often carry quite different endonucleases. There are differences in the details of insertion; for example, the endonuclease coded by the phage T4 *td* intron cleaves a target site that is 24 bp upstream of the site at which the intron is itself inserted. The dissociation between the intron sequence and the endonuclease sequence is emphasized by the fact that the same endonuclease sequences are found in inteins (sequences that code for self-splicing proteins; see *Section 23.12, Protein Splicing Is Autocatalytic*).

The variation in the endonucleases means that there is no homology between the sequences of their target sites. The target sites are among the longest and therefore the most specific known for any endonucleases (with a range of 14 to 40 bp). The specificity ensures that the intron perpetuates itself only by insertion into a single target site and not elsewhere in the genome. This is called **intron homing**.

Introns carrying sequences that code for endonucleases are found in a variety of bacteria and unicellular/oligocellular eukaryotes. These results strengthen the view that introns carrying coding sequences originated as independent elements.

23.6 Group II Introns May Code for Multifunction Proteins

Key concepts

- Group II introns can autosplice *in vitro*, but are usually assisted by protein activities encoded in the intron.
- A single coding frame specifies a protein with reverse transcriptase activity, maturase activity, a DNA-binding motif, and a DNA endonuclease.
- The endonuclease cleaves target DNA to allow insertion of the intron at a new site.
- The reverse transcriptase generates a DNA copy of the inserted RNA intron sequence.

The best characterized mobile group II introns code for a single protein in a region of the intron beyond its catalytic core. The typical protein contains an N-terminal reverse transcriptase activity, a central domain associated with an ancillary activity that assists folding of the intron into its active structure (called the *maturase*; see *Section 23.7, Some Autosplicing Introns Require Maturases*), a DNA-binding domain, and a C-terminal endonuclease domain.

The endonuclease initiates the transposition reaction and plays the same role in homing as its counterpart in a group I intron. The reverse transcriptase generates a DNA copy of the intron that is inserted at the homing site. The endonuclease also cleaves target sites that resemble, but are not identical to, the homing site at much lower frequency, leading to insertion of the intron at new locations.

FIGURE 23.12 illustrates the transposition reaction for a typical group II intron. First, the endonuclease makes a single-strand break in the antisense strand. Cleavage of the sense strand is achieved by a reverse splicing reaction, with the RNA intron inserting itself into the DNA between the DNA exons. This newly inserted RNA intron can now act as a template for the reverse transcriptase. Almost all group II introns have a reverse transcriptase activity that is specific for the intron. The reverse transcriptase generates a DNA copy of the intron, with the end result being the insertion of the intron into the target site as a duplex DNA.

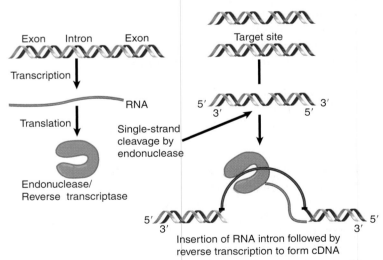

FIGURE 23.12 Reverse transcriptase/endonuclease coded by an intron allows a copy of the RNA to be inserted into a target site.

23.7 Some Autosplicing Introns Require Maturases

Key concept

- Autosplicing introns may require maturase activities encoded within the intron to assist folding into the active catalytic structure.

Although group I and group II introns both have the capacity to autosplice *in vitro*, under physiological conditions they usually require assistance from proteins. Both types of intron may code for **maturase** activities that are required to assist the splicing reaction.

The maturase activity is part of the single open reading frame coded by the intron. In the example of introns that code for homing endonucleases, the single protein product has both endonuclease and maturase activity. Mutational analysis shows that the two activities are independent. Structural analysis confirms the mutational data and shows that the endonuclease and maturase activities are provided by different active sites in the protein, each coded by a separate domain. The coexistence of endonuclease and maturase activities in the same protein suggests a route for the evolution of the intron. **FIGURE 23.13** suggests that the intron originated in an independent autosplicing element. While Figure 23.13 depicts a group I intron, the process for group II introns is presumed to be similar. The insertion into this element of a sequence coding for an endonuclease gave it mobility. The insertion, however, might well disrupt the ability of the RNA sequence to fold into the active structure. This would create pressure for assistance from proteins that could restore folding ability. The incorporation of such a sequence into the intron would maintain its independence.

Some group II introns, however, do not code for maturase activity. These group II introns may use proteins (comparable to intron-encoded maturases) that are instead encoded by sequences in the host genome. This suggests a possible route for the evolution of general splicing factors. The factor may have originated as a maturase that specifically assisted the splicing of a particular intron. The coding sequence

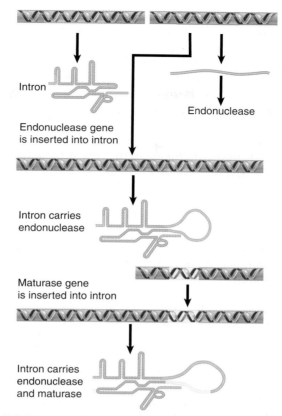

FIGURE 23.13 The intron originated as independent sequence coding for a self-splicing RNA. The insertion of the endonuclease sequence created a homing intron that was mobile. The insertion of the maturase sequence then enhanced the ability of the intron sequences to fold into the active structure for splicing.

became isolated from the intron in the host genome and then it evolved to function with a wider range of substrates that the original intron sequence. The catalytic core of the intron could have evolved into an snRNA.

23.8 The Catalytic Activity of RNase P Is Due to RNA

Key concepts

- Ribonuclease P (RNase P) is a ribonucleoprotein in which the RNA has catalytic activity.
- RNase P is essential for bacteria, archaea, and eukaryotes.
- RNase MRP in eukaryotes is related to RNase P and is involved in rRNA processing and degradation of cyclin B mRNA.

One of the first demonstrations of the catalytic capabilities of RNA was provided by the dissection of ribonuclease P (RNase P) from *E. coli*. While originally identified in bacteria, RNase P has been identified as an essential endonuclease involved in tRNA processing in most, if not all, bacterial, archaeal, and eukaryotic organisms.

In its simplest form, bacterial RNase P can be dissociated into two components: a base RNA of 350 to 400 nucleotides and a single protein subunit. The RNA subunit from bacteria, when isolated *in vitro*, displays catalytic activity. RNase P from archaea and eukaryotes consists of a single RNA structurally related to that found in bacteria, but it has a higher protein content and the RNA has little if any catalytic activity when examined *in vitro*. Typically, archaeal RNase P is associated with four proteins, whereas the yeast version is associated with nine proteins and the human version with ten proteins. In all cases, the protein component is required to support RNase P activity *in vivo*. Mutations in either the gene for the RNA or the gene for the protein can inactivate RNase P *in vivo*, so we know that both components are necessary for natural enzyme activity. Originally it had been assumed that the protein provided the catalytic activity, while the RNA filled some subsidiary role—for example, assisting in the binding of substrate, since it has some short sequences complementary to exposed regions of tRNA. These roles, however, are reversed, with the RNA actually providing the catalytic activity while the protein provides structural support.

Analyzing the results as though the RNA were an enzyme, each "enzyme" catalyzes the cleavage of multiple substrates. Although the catalytic activity resides in the RNA, the protein component greatly increases the speed of the reaction, as seen in the increase in turnover number (see Figure 23.9).

In addition to RNase P, eukaryotes have another essential RNA-based endonuclease, RNase MRP (*mitochondrial RNA processing*). This endonuclease is composed of a structurally related catalytic RNA and shares many of the same protein subunits that are found in RNase P. While originally identified for its role in processing mitochondrial RNAs, RNase MRP functions mainly in the nucleus, processing precursor ribosomal RNA. RNase MRP may also play an important role in cell cycle regulation as it is involved in degradation of cyclin B mRNA. Identification of RNase MRP is provocative, as it appears that the protein component is largely conserved between RNase P and RNase MRP,

with the change in substrate specificity provided by exchanging the catalytic RNA.

23.9 Viroids Have Catalytic Activity

Key concepts

- Viroids and virusoids form a hammerhead structure that has a self-cleaving activity.
- Similar structures can be generated by pairing a substrate strand that is cleaved by an enzyme strand.
- When an enzyme strand is introduced into a cell, it can pair with a substrate strand target that is then cleaved.

Another example of the ability of RNA to function as an endonuclease is provided by some small (~350 nt) plant RNAs that undertake a self-cleavage reaction. As with the case of the *Tetrahymena* group I intron, however, it is possible to engineer constructs that can function on external substrates.

These small plant RNAs fall into two general groups: viroids and virusoids. The **viroids** are infectious RNA molecules that function independently without encapsidation by any protein coat. The **virusoids** (which are sometimes called **satellite RNAs**) are similar in organization but are encapsidated by plant viruses, being packaged together with a viral genome. The virusoids cannot replicate independently, as they require assistance from the virus.

Viroids and virusoids both replicate via rolling circles (see Figure 12.6). The strand of RNA that is packaged into the virus is called the *plus strand*. The complementary strand, generated during replication of the RNA, is called the *minus strand*. Multimers of both plus and minus strands are found. Both types of monomer are generated by cleaving the tail of a rolling circle; circular plus-strand monomers are generated by ligating the ends of the linear monomer.

Both plus and minus strands of viroids and virusoids undergo self-cleavage *in vitro*. Some of the RNAs cleave *in vitro* under physiological conditions. Others do so only after a cycle of heating and cooling; this suggests that the isolated RNA has an inappropriate conformation, but can generate an active conformation when it is denatured and renatured.

The viroids and virusoids that undergo self-cleavage form a "hammerhead" secondary structure at the cleavage site, as drawn in the upper part of **FIGURE 23.14**. Hammerhead ribozymes belong to a family of ribozymes that

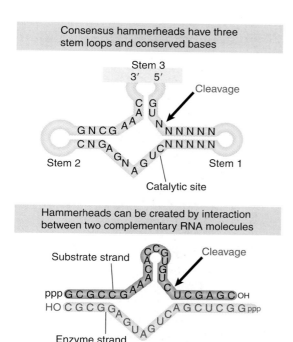

Consensus hammerheads have three stem loops and conserved bases

FIGURE 23.14 Self-cleavage sites of viroids and virusoids have a consensus sequence and form a hammerhead secondary structure by intramolecular pairing. Hammerheads can also be generated by pairing between a substrate strand and an "enzyme" strand.

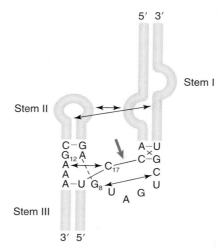

FIGURE 23.15 The hammerhead ribozyme structure is held in an active tertiary conformation by interactions between stem loops, indicated by arrows. The site of cleavage is marked with a red arrow. Adapted from M. Martick and W. G. Scott, *Cell* (126): 309–320.

include hepatitis delta virus (HDV), hairpin ribozymes, and Varkud satellite (VS) ribozyme. Functionally, HDV requires divalent metal cations to promote cleavage, while hammerhead and hairpin ribozymes do not require metal. The importance of metal for VS ribozyme cleavage is still ambiguous. All of these ribozymes, however, generate a cleavage that leaves 5'–OH and 2'–3'–cyclic phosphodiester termini.

Unlike all other ribozymes identified to date, hammerhead ribozymes and other members of the family do not require a protein component to function *in vivo*, as the sequence of this structure is sufficient for cleavage. Minimally, for hammerhead ribozymes the active site is a sequence of only 58 nucleotides. The hammerhead contains three stem-loop regions whose position and size are constant and 13 conserved nucleotides, mostly in the regions connecting the center of the structure. The conserved bases and duplex stems generate an RNA with the intrinsic ability to cleave.

An active hammerhead can also be generated by pairing an RNA representing one side of the structure with an RNA representing the other side. The lower part of Figure 23.14 shows an example of a hammerhead generated by hybridizing a 19 nt molecule with a 24 nt molecule. The hybrid mimics the hammerhead

structure, with the omission of loops I and III. When the 19 nt RNA is added to the 24 nt RNA, cleavage occurs at the appropriate position in the hammerhead. We may regard the top (24 nt) strand of this hybrid as comprising the "substrate" and the bottom (19 nt) strand as comprising the "enzyme." When the 19 nt RNA is mixed with an excess of the 24 nt RNA, multiple copies of the 24 nt RNA are cleaved. This suggests that there is a cycle of 19 nt–24 nt pairing, cleavage, dissociation of the cleaved fragments from the 19 nt RNA, and pairing of the 19 nt RNA with a new 24 nt substrate. The 19 nt RNA is therefore a ribozyme with endonuclease activity. The parameters of the reaction are similar to those of other RNA-catalyzed reactions (see Figure 23.9).

Previously, the crystal structure of a minimal hammerhead ribozyme was solved. In the minimal structure, however, the architecture of the active site was such that it was unclear how catalysis could proceed. Recently, the crystal structure of the full-length hammerhead ribozyme from *Schistosoma mansoni*, a nonviral species, has been solved, and it gives insight into catalysis. This structure, schematically illustrated in FIGURE 23.15, reveals a critical tertiary interaction between a bulge in stem I and the loop of stem II. This interaction stabilizes the active site in a conformation such that G12 can deprotonate the 2'-OH of C17, the scissile bond, and create the 2'-attacking oxygen. G8, in turn, provides the hydrogen to stabilize the newly formed 5'–OH end of the 3' cleavage product.

It is possible to design enzyme-substrate combinations that can form minimal hammerhead structures. These structures have been used to demonstrate that introduction of the appropriate RNA molecules into a cell can allow the enzymatic reaction to occur *in vivo*. A ribozyme designed in this way essentially provides a highly specific restriction endonuclease-like activity directed against an RNA target. By placing the ribozyme under control of a regulated promoter, it can be used in the same way as, for example, antisense constructs to specifically turn off expression of a target gene under defined circumstances.

23.10 RNA Editing Occurs at Individual Bases

Key concept

• Apolipoprotein-B and glutamate receptors have site-specific deaminations catalyzed by cytidine and adenosine deaminases that change the coding sequence.

A prime axiom of molecular biology is that the sequence of an mRNA can only represent what is coded in the DNA. The central dogma envisaged a linear relationship in which a continuous sequence of DNA is transcribed into a sequence of mRNA that is, in turn, directly translated into polypeptide. The occurrence of interrupted genes and the removal of introns by RNA splicing introduce an additional step into the process of gene expression (see Chapter 21, *RNA Splicing and Processing*, for details). Briefly, splicing occurs at the RNA level, and it results in removal of noncoding sequences (introns) that interrupt the coding sequences (exons) that are encoded in the DNA sequence. The process remains one of information transfer, though, in which the actual coding sequence in DNA remains unchanged.

Changes in the information encoded by DNA occur in some exceptional circumstances, most notably in the generation of new sequences coding for immunoglobulins in mammals and birds. These changes occur specifically in the somatic cells (B lymphocytes) in which immunoglobulins are synthesized (see Chapter 18, *Recombination in the Immune System*). New information is generated in the DNA of an individual during the process of reconstructing an immunoglobulin gene and information coded in the DNA is changed by somatic mutation. The information in DNA continues to be faithfully transcribed into RNA.

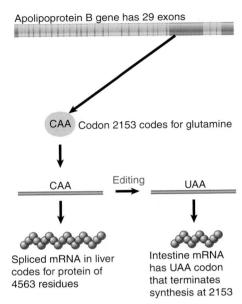

Apolipoprotein B gene has 29 exons

CAA Codon 2153 codes for glutamine

CAA → Editing → UAA

Spliced mRNA in liver codes for protein of 4563 residues

Intestine mRNA has UAA codon that terminates synthesis at 2153

FIGURE 23.16 The sequence of the apo-B gene is the same in intestine and liver, but the sequence of the mRNA is modified by a base change that creates a termination codon in intestine.

RNA editing is a process in which *information changes at the level of mRNA*. It is revealed by situations in which the coding sequence in an RNA differs from the sequence of DNA from which it was transcribed. RNA editing occurs in two different situations, each with different causes. In mammalian cells there are cases in which a substitution occurs in an individual base in mRNA that can cause a change in the sequence of the polypeptide that is encoded. This base substitution is the result of deamination of either adenosine to become inosine, or cytidine to become uridine. In trypanosome mitochondria, more widespread changes occur in transcripts of several genes, when bases are systematically added or deleted.

FIGURE 23.16 summarizes the sequences of the apolipoprotein-B (*apo-B*) gene and mRNA in mammalian intestine and liver. The genome contains a single interrupted gene whose sequence is identical in all tissues, with a coding region of 4563 codons. This gene is transcribed into an mRNA that is translated into a protein of 512 kDa representing the full coding sequence in the liver. A shorter form of the protein (~250 kDa) is synthesized in the intestine. This protein consists of the N-terminal half of the full-length protein. It is translated from an mRNA whose sequence is identical with that of liver except for a change from C to U at codon 2153. This substitution changes the codon CAA for glutamine into the ochre codon UAA for termination. Given that no alternative gene or exon is available in the genome to code for the new sequence and no change in the pattern of splicing can be discovered, we are forced to conclude that a change has been made directly in the sequence of the RNA transcript.

Another example is provided by glutamate receptors in rat brain. Editing at one position changes a glutamine codon in DNA into a codon for arginine in the mRNA. The change from glutamine to arginine affects the conductivity of the channel and therefore has an important effect on controlling ion flow through the neurotransmitter. At another position in the receptor, an arginine codon is converted to a glycine codon.

The events outlined for apo-B and glutamate receptors are the result of *deaminations* in which the amino group on the nucleotide ring is removed. The editing event in apo-B causes C_{2153} to be changed to U and both changes in the glutamate receptor are from A to I (inosine). Deaminations in apolipoprotein B are catalyzed by the cytidine deaminase APOBEC (*apo*lipoprotien *B* mRNA *e*diting *e*nzyme *c*omplex), whereas deaminations in the glutamate receptor are performed by *a*denosine *d*eaminases *a*cting on *R*NA (termed ADARs). This type of editing appears to occur largely in the nervous system. There are 16 (potential) targets for ADARs in *Drosophila melanogaster* and all are genes involved in neurotransmission. In many cases, the editing event changes an amino acid at a functionally important position in the protein.

Enzymes that undertake deamination as such often have broad specificity—for example, the best-characterized adenosine deaminase acts on any A residues in a duplexed RNA region. Deamination of adenosine and cytidine in RNA, however, displays specificity. Editing enzymes are related to the general deaminases, but have other regions or additional subunits that control their specificity. In the case of apo-B editing, the catalytic subunit of an editing complex is related to bacterial cytidine deaminase, but has an additional RNA-binding region that helps to recognize the specific target site for editing. A special adenosine deaminase enzyme recognizes the target sites in the glutamate receptor RNA, and similar events occur in a serotonin receptor RNA. The complex may recognize a particular region of secondary structure in a manner analogous to tRNA-modifying enzymes, or could directly recognize a nucleotide sequence. The develop-

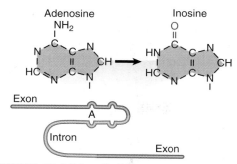

FIGURE 23.17 Editing of mRNA occurs when a deaminase acts on an adenine in an imperfectly paired RNA duplex region.

ment of an *in vitro* system for the apo-B editing event suggests that a relatively small sequence (~26 nucleotides) surrounding the editing site provides a sufficient target. **FIGURE 23.17** shows that in the case of the RNA for the glutamate receptor, GluR-B, a base-paired region that is necessary for recognition of the target site, is formed between the edited region in the exon and a complementary sequence in the downstream intron. A pattern of mispairing within the duplex region is necessary for specific recognition. Thus different editing systems may have different requirements for sequence specificity in their substrates.

23.11 RNA Editing Can Be Directed by Guide RNAs

Key concepts
- Extensive RNA editing in trypanosome mitochondria occurs by insertions or deletions of uridine.
- The substrate RNA base pairs with a guide RNA on both sides of the region to be edited.
- The guide RNA provides the template for addition (or less often, deletion) of uridines.
- Editing is catalyzed by the editosome, a complex of endonuclease, exonuclease, terminal uridyltransferase activity, and RNA ligase.

Another type of editing is revealed by dramatic changes in sequence in the products of several genes of trypanosome mitochondria. In the first case to be discovered, the sequence of the cytochrome oxidase subunit II protein has an internal frameshift that is not predicted based on the nucleotide sequence of the *coxII* gene. The sequences of the gene and protein given in **FIGURE 23.18** are conserved in several trypanosome species, so the method of RNA editing is not unique to a single organism.

The discrepancy between the sequence of the *coxII* gene and the protein product is due

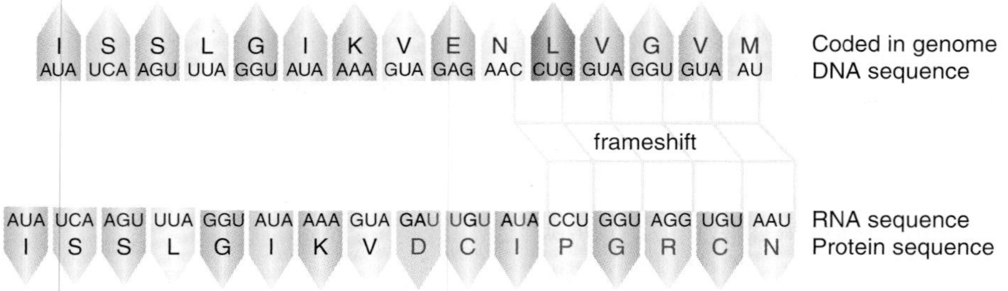

I	S	S	L	G	I	K	V	E	N	L	V	G	V	M	Coded in genome
AUA	UCA	AGU	UUA	GGU	AUA	AAA	GUA	GAG	AAC	CUG	GUA	GGU	GUA	AU	DNA sequence

frameshift

AUA	UCA	AGU	UUA	GGU	AUA	AAA	GUA	GAU	UGU	AUA	CCU	GGU	AGG	UGU	AAU	RNA sequence
I	S	S	L	G	I	K	V	D	C	I	P	G	R	C	N	Protein sequence

FIGURE 23.18 The mRNA for the trypanosome *coxII* gene has a frameshift relative to the DNA; the correct reading frame observed in the protein is created by the insertion of four uridines.

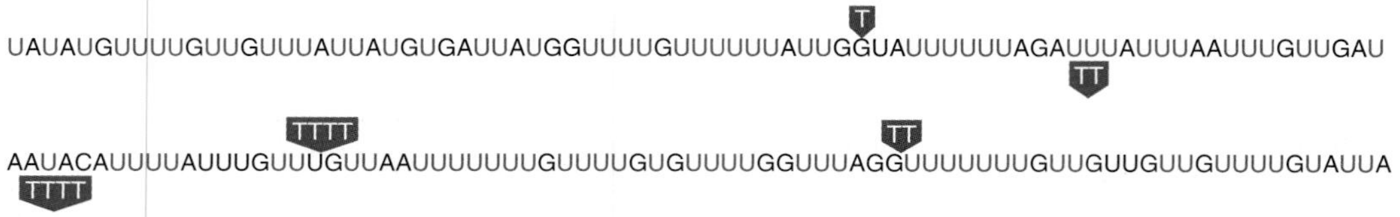

UAUAUGUUUUGUUGUUUAUUAUGUGAUUAUGGUUUUGUUUUUUAUUGGUAUUUUUUAGAUUUAUUUAAUUUGUUGAU

AAUACAUUUUAUUUGUUUGUUAAUUUUUUUGUUUUGUGUUUUGGUUUAGGUUUUUUUGUUGUUGUUGUUUUGUAUUA

FIGURE 23.19 Part of the mRNA sequence of *T. brucei coxIII* shows many uridines that are not coded in the DNA (shown in red) or that are removed from the RNA (shown as T).

to an RNA editing event. The *coxII* mRNA has an insert of an additional four nucleotides (all uridines) around the site of frameshift. The insertion establishes the proper reading frame for the protein. No second *coxII* gene carrying the frameshift sequence can be discovered; thus we are forced to conclude that the extra bases are inserted during or after transcription. A similar discrepancy between mRNA and genomic sequences is found in genes of the SV5 and measles paramyxoviruses, in these cases involving the addition of G residues in the mRNA.

Similar editing of RNA sequences occurs for other genes and includes deletions as well as additions of uridine. The extraordinary case of the cytochrome c oxidase III (*coxIII*) gene of *Trypanosoma brucei* is summarized in **FIGURE 23.19**. *More than half of the residues in the mRNA consist of uridines that are not encoded in the gene.* Comparison between the genomic DNA and the mRNA shows that no stretch longer than seven nucleotides is represented in the mRNA without alteration and runs of uridine up to seven bases long are inserted. The information for the specific insertion of uridines is provided by a **guide RNA**.

Guide RNA contains a sequence that is complementary to the correctly edited mRNA. **FIGURE 23.20** shows a model for its action in the cytochrome b gene of another trypanosome, *Leishmania*. The sequence at the top of the figure shows the original transcript, or preedited

RNA. Gaps show where bases will be inserted in the editing process. Eight uridines must be inserted into this region to create the valid mRNA sequence. The guide RNA is complementary to the mRNA for a significant distance, including and surrounding the edited region. Typically the complementarity is more extensive on the 3' side of the edited region and is rather short on the 5' side. Pairing between the guide RNA and the preedited RNA leaves gaps where unpaired A residues in the guide RNA do not find complements in the preedited RNA. The guide RNA provides a template that allows the missing U residues to be inserted at these positions in a process described below. When the reaction is completed the guide RNA separates from the mRNA, which becomes available for translation.

Specification of the final edited sequence can be quite complex. In the example of *Leishmania* cytochrome b, a lengthy stretch of the transcript is edited by the insertion of a total of 39 U residues, which appears to require two guide RNAs acting at adjacent sites. The first guide RNA pairs at the 3'–most site and the edited sequence then becomes a substrate for further editing by the next guide RNA. The guide RNAs are encoded as independent transcription units. **FIGURE 23.21** shows a map of the relevant region of the *Leishmania* mitochondrial DNA. It includes the gene for cytochrome b, which codes for the preedited sequence and two regions that specify guide RNAs. Genes for the

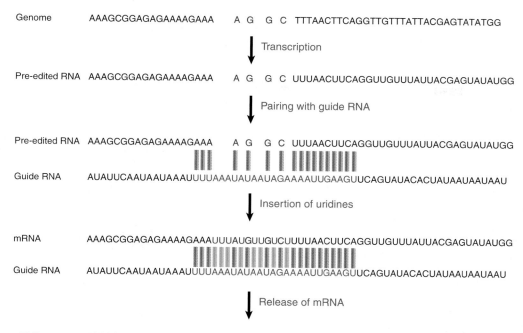

FIGURE 23.20 Pre-edited RNA base pairs with a guide RNA on both sides of the region to be edited. The guide RNA provides a template for the insertion of uridines. The mRNA produced by the insertions is complementary to the guide RNA.

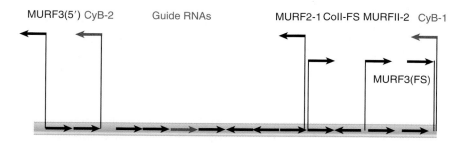

FIGURE 23.21 The *Leishmania* genome contains genes coding for pre-edited RNAs interspersed with units that code for the guide RNAs required to generate the correct mRNA sequences. Some genes have multiple guide RNAs. *CyB* is the gene for pre-edited cytochrome b, and *CyB-1* and *CyB-2* are genes for the guide RNAs involved in its editing.

major coding regions and for their guide RNAs are interspersed.

In principle, a mutation in either the gene or one of its guide RNAs could change the primary sequence of the mRNA and thus the primary sequence of the protein. By genetic criteria, each of these units could be considered to comprise part of the gene. The units are independently expressed, and as a result they should of course complement in *trans*. If mutations were available, we should therefore find that three complementation groups were needed to code for the primary sequence of a single protein.

The characterization of intermediates that are partially edited suggests that the reaction proceeds along the preedited RNA in the 3'–5' direction. The guide RNA determines the specificity of uridine insertions by its pairing with the preedited RNA.

Editing of uridines is catalyzed by a 20S enzyme complex called the *editosome* that is composed of about 20 proteins and contains an endonuclease, a terminal uridyltransferase (TUTase), a 3'–5' U-specific exonuclease (exoUase), and an RNA ligase. As illustrated in **FIGURE 23.22**, the editosome binds the guide RNA and uses it to pair with the preedited

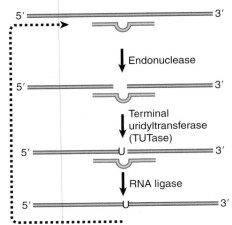

FIGURE 23.22 Addition or deletion of U residues occurs by cleavage of the RNA, removal or addition of the U, and ligation of the ends. The reactions are catalyzed by a complex of enzymes under the direction of guide RNA.

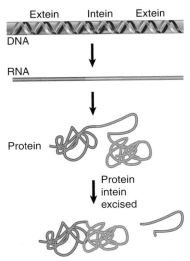

FIGURE 23.23 In protein splicing the exteins are connected by removing the intein from the protein.

mRNA. The substrate RNA is cleaved at a site that is (presumably) identified by the absence of pairing with the guide RNA; a uridine is inserted or deleted to base pair with the guide RNA and then the substrate RNA is ligated. Uridine triphosphate (UTP) provides the source for the uridyl residue. It is added by the TUTase activity. Deletion of U residues is mediated by an exoUase which functions in concert with a 3′ phosphatase to allow the newly edited RNA construct to religate.

The structures of partially edited molecules suggest that the U residues are added one at a time rather than in groups. It is possible that the reaction proceeds through successive cycles in which U residues are added, tested for complementarity with the guide RNA, and retained if acceptable and removed if not, so that the construction of the correct edited sequence occurs gradually. We do not know whether the same types of reaction are involved in editing reactions that add C residues.

23.12 Protein Splicing Is Autocatalytic

Key concepts

- An intein has the ability to catalyze its own removal from a protein in such a way that the flanking exteins are connected.
- Protein splicing is catalyzed by the intein.
- Most inteins have two independent activities: protein splicing and a homing endonuclease.

Protein splicing has the same effect as RNA splicing: a sequence that is represented within the gene fails to be represented in the protein.

The parts of the protein are named by analogy with RNA splicing: **exteins** are the sequences that are represented in the mature protein, and **inteins** are the sequences that are removed. The mechanism of removing the intein is completely different from that of RNA splicing. **FIGURE 23.23** shows that the gene is translated into a protein precursor that contains the intein and then the intein is excised from the protein. Over 350 examples of protein splicing are known and are spread throughout all classes of organisms. The typical gene whose product undergoes protein splicing has a single intein.

The first intein was discovered in an archaeal DNA polymerase gene in the form of an intervening sequence in the gene that does not conform to the rules for introns. It was then demonstrated that the purified protein can splice this sequence out of itself in an autocatalytic reaction. The reaction does not require input of energy and occurs through the series of bond rearrangements shown in **FIGURE 23.24**. The reaction is a function of the intein, although its efficiency can be influenced by the exteins.

The first reaction is an attack by an –OH or –SH side chain of the first amino acid in the intein on the peptide bond that connects it to the first extein. This transfers the extein from the amino-terminal group of the intein to an N-O or N-S acyl connection. This bond is then attacked by the –OH or –SH side chain of the first amino acid in the second extein. The result is to transfer extein1 to the side chain of the amino-terminal acid of extein2. Finally, the C-terminal asparagine of the intein cyclizes, and the terminal NH of extein2 attacks the acyl bond to replace it with a conventional peptide bond. Each of these reactions can occur spontaneously at very low rates, but their occurrence in a coordinated manner that is rapid enough to achieve protein splicing requires catalysis by the intein.

Inteins have characteristic features. They are found as in-frame insertions into coding sequences. They can be recognized as such because of the existence of homologous genes that lack the insertion. They have an N-terminal serine or cysteine (to provide the –XH side chain) and a C-terminal asparagine. A typical intein has a sequence of ~150 amino acids at the N-terminal end and ~50 amino acids at the C-terminal end that are involved in catalyzing the protein splicing reaction. The sequence in the center of the intein can have other functions.

An extraordinary feature of many inteins is that they have homing endonuclease activity. A homing endonuclease cleaves a target DNA to create a site into which the DNA sequence coding for the intein can be inserted (see Figure 23.11 in *Section 23.5, Some Group I Introns Code for Endonucleases That Sponsor Mobility*). The protein splicing and homing endonuclease activities of an intein are independent.

We do not really understand the connection between the presence of both these activities in an intein, but two types of model have been suggested. One is to suppose that there was originally some sort of connection between the activities, but that they have since become independent and some inteins have lost the homing endonuclease. The other is to suppose that inteins may have originated as protein splicing units, most of which (for unknown reasons) were subsequently invaded by homing endonucleases. This is consistent with the fact that homing endonucleases appear to have invaded other types of units as well, including, most notably, group I introns.

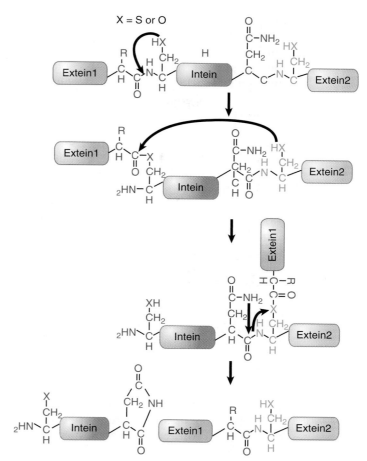

FIGURE 23.24 Bonds are rearranged through a series of transesterifications involving the –OH groups of serine or threonine or the –SH group of cysteine until the exteins are connected by a peptide bond and the intein is released with a circularized C-terminus.

23.13 Summary

Self-splicing is a property of two groups of introns, which are widely dispersed in unicellular/oligocellular eukaryotes, prokaryotic systems, and mitochondria. The information necessary for the reaction resides in the intron sequence, although the reaction is actually assisted by proteins *in vivo*. For both group I and group II introns, the reaction requires formation of a specific secondary/tertiary structure involving short consensus sequences. Group I intron RNA creates a structure in which the substrate sequence is held by the IGS region of the intron and other conserved sequences generate a guanine nucleotide binding site. It occurs by a transesterification involving a guanosine residue as cofactor. No input of energy is required. The guanosine breaks the bond at the 5′ exon–intron junction and becomes linked to the intron; the hydroxyl at the free end of the exon then attacks the 3′ exon–intron junction. The intron cyclizes and loses the guanosine and the terminal 15 bases. A series of related reactions can be catalyzed via attacks by the terminal G-OH residue of the intron on internal phosphodiester bonds. By providing appropriate substrates, it has been possible to engineer ribozymes that perform a variety of catalytic reactions, including nucleotidyl transferase activities.

Some group I and group II mitochondrial introns have open reading frames. The proteins coded by group I introns are endonucleases that make double-stranded cleavages in target sites in DNA. The endonucleolytic cleavage initiates a gene conversion process in which the sequence of the intron itself is copied into the target site. The proteins coded by group II introns include an endonuclease activity that initiates the transposition process and a reverse transcriptase that enables an RNA copy of the intron to be copied into the target site. These types of introns probably originated by insertion events. The proteins encoded by both groups of introns may include maturase activities that assist splicing of the intron by stabilizing the formation of the secondary/tertiary structure of the active site.

Catalytic reactions are undertaken by the RNA component of the RNAase P ribonucleoprotein. Virusoid RNAs can undertake self-cleavage at a "hammerhead" structure. Hammerhead structures can form between a substrate RNA and a ribozyme RNA, which allows cleavage to be directed at highly specific sequences. These reactions support the view that RNA can form specific active sites that have catalytic activity.

RNA editing changes the sequence of an RNA during or after its transcription. The changes are required to create a meaningful coding sequence. Substitutions of individual bases occur in mammalian systems; they take the form of deaminations in which C is converted to U or A is converted to I. A catalytic subunit related to cytidine or adenosine deaminase functions as part of a larger complex that has specificity for a particular target sequence.

Additions and deletions (most often of uridine) occur in trypanosome mitochondria and in paramyxoviruses. Extensive editing reactions occur in trypanosomes in which as many as half of the bases in an mRNA are derived from editing. The editing reaction uses a template consisting of a guide RNA that is complementary to the mRNA sequence. The reaction is catalyzed by the editosome, an enzyme complex that includes an endonuclease, exonuclease terminal uridyltransferase, and RNA ligase, using free nucleotides as the source for additions, or releasing cleaved nucleotides following deletion.

Protein splicing is an autocatalytic reaction that occurs by bond transfer reactions and input of energy is not required. The intein catalyzes its own splicing out of the flanking exteins. Many inteins have a homing endonuclease activity that is independent of the protein splicing activity.

References

23.2 Group I Introns Undertake Self-Splicing by Transesterification

Reviews

Cech, T. R. (1985). Self-splicing RNA: implications for evolution. *Int. Rev. Cytol.* 93, 3–22.

Cech, T. R. (1987). The chemistry of self-splicing RNA and RNA enzymes. *Science* 236, 1532–1539.

Vicens, Q. and Cech, T. T. (2006). Atomic level architecture of group I introns revealed. *Trends Biochem. Sci.* 31, 41–51.

Research

Been, M. D. and Cech, T. R. (1986). One binding site determines sequence specificity of *Tetrahymena* pre-rRNA self-splicing, *trans*-splicing, and RNA enzyme activity. *Cell* 47, 207–216.

Belfort, M., Pedersen-Lane, J., West, D., Ehrenman, K., Maley, G., Chu, F., and Maley, F. (1985). Processing of the intron-containing thymidylate synthase (td) gene of phage T4 is at the RNA level. *Cell* 41, 375–382.

Cech, T. R., Zaug, A. J., and Grabowski, P. J. (1981). *In vitro* splicing of the rRNA precursor of *Tetrahymena:* involvement of a guanosine nucleotide in the excision of the intervening sequence. *Cell* 27, 487–496.

Kruger, K., Grabowski, P. J., Zaug, A. J., Sands, J., Gottschling, D. E., and Cech, T. R. (1982). Self-splicing RNA: autoexcision and autocyclization of the ribosomal RNA intervening sequence of *Tetrahymena. Cell* 31, 147–157.

Myers, C. A., Kuhla, B., Cusack, S., and Lambowitz, A. M. (2002). tRNA-like recognition of group I introns by a tyrosyl-tRNA synthetase. *Proc. Natl. Acad. Sci. USA* 99, 2630–2635.

23.3 Group I Introns Form a Characteristic Secondary Structure

Research

Burke, J. M., Irvine, K. D., Kaneko, K. J., Kerker, B. J., Oettgen, A. B., Tierney, W. M., Williamson, C. L., Zaug, A. J., and Cech, T. R. (1986). Role of conserved sequence elements 9L and 2 in self-splicing of the *Tetrahymena* ribosomal RNA precursor. *Cell* 45, 167–176.

Michel, F. and Wetshof, E. (1990). Modeling of the three-dimensional architecture of group I catalytic introns based on comparative sequence analysis. *J. Mol. Biol.* 216, 585–610.

23.4 Ribozymes Have Various Catalytic Activities

Review
Cech, T. R. (1990). Self-splicing of group I introns. *Annu. Rev. Biochem.* 59, 543–568.

Research
Edwards, T. E., Klein, D. J., and Ferre-D'Amare, A. R. (2007). Riboswitches: small-molecule recognition by gene regulatory RNAs. *Curr. Opin. Struct. Biol.* 17: 273–279.

Serganov, A. and Patel, D. J. (2007). Ribozymes, riboswitches and beyond: regulation of gene expression without proteins. *Nat. Rev. Genet.* 8: 776–790.

Winkler, W. C., Nahvi, A., Roth, A., Collins, J. A., and Breaker, R. R. (2004). Control of gene expression by a natural metabolite-responsive ribozyme. *Nature* 428, 281–286.

23.5 Some Group I Introns Code for Endonucleases That Sponsor Mobility

Reviews
Belfort, M. and Roberts, R. J. (1997). Homing endonucleases: keeping the house in order. *Nucleic Acids Res.* 25, 3379–3388.

Haugen, P., Reeb, V., Lutzoni, F., and Bhatacharya, D. (2004). The evolution of homing endonuclease genes and group I introns in nuclear rDNA. *Mol. Biol. Evol.* 21: 129–140.

23.6 Group II Introns May Code for Multifunction Proteins

Reviews
Lambowitz, A. M. and Belfort, M. (1993). Introns as mobile genetic elements. *Annu. Rev. Biochem.* 62, 587–622.

Lambowitz, A. M. and Zimmerly, S. (2004). Mobile group II introns. *Annu. Rev. Genet.* 38, 1–35.

Research
Dickson, L., Huang, H. R., Liu, L., Matsuura, M., Lambowitz, A. M., and Perlman, P. S. (2001). Retrotransposition of a yeast group II intron occurs by reverse splicing directly into ectopic DNA sites. *Proc. Natl. Acad. Sci. USA* 98, 13207–13212.

Zimmerly, S., Guo, H., Perlman, P. S., and Lambowitz, A. M. (1995). Group II intron mobility occurs by target DNA-primed reverse transcription. *Cell* 82, 545–554.

Zimmerly, S., Guo, H., Eskes, R., Yang, J., Perlman, P. S., and Lambowitz, A. M. (1995). A group II intron is a catalytic component of a DNA endonuclease involved in intron mobility. *Cell* 83, 529–538.

23.7 Some Autosplicing Introns Require Maturases

Research
Bolduc, J. M., Spiegel, P. C., Chatterjee, P., Brady, K. L., Downing, M. E., Caprara, M. G., Waring, R. B., and Stoddard, B. L. (2003). Structural and biochemical analyses of DNA and RNA binding by a bifunctional homing endonuclease and group I splicing factor. *Genes. Dev.* 17, 2875–2888.

Carignani, G., Groudinsky, O., Frezza, D., Schiavon, E., Bergantino, E., and Slonimski, P. P. (1983). An RNA maturase is encoded by the first intron of the mitochondrial gene for the subunit I of cytochrome oxidase in *S. cerevisiae*. *Cell* 35, 733–742.

Henke, R. M., Butow, R. A., and Perlman, P. S. (1995). Maturase and endonuclease functions depend on separate conserved domains of the bifunctional protein encoded by the group I intron aI4 alpha of yeast mitochondrial DNA. *EMBO J.* 14, 5094–5099.

Matsuura, M., Noah, J. W., and Lambowitz, A. M. (2001). Mechanism of maturase-promoted group II intron splicing. *EMBO J.* 20, 7259–7270.

23.8 The Catalytic Activity of RNase P Is Due to RNA

Reviews
Altman, S. (2007). A view of RNase P. *Mol. BioSyst.* 3: 604–607.

Walker, S. C. and Engelke, D. R. (2006). Ribonuclease P: the evolution of an ancient RNA enzyme. *Crit. Rev. Biochem. Mol. Biol.* 41: 77–102.

23.9 Viroids Have Catalytic Activity

Reviews
Cochrane, J. C. and Strobel, S. A. (2008). Catalytic strategies of self-cleaving ribozymes. *Acc. Chem. Res.* 41: 1027–1035.

Doherty, E. A. and Doudna, J. A. (2000). Ribozyme structures and mechanisms. *Annu. Rev. Biochem.* 69, 597–615.

Symons, R. H. (1992). Small catalytic RNAs. *Annu. Rev. Biochem.* 61, 641–671.

Research
Forster, A. C. and Symons, R. H. (1987). Self-cleavage of virusoid RNA is performed by the proposed 55-nucleotide active site. *Cell* 50, 9–16.

Guerrier-Takada, C., Gardiner, K., Marsh, T., Pace, N., and Altman, S. (1983). The RNA moiety of ribonuclease P is the catalytic subunit of the enzyme. *Cell* 35, 849–857.

Martick, M. and Scott, W. G. (2006). Tertiary contacts distant from the active site prime a ribozyme for catalysis. *Cell* 126: 309–320.

Scott, W. G., Finch, J. T., and Klug, A. (1995). The crystal structure of an all-RNA hammerhead ribozyme: a proposed mechanism for RNA catalytic cleavage. *Cell* 81, 991–1002.

23.10   RNA Editing Occurs at Individual Bases

Review

Hoopengardner, B. (2006). Adenosine-to-inosine RNA editing: perspectives and predictions. *Mini-Rev. Med. Chem.* 6: 1213–1216.

Research

Higuchi, M., Single, F. N., Köhler, M., Sommer, B., Sprengel, R., and Seeburg, P. H. (1993). RNA editing of AMPA receptor subunit GluR-B: a base-paired intron-exon structure determines position and efficiency. *Cell* 75, 1361–1370.

Hoopengardner, B., Bhalla, T., Staber, C., and Reenan, R. (2003). Nervous system targets of RNA editing identified by comparative genomics. *Science.* 301: 832–836.

Navaratnam, N., Bhattacharya, S., Fujino, T., Patel, D., Jarmuz, A. L., and Scott, J. (1995). Evolutionary origins of apoB mRNA editing: catalysis by a cytidine deaminase that has acquired a novel RNA-binding motif at its active site. *Cell* 81, 187–195.

Powell, L. M., Wallis, S. C., Pease, R. J., Edwards, Y. H., Knott, T. J., and Scott, J. (1987). A novel form of tissue-specific RNA processing produces apolipoprotein-B48 in intestine. *Cell* 50, 831–840.

Sommer, B., Köhler, M., Sprengel, R., and Seeburg, P. H. (1991). RNA editing in brain controls a determinant of ion flow in glutamate-gated channels. *Cell* 67, 11–19.

23.11   RNA Editing Can Be Directed by Guide RNAs

Reviews

Aphasizhev, R. (2005). RNA uridylyltransferases. *Cell. Mol. Life Sci.* 62: 2194–2203.

Stuart, K. D., Schnaufer, A., Ernst, N. L., and Panigrahi, A. K. (2005). Complex management: RNA editing in trypanosomes. *Trends Biochem. Sci.* 30: 97–105.

Research

Aphasizhev, R., Sbicego, S., Peris, M., Jang, S. H., Aphasizheva, I., Simpson, A. M., Rivlin, A., and Simpson, L. (2002). Trypanosome mitochondrial 3′ terminal uridylyl transferase (TUTase): the key enzyme in U-insertion/deletion RNA editing. *Cell* 108, 637–648.

Benne, R., Van den Burg J., Brakenhoff, J. P., Sloof, P., Van Boom, J. H., and Tromp, M. C. (1986). Major transcript of the frameshifted coxII gene from trypanosome mitochondria contains four nucleotides that are not encoded in the DNA. *Cell* 46, 819–826.

Blum, B., Bakalara, N., and Simpson, L. (1990). A model for RNA editing in kinetoplastid mitochondria: "guide" RNA molecules transcribed from maxicircle DNA provide the edited information. *Cell* 60, 189–198.

Feagin, J. E., Abraham, J. M., and Stuart, K. (1988). Extensive editing of the cytochrome c oxidase III transcript in *Trypanosoma brucei. Cell* 53, 413–422.

Niemann, M., Kaibel, H., Schluter, E., Weitzel, K., Brecht, M., and Goringer, H. U. (2009). Kinetoplastid RNA editing involves a 3′ nucleotidyl phosphatase activity. *Nucleic Acids Res.* advanced access, published on February 3, 2009.

Seiwert, S. D., Heidmann, S., and Stuart, K. (1996). Direct visualization of uridylate deletion *in vitro* suggests a mechanism for kinetoplastid editing. *Cell* 84, 831–841.

23.12   Protein Splicing Is Autocatalytic

Reviews

Paulus, H. (2000). Protein splicing and related forms of protein autoprocessing. *Annu. Rev. Biochem.* 69, 447–496.

Saleh, L. and Perler, F. B. (2006). Protein splicing in *cis* and in *trans. Chem. Rec.* 6: 183–193.

Research

Derbyshire, V., Wood, D. W., Wu, W., Dansereau, J. T., Dalgaard, J. Z., and Belfort, M. (1997). Genetic definition of a protein-splicing domain: functional mini-inteins support structure predictions and a model for intein evolution. *Proc. Natl. Acad. Sci. USA* 94, 11466–11471.

Perler, F. B. et al. (1992). Intervening sequences in an Archaea DNA polymerase gene. *Proc. Natl. Acad. Sci. USA* 89, 5577–5581.

Xu, M. Q., Southworth, M. W., Mersha, F. B., Hornstra, L. J., and Perler, F. B. (1993). *in vitro* protein splicing of purified precursor and the identification of a branched intermediate. *Cell* 75, 1371–1377.

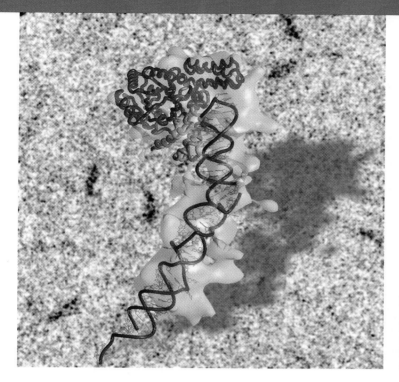

Translation

Edited by Cheryl Keller Capone

CHAPTER OUTLINE

24.8 Small Subunits Scan for Initiation Sites on Eukaryotic mRNA

- Eukaryotic 40S ribosomal subunits bind to the 5′ end of mRNA and scan the mRNA until they reach an initiation site.
- A eukaryotic initiation site consists of a ten-nucleotide sequence that includes an AUG codon.
- 60S ribosomal subunits join the complex at the initiation site.

24.9 Eukaryotes Use a Complex of Many Initiation Factors

- Initiation factors are required for all stages of initiation, including binding the initiator tRNA, 40S subunit attachment to mRNA, movement along the mRNA, and joining of the 60S subunit.
- Eukaryotic initiator tRNA is a Met-tRNA that is different from the Met-tRNA used in elongation, but the methionine is not formylated.
- eIF2 binds the initiator Met-tRNA$_i$ and GTP forming a ternary complex that binds to the 40S subunit before it associates with mRNA.
- A cap-binding complex binds to the 5′ end of mRNA prior to association of the mRNA with the 40S subunit.

24.10 Elongation Factor Tu Loads Aminoacyl-tRNA into the A Site

- EF-Tu is a monomeric G protein whose active form (bound to GTP) binds to aminoacyl-tRNA.
- The EF-Tu-GTP-aminoacyl-tRNA complex binds to the ribosome A site.

24.11 The Polypeptide Chain Is Transferred to Aminoacyl-tRNA

- The 50S subunit has peptidyl transferase activity as provided by an rRNA ribozyme.
- The nascent polypeptide chain is transferred from peptidyl-tRNA in the P site to aminoacyl-tRNA in the A site.
- Peptide bond synthesis generates deacylated tRNA in the P site and peptidyl-tRNA in the A site.

24.12 Translocation Moves the Ribosome

- Ribosomal translocation moves the mRNA through the ribosome by three bases.
- Translocation moves deacylated tRNA into the E site and peptidyl-tRNA into the P site, and empties the A site.
- The hybrid state model proposes that translocation occurs in two stages, in which the 50S moves relative to the 30S, and then the 30S moves along mRNA to restore the original conformation.

24.13 Elongation Factors Bind Alternately to the Ribosome

- Translocation requires EF-G, whose structure resembles the aminoacyl-tRNA-EF-Tu-GTP complex.
- Binding of EF-Tu and EF-G to the ribosome is mutually exclusive.
- Translocation requires GTP hydrolysis, which triggers a change in EF-G, which in turn triggers a change in ribosome structure.

24.14 Three Codons Terminate Translation

- The codons UAA (ochre), UAG (amber), and UGA (sometimes called opal) terminate translation.
- In bacteria, they are used most often with relative frequencies UAA>UGA>UAG.

24.15 Termination Codons Are Recognized by Protein Factors

- Termination codons are recognized by protein release factors, not by aminoacyl-tRNAs.
- The structures of the class 1 release factors resemble aminoacyl-tRNA-EF-Tu and EF-G.
- The class 1 release factors respond to specific termination codons and hydrolyze the polypeptide-tRNA linkage.
- The class 1 release factors are assisted by class 2 release factors that depend on GTP.
- The mechanism is similar in bacteria (which have two types of class 1 release factors) and eukaryotes (which have only one class 1 release factor).

24.16 Ribosomal RNA Pervades Both Ribosomal Subunits

- Each rRNA has several distinct domains that fold independently.
- Virtually all ribosomal proteins are in contact with rRNA.
- Most of the contacts between ribosomal subunits are made between the 16S and 23S rRNAs.

24.17 Ribosomes Have Several Active Centers

- Interactions involving rRNA are a key part of ribosome function.
- The environment of the tRNA-binding sites is largely determined by rRNA.

24.18 16S rRNA Plays an Active Role in Translation

- 16S rRNA plays an active role in the functions of the 30S subunit. It interacts directly with mRNA, with the 50S subunit, and with the anticodons of tRNAs in the P and A sites.

24.19 23S rRNA Has Peptidyl Transferase Activity

- Peptidyl transferase activity resides exclusively in the 23S rRNA.

24.20 Ribosomal Structures Change When the Subunits Come Together

- The head of the 30S subunit swivels around the neck when complete ribosomes are formed.
- The peptidyl transferase active site of the 50S subunit is more active in complete ribosomes than in individual 50S subunits.
- The interface between the 30S and 50S subunits is very rich in solvent contacts.

24.21 Summary

24.1 Introduction

An mRNA contains a series of codons that interact with the anticodons of aminoacyl-tRNAs so that a corresponding series of amino acids is incorporated into a polypeptide chain. The ribosome provides the environment for controlling the interaction between mRNA and aminoacyl-tRNA. The ribosome behaves like a small migrating factory that travels along the template, engaging in rapid cycles of peptide bond synthesis. Aminoacyl-tRNAs shoot in and out of the particle at an incredibly fast rate while depositing amino acids, and elongation factors cyclically associate with and dissociate from the ribosome. Together with its accessory factors, the ribosome provides the full range of activities required for all the steps of translation.

FIGURE 24.1 shows the relative dimensions of the components of the protein synthetic apparatus. The ribosome consists of two subunits that have specific roles in translation. Messenger RNA is associated with the small subunit; ~35 bases of the mRNA are bound at any time. The mRNA threads its way along the surface close to the junction of the subunits. Two tRNA molecules are active in translation at any moment, so polypeptide elongation involves reactions taking place at just two of the (roughly) ten codons covered by the ribosome. The two tRNAs are inserted into internal sites that stretch across the subunits. A third tRNA may remain on the ribosome after it has been used in translation before being recycled.

The basic form of the ribosome has been conserved in evolution, but there are appreciable variations in the overall size and proportions of RNA and protein in the ribosomes of bacteria, eukaryotic cytoplasm, and organelles. FIGURE 24.2 compares the components of bacterial and mammalian ribosomes. Both are ribonucleoprotein particles that contain more RNA than protein. The ribosomal proteins are known as *r-proteins*.

Each of the ribosome subunits contains a major rRNA and a number of small proteins. The large subunit may also contain smaller RNA(s). In *E. coli*, the small (30S) subunit consists of the 16S rRNA and 21 r-proteins. The large (50S) subunit contains 23S rRNA, the small 5S RNA, and 31 proteins. With the exception of one protein present at four copies per ribosome, there is one copy of each protein. The major RNAs constitute the major part of the mass of the bacterial ribosome. Their presence is pervasive, and probably most or all of the ribosomal proteins actually contact rRNA. So the major rRNAs form what is sometimes thought of as the backbone of each subunit—a continuous thread whose presence dominates the structure and which determines the positions of the ribosomal proteins.

The ribosomes in the cytosol of eukaryotes are larger than those of bacteria. The total content of both RNA and protein is greater; the major RNA molecules are longer (called 18S and 28S rRNAs), and there are more proteins. RNA is still the predominant component by mass.

The ribosomes of mitochondria and chloroplasts are distinct from the ribosomes of the cytosol and take varied forms. In some cases, they are almost the size of bacterial ribosomes and have 70% RNA; in other cases, they are only 60S and have <30% RNA.

The ribosome possesses several active centers, each of which is constructed from a group of proteins associated with a region of ribosomal RNA. The active centers require the direct participation of rRNA in a structural or even catalytic role. Some catalytic functions require individual proteins, but none of the activities can be reproduced by isolated proteins

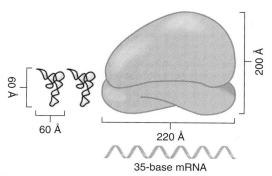

FIGURE 24.1 Size comparisons show that the ribosome is large enough to bind tRNAs and mRNA.

Ribosomes		rRNAs	r-proteins
Bacterial (70S) mass: 2.5 MDa 66% RNA	50S	23S = 2904 bases 5S = 120 bases	31
	30S	16S = 1542 bases	21
Mammalian (80S) mass: 4.2 MDa 60% RNA	60S	28S = 4718 bases 5.8S = 160 bases 5S = 120 bases	49
	40S	18S = 1874 bases	33

FIGURE 24.2 Ribosomes are large ribonucleoprotein particles that contain more RNA than protein and dissociate into large and small subunits.

or groups of proteins; they function only in the context of the ribosome.

In analyzing the functions of structural components of the ribosome, there are two experimental approaches. Mutational analysis implicates specific ribosomal proteins or bases in rRNA in participating in particular reactions. Structural analysis, including direct modification of components of the ribosome and comparisons to identify conserved features in rRNA, identifies the physical locations of components involved in particular functions.

24.2 Translation Occurs by Initiation, Elongation, and Termination

Key concepts

- The ribosome has three tRNA-binding sites.
- An aminoacyl-tRNA enters the A site.
- Peptidyl-tRNA is bound in the P site.
- Deacylated tRNA exits via the E site.
- An amino acid is added to the polypeptide chain by transferring the polypeptide from peptidyl-tRNA in the P site to aminoacyl-tRNA in the A site.

An amino acid is brought to the ribosome by an aminoacyl-tRNA. Its addition to the growing protein chain occurs by an interaction with the tRNA that brought the previous amino acid. Each of these tRNA lies in a distinct site on the ribosome. **FIGURE 24.3** shows that the two sites have different features:

- An incoming aminoacyl-tRNA binds to the **A site**. Prior to the entry of aminoacyl-tRNA, the site exposes the codon representing the next amino acid to be added to the chain.
- The codon representing the most recent amino acid to have been added to the nascent polypeptide chain lies in the **P site**. This site is occupied by **peptidyl-tRNA**, a tRNA carrying the nascent polypeptide chain.

FIGURE 24.4 shows that the aminoacyl end of the tRNA is located on the large subunit, whereas the anticodon at the other end interacts with the mRNA bound by the small subunit. So the P and A sites each extend across both ribosomal subunits.

For a ribosome to synthesize a peptide bond, it must be in the state shown in step 1 in Figure 24.3, when peptidyl-tRNA is in the P site and aminoacyl-tRNA is in the A site. Peptide bond formation occurs when the polypeptide

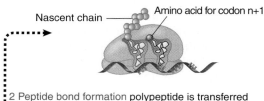

Codon "n"
P site holds
peptidyl-tRNA

Codon "n+1"
A site is entered
by aminoacyl-tRNA

Ribosome movement

5′   3′

1 Before peptide bond formation peptidyl-tRNA occupies P site; aminoacyl-tRNA occupies A site

Nascent chain — Amino acid for codon n+1

2 Peptide bond formation polypeptide is transferred from peptidyl-tRNA in P site to aminoacyl-tRNA in A site

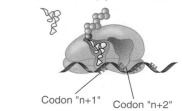

3 Translocation moves ribosome one codon; places peptidyl-tRNA in P site; deacylated tRNA leaves via E site; A site is empty for next aa-tRNA

Codon "n+1"   Codon "n+2"

FIGURE 24.3 The ribosome has two sites for binding charged tRNA.

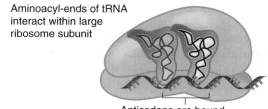

Aminoacyl-ends of tRNA interact within large ribosome subunit

Anticodons are bound to adjacent triplets on mRNA in small ribosome subunit

FIGURE 24.4 The P and A sites position the two interacting tRNAs across both ribosome subunits.

carried by the peptidyl-tRNA is transferred to the amino acid carried by the aminoacyl-tRNA. This step requires correct positioning of the aminoacyl-ends of the two tRNAs within the large subunit. This reaction is catalyzed by the large subunit of the ribosome.

Transfer of the polypeptide generates the ribosome shown in step 2, in which the **deacylated tRNA**, lacking any amino acid, lies in the P site and a new peptidyl-tRNA has been created in the A site. This peptidyl-tRNA is one amino acid residue longer than the peptidyl-tRNA that had been in the P site in step 1.

The ribosome now moves one triplet along the messenger RNA. This stage is called **translocation**. The movement transfers the deacylated tRNA out of the P site and moves the peptidyl-tRNA into the P site (see step 3 in Figure 24.3). The next codon to be translated now lies in the A site, ready for a new aminoacyl-tRNA to enter, at which point the cycle will be repeated. **FIGURE 24.5** summarizes the interaction between tRNAs and the ribosome.

The deacylated tRNA leaves the ribosome via another tRNA-binding site, the E site. This site is transiently occupied by the tRNA *en route* between leaving the P site and being released from the ribosome into the cytosol. Thus the flow of tRNA is into the A site, through the P site, and out through the E site (see also Figure 24.28 in *Section 24.12, Translocation Moves the Ribosome*). **FIGURE 24.6** compares the movement of tRNA and mRNA, which may be thought of as a sort of ratchet in which the reaction is driven by the codon–anticodon interaction.

Translation is divided into the three stages shown in **FIGURE 24.7**:

- **Initiation** involves the reactions that precede formation of the peptide bond between the first two amino acids of the protein. It requires the ribosome to bind to the mRNA, which forms an initiation complex that contains the first aminoacyl-tRNA. This is a relatively slow step in translation and usually determines the rate at which an mRNA is translated.
- **Elongation** includes all the reactions from synthesis of the first peptide bond to addition of the last amino acid. Amino acids are added to the chain one at a time; the addition of an amino acid is the most rapid step in translation.
- **Termination** encompasses the steps that are needed to release the completed polypeptide chain; at the same time, the ribosome dissociates from the mRNA.

Different sets of accessory factors assist the ribosome at each stage. Energy is provided at various stages by the hydrolysis of guanine triphosphate (GTP).

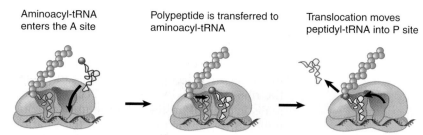

FIGURE 24.5 Aminoacyl-tRNA enters the A site, receives the polypeptide chain from peptidyl-tRNA, and is transferred into the P site for the next cycle of elongation.

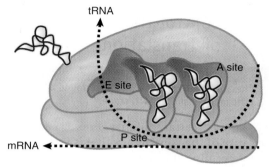

FIGURE 24.6 tRNA and mRNA move through the ribosome in the same direction.

Initiation small subunit on mRNA binding site is joined by large subunit and aminoacyl-tRNA binds

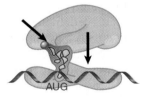

Elongation Ribosome moves along mRNA, extending protein by transfer from peptidyl-tRNA to aminoacyl-tRNA

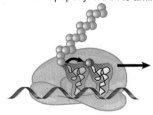

Termination Polypeptide chain is released from tRNA, and ribosome dissociates from mRNA

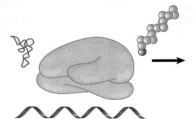

FIGURE 24.7 Translation falls into three stages.

During initiation, the small ribosomal subunit binds to mRNA and then is joined by the large subunit. During elongation, the mRNA moves through the ribosome and is translated in triplets. (Although we usually talk about the ribosome moving along mRNA, it is more realistic to think in terms of the mRNA being pulled through the ribosome.) At termination the protein is released, mRNA is released, and the individual ribosomal subunits dissociate in order to be used again.

24.3 Special Mechanisms Control the Accuracy of Translation

Key concept

- The accuracy of translation is controlled by specific mechanisms at each stage.

We know that translation is generally accurate, because of the consistency that is found when we determine the sequence of a polypeptide. There are few detailed measurements of the error rate *in vivo*, but it is generally thought to lie in the range of one error for every 10^4 to 10^5 amino acids incorporated. Considering that most proteins are produced in large quantities, this means that the error rate is too low to have any effect on the phenotype of the cell.

It is not immediately obvious how such a low error rate is achieved. In fact, the nature of discriminatory events is a general issue raised by several steps in gene expression:

- How do the enzymes that synthesize RNA recognize only the base complementary to the template?
- How do synthetases recognize just the corresponding tRNAs and amino acids?
- How does a ribosome recognize only the tRNA corresponding to the codon in the A site?

Each case poses a similar problem: how to distinguish one particular member from the entire set, all of which share the same general features.

Probably any substrate initially can contact the active center by a random-hit process, but then the wrong substrates are rejected and only the appropriate one is accepted. The appropriate substrate is always in a minority (one of four bases, one of twenty amino acids, one of ~30 to 50 tRNAs), so the criteria for discrimination must be strict. The point is that the enzyme must have some mechanism for increasing discrimination from the level that would be

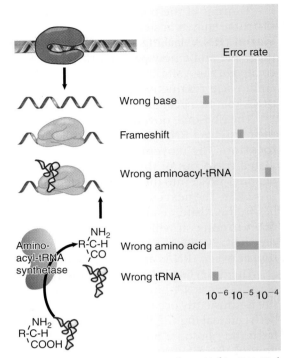

FIGURE 24.8 Errors occur at rates from 10^{-6} to 5×10^{-4} at different stages of translation.

achieved merely by making contacts with the available surfaces of the substrates.

FIGURE 24.8 summarizes the error rates at the steps that can affect the accuracy of translation.

Errors in transcribing mRNA are rare—probably $<10^{-6}$. This is an important stage for accuracy, because a single mRNA molecule is translated into many protein copies. The mechanisms that ensure transcriptional accuracy are discussed in Chapter 19, *Prokaryotic Transcription*.

The ribosome can make two types of errors in translation. It may cause a frameshift by skipping a base when it reads the mRNA (or in the reverse direction by reading a base twice—once as the last base of one codon and then again as the first base of the next codon). These errors are rare, occurring at ~10^{-5}. Or it may allow an incorrect aminoacyl-tRNA to (mis)pair with a codon, so that the wrong amino acid is incorporated. This is probably the most common error in translation, occurring at ~ 5×10^{-4}. It is controlled by ribosome structure and dissociation kinetics (see *Section 25.15, The Ribosome Influences the Accuracy of Translation*).

A tRNA synthetase can make two types of errors: It can place the wrong amino acid on its tRNA, or it can charge its amino acid with the wrong tRNA (see *Section 25.9, tRNAs Are Selectively Paired with Amino Acids by Ami-*

noacyl-tRNA Synthetases. The incorporation of the wrong amino acid is more common, probably because the tRNA offers a larger surface with which the enzyme can make many more contacts to ensure specificity. Aminoacyl-tRNA synthetases have specific mechanisms to correct errors before a mischarged tRNA is released (see *Section 25.11, Synthetases Use Proofreading to Improve Accuracy*).

24.4 Initiation in Bacteria Needs 30S Subunits and Accessory Factors

Key concepts

- Initiation of translation requires separate 30S and 50S ribosome subunits.
- Initiation factors (IF-1, -2, and -3), which bind to 30S subunits, are also required.
- A 30S subunit carrying initiation factors binds to an initiation site on mRNA to form an initiation complex.
- IF-3 must be released to allow 50S subunits to join the 30S-mRNA complex.

Bacterial ribosomes engaged in elongating a polypeptide chain exist as 70S particles. At termination, they are released from the mRNA as free ribosomes or ribosomal subunits. In growing bacteria, the majority of ribosomes are synthesizing proteins; the free pool is likely to contain ~20% of the ribosomes.

Ribosomes in the free pool can dissociate into separate subunits; this means that 70S ribosomes are in dynamic equilibrium with 30S and 50S subunits. *Initiation of translation is not a function of intact ribosomes, but is undertaken by the separate subunits*, which reassociate during the initiation reaction. **FIGURE 24.9** summarizes the ribosomal subunit cycle during translation in bacteria.

Initiation occurs at a special sequence on mRNA called the **ribosome-binding site** (including the **Shine–Dalgarno** sequence, which will be discussed in the next section). This is a short sequence of bases that precedes the coding region (see Figure 24.12) and is complementary to a portion of the 16S rRNA (see *Section 24.18, 16S rRNA Plays an Active Role in Translation*). The small and large subunits associate at the ribosome-binding site to form an intact ribosome. The reaction occurs in two steps:

- Recognition of mRNA occurs when a small subunit binds to form an *initiation complex* at the ribosome-binding site.

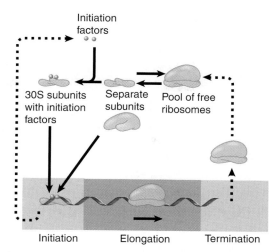

FIGURE 24.9 Initiation requires free ribosome subunits. When ribosomes are released at termination, the 30S subunits bind initiation factors and dissociate to generate free subunits. When subunits reassociate to give a functional ribosome at initiation, they release the factors.

- A large subunit then joins the complex to generate a complete ribosome.

Although the 30S subunit is involved in initiation, it is not by itself competent to undertake the reactions of binding mRNA and tRNA. It requires additional proteins called **initiation factors (IF)**. These factors are found only on 30S subunits, and they are released when the 30S subunits associate with 50S subunits to generate 70S ribosomes. This action distinguishes initiation factors from the structural proteins of the ribosome. The initiation factors are concerned solely with formation of the initiation complex; they are absent from 70S ribosomes, and they play no part in the stages of elongation. **FIGURE 24.10** summarizes the stages of initiation.

Bacteria use three initiation factors, numbered **IF-1**, **IF-2**, and **IF-3**. They are needed for both mRNA and tRNA to enter the initiation complex:

- IF-3 has multiple functions: it is needed to stabilize (free) 30S subunits, and to inhibit the premature binding of the 50S subunit; it enables 30S subunits to bind to initiation sites in mRNA; and as part of the 30S-mRNA complex, it checks the accuracy of recognition of the first aminoacyl-tRNA.
- IF-2 binds a special initiator tRNA and controls its entry into the ribosome.
- IF-1 binds to 30S subunits as a part of the complete initiation complex. It binds in the vicinity of the A site and prevents aminoacyl-tRNA from

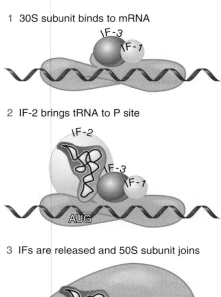

1 30S subunit binds to mRNA

2 IF-2 brings tRNA to P site

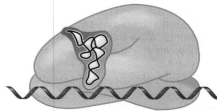

3 IFs are released and 50S subunit joins

FIGURE 24.10 Initiation factors stabilize free 30S subunits and bind initiator tRNA to the 30S-mRNA complex.

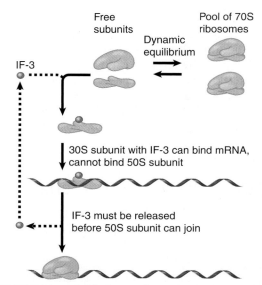

FIGURE 24.11 Initiation requires 30S subunits that carry IF-3.

30S subunit with IF-3 can bind mRNA, cannot bind 50S subunit

IF-3 must be released before 50S subunit can join

entering. Its location also may impede the 30S subunit from binding to the 50S subunit.

Numerous structural studies indicate that IF-3 has two distinct, largely globular domains with the C-terminal domain at the 50S contact site on the 30S subunit, and the N-terminal domain in the vicinity of the 30S E site. This broad positioning of IF-3 on the 30S subunit is consistent with its multiple functions.

The first function of IF-3 controls the equilibrium between ribosomal states, as shown in **FIGURE 24.11**. IF-3 binds to free 30S subunits that are released from the pool of 70S ribosomes. The presence of IF-3 prevents the 30S subunit from reassociating with a 50S subunit. IF-3 can interact with directly with 16S rRNA, and there is significant overlap between the bases in 16S rRNA protected by IF-3 and those protected by binding of the 50S subunit, suggesting that it physically prevents junction of the subunits. IF-3 therefore behaves as an anti-association factor that causes a 30S subunit to remain in the pool of free subunits. The reaction between IF-3 and the 30S subunit is stoichiometric: one

molecule of IF-3 binds per subunit. There is a relatively small amount of IF-3, so its availability determines the number of free 30S subunits.

The second function of IF-3 controls the ability of 30S subunits to bind to mRNA. Small subunits must have IF-3 in order to form initiation complexes with mRNA. IF-3 must be released from the 30S-mRNA complex in order to enable the 50S subunit to join. On its release, IF-3 immediately recycles by finding another 30S subunit.

Finally, IF-3 checks the accuracy of recognition of the first aminoacyl-tRNA and helps to direct it to the P site of the 30S subunit. The former has been attributed to the C-terminal domain of IF-3 (see *Section 24.7, Use of fMet-tRNAf Is Controlled by IF-2 and the Ribosome*). By comparison, the N-terminal domain of IF-3 is positioned to help direct the aminoacyl-tRNA into the P site of the 30S subunit by blocking the E site at the same time that IF-1 is blocking the A site.

IF-2 has a ribosome-dependent GTPase activity: It sponsors the hydrolysis of GTP in the presence of ribosomes, releasing the energy stored in the high-energy bond. The GTP is hydrolyzed when the 50S subunit joins to generate a complete ribosome. The GTP cleavage could be involved in changing the conformation of the ribosome, so that the joined subunits are converted into an active 70S ribosome.

Initiation Involves Base Pairing between mRNA and rRNA

Key concepts

- An initiation site on bacterial mRNA consists of the AUG initiation codon preceded with a gap of ~10 bases by the Shine–Dalgarno polypurine hexamer.
- The rRNA of the 30S bacterial ribosomal subunit has a complementary sequence that base pairs with the Shine–Dalgarno sequence during initiation.

The signal for initiating a polypeptide chain is a special initiation codon that marks the start of the reading frame. Usually the initiation codon is the triplet AUG, but in bacteria GUG or UUG are also used.

An mRNA contains many AUG triplets: How is the initiation codon recognized as providing the starting point for translation? The sites on mRNA where translation is initiated can be identified by binding the ribosome to mRNA under conditions that block elongation. Then the ribosome remains at the initiation site. When ribonuclease is added to the blocked initiation complex, all the regions of mRNA outside the ribosome are degraded. Those actually bound to it are protected, though, as illustrated in **FIGURE 24.12**. The protected fragments can be recovered and characterized.

The initiation sequences protected by bacterial ribosomes are ~30 bases long. The ribosome-binding sites of different bacterial mRNAs display two common features:

- The AUG (or less often, GUG or UUG) initiation codon is always included within the protected sequence.
- Within ten bases upstream of the AUG is a sequence that corresponds to part or all of the hexamer:

$$5'\ldots A G G A G G\ldots 3'$$

This polypurine stretch is known as the *Shine–Dalgarno sequence*. It is complementary to a highly conserved sequence close to the 3' end of 16S rRNA. (The extent of complementarity differs with individual mRNAs, and may extend from a four-base core sequence GAGG to a nine-base sequence extending beyond each end of the hexamer.) Written in reverse direction, the rRNA sequence is the hexamer:

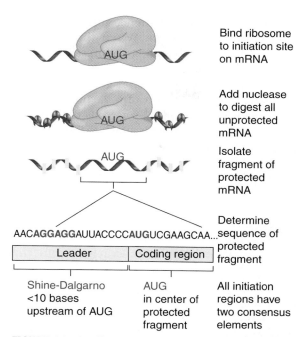

FIGURE 24.12 Ribosome-binding sites on mRNA can be recovered from initiation complexes. They include the upstream Shine–Dalgarno sequence and the initiation codon.

$$3'\ldots U C C U C C\ldots 5'$$

Does the Shine–Dalgarno sequence pair with its rRNA complement during mRNA-ribosome binding? Mutations of either sequence demonstrate its importance in initiation. Point mutations in the Shine–Dalgarno sequence can prevent an mRNA from being translated. In addition, the introduction of mutations into the complementary sequence in rRNA is deleterious to the cell and changes the pattern of translation. The decisive confirmation of the base-pairing reaction is that a mutation in the Shine–Dalgarno sequence of an mRNA can be suppressed by a mutation in the rRNA that restores base pairing.

The sequence at the 3' end of rRNA is conserved between prokaryotes and eukaryotes, except that in all eukaryotes there is a deletion of the five-base sequence CCUCC that is the principal complement to the Shine–Dalgarno sequence. There does not appear to be base pairing between eukaryotic mRNA and 18S rRNA. This is a significant difference in the mechanism of initiation.

In bacteria, a 30S subunit binds directly to a ribosome-binding site. As a result, the initiation complex forms at a sequence surrounding the AUG initiation codon. When the mRNA is polycistronic, each coding region starts with a ribosome-binding site.

The nature of bacterial gene expression means that translation of a bacterial mRNA proceeds sequentially through its cistrons. At the time when ribosomes attach to the first coding region, the subsequent coding regions have not yet even been transcribed. By the time the second ribosome site is available, translation is well under way through the first cistron.

What happens between the coding regions depends on the individual mRNA. In most cases, the ribosomes probably bind independently at the beginning of each cistron. The most common series of events is illustrated in **FIGURE 24.13**. When synthesis of the first protein terminates, the ribosomes leave the mRNA and dissociate into subunits. Then a new ribosome must assemble at the next coding region and set out to translate the next cistron.

In some bacterial mRNAs, translation between adjacent cistrons is directly linked, because ribosomes gain access to the initiation codon of the second cistron as they complete translation of the first cistron. This effect requires the space between the two coding regions to be small. It may depend on the high local density of ribosomes, or the juxtaposition of termination and initiation sites could allow some of the usual intercistronic events to be bypassed. A ribosome physically spans ~30 bases of mRNA, so that it could simultaneously contact a termination codon and the next initiation site if they are separated by only a few bases.

A Special Initiator tRNA Starts the Polypeptide Chain

Key concepts

- Translation starts with a methionine amino acid usually coded by AUG.
- Different methionine tRNAs are involved in initiation and elongation.
- The initiator tRNA has unique structural features that distinguish it from all other tRNAs.
- The NH_2 group of the methionine bound to bacterial initiator tRNA is formylated.

Synthesis of all proteins starts with the same amino acid: methionine. tRNAs recognizing the AUG codon carry methionine, and two types of tRNA can carry this amino acid. One is used for initiation, the other for recognizing AUG codons during elongation.

In bacteria, mitochondria, and chloroplasts, the initiator tRNA carries a methionine residue that has been formylated on its amino group, forming a molecule of **N-formyl-methionyl-tRNA**. The tRNA is known as **$tRNA_f^{Met}$**. The name of the aminoacyl-tRNA is usually abbreviated to fMet-tRNA$_f$.

The initiator tRNA gains its modified amino acid in a two-stage reaction. First, it is charged with the amino acid to generate Met-tRNA$_f$; and then the formylation reaction shown in **FIGURE 24.14** blocks the free NH_2 group. Although the blocked amino acid group would prevent the initiator from participating in chain elongation, it does not interfere with the ability to initiate a protein.

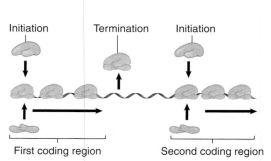

FIGURE 24.13 Initiation occurs independently at each cistron in a polycistronic mRNA. When the intercistronic region is longer than the span of the ribosome, dissociation at the termination site is followed by independent reinitiation at the next cistron.

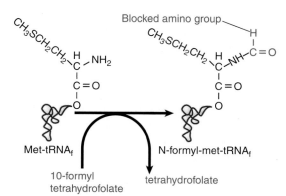

FIGURE 24.14 The initiator N-formyl-methionyl-tRNA (fMet-tRNA$_f$) is generated by formylation of methionyl-tRNA, using formyl-tetrahydrofolate as cofactor.

This tRNA is used only for initiation. It recognizes the codons AUG or GUG (occasionally UUG). The codons are not recognized equally well: the extent of initiation declines by about half when AUG is replaced by GUG, and declines by about half again when UUG is employed.

The tRNA species responsible for recognizing AUG codons in internal locations is **tRNA$_m^{Met}$**. This tRNA responds only to internal AUG codons. Its methionine cannot be formylated.

What features distinguish the fMet-tRNA$_f$ initiator and the Met-tRNA$_m$ elongator? Some characteristic features of the tRNA sequence are important, as summarized in **FIGURE 24.15**. Some of these features are needed to prevent the initiator from being used in elongation, whereas others are necessary for it to function in initiation:

- Formylation is not strictly necessary, because nonformylated Met-tRNA$_f$ can function as an initiator. Formylation improves the efficiency with which the Met-tRNA$_f$ is used, though, because it is one of the features recognized by the factor IF-2 that binds the initiator tRNA.
- The bases that face one another at the last position of the stem to which the amino acid is connected are paired in all tRNAs except tRNA$_f^{Met}$. Mutations that create a base pair in this position of tRNA$_f^{Met}$ allow it to function in elonga-

tion. The absence of this pair is therefore important in preventing tRNA$_f^{Met}$ from being used in elongation. It is also needed for the formylation reaction.

- A series of 3 G-C pairs in the stem that precedes the loop containing the anticodon is unique to tRNA$_f^{Met}$. These base pairs are required to allow the fMet-tRNA$_f$ to be inserted directly into the P site.

In bacteria and mitochondria, the formyl residue on the initiator methionine is removed by a specific deformylase enzyme to generate a normal NH$_2$ terminus. If methionine is to be the N-terminal amino acid of the protein, this is the only necessary step. In about half the proteins, the methionine at the terminus is removed by an aminopeptidase, which creates a new terminus from R$_2$ (originally the second amino acid incorporated into the chain). When both steps are necessary, they occur sequentially. The removal reaction(s) occur rather rapidly, probably when the nascent polypeptide chain has reached a length of 15 amino acids.

24.7 Use of fMet-tRNA$_f$ Is Controlled by IF-2 and the Ribosome

Key concept
- IF-2 binds the initiator fMet-tRNA$_f$ and allows it to enter the partial P site on the 30S subunit.

In bacterial translation, the meaning of the AUG and GUG codons depends on their **context**. When the AUG codon is used for initiation, it is read as formyl-methionine; when used within the coding region, methionine is added to the polypeptide. The meaning of the GUG codon is even more dependent on its location. When present as the first codon, formyl-methionine is added, but when present within a gene, it is bound by Val-tRNA, one of the regular members of the tRNA set, to provide valine as specified by the genetic code.

How is the context of AUG and GUG codons interpreted? **FIGURE 24.16** illustrates the decisive role of the ribosome when acting in conjunction with accessory factors.

In an initiation complex, the small subunit alone is bound to mRNA. The initiation codon

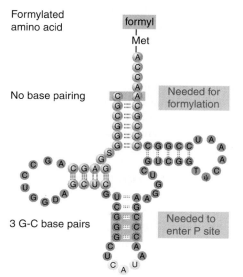

FIGURE 24.15 fMet-tRNA$_f$ has unique features that distinguish it as the initiator tRNA.

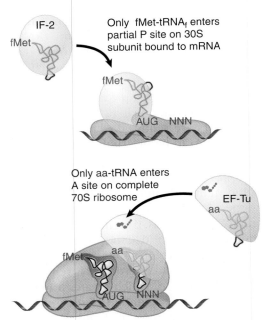

FIGURE 24.16 Only fMet-tRNA_f can be used for initiation by 30S subunits; other aminoacyl-tRNAs (aa-tRNA) must be used for elongation by 70S ribosomes.

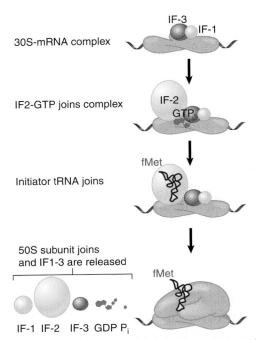

FIGURE 24.17 IF-2 is needed to bind fMet-tRNA_f to the 30S-mRNA complex. After 50S binding, all IF factors are released and GTP is cleaved.

lies within the part of the P site carried by the small subunit. The only aminoacyl-tRNA that can become part of the initiation complex is the initiator, which has the unique property of being able to enter directly into the partial P site to recognize its codon.

When the large subunit joins the complex, the partial tRNA-binding sites are converted into the intact P and A sites. The initiator fMet-tRNA_f occupies the P site, and the A site is available for entry of the aminoacyl-tRNA complementary to the second codon of the gene. The first peptide bond forms between the initiator and the next aminoacyl-tRNA.

Initiation prevails when an AUG (or GUG) codon lies within a ribosome-binding site, because only the initiator tRNA can enter the partial P site generated when the 30S subunit binds *de novo* to the mRNA. Internal reading prevails subsequently, when the codons are encountered by a ribosome that is continuing to translate an mRNA, because only the regular aminoacyl-tRNAs can enter the (complete) A site.

Accessory factors are critical in controlling the usage of aminoacyl-tRNAs. All aminoacyl-tRNAs associate with the ribosome by binding to an accessory factor. The factor used in initiation is IF-2 (see *Section 24.4, Initiation in Bacteria Needs 30S Subunits and Accessory Factors*), and the corresponding factor used at elongation is EF-Tu (see *Section 24.10, Elongation Factor Tu Loads Aminoacyl-tRNA into the A Site*).

The initiation factor IF-2 places the initiator tRNA into the P site. By forming a complex specifically with fMet-tRNA_f, IF-2 ensures that only the initiator tRNA, and none of the regular aminoacyl-tRNAs, participates in the initiation reaction. Conversely, EF-Tu, which places aminoacyl-tRNAs in the A site, cannot bind fMet-tRNA_f, which is therefore excluded from use during elongation.

An additional check on accuracy is made by IF-3, which stabilizes binding of the initiator tRNA by recognizing correct base pairing with the second and third bases of the AUG initiation codon.

FIGURE 24.17 details the series of events by which IF-2 places the fMet-tRNA_f initiator in the P site. IF-2, bound to GTP, associates with the P site of the 30S subunit. At this point, the 30S subunit carries all the initiation factors. fMet-tRNA_f binds to the IF-2 on the 30S subunit, and then IF-2 transfers the tRNA into the partial P site.

24.8 Small Subunits Scan for Initiation Sites on Eukaryotic mRNA

Key concepts

- Eukaryotic 40S ribosomal subunits bind to the 5′ end of mRNA and scan the mRNA until they reach an initiation site.
- A eukaryotic initiation site consists of a ten-nucleotide sequence that includes an AUG codon.
- 60S ribosomal subunits join the complex at the initiation site.

Initiation of translation in eukaryotic cytoplasm resembles the process in bacteria, but the order of events is different and the number of accessory factors is greater. Some of the differences in initiation are related to a difference in the way that bacterial 30S and eukaryotic 40S subunits find their binding sites for initiating translation on mRNA. In eukaryotes, small subunits first recognize the 5′ end of the mRNA and then move to the initiation site, where they are joined by large subunits. (In prokaryotes, small subunits bind directly to the initiation site.)

Virtually all eukaryotic mRNAs are monocistronic, but each mRNA usually is substantially longer than necessary just to code for its protein. The average mRNA in eukaryotic cytoplasm is 1000 to 2000 bases long, has a methylated cap at the 5′ terminus, and carries 100 to 200 bases of poly(A) at the 3′ terminus.

The nontranslated 5′ leader is relatively short, usually <100 bases. The length of the coding region is determined by the size of the protein. The nontranslated 3′ trailer is often rather long, at times reaching lengths of up to ~1000 bases.

The first feature to be recognized during translation of a eukaryotic mRNA is the methylated cap that marks the 5′ end. Messenger RNAs whose caps have been removed are not translated efficiently *in vitro*. Binding of 40S subunits to mRNA requires several initiation factors, including proteins that recognize the structure of the cap.

Modification at the 5′ end occurs to almost all cellular or viral mRNAs and is essential for their translation in eukaryotic cytoplasm (although it is not needed in organelles). The sole exception to this rule is provided by a few viral mRNAs (such as poliovirus) that are not

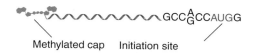

1 Small subunit binds to methylated cap

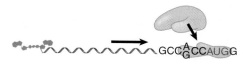

2 Small subunit migrates to initiation site

3 If leader is long, subunits may form queue

FIGURE 24.18 Eukaryotic ribosomes migrate from the 5′ end of mRNA to the initiation site, which includes an AUG initiation codon.

capped; only these exceptional viral mRNAs can be translated *in vitro* without caps. They use an alternative pathway that bypasses the need for the cap.

We have dealt with the process of initiation as though the initiation site is always freely available. Its availability may, however, be impeded by secondary structure. The recognition of mRNA requires several additional factors; an important part of their function is to remove any secondary structure in the mRNA (see Figure 24.22).

Sometimes the AUG initiation codon lies within 40 bases of the 5′ terminus of the mRNA, so that both the cap and AUG lie within the span of ribosome binding. In many mRNAs, however, the cap and AUG are farther apart—in extreme cases, they can be as much as 1000 bases away from each other. Yet the presence of the cap still is necessary for a stable complex to be formed at the initiation codon. How can the ribosome rely on two sites so far apart?

FIGURE 24.18 illustrates the "scanning" model, which supposes that the 40S subunit initially recognizes the 5′ cap and then "migrates" along the mRNA. Scanning from the 5′ end is a linear process. When 40S subunits scan the leader region, they can melt secondary structure hairpins with stabilities <–30 kcal, but hairpins of greater stability impede or prevent migration.

Migration stops when the 40S subunit encounters the AUG initiation codon. Usually, although not always, the first AUG triplet sequence to be encountered will be the initiation codon. The AUG triplet by itself, however, is not sufficient to halt migration; it is recognized efficiently as an initiation codon only when it is in the right context. The most important determinants of context are the bases in positions –4 and +1. An initiation codon may be recognized in the sequence NNNPuNNAUGG. The purine (A or G) three bases before the AUG codon, and the G immediately following it, can influence the efficiency of translation by 10×. When the leader sequence is long, further 40S subunits can recognize the 5′ end before the first has left the initiation site, creating a queue of subunits proceeding along the leader to the initiation site.

It is probably true that the initiation codon is the first AUG to be encountered in the most efficiently translated mRNAs. What happens, though, when there is an AUG triplet in the 5′ nontranslated region? There are two possible escape mechanisms for a ribosome that starts scanning at the 5′ end. The most common is that scanning is leaky; that is, a ribosome may continue past a noninitiation AUG because it is not in the right context. In the rare case that it does recognize the AUG, it may initiate translation but terminate before the proper initiation codon, after which it resumes scanning.

The majority of eukaryotic initiation events involve scanning from the 5′ cap, but there is an alternative means of initiation, used especially by certain viral RNAs, in which a 40S subunit associates directly with an internal site called an **IRES (internal ribosome entry site)**. In this case, any AUG codons that may be in the 5′ nontranslated region are bypassed entirely. There are few sequence homologies between known IRES elements. We can distinguish three types on the basis of their interaction with the 40S subunit:

- One type of IRES includes the AUG initiation codon at its upstream boundary. The 40S subunit binds directly to it, using a subset of the same factors that are required for initiation at 5′ ends.
- Another is located as much as 100 nucleotides upstream of the AUG, requiring a

40S subunit to migrate, again probably by a scanning mechanism.

- An exceptional type of IRES in hepatitis C virus can bind a 40S subunit directly, without requiring any initiation factors. The order of events is different from all other eukaryotic initiation. Following 40S-mRNA binding, a complex containing initiator factors and the initiator tRNA binds.

Use of the IRES is especially important in picornavirus infection, where it was first discovered, because the virus inhibits host translation by destroying cap structures and inhibiting the initiation factors that bind them. One such target is subunit eIF4G (see *Section 24.9, Eukaryotes Use a Complex of Many Initiation Factors*), which binds the 5′ end of mRNA. Thus, infection prevents translation of host mRNAs, but allows viral mRNAs to be translated because they use the IRES.

Binding is stabilized at the initiation site. When the 40S subunit is joined by a 60S subunit, the intact ribosome is located at the site identified by the protection assay. A 40S subunit protects a region of up to 60 bases; when the 60S subunits join the complex, the protected region contracts to about the same length of 30 to 40 bases seen in prokaryotes.

24.9 Eukaryotes Use a Complex of Many Initiation Factors

Key concepts

- Initiation factors are required for all stages of initiation, including binding the initiator tRNA, 40S subunit attachment to mRNA, movement along the mRNA, and joining of the 60S subunit.
- Eukaryotic initiator tRNA is a Met-tRNA that is different from the Met-tRNA used in elongation, but the methionine is not formylated.
- eIF2 binds the initiator Met-tRNA$_i$ and GTP forming a ternary complex that binds to the 40S subunit before it associates with mRNA.
- A cap-binding complex binds to the 5′ end of mRNA prior to association of the mRNA with the 40S subunit.

Initiation in eukaryotes has the same general features as in bacteria in using a specific initiation codon and initiator tRNA. Initiation in eukaryotic cytoplasm uses AUG as the initiator. The initiator tRNA is a distinct species, but its methionine does not become formylated. It is called $tRNA_i^{Met}$. Thus the difference between the initiating and elongating Met-tRNAs lies solely in the tRNA moiety, with Met-tRNA$_i$ used for initiation and Met-tRNA$_m$ used for elongation.

At least two features are unique to the initiator tRNA$_i^{Met}$ in yeast: It has an unusual tertiary structure, and it is modified by phosphorylation of the 2′ ribose position on base 64 (if this modification is prevented, the initiator can be used in elongation). Thus the principle of a distinction between initiator and elongator Met-tRNAs is maintained in eukaryotes, but its structural basis is different from that in bacteria (for comparison, see Figure 24.15).

Eukaryotic cells have more initiation factors than bacteria—the current list includes 12 factors that are directly or indirectly required for initiation. The factors are named similarly to those in bacteria (sometimes by analogy with the bacterial factors) and are given the prefix "e" to indicate their eukaryotic origin. They act at all stages of the process, including:

- forming an initiation complex with the 5′ end of mRNA;
- forming a complex with Met-tRNA$_i$;
- binding the mRNA-factor complex to the Met-tRNA$_i$-factor complex;
- enabling the ribosome to scan mRNA from the 5′ end to the first AUG;
- detecting binding of initiator tRNA to AUG at the start site; and
- mediating joining of the 60S subunit.

FIGURE 24.19 summarizes the stages of initiation and shows which initiation factors are involved at each stage. eIF2, together with Met-tRNA$_i$, eIF3, eIF1, and eIF1A, binds to the 40S ribosome subunit to form the 43S preinitiation complex. eIF4A, eIF4B, eIF4E, and eIF4G bind to the 5′ end of the mRNA to form the cap-binding complex. This complex associates with 3′ end of the mRNA via eIF4G, which interacts with poly(A)-binding protein (PABP). The 43S complex binds the initiation factors at the 5′ end of the mRNA and scans for the initiation codon. It can be isolated as the 48S initiation complex.

The subunit eIF2 is the key factor in binding Met-tRNA$_i$. Unlike bacterial IF2, which is a monomeric GTP-binding protein, eIF2 is a hetero-trimeric GTP-binding protein consisting of α, β, and γ subunits, none of which is homologous to bacterial IF2 (see Figure 24.36 in *Section 24.15, Termination Codons Are Recognized by Protein Factors*). eIF2 is active when bound to GTP and inactive when bound to guanine diphosphate (GDP). **FIGURE 24.20** shows that the

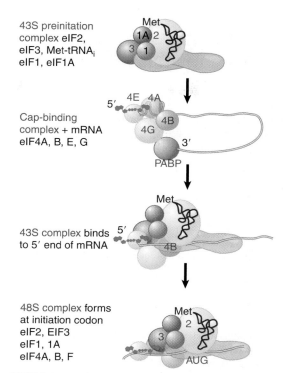

FIGURE 24.19 Some initiation factors bind to the 40S ribosome subunit to form the 43S preinitiation complex; others bind to mRNA. When the 43S complex binds to mRNA, it scans for the initiation codon and can be isolated as the 48S complex.

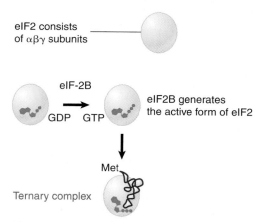

FIGURE 24.20 In eukaryotic initiation, eIF-2 forms a ternary complex with Met-tRNA$_i$ and GTP. The ternary complex binds to free 40S subunits, which attach to the 5′ end of mRNA. Later in the reaction, GTP is hydrolyzed and eIF2 is released in the form of eIF2-GDP. eIF2B regenerates the active form.

eIF2-GTP binds to Met-tRNA$_i$. The product is sometimes called the ternary complex (after its three components, eIF2, GTP, and Met-tRNA$_i$). Assembly of the ternary complex is regulated by the guanine nucleotide exchange factor (GEF) eIF2B, which exchanges GDP for GTP following hydrolysis of GTP by eIF2.

eIF3 maintains free 40S subunits

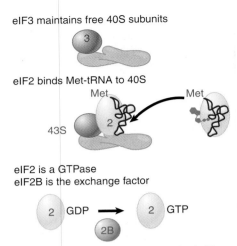

eIF2 binds Met-tRNA to 40S

43S

eIF2 is a GTPase
eIF2B is the exchange factor

FIGURE 24.21 Initiation factors bind the initiator Met-tRNA to the 40S subunit to form a 43S complex. Later in the reaction, GTP is hydrolyzed and eIF2 is released in the form of eIF2-GDP. eIF2B regenerates the active form.

eIF4F is a heterotrimer consisting of:

eIF4G is a scaffold protein
eiF4E binds the 5' methyl cap
eIF4A is a helicase that unwinds the 5' structure

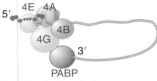

eIF4G binds two further factors
eIF4B stimulates eIF4A helicase
PABP binds 3' poly(A)

FIGURE 24.22 The heterotrimer eIF4F binds the 5' end of mRNA as well as other factors.

FIGURE 24.21 shows that the ternary complex places Met-tRNA$_i$ onto the 40S subunit. Along with factors eIF1, eIF1A, and eIF3, this generates the 43S preinitiation complex. The reaction is independent of the presence of mRNA. In fact, the Met-tRNA$_i$ initiator must be present in order for the 40S subunit to bind to mRNA. eIF3, which is required to maintain 40S subunits in their dissociated state, is a very large factor, with eight to ten subunits. eIF1 and eIF1A, which is homologous to bacterial IF1, appear to enhance eIF3's dissociation activity.

FIGURE 24.22 shows the group of factors that bind to the 5' end of mRNA. The factor eIF4F is a protein complex that contains three of the initiation factors. It is not clear whether it preassembles as a complex before binding to mRNA or whether the subunits are added individually to form the complex on mRNA. It includes the cap-binding subunit eIF4E, the helicase eIF4A, and the "scaffolding" subunit eIF4G.

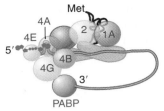

Possible interactions:
eIF4G binds to eIF3
mRNA binds eIF4G, eIF3, and 40S subunit

FIGURE 24.23 Interactions involving initiation factors are important when mRNA binds to the 43S complex.

After eIF4E binds the cap, eIF4A unwinds any secondary structure that exists in the first 15 bases of the mRNA. Energy for the unwinding is provided by hydrolysis of ATP. Unwinding of structure further along the mRNA is accomplished by eIF4A together with another factor, eIF4B. The main role of eIF4G is to link other components of the initiation complex.

The subunit eIF4E is a focus for regulation. Its activity is increased by phosphorylation, which is triggered by stimuli that increase translation and reversed by stimuli that repress translation. The subunit eIF4F has a kinase activity that phosphorylates eIF4E. The availability of eIF4E is also controlled by proteins that bind to it (called 4E-BP1, -2, and -3), to prevent it from functioning in initiation.

The presence of poly(A) on the 3' tail of the mRNA stimulates the formation of the initiation complex at the 5' end. PABP binds to the eIF4G scaffolding protein, bringing about a circular organization of the mRNA with both the 5' and 3' ends held in this complex (see Figure 24.22). The formation of this closed loop stimulates transcription, and PABP is required for this effect. The PABP/eIF4G interaction on the mRNA promotes recruitment of the 43S complex to the mRNA, as well as joining of the 60S subunit.

FIGURE 24.23 shows that the interactions involved in binding the mRNA to the 43S complex are not completely defined, but appear to involve eIF4G and eIF3 as well as the mRNA and 40S subunit. The subunit eIF4G binds to eIF3. This provides the means by which the 40S ribosomal subunit binds to eIF4F, and thus is recruited to the complex. In effect, eIF4F functions to get eIF4G in place so that it can attract the small ribosomal subunit.

When the small subunit has bound mRNA, it migrates to (usually) the first AUG codon. Scanning is assisted by the factors eIF1 and eIF1A. This process requires expenditure of

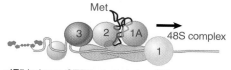

eIF1 and eIF1A enable scanning

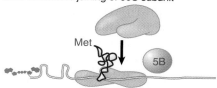

eIF5 induces GTP hydrolysis by eIF2
eIF2 and eIF3 are released

eIF5B mediates joining of 60S subunit

FIGURE 24.24 eIF1 and eIF1A help the 43S initiation complex to scan the mRNA until it reaches an AUG codon. eIF2 hydrolyzes its GTP to enable its release together with IF3. eIF5B mediates 60S–40S joining.

energy in the form of ATP, and thus factors associated with ATP hydrolysis (eIF4A, IF4B, and eIF4F) also play a role in this step. **FIGURE 24.24** shows that the small subunit stops when it reaches the initiation site, at which point the initiator tRNA base-pairs with the AUG initiation codon, forming a stable 48S complex.

Junction of the 60S subunits with the initiation complex cannot occur until eIF2 and eIF3 have been released from the initiation complex. This is mediated by eIF5 and causes eIF2 to hydrolyze its GTP. The reaction occurs on the small ribosome subunit and requires the base-pairing of the initiator tRNA with the initiation codon. All of the remaining factors likely are released when the complete 80S ribosome is formed.

Finally, the factor eIF5B enables the 60S subunit to join the complex, forming an intact ribosome that is ready to start elongation. The subunit eIF5B has a similar sequence to the prokaryotic factor IF2, which has a similar role in hydrolyzing GTP (in addition to its role in binding the initiator tRNA).

Once the factors have been released, they can associate with the initiator tRNA and ribosomal subunits in another initiation cycle. The subunit eIF2 has hydrolyzed its GTP; as a result, the active form must be regenerated. This is accomplished by the GEF (guanosine exchange factor), eIF2B, which displaces the GDP so that it can be replaced by GTP.

The subunit eIF2 is a target for regulation. Several regulatory kinases act on the α subunit of eIF2. Phosphorylation prevents eIF2B from

regenerating the active form. This limits the action of eIF2B to one cycle of initiation, and thereby inhibits translation.

24.10 Elongation Factor Tu Loads Aminoacyl-tRNA into the A Site

Key concepts

- EF-Tu is a monomeric G protein whose active form (bound to GTP) binds to aminoacyl-tRNA.
- The EF-Tu-GTP-aminoacyl-tRNA complex binds to the ribosome A site.

Once the complete ribosome is formed at the initiation codon, the stage is set for a cycle in which aminoacyl-tRNA enters the A site of a ribosome whose P site is occupied by peptidyl-tRNA. Any aminoacyl-tRNA except the initiator can enter the A site. Its entry is mediated by an **elongation factor** (**EF-Tu** in bacteria). The process is similar in eukaryotes. EF-Tu is a highly conserved protein throughout bacteria and mitochondria and is homologous to its eukaryotic counterpart.

Just like its counterpart in initiation (IF-2), EF-Tu is associated with the ribosome only during the process of aminoacyl-tRNA entry. Once the aminoacyl-tRNA is in place, EF-Tu leaves the ribosome, to work again with another aminoacyl-tRNA. Thus it displays the cyclic association with, and dissociation from, the ribosome that is the hallmark of the accessory factors.

FIGURE 24.25 depicts the role of EF-Tu in bringing aminoacyl-tRNA to the A site. EF-Tu is a monomeric GTP-binding protein that is active when bound to GTP and inactive when bound to guanine diphosphate (GDP). The binary complex of EF-Tu-GTP binds aminoacyl-tRNA to form a ternary complex of aminoacyl-tRNA-EF-Tu-GTP. The ternary complex binds only to the A site of ribosomes whose P site is already occupied by peptidyl-tRNA. This is the critical reaction in ensuring that the aminoacyl-tRNA and peptidyl-tRNA are correctly positioned for peptide bond formation.

Aminoacyl-tRNA is loaded into the A site in two stages. First, the anticodon end binds to the A site of the 30S subunit. Then, codon–anticodon recognition triggers a change in the conformation of the ribosome. This stabilizes tRNA binding and causes EF-Tu to hydrolyze its GTP. The CCA end of the tRNA now moves into the A site on the 50S subunit. The binary complex EF-Tu-GDP is released. This form of

EF-Tu is inactive and does not bind aminoacyl-tRNA effectively.

The guanine nucleotide exchange factor, EF-Ts, mediates the regeneration of the used form, EF-Tu·GDP, into the active form EF-Tu·GTP. First, EF-Ts displaces the GDP from EF-Tu, forming the combined factor EF-Tu·EF-Ts. Then the EF-Ts is in turn displaced by GTP, reforming EF-Tu·GTP. The active binary complex binds aminoacyl-tRNA, and the released EF-Ts can recycle.

There are ~70,000 molecules of EF-Tu per bacterium (~5% of the total bacterial protein), which approaches the number of aminoacyl-tRNA molecules. This implies that most aminoacyl-tRNAs are likely to be present in ternary complexes. There are only ~10,000 molecules of EF-Ts per cell (about the same as the number of ribosomes). The kinetics of the interaction between EF-Tu and EF-Ts suggest that the EF-Tu·EF-Ts complex exists only transiently, so that the EF-Tu is very rapidly converted to the GTP-bound form, and then to a ternary complex.

The role of GTP in the ternary complex has been studied by substituting an analog that cannot be hydrolyzed. The compound **GMP-PCP**

has a methylene bridge in place of the oxygen that links the β and γ phosphates in GTP. In the presence of GMP-PCP, a ternary complex can be formed that binds aminoacyl-tRNA to the ribosome. The peptide bond cannot be formed, though, so the presence of GTP is needed for aminoacyl-tRNA to be bound at the A site. The hydrolysis is not required until later.

Kirromycin is an antibiotic that inhibits the function of EF-Tu. When EF-Tu is bound by kirromycin, it remains able to bind aminoacyl-tRNA to the A site. The EF-Tu·GDP complex cannot be released from the ribosome, though. Its continued presence prevents formation of the peptide bond between the peptidyl-tRNA and the aminoacyl-tRNA. As a result, the ribosome becomes "stalled" on mRNA, bringing translation to a halt.

This effect of kirromycin demonstrates that inhibiting one step in translation blocks the next step. The reason is that the continued presence of EF-Tu prevents the aminoacyl end of aminoacyl-tRNA from entering the A site on the 50S subunit (see Figure 24.30). Thus the release of EF-Tu·GDP is needed for the ribosome to undertake peptide bond formation. The same principle is seen at other stages of translation: One reaction must be completed properly before the next can occur.

The interaction with EF-Tu also plays a role in quality control. Aminoacyl-tRNAs are brought into the A site without knowing whether their anticodons will fit the codon. The hydrolysis of EF-Tu·GTP is relatively slow: it takes longer than the time required for an incorrect aminoacyl-tRNA to dissociate from the A site; therefore, most incorrect species are removed at this stage. The release of EF-Tu·GDP after hydrolysis also is slow, so any surviving incorrect aminoacyl-tRNAs may dissociate at this stage. The basic principle is that the reactions involving EF-Tu occur slowly enough to allow incorrect aminoacyl-tRNAs to dissociate before they become trapped in translation.

In eukaryotes, the factor eEF1α is responsible for bringing aminoacyl-tRNA to the ribosome, again in a reaction that involves cleavage of a high-energy bond in GTP. Like its prokaryotic homolog (EF-Tu), it is an abundant protein. After hydrolysis of GTP, the active form is regenerated by the factor eEF1βγ, a counterpart to EF-Ts.

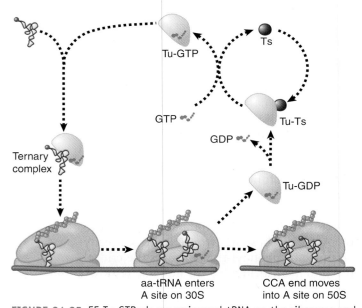

Tu-GTP

Ts

GTP

Tu-Ts

GDP

Ternary complex

Tu-GDP

aa-tRNA enters A site on 30S

CCA end moves into A site on 50S

FIGURE 24.25 EF-Tu·GTP places aminoacyl-tRNA on the ribosome and then is released as EF-Tu·GDP. EF-Ts is required to mediate the replacement of GDP by GTP. The reaction consumes GTP and releases GDP. The only aminoacyl-tRNA that cannot be recognized by EF-Tu·GTP is fMet-tRNA$_f$, whose failure to bind prevents it from responding to internal AUG or GUG codons.

24.11 The Polypeptide Chain Is Transferred to Aminoacyl-tRNA

Key concepts

- The 50S subunit has peptidyl transferase activity as provided by an rRNA ribozyme.
- The nascent polypeptide chain is transferred from peptidyl-tRNA in the P site to aminoacyl-tRNA in the A site.
- Peptide bond synthesis generates deacylated tRNA in the P site and peptidyl-tRNA in the A site.

The ribosome remains in place while the polypeptide chain is elongated by transferring the polypeptide attached to the tRNA in the P site to the aminoacyl-tRNA in the A site. The reaction is shown in **FIGURE 24.26**. The activity responsible for synthesis of the peptide bond is called **peptidyl transferase**.

Peptidyl transferase is a function of the large (50S or 60S) ribosomal subunit. The reaction is triggered when EF-Tu releases the aminoacyl end of its tRNA. The aminoacyl end then swings into a location close to the end of the peptidyl-tRNA. This site has a peptidyl transferase activity that essentially ensures a rapid transfer of the peptide chain to the aminoacyl-tRNA. Both rRNA and 50S subunit proteins are necessary for this activity, but the actual act of catalysis is a property of the ribosomal RNA of the 50S subunit (see *Section 24.19, 23S rRNA Has Peptidyl Transferase Activity*).

The nature of the transfer reaction is revealed by the ability of the antibiotic **puromycin** to inhibit translation. Puromycin resembles an amino acid attached to the terminal adenosine of tRNA. **FIGURE 24.27** shows that puromycin has an N instead of the O that joins an amino acid to tRNA. The antibiotic is treated by the ribosome as though it were an incoming aminoacyl-tRNA, after which the polypeptide attached to peptidyl-tRNA is transferred to the NH₂ group of the puromycin.

The puromycin moiety is not anchored to the A site of the ribosome, and as a result the polypeptidyl-puromycin adduct is released from the ribosome in the form of polypeptidyl-puromycin. This premature termination of translation is responsible for the lethal action of the antibiotic.

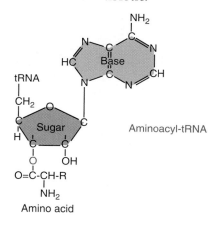

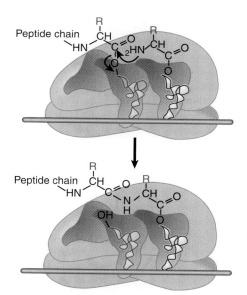

FIGURE 24.26 Peptide bond formation takes place by reaction between the polypeptide of peptidyl-tRNA in the P site and the amino acid of aminoacyl-tRNA in the A site.

FIGURE 24.27 Puromycin mimics aminoacyl-tRNA because it resembles an aromatic amino acid linked to a sugar-base moiety.

24.12 Translocation Moves the Ribosome

Key concepts

- Ribosomal translocation moves the mRNA through the ribosome by three bases.
- Translocation moves deacylated tRNA into the E site and peptidyl-tRNA into the P site, and empties the A site.
- The hybrid state model proposes that translocation occurs in two stages, in which the 50S moves relative to the 30S, and then the 30S moves along mRNA to restore the original conformation.

The cycle of addition of amino acids to the growing polypeptide chain is completed by *translocation*, when the ribosome advances three nucleotides along the mRNA. **FIGURE 24.28** shows that translocation expels the uncharged tRNA from the P site, so that the new peptidyl-tRNA can enter. The ribosome then has an empty A site ready for entry of the aminoacyl-tRNA corresponding to the next codon. As the figure shows, in bacteria the discharged tRNA is transferred from the P site to the E site (from which it is then expelled directly into the cytosol). In eukaryotes it is expelled directly into the cytosol without the presence of an E site. The A and P sites straddle both the large and small subunits; the E site (in bacteria) is located largely on the 50S subunit, but has some contacts in the 30S subunit.

Most thinking about translocation follows the *hybrid state model*, which has translocation occurring in two stages. **FIGURE 24.29** shows that first there is a shift of the 50S subunit relative to the 30S subunit, followed by a second shift that occurs when the 30S subunit moves along mRNA to restore the original conformation. The basis for this model was the observation that the pattern of contacts that tRNA makes with the ribosome (measured by chemical footprinting) changes in two stages. When puromycin is added to a ribosome that has an aminoacylated tRNA in the P site, the contacts of tRNA on the 50S subunit change from the P site to the

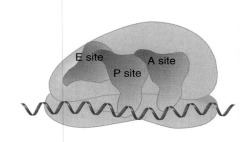

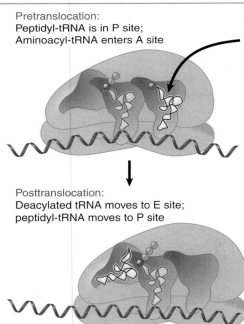

Pretranslocation:
Peptidyl-tRNA is in P site;
Aminoacyl-tRNA enters A site

Posttranslocation:
Deacylated tRNA moves to E site;
peptidyl-tRNA moves to P site

FIGURE 24.28 A bacterial ribosome has three tRNA-binding sites. Aminoacyl-tRNA enters the A site of a ribosome that has peptidyl-tRNA in the P site. Peptide bond synthesis deacylates the P site tRNA and generates peptidyl-tRNA in the A site. Translocation moves the deacylated tRNA into the E site and moves peptidyl-tRNA into the P site.

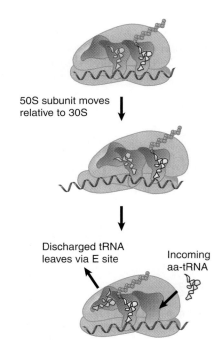

50S subunit moves relative to 30S

Discharged tRNA leaves via E site

Incoming aa-tRNA

FIGURE 24.29 Models for translocation involve two stages. First, at peptide bond formation the aminoacyl end of the tRNA in the A site becomes relocated in the P site. Second, the anticodon end of the tRNA becomes relocated in the P site.

E site, but the contacts on the 30S subunit do not change. This suggests that the 50S subunit has moved to a posttransfer state, but the 30S subunit has not changed.

The interpretation of these results is that first the aminoacyl ends of the tRNAs (located in the 50S subunit) move into the new sites (while the anticodon ends remain bound to their anticodons in the 30S subunit). At this stage, the tRNAs are effectively bound in hybrid sites, consisting of the 50SE/30S P and the 50SP/30S A sites. Then movement is extended to the 30S subunits, so that the anticodon–codon pairing region finds itself in the right site. The most likely means of creating the hybrid state is by a movement of one ribosomal subunit relative to the other, so that translocation in effect involves two stages, with the normal structure of the ribosome being restored by the second stage.

The ribosome faces an interesting dilemma at translocation. It needs to break many of its contacts with tRNA in order to allow movement. At the same time, however, it must maintain pairing between tRNA and the anticodon (breaking the pairing of the deacylated tRNA only at the right moment). One possibility is that the ribosome switches between alternative, discrete conformations. The switch could consist of changes in rRNA base pairing. The accuracy of translation is influenced by certain mutations that influence alternative base pairing arrangements. The most likely interpretation is that the effect is mediated by the tightness of binding to tRNA of the alternative conformations.

24.13 Elongation Factors Bind Alternately to the Ribosome

Key concepts

- Translocation requires EF-G, whose structure resembles the aminoacyl-tRNA-EF-Tu-GTP complex.
- Binding of EF-Tu and EF-G to the ribosome is mutually exclusive.
- Translocation requires GTP hydrolysis, which triggers a change in EF-G, which in turn triggers a change in ribosome structure.

Translocation requires GTP and another elongation factor, EF-G. This factor is a major constituent of the cell: it is present at a level of ~1 copy per ribosome (20,000 molecules per cell).

Ribosomes cannot bind EF-Tu and EF-G simultaneously, so translation follows the cycle illustrated in FIGURE 24.30, in which the factors are alternately bound to, and released from, the ribosome. Thus EF-Tu-GDP must be released before EF-G can bind; and then EF-G must be released before aminoacyl-tRNA-EF-Tu-GTP can bind.

Does the ability of each elongation factor to exclude the other rely on an allosteric effect on the overall conformation of the ribosome or on direct competition for overlapping binding sites? FIGURE 24.31 shows an extraordinary similarity between the structures of the ternary complex of aminoacyl-tRNA-EF-Tu-GDP and EF-G. The structure of EF-G mimics the overall structure of EF-Tu bound to the amino acceptor stem of aminoacyl-tRNA. This creates the immediate assumption that they compete for the same binding site (presumably in the vicinity of the A site). The need for each factor to be released before the other can bind ensures that

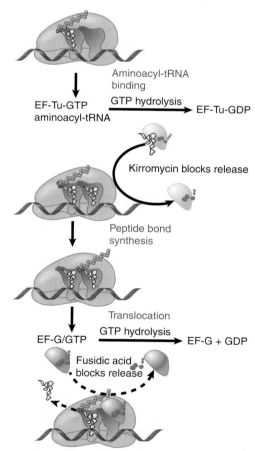

FIGURE 24.30 Binding of factors EF-Tu and EF-G alternates as ribosomes accept new aminoacyl-tRNA, form peptide bonds, and translocate.

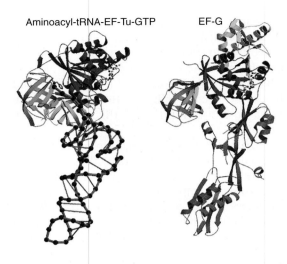

FIGURE 24.31 The structure of the ternary complex of aminoacyl-tRNA-EF-Tu-GTP (left) resembles the structure of EF-G (right). Structurally conserved domains of EF-Tu and EF-G are in red and green; the tRNA and the domain resembling it in EF-G are in purple. Photo courtesy of Poul Nissen, University of Aarhus, Denmark.

the events of translation proceed in an orderly manner.

Both elongation factors are monomeric GTP-binding proteins that are active when bound to GTP, but inactive when bound to GDP. The triphosphate form is required for binding to the ribosome, which ensures that each factor obtains access to the ribosome only in the company of the GTP that it needs to fulfill its function.

EF-G binds to the ribosome to sponsor translocation, and then is released following ribosome movement. EF-G can still bind to the ribosome when GMP-PCP is substituted for GTP; thus the presence of a guanine nucleotide is needed for binding, but its hydrolysis is not absolutely essential for translocation (although translocation is much slower in the absence of GTP hydrolysis). The hydrolysis of GTP is needed to release EF-G.

The need for EF-G release was discovered by the effects of the steroid antibiotic fusidic acid, which "jams" the ribosome in its post-translocation state (see Figure 24.30). In the presence of fusidic acid, one round of translocation occurs: EF-G binds to the ribosome, GTP is hydrolyzed, and the ribosome moves three nucleotides. Fusidic acid stabilizes the ribosome-EF-G-GDP complex, though, so that EF-G and GDP remain on the ribosome instead of being released. As a result, the ribosome

cannot bind aminoacyl-tRNA, and no further amino acids can be added to the chain.

Translocation is an intrinsic property of the ribosome that requires a major change in structure (see *Section 24.17, Ribosomes Have Several Active Centers*). This intrinsic translocation is activated by EF-G in conjunction with GTP hydrolysis, which occurs before translocation and accelerates the ribosome movement. The most likely mechanism is that GTP hydrolysis causes a change in the structure of EF-G, which in turn forces a change in the ribosome structure. An extensive reorientation of EF-G occurs at translocation. Before translocation, it is bound across the two ribosomal subunits. Most of its contacts with the 30S subunit are made by a region called domain 4, which is inserted into the A site. This domain could be responsible for displacing the tRNA. After translocation, domain 4 is instead oriented toward the 50S subunit.

The eukaryotic counterpart to EF-G is the protein eEF2, which functions in a similar manner as a translocase dependent on GTP hydrolysis. Its action also is inhibited by fusidic acid. A stable complex of eEF2 with GTP can be isolated, and the complex can bind to ribosomes with consequent hydrolysis of its GTP.

A unique reaction of eEF2 is its susceptibility to diphtheria toxin. The toxin uses nicotinamide adenine dinucleotide (NAD) as a cofactor to transfer an adenosine diphosphate ribosyl (ADPR) moiety onto the eEF2. The ADPR-eEF2 conjugate is inactive in translation. The substrate for the attachment is an unusual amino acid that is produced by modifying a histidine; it is common to the eEF2 of many species.

The ADP-ribosylation is responsible for the lethal effects of diphtheria toxin. The reaction is extremely effective: A single molecule of toxin can modify sufficient eEF2 molecules to kill a cell.

24.14 Three Codons Terminate Translation

Key concepts

- The codons UAA (ochre), UAG (amber), and UGA (sometimes called opal) terminate translation.
- In bacteria, they are used most often with relative frequencies UAA>UGA>UAG.

Only 61 triplets specify amino acids. The other three triplets are termination codons (or **stop**

codons), which end translation. They have casual names from the history of their discovery. The UAG triplet is called the **amber codon**, UAA is the **ochre codon**, and UGA is sometimes called the **opal codon**.

The nature of these triplets was originally shown by a genetic test that distinguished two types of point mutation:

- A point mutation that changes a codon to represent a different amino acid is called a *missense* mutation. One amino acid replaces the other in the protein; the effect on protein function depends on the site of mutation and the nature of the amino acid replacement.

- A point mutation that changes a codon to create one of the three termination codons is called a *nonsense* mutation. It causes **premature termination** of translation at the mutant codon. Only the first part of the protein is made in the mutant cell. This is likely to abolish protein function (depending, of course, on how far along the protein the mutant site is located).

(Note that the term *nonsense codon* sometimes is used to describe the termination triplets. "Nonsense" is really a term that describes the effect of a mutation in a gene rather than the meaning of the codon for translation. *Stop codon* is a better term.)

In every gene that has been sequenced, one of the termination codons lies immediately after the codon representing the C-terminal amino acid of the wild-type sequence. Nonsense mutations show that any one of the three codons is sufficient to terminate translation within a gene. The UAG, UAA, and UGA triplet sequences are therefore necessary and sufficient to end translation, whether occurring naturally at the end of a gene or created by mutation within a coding sequence.

In bacterial genes, UAA is the most commonly used termination codon. UGA is used more heavily than UAG, although there appear to be more errors reading UGA. (An error in reading a termination codon, when an aminoacyl-tRNA improperly responds to it, results in the continuation of translation until another termination codon is encountered or the ribosome reaches the 3' end of the mRNA, which may result in other problems. For this circumstance, bacteria have a special RNA.)

24.15 Termination Codons Are Recognized by Protein Factors

Key concepts

- Termination codons are recognized by protein release factors, not by aminoacyl-tRNAs.
- The structures of the class 1 release factors resemble aminoacyl-tRNA-EF-Tu and EF-G.
- The class 1 release factors respond to specific termination codons and hydrolyze the polypeptide-tRNA linkage.
- The class 1 release factors are assisted by class 2 release factors that depend on GTP.
- The mechanism is similar in bacteria (which have two types of class 1 release factors) and eukaryotes (which have only one class 1 release factor).

Two stages are involved in ending translation. The *termination reaction* itself involves release of the protein chain from the last tRNA. The *post-termination reaction* involves release of the tRNA and mRNA and dissociation of the ribosome into its subunits.

None of the termination codons is represented by a tRNA. They function in an entirely different manner from other codons and are recognized directly by protein factors. (The reaction does not depend on codon–anticodon recognition, so there seems to be no particular reason why it should require a triplet sequence. Presumably this reflects the evolution of the genetic code.)

Termination codons are recognized by class 1 **release factors (RF)**. In *E. coli*, two class 1 release factors are specific for different sequences. **RF1** recognizes UAA and UAG; **RF2** recognizes UGA and UAA. The factors act at the ribosomal A site and require polypeptidyl-tRNA in the P site. The RFs are present at much lower levels than initiation or elongation factors; there are ~600 molecules of each per cell, equivalent to one RF per ten ribosomes. At one time there probably was only a single release factor that recognized all termination codons, which later evolved into two factors with specificities for particular codons. In eukaryotes, there is only a single class 1 release factor, called eRF. The efficiency with which the bacterial factors recognize their target codons is influenced by the bases on the 3' side.

The class 1 release factors are assisted by class 2 release factors, which are not codon-specific. The class 2 factors are GTP-binding

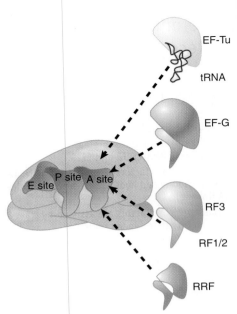

FIGURE 24.32 Molecular mimicry enables the elongation factor Tu-tRNA complex, the translocation factor EF-G, and the release factors RF1/2-RF3 to bind to the same ribosomal site. RRF is the ribosome recycling factor (see Figure 24.35).

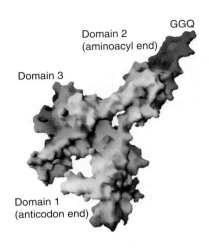

FIGURE 24.33 The eukaryotic termination factor eRF1 has a structure that mimics tRNA. The motif GGQ at the tip of domain 2 is essential for hydrolyzing the polypeptide chain from tRNA. Photo courtesy of David Barford, The Institute of Cancer Research.

proteins. In *E. coli*, the role of the class 2 factor, **RF3**, is to release the class 1 factor from the ribosome. RF3 is a GTP-binding protein that is related to the elongation factors.

Although the general mechanism of termination is similar in prokaryotes and eukaryotes, the interactions between the class 1 and class 2 factors have some differences.

The class 1 factors RF1 and RF2 recognize the termination codons and activate the ribosome to hydrolyze the peptidyl tRNA. Cleavage of polypeptide from tRNA takes place by a reaction analogous to the usual peptidyl transfer, except that the acceptor is H_2O instead of aminoacyl-tRNA (see Figure 24.34).

At this point RF1 or RF2 is released from the ribosome by the class 2 factor RF3, which is related to EF-G. RF3-GDP binds to the ribosome before the termination reaction occurs, and the GDP is replaced by GTP. This enables RF3 to contact the ribosome GTPase center, where it causes RF1/2 to be released when the polypeptide chain is terminated.

RF3 resembles the GTP-binding domains of EF-Tu and EF-G, and RF1 and RF2 resemble the C-terminal domain of EF-G, which mimics tRNA. This suggests that the release factors utilize the same site that is used by the elongation factors. FIGURE 24.32 illustrates the basic idea that these factors all have the same general shape and bind to the ribosome successively at the same site (basically the A site or a region extensively overlapping with it).

The eukaryotic class 1 release factor, eRF1, is a single protein that recognizes all three termination codons. Its sequence is unrelated to the bacterial factors. It can terminate translation *in vitro* without the class 2 factor, eRF2, although eRF2 is essential in yeast *in vivo*. The structure of eRF1 follows a familiar theme: FIGURE 24.33 shows that it consists of three domains that mimic the structure of tRNA.

An essential motif of three amino acids, GGQ, is exposed at the top of domain 2. Its position in the A site corresponds to the usual location of an amino acid on an aminoacyl-tRNA. This positions it to use the glutamine (Q) to position H_2O to substitute for the amino acid of aminoacyl-tRNA in the peptidyl transfer reaction. FIGURE 24.34 compares the termination reaction with the usual peptide transfer reaction. Termination transfers a hydroxyl group from H_2O, thus effectively hydrolyzing the peptide-tRNA bond (see Figure 24.48 for discussion of how the peptidyl transferase center works).

Mutations in the RF genes reduce the efficiency of termination, as seen by an increased ability to continue translation past the termination codon. Overexpression of RF1 or RF2 increases the efficiency of termination at the codons on which it acts. This suggests that codon recognition by RF1 or RF2 competes with

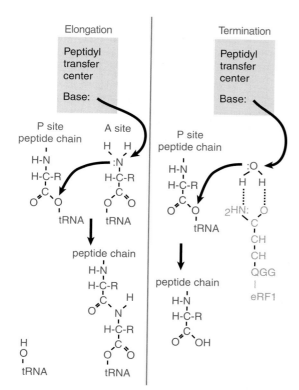

FIGURE 24.34 Peptide transfer and termination are similar reactions in which a base in the peptidyl transfer center triggers a transesterification reaction by attacking an N–H or O–H bond, releasing the N or O to attack the link to tRNA.

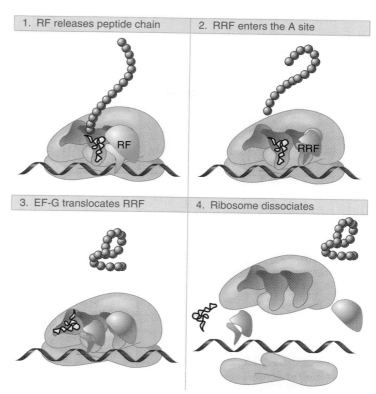

FIGURE 24.35 The RF (release factor) terminates translation by releasing the protein chain. The RRF (ribosome recycling factor) releases the last tRNA, and EF-G releases RRF, causing the ribosome to dissociate.

aminoacyl-tRNAs that erroneously recognize the termination codons. The release factors recognize their target sequences very efficiently.

The termination reaction involves release of the completed polypeptide, but leaves a deacylated tRNA and the mRNA still associated with the ribosome. FIGURE 24.35 shows that the dissociation of the remaining components (tRNA, mRNA, 30S, and 50S subunits) requires *ribosome recycling factor (RRF)*. RRF acts together with EF-G in a reaction that uses hydrolysis of GTP. As for the other factors involved in release, RRF has a structure that mimics tRNA, except that it lacks an equivalent for the 3′ amino acid-binding region. IF-3 is also required, which brings the wheel full circle to its original discovery, when it was proposed to be a dissociation factor! RRF acts on the 50S subunit, and IF-3 acts to remove deacylated tRNA from the 30S subunit. Once the subunits have separated, IF-3 remains necessary, of course, to prevent their reassociation.

FIGURE 24.36 compares the functional and sequence homologies of the prokaryotic and eukaryotic translation factors.

Ribosomal RNA Pervades Both Ribosomal Subunits

Key concepts

- Each rRNA has several distinct domains that fold independently.
- Virtually all ribosomal proteins are in contact with rRNA.
- Most of the contacts between ribosomal subunits are made between the 16S and 23S rRNAs.

Two thirds of the mass of the bacterial ribosome is made up of rRNA. The most penetrating approach to analyzing secondary structure of large RNAs is to compare the sequences of corresponding rRNAs in related organisms. Those regions that are important in the secondary structure retain the ability to interact by base pairing. Thus if a base pair is required, it can form at the same relative position in each rRNA. This approach has enabled detailed models of both 16S and 23S rRNA to be constructed.

Each of the major rRNAs can be drawn in a secondary structure with several discrete domains. Four general domains are formed

Initiation Factors			
Prokaryotic	Eukaryotic	General Function	Notes
IF-1	eIF1A	Blocks A site	eIF1A assists eIF2 in promoting Met-tRNA$_i^{Met}$ to binding to 40S; also promotes subunit dissociation
IF-2*†	eIF2, eIF3, eIF5B*	Entry of initiator tRNA	eIF2 is a GTPase
			eIF3 stimulates formation of the ternary complex, its binding to 40S, and binding and scanning of mRNA
			eIF5B is involved in initiator tRNA entry and is a GTPase
IF-3	eIF1, eIF4 complex, eIF3	Small subunit binding to mRNA	eIF4 complex functions in cap binding

Elongation Factors			
Prokaryotic	Eukaryotic	General Function	
EF-Tu††‡, EF-G†	eEF1α‡	GTP-binding	
EF-Ts	eEF1β, eEF1γ	GDP-exchanging	
EF-G§	eEF2§	Ribosome translocation	

Release Factors			
Prokaryotic	Eukaryotic	General Function	
RF1	eRF1	UAA/UAG recognition	
RF2	eRF1	UAA/UGA recognition	
RF3†	eRF3	Stimulation of other RF(s)	

* IF-2 and eIF5B have sequence homology.
† IF-2, EF-Tu, EF-G, and RF3 have sequence homology.
‡ EF-Tu and eEF1α have sequence homology.
§ EF-G and eEF2 have sequence homology.

FIGURE 24.36 Functional homologies of prokaryotic and eukaryotic translation factors.

by 16S rRNA, in which just under half of the sequence is base paired (see Figure 24.46). Six general domains are formed by 23S rRNA. The individual double-helical regions tend to be short (<8 bp). Often the duplex regions are not perfect and contain bulges of unpaired bases. Comparable models have been drawn for mitochondrial rRNAs (which are shorter and have fewer domains) and for eukaryotic cytosolic rRNAs (which are longer and have more domains). The greater length of eukaryotic rRNAs is due largely to the acquisition of sequences representing additional domains. The crystal structure of the ribosome shows that in each subunit the domains of the major rRNA fold independently and have discrete locations.

Differences in the ability of 16S rRNA to react with chemical agents are found when 30S subunits are compared with 70S ribosomes; there also are differences between free ribosomes and those engaged in translation. Changes in the reactivity of the rRNA occur when mRNA is bound, when the subunits associate, or when tRNA is bound. Some changes reflect a direct interaction of the rRNA with mRNA or tRNA, whereas others are caused indirectly by other changes in ribosome structure.

The main point is that ribosome conformation is flexible during translation, particularly that of the small subunit as it must physically check the accuracy of codon–anticodon pairing.

A feature of the primary structure of rRNA is the presence of methylated residues. There are ~10 methyl groups in 16S rRNA (located mostly toward the 3' end of the molecule) and ~20 in 23S rRNA. In mammalian cells, the 18S and 28S rRNAs carry 43 and 74 methyl groups, respectively, so ~2% of the nucleotides are methylated (about three times the proportion methylated in bacteria).

The large ribosomal subunit also contains a molecule of a 120-base *5S RNA* (in all ribosomes except those of mitochondria). The sequence of 5S RNA is less well conserved than those of the major rRNAs. All 5S RNA molecules display a highly base-paired structure.

In eukaryotic cytosolic ribosomes, another small RNA is present in the large subunit. This is the *5.8S RNA*. Its sequence corresponds to the 5' end of the prokaryotic 23S rRNA.

Some ribosomal proteins bind strongly to isolated rRNA. Others do not bind to free rRNA, but can bind after other proteins have bound. This suggests that the conformation of the rRNA

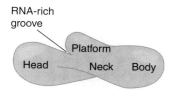

RNA-rich groove

FIGURE 24.37 The 30S subunit has a head separated by a neck from the body, with a protruding platform.

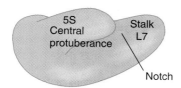

FIGURE 24.38 The 50S subunit has a central protuberance where 5S rRNA is located, separated by a notch from a stalk made of copies of the protein L7.

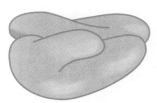

FIGURE 24.39 The platform of the 30S subunit fits into the notch of the 50S subunit to form the 70S ribosome.

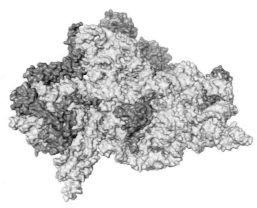

FIGURE 24.40 The 30S ribosomal subunit is a ribonucleoprotein particle. Proteins are in yellow. Photo courtesy of V. Ramakrishnan, Medical Research Council (UK).

is important in determining whether binding sites exist for some proteins. As each protein binds, it induces conformational changes in the rRNA that make it possible for other proteins to bind. In *E. coli*, virtually all the 30S ribosomal proteins interact (albeit to varying degrees) with 16S rRNA. The binding sites on the proteins show a wide variety of structural features, suggesting that protein–RNA recognition mechanisms may be diverse.

The 70S ribosome has an asymmetric construction. **FIGURE 24.37** shows a schematic of the structure of the 30S subunit, which is divided into four regions: the head, neck, body, and platform. **FIGURE 24.38** shows a similar representation of the 50S subunit, where two prominent features are the central protuberance (where 5S rRNA is located) and the stalk (made of multiple copies of protein L7). **FIGURE 24.39** shows that the platform of the small subunit fits into the notch of the large subunit. There is a cavity between the subunits that contains some of the important sites.

The structure of the 30S subunit follows the organization of 16S rRNA, with each structural feature corresponding to a domain of the

rRNA. The body is based on the 5′ domain, the platform on the central domain, and the head on the 3′ region. **FIGURE 24.40** shows that the 30S subunit has an asymmetrical distribution of RNA and protein. One important feature is that the platform of the 30S subunit that provides the interface with the 50S subunit is composed almost entirely of RNA. At most two proteins (a small part of S7 and possibly part of S12) lie near the interface. This means that the association and dissociation of ribosomal subunits must depend on interactions with the 16S rRNA. Subunit association is affected by a mutation in a loop of 16S rRNA (at position 791) that is located at the subunit interface, and other nucleotides in 16S rRNA have been shown to be involved by modification/interference experiments. This observation supports the idea that the evolutionary origin of the ribosome may have been as a particle consisting of RNA rather than protein.

The 50S subunit has a more even distribution of components than the 30S, with long rods of double-stranded RNA crisscrossing the structure. The RNA forms a mass of tightly packed helices. The exterior surface largely consists of protein, except for the peptidyl transferase center (see *Section 24.19, 23S rRNA Has Peptidyl Transferase Activity*). Almost all segments of the 23S rRNA interact with protein, but many of the proteins are relatively unstructured.

The junction of subunits in the 70S ribosome involves contacts between 16S rRNA (many in the platform region) and 23S rRNA. There are also some interactions between rRNA of each subunit with proteins in the other, and a few protein–protein contacts. **FIGURE 24.41** identifies the contact points on the rRNA structures. **FIGURE 24.42** opens out the structure (imagine the

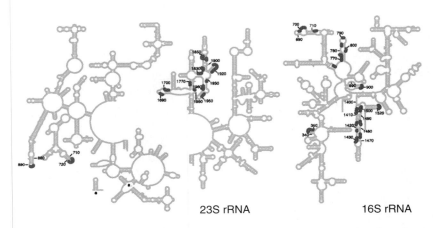

FIGURE 24.41 Contact points between the rRNAs are located in two domains of 16S rRNA and one domain of 23S rRNA. Reproduced from M. M. Yusupov, et al., *Science* 292 (2001): 883–896 [http://www.sciencemag.org]. Reprinted with permission from AAAS. Photo courtesy of Harry Noller, University of California, Santa Cruz.

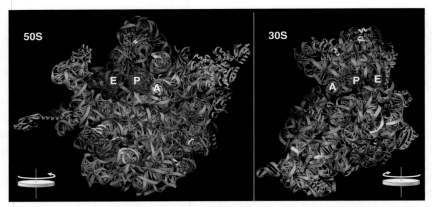

FIGURE 24.42 Contacts between the ribosomal subunits are mostly made by RNA (shown in purple). Contacts involving proteins are shown in yellow. The two subunits are rotated away from one another to show the faces where contacts are made; from a plane of contact perpendicular to the screen, the 50S subunit is rotated 90° counterclockwise, and the 30S is rotated 90° clockwise (this shows it in the reverse of the usual orientation). Photo courtesy of Harry Noller, University of California, Santa Cruz.

50S subunit rotated counterclockwise and the 30S subunit rotated clockwise around the axis shown in the figure) to show the locations of the contact points on the face of each subunit.

24.17 Ribosomes Have Several Active Centers

Key concepts

- Interactions involving rRNA are a key part of ribosome function.
- The environment of the tRNA-binding sites is largely determined by rRNA.

The basic ribosomal feature to remember is that it is a cooperative structure that depends on changes in the relationships among its active sites during translation. The active sites are not small, discrete regions like the active centers of enzymes. They are large regions whose construction and activities may depend just as much on the rRNA as on the ribosomal proteins. The crystal structures of the individual subunits and bacterial ribosomes give us a good impression of the overall organization and emphasize the role of the rRNA. The most recent structure, at 3.5 Å resolution, clearly identifies the locations of the tRNAs and the functional sites. We can now account for many ribosomal functions in terms of its structure.

Ribosomal functions are centered around the interaction with tRNAs. **FIGURE 24.43** shows the 70S ribosome with the positions of tRNAs in the three binding sites. The tRNAs in the A and

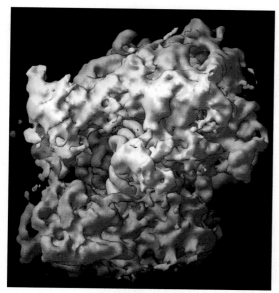

FIGURE 24.43 The 70S ribosome consists of the 50S subunit (white) and the 30S subunit (purple) with three tRNAs located superficially: yellow in the A site, blue in the P site, and green in the E site. Photo courtesy of Harry Noller, University of California, Santa Cruz.

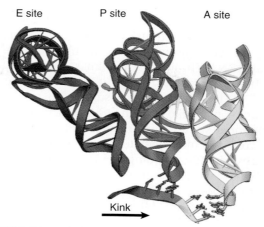

FIGURE 24.44 Three tRNAs have different orientations on the ribosome. mRNA turns between the P and A sites to allow aminoacyl-tRNAs to bind adjacent codons. Photo courtesy of Harry Noller, University of California, Santa Cruz.

P sites are nearly parallel to one another. All three tRNAs are aligned with their anticodon loops bound to the mRNA in the groove on the 30S subunit. The rest of each tRNA is bound to the 50S subunit. The environment surrounding each tRNA is mostly provided by rRNA. In each site, the rRNA contacts the tRNA at parts of the structure that are universally conserved.

It has always been a big puzzle to understand how two bulky tRNAs can fit next to one another in reading adjacent codons. The crystal structure shows a 45° kink in the mRNA between the P and A sites, which allows the tRNAs to fit as shown in the expansion of **FIGURE 24.44**. The tRNAs in the P and A sites are angled at 26° relative to each other at their anticodons. The closest approach between the backbones of the tRNAs occurs at the 3' ends, where they converge to within 5 Å (perpendicular to the plane of the page). This allows the peptide chain to be transferred from the peptidyl-tRNA in the P site to the aminoacyl-tRNA in the A site.

Aminoacyl-tRNA is inserted into the A site by EF-Tu, and its pairing with the codon is necessary for EF-Tu to hydrolyze GTP and be released from the ribosome (see *Section 24.10, Elongation Factor Tu Loads Aminoacyl-tRNA into the A Site*). EF-Tu initially places the aminoacyl-tRNA into the small subunit, where the anticodon pairs with the codon. Movement of the tRNA is required to bring it fully into the A site, when its 3' end enters the peptidyl transferase center on the large subunit. There

are different models for how this process may occur. One calls for the entire tRNA to swivel, so that the elbow in the L-shaped structure made by the D and TΨC arms moves into the ribosome, enabling the TΨC arm to pair with rRNA. Another calls for the internal structure of the tRNA to change, using the anticodon loop as a hinge, with the rest of the tRNA rotating from a position in which it is stacked on the 3' side of the anticodon loop to one in which it is stacked on the 5' side. Following the transition, EF-Tu hydrolyzes GTP, allowing peptide synthesis to proceed.

Translocation involves large movements in the positions of the tRNAs within the ribosome. The anticodon end of tRNA moves ~28 Å from the A site to the P site, and then moves an additional 20 Å from the P site to the E site. As a result of the angle of each tRNA relative to the anticodon, the bulk of the tRNA moves much larger distances: 40 Å from the A site to the P site, and 55 Å from the P site to the E site. This suggests that translocation requires a major reorganization of structure.

For many years, it was thought that translocation could occur only in the presence of the factor EF-G. The antibiotic sparsomycin (which inhibits the peptidyl transferase activity), however, triggers translocation. This suggests that the energy to drive translocation actually is stored in the ribosome after peptide bond formation has occurred. Usually EF-G acts on the ribosome to release this energy and enable it to drive translocation, but sparsomycin can have the same role. Sparsomycin inhibits peptidyl transferase by binding to the peptidyl-tRNA, blocking its interaction with aminoacyl-tRNA. It probably creates a conformation that resembles the usual posttranslocation conformation, which in turn promotes movement of the peptidyl-tRNA. The important point is that translocation is an intrinsic property of the ribosome.

The hybrid states model suggests that translocation may take place in two stages, with one ribosomal subunit moving relative to the other to create an intermediate stage in which there are hybrid tRNA-binding sites (50S E/30S P and 50SP/30S A) (see Figure 24.29). Comparisons of the ribosome structure between pre- and posttranslocation states, and comparisons in 16S rRNA conformation between free 30S subunits and 70S ribosomes, suggest that mobility of structure is especially marked in the head and platform regions of the 30S subunit. An interesting insight on the hybrid states model is cast by the fact that many bases in rRNA involved in subunit association are close to bases involved in interacting with tRNA. This suggests that tRNA-binding sites are close to the interface between subunits, and carries the implication that changes in subunit interaction could be connected with movement of tRNA.

Much of the structure of the ribosome is occupied by its active centers. The schematic view of the ribosomal sites in **FIGURE 24.45** shows they comprise about two-thirds of the ribosomal structure. A tRNA enters the A site, is transferred by translocation into the P site, and then leaves the ribosome by the E site. The A and P sites extend across both ribosome subunits; tRNA is paired with mRNA in the 30S subunit, but peptide transfer takes place in the 50S subunit. The A and P sites are adjacent,

enabling translocation to move the tRNA from one site into the other. The E site is located near the P site (representing a position *en route* to the surface of the 50S subunit). The peptidyl transferase center is located on the 50S subunit, close to the aminoacyl ends of the tRNAs in the A and P sites (see *Section 24.18, 16S rRNA Plays an Active Role in Translation*).

All of the GTP-binding proteins that function in translation (EF-Tu, EF-G, IF-2, RF1, RF2, and RF3) bind to the same factor-binding site (sometimes called the GTPase center), which probably triggers their hydrolysis of GTP. This site is located at the base of the stalk of the large subunit, which consists of the proteins L7 and L12. (L7 is a modification of L12 and has an acetyl group on the N terminus.) In addition to this region, the complex of protein L11 with a 58-base stretch of 23S rRNA provides the binding site for some antibiotics that affect GTPase activity. Neither of these ribosomal structures actually possesses GTPase activity, but they are both necessary for it. The role of the ribosome is to trigger GTP hydrolysis by factors bound in the factor-binding site.

Initial binding of 30S subunits to mRNA requires protein S1, which has a strong affinity for single-stranded nucleic acid. It is responsible for maintaining the single-stranded state in mRNA that is bound to the 30S subunit. This action is necessary to prevent the mRNA from taking up a base-paired conformation that would be unsuitable for translation. S1 has an extremely elongated structure and associates with S18 and S21. The three proteins constitute a domain that is involved in the initial binding of mRNA and in binding initiator tRNA. This locates the mRNA-binding site in the vicinity of the cleft of the small subunit (see Figure 24.3). The 3' end of rRNA, which pairs with the mRNA initiation site, is located in this region.

The initiation factors bind in the same region of the ribosome. IF-3 can be crosslinked to the 3' end of the rRNA, as well as to several ribosomal proteins, including those probably involved in binding mRNA. The role of IF-3 could be to stabilize mRNA-30S subunit binding; then it would be displaced when the 50S subunit joins.

The incorporation of 5S RNA into 50S subunits that are assembled *in vitro* depends on the ability of three proteins—L5, L8, and L25—to form a stoichiometric complex with it. The complex can bind to 23S rRNA, although none of the isolated components can do so. It lies in the vicinity of the P and A sites.

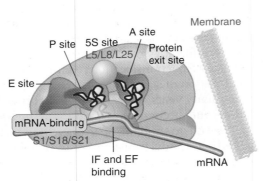

FIGURE 24.45 The ribosome has several active centers. It may be associated with a membrane. mRNA takes a turn as it passes through the A and P sites, which are angled with regard to each other. The E site lies beyond the P site. The peptidyl transferase site (not shown) stretches across the tops of the A and P sites. Part of the site bound by EF-Tu/G lies at the base of the A and P sites.

A nascent polypeptide extends through the ribosome, away from the active sites, into the region in which ribosomes may be attached to membranes. A polypeptide chain emerges from the ribosome through an exit channel, which leads from the peptidyl transferase site to the surface of the 50S subunit. The tunnel is composed mostly of rRNA. It is quite narrow—only 1 to 2 nm wide—and is ~10 nm long. The nascent polypeptide emerges from the ribosome ~15 Å away from the peptidyl transferase site. The tunnel can hold ~50 amino acids, and probably constrains the polypeptide chain so that it cannot fold until it leaves the exit domain, though some limited secondary structures may form.

24.18 16S rRNA Plays an Active Role in Translation

Key concept

- 16S rRNA plays an active role in the functions of the 30S subunit. It interacts directly with mRNA, with the 50S subunit, and with the anticodons of tRNAs in the P and A sites.

The ribosome was originally viewed as a collection of proteins with various catalytic activities held together by protein–protein interactions and RNA-protein interactions. The discovery of RNA molecules with catalytic activities (see Chapter 21, *RNA Splicing and Processing*) immediately suggests, however, that rRNA might play a more active role in ribosome function. There is now evidence that rRNA interacts with mRNA or tRNA at each stage of translation, and that the proteins are necessary to maintain the rRNA in a structure in which it can perform the catalytic functions. Several interactions involve specific regions of rRNA:

- The 3' terminus of the rRNA interacts directly with mRNA at initiation.
- Specific regions of 16S rRNA interact directly with the anticodon regions of tRNAs in both the A site and the P site. Similarly, 23S rRNA interacts with the CCA terminus of peptidyl-tRNA in both the P site and A site.
- Subunit interaction involves interactions between 16S and 23S rRNAs (see *Section 24.16, Ribosomal RNA Pervades Both Ribosomal Subunits*).

Much information about the individual steps of bacterial translation has been obtained by using antibiotics that inhibit the process at particular stages. The target for the antibiotic can be identified by the component in which resistant mutations occur. Some antibiotics act on individual ribosomal proteins, but several act on rRNA, which suggests that the rRNA is involved with many or even all of the functions of the ribosome.

The functions of rRNA have been investigated by two types of approach. Structural studies show that particular regions of rRNA are located in important sites of the ribosome, and that chemical modifications of these bases impede particular ribosomal functions. In addition, mutations identify bases in rRNA that are required for particular ribosomal functions. **FIGURE 24.46** summarizes the sites in 16S rRNA that have been identified by these means.

An indication of the importance of the 3' end of 16S rRNA is given by its susceptibility to the lethal agent colicin E3. Produced by some bacteria, the colicin cleaves ~50 nucleotides from the 3' end of the 16S rRNA of *E. coli*. The

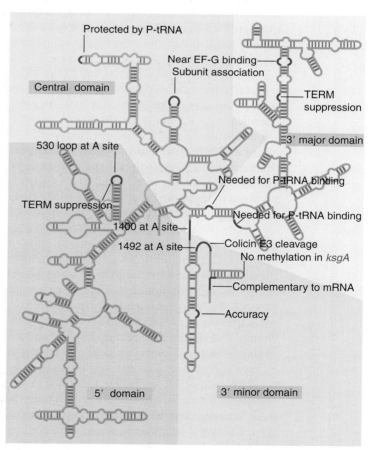

FIGURE 24.46 Some sites in 16S rRNA are protected from chemical probes when 50S subunits join 30S subunits or when aminoacyl-tRNA binds to the A site. Others are the sites of mutations that affect translation. TERM suppression sites may affect termination at some or several termination codons. The large colored blocks indicate the four domains of the rRNA.

cleavage entirely abolishes initiation of translation. Several important functions require the region that is cleaved: binding the factor IF-3, recognition of mRNA, and binding of tRNA.

The 3' end of the 16S rRNA is directly involved in the initiation reaction by pairing with the Shine–Dalgarno sequence in the ribosome-binding site of mRNA (see Figure 24.12). Another direct role for the 3' end of 16S rRNA in translation is shown by the properties of kasugamycin-resistant mutants, which lack certain modifications in 16S rRNA. Kasugamycin blocks initiation of translation. Resistant mutants of the type *ksgA* lack a methylase enzyme that introduces four methyl groups into two adjacent adenines at a site near the 3' terminus of the 16S rRNA. The methylation generates the highly conserved sequence $G–m_2^6A–m_2^6A$, found in both prokaryotic and eukaryotic small rRNA. The methylated sequence is involved in the joining of the 30S and 50S subunits, which in turn is connected also with the retention of initiator tRNA in the complete ribosome. Kasugamycin causes fMet-tRNA$_f$ to be released from the sensitive (methylated) ribosomes, but the resistant ribosomes are able to retain the initiator.

Changes in the structure of 16S rRNA occur when ribosomes are engaged in translation, as seen by protection of particular bases against chemical attack. The individual sites fall into a few groups that are concentrated in the 3' minor and central domains. Although the locations are dispersed in the linear sequence of 16S rRNA, it seems likely that base positions involved in the same function are actually close together in the tertiary structure.

Some of the changes in 16S rRNA are triggered by joining with 50S subunits, binding of mRNA, or binding of tRNA. They indicate that these events are associated with changes in ribosome conformation that affect the exposure of rRNA. They do not necessarily indicate direct participation of rRNA in these functions. One change that occurs during translation is shown in **FIGURE 24.47**; it involves a local movement to change the nature of a short duplex sequence.

The 16S rRNA is involved in both A site and P site function, and significant changes in its structure occur when these sites are occupied. Certain distinct regions are protected by tRNA bound in the A site (see Figure 24.46). One is the 530 loop (which also is the site of a mutation that prevents termination at the UAA, UAG, and UGA codons). The other is the 1400 to 1500 region (so called because bases

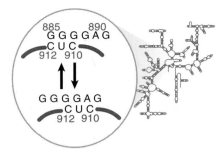

FIGURE 24.47 A change in conformation of 16S rRNA may occur during translation.

1399 to 1492 and the adenines at 1492 and 1493 are two single-stranded stretches that are connected by a long hairpin). All of the effects that tRNA binding has on 16S rRNA can be produced by the isolated oligonucleotide of the anticodon stem–loop, so that tRNA-30S subunit binding must involve this region.

The adenines at 1492 and 1493 provide a mechanism for detecting properly paired codon–anticodon complexes. The principle of the interaction is that the structure of the 16S rRNA responds to the structure of the first two bases pairs in the minor groove of the duplex formed by the codon–anticodon interaction. Modification of the N1 position of either base 1492 or 1493 in rRNA prevents tRNA from binding in the A site. Mutations at 1492 or 1493, however, can be suppressed by the introduction of fluorine at the 2' position of the corresponding bases in mRNA (which restores the interaction). **FIGURE 24.48** shows that codon–anticodon pairing allows the N1 of each adenine to interact with the 2'–OH in the mRNA backbone. The interaction stabilizes the association of tRNA with the A site. When an incorrect tRNA enters the A site, the structure of the codon–anticodon complex is distorted and this interaction cannot occur.

A variety of bases in different positions of 16S rRNA are protected by tRNA in the P site; most likely the bases lie near one another in the tertiary structure. In fact, there are more contacts with tRNA when it is in the P site than when it is in the A site. This may be responsible for the increased stability of peptidyl-tRNA compared with aminoacyl-tRNA. This makes sense: Once the tRNA has reached the P site, the ribosome has decided that it is correctly bound, whereas in the A site, the assessment of binding is being made. The 1400 region can be directly crosslinked to peptidyl-tRNA, which suggests that this region is a structural component of the P site.

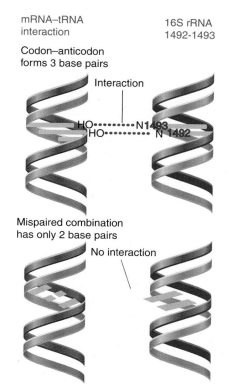

mRNA–tRNA interaction

Codon–anticodon forms 3 base pairs

16S rRNA 1492-1493

Interaction

HO·······N1493
HO·······N 1492

Mispaired combination has only 2 base pairs

No interaction

FIGURE 24.48 Codon–anticodon pairing supports interaction with adenines 1492–1493 of 16S rRNA, but mispaired tRNA–mRNA cannot interact.

The basic conclusion to be drawn from these results is that rRNA has many interactions with both tRNA and mRNA, and that these interactions recur in each cycle of peptide bond formation.

24.19 23S rRNA Has Peptidyl Transferase Activity

Key concept

- Peptidyl transferase activity resides exclusively in the 23S rRNA.

The sites involved in the functions of 23S rRNA are less well identified than those of 16S rRNA, but the same general pattern is observed: bases at certain positions affect specific functions. Bases at some positions in 23S rRNA are affected by the conformation of the A site or the P site. In particular, oligonucleotides derived from the 3′ CCA terminus of tRNA protect a set of bases in 23S rRNA that essentially are the same as those protected by peptidyl-tRNA. This suggests that the major interaction of 23S rRNA with peptidyl-tRNA in the P site involves the 3′ end of the tRNA.

The tRNA makes contacts with the 23S rRNA in both the P and A sites. At the P site, G2552 of 23S rRNA base pairs with C74 of the peptidyl tRNA. A mutation in the G in the rRNA prevents interaction with tRNA, but interaction is restored by a compensating mutation in the C of the amino acceptor end of the tRNA. At the A site, G2553 of the 23S rRNA base pairs with C75 of the aminoacyl-tRNA. Thus there is a close role for rRNA in both the tRNA-binding sites. As structural studies continue to emerge, the movements of tRNA between the A and P sites in terms of making and breaking contacts with rRNA will be elucidated.

Another site that binds tRNA is the E site, which is localized almost exclusively on the 50S subunit. Bases affected by its conformation can be identified in 23S rRNA.

What is the nature of the site on the 50S subunit that provides peptidyl transferase function? A long search for ribosomal proteins that might possess the catalytic activity was unsuccessful, and led to the discovery that the ribosomal RNA of the large subunit can catalyze the formation of a peptide bond between peptidyl-tRNA and aminoacyl-tRNA. The involvement of rRNA was first indicated because a region of the 23S rRNA is the site of mutations that confer resistance to antibiotics that inhibit peptidyl transferase. Extraction of almost all the protein content of 50S subunits leaves the 23S rRNA associated largely with fragments of proteins, amounting to <5% of the mass of the ribosomal proteins. This preparation retains peptidyl transferase activity. Treatments that damage the RNA abolish the catalytic activity.

Following from these results, 23S rRNA prepared by transcription *in vitro* can catalyze the formation of a peptide bond between Ac-Phe-tRNA and Phe-tRNA. The yield of Ac-Phe-Phe is very low, suggesting that the 23S rRNA requires proteins in order to function at a high efficiency. Given that the rRNA has the basic catalytic activity, though, the role of the proteins must be indirect, serving to fold the rRNA properly or to present the substrates to it. The reaction also works, although less effectively, if the domains of 23S rRNA are synthesized separately and then combined. In fact, some activity is shown by domain V alone, which has the catalytic center. Activity is abolished by mutations in position 2252 of domain V that lies in the P site.

The crystal structure of an archaeal 50S subunit shows that the peptidyl transferase site basically consists of 23S rRNA. There is no

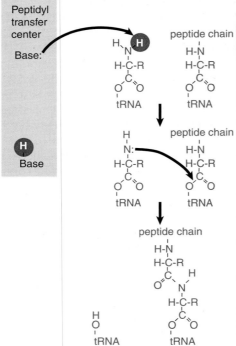

FIGURE 24.49 Peptide bond formation requires acid-base catalysis in which an H atom is transferred to a basic residue.

protein within 18 Å of the active site where the transfer reaction occurs between peptidyl-tRNA and aminoacyl-tRNA!

Peptide bond synthesis requires an attack by the amino group of one amino acid on the carboxyl group of another amino acid. Catalysis requires a basic residue to accept the hydrogen atom that is released from the amino group, as shown in **FIGURE 24.49**. If rRNA is the catalyst it must provide this residue, but we do not know how this happens. The purine and pyrimidine bases are not basic at physiological pH. A highly conserved base (at position 2451 in *E. coli*) had been implicated, but appears now neither to have the right properties nor to be crucial for peptidyl transferase activity.

The catalytic activity of isolated rRNA is quite low, and proteins that are bound to the 23S rRNA outside of the peptidyl transfer region are almost certainly required to enable the rRNA to form the proper structure *in vivo*. The idea that rRNA is the catalytic component is consistent with the results discussed in Chapter 21, *RNA Splicing and Processing*, which identify catalytic properties in RNA that are involved with several RNA processing reactions. It fits with the notion that the ribosome evolved from a prototype originally composed of RNA.

24.20 Ribosomal Structures Change When the Subunits Come Together

Key concepts
- The head of the 30S subunit swivels around the neck when complete ribosomes are formed.
- The peptidyl transferase active site of the 50S subunit is more active in complete ribosomes than in individual 50S subunits.
- The interface between the 30S and 50S subunits is very rich in solvent contacts.

Much indirect evidence suggests that the structures of the individual subunits change significantly when they join together to form a complete ribosome. Differences in the susceptibilities of the rRNAs to outside agents are one of the strongest indicators (see *Section 24.18, 16S rRNA Plays an Active Role in Translation*). More directly, comparisons of the high resolution crystal structures of the individual subunits with the lower resolution structure of the intact ribosome suggests the existence of significant differences. These ideas have been confirmed by a crystal structure of the *E. coli* ribosome at 3.5 Å, which furthermore identifies two different conformations of the ribosome, possibly representing different stages in translation.

The crystal contains two ribosomes per unit, each with a different conformation. The differences are due to changes in the positioning of domains within each subunit, the most important being that in one conformation the head of the small subunit has swiveled 6° around the neck region toward the E site. Also, a 6° rotation in the opposite direction is seen in the (low resolution) structures of *Thermus thermophilus* ribosomes that are bound to mRNA and have tRNAs in both A and P sites, suggesting that the head may swivel overall by 12° depending on the stage of translation. The rotation of the head follows the path of tRNAs through the ribosome, raising the possibility that its swiveling controls movement of mRNA and tRNA.

The changes in conformation that occur when subunits join together are much more marked in the 30S subunit than in the 50S subunit. The changes are probably concerned with controlling the position and movement of mRNA. The most significant change in the 50S subunit concerns the peptidyl transferase center. 50S subunits are ~1000× less effective in catalyzing peptide bond synthesis than complete ribosomes; the reason may be a change in structure that positions the substrate more effectively in the active site in the complete ribosome.

One of the main features emerging from the structure of the complete ribosome is the very high density of solvent contacts at their interface; this may help the making and breaking of contacts that is essential for subunit association and dissociation, and may also be involved in structural changes that occur during translocation.

24.21 Summary

A codon in mRNA is recognized by an amino-acyl-tRNA, which has an anticodon complementary to the codon and carries the amino acid corresponding to the codon. A special initiator tRNA (fMet-tRNA$_f$ in prokaryotes or Met-tRNA$_i$ in eukaryotes) recognizes the AUG codon, which is used to start all coding sequences. In prokaryotes, GUG is also used. Only the termination (nonsense) codons, UAA, UAG, and UGA, are not recognized by aminoacyl-tRNAs.

Ribosomes are released from translation to enter a pool of free ribosomes that are in equilibrium with separate small and large subunits. Small subunits bind to mRNA and then are joined by large subunits to generate an intact ribosome that undertakes translation. Recognition of a prokaryotic initiation site involves binding of a sequence at the 3' end of rRNA to the Shine–Dalgarno motif, which precedes the AUG (or GUG) codon in the mRNA. Recognition of a eukaryotic mRNA involves binding to the 5' cap; the small subunit then migrates to the initiation site by scanning for AUG codons. When it recognizes an appropriate AUG codon (usually, but not always, the first it encounters), it is joined by a large subunit.

A ribosome can carry at least two aminoacyl-tRNAs simultaneously: Its P site is occupied by a polypeptidyl-tRNA, which carries the polypeptide chain synthesized so far, whereas the A site is used for entry by an aminoacyl-tRNA carrying the next amino acid to be added to the chain. Bacterial ribosomes also have an E site, through which deacylated tRNA passes before it is released after being used in translation. The polypeptide chain in the P site is transferred to the aminoacyl-tRNA in the A site, creating a deacylated tRNA in the P site and a peptidyl-tRNA in the A site.

Following peptide bond synthesis, the ribosome translocates one codon along the mRNA, moving deacylated tRNA into the E site and peptidyl tRNA from the A site into the P site.

Translocation is catalyzed by the elongation factor EF-G and, like several other stages of ribosome function, requires hydrolysis of GTP. During translocation, the ribosome passes through a hybrid stage in which the 50S subunit moves relative to the 30S subunit.

Translation is an expensive process. ATP is used to provide energy at several stages, including the charging of tRNA with its amino acid and the unwinding of mRNA. It has been estimated that up to 90% of all the ATP molecules synthesized in a rapidly growing bacterium are consumed in assembling amino acids into protein!

Additional factors are required at each stage of translation. They are defined by their cyclic association with, and dissociation from, the ribosome. Initiation factors are involved in prokaryotic initiation. IF-3 is needed for 30S subunits to bind to mRNA, and also is responsible for maintaining the 30S subunit in a free form. IF-2 is needed for fMet-tRNA$_f$ to bind to the 30S subunit and is responsible for excluding other aminoacyl-tRNAs from the initiation reaction. GTP is hydrolyzed after the initiator tRNA has been bound to the initiation complex. The initiation factors must be released in order to allow a large subunit to join the initiation complex.

Eukaryotic initiation involves a greater number of factors. Some of them are involved in the initial binding of the 40S subunit to the capped 5' end of the mRNA, at which point the initiator tRNA is bound by another group of factors. After this initial binding, the small subunit scans the mRNA until it recognizes the correct AUG codon. At this point, initiation factors are released and the 60S subunit joins the complex.

Prokaryotic elongation factors are involved in elongation. EF-Tu binds aminoacyl-tRNA to the 70S ribosome. GTP is hydrolyzed when EF-Tu is released, and EF-Ts is required to regenerate the active form of EF-Tu. EF-G is required for translocation. Binding of the EF-Tu and EF-G factors to ribosomes is mutually exclusive, which ensures that each step must be completed before the next can be started.

Termination occurs at any one of the three special codons, UAA, UAG, and UGA. Class 1 release factors that specifically recognize the termination codons activate the ribosome to hydrolyze the peptidyl-tRNA. A class 2 RF factor is required to release the class 1 RF factor from the ribosome. The GTP-binding factors IF-2, EF-Tu, EF-G, and RF3 all have similar structures, with the latter two mimicking the

RNA-protein structure of the first two when they are bound to tRNA. They all bind to the same ribosomal site, the G-factor binding site.

Ribosomes are ribonucleoprotein particles in which a majority of the mass is provided by rRNA. The shapes of all ribosomes are generally similar, but only those of bacteria (70S) have been characterized in detail. The small (30S) subunit has a squashed shape, with a "body" containing about two thirds of the mass divided from the "head" by a cleft. The large (50S) subunit is more spherical, with a prominent "stalk" on the right and a "central protuberance." Approximate locations of all proteins in the small subunit are known.

Each subunit contains a single major rRNA, 16S and 23S in prokaryotes, and 18S and 28S in eukaryotic cytosol. There are also minor rRNAs, most notably 5S rRNA in the large subunit. Both major rRNAs have extensive base pairing, mostly in the form of short, imperfectly paired duplex stems with single-stranded loops. Conserved features in the rRNA can be identified by comparing sequences and the secondary structures that can be drawn for rRNA of a variety of organisms. The 16S rRNA has four distinct domains; the 23S rRNA has six distinct domains. Eukaryotic rRNAs have additional domains.

The crystal structure shows that the 30S subunit has an asymmetrical distribution of RNA and protein. RNA is concentrated at the interface with the 50S subunit. The 50S subunit has a surface of protein, with long rods of double-stranded RNA crisscrossing the structure. 30S-to-50S joining involves contacts between 16S rRNA and 23S rRNA. The interface between the subunits is very rich in contacts for solvent. Structural changes occur in both subunits when they join to form a complete ribosome.

Each subunit has several active centers, which are concentrated in the translational domain of the ribosome where proteins are synthesized. Polypeptides leave the ribosome through the exit domain, which can associate with a membrane. The major active sites are the P and A sites, the E site, the EF-Tu and EF-G binding sites, peptidyl transferase, and the mRNA-binding site. Ribosome conformation may change at stages during translation; differences in the accessibility of particular regions of the major rRNAs have been detected.

The tRNAs in the A and P sites are parallel to one another. The anticodon loops are bound to mRNA in a groove on the 30S subunit. The rest of each tRNA is bound to the 50S subunit.

A conformational shift of tRNA within the A site is required to bring its aminoacyl end into juxtaposition with the end of the peptidyl-tRNA in the P site. The peptidyl transferase site that links the P- and A-binding sites is made of 23S rRNA, which has the peptidyl transferase catalytic activity, although proteins are probably needed to acquire the correct structure.

An active role for the rRNAs in translation is indicated by mutations that affect ribosomal function, interactions with mRNA or tRNA that can be detected by chemical crosslinking, and the requirement to maintain individual base pairing interactions with the tRNA or mRNA. The 3' terminal region of the rRNA base pairs with mRNA at initiation. Internal regions make individual contacts with the tRNAs in both the P and A sites. Ribosomal RNA is the target for some antibiotics or other agents that inhibit translation.

References

24.4 Initiation in Bacteria Needs 30S Subunits and Accessory Factors

Reviews

Maitra, U. (1982). Initiation factors in protein biosynthesis. *Annu. Rev. Biochem.* 51, 869–900.

Noller, H. F. (2007). Structure of the bacterial ribosome and some implications for translational regulation. *In* "Translational Control in Biology and Medicine." (Mathews, M. B., Sonenberg, N., and Hershey, J. W. B., Eds.), pp. 87–128. Cold Spring Harbor Laboratory Press, New York.

Research

Carter, A. P., Clemons, W. M., Brodersen, D. E., Morgan-Warren, R. J., Hartsch, T., Wimberly, B. T., and Ramakrishnan, V. (2001). Crystal structure of an initiation factor bound to the 30S ribosomal subunit. *Science* 291, 498–501.

Dallas, A. and Noller, H. F. (2001). Interaction of translation initiation factor 3 with the 30S ribosomal subunit. *Mol. Cell* 8, 855–864.

Moazed, D., Samaha, R. R., Gualerzi, C., and Noller, H. F. (1995). Specific protection of 16S rRNA by translational initiation factors. *J. Mol. Biol.* 248, 207–210.

24.6 A Special Initiator tRNA Starts the Polypeptide Chain

Research

Lee, C. P., Seong, B. L., and RajBhandary, U. L. (1991). Structural and sequence elements important for recognition of *E. coli* formylmethionine tRNA by methionyl-tRNA transformylase are clustered in the acceptor stem. *J. Biol. Chem.* 266, 18012–18017.

Marcker, K. and Sanger, F. (1964). N-Formyl-methionyl-S-RNA. *J. Mol. Biol.* 8, 835–840.

Sundari, R. M., Stringer, E. A., Schulman, L. H., and Maitra, U. (1976). Interaction of bacterial initiation factor 2 with initiator tRNA. *J. Biol. Chem.* 251, 3338–3345.

24.8 Small Subunits Scan for Initiation Sites on Eukaryotic mRNA

Reviews

Hellen, C. U. and Sarnow, P. (2001). Internal ribosome entry sites in eukaryotic mRNA molecules. *Genes Dev.* 15, 1593–1612.

Kozak, M. (1978). How do eukaryotic ribosomes select initiation regions in mRNA? *Cell* 15, 1109–1123.

Kozak, M. (1983). Comparison of initiation of protein synthesis in prokaryotes, eukaryotes, and organelles. *Microbiol. Rev.* 47, 1–45.

Research

Kaminski, A., Howell, M. T., and Jackson, R. J. (1990). Initiation of encephalomyocarditis virus RNA translation: the authentic initiation site is not selected by a scanning mechanism. *EMBO J.* 9, 3753–3759.

Pelletier, J. and Sonenberg, N. (1988). Internal initiation of translation of eukaryotic mRNA directed by a sequence derived from poliovirus RNA. *Nature* 334, 320–325.

Pestova, T. V., Hellen, C. U., and Shatsky, I. N. (1996). Canonical eukaryotic initiation factors determine initiation of translation by internal ribosomal entry. *Mol. Cell Biol.* 16, 6859–6869.

Pestova, T. V., Shatsky, I. N., Fletcher, S. P., Jackson, R. J., and Hellen, C. U. (1998). A prokaryotic-like mode of cytoplasmic eukaryotic ribosome binding to the initiation codon during internal translation initiation of hepatitis C and classical swine fever virus RNAs. *Genes Dev.* 12, 67–83.

24.9 Eukaryotes Use a Complex of Many Initiation Factors

Reviews

Dever, T. E. (2002). Gene-specific regulation by general translation factors. *Cell* 108, 545–556.

Gingras, A. C., Raught, B., and Sonenberg, N. (1999). eIF4 initiation factors: effectors of mRNA recruitment to ribosomes and regulators of translation. *Annu. Rev. Biochem.* 68, 913–963.

Gebauer, F. and Hentze, M. W. (2004). Molecular mechanisms of translational control. *Nat. Rev. Cell. Mol. Biol.* 5, 827–835.

Hershey, J. W. B. (1991). Translational control in mammalian cells. *Annu. Rev. Biochem.* 60, 717–755.

Lackner, D. H. and Bähler, J. (2008). Translational control of gene expression from transcripts to transcriptomes. *Int. Rev. Cell. Mol. Biol.* 271, 199–251.

Merrick, W. C. (1992). Mechanism and regulation of eukaryotic protein synthesis. *Microbiol. Rev.* 56, 291–315.

Pestova, T. V., Kolupaeva, V. G., Lomakin, I. B., Pilipenko, E. V., Shatsky, I. N., Agol, V. I., and Hellen, C. U. (2001). Molecular mechanisms of translation initiation in eukaryotes. *Proc. Natl. Acad. Sci. USA* 98, 7029–7036.

Pestova, T. V., Lorsch, J. R., and Hellen, C. U. T. (2007). The mechanism of translation initiation in eukaryotes. *In* "Translational Control in Biology and Medicine." (M. B. Mathews, N. Sonenberg, and J. W. B. Hershey, Eds.), pp. 87–128. Cold Spring Harbor Laboratory Press, New York.

Sachs, A., Sarnow, P., and Hentze, M. W. (1997). Starting at the beginning, middle, and end: translation initiation in eukaryotes. *Cell* 89, 831–838.

Research

Asano, K., Clayton, J., Shalev, A., and Hinnebusch, A. G. (2000). A multifactor complex of eukaryotic initiation factors, eIF1, eIF2, eIF3, eIF5, and initiator tRNA(Met) is an important translation initiation intermediate *in vitro*. *Genes Dev.* 14, 2534–2546.

Huang, H. K., Yoon, H., Hannig, E. M., and Donahue, T. F. (1997). GTP hydrolysis controls stringent selection of the AUG start codon during translation initiation in *S. cerevisiae*. *Genes Dev.* 11, 2396–2413.

Kahvejian, A., Svitkin, Y. V., Sukarieh, R., M'Boutchou, M.-N., Sonenberg, N. (2005). Mammalian poly(A)-binding is a eukaryotic translation initiation factor, which acts via multiple mechanisms. *Genes Dev.* 19, 104–113.

Pestova, T. V. and Kolupaeva, V. G. (2002). The roles of individual eukaryotic translation initiation factors in ribosomal scanning and initiation codon selection. *Genes Dev.* 16, 2906–2922.

Pestova, T. V., Lomakin, I. B., Lee, J. H., Choi, S. K., Dever, T. E., and Hellen, C. U. (2000). The joining of ribosomal subunits in eukaryotes requires eIF5B. *Nature* 403, 332–335.

Tarun, S. Z. and Sachs, A. B. (1996). Association of the yeast poly(A) tail binding protein with translation initiation factor eIF-4G. *EMBO J.* 15, 7168–7177.

24.12 Translocation Moves the Ribosome

Reviews

Ramakrishnan, V. (2002). Ribosome structure and the mechanism of translation. *Cell* 108, 557–572.

Wilson, K. S. and Noller, H. F. (1998). Molecular movement inside the translational engine. *Cell* 92, 337–349.

Research

Moazed, D. and Noller, H. F. (1986). Transfer RNA shields specific nucleotides in 16S ribosomal RNA from attack by chemical probes. *Cell* 47, 985–994.

Moazed, D. and Noller, H. F. (1989). Intermediate states in the movement of tRNA in the ribosome. *Nature* 342, 142–148.

24.13 Elongation Factors Bind Alternately to the Ribosome

Research

Nissen, P., Kjeldgaard, M., Thirup, S., Polekhina, G., Reshetnikova, L., Clark, B. F., and Nyborg, J. (1995). Crystal structure of the ternary complex of Phe-tRNAPhe, EF-Tu, and a GTP analog. *Science* 270, 1464–1472.

Stark, H., Rodnina, M. V., Wieden, H. J., van Heel, M., and Wintermeyer, W. (2000). Large-scale movement of elongation factor G and extensive conformational change of the ribosome during translocation. *Cell* 100, 301–309.

24.15 Termination Codons Are Recognized by Protein Factors

Reviews

Eggertsson, G. and Soll, D. (1988). Transfer RNA-mediated suppression of termination codons in *E. coli. Microbiol. Rev.* 52, 354–374.

Frolova, L. et al. (1994). A highly conserved eukaryotic protein family possessing properties of polypeptide chain release factor. *Nature* 372, 701–703.

Nissen, P., Kjeldgaard, M., and Nyborg, J. (2000). Macromolecular mimicry. *EMBO J.* 19, 489–495.

Research

Freistroffer, D. V., Kwiatkowski, M., Buckingham, R. H., and Ehrenberg, M. (2000). The accuracy of codon recognition by polypeptide release factors. *Proc. Natl. Acad. Sci. USA* 97, 2046–2051.

Ito, K., Ebihara, K., Uno, M., and Nakamura, Y. (1996). Conserved motifs in prokaryotic and eukaryotic polypeptide release factors: tRNA-protein mimicry hypothesis. *Proc. Natl. Acad. Sci. USA* 93, 5443–5448.

Klaholz, B. P., Myasnikov, A. G., and van Heel, M. (2004). Visualization of release factor 3 on the ribosome during termination of protein synthesis. *Nature* 427, 862–865.

Mikuni, O., Ito, K., Moffat, J., Matsumura, K., McCaughan, K., Nobukuni, T., Tate, W., and Nakamura, Y. (1994). Identification of the *prfC* gene, which encodes peptide-chain-release factor 3 of *E. coli. Proc. Natl. Acad. Sci. USA* 91, 5798–5802.

Milman, G., Goldstein, J., Scolnick, E., and Caskey, T. (1969). Peptide chain termination.
3. Stimulation of *in vitro* termination. *Proc. Natl. Acad. Sci. USA* 63, 183–190.

Scolnick, E. et al. (1968). Release factors differing in specificity for terminator codons. *Proc. Natl. Acad. Sci. USA* 61, 768–774.

Selmer, M., Al-Karadaghi, S., Hirokawa, G., Kaji, A., and Liljas, A. (1999). Crystal structure of *Thermotoga maritima* ribosome recycling factor: a tRNA mimic. *Science* 286, 2349–2352.

Song, H., Mugnier, P., Das, A. K., Webb, H. M., Evans, D. R., Tuite, M. F., Hemmings, B. A., and Barford, D. (2000). The crystal structure of human eukaryotic release factor eRF1—mechanism of stop codon recognition and peptidyl-tRNA hydrolysis. *Cell* 100, 311–321.

24.16 Ribosomal RNA Pervades Both Ribosomal Subunits

Reviews

Hill, W. E. et al. (1990). *The Ribosome.* Washington, DC: American Society for Microbiology.

Noller, H. F. (1984). Structure of ribosomal RNA. *Annu. Rev. Biochem.* 53, 119–162.

Noller, H. F. (2005). RNA structure: reading the ribosome. *Science* 309, 1508–1514.

Noller, H. F. and Nomura, M. (1987). *E. coli and S. typhimurium.* Washington, DC: American Society for Microbiology.

Wittman, H. G. (1983). Architecture of prokaryotic ribosomes. *Annu. Rev. Biochem.* 52, 35–65.

Research

Ban, N., Nissen, P., Hansen, J., Capel, M., Moore, P. B., and Steitz, T. A. (1999). Placement of protein and RNA structures into a 5 Å-resolution map of the 50S ribosomal subunit. *Nature* 400, 841–847.

Ban, N., Nissen, P., Hansen, J., Moore, P. B., and Steitz, T. A. (2000). The complete atomic structure of the large ribosomal subunit at 2.4 Å resolution. *Science* 289, 905–920.

Clemons, W. M. et al. (1999). Structure of a bacterial 30S ribosomal subunit at 5.5 Å resolution. *Nature* 400, 833–840.

Wimberly, B. T., Brodersen, D. E., Clemons, W. M. Jr., Morgan-Warren, R. J., Carter, A. P., Vonrhein, C., Hartsch, T., and Ramakrishnan, V. (2000). Structure of the 30S ribosomal subunit. *Nature* 407, 327–339.

Yusupov, M. M., Yusupova, G. Z., Baucom, A., Lieberman, A., Earnest, T. N., Cate, J. H. D., and Noller, H. F. (2001). Crystal structure of the ribosome at 5.5 Å resolution. *Science* 292, 883–896.

24.17 Ribosomes Have Several Active Centers

Reviews

Lafontaine, D. L. and Tollervey, D. (2001). The function and synthesis of ribosomes. *Nat. Rev. Mol. Cell Biol.* 2, 514–520.

Moore, P. B. and Steitz, T. A. (2003). The structural basis of large ribosomal subunit function. *Annu. Rev. Biochem.* 72, 813–850.

Ramakrishnan, V. (2002). Ribosome structure and the mechanism of translation. *Cell* 108, 557–572.

Research

Cate, J. H., Yusupov, M. M., Yusupova, G. Z., Earnest, T. N., and Noller, H. F. (1999). X-ray crystal structures of 70S ribosome functional complexes. *Science* 285, 2095–2104.

Fredrick, K. and Noller, H. F. (2003). Catalysis of ribosomal translocation by sparsomycin. *Science* 300, 1159–1162.

Sengupta, J., Agrawal, R. K., and Frank, J. (2001). Visualization of protein S1 within the 30S ribosomal subunit and its interaction with messenger RNA. *Proc. Natl. Acad. Sci. USA* 98, 11991–11996.

Simonson, A. B. and Simonson, J. A. (2002). The transorientation hypothesis for codon recognition during protein synthesis. *Nature* 416, 281–285.

Valle, M., Sengupta, J., Swami, N. K., Grassucci, R. A., Burkhardt, N., Nierhaus, K. H., Agrawal, R. K., and Frank, J. (2002). Cryo-EM reveals an active role for aminoacyl-tRNA in the accommodation process. *EMBO J.* 21, 3557–3567.

Yusupov, M. M., Yusupova, G. Z., Baucom, A., Lieberman, A., Earnest, T. N., Cate, J. H. D., and Noller, H. F. (2001). Crystal structure of the ribosome at 5.5 Å resolution. *Science* 292, 883–896.

24.18 16S rRNA Plays an Active Role in Translation

Reviews

Noller, H. F. (1991). Ribosomal RNA and translation. *Annu. Rev. Biochem.* 60, 191–227.

Yonath, A. (2005). Antibiotics targeting ribosomes: resistance, selectivity, synergism and cellular regulation. *Annu. Rev. Biochem.* 74, 649–679.

Research

Lodmell, J. S. and Dahlberg, A. E. (1997). A conformational switch in *E. coli* 16S rRNA during decoding of mRNA. *Science* 277, 1262–1267.

Moazed, D. and Noller, H. F. (1986). Transfer RNA shields specific nucleotides in 16S ribosomal RNA from attack by chemical probes. *Cell* 47, 985–994.

Yoshizawa, S., Fourmy, D., and Puglisi, J. D. (1999). Recognition of the codon-anticodon helix by rRNA. *Science* 285, 1722–1725.

24.19 23S rRNA Has Peptidyl Transferase Activity

Research

Ban, N., Nissen, P., Hansen, J., Moore, P. B., and Steitz, T. A. (2000). The complete atomic structure of the large ribosomal subunit at 2.4 Å resolution. *Science* 289, 905–920.

Bayfield, M. A., Dahlberg, A. E., Schulmeister, U., Dorner, S., and Barta, A. (2001). A conformational change in the ribosomal peptidyl transferase center upon active/inactive transition. *Proc. Natl. Acad. Sci. USA* 98, 10096–10101.

Noller, H. F., Hoffarth, V., and Zimniak, L. (1992). Unusual resistance of peptidyl transferase to protein extraction procedures. *Science* 256, 1416–1419.

Samaha, R. R., Green, R., and Noller, H. F. (1995). A base pair between tRNA and 23S rRNA in the peptidyl transferase center of the ribosome. *Nature* 377, 309–314.

Thompson, J., Thompson, D. F., O'Connor, M., Lieberman, K. R., Bayfield, M. A., Gregory, S. T., Green, R., Noller, H. F., and Dahlberg, A. E. (2001). Analysis of mutations at residues A2451 and G2447 of 23S rRNA in the peptidyltransferase active site of the 50S ribosomal subunit. *Proc. Natl. Acad. Sci. USA* 98, 9002–9007.

24.20 Ribosomal Structures Change When the Subunits Come Together

Reference

Schuwirth, B. S., Borovinskaya, M. A., Hau, C. W., Zhang, W., Vila-Sanjurjo, A., Holton, J. M., and Cate, J. H. (2005). Structures of the bacterial ribosome at 3.5 Å resolution. *Science* 310, 827–834.

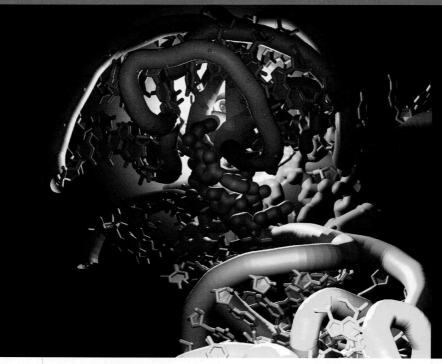

Using the Genetic Code

Edited by John Perona

CHAPTER OUTLINE

25.1 Introduction

The sequence of a coding strand of DNA, read in the direction from 5' to 3', consists of nucleotide triplets (codons) corresponding to the amino acid sequence of a protein read from N-terminus to C-terminus. Sequencing of DNA and proteins makes it possible to compare corresponding nucleotide and amino acid sequences directly. There are sixty-four codons (each of four possible nucleotides can occupy each of the three positions of the codon, making $4^3 = 64$ possible trinucleotide sequences). Each of these codons has a specific meaning in translation: sixty-one codons represent amino acids; three codons cause the termination of translation.

The meaning of a codon that represents an amino acid is determined by the tRNA that corresponds to it; the meaning of the termination codons is determined directly by protein factors.

The breaking of the genetic code originally showed that genetic information is stored in the form of nucleotide triplets, but did not reveal which amino acid is specified by each triplet codon. Before the advent of sequencing, codon assignments were deduced on the basis of two types of *in vitro* studies. A system involving the translation of synthetic polynucleotides was introduced in 1961, when Nirenberg showed that polyuridylic acid [poly(U)] directs the assembly of phenylalanine into polyphenylalanine. This result means that UUU must be a codon for phenylalanine. A second system was later introduced in which a trinucleotide was used to mimic a codon, thus causing the corresponding aminoacyl-tRNA to bind to a ribosome. By identifying the amino acid component of the aminoacyl-tRNA, the meaning of the codon can be found. The two techniques together assigned meaning to all of the codons that represent amino acids.

Sixty-one of the sixty-four codons represent amino acids. The other three cause termination of translation. The assignment of amino acids to codons is not random, but shows relationships in which the third base has less effect on codon meaning. In addition, chemically similar amino acids are often represented by related codons.

25.2 Related Codons Represent Chemically Similar Amino Acids

Key concepts

- Sixty-one of the sixty-four possible triplets code for twenty amino acids.
- Three codons do not represent amino acids and cause termination.
- The genetic code was frozen at an early stage of evolution and is universal.
- Most amino acids are represented by more than one codon.
- The multiple codons for an amino acid are usually related.
- Chemically similar amino acids often have related codons, minimizing the effects of mutation.

The code is summarized in **FIGURE 25.1**. There are more codons than there are amino acids,

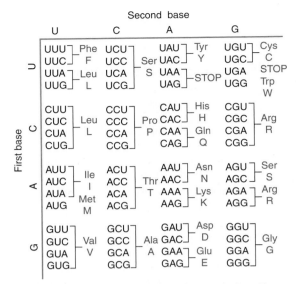

FIGURE 25.1 All the triplet codons have meaning: Sixty-one represent amino acids, and three cause termination (STOP).

and as a result almost all amino acids are represented by more than one codon. The only exceptions are methionine and tryptophan. Codons that have the same meaning are said to be **synonymous**. The genetic code is actually read on the mRNA, and thus it is usually described in terms of the four bases present in RNA: U, C, A, and G.

Codons representing the same or chemically similar amino acids tend to be similar in sequence. Often the base in the third position of a codon is not significant, because the four codons differing only in the third base represent the same amino acid. Sometimes a distinction is made only between a purine versus a pyrimidine in this position. The reduced specificity at the last position is known as **third-base degeneracy**.

To be interpreted, a codon in mRNA must first base-pair with the anticodon of the corresponding aminoacyl-tRNA. This pairing occurs within the ribosome, where the interaction between complementary trinucleotides is stabilized by highly conserved 16S rRNA nucleotides in the A site. Stringent monitoring of the overall base-pair shape by rRNA permits only conventional A-U and G-C pairing to occur at the first two positions of the codon, but additional pairs are permitted at the third codon base, where rRNA contacts are less complementary. As a result, a single aminoacyl-tRNA may recognize more than one codon, by means of the additional, noncanonical pairs permitted at the third position. Furthermore, pairing

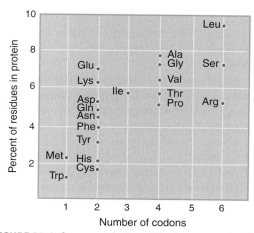

FIGURE 25.2 Some correlation of the frequency of amino acid use in proteins with the number of codons specifying the amino acid is observed. An exception is found for amino acids specified by two codons, which occur with a wide variety of frequencies.

interactions may also be influenced by the post-transcriptional modification of tRNA, especially within or directly adjacent to the anticodon.

The tendency for identical or chemically similar amino acids to be represented by related codons minimizes the effects of mutations. It increases the probability that a single random base change will result in no amino acid substitution or in one involving amino acids of similar character. For example, a mutation of CUC to CUG has no effect, because both codons represent leucine. Mutation of CUU to AUU results in replacement of leucine with isoleucine; both of these amino acids are hydrophobic and are likely to play similar roles in the encoded protein.

FIGURE 25.2 plots the number of codons representing each amino acid against the frequency with which the amino acid is used in proteins (in *E. coli*). In general, amino acids that are more common are represented by more codons. This suggests that there has been some optimization of the genetic code with regard to the utilization of amino acids.

The three codons (UAA, UAG, and UGA) that do not represent amino acids are used specifically to terminate translation. One of these **stop codons** marks the end of every open reading frame.

Comparisons of DNA sequences with the corresponding polypeptide sequences reveal that the identical set of codon assignments is used in bacteria and in eukaryotic cytoplasm. As a result, mRNA from one species usually can be translated correctly *in vitro* or *in vivo* by the translation apparatus of another species. Thus the codons used in the mRNA of one species have the same meaning for the ribosomes and tRNAs of other species.

The universality (with minor exceptions) of the code argues that it must have been established very early in evolution. Perhaps the code started in a primitive form in which a small number of codons were used to represent comparatively few amino acids, possibly even with one codon corresponding to any member of a group of amino acids. More precise codon meanings and additional amino acids could have been introduced later. One possibility is that at first only two of the three bases in each codon were used; discrimination at the third position could have evolved later.

Evolution of the code could have become "frozen" at a point at which the system had become so complex that any changes in codon meaning would disrupt existing proteins by substituting unacceptable amino acids. Its universality implies that this must have happened at such an early stage that all living organisms are descended from a single pool of primitive cells in which this occurred.

Exceptions to the universal genetic code are rare. Changes in meaning in the principal genome of a species usually concern the termination codons. For example, in a *Mycoplasma*, UGA codes for tryptophan; in certain species of the ciliates *Tetrahymena* and *Paramecium*, UAA and UAG code for glutamine. Systematic alterations of the code have occurred only in mitochondrial DNA (see *Section 25.7, There Are Sporadic Alterations of the Universal Code*).

25.3 Codon–Anticodon Recognition Involves Wobbling

Key concepts

- Multiple codons that represent the same amino acid most often differ at the third base position.
- The wobble in pairing between the first base of the anticodon and the third base of the codon results from looser monitoring of the pairing by rRNA nucleotides in the ribosomal A site.

The function of tRNA in translation is fulfilled when it recognizes the codon in the ribosomal A site. The interaction between anticodon and codon takes place by base pairing, but under rules that extend pairing beyond the usual G-C and A-U partnerships.

The genetic code itself yields some important clues about the process of codon recognition.

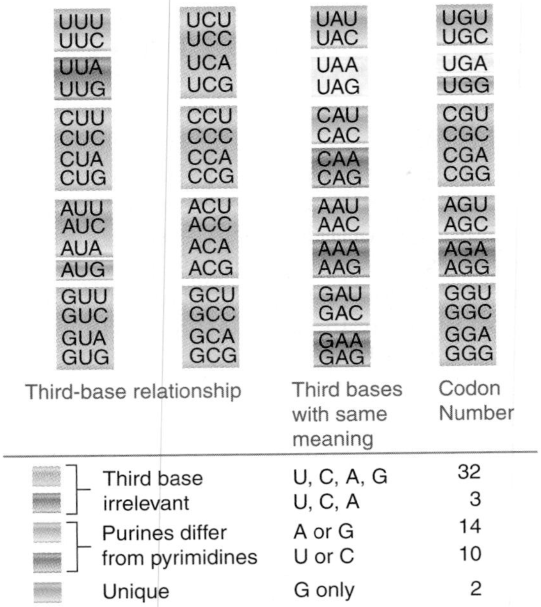

Third-base relationship		Third bases with same meaning	Codon Number
	Third base irrelevant	U, C, A, G	32
		U, C, A	3
	Purines differ from pyrimidines	A or G	14
		U or C	10
	Unique	G only	2

FIGURE 25.3 Third bases have the least influence on codon meanings. Boxes indicate groups of codons within which third-base degeneracy ensures that the meaning is the same.

The pattern of third-base degeneracy is drawn in **FIGURE 25.3**, which shows that in almost all cases either the third base is irrelevant or a distinction is made only between purines and pyrimidines.

There are eight codon families in which all four codons sharing the same first two bases have the same meaning, so that the third base has no role at all in specifying the amino acid. There are seven codon pairs in which the meaning is the same regardless of which pyrimidine is present at the third position, and there are five codon pairs in which either purine may be present without changing the amino acid that is coded.

There are only three cases in which a unique meaning is conferred by the presence of a particular base at the third position: AUG (for methionine), UGG (for tryptophan), and UGA (termination). So C and U never have a unique meaning in the third position, and A never signifies a unique amino acid.

The anticodon is complementary to the codon; thus it is the first base in the anticodon sequence written conventionally in the direction from 5' to 3' that pairs with the third base in the codon sequence written by the same convention. So the combination

Codon	5' A C G 3'
Anticodon	3' U G C 5'

is usually written as codon ACG/anticodon CGU, where the anticodon sequence must be read backward for complementarity with the codon.

To avoid confusion, we shall retain the usual convention in which all sequences are written 5'–3', but indicate anticodon sequences with a backward arrow as a reminder of the relationship with the codon. Thus the codon/anticodon pair shown above will be written as ACG and CGU$^{\leftarrow}$, respectively.

Does each triplet codon demand its own tRNA with a complementary anticodon? Or can a single tRNA respond to both members of a codon pair and to all (or at least some) of the four members of a codon family?

Often one tRNA can recognize more than one codon. All codons that a particular tRNA recognizes must be identical at their first two positions. By contrast, the base in the first position of the tRNA anticodon is able to partner alternative bases in the corresponding third position of the codon. Base pairing at this position is not limited to the usual G-C and A-U partnerships.

The rules governing the recognition patterns are summarized in the **wobble hypothesis**, which states that the pairing between codon and anticodon at the first two codon positions always follows the usual rules, but that exceptional "wobbles" occur at the third position. Wobbling occurs because the structure of the ribosomal A site, in which the codon–anticodon pairing occurs, permits increased flexibility at the first base of the anticodon. The most common nonconventional pair that is found at this position is G-U (**FIGURE 25.4**). For example, the anticodon UUG in tRNA^{Gln} recognizes both the CAA and CAG glutamine codons, and the anticodon GUG in tRNA^{His} recognizes both the CAU and CAC histidine codons. Other nonconventional pairs that are tolerated at the third codon position involve modified bases (see *Section 25.6, Modified Bases Affect Anticodon–Codon Pairing*).

This capacity of the third codon position to tolerate G-U pairs creates a pattern of base pairing in which A can no longer have a unique meaning in the codon (because the U that recognizes it must also recognize G). Similarly, C also no longer has a unique meaning (because the G that recognizes it also must recognize U). **FIGURE 25.5** summarizes the pattern of recognition. It is therefore possible to recognize unique codons only when the third bases are G or U. Only UGG and AUG, however, provide examples of such unique recognition.

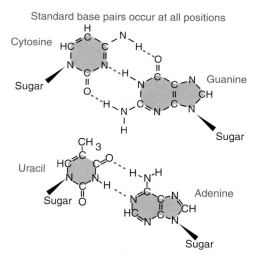

Standard base pairs occur at all positions

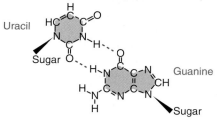

G-U wobble pairing occurs only at third codon position

FIGURE 25.4 Wobble in base pairing allows G-U pairs to form between the third base of the codon and the first base of the anticodon.

Base in first position of anticodon	Base(s) recognized in third position of codon
U	A or G
C	G only
A	U only
G	C or U

FIGURE 25.5 Codon–anticodon pairing involves wobbling at the third position.

25.4 tRNAs Are Processed from Longer Precursors

Key concepts

- A mature tRNA is generated by processing a precursor.
- The 5′ end is generated by cleavage by the endonuclease RNAase P.
- The 3′ end is generated by multiple endonucleolytic and exonucleolytic cleavages, followed by addition of the common terminal trinucleotide CCA.

tRNAs are commonly synthesized as precursor chains with additional material at one or both ends. **FIGURE 25.6** shows that the extra sequences are removed by combinations of endonucleolytic and exonucleolytic activities. The three

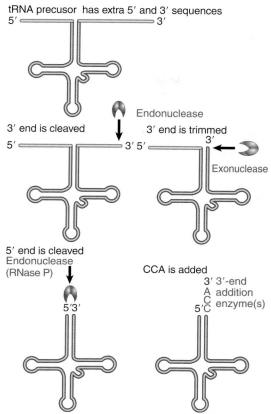

FIGURE 25.6 The tRNA 3′ end is generated by cutting (endonucleolytic) and trimming (exonucleolytic) reactions, followed by addition of CCA when this sequence is not coded; the 5′ end is generated by a precise endonucleolytic cleavage.

nucleotides at the 3′ terminus, which are always present as the triplet sequence CCA, are sometimes not coded in the genome. In such cases, they are added as part of the tRNA processing.

The 5′ end of tRNA is generated by a cleavage action catalyzed by the ribonucleoprotein enzyme ribonuclease P. This enzyme recognizes the global L-shaped tRNA structure, and specifically hydrolyzes the phosphodiester linkage that liberates the mature 5′-end of the molecule, leaving a 5′-phosphate group. In *E. coli*, RNase P consists of a 377 nucleotide RNA and 17.5 kD protein, and its active site is composed of RNA. *In vitro*, the RNA component alone is able to catalyze the tRNA processing reaction. (This is an example of a *ribozyme*; see Chapter 23, *Catalytic RNA*.) The function of the protein subunit is to stabilize a conformation of the RNA active site that is complementary to the tRNA precursor. This is discussed further in *Section 23.8, The Catalytic Activity of RNase P Is Due to RNA.*

The enzymes that process the 3' end are best characterized in *E. coli*, where an endonuclease triggers the reaction by cleaving the precursor downstream, and several exonucleases then trim the end by degradation in the 3'–5' direction. tRNA 3'-end processing also involves several enzymes in eukaryotes. The addition of the 3'-CCA is catalyzed by the enzyme tRNA nucleotidyltransferase, which functions as a nontemplate-directed RNA polymerase. That is, the enzyme specifically adds C, C, and A in sequence, without pairing the cytosine and adenine to complementary guanine and uracil bases on a template. Instead, the enzyme structure itself is sufficient to form sequential complementary binding sites for C, C, and A. As the nucleotides are added, the enzyme–tRNA complex changes conformation to become complementary to each successive nucleotide.

All three nucleotides are added by tRNA nucleotidyltransferase when they are not encoded in the tRNA gene sequence. Interestingly, the enzyme also plays an essential role in repairing damaged tRNA 3'-ends in organisms such as *E. coli* that *do* encode CCA. In these organisms, three different tRNA substrates are recognized: those lacking CCA, those possessing a 3'-C, and those possessing a 3'-CC.

tRNA nucleotidyltransferase enzymes are divided into two classes that retain significant amino acid similarity only in their active-site regions. Class I enzymes are found in Archaea, while bacterial and eukaryotic enzymes together make up a second class. In some very ancient bacteria, CCA addition is catalyzed by two closely related class II enzymes; one of these enzymes adds –CC while the other adds the 3'-terminal A.

25.5 tRNA Contains Modified Bases

Key concepts

- tRNAs contain over 90 modified bases.
- Modification usually involves direct alteration of the primary bases in tRNA, but there are some exceptions in which a base is removed and replaced by another base.
- Known functions of modified bases are to confer increased stability to tRNAs, and to modulate their recognition by proteins and other RNAs in the translational apparatus.

Transfer RNA is unique among nucleic acids in its content of modified bases. A modified base is any purine or pyrimidine ring except the usual A, G, C, and U from which all RNAs are synthesized. All other bases are produced by posttranscriptional **modification** of one of the four bases after it has been incorporated into the polyribonucleotide chain. The ribose sugar of some tRNA nucleotides is also methylated on the 2'-hydroxyl to produce the 2'-O-methyl modification.

While all classes of RNA display some degree of modification, the range of chemical alterations to the bases is much greater in tRNA. The modifications range from simple methylation to wholesale restructuring of the base. Modifications occur in all parts of the tRNA molecule. They vary considerably in their extent of conservation among tRNA species, and in the location within the molecule at which they are found. Modifications specific for particular tRNAs or small subgroups of tRNAs are generally less common than those present more broadly. There are also some species-specific patterns. In all there are over 70 different types of modified bases in tRNA. Each tRNA is modified, on average, at about 15% to 20% of its bases.

The modified nucleosides are synthesized by specific tRNA-modifying enzymes. The original nucleoside present at each position can be determined either by comparing the sequence of tRNA with that of its gene or (less efficiently) by isolating precursor molecules that lack some or all of the modifications. The sequences of precursors show that different modifications are introduced at different stages during the maturation of tRNA.

The many tRNA-modifying enzymes vary greatly in specificity. In some cases, a single enzyme acts to make a particular modification at a single position. In other cases, an enzyme can modify bases at several different target positions. Some enzymes undertake single reactions with individual tRNAs; others have a range of substrate molecules. Some modifications require the successive actions of more than one enzyme.

Details of the structural basis for tRNA modification by enzymes are just beginning to emerge. One striking example is the mechanism by which archaeosine, a modified G, is introduced into the D loop of certain archaeal tRNAs. To access the base to be modified, which is normally buried within the tRNA tertiary core, the tRNA guanine transglycosylase enzyme facilitates a dramatic induced-fit rearrangement of the tRNA to produce an alternative tertiary structure termed the lambda form. Induced-fit

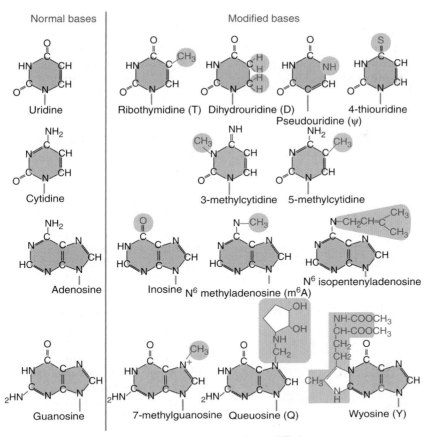

FIGURE 25.7 Each of the four bases in tRNA can be modified.

rearrangements of the tRNA structure have also been observed for other modifying enzymes, and constitute a common theme in recognition.

Known functions of modified bases are to confer increased stability to tRNAs, and to modulate their recognition by proteins and other RNAs in the translational apparatus. Roles for modified bases in recognition by aminoacyl-tRNA synthetases, for example, have been clearly defined in a number of cases (see *Sections 25.9 to 25.11*). In many cases, however, the biological role of the tRNA modification remains unknown.

FIGURE 25.7 shows some of the more common modified bases. Modifications of pyrimidines (C and U) are generally less complex than those of purines (A and G).

The most common modification made to uridine and cytosine is methylation, which may occur at several different positions on the ring. Methylation at position 5 of uracil creates ribothymidine (T). The thymidine base is identical to that found in DNA, but in tRNA is attached to ribose rather than deoxyribose. This thymidine is found in nearly all tRNA molecules at position 54 in the TψC loop. Pseudouridine is a striking uridine modification that is generated by cleavage of the glycosidic bond, followed by constrained rotation of the liberated ring and rejoining of the C5 carbon to the C1 carbon of the ribose. Thus, pseudouridine lacks an N-glycosidic linkage. Nearly all tRNAs possess pseudouridine at position 55 of the TψC loop. Position 56 is also very highly conserved as cytosine; together, the TψC sequence at positions 54–56 provides the basis for naming this portion of the tRNA molecule.

The dihydrouridine (D) modification, which is generated by saturation of the double bond joining C5 and C6 of uracil, is nearly universally found in the D loop of tRNAs. As for the TψC sequence, this D modification provides the basis for naming the D stem-loop of the tRNA. The removal of the double bond in D destroys the aromaticity and planarity of the uracil ring, generating an unusual structure that subtly modifies the shape of the globular core of the tRNA.

The nucleoside inosine (I) is found normally in the cell as an intermediate in the purine biosynthetic pathway. It is not, however, incorporated directly into RNA. Instead, its existence depends on modification of A to create I. The incorporation of I at the 5'-anticodon position contributes importantly to wobble base-pairing at the third codon position of mRNA (see *Section 25.6, Modified Bases Affect Anticodon–Codon Pairing*).

Modifications of A and G often generate dramatic new structures (Figure 25.7). For example, two complex series of nucleotides depend on modification of G. The Q bases, such as queuosine, have an additional pentenyl ring added via an NH linkage to the methyl group of 7-methylguanosine. The pentenyl ring may carry various further groups. The Y bases, such as wyosine, have an additional ring fused with the purine ring itself. This extra ring carries a long carbon chain; again, it is a chain to which further groups are added in different cases.

25.6 Modified Bases Affect Anticodon–Codon Pairing

Key concept

- Modifications in the anticodon affect the pattern of wobble pairing and therefore are important in determining tRNA specificity.

tRNA modifications in and adjacent to the anticodon influence its ability to pair with the mRNA codon. Most such modifications are present at positions 34 and 37 of the anticodon loop, and they generally function by constraining the range of available motion in the anticodon. In turn, this facilitates docking of the tRNA into the A site of the ribosome. These modifications influence codon pairing, and as a result they function directly to help determine how the cell assigns the meaning of the tRNA. Modified bases permit further pairing patterns in addition to those involving regular and wobble pairing of A, C, U, and G.

Inosine is particularly important when present at the first anticodon position (nucleotide 34 in the sequence), because it is able to pair with any one of three bases U, C, and A (FIGURE 25.8). The role of inosine is well illustrated in the decoding of isoleucine codons. Here AUA codes for isoleucine, whereas AUG codes for methionine. To read the A at the third codon position, a tRNA would require U at the first anticodon position—but this U in the wobble position would necessarily also pair with G. Thus, any tRNA with a 5' U in its anticodon would recognize both AUG and

AUA. This problem is resolved by synthesis of an isoleucine tRNA possessing A34, followed by modification of A34 to I34 by the enzyme tRNA adenosine deaminase. I34 then is able to recognize all three codons of the isoleucine set: AUU, AUC, and AUA.

In most cases, U at the first position of the anticodon is also converted to a modified form that has altered pairing properties. Derivatives of U possessing the 2-thio group in place of oxygen show improved selectivity in pairing to A as compared with G (FIGURE 25.9). Anticodons with uridine-5-oxyacetic acid and related modifications in the first position have the remarkable property of permitting the single tRNA to read three and sometimes all four of the synonymous codons NNA, NNC, NNU, and NNG.

These and other pairing relationships show that there are multiple ways to construct a set of tRNAs able to recognize all the sixty-one codons representing amino acids. No particular pattern predominates in any given organism, although the absence of a certain pathway for modification can prevent the use of some recognition patterns. Thus, a particular codon family is read by tRNAs with different anticodons in different organisms.

Often the tRNAs will have overlapping capacities to read certain codons, so that a particular codon is read by more than one tRNA. In such cases there may be differences in the efficiencies of the alternative recognition reactions (as a general rule, codons that are commonly used tend to be more efficiently read.)

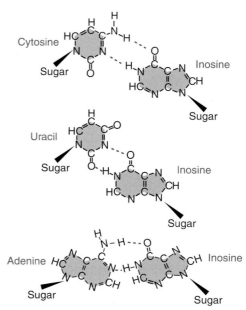

FIGURE 25.8 Inosine can pair with U, C, or A.

The predictions of wobble pairing accord very well with experimental evidence for almost all tRNAs. There are, however, exceptions in which the codons recognized by a tRNA differ from those predicted by the wobble rules. Such effects probably result from the influence of neighboring bases and/or the conformation of the anticodon loop in the overall tertiary structure of the tRNA. Further support for the influence of the surrounding structure is provided by the isolation of occasional mutants in which a change in a base in some other region of the molecule alters the ability of the anticodon to recognize codons.

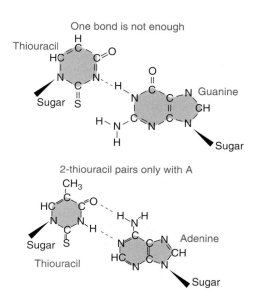

FIGURE 25.9 Modification to 2-thiouridine restricts pairing to A alone because only one H-bond can form with G.

25.7 There Are Sporadic Alterations of the Universal Code

Key concepts

- Changes in the universal genetic code have occurred in some species.
- These changes are more common in mitochondrial genomes, where a phylogenetic tree can be constructed for the changes.
- In nuclear genomes, the changes usually affect only termination codons.

The universality of the genetic code is striking, but some exceptions exist. They tend to affect the codons involved in initiation or termination. The changes found in principal (bacterial or nuclear) genomes are summarized in **FIGURE 25.10**.

Almost all of the changes in nuclear genomes that allow a codon to represent an amino acid affect termination codons:

- In the prokaryote *Mycoplasma capricolum*, UGA is not used for termination, but instead codes for tryptophan (Trp). In fact, it is the predominant Trp codon, and UGG is used only rarely. Two Trp-tRNA species exist, which have the anticodons UCA← (reads UGA and UGG) and CCA← (reads only UGG).
- Some ciliates (unicellular protozoa) read UAA and UAG as glutamine instead of termination signals. *Tetrahymena thermophila*, which is one of the ciliates, contains three tRNA^{Gln} species. One tRNA^{Gln} with UUG anticodon

UUU	Phe	F	UCU			UAU	Tyr	Y	UGU	Cys	C
UUC			UCC	Ser S		UAC			UGC	
UUA	Leu	L	UCA			UAA	STOP→Gln Q		UGA	STOP→Trp, Cys, Sel W C S
UUG			UCG			UAG			UGG	Trp	W
CUU			CCU			CAU	His	H	CGU	
CUC	Leu	L	CCC	Pro P		CAC			CGC	Arg	R
CUA			CCA			CAA	Gln	Q	CGA	
CUG	Leu→Ser S		CCG			CAG			CGG	Arg→NONE
AUU			ACU			AAU	Asn	N	AGU	Ser	S
AUC	Ile	I	ACC	Thr T		AAC			AGC	
AUA	Ile→NONE		ACA			AAA	Lys	K	AGA	Arg→NONE
AUG	Met	M	ACG			AAG			AGG	Arg	R
GUU			GCU			GAU	Asp	D	GGU	
GUC	Val	V	GCC	Ala A		GAC			GGC	Gly	G
GUA			GCA			GAA	Glu	E	GGA	
GUG			GCG			GAG			GGG	

FIGURE 25.10 Changes in the genetic code in bacterial or eukaryotic nuclear genomes usually assign amino acids to stop codons or change a codon so that it no longer specifies an amino acid. A change in meaning from one amino acid to another is unusual.

recognizes the usual codons CAA and CAG for glutamine, a second species with anticodon UUA recognizes both UAA and UAG (in accordance with the wobble hypothesis), and the third with anticodon CUA recognizes only UAG. Restriction of the specificity of the release factor eRF, so that it recognizes the UGA stop codon only, is also necessary to prevent premature termination at the newly reassigned glutamine codons.

- In another ciliate (*Euplotes octacarinatus*), the UGA stop codon is reassigned to cysteine. Only UAA is used as a termination codon, and UAG is not found. The change in meaning of UGA might be accomplished by modifying the anticodon of tRNA^Cys with I34, so that it is able to read UGA together with the usual codons UGU and UGC. UGA has dual meaning in *Euplotes crassus* (see Section 25.8).
- In a yeast (*Candida*), CUG is reassigned to serine instead of leucine. This is a rare example of reassignment from one sense codon to another.

In general, acquisition of a coding function by a termination codon requires two types of change: a tRNA must be mutated so as to recognize the codon, and the class I release factor must be mutated so that it does not terminate at this codon. The other common type of change is loss of the tRNA that responds to a codon, so that the codon no longer specifies any amino acid.

All of these changes are sporadic, meaning that they appear to have occurred independently in specific evolutionary lineages. They may be concentrated in termination codons, because at these positions there is no substitution of one amino acid for another. Once the genetic code was established, early in evolution, any general change in the meaning of a codon would cause a substitution in all the proteins that contain that amino acid. It seems likely that the change would be deleterious in at least some of these proteins, with the result that it would be strongly selected against. The divergent uses of the termination codons could represent their "capture" for normal coding purposes. If some termination codons were used only rarely, their recruitment to coding purposes, by way of changes in tRNAs that permit reassignment, would have been more likely.

Exceptions to the universal genetic code also occur in the mitochondria from several spe-

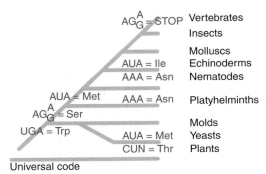

FIGURE 25.11 Changes in the genetic code in mitochondria can be traced in phylogeny. The minimum number of independent changes is generated by supposing that the AUA=Met and the AAA=Asn changes each occurred independently twice, and that the early AUA=Met change was reversed in echinoderms.

cies. **FIGURE 25.11** constructs a phylogeny for the changes. The ability to construct a phylogeny suggests that there was a universal code that was changed at various points in mitochondrial evolution. The earliest change was the employment of UGA to code for tryptophan, which is common to all nonplant mitochondria.

Some of the mitochondrial changes make the code simpler by replacing two codons that had different meanings with a pair that has a single meaning. Pairs treated this way include UGG and UGA (both Trp instead of one Trp and one termination) and AUG and AUA (both Met instead of one Met and the other Ile).

Why have changes to the code been able to evolve more readily in mitochondria, as compared with the nucleus? The mitochondrion synthesizes only a small number of proteins (~10), and as a result the problem of disruption by changes in meaning is much less severe. It is likely that the altered codons were not used extensively in locations where amino acid substitutions would have been deleterious.

According to the wobble hypothesis, a minimum of 31 tRNAs (excluding the initiator) are required to recognize all sixty-one codons (at least two tRNAs are required for each four-codon family and one tRNA is needed per codon pair or single codon). The streamlined mammalian mitochondrial genome, however, encodes only 22 tRNAs. tRNAs encoded in the nuclear genome are not imported into the mitochondrion in mammals; thus it can be inferred there must be some modification to the wobble rules for translation on the mitochondrial ribosome. Interestingly, in mitochondria an unmodified uridine at the first position of the anticodon is able to pair with all four bases at the third codon position. Such an unmodified uridine exists for the tRNAs representing all eight 4-codon families: Pro, Thr, Ala, Ser, Leu, Val, Gly, and Arg. This reduces the total number of tRNAs required in mitochondria by eight. The conversion of AGA and AGG to stop codons in mammalian mitochondria (see Figure 25.11) eliminates the need for one additional tRNA, bringing the total required number of tRNAs to just 22. The conversion of AUA to methionine further eliminates the need for inosine modification at position 34 of tRNA^{Ile} (see *Section 25.6, Modified Bases Affect Anticodon–Codon Pairing*).

The different wobble rules for mitochondrial and nuclear translation very likely arise from differences in the detailed structure of the respective mitochondria that translate the two genomes. In cytoplasmic ribosomes, modifications to U34 are used to expand the decoding capacities of certain tRNAs (see *Section 25.6, Modified Bases Affect Anticodon–Codon Pairing*). On mitochondrial ribosomes, modifications to U34 are instead used to restrict pairing to codons containing A or G at the third position, according to the usual wobble rules. Modifications to U34 are indeed found in mitochondrial tRNAs representing amino acids for two-codon sets, to avoid the misreading that would otherwise occur.

 ## 25.8 Novel Amino Acids Can Be Inserted at Certain Stop Codons

Key concepts

- The insertion of selenocysteine at some UGA codons requires the action of an unusual tRNA in combination with several proteins.
- The unusual amino acid pyrrolysine can be inserted at certain UAG codons.
- The UGA codon specifies both selenocysteine and cysteine in the ciliate *Euplotes crassus*.

There are two known instances in which a stop codon is used to specify an unusual amino acid apart from the classical twenty. Only particular stop codons are reinterpreted in this way by the translational apparatus. This demonstrates that the meaning of the codon triplet is influenced by the identity of other bases in the mRNA. Such a dual meaning for a particular codon in a genome should be distinguished from the context-independent complete reassignment of codons in some organisms or in mitochondria, as described in *Section 25.7, There Are Sporadic Alterations of the Universal Code*.

Selenocysteine, in which the sulfur of cysteine is replaced by selenium, is incorporated at certain UGA codons within genes coding for selenoproteins in all three domains of life. Usually these proteins catalyze oxidation-reduction reactions. The selenocysteine residue is typically located in the active site, where it directly facilitates the reaction chemistry. For example, the UGA codon specifies selenocysteine in three *E. coli* genes coding for formate dehydrogenase isozymes; the incorporated selenium directly ligates a catalytic molybdenum ion in the active site.

Organisms capable of coding for selenocysteine possess an unusual tRNA, tRNA^{Sec}, which is over 90 nucleotides long and contains acceptor and T stems of nonstandard length. Instead of seven base pairs in the acceptor stem and five in the T stem (a 7/5 structure), bacterial tRNA^{Sec} possesses an 8/5 structure, while archaeal and eukaryotic tRNA^{Sec} likely possess a 9/4 structure. These tRNAs also possess the 5′-UCA anticodon, allowing them to read UGA. In all organisms, tRNA^{Sec} is first aminoacylated with serine by seryl-tRNA synthetase (SerRS) to produce seryl-tRNA^{Sec}. In bacteria, the enzyme selenocysteine synthase next converts Ser-tRNA^{Sec} directly to selenocysteinyl (Sec)-tRNA^{Sec} using selenophosphate as the selenium donor. In Archaea and eukaryotes, Ser-tRNA^{Sec} is first phosphorylated by the kinase PSTK to produce phosphoseryl (Sep)-tRNA^{Sec}. In a second step, Sep-tRNA^{Sec} is converted to Sec-tRNA^{Sec} by the enzyme SepSecS. The exquisite specificity of PSTK is notable: It is capable of efficiently phosphorylating Ser-tRNA^{Sec} while excluding the standard Ser-tRNA^{Ser}. Improper phosphorylation of Ser-tRNA^{Ser} by PSTK could result in the incorporation of selenocysteine in response to serine codons.

The choice of which UGA codons are to be interpreted as selenocysteine is determined by the local secondary structure of the mRNA. A hairpin loop downstream of the UGA codon, termed the *SECIS element*, is required for incorporation of selenocysteine and exclusion of release factor binding. The SECIS element is directly adjacent to the UGA codon in bacteria, but is located in the 3′-untranslated region of the mRNA in Archaea and eukaryotes. In *E. coli*, a specialized translation elongation factor, SelB, interacts solely with Sec-tRNA^{Sec} and not with any other aminoacylated tRNA, including the precursor Ser-tRNA^{Sec}. SelB also binds directly to the SECIS element. The consequence of the action of SelB is that only those UGA codons that also possess a properly juxtaposed SECIS

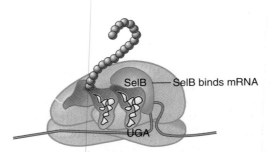

FIGURE 25.12 SelB is an elongation factor that specifically binds tRNA^Sec to a UGA codon that is followed by a stem-loop structure in mRNA.

site will be able to productively bind Sec-tRNA^Sec in the ribosomal A site (**FIGURE 25.12**). Archaea and eukaryotes possess a homolog to SelB, but also require the presence of an additional protein, SBP2, to permit the ribosome to insert selenocysteine.

Another example of the insertion of a special amino acid is the placement of pyrrolysine at certain UAG codons in the archaeal genus *Methanosarcina*, as well as in a few bacteria. In *Methanosarcina*, pyrrolysine is found in the active site of methylamine methyltransferases, where it plays an important role in the reaction chemistry. The incorporation of pyrrolysine requires a specialized aminoacyl-tRNA synthetase, pyrrolysyl-tRNA synthetase (PylRS), which aminoacylates a specialized tRNA^Pyl with pyrrolysine. tRNA^Pyl possesses the 5′-CUA anticodon, enabling it to read UAG. As found for tRNA^Sec, tRNA^Pyl also possess unusual structural features not found in other tRNAs; for example, it lacks the otherwise invariant U8 nucleotide and features atypically short D and variable loops. The mechanism by which particular UAG codons are read as pyrrolysine is not yet resolved, because it has not been possible to unambiguously identify a secondary structure element in all mRNAs that incorporate the amino acid. Further, no specific elongation factor targeting Pyl-tRNA^Pyl to the ribosome has been identified.

Very recently, it was found that the UGA codon specifies insertion of either cysteine or selenocysteine in the ciliate *Euplotes crassus*. Dual use of UGA was found to occur even within the same gene, and the choice of which amino acid is inserted depends on the structure of the 3′-untranslated region of the mRNA. UGA specifies Cys generally in *Euplodes*, and does not function as a stop codon. As a result, this work shows that position-specific dual use can occur within the context of a codon that is not otherwise used for termination in that organism.

tRNAs Are Selectively Paired with Amino Acids by Aminoacyl-tRNA Synthetases

Key concepts

- Aminoacyl-tRNA synthetases are a family of enzymes that attach amino acid to tRNA, generating aminoacyl-tRNA in a two-step reaction that uses energy from ATP.
- Each tRNA synthetase aminoacylates all the tRNAs in an isoaccepting group, representing a particular amino acid.
- Recognition of tRNA by tRNA synthetases is based on a particular set of nucleotides, the tRNA "identity set," that often are concentrated in the acceptor stem and anticodon loop regions of the molecule.

Amino acids enter the translation pathway through the action of aminoacyl-tRNA synthetases, which provide the essential decoding step converting the information in nucleic acids into the polypeptide sequence. All synthetases function by the two-step mechanism depicted in **FIGURE 25.13**:

- The amino acid first reacts with ATP to form an aminoacyl adenylate intermediate, releasing pyrophosphate. Part of the energy released in ATP hydrolysis is trapped as a high-energy mixed anhydride linkage in the adenylate.
- Next, either the 2′-OH or 3′-OH group located on the 3′-A76 nucleotide of tRNA attacks the carbonyl carbon atom of the mixed anhydride, generating aminoacyl-tRNA with concomitant release of AMP.

A subset of four tRNA synthetases—those specific to glutamine, glutamate, arginine, and lysine—require the presence of tRNA to synthesize the aminoacyl adenylate intermediate. For these enzymes, the tRNA synthetase is properly considered as a ribonucleoprotein particle (RNP), in which the RNA subunit functions to assist the protein in attaining a catalytically competent conformation. In the second step of aminoacylation, the amino acid portion of the aminoacyl adenylate is then transferred to the RNA component of the RNP (the tRNA).

Each tRNA synthetase is selective for a single amino acid among all the amino acids in the cellular pool. It also discriminates among all tRNAs in the cell. Usually, each amino acid is represented by more than one tRNA. Several

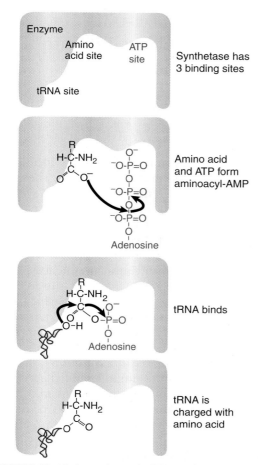

Enzyme

Amino acid site   ATP site

Synthetase has 3 binding sites

tRNA site

Amino acid and ATP form aminoacyl-AMP

Adenosine

tRNA binds

Adenosine

tRNA is charged with amino acid

FIGURE 25.13 An aminoacyl-tRNA synthetase charges tRNA with an amino acid.

tRNAs may be needed to respond to synonym codons, and sometimes there are multiple species of tRNA that base-pair with the same codon. Multiple tRNAs representing the same amino acid are called **isoaccepting tRNAs**; because they are all recognized by the same synthetase, they are also described as its **cognate tRNAs**.

All tRNAs possess the canonical L-shaped tertiary structure (see Chapter 24, *Translation*). The tRNA folds such that the acceptor and T stems form one coaxial stack, while the D and anticodon stems together form the perpendicular arm of the L-shape. The anticodon loop and CCA acceptor end are located at opposite ends of the molecule and are separated by approximately 40 Å. The globular hinge region of the tRNA, which connects the two perpendicular stacks, is composed of the D loop, T loop, variable arm, and two-nucleotide spacer between the acceptor and D stems. Most tRNAs possess small variable regions consisting of a four to five nucleotide loop, whereas a few isoaccepting groups feature a larger variable arm including a base-paired stem, which protrudes from the globular core. The common tRNA L-shape is essential for the interaction of all tRNAs with elongation factors and with the ribosome.

Within the context of this common L-shaped structure, enforced by the presence of conserved tertiary interactions within the globular core, tRNA sequences are found to diverge at a majority of positions in all four arms of the molecule. This sequence diversity can generate subtle differences in the angle between the two arms of the L-shape, and, more importantly, leads to variations in the detailed path of the polynucleotide backbone throughout the molecule. It is this structural diversity that forms the basis for discrimination by the tRNA synthetases.

tRNA synthetases discriminate among tRNAs by means of two general mechanisms: *direct readout* and *indirect readout*. In direct readout, the enzyme recognizes base-specific functional groups directly—for example, a surface amino acid of a tRNA synthetase may accept a hydrogen bond from the exocyclic amine group of guanine (the N2 of G), a minor-groove group not found on the other three bases. By contrast, in indirect readout, the enzyme directly binds nonspecific portions of the tRNA: the sugar–phosphate backbone and nonspecific portions of the nucleotide bases. For example, sequences in the variable and D arms of a tRNA may produce a distinctively shaped surface that is complementary to the cognate tRNA synthetase, but not to other tRNA synthetases. In this way nucleotides distant from the enzyme-tRNA interface create an interface structure that is in turn directly bound. Both direct and indirect readout usually function within the context of mutual induced fit: Conformational changes in both the tRNA and enzyme occur after initial binding, to form a productive catalytic complex. Both these mechanisms also often involve the participation of bound water molecules at the interface between the tRNA and enzyme. For example, when glutaminyl-tRNA synthetase (GlnRS) binds tRNA^Gln, two domains of the enzyme rotate with respect to each other; simultaneously, the 3'-single-stranded end and the anticodon loop of the tRNA undergo substantial conformational changes as compared with their presumed structures in the unliganded state.

In many cases the determinants in tRNA that are needed for specific recognition are located at the extremities of the molecule, in the acceptor stem and the anticodon loop. There are also a number of examples, however, where nucleotides in the tertiary core provide the identity signals. Another commonly used identity nucleotide is the "discriminator base" at position 73 in the tRNA, which is located directly 5′ to the 3′-terminal CCA sequence. Interestingly, the anticodon sequence of the tRNA is not necessarily required for specific tRNA synthetase recognition. In general, the tRNA identity set is idiosyncratic to each tRNA synthetase.

The identity determinants vary in their importance, and are sometimes conserved in evolution. The conservation in tRNA identity elements is demonstrated by the capacities of many tRNA synthetases to aminoacylate tRNAs that are derived from different organisms. Hypotheses regarding the set of tRNA identity elements necessary for selectivity by a tRNA synthetase are derived from X-ray cocrystal structures of tRNA synthetase complexes, from classical genetics, and from *in vitro* mutagenesis. Final proof that a tRNA identity set has been well defined is obtained from transplantation experiments, in which the hypothesized set of nucleotides is incorporated into a tRNA from a different isoaccepting group. For example, replacement of 15 nucleotides in the acceptor stem and anticodon loop of tRNA^{Asp}, with the corresponding nucleotides in tRNA^{Gln}, allowed glutaminyl-tRNA synthetase (GlnRS) to aminoacylate the modified tRNA^{Asp} with glutamine, with an efficiency and selectivity comparable to that of the cognate GlnRS reaction.

Many tRNA synthetases can specifically aminoacylate a tRNA "minihelix," which consists only of the acceptor and TψC arms of the molecule. In some cases, a tRNA microhelix, consisting of the acceptor stem alone closed at its distal end by a stable tetraloop, can serve as a substrate. For both minihelices and microhelices, the efficiency of aminoacylation is very substantially weaker than in the case of the intact tRNA. These experiments have some significance, though, to the evolutionary development of tRNA synthetase complexes. At an early evolutionary stage, tRNAs may have consisted solely of the acceptor arm of the contemporary molecule.

25.10 Aminoacyl-tRNA Synthetases Fall into Two Families

Key concept

- Aminoacyl-tRNA synthetases are divided into class I and class II families based on mutually exclusive sets of sequence motifs and structural domains.

In spite of their common function, synthetases are a very diverse group of enzymes. They are divisible into two families. *Class I tRNA synthetases* are primarily monomeric, and feature structurally similar active-site Rossmann fold domains at or near their N-termini. The Rossmann fold consists of a five- or six-stranded parallel β-sheet with connecting helices. This domain is homologous to the active-site domain of dehydrogenases, and is responsible for binding the ATP, the amino acid, and the 3′-terminus of tRNA. All class I tRNA synthetases contain an "acceptor-binding" domain that is inserted into the Rossmann fold at a common location, which also binds the single-stranded acceptor end of the tRNA, and which contains an editing active site in some of the enzymes (see *Section 25.11, Synthetases Use Proofreading to Improve Accuracy*). The C-terminal domains of class I synthetases bind the inner corner of the L-shaped tRNA and the anticodon arm, and function to discriminate among tRNAs. There are two short common sequence motifs found in the active site Rossmann fold, which are involved in ATP binding. Aside from some limited homology among a few of the enzymes, there are no significant structural or sequence similarities among class I enzymes, outside of the Rossmann fold.

Class II tRNA synthetases are similarly diverse. Their quaternary structures are generally dimeric, but in some cases form homotetramers or α$_2$β$_2$ heterotetramers. Like class I enzymes, class II tRNA synthetases also possess a structurally conserved active site domain—in this case a mixed α/β domain dissimilar to the Rossmann fold. The active sites of class II tRNA synthetases are located toward the C-terminal end of the polypeptides. Three short sequence motifs in the active site domain are conserved in this family; one of these motifs functions in multimerization, whereas the other two have catalytic roles.

Aminoacyl-tRNA synthetases

Class I		Class II	
Gln	(α)	Asn	(α_2)
Glu	(α)	Asp	(α_2)
Arg	(α)	Ser	(α_2)
Lys	(α)	His	(α_2)
Val	(α)	Lys	(α_2)
Ile	(α)	Thr	(α_2)
Leu	(α)	Pro	(α_2)
Met	(α, α_2)	Phe	$(\alpha, \alpha_2\beta_2)$
Cys	(α, α_2)	Ala	(α_2, α_4)
Tyr	(α_2)	Gly	$(\alpha_2, \alpha_2\beta_2)$
Trp	(α_2)	Sep	(α_4)
		Pyl	(?)

FIGURE 25.14 Separation of tRNA synthetases into two classes possessing mutually exclusive sets of sequence motifs and active-site structural domains. The quaternary structure of the enzyme is noted. Multiple designations indicate that the quaternary structure differs in different organisms. The quaternary structure of PylRS has not been clearly established.

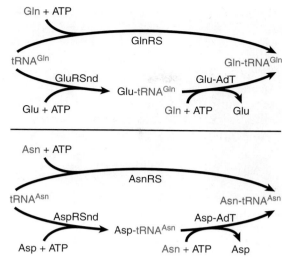

FIGURE 25.15 Mechanisms for the synthesis of Gln-tRNA^{Gln} and Asn-tRNA^{Asn}. The top route in each case indicates the one-step pathway catalyzed by the conventional tRNA synthetase. The bottom, two-step pathways are found in most organisms. They consist of a nondiscriminating tRNA synthetase followed by the action of a tRNA-dependent amidotransferase enzyme (AdT).

There are 23 phylogenetically distinct families of tRNA synthetases. Eleven of these families fall into class I, and the remaining 12 are class II enzymes (**FIGURE 25.14**). Interestingly, there are two distinct types of LysRS enzymes that fall into separate classes. Two noncanonical tRNA synthetase families with limited phylogenetic scope have also recently been discovered. These enzymes are the class II pyrrolysyl-tRNA synthetase (PylRS) (discussed in *Section 25.8, Novel Amino Acids Can Be Inserted at Certain Stop Codons*), and the class II phosphoseryl-tRNA synthetase (SepRS). SepRS is restricted to methanogens (a subclass of Archaea) and the closely related *Archaeoglobus fulgidus*. It attaches phosphoserine (Sep) onto tRNA^{Cys} acceptors to produce a misacylated Sep-tRNA^{Cys} species. All organisms possessing SepRS also possess a pyridoxal phosphate-dependent companion enzyme, SepCysS, which converts Sep-tRNA^{Cys} to Cys-tRNA^{Cys}. The sulfur donor used by SepCysS *in vivo* is unknown. Interestingly, some methanogens possess both the SepRS/SepCysS two-step pathway and, in parallel, the canonical CysRS enzyme.

Although there are 23 phylogenetically distinct tRNA synthetase families, most organisms possess only 18 of the enzymes. Typically missing from the repertoire are GlnRS and asparaginyl-tRNA synthetase (AsnRS). To syn-thesize Gln-tRNA^{Gln} and Asn-tRNA^{Asn}, these organisms possess distinct glutamyl-tRNA synthetase (GluRS) and aspartyl-tRNA synthetase (AspRS) enzymes that are nondiscriminating (ND). GluRSND synthesizes both Glu-tRNA^{Glu} as well as misacylated Glu-tRNA^{Gln}, whereas AspRSND synthesizes both Asp-tRNA^{Asp} and misacylated Asp-tRNA^{Asn}. The misacylated tRNAs are then converted to Gln-tRNA^{Gln} and Asn-tRNA^{Asn} by the action of a tRNA-dependent amidotransferase enzyme (AdT). AdTs are remarkable multimeric enzymes possessing three distinct activities (**FIGURE 25.15**). They first generate ammonia in one active site by deamidation of a nitrogen donor such as glutamine or asparagine. The ammonia is then shuttled through an intramolecular tunnel in the enzyme to emerge in a second site that binds the 3'-end of the misacylated tRNA. In the second active site, a kinase activity γ-phosphorylates the side-chain amino acid carboxylate of Glu-tRNA^{Gln} or Asp-tRNA^{Asn}. Finally, the ammonia reacts to displace phosphate, forming Gln-tRNA^{Gln} or Asn-tRNA^{Asn}. Distinct AdT families exist that function on both misacylated tRNAs, or that are restricted to Gln-tRNA^{Gln} formation only.

Class I and class II synthetases are functionally differentiated in a number of ways.

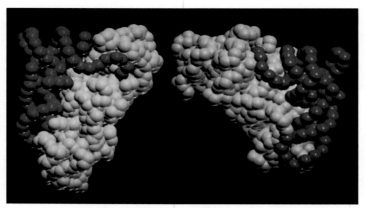

FIGURE 25.16 Crystal structures show that class I and class II aminoacyl-tRNA synthetases bind the opposite faces of their tRNA substrates. The tRNA is shown in red and the protein in blue. Photo courtesy of Dino Moras, Institute of Genetics and Molecular and Cellular Biology (IGBMC).

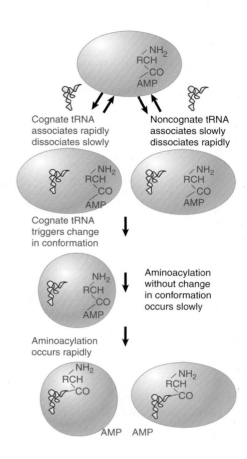

FIGURE 25.17 Aminoacylation of cognate tRNAs by synthetase is based in part on greater affinities for these species, coupled with weak affinities for noncognate species. In addition, noncognate tRNAs are unable to fully undergo the induced-fit conformational changes required for the later catalytic steps.

First, class I enzymes aminoacylate tRNA at the 2'-OH position of A76, whereas class II enzymes generally aminoacylate tRNA on the 3'-OH. The position of initial aminoacylation is related to the binding orientation of the tRNA on the enzyme. Class I synthetases bind tRNA on the minor groove side of the acceptor stem, and require that the single-stranded 3'-terminus form a hairpin structure for proper juxtaposition with the amino acid and ATP in the active site (**FIGURE 25.16**). Class II synthetases instead bind the major groove side of the tRNA acceptor stem, and do not require hairpinning of the tRNA 3'-end into the active site. There is also a mechanistic distinction that has recently emerged: the reaction rates of class I synthetases are limited by release of aminoacylated tRNA product, whereas class II synthetases are limited by earlier chemical steps and/or physical rearrangements in the active sites.

25.11 Synthetases Use Proofreading to Improve Accuracy

Key concept

• Specificity of amino acid-tRNA pairing is controlled by proofreading reactions that hydrolyze incorrectly formed aminoacyl adenylates and aminoacyl-tRNAs.

Aminoacyl-tRNA synthetases must distinguish one specific amino acid from the cellular pool of amino acids and related molecules, and must also differentiate cognate tRNAs in a particular isoaccepting group (typically one to three) from the total set of tRNAs. tRNA discrimina-

tion can be successfully accomplished based on detailed differences in the L-shaped structures (see *Section 25.9, tRNAs Are Selectively Paired with Amino Acids by Aminoacyl-tRNA Synthetases*). This occurs at both the initial binding step, and at the level of induced fit; noncognate tRNAs derived from other isoaccepting groups lack the full identity set of nucleotides, and are consequently unable to rearrange their structure to adopt an enzyme-bound conformation in which the reactive CCA terminus is properly aligned with the amino acid carboxylate group and the ATP α-phosphate. This rejection of noncognate tRNAs, at a stage of the reaction that precedes the synthesis of misacylated tRNA, is sometimes referred to as **kinetic proofreading**. The inability of noncognate tRNAs to proceed through the chemical steps of aminoacylation arises because the tRNA dissociates from the enzyme much faster than it can react (**FIGURE 25.17**).

In contrast, tRNA synthetases are unable to distinguish between some structurally similar amino acids in the course of the two-step aminoacyl-tRNA synthesis reaction alone. It is especially difficult for the enzymes to distinguish between two amino acids that differ only in the length of the carbon backbone (that is, by one CH_2 group), or between amino acids of the same size that differ at only one atomic position. For example, the amino acid binding pocket of isoleucyl-tRNA synthetase (IleRS) cannot distinguish isoleucine from valine sufficiently well to prevent synthesis of a significant amount of Val-tRNA^{Ile}. Similarly, valyl-tRNA synthetase (ValRS) synthesizes Thr-tRNA^{Val} to a significant extent.

IleRS, ValRS, and at least seven additional tRNA synthetases (those specific to leucine, methionine, alanine, proline, phenylalanine, threonine, and lysine) are able to correct, or proofread, the aminoacyl adenylates and aminoacyl-tRNA formed in their active sites, by means of additional activities that either hydrolyze the aminoacyl-AMP to yield free amino acid and AMP, or that hydrolyze the misacylated tRNA to yield free amino acid and deacylated tRNA. The hydrolysis of aminoacyl-AMP is referred to as *pretransfer editing*, while the hydrolysis of aminoacyl-tRNA is referred to as *posttransfer editing* (**FIGURE 25.18**). In the case of pretransfer editing, it is also possible that some of the incorrectly formed aminoacyl-AMP dissociates from the active site, after which it is hydrolyzed nonenzymatically in solution (the aminoacyl ester bond is relatively unstable). This type of editing reaction can also be considered as a form of kinetic proofreading. In contrast, pretransfer hydrolysis of noncognate aminoacyl adenylate when bound by the enzyme, as well as enzyme-catalyzed posttransfer editing, are each known as **chemical proofreading**. Although pretransfer editing reactions may sometimes occur in the absence of tRNA (that is, before tRNA binding), the presence of tRNA generally substantially improves the efficiency of the hydrolytic reaction. The extent to which pretransfer versus posttransfer editing predominates varies with the individual synthetase.

A general way to think of the editing reaction is in terms of the classic double-sieve mechanism, illustrated for IleRS in **FIGURE 25.19**, in which the size of the amino acid is used as the basis for discrimination. IleRS possesses two active sites: the synthetic (or activation) site located in the common class I Rossmann

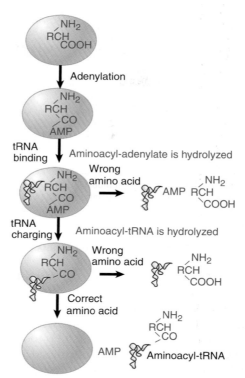

FIGURE 25.18 Proofreading by aminoacyl-tRNA synthetases may take place at the stage prior to aminoacylation (pretransfer editing), in which the noncognate aminoacyl adenylate is hydrolyzed. Alternatively or additionally, hydrolysis of incorrectly formed aminoacyl-tRNA may occur after its synthesis (posttransfer editing).

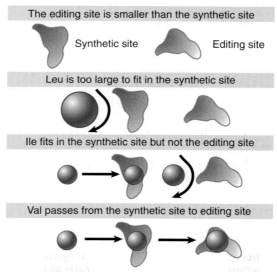

FIGURE 25.19 Isoleucyl-tRNA synthetase has two active sites. Amino acids larger than Ile cannot be activated because they do not fit in the synthetic site. Amino acids smaller than Ile are removed because they are able to enter the editing site.

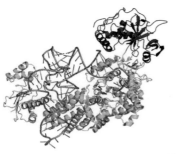

FIGURE 25.20 Cocrystal structure of leucyl-tRNA synthetase (LeuRS) bound to tRNA^{Leu} in the posttransfer editing conformation (left) and in the aminoacylation conformation (right). The editing domain in dark blue rotates substantially between the two conformations, and the acceptor end of the tRNA converts between an extended helical (left) and hairpinned conformation (right). Reprinted with permission from J. J. Perona, et al., *Biochemistry* 46 (2007): 10419–10432. Copyright 2007, American Chemical Society. Photos courtesy of John J. Perona, University of California, Santa Barbara.

fold domain, and the editing (or hydrolytic) site located in the acceptor-binding domain (see *Section 25.10, Aminoacyl-tRNA Synthetases Fall into Two Families*). The crystal structure of IleRS shows that the synthetic site is too small to allow leucine to enter (the leucine side-chain is branched at a different position as compared with isoleucine). Indeed, all amino acids larger than isoleucine are excluded from activation because they cannot enter the synthetic site. Some smaller amino acids that retain sufficient capacity to bind, though—such as valine—can enter the synthetic site and become attached to tRNA. The synthetic site functions as the first sieve. The editing site is smaller than the synthetic site, and cannot accommodate the cognate isoleucine, but it does bind valine. Thus, Val-tRNA^{Ile} can be hydrolyzed in the editing site, functioning as the second sieve, while Ile-tRNA^{Ile} is not hydrolyzed.

The double-sieve model functions as a convenient and generally accurate way to think of posttransfer editing. In IleRS, as well as in other editing tRNA synthetases from both class I and class II, the synthetic and editing sites are located a considerable distance apart, on the order of 10–40 Å. For posttransfer hydrolysis (editing) to occur, the misacylated aminoacyl-tRNA acceptor end is translocated across the surface of the enzyme (**FIGURE 25.20**), moving from the synthetic site to the editing site. This involves a change in the conformation of the acceptor end of the tRNA. In class I tRNA synthetases, the acceptor end adopts a hairpinned conformation when bound in the synthetic site (see *Section 25.10, Aminoacyl-tRNA Synthetases Fall into Two Families*), and an extended structure when bound in the editing site.

Translocation of the incorrect amino acid across the tRNA synthetase surface is possible

in the posttransfer editing mechanism, because the amino acid is covalently bound to the tRNA. In contrast, pretransfer editing occurs before a covalent linkage of the amino acid to the tRNA is formed. For this reason, it is unlikely that pretransfer editing occurs within the editing domain of the enzyme, because there is no mechanism by which the noncognate aminoacyl adenylate could be prevented from dissociating *en route* from the synthetic site to the editing site. Instead, when enzyme-mediated, the pretransfer editing reaction likely occurs within the confines of the synthetic active site.

25.12 Suppressor tRNAs Have Mutated Anticodons That Read New Codons

Key concepts

- A suppressor tRNA typically has a mutation in the anticodon that changes the codons to which it responds.
- When the new anticodon corresponds to a termination codon, an amino acid is inserted and the polypeptide chain is extended beyond the termination codon. This results in nonsense suppression at a site of nonsense mutation, or in readthrough at a natural termination codon.
- Missense suppression occurs when the tRNA recognizes a different codon from usual, so that one amino acid is substituted for another.

Isolation of mutant tRNAs has been one of the most potent tools for analyzing the ability of a tRNA to respond to its codon(s) in mRNA, and for determining the effects that different parts of the tRNA molecule have on codon–anticodon recognition.

Mutant tRNAs are isolated by virtue of their ability to overcome the effects of mutations in genes coding for polypeptides. In general genetic terminology, a mutation that is able to overcome the effects of another mutation is called a **suppressor**.

In tRNA suppressor systems, the primary mutation changes a codon in an mRNA so that the polypeptide product is no longer functional. The secondary suppressor mutation changes the anticodon of a tRNA, so that it recognizes the mutant codon instead of (or as well as) its original target codon. The amino acid that is now inserted restores polypeptide function. The suppressors are described as **nonsense suppressors** or **missense suppressors**, depending on the nature of the original mutation.

A nonsense mutation converts a codon that specifies an amino acid to one of the three stop

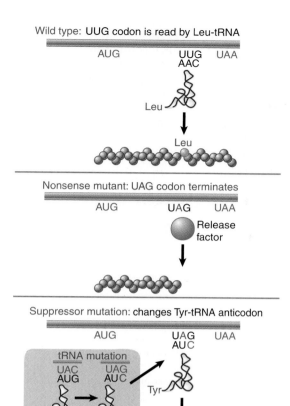

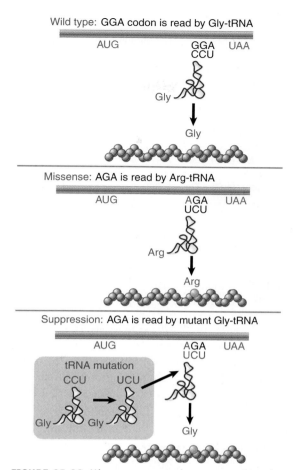

FIGURE 25.21 Nonsense mutations can be suppressed by a tRNA with a mutant anticodon, which inserts an amino acid at the mutant codon, producing a full length protein in which the original Leu residue has been replaced by Tyr.

FIGURE 25.22 Missense suppression occurs when the anticodon of tRNA is mutated so that it responds to the wrong codon.

codons. In a wild-type cell, such a nonsense mutation is recognized only by a release factor, which terminates protein synthesis. The second suppressor mutation in the tRNA anticodon, however, creates an aminoacyl-tRNA that can recognize the termination codon. By inserting an amino acid, the second-site suppressor allows translation to continue beyond the site of nonsense mutation. This new capacity of the translation system allows a full-length polypeptide to be synthesized, as illustrated in **FIGURE 25.21**. If the amino acid inserted by suppression is different from the amino acid that was originally present at this site in the wild-type polypeptide, the activity of the polypeptide may be altered.

Missense mutations change a codon representing one amino acid into a codon representing another amino acid—one that cannot function in the polypeptide in place of the original residue. (Formally, any substitution of amino acids constitutes a missense mutation, but in practice it is detected only if it changes

the activity of the polypeptide.) The mutation can be suppressed by the insertion either of the original amino acid or of some other amino acid that restores the function of the polypeptide.

FIGURE 25.22 demonstrates that missense suppression can be accomplished in the same way as nonsense suppression, by mutating the anticodon of a tRNA carrying an acceptable amino acid so that it responds to the mutant codon. So missense suppression involves a change in the meaning of the codon from one amino acid to another.

25.13 There Are Nonsense Suppressors for Each Termination Codon

Key concepts

- Each type of nonsense codon is suppressed by a tRNA with a mutated anticodon.
- Some rare suppressor tRNAs have mutations in other parts of the molecule.

Locus	tRNA	Wild Type		Suppressor	
		Codon/Anti		Anti/Codon	
supD (su1)	Ser	UCG	CGA	CUA	UAG
supE (su2)	Gln	CAG	CUG	CUA	UAG
supF (su3)	Tyr	UA$_U^C$	GUA	CUA	UAG
supC (su4)	Tyr	UA$_U^C$	GUA	UUA	UA$_G^A$
supG (su5)	Lys	AA$_G^A$	UUU	UUA	UA$_G^A$
supU (su7)	Trp	UGG	CCA	UCA	UG$_G^A$

FIGURE 25.23 Nonsense suppressor tRNAs are generated by mutations in the anticodon.

Nonsense suppressors fall into three classes, one for each type of termination codon. **FIGURE 25.23** describes the properties of some of the best characterized suppressors.

The easiest to characterize have been amber suppressors. In *E. coli*, at least six tRNAs have been mutated to recognize UAG codons. All of the amber suppressor tRNAs have the anticodon CUA←, in each case derived from wild type by a single base change. The site of mutation can be any one of the three bases of the anticodon, as seen from *supD*, *supE*, and *supF*. Each suppressor tRNA recognizes only the UAG codon, instead of its former codon(s). The amino acids inserted are serine, glutamine, or tyrosine, the same as those carried by the corresponding wild-type tRNAs.

Ochre suppressors also arise by mutations in the anticodon. The best known are *supC* and *supG*, which insert tyrosine or lysine in response to both ochre (UAA) and amber (UAG) codons. This is consistent with the prediction of the wobble hypothesis that UAA cannot be recognized alone.

A UGA suppressor has an unexpected property. It is derived from tRNA^{Trp}, but its only mutation is the substitution of A in place of G at position 24. This change replaces a G-U pair in the D stem with an A-U pair, increasing the stability of the helix. The sequence of the anticodon remains the same as the wild type, CCA←. So the mutation in the D stem must in some way alter the conformation of the anticodon loop, allowing CCA← to pair with UGA in an unusual wobble pairing of C with A. The suppressor tRNA continues to recognize its usual codon, UGG.

A related situation is seen in the case of a particular eukaryotic tRNA. Bovine liver contains a tRNA^{Ser} with the anticodon ^{m}CCA←. The wobble rules predict that this tRNA should respond to the tryptophan codon UGG, but in fact it responds to the termination codon UGA. So it is possible that UGA is suppressed naturally in this situation.

The general importance of these observations lies in the demonstration that codon–anticodon recognition of either wild-type or mutant tRNA cannot be predicted entirely from the relevant triplet sequences, but may in some cases be influenced by other features of the molecule.

25.14 Suppressors May Compete with Wild-Type Reading of the Code

Key concepts

- Suppressor tRNAs compete with wild-type tRNAs that have the same anticodon to read the corresponding codon(s).
- Efficient suppression is deleterious because it results in readthrough past normal termination codons.
- The UGA codon is leaky and is misread by Trp-tRNA at 1% to 3% frequency.

There is an interesting difference between the usual recognition of a codon by its proper aminoacyl-tRNA, and the situation in which mutation allows a suppressor tRNA to recognize a new codon. In the wild-type cell, only one meaning can be attributed to a given codon, which represents either a particular amino acid or a signal for termination. In a cell carrying a suppressor mutation, however, the mutant codon may either be recognized by the suppressor tRNA, or be read with its usual meaning.

A nonsense suppressor tRNA must compete with the release factors that recognize the termination codon(s). A missense suppressor tRNA must compete with the tRNAs that respond properly to its new codon. In each case, the extent of competition influences the efficiency of suppression; thus the effectiveness of a particular suppressor depends not only on the affinity between its anticodon and the target codon, but also on its concentration in the cell, and on the parameters governing the competing termination or insertion reactions.

The efficiency with which any particular codon is read is influenced by its location. Thus the extent of nonsense suppression by a given tRNA can vary quite widely, depending on the context of the codon. The effect that neighboring bases in mRNA have on codon–anticodon recognition is poorly understood, but the context can change the frequency with which a

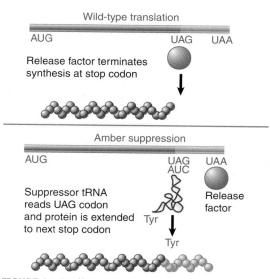

Wild-type translation

AUG   UAG   UAA

Release factor terminates
synthesis at stop codon

Amber suppression

AUG   UAG   UAA
             AUC

Suppressor tRNA
reads UAG codon   Release
and protein is extended   factor
to next stop codon   Tyr

                      Tyr

FIGURE 25.24 Nonsense suppressors also read through natural termination codons, synthesizing proteins that are longer than wild-type.

codon is recognized by a particular tRNA by more than an order of magnitude.

A nonsense suppressor is isolated by its ability to respond to a mutant nonsense codon. The same triplet sequence, however, constitutes one of the normal termination signals of the cell. The mutant tRNA that suppresses the nonsense mutation must in principle be able to suppress natural termination at the end of any gene that uses this codon. **FIGURE 25.24** shows that this **readthrough** results in the synthesis of a longer polypeptide, with additional C-terminal material. The extended polypeptide will end at the next termination triplet sequence found in the phase of the reading frame. Any extensive suppression of termination is likely to be deleterious to the cell by producing extended polypeptides whose functions are thereby altered.

Amber suppressors tend to be relatively efficient, usually in the range of 10% to 50%, depending on the system. This efficiency is possible because amber codons are used relatively infrequently to terminate translation in *E. coli*. In contrast, ochre suppressors are difficult to isolate. They are always much less efficient, usually with activities below 10%. All ochre suppressors grow rather poorly, which indicates that suppression of both UAA and UAG is damaging to *E. coli*, probably because the UAA ochre codon is used most frequently as a natural termination signal. Finally, UGA is the least efficient of the termination codons in its natural function; it is misread by Trp-tRNA as

frequently as 1% to 3% in wild-type situations. In spite of this deficiency, however, UGA is used more commonly than the amber triplet UAG to terminate bacterial genes.

A missense suppressor tRNA that compensates for a mutated codon at one position may have the effect of introducing an unwanted mutation in another gene. A suppressor corrects a mutation by substituting one amino acid for another at the mutant site. In other locations, though, the same substitution will replace the wild-type amino acid with a new amino acid. The change may inhibit normal polypeptide function. This poses a dilemma for the cell: it must suppress what is a mutant codon at one location while failing to change too extensively its normal meaning at other locations. The absence of any strong missense suppressors is therefore explained by the damaging effects that would be caused by a general and efficient substitution of amino acids.

A mutation that creates a suppressor tRNA can have two consequences. First, it allows the tRNA to recognize a new codon. Second, it sometimes prevents the tRNA from recognizing the codons to which it previously responded. It is significant that all the high-efficiency amber suppressors are derived by mutation of one copy of a redundant tRNA set. In these cases, the cell has several tRNAs able to respond to the codon originally recognized by the wild-type tRNA. Thus the mutation does not abolish recognition of the old codons, which continue to be served adequately by the tRNAs of the set. In the unusual situation in which there is only a single tRNA that responds to a particular codon, any mutation that prevents the response is lethal.

Suppression is most often considered in the context of a mutation that changes the reading of a codon. There are, however, some situations in which a stop codon is read as an amino acid at a low frequency in the wild-type situation. The first example to be discovered was the coat protein gene of the RNA phage Qβ. The formation of infective Qβ particles requires that the stop codon at the end of this gene is suppressed at a low frequency to generate a small proportion of coat proteins with a C-terminal extension. In effect, this stop codon is leaky. The reason is that Trp-tRNA recognizes the codon at a low frequency.

Readthrough past stop codons also occurs in eukaryotes, where it is employed most often by RNA viruses. This may involve the suppression of UAG/UAA by Tyr-tRNA, Gln-tRNA, or

Leu-tRNA, or the suppression of UGA by Trp-tRNA or Arg-tRNA. The extent of partial suppression is dictated by the context surrounding the codon.

25.15 The Ribosome Influences the Accuracy of Translation

Key concept

- The structure of the 16S rRNA at the P and A sites of the ribosome influences the accuracy of translation.

The error rate for incorporation of amino acids into polypeptides must be kept low, in the range of one misincorporation per 10,000 amino acids, to ensure that the functional properties of the encoded polypeptides are not altered in such a way as to be deleterious to the cell. There are three general stages in translation at which errors might be made (see Figure 24.8 in *Section 24.3, Special Mechanisms Control the Accuracy of Translation*):

- Charging a tRNA only with its correct amino acid clearly is critical. This is a function of the aminoacyl-tRNA synthetase. The error rate varies with the particular enzyme, in the range of one misincorporation per 10^5–10^7 aminoacylations (see *Sections 25.9 to 25.11*).

- Transporting only correctly aminoacylated tRNA to the ribosome, the function of initiation or elongation factors, can provide a mechanism for enhancing overall selectivity. In addition, these factors assist in the process of docking aminoacyl-tRNA to the ribosomal P and A sites.

- The specificity of codon–anticodon recognition is also crucial. Although binding constants vary with the individual codon–anticodon pairing, the intrinsic specificity associated with formation of a cognate versus noncognate three base-pair sequence (about 10^{-1} to 10^{-2}) is far too low to provide an error rate of $<10^{-5}$.

It had long been assumed that the bacterial elongation factor EF-Tu is a nonsequence-specific RNA binding protein, given that it must transport all aminoacyl-tRNAs (except for the initiator tRNA) to the ribosome. EF-Tu recognizes both the amino acid portion of the aminoacyl-tRNA bond and the tRNA body, however, where it primarily binds to the sugar–phosphate backbone in the acceptor and T stems. Studies in which EF-Tu binding affinity to correctly and incorrectly aminoacylated tRNA was measured have shown that the strength of binding to the amino acid is inversely correlated with the strength of binding to the tRNA body. That is, weakly bound amino acids are correctly esterified to tightly bound tRNA bodies, and tightly bound amino acids are correctly esterified to weakly bound tRNA bodies. As a result, correctly acylated aminoacyl-tRNAs bind EF-Tu with quite similar affinities. Selectivity in overall translation can then come about because misacylation of a weakly bound amino acid to a weakly bound tRNA body produces a noncognate aminoacyl-tRNA that interacts very poorly with EF-Tu. It is also possible that a misacylated aminoacyl-tRNA that binds more tightly to EF-Tu may be discriminated against because it is more difficult to properly release this species upon docking to the ribosome.

It has been found that mutations in EF-Tu are able to suppress frameshifting errors (see *Section 25.16, Frameshifting Occurs at Slippery Sequences*, for a discussion of frameshifting). This implies that EF-Tu does not merely bring aminoacyl-tRNA to the A site, but also is involved in positioning the incoming aminoacyl-tRNA relative to the peptidyl-tRNA in the P site. Similarly, mutations in the yeast initiation factor eIF2 allow the initiation of translation at a start codon that is mutated from AUG to UUG. This implies a role for eIF2 in assisting the docking of tRNA$_i$^{Met} to the P site.

Proofreading on the ribosome, to enhance the intrinsically low level of specificity achievable from codon–anticodon base pairing alone, requires additional interactions provided by the local environment in the 30S subunit. In its function as a proofreader, the ribosome amplifies the modest intrinsic selectivity of trinucleotide pairing by as much as 1000-fold (**FIGURE 25.25**).

Aminoacyl-tRNA selection by the ribosome occurs at several stages along the pathway by which the EF-Tu:GTP:aminoacyl-tRNA ternary complex formed after aminoacylation delivers aminoacyl-tRNA to the ribosomal A site. First, a rather unstable initial binding complex forms with the ribosome. Next, there is a codon recognition step, by which the initial complex is rearranged to permit codon–anticodon pairing in the A site. Recall that the adjacent P site accommodates peptidyl-tRNA (see Chapter 24, *Translation*). Both the initial binding step and

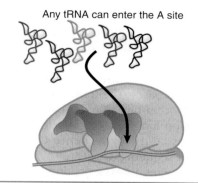

Any tRNA can enter the A site

The correct tRNA interacts with rRNA

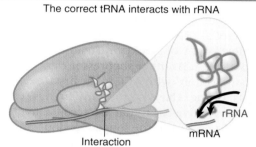

rRNA

mRNA

Interaction

An incorrect tRNA diffuses out

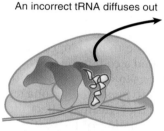

FIGURE 25.25 Any aminoacyl-tRNA can be placed in the A site (by EF-Tu), but only one that pairs with the anticodon can make stabilizing contacts with rRNA. In the absence of these contacts, the aminoacyl-tRNA diffuses out of the A site.

the subsequent codon recognition step are reversible. Mispaired aminoacyl-tRNAs can be rejected at these stages by a combination of increased dissociation rates and/or lowered association rates for mispaired complexes.

After codon–anticodon recognition, a further conformational change triggers hydrolysis of GTP. Release of phosphate from the GDP-bound EF-Tu then occurs: This release triggers another extensive conformational rearrangement, whereby EF-Tu:GDP dissociates from the aminoacyl-tRNA:ribosome complex. Only after EF-Tu dissociation do final conformational rearrangements associated with docking of the aminoacyl moiety into the 50S peptidyl transfer site, and the subsequent peptidyl transfer reaction, occur. In addition to selection at the early binding stage, rejection of mispaired aminoacyl-tRNA can also take place after the GTP hydrolysis step. Here the rejection occurs

because the rate of the final conformational transition is very slow in the case of a misacylated complex. Thus, the overall specificity is enhanced because the tRNA must pass through two selection steps before peptide bond formation can occur.

The precision of codon–anticodon pairing in the A site is maintained by close monitoring of the steric and electrostatic properties of the trinucleotide. Three conserved bases in the 16S ribosomal RNA (A1492, A1493, and G530) interact closely with the minor groove of the codon–anticodon helix at the first two base-pairs, and are able to accurately sense the presence of canonical Watson–Crick pairs at these position. At the third (wobble) position, some noncanonical pairs can be accommodated because the ribosomal RNA does not monitor the pairing as closely. Ultimately, it is the failure of misacylated tRNA to fully meet the scrutiny of the ribosome at the codon–anticodon helix, and perhaps other positions, that leads to its rejection either before or after the GTP hydrolysis step.

Recently, an additional mechanism that contributes to the specificity of translation has been discovered: The ribosome is able to exert quality control after the synthesis of the peptide bond. In this mechanism, the synthesis of a peptide bond that arises from a mismatched aminoacyl-tRNA in the A site leads to a more general loss in specificity in the A site. In turn, this gives rise to the early termination of translation.

The mechanism by which the ribosome recognizes errors after peptide bond synthesis is by monitoring the precise complementarity of the codon–anticodon helix in the peptidyl (P) site. The consequence of the misincorporation is the increased capacity of release factors to bind in the A site to cause premature termination, even when a stop codon is not present. Additionally, there is an increased rate of improper coding in the adjacent A site. The resulting propagation of errors ultimately leads to premature termination.

The cost of translation, as calculated by the number of high-energy bonds that must be hydrolyzed, is clearly increased by proofreading processes. The extent of the increased energetic cost depends the stage at which the misacylated tRNA is rejected. The cost associated with rejection before GTP hydrolysis is associated only with the production of the misacylated tRNA by the tRNA synthetase. If, however, GTP is hydrolyzed before the mismatched aminoacyl-tRNA dissociates, the energetic cost will be greater.

The greatest cost, of course, is associated with the premature termination of translation to give a nonfunctional product, in postpeptidyl transfer quality control. In that case, the full energetic payment associated with synthesis of the polypeptide to the point of premature release must be paid.

25.16 Frameshifting Occurs at Slippery Sequences

Key concepts

- The reading frame may be influenced by the sequence of mRNA and the ribosomal environment.
- Slippery sequences allow a tRNA to shift by one base after it has paired with its anticodon, thereby changing the reading frame.
- Translation of some genes depends upon the regular occurrence of programmed frameshifting.

Recoding events usually involve changes to the meaning of a single codon. Examples include the phenomenon of tRNA suppression (*Section 25.12, Suppressor tRNAs Have Mutated Anticodons That Read New Codons*), and the covalent modification of an aminoacyl-tRNA (*Section 25.8, Novel Amino Acids Can Be Inserted at Certain Stop Codons*). Three other types of recoding, however, cause more global changes in the resulting polypeptide product. These are frameshifting (considered in this section), bypassing, and the use of two mRNAs to synthesize one polypeptide (both are discussed in *Section 25.17, Other Recoding Events: Translational Bypassing and the tmRNA Mechanism to Free Stalled Ribosomes*).

Frameshifting is associated with specific tRNAs in two circumstances:

- Some mutant tRNA suppressors recognize a "codon" for four bases instead of the usual three bases.
- Certain "slippery" sequences allow a tRNA to move along the mRNA in the A site by one base in either the 5' or 3' direction.

Frameshift mutants in a polypeptide result from an aberrant reading of the mRNA codon. Instead of reading a codon triplet, the ribosome reads either a doublet or a quadruplet set of nucleotides. In either case, resumption of triplet reading following this event results in a polypeptide that is out of frame. A frameshift can be suppressed by means of a tRNA that is capable of reading a two- or four-base codon. In the case of four-base codons, the tRNA possesses an expanded anticodon loop consisting of eight nucleotides instead of the normal seven. For example, a G may be inserted in a run of several contiguous G bases. The frameshift suppressor is a tRNA^{Gly} that has an extra base inserted in its anticodon loop, converting the anticodon from the usual triplet sequence CCC$^{\leftarrow}$ to the quadruplet sequence CCCC$^{\leftarrow}$. The suppressor tRNA recognizes a 4-base "codon."

Some frameshift suppressors can recognize more than one 4-base "codon." For example, a bacterial tRNA^{Lys} suppressor can respond to either AAAA or AAAU, instead of the usual codon AAA. Another suppressor can read any 4-base "codon" with ACC in the first three positions; the next base is irrelevant. In these cases, the alternative bases that are acceptable in the fourth position of the longer "codon" are not related by the usual wobble rules. The suppressor tRNA probably recognizes a 3-base codon, but for some other reason—most likely steric hindrance—the adjacent base is blocked. This forces one base to be skipped before the next tRNA can find a codon.

Situations in which frameshifting is a normal event are presented by phages and other viruses. Such events may affect the continuation or termination of translation, and result from the intrinsic properties of the mRNA.

In retroviruses, translation of the first gene is terminated by a nonsense codon in phase with the reading frame. The second gene lies in a different reading frame, and (in some viruses) is translated by a frameshift that allows a shift into the second reading frame and therefore bypasses the termination codon (see **FIGURE 25.26** and also *Section 17.12, Retroviral Genes Code for Polyproteins*). The efficiency of the frameshift is low, typically ~5%. The low efficiency is important in the biology of the virus; an increase in efficiency can be damaging. **FIGURE 25.27** illustrates the similar situation of the yeast Ty element, in which the termination codon of *tya* must be bypassed by a frameshift in order to read the subsequent *tyb* gene.

Such situations make the important point that the rare (but predictable) occurrence of "misreading" events can be relied on as a necessary step in natural translation. This is called **programmed frameshifting**. It occurs at particular sites at frequencies that are 100 to

−1 frameshift in HIV retrovirus

NNNNUUUUUUAGGNNNNNNNN

Last codon read in initial reading frame

First codon read in new reading frame

Reading without frameshift

NNNNUUUUUUAGGNNNNNNNN

Reading after frameshift

NNNNUUUUUU UAGGNNNNNNNN

FIGURE 25.26 A tRNA that slips one base in pairing with a codon causes a frameshift that can suppress termination. The efficiency is usually ~5%.

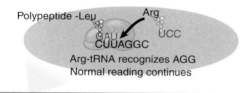

Polypeptide -Leu   Arg
                        UCC
    GAU
    CUUAGGC
Arg-tRNA recognizes AGG
Normal reading continues

Alternative modes of translation give Tya or Tya-Tyb

Tya protein

Initiation   Termination

AUG   tya   UAG tyb   UAA

Initiation   Frameshift   Termination

Tya-Tyb fusion protein

In absence of Arg-tRNA, Leu-tRNA slips 1 base
Gly-tRNA recognizes GGC

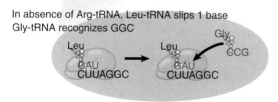

Leu   Leu   Gly
                          CCG
    GAU   GAU
    CUUAGGC   CUUAGGC

FIGURE 25.27 Bypassing occurs when the ribosome moves along mRNA so that the peptidyl-tRNA in the P site is released from pairing with its codon and then repairs with another codon farther along.

1000× greater than the rate at which errors are made at nonprogrammed sites (~3 × 10^{-5} per codon).

There are two common features in this type of frameshifting:

- A "slippery" sequence allows an aminoacyl-tRNA to pair with its codon and then to move +1 or −1 base to pair with an overlapping triplet sequence that can also pair with its anticodon.
- The ribosome is delayed at the frameshifting site to allow time for the aminoacyl-tRNA to rearrange its pairing. The cause of the delay can be an adjacent codon that requires a scarce aminoacyl-tRNA, a termination codon that is recognized slowly by its release factor, or a structural impediment in mRNA (for example, a "pseudoknot," a particular conformation of RNA) that impedes the ribosome.

Slippery events can involve movement in either direction; a −1 frameshift is caused when the tRNA moves backward, and a +1 frameshift is caused when it moves forward. In either case, the result is to expose an out-of-phase triplet in the A site for the next aminoacyl-tRNA. The frameshifting event occurs before peptide bond synthesis. In the most common type of case, when it is triggered by a slippery sequence in conjunction with a downstream hairpin in mRNA, the surrounding sequences influence its efficiency.

The frameshifting in Figure 25.27 shows the behavior of a typical slippery sequence. The seven-nucleotide sequence CUUAGGC is usually recognized by Leu-tRNA at CUU, followed by Arg-tRNA at AGG. The Arg-tRNA is scarce, though, and when its scarcity results in a delay, the Leu-tRNA slips from the CUU codon to the overlapping UUA triplet. This causes a frameshift, because the next triplet in phase with the new pairing (GGC) is read by Gly-tRNA. Slippage usually occurs in the P site (when the Leu-tRNA actually has become peptidyl-tRNA, carrying the nascent chain).

Frameshifting at a stop codon causes readthrough of the protein. The base on the 3′ side of the stop codon influences the relative frequencies of termination and frameshifting, and thus affects the efficiency of the termination signal. This helps to explain the significance of context on termination.

25.17 Other Recoding Events: Translational Bypassing and the tmRNA Mechanism to Free Stalled Ribosomes

Key concepts

- Bypassing involves the capacity of the ribosome to stop translation, release from mRNA, and resume translation some 50 nucleotides downstream.
- Ribosomes that are stalled on mRNA after partial synthesis of a protein may be freed by the action of tmRNA, a unique RNA that incorporates features of both tRNA and mRNA.

Bypassing involves a movement of the ribosome to change the codon that is paired with the peptidyl-tRNA in the P site. The sequence between the two codons is skipped over and is not represented in the polypeptide product. As shown in **FIGURE 25.28**, this allows translation to continue past any termination codons in the intervening region. The most dramatic example of bypassing is in gene *60* of phage T4, where the ribosome moves 50 nucleotides along the mRNA.

The key to the bypass system is that there are identical (or synonymous) codons at either end of the sequence that is skipped. These are sometimes referred to as the "take-off" and "landing" sites. Before bypass, the ribosome is positioned with a peptidyl-tRNA paired with the take-off codon in the P site, with an empty A site waiting for an aminoacyl-tRNA to enter. **FIGURE 25.29** shows that the ribosome slides along mRNA in this condition until the peptidyl-tRNA can become paired with the codon in the landing site.

The sequence of the mRNA triggers the bypass. The important features are the two GGA codons for take-off and landing, the spacing between them, a stem-loop structure that includes the take-off codon, and a stop codon positioned adjacent to the take-off codon.

The take-off stage requires the peptidyl-tRNA to unpair from its codon. This is followed by a movement of the mRNA that prevents it from repairing. Then the ribosome scans the mRNA until the peptidyl-tRNA can repair with the codon in the landing reaction. This is followed by the resumption of protein synthesis when aminoacyl-tRNA enters the A site in the usual way.

Like frameshifting, the bypass reaction depends on a pause by the ribosome. The probability that peptidyl-tRNA will dissociate from its codon in the P site is increased by delays in the entry of aminoacyl-tRNA into the A site. Starvation for an amino acid can trigger bypassing in bacterial genes because of the delay that occurs when there is no aminoacyl-tRNA available to enter the A site. In phage T4 gene *60*,

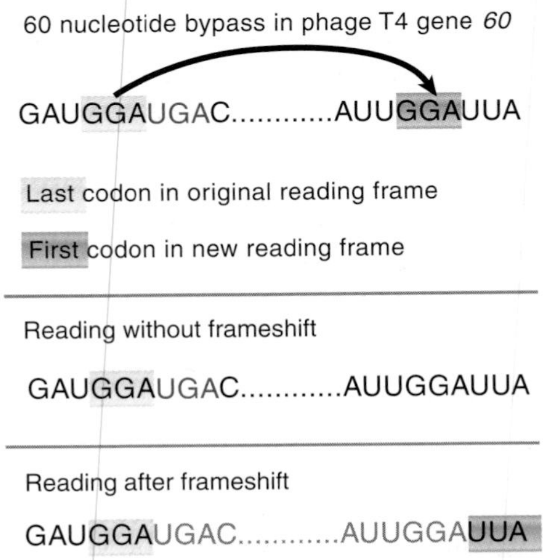

60 nucleotide bypass in phage T4 gene *60*

GAUGGAUGAC............AUUGGAUUA

Last codon in original reading frame

First codon in new reading frame

Reading without frameshift

GAUGGAUGAC............AUUGGAUUA

Reading after frameshift

GAUGGAUGAC............AUUGGAUUA

FIGURE 25.28 A +1 frameshift is required for expression of the *tyb* gene of the yeast Ty element. The shift occurs at a 7-base sequence at which two Leu codon(s) are followed by a scarce Arg codon.

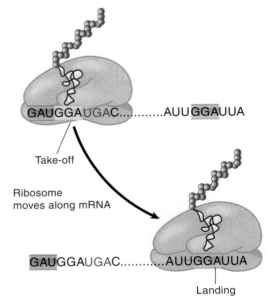

GAUGGAUGAC............AUUGGAUUA

Take-off

Ribosome moves along mRNA

GAUGGAUGAC............AUUGGAUUA

Landing

Peptidyl-tRNA re-pairs with new codon

FIGURE 25.29 In bypass mode, a ribosome with its P site occupied can stop translation. It slides along mRNA to a site where peptidyl-tRNA pairs with a new codon in the P site. Then translation is resumed.

one role of mRNA structure may be to reduce the efficiency of termination, thus creating the delay that is needed for the take-off reaction.

The rescue of stalled ribosomes in bacteria and some mitochondria is accomplished by means of a unique mRNA-tRNA hybrid, termed **tmRNA**, which contains two functional domains. One domain mimics part of tRNA^{Ala}, whereas the second domain encodes a short polypeptide. tmRNA is first aminoacylated by alanyl-tRNA synthetase (AlaRS). It is then bound by EF-Tu and subsequently as a ternary complex at the A site of stalled ribosomes. Peptidyl transfer occurs on the ribosome to join alanine to the C-terminal end of the stalled nascent protein; simultaneously, the mRNA present on the ribosome is replaced by the second domain of tmRNA. tmRNA then functions as template for the synthesis of ten additional amino acids, after which a stop codon is present to terminate translation and release the protein. The newly added C-terminal sequence then acts as a tag for subsequent recognition by proteases, which degrade the truncated protein. tmRNA thus functions as a quality control mechanism to recycle stalled ribosomes and to remove truncated proteins which might otherwise accumulate.

25.18 Summary

The sequence of mRNA read in triplets in the 5'→3' direction is related by the genetic code to the amino acid sequence of polypeptides read from N- to C-terminus. Of the sixty-four triplets, sixty-one encode amino acids and three provide termination signals. Synonymous codons that represent the same amino acids are related in sequence, often by a change in the third base of the codon. This third-base degeneracy, coupled with a pattern in which chemically similar amino acids tend to be coded by related codons, minimizes the effects of mutations. The genetic code is universal and must have been established very early in evolution. Changes in nuclear genomes are rare, but some changes have occurred during mitochondrial evolution.

Multiple tRNAs may respond to a particular codon. The set of tRNAs responding to the various codons for each amino acid is distinctive for each organism. Codon–anticodon recognition involves wobbling at the first position of the anticodon (third position of the codon), which allows some tRNAs to recognize multiple codons. All tRNAs have modified bases, introduced by enzymes that recognize target bases in the tRNA structure. Codon–anticodon pairing is influenced by modifications of the anticodon itself and also by the context of adjacent bases, especially on the 3' side of the anticodon. Taking advantage of codon–anticodon wobble allows vertebrate mitochondria to use only twenty-two tRNAs to recognize all codons, compared with the usual minimum of thirty-one tRNAs; this is assisted by the changes in the mitochondrial code.

Each amino acid is recognized by a particular aminoacyl-tRNA synthetase, which also recognizes all of the tRNAs coding for that amino acid. Some aminoacyl-tRNA synthetases have a proofreading function that scrutinizes the aminoacyl-tRNA products and hydrolyzes incorrectly joined aminoacyl-tRNAs.

Aminoacyl-tRNA synthetases vary widely, but fall into two general groups featuring mutually exclusive sequence motifs and protein structures in their catalytic domains. The two groups of synthetases are also distinguished by the initial site of aminoacylation on the 3'-terminal tRNA ribose, by the orientation of binding of the tRNA acceptor helix, and by the rate-limiting step in aminoacylation. A defined set of nucleotides in the tRNA, termed the identity set, is selectively recognized by the synthetase using a combination of direct and indirect readout mechanisms. In many case the identity set is localized at the anticodon and 3'-acceptor ends of the molecule.

Mutations may allow a tRNA to read different codons; the most common form of such mutations occurs in the anticodon itself. Alteration of the anticodon may allow a tRNA to suppress a mutation in a gene coding for protein. A tRNA that recognizes a termination codon provides a nonsense suppressor, whereas a tRNA that changes the amino acid responding to a codon is a missense suppressor. Suppressors of UAG codons are more efficient than those of UAA codons; this is explained by the fact that UAA is the most commonly used natural termination codon. The efficiency of all suppressors, however, depends on the context of the individual target codon.

Frameshifts of either +1 or −1 may be caused by slippery sequences in mRNA that allow a peptidyl-tRNA to slip from its codon by one base in either the 5' or 3' direction. Certain programmed frameshifts determined by the mRNA sequence are required for expression of natural genes. Bypassing occurs when a ribosome stops translation and moves along mRNA with its peptidyl-tRNA in the P site until the peptidyl-tRNA pairs with an appropriate codon; then translation resumes. The use of tmRNA provides a quality control mechanism to recycle stalled ribosome and to remove undesirable truncated protein products.

References

25.1   Introduction

Research

Nirenberg, M. W. and Leder, P. (1964). The effect of trinucleotides upon the binding of sRNA to ribosomes. *Science* 145, 1399–1407.

Nirenberg, M. W. and Matthaei, H. J. (1961). The dependence of cell-free protein synthesis in *E. coli* upon naturally occurring or synthetic polyribonucleotides. *Proc. Natl. Acad. Sci. USA* 47, 1588–1602.

25.3   Codon–Anticodon Recognition Involves Wobbling

Research

Crick, F. H. C. (1966). Codon-anticodon pairing: the wobble hypothesis. *J. Mol. Biol.* 19, 548–555.

25.4   tRNAs Are Processed from Longer Precursors

Review

Hopper, A. K. and Phizicky, E. M. (2003). tRNA transfers to the limelight. *Genes Dev.* 17, 162–180.

25.5   tRNA Contains Modified Bases

Review

Hopper, A. K. and Phizicky, E. M. (2003). tRNA transfers to the limelight. *Genes Dev.* 17, 162–180.

25.6   Modified Bases Affect Anticodon–Codon Pairing

Review

Agris, P. F. (2008). Bringing order to translation: the contributions of transfer RNA anticodon-domain modifications. *EMBO Reports* 9, 629–635.

25.7   There Are Sporadic Alterations of the Universal Code

Reviews

Osawa, S. et al. (1992). Recent evidence for evolution of the genetic code. *Microbiol. Rev.* 56, 229–264.

Santos, M. A. S., Moura, G., Massey, S. E., and Tuite, M. F. (2004). Driving change: the evolution of alternative genetic codes. *Trends in Genetics* 20, 95–102.

25.8   Novel Amino Acids Can Be Inserted at Certain Stop Codons

Reviews

Krzycki, J. (2005). The direct genetic encoding of pyrrolysine. *Curr. Opin. Microbiol.* 8, 706–712.

Ambrogelly, A., Palioura, S., and Söll, D. (2007). Natural expansion of the genetic code. *Nat. Chem. Biol.* 3, 29–35.

Research

Srinivasan, G., James, C. M., and Krzycki, J. A. (2002). Pyrrolysine encoded by UAG in Archaea: charging of a UAG-decoding specialized tRNA. *Science* 296, 1459–1462.

Turanov, A. A., Lobanov, A. V., Fomenko, E. D., Morrison, H. G., Sogin, M. L., Klobutcher, L. A., Hatfield, D. L., and Gladyshev, V. N. (2009). Genetic code supports targeted insertion of two amino acids by one codon. *Science* 323, 259–261.

25.9   tRNAs Are Selectively Paired with Amino Acids by Aminoacyl-tRNA Synthetases

Reviews

Giege, R., Sissler, M., and Florentz, C. (1998). Universal rules and idiosyncratic features in tRNA identity. *Nucleic Acids Res.* 26, 5017–5035.

Ibba, M. and Söll, D. (2000). Aminoacyl-tRNA synthesis. *Annu. Rev. Biochem.* 69, 617–650.

Perona, J. J. and Hou, Y-M. (2007). Indirect readout of tRNA for aminoacylation. *Biochemistry* 46, 10419–10432.

25.10   Aminoacyl-tRNA Synthetases Fall into Two Families

Review

Ibba, M. and Söll, D. (2004). Aminoacyl-tRNAs: setting the limits of the genetic code. *Genes Dev.* 18, 731–738.

Research

Eriani, G., Delarue, M., Poch, O., Gangloff, J., and Moras, D. (1990). Partition of tRNA synthetases into two classes based on mutually exclusive sets of sequence motifs. *Nature* 347, 203–206.

Rould, M. A. et al. (1989). Structure of *E. coli* glutaminyl-tRNA synthetase complexed with tRNA^{Gln} and ATP at 28Å resolution. *Science* 246, 1135–1142.

Ruff, M. et al. (1991). Class II aminoacyl tRNA synthetases: crystal structure of yeast aspartyl-tRNA synthetase complexes with tRNA^{Asp}. *Science* 252, 1682–1689.

Sauerwald, A., Zhu, W., Major, T. A., Roy, H., Palioura, S., Jahn, D., Whitman, W. B., Yates, J. R. 3d., Ibba, M., and Söll, D. (2005). RNA-dependent cysteine biosynthesis in archaea. *Science* 307, 1969–1972.

25.11 Synthetases Use Proofreading to Improve Accuracy

Research

Dock-Bregeon, A., Sankaranarayanan, R., Romby, P., Caillet, J., Springer, M., Rees, B., Francklyn, C. S., Ehresmann, C., and Moras, D. (2000). Transfer RNA-mediated editing in threonyl-tRNA synthetase. The class II solution to the double discrimination problem. *Cell* 103, 877–884.

Hopfield, J. J. (1974). Kinetic proofreading: a new mechanism for reducing errors in biosynthetic processes requiring high specificity. *Proc. Natl. Acad. Sci. USA* 71, 4135–4139.

Silvian, L. F., Wang, J., and Steitz, T. A. (1999). Insights into editing from an Ile-tRNA synthetase structure with tRNA^{Ile} and mupirocin. *Science* 285, 1074–1077.

Tukalo, M., Yaremchuk, A., Fukunaga, R., Yokoyama, S., and Cusack, S. (2005). The crystal structure of leucyl-tRNA synthetase complexed with tRNA^{Leu} in the post-transfer-editing conformation. *Nat. Struct. Mol. Biol.* 12, 923–930.

25.14 Suppressors May Compete with Wild-Type Reading of the Code

Reviews

Beier, H. and Grimm, M. (2001). Misreading of termination codons in eukaryotes by natural nonsense suppressor tRNAs. *Nucleic Acids Res.* 29, 4767–4782.

Eggertsson, G. and Soll, D. (1988). Transfer RNA-mediated suppression of termination codons in *E. coli. Microbiol. Rev.* 52, 354–374.

Murgola, E. J. (1985). tRNA, suppression, and the code. *Annu. Rev. Genet.* 19, 57–80.

Research

Ruan, B., Palioura, S., Sabina, J., Marvin-Guy, L., Kochhar, S., LaRossa, R. A., and Söll, D. (2009). Quality control despite mistranslation caused by an ambiguous genetic code. *Proc. Natl. Acad. Sci. USA* 105, 16502–16507.

25.15 The Ribosome Influences the Accuracy of Translation

Reviews

Daviter, T., Gromadski, K. B., and Rodnina, M. V. (2006). The ribosome's response to codon-anticodon mismatches. *Biochimie* 88, 1001–1011.

Ogle, J. M. and Ramakrishnan, V. (2005). Structural insights into translational fidelity. *Annu. Rev. Biochem.* 74, 129–177.

Research

LaRiviere, F. J., Wolfson, A. D., and Uhlenbeck, O. C. (2001). Uniform binding of aminoacyl-tRNAs to elongation factor Tu by thermodynamic compensation. *Science* 294, 165–168.

Ogle, J. M., Brodersen, D. E., Clemons, W. M., Tarry, M. J., Carter, A. P., and Ramakrishnan, V. (2001). Recognition of cognate transfer RNA by the 30S ribosomal subunit. *Science* 292, 897–902.

Zaher, H. S. and Green, R. (2009). Quality control by the ribosome following peptide bond formation. *Nature* 457, 161–166.

25.16 Frameshifting Occurs at Slippery Sequences

Reviews

Baranov, P. B., Gesteland, R. F., and Atkins, J. F. (2002). Recoding: translational bifurcations in gene expression. *Gene* 286, 187–202.

Gesteland, R. F. and Atkins, J. F. (1996). Recoding: dynamic reprogramming of translation. *Annu. Rev. Biochem.* 65, 741–768.

Research

Jacks, T., Power, M. D., Masiarz, F. R., Luciw, P. A., Barr, P. J., and Varmus, H. E. (1988). Characterization of ribosomal frameshifting in HIV-1 gag-pol expression. *Nature* 331, 280–283.

25.17 Other Recoding Events: Translational Bypassing and the tmRNA Mechanism to Free Stalled Ribosomes

Review

Herr, A. J., Atkins, J. F., and Gesteland, R. F. (2000). Coupling of open reading frames by translational bypassing. *Annu. Rev. Biochem.* 69, 343–372.

Research

Gallant, J. A. and Lindsley, D. (1998). Ribosomes can slide over and beyond "hungry" codons, resuming protein chain elongation many nucleotides downstream. *Proc. Natl. Acad. Sci. USA* 95, 13771–13776.

Huang, W. M., Ao, S. Z., Casjens, S., Orlandi, R., Zeikus, R., Weiss, R., Winge, D., and Fang, M. (1988). A persistent untranslated sequence within bacteriophage T4 DNA topoisomerase gene 60. *Science* 239, 1005–1012.

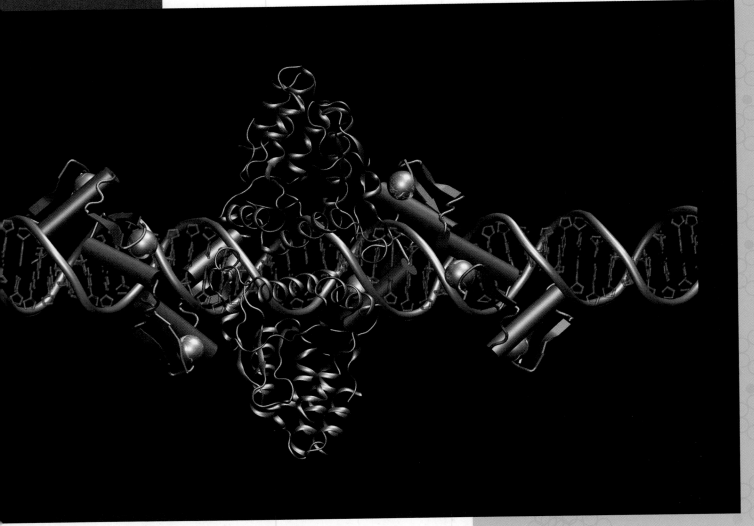

© 2007, Sangamo BioSciences, Inc. (www.sangamo.com)

26

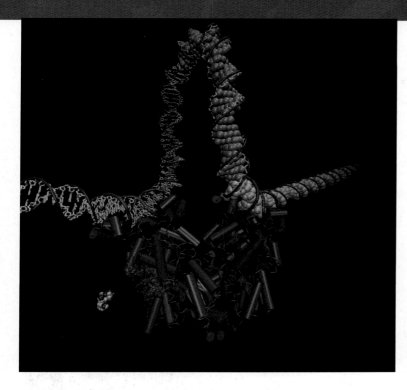

Figure created by Elizabeth Villa using VMD and under copyright of the Theoretical and Computational Biophysics Group, NIH Resource for Macromolecular Modeling and Bioinformatics, at the Beckman Institute, University of Illinois at Urbana-Champaign. The work was partially supported by the National Center for Supercomputing Applications under National Institutes Health Grant PHS5P41RR05969-04 and National Science Foundation Grant MCB-9982629. Computer time was provided by National Resource Allocations Committee Grant MCA93S028.

The Operon

Edited by Liskin Swint-Kruse

CHAPTER OUTLINE

- These mutations are *cis*-acting and affect only those genes on the contiguous stretch of DNA.
- Mutations in the promoter prevent expression of *lacZYA* are uninducible and *cis*-acting.

26.6 *trans*-Acting Mutations Identify the Regulator Gene

- Mutations in the *lacI* gene are *trans*-acting and affect expression of all *lacZYA* clusters in the bacterium.
- Mutations that eliminate *lacI* function cause constitutive expression and are recessive (*lacI⁻*).
- Mutations in the DNA-binding site of the repressor are constitutive because the repressor cannot bind the operator.
- Mutations in the inducer-binding site of the repressor prevent it from being inactivated and cause uninducibility.
- When mutant and wild-type subunits are present, a single *lacI⁻ᵈ* mutant subunit can inactivate a tetramer whose other subunits are wild-type.
- *lacI⁻ᵈ* mutations occur in the DNA-binding site. Their effect is explained by the fact that repressor activity requires all DNA-binding sites in the tetramer to be active.

26.7 *lac* Repressor Is a Tetramer Made of Two Dimers

- A single repressor subunit can be divided into the N-terminal DNA-binding domain, a hinge, and the core of the protein.
- The DNA-binding domain contains two short α-helical regions that bind the major groove of DNA.
- The inducer-binding site and the regions responsible for multimerization are located in the core.
- Monomers form a dimer by making contacts between core subdomains 1 and 2.
- Dimers form a tetramer by interactions between the tetramerization helices.
- Different types of mutations occur in different domains of the repressor protein.

26.8 *lac* Repressor Binding to the Operator Is Regulated by an Allosteric Change in Conformation

- *lac* repressor protein binds to the double-stranded DNA sequence of the operator.
- The operator is a palindromic sequence of 26 bp.
- Each inverted repeat of the operator binds to the DNA-binding site of one repressor subunit.
- Inducer binding causes a change in repressor conformation that reduces its affinity for DNA and releases it from the operator.

26.9 *lac* Repressor Binds to Three Operators and Interacts with RNA Polymerase

- Each dimer in a repressor tetramer can bind an operator, so that the tetramer can bind two operators simultaneously.
- Full repression requires the repressor to bind to an additional operator downstream or upstream as well as to the primary operator at the *lacZ* promoter.

- Binding of repressor at the operator stimulates binding of RNA polymerase at the promoter but precludes transcription.

26.10 The Operator Competes with Low-Affinity Sites to Bind Repressor

- Proteins that have a high affinity for a specific DNA sequence also have a low affinity for other DNA sequences.
- Every base pair in the bacterial genome is the start of a low-affinity binding site for repressor.
- The large number of low-affinity sites ensures that all repressor protein is bound to DNA.
- Repressor binds to the operator by moving from a low-affinity site rather than by equilibrating from solution.
- In the absence of inducer, the operator has an affinity for repressor that is 10^7 times that of a low-affinity site.
- The level of 10 repressor tetramers per cell ensures that the operator is bound by repressor 96% of the time.
- Induction reduces the affinity for the operator to 10^4 times that of low-affinity sites, so that operator is bound only 3% of the time.

26.11 The *lac* Operon Has a Second Layer of Control: Catabolite Repression

- CRP is an activator protein that binds to a target sequence at a promoter.
- A dimer of CRP is activated by a single molecule of cAMP.
- cAMP is controlled by the level of glucose in the cell; a low glucose level allows cAMP to be made.
- CRP interacts with the C-terminal domain of the α subunit of RNA polymerase to activate it.

26.12 The *trp* Operon Is a Repressible Operon with Three Transcription Units

- The *trp* operon is negatively controlled by the level of its product, the amino acid tryptophan.
- The amino acid tryptophan activates an inactive repressor encoded by *trpR*.
- A repressor (or activator) will act on all loci that have a copy of its target operator sequence.

26.13 The *trp* Operon Is Also Controlled by Attenuation

- An attenuator (intrinsic terminator) is located between the promoter and the first gene of the *trp* cluster.
- The absence of Trp-tRNA suppresses termination and results in a 10× increase in transcription.

26.14 Attenuation Can Be Controlled by Translation

- The leader region of the *trp* operon has a fourteen-codon open reading frame that includes two codons for tryptophan.
- The structure of RNA at the attenuator depends on whether this reading frame is translated.

- In the presence of Trp-tRNA, the leader is translated, and the attenuator is able to form the hairpin that causes termination.
- In the absence of Trp-tRNA, the ribosome stalls at the tryptophan codons and an alternative secondary structure prevents formation of the hairpin, so that transcription continues.

26.15 Translation Can Be Regulated

- Translation can be regulated by the 5' UTR of the mRNA.
- Translation may be regulated by the abundance of various tRNAs.

- A repressor protein can regulate translation by preventing a ribosome from binding to an initiation codon.
- Accessibility of initiation codons in a polycistronic mRNA can be controlled by changes in the structure of the mRNA that occur as the result of translation.

26.16 r-Protein Synthesis Is Controlled by Autoregulation

- Translation of an r-protein operon can be controlled by a product of the operon that binds to a site on the polycistronic mRNA.

26.17 Summary

26.1 Introduction

Key concepts

- In negative regulation, a repressor protein binds to an operator to prevent a gene from being expressed.
- In positive regulation, a transcription factor is required to bind at the promoter in order to enable RNA polymerase to initiate transcription.
- In inducible regulation, the gene is regulated by the presence of its substrate.
- In repressible regulation, the gene is regulated by the product of its enzyme pathway.
- Gene regulation *in vivo* can utilize any of these mechanisms, resulting in all four combinations: negative inducible, negative repressible, positive inducible, and positive repressible.

Gene expression can be controlled at any of several stages, which we divide broadly into transcription, processing, and translation:

- Transcription often is controlled at the stage of initiation. Transcription is not usually controlled at elongation, but may be controlled at termination to determine whether RNA polymerase is allowed to proceed past a terminator to the gene(s) beyond.
- In bacteria, an mRNA is typically available for translation while it is being synthesized; this is called **coupled transcription/translation**. (In eukaryotic cells, processing of the RNA product may be regulated at the stages of modification, splicing, transport, or stability.)
- Translation in bacteria may also be directly regulated, but more commonly it is passively modulated. The coding portion or open reading frame of a gene can be assembled either with common or rare codons, which correspond to common or rare tRNAs. mRNAs containing a number of rare codons are more difficult to translate.

The basic concept for the way transcription is controlled in bacteria is called the **operon** model and was proposed by François Jacob and Jacques Monod in 1961. They distinguished between two types of sequences in DNA: sequences that code for *trans*-acting products (usually proteins) and *cis*-acting DNA sequences. Gene activity is regulated by the specific interactions of the *trans*-acting products with the *cis*-acting sequences (see *Section 2.12, Proteins Are* trans-*acting, but Sites on DNA Are* cis-*acting*). In more formal terms:

- A gene is a sequence of DNA that codes for a diffusible product, either RNA or a protein. The crucial feature is that the product diffuses away from its site of synthesis to act elsewhere. Any gene product that is free to diffuse to find its target is described as *trans*-acting.
- The description *cis*-acting applies to any sequence of DNA that functions exclusively as a DNA sequence, affecting only the DNA to which it is physically linked.

To help distinguish between the components of regulatory circuits and the genes that they regulate, we sometimes use the terms *structural gene* and *regulator gene*. A **structural gene** is simply any gene that codes for a protein (or RNA) product. Protein structural genes represent an enormous variety of structures and functions, including structural proteins, enzymes with catalytic activities, and regulatory proteins. A type of structural gene is a **regulator gene**, which simply describes a gene that codes for a protein or an RNA involved in regulating the expression of other genes.

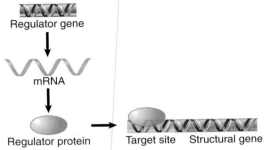

Regulator gene

mRNA

Regulator protein   Target site Structural gene

FIGURE 26.1 A regulator gene codes for a protein that acts at a target site on DNA.

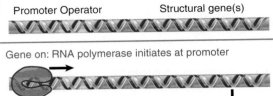

cis-acting operator/promoter precedes structural gene(s)

Promoter Operator   Structural gene(s)

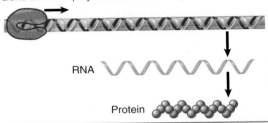

Gene on: RNA polymerase initiates at promoter

RNA

Protein

Gene is turned off when repressor binds to operator
    Repressor

FIGURE 26.2 In negative control, a *trans*-acting repressor binds to the *cis*-acting operator to turn off transcription.

GENE OFF BY DEFAULT

Startpoint Gene

Promoter

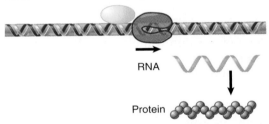

GENE TURNED ON BY ACTIVATORS

Factors interact with RNA polymerase

RNA

Protein

FIGURE 26.3 In positive control, a *trans*-acting factor must bind to *cis*-acting site in order for RNA polymerase to initiate transcription at the promoter.

The simplest form of the regulatory model is illustrated in **FIGURE 26.1**: *a regulator gene codes for a protein that controls transcription by binding to particular site(s) on DNA.* This interaction can regulate a target gene in either a positive manner (the interaction turns the gene on) or a negative manner (the interaction turns the gene off). The sites on DNA are usually (but not exclusively) located just upstream of the target gene.

The sequences that mark the beginning and end of the transcription unit—the promoter and terminator—are examples of *cis*-acting sites. *A promoter serves to initiate transcription only of the gene or genes physically connected to it on the same stretch of DNA.* In the same way, a terminator can terminate transcription only by an RNA polymerase that has traversed the preceding gene(s). In their simplest forms, promoters and terminators are *cis*-acting elements that are recognized by the same *trans*-acting species; that is, by RNA polymerase (although other factors also participate at each site).

Additional *cis*-acting regulatory sites are often combined with the promoter. A bacterial promoter may have one or more such sites located close by; that is, in the immediate vicinity of the startpoint. A eukaryotic promoter is likely to have a greater number of sites that are spread out over a longer distance, as we will see in *Section 28.5, Activators Interact with the Basal Apparatus.*

A classic mode of transcription control in bacteria is **negative control**: a repressor protein prevents a gene from being expressed. **FIGURE 26.2** shows that in the absence of the negative regulator, the gene is expressed. Close to the promoter is another *cis*-acting site called the **operator**, which is the binding site for the repressor protein. When the repressor binds to the operator, RNA polymerase is prevented from initiating transcription, and *gene expression is therefore turned off.* An alternative mode

of control is **positive control**. This is used in bacteria (probably) with about equal frequency to negative control, and it is the most common mode of control in eukaryotes. *A transcription factor is required to assist RNA polymerase in initiating at the promoter.* **FIGURE 26.3** shows that in the absence of the positive regulator, the gene is inactive: RNA polymerase cannot by itself initiate transcription at the promoter.

In addition to negative and positive control, a gene that encodes an enzyme may be regulated by the concentration of its substrate or product (or a chemical derivative of either). Bacteria need to respond swiftly to changes in their environment. Fluctuations in the supply of nutrients (such as the sugars glucose or lactose) can occur at any time, and survival depends on the ability to switch from metabolizing one substrate to another. Yet economy is important, too: a bacterium that indulges in energetically expensive ways to meet the demands of the environment is likely to be at a disadvantage. Thus a bacterium avoids synthesizing the enzymes of a pathway in the absence of the substrate, but is ready to produce the enzymes if the substrate should appear. *The synthesis of enzymes in response to the appearance of a specific substrate is called* **induction** *and the gene is an* **inducible gene***.*

The opposite of induction is **repression**, where *the* **repressible gene** *is controlled by the amount of the product made by the enzyme.* For example, *E. coli* synthesizes the amino acid tryptophan through the actions of an enzyme complex containing tryptophan synthetase and four other enzymes. If, however, tryptophan is provided in the medium on which the bacteria are growing, the production of the enzyme is immediately halted. This allows the bacterium to avoid devoting its resources to unnecessary synthetic activities.

Induction and repression represent similar phenomena. In one case the bacterium adjusts its ability to use a given substrate (such as lactose) for growth; in the other it adjusts its ability to synthesize a particular metabolic intermediate (such as an essential amino acid). The trigger for either type of adjustment is a small molecule that is the substrate (or related to the substrate) for the enzyme, or the product of the enzyme activity, respectively. Small molecules that cause the production of enzymes that are able to metabolize them (or their analogues) are called **inducers**. Those that prevent the production of enzymes that are able to synthesize them are called **corepressors**.

These two ways of looking at regulation— negative versus positive control and inducible versus repressible control—are typically combined to give four different patterns of gene regulation: **negative inducible**, **negative repressible**, **positive inducible**, and **positive repressible**, as shown in **FIGURE 26.4**. This enables a bacterium to perform the ultimate in inventory control of its metabolism to allow survival in rapidly changing environments.

The unifying theme is that regulatory proteins are *trans*-acting factors that recognize *cis*-

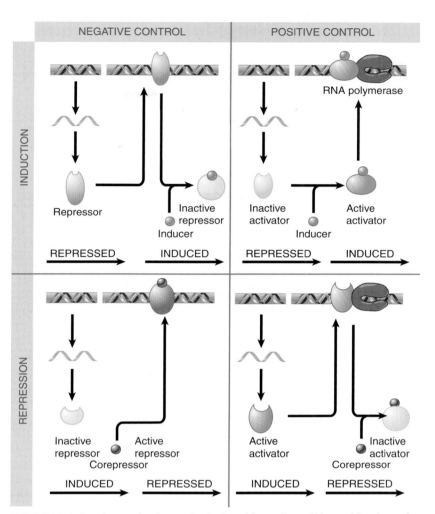

FIGURE 26.4 Regulatory circuits can be designed from all possible combinations of positive and negative control with inducible and repressible control.

acting elements (usually) upstream of the gene. The consequences of this recognition are either to activate or to repress the gene, depending on the individual type of regulatory protein. A typical feature is that the protein functions by recognizing a very short sequence in DNA, usually <10 bp in length, although the protein actually binds over a somewhat greater distance of DNA. The bacterial promoter is an example: RNA polymerase covers >70 bp of DNA at initiation, but the crucial sequences that it recognizes are the hexamers centered at −35 and −10.

A significant difference in gene organization between prokaryotes and eukaryotes is that structural genes in bacteria are organized in operons that are coordinately controlled by means of interactions at a single regulator. In contrast, genes in eukaryotes are controlled individually. As a result, an entire related set of bacterial genes is either transcribed or not transcribed. In this chapter, we discuss this mode of control and its use by bacteria. The means employed to coordinate control of dispersed eukaryotic genes are discussed in Chapter 20, *Eukaryotic Transcription*.

26.2 Structural Gene Clusters Are Coordinately Controlled

Key concept

- Genes coding for proteins that function in the same pathway may be located adjacent to one another and controlled as a single unit that is transcribed into a polycistronic mRNA.

Bacterial genes are often organized into operons that include genes coding for proteins whose functions are related. The genes coding for the enzymes of a metabolic pathway are commonly organized into such a cluster. In addition to the enzymes actually involved in the pathway, other related activities may be included in the unit of coordinated control, such as the protein responsible for transporting the small molecule substrate into the cell.

The cluster of the *lac* operon containing the three *lac* structural genes, *lacZ*, *lacY*, and *lacA*, is typical. **FIGURE 26.5** summarizes the organization of the structural genes, their associated *cis*-acting regulatory elements, and the *trans*-acting regulatory gene. *The key feature is that the structural gene cluster is transcribed into a single **polycistronic mRNA** from a promoter where initiation of transcription is regulated.*

The protein products enable cells to take up and metabolize β-galactoside sugars, such as lactose. The roles of the three structural genes are:

- *lacZ* codes for the enzyme β-galactosidase, whose active form is a tetramer of ~500 kD. The enzyme breaks the complex β-galactoside into its component sugars. For example, lactose is cleaved into glucose and galactose (which are then further metabolized). This enzyme also produces an important by-product, β-1,6-allolactase, which we will see below has a role in regulation.
- *lacY* codes for the β-galactoside permease, a 30-kD membrane-bound protein constituent of the transport system. This transports β-galactosides into the cell.
- *lacA* codes for β-galactoside transacetylase, an enzyme that transfers an acetyl group from acetyl-CoA to β-galactosides.

Mutations in either *lacZ* or *lacY* can create the *lac* genotype, in which cells cannot utilize lactose. (The genotypic description "lac" without a qualifier indicates loss-of-function.) The *lacZ* mutations abolish enzyme activity, directly preventing metabolism of lactose. The *lacY* mutants cannot take up lactose efficiently from the medium. (No defect is identifiable in *lacA* cells, which is puzzling. The acetylation reaction might give an advantage when the bacteria grow in the presence of certain analogs of β-galactosides that cannot be metabolized, because the modification results in detoxification and excretion.)

The entire system, including structural genes and the elements that control their expression, forms a common unit of regulation called an operon. The activity of the operon is controlled by regulator gene(s) whose protein products interact with the *cis*-acting control elements.

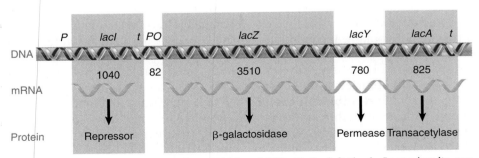

FIGURE 26.5 The *lac* operon occupies ~6000 bp of DNA. At the left the *lacI* gene has its own promoter and terminator. The end of the *lacI* region is adjacent to the *lacZYA* promoter, *P*. Its operator, *O*, occupies the first 26 bp of the transcription unit. The long *lacZ* gene starts at base 39, and is followed by the *lacY* and *lacA* genes and a terminator.

26.3 The *lac* Operon Is Negative Inducible

Key concepts

- Transcription of the *lacZYA* operon is controlled by a repressor protein that binds to an operator that overlaps the promoter at the start of the cluster.
- In the absence of β-galactosides, the *lac* operon is expressed only at a very low (basal) level.
- The repressor protein is a tetramer of identical subunits coded by the *lacI* gene.
- β-galactoside sugars, the substrates of the *lac* operon, are its inducer.
- Addition of specific β-galactosides induces transcription of all three genes of the *lac* operon.
- The *lac* mRNA is extremely unstable; as a result, induction can be rapidly reversed.

We can distinguish between structural genes and regulator genes by the effects of mutations. A mutation in a structural gene deprives the cell of the particular protein for which the gene codes. A mutation in a regulator gene, however, influences the expression of all the structural genes that it controls. The consequences of a regulatory mutation reveal the type of regulation.

Transcription of the *lacZYA* genes is controlled by a regulator protein encoded by the *lacI* gene. Although adjacent to the structural genes, *lacI* comprises an independent transcription unit with its own promoter and terminator. In principle, *lacI* need not be located near the structural genes because it specifies a diffusible product. The *lacI* gene can function equally well if moved elsewhere, or can be carried on a separate DNA molecule (the classic test for a *trans*-acting regulator).

The *lacZYA* genes are negatively regulated: *they are transcribed unless turned off by the regulator protein*. Note that repression is not an absolute phenomenon; turning off a gene is not like turning off a lightbulb. Repression can often be a reduction in transcription by five-fold or 100-fold. A mutation that inactivates the regulator causes the structural genes to be continually expressed, a condition called **constitutive expression**. The product of *lacI* is called the ***lac* repressor**, because its function is to prevent the expression of the *lacZYA* structural genes.

The repressor is a tetramer of identical subunits of 38 kD each. A wild-type cell contains ~10 tetramers. The repressor gene is not controlled; it is an unregulated gene. It is transcribed into a monocistronic mRNA at a rate that appears to be governed simply by the affinity of its (poor) promoter for RNA poly-

merase. In addition, *lacI* is transcribed into a poor mRNA. This is a common way to restrict the amount of protein made. In this case, the mRNA has virtually no 5′ UTR, which restricts the ability of a ribosome to start translation. These two features account for the low abundance of *lac* repressor protein in the cell.

The repressor functions by binding to an operator (formally denoted O$_{lac}$) at the start of the *lacZYA* cluster. The sequence of the operator includes an inverted repeat. The operator lies between the promoter (P$_{lac}$) and the structural genes (*lacZYA*). *When the repressor binds at the operator, it prevents RNA polymerase from initiating transcription at the promoter.* FIGURE 26.6 expands our view of the region at the start of the *lac* structural genes. The operator extends from position −5 just upstream of the mRNA startpoint to position +21 within the transcription unit; thus it overlaps the 3′, right end of the promoter. A mutation that inactivates the operator also causes constitutive expression.

When cells of *E. coli* are grown in the absence of a β-galactoside they have no need for β-galactosidase, and they contain very few molecules of the enzyme, about five per cell. When a suitable substrate is added, the enzyme activity appears very rapidly in the bacteria. Within two to three minutes some enzyme is present, and soon each bacterium accumulates ~5000 molecules of enzyme. (Under suitable conditions, β-galactosidase can account for 5% to 10% of the total soluble protein of the bacterium.) If the substrate is removed from the medium, the synthesis of enzyme stops as rapidly as it started.

FIGURE 26.7 summarizes the essential features of this induction. Control of transcription of the *lac* operon responds very rapidly to the inducer, as shown in the upper part of the figure. In the absence of inducer, the operon

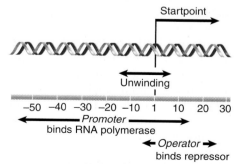

FIGURE 26.6 *lac* repressor and RNA polymerase bind at sites that overlap around the transcription startpoint of the *lac* operon.

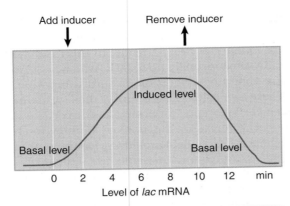

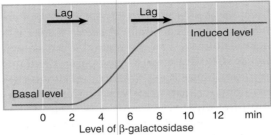

FIGURE 26.7 Addition of inducer results in rapid induction of *lac* mRNA, and is followed after a short lag by synthesis of the enzymes; removal of inducer is followed by rapid cessation of synthesis.

is transcribed at a very low basal level (this is an important concept; see the next section). Transcription is stimulated as soon as inducer is added; the amount of *lac* mRNA increases rapidly to an induced level that reflects a balance between synthesis and degradation of the mRNA.

The *lac* mRNA (as most mRNA is in bacteria) is extremely unstable and decays with a half-life of only ~3 minutes. This feature allows induction to be reversed rapidly by repressing transcription as soon as the inducer is removed. In a very short time all the *lac* mRNA is destroyed and enzyme synthesis ceases.

The production of protein is followed in the lower part of the figure. Translation of the *lac* mRNA produces β-galactosidase (and the products of the other *lac* genes). A short lag occurs between the appearance of *lac* mRNA and appearance of the first completed enzyme molecules (~2 minutes lapses between the rise of mRNA from basal level and increased protein level). There is a similar lag between reaching maximal induced levels of mRNA and protein. When inducer is removed, synthesis of enzyme ceases almost immediately (as the *lacZYA* mRNA is quickly degraded), but the β-galactosidase in the cell is more stable, so that the enzyme activity remains at the induced level for longer.

26.4 *lac* Repressor Is Controlled by a Small-Molecule Inducer

Key concepts

- An inducer functions by converting the repressor protein into a form with lower operator affinity.
- Repressor has two binding sites, one for the operator DNA and another for the inducer.
- Repressor is inactivated by an allosteric interaction in which binding of inducer at its site changes the properties of the DNA-binding site.
- The true inducer is allolactose, not the actual substrate of β-galactosidase.

The ability to act as inducer or corepressor is highly specific. Only the substrate/product of the regulated enzymes or a closely related molecule can serve. In most cases, though, the activity of the small molecule does not depend on its interaction with the target enzyme. For the *lac* system the natural inducer is not lactose, but a byproduct of the lacZ enzyme, **allolactose**. Allolactose is also a substrate of the *lacZ* enzyme, so it does not persist in the cell. Some inducers resemble the natural inducers of the *lac* operon but cannot be metabolized by the enzyme. The example *par excellence* is isopropylthiogalactoside (IPTG), one of several thiogalactosides with this property. IPTG is not metabolized by β-galactosidase; even so, it is a very efficient inducer of the *lac* genes.

Molecules that induce enzyme synthesis but are not metabolized are called **gratuitous inducers**. The existence of gratuitous inducers reveals an important point. The system must possess some component, distinct from the target enzyme, that recognizes the appropriate substrate, and its ability to recognize related potential substrates is different from that of the enzyme. The separate component that represses the *lac* operon is the *lac* repressor protein, which is encoded by the *lacI* gene. The *lac* repressor protein is induced by allolactose and IPTG to allow expression of *lacZYA*. The LacZ enzyme (β-galactosidase) utilizes allolactose and lactose as substrates. *lacI* is not induced by lactose, and the LacZ enzyme does not metabolize IPTG.

The component that responds to the inducer is the repressor protein encoded by *lacI*. Its target, the *lacZYA* structural genes, is transcribed into a single mRNA from the promoter just upstream of *lacZ*. The state of the repressor determines whether this promoter is turned off or on.

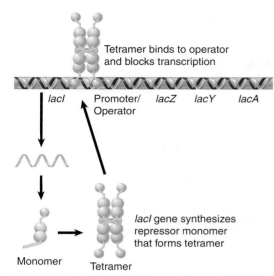

FIGURE 26.8 *lac* repressor maintains the *lac* operon in the inactive condition by binding to the operator. The shape of the repressor is represented as a series of connected domains as revealed by its crystal structure (see Figure 26.14).

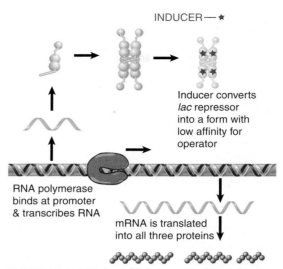

FIGURE 26.9 Addition of inducer converts repressor to a form with low affinity for the operator. This allows RNA polymerase to initiate transcription.

- **FIGURE 26.8** shows that in the absence of an inducer the genes are not transcribed, because repressor protein is in an active form that is bound to the operator.
- **FIGURE 26.9** shows that when an inducer is added, the repressor is converted into either a form with lower affinity for operator or a lower affinity form that leaves the operator. Transcription then starts at the promoter and proceeds through the genes to a terminator located beyond the 3' end of *lacA*.

The crucial features of the control circuit reside in the dual properties of the repressor: it can prevent transcription, and it can recognize the small-molecule inducer. The repressor has two types of binding site: one type for the operator DNA and one type for the inducer. When the inducer binds at its site, it changes the structure of the protein in such a way as to influence the activity of the operator-binding site. The ability of one site in the protein to control the activity of another is called **allosteric control**.

Induction accomplishes a coordinate regulation: all the genes are expressed (or not expressed) in unison. The mRNA is translated sequentially from its 5' end, which explains why induction always causes the appearance of β-galactosidase, β-galactoside permease, and β-galactoside transacetylase, in that order. Translation of a common mRNA explains why the relative amounts of the three enzymes always remain the same under varying conditions of induction. Usually, the most important enzyme is first in the operon.

The constitution of the *lac* operon has several potential paradoxes. First, the *lac* operon contains the structural gene (*lacZ*) coding for the β-galactosidase activity needed to metabolize the sugar; it also includes the gene (*lacY*) that codes for the protein needed to transport the substrate into the cell. If the operon is in a repressed state, how does the inducer enter the cell to start the process of induction? The second paradox is that β-galactosidase (encoded by *lacZ*) is required to make the inducer allolactose to induce the synthesis of β-galactosidase. How is allolactose synthesized to allow induction of the gene? (An operon with a mutant *lacZ* gene cannot be induced.)

Two features ensure induction of the *lac* operon. First, the operon has a basal level of expression, ensuring that a minimal amount of LacZ and LacY proteins are present in the cell—enough to start the process. Even when the *lac* operon is not induced, it is expressed at a residual level (0.1% of the induced level). In addition, some inducer enters the cell via another uptake system. The basal level of β-galactosidase then converts some lactose to allolactose, leading to induction of the *lac* operon.

26.5 cis-Acting Constitutive Mutations Identify the Operator

Key concepts

- Mutations in the operator cause constitutive expression of all three *lac* structural genes.
- These mutations are *cis*-acting and affect only those genes on the contiguous stretch of DNA.
- Mutations in the promoter prevent expression of *lacZYA* are uninducible and *cis*-acting.

Mutations in the regulatory circuit may either abolish expression of the operon or cause constitutive expression. Mutants that cannot be expressed at all are called **uninducible**. Mutants that are continuously expressed are called *constitutive mutants.*

Components of the regulatory circuit of the operon can be identified by mutations that (1) affect the expression of all the regulated structural genes, and (2) map outside them. They fall into two classes, *cis*-acting and *trans*-acting. The promoter and the operator are identified as targets for the regulatory proteins (RNA polymerase and repressor, respectively) by *cis*-acting mutations. The locus *lacI* is identified to code for the repressor protein by mutations that eliminate the *trans*-acting product.

The operator was originally identified by constitutive mutations, denoted O^c, whose distinctive properties provided the first evidence for an element that functions without being represented in a diffusible product. The structural genes contiguous with an O^c mutation are expressed constitutively because the mutation changes the operator so that the repressor no longer binds to it. Thus the repressor cannot prevent RNA polymerase from initiating transcription. The operon is transcribed constitutively, as illustrated in **FIGURE 26.10**.

The operator can control only the *lac* genes that are adjacent to it. If a second *lac* operon is introduced into the bacterium on an independent molecule of DNA, it has its own operator. Neither operator is influenced by the other. Thus if one operon has a wild-type operator it will be repressed under the usual conditions, whereas a second operon with an O^c mutation will be expressed in its characteristic fashion.

Promoter mutations are also *cis*-acting. If they prevent RNA polymerase from binding at P$_{lac}$, the structural genes are never transcribed. These mutations are described as being uninducible. Like O^c mutations, mutations in the promoter only affect contiguous structural

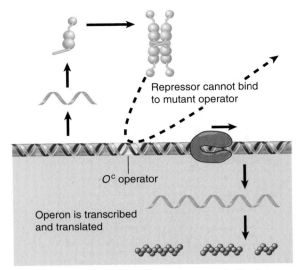

FIGURE 26.10 Operator mutations are constitutive because the operator is unable to bind repressor protein; this allows RNA polymerase to have unrestrained access to the promoter. The O^c mutations are *cis*-acting, because they affect only the contiguous set of structural genes.

genes and cannot be substituted with another promoter that is present on an independent molecule of DNA.

These properties define the operator as a typical *cis*-acting site, whose function depends upon recognition of its DNA sequence by some *trans*-acting factor. The operator controls the adjacent genes irrespective of the presence in the cell of other alleles of the site. A mutation in such a site—for example, the O^c mutation—is formally described as **cis-dominant**.

26.6 trans-Acting Mutations Identify the Regulator Gene

Key concepts

- Mutations in the *lacI* gene are *trans*-acting and affect expression of all *lacZYA* clusters in the bacterium.
- Mutations that eliminate *lacI* function cause constitutive expression and are recessive (*lacI*⁻).
- Mutations in the DNA-binding site of the repressor are constitutive because the repressor cannot bind the operator.
- Mutations in the inducer-binding site of the repressor prevent it from being inactivated and cause uninducibility.
- When mutant and wild-type subunits are present, a single *lacI*$^{-d}$ mutant subunit can inactivate a tetramer whose other subunits are wild-type.
- *lacI*$^{-d}$ mutations occur in the DNA-binding site. Their effect is explained by the fact that repressor activity requires all DNA-binding sites in the tetramer to be active.

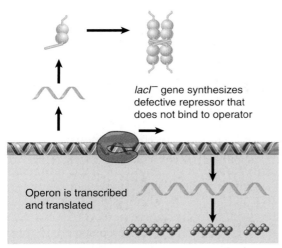

lacI⁻ gene synthesizes defective repressor that does not bind to operator

Operon is transcribed and translated

FIGURE 26.11 Mutations that inactivate the *lacI* gene cause the operon to be constitutively expressed, because the mutant repressor protein cannot bind to the operator.

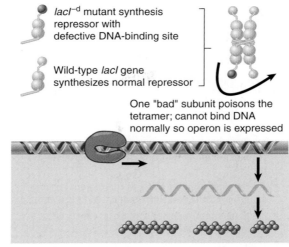

lacI⁻ᵈ mutant synthesis repressor with defective DNA-binding site

Wild-type *lacI* gene synthesizes normal repressor

One "bad" subunit poisons the tetramer; cannot bind DNA normally so operon is expressed

FIGURE 26.12 A *lacI*⁻ᵈ mutant gene makes a monomer that has a damaged DNA binding (shown by the red circle). When it is present in the same cell as a wild type gene, multimeric repressors are assembled at random from both types of subunits. It only requires one of the subunits of the multimer to be of the *lacI*⁻ᵈ type to block repressor function. This explains the dominant negative behavior of the *lacI*⁻ᵈ mutation.

Two types of constitutive mutations can be distinguished genetically. O^c mutants are *cis*-dominant, whereas *lacI*⁻ mutants are recessive. This means that the introduction of a normal, *lacI*⁺ gene can restore control, even in the presence of a defective *lacI*⁻ gene. The *lac* repressor protein is diffusible; thus the normal *lacI* gene can be placed on an independent molecule of DNA. Other *lacI* mutations can cause the operon to be uninducible (unable to be turned on, denoted *lacI*ˢ), similar to mutations in the promoter.

Constitutive transcription is caused by mutations of the *lacI*⁻ type, which are caused by loss of DNA-binding function (including deletions of the gene). When the repressor is inactive or absent, transcription of the lac operon can initiate at the lac operon promoter. **FIGURE 26.11** shows that the *lacI*⁻ mutants express the structural genes all the time (constitutively), irrespective of whether the inducer is present or absent, because the repressor is inactive. One important subset of *lacI*⁻ mutations (called *lacI*⁻ᵈ; see below) is localized in the DNA-binding site of the repressor. The *lacI*⁻ᵈ mutations abolish the ability to turn off the gene by damaging the site that the repressor uses to contact the operator. They are dominant mutations because a mixed tetramer with both normal and mutant repressor subunits cannot bind the operator (see below).

Uninducible mutants are caused by mutations that abolish the ability of repressor to bind or to respond to the inducer. They are described as *lacI*ˢ. The repressor is "locked in" to the active form that recognizes the operator and prevents transcription. These mutations identify the inducer-binding site and other positions involved in allosteric control of the

DNA-binding site. The mutant repressor binds to all *lac* operators in the cell to prevent their transcription, and cannot be removed from the operator, even if wild-type protein is present.

An important feature of the repressor protein is that it is multimeric. Repressor subunits associate at random in the cell to form the active tetramer. When two different alleles of the *lacI* gene are present, the subunits made by each can associate to form a heterotetramer, whose properties differ from those of either homotetramer. This type of interaction between subunits is a characteristic feature of multimeric proteins and is described as **interallelic complementation**.

Most *lacI*⁻ mutations inactivate the repressor. Thus these genes are recessive when coexpressed with the wild-type repressor and the *lac* operon is normally regulated. Combinations of certain repressor mutants, however, display a form of interallelic complementation called **negative complementation**. As mentioned above, *lacI*⁻ᵈ mutations are dominant when paired with a wild-type allele. Such mutations are called **dominant negative** as seen in **FIGURE 26.12**. The reason for their behavior is that one mutant subunit in a tetramer can antagonize the function of the wild-type subunits, as discussed in the next section. The *lacI*⁻ᵈ mutation alone results in the production of a repressor that cannot bind the operator, and it is therefore constitutive like other *lacI*⁻ alleles.

26.7 *lac* Repressor Is a Tetramer Made of Two Dimers

Key concepts

- A single repressor subunit can be divided into the N-terminal DNA-binding domain, a hinge, and the core of the protein.
- The DNA-binding domain contains two short α-helical regions that bind the major groove of DNA.
- The inducer-binding site and the regions responsible for multimerization are located in the core.
- Monomers form a dimer by making contacts between core subdomains 1 and 2.
- Dimers form a tetramer by interactions between the tetramerization helices.
- Different types of mutations occur in different domains of the repressor protein.

The repressor protein has several domains, as shown in the crystal structure illustrated in **FIGURE 26.13**. A major feature is that the DNA-binding domain is separate from the rest of the protein.

The DNA-binding domain occupies residues 1–59. It contains two α-helices separated by a turn. This is a common DNA-binding motif known as the HTH (helix-turn-helix); the two α-helices fit into the major groove of DNA, where they make contacts with specific bases (see *Section 27.11, Lambda Repressor Uses a Helix-Turn-Helix Motif to Bind DNA*). This region is connected by a hinge sequence to the main body of the protein. In the DNA-binding form of repressor, the hinge forms a small α-helix (as shown in Figure 26.13); but when the repressor is not bound to DNA, this region is disordered. The HTH and hinge are sometimes referred to as the *headpiece*.

The remainder of the protein is called the "core." The bulk of the core consists of two interconnected regions with similar structures (core subdomains 1 and 2). Each has a six-stranded parallel β-sheet sandwiched between two α-helices on either side. The inducer binds in a cleft between the two regions. Two monomer core domains can associate to form a dimeric version of LacI. Dimeric LacI tightly binds operator DNA because it recognizes both halves of the operator sequence, which is an inverted repeat (see below).

The C-terminus of the monomer contains an α-helix with two leucine heptad repeats. This is the tetramerization domain. The tetramerization helices of four monomers associate to maintain the tetrameric structure. **FIGURE 26.14** shows the structure of the tetrameric core (using a different modeling system from Figure 26.13). It consists, in effect, of two dimers. The body of the dimer contains an interface between the subdomains of the two core monomers and two clefts in which two inducers bind (top). The C-terminal regions of each monomer protrude as helices. (The headpiece would join with the N-terminal regions at the top.) Together, two dimers form a tetramer (center) that is held together by a C-terminal bundle of four helices.

FIGURE 26.15 shows a schematic for how the monomers are organized into the tetramer. Two monomers form a dimer by means of contacts at core subdomains 1 and 2; other contacts occur between their respective tetramerization helices. The dimer has two DNA-binding domains at one end of the structure and the tetramerization helices at the other end. Two dimers then form a tetramer by interactions at the tetramerization interface. Each tetramer has four inducer-binding sites and two DNA-binding sites.

Mutations in the *lac* repressor identified the existence of different domains even before the structure was known. We can now explain the nature of the mutations more fully by reference to the structure, as summarized in **FIGURE 26.16**. Recessive mutations of the *lacI⁻* type can occur anywhere in the bulk of the protein. Basically, any mutation that inactivates the

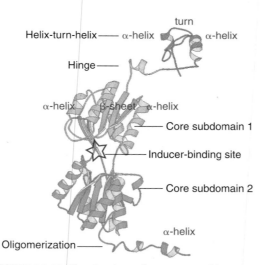

FIGURE 26.13 The structure of a monomer of *lac* repressor identifies several independent domains. Structure from Protein Data Bank 1LBG M. Lewis, et al., *Science* 271 (1996): 1247–1254. Photo courtesy of Hongli Zhan and Kathleen S. Matthews, Rice University.

protein will have this phenotype. The more detailed mapping of mutations on to the crystal structure in Figure 26.14 identifies specific impairments for some of these mutations—for example, those that affect oligomerization.

The special class of dominant-negative *lacI⁻ᵈ* mutations lies in the DNA-binding site of the repressor subunit (see *Section 26.6*, trans-*Acting Mutations Identify Regulator Gene*). This explains their ability to prevent mixed tetramers from binding to the operator; reducing the number of binding sites reduces the specific affinity for the operator. The role of the N-terminal region in specifically binding DNA is also shown by the occurrence of "tight binding" mutations in this region. These rare mutations increase the affinity of the repressor for the operator, sometimes so much that it cannot be released by inducer.

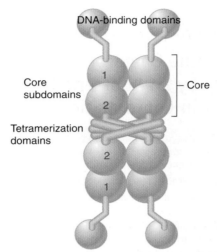

FIGURE 26.15 The repressor tetramer consists of two dimers. Dimers are held together by contacts involving core subdomains 1 and 2 as well as by the tetramerization helix. The dimers are linked into the tetramer by the tetramerization interface.

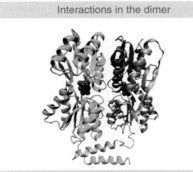

Interactions in the dimer

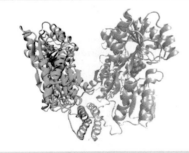

Two dimers make a tetramer

Mutations identify functional sites

FIGURE 26.14 The crystal structure of the core region of *lac* repressor identifies the interactions between monomers in the tetramer. Each monomer is identified by a different color. Mutations are colored as: dimer interface = yellow; inducer-binding = blue; oligomerization = white and purple. The protein orientation in the middle panel is rotated ~90 degrees along the Z axis relative to the top panel. Photos courtesy of Benjamin Wieder and Ponzy Lu, University of Pennsylvania.

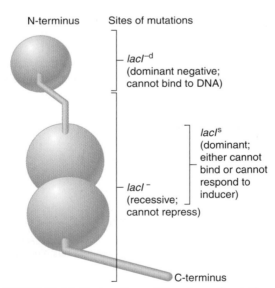

FIGURE 26.16 The locations of three type of mutations in lactose repressor are mapped on the domain structure of the protein. Recessive *lacI⁻* mutants that cannot repress can map anywhere in the protein. Dominant negative *lacI⁻ᵈ* mutants that cannot repress map to the DNA-binding domain. Dominant *lacIˢ* mutants that cannot induce because they do not bind inducer or cannot undergo the allosteric change map to core subdomain 1.

Uninducible *lacI^s* mutations map largely in a region of the core subdomain 1, extending from the inducer-binding site to the hinge. One group lies in amino acids that contact the inducer, and these mutations prevent binding of inducer. The remaining mutations lie at sites that must be involved in transmitting the allosteric change in conformation to the hinge when inducer binds.

26.8 *lac* Repressor Binding to the Operator Is Regulated by an Allosteric Change in Conformation

Key concepts

- *lac* repressor protein binds to the double-stranded DNA sequence of the operator.
- The operator is a palindromic sequence of 26 bp.
- Each inverted repeat of the operator binds to the DNA-binding site of one repressor subunit.
- Inducer binding causes a change in repressor conformation that reduces its affinity for DNA and releases it from the operator.

How does the repressor recognize the specific sequence of operator DNA? The operator has a feature common to many recognition sites for regulator proteins: it is a type of **palindrome** known as an inverted repeat. The inverted repeats are highlighted in **FIGURE 26.17**. Each repeat can be regarded as a half-site of the operator. The symmetry of the operator matches the symmetry of the repressor protein dimer. Each DNA-binding domain of the identical subunits in a repressor can bind one half-site of the operator; two DNA-binding domains of a dimer are required to bind the full-length operator. **FIGURE 26.18** shows that the two DNA-binding domains in a dimeric unit contact DNA by inserting into successive turns of the major groove. This enormously increases affinity for the operator. Note that the *lac* operator is not a perfectly symmetrical sequence; it contains a single central base pair and the sequence of the left side binds to the repressor more strongly than the sequence of the right side. An artificial, perfectly palindromic operator sequence binds to the *lac* repressor protein 10× more tightly than the natural sequence!

The importance of particular bases within the operator sequence can be determined by identifying those that contact the repressor protein or in which mutations change the binding of repressor. The *lac* repressor dimer contacts the operator in such a way that each inverted repeat of the operator makes the same pattern of contacts with a repressor monomer. This is shown by symmetry in the contacts that repressor makes with the operator (the pattern between +1 and +6 is identical with that between +21 and +16) and by matching constitutive mutations in each inverted repeat, as shown in **FIGURE 26.19**. The region of DNA contacted by protein extends for 26 bp, and within this region are eight sites at which constitutive mutations occur. This emphasizes the

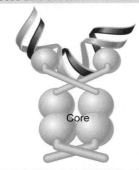

Headpieces bind successive turns in major groove

Core

Inducer binding changes conformation

Inducer

FIGURE 26.18 Inducer changes the structure of the core so that the headpieces of a repressor dimer are no longer in an orientation with high affinity for operator.

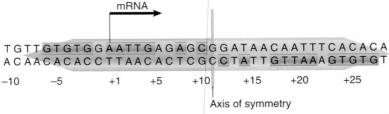

mRNA

T G T T G T G T G G A A T T G A G A G C G G A T A A C A A T T T C A C A C A
A C A A C A C A C C T T A A C A C T C G C C T A T T G T T A A A G T G T G T

−10   −5   +1   +5   +10   +15   +20   +25

Axis of symmetry

FIGURE 26.17 The *lac* operator has a symmetrical sequence. The sequence is numbered relative to the startpoint for transcription at +1. The pink arrows to the left and to the right identify the two dyad repeats. The green blocks indicate the positions of identity.

same point made by promoter mutations: *A small number of essential specific contacts within a larger region can be responsible for sequence-specific association of a protein binding to DNA.*

Figure 26.18 shows another key element of repressor-operator binding: the insertion of the hinge helix into the minor groove of operator DNA, which bends the DNA by ~45°. This bend orients the major groove for HTH binding. DNA-bending is commonly seen when a sequence is bound to a regulatory protein, and illustrates the principle that the structure of DNA is more complicated than the canonical double helix.

The interaction between the *lac* repressor protein and the operator DNA is altered when the repressor is induced as shown in **FIGURE 26.20**. Binding of inducer (e.g., allolactose or IPTG) causes an immediate conformational change in the repressor protein. The change probably disrupts the hinge helices, changing the orientation of the headpieces relative to the core, with the result that repressor's affinity for DNA is lowered dramatically. Although the repressor has weak affinity for operator DNA, other sequences of genomic DNA can bind to the repressor with similar affinity. Thus, the operator and other DNA are in competition for the repressor protein. A cell contains much more genomic DNA than the single copy of the operator sequence; as a result, the genomic DNA "wins" the repressor protein, and the operator is vacant.

Some structural and molecular details of induction process remain the subject of active research. The number of inducers that must be bound to a dimer (within the tetramer) in order to cause induction is under debate. The nature of the conformational change caused in *lac* repressor by binding to inducer is also not completely known, because no high-resolution structure has been obtained for the repressor-operator-inducer complex. In the absence of DNA, inducer binding causes a change in the orientation of the core subdomains that are closest to the hinge helices. A similar change might occur when inducer binds to the repressor-operator complex. Such a change could disrupt the relative orientations of the hinge helices, lowering affinity for DNA. Low-resolution structural information of the low affinity repressor-operator-inducer complex shows that conformational changes in induced *lac* repressor are probably not very large.

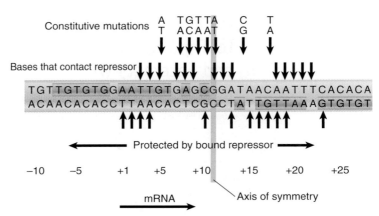

FIGURE 26.19 Bases that contact the repressor can be identified by chemical crosslinking or by experiments to see whether modifications prevent binding. They identify positions on both strands of DNA extending from +1 to +23. Constitutive mutations occur at eight positions in the operator between +5 and +17.

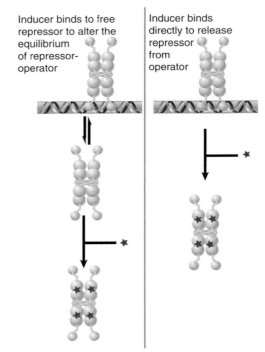

FIGURE 26.20 Does the inducer bind to the free repressor to upset an equilibrium (left) or directly to repressor bound at the operator (right)?

lac Repressor Binds to Three Operators and Interacts with RNA Polymerase

- Each dimer in a repressor tetramer can bind an operator, so that the tetramer can bind two operators simultaneously.
- Full repression requires the repressor to bind to an additional operator downstream or upstream as well as to the primary operator at the *lacZ* promoter.
- Binding of repressor at the operator stimulates binding of RNA polymerase at the promoter but precludes transcription.

The repressor dimer is sufficient to bind the entire operator sequence. Why, then, is a tetramer required to establish full repression?

Each dimer can bind an operator sequence. This enables the intact tetrameric repressor to bind to two operator sites simultaneously. In fact, there are two additional operator sites in the initial region of the *lac* operon. The original operator, *O1*, is located just at the start of the *lacZ* gene. It has the strongest affinity for repressor. Weaker operator sequences are located on either side; *O2* is 410 bp downstream of the startpoint in *lacZ* and *O3* is 88 bp upstream of *lacO¹*, within the *lacI* gene.

FIGURE 26.21 predicts what happens when a DNA-binding protein simultaneously binds to two separated sites on DNA. The DNA between the two sites forms a loop from a base where the protein has bound the two sites. The length of the loop depends on the distance between the two binding sites. When *lac* repressor binds simultaneously to *O1* and to one of the other operators, it causes the DNA between them to form a rather short loop, significantly constraining the DNA structure. A scale model for binding of tetrameric repressor to two operators is shown in **FIGURE 26.22**. Low resolution, looped complexes have been directly visualized with single-molecule experiments.

Binding at the additional operators affects the level of repression. Elimination of either the downstream operator (*O2*) or the upstream operator (*O3*) reduces the efficiency of repression by 2× to 4×. If, however, both *O2* and *O3* are eliminated, repression is reduced more than 50×. *This suggests that the ability of the repressor to bind to one of the two other operators, as well as to O1, is important for establishing strong repression.* *In vitro* experiments with supercoiled plasmids containing multiple operators demonstrate significant stabilization of the lacI-DNA complex. Nonetheless, these looped DNAs are released rapidly when *lac* repressor binds to IPTG.

We have several lines of evidence as to how binding of repressor to the operator (*O1*) inhibits transcription initiation by polymerase. It was originally thought that repressor binding would occlude RNA polymerase from binding to the promoter. We now know that the two proteins may be bound to DNA simultaneously, and that *the binding of repressor actually enhances the binding of RNA polymerase*! The bound enzyme is prevented from initiating transcription, though. The repressor in effect causes RNA polymerase to be stored at the promoter. When inducer is

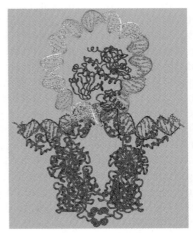

FIGURE 26.22 When a repressor tetramer binds to two operators, the stretch of DNA between them is forced into a tight loop. (The blue structure in the center of the looped DNA represents CRP, which is another regulator protein that binds in this region.) Reproduced from M. Lewis et al., *Science* 271 (1996): 1247–1254 [http://www.sciencemag.org]. Reprinted with permission from AAAS. Photo courtesy of Ponzy Lu, University of Pennsylvania.

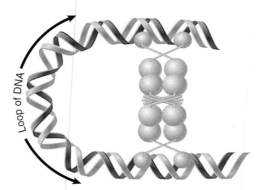

FIGURE 26.21 If both dimers in a repressor tetramer bind to DNA, the DNA between the two binding sites is held in a loop.

added, the repressor is released, and RNA polymerase can initiate transcription immediately. The overall effect of repressor is to speed up the induction process.

Does this model apply to other systems? The interaction between RNA polymerase, repressor, and the promoter/operator region is distinct in each system, because the operator does not always overlap with the same region of the promoter (see Figure 26.24). For example, in phage lambda, the operator lies in the upstream region of the promoter, and binding of lambda repressor occludes the binding of RNA polymerase (see Chapter 27, *Phage Strategies*). Thus a bound repressor does not interact with RNA polymerase in the same way in all systems.

26.10 The Operator Competes with Low-Affinity Sites to Bind Repressor

Key concepts

- Proteins that have a high affinity for a specific DNA sequence also have a low affinity for other DNA sequences.
- Every base pair in the bacterial genome is the start of a low-affinity binding site for repressor.
- The large number of low-affinity sites ensures that all repressor protein is bound to DNA.
- Repressor binds to the operator by moving from a low-affinity site rather than by equilibrating from solution.
- In the absence of inducer, the operator has an affinity for repressor that is 10^7 times that of a low-affinity site.
- The level of 10 repressor tetramers per cell ensures that the operator is bound by repressor 96% of the time.
- Induction reduces the affinity for the operator to 10^4 times that of low-affinity sites, so that operator is bound only 3% of the time.

Probably all proteins that have a high affinity for a specific sequence also possess a low affinity for any random DNA sequence. A large number of low-affinity sites will compete just as well for a repressor as a small number of high-affinity sites. The *E. coli* genome contains only one *lac* operon, which contains the only high-affinity sites. The remainder of the DNA provides low-affinity binding sites. Every base pair in the genome starts a new low-affinity binding site. Simply moving one base pair from the operator creates a low-affinity site! That means that there are 4.2×10^6 low-affinity sites in the *E. coli* genome.

The large number of low-affinity sites means that even in the absence of a specific binding site, almost all of the repressor is bound to DNA, and very little remains free in solution. LacI binding to nonspecific genomic sites has been visualized *in vivo* by single molecule experiments. Using the binding affinities, we can deduce that *all but 0.01% of repressors are bound to random DNA*. There are only about 10 molecules of repressor tetramer per wild-type cell; this says that there is no free repressor protein. Thus, the critical factor of the repressor–operator interaction is the partitioning of the repressor on DNA; the single high-affinity site of the operator must compete with a large number of low-affinity sites.

The efficiency of repression therefore depends on the relative affinity of the repressor for its operator compared with other random DNA sequences. The affinity must be great enough to overcome the large number of random sites. We can see how this works by comparing the equilibrium constants for *lac* repressor/operator binding with repressor/general DNA binding. FIGURE 26.23 shows that the ratio is 10^7 for an active repressor, enough to ensure that the operator is bound by repressor 96% of the time so that transcription is effectively—but not completely—repressed. (Remember that because allolactose is the inducer and not lactose, we always need a little β-galactosidase in the cell.) When inducer is added, the ratio is reduced to 10^4. At this level, only 3% of the operators are bound and the operon is effectively induced.

The consequence of these affinities is that in an uninduced cell, one tetramer of repressor usually is bound to the operator. All, or almost all, of the remaining tetramers are bound at random to other regions of DNA, as illustrated in FIGURE 26.24. There are likely to be very few or no repressor tetramers free within the cell.

The addition of inducer abolishes the ability of repressor to bind specifically at the operator.

DNA	Repressor	Repressor + inducer
Operator	2×10^{13}	2×10^{10}
Other DNA	2×10^6	2×10^6
Specificity	10^7	10^4
Operators bound	96%	3%
Operon is:	repressed	induced

FIGURE 26.23 *lac* repressor binds strongly and specifically to its operator, but is released by inducer. All equilibrium constants are in M^{-1}.

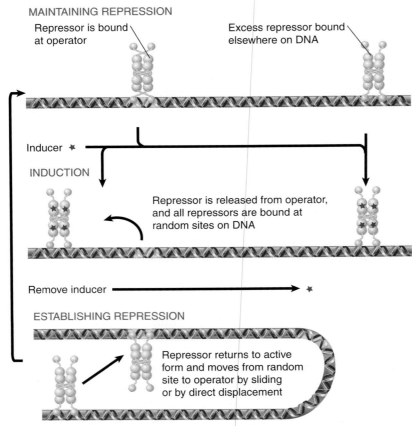

MAINTAINING REPRESSION

Repressor is bound at operator

Excess repressor bound elsewhere on DNA

Inducer ✳

INDUCTION

Repressor is released from operator, and all repressors are bound at random sites on DNA

Remove inducer ⟶ ✳

ESTABLISHING REPRESSION

Repressor returns to active form and moves from random site to operator by sliding or by direct displacement

FIGURE 26.24 Virtually all the repressor in the cell is bound to DNA.

Those repressors bound at the operator are released and bind to random (low-affinity) sites. Thus in an induced cell, the repressor tetramers are "stored" on random DNA sites. In a noninduced cell a tetramer is bound at the operator, whereas the remaining repressor molecules are bound to nonspecific sites. The effect of induction is therefore to change the distribution of repressor on DNA, rather than to generate free repressor. In the same way that RNA polymerase probably moves between promoters and other DNA by swapping one sequence for another, the repressor also may directly displace one bound DNA sequence with another in order to move between sites. We can define the parameters that influence the ability of a regulator protein to saturate its target site by comparing the equilibrium equations for specific and nonspecific binding. As might be expected, the important parameters are:

- The size of the genome dilutes the ability of a protein to bind specific target sites (recall how large eukaryote genomes are).
- The specificity of a protein counters the effect of the mass of the DNA.

- The amount of the protein that is required increases with the total amount of DNA in the genome and decreases the specificity of DNA binding.
- The amount of the protein also must be in reasonable excess of the total number of specific target sites, so we expect regulators with many targets to be found in greater quantities than regulators with fewer targets.

26.11 The *lac* Operon Has a Second Layer of Control: Catabolite Repression

Key concepts

- CRP is an activator protein that binds to a target sequence at a promoter.
- A dimer of CRP is activated by a single molecule of cAMP.
- cAMP is controlled by the level of glucose in the cell; a low glucose level allows cAMP to be made.
- CRP interacts with the C-terminal domain of the α subunit of RNA polymerase to activate it.

The *E. coli lac* operon is negative inducible. Transcription is turned on by the presence of lactose by removing the *lac* repressor. This operon, however, is also under a second layer of control and cannot be turned on by lactose if the bacterium has is a sufficient supply of glucose. The rationale for this is that glucose is a better energy source than lactose, so there is no need to turn on the operon if there is glucose available. This system is part of a global network called **catabolite repression** that affects about 20 genes in *E. coli*. Catabolite repression is exerted through a second messenger called **cyclic AMP (cAMP)** and the positive regulator protein called the **catabolite repressor protein** or **CRP** (also called *catabolite activator protein*, or CAP). The *lac* operon is therefore under dual control.

Thus far we have dealt with the promoter as a DNA sequence that is competent to bind RNA polymerase, which then initiates transcription. Some promoters, though, do not allow RNA polymerase to initiate transcription without assistance from an ancillary protein. Such proteins are positive regulators, because their presence is necessary to switch on the transcription unit. Typically, the activator overcomes a deficiency in the promoter; for example, a poor consensus sequence at −35 or −10.

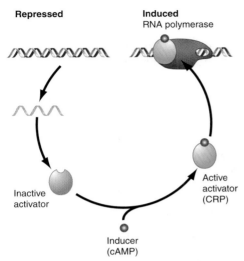

FIGURE 26.25 A small molecule inducer, cAMP, converts an activator protein CRP to a form that binds the promoter and assists RNA polymerase in initiating transcription.

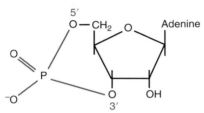

FIGURE 26.26 Cyclic AMP has a single phosphate group connected to both the 3' and 5' positions of the sugar ring.

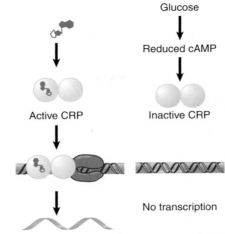

FIGURE 26.27 By reducing the level of cyclic AMP, glucose inhibits the transcription of operons that require CRP activity.

FIGURE 26.28 The consensus sequence for CRP contains the well-conserved pentamer TGTGA and (sometimes) an inversion of this sequence (TCANA).

One of the most widely acting activators is CRP. This protein is a positive regulator whose presence is necessary to initiate transcription at dependent promoters. CRP is active *only in the presence of cAMP*, which behaves as a classic small-molecule inducer for positive control (see **FIGURE 26.25**).

cAMP is synthesized by the enzyme adenylate cyclase. The reaction uses ATP as substrate and introduces an internal 3'–5' link via a phosphodiester bond, which generates the structure drawn in **FIGURE 26.26**. Adenylate cylase activity is repressed by high glucose as shown in **FIGURE 26.27**. Thus, the level of cAMP is inversely related to the level of glucose. Only with low levels of glucose is the enzyme active and able to synthesize cAMP. In turn, cAMP binding is required for CRP to bind DNA and activate transcription. Thus, transcription activation by CRP only occurs when cellular glucose levels are low.

CRP is a dimer of two identical subunits of 22.5 kD, which can be activated by a single molecule of cAMP. A CRP monomer contains a DNA-binding region and a transcription-activating region. A CRP dimer binds to a site of ~22 bp at a responsive promoter. The binding sites include variations of the 5-bp consensus sequence given in **FIGURE 26.28**. Mutations preventing CRP action usually are located within the well-conserved pentamer, which appears to be the essential element in recognition. CRP binds most strongly to sites that contain two (inverted) versions of the pentamer, because this enables both subunits of the dimer to bind to the DNA.

CRP introduces a large bend when it binds DNA. In the *lac* promoter, this point lies at the center of dyad symmetry. The bend is quite severe, >90°, as illustrated in the model of **FIGURE 26.29**. There is, therefore, a dramatic change in the organization of the DNA double helix when CRP protein binds. The mechanism of bending is to introduce a sharp kink within the TGTGA consensus sequence. When there are inverted repeats of the consensus, the two kinks in each copy present in a palindrome cause the overall 90° bend. It is possible that the bend has some direct effect upon transcription, but it could be the case that it is needed simply to allow CRP to contact RNA polymerase at the promoter.

The action of CRP has the curious feature that its binding sites lie at different locations relative to the startpoint in the various operons that it regulates. The TGTGA pentamer may lie in either orientation. The three examples summarized in **FIGURE 26.30** encompass the range of locations:

- The CRP-binding site is adjacent to the promoter, as in the *lac* operon, in which the region of DNA protected by CRP is centered on –61. It is possible that two dimers of CRP are bound. The binding pattern is consistent with the presence of CRP largely on one face of DNA, which is the same face that is bound by RNA polymerase. This location would place the two proteins just about in reach of each other.
- Sometimes the CRP-binding site lies within the promoter, as in the *gal* locus, where the CRP-binding site is centered on –41. It is likely that only a single CRP dimer is bound, probably in quite intimate contact with RNA polymerase,

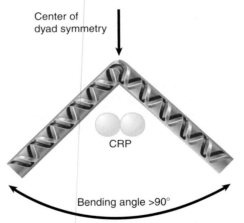

Center of dyad symmetry

CRP

Bending angle >90°

FIGURE 26.29 CRP bends DNA >90° around the center of symmetry.

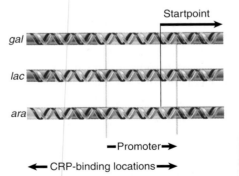

Startpoint

gal

lac

ara

Promoter

CRP-binding locations

FIGURE 26.30 The CRP protein can bind at different sites relative to RNA polymerase.

because the CRP-binding site extends well into the region generally protected by the RNA polymerase.
- In other operons, the CRP-binding site lies well upstream of the promoter. In the *ara* region, the binding site for a single CRP is the farthest from the startpoint, centered at –92.

Dependence on CRP is related to the intrinsic efficiency of the promoter. No CRP-dependent promoter has a good –35 sequence and some also lack good –10 sequences. In fact, we might argue that effective control by CRP would be difficult if the promoter had effective –35 and –10 regions that interacted independently with RNA polymerase.

There are in principle two ways in which CRP might activate transcription: it could interact directly with RNA polymerase, or it could act upon DNA to change its structure in some way that assists RNA polymerase to bind. In fact, CRP has effects upon both RNA polymerase and DNA.

Binding sites for CRP at most promoters resemble either *lac* (centered at –61) or *gal* (centered at –41 bp). The basic difference between them is that in the first type (called class I) the CRP-binding site is entirely upstream of the promoter, whereas in the second type (called class II) the CRP-binding site overlaps the binding site for RNA polymerase. (The interactions at the *ara* promoter may be different.)

In both types of promoter, the CRP binding site is centered an integral number of turns of the double helix from the startpoint. This suggests that CRP is bound to the same face of DNA as RNA polymerase. The nature of the interaction between CRP and RNA polymerase is, however, different at the two types of promoter.

When the α subunit of RNA polymerase has a deletion in the C-terminal end, transcription appears normal except for the loss of ability to be activated by CRP. CRP has an "activating region" that is required for activating both types of its promoters. This activating region, which consists of an exposed loop of ~10 amino acids, is a small patch that interacts directly with the α subunit of RNA polymerase to stimulate the enzyme. At class I promoters, this interaction is sufficient. At class II promoters, a second interaction is required, which involves another region of CRP and the N-terminal region of the RNA polymerase α subunit.

Experiments using CRP dimers in which only one of the subunits has a functional transcription-activating region shows that, when

CRP is bound at the *lac* promoter, only the activating region of the subunit nearer the startpoint is required, presumably because it touches RNA polymerase. This offers an explanation for the lack of dependence on the orientation of the binding site: the dimeric structure of CRP ensures that one of the subunits is available to contact RNA polymerase, no matter which subunit binds to DNA and in which orientation.

The effect upon RNA polymerase binding depends on the relative locations of the two proteins. At class I promoters, where CRP binds adjacent to the promoter, it increases the rate of initial binding to form a closed complex. At class II promoters, where CRP binds within the promoter, it increases the rate of transition from the closed to open complex.

26.12 The *trp* Operon Is a Repressible Operon with Three Transcription Units

Key concepts

- The *trp* operon is negatively controlled by the level of its product, the amino acid tryptophan.
- The amino acid tryptophan activates an inactive repressor encoded by *trpR*.
- A repressor (or activator) will act on all loci that have a copy of its target operator sequence.

The *lac* repressor acts only on the operator of the *lacZYA* cluster. Some repressors, however, control dispersed structural genes by binding at more than one operator. An example is the *trp* repressor (a small 25 kD dimeric protein), which controls three unlinked sets of genes:

- An operator at the cluster of structural genes *trpEDCBA* controls coordinate synthesis of the enzymes that synthesize tryptophan. This is an example of a *repressible operon,* one that is controlled by the product of the operon: tryptophan (see below).

- The *trpR* regulator gene is repressed by its own product, the *trp* repressor. Thus the repressor protein acts to reduce its own synthesis: it is **autoregulated**. (Remember, the *lacI* regulator gene is unregulated.) Such circuits are quite common in regulatory genes and may be either negative or positive (see *Section 26.15, Translation Can Be Regulated,* and *Section 27.13, Lambda Repressor Maintains an Autoregulatory Circuit*).

- An operator at a third locus controls the *aroH* gene, which codes for one of the three isoenzymes that catalyzes the initial reaction in the common pathway of aromatic amino acid biosynthesis leading to the synthesis of tryptophan, phenylalanine, and tyrosine.

A related 21-bp operator sequence is present at each of the three loci at which the *trp* repressor acts. The conservation of sequence is indicated in **FIGURE 26.31**. Each operator contains appreciable (but not identical) dyad symmetry. The features conserved at all three operators include the important points of contact for *trp* repressor. This explains how one repressor protein acts on several loci: each locus has a copy of a specific DNA-binding sequence recognized by the repressor (just as each promoter shares consensus sequences with other promoters).

FIGURE 26.32 summarizes the variety of relationships between operators and promoters. A notable feature of the dispersed operators recognized by TrpR is their presence at different locations within the promoter in each locus. In *trpR* the operator lies between positions −12 and +9, whereas in the *trp* operon it occupies positions −23 to −3. In another gene system, the *aroH* locus, it lies farther upstream, between −49 and −29. In other cases, the operator can lie either downstream from the promoter (as in *lac*), or just upstream of the promoter (as in *gal*, for which the nature of the repressive effect is not quite clear). The ability of the repressors

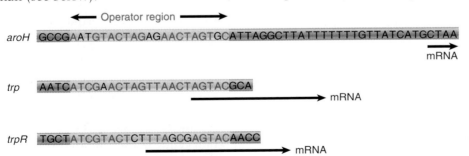

FIGURE 26.31 The *trp* repressor recognizes operators at three loci. Conserved bases are shown in red. The location of the startpoint and mRNA varies, as indicated by the black arrows.

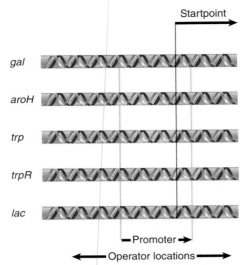

FIGURE 26.32 Operators may lie at various positions relative to the promoter.

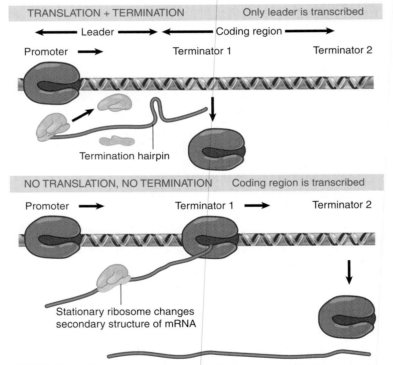

FIGURE 26.33 Termination can be controlled via changes in RNA secondary structure that are determined by ribosome movement.

to act at operators whose positions are different in each target promoter suggests possible differences in the exact mode of repression: the common feature is prevention of RNA polymerase from binding and initiating transcription at the promoter.

The *trp* operon itself is under negative repressible control. This means that the *trpR* gene product, the *trp* repressor, is made as an inactive negative regulator. Repression means

that that the product of the *trp* operon, the amino acid tryptophan is a coregulator for the *trp* repressor. When the level of the amino acid tryptophan builds up, two molecules bind to the dimeric *trp* repressor, changing its conformation to the active DNA-binding conformation and its binding to the operator. This precludes RNA polymerase binding to the overlapping promoter. Up to three *trp* repressor dimers can bind to the operator, depending on the tryptophan concentration and the concentration of repressor. The central dimer binds the tightest.

As we will see in the next section, the *trp* operon is also under dual control (like the *lac* operon above), but the second level is quite different.

26.13 The *trp* Operon Is Also Controlled by Attenuation

Key concepts

- An attenuator (intrinsic terminator) is located between the promoter and the first gene of the *trp* cluster.
- The absence of Trp-tRNA suppresses termination and results in a 10× increase in transcription.

A complex regulatory system of repression and **attenuation** is used in the *E. coli trp* operon (where attenuation was originally discovered). As discussed in the previous section (*Section 26.12, The* trp *Operon Is a Repressible Operon with Three Transcription Units*), the first level of control of gene expression is that the operon is *negative repressible*, which means that it is prevented from initiating transcription by its product, the free amino acid tryptophan. Attenuation is the second level of control. There is a region in the 5' leader of the mRNA called the **attenuator** that contains a small ORF. Attenuation in the *E. coli trp* operon means that *transcription termination is controlled by the rate of translation of the attenuator ORF*. This allows *E. coli* to also monitor the second pool of tryptophan, that of Trp-tRNA. High levels of Trp-tRNA will attenuate or terminate transcription, whereas low levels will allow the *trpEDCBA* operon to be transcribed. This is accomplished by changes in secondary structure of the attenuator RNA that are determined by the position of the ribosome on mRNA. **FIGURE 26.33** shows that termination requires that the ribosome can translate the attenuator. When the ribosome translates the leader region, a termination hairpin forms

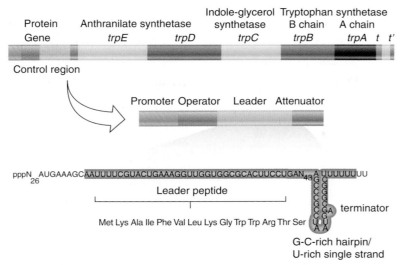

FIGURE 26.34 An attenuator controls the progression of RNA polymerase into the *trp* genes. RNA polymerase initiates at the promoter and then proceeds to position 90, where it pauses before proceeding to the attenuator at position 140. In the absence of tryptophan, the polymerase continues into the structural genes (*trpE* starts at +163). In the presence of tryptophan there is ~90% probability of termination to release the 140-base leader RNA.

at terminator 1. When the ribosome is prevented from translating the leader, though, the termination hairpin does not form, and RNA polymerase transcribes the coding region. *This mechanism of antitermination therefore depends upon the level of Trp-tRNA to influence the rate of ribosome movement in the leader region.*

Attenuation was first revealed by the observation that deleting a sequence between the operator and the *trpE* coding region can increase the expression of the structural genes. This effect is independent of repression: both the basal and derepressed levels of transcription are increased. Thus this site influences events that occur after RNA polymerase has set out from the promoter (irrespective of the conditions prevailing at initiation).

Termination at the attenuator responds to the level of Trp-tRNA, as illustrated in **FIGURE 26.34**. In the presence of adequate amounts of Trp-tRNA, termination is efficient. With low levels of Trp-tRNA, however, RNA polymerase can continue into the structural genes.

Repression and attenuation respond in the same way to the levels of the two pools of tryptophan. When free amino acid tryptophan is present, the operon is repressed. When tryptophan is removed, RNA polymerase has free access to the promoter, and can start transcribing the operon. When Trp-tRNA is present, the operon is attenuated and transcription terminates. When the pool of tryptophan bound to its tRNA is depleted, the

RNA polymerase can continue to transcribe the operon. Note the pool of free tryptophan may be low and allow transcription to begin, but if the Trp-tRNA is fully charged, transcription will terminate.

Attenuation has ~10× effect on transcription. When tryptophan is present termination is effective, and the attenuator allows only ~10% of the RNA polymerases to proceed. In the absence of tryptophan, attenuation allows virtually all of the polymerases to proceed. Together with the ~70× increase in initiation of transcription that results from the release of repression, this allows an ~700-fold range of regulation of the operon.

26.14 Attenuation Can Be Controlled by Translation

Key concepts

- The leader region of the *trp* operon has a fourteen-codon open reading frame that includes two codons for tryptophan.
- The structure of RNA at the attenuator depends on whether this reading frame is translated.
- In the presence of Trp-tRNA, the leader is translated, and the attenuator is able to form the hairpin that causes termination.
- In the absence of Trp-tRNA, the ribosome stalls at the tryptophan codons and an alternative secondary structure prevents formation of the hairpin, so that transcription continues.

TRANSCRIPTION OF LEADER REGION

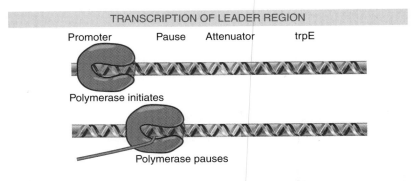

Promoter   Pause   Attenuator   trpE

Polymerase initiates

Polymerase pauses

TRYPTOPHAN ABSENT: TRANSCRIPTION CONTINUES INTO OPERON

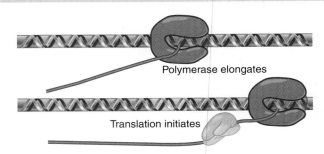

Polymerase elongates

Translation initiates

TRYPTOPHAN PRESENT: TRANSCRIPTION TERMINATES AT ATTENUATOR

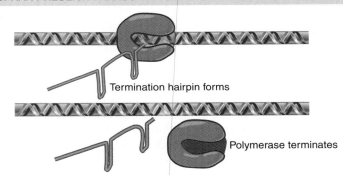

Termination hairpin forms

Polymerase terminates

FIGURE 26.35 The *trp* operon has a short sequence coding for a leader peptide that is located between the operator and the attenuator.

How can termination of transcription at the attenuator respond to the level of Trp-tRNA? The sequence of the leader region suggests a mechanism. It has a short coding sequence that could represent a **leader peptide** of fourteen amino acids. **FIGURE 26.35** shows that it contains a ribosome-binding site whose AUG codon is followed by a short coding region that contains two successive codons for tryptophan. When the cell has a low level of Trp-tRNA, ribosomes initiate translation of the leader peptide but stop when they reach the Trp codons. The sequence of the mRNA suggests that this **ribosome stalling** influences termination at the attenuator.

The leader sequence can be written in alternative base-paired structures. The ability of the ribosome to proceed through the leader

region controls transitions between these structures. The structure determines whether the mRNA can provide the features needed for termination.

FIGURE 26.36 shows these structures. In the first, region 1 pairs with region 2 and region 3 pairs with region 4. The pairing of regions 3 and 4 generates the hairpin that precedes the U_8 sequence: this is the essential signal for intrinsic termination. It is likely that the RNA would form this structure automatically.

A different structure is formed if region 1 is prevented from pairing with region 2. In this case, region 2 is free to pair with region 3. Region 4 then has no available pairing partner, so it is compelled to remain single-stranded. Thus the terminator hairpin cannot be formed.

FIGURE 26.37 shows that the position of the ribosome can determine which structure is formed in such a way that termination is attenuated only in the absence of tryptophan. The crucial feature is the position of the Trp codons in the leader peptide coding sequence.

When Trp-tRNA is abundant, ribosomes are able to synthesize the leader peptide. They continue along the leader section of the mRNA to the UGA codon, which lies between regions 1 and 2. As shown in the lower part of the figure, by progressing to this point, the ribosomes extend over region 2 and prevent it from base pairing. The result is that region 3 is available to base pair with region 4, which generates the terminator hairpin. Under these conditions, therefore, RNA polymerase terminates at the attenuator.

When Trp-tRNA is not abundant, ribosomes stall at the Trp codons, which are part of region 1, as shown in the upper part of the figure. Thus region 1 is sequestered within the ribosome and cannot base pair with region 2. This means that regions 2 and 3 become base-paired before region 4 has been transcribed. This compels region 4 to remain in a single-stranded form. In the absence of the terminator hairpin, RNA polymerase continues transcription past the attenuator.

Control by attenuation requires a precise timing of events. For ribosome movement to determine formation of alternative secondary structures that control termination, translation of the leader must occur at the same time when RNA polymerase approaches the terminator site. A critical event in controlling the timing is the presence of a site that causes the RNA polymerase to pause at base 90 along the leader. The RNA polymerase remains

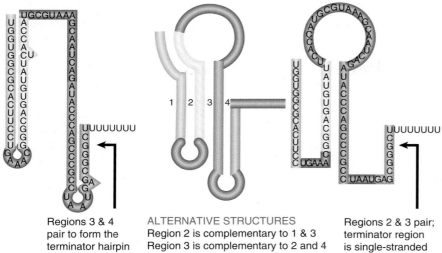

Regions 3 & 4
pair to form the
terminator hairpin

ALTERNATIVE STRUCTURES
Region 2 is complementary to 1 & 3
Region 3 is complementary to 2 and 4

Regions 2 & 3 pair;
terminator region
is single-stranded

FIGURE 26.36 The *trp* leader region can exist in alternative base-paired conformations. The center shows the four regions that can base pair. Region 1 is complementary to region 2, which is complementary to region 3, which is complementary to region 4. On the left is the conformation produced when region 1 pairs with region 2 and region 3 pairs with region 4. On the right is the conformation when region 2 pairs with region 3, leaving regions 1 and 4 unpaired.

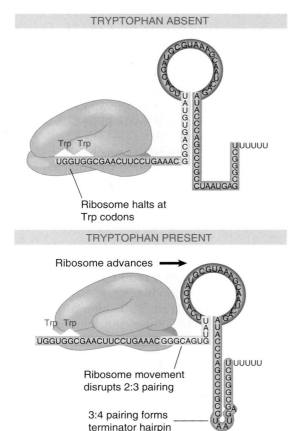

TRYPTOPHAN ABSENT

Ribosome halts at
Trp codons

TRYPTOPHAN PRESENT

Ribosome advances

Ribosome movement
disrupts 2:3 pairing

3:4 pairing forms
terminator hairpin

FIGURE 26.37 The alternatives for RNA polymerase at the attenuator depend on the location of the ribosome, which determines whether regions 3 and 4 can pair to form the terminator hairpin.

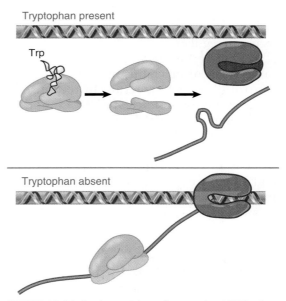

Tryptophan present

Trp

Tryptophan absent

FIGURE 26.38 In the presence of tryptophan tRNA, ribosomes translate the leader peptide and are released. This allows hairpin formation, so that RNA polymerase terminates. In the absence of tryptophan tRNA, the ribosome is blocked, the termination hairpin cannot form, and RNA polymerase continues.

paused until a ribosome translates the leader peptide. The polymerase is then released and moves off toward the attenuation site. By the time it arrives there, the secondary structure of the attenuation region has been determined.

FIGURE 26.38 summarizes the role of Trp-tRNA in controlling expression of the operon.

By providing a mechanism to sense the abundance of Trp-tRNA, attenuation responds directly to the need of the cell for tryptophan in protein synthesis.

How widespread is the use of attenuation as a control mechanism for bacterial operons? It is used in at least six operons that code for enzymes concerned with the biosynthesis of amino acids. Thus a feedback from the level of the amino acid available for protein synthesis (as represented by the availability of aminoacyl-tRNA) to the production of the enzymes may be common.

The use of the ribosome to control RNA secondary structure in response to the availability of an aminoacyl-tRNA establishes an inverse relationship between the presence of aminoacyl-tRNA and the transcription of the operon, which is equivalent to a situation in which aminoacyl-tRNA functions as a corepressor of transcription. The regulatory mechanism is mediated by changes in the formation of duplex regions; thus attenuation provides a striking example of the importance of secondary structure in the termination event and of its use in regulation.

E. coli and *B. subtilis*, therefore, use the same types of mechanisms, which involve control of mRNA structure in response to the presence or absence of a tRNA, but they have combined the individual interactions in different ways. The end result is the same: to inhibit production of the enzymes when there is an excess supply of the amino acid, and to activate production when a shortage is indicated by the accumulation of uncharged tRNA^{Trp}.

26.15 Translation Can Be Regulated

Key concepts

- Translation can be regulated by the 5' UTR of the mRNA.
- Translation may be regulated by the abundance of various tRNAs.
- A repressor protein can regulate translation by preventing a ribosome from binding to an initiation codon.
- Accessibility of initiation codons in a polycistronic mRNA can be controlled by changes in the structure of the mRNA that occur as the result of translation.

Control over which and how much protein is made occurs first at the level of transcription control (as we have just discussed), then through RNA processing control (rare in bacteria, but common in eukaryotes), and then finally translation level control, which we will examine here.

The *lac* repressor is encoded by the *lacI* gene; this is an unregulated gene that is continuously transcribed, but from a poor promoter. Also, the coding region of the *lac* repressor is in a very poor mRNA. This simply means that the 5' UTR (untranslated region) of the mRNA has a poor sequence context that does not allow rapid ribosome binding or movement onto the ORF. Just as we have seen that promoters can be "good" or "poor," so can mRNAs. Together, this means that ribosomes do not translate the small amount of mRNA at the same level as the *LacZYA* polycistronic mRNA. Thus we find very little *lac* repressor in a cell—only about 10 tetramers.

A second way that translation can be modulated is by **codon usage**. Multiple codons exist for most of the amino acids. These codons are not decoded equally by tRNAs. Some have abundant tRNAs and some do not. An ORF constructed from codons with abundant tRNAs can be rapidly translated, whereas another ORF that contains codons with less abundant tRNAs will be translated much more slowly.

Additional, more active mechanisms exist for translation-level control. One mechanism for controlling gene expression at the level of translation is a parallel to the use of a repressor to prevent transcription. Translational repression occurs when a protein binds to a target region on mRNA to prevent ribosomes from recognizing the initiation region. Formally, protein-mRNA binding is equivalent to a repressor protein binding to DNA to prevent polymerase from utilizing a promoter. Polycistronic RNA allows coordinate regulation of translation, analogous to transcription repression of an operon. **FIGURE 26.39** illustrates the most common form of this interaction, in which the regulator protein binds directly to a sequence that includes the AUG initiation codon, thereby preventing the ribosome from binding.

Some examples of translational repressors and their targets are summarized in **FIGURE 26.40**. A classic example of how the product of translation can directly control the translation of its mRNA is the coat protein of the RNA phage R17; it binds to a hairpin that encompasses the ribosome-binding site in the phage mRNA. Similarly, the phage T4 RegA protein binds to a consensus sequence that includes the AUG initiation codon in several T4 early mRNAs, and T4 DNA polymerase binds to a sequence in its own mRNA that includes the Shine–Dalgarno element needed for ribosome binding.

Another form of translational control occurs when translation of one gene requires changes in secondary structure that depend on translation of an immediately preceding gene. This happens during translation of the RNA phages, whose genes always are expressed in a set order. **FIGURE 26.41** shows that the phage RNA takes up a secondary structure in which only one initiation sequence is accessible; the second cannot be recognized by ribosomes because it is base-paired with other regions of the RNA. Translation of the first gene, however, disrupts the secondary structure, allowing ribosomes to bind to the initiation site of the next gene. In this mRNA, secondary structure controls translatability.

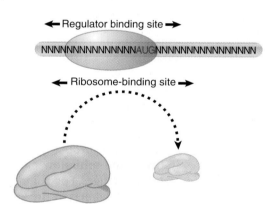

FIGURE 26.39 A regulator protein may block translation by binding to a site on mRNA that overlaps the ribosome-binding site at the initiation codon.

26.16 r-Protein Synthesis Is Controlled by Autoregulation

Key concept

- Translation of an r-protein operon can be controlled by a product of the operon that binds to a site on the polycistronic mRNA.

About seventy or so proteins constitute the apparatus for bacterial gene expression. The

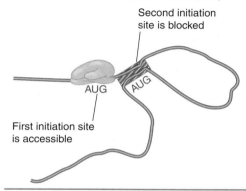

Only one initiation site is available initially

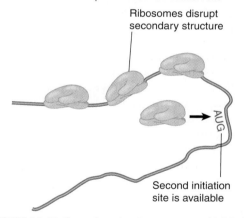

Translation exposes second initiation site

FIGURE 26.41 Secondary structure can control initiation. Only one initiation site is available in the RNA phage, but translation of the first cistron changes the conformation of the RNA so that other initiation site(s) become available.

Repressor	Target Gene	Site of Action
R17 coat protein	R17 replicase	hairpin that includes ribosome binding site
T4 RegA	early T4 mRNAs	various sequences including initiation codon
T4 DNA polymerase	T4 DNA polymerase	Shine-Dalgarno sequence
T4 p32	gene 32	single-stranded 5′ leader

FIGURE 26.40 Proteins that bind to sequences within the initiation regions of mRNAs may function as translational repressors.

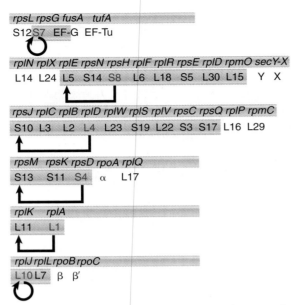

FIGURE 26.42 Genes for ribosomal proteins, protein synthesis factors, and RNA polymerase subunits are interspersed in a small number of operons that are autonomously regulated. The regulator is named in blue; the proteins that are regulated are shaded in pink.

ribosomal proteins are the major component, together with the ancillary proteins involved in protein synthesis. The subunits of RNA polymerase and its accessory factors make up the remainder. The genes coding for ribosomal proteins, protein-synthesis factors, and RNA polymerase subunits all are intermingled and organized into a small number of operons. Most of these proteins are represented only by single genes in *E. coli*.

Coordinate controls ensure that these proteins are synthesized in amounts appropriate for the growth conditions: when bacteria grow more rapidly, they devote a greater proportion of their efforts to the production of the apparatus for gene expression. An array of mechanisms is used to control the expression of the genes coding for this apparatus and to ensure that the proteins are synthesized at comparable levels that are related to the levels of the rRNAs.

The organization of six operons is summarized in **FIGURE 26.42**. About half of the genes for ribosomal proteins (r-proteins) map to four operons that lie close together (named *str*, *spc*, *S10*, and α simply for the first one of the functions to have been identified in each case). The *rif* and *L11* operons lie together at another location.

Each operon codes for a variety of functions. The *str* operon has genes for small subunit ribosomal proteins as well as for EF-Tu

and EF-G. The *spc* and *S10* operons have genes interspersed for both small and large ribosomal subunit proteins. The α operon has genes for proteins of both ribosomal subunit, as well as for the α subunit of RNA polymerase. The *rif* locus has genes for large subunit ribosomal proteins and for the β and β' subunits of RNA polymerase.

All except one of the ribosomal proteins are needed in equimolar amounts, which must be coordinated with the level of rRNA. The dispersion of genes whose products must be equimolar, and their intermingling with genes whose products are needed in different amounts, pose some interesting problems for coordinate regulation.

A feature common to all of the operons described in Figure 26.42 is regulation of some of the genes by one of the products. In each case, the gene coding for the regulatory product is itself one of the targets for regulation. Autoregulation occurs whenever a protein (or RNA) regulates its own production. In the case of the r-protein operons, the regulatory protein inhibits expression of a contiguous set of genes within the operon, so this is an example of negative autoregulation.

In each case, *accumulation of the protein inhibits further synthesis of itself and of some other gene products.* The effect often is exercised at the level of translation of the polycistronic mRNA. Each of the regulators is a ribosomal protein that binds directly to rRNA. *Its effect on translation is a result of its ability also to bind to its own mRNA.* The sites on mRNA at which these proteins bind either overlap the sequence where translation is initiated or lie nearby and probably influence the accessibility of the initiation site by inducing conformational changes. For example, in the S10 operon, protein L4 acts at the very start of the mRNA to inhibit translation of S10 and the subsequent genes. The inhibition may result from a simple block to ribosome access, as illustrated previously in Figure 26.39, or it may prevent a subsequent stage of translation. In two cases (including S4 in the α operon), the regulatory protein stabilizes a particular secondary structure in the mRNA that prevents the initiation reaction from continuing after the 30S subunit has bound.

The use of r-proteins that bind rRNA to establish autogenous regulation immediately suggests that this provides a mechanism to link r-protein synthesis to rRNA synthesis. A generalized model is depicted in **FIGURE 26.43**. Suppose that the binding sites for the autogenous regula-

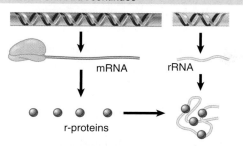

When rRNA is available, the r-proteins associate with it. Translation of mRNA continues

mRNA   rRNA

r-proteins

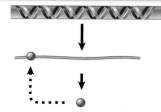

When no rRNA is availale, r-proteins accumulate. An r-protein binds to mRNA and prevents translation

FIGURE 26.43 Translation of the r-protein operons is autogenously controlled and responds to the level of rRNA.

tor r-proteins on rRNA are much stronger than those on the mRNAs. As long as any free rRNA is available, the newly synthesized r-proteins will associate with it to start ribosome assembly. There will be no free r-protein available to bind to the mRNA, so its translation will continue. As soon as the synthesis of rRNA slows or stops, though, free r-proteins begin to accumulate. They are then available to bind their mRNAs and thus repress further translation. This circuit ensures that each r-protein operon responds in the same way to the level of rRNA: as soon as there is an excess of r-protein relative to rRNA, synthesis of the protein is repressed.

26.17 Summary

Transcription is regulated by the interaction between *trans*-acting factors and *cis*-acting sites. A *trans*-acting factor is the product of a regulator gene. It is usually protein but also can be RNA. It diffuses in the cell, and as a result it can act on any appropriate target gene. A *cis*-acting site in DNA (or RNA) is a sequence that functions by being recognized *in situ*. It has no coding function and can regulate only those sequences with which it is physically contiguous. Bacterial genes coding for proteins whose functions are related, such as successive enzymes in a pathway, may be organized in a cluster that is transcribed into a polycistronic mRNA from a single promoter. Control of this promoter regulates expression of the entire pathway. The unit of regulation, which contains structural genes and *cis*-acting elements, is called the operon.

Initiation of transcription is regulated by interactions that occur in the vicinity of the promoter. The ability of RNA polymerase to initiate at the promoter is prevented or activated by other proteins. Genes that are active unless they are turned off by binding the regulator are said to be under negative control. Genes that are active only when the regulator is bound to them are said to be under positive control. The type of control can be determined by the dominance relationships between wild-type genes and mutants that are constitutive/derepressed (permanently on) or uninducible/super-repressed (permanently off).

A repressor or activator can control multiple targets that have copies of an operator, or its consensus sequence. A repressor protein prevents RNA polymerase from either binding to the promoter or activating transcription. The repressor binds to a target sequence, the operator, which is usually located around or upstream of the transcription startpoint. Operator sequences are short and often are palindromic. The repressor is often a homomultimer whose symmetry reflects that of its target.

The ability of the repressor protein to bind to its operator is often regulated by small molecules, which provide a second level of gene regulation. If the repressor regulates genes that code for enzymes, the system may be induced by enzyme substrates or repressed by enzyme products. In a negative inducible gene, the substrate (an inducer) prevents a repressor from binding the operator. In a negative repressible gene, the product or corepressor enables the regulator to bind the operator and turn off gene expression. Binding of the inducer or corepressor to its site on the regulator protein produces a change in the structure of the DNA-binding site of the protein. This allosteric reaction occurs both in free repressor proteins and directly in repressor proteins already bound to DNA.

The lactose pathway in *E. coli* operates by negative induction. When an inducer, the substrate β-galactoside, diminishes the ability of repressor to bind its operator, transcription and translation of the *lacZ* gene then produce β-galactosidase, the enzyme that metabolizes β-galactosides.

A protein with a high affinity for a particular target sequence in DNA has a lower affinity for all DNA. The ratio defines the specificity of the protein. There are many more nonspecific sites (any DNA sequence) than specific target sites in a genome; as a result, a DNA-binding protein such as a repressor or RNA polymerase is "stored" on DNA. (It is likely that none, or very little, is free.) The specificity for the target sequence must be great enough to counterbalance the excess of nonspecific sites over specific sites. The balance for bacterial proteins is adjusted so that the amount of protein and its specificity allow specific recognition of the target in "on" conditions, but allow almost complete release of the target in "off" conditions.

Some promoters cannot be recognized by RNA polymerase, or are recognized only poorly unless a specific activator protein (a positive regulator) is present. Activator proteins may also be regulated by small molecules. The CRP activator is only able to bind to target sequences when complexed with cAMP, which only happens in conditions of low glucose. All promoters that are controlled by catabolite repression have at least one copy of the CRP-binding site. Direct contact between CRP and RNA polymerase occurs through the C-terminal domain of the α subunits.

The tryptophan pathway operates by negative repression. The corepressor tryptophan, the product of the pathway, activates the repressor protein so that it binds to the operator and prevents expression of the genes that code for the enzymes that synthesize tryptophan. The *trp* operon is also controlled by attenuation.

Gene expression may also be modulated at the level of translation by the ability of an mRNA to attract a ribosome and by the abundance of specific tRNAs that recognize different codons. More active mechanisms that regulate at the level of translation are also found. Translation may be regulated by a protein that can bind to the mRNA to prevent the ribosome from binding. Most proteins that repress translation possess this capacity in addition to other functional roles; in particular, translation is controlled in some cases of autoregulation, when a gene product regulates translation of the mRNA containing its own open reading frame.

References

26.1 Introduction

Review

Miller, J. and Reznikoff, W., eds. (1980). *The Operon*, 2nd ed. Woodbury, NY: Cold Spring Harbor Laboratory Press.

Research

Jacob, F. and Monod, J. (1961). Genetic regulatory mechanisms in the synthesis of proteins. *J. Mol. Biol.* 3, 318–389.

26.3 The *lac* Operon Is Negative Inducible

Reviews

Beckwith, J. (1978). *lac*: the genetic system. In Miller, J. H. and Reznikoff, W., eds. *The Operon*. New York: Cold Spring Harbor Laboratory, pp. 11–30.

Beyreuther, K. (1978). Chemical structure and functional organization of the *lac* repressor from *E. coli*. In Miller, J. H. and Reznikoff, W., eds. *The Operon*. New York: Cold Spring Harbor Laboratory, pp. 123–154.

Miller, J. H. (1978). The *lacI* gene: its role in *lac* operon control and its use as a genetic system. In Miller, J. H. and Reznikoff, W., eds. *The Operon*. New York: Cold Spring Harbor Laboratory, pp. 31–88.

Weber, K. and Geisler, N. (1978). Lac repressor fragments produced *in vivo* and *in vitro*: an approach to the understanding of the interaction of repressor and DNA. In Miller, J. H. and Reznikoff, W., eds. *The Operon*. New York: Cold Spring Harbor Laboratory, pp. 155–176.

Wilson, C. J., Zahn, H., Swint-Kruse, L., and Matthews, K. S. (2007). The lactose repressor system: paradigms for regulation, allosteric behavior and protein folding. *Cell. Mol. Life Sci.* 64, 3–16.

Research

Jacob, F. and Monod, J. (1961). Genetic regulatory mechanisms in the synthesis of proteins. *J. Mol. Biol.* 3, 318–389.

26.7 *lac* Repressor Is a Tetramer Made of Two Dimers

Research

Friedman, A. M., Fischmann, T. O., and Steitz, T. A. (1995). Crystal structure of *lac* repressor core tetramer and its implications for DNA looping. *Science* 268, 1721–1727.

Lewis, M. et al. (1996). Crystal structure of the lactose operon repressor and its complexes with DNA and inducer. *Science* 271, 1247–1254.

26.8 *lac* Repressor Binding to the Operator Is Regulated by an Allosteric Change in Conformation

Reviews

Markiewicz, P., Kleina, L. G., Cruz, C., Ehret, S., and Miller, J. H. (1994). Genetic studies of the lac repressor. XIV. Analysis of 4000 altered *E. coli lac* repressors reveals essential and nonessential residues, as well as spacers which do not require a specific sequence. *J. Mol. Biol.* 240, 421–433.

Pace, H. C., Kercher, M. A., Lu, P., Markiewicz, P., Miller, J. H., Chang, G., and Lewis, M. (1997). Lac repressor genetic map in real space. *Trends Biochem. Sci.* 22, 334–339.

Suckow, J., Markiewicz, P., Kleina, L. G., Miller, J., Kisters-Woike, B., and Müller-Hill, B. (1996). Genetic studies of the Lac repressor. XV: 4000 single amino acid substitutions and analysis of the resulting phenotypes on the basis of the protein structure. *J. Mol. Biol.* 261, 509–523.

Research

Gilbert, W. and Müller-Hill, B. (1966). Isolation of the *lac* repressor. *Proc. Natl. Acad. Sci. USA* 56, 1891–1898.

Gilbert, W. and Müller-Hill, B. (1967). The *lac* operator is DNA. *Proc. Natl. Acad. Sci. USA* 58, 2415–2421.

Taraban, M., Zhan, H., Whitten, A. E., Langley, D. B., Matthews, K. S., Swint-Kruse, L., Trewhella, J. (2008). Ligand-induced conformational changes and conformational dynamics in the solution structure of the lactose repressor protein. *J. Mol. Biol.* 376, 466–481.

Yu, H. and Gertstein, M. (2006). Genomic analysis of the hierarchical structure of regulatory networks. *Proc. Natl. Acad. Sci.* 103, 14724–14731.

26.9 *lac* Repressor Binds to Three Operators and Interacts with RNA Polymerase

Research

Oehler, S. et al. (1990). The three operators of the lac operon cooperate in repression. *EMBO J.* 9, 973–979.

Swigon, D., Coleman, B. D., and Olson, W. K. (2006). Modeling the *lac* Repressor-operator assembly: the influence of DNA looping on *lac* Repressor conformation. *Proc. Natl. Acad. Sci.* 103, 9879–9884.

Wong, O. K., Guthold, M., Erie, D. A., Gelles, J. (2008). Interconvertible *lac* repressor-DNA loops revealed by single-molecule experiments. *PLoS Biol.* 6:e232.

26.10 The Operator Competes with Low-Affinity Sites to Bind Repressor

Research

Cronin, C. A., Gluba, W., and Scrable, H. (2001). The *lac* operator-repressor system is functional in the mouse. *Genes Dev.* 15, 1506–1517.

Elf, J., Li, G.-W., and Xie, X. S. (2007). Probing transcription factor dynamics at the single-molecule level in a living cell. *Science* 316, 1191–1194.

Hildebrandt, E. R. et al. (1995). Comparison of recombination *in vitro* and in *E. coli* cells: measure of the effective concentration of DNA *in vivo*. *Cell* 81, 331–340.

Lin, S.-Y. and Riggs, A. D. (1975). The general affinity of lac repressor for *E. coli* DNA: impli-

cations for gene regulation in prokaryotes and eukaryotes. *Cell* 4, 107–111.

26.11 The *lac* Operon Has a Second Layer of Control: Catabolite Repression

Reviews

Botsford, J. L. and Harman, J. G. (1992). Cyclic AMP in prokaryotes. *Microbiol. Rev.* 56, 100–122.

Kolb, A. (1993). Transcriptional regulation by cAMP and its receptor protein. *Annu. Rev. Biochem.* 62, 749–795.

Research

Niu, W., Kim, Y., Tau, G., Heyduk, T., and Ebright, R. H. (1996). Transcription activation at class II CAP-dependent promoters: two interactions between CAP and RNA polymerase. *Cell* 87, 1123–1134.

Zhou, Y., Busby, S., and Ebright, R. H. (1993). Identification of the functional subunit of a dimeric transcription activator protein by use of oriented heterodimers. *Cell* 73, 375–379.

Zhou, Y., Merkel, T. J., and Ebright, R. H. (1994). Characterization of the activating region of *E. coli* catabolite gene activator protein (CAP). II. Role at class I and class II CAP-dependent promoters. *J. Mol. Biol.* 243, 603–610.

26.12 The *trp* Operon Is a Repressible Operon with Three Transcription Units

Research

Tabaka, M., Cybutski, O., and Holyst, R. (2008). Accurate genetic switch in *E. coli*: novel mechanism of regulation corepressor *J. Mol. Biol.* 377, 1002–1014.

26.13 The *trp* Operon Is Also Controlled by Attenuation

Review

Yanofsky, C. (1981). Attenuation in the control of expression of bacterial operons. *Nature* 289, 751–758.

26.14 Attenuation Can Be Controlled by Translation

Reviews

Bauer, C. E., Carey, J., Kasper, L. M., Lynn, S. P., Waechter, D. A., and Gardner, J. F. (1983). Attenuation in bacterial operons. In Beckwith, J., Davies, J., and Gallant, J. A., eds. *Gene Function in Prokaryotes*. Cold Spring Harbor, NY: Cold Spring Harbor Press, pp. 65–89.

Landick, R. and Yanofsky, C. (1987). In Neidhardt, F. C., ed. *E. coli* and *S. typhimurium Cellular and Molecular Biology*. Washington, DC: American Society for Microbiology, pp. 1276–1301.

Yanofsky, C. and Crawford, I. P. (1987). In Ingraham, J. L. et al., eds. Escherichia coli *and*

Salmonella typhimurium. Washington, DC: American Society for Microbiology, pp. 1453–1472.

Research

Lee, F. and Yanofsky, C. (1977). Transcription termination at the *trp* operon attenuators of *E. coli* and *S. typhimurium*: RNA secondary structure and regulation of termination. *Proc. Natl. Acad. Sci. USA* 74, 4365–4368.

Zurawski, G. et al. (1978). Translational control of transcription termination at the attenuator of the *E. coli* tryptophan operon. *Proc. Natl. Acad. Sci. USA* 75, 5988–5991.

26.16 r-Protein Synthesis Is Controlled by Autoregulation

Review

Nomura, M. et al. (1984). Regulation of the synthesis of ribosomes and ribosomal components. *Annu. Rev. Biochem.* 53, 75–117.

Research

Baughman, G. and Nomura, M. (1983). Localization of the target site for translational regulation of the L11 operon and direct evidence for translational coupling in *E. coli*. *Cell* 34, 979–988.

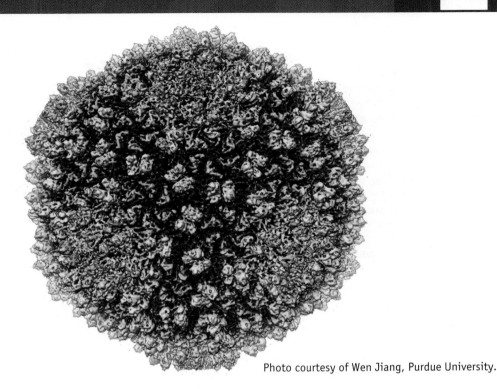

Photo courtesy of Wen Jiang, Purdue University.

Phage Strategies

CHAPTER OUTLINE

- pQ is the product of a delayed early gene and is an antiterminator that allows RNA polymerase to transcribe the late genes.
- Lambda DNA circularizes after infection; as a result, the late genes form a single transcription unit.

27.8 Lysogeny Is Maintained by the Lambda Repressor Protein

- The lambda repressor, encoded by the *cI* gene, is required to maintain lysogeny.
- The lambda repressor acts at the O_L and O_R operators to block transcription of the immediate early genes.
- The immediate early genes trigger a regulatory cascade; as a result, their repression prevents the lytic cycle from proceeding.

27.9 The Lambda Repressor and Its Operators Define the Immunity Region

- Several lambdoid phages have different immunity regions.
- A lysogenic phage confers immunity to further infection by any other phage with the same immunity region.

27.10 The DNA-Binding Form of the Lambda Repressor Is a Dimer

- A repressor monomer has two distinct domains.
- The N-terminal domain contains the DNA-binding site.
- The C-terminal domain dimerizes.
- Binding to the operator requires the dimeric form so that two DNA-binding domains can contact the operator simultaneously.
- Cleavage of the repressor between the two domains reduces the affinity for the operator and induces a lytic cycle.

27.11 Lambda Repressor Uses a Helix-Turn-Helix Motif to Bind DNA

- Each DNA-binding region in the repressor contacts a half-site in the DNA.
- The DNA-binding site of the repressor includes two short α-helical regions that fit into the successive turns of the major groove of DNA.
- A DNA-binding site is a (partially) palindromic sequence of 17 bp.
- The amino acid sequence of the recognition helix makes contacts with particular bases in the operator sequence that it recognizes.

27.12 Lambda Repressor Dimers Bind Cooperatively to the Operator

- Repressor binding to one operator increases the affinity for binding a second repressor dimer to the adjacent operator.
- The affinity is $10\times$ greater for O_L1 and O_R1 than other operators, so they are bound first.
- Cooperativity allows repressor to bind the O_L2/O_R2 sites at lower concentrations.

27.13 Lambda Repressor Maintains an Autoregulatory Circuit

- The DNA-binding region of repressor at O_R2 contacts RNA polymerase and stabilizes its binding to P_{RM}.
- This is the basis for the autoregulatory control of repressor maintenance.
- Repressor binding at O_L blocks transcription of gene *N* from P_L.
- Repressor binding at O_R blocks transcription of cro, but also is required for transcription of *cI*.
- Repressor binding to the operators therefore simultaneously blocks entry to the lytic cycle and promotes its own synthesis.

27.14 Cooperative Interactions Increase the Sensitivity of Regulation

- Repressor dimers bound at O_L1 and O_L2 interact with dimers bound at O_R1 and O_R2 to form octamers.
- These cooperative interactions increase the sensitivity of regulation.

27.15 The *cII* and *cIII* Genes Are Needed to Establish Lysogeny

- The delayed early gene products cII and cIII are necessary for RNA polymerase to initiate transcription at the promoter P_{RE}.
- cII acts directly at the promoter and cIII protects cII from degradation.
- Transcription from P_{RE} leads to synthesis of repressor and also blocks the transcription of *cro*.

27.16 A Poor Promoter Requires cII Protein

- P_{RE} has atypical sequences at −10 and −35.
- RNA polymerase binds the promoter only in the presence of cII.
- cII binds to sequences close to the −35 region.

27.17 Lysogeny Requires Several Events

- cII and cIII cause repressor synthesis to be established and also trigger inhibition of late gene transcription.
- Establishment of repressor turns off immediate and delayed early gene expression.
- Repressor turns on the maintenance circuit for its own synthesis.
- Lambda DNA is integrated into the bacterial genome at the final stage in establishing lysogeny.

27.18 The Cro Repressor Is Needed for Lytic Infection

- Cro binds to the same operators as the lambda repressor, but with different affinities.
- When Cro binds to O_R3, it prevents RNA polymerase from binding to P_{RM} and blocks the maintenance of repressor promoter.
- When Cro binds to other operators at O_R or O_L, it prevents RNA polymerase from expressing immediate early genes, which (indirectly) blocks repressor establishment.

CHAPTER OUTLINE, CONTINUED

27.1 Introduction

A virus consists of a nucleic acid genome contained in a protein coat. In order to reproduce, the virus must infect a host cell. The typical pattern of an infection is to subvert the functions of the host cell for the purpose of producing a large number of progeny viruses. Viruses that infect bacteria are generally called **bacteriophages**, often abbreviated to *phage* or simply φ. Usually a **phage** infection kills the bacterium. The process by which a phage infects a bacterium, reproduces itself, and then kills its host is called **lytic infection**. In the typical lytic cycle, the phage DNA (or RNA) enters the host bacterium, its genes are transcribed in a set order, the phage genetic material is replicated, and the protein components of the phage particle are produced. Finally, the host bacterium is broken open (lysed) to release the assembled progeny particles by the process of **lysis**. For some phages, called **virulent phages**, this is their only strategy for survival.

Other phages have a dual existence. They are able to perpetuate themselves via the same sort of lytic cycle in what amounts to an open strategy for producing as many copies of the phage as rapidly as possible. They also have an alternative form of existence, though, in which the phage genome is present in the bacterium in a latent form known as a **prophage**. This form of propagation is called **lysogeny** and the infected bacteria are known as lysogens. Phages that follow this pathway are called **temperate phages**.

In a lysogenic bacterium, the prophage is inserted, or recombined, into the bacterial genome and is inherited in the same way as bacterial genes. The process by which it is converted from an independent phage genome into a prophage that is a linear part of the bacterial genome is described as **integration**. By virtue of its possession of a prophage, a lysogenic bac-

terium has immunity against infection by other phage particles of the same type. Immunity is established by a single integrated prophage, so in general a bacterial genome contains only one copy of a prophage of any particular type.

There are transitions between the lysogenic and lytic modes of existence. **FIGURE 27.1** shows that when a temperate phage produced by a

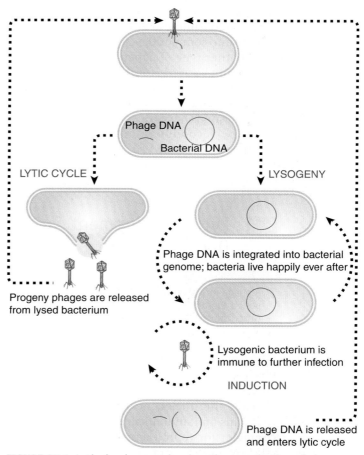

Phage DNA

Bacterial DNA

LYTIC CYCLE

LYSOGENY

Phage DNA is integrated into bacterial genome; bacteria live happily ever after

Progeny phages are released from lysed bacterium

Lysogenic bacterium is immune to further infection

INDUCTION

Phage DNA is released and enters lytic cycle

FIGURE 27.1 Lytic development involves the reproduction of phage particles with destruction of the host bacterium, but lysogenic existence allows the phage genome to be carried as part of the bacterial genetic information.

lytic cycle enters a new bacterial host cell, it either repeats the lytic cycle or enters the lysogenic state. The outcome depends on the conditions of infection and the genotypes of phage and bacterium.

A prophage is freed from the restrictions of lysogeny by a process called **induction**. First the phage DNA is released from the bacterial chromosome by another recombination event called **excision**; then the free DNA proceeds through the lytic pathway.

The alternative forms in which these phages are propagated are determined by the regulation of transcription. Lysogeny is maintained by the interaction of a phage repressor with an operator. The lytic cycle requires a cascade of transcriptional controls. The transition between the two lifestyles is accomplished by the establishment of repression (lytic cycle to lysogeny) or by the relief of repression (induction of lysogen to lytic phage). These regulatory processes provide a wonderful example of how a series of relatively simple regulatory actions can be built up into complex developmental pathways.

Another type of genetic element that can exist within bacteria is a **plasmid**. Plasmids are autonomous units that exist in the cell as extrachromosomal genomes that are self-replicating. Some plasmids have the ability to insert themselves by recombination into the bacterial chromosome (see *Section 12.6, The F Plasmid Is Transferred by Conjugation between Bacteria.*) This class of plasmid is called an **episome**.

27.2 Lytic Development Is Divided into Two Periods

Key concepts

- A phage infective cycle is divided into the early period (before replication) and the late period (after the onset of replication).
- A phage infection generates a pool of progeny phage genomes that replicate and recombine.

Phage genomes by necessity are small. As with all viruses, they are restricted by the need to package the nucleic acid within the protein coat. This limitation dictates many of the viral strategies for reproduction. Typically a virus takes over the apparatus of the host cell, which then replicates and expresses phage genes instead of the bacterial genes.

Usually the phage has genes whose function is to ensure preferential replication of phage DNA. These genes are concerned with the initiation of replication and may even

include a new DNA polymerase. Changes are introduced in the capacity of the host cell to engage in transcription. They involve replacing the RNA polymerase or modifying its capacity for initiation or termination. The result is always the same: phage mRNAs are preferentially transcribed. As far as protein synthesis is concerned, the phage is, for the most part, content to use the host apparatus, redirecting its activities principally by replacing bacterial mRNA with phage mRNA.

Lytic development is accomplished by a pathway in which the phage genes are expressed in a particular order. This ensures that the right amount of each component is present at the appropriate time. The cycle can be divided into the two general parts illustrated in **FIGURE 27.2**:

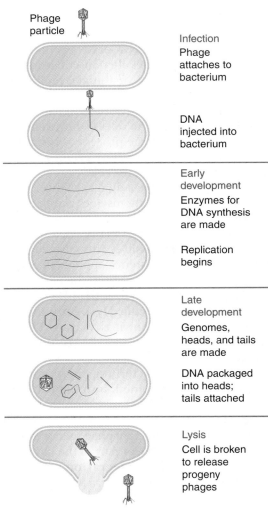

FIGURE 27.2 Lytic development takes place by producing phage genomes and protein particles that are assembled into progeny phages.

- **Early infection** describes the period from entry of the DNA to the start of its replication.
- **Late infection** defines the period from the start of replication to the final step of lysing the bacterial cell to release progeny phage particles.

The early phase is devoted to the production of enzymes involved in the reproduction of DNA. These include the enzymes concerned with DNA synthesis, recombination, and sometimes modification. Their activities cause a pool of phage genomes to accumulate. In this pool, genomes are continually replicating and recombining, so that *the events of a single lytic cycle concern a population of phage genomes.*

During the late phase, the protein components of the phage particle are synthesized. Often many different proteins are needed to make up head and tail structures, so the largest part of the phage genome consists of late functions. In addition to the structural proteins, "assembly proteins" are needed to help construct the particle, although they are not incorporated into it themselves. By the time the structural components are assembling into heads and tails, replication of DNA has reached its maximum rate. The genomes then are inserted into the empty protein heads, tails are added, and the host cell is lysed to allow release of new viral particles.

27.3 Lytic Development Is Controlled by a Cascade

Key concepts
- The early genes transcribed by host RNA polymerase following infection include, or comprise, regulators required for expression of the middle set of phage genes.
- The middle group of genes includes regulators to transcribe the late genes.
- This results in the ordered expression of groups of genes during phage infection.

The organization of the phage genetic map often reflects the sequence of lytic development. The concept of the operon is taken to somewhat of an extreme, in which the genes coding for proteins with related functions are clustered to allow their control with the maximum economy. This allows the pathway of lytic development to be controlled with a small number of regulatory switches.

The lytic cycle is under positive control, so that each group of phage genes can be expressed only when an appropriate signal is given. **FIGURE 27.3** is an overview showing that the regulatory genes function in a **cascade**, in which a gene expressed at one stage is necessary for synthesis of the genes that are expressed at the next stage.

The early part of the first stage of gene expression necessarily relies on the transcription apparatus of the host cell. In general, only a few genes are expressed at this time. Their promoters are indistinguishable from those of host genes. The name of this class of genes depends on the phage. In most cases, they are known as the **early genes**. In phage lambda, they are given the evocative description of **immediate early genes**. Irrespective of the name, they constitute only a preliminary set of genes, representing just the initial part of the early period. Sometimes they are exclusively occupied with the transition to the next period. At all cases, *one of these genes always codes for a protein, a gene regulator that is necessary for transcription of the next class of genes.*

This next class of genes in the early stage is known variously as the **delayed early** or

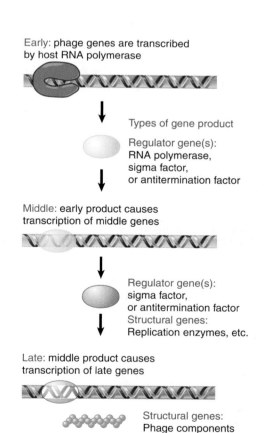

Early: phage genes are transcribed by host RNA polymerase

Types of gene product

Regulator gene(s):
RNA polymerase,
sigma factor,
or antitermination factor

Middle: early product causes transcription of middle genes

Regulator gene(s):
sigma factor,
or antitermination factor
Structural genes:
Replication enzymes, etc.

Late: middle product causes transcription of late genes

Structural genes:
Phage components

FIGURE 27.3 Phage lytic development proceeds by a regulatory cascade, in which a gene product at each stage is needed for expression of the genes at the next stage.

middle gene group. Its expression typically starts as soon as the regulator protein coded by the early gene(s) is available. Depending on the nature of the control circuit, the initial set of early genes may or may not continue to be expressed at this stage. If control is at transcription initiation, the two events are independent (as seen in **FIGURE 27.4**) and early genes can be switched off when middle genes are transcribed. If control is at transcription termination, the early genes must continue to be expressed, as seen in **FIGURE 27.5**. Often, the expression of host genes is reduced. Together the two sets

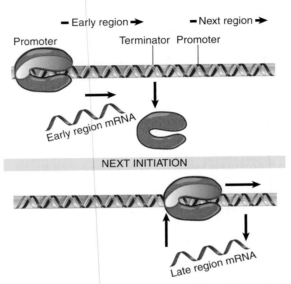

FIGURE 27.4 Control at initiation utilizes independent transcription units, each with its own promoter and terminator, which produce independent mRNAs. The transcription units need not be located near one another.

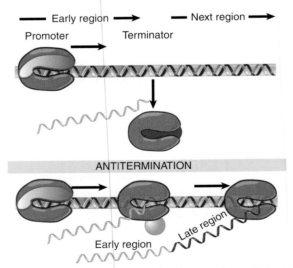

FIGURE 27.5 Control at termination requires adjacent units, so that transcription can read from the first gene into the next gene. This produces a single mRNA that contains both sets of genes.

of early genes account for all necessary phage functions except those needed to assemble the particle coat itself and to lyse the cell.

When the replication of phage DNA begins, it is time for the **late genes** to be expressed. Their transcription at this stage usually is arranged by embedding an additional regulator gene within the previous (delayed early or middle) set of genes. This regulator may be another antitermination factor (as in lambda) or it may be another sigma factor (such as the *B. subtilis* SPO1 factor).

A lytic infection often falls into the stages described above, beginning with the early genes transcribed by host RNA polymerase (sometimes the regulators are the only products at this stage). This stage is followed by those genes transcribed under the direction of the regulator produced in the first stage (most of these genes code for enzymes needed for replication of phage DNA). The final stage consists of genes for phage components, which are transcribed under the direction of a regulator synthesized in the second stage.

The use of these successive controls, in which each set of genes contains a regulator that is necessary for expression of the next set, creates a cascade in which groups of genes are turned on (and sometimes off) at particular times. The means used to construct each phage cascade are different but the results are similar.

27.4 Two Types of Regulatory Events Control the Lytic Cascade

Key concept

• Regulator proteins used in phage cascades may sponsor initiation at new (phage) promoters or cause the host polymerase to read through transcription terminators.

At every stage of phage expression, one or more of the active genes is a regulator that is needed for the subsequent stage. The regulator may take the form of a new sigma factor that redirects the specificity of the host RNA polymerase (see *Section 19.8, Sigma Factors Control Binding to DNA by Recognizing Specific Sequences in Promoters DNA*) or an antitermination factor that allows it to read a new group of genes (see *Section 19.22, Antitermination Can Be a Regulatory Event*). Now, let's compare the use of switching at initiation or termination to control gene expression.

One mechanism for recognizing new phage promoters is to replace the sigma factor of the host enzyme with another factor that redirects its specificity in initiation as seen in **FIGURE 27.6**. An alternative is to synthesize a new phage RNA polymerase. In either case, the critical feature that distinguishes the new set of genes is their possession of *different promoters from those originally recognized by host RNA polymerase*. Figure 27.4 shows that the two sets of transcripts are independent; as a consequence, early gene expression can cease after the new sigma factor or polymerase has been produced.

Antitermination provides an alternative mechanism for phages to control the switch from early genes to the next stage of expression. The use of antitermination depends on a particular arrangement of genes. Figure 27.5 shows that the early genes lie adjacent to the genes that are to be expressed next, but are separated from them by terminator sites. *If termination is prevented at these sites, the polymerase reads through into the genes on the other side.* So in antitermination, the *same promoters* continue to be recognized by RNA polymerase. The new genes are expressed only by extending the RNA chain to form molecules that contain the early gene sequences at the 5′ end and the new gene sequences at the 3′ end. The two types

of sequences remain linked; thus early gene expression inevitably continues.

The regulator gene that controls the switch from immediate early to delayed early expression in phage lambda is identified by mutations in gene *N* that can transcribe *only* the immediate early genes; they proceed no further into the infective cycle (see Figure 27.10). From the genetic point of view, the mechanisms of new initiation and antitermination are similar. *Both are positive controls in which an early gene product must be made by the phage in order to express the next set of genes.* By employing either sigma factor or antitermination proteins with different specifications, a cascade for gene expression can be constructed.

27.5 The Phage T7 and T4 Genomes Show Functional Clustering

Key concepts

- Genes concerned with related functions are often clustered.
- Phages T7 and T4 are examples of regulatory cascades in which phage infection is divided into three periods.

The genome of phage T7 has three classes of genes, each of which constitutes a group of adjacent loci. As **FIGURE 27.7** shows, the class I genes are the immediate early type and are expressed by host RNA polymerase as soon as the phage DNA enters the cell. Among the products of these genes are a phage RNA polymerase and enzymes that interfere with host gene expression. The phage RNA polymerase is responsible for expressing the class II genes (which are concerned principally with DNA synthesis functions) and the class III genes

Holoenzyme with σ70 recognizes one set of promoters

Phage synthesizes new sigma or RNA polymerase

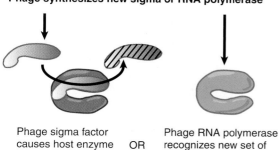

Phage sigma factor causes host enzyme to recognize new promoters   OR   Phage RNA polymerase recognizes new set of promoters

FIGURE 27.6 A phage may control transcription at initiation either by synthesizing a new sigma factor that replaces the host sigma factor or by synthesizing a new RNA polymerase.

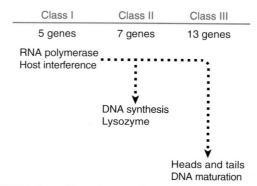

Class I	Class II	Class III
5 genes	7 genes	13 genes

RNA polymerase
Host interference

DNA synthesis
Lysozyme

Heads and tails
DNA maturation

FIGURE 27.7 Phage T7 contains three classes of genes that are expressed sequentially. The genome is ~38 kb.

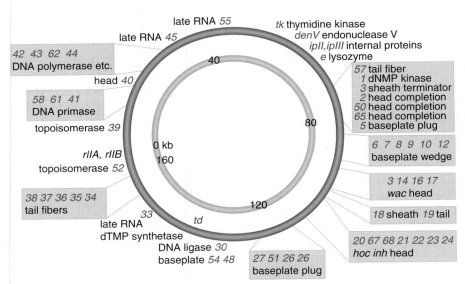

FIGURE 27.8 The map of T4 is circular. There is extensive clustering of genes coding for components of the phage and processes such as DNA replication, but there is also dispersion of genes coding for a variety of enzymatic and other functions. Essential genes are indicated by numbers. Nonessential genes are identified by letters. Only some representative T4 genes are shown on the map.

(which are concerned with assembling the mature phage particle).

Phage T4 has one of the larger phage genomes (165 kb), which is organized with extensive functional grouping of genes. **FIGURE 27.8** presents the genetic map. Essential genes are numbered: a mutation in any one of these loci prevents successful completion of the lytic cycle. Nonessential genes are indicated by three-letter abbreviations. (They are defined as nonessential under the usual conditions of infection. We do not really understand the inclusion of many nonessential genes, but presumably they confer a selective advantage in some of T4's habitats. In smaller phage genomes, most or all of the genes are essential.)

There are three phases of gene expression. A summary of the functions of the genes expressed at each stage is given in **FIGURE 27.9**. The early genes are transcribed by host RNA polymerase. The middle genes are also transcribed by host RNA polymerase, but two phage-encoded products, MotA and AsiA, are also required. The middle promoters lack a consensus −35 sequence and instead have a binding sequence for MotA. The phage protein is an activator that compensates for the deficiency in the promoter by assisting host RNA polymerase to bind. (This is similar to a mechanism employed by phage lambda with its cII gene, which is illustrated later in Figure 27.30.) The early and middle genes account for virtually all of the phage functions concerned with the

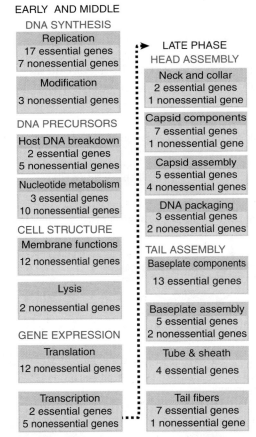

FIGURE 27.9 The phage T4 lytic cascade falls into two parts; early functions are concerned with DNA synthesis; late functions with particle assembly.

synthesis of DNA, modifying cell structure, and transcribing and translating phage genes.

The two essential genes in the "transcription" category fulfill a regulatory function: their products are necessary for late gene expression. Phage T4 infection depends on a mechanical link between replication and late gene expression. Only actively replicating DNA can be used as a template for late gene transcription. The connection is generated by introducing a new sigma factor and also by making other modifications in the host RNA polymerase so that it is active only with a template of replicating DNA. This link establishes a correlation between the synthesis of phage protein components and the number of genomes available for packaging.

27.6 Lambda Immediate Early and Delayed Early Genes Are Needed for Both Lysogeny and the Lytic Cycle

Key concepts

- Lambda has two immediate early genes, *N* and *cro*, which are transcribed by host RNA polymerase.
- The product of the *N* gene is required to express the delayed early genes.
- Three of the delayed early gene products are regulators.
- Lysogeny requires the delayed early genes *cII-cIII*.
- The lytic cycle requires the immediate early gene *cro* and the delayed early gene *Q*.

One of the most intricate cascade circuits is provided by phage lambda. Actually, the cascade for lytic development itself is straightforward, with two regulators controlling the successive stages of development. The circuit for the lytic cycle, though, is interlocked with the circuit for establishing lysogeny, as summarized in **FIGURE 27.10**.

When lambda DNA enters a new host cell, the lytic and lysogenic pathways start off the same way. Both require expression of the immediate early and delayed early genes, but then they diverge: lytic development follows if the late genes are expressed, and lysogeny ensues if synthesis of a gene regulator called the lambda repressor is established by turning on its gene, the *cI* gene. Lambda has only two immediate early genes, transcribed independently by host RNA polymerase:

- The *N* gene codes for an antitermination factor whose action at *nut* (N utilization)

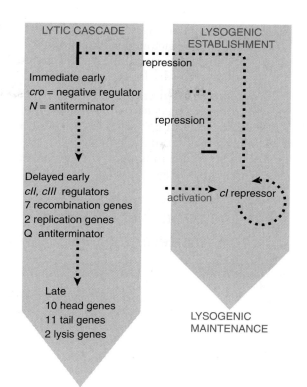

FIGURE 27.10 The lambda lytic cascade is interlocked with the circuitry for lysogeny.

sites allows transcription to proceed into the delayed early genes (see *Section 19.22, Antitermination Can Be a Regulated Event*). The *N* gene is required for both the lytic and lysogenic pathways.

- The *cro* gene codes for a repressor that prevents expression of the *cI* gene coding for the lambda repressor (essentially derepressing the late genes, a necessary action if the lytic cycle is to proceed). It also turns off expression of the immediate early genes (which are not needed later in the lytic cycle). The lambda repressor is the major regulator required for lysogenic development.

The delayed early genes, turned on by the product of the *N* gene, include two replication genes (needed for lytic infection), seven recombination genes (some involved in recombination during lytic infection, and two necessary to integrate lambda DNA into the bacterial chromosome for lysogeny), and three regulator genes. These regulator genes have opposing functions:

- The *cII-cIII* pair of regulator genes is needed to establish the synthesis of the lambda repressor for the lysogenic pathway.

- The *Q* regulator gene codes for an anti-termination factor that allows host RNA polymerase to transcribe the late genes and is necessary for the lytic cycle.

Thus the delayed early genes serve two masters: some are needed for the phage to enter lysogeny, and the others are concerned with controlling the order of the lytic cycle. At this point, lambda is keeping open the option to choose either pathway.

27.7 The Lytic Cycle Depends on Antitermination by pN

Key concepts

- pN is an antitermination factor that allows RNA polymerase to continue transcription past the ends of the two immediate early genes.
- pQ is the product of a delayed early gene and is an antiterminator that allows RNA polymerase to transcribe the late genes.
- Lambda DNA circularizes after infection; as a result, the late genes form a single transcription unit.

To disentangle the lytic and lysogenic pathways, let's first consider just the lytic cycle. **FIGURE 27.11** gives the map of lambda phage DNA. A group of genes concerned with regulation is surrounded by genes needed for recombination and replication. The genes coding for structural components of the phage are clustered. All of the genes necessary for the lytic cycle are expressed in polycistronic transcripts from three promoters.

FIGURE 27.12 shows that the two immediate early genes, *N* and *cro*, are transcribed by host RNA polymerase. *N* is transcribed toward the left and *cro* toward the right. Each transcript is terminated at the end of the gene. The protein

pN is the regulator, the antitermination factor that allows transcription to continue into the delayed early genes by suppressing use of the terminators t_L and t_R (see *Section 19.22, Antitermination Can Be a Regulated Event*). In the presence of pN, transcription continues to the left of the *N* gene into the recombination genes and to the right of the *cro* gene into the replication genes.

The map in Figure 27.11 gives the organization of the lambda DNA as it exists in the phage particle. Shortly after infection, though, the ends of the DNA join to form a circle. **FIGURE 27.13** shows the true state of lambda DNA during infection. The late genes are welded into a single group, which contains the lysis genes *S-R* from the right end of the linear DNA and the head and tail genes *A-J* from the left end.

The late genes are expressed as a single transcription unit, starting from a promoter $P_{R'}$ that lies between *Q* and *S*. The late promoter is used constitutively. In the absence of the product of gene *Q* (which is the last gene in the rightward delayed early unit), however,

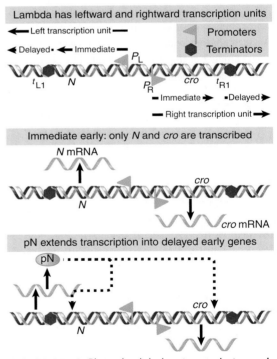

FIGURE 27.12 Phage lambda has two early transcription units. In the "leftward" unit, the "upper" strand is transcribed toward the left; in the "rightward" unit, the "lower" strand is transcribed toward the right. Genes *N* and *cro* are the immediate early functions and are separated from the delayed early genes by the terminators. Synthesis of N protein allows RNA polymerase to pass the terminators t_{L1} to the left and t_{R1} to the right.

Promoters for the lytic cycle	$P_L P_R$				$P_{R'}$	
Head genes Tail genes		Recombination	Regulation	Replication	Lysis	

*AWBCNu3DEF$_I$F$_{II}$ZUVGTHMLKIJ att int xis*αβγ*cIII N cI cro cII O P QSR*

Required for:

lysogeny	*cIII* maintains *cII*
lysogeny and lysis	*N* turns on delayed early
lysogeny	*cI* is lysogenic repressor
lysis	*cro* turns off repressor
lysogeny	*cII* turns on repressor
lysis	*Q* turns on late

FIGURE 27.11 The lambda map shows clustering of related functions. The genome is 48,514 bp.

late transcription terminates at a site t_{R3}. The transcript resulting from this termination event is 194 bases long; it is known as 6S RNA. When pQ becomes available, it suppresses termination at t_{R3} and the 6S RNA is extended, with the result that the late genes are expressed.

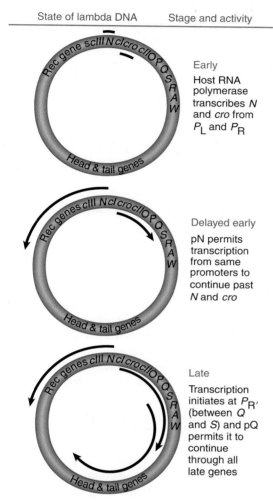

FIGURE 27.13 Lambda DNA circularizes during infection, so that the late gene cluster is intact in one transcription unit.

State of lambda DNA — Stage and activity

Early
Host RNA polymerase transcribes *N* and *cro* from P_L and P_R

Delayed early
pN permits transcription from same promoters to continue past *N* and *cro*

Late
Transcription initiates at $P_{R'}$ (between *Q* and *S*) and pQ permits it to continue through all late genes

27.8 Lysogeny Is Maintained by the Lambda Repressor Protein

Key concepts

- The lambda repressor, encoded by the *cI* gene, is required to maintain lysogeny.
- The lambda repressor acts at the O_L and O_R operators to block transcription of the immediate early genes.
- The immediate early genes trigger a regulatory cascade; as a result, their repression prevents the lytic cycle from proceeding.

Looking at the lambda lytic cascade, we see that the entire program is set in motion by the initiation of transcription at the two promoters P_L and P_R for the immediate early genes *N* and *cro*. Lambda uses antitermination to proceed to the next stage of (delayed early) expression; therefore, the same two promoters continue to be used throughout the early period.

The expanded map of the regulatory region drawn in **FIGURE 27.14** shows that the promoters P_L and P_R lie on either side of the *cI* gene. Associated with each promoter is an operator (O_L, O_R) at which repressor protein binds to prevent RNA polymerase from initiating transcription. The sequence of each operator overlaps with the promoter that it controls, and because this occurs so often these sequences are described as the P_L/O_L and P_R/O_R control regions.

The sequential nature of the lytic cascade; as a result, the control regions provide a pressure point at which entry to the entire cycle can be controlled. *By denying RNA polymerase access to these promoters, the lambda repressor protein prevents the phage genome from entering the lytic cycle.* The lambda repressor functions in the same way as repressors of bacterial operons: it binds to specific operators.

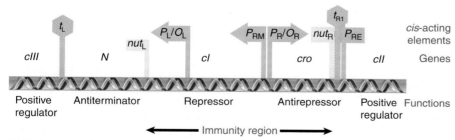

FIGURE 27.14 The lambda regulatory region contains a cluster of *trans*-acting functions and *cis*-acting elements.

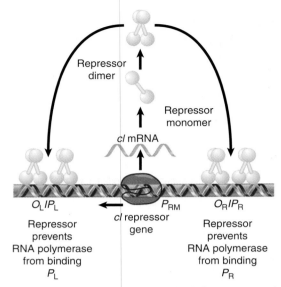

FIGURE 27.15 Repressor acts at the left operator and right operator to prevent transcription of the immediate early genes (*N* and *cro*). It also acts at the promoter P_{RM} to activate transcription by RNA polymerase of its own gene.

The lambda repressor protein is encoded by the *cI* gene. Note in Figure 27.14 that the *cI* gene has two promoters, P_{RM} (promoter right maintenance) and P_{RE} (promoter right establishment). Mutants in this gene cannot maintain lysogeny, but always enter the lytic cycle. In the time since the original isolation of the lambda repressor protein, the characterization of the repressor protein has shown how it both maintains the lysogenic state and provides immunity for a lysogen against superinfection by new phage lambda genomes.

The lambda repressor binds independently to the two operators, O_L and O_R. Its ability to repress transcription at the associated promoters is illustrated in **FIGURE 27.15**.

At O_L the lambda repressor has the same sort of effect that we have already discussed for several other systems: it prevents RNA polymerase from initiating transcription at P_L. This stops the expression of gene *N*. P_L is used for all leftward early gene transcription; thus this action prevents expression of the entire leftward early transcription unit. So the lytic cycle is blocked before it can proceed beyond early stages.

At O_R, repressor binding prevents the use of P_R so *cro* and the other rightward early genes cannot be expressed. The lambda repressor protein binding at O_R also stimulates transcription of *cI*, its own gene from P_{RM}.

The nature of this control circuit explains the biological features of lysogenic existence. Lysogeny is stable because the control circuit ensures that, so long as the level of lambda repressor is adequate, there is continued expression of the *cI* gene. The result is that O_L and O_R remain occupied indefinitely. By repressing the entire lytic cascade, this action maintains the prophage in its inert form.

27.9 The Lambda Repressor and Its Operators Define the Immunity Region

Key concepts

- Several lambdoid phages have different immunity regions.
- A lysogenic phage confers immunity to further infection by any other phage with the same immunity region.

The presence of lambda repressor explains the phenomenon of **immunity**. If a second lambda phage DNA enters a lysogenic cell, repressor protein synthesized from the resident prophage genome will immediately bind to O_L and O_R in the new genome. This prevents the second phage from entering the lytic cycle.

The operators were originally identified as the targets for repressor action by **virulent mutations** (λvir). These mutations prevent the repressor from binding at O_L or O_R, with the result that the phage inevitably proceeds into the lytic pathway when it infects a new host bacterium. Note that λvir mutants can grow on lysogens because the virulent mutations in O_L and O_R allow the incoming phage to ignore the resident repressor and thus enter the lytic cycle. Virulent mutations in phages are the equivalent of operator-constitutive mutations in bacterial operons.

A prophage is induced to enter the lytic cycle when the lysogenic circuit is broken. This happens when the repressor is inactivated (see Section *27.10, The DNA-Binding Form of the Lambda Repressor Is a Dimer*). The absence of repressor allows RNA polymerase to bind at P_L and P_R, starting the lytic cycle as shown in **FIGURE 27.16**.

The autoregulatory nature of the repressor-maintenance circuit creates a sensitive response. The presence of the lambda repressor is necessary for its own synthesis; therefore, expres-

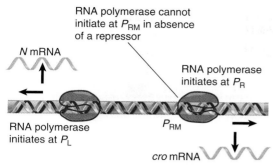

FIGURE 27.16 In the absence of repressor, RNA polymerase initiates at the left and right promoters. It cannot initiate at P_{RM} in the absence of repressor.

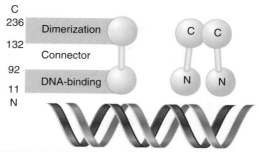

FIGURE 27.17 The N-terminal and C-terminal regions of repressor form separate domains. The C-terminal domains associate to form dimers; the N-terminal domains bind DNA.

sion of the *cI* gene stops as soon as the existing repressor is destroyed. Thus no repressor is synthesized to replace the molecules that have been damaged. This enables the lytic cycle to start without interference from the circuit that maintains lysogeny.

The region including the left and right operators, the *cI* gene, and the *cro* gene determines the immunity of the phage. Any phage that possesses this region has the same type of immunity, because it specifies both the repressor protein and the sites on which the repressor acts. Accordingly, this is called the **immunity region** (as marked in Figure 27.14). Each of the four lambdoid phages φ80, *21*, *434*, and λ has a unique immunity region. When we say that a lysogenic phage confers immunity to any other phage of the same type, we mean more precisely that the immunity is to any other phage that has the same immunity region (irrespective of differences in other regions).

27.10 The DNA-Binding Form of the Lambda Repressor Is a Dimer

Key concepts

- A repressor monomer has two distinct domains.
- The N-terminal domain contains the DNA-binding site.
- The C-terminal domain dimerizes.
- Binding to the operator requires the dimeric form so that two DNA-binding domains can contact the operator simultaneously.
- Cleavage of the repressor between the two domains reduces the affinity for the operator and induces a lytic cycle.

The lambda repressor subunit is a polypeptide of 27 kD with the two distinct domains summarized in **FIGURE 27.17**.

- The N-terminal domain, residues 1–92, provides the operator-binding site.
- The C-terminal domain, residues 132–236, is responsible for dimerization.

The two domains are joined by a connector of forty residues. When repressor is digested by a protease, each domain is released as a separate fragment.

Each domain can exercise its function independently of the other. The C-terminal fragment can form oligomers. The N-terminal fragment can bind the operators, although with a lower affinity than the intact lambda repressor. Thus the information for specifically contacting DNA is contained within the N-terminal domain, but the efficiency of the process is enhanced by the attachment of the C-terminal domain.

The dimeric structure of the lambda repressor is crucial in maintaining lysogeny. The induction of a lysogenic prophage to enter the lytic cycle is caused by cleavage of the repressor subunit in the connector region, between residues 111 and 113. (This is a counterpart to the allosteric change in conformation that results when a small-molecule inducer inactivates the repressor of a bacterial operon, a capacity that the lysogenic repressor does not have.) Induction occurs under certain adverse conditions, such as exposure of lysogenic bacteria to UV irradiation, which leads to proteolytic inactivation of the repressor.

In the intact state, dimerization of the C-terminal domains ensures that when the repressor binds to DNA, its two N-terminal domains each contact DNA simultaneously.

Cleavage releases the C-terminal domains from the N-terminal domains, though. As illustrated in **FIGURE 27.18**, this means that the N-terminal domains can no longer dimerize, which upsets the equilibrium between monomers and dimers. As a result, they do not have sufficient affinity for the lambda repressor to remain bound to DNA, which allows the lytic cycle to start. Also, two dimers usually cooperate to bind at an operator, and the cleavage destabilizes this interaction.

The balance between lysogeny and the lytic cycle depends on the concentration of repressor. Intact repressor is present in a lysogenic cell at a concentration sufficient to ensure that the operators are occupied. If the repressor is cleaved, however, this concentration is inadequate, because of the lower affinity of the separate N-terminal domain for the operator. A concentration of repressor that is too high would make it impossible to induce the lytic cycle in this way; a level that is too low, of course, would make it impossible to maintain lysogeny.

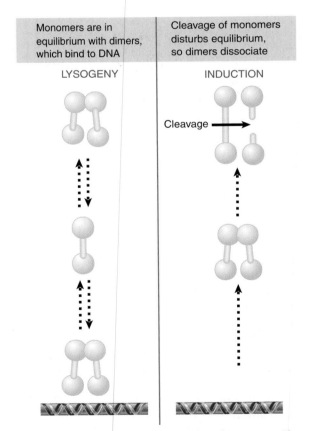

FIGURE 27.18 Repressor dimers bind to the operator. The affinity of the N-terminal domains for DNA is controlled by the dimerization of the C-terminal domains.

27.11 Lambda Repressor Uses a Helix-Turn-Helix Motif to Bind DNA

Key concepts

- Each DNA-binding region in the repressor contacts a half-site in the DNA.
- The DNA-binding site of the repressor includes two short α-helical regions that fit into the successive turns of the major groove of DNA.
- A DNA-binding site is a (partially) palindromic sequence of 17 bp.
- The amino acid sequence of the recognition helix makes contacts with particular bases in the operator sequence that it recognizes.

A repressor dimer is the unit that binds to DNA. It recognizes a sequence of 17 bp displaying partial symmetry about an axis through the central base pair. **FIGURE 27.19** shows an example of a binding site. The sequence on each side of the central base pair is sometimes called a "half-site." Each individual N-terminal region contacts a half-site. Several DNA-binding proteins that regulate bacterial transcription share a similar mode of holding DNA, in which the active domain contains two short regions of α-helix that contact DNA. (Some transcription factors in eukaryotic cells use a similar motif; see *Section 28.6, There Are Many Types of DNA-Binding Domains*.)

The N-terminal domain of lambda repressor contains several stretches of α-helix, which are arranged as illustrated diagrammatically in **FIGURE 27.20**. Two of the helical regions are responsible for binding DNA. The **helix-turn-helix** model for contact is illustrated in **FIGURE 27.21**. Looking at a single monomer, α-helix-3 consists of nine amino acids, each of which lies at an angle to the preceding region of seven amino acids that forms α-helix-2. In the dimer, the two apposed helix-3 regions lie 34 Å apart, enabling them to fit into successive major grooves of DNA. The helix-2 regions lie at an angle that would place them across the

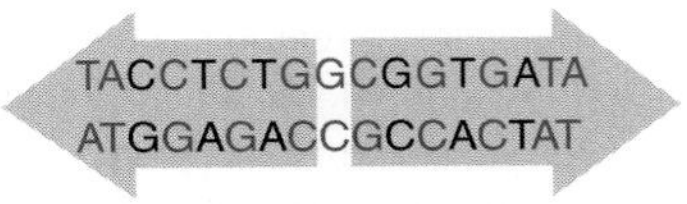

TACCTCTGGCGGTGATA
ATGGAGACCGCCACTAT

FIGURE 27.19 The operator is a 17-bp sequence with an axis of symmetry through the central base pair. Each half-site is marked in light blue. Base pairs that are identical in each operator half are in dark blue.

groove. The symmetrical binding of dimer to the site means that each N-terminal domain of the dimer contacts a similar set of bases in its half-site.

Related forms of the α-helical motifs employed in the helix-turn-helix of the lambda repressor are found in several DNA-binding proteins, including catabolite repressor protein (CRP), the *lac* repressor, and several other phage repressors. By comparing the abilities of these proteins to bind DNA, we can define the roles of each helix:

- Contacts between helix-2 and helix-3 are maintained by interactions between hydrophobic amino acids.

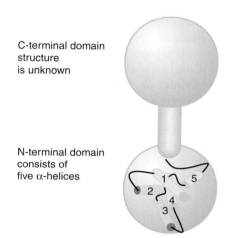

FIGURE 27.20 Lambda repressor's N-terminal domain contains five stretches of α-helix; helices 2 and 3 bind DNA.

- Contacts between helix-3 and DNA rely on hydrogen bonds between the amino acid side chains and the exposed positions of the base pairs. This helix is responsible for recognizing the specific target DNA sequence and is therefore also known as the **recognition helix**. By comparing the contact patterns summarized in **FIGURE 27.22**, we see that the lambda repressor and Cro select different sequences in the DNA as their most favored targets because they have different amino acids in the corresponding positions in helix-3.

- Contacts from helix-2 to the DNA take the form of hydrogen bonds connecting with the phosphate backbone. These interactions are necessary for binding, but do not control the specificity of target recognition. In addition to these contacts, a large part of the overall energy of interaction with DNA is provided by ionic interactions with the phosphate backbone.

What happens if we manipulate the coding sequence to construct a new protein by substituting the recognition helix in one repressor with the corresponding sequence from a closely related repressor? The specificity of the hybrid protein is that of its new recognition helix. *The amino acid sequence of this short region determines the sequence specificities of the individual proteins*

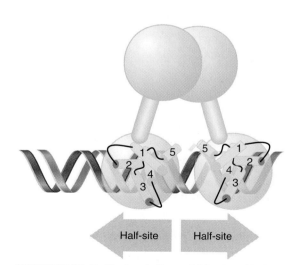

FIGURE 27.21 In the two-helix model for DNA binding, helix-3 of each monomer lies in the wide groove on the same face of DNA, and helix-2 lies across the groove.

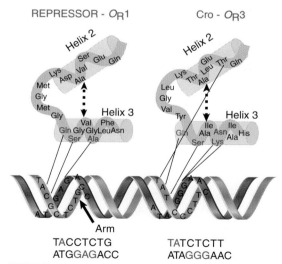

FIGURE 27.22 Two proteins that use the two-helix arrangement to contact DNA recognize lambda operators with affinities determined by the amino acid sequence of helix-3.

and is able to act in conjunction with the rest of the polypeptide chain.

The bases contacted by helix-3 lie on one face of the DNA, as can be seen from the positions indicated on the helical diagram in Figure 27.22. Repressor makes an additional contact with the other face of DNA, though. The last six N-terminal amino acids of the N-terminal domain form an "arm" extending around the back. **FIGURE 27.23** shows the view from the back. Lysine residues in the arm make contact with G residues in the major groove, and also with the phosphate backbone. The interaction between the arm and DNA contributes heavily to DNA binding; the binding affinity of a mutant armless repressor is reduced by ~1000 fold.

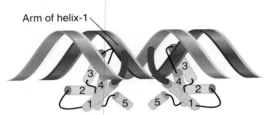

FIGURE 27.23 A view from the back shows that the bulk of the repressor contacts one face of DNA, but its N-terminal arms reach around to the other face.

27.12 Lambda Repressor Dimers Bind Cooperatively to the Operator

Key concepts

- Repressor binding to one operator increases the affinity for binding a second repressor dimer to the adjacent operator.
- The affinity is 10× greater for O_L1 and O_R1 than other operators, so they are bound first.
- Cooperativity allows repressor to bind the O_L2/O_R2 sites at lower concentrations.

Each operator contains three repressor-binding sites. As can be seen from **FIGURE 27.24**, no two of the six individual repressor-binding sites are identical, but they all conform to a consensus sequence. The binding sites within each operator are separated by spacers of 3 to 7 bp that are rich in A-T base pairs. The sites at each operator are numbered so that O_R consists of the series of binding sites O_R1-O_R2-O_R3, whereas O_L consists of the series O_L1-O_L2-O_L3. In each case, site 1 lies closest to the startpoint for transcription in the promoter, and sites 2 and 3 lie farther upstream.

Faced with the triplication of binding sites at each operator, how does the lambda repres-

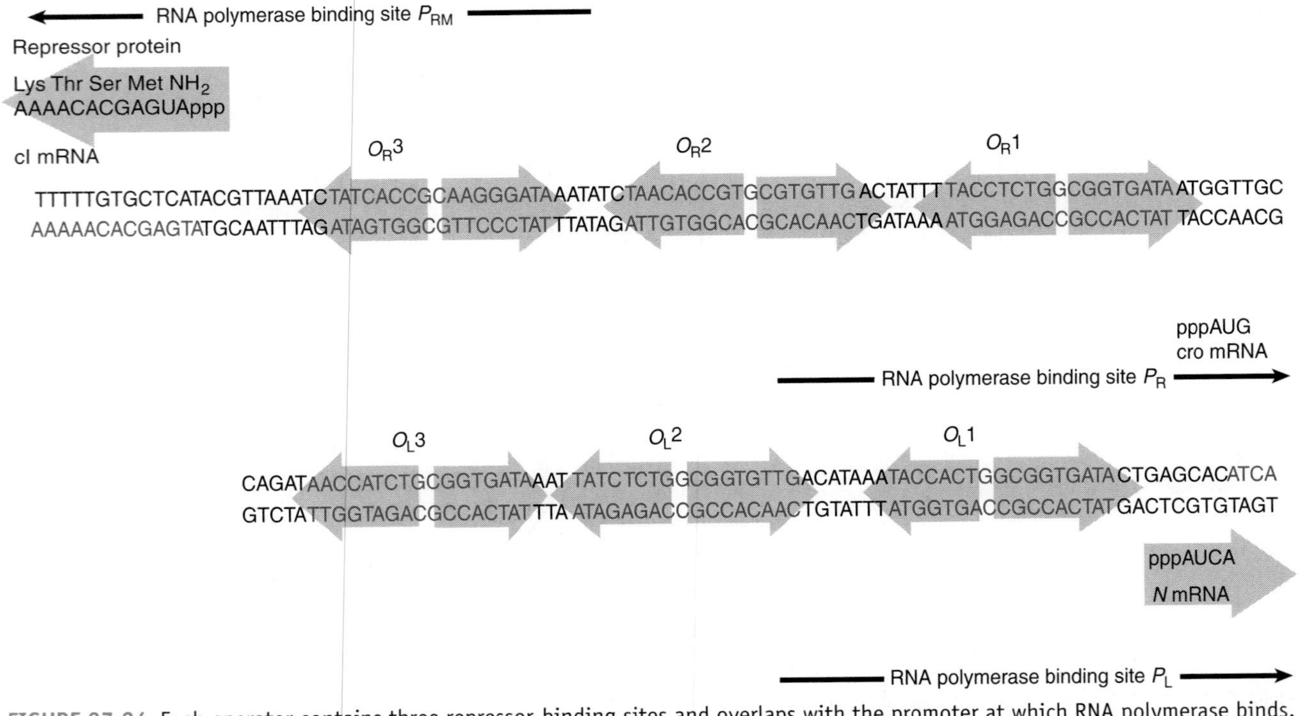

FIGURE 27.24 Each operator contains three repressor-binding sites and overlaps with the promoter at which RNA polymerase binds. The orientation of O_l has been reversed from usual to facilitate comparison with O_r.

sor decide where to start binding? At each operator, site 1 has a greater affinity (roughly tenfold) than the other sites for the lambda repressor. Thus it always binds first to O_L1 and O_R1.

Lambda repressor binds to subsequent sites within each operator in a cooperative manner. The presence of a dimer at site 1 greatly increases the affinity with which a second dimer can bind to site 2. When both sites 1 and 2 are occupied, this interaction does *not* extend farther, to site 3. At the concentrations of the lambda repressor usually found in a lysogen, both sites 1 and 2 are filled at each operator, but site 3 is not occupied.

The C-terminal domain is responsible for the cooperative interaction between dimers, as well as for the dimer formation between subunits. **FIGURE 27.25** shows that it involves both subunits of each dimer; that is, each subunit contacts its counterpart in the other dimer, forming a tetrameric structure.

A result of cooperative binding is to increase the effective affinity of repressor for the operator at physiological concentrations. This enables a lower concentration of repressor to achieve occupancy of the operator. This is an important consideration in a system in which release of repression has irreversible consequences. In an operon coding for metabolic enzymes, after all, failure to repress will merely allow unnecessary synthesis of enzymes. Failure to repress lambda prophage, however, will lead to induction of phage and lysis of the cell.

From the sequences shown in Figure 27.21, we see that O_L1 and O_R1 lie more or less in the center of the RNA polymerase binding sites of P_L and P_R, respectively. Occupancy of O_L1-O_L2 and O_R1-O_R2 thus physically blocks access of RNA polymerase to the corresponding promoters.

FIGURE 27.25 When two lambda repressor dimers bind cooperatively, each of the subunits of one dimer contacts a subunit in the other dimer.

27.13 Lambda Repressor Maintains an Autoregulatory Circuit

Key concepts

- The DNA-binding region of repressor at O_R2 contacts RNA polymerase and stabilizes its binding to P_{RM}.
- This is the basis for the autoregulatory control of repressor maintenance.
- Repressor binding at O_L blocks transcription of gene *N* from P_L.
- Repressor binding at O_R blocks transcription of *cro*, but also is required for transcription of *cI*.
- Repressor binding to the operators therefore simultaneously blocks entry to the lytic cycle and promotes its own synthesis.

Once lysogeny has been established, the *cI* gene is transcribed from the P_{RM} promoter (see Figure 27.14) that lies to its right, close to P_R/O_R. Transcription terminates at the left end of the gene. The mRNA starts with the AUG initiation codon; because of the absence of a 5′ UTR containing a ribosome binding site, this is a very poor message that is translated inefficiently, producing only a low level of protein. Note that we have not yet described how transcription for the *cI* gene is established (see *Section 27.18, The Cro Repressor Is Needed for Lytic Infection*).

The presence of the lambda repressor at O_R has dual effects as noted above (*Section 27.8, Lysogeny Is Maintained by the Lambda Repressor Protein*). It blocks expression from P_R, but it assists transcription from P_{RM}. *RNA polymerase can initiate efficiently at P_{RM} only when the lambda repressor is bound at O_R.* The lambda repressor thus behaves as a positive regulator protein that is necessary for transcription of its own gene, *cI*. This is the definition of an autoregulatory circuit.

At O_L, the repressor has the same sort of effect that we see above. It prevents RNA polymerase from initiating transcription at P_L; this stops the expression of gene *N*. P_L is used for all leftward early gene transcription. As a result, this action prevents expression of the entire leftward early transcription unit. *Thus the lytic cycle is blocked before it can proceed beyond early stages.* Its actions at O_R and O_L are summarized in **FIGURE 27.26**.

The RNA polymerase binding site at P_{RM} is adjacent to O_R2. This explains how the lambda repressor autoregulates its own synthesis. When two dimers are bound at O_R1-O_R2, the amino terminal domain of the dimer at O_R2

interacts with RNA polymerase. The nature of the interaction is identified by mutations in the repressor that abolish positive control because they cannot stimulate RNA polymerase to transcribe from P_{RM}. They map within a small group of amino acids, located on the outside of helix-2 or in the turn between helix-2 and helix-3. The mutations reduce the negative charge of the region; conversely, mutations that increase the negative charge enhance the activation of RNA polymerase. This suggests that the group of amino acids constitutes an "acidic patch" that functions by an electrostatic interaction with a basic region on RNA polymerase to activate it.

The location of these "positive control mutations" in the repressor is indicated in **FIGURE 27.27**. They lie at a site on repressor that is close to a phosphate group on DNA, which is also close to RNA polymerase. Thus the group of amino acids on repressor that is involved in positive control is in a position to contact the polymerase. The important principle is that protein–protein interactions can release energy that is used to help to initiate transcription.

The target site on RNA polymerase that the repressor contacts is in the σ^{70} subunit, which is within the region that contacts the −35 region of the promoter. The interaction between repressor and polymerase is needed for the polymerase to make the transition from a closed complex to an open complex.

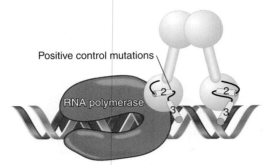

FIGURE 27.26 Lysogeny is maintained by an autoregulatory circuit.

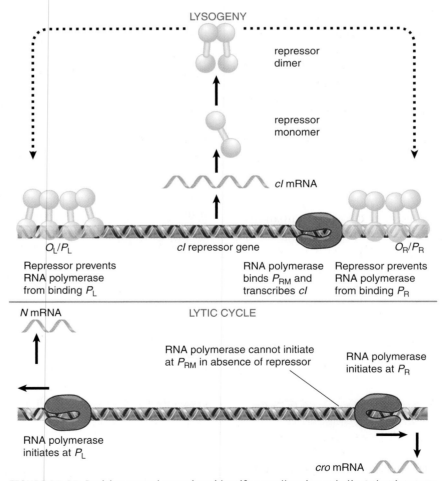

FIGURE 27.27 Positive control mutations identify a small region at helix-2 that interacts directly with RNA polymerase.

This explains how low levels of repressor positively regulate its own synthesis. As long as enough repressor is available to fill O_R2, RNA polymerase will continue to transcribe the cI gene from P_{RM}.

27.14 Cooperative Interactions Increase the Sensitivity of Regulation

Key concepts

- Repressor dimers bound at O_L1 and O_L2 interact with dimers bound at O_R1 and O_R2 to form octamers.
- These cooperative interactions increase the sensitivity of regulation.

Lambda repressor dimers interact cooperatively at both the left and right operators, so that their normal condition when occupied by repressor is to have dimers at both the 1 and 2 binding sites. In effect, each operator has a tetramer of repressor. This is not the end of the story, though. The two dimers interact with one another through their C-terminal domains to form an octamer as depicted in **FIGURE 27.28**, which shows the distribution of repressors at the operator sites that are occupied in a lysogen. Repressors are occupying O_L1, O_L2, O_R1, and O_R2, and the repressor at the last of these sites is interacting with RNA polymerase, which is initiating transcription at P_{RM}.

The interaction between the two operators has several consequences. It stabilizes repressor binding, thereby making it possible for repressor to occupy operators at lower concentrations. Binding at O_R2 stabilizes RNA polymerase binding at P_{RM}, which enables low concentrations of repressor to autogenously stimulate their own production.

The DNA between the O_L and O_R sites (that is, the gene cI) forms a large loop, which is held together by the repressor octamer. The octamer brings the sites O_L3 and O_R3 into proximity. As a result, two repressor dimers can bind to these sites and interact with one another, as shown in **FIGURE 27.29**. The occupation of O_R3 prevents RNA polymerase from binding to P_{RM} and therefore turns off expression of repressor.

This shows us how the expression of the cI gene becomes exquisitely sensitive to repressor concentration. At the lowest concentrations, it forms the octamer and activates RNA polymerase in a positive autogenous regulation. An increase in concentration allows binding

to O_L3 and O_R3 and turns off transcription in a negative autogenous regulation. The threshold levels of repressor that are required for each of these events are reduced by the cooperative interactions, which make the overall regulatory system much more sensitive. Any change in repressor level triggers the appropriate regulatory response to restore the lysogenic level.

The overall level of repressor has been reduced (about threefold from the level that would be required if there were no cooperative effects), and thus there is less repressor that has to be eliminated when it becomes necessary to induce the phage. This increases the efficiency of induction.

27.15 The cII and $cIII$ Genes Are Needed to Establish Lysogeny

Key concepts

- The delayed early gene products cII and cIII are necessary for RNA polymerase to initiate transcription at the promoter P_{RE}.
- cII acts directly at the promoter and cIII protects cII from degradation.
- Transcription from P_{RE} leads to synthesis of repressor and also blocks the transcription of cro.

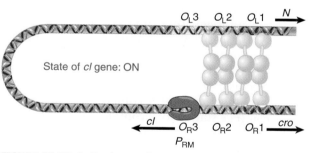

FIGURE 27.28 In the lysogenic state, the repressors bound at O_l1 and O_l2 interact with those bound at O_l1 and O_r2. RNA polymerase is bound at P_{rm} (which overlaps with O_r3) and interacts with the repressor bound at O_r2.

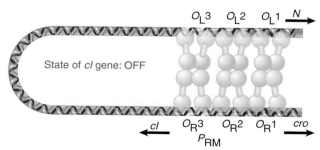

FIGURE 27.29 O_l3 and O_r3 are brought into proximity by formation of the repressor octamer, and an increase in repressor concentration allows dimers to bind at these sites and to interact.

The control circuit for maintaining lysogeny presents a paradox. *The presence of repressor protein is necessary for its own synthesis.* This explains how the lysogenic condition is perpetuated. How, though, is the synthesis of repressor established in the first place?

When a lambda DNA enters a new host cell, RNA polymerase cannot transcribe *cI* because there is no repressor present to aid its binding at P_{RM}. This same absence of repressor, however, means that P_R and P_L are available. Thus the first event after lambda DNA infects a bacterium is when genes *N* and *cro* are transcribed. After this, pN allows transcription to be extended farther. This allows *cIII* (and other genes) to be transcribed on the left, whereas *cII* (and other genes) are transcribed on the right (see Figure 27.14).

The *cII* and *cIII* genes share with *cI* the property that mutations in them hinder lytic development. There is, however, a difference. The *cI* mutants can neither establish nor maintain lysogeny. The *cII* or *cIII* mutants have some difficulty in establishing lysogeny, but once it is established they are able to maintain it by the *cI* autoregulatory circuit.

This implicates the *cII* and *cIII* genes as positive regulators whose products are needed for an alternative system for repressor synthesis. The system is needed only to initiate the expression of *cI* in order to circumvent the inability of the autoregulatory circuit to engage in *de novo* synthesis. They are not needed for continued expression.

The cII protein acts directly on gene expression as a positive regulator. Between the *cro* and *cII* genes is the second *cI* promoter, called P_{RE} (P_{RE} stands for promoter right establishment). This promoter can be recognized by RNA polymerase only in the presence of cII protein, whose action is illustrated in **FIGURE 27.30**. The

cII protein is extremely unstable *in vivo*, because it is degraded as the result of the activity of a host protein called HflA ("hfl" stands for *high frequency lysogenization*). The role of cIII is to protect cII against this degradation.

Transcription from P_{RE} promotes lysogeny in two ways. Its direct effect is that *cI* mRNA is translated into repressor protein. An indirect effect is that transcription proceeds through the *cro* gene in the "wrong" direction. Thus the 5' part of the RNA corresponds to an antisense transcript of *cro*; in fact, it hybridizes to authentic *cro* mRNA, which inhibits its translation. This is important because *cro* expression is needed to enter the lytic cycle (see *Section 27.18, The Cro Repressor Is Needed for Lytic Infection*).

The *cI* coding region on the P_{RE} transcript is very efficiently translated, in contrast with the weak translation of the P_{RM} transcript. In fact, repressor is synthesized approximately seven to eight times more effectively via expression from P_{RE} than from P_{RM}. This reflects the fact that the P_{RE} transcript has an efficient 5' UTR containing a strong ribosome-binding site, whereas the P_{RM} transcript (as noted in *Section 27.13, Lambda Repressor Maintains an Autoregulatory Circuit*) is a very poor mRNA.

27.16 A Poor Promoter Requires cII Protein

Key concepts

- P_{RE} has atypical sequences at −10 and −35.
- RNA polymerase binds the promoter only in the presence of cII.
- cII binds to sequences close to the −35 region.

The P_{RE} promoter has a poor fit with the consensus at −10 and lacks a consensus sequence at −35. This deficiency explains its dependence on the positive regulator *cII*. The promoter cannot be transcribed by RNA polymerase alone *in vitro*, but can be transcribed when cII is added. The regulator binds to a region extending from about −25 to −45. When RNA polymerase is added, an additional region, which extends from −12 to 13, is protected. As summarized in **FIGURE 27.31**, the two proteins bind to overlapping sites.

The importance of the −35 and −10 regions for promoter function, in spite of their lack of resemblance with the consensus, is indicated by the existence of *cy* mutations. These have

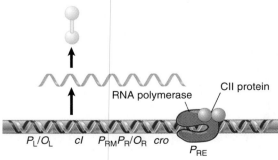

FIGURE 27.30 Repressor synthesis is established by the action of cII and RNA polymerase at P_{RE} to initiate transcription that extends from the antisense strand of *cro* through the *cI* gene.

effects similar to those of *cII* and *cIII* mutations in preventing the establishment of lysogeny, but they are *cis*-acting instead of *trans*-acting. They fall into two groups, *cyL* and *cyR*, which are localized at the consensus operator positions of −10 and −35.

The *cyL* mutations are located around −10 and probably prevent RNA polymerase from recognizing the promoter.

The *cyR* mutations are located around −35 and fall into two types, which affect either RNA polymerase or cII binding. Mutations in the center of the region do not affect cII binding; presumably they prevent RNA polymerase binding. On either side of this region, mutations in short tetrameric repeats, TTGC, prevent cII from binding. Each base in the tetramer is 10 bp (one helical turn) separated from its homolog in the other tetramer, so that when cII recognizes the two tetramers, it lies on one face of the double helix.

Positive control of a promoter implies that an accessory protein has increased the efficiency with which RNA polymerase initiates transcription. **FIGURE 27.32** reports that either or both stages of the interaction between promoter and polymerase can be the target for regulation. Initial binding to form a closed complex or its conversion into an open complex can be enhanced.

27.17 Lysogeny Requires Several Events

Key concepts

- cII and cIII cause repressor synthesis to be established and also trigger inhibition of late gene transcription.
- Establishment of repressor turns off immediate and delayed early gene expression.
- Repressor turns on the maintenance circuit for its own synthesis.
- Lambda DNA is integrated into the bacterial genome at the final stage in establishing lysogeny.

Now we can see how lysogeny is established during an infection. **FIGURE 27.33** recapitulates the early stages and shows what happens as the result of expression of *cIII* and *cII*. cIII protects cII. The presence of cII allows P_{RE} to be used for transcription extending through *cI*. Lambda repressor protein is synthesized in high amounts from this transcript and immediately binds to O_L and O_R.

By directly inhibiting any further transcription from P_L and P_R, repressor binding turns off the expression of all phage genes. This halts the synthesis of cII and cIII proteins, which are unstable; they decay rapidly, with the result that P_{RE} can no longer be used. Thus the synthesis of repressor via the establishment circuit is brought to a halt.

The lambda repressor is now present at O_{R2}, though. Acting as a positive regulator, it switches on the maintenance circuit for expression from P_{RM} by making contact with the RNA polymerase sigma factor. This may be a redundant mechanism, simply to ensure the switch. Repressor continues to be synthesized, although at the lower level typical of P_{RM} function. So the establishment circuit starts off repressor synthesis at a high level; then repressor turns off all other functions, while at the same time turning on the maintenance circuit, which functions at the low level adequate to sustain lysogeny. At even higher levels of lambda repressor, with occupancy of O_{R3}, lambda repressor turns off its own synthesis.

We shall not at this point deal in detail with the other functions needed to establish

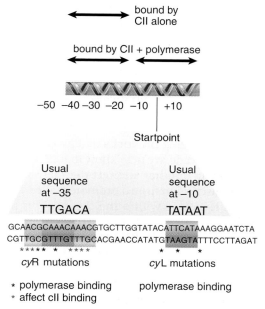

FIGURE 27.31 RNA polymerase binds to P_{RE} only in the presence of cII, which controls the region around -35.

Promoter	Regulator	Polymerase Binding (equilibrium constant, K_B)	Closed-Open Conversion (rate constant, k_2)
P_{RM}	repressor	no effect	11X
P_{RE}	cII	100X	100X

FIGURE 27.32 Positive regulation can influence RNA polymerase at either stage of initiating transcription.

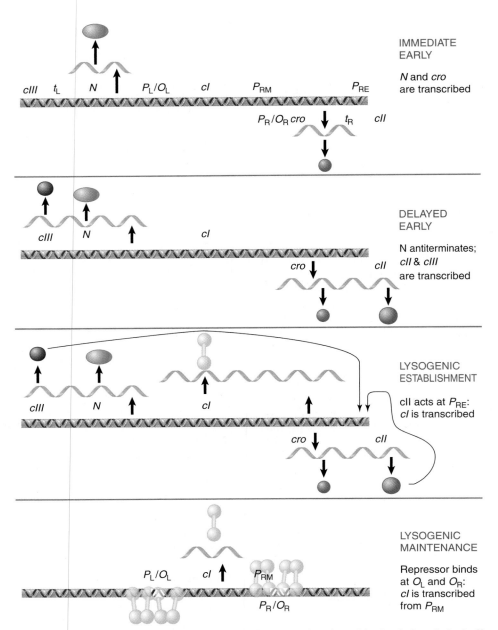

IMMEDIATE
EARLY

N and *cro*
are transcribed

cIII   *t*$_L$ *N*   *P*$_L$/*O*$_L$ *cI*   *P*$_{RM}$ *P*$_{RE}$

P$_R$/*O*$_R$ *cro*   *t*$_R$ *cII*

DELAYED
EARLY

N antiterminates;
cII & *cIII*
are transcribed

cIII   *N*   *cI*

cro   *cII*

LYSOGENIC
ESTABLISHMENT

cII acts at *P*$_{RE}$:
cI is transcribed

cIII   *N*   *cI*

cro   *cII*

LYSOGENIC
MAINTENANCE

Repressor binds
at *O*$_L$ and *O*$_R$:
cI is transcribed
from *P*$_{RM}$

P$_L$/*O*$_L$ *cI*   *P*$_{RM}$

P$_R$/*O*$_R$

FIGURE 27.33 A cascade is needed to establish lysogeny, but then this circuit is switched off and replaced by the autogenous repressor-maintenance circuit.

lysogeny, but we can just briefly remark that the infecting lambda DNA must be inserted into the bacterial genome (see *Section 15.16, Specialized Recombination Involves Specific Sites*). The insertion requires the product of gene *int*, which is expressed from its own promoter *P*$_I$, at which the cII positive regulator also is necessary. The functions necessary for establishing the lysogenic control circuit are therefore under the same control as the function needed to integrate the phage DNA into the bacterial genome. Thus the establishment of lysogeny is under a control that ensures all the necessary events occur with the same timing.

Emphasizing the tricky quality of lambda's intricate cascade, we now know that *cII* promotes lysogeny in another, indirect manner. It sponsors transcription from a promoter called *P*$_{anti-Q}$, which is located within the *Q* gene. This transcript is an antisense version of the *Q* region, and it hybridizes with *Q* mRNA to prevent translation of Q protein, whose synthesis is essential for lytic development. Thus the same mechanisms that directly promote lysogeny by causing transcription of the *cI* repressor gene also indirectly help lysogeny by inhibiting the expression of *cro* (see above) and *Q*, the regulator genes needed for the antagonistic lytic pathway.

27.18 The Cro Repressor Is Needed for Lytic Infection

Key concepts

- Cro binds to the same operators as the lambda repressor, but with different affinities.
- When Cro binds to O_R3, it prevents RNA polymerase from binding to P_{RM} and blocks the maintenance of repressor promoter.
- When Cro binds to other operators at O_R or O_L, it prevents RNA polymerase from expressing immediate early genes, which (indirectly) blocks repressor establishment.

Lambda is a temperate virus; thus it has the alternatives of entering either the lysogenic pathway or the lytic pathway. Lysogeny is initiated by establishing an autoregulatory maintenance circuit that inhibits the entire lytic cascade through applying pressure at two points, P_LO_L and P_RO_R. The two pathways begin exactly the same—with the immediate early gene expression of the N gene and the cro gene, followed by the pN-directed delayed early transcription. We now face a problem. How does the phage enter the lytic cycle?

The key requirement on the lytic cycle is the role of gene cro, which codes for another repressor protein. *Cro is responsible for preventing the synthesis of the lambda repressor protein cI;* this action shuts off the possibility of establishing lysogeny. Cro mutants usually establish lysogeny rather than entering the lytic pathway, because they lack the ability to switch events away from the expression of repressor.

Cro forms a small dimer (the monomer is 9 kD) that acts within the immunity region. It has two effects:

- It prevents the synthesis of the lambda repressor via the maintenance circuit; that is, it prevents transcription via P_{RM}.
- It also inhibits the expression of early genes from both P_L and P_R.

This means that when a phage enters the lytic pathway, Cro has responsibility both for preventing the synthesis of the lambda repressor and subsequently for turning down the expression of the early genes once there has been enough product made.

Note that Cro achieves its function by binding to the same operators as the lambda repressor protein, cI. Cro contains a region with the same general structure as the lambda repressor; a helix-2 is offset at an angle from the recogni-tion helix-3. The remainder of the structure is different, which demonstrates that the helix-turn-helix motif can operate within various contexts. As does the lambda repressor, Cro binds symmetrically at the operators.

The sequence of Cro and the lambda repressor in the helix-turn-helix region are related, which explains their ability to contact the same DNA sequence (see Figure 27.22). Cro makes similar contacts to those made by the lambda repressor, but binds to only one face of DNA; it lacks the N-terminal arms by which the lambda repressor reaches around to the other side.

How can two proteins have the same sites of action, yet have such opposite effects? The answer lies in the different affinities that each protein has for the individual binding sites within the operators. Let us just consider O_R, about which more is known, and where Cro exerts both its effects. The series of events is illustrated in **FIGURE 27.34**. (Note that the first two stages are identical to those of the lysogenic circuit shown in Figure 27.33.)

The affinity of Cro for O_R3 is greater than its affinity for O_R2 or O_R1. Thus it binds first to O_R3. This inhibits RNA polymerase from binding to P_{RM}. As a result, Cro's first action is to prevent the maintenance circuit for lysogeny from coming into play.

Cro then binds to O_R2 or O_R1. Its affinity for these sites is similar, and there is no cooperative effect. Its presence at either site is sufficient to prevent RNA polymerase from using P_R. This in turn stops the production of the early functions (including Cro itself). As a result of cII's instability, any use of P_{RE} is brought to a halt. Thus the two actions of Cro together block all production of the lambda repressor.

As far as the lytic cycle is concerned, Cro turns down (although it does not completely eliminate) the expression of the early genes. Its incomplete effect is explained by its affinity for O_R1 and O_R2, which is about eight times lower than that of the lambda repressor. This effect of Cro does not occur until the early genes have become more or less superfluous, because the pQ protein is present; by this time, the phage has started late gene expression and is concentrating on the production of progeny phage particles.

Note that in the early stages of the infection, Cro is given a head start over the lambda repressor, so it would seem that the lytic pathway is favored. Ultimately, the outcome will be determined by the concentration of the two proteins and their intrinsic DNA binding affinities.

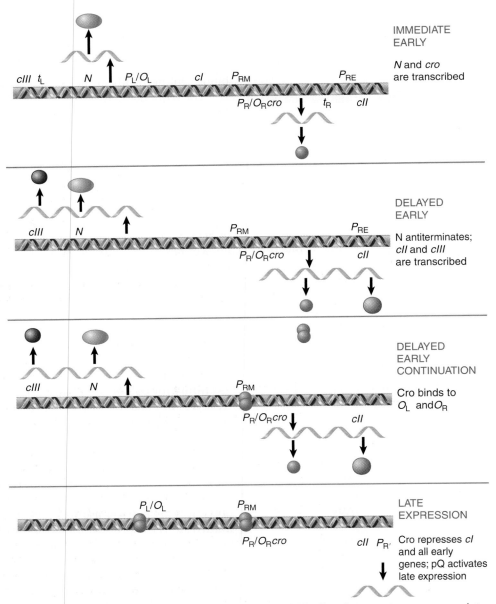

FIGURE 27.34 The lytic cascade requires Cro protein, which directly prevents repressor maintenance via P_{RM}, as well as turning off delayed early gene expression, indirectly preventing repressor establishment.

Labels within figure, top panel:

IMMEDIATE EARLY

N and *cro* are transcribed

cIII t_L *N* P_L/O_L *cI* P_{RM} P_{RE}

$P_R/O_R cro$ t_R *cII*

Second panel:

DELAYED EARLY

N antiterminates; *cII* and *cIII* are transcribed

cIII   *N* P_{RM} P_{RE}

$P_R/O_R cro$ *cII*

Third panel:

DELAYED EARLY CONTINUATION

Cro binds to O_L and O_R

cIII   *N* P_{RM}

$P_R/O_R cro$ *cII*

Fourth panel:

LATE EXPRESSION

Cro represses *cI* and all early genes; pQ activates late expression

P_L/O_L P_{RM}

$P_R/O_R cro$ *cII* $P_{R'}$

27.19 What Determines the Balance between Lysogeny and the Lytic Cycle?

Key concepts

• The delayed early stage when both Cro and repressor are being expressed is common to lysogeny and the lytic cycle.
• The critical event is whether cII causes sufficient synthesis of repressor to overcome the action of Cro.

The programs for the lysogenic and lytic pathways are so intimately related that it is impossible to predict the fate of an individual phage genome when it enters a new host bacterium. Will the antagonism between the lambda repressor and Cro be resolved by establishing the autoregulatory maintenance circuit shown in Figure 27.33, or by turning off lambda repressor synthesis and entering the late stage of development shown in Figure 27.34?

The same pathway is followed in both cases right up to the brink of decision. Both involve the expression of the immediate early genes

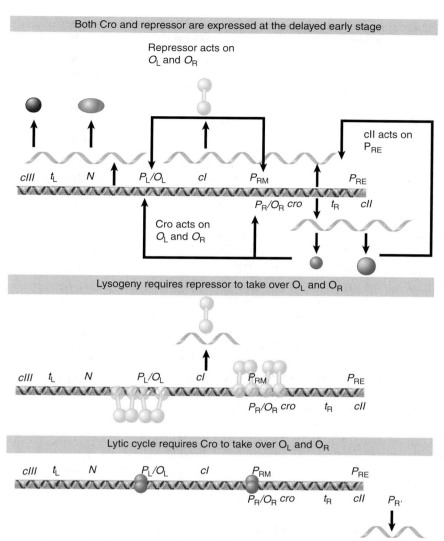

FIGURE 27.35 The critical stage in deciding between lysogeny and lysis is when delayed early genes are being expressed. If cII causes sufficient synthesis of repressor, lysogeny will result because repressor occupies the operators. Otherwise Cro occupies the operators, resulting in a lytic cycle.

and extension into the delayed early genes. The difference between them comes down to the question of whether the lambda repressor or Cro will obtain occupancy of the two operators O_L and P_L.

The early phase during which the decision is made is limited in duration in either case. No matter which pathway the phage follows, expression of all early genes will be prevented as P_L and P_R are repressed and, as a consequence of the disappearance of cII and cIII, production of repressor via P_{RE} will cease.

The critical question comes down to whether the cessation of transcription from P_{RE} is followed by activation of P_{RM} and the establishment of lysogeny, or whether P_{RM} fails to

become active and the pQ regulator commits the phage to lytic development. **FIGURE 27.35** shows the critical stage at which both repressor and Cro are being synthesized. This will be determined by how much lambda repressor was made. This in turn will be determined by how much cII transcription factor was made. Finally, this in turn will be—at least partly—determined by how much cIII protein was made.

The initial event in establishing lysogeny is the binding of lambda repressor at O_L1 and O_R1. Binding at the first sites is rapidly succeeded by cooperative binding of further repressor dimers at O_L2 and O_R2. This shuts off the synthesis of Cro and starts up the synthesis of lambda repressor via P_{RM}.

The initial event in entering the lytic cycle is the binding of Cro at O_R3. This stops the lysogenic-maintenance circuit from starting up at P_{RM}. Cro must then bind to O_R1 or O_R2, and to O_L1 or O_L2, to turn down early gene expression. By halting production of cII and cIII, this action leads to the cessation of lambda repressor synthesis via P_{RE}. The shutoff of lambda repressor establishment occurs when the unstable cII and cIII proteins decay.

The critical influence over the switch between lysogeny and lysis is how much cII protein is made. If cII is abundant, synthesis of repressor via the establishment promoter is effective, and, as a result, the lambda repressor gains occupancy of the operators. If cII is not abundant, lambda repressor establishment fails, and Cro binds to the operators.

The level of cII protein under any particular set of circumstances determines the outcome of an infection. Mutations that increase the stability of cII increase the frequency of lysogenization. Such mutations occur in *cII* itself or in other genes. The cause of cII's instability is its susceptibility to degradation by host proteases. Its level in the cell is influenced by cIII as well as by host functions.

The effect of the lambda protein cIII is secondary: it helps to protect cII against degradation. The presence of cIII does not guarantee the survival of cII; however, in the absence of cIII, cII is virtually always inactivated.

Host gene products act on this pathway. Mutations in the host genes *hflA* and *hflB* increase lysogeny. The mutations stabilize cII because they inactivate host protease(s) that degrade it.

The influence of the host cell on the level of cII provides a route for the bacterium to interfere with the decision-taking process. For example, host proteases that degrade cII are activated by growth on rich medium. Thus lambda tends to lyse cells that are growing well, but is more likely to enter lysogeny on cells that are starving (and that lack components necessary for efficient lytic growth).

27.20 Summary

Virulent phages follow a lytic life cycle, in which infection of a host bacterium is followed by production of a large number of phage particles, lysis of the cell, and release of the viruses. Temperate phages can follow the lytic pathway or the lysogenic pathway, in which the phage genome is integrated into the bacterial chromosome and is inherited in this inert, latent form like any other bacterial gene.

In general, lytic infection can be described as falling into three phases. In the first phase a small number of phage genes are transcribed by the host RNA polymerase. One or more of these genes is a regulator that controls expression of the group of genes expressed in the second phase. The pattern is repeated in the second phase, when one or more genes is a regulator needed for expression of the genes of the third phase. Genes active during the first two phases code for enzymes needed to reproduce phage DNA; genes of the final phase code for structural components of the phage particle. It is common for the very early genes to be turned off during the later phases.

In phage lambda, the genes are organized into groups whose expression is controlled by individual regulatory events. The immediate early gene N codes for an antiterminator that allows transcription of the leftward and rightward groups of delayed early genes from the early promoters P_R and P_L. The delayed early gene Q has a similar antitermination function that allows transcription of all late genes from the promoter $P_{R'}$. The lytic cycle is repressed, and the lysogenic state maintained, by expression of the *cI* gene, whose product is a repressor protein, the lambda repressor, that acts at the operators O_R and O_L to prevent use of the promoters P_R and P_L, respectively. A lysogenic phage genome expresses only the *cI* gene from its promoter, P_{RM}. Transcription from this promoter involves positive autoregulation, in which repressor bound at O_R activates RNA polymerase at P_{RM}.

Each operator consists of three binding sites for the lambda repressor. Each site is palindromic, consisting of symmetrical half-sites. Lambda repressor functions as a dimer. Each half-binding site is contacted by a repressor monomer. The N-terminal domain of repressor contains a helix-turn-helix motif that contacts DNA. Helix-3 is the recognition helix and is responsible for making specific contacts with base pairs in the operator. Helix-2 is involved in positioning helix-3; it is also involved in contacting RNA polymerase at P_{RM}. The C-terminal domain is required for dimerization. Induction is caused by cleavage between the N- and C-terminal domains, which prevents the DNA-binding regions from functioning in dimeric form, thereby reducing their affinity for DNA and making it impossible to maintain lysogeny. Lambda repressor–operator binding is cooperative, so that once one dimer has bound to the first site, a second dimer binds more readily to the adjacent site.

The helix-turn-helix motif is used by other DNA-binding proteins, including lambda Cro. Cro binds to the same operators but has a different affinity for the individual operator sites, which are determined by the sequence of helix-3. Cro binds individually to operator sites, starting with O_R3, in a noncooperative manner. It is needed for progression through the lytic cycle. Its binding to O_R3 first prevents synthesis of repressor from P_{RM}, and then its binding to O_R2 and O_R1 prevents continued expression of early genes, an effect also seen in its binding to O_L1 and O_L2.

Establishment of lambda repressor synthesis requires use of the promoter P_{RE}, which is activated by the product of the *cII* gene. The product of *cIII* is required to stabilize the *cII* product against degradation. By turning off *cII* and *cIII* expression, Cro acts to prevent lysogeny. By turning off all transcription except that of its own gene, the repressor acts to prevent the lytic cycle. The choice between lysis and lysogeny depends on whether repressor or Cro gains occupancy of the operators in a particular infection. The stability of cII protein in the infected cell is a primary determinant of the outcome.

References

27.4 Two Types of Regulatory Event Control the Lytic Cascade

Review

Greenblatt, J., Nodwell, J. R., and Mason, S. W. (1993). Transcriptional antitermination. *Nature* 364, 401–406.

27.6 Lambda Immediate Early and Delayed Early Genes Are Needed for Both Lysogeny and the Lytic Cycle

Review

Ptashne, M. (2004). *The genetic switch: Phage lambda revisited*. Cold Spring Harbor, NY: Cold Spring Harbor Press.

27.8 Lysogeny Is Maintained by the Lambda Repressor Protein

Research

Pirrotta, V., Chadwick, P., and Ptashne, M. (1970). Active form of two coliphage repressors. *Nature* 227, 41–44.

Ptashne, M. (1967). Isolation of the lambda phage repressor. *Proc. Natl. Acad. Sci. USA* 57, 306–313.

Ptashne, M. (1967). Specific binding of the lambda phage repressor to lambda DNA. *Nature* 214, 232–234.

27.9 The Lambda Repressor and Its Operators Define the Immunity Region

Review

Friedman, D. I. and Gottesman, M. (1982). *Lambda II*. Cambridge, MA: Cell Press.

27.10 The DNA-Binding Form of the Lambda Repressor Is a Dimer

Research

Pabo, C. O. and Lewis, M. (1982). The operator-binding domain of lambda repressor: structure and DNA recognition. *Nature* 298, 443–447.

27.11 Lambda Repressor Uses a Helix-Turn-Helix Motif to Bind DNA

Research

Brennan, R. G. et al. (1990). Protein-DNA conformational changes in the crystal structure of a lambda Cro-operator complex. *Proc. Natl. Acad. Sci. USA* 87, 8165–8169.

Sauer, R. T. et al. (1982). Homology among DNA-binding proteins suggests use of a conserved super-secondary structure. *Nature* 298, 447–451.

Wharton, R. L., Brown, E. L., and Ptashne, M. (1984). Substituting an α-helix switches the sequence specific DNA interactions of a repressor. *Cell* 38, 361–369.

27.12 Lambda Repressor Dimers Bind Cooperatively to the Operator

Research

Bell, C. E., Frescura, P., Hochschild, A., and Lewis, M. (2000). Crystal structure of the lambda repressor C-terminal domain provides a model for cooperative operator binding. *Cell* 101, 801–811.

Johnson, A. D., Meyer, B. J., and Ptashne, M. (1979). Interactions between DNA-bound repressors govern regulation by the phage lambda repressor. *Proc. Natl. Acad. Sci. USA* 76, 5061–5065.

27.13 Lambda Repressor Maintains an Autoregulatory Circuit

Research

Hochschild, A., Irwin, N., and Ptashne, M. (1983). Repressor structure and the mechanism of positive control. *Cell* 32, 319–325.

Li, M., Moyle, H., and Susskind, M. M. (1994). Target of the transcriptional activation function of phage lambda cI protein. *Science* 263, 75–77.

Michalowski, C. B. and Little, J. W. (2005). Positive autoregulation of CI is a dispensable feature of the phage lambda gene regulatory circuitry. *J. Bact.* 187, 6430–6442.

27.14 Cooperative Interactions Increase the Sensitivity of Regulation

Review

Ptashne, M. (2004). *The genetic switch: Phage lambda revisited*. Cold Spring Harbor, NY: Cold Spring Harbor Press.

Research

Anderson and Yang (2008). DNA looping can enhance lysogenic CI transcription in phage lambda. *Proc. Natl. Acad. Sci. USA* 105, 5827–5832.

Bell, C. E. and Lewis, M. (2001). Crystal structure of the lambda repressor C-terminal domain octamer. *J. Mol. Biol.* 314, 1127–1136.

Dodd, I. B., Perkins, A. J., Tsemitsidis, D., and Egan, J. B. (2001). Octamerization of lambda CI repressor is needed for effective repression of P(RM) and efficient switching from lysogeny. *Genes Dev.* 15, 3013–3022.

27.19 What Determines the Balance between Lysogeny and the Lytic Cycle?

Review

Oppenheim, A. B., Kobiler, O., Stavans, J., Court, D. L., and Adhya, S. (2005). Switches in bacteriophage lambda development. *Annu. Rev. Gen.* 39, 409–429.

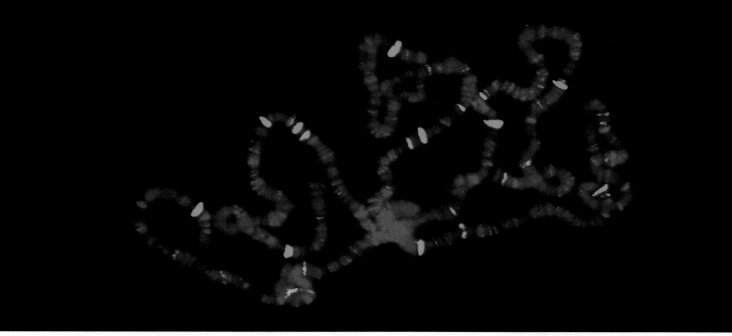

Reproduced from D. Karachentsev, et al., *Genes Dev.* 19 (2005): 431–435. Used with permission of Cold Spring Harbor Laboratory Press. Photo courtesy of Ruth Steward, Waksman Institute, Rutgers University.

Eukaryotic Transcription Regulation

CHAPTER OUTLINE

28.1 Introduction
- Eukaryotic gene expression is usually controlled at the level of initiation of transcription by opening the chromatin.

28.2 Mechanism of Action of Activators and Repressors
- Activators determine the frequency of transcription.
- Activators work by making protein–protein contacts with the basal factors.
- Activators may work via coactivators.
- Activators are regulated in many different ways.
- Some components of the transcriptional apparatus work by changing chromatin structure.
- Repression is achieved by affecting chromatin structure or by binding to and masking activators.

28.3 Independent Domains Bind DNA and Activate Transcription
- DNA-binding and transcription-activation activities are carried by independent domains of an activator.
- The role of the DNA-binding domain is to bring the transcription-activation domain into the vicinity of the promoter.

28.4 The Two-Hybrid Assay Detects Protein–Protein Interactions
- The two-hybrid assay works by requiring an interaction between two proteins, where one has a DNA-binding domain and the other has a transcription-activation domain.

28.5 Activators Interact with the Basal Apparatus
- The principle that governs the function of all activators is that a DNA-binding domain determines specificity for the target promoter or enhancer.
- The DNA-binding domain is responsible for localizing a transcription-activating domain in the proximity of the basal apparatus.
- An activator that works directly has a DNA-binding domain and an activating domain.
- An activator that does not have an activating domain may work by binding a coactivator that has an activating domain.
- Several factors in the basal apparatus are targets with which activators or coactivators interact.
- RNA polymerase may be associated with various alternative sets of transcription factors in the form of a holoenzyme complex.

28.1 Introduction

Key concept

- Eukaryotic gene expression is usually controlled at the level of initiation of transcription by opening the chromatin.

The phenotypic differences that distinguish the various kinds of cells in a higher eukaryote are largely due to differences in the expression of genes that code for proteins, that is, those transcribed by RNA polymerase II. In principle, the expression of these genes might be regulated at any one of several stages. In **FIGURE 28.1**, we can distinguish (at least) six potential control points, which form the following series:

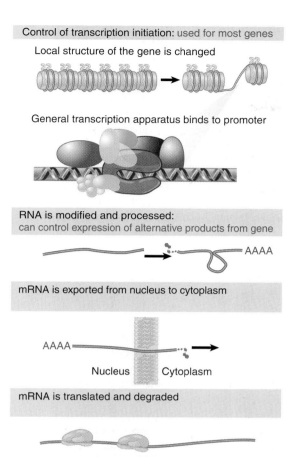

Control of transcription initiation: used for most genes

Local structure of the gene is changed

General transcription apparatus binds to promoter

RNA is modified and processed:
can control expression of alternative products from gene

AAAA

mRNA is exported from nucleus to cytoplasm

AAAA

Nucleus   Cytoplasm

mRNA is translated and degraded

FIGURE 28.1 Gene expression is controlled principally at the initiation of transcription. Control of processing may be used to determine which form of a gene is represented in mRNA. The mRNA may be regulated during transport to the cytoplasm, during translation, and by degradation.

Activation of gene structure: open chromatin
↓
Initiation of transcription and elongation
↓
Processing the transcript
↓
Transport to the cytoplasm from the nucleus
↓
Translation of mRNA
↓
Degradation and turnover of mRNA

The determination of whether a gene is expressed depends on the structure of chromatin both locally (at the promoter) and in the surrounding domain. Chromatin structure correspondingly can be regulated by individual activation events or by changes that affect a wide chromosomal region. The most localized events concern an individual target gene, where changes in nucleosomal structure and organiza-tion occur in the immediate vicinity of the pro-moter. Many genes have multiple promoters; the choice of the promoter can influence how the mRNA is used because it will change the 5′ UTR. More general changes may affect regions as large as a whole chromosome. Activation of a gene requires changes in the state of chromatin. The essential issue is how the transcription fac-tors gain access to the promoter DNA.

Local chromatin structure is an integral part of controlling gene expression. Genes may exist in either of two structural conditions. Genes are found in an "active" state only in the cells in which they are expressed. The change of structure precedes the act of transcription and indicates that the gene is "transcribable." This suggests that acquisition of the "active" structure must be the first step in gene expres-sion. Active genes are found in domains of euchromatin with a preferential susceptibility to nucleases, and hypersensitive sites are cre-ated at promoters before a gene is activated (see *Section 10.11, DNase Sensitivity Detects Changes in Chromatin Structure*).

There is an intimate and continuing con-nection between initiation of transcription and chromatin structure. Some activators of gene transcription directly modify histones; in par-ticular, acetylation of histones is associated with gene activation. Conversely, some repressors of transcription function by deacetylating his-tones. Thus a reversible change in histone struc-ture in the vicinity of the promoter is involved in the control of gene expression. These changes influence the association of histone octamers with DNA, and are responsible for controlling the presence and structure of nucleosomes at specific sites. This is an important aspect of the mechanism by which a gene is maintained in an active or inactive state.

The mechanisms by which regions of chro-matin are maintained in an inactive (silent) state are related to the means by which an individual promoter is repressed. The proteins involved in the formation of heterochromatin act on chromatin via the histones, and modi-fications of the histones are an important fea-ture in the interaction. Once established, such changes in chromatin can persist through cell divisions, creating an **epigenetic** state in which the properties of a gene are determined by the self-perpetuating structure of chromatin. The name *epigenetic* reflects the fact that a gene may have an inherited condition (it may be active or inactive) that does not depend on its sequence (see Chapter 29, *Epigenetic Effects Are Inherited*).

Once transcription begins, regulation during the elongation phase of transcription is less likely. Attenuation as we saw in bacteria (see *Section 26.13, The* trp *Operon Is Also Controlled by Attenuation*) cannot occur in eukaryotes because of the separation of chromosomes from the cytoplasm by the nuclear membrane. However, control of transcription elongation does occur. The primary transcript is modified by capping at the 5′ end, and in general also is modified by polyadenylation at the 3′ end (see Chapter 21, *RNA Splicing and Processing*). Many genes also have multiple termination sites, which can alter the 3′ UTR and thus mRNA function and behavior.

Introns must be excised from the transcripts of interrupted genes. The mature RNA must then be exported from the nucleus to the cytoplasm. Regulation of gene expression at the level of nuclear RNA processing might involve any or all of these stages, but the one for which we have most evidence concerns changes in splicing; some genes are expressed by means of alternative splicing patterns whose regulation controls the type of protein product (see *Section 21.12, Alternative Splicing Is a Rule, Rather Than an Exception, in Multicellular Eukaryotes*).

The translation of an mRNA in the cytoplasm can be specifically controlled, as can the turnover rate of the mRNA. While translation level control is uncommon in adult somatic cells, it does occur in some embryonic situations. This can also involve the localization of the mRNA to specific sites where it is expressed; in addition, the blocking of initiation of translation by specific protein factors may occur. Different mRNAs may have different intrinsic half-lives determined by specific sequence elements.

Regulation of tissue-specific gene transcription lies at the heart of eukaryotic differentiation. It is also important for control of metabolic and catabolic pathways. A regulatory transcription factor serves to provide common control of a large number of target genes, and we seek to answer two questions about this mode of regulation: "How does the transcription factor identify its group of target genes?" and "How is the activity of the transcription factor itself regulated in response to intrinsic or extrinsic signals?"

28.2 Mechanism of Action of Activators and Repressors

Key concepts

- Activators determine the frequency of transcription.
- Activators work by making protein–protein contacts with the basal factors.
- Activators may work via coactivators.
- Activators are regulated in many different ways.
- Some components of the transcriptional apparatus work by changing chromatin structure.
- Repression is achieved by affecting chromatin structure or by binding to and masking activators.

Initiation of transcription involves many protein–protein interactions between transcription factors bound at enhancers with the basal apparatus that assembles at the promoter, including RNA polymerase. We can divide these transcription factors into two opposing classes: positive **activators** and negative **repressors**.

We saw in Chapter 26 that **positive control** in bacteria entails a regulator that aids the RNA polymerase in the transition from the closed complex to the open complex. Transcription factors like the *E. coli* CRP typically bind close to the promoter to allow the CTD of the α subunit of RNA polymerase to make direct physical contact. This usually occurs in a gene having a poor promoter sequence. The activator functions to overcome the inability of the RNA polymerase to open the promoter. Positive control in eukaryotes is quite different. We can identify three classes of activators that differ by function.

The first class is the **true activators** (see *Section 20.9, Enhancers Contain Bidirectional Elements That Assist Initiation*). These are the classical transcription factors that function by making direct physical contact with the basal apparatus at the promoter (see *Section 28.3, Independent Domains Bind DNA and Activate Transcription*) either directly, or indirectly, through a coactivator. These transcription factors function on DNA or chromatin templates.

The activity of a true activator may be regulated in any one of several ways, as illustrated schematically in **FIGURE 28.2**:

- A factor is tissue-specific because it is synthesized only in a particular type of

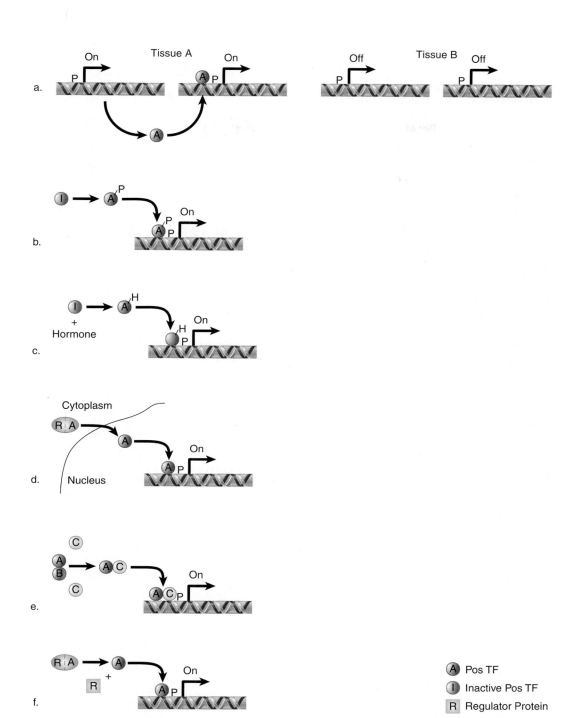

FIGURE 28.2 The activity of a positive regulatory transcription factor may be controlled by synthesis of protein, covalent modification of protein, ligand binding, or binding of inhibitors that sequester the protein or affect its ability to bind to DNA.

cell. This is typical of factors that regulate development, such as homeodomain proteins.

- The activity of a factor may be directly controlled by modification. HSF (*heat shock transcription factor*) is converted to the active form by phosphorylation.

- A factor is activated or inactivated by binding a ligand. The steroid receptors are prime examples. Ligand binding may influence the localization of the protein (causing transport from cytoplasm to nucleus), as well as determine its ability to bind to DNA.

- Availability of a factor may vary; for example, the factor NF-κB (which activates immunoglobulin κ genes in B lymphocytes) is present in many cell types. It is sequestered or masked in the cytoplasm, however, by the inhibitory protein I-κB. In B lymphocytes, NF-κB is released from I-κB and moves to the nucleus, where it activates transcription.

- A dimeric factor may have alternative partners. One partner may cause it to be inactive; synthesis of the active partner may displace the inactive partner. Such situations may be amplified into networks in which various alternative partners pair with one another, especially among the HLH proteins.

- The factor may be cleaved from an inactive precursor. One activator is produced as a protein bound to the nuclear envelope and endoplasmic reticulum. The absence of sterols (such as cholesterol) causes the cytosolic domain to be cleaved; it then translocates to the nucleus and provides the active form of the activator.

The second class includes the **antirepressors**. When one of these activators is bound to its enhancer, it recruits the histone modifier enzymes and/or the chromatin remodeler complexes to convert the chromatin from the closed state to the open state. This class has no activity on a DNA template; it only functions on chromatin templates (described below in *Section 28.7, Chromatin Remodeling Is an Active Process*).

The third class includes **architectural proteins** such as Yin-Yang; these proteins function to bend the DNA, either bringing bound proteins together to facilitate forming a cooperative complex, or bending the DNA the other way to

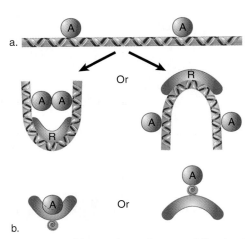

FIGURE 28.3 Architectural proteins control the structure of DNA and thus control whether bound proteins can contact each other.

prevent complex formation, as shown in **FIGURE 28.3**. Note that a strand of DNA may thus be bent in two different directions depending on whether the regulator binds to the top or to the bottom. This is a difference of one half of a turn of the helix, which is ~5 base pairs (10.5 bp per turn).

We have seen several examples of **negative control** in bacteria, in the *lac* operon and in the *trp* operon in Chapter 26. Repression can occur in bacteria when the repressor prevents the RNA polymerase from converting from the closed complex to the open complex as in the *lac* operon, or binds to the promoter sequence to prevent polymerase from binding as in the *trp* operon. There are many more mechanisms by which repressors act in eukaryotes, which are illustrated in **FIGURE 28.4**.

- One mechanism of action by which a eukaryote repressor can prevent gene expression is to *sequester an activator* in the cytoplasm. Eukaryotic proteins are synthesized in the cytoplasm. Proteins that function in the nucleus have a domain that directs their transport through the nuclear membrane. A repressor can bind to that domain and mask it.

- Several variations of that mechanism are possible. One that takes place in the nucleus occurs when the repressor binds to an activator that is already bound to an enhancer and *masks its activation domain*, thus preventing it from functioning (such as with the Gal80 repressor; see *Section 28.14, Yeast GAL Genes: A Model for Activation and Repression*).

- Alternatively, the repressor can be *masked and held in the cytoplasm* until it is released to enter the nucleus.
- A fourth mechanism is simple *competition for an enhancer*, where either the repressor and activator have the same binding site sequence or have overlapping but different binding site sequences. This is a very versatile mechanism for a cell because there are two variables at work here: one is strength of factor binding to DNA and the second is factor concentration. By only slightly varying the concentration of a factor, a cell can dramatically alter its developmental path.

The transcription factors that recruit the histone modifiers and chromatin remodelers have their counterparts as repressors that recruit the complexes that undo the modifications and remodeling. The same is true for the architectural proteins, where, in fact, the same protein bound to a different site prevents activator complexes from forming.

28.3 Independent Domains Bind DNA and Activate Transcription

Key concepts

- DNA-binding and transcription-activation activities are carried by independent domains of an activator.
- The role of the DNA-binding domain is to bring the transcription-activation domain into the vicinity of the promoter.

We know the most about the activator class of transcription factors. Activators require protein domains with multiple functions:

- They recognize specific DNA target sequences located in enhancers that affect a particular target gene.
- Having bound to DNA, an activator exercises its function by binding to components of the basal transcription apparatus.
- Many require a dimerization domain to form complexes with other proteins.

Can we characterize domains in the activator that are responsible for these activities? Often an activator has a separate domain that binds DNA and a separate domain that activates transcription. Each domain behaves as a separate module that functions independently

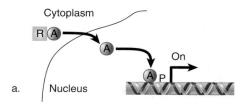
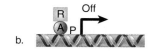
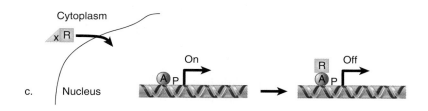
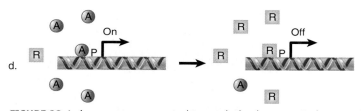

FIGURE 28.4 A repressor may control transcription by sequestering an activator in the cytoplasm, by binding an activator and masking its activation domain, by being held in the cytoplasm until it is needed, or by competition with an activator for a binding site.

when it is linked to a domain of the other type. The geometry of the overall transcription complex must allow the activating domain to contact the basal apparatus irrespective of the exact location and orientation of the DNA-binding domain.

Enhancer elements near the promoter may still be an appreciable distance from the startpoint, and in many cases may be oriented in either direction. Enhancers may even be farther away and always show orientation independence. This organization has implications for both the DNA and proteins. The DNA may be looped or condensed in some way to allow the formation of the transcription complex. In addition, the domains of the activator may be connected in a flexible way, as illustrated diagrammatically in **FIGURE 28.5**. The main point here is that the DNA-binding and activating domains are independent, and are connected

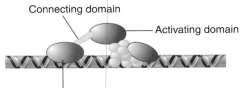

FIGURE 28.5 DNA-binding and activating functions in a transcription factor may comprise independent domains of the protein.

in a way that allows the activating domain to interact with the basal apparatus irrespective of the orientation and exact location of the DNA-binding domain.

Binding to DNA is usually necessary for activating transcription, but there are transcription factors that function without a DNA-binding domain by virtue of protein–protein dimerization. Does activation depend on the particular DNA-binding domain? This question has been answered by making hybrid proteins that consist of the DNA-binding domain of one activator linked to the activation domain of another activator. The hybrid functions in transcription at sites dictated by its DNA-binding domain, but in a way determined by its activation domain.

This result fits the modular view of transcription activators. *The function of the DNA-binding domain is to bring the activation domain to the basal apparatus at the promoter.* Precisely how or where it is bound to DNA is irrelevant, but once it is there, the activation domain can play its role. This explains why the exact locations of DNA-binding sites can vary. The ability of the two types of module to function in hybrid proteins suggests that each domain of the protein folds independently into an active structure that is not influenced by the rest of the protein.

28.4 The Two-Hybrid Assay Detects Protein–Protein Interactions

Key concept

• The two-hybrid assay works by requiring an interaction between two proteins, where one has a DNA-binding domain and the other has a transcription-activation domain.

The model of domain independence is the basis for an extremely useful assay for detecting protein interactions. The principle is illustrated in **FIGURE 28.6**. We fuse one of the proteins to be tested to a DNA-binding domain. We fuse the other protein to a transcription-activating

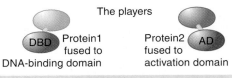
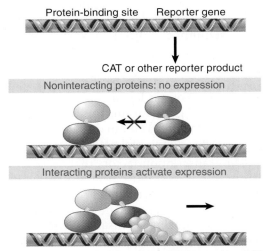

FIGURE 28.6 The two-hybrid technique tests the ability of two proteins to interact by incorporating them into hybrid proteins where one has a DNA-binding domain and the other has a transcription-activating domain.

domain. (This is done by linking the appropriate coding sequences in each case and making synthetic proteins by expressing each hybrid gene.)

If the two proteins that are being tested can interact with one another, the two hybrid proteins will interact. This is reflected in the name of the technique: the two-hybrid assay. The protein with the DNA-binding domain binds to a reporter gene that has a simple promoter containing its target site. It cannot, however, activate the gene by itself. Activation occurs only if the second hybrid binds to the first hybrid to bring the activation domain to the promoter. Any reporter gene can be used where the product is readily assayed, and this technique has given rise to several automated procedures for rapidly testing protein–protein interactions.

The effectiveness of the technique dramatically illustrates the modular nature of proteins. Even when fused to another protein, the DNA-binding domain can bind to DNA and the transcription-activating domain can activate transcription. Correspondingly, the interaction ability of the two proteins being tested is not inhibited by the attachment of the DNA-binding or transcription-activating domains. (Of course, there are some exceptions for which these simple rules do not apply and interference between the domains of the hybrid protein prevents the technique from working.)

The power of this assay is that it requires only that the two proteins being tested can interact with each other. They need not have anything to do with transcription. As a result of the independence of the DNA-binding and transcription-activating domains, all we require is that they are brought together. This will happen so long as the two proteins being tested can interact in the environment of the nucleus.

28.5 Activators Interact with the Basal Apparatus

Key concepts

- The principle that governs the function of all activators is that a DNA-binding domain determines specificity for the target promoter or enhancer.
- The DNA-binding domain is responsible for localizing a transcription-activating domain in the proximity of the basal apparatus.
- An activator that works directly has a DNA-binding domain and an activating domain.
- An activator that does not have an activating domain may work by binding a coactivator that has an activating domain.
- Several factors in the basal apparatus are targets with which activators or coactivators interact.
- RNA polymerase may be associated with various alternative sets of transcription factors in the form of a holoenzyme complex.

The true activator class of transcription factors may work directly when it consists of a DNA-binding domain linked to a transcription-activating domain, as illustrated in Figure 28.4. In other cases, the activator does not itself have a transcription-activating domain (or contains only a weak activation domain), but binds another protein—a coactivator—that has the transcription-activating activity. **FIGURE 28.7** shows the action of such an activator. We may regard **coactivators** as transcription factors whose specificity is conferred by the ability to bind to proteins that bind to DNA instead of directly to DNA. A particular activator may require a specific coactivator.

Although the protein components are organized differently, the mechanism is the same. An activator that contacts the basal apparatus directly has an activation domain covalently connected to the DNA-binding domain. When an activator works through a coactivator, the connections involve noncovalent binding between protein subunits (compare Figure 28.4

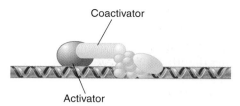

FIGURE 28.7 An activator may bind a coactivator that contacts the basal apparatus.

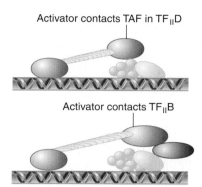

FIGURE 28.8 Activators may work at different stages of initiation by contacting the TAFs of TF$_{II}$D or by contacting TF$_{II}$B.

and Figure 28.5). The same interactions are responsible for activation, irrespective of whether the various domains are present in the same protein subunit or divided into multiple protein subunits. In addition, many coactivators also contain additional enzymatic activities that promote transcription activation, such as activities that modify chromatin structure (see *Section 28.9, Histone Acetylation Is Associated with Transcription Activation*).

An activation domain works by making protein–protein contacts with general transcription factors that promote assembly of the **basal apparatus**. Contact with the basal apparatus may be made with any one of several basal factors, but typically occurs with TF$_{II}$D, TF$_{II}$B, or TF$_{II}$A. All of these factors participate in early stages of assembly of the basal apparatus (see Figure 20.11). **FIGURE 28.8** illustrates the situation when such a contact is made. The major effect of the activators is to influence the assembly of the basal apparatus.

TF$_{II}$D may be the most common target for activators, which may contact any one of several TAFs. In fact, a major role of the TAFs is to provide the connection from the basal apparatus to activators. This explains why TBP alone can support basal-level transcription, whereas the TAFs of TF$_{II}$D are required for the higher levels of transcription that are stimulated by activators. Different TAFs in TF$_{II}$D may provide surfaces that interact with different activators. Some activators interact only with individual TAFs; others interact with multiple TAFs. We assume that the interaction assists the binding of TF$_{II}$D to the TATA box, assists the binding of other basal apparatus components around the TF$_{II}$D-TATA box complex, or controls the phosphorylation of the CTD. In any case, the interaction stabilizes the basal transcription complex, speeds the process of initiation, and thereby increases use of the promoter.

The activating domains of the yeast activators Gal4 (see *Section 28.14, Yeast GAL Genes: A Model for Activation and Repression*) and others have multiple negative charges, giving rise to their description as "acidic activators." Acidic activators function by enhancing the ability of TF$_{II}$B to join the basal initiation complex. Experiments *in vitro* show that binding of TF$_{II}$B to an initiation complex at an adenovirus promoter is stimulated by the presence of Gal4 or other acid activators, and that the activator can bind directly to TF$_{II}$B. Assembly of TF$_{II}$B into the complex at this promoter is therefore a rate-limiting step that is stimulated by the presence of an acidic activator.

The resilience of an RNA polymerase II promoter to the rearrangement of elements, and its indifference even to the particular elements present, suggests that the events by which it is activated are relatively general in nature. Any activators whose activating region is brought within range of the basal initiation complex may be able to stimulate its formation. Some striking illustrations of such versatility have been accomplished by constructing promoters consisting of new combinations of elements.

How does an activator stimulate transcription? We can imagine two general types of model:

- The recruitment model argues that the activator's sole effect is to increase the binding of RNA polymerase to the promoter.
- An alternative model is to suppose that the activator induces some change in the transcriptional complex—for example, in the conformation of enzymes such as protein kinases, which increases its efficiency.

When we add up all the components required for efficient transcription—basal factors, RNA polymerase, activators, and coactivators—we get a very large apparatus that consists of ~40 proteins. Is it feasible for this apparatus to assemble step by step at the promoter? Some activators,

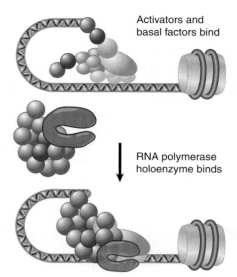

FIGURE 28.9 RNA polymerase exists as a holoenzyme containing many activators.

coactivators, and basal factors may assemble stepwise at the promoter, but then they may be joined by a very large complex consisting of RNA polymerase preassembled with further activators and coactivators, as illustrated in **FIGURE 28.9**.

Several forms of RNA polymerase in which the enzyme is associated with various transcription factors have been found. The most prominent "holoenzyme complex" in yeast (defined as being capable of initiating transcription without additional components) consists of RNA polymerase associated with a 20-subunit complex called **Mediator**. Mediator includes products of several genes in which mutations block transcription, including some *SRB* loci (so named because many of their genes were originally identified as suppressors of mutations in RNA polymerase B). The name was suggested by its ability to mediate the effects of activators. Mediator is necessary for transcription of most yeast genes. Homologous complexes are required for the transcription of most genes in multicellular eukaryotes as well. Mediator undergoes a conformational change when it interacts with the CTD of RNA polymerase. It can transmit either activating or repressing effects from upstream components to the RNA polymerase. It is probably released when a polymerase starts elongation. Some transcription factors influence transcription directly by interacting with RNA polymerase or the basal apparatus, whereas others work by manipulating the structure of chromatin (see *Section 28.7 Chromatin Remodeling Is an Active Process*).

28.6 There Are Many Types of DNA-Binding Domains

Key concepts

- Activators are classified according to the type of DNA-binding domain.
- Members of the same group have sequence variations of a specific motif that confer specificity for individual DNA target sites.

It is common for an activator to have a modular structure in which different domains are responsible for binding to DNA and for activating transcription. Factors are often classified according to the type of DNA-binding domain. In general, a relatively short motif in this domain is responsible for binding to DNA:

- The **zinc finger** motif comprises a DNA-binding domain. It was originally recognized in factor $TF_{III}A$, which is required for RNA polymerase III to transcribe 5S rRNA genes.

  The consensus sequence of a single finger is:

  $$Cys-X_{2-4}-Cys-X_3-Phe-X_5-Leu-X_2-His-X_3-His$$

  The motif takes its name from the loop of ~23 amino acids that protrudes from the zinc-binding site and is described as the Cys_2/His_2 finger. The zinc is held in a tetrahedral structure formed by the conserved Cys and His residues. This motif has since been identified in numerous other transcription factors (and presumed transcription factors). Proteins often contain multiple zinc fingers, such as the three shown in **FIGURE 28.10**. Some zinc finger proteins can bind to RNA.

- **Steroid receptors** (and some other proteins) have another type of zinc finger that is different from the Cys_2/His_2 finger. Its structure is based on a sequence with the zinc-binding consensus:

  $$Cys-X_2-Cys-X_{13}-Cys-X_2-Cys$$

  These sequences are called Cys_2/Cys_2 fingers. The steroid receptors are defined as a group by a functional relationship: each receptor is activated by binding a particular steroid, such as glucocorticoid binding to the glucocorticoid receptor. Together with other receptors, such as the thyroid hormone receptor or the retinoic acid receptor, the steroid

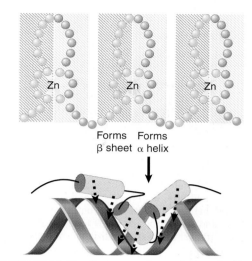

FIGURE 28.10 Zinc fingers may form α helices that insert into the major groove, which is associated with β sheets on the other side.

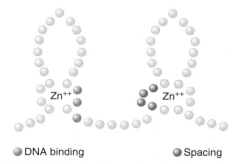

DNA binding   Spacing

FIGURE 28.11 The first finger of a steroid receptor controls which DNA sequence is bound (positions shown in purple); the second finger controls spacing between the sequences (positions shown in blue).

receptors are members of the superfamily of ligand-activated activators with the same general modus operandi: the protein factor is inactive until it binds a small ligand, as shown in **FIGURE 28.11**. The steroid receptors bind to DNA as dimers, either homodimers or heterodimers. Each monomer of the dimer binds to a half-site that may be palindromic or directly repeated.

- The **helix-turn-helix** motif was originally identified as the DNA-binding domain of phage repressors. The C-terminal α-helix lies in the major groove of DNA and is the recognition helix; the middle α-helix lies at an angle across DNA. The N-terminal arm lies in the minor groove and makes additional contacts. A related form of the motif is present in the **homeodomain**, a sequence first characterized in

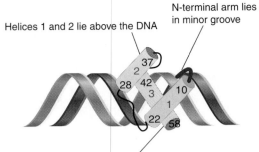

Helices 1 and 2 lie above the DNA

N-terminal arm lies in minor groove

37
2
28 42
3 10
1
22 58

Helix 3 lies in the major groove

FIGURE 28.12 Helix 3 of the homeodomain binds in the major groove of DNA, with helices 1 and 2 lying outside the double helix. Helix 3 contacts both the phosphate backbone and specific bases. The N-terminal arm lies in the minor groove, and makes additional contacts.

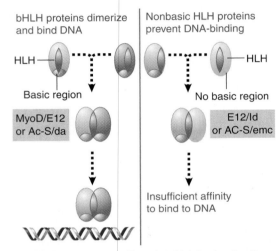

bHLH proteins dimerize and bind DNA

Nonbasic HLH proteins prevent DNA-binding

HLH

HLH

Basic region

No basic region

MyoD/E12 or Ac-S/da

E12/Id or AC-S/emc

Insufficient affinity to bind to DNA

FIGURE 28.13 An HLH dimer in which both subunits are of the bHLH type can bind DNA, but a dimer in which one subunit lacks the basic region cannot bind DNA.

several proteins encoded by *Homeobox* genes involved in developmental regulation in *Drosophila,* and by the comparable human *Hox* genes shown in **FIGURE 28.12**. Homeodomain proteins can be activators or repressors.

- The amphipathic **helix-loop-helix (HLH)** motif has been identified in some developmental regulators and in genes coding for eukaryotic DNA-binding proteins. Each *amphipathic helix* presents a face of hydrophobic residues on one side and charged residues on the other side. The length of the connecting loop varies from 12 to 28 amino acids. The motif enables proteins to dimerize, either homodimers or heterodimers, and a basic region near this motif contacts DNA as seen in **FIGURE 28.13**. Not all of the HLH proteins contain a DNA-

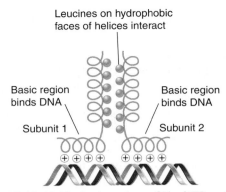

Leucines on hydrophobic faces of helices interact

Basic region binds DNA

Basic region binds DNA

Subunit 1

Subunit 2

FIGURE 28.14 The basic regions of the bZIP motif are held together by the dimerization at the adjacent zipper region when the hydrophobic faces of two leucine zippers interact in parallel orientation.

binding domain, but rather rely on their partner for sequence specificity. Partners may change during development to provide additional combinations.

- **Leucine zippers** consist of an amphipathic α-helix with a leucine residue in every seventh position. The hydrophobic groups, including leucine, face one side while the charged groups face the other side. A leucine zipper domain in one polypeptide interacts with a leucine zipper domain in another polypeptide to form a protein dimer. There are rules for which zippers may dimerize. Adjacent to each zipper is another domain containing positively charged residues that is involved in binding to DNA; this is known as the **bZIP ("basic zipper")** structural motif shown in **FIGURE 28.14**.

28.7 Chromatin Remodeling Is an Active Process

Key concepts

- There are numerous chromatin-remodeling complexes that use energy provided by hydrolysis of ATP.
- All remodeling complexes contain a related ATPase catalytic subunit, and are grouped into subfamilies containing more closely related ATPase subunits.
- Remodeling complexes can alter, slide, or displace nucleosomes.
- Some remodeling complexes can exchange one histone for another in a nucleosome.

Transcriptional activators face a challenge when trying to bind to their recognition sites in eukaryotic chromatin. **FIGURE 28.15** illustrates two general states that can exist at a eukaryotic

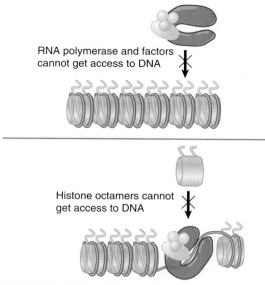

RNA polymerase and factors cannot get access to DNA

Histone octamers cannot get access to DNA

FIGURE 28.15 If nucleosomes form at a promoter, transcription factors (and RNA polymerase) cannot bind. If transcription factors (and RNA polymerase) bind to the promoter to establish a stable complex for initiation, histones are excluded.

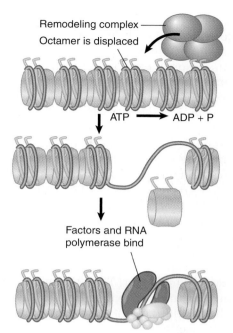

Remodeling complex
Octamer is displaced

ATP → ADP + P

Factors and RNA polymerase bind

FIGURE 28.16 The dynamic model for transcription of chromatin relies upon factors that can use energy provided by hydrolysis of ATP to displace nucleosomes from specific DNA sequences.

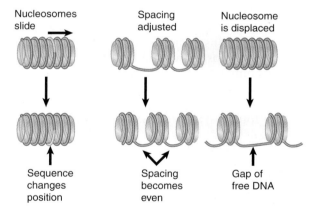

Nucleosomes slide

Spacing adjusted

Nucleosome is displaced

Sequence changes position

Spacing becomes even

Gap of free DNA

FIGURE 28.17 Remodeling complexes can cause nucleosomes to slide along DNA, can displace nucleosomes from DNA, or can reorganize the spacing between nucleosomes.

promoter. In the inactive state, nucleosomes are present, and they prevent basal factors and RNA polymerase from binding. In the active state, the basal apparatus occupies the promoter, and histone octamers cannot bind to it. Each type of state is stable. In order to convert a promoter from the inactive state to the active state, the chromatin structure must be perturbed in order to allow binding of the basal factors.

The general process of inducing changes in chromatin structure is called **chromatin remodeling**. This consists of mechanisms for displacing histones that depend on the input of energy. Many protein–protein and protein–DNA contacts need to be disrupted to release histones from chromatin. There is no free ride: energy must be provided to disrupt these contacts. **FIGURE 28.16** illustrates the principle of a dynamic model by a factor that hydrolyzes ATP. When the histone octamer is released from DNA, other proteins (in this case transcription factors and RNA polymerase) can bind.

There are several alternative outcomes of chromatin remodeling, summarized in **FIGURE 28.17**:

- Histone octamers may *slide* along DNA, changing the relationship between the nucleic acid and the protein. This can alter both the rotational and the translational position of a particular sequence on the nucleosome.

- The *spacing* between histone octamers may be changed, again with the result that the positions of individual sequences are altered relative to protein.
- The most extensive change is that an octamer(s) may be *displaced entirely* from DNA to generate a nucleosome-free gap. Alternatively, one or both H2A-H2B dimers can be displaced.

A major role of chromatin remodeling is to change the organization of nucleosomes at the promoter of a gene that is to be transcribed. This is required to allow the transcription apparatus to gain access to the promoter. Remodeling can also act to prevent transcription by moving nucleosomes onto, rather than away from, essential promoter sequences. Remodeling is also required to enable other manipulations of chromatin, including repair of damaged DNA (see Chapter 16, *Repair Systems*).

Remodeling often takes the form of displacing one or more histone octamers. This can result in the creation of a site that is hypersensitive to cleavage with DNase I (see *Section 10.11, DNase Sensitivity Detects Changes in Chromatin Structure*). Sometimes there are less dramatic changes; for example, alteration of the rotational positioning of a single nucleosome, detectable by loss or change of the DNase I 10 bp ladder. Thus changes in chromatin structure can extend from subtly altering the positions of nucleosomes to removing them altogether.

Chromatin remodeling is undertaken by **ATP-dependent chromatin remodeling complexes**, which use ATP hydrolysis to provide the energy for remodeling. The heart of the remodeling complex is its *ATPase subunit*. The ATPase subunits of all remodeling complexes are related members of a large *superfamily* of proteins, which is divided into *subfamilies* of more closely related members. Remodeling complexes are classified according to the subfamily of ATPase that they contain as their catalytic subunit. There are many subfamilies; the four major ones (SWI/SNF, ISWI, CHD,

and INO80/SWR1) are shown in **FIGURE 28.18**. The first remodeling complex described was the SWI/SNF ("switch sniff") complex in yeast, which has homologs in all eukaryotes. The chromatin remodeling superfamily is large and diverse, and most species have multiple complexes in different subfamilies. Yeast has two SWI/SNF-related complexes and three ISWI complexes. Eight different ISWI complexes have been identified thus far in mammals. Remodeling complexes range from small heterodimeric complexes (the ATPase subunit plus a single partner) to massive complexes of ten or more subunits. Each type of complex may undertake a different range of remodeling activities.

SWI/SNF is the prototypic remodeling complex. Its name reflects the fact that many of its subunits are encoded by genes originally identified by *swi* or *snf* mutations in *Saccharomyces cerevisiae*. (*swi* mutants cannot *swi*tch mating type, and *snf*—sucrose *n*on*f*ermenting—mutants cannot use sucrose as a carbon source.) Mutations in these loci are pleiotropic, and the range of defects is similar to those shown by mutants that have lost part of the carboxyl-terminal domain (CTD) of RNA polymerase II. Early hints that these genes might be linked to chromatin came from evidence that these mutations show genetic interactions with mutations in genes that code for components of chromatin: *SIN1*, which codes for a nonhistone chromatin protein, and *SIN2*, which codes for histone H3. The *SWI* and *SNF* genes are required for expression of a variety of individual loci (~120, or 2%, of *S. cerevisiae* genes require SWI/SNF for normal expression). Expression of these loci may require the SWI/SNF complex to remodel chromatin at their promoters.

SWI/SNF acts catalytically *in vitro*, and there are only ~150 complexes per yeast cell. All of the genes encoding the SWI/SNF subunits are nonessential, which implies that yeast must also have other ways of remodeling chromatin. The related RSC (*r*emodels the *s*tructure of *c*hromatin) complex is more abundant and also is essential. It acts at ~700 target loci.

Different subfamilies of remodeling complexes have distinct modes of remodeling, reflecting differences in their ATPase subunits as well as effects of other proteins in individual remodeling complexes. SWI/SNF complexes can remodel chromatin *in vitro* without overall

Type of Complex	SWI/SNF	ISWI	CHD	INO80/SWRI
Yeast	SWI/SNF RSC	ISW1a, ISWb ISW2	CHDI	INO80 SWRI
Fly	dSWI/SNF (brahma)	NURF CHRAC ACF	JMIZ	Tip60
Human	hSWI/SNF	RSF hACF/WCFR hCHRAC WICH	NuRD	INO80 SRCAP
Frog		WICH CHRAC ACF	Mi-2	

FIGURE 28.18 Remodeling complexes can be classified by their ATPase subunits.

loss of histones or can displace histone octamers. These reactions likely pass through the same intermediate in which the structure of the target nucleosome is altered, leading either to reformation of a (remodeled) nucleosome on the original DNA or to displacement of the histone octamer to a different DNA molecule. In contrast, the ISWI family primarily affects nucleosome positioning *without* displacing octamers, in a sliding reaction in which the octamer moves along DNA. The activity of ISWI requires the histone H4 tail as well as binding to linker DNA.

There are many contacts between DNA and a histone octamer; fourteen are identified in the crystal structure. All of these contacts must be broken for an octamer to be released or for it to move to a new position. How is this achieved? The ATPase subunits are distantly related to helicases (enzymes that unwind double-stranded nucleic acids), but remodeling complexes do not have any unwinding activity. Present thinking is that remodeling complexes in the SWI/SNF and ISWI classes use the hydrolysis of ATP to *twist* DNA on the nucleosomal surface. This twisting creates a mechanical force that allows a small region of DNA to be released from the surface and then repositioned. This mechanism creates transient loops of DNA on the surface of the octamer; these loops are themselves accessible to interact with other factors, or they can propagate along the nucleosome, ultimately resulting in nucleosome sliding.

Different remodeling complexes have different roles in the cell. SWI/SNF complexes are generally involved in transcriptional activation, whereas some ISWI complexes act as repressors, using their remodeling activity to slide nucleosomes *onto* promoter regions to prevent transcription. Members of the CHD (*c*hromodomain *h*elicase *D*NA-binding) family have also been implicated in repression, particularly the Mi-2/NuRD complexes, which contain both chromatin remodeling and histone deacetylase activities. Remodelers in the SWR1/INO80 class have a unique activity: in addition to their normal remodeling capabilities, some members of this class also have *histone exchange* capability, in which individual histones (usually H2A/H2B dimers) can be replaced in a nucleosome, typically with a histone variant (see *Section 10.5, Histone Variants Produce Alternative Nucleosomes*).

28.8 Nucleosome Organization or Content May Be Changed at the Promoter

Key concepts

- A remodeling complex does not itself have specificity for any particular target site, but must be recruited by a component of the transcription apparatus.
- Remodeling complexes are recruited to promoters by sequence-specific activators.
- The factor may be released once the remodeling complex has bound.
- Transcription activation often involves nucleosome displacement at the promoter.
- Promoters contain nucleosome-free regions flanked by nucleosomes containing the H2A variant H2AZ (Htz1 in yeast).
- The MMTV promoter requires a change in rotational positioning of a nucleosome to allow an activator to bind to DNA on the nucleosome.

How are remodeling complexes targeted to specific sites on chromatin? They do not themselves contain subunits that bind specific DNA sequences. This suggests the model shown in **FIGURE 28.19**, in which they are recruited by activators or (sometimes) by repressors.

The interaction between transcription factors and remodeling complexes gives a key insight into their modus operandi. The transcription factor Swi5 activates the *HO* gene in yeast, a gene involved in mating-type switching. (Note that despite its name, Swi5 is not a member of the SWI/SNF complex.) Swi5 enters the nucleus near the end of mitosis and binds to the *HO* promoter. It then recruits SWI/SNF to the promoter. Swi5 is then released, leaving SWI/SNF at the promoter. This means that a transcription factor can activate a promoter by a "hit and run" mechanism, in which its function is fulfilled once the remodeling complex has bound.

The involvement of remodeling complexes in gene activation was discovered because the complexes are necessary to enable certain transcription factors to activate their target genes. One of the first examples was the GAGA factor, which activates the *Drosophila hsp70* promoter. Binding of GAGA to four $(CT)_n$-rich sites near the promoter disrupts the nucleosomes, creates a hypersensitive region, and causes the adjacent nucleosomes to be rearranged so that they occupy preferential instead of random positions. Disruption is an energy-dependent

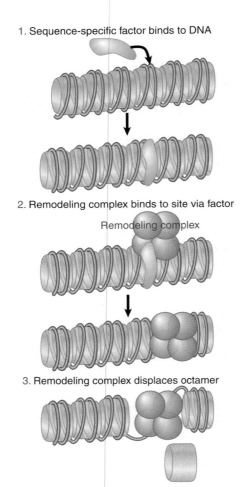

1. Sequence-specific factor binds to DNA

2. Remodeling complex binds to site via factor

Remodeling complex

3. Remodeling complex displaces octamer

FIGURE 28.19 A remodeling complex binds to chromatin via an activator (or repressor).

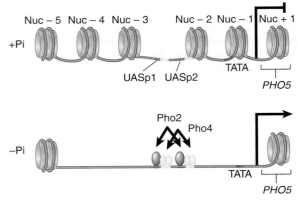

FIGURE 28.20 Nucleosomes are displaced from promoters during activation. The *PHO5* promoter contains nucleosomes positioned over the TATA box and one of the binding sites for the Pho4 and Pho2 activators. When *PHO5* is induced by phosphate starvation (−Pi), promoter nucleosomes are displaced.

process that requires the NURF remodeling complex, a complex in the ISWI subfamily. The organization of nucleosomes is altered so as to create a boundary that determines the positions of the adjacent nucleosomes. During this process, GAGA binds to its target sites and DNA, and its presence fixes the remodeled state.

The *PHO* system was one of the first in which it was shown that a change in nucleosome organization is involved in gene activation. At the *PHO5* promoter, the bHLH activator Pho4 responds to phosphate starvation by inducing the disruption of four precisely positioned nucleosomes, as depicted in **FIGURE 28.20**. This event is independent of transcription (it occurs in a *TATA⁻* mutant) and independent of replication. There are two binding sites for Pho4 (and another activator, Pho2) at the promoter. One is located between nucleosomes, which can be bound by the isolated DNA-binding domain of Pho4; the other lies within a nucleosome, which cannot be recognized. Disruption

of the nucleosome to allow DNA binding at the second site is necessary for gene activation. This action requires the presence of the transcription-activating domain, and appears to involve at least two remodelers: SWI/SNF and INO80. In addition, chromatin disassembly at *PHO5* also requires a histone chaperone, Asf1, which may assist in nucleosome removal or act as a recipient of displaced histones.

A survey of nucleosome positions in a large region of the yeast genome shows that most sites that bind transcription factors are free of nucleosomes. Promoters for RNA polymerase II typically have a nucleosome-free region (NFR) ~200 bp upstream of the startpoint, which is flanked by positioned nucleosomes on either side. These positioned nucleosomes typically contain the histone variant H2AZ (called Htz1 in yeast); the deposition of H2AZ requires the SWR1 remodeling complex. This organization appears to be present in many human promoters as well. It has been suggested that H2AZ-containing nucleosomes are more easily evicted during transcription activation, thus "poising" promoters for activation; however, the actual effects of H2AZ on nucleosome stability *in vivo* are controversial.

It is not always the case, though, that nucleosomes must be excluded in order to permit initiation of transcription. Some activators can bind to DNA on a nucleosomal surface. Nucleosomes appear to be precisely positioned at some steroid hormone response elements in such a way that receptors can bind. Receptor binding may alter the interaction of DNA with histones, and may even lead to expo-

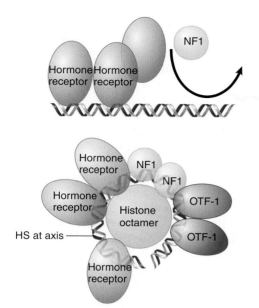

FIGURE 28.21 Hormone receptor and NF1 cannot bind simultaneously to the MMTV promoter in the form of linear DNA, but can bind when the DNA is presented on a nucleosomal surface.

sure of new binding sites. The exact positioning of nucleosomes could be required either because the nucleosome "presents" DNA in a particular rotational phase or because there are protein–protein interactions between the activators and histones or other components of chromatin. Thus we have now moved some way from viewing chromatin exclusively as a repressive structure to considering which interactions between activators and chromatin can be required for activation.

The MMTV promoter presents an example of the need for specific nucleosomal organization. It contains an array of six partly palindromic sites that constitute the HRE (hormone response element). Each site is bound by one dimer of hormone receptor (HR). The MMTV promoter also has a single binding site for the factor NF1, and two adjacent sites for the factor OTF. HR and NF1 cannot bind simultaneously to their sites in free DNA. **FIGURE 28.21** shows how the nucleosomal structure controls binding of the factors.

The HR protects its binding sites at the promoter when hormone is added, but does not affect the micrococcal nuclease-sensitive sites that mark either side of the nucleosome. This suggests that HR is binding to the DNA on the nucleosomal surface; however, the rotational positioning of DNA on the nucleosome prior to hormone addition allows access to only two

of the four sites. Binding to the other two sites requires a change in rotational positioning on the nucleosome. This can be detected by the appearance of a sensitive site at the axis of dyad symmetry (which is in the center of the binding sites that constitute the HRE). NF1 can be detected on the nucleosome after hormone induction, so these structural changes may be necessary to allow NF1 to bind, perhaps because they expose DNA and abolish the steric hindrance by which HR blocks NF1 binding to free DNA.

28.9 Histone Acetylation Is Associated with Transcription Activation

Key concepts

- Newly synthesized histones are acetylated at specific sites, then deacetylated after incorporation into nucleosomes.
- Histone acetylation is associated with activation of gene expression.
- Transcription activators are associated with histone acetylase activities in large complexes.
- Histone acetyltransferases vary in their target specificity.
- Deacetylation is associated with repression of gene activity.
- Deacetylases are present in complexes with repressor activity.

All of the core histones are subject to multiple covalent modifications, as discussed in *Section 10.4, Nucleosomes Are Covalently Modified*. Different modifications result in different functional outcomes. The most extensively studied modification (and the first to be characterized in detail) is lysine acetylation. All core histones dynamically acetylated on lysine residues in the tails (and occasionally within the globular core). As described in *Section 10.4*, certain patterns of acetylation are associated with newly synthesized histones that are deposited during DNA synthesis in S phase. This specific acetylation pattern is then erased after histones are incorporated into nucleosomes.

Outside of S phase, acetylation of histones in chromatin is generally correlated with the state of gene expression. The correlation was first noticed because histone acetylation is increased in a domain containing active genes, and acetylated chromatin is more sensitive to DNase I. We now know that this occurs largely because of acetylation of the nucleosomes (on specific lysines) in the vicinity of the promoter when a gene is activated.

The range of nucleosomes targeted for modification can vary. Modification can be a local event—for example, restricted to nucleosomes at a promoter. It can also be a general event, extending over large domains or even to an entire chromosome. Global changes in acetylation occur on sex chromosomes. This is part of the mechanism by which the activities of genes on the X chromosome are altered to compensate for the presence of two X chromosomes in one sex but only one X chromosome (in addition to the Y chromosome) in other sex (see *Section 29.5, X Chromosomes Undergo Global Changes*). The inactive X chromosome in female mammals has underacetylated histones. The super-active X chromosome in *Drosophila* males has increased acetylation of H4. This suggests that the presence of acetyl groups may be a prerequisite for a less condensed, active structure. In male *Drosophila*, the X chromosome is acetylated specifically at K16 of histone H4. The enzyme responsible for this acetylation is called MOF; MOF is recruited to the chromosome as part of a large protein complex. This "dosage compensation" complex is responsible for introducing general changes in the X chromosome that enable it to be more highly expressed. The increased acetylation is only one of its activities.

Acetylation is reversible. Each direction of the reaction is catalyzed by a specific type of enzyme. Enzymes that can acetylate lysine residues in proteins are called **lysine (K) acetyltransferases** or **KATs**; when these enzymes target lysines in histones they are also known as *histone acetyltransferases* or *HATs*. The acetyl groups are removed by **histone deacetylases** or **HDACs**. There are two classes of HAT enzymes: those in group A act on histones in chromatin and are involved with the control of transcription; those in group B act on newly synthesized histones in the cytosol, and are involved with nucleosome assembly.

Two inhibitors have been useful in analyzing acetylation. Trichostatin and butyric acid inhibit histone deacetylases, and cause acetylated nucleosomes to accumulate. The use of these inhibitors has supported the general view that acetylation is associated with gene expression; in fact, the ability of butyric acid to cause changes in chromatin resembling those found upon gene activation was one of the first indications of the connection between acetylation and gene activity.

The breakthrough in analyzing the role of histone acetylation was provided by the characterization of the acetylating and deacetylating enzymes, and their association with other proteins that are involved in specific events of activation and repression. A basic change in our view of histone acetylation was caused by the discovery that previously identified activators of transcription turned out to also have HAT activity.

The connection was established when the catalytic subunit of a group A HAT was identified as a homolog of the yeast regulator protein Gcn5. It then was shown that yeast Gcn5 itself has HAT activity, with histones H3 and H2B as its preferred substrates *in vivo*. Gcn5 had previously been identified as part of an adaptor complex required for the function of certain enhancers and their target promoters. It is now known that Gcn5's HAT activity is required for activation of a number of target genes.

Gcn5 was the prototypic HAT that opened the way to the identification of a large family of related acetyltransferase complexes conserved from yeast to mammals. In yeast, Gcn5 is the catalytic HAT subunit of the 1.8 MDa Spt-Ada-Gcn5-acetyltransferase (SAGA) complex, which contains several proteins that are involved in transcription. Among these proteins are several TAF$_{II}$s. In addition, the Taf1 subunit of TF$_{II}$D is itself an acetyltransferase. There are some functional overlaps between TF$_{II}$D and SAGA, most notably that yeast can survive the loss of either Taf1 or Gcn5, but cannot tolerate the deletion of both. This suggests that an acetyltransferase activity is essential for gene expression, but can be provided by either TF$_{II}$D or SAGA. As might be expected from the size of the SAGA complex, acetylation is only one of its functions. The SAGA complex has histone H2B deubiquitylation activity (dynamic H2B ubiquitylation/deubiquitylation is also associated with transcription), and also contains subunits possessing bromodomains and chromodomains, allowing this complex to interact with acetylated and methylated histones.

One of the first general activators to be characterized as HAT was p300/CREB-binding protein (CBP). (Actually, p300 and CBP are different proteins, but they are so closely related that they are often referred to as a single type of activity.) p300/CBP is a coactivator that links an activator to the basal apparatus (see Figure 28.7). p300/CBP interacts with various activators, including hormone receptors, AP-1 (c-Jun and c-Fos), and MyoD. p300/CBP acetylates multiple histone targets, with a preference for the H4 tail. p300/CBP interacts with another coactivator, PCAF, which is related to Gcn5 and preferentially acetylates H3 in nucleosomes. p300/CBP and PCAF form a complex that functions in transcriptional activation. In some cases yet another HAT can be involved, such as the hormone receptor coactivator ACTR, which is itself a HAT that acts on H3 and H4. One explanation for the presence of multiple HAT activities in a coactivating complex is that each HAT has a different specificity, and that multiple different acetylation events are required for activation. This enables us to redraw our picture for the action of coactivators as shown in FIGURE 28.22, where RNA polymerase II is bound at a hypersensitive site and coactivators are acetylating histones in the nucleosomes in the vicinity.

Group A HATs, like ATP-dependent remodeling enzymes, are typically found in large complexes. FIGURE 28.23 shows a simplified model for their behavior. HAT complexes can be targeted to DNA by interactions with DNA-binding factors. The complex also contains effector subunits that affect chromatin structure or act directly on transcription. It is likely that at least some of the effectors require the acetylation event in order to act (such as the deubiquitylation activity of SAGA).

The effect of acetylation may be both quantitative and qualitative. In cases where the effect of charge neutralization on chromatin structure is key, a certain minimal number of acetyl groups should be required to have an effect, and the exact positions at which they occur are largely irrelevant. In the case where the role of acetylation is primarily in the creation of a binding site (for a bromodomain-containing factor, for example), the specific position of the acetylation event will be critical. We might interpret the existence of complexes containing multiple HAT activities in either way—if individual enzymes have different specificities, we may need multiple activities either to acetylate a sufficient number of different positions or because the individual events are necessary for different effects upon transcription. At replication, it appears (at least with respect to histone H4) that acetylation at any two of three particular positions is adequate, favoring a quantitative model in this case. Where chromatin structure is changed to affect transcription, acetylation at specific positions is important (for example, see *Section 29.3, Heterochromatin Depends on Interactions with Histones*).

As acetylation is linked to activation, deacetylation is linked to transcriptional repression. Whereas site-specific activators recruit coactivators with HAT activity, site-specific repressor proteins can recruit corepressor complexes, which often contain HDAC activity.

In yeast, mutations in *SIN3* and *RPD3* result in increased expression of a variety of genes, indicating that Sin3 and Rpd3 proteins act as repressors of transcription. Sin3 and Rpd3 are recruited to a number of genes by interacting with the DNA-binding protein Ume6, which binds to the *URS1* (upstream repressive sequence) element. The complex represses transcription at the promoters containing *URS1*,

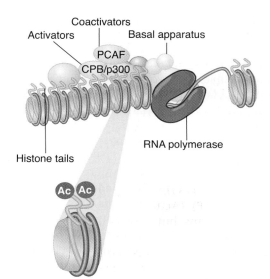

FIGURE 28.22 Coactivators may have HAT activities that acetylate the tails of nucleosomal histones.

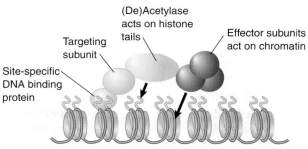

FIGURE 28.23 Complexes that control acetylation levels have targeting subunits that determine their sites of action (usually subunits that interact with site-specific DNA binding proteins), HAT or HDAC enzymes that acetylate or deacetylate histones, and effector subunits that have other actions on chromatin or DNA.

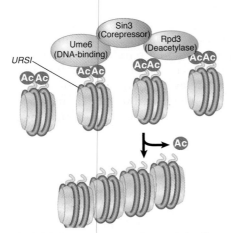

FIGURE 28.24 A repressor complex contains three components: a DNA-binding subunit, a corepressor, and a histone deacetylase.

as illustrated in **FIGURE 28.24**. Rpd3 is a histone deacetylase, and its recruitment leads to deacetylation of nucleosomes at the promoter. Rpd3 and its homologs are present in multiple HDAC complexes found in eukaryotes from yeast to humans; these large complexes are typically built around Sin3 and its homologs.

In mammalian cells, Sin3 is part of a repressive complex that includes histone-binding proteins and the Rpd3 homologs HDAC1 and HDAC2. This corepressor complex can be recruited by a variety of repressors to specific gene targets. The bHLH family of transcription regulators includes activators that function as heterodimers, including MyoD. This family also includes repressors, in particular the heterodimer Mad:Max, where Mad can be any one of a group of closely related proteins. The Mad:Max heterodimer (which binds to specific DNA sites) interacts with Sin3/HDAC1/2 complex, and requires the deacetylase activity of this complex for repression. Similarly, the SMRT corepressor (which enables retinoid hormone receptors to repress certain target genes) binds mSin3, which in turns brings the HDAC activities to the site. Another means of bringing HDAC activities to a DNA site can be an interaction with MeCP2, a protein that binds to methylated cytosines, a mark of transcriptional silencing (see *Section 20.12, CpG Islands Are Regulatory Targets* and *Section 29.7, CpG Islands Are Subject to Methylation*).

Absence of histone acetylation is also a feature of heterochromatin. This is true of both constitutive heterochromatin (typically involving regions of centromeres or telomeres) and facultative heterochromatin (regions that are inactivated in one cell although they may be active in another). Typically the N-terminal tails of histones H3 and H4 are not acetylated in heterochromatic regions (see *Section 29.3, Heterochromatin Depends on Interactions with Histones*).

28.10 Methylation of Histones and DNA Is Connected

Key concepts

- Methylation of both DNA and specific sites on histones is a feature of inactive chromatin.
- The SET domain is part of the catalytic site of protein methyltransferases.
- The two types of methylation event are connected.

DNA methylation is associated with transcriptional inactivity, whereas histone methylation can be linked to either active or inactive regions, depending on the specific site of methylation. There are numerous sites of lysine methylation in the tail and core of histone H3 (a few of which occur only in some species), and a single lysine in the tail of H4. In addition, three arginines in H3 and one in H4 are also methylated.

Di- or trimethylation of H3K4 is associated with transcriptional activation, and trimethylated H3K4 occurs around the start sites of active genes. In contrast, H3 methylated at K9 or K27 is a feature of transcriptionally silent regions of chromatin, including heterochromatin and smaller regions containing one or more silent genes. Whole genome studies have begun to uncover general patterns of modifications linked to different transcriptional states, as shown in **FIGURE 28.25**.

Histone lysine methylation is catalyzed by lysine methyltransferases (KMTs or HMTs), most of which contain a conserved region called the SET domain. Like acetylation, methylation is reversible, and two different families of lysine demethylases (KDMs) have been identified: the LSD1 (lysine-specific demethylase 1, also known as KDM1) family and the Jumonji family. Different classes of enzymes demethylate arginines.

In silent or heterochromatic regions, the methylation of H3 at K9 is linked to DNA methylation. The enzyme that targets this lysine is a SET-domain containing enzyme called Suv39h1. Deacetylation of H3K9 by HDACs must occur before this lysine can be methylated. H3K9 methylation then recruits a protein called HP1 (heterochromatin protein 1), which binds H3K9me via its chromodomain. HP1 then targets the activity of DNA methyltransferases

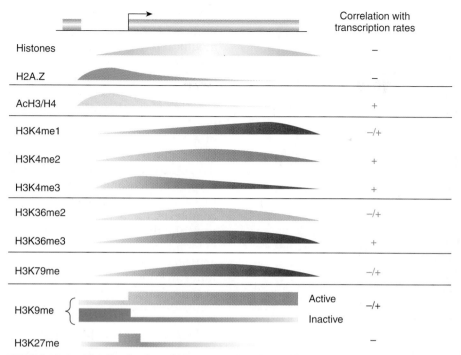

FIGURE 28.25 The distribution of histones and their modifications are mapped on an arbitrary gene relative to its promoter. The curves represent the patterns that are determined via genome-wide approaches. The location of the histone variant H2A.Z is also shown. With the exception of the data on K9 and K27 methylation, most of the data are based on yeast genes. Reprinted from *Cell*, vol. 128, B. Li, M. Carey, and J. L. Workman, The Role of Chromatin during Transcription, pp. 707–719. Copyright 2007, with permission from Elsevier [http://www.sciencedirect.com/science/journal/00928674].

(DNMTs). Most of the methylation sites in DNA are CpG islands (see *Section 29.7, CpG Islands Are Subject to Methylation*). CpG sequences in heterochromatin are usually methylated. Conversely, it is necessary for the CpG islands located in promoter regions to be unmethylated in order for a gene to be expressed.

Methylation of DNA and methylation of histones is connected in a mutually reinforcing circuit. In addition to the recruitment of DNMTs via HP1 binding to H3K4me, DNA methylation can in turn result in histone methylation. Some histone methyltransferase complexes (as well as some HDAC complexes) contain binding domains that recognize the methylated CpG doublet, so the DNA methylation reinforces the circuit by providing a target for the histone deacetylases and methyltransferases to bind. The important point is that one type of modification can be the trigger for another. These systems are widespread, as can be seen by evidence for these connections in fungi, plants, and animal cells, and for regulating transcription at promoters used by both RNA polymerases I and II, as well as maintaining heterochromatin in an inert state.

28.11 Promoter Activation Involves Multiple Changes to Chromatin

Key concepts

- Remodeling complexes can facilitate binding of acetyltransferase complexes, and vice versa.
- Histone methylation can also recruit chromatin-modifying complexes.
- Different modifications and complexes facilitate transcription elongation.

FIGURE 28.26 summarizes three general differences between active chromatin and inactive chromatin:

- Active chromatin is acetylated on the tails of histones H3 and H4.
- Inactive chromatin is methylated on specific lysines (such as K9) of histone H3.
- Inactive chromatin is methylated on cytosines of CpG doublets.

The reverse events occur if we compare the activation of a promoter with the generation of heterochromatin. The actions of the enzymes

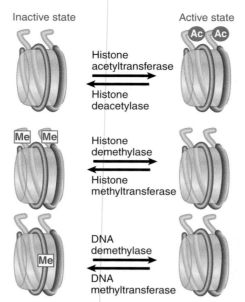

Inactive state | Active state

Histone acetyltransferase → Histone deacetylase

Histone demethylase → Histone methyltransferase

DNA demethylase → DNA methyltransferase

FIGURE 28.26 Acetylation of histones activates chromatin, and methylation of DNA and specific sites on histones inactivates chromatin.

that modify chromatin ensure that activating events are mutually exclusive with inactivating events. For example, the silencing methylation of H3 K9 and the activating acetylation of H3 at K9 and K14 are mutually antagonistic.

How are histone-modifying enzymes such as acetyltransferases or deacetylases recruited to their specific targets? As we have seen with remodeling complexes, the process is likely to be indirect. A sequence-specific activator (or repressor) may interact with a component of the acetyltransferase (or deacetylase) complex to recruit it to a promoter.

There can also be direct interactions between remodeling complexes and histone-modifying complexes. Binding by the SWI/SNF remodeling complex may lead in turn to binding by the SAGA acetyltransferase complex. Acetylation of histones may then stabilize the association with the SWI/SNF complex, making a mutual reinforcement of the changes in the components at the promoter. Some of these events result in displacement of nucleosomes from the promoter. Methylation of histone H3 on K4 also results in recruitment of numerous factors, including the chromodomain-containing remodeler Chd1, which also associates with SAGA. H3K4me also directly recruits another acetyltransferase complex, NuA3, which recognizes H3K4me via a PHD domain in one of its subunits. These are just a few of the interactions that occur during transcription activation in yeast; similar complex networks of interactions also facilitate transcription in multicellular eukaryotes. A further set of dynamic modifications serves to facilitate transcriptional elongation, and to "reset" the chromatin behind the elongating polymerase.

We can connect all of the events at the promoter into the series summarized in **FIGURE 28.27**. The initiating event is the binding of a sequence-specific component, which is either able to find its target DNA sequence in the context of chromatin or which binds to a site in a nucleosome-free region. This activator recruits remodeling and/or acetyltransferase complexes. Changes occur in nucleosome structure, and the acetylation of target histones provides a covalent mark that the locus has been activated. Initiation complex assembly follows (after any other necessary activators bind), and at some point histones are typically displaced.

28.12 Histone Phosphorylation Affects Chromatin Structure

Key concept
- Histone phosphorylation is linked to transcription, repair, chromosome condensation, and cell cycle progression.

All histones can be phosphorylated *in vivo* in different contexts. Histones are phosphorylated in three circumstances:

- cyclically during the cell cycle,
- in association with chromatin remodeling during transcription, and
- during DNA repair.

It has been known for a long time that the linker histone H1 is phosphorylated at mitosis, and more recently it was discovered that H1 is an extremely good substrate for the Cdc2 kinase that controls cell division. This led to speculation that the phosphorylation might be connected with the condensation of chromatin, but so far no direct effect of this phosphorylation event has been demonstrated, and we do not know whether it plays a role in cell division. In *Tetrahymena*, it is possible to delete all the genes for H1 without significantly affecting the overall properties of chromatin. There is a relatively small effect on the ability of chromatin to condense at mitosis. Some genes are activated and others are repressed by this change, which suggests that there are alterations in local structure. Mutations that eliminate sites of phosphorylation in H1 have no effect, but mutations that

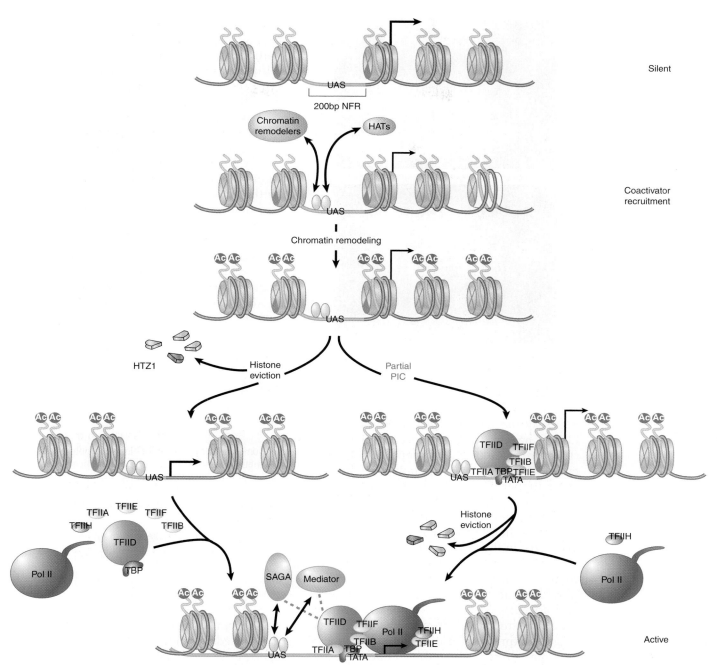

FIGURE 28.27 Htz1-containing nucleosomes flank a 200 bp NFR on both sides of a promoter. Upon targeting to the upstream-activation sequence (UAS), activators recruit various coactivators (such as Swi/Snf or SAGA). This recruitment further increases the binding of activators, particularly for those bound within nucleosomal regions. More importantly, histones are acetylated at promoter-proximal regions, and these nucleosomes become much more mobile. In one model (left), a combination of acetylation and chromatin remodeling directly results in the loss of Htz1-containing nucleosome, thereby exposing the entire core promoter to the GTFs and Pol II. SAGA and mediator then facilitate PIC formation through direct interactions. In the other model (right), which represents the remodeled state, partial PICs could be assembled at the core promoter without loss of Htz1. It is the binding of Pol II and TFIIH that leads to the displacement of Htz1-containing nucleosomes and the full assembly of PIC. Reprinted from *Cell*, vol. 128, B. Li, M. Carey, and J. L. Workman, The Role of Chromatin during Transcription, pp. 707–719. Copyright 2007, with permission from Elsevier [http://www.sciencedirect.com/science/journal/00928674].

mimic the effects of phosphorylation produce a phenotype that resembles the deletion. This suggests that the effect of phosphorylating H1 is to eliminate its effects on local chromatin structure.

Phosphorylation of serine 10 of histone H3 is linked to transcriptional activation (where it promotes acetylation of K14 in the same tail), as well to chromosome condensation and mitotic progression. In *Drosophila melanogaster*, loss of

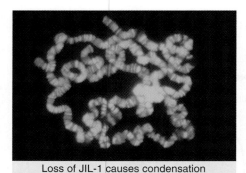

Loss of JIL-1 causes condensation

FIGURE 28.28 Flies that have no JIL-1 kinase have abnormal polytene chromosomes that are condensed instead of extended. Photos courtesy of Jorgen Johansen and Kristen M. Johansen, Iowa State University.

a kinase that phosphorylates histone H3S10 (JIL-1) has devastating effects on chromatin structure. **FIGURE 28.28** compares the usual extended structure of the polytene chromosome (upper photograph) with the structure that is found in a null mutant that has no JIL-1 kinase (lower photograph). The absence of JIL-1 is lethal, but the chromosomes can be visualized in the larvae before they die.

This suggests that H3 phosphorylation is required to generate the more extended chromosome structure of euchromatic regions. JIL-1 also associates with the complex of proteins that binds to the X chromosome to increase its gene expression in males (see *Section 29.5, X Chromosomes Undergo Global Changes*), and JIL-1-dependent H3S10 phosphorylation also antagonizes H3K9 dimethylation, a heterochromatic mark. These results are consistent with a role for JIL-1 in promoting an active chromatin conformation. Interestingly, H3S10 phosphorylation by JIL-1 is itself promoted by acetylation of H4K12 by the ATAC acetyltransferase complex; these complicated interactions make it challenging to determine whether one single modification is key for the transitions in chromatin structure, or whether several modifications must occur together. It is also not clear

how this role of H3 phosphorylation in promoting transcriptionally active chromatin is related to the requirement for H3 phosphorylation to initiate chromosome condensation in at least some species (including mammals and the ciliate *Tetrahymena*).

This leaves us with somewhat conflicting impressions of the roles of histone phosphorylation. Where it is important in the cell cycle, it is likely to be as a signal for condensation. Its effect in transcription and repair appears to be the opposite, where it contributes to open chromatin structures compatible with transcription activation and repair processes. (Histone phosphorylation during repair is discussed in *Section 10.5, Histone Variants Produce Alternative Nucleosomes* and *Section 16.12, DNA Repair in Eukaryotes Occurs in the Context of Chromatin*.)

It is possible, of course, that phosphorylation of different histones, or even of different amino acid residues in one histone, has opposite effects on chromatin structure.

28.13 How Is a Gene Turned On?

Key concepts

- Some transcription factors may compete with histones for DNA after passage of a replication fork.
- Some transcription factors can recognize their targets in closed chromatin to initiate activation.
- The genome is divided into domains by boundary elements (insulators).
- Insulators can block the spreading of chromatin modifications from one domain to another.

Multicellular eukaryotes typically begin life through the fertilization of an egg by a sperm. In both of these haploid gametes, but especially the sperm, the chromosomes are in supercondensed modified chromatin. Males of some species use positively charged *polyamines* like spermines and spermidines to replace the histones in sperm chromatin; others include sperm-specific histone variants. Once the process of fusion of the two haploid nuclei is complete in the egg, genes are then activated in a cascade of regulatory events. The general question of how a gene in closed chromatin is turned on can be broken down into (at least) two parts. How do we identify and target for activation an individual gene that is wrapped up in condensed chromatin? Furthermore, when we begin to modify the histones and remodel the chromatin, how do we prevent that from spreading to genes we do not wish to turn on?

First of all, we can imagine that replication is one mechanism by which closed chromatin can be disrupted in order to allow DNA-binding sequences to become accessible. Replication opens higher-order chromatin structure by temporarily displacing histone octamers. The occupation of enhancer DNA sites on daughter strands subsequently can be viewed as competition. Chromatin can be opened if transcription factors are present in high enough concentration, as shown in **FIGURE 28.29**. If transcription factor concentration is low, then nucleosomes can bind and condense the region. This occurs in *Xenopus* embryos as oocyte 5S ribosomal genes are repressed in the embryo after fertilization.

Second, it is clear that some transcription factors can bind to their DNA target sequence in closed chromatin. The DNA exposed on the surface of the histone octamer is potentially accessible. These transcription factors can then recruit the histone modifiers and chromatin remodelers to begin the process of opening the gene region and clearing the promoter (see Figure 28.17). Recently described examples of antisense transcription through a gene region can facilitate this process; these are described in Chapter 30, *Regulatory RNA* (see Figures 30.4 and 30.5).

Chromatin modification typically originates from a point source (an enhancer) and then spreads, in most cases unidirectionally. (In those cases where modification spreads in a unidirectional fashion, we can ask why it is not spread bidirectionally.) The next question is, what prevents chromatin modification from spreading into distant gene regions?

Activation is limited by boundaries called **insulators** or **boundary elements** (see *Section 10.12, Insulators Define Transcriptionally Independent Domains*). Very few of these insulators have been described in detail, and their mechanisms of action are still poorly understood. In one sense, they are very much like enhancers. They are modular, compact sequence sets that bind specific proteins. Insulators can also function within complex loci to separate multiple temporal and tissue-specific enhancers so that only one can function at a time. Boundary elements are also required to prevent the heterochromatin at centromeres and telomeres from spreading into euchromatin.

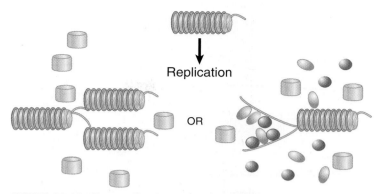

FIGURE 28.29 When replication disrupts chromatin structure, after the Y fork has passed, either chromatin can reform or transcription factors can bind and prevent chromatin formation.

28.14 Yeast *GAL* Genes: A Model for Activation and Repression

Key concepts

- *GAL1/10* genes are positively regulated by the activator Gal4.
- *GAL1/10* genes are negatively regulated by a non-coding RNA synthesized from a cryptic promoter that controls chromatin structure.
- Gal4 is negatively regulated by Gal80, which shuttles between the nucleus and the cytoplasm.
- Gal80 is negatively regulated in the cytoplasm by Gal3, which is activated by the inducer, galactose.
- Activated Gal4 recruits the machinery necessary to alter the chromatin and recruit RNA polymerase.

Yeast, like bacteria, need to be able to rapidly respond to their environment (see *Section 26.3, The* lac *Operon Is Negative Inducible*). In the yeast *Saccharomyces cerevisiae*, the *GAL* genes serve a similar function to the *lac* operon in *E. coli*. In an emergency, when there is little or no glucose as an energy source and only galactose (or in *E. coli*, lactose) is available, then the cell will survive because it can catabolize the alternate sugar to generate ATP. The *GAL* system in *S. cerevisiae* has been a model system to investigate gene regulation in eukaryotes for many years. We will focus on two of the genes, *GAL1* and *GAL10*, which are shown in **FIGURE 28.30**. Like most eukaryotic genes, the *GAL* genes are monocistronic. These two genes are divergently transcribed and regulated from a central control region called the **UAS (*upstream activating sequence*)**, which is similar to an enhancer. Like the *lac* operon in *E. coli*, the *GAL* genes are

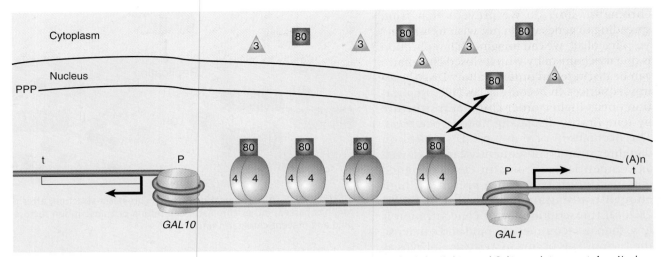

FIGURE 28.30 The yeast *GAL1/GAL10* locus highlighting the UAS and showing the Gal4, Gal80, and Gal3 regulatory proteins. Nucleosomes are positioned at the promoters when the genes are not being transcribed.

induced by their substrate, galactose. For the same reason as in *E. coli*, the *GAL* genes are also under a second level of control described below: catabolite repression. They cannot be activated by the substrate galactose when there is a sufficient supply of glucose, the preferred energy source.

The *GAL* genes are under five different levels of control. The first level is chromatin structure. Mutations in any of the subunits of SWI/SNF and in the acetyltransferase complex SAGA will result in reduced expression of the *GAL* genes. Second, in the UAS there are both general enhancer and Mig1 repressor binding sites. The third level is through a noncoding RNA transcript that assists in maintaining repressed chromatin over the open reading frames. The fourth level is the *GAL* gene-specific, galactose induction mechanism. The fifth level is catabolite (glucose) repression.

GAL1 is an unusual gene in that it lacks the typical nucleosome-free region present at the start sites of most yeast genes. Instead, the start site is contained in a well-positioned nucleosome, whereas the ~170 bp UAS region is held in a nucleosome-free state, which may be partly dependent on the chromatin remodeler SWI/SNF. This DNA region has an unusual base composition, short-phased AT repeats every 10 base pairs, which causes the DNA to bend. Nucleosomes containing the histone variant H2A.Z (Htz in yeast) are positioned over the promoters of both *GAL1* and *GAL10*, presumably aided in their positioning by the bent DNA.

The *GAL10* gene is also an unusual gene in that it has a cryptic promoter in open chromatin at its 3′ end. This promoter transcribes a noncoding RNA that is antisense to *GAL10* and extends through and includes *GAL1* (see *Section 30.3 Noncoding RNAs Can Be Used to Regulate Gene Expression*). Transcription is very inefficient and the RNA abundance is extremely low (less than one copy per cell), due in part to rapid degradation. Under repressed conditions this promoter is stimulated by the Reb1 transcription factor, usually thought to be an RNA polymerase I transcription factor. The noncoding transcript represses transcription of the *GAL1/10* pair of genes by recruiting the Set2 methyltransferase, which leads to H3K36 di- and trimethylation. H3K36 di- and trimethylation lead to the recruitment of HDAC to deacetylate the chromatin, which in turn leads to repressed chromatin structure.

The *GAL* genes are ultimately controlled by the positive regulator Gal4, which binds as a dimer to four binding sites in the UAS region, as shown in Figure 28.30 and **FIGURE 28.31**. Its activation domain consists of two acidic patch domains. Gal4 in turn is regulated by Gal80, a negative regulator that binds to Gal4 and masks its activation domain, preventing it from activating transcription. This is the normal state for the *GAL* genes: turned off and waiting to be induced. Gal80 normally shuttles back and forth between the cytoplasm and the nucleus, reentering the nucleus because of a nuclear localization domain. Gal80 in turn is regulated in the cytoplasm by the negative regulator Gal3, which is itself controlled by the inducer galactose. More recent data indicate that Gal3 may also function in the nucleus.

Gal3 is an interesting protein, having very high homology to Gal1, which is a galactokinase enzyme whose function is to phosphorylate galactose. Gal3 has no enzymatic activity, but retains the ability to bind galactose and ATP. This changes the structure of Gal3 to enable it to bind to Gal80 in the presence of NADP. When it does, Gal3 masks the nuclear localization signal of Gal80, preventing it from shuttling back into the nucleus. Gal3 is thus a negative regulator of a negative regulator, which makes it a positive regulator of Gal4. This depletes the nuclear level of Gal80, unmasking Gal4 and allowing activation of the genes. NADP is thought to be a "second messenger" metabolic sensor.

Unmasked Gal4 is now able to begin the process of turning on the *GAL1/10* genes through direct contact with a number of proteins at the promoter. During induction, Reb1 no longer binds to the cryptic promoter in *GAL10*. Gal4 recruits an H2B histone ubiquitylation factor (Rad6), which then stimulates histone di- and trimethylation of histone H3K4 by Set1. Next, the SAGA acetyltransferase complex is recruited by Gal4 and both deubiquitylates H2B and acetylates histone H3, ultimately resulting in the eviction of the poised nucleosomes from the two promoters. The removal is facilitated by the remodeler SWI/SNF and the chaperones Hsp90/70. SWI/SNF is not absolutely required but speeds the process. This allows the recruitment of TBP/TF$_{II}$D, which then recruits RNA polymerase II and the coactivator complex Mediator. The elongation control factor TF$_{II}$S is also recruited, which actually plays a role in initiation for at least some genes.

During the elongation phase of transcription, nucleosomes are disrupted (see *Section 20.8, Initiation Is Followed by Promoter Clearance and Elongation*). In order to prevent spurious transcription from internal cryptic promoters on either strand, histone octamers must reform as RNA polymerase II passes. A number of histone chaperones and the FACT (*f*acilitating *c*hromatin *t*ranscription) complex play a role in the dynamics of octamer disassembly and assembly during elongation.

This system is also poised to rapidly repress transcription when the supply of galactose is used up or glucose becomes available. As Gal4 is activating transcription by RNA polymerase II, protein kinases associated with the activation of the polymerase also phosphorylate Gal4. This phosphorylation then leads to ubiquitination and destruction of Gal4. This turnover may be essential for RNA polymerase clearance and

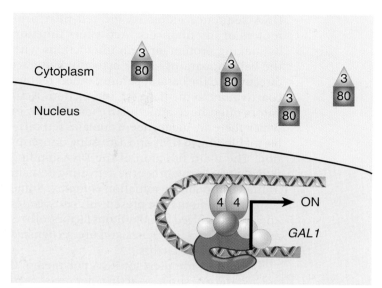

FIGURE 28.31 The yeast *GAL1* gene as it is being activated. Gal3 is holding Gal80 in the cytoplasm, allowing Gal4 to recruit the transcription machinery and activate transcription.

elongation. This is a dynamic system in which there must be a continuous positive signal, the presence of galactose.

Although catabolite repression in eukaryotes is used for the same purpose as in *E. coli* (which uses cAMP as a positive coregulator), it has a completely different mechanism. Glucose is a preferred sugar source compared to galactose. If the cell has both sugars, it will preferentially use the best source, glucose, and repress the genes for galactose utilization. Glucose repression of the yeast *GAL* genes is multifaceted. The glucose-dependent switch is the protein kinase Snf1. In low glucose, the *GAL* genes are transcribed because the general glucose-dependent repressor Mig1 has been inactivated, phosphorylated by Snf1. Glucose repression inactivates Snf1, which allows Mig1 to be active.

A number of other genes involving galactose usage are also downregulated in glucose, including the galactose transporter and Gal4 itself. Glucose inactivates Snf1, which leads to the activation of Mig1 at the *GAL* locus. Mig1 interacts at the *GAL* locus with the Cyc8-Tup1 corepressor, which is known to recruit histone deacetylases.

28.15 Summary

Transcription factors include basal factors, activators, and coactivators. Basal factors interact with RNA polymerase at the startpoint within the promoter. Activators bind specific short

DNA sequence elements located near promoters or in enhancers. Activators function by making protein–protein interactions with the basal apparatus. Some activators interact directly with the basal apparatus; others require coactivators to mediate the interaction. Activators often have a modular construction, in which there are independent domains responsible for binding to DNA and activating transcription. The main function of the DNA-binding domain may be to tether the activating domain in the vicinity of the initiation complex. Some response elements are present in many genes and are recognized by ubiquitous factors; others are present in a few genes and are recognized by tissue-specific factors.

Near the promoters for RNA polymerase II are a variety of short *cis*-acting elements, each of which is recognized by a *trans*-acting factor. The *cis*-acting elements can be located upstream of the TATA box and may be present in either orientation and at a variety of distances with regard to the startpoint or downstream within an intron. These elements are recognized by activators or repressors that interact with the basal transcription complex to determine the efficiency with which the promoter is used. Some activators interact directly with components of the basal apparatus; others interact via intermediaries called coactivators. The targets in the basal apparatus are the TAFs of $TF_{II}D$, $TF_{II}B$, or $TF_{II}A$. The interaction stimulates assembly of the basal apparatus.

Several groups of transcription factors have been identified by sequence homology. The homeodomain is a 60-amino-acid sequence that regulates development in insects, worms, and humans. It is related to the prokaryotic helix-turn-helix motif and is the DNA-binding motif for these transcription factors.

Another motif involved in DNA binding is the zinc finger, which is found in proteins that bind DNA or RNA (or sometimes both). A zinc finger has cysteine and histidine residues that bind zinc. One type of finger is found in multiple repeats in some transcription factors; another is found in single or double repeats in others.

The leucine zipper contains a stretch of amino acids rich in leucine that are involved in dimerization of transcription factors. An adjacent basic region is responsible for binding to DNA in the bZIP transcription factors.

Steroid receptors were the first members identified of a group of transcription factors in which the protein is activated by binding a small hydrophobic hormone. The activated factor becomes localized in the nucleus and binds to its specific response element, where it activates transcription. The DNA-binding domain has zinc fingers.

HLH (helix-loop-helix) proteins have amphipathic helices that are responsible for dimerization, which are adjacent to basic regions that bind to DNA. bHLH proteins have a basic region that binds to DNA. They fall into two groups: ubiquitously expressed and tissue-specific. An active protein is usually a heterodimer between two subunits, one from each group. When a dimer has one subunit that does not have the basic region, it fails to bind DNA; thus such subunits can prevent gene expression. Combinatorial associations of subunits form regulatory networks.

Many transcription factors function as dimers, and it is common for there to be multiple members of a family that form homodimers and heterodimers. This creates the potential for complex combinations to govern gene expression. In some cases, a family includes inhibitory members whose participation in dimer formation prevents the partner from activating transcription.

Genes whose control regions are organized in nucleosomes usually are not expressed. In the absence of specific regulatory proteins, promoters and other regulatory regions are organized by histone octamers into a state in which they cannot be activated. This may explain the need for nucleosomes to be precisely positioned in the vicinity of a promoter, so that essential regulatory sites are appropriately exposed. Some transcription factors have the capacity to recognize DNA on the nucleosomal surface, and a particular positioning of DNA may be required for initiation of transcription.

Chromatin remodeling complexes have the ability to slide or displace histone octamers by a mechanism that involves hydrolysis of ATP. Remodeling complexes range from small to extremely large and are classified according to the type of the ATPase subunit. Common types are SWI/SNF, ISWI, CHD, and SWR1/INO80. A typical form of this chromatin remodeling is to displace one or more histone octamers from specific sequences of DNA, creating a boundary that results in the precise or preferential positioning of adjacent nucleosomes. Chromatin remodeling may also involve changes in the positions of nucleosomes, sometimes involving sliding of histone octamers along DNA.

Extensive covalent modifications occur on histone tails, all of which are reversible. Acetylation of histones occurs at both replication and transcription and facilitates formation of a less compact chromatin structure. Some coactivators, which connect transcription factors to the basal apparatus, have histone acetylase activity. Conversely, repressors may be associated with deacetylases. The modifying enzymes are usually specific for particular amino acids in particular histones. Some histone modifications may be exclusive or synergistic with others.

Large activating (or repressing) complexes often contain several activities that undertake different modifications of chromatin. Some common motifs found in proteins that modify chromatin are the chromodomain (which binds methylated lysine), the bromodomain (which targets acetylated lysine), and the SET domain (which is part of the active sites of histone methyltransferases).

References

28.2 Mechanisms of Action of Activators and Repressors

Reviews

Guarente, L. (1987). Regulatory proteins in yeast. *Annu. Rev. Genet.* 21, 425–452.

Lee, T. I. and Young, R. A. (2000). Transcription of eukaryotic protein-coding genes. *Annu. Rev. Genet.* 34, 77–137.

Lemon, B. and Tjian, R. (2000). Orchestrated response: a symphony of transcription factors for gene control. *Genes Dev.* 14, 2551–2569.

Ptashne, M. (1988). How eukaryotic transcriptional activators work. *Nature* 335, 683–689.

28.3 Independent Domains Bind DNA and Activate Transcription

Reviews

Guarente, L. (1987). Regulatory proteins in yeast. *Annu. Rev. Genet.* 21, 425–452.

Ptashne, M. (1988). How eukaryotic transcriptional activators work. *Nature* 335, 683–689.

28.4 The Two-Hybrid Assay Detects Protein–Protein Interactions

Research

Fields, S. and Song, O. (1989). A novel genetic system to detect protein-protein interactions. *Nature* 340, 245–246.

28.5 Activators Interact with the Basal Apparatus

Reviews

Lemon, B. and Tjian, R. (2000). Orchestrated response: a symphony of transcription factors for gene control. *Genes Dev.* 14, 2551–2569.

Maniatis, T., Goodbourn, S., and Fischer, J. A. (1987). Regulation of inducible and tissue-specific gene expression. *Science* 236, 1237–1245.

Mitchell, P. and Tjian, R. (1989). Transcriptional regulation in mammalian cells by sequence-specific DNA-binding proteins. *Science* 245, 371–378.

Myers, L. C. and Kornberg, R. D. (2000). Mediator of transcriptional regulation. *Annu. Rev. Biochem.* 69, 729–749.

Research

Asturias, F. J., Jiang, Y. W., Myers, L. C., Gustafsson, C. M., and Kornberg, R. D. (1999). Conserved structures of mediator and RNA polymerase II holoenzyme. *Science* 283, 985–987.

Chen, J.-L. et al. (1994). Assembly of recombinant TFIID reveals differential coactivator requirements for distinct transcriptional activators. *Cell* 79, 93–105.

Dotson, M. R., Yuan, C. X., Roeder, R. G., Myers, L. C., Gustafsson, C. M., Jiang, Y. W., Li, Y., Kornberg, R. D., and Asturias, F. J. (2000). Structural organization of yeast and mammalian mediator complexes. *Proc. Natl. Acad. Sci. USA* 97, 14307–14310.

Dynlacht, B. D., Hoey, T., and Tjian, R. (1991). Isolation of coactivators associated with the TATA-binding protein that mediate transcriptional activation. *Cell* 66, 563–576.

Kim, Y. J., Bjorklund, S., Li, Y., Sayre, M. H., and Kornberg, R. D. (1994). A multiprotein mediator of transcriptional activation and its interaction with the C-terminal repeat domain of RNA polymerase II. *Cell* 77, 599–608.

Ma, J. and Ptashne, M. (1987). A new class of yeast transcriptional activators. *Cell* 51, 113–119.

Pugh, B. F. and Tjian, R. (1990). Mechanism of transcriptional activation by Sp1: evidence for coactivators. *Cell* 61, 1187–1197.

28.6 There Are Many Types of DNA-Binding Domains

Reviews

Harrison, S. C. (1991). A structural taxonomy of DNA-binding proteins. *Nature* 353, 715–719.

Pabo, C. T. and Sauer, R. T. (1992). Transcription factors: structural families and principles of DNA recognition. *Annu. Rev. Biochem.* 61, 1053–1095.

28.7 Chromatin Remodeling Is an Active Process

Reviews

Becker, P. B. and Horz, W. (2002). ATP-dependent nucleosome remodeling. *Annu. Rev. Biochem.* 71, 247–273.

Cairns, B. (2005). Chromatin remodeling complexes: strength in diversity, precision through specialization. *Curr. Op. Genet. Dev.* 15, 185–190.

Felsenfeld, G. (1992). Chromatin as an essential part of the transcriptional mechanism. *Nature* 355, 219–224.

Grunstein, M. (1990). Histone function in transcription. *Annu. Rev. Cell Biol.* 6, 643–678.

Narlikar, G. J., Fan, H. Y., and Kingston, R. E. (2002). Cooperation between complexes that regulate chromatin structure and transcription. *Cell* 108, 475–487.

Peterson, C. L. and Côté, J. (2004). Cellular machineries for chromosomal DNA repair. *Genes Dev.* 18, 602–616.

Schnitzler, G. R. (2008). Control of nucleosome positions by DNA sequence and remodeling machines. *Cell Biochem. Biophys.* 51, 67–80.

Tsukiyama, T. (2002). The in vivo functions of ATP-dependent chromatin-remodelling factors. *Nat. Rev. Mol. Cell Biol.* 3, 422–429.

Vignali, M., Hassan, A. H., Neely, K. E., and Workman, J. L. (2000). ATP-dependent chromatin-remodeling complexes. *Mol. Cell Biol.* 20, 1899–1910.

Research

Cairns, B. R., Kim, Y.-J., Sayre, M. H., Laurent, B. C., and Kornberg, R. (1994). A multisubunit complex containing the SWI/ADR6, SWI2/1, SWI3, SNF5, and SNF6 gene products isolated from yeast. *Proc. Natl. Acad. Sci. USA* 91, 1950–1954.

Côte, J., Quinn, J., Workman, J. L., and Peterson, C. L. (1994). Stimulation of GAL4 derivative binding to nucleosomal DNA by the yeast SWI/SNF complex. *Science* 265, 53–60.

Gavin, I., Horn, P. J., and Peterson, C. L. (2001). SWI/SNF chromatin remodeling requires changes in DNA topology. *Mol. Cell* 7, 97–104.

Hamiche, A., Kang, J. G., Dennis, C., Xiao, H., and Wu, C. (2001). Histone tails modulate nucleosome mobility and regulate ATP-dependent nucleosome sliding by NURF. *Proc. Natl. Acad. Sci. USA* 98, 14316–14321.

Kingston, R. E. and Narlikar, G. J. (1999). ATP-dependent remodeling and acetylation as regulators of chromatin fluidity. *Genes Dev.* 13, 2339–2352.

Kwon, H., Imbaizano, A. N., Khavari, P. A., Kingston, R. E., and Green, M. R. (1994). Nucleosome disruption and enhancement of activator binding of human SWI/SNF complex. *Nature* 370, 477–481.

Logie, C. and Peterson, C. L. (1997). Catalytic activity of the yeast SWI/SNF complex on reconstituted nucleosome arrays. *EMBO J.* 16, 6772–6782.

Lorch, Y., Cairns, B. R., Zhang, M., and Kornberg, R. D. (1998). Activated RSC-nucleosome complex and persistently altered form of the nucleosome. *Cell* 94, 29–34.

Lorch, Y., Zhang, M., and Kornberg, R. D. (1999). Histone octamer transfer by a chromatin-remodeling complex. *Cell* 96, 389–392.

Peterson, C. L. and Herskowitz, I. (1992). Characterization of the yeast SWI1, SWI2, and SWI3 genes, which encode a global activator of transcription. *Cell* 68, 573–583.

Robert, F., Young, R. A., and Struhl, K. (2002). Genome-wide location and regulated recruitment of the RSC nucleosome remodeling complex. *Genes Dev.* 16, 806–819.

Schnitzler, G., Sif, S., and Kingston, R. E. (1998). Human SWI/SNF interconverts a nucleosome between its base state and a stable remodeled state. *Cell* 94, 17–27.

Tamkun, J. W., Deuring, R., Scott, M. P., Kissinger, M., Pattatucci, A. M., Kaufman, T. C., and Kennison, J. A. (1992). Brahma: a regulator of Drosophila homeotic genes structurally related to the yeast transcriptional activator SNF2/SWI2. *Cell* 68, 561–572.

Tsukiyama, T., Daniel, C., Tamkun, J., and Wu, C. (1995). ISWI, a member of the SWI2/SNF2 ATPase family, encodes the 140 kDa subunit of the nucleosome remodeling factor. *Cell* 83, 1021–1026.

Tsukiyama, T., Palmer, J., Landel, C. C., Shiloach, J., and Wu, C. (1999). Characterization of the imitation switch subfamily of ATP-dependent chromatin-remodeling factors in *S. cerevisiae*. *Genes Dev.* 13, 686–697.

Whitehouse, I., Flaus, A., Cairns, B. R., White, M. F., Workman, J. L., and Owen-Hughes, T. (1999). Nucleosome mobilization catalysed by the yeast SWI/SNF complex. *Nature* 400, 784–787.

28.8 Nucleosome Organization or Content May Be Changed at the Promoter

Review

Lohr, D. (1997). Nucleosome transactions on the promoters of the yeast *GAL* and *PHO* genes. *J. Biol. Chem.* 272, 26795–26798.

Research

Cosma, M. P., Tanaka, T., and Nasmyth, K. (1999). Ordered recruitment of transcription and chromatin remodeling factors to a cell cycle and developmentally regulated promoter. *Cell* 97, 299–311.

Kadam, S., McAlpine, G. S., Phelan, M. L., Kingston, R. E., Jones, K. A., and Emerson, B. M. (2000). Functional selectivity of recombinant mammalian SWI/SNF subunits. *Genes Dev.* 14, 2441–2451.

McPherson, C. E., Shim, E.-Y., Friedman, D. S., and Zaret, K. S. (1993). An active tissue-specific enhancer and bound transcription factors existing in a precisely positioned nucleosomal array. *Cell* 75, 387–398.

Schmid, V. M., Fascher, K.-D., and Horz, W. (1992). Nucleosome disruption at the yeast *PHO5* promoter upon *PHO5* induction occurs in the absence of DNA replication. *Cell* 71, 853–864.

Truss, M., Barstch, J., Schelbert, A., Hache, R. J. G., and Beato, M. (1994). Hormone induces binding of receptors and transcription factors to a rearranged nucleosome on the MMTV promoter in vitro. *EMBO J.* 14, 1737–1751.

Tsukiyama, T., Becker, P. B., and Wu, C. (1994). ATP-dependent nucleosome disruption at a heat shock promoter mediated by binding of GAGA transcription factor. *Nature* 367, 525–532.

Yudkovsky, N., Logie, C., Hahn, S., and Peterson, C. L. (1999). Recruitment of the SWI/SNF chromatin remodeling complex by transcriptional activators. *Genes Dev.* 13, 2369–2374.

28.9 Histone Acetylation Is Associated with Transcription Activation

Review

Jenuwein, T. and Allis, C. D. (2001). Translating the histone code. *Science* 293, 1074–1080.

Lee, K. K. and Workman, J. L. (2007). Histone acetyltransferase complexes: one size doesn't fit all. *Nat. Rev. Mol. Cell Biol.* 8, 284–295.

Ruthenburg, A. J., Li, H., Patel, D. J., and Allis, C. D. (2007). Multivalent engagement of chromatin modifications by linked binding modules. *Nat. Rev. Mol. Cell Biol.* 8, 983–994.

Research

Akhtar, A. and Becker, P. B. (2000). Activation of transcription through histone H4 acetylation by MOF, an acetyltransferase essential for dosage compensation in Drosophila. *Mol. Cell* 5, 367–375.

Ayer, D. E., Lawrence, Q. A., and Eisenman, R. N. (1995). Mad-Max transcriptional repression is mediated by ternary complex formation with mammalian homologs of yeast repressor Sin3. *Cell* 80, 767–776.

Brownell, J. E. et al. (1996). Tetrahymena histone acetyltransferase A: a homologue to yeast Gcn5p linking histone acetylation to gene activation. *Cell* 84, 843–851.

Chen, H. et al. (1997). Nuclear receptor coactivator ACTR is a novel histone acetyltransferase and forms a multimeric activation complex with P/CAF and CP/p300. *Cell* 90, 569–580.

Grant, P. A. et al. (1998). A subset of TAF$_{II}$s are integral components of the SAGA complex required for nucleosome acetylation and transcriptional stimulation. *Cell* 94, 45–53.

Jackson, V., Shires, A., Tanphaichitr, N., and Chalkley, R. (1976). Modifications to histones immediately after synthesis. *J. Mol. Biol.* 104, 471–483.

Kadosh, D. and Struhl, K. (1997). Repression by Ume6 involves recruitment of a complex containing Sin3 corepressor and Rpd3 histone deacetylase to target promoters. *Cell* 89, 365–371.

Kingston, R. E. and Narlikar, G. J. (1999). ATP-dependent remodeling and acetylation as regulators of chromatin fluidity. *Genes Dev.* 13, 2339–2352.

Krebs, J. E., Kuo, M. H., Allis, C. D. and Peterson, C. L. (1999). Cell-cycle regulated histone acetylation required for expression of the yeast *HO* gene. *Genes Dev.* 13, 1412–1421.

Lee, T. I., Causton, H. C., Holstege, F. C., Shen, W. C., Hannett, N., Jennings, E. G., Winston, F., Green, M. R., and Young, R. A. (2000). Redundant roles for the TF$_{II}$D and SAGA complexes in global transcription. *Nature* 405, 701–704.

Ling, X., Harkness, T. A., Schultz, M. C., Fisher-Adams, G., and Grunstein, M. (1996). Yeast histone H3 and H4 amino termini are important for nucleosome assembly in vivo and in vitro: redundant and position-independent functions in assembly but not in gene regulation. *Genes Dev.* 10, 686–699.

Osada, S., Sutton, A., Muster, N., Brown, C. E., Yates, J. R., Sternglanz, R., and Workman, J. L. (2001). The yeast SAS (something about silencing) protein complex contains a MYST-type putative acetyltransferase and functions with chromatin assembly factor ASF1. *Genes Dev.* 15, 3155–3168.

Schreiber-Agus, N., Chin, L., Chen, K., Torres, R., Rao, G., Guida, P., Skoultchi, A. I., and DePinho, R. A. (1995). An amino-terminal domain of Mxi1 mediates anti-Myc oncogenic activity and interacts with a homolog of the yeast transcriptional repressor SIN3. *Cell* 80, 777–786.

Shibahara, K., Verreault, A., and Stillman, B. (2000). The N-terminal domains of histones H3 and H4 are not necessary for chromatin assembly factor-1-mediated nucleosome assembly onto replicated DNA in vitro. *Proc. Natl. Acad. Sci. USA* 97, 7766–7771.

Turner, B. M., Birley, A. J., and Lavender, J. (1992). Histone H4 isoforms acetylated at specific lysine residues define individual chromosomes and chromatin domains in Drosophila polytene nuclei. *Cell* 69, 375–384.

28.10 Methylation of Histones and DNA Is Connected

Reviews

Bannister, A. J. and Kouzarides, T. (2005). Reversing histone methylation. *Nature* 436, 1103–1106.

Richards, E. J., Elgin, S. C., and Richards, S. C. (2002). Epigenetic codes for heterochromatin formation and silencing: rounding up the usual suspects. *Cell* 108, 489–500.

Zhang, Y. and Reinberg, D. (2001). Transcription regulation by histone methylation: interplay between different covalent modifications of the core histone tails. *Genes Dev.* 15, 2343–2360.

Research

Cuthbert, G. L., Daujat, S., Snowden, A. W., Erdjument-Bromage, H., Hagiwara, T., Yamada, M., Schneider, R., Gregory, P. D., Tempst, P., Bannister, A. J., and Kouzarides, T. (2004). Histone deimination antagonizes arginine methylation. *Cell* 118, 545–553.

Fuks, F., Hurd, P. J., Wolf, D., Nan, X., Bird, A. P., and Kouzarides, T. (2003). The methyl-CpG-binding protein MeCP2 links DNA methylation to histone methylation. *J. Biol. Chem.* 278, 4035–4040.

Gendrel, A. V., Lippman, Z., Yordan, C., Colot, V., and Martienssen, R. A. (2002). Dependence of heterochromatic histone H3 methylation patterns on the Arabidopsis gene DDM1. *Science* 297, 1871–1873.

Johnson, L., Cao, X., and Jacobsen, S. (2002). Interplay between two epigenetic marks: DNA methylation and histone H3 lysine 9 methylation. *Curr. Biol.* 12, 1360–1367.

Lawrence, R. J., Earley, K., Pontes, O., Silva, M., Chen, Z. J., Neves, N., Viegas, W., and Pikaard, C. S. (2004). A concerted DNA methylation/histone methylation switch regulates rRNA gene dosage control and nucleolar dominance. *Mol. Cell* 13, 599–609.

Ng, H. H., Feng, Q., Wang, H., Erdjument-Bromage, H., Tempst, P., Zhang, Y., and Struhl, K. (2002). Lysine methylation within the globular domain of histone H3 by Dot1 is important for telomeric silencing and Sir protein association. *Genes Dev.* 16, 1518–1527.

Rea, S., Eisenhaber, F., O'Carroll, D., Strahl, B. D., Sun, Z. W., Sun, M., Opravil, S., Mechtler, K., Ponting, C. P., Allis, C. D., and Jenuwein, T. (2000). Regulation of chromatin structure by site-specific histone H3 methyltransferases. *Nature* 406, 593–599.

Shi, Y., Lan, F., Matson, C., Mulligan, P., Whetstine, J. R., Cole, P. A., and Casero, R. A. (2004). Histone demethylation mediated by the nuclear amine oxidase homolog LSD1. *Cell* 119, 941–953.

Tamaru, H. and Selker, E. U. (2001). A histone H3 methyltransferase controls DNA methylation in *Neurospora crassa*. *Nature* 414, 277–283.

Tamaru, H., Zhang, X., McMillen, D., Singh, P. B., Nakayama, J., Grewal, S. I., Allis, C. D., Cheng, X., and Selker, E. U. (2003). Trimethylated lysine 9 of histone H3 is a mark for DNA methylation in *Neurospora crassa*. *Nat. Genet.* 34, 75–79.

Wang, Y., Wysocka, J., Sayegh, J., Lee, Y. H., Perlin, J. R., Leonelli, L., Sonbuchner, L. S., McDonald, C. H., Cook, R. G., Dou, Y., et al. (2004). Human PAD4 regulates histone arginine methylation levels via demethylimination. *Science* 306, 279–283.

28.11 Promoter Activation Involves Multiple Changes to Chromatin

Review

Li, B., M. Carey and Workman, J. L. (2007). The role of chromatin during transcription. *Cell* 128, 707–719.

Orphanides, G. and Reinberg, D. (2000). RNA polymerase II elongation through chromatin. *Nature* 407, 471–475.

Research

Bortvin, A. and Winston, F. (1996). Evidence that Spt6p controls chromatin structure by a direct interaction with histones. *Science* 272, 1473–1476.

Cosma, M. P., Tanaka, T., and Nasmyth, K. (1999). Ordered recruitment of transcription and chromatin remodeling factors to a cell cycle and developmentally regulated promoter. *Cell* 97, 299–311.

Hassan, A. H., Neely, K. E., and Workman, J. L. (2001). Histone acetyltransferase complexes stabilize SWI/SNF binding to promoter nucleosomes. *Cell* 104, 817–827.

Krebs, J. E., Kuo, M. H., Allis, C. D., and Peterson, C. L. (1999). Cell-cycle regulated histone acetylation required for expression of the yeast *HO* gene. *Genes Dev.* 13, 1412–1421.

Orphanides, G., LeRoy, G., Chang, C. H., Luse, D. S., and Reinberg, D. (1998). FACT, a factor that facilitates transcript elongation through nucleosomes. *Cell* 92, 105–116.

Wada, T., Takagi, T., Yamaguchi, Y., Ferdous, A., Imai, T., Hirose, S., Sugimoto, S., Yano, K., Hartzog, G. A., Winston, F., Buratowski, S., and Handa, H. (1998). DSIF, a novel transcription elongation factor that regulates RNA polymerase II processivity, is composed of human Spt4 and Spt5 homologs. *Genes Dev.* 12, 343–356.

28.12 Histone Phosphorylation Affects Chromatin Structure

Research

Ciurciu, A., Komonyi, O., and Boros, I. M. (2008). Loss of ATAC-specific acetylation of histone H4 at Lys12 reduces binding of JIL-1 to chromatin and phosphorylation of histone H3 and Ser10. *J. Cell Sci.* 121, 3366–3372.

Wang, Y., Zhang, W., Jin, Y., Johansen, J., and Johansen, K. M. (2001). The JIL-1 tandem kinase mediates histone H3 phosphorylation and is required for maintenance of chromatin structure in Drosophila. *Cell* 105, 433–443.

28.14 Yeast *GAL* Genes: A Model for Activation and Repression

Review

Armstrong, J. A. (2007). Negotiating the nucleosome: factors that allow RNA polymerase II to elongate through chromatin. *Biochem. Cell Biol.* 85, 426–434.

Peng, G. and Hopper, J. E. (2002). Gene activation by interaction of an inhibitor with a cytoplasmic signaling protein. *Proc. Natl. Acad. Sci. USA* 99, 8548–8553.

Research

Ahuatzi, D., Riera, A., Pelaez, R., Herrero, P., and Moreno, F. (2007). Hxk2 regulates the phosphorylation state of Mig1 and therefore its nucleocytplasmic distribution. *J. Biol. Chem.* 282, 4485–4493.

Bryant, G. O., Prabhu, V., Floer, M., Wang, X., Spagna, D., Schreiber, D., and Ptashne, M. (2008). Activator control of nucleosome occupancy in activation and repression of transcription. *PLoS* 6, 2928–2938.

Floer, M., Bryant, G. O., and Ptashne, M. (2008). HSP90/70 chaperones are required for rapid nucleosome removal upon induction of the *GAL* genes in yeast. *Proc. Natl. Acad. Sci. USA* 105, 2975–2980.

Houseley, J., Rubbi, L., Grunstein, M., Tollervey, D., and Vogelauer, M. (2008). A ncRNA modulates histone modification and mRNA induction in the yeast *GAL* gene cluster. *Mol. Cell* 32, 685–695.

Imbeault, D., Gamar, L., Rufiange, A., Paquet, E., and Nourani, A. (2008). The Rtt106 histone chaperone is functionally linked to transcription elongation and is involved in the regulation of spurious transcription from cryptic promoters in yeast. *J. Biol. Chem.* 283, 27350–27354.

Ingvarsdottir, K., Edwards, C., Lee, M. G., Lee, J. S., Schultz, D.C., Shilatifard, A., Shiekhattar, R., and Berger, S. (2007). Histone H3 K4 demethylation during activation and attenuation of *GAL1* transcription in *Saccharomyces cerevisiae*. *Mol. Cell Biol.* 27, 7856–7864.

Kumar, P. R., Yu, Y., Sternglanz, R., Johnston, S. A., and Joshua-Tor, L. (2008). NADP regulates the yeast *GAL* system. *Science* 319, 1090–1092.

Muratani, M., Kung, C., Shokat, K. M., and Tansey, W. P. (2005). The F box protein Dsg1/Mdm30 is a transcriptional Coactivator that stimulates Gal4 turnover and cotranscriptional mRNA processing. *Cell* 120, 887–899.

Westergaard, S. L., Oliveira, A. P., Bro, C., Olsson, L., and Nielsen. (2007). A systems approach to study glucose repression in the yeast *Saccharomyces cerevisiae*. *Biotech. And Bioeng.* 96, 134–141

Wightman, R., Bell, R., and Reece, R. J. (2008). Localization and interaction of the proteins constituting the *GAL* genetic switch in *Saccharomyces cerevisiae*. *Euk. Cell* 7, 2061–2068.

29

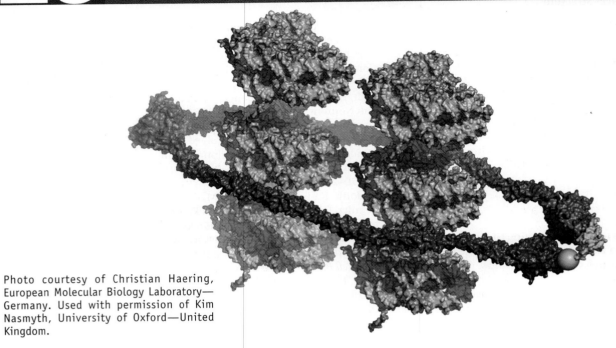

Photo courtesy of Christian Haering, European Molecular Biology Laboratory—Germany. Used with permission of Kim Nasmyth, University of Oxford—United Kingdom.

Epigenetic Effects Are Inherited

Edited by Trygve Tollefsbol

29.1 Introduction

Key concepts
- Epigenetic effects can result from modification of a nucleic acid after it has been synthesized or by the perpetuation of protein structures.

Epigenetic inheritance describes the ability of different states, which may have different phe-notypic consequences, to be inherited without any change in the sequence of DNA. This means that two individuals with the same DNA sequence at the locus that controls the effect may show different phenotypes. The basic cause of this phenomenon is the existence of a self-perpetuating structure in one of the individuals

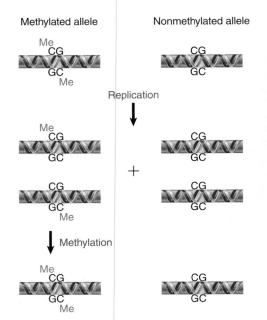

Methylated allele   Nonmethylated allele

FIGURE 29.1 Replication of a methylated site produces hemimethylated DNA, in which only the parental strand is methylated. A perpetuation methylase recognizes hemimethylated sites and adds a methyl group to the base on the daughter strand. This restores the original situation, in which the site is methylated on both strands. An unmethylated site remains unmethylated after replication.

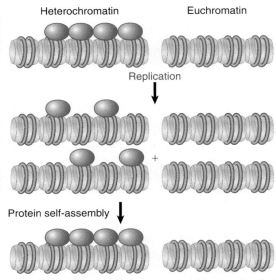

Heterochromatin   Euchromatin

FIGURE 29.2 Heterochromatin is created by proteins that associate with histones. Perpetuation through division requires that the proteins associate with each daughter duplex and then recruit new subunits to reassemble the repressive complexes.

that does not depend on DNA sequence. Several different types of structures have the ability to sustain epigenetic effects:

- A covalent modification of DNA (methylation of a base).
- A proteinaceous structure that assembles on DNA.
- A protein aggregate that controls the conformation of new subunits as they are synthesized.

In each case the epigenetic state results from a difference in function (typically inactivation) that is determined by the structure.

In the case of DNA methylation, a DNA sequence methylated in its control region may fail to be transcribed, whereas the unmethylated sequence will be expressed (this idea was introduced in *Section 20.12, CpG Islands Are Regulatory Targets*). **FIGURE 29.1** shows how this situation is inherited. One allele has a sequence that is methylated on both strands of DNA, whereas the other allele has an unmethylated sequence. Replication of the methylated allele creates hemimethylated daughters that are restored to the methylated state by a constitutively active methylase enzyme. Replication does not affect the state of the unmethylated allele. If the state of methylation affects transcription, the two

alleles differ in their state of gene expression, even though their sequences are identical.

Self-perpetuating structures that assemble on DNA usually have a repressive effect by forming heterochromatic regions that prevent the expression of genes within them. Their perpetuation depends on the ability of proteins in a heterochromatic region to remain bound to those regions after replication, and then to recruit more protein subunits to sustain the complex. If individual subunits are distributed at random to each daughter duplex at replication, the two daughters will continue to be marked by the protein, although its density will be reduced to half of the level before replication. **FIGURE 29.2** shows that the existence of epigenetic effects forces us to the view that a protein responsible for such a situation must have some sort of self-templating or self-assembling capacity to restore the original complex.

It can be the state of protein modification, rather than the presence of the protein *per se*, that is responsible for an epigenetic effect. Usually the tails of histones H3 and H4 are not acetylated in constitutive heterochromatin. If centromeric heterochromatin is acetylated, though, silenced genes may become active. The effect may be perpetuated through mitosis and meiosis, which suggests that an epigenetic effect has been created by changing the state of histone acetylation.

Independent protein aggregates that cause epigenetic effects (called **prions**) work by sequestering the protein in a form in which its normal function cannot be displayed. Once the protein aggregate has formed, it forces newly synthesized protein subunits to join it in the inactive conformation.

29.2 Heterochromatin Propagates from a Nucleation Event

Key concepts

- Heterochromatin is nucleated at a specific sequence and the inactive structure propagates along the chromatin fiber.
- Genes within regions of heterochromatin are inactivated.
- The length of the inactive region varies from cell to cell; as a result, inactivation of genes in this vicinity causes position effect variegation.
- Similar spreading effects occur at telomeres and at the silent cassettes in yeast mating type.

An interphase nucleus contains both euchromatin and heterochromatin. The condensation state of heterochromatin is close to that of mitotic chromosomes. Heterochromatin is inert. It remains condensed in interphase, is transcriptionally repressed, replicates late in S phase, and may be localized to the nuclear periphery. Centromeric heterochromatin typically consists of satellite DNAs; however, the formation of heterochromatin is not rigorously defined by sequence. When a gene is transferred, either by a chromosomal translocation or by transfection and integration, into a position adjacent to heterochromatin, it may become inactive as the result of its new location, implying that it has become heterochromatic.

Such inactivation is the result of an epigenetic effect (see *Section 29.10, Epigenetic Effects Can Be Inherited*). It may differ between individual cells in an animal, and results in the phenomenon of **position effect variegation (PEV)**, in which genetically identical cells have different phenotypes. This has been well characterized in *Drosophila*. **FIGURE 29.3** shows an example of position effect variegation in the fly eye. Some of the regions in the eye lack color, whereas others are red. This is because the *white* gene (required to develop red pigment) was inactivated by adjacent heterochromatin in some cells, but remained active in others.

The explanation for this effect is shown in **FIGURE 29.4**. Inactivation spreads from heterochromatin into the adjacent region for a variable distance. In some cells it goes far enough to inactivate a nearby gene, whereas in others it does

FIGURE 29.3 Position-effect variegation in eye color results when the *white* gene is integrated near heterochromatin. Cells in which *white* is inactive give patches of white eye, whereas cells in which *white* is active give red patches. The severity of the effect is determined by the closeness of the integrated gene to heterochromatin. Photo courtesy of Steven Henikoff, Fred Hutchinson Cancer Research Center.

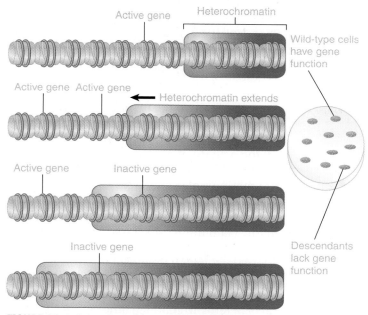

FIGURE 29.4 Extension of heterochromatin inactivates genes. The probability that a gene will be inactivated depends on its distance from the heterochromatin region.

not. This happens at a certain point in embryonic development, and after that point the state of the gene is stably inherited by all the progeny cells. Cells descended from an ancestor in which the gene was inactivated form patches corresponding to the phenotype of loss-of-function (in the case of *white*, the absence of color).

The closer a gene lies to heterochromatin, the higher the probability that it will be inactivated. This suggests that the formation of heterochromatin may be a two-stage process: A *nucleation* event occurs at a specific sequence (triggered by binding of a protein that recognizes this sequence), and then the inactive structure *propagates* along the chromatin fiber. The distance for which the inactive structure extends is not precisely determined and may be stochastic, being influenced by parameters such as the quantities of limiting protein components. One factor that may affect the spreading process is the activation of promoters in the region; an active promoter may inhibit spreading. Genes near heterochromatin are more likely to be inactivated; however, insulators can protect a transcriptionally active region by preventing heterochromatin from spreading (see *Section 10.12, Insulators Define Transcriptionally Independent Domains*).

The effect of **telomeric silencing** in yeast is analogous to position effect variegation in *Drosophila*; genes translocated to a telomeric location show the same sort of variable loss of activity. This results from a spreading effect that propagates from the telomeres. In this case, the binding of the Rap1 protein to telomeric repeats triggers the nucleation event, which results in the recruitment of heterochromatin proteins, as described next in *Section 29.3, Heterochromatin Depends on Interactions with Histones*.

In addition to the telomeres, there are two other sites at which heterochromatin is nucleated in yeast. Yeast mating type is determined by the activity of a single active locus (*MAT*), but the genome contains two other copies of the mating type sequences (*HML* and *HMR*), which are maintained in an inactive form. The silent loci *HML* and *HMR* nucleate heterochromatin via binding of several proteins (rather than the single protein, Rap1, required at telomeres), which then lead to propagation of heterochromatin similar to that at telomeres. Heterochromatin in yeast exhibits features typical of heterochromatin in other species, such as transcriptional inactivity and self-perpetuating protein structures superimposed on nucleosomes (which are generally deacetylated). The only notable difference between yeast heterochromatin and that of most other species is that histone methylation in yeast is not associated with silencing, whereas specific sites of histone

methylation are a key feature of heterochromatin formation in most multicellular eukaryotes.

29.3 Heterochromatin Depends on Interactions with Histones

Key concepts

- HP1 is the key protein in forming mammalian heterochromatin, and acts by binding to methylated histone H3.
- Rap1 initiates formation of heterochromatin in yeast by binding to specific target sequences in DNA.
- The targets of Rap1 include telomeric repeats and silencers at *HML* and *HMR*.
- Rap1 recruits Sir3 and Sir4, which interact with the N-terminal tails of H3 and H4.
- Sir2 deacetylates the N-terminal tails of H3 and H4 and promotes spreading of Sir3 and Sir4.
- RNAi pathways promote heterochromatin formation at centromeres.

Inactivation of chromatin occurs by the addition of proteins to the nucleosomal fiber. The inactivation may be due to a variety of effects, including condensation of chromatin to make it inaccessible to the apparatus needed for gene expression, addition of proteins that directly block access to regulatory sites, or proteins that directly inhibit transcription.

Two systems that have been characterized at the molecular level involve HP1 in mammals and the SIR complex in yeast. Although many of the proteins involved in each system are not evolutionarily related, the general mechanism of reaction is similar: The points of contact in chromatin are the N-terminal tails of the histones.

Our insights into the molecular mechanisms for regulating the formation of heterochromatin originated with mutants that affect position effect variegation. Some 30 genes have been identified in *Drosophila*. They are named systematically as *Su(var)* for genes whose products act to suppress variegation and *E(var)* for genes whose products enhance variegation. These genes were named for the behavior of the *mutant* loci; thus *Su(var)* mutations lie in genes whose products are needed for the formation of heterochromatin. They include enzymes that act on chromatin, such as histone deacetylases, and proteins that are localized to heterochromatin. In contrast, *E(var)* mutations lie in genes whose products are needed to activate gene expression. They include members of the SWI/SNF complex (see *Section 28.7, Chromatin Remodeling Is an Active Process*).

HP1 (heterochromatin protein 1) is one of the most important Su(var) proteins. It was originally identified as a protein that is localized to heterochromatin by staining polytene chromosomes with an antibody directed against the protein. It was later shown to be the product of the gene *Su(var)2–5*. Its homolog in the yeast *Schizosaccaromyces pombe* is encoded by *swi6*. HP1 is now called HP1α because two related proteins, HP1β and HP1γ, have since been found.

HP1 contains a chromodomain near the N-terminus, and another domain that is related to it (the chromo-shadow domain) at the C-terminus (see Figure 29.6). The HP1 chromodomain binds to histone H3 that is trimethylated at lysine 9 (H3K9me3).

Mutation of a deacetylase that acts on H3K14Ac prevents the methylation at K9. H3 that is trimethylated at K9 binds the protein HP1 via the chromodomain. This suggests the model for initiating formation of heterochromatin shown in **FIGURE 29.5**. First the deacetylase acts to remove the modification at K14, and this allows the SUV39H1 methyltransferase (also known as KMT1A) to methylate H3K9 to create the methylated signal to which HP1 will bind. **FIGURE 29.6** expands the reaction to show that the interaction occurs between the chromodomain and the methylated lysine. This is a trigger for forming inactive chromatin. **FIGURE 29.7** shows that the inactive region may then be extended by the ability of further HP1 molecules to interact with one another.

The state of histone methylation is important in the control of heterochromatin or euchromatin states. Methylation of histone H3 lysine 9 [H3K9] demarcates heterochromatin while H3K4 methylation demarcates euchromatin. A trimethyl H3K4 demethylase found in *S. pombe* referred to as Lid2 interacts with the Clr4 H3K9 methyltransferase, resulting in H3K4 hypomethylation and heterochromatin formation. The link between H3K4 demethylation and H3K9 methylation suggests that the two reactions act in a coordinated manner to control the relative state of heterochromatin or euchromatin of a specific region.

Heterochromatin formation at telomeres and silent mating-type loci in yeast relies on an overlapping set of genes known as *silent information regulators* (*SIR* genes). Mutations in *SIR2*, *SIR3*, or *SIR4* cause *HML* and *HMR* to become activated, and also relieve the inactivation of genes that have been integrated near telomeric

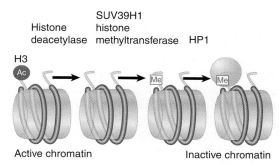

FIGURE 29.5 SUV39H1 is a histone methyltransferase that acts on K9 of histone H3. HP1 binds to the methylated histone.

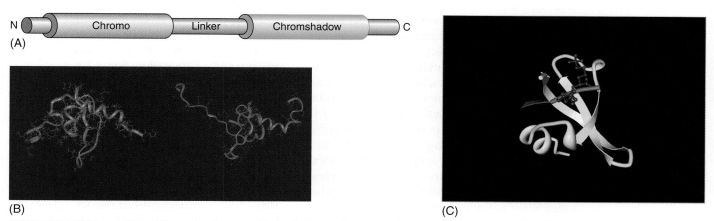

FIGURE 29.6 HP1 contains a chromodomain and a chromoshadow domain. Methylation of histone H3 creates a binding site for HP1. (A & B) Photo reproduced from G. Lomberk, L. Wallrath, and R. Urrutia, *Genome Biol.* 7 (2006): p. 228. Used with permission of Raul A. Urrutia and Gwen Lamberk, Mayo Clinic. (C) Structure from Protein Data Bank 1KNE. S. A. Jacobs and S. Khorasanizadeh, *Science* 295 (2002): 2080–2083.

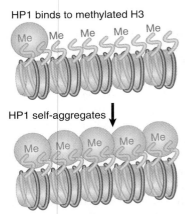

HP1 binds to methylated H3

FIGURE 29.7 Binding of HP1 to methylated histone H3 forms a trigger for silencing because further molecules of HP1 aggregate along the methylation chromatin domain.

heterochromatin. The products of these loci therefore function to maintain the inactive state of both types of heterochromatin.

FIGURE 29.8 shows a model for the actions of these proteins. Only one of them—Rap1—is a sequence-specific DNA-binding protein. It binds to the $C_{1-3}A$ repeats at the telomeres, and also binds to the *cis*-acting silencer elements that are needed for repression of *HML* and *HMR*. The proteins Sir3 and Sir4 interact with Rap1 and also with one another (they may function as a heteromultimer). Sir3 and Sir4 interact with the N-terminal tails of the histones H3 and H4, with a preference for unacetylated tails. Another SIR protein, Sir2, is a deacetylase, and its activity is necessary to maintain binding of the Sir3/Sir4 complex to chromatin.

Rap1 has the crucial role of identifying the DNA sequences at which heterochromatin forms. It recruits Sir4, which in turn recruits both its binding partner Sir3 and the HDAC Sir2. Sir3 and Sir4 then interact directly with histones H3 and H4. Once Sir3 and Sir4 have bound to histones H3 and H4, the complex (including Sir2) can polymerize further and spread along the chromatin fiber. This may inactivate the region, either because coating with the Sir3/Sir4 complex itself has an inhibitory effect, or because Sir2-dependent deacetylation represses transcription. We do not know what limits the spreading of the complex. The C-terminus of Sir3 has a similarity to nuclear lamin proteins (constituents of the nuclear matrix) and may be responsible for tethering heterochromatin to the nuclear periphery.

A similar series of events forms the silenced regions at *HMR* and *HML*. Three sequence-specific factors are involved in triggering formation of the complex: Rap1, Abf1 (a transcription factor),

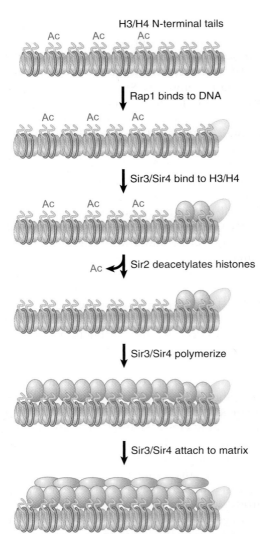

FIGURE 29.8 Formation of heterochromatin is initiated when Rap1 binds to DNA. Sir3/4 bind to Rap1 and also to histones H3/H4. Sir2 deacetylates histones. The SIR complex polymerizes along chromatin and may connect telomeres to the nuclear matrix.

and ORC (the origin replication complex). In this case, Sir1 binds to a sequence-specific factor and recruits Sir2, –3, and –4 to form the repressive structure. As at the telomeres, Sir2-dependent deacetylation is necessary to maintain binding of the Sir complex to chromatin.

Formation of heterochromatin in the yeast *S. pombe* utilizes an RNAi-dependent pathway (see *Section 30.6, How Does RNA Interference Work?*). This pathway is initiated by the production of siRNA molecules resulting from transcription of centromeric repeats. These siRNAs result in formation of the RNA-induced transcriptional gene silencing (RITS) complex. The siRNA components are responsible for localizing the complex at centromeres. The complex contains proteins that are homologs of those involved in heterochromatin formation in other organisms such as plants, *Caenorhabditis elegans*, and *D. melanogaster*. This complex includes Argonaute, which is involved in targeting RNA-induced silencing complex (RISC) remodeling complexes to chromatin. The siRNA complex promotes methylation of histone H3K9 by the Clr4 methyltransferase (also known as KMT1, a homolog of *Drosophila Su(Var)3–9*). H3K9 methylation recruits the *S. pombe* homolog of HP1, Swi6.

How does a silencing complex repress chromatin activity? It could condense chromatin so that regulator proteins cannot find their targets. The simplest case would be to suppose that the presence of a silencing complex is mutually incompatible with the presence of transcription factors and RNA polymerase. The cause could be that silencing complexes block remodeling (and thus indirectly prevent factors from binding) or that they directly obscure the binding sites on DNA for the transcription factors. The situation may not be that simple, though, because transcription factors and RNA polymerase can be found at promoters in silenced chromatin. This could mean that the silencing complex prevents the factors from working rather than from binding as such. In fact, there may be competition between gene activators and the repressing effects of chromatin, so that activation of a promoter inhibits spread of the silencing complex.

Centromeric heterochromatin is particularly interesting, as it is not necessarily nucleated by simple sequences (as is the case for telomeres and the mating type loci in yeast), but instead depends on more complex mechanisms, some of which are RNAi-dependent. The specialized chromatin structure that forms at the centromere may be associated with the formation of heterochromatin in the region. The unique centromeric chromatin structure, and the centromere-specific histone H3 variant, were discussed in *Section 9.15, The* S. cerevisiae *Centromere Binds a Protein Complex*, and *Section 10.5, Histone Variants Produce Alternative Nucleosomes*. In human cells, the centromere-specific protein CENP-B is required to initiate modifications of histone H3 (deacetylation of K9 and K14, followed by methylation of K9) that trigger an association with HP1 that leads to the formation of heterochromatin in the region. Moreover, heterochromatin and RNAi are required to establish the human CenH3 homolog, CENP-A, at centromeres. Heterochromatin is often present near CENP-A chromatin and

the RNAi-directed heterochromatin flanking the central kinetochore domain is required for kinetochore assembly. Several factors, such as the Suv39 methyltransferase, HP1, and components of the RNAi pathway (see *Section 30.6, How Does RNA Interference Work?*), are required to form the CENP-A chromatin.

Studies of the propagation of the pathogenic yeast, *Candida albicans*, have shown that naked centromeric DNA that can confer centromeric activity *in vivo* is not able to assemble functional centromeric chromatin *de novo* when reintroduced into cells. This suggests that *C. albicans* centromeres are dependent on their preexisting chromatin state and provides an example of epigenetic propagation of a centromere.

29.4 Polycomb and Trithorax Are Antagonistic Repressors and Activators

Key concepts

- Polycomb group proteins (Pc-G) perpetuate a state of repression through cell divisions.
- The PRE is a DNA sequence that is required for the action of Pc-G.
- The PRE provides a nucleation center from which Pc-G proteins propagate an inactive structure.
- Trithorax group proteins (trxG) antagonize the actions of the Pc-G.
- Pc-G and trxG can bind to the same PRE with opposing effects.

Regions of constitutive heterochromatin, such as at telomeres and centromeres, provide one example of the specific repression of chromatin. Another is provided by the genetics of homeotic genes (which affect the identity of body segments) in *Drosophila*, which has led to the identification of a protein complex that may *maintain* certain genes in a repressed state. *Polycomb (Pc)* mutants show transformations of cell type that are equivalent to gain-of-function mutations in the genes *Antennapedia (Antp)* or *Ultrabithorax*, because these genes are expressed in tissues in which usually they are repressed. This implicates *Pc* in regulating transcription. Furthermore, *Pc* is the prototype for a class of ~15 loci called the *Pc-group (Pc-G)*; mutations in these genes generally have the same result of derepressing homeotic genes, which suggests the possibility that the group of proteins has some common regulatory role.

The Pc proteins function in large complexes. The PRC1 (Polycomb-repressive complex)

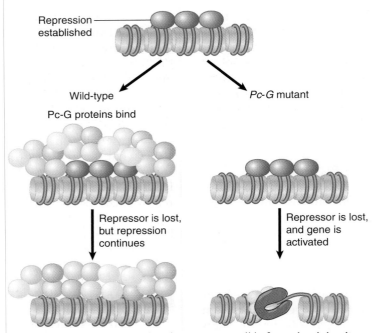

Repression established

Wild-type
Pc-G proteins bind

Pc-G mutant

Repressor is lost, but repression continues

Repressor is lost, and gene is activated

FIGURE 29.9 Pc-G proteins do not initiate repression, but are responsible for maintaining it.

contains Pc itself, several other Pc-G proteins, and five general transcription factors. The Esc-E(z) complex contains Esc, E(z), other Pc-G proteins, a histone-binding protein, and a histone deacetylase. Pc itself has a chromodomain that binds to methylated H3, and E(z) is a methyltransferase that acts on H3. These properties directly support the connection between chromatin remodeling and repression that was initially suggested by the properties of *brahma*, a fly counterpart to *SWI2*. *brahma* codes for a component of the SWI/SNF remodeling complex (see *Section 28.7, Chromatin Remodeling Is an Active Process*), and loss of *brahma* function suppresses mutations in *Polycomb*.

Consistent with the pleiotropy of *Pc* mutations, Pc is a nuclear protein that can be visualized at ~80 sites on polytene chromosomes. These sites include the *Antp* gene. Another member of the *Pc-G, polyhomeotic*, is visualized at a set of polytene chromosome bands that are identical with those bound by Pc. The two proteins coimmunoprecipitate in a complex of ~2.5 × 10⁶ D that contains 10 to 15 polypeptides. The relationship between these proteins and the products of the ~30 *Pc-G* genes remains to be established. One possibility is that some of these gene products form a general repressive complex, and then some of the other proteins associate with it to determine its specificity.

The Pc-G proteins are not conventional repressors. They are not responsible for determining the initial pattern of expression of the genes on which they act. In the absence of Pc-G proteins, these genes are initially repressed as usual, but later in development the repression is lost without Pc-G group functions. This suggests that *the Pc-G proteins in some way recognize the state of repression when it is established, and they then act to perpetuate it through cell division of the daughter cells*. **FIGURE 29.9** shows a model in which Pc-G proteins bind in conjunction with a repressor, but the Pc-G proteins remain bound after the repressor is no longer available. This is necessary to maintain repression; otherwise, the gene becomes activated if Pc-G proteins are absent.

A region of DNA that is sufficient to enable the response to the *Pc-G* genes is called a PRE (*Polycomb* response element). It can be defined operationally by the property that it maintains repression in its vicinity throughout development. The assay for a PRE is to insert it close to a reporter gene that is controlled by an enhancer that is repressed in early development, and then to determine whether the reporter becomes expressed subsequently in the descendants. An effective PRE will prevent such reexpression.

The PRE is a complex structure that measures ~10 kb. Several proteins with DNA-binding activity for sites within the PRE, including Pho, Pho1, and GAGA factor (GAF), have been identified, but there could be others. When a locus is repressed by Pc-G, however, the Pc-G proteins occupy a much larger length of DNA than the PRE itself. Pc is found locally over a few kilobases of DNA surrounding a PRE. This suggests that the PRE may provide a nucleation center, from which a structural state depending on Pc-G proteins may propagate. This model is supported by the observation of effects related to position effect variegation (see Figure 29.4); that is, a gene near to a locus whose repression is maintained by Pc-G may become heritably inactivated in some cells but not others. In one typical situation, crosslinking experiments *in vivo* showed that Pc protein is found over large regions of the *bithorax* complex that are inactive, but the protein is excluded from regions that contain active genes. The idea that this could be due to cooperative interactions within a multimeric complex is supported by the existence of mutations in *Pc* that change its nuclear distribution and abolish the ability of other *Pc-G* members to localize in the nucleus. The role of Pc-G proteins in maintaining, as opposed to establishing, repression must mean that the formation of the complex at the PRE also depends on the local state of gene expression.

The effects of Pc-G proteins are vast in that hundreds of potential Pc-G targets in plants, insects, and mammals have been identified. A working model for Pc-G binding at a PRE is suggested by the properties of the individual proteins. First Pho and Pho1 bind to specific sequences within the PRE. Esc-E(z) is recruited to Pho/Pho1; it then uses its methyltransferase activity to methylate K27 of histone H3. This creates the binding site for the PRC, because the chromodomain of Pc binds to the methylated lysine. The Polycomb complex induces a more compact structure in chromatin; each PRC1 complex causes about three nucleosomes to become less accessible.

In fact, the chromodomain was first identified as a region of homology between Pc and the protein HP1 found in heterochromatin. Binding of the chromodomain of Pc to K27 on H3 is analogous to HP1's use of its chromodomain to bind to methylated K9. Variegation is caused by the spreading of inactivity from constitutive heterochromatin, and as a result it is likely that the chromodomain is used by Pc and HP1 in a similar way to induce the formation of heterochromatic or inactive structures. This model implies that similar mechanisms are used to repress individual loci or to create heterochromatin.

The *trithorax* group (*trxG*) of proteins have the opposite effect to the Pc-G proteins: They act to maintain genes in an active state. trxG proteins are quite diverse; some comprise subunits of chromatin remodeling enzymes such as

SWI/SNF, whereas others also possess important histone modification activities (such as histone **demethylases**) which could oppose the activities of Pc-G proteins. There may be some similarities in the actions of the two groups: mutations in some loci prevent both Pc-G and trxG from functioning, suggesting that they could rely on common components. The GAGA factor, which is encoded by the *trithorax-like* gene, has binding sites in the PRE. In fact, the sites where Pc binds to DNA coincide with the sites where GAGA factor binds. What does this mean? GAGA is probably needed for activating factors, including trxG members, to bind to DNA. Is it also needed for Pc-G proteins to bind and exercise repression? This is not yet clear, but such a model would demand that something other than GAGA determines which of the alternative types of complex subsequently assemble at the site.

The trxG proteins act by making chromatin continuously accessible to transcription factors. Although PcG and trxG proteins promote opposite outcomes, they bind to the same PREs, which can regulate homeotic gene promoters some distance away from the PRE through looping of DNA.

29.5 X Chromosomes Undergo Global Changes

Key concepts

- One of the two X chromosomes is inactivated at random in each cell during embryogenesis of eutherian mammals.
- In exceptional cases where there are >2 X chromosomes, all but one are inactivated.
- The *Xic* (X inactivation center) is a *cis*-acting region on the X chromosome that is necessary and sufficient to ensure that only one X chromosome remains alive.
- *Xic* includes the *Xist* gene, which codes for an RNA that is found only on inactive X chromosomes.
- *Xist* recruits Polycomb complexes, which modify histones on the inactive X.
- The mechanism that is responsible for preventing *Xist* RNA from accumulating on the active chromosome is unknown.

For species with chromosomal sex determination, the sex of the individual presents an interesting problem for gene regulation, because of the variation in the number of X chromosomes. If X-linked genes were expressed equally well in each sex, females would have twice as much of each product as males. The importance of avoiding this situation is shown by the existence of

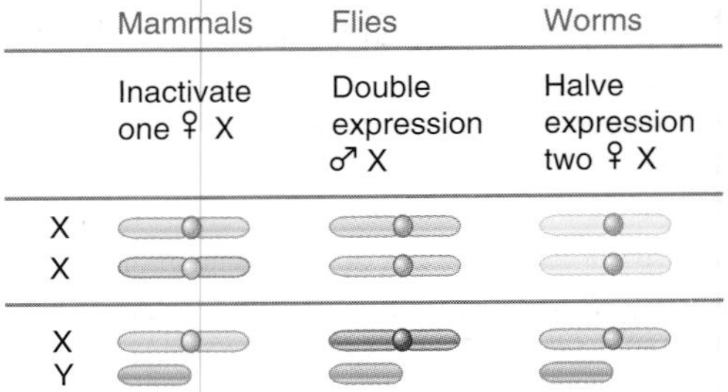

	Mammals	Flies	Worms
	Inactivate one ♀ X	Double expression ♂ X	Halve expression two ♀ X
X X			
X Y			

FIGURE 29.10 Different means of dosage compensation are used to equalize X chromosome expression in male and female.

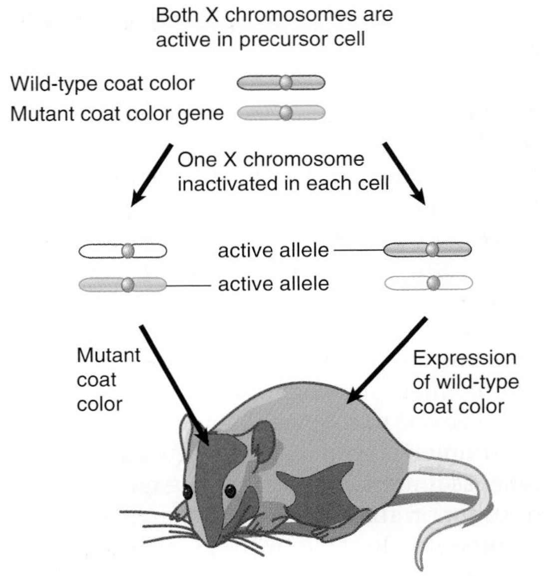

FIGURE 29.11 X-linked variegation is caused by the random inactivation of one X chromosome in each precursor cell. Cells in which the + allele is on the active chromosome have wild phenotype; cells in which the − allele is on the active chromosome have mutant phenotype.

dosage compensation, which equalizes the level of expression of X-linked genes in the two sexes. Mechanisms used in different species are summarized in **FIGURE 29.10**:

- In mammals, one of the two female X chromosomes is inactivated completely. The result is that females have only one active X chromosome, which is the same situation found in males. The active X chromosome of females and the single X chromosome of males are expressed at the same level.
- In *Drosophila*, the expression of the single male X chromosome is doubled relative to the expression of each female X chromosome.
- In *Caenorhabditis elegans*, the expression of each female (hermaphrodite) X chromosome is halved relative to the expression of the single male X chromosome.

The common feature in all these mechanisms of dosage compensation is that *the entire chromosome is the target for regulation*. A global change occurs that quantitatively affects almost all of the promoters on the chromosome. We know the most about the inactivation of the X chromosome in mammalian females, where the entire chromosome becomes heterochromatic.

The twin properties of heterochromatin are its condensed state and associated inactivity. It can be divided into two types:

- **Constitutive heterochromatin** contains specific sequences that have no coding function. These include satellite DNAs, which are often found at the centromeres. These regions are invariably heterochromatic because of their intrinsic nature.
- **Facultative heterochromatin** takes the form of chromosome segments or entire chromosomes that are inactive in one cell lineage, although they

can be expressed in other lineages. The example *par excellence* is the mammalian X chromosome. The inactive X chromosome is perpetuated in a heterochromatic state, whereas the active X chromosome is euchromatic. Thus identical DNA sequences are involved in both states. Once the inactive state has been established, it is inherited by descendant cells. This is an example of epigenetic inheritance, because it does not depend on the DNA sequence.

Our basic view of the situation of the female mammalian X chromosomes was formed by the **single X hypothesis** in 1961. Female mice that are heterozygous for X-linked coat color mutations have a variegated phenotype in which some areas of the coat are wild-type but others are mutant. **FIGURE 29.11** shows that this can be explained if one of the two X chromosomes is inactivated at random in each cell of a small precursor population. Cells in which the X chromosome carrying the wild-type gene is inactivated give rise to progeny that express only the mutant allele on the active chromosome. Cells derived from a precursor where the other chromosome was inactivated have an active wild-type gene. In the case of coat color, cells descended from a particular precursor stay together and thus form a patch of the same color, creating the pattern of visible variegation. In other cases, individual cells in a population will express one or the other of X-linked alleles; for example, in heterozygotes for the X-linked locus *G6PD*, any particular red blood cell will express only one of the two allelic forms. (Random inactivation of one X chromosome occurs in eutherian mammals. In marsupials, the choice is directed: It is always the X chromosome inherited from the father that is inactivated.)

Inactivation of the X chromosome in females is governed by the **n–1 rule**: Regardless of how many X chromosomes are present, all but one will be inactivated. In normal females there are of course two X chromosomes, but in rare cases where nondisjunction has generated a 3X or greater genotype, only one X chromosome remains active. This suggests a general model in which a specific event is limited to one X chromosome and protects it from an inactivation mechanism that applies to all the others.

A single locus on the X chromosome is sufficient for inactivation. When a translocation occurs between the X chromosome and an autosome, this locus is present on only one of the reciprocal products, and only that product can be inactivated. By comparing different translocations, it is possible to map this locus, which is called the *Xic* (X-inactivation center). A cloned region of 450 kb contains all the properties of the *Xic*. When this sequence is inserted as a transgene onto an autosome, the autosome becomes subject to inactivation (at least in a cell culture system). Pairing of *Xic* loci on the two X chromosomes has been implicated in the mechanism for the random choice of X inactivation. Moreover, differences in the sister chromatid cohesion correlates with the outcome of the choice of the X chromosome to be inactivated, indicating that alternate states present before the inactivation process may direct the choice of which X chromosome will become inactivated.

Xic is a *cis*-acting locus that contains the information necessary to count X chromosomes and inactivate all copies but one. Inactivation spreads from *Xic* along the entire X chromosome. When *Xic* is present on an X chromosome-autosome translocation, inactivation spreads into the autosomal regions (although the effect is not always complete).

Xic is a complex genetic locus that expresses several long noncoding RNAs (ncRNAs). The most important of these is a gene called *Xist* (X inactive specific transcript), which is stably expressed only on the *inactive* X chromosome. The behavior of this gene is effectively the

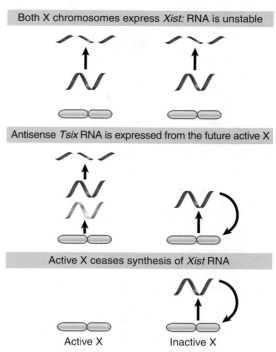

FIGURE 29.12 X-inactivation involves stabilization of *Xist* RNA, which coats the inactive chromosome. *Tsix* prevents *Xist* expression on the future active X.

opposite from all other loci on the chromosome, which are turned off. Deletion of *Xist* prevents an X chromosome from being inactivated. It does not, however, interfere with the counting mechanism (because other X chromosomes can be inactivated). Thus we can distinguish two features of *Xic*: an unidentified element(s) required for counting, and the *Xist* gene required for inactivation.

The n–1 rule suggests that stabilization of *Xist* RNA is the "default," and that some blocking mechanism prevents stabilization at one X chromosome (which will be the active X). This means that, although *Xic* is necessary and sufficient for a chromosome to be *inactivated*, the products of other loci are necessary for the establishment of an *active* X chromosome.

The *Xist* transcript is regulated in a negative manner by *Tsix*, its antisense partner. Loss of *Tsix* expression on the future inactive X chromosome permits *Xist* to become upregulated and stabilized, and persistence of *Tsix* on the future active X chromosome prevents *Xist* upregulation. *Tsix* is regulated by *Xite*, which has a *Tsix*-specific enhancer and is located 10 kb upstream of *Tsix*.

FIGURE 29.12 illustrates the role of *Xist* RNA in X-inactivation. *Xist* codes for an ncRNA that lacks open reading frames. The *Xist* RNA "coats" the X chromosome from which it is synthesized,

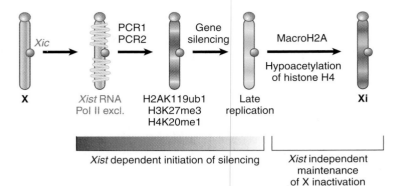

FIGURE 29.13 *Xist* RNA produced from the *Xic* locus accumulates on the future inactive X (Xi). This excludes transcription machinery, such as RNA polymerase II (Pol II). Polycomb group complexes are recruited to the *Xist*-covered chromosome and establish chromosome-wide histone modifications. Histone macroH2A becomes enriched on the Xi and promoters of genes on the Xi are methylated. In this phase X inactivation is irreversible and *Xist* is not required for maintenance of the silent state. Adapted from A. Wutz and J. Gribnau, *Curr. Opin. Genet. Dev.* 17 (2007): 387–393.

which suggests that it has a structural role. Prior to X-inactivation, it is synthesized by both female X chromosomes. Following inactivation, the RNA is found only on the inactive X chromosome. The transcription rate remains the same before and after inactivation, so the transition depends on posttranscriptional events.

Prior to X-inactivation, *Xist* RNA decays with a half-life of ~2 hours. X-inactivation is mediated by stabilizing the *Xist* RNA on the inactive X chromosome. The *Xist* RNA shows a punctate distribution along the X chromosome, which suggests that association with proteins to form particulate structures may be the means of stabilization. We do not know yet what other factors may be involved in this reaction and how the *Xist* RNA is limited to spreading in *cis* along the chromosome.

Accumulation of *Xist* on the future inactive X results in exclusion of transcription machinery (such as RNA polymerase II), and leads to the recruitment of Polycomb repressor complexes (PRC1 and PRC2), which trigger a series of chromosome-wide histone modifications (H2AK119 ubiquitination, H3K27 methylation, H4K20 methylation, and H4 deacetylation). Late in the process, an inactive X-specific histone variant, macroH2A, is incorporated into the chromatin, and promoter DNA is methylated. These changes are summarized in **FIGURE 29.13**. (The repressive effects of promoter methylation are discussed in the following sections.) At this point, the heterochromatic state of the inactive X is stable, and *Xist* is not required to maintain the silent state of the chromosome.

Despite these findings, none of the chromatin components or modifications found have been shown on their own to be essential for X chromosome silencing, indicating potential redundancy among them or the existence of pathways that are yet to be identified.

Global changes also occur in other types of dosage compensation. In *Drosophila*, a large ribonucleoprotein complex, MSL, is found only in males, where it localizes on the X chromosome. This complex contains two noncoding RNAs, which appear to be needed for localization to the male X (perhaps analogous to the localization of *Xist* to the inactive mammalian X), and a histone acetyltransferase that acetylates histone H4 on K16 throughout the male X. The net result of the action of this complex is the twofold increase in transcription of all genes on the male X. In the next section, we will discuss a third mechanism for dosage compensation, a global reduction in X-linked gene expression in XX (hermaphrodite) nematodes.

29.6 Chromosome Condensation Is Caused by Condensins

Key concepts

- SMC proteins are ATPases that include condensins and cohesins.
- A heterodimer of SMC proteins associates with other subunits.
- Condensins cause chromatin to be more tightly coiled by introducing positive supercoils into DNA.
- Condensins are responsible for condensing chromosomes at mitosis.
- Chromosome-specific condensins are responsible for condensing inactive X chromosomes in *C. elegans*.

The structures of entire chromosomes are influenced by interactions with proteins of the **SMC (structural maintenance of chromosome)** family. They are ATPases that fall into two functional groups. **Condensins** are involved with the control of overall structure, and are responsible for the condensation into compact chromosomes at mitosis. **Cohesins** are concerned with connections between sister chromatids that concatenate through a cohesion ring, which must be released at mitosis. Both consist of dimers formed by SMC proteins. Condensins form complexes that have a core of the heterodimer SMC2-SMC4 associated with other (non-SMC) proteins. Cohesins have a similar organization but consist of SMC1 and SMC3,

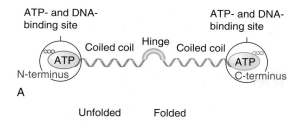

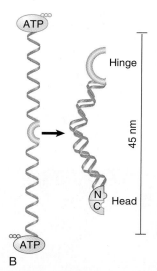

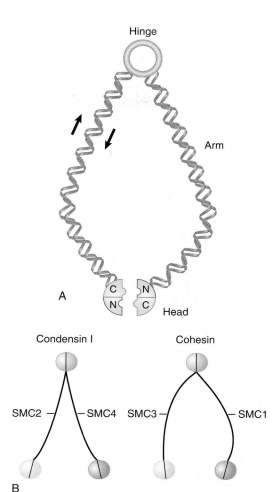

FIGURE 29.14 (A) An SMC protein has a "Walker module" with an ATP-binding motif and DNA-binding site at each end, which are connected by coiled coils that are linked by a hinge region. (B) SMC monomers fold at the hinge regions and interact along the length of the coiled coils. The N- and C-termini interact to form a head domain. Adapted from I. Onn, et al., *Annu. Rev. Cell Dev. Biol.* 24 (2008): 105–129.

FIGURE 29.15 (A) The basic architecture of condensin and cohesin complexes. (B) Condensin and cohesin consist of V-shaped dimers of two SMC proteins interacting through their hinge domains. The two monomers in a condensin dimer tend to exhibit a very small separation between the two arms of the V, while cohesins have a much larger angle of separation between the arms. Part A and B adapted from T. Hirano, *Nat. Rev. Mol. Cell Biol.* 7 (2006): 311–322.

and also interact with smaller non-SMC subunits termed Scc1/Rad21 and Scc3/SA.

FIGURE 29.14 shows that an SMC protein has a coiled-coil structure in its center that is interrupted by a flexible hinge region. Both the amino and carboxyl termini have ATP- and DNA-binding motifs. The ATP-binding motif is also known as a "Walker module." SMC monomers fold at the hinge region, forming an antiparallel interaction between the two halves of each coiled coil. This allows the amino and carboxyl termini to interact to form a "head" domain. Different models have been proposed for the actions of these proteins depending on whether they dimerize by intra- or intermolecular interactions.

Folded SMC proteins form dimers via several different interactions. The most stable association occurs between hydrophobic domains in the hinge regions. **FIGURE 29.15** shows that these

hinge–hinge interactions result in V-shaped structures. Electron microscopy shows that in solution, cohesins tend to form V's with the arms separated by a large angle, whereas condensins form more linear structures, with only a small angle between the arms. In addition, the heads of the two monomers can interact, closing the V, and the coiled coils of the individual monomers may also interact with each other. Various non-SMC proteins interact with SMC dimers and can influence the final structure of the dimer.

The function of cohesins is to hold sister chromatids together, but it is not yet clear

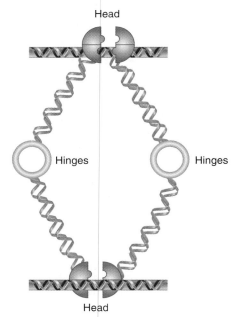

Head

Hinges   **Hinges**

Head

FIGURE 29.16 One model for DNA linking by cohesins. Cohesins may form an extended structure in which each monomer binds DNA and connects via the hinge region, allowing two different DNA molecules to be linked. Head domain interactions can result in binding by two cohesin dimers. Adapted from I. Onn, et al., *Annu. Rev. Cell Dev. Biol.* 24 (2008): 105–129.

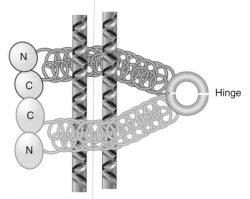

N
C
C
N

Hinge

FIGURE 29.17 Cohesins may dimerize by intramolecular connections, and then form multimers that are connected at the heads and at the hinge. Such a structure could hold two molecules of DNA together by surrounding them.

how this is achieved. There are several different models for cohesin function. **FIGURE 29.16** shows one model in which a cohesin could take the form of extended dimers, interacting hinge-to-hinge, that crosslink two DNA molecules. Head–head interactions would create tetrameric structures, adding to the stability of cohesion. An alternative "ring" model is shown in **FIGURE 29.17**. In this model, dimers interact at both their head and hinge regions to form a circular structure. Instead of binding directly to DNA, a structure of this type could hold DNA molecules together by encircling them.

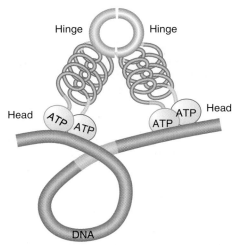

Hinge   **Hinge**

Head   ATP   ATP   ATP   ATP   **Head**

DNA

FIGURE 29.18 Condensins may form a compact structure by bending at the hinge, causing DNA to become compacted.

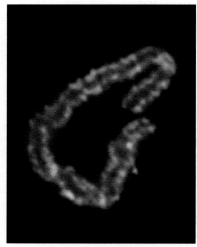

FIGURE 29.19 Condensins are located along the entire length of a mitotic chromosome. DNA is red; condensins are yellow. Photo courtesy of Ana Losada and Tatsuya Hirano.

While cohesins act to hold separate sister chromatids together, condensins are responsible for chromatin condensation. **FIGURE 29.18** shows that a condensin could take the form of a V-shaped dimer, interacting via the hinge domains, that pulls together distant sites on the same DNA molecule, causing it to condense. It is thought that dynamic head–head interactions could act to promote the ordered assembly of condensed loops, but the details of condensin action are still far from clear.

Visualization of mitotic chromosomes shows that condensins are located all along the length of the chromosome, as can be seen in **FIGURE 29.19**. (By contrast, cohesins are found at discrete locations in a focal nonrandom pattern with an average spacing of about 10 kb.) The condensin complex was named for its ability to cause chromatin to condense *in vitro*. It has an ability to introduce positive supercoils into DNA in an action that uses hydrolysis of ATP and depends on the presence of topoisomerase I. This ability is controlled by the phosphorylation of the non-SMC subunits, which occurs at mitosis. We do not know yet how this connects with other modifications of chromatin—for example, the phosphorylation of histones. The activation of the condensin complex specifically at mitosis makes it questionable whether it is also involved in the formation of interphase heterochromatin.

We discussed in the previous section the dramatic chromosomal changes that occur during X inactivation in female mammals and in X chromosome upregulation in male flies. In the nematode *C. elegans*, a third approach is used: twofold reduction of X-chromosome transcription in XX hermaphrodites relative to XO males. A dosage compensation complex (DCC) is maternally provided to both XX and XO embryos, but it then associates with both X chromosomes in only in XX animals, while remaining diffusely distributed in the nuclei of XO animals. The protein complex contains an SMC core, and is similar to the condensin complexes that are associated with mitotic chromosomes in other species. This suggests that it has a structural role in causing the chromosome to take up a more condensed, inactive state. Recent studies have shown, though, that SMC-related proteins may also have roles in dosage compensation in mammals: the protein SmcHD1 (SMC-hinge domain 1) may actually contribute to the deposition of DNA methylation on the inactive X. SMCs could recruit DNA methyltransferase via a component of the SMC core that is involved in RNAi-directed DNA methylation, such as occurs in *Arabidopsis* via the DMS3 protein (another SMC-related protein).

Whatever the mechanism of transcriptional downregulation, multiple sites on the X chromosome appear to be needed for the DCC to be fully distributed along it, and short DNA sequence motifs have been identified that appear to be key for localization of DCC. The complex binds to these sites, and then spreads along the chromosome to cover it more thoroughly.

Changes affecting all the genes on a chromosome, either negatively (mammals and *C. elegans*) or positively (*Drosophila*), are there-fore a common feature of dosage compensation. The components of the dosage compensation apparatus may vary, however, as well as the means by which it is localized to the chromosome. Dosage compensation in mammals and *Drosophila* both entail chromosome-wide changes in histone acetylation, and involve noncoding RNAs that play central roles in targeting X chromosomes for global change. In *C. elegans*, chromosome condensation by condensin homologs is used to accomplish dosage compensation. It remains to be seen whether there are also global changes in histone acetylation or other modifications in XX *C. elegans* that reflect the twofold reduction in transcription of the X chromosomes.

29.7 CpG Islands Are Subject to Methylation

Key concepts

- Most methyl groups in DNA are found on cytosine on both strands of the CpG doublet.
- Replication converts a fully methylated site to a hemimethylated site.
- Hemimethylated sites are converted to fully methylated sites by a maintenance methyltransferase.

Methylation of DNA occurs at specific sites. In bacteria, it is associated with identifying the bacterial restriction-methylation system used for phage defense, and also with distinguishing replicated and nonreplicated DNA. In eukaryotes, its principal known function is connected with the control of transcription; methylation of a control region is usually associated with gene inactivation. Methylation in eukaryotes principally occurs at **CpG islands** in the 5′ regions of some genes; these islands are defined by the presence of an increased density of the dinucleotide sequence, CpG (see *Section 20.12, CpG Islands Are Regulatory Targets*).

From 2% to 7% of the cytosines of animal cell DNA are methylated (the value varies with the species). The methylation occurs at the 5 position of cytosine producing 5-methyl-cytosine. Most of the methyl groups are found in CG dinucleotides in CpG islands, where the C residues on both strands of this short palindromic sequence are methylated.

Such a site is described as **fully methylated**. Consider, though, the consequences of replicating this site. **FIGURE 29.20** shows that each daughter duplex has one methylated strand and one unmethylated strand. Such a site is called **hemimethylated**.

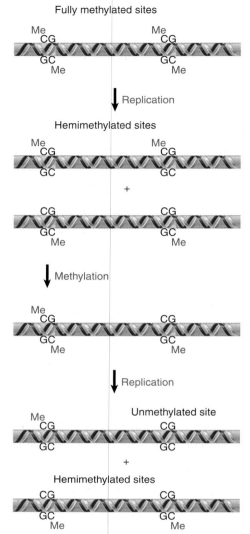

Fully methylated sites

↓ Replication

Hemimethylated sites

+

↓ Methylation

↓ Replication

Unmethylated site

+

Hemimethylated sites

FIGURE 29.20 The state of methylated sites could be perpetuated by an enzyme (Dnmt1) that recognizes only hemimethylated sites as substrates.

The perpetuation of the methylated site now depends on what happens to hemimethylated DNA. If methylation of the unmethylated strand occurs, the site is restored to the fully methylated condition. If replication occurs first, though, the hemimethylated condition will be perpetuated on one daughter duplex, but the site will become unmethylated on the other daughter duplex. **FIGURE 29.21** shows that the state of methylation of DNA is controlled by **DNA methyltransferases** (often shortened to *methylases*), or Dnmts, which add methyl groups to the 5 position of cytosine, and **demethylases**, which remove the methyl groups.

There are two types of DNA methyltransferase, whose actions are distinguished by the state of the methylated DNA. To modify DNA

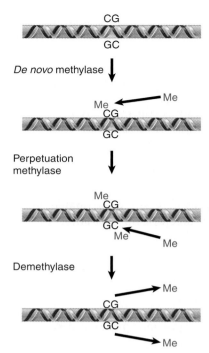

De novo methylase

Perpetuation methylase

Demethylase

FIGURE 29.21 The state of methylation is controlled by three types of enzyme. *De novo* and perpetuation methylases are known, but demethylases gave not been identified.

at a new position requires the action of the *de novo* **methyltransferase**, which recognizes DNA by virtue of a specific sequence. It acts *only* on unmethylated DNA, to add a methyl group to one strand. There are two *de novo* methyltransferases (Dnmt3A and Dnmt3B) in mouse; they have different target sites, and both are essential for development.

A **maintenance methyltransferase** acts constitutively *only on hemimethylated sites* to convert them to fully methylated sites. Its existence means that any methylated site is perpetuated after replication. There is one maintenance methyltransferase (Dnmt1) in mouse, and it is essential: mouse embryos in which its gene has been disrupted do not survive past early embryogenesis.

Maintenance methylation is almost 100% efficient. The result is that, if a *de novo* methylation occurs on one allele but not on the other, this difference will be perpetuated through ensuing cell divisions, maintaining a difference between the alleles that does not depend on their sequences.

How does a maintenance methyltransferase such as Dnmt1 target methylated CpG sites to preserve DNA methylation patterns with each cell replication? One possibility is that Dnmt1 is brought to hemimethylated sites by factors that recognize methylated CpG sites. Consistent with this concept, a protein has been identified,

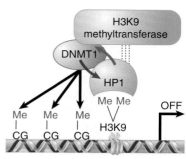

FIGURE 29.22 Mammalian HP1 is recruited to regions where lysine 9 of histone H3 (H3K9) has been methylated by a histone methyltransferase. HP1 then binds to DNMT1 and potentiates its DNA methyltransferase activity (blue arrow), thereby enhancing cytosine methylation (meCG) on nearby DNA. DNMT1 could in turn assist HP1 loading onto chromatin (red arrow). Furthermore, association of DNMT1 with the histone methyltransferase could allow a positive feedback loop to stablize inactive chromatin.

UHRF1, that is important for the maintenance of methylation both locally and globally through its association with Dnmt1. This protein is able to recognize CpG dinucleotides and to preferentially bind to hemimethylated DNA. Most importantly, however, UHRF1 binds to Dnmt1 and appears to increase the efficacy of Dnmt1 for maintenance methylation at hemimethylated CpG dinucleotides. Thus UHRF1 has dual functions in recognizing sites for maintenance methylation as well as in recruitment of the maintenance methyltransferase to these sites for methylation of the unmethylated CpG on the newly synthesized strand, thereby preserving methylation patterns with each cell replication.

Strikingly, UHRF1 also interacts with methylated histone H3, which connects the maintenance of DNA methylation with the stabilization of heterochromatin structure (see *Section 28.10, Methylation of Histones and DNA Is Connected*). DNA methylation and heterochromatin are in fact mutually reinforcing in several ways, such as the one depicted in **FIGURE 29.22**. Recall that HP1 is recruited to regions in which histone H3 has been methylated at lysine 9, a modification involved in heterochromatin formation. It turns out that HP1 can also interact with Dnmt1, which can promote DNA methylation in the vicinity of HP1 binding. Furthermore, Dnmt1 can directly interact with the methyltransferase responsible for H3K9 methylation, creating a positive feedback loop to ensure continued DNA and histone methylation. These interactions (and other similar networks of interactions) contribute to the stability of epigenetic states, allowing a heterochromatin region to be maintained through many cell divisions.

Methylation has various functional targets. Gene promoters are the most common target. The promoter may be methylated when a gene is inactive, and is always unmethylated when it is active. The absence of Dnmt1 in mouse causes widespread demethylation at promoters; we assume this is lethal because of the uncontrolled gene expression. Satellite DNA is another target. Mutations in Dnmt3B prevent methylation of satellite DNA, which causes centromere instability at the cellular level. Mutations in the corresponding human gene cause a disease called ICF (immunodeficiency/centromere instability, facial anomalies). The importance of methylation is emphasized by another human disease, Rett syndrome, which is caused by mutation of the gene for the protein MeCP2 that binds methylated CpG sequences. Patients with Rett syndrome exhibit autism-like symptoms that appear to be the result of a failure of normal gene silencing in the brain.

How are demethylated regions established and maintained? If a DNA site has not been methylated, a protein that recognizes the unmethylated sequence could protect it against methylation. Once a site has been methylated, there are several possible ways to generate demethylated sites. One is the loss of methylation at that site due to incomplete fidelity of Dnmt1 during maintenance methylation. Another mechanism is to block the maintenance methylase from acting on the site when it is replicated. After a second replication cycle, one of the daughter duplexes will be unmethylated. A third mechanism is to actively demethylate the site, either by removing the methyl group directly from cytosine, or by excising the methylated cytosine or cytidine from DNA for replacement by a repair system.

We know that active demethylation can occur to the paternal genome soon after fertilization, but we do not know what mechanism is used. One interesting possibility is that the cytidine deaminase AID may be involved; it can deaminate methylated C residues, creating a mismatched base pair that a repair system might then correct to a standard (unmethylated) C-G pair.

Plants transmit genomic methylation patterns through each generation, although methylation is removed from repeated sequences to prevent interference with nearby gene expression. Plants therefore can easily remove DNA methylation. This occurs through removal of 5-methylcytosine by DEMETER, followed by cleavage of the DNA backbone phosphodiester bond by AP endonuclease and insertion of the unmethylated dCMP base through base excision repair. In mammals, however, the genomic methylation patterns are erased in primordial germ cells—the cells that ultimately give rise to the germ line (see *Section 29.8, DNA Methylation Is Responsible for Imprinting*). Primordial germ cells have low levels of Dnmt1, thereby eliminating the need for demethylation on larger scales, as seen in plants. This reduced need for DNA demethylation in mammals relative to plants may explain the challenges in characterizing their mechanisms for DNA demethylation. DNMT3A and DNMT3B (*de novo* methyltransferases) may paradoxically participate in active DNA demethylation in mammals, though. DNMT3A and DNMT3B may possess deaminase activity and are involved not only in gene demethylation, but also cyclical demethylation and remethylation within the cell cycle. These enzymes appear to mediate oxidative deamination at cytosine C4 in the absence of the methyl donor (S-adenosylmethionine) to convert 5-methylcytosine to thymine. The resulting guanine-thymine (G-T) mismatch is repaired by base excision, thereby returning the mismatch to a guanine-cytosine (G-C) pair and leading to demethylation of a previously methylated CpG site.

29.8 DNA Methylation Is Responsible for Imprinting

Key concepts

- Paternal and maternal alleles may have different patterns of methylation at fertilization.
- Methylation is usually associated with inactivation of the gene.
- When genes are differentially imprinted, survival of the embryo may require that the functional allele is provided by the parent with the unmethylated allele.
- Survival of heterozygotes for imprinted genes is different, depending on the direction of the cross.
- Imprinted genes occur in clusters and may depend on a local control site where *de novo* methylation occurs unless specifically prevented.

The pattern of methylation of germ cells is established in each sex during gametogenesis by a two-stage process: First the existing pattern is erased by a genome-wide demethylation in primordial germ cells, and then the pattern specific for each sex is imposed during meiosis.

All allelic differences are lost when primordial germ cells develop in the embryo; irrespective of sex, the previous patterns of methylation are erased, and a typical gene is then unmethylated. In males, the pattern develops in two stages. The methylation pattern that is characteristic of mature sperm is established in the spermatocyte, but further changes are made in this pattern after fertilization. In females, the maternal pattern is imposed during oogenesis, when oocytes mature through meiosis after birth.

As may be expected from the inactivity of genes in gametes, the typical state is to be methylated. There are cases of differences between the two sexes, though, for which a locus is unmethylated in one sex. A major question is how the specificity of methylation is determined in the male and female gametes.

Systematic changes occur in early embryogenesis. Some sites will continue to be methylated, whereas others will be specifically unmethylated in cells in which a gene is expressed. From the pattern of changes, we may infer that individual sequence-specific demethylation events occur during somatic development of the organism as particular genes are activated.

The specific pattern of methyl groups in germ cells is responsible for the phenomenon of **imprinting**, which describes a difference in behavior between the alleles inherited from each parent. The expression of certain genes in mouse embryos depends upon the sex of the parent from which they were inherited. For example, the allele coding for IGF-II (insulin-like growth factor II) that is inherited from the father is expressed, but the allele that is inherited from the mother is not expressed. The IGF-II gene of oocytes is methylated in its promoter, whereas the IGF-II gene of sperm is not, so that the two alleles behave differently in the zygote. This is the most common pattern, but the dependence on sex is reversed for some genes. In fact, the opposite pattern (expression of maternal copy) is shown for IGF-IIR, a receptor that causes the rapid turnover of IGF-II.

This sex-specific mode of inheritance requires that the pattern of methylation is established specifically during each gametogenesis. The fate of a hypothetical locus in a mouse is illustrated in **FIGURE 29.23**. In the early embryo, the paternal allele is unmethylated and expressed, and the maternal allele is methylated and silent. What happens when this mouse itself forms gametes? If it is a male, the allele contributed to the sperm must be nonmethyl-

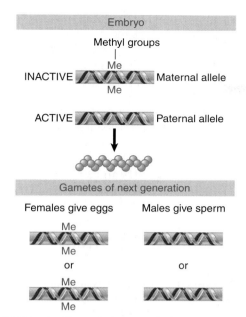

FIGURE 29.23 The typical pattern for imprinting is that a methylated locus is inactive. If this is the maternal allele, only the paternal allele is active, and will be essential for viability. The methylation pattern is reset when gametes are formed, so that all sperm have the paternal type and all oocytes have the maternal type.

ated, irrespective of whether it was originally methylated or not. Thus when the maternal allele finds itself in a sperm, it must be demethylated. If the mouse is a female, the allele contributed to the egg must be methylated; if it was originally the paternal allele, methyl groups must be added.

The consequence of imprinting is that an embryo is *hemizygous* for any imprinted gene. Thus in the case of a heterozygous cross where the allele of one parent has an inactivating mutation, the embryo will survive if the wild-type allele comes from the parent in which this allele is active, but will die if the wild-type allele is the imprinted (silenced) allele. This type of dependence on the directionality of the cross (in contrast with Mendelian genetics) is an example of epigenetic inheritance, where some factor other than the sequences of the genes themselves influences their effects. Although the paternal and maternal alleles have identical sequences, they display different properties, depending on which parent provided them. These properties are inherited through meiosis and the subsequent somatic mitoses.

Although imprinted genes are estimated to comprise 1%–2% of the mammalian transcriptome, these genes are sometimes clustered. More than half of the ~25 known imprinted genes in mouse are contained in two particular regions, each containing both maternally and paternally expressed genes. This suggests the possibility that imprinting mechanisms may function over long distances. Some insights into this possibility come from deletions in the human population that cause the Prader–Willi and Angelman diseases. Most cases of these neurodevelopmental disorders involving the proximal long arm of chromosome 15 are caused by the same 4 Mb deletion, but the syndromes are different, depending on which parent contributed the deletion. The reason is that the deleted region contains at least one gene that is paternally imprinted and at least one that is maternally imprinted. There are some rare cases, however, with much smaller deletions. Prader–Willi syndrome can be caused by a 20 kb deletion that silences distant genes on either side of the deletion. The basic effect of the deletion is to prevent a father from resetting the paternal mode to a chromosome inherited from his mother. The result is that these genes remain in maternal mode, so that the paternal as well as maternal alleles are silent in the offspring. The inverse effect is found in some small deletions that cause Angelman's syndrome. These mutations have led to the identification of a Prader–Willi/Angelman's syndrome "imprint center" (PW/AS IC) that acts at a distance to regulate imprinting in either sex across the entire region.

A microdeletion resulting in removal of a cluster of small nucleolar RNAs (snoRNAs) that is paternally derived may result in the key aspects of Prader–Willi syndrome. Mutations that separate the snoRNA HBII-85 cluster from its promoter cause Prader–Willi syndrome, although other genes in the region could also contribute to the syndrome.

Imprinting may also regulate alternative polyadenylation. A number of mammalian genes utilize multiple polyadenylation (polyA) sites to confer diversity on gene transcription. The *H13* murine gene undergoes alternative polyadenylation in an allele-specific manner, in that polyA sites are differentially methylated in the maternal and paternal genome of this imprinted gene. Elongation proceeds to downstream polyadenylation sites when the allele is methylated, indicating that epigenetic processes may influence alternative polyadenylation contributing to the diversity of gene transcription in mammals.

29.9 Oppositely Imprinted Genes Can Be Controlled by a Single Center

Key concepts

- Imprinted genes are controlled by methylation of *cis*-acting sites.
- Methylation may be responsible for either inactivating or activating a gene.

Imprinting is determined by the state of methylation of a *cis*-acting site near a target gene or genes. These regulatory sites are known as differentially methylated domains (DMDs) or imprinting control regions (ICRs). Deletion of these sites removes imprinting, and the target loci then behave the same in both maternal and paternal genomes.

The behavior of a region containing two genes, *Igf2* and *H19*, illustrates the ways in which methylation can control gene activity. FIGURE 29.24 shows that these two genes react oppositely to the state of methylation at the ICR located between them. The ICR is methylated on the paternal allele. *H19* shows the typical response of inactivation. Note, however, that *Igf2* is expressed. The reverse situation is found on a maternal allele, where the ICR is not methylated. *H19* now becomes expressed, but *Igf2* is inactivated.

The control of *Igf2* is exercised by an insulator contained within the ICR (see *Section 10.12, Insulators Define Transcriptionally Independent Domains*). FIGURE 29.25 shows that when the ICR is unmethylated, it binds the protein CTCF. This creates a functional insulator that blocks an enhancer from activating the *Igf2* promoter. This is an unusual effect in which methylation indirectly activates a gene by blocking an insulator.

The regulation of *H19* shows the more usual direction of control in which methylation creates an inactive imprinted state. This could reflect a direct effect of methylation on promoter activity, though the effect could also be due to additional factors. CTCF regulates chromatin by repressing H3K27 trimethylation at the *Igf2* locus independently of repression by DNA hypermethylation. As a result, the effects of CTCF on chromatin, as well as on DNA methylation, likely contribute to the imprinting of *H19* and *Igf2*.

29.10 Epigenetic Effects Can Be Inherited

Key concepts

- Epigenetic effects can result from modification of a nucleic acid after it has been synthesized or by the perpetuation of protein structures.
- Epigenetic effects may be inherited through generations.

Epigenetic inheritance describes the ability of different states, which may have different phenotypic consequences, to be inherited without any change in the sequence of DNA. How can this occur? We can divide epigenetic mechanisms into two general classes:

- DNA may be modified by the covalent attachment of a moiety that is then perpetuated. Two alleles with the same sequence may have different states of methylation that confer different properties.
- A self-perpetuating protein state may be established. This might involve assembly of a protein complex, modification of specific protein(s), or establishment of an alternative protein conformation.

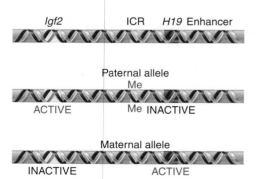

FIGURE 29.24 The ICR is methylated on the paternal allele, where *Igf2* is active and *H19* is inactive. ICR is unmethylated on the maternal allele, where *Igf2* is inactive and *H19* is active.

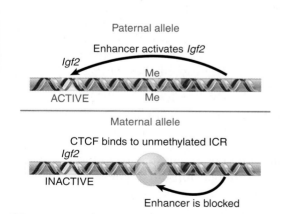

FIGURE 29.25 The ICR contains an insulator that prevents an enhancer from activating *Igf2*. The insulator functions only when CTCF binds to unmethylated DNA.

Methylation establishes epigenetic inheritance so long as the maintenance methyltransferase acts constitutively to restore the methylated state after each cycle of replication, as shown in Figure 29.20. A state of methylation can be perpetuated through an indefinite series of somatic mitoses. This is probably the "default" situation. Methylation can also be perpetuated through meiosis: for example, in the fungus *Ascobolus* there are epigenetic effects that can be transmitted through both mitosis and meiosis by maintaining the state of methylation. In mammalian cells, epigenetic effects are first erased in primordial germ cells and then created by resetting the state of methylation differently in male and female meioses during gametogenesis, as described in the previous sections.

Situations in which epigenetic effects appear to be maintained by means of protein states are less well understood in molecular terms. Position effect variegation shows that constitutive heterochromatin may extend for a variable distance, and the structure is then perpetuated through somatic divisions. There is no methylation of DNA in *Saccharomyces* and a vanishingly small amount in *Drosophila*, and as a result the inheritance of epigenetic states of position effect variegation or telomeric silencing in these organisms is likely to be due to the perpetuation of protein structures.

FIGURE 29.26 considers two extreme possibilities for the fate of a protein complex at replication.

- A complex could perpetuate itself if it splits symmetrically, so that half complexes associate with each daughter duplex. If the half complexes have the capacity to nucleate formation of full complexes, the original state will be restored. This is basically analogous to the maintenance of methylation. The problem with this model is that there is no evident reason why protein complexes should behave in this way.
- A complex could be maintained as a unit and segregate to one of the two daughter duplexes. The problem with this model is that it requires a new complex to be assembled *de novo* on the other daughter duplex, and it is not evident why this should happen.

Consider now the need to perpetuate a heterochromatic structure consisting of protein complexes. Suppose that a protein is distributed more or less continuously along a stretch of heterochromatin, as implied in Figure 29.4. If individual subunits are distributed at random to each daughter duplex at replication, the two daughters will continue to be marked by the protein, although its density will be reduced to half of the level before replication. If the protein has a self-assembling property that causes new subunits to associate with it, the original situation may be restored. *Basically, the existence of epigenetic effects forces us to the view that a protein responsible for such a situation must have some sort of self-templating or self-assembling capacity.*

In some cases, it may be the state of protein modification, rather than the presence of the protein *per se*, that is responsible for an epigenetic effect. There is a general correlation between the activity of chromatin and the state of acetylation of the histones, in particular the acetylation of the N-terminal tails of histones H3 and H4. Activation of transcription is associated with acetylation in the vicinity of the promoter; and repression of transcription is associated with deacetylation (see *Section 28.9, Histone Acetylation Is Associated with Transcription Activation*). The most dramatic correlation is that the inactive X chromosome in mammalian female cells is underacetylated.

The inactivity of constitutive heterochromatin may require that the histones are not acetylated. If a histone acetyltransferase is tethered to a region of telomeric heterochromatin

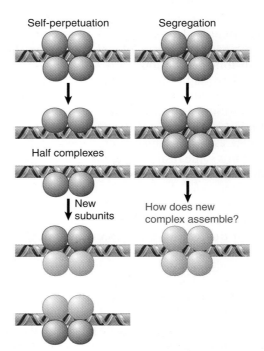

FIGURE 29.26 What happens to protein complexes on chromatin during replication?

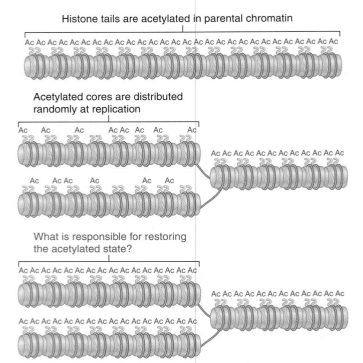

Histone tails are acetylated in parental chromatin

Acetylated cores are distributed randomly at replication

What is responsible for restoring the acetylated state?

FIGURE 29.27 Acetylated histones are conserved and distributed at random to the daughter chromatin fibers at replication. Each daughter fiber has a mixture of old (acetylated) cores and new (unacetylated) histones.

in yeast, silenced genes become active. When yeast is exposed to trichostatin (an inhibitor of deacetylation), centromeric heterochromatin becomes acetylated, and silenced genes in centromeric regions may become active. *The effect may persist even after trichostatin has been removed.* In fact, it may be perpetuated through mitosis and meiosis. This suggests that an epigenetic effect has been created by changing the state of histone acetylation.

How might the state of acetylation be perpetuated? Suppose that the $H3_2$-$H4_2$ tetramer is distributed at random to the two daughter duplexes. This creates the situation shown in **FIGURE 29.27**, in which each daughter duplex contains some histone octamers that are acetylated on the H3 and H4 tails, whereas others are unacetylated. To account for the epigenetic effect, we could suppose that the presence of some acetylated histone octamers provides a signal that causes the unacetylated octamers to be acetylated.

We do not yet fully understand how epigenetic changes are inherited mitotically in somatic cells, but it is clear that this occurs. Surprisingly, several lines of evidence indicate that epigenetic effects may also be transmitted *across generations* in a process referred to as **transgenerational epigenetics**. Evidence that

DNA methylation is a central coordinator that secures stable transgenerational inheritance in plants come from studies of an *Arabidopsis thaliana* mutant deficient in maintaining DNA methylation. The loss of DNA methylation triggers genome-wide activation of alternative epigenetic mechanisms such as RNA-directed DNA methylation, DNA demethylase inhibition, and retargeting of histone H3K9 methylation. In the absence of maintenance methylation, new and aberrant patterns of epigenetic marks accumulate over several generations, leaving these plants dwarfed and sterile. As a result—at least in plants—the case is strong that intact maintenance methylation plays a major role in transgenerational epigenetics.

In mammals, support for transgenerational epigenetics is less strong, but several lines of evidence indicate that this process occurs in mammals as well. *Metastable epialleles* are dependent upon the epigenetic state for their transcription. This state can vary not only between cells, but also between tissues. Although the epigenetic state of the genome undergoes reprogramming in the parental genomes and during early embryogenesis, some loci may transmit the epigenetic state through the gametes to the next generation (transgenerational epigenetics). For example, in mice there is a dominant mutation of the *agouti* locus (a coat color gene) known as *agouti viable yellow*, which is caused by the insertion of a retrotransposon upstream of the *agouti* coding region. This allele shows variegation, resulting in coat colors ranging from solid yellow, to mottled, to completely agouti (dark). It has been observed that agouti females are more likely to produce agouti offspring and yellow females are more likely to produce yellow offspring—in other words, the variable level of expression of *agouti* in the mother appears to be transmitted to the offspring (while the color of the father is irrelevant). It turns out that DNA methylation of the inserted retrotransposon determines the coat color of the agouti mice, indicating transgenerational conservation of expression levels due to incomplete erasure of the epigenetic mark between generations.

Metastable alleles may also play a role in transgenerational epigenetic inheritance in humans, as suggested by the high degree of copy-number variation within monozygotic twins. Moreover, in some cases of Prader–Willi syndrome there is no apparent mutation but, rather, an *epimutation* involving aberrant DNA methylation. The cause for the epimutation may be due to an allele that has passed through the male germ line without erasure of the silent epigenetic state established in the grandmother. Thus the evidence for transgenerational epigenetic inheritance is emerging not only in plants and mammals, but also as a potential cause for gene control or diseases due to aberrant epigenetic control of transcription in humans.

29.11 Yeast Prions Show Unusual Inheritance

Key concepts

- The Sup35 protein in its wild-type soluble form is a termination factor for translation.
- Sup35 can also exist in an alternative form of oligomeric aggregates, in which it is not active in protein synthesis.
- The presence of the oligomeric form causes newly synthesized protein to acquire the inactive structure.
- Conversion between the two forms is influenced by chaperones.
- The wild-type form has the recessive genetic state *psi*⁻ and the mutant form has the dominant genetic state *PSI*⁺.

One of the clearest cases of the dependence of epigenetic inheritance on the condition of a protein is provided by the behavior of *prions*. They have been characterized in two circumstances: by genetic effects in yeast, and as the causative agents of neurological diseases in mammals, including humans. A striking epigenetic effect is found in yeast, where two different states can be inherited that map to a single genetic locus, *although the sequence of the gene is the same in both states*. The two different states are [*psi*⁻] and [*PSI*⁺]. A switch in condition occurs at a low frequency as the result of a spontaneous transition between the states.

The [*psi*] genotype maps to the locus *SUP35*, which codes for a translation termination factor. **FIGURE 29.28** summarizes the effects of the Sup35 protein in yeast. In wild-type cells, which are characterized as [*psi*⁻], the gene is active, and Sup35 protein terminates protein synthesis. In cells of the mutant [*PSI*⁺] type, the oligomerized factor does not function, which causes a failure to terminate protein synthesis properly. (This was originally detected by the lethal effects of the enhanced efficiency of suppressors of ochre codons in [*PSI*⁺] strains.)

[*PSI*⁺] strains have unusual genetic properties. When a [*psi*⁻] strain is crossed with a [*PSI*⁺] strain, *all of the progeny are* [*PSI*⁺]. This is

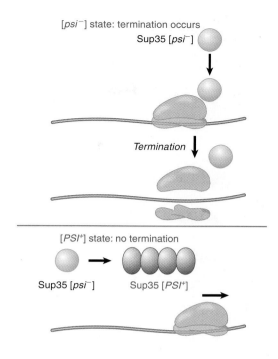

FIGURE 29.28 The state of the Sup35 protein determines whether termination of translation occurs.

a pattern of inheritance that would be expected of an extrachromosomal agent, but the [*PSI*⁺] trait cannot be mapped to any such nucleic acid. The [*PSI*⁺] trait is metastable, which means that, although it is inherited by most progeny, it is lost at a higher rate than is consistent with mutation. Similar behavior also is shown by the locus *URE2*, which codes for a protein required for nitrogen-mediated repression of certain catabolic enzymes. When a yeast strain is converted into an alternative state called [*URE3*], the Ure2 protein is no longer functional.

The [*PSI*⁺] state is determined by the conformation of the Sup35 protein. In a wild-type [*psi*⁻] cell, the protein displays its normal function. In a [*PSI*⁺] cell, though, the protein is present in an alternative conformation in which its normal function has been lost. To explain the unilateral dominance of [*PSI*⁺] over [*psi*⁻] in genetic crosses, we must suppose that *the presence of protein in the* [*PSI*⁺] *state causes all the protein in the cell to enter this state*. This requires an interaction between the [*PSI*⁺] protein and newly synthesized protein, which probably reflects the generation of an oligomeric state in which the [*PSI*⁺] protein has a nucleating role, as illustrated in **FIGURE 29.29**.

A feature common to both the Sup35 and Ure2 proteins is that each consists of two domains that function independently. The

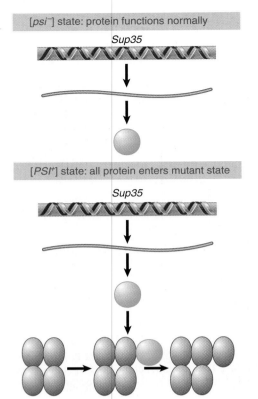

FIGURE 29.29 Newly synthesized Sup35 protein is converted into the [*PSI*⁺] state by the presence of preexisting [*PSI*⁺] protein.

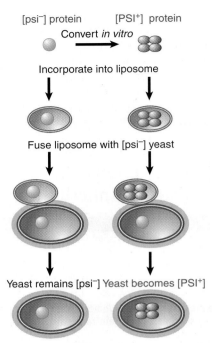

FIGURE 29.30 Purified protein can convert the [*psi*⁻] state of yeast to [*PSI*⁺].

C-terminal domain is sufficient for the activity of the protein. The N-terminal domain is sufficient for formation of the structures that make the protein inactive. Thus yeast in which the N-terminal domain of Sup35 has been deleted cannot acquire the [*PSI*⁺] state, and the presence of a [*PSI*⁺] N-terminal domain is sufficient to maintain Sup35 protein in the [*PSI*⁺] condition. The critical feature of the N-terminal domain is that it is rich in glutamine and asparagine residues.

Loss of function in the [*PSI*⁺] state is due to the sequestration of the protein in an oligomeric complex. Sup35 protein in [*PSI*⁺] cells is clustered in discrete foci, whereas the protein in [*psi*⁻] cells is diffused in the cytosol. Sup35 protein from [*PSI*⁺] cells forms **amyloid fibers** *in vitro*—these have a characteristic high content of β-sheet structures. These amyloid fibers consist of a parallel in-register β-sheet structure, which allows the prion amyloid to induce a "templating" action at the end of filaments. This templating action provides the faithful transmission of variant differences in these molecules and allows self-reproduction encod-

ing heritable information reminiscent of the behavior of genes.

The involvement of protein conformation (rather than covalent modification) is suggested by the effects of conditions that affect protein structure. Denaturing treatments cause loss of the [*PSI*⁺] state. In particular, the chaperone Hsp104 is involved in inheritance of [*PSI*⁺]. Its effects are paradoxical. Deletion of *HSP104* prevents maintenance of the [*PSI*⁺] state, and overexpression of Hsp104 also causes loss of the [*PSI*⁺] state through elimination of Sup35 proteins. The Ssa and Ssb components of the Hsp70 chaperone system affect Sup35 priongenesis directly through cooperation with Hsp104. Ssa and Ssb binding are facilitated by Hsp40 chaperones through interactions with Sup35 oligomers. At high concentrations, Hsp104 eliminates Sup35 prions while low levels of Hsp104 stimulate priongenesis and alleviate some Hsp70:Hsp40 pairs. Thus the interplay between Hsp104, Hsp70, and Hsp40 regulates the formation, growth, and elimination of Sup35 prions.

Using the ability of Sup35 to form the inactive structure *in vitro*, it is possible to provide biochemical proof for the role of the protein. **FIGURE 29.30** illustrates a striking experiment in which the protein was converted to the inactive form *in vitro*, put into liposomes (where in effect the protein is surrounded by an artificial membrane), and then introduced directly into cells by fusing the liposomes with [*psi*⁻] yeast. The yeast cells were converted to [*PSI*⁺]! This experiment refutes all of the objections that were raised to the conclusion that the protein has the ability to confer the epigenetic state. Experiments in which cells are mated, or in which extracts are taken from one cell to treat another cell, always are susceptible to the possibility that a nucleic acid has been transferred. When the protein by itself does not convert target cells, though (even though protein converted to the inactive state can do so), the only difference is the treatment of the protein—which must therefore be responsible for the conversion.

The ability of yeast to form the [*PSI*⁺] prion state depends on the yeast's genetic background. The yeast must be [*PIN*⁺] in order for the [*PSI*⁺] state to form. The [*PIN*⁺] condition itself is an epigenetic state. It can be created by the formation of prions from any one of several different proteins. These proteins share a key characteristic of Sup35, which is that they have Gln/Asn-rich domains. Overexpression of these domains in yeast stimulates formation of the [*PSI*⁺] state. This suggests that there is a common model for the formation of the prion state that involves aggregation of the Gln/Asn domains into self-propagating amyloid structure.

How does the presence of one Gln/Asn protein influence the formation of prions by another? We know that the formation of Sup35 prions is specific to Sup35 protein; that is, it does not occur by cross-aggregation with other proteins. This suggests that the yeast cell may contain soluble proteins that antagonize prion formation. These proteins are not specific for any one prion. As a result, the introduction of any Gln/Asn domain protein that interacts with these proteins will reduce the concentration. This will allow other Gln/Asn proteins to aggregate more easily.

Prions have recently been linked to chromatin remodeling factors. Swi1 is a subunit of the SWI/SNF chromatin-remodeling complex (see *Section 28.7, Chromatin Remodeling Is an Active Process*), and this protein can become a prion. Swi1 aggregates in [*SWI*⁺] cells but not in nonprion cells, and is dominantly and cytoplasmically transmitted. This suggests that inheritance through proteins can impact chromatin remodeling and potentially affect gene regulation throughout the genome.

29.12 Prions Cause Diseases in Mammals

Key concepts

- The protein responsible for scrapie exists in two forms: the wild-type noninfectious form PrP^C, which is susceptible to proteases, and the disease-causing PrP^{Sc}, which is resistant to proteases.
- The neurological disease can be transmitted to mice by injecting the purified PrP^{Sc} protein into mice.
- The recipient mouse must have a copy of the *PrP* gene coding for the mouse protein.
- The PrP^{Sc} protein can perpetuate itself by causing the newly synthesized PrP protein to take up the PrP^{Sc} form instead of the PrP^C form.
- Multiple strains of PrP^{Sc} may have different conformations of the protein.

Prion diseases have been found in humans, sheep, cows, and more recently in wild deer and elk. The basic phenotype is an ataxia—a neurodegenerative disorder that is manifested by an inability to remain upright. The name of the disease in sheep, **scrapie**, reflects the phenotype: The sheep rub against walls in order to stay upright. Scrapie can be perpetuated by inoculating sheep with tissue extracts from infected animals. In humans, the disease **kuru** was found in New Guinea, where it appeared to be perpetuated by cannibalism, in particular the eating of brains. Related diseases in Western populations with a pattern of genetic transmission include Gerstmann–Straussler syndrome and the related Creutzfeldt–Jakob disease (CJD), which occurs sporadically. A disease resembling CJD appears to have been transmitted by consumption of meat from cows suffering from "mad cow" disease.

When tissue from scrapie-infected sheep is inoculated into mice, the disease occurs in a period ranging from 75 to 150 days. The active component is a protease-resistant protein. The protein is coded by a gene that is normally expressed in the brain. The form of the protein in normal brain, called PrP^C, is sensitive to proteases. Its conversion to the resistant form, called Prp^{Sc}, is associated with occurrence of the disease. Neurotoxicity is mediated by PrP^L, which is catalyzed by PrP^{Sc} and occurs when PrP^L concentration becomes too high. Rapid propagation results in severe neurotoxicity and eventual death. The infectious preparation has no detectable nucleic acid, is sensitive to UV irradiation at wavelengths that damage protein, and has a low infectivity (1 infectious

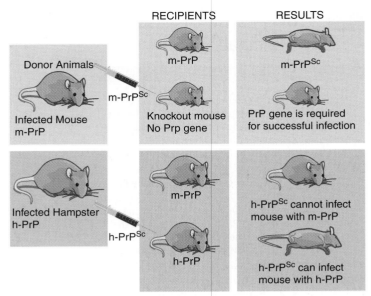

FIGURE 29.31 A PrP^{Sc} protein can only infect an animal that has the same type of endogenous PrP^C protein.

unit/10^5 PrP^{Sc} proteins). This corresponds to an epigenetic inheritance in which there is no change in genetic information (because normal and diseased cells have the same *PrP* gene sequence), but the PrP^{Sc} form of the protein is the infectious agent (whereas PrP^C is harmless). The PrP^{Sc} form has a high content of β-sheets, which form an amyloid fibrillous structure that is absent from the PrP^C form. The basis for the difference between the PrP^{Sc} and PrP^C forms appears to lie with a change in conformation rather than with any covalent alteration. Both proteins are glycosylated and linked to the membrane by a GPI-linkage.

The assay for infectivity in mice allows the dependence on protein sequence to be tested. **FIGURE 29.31** illustrates the results of some critical experiments. In the normal situation, PrP^{Sc} protein extracted from an infected mouse will induce disease (and ultimately kill) when it is injected into a recipient mouse. If the *PrP* gene is "knocked out," a mouse becomes resistant to infection. This experiment demonstrates two things. First, the endogenous protein is necessary for an infection, presumably because it provides the raw material that is converted into the infectious agent. Second, the cause of disease is not the removal of the PrP^C form of the protein, because a mouse with no PrP^C survives normally: The disease is caused by a gain-of-function in PrP^{Sc}. If the PrP gene is altered to prevent the GPI-linkage from occur-

ring, mice infected with PrP^{Sc} do not develop disease, which suggests that the gain of function involves an altered signaling function for which the GPI-linkage is required.

The existence of species barriers allows hybrid proteins to be constructed to delineate the features required for infectivity. The original preparations of scrapie were perpetuated in several types of animal, but these cannot always be transferred readily. For example, mice are resistant to infection from prions of hamsters. This means that hamster-PrP^{Sc} cannot convert mouse-PrP^C to PrP^{Sc}. The situation changes, though, if the mouse *PrP* gene is replaced by a hamster *PrP* gene. (This can be done by introducing the hamster *PrP* gene into the *PrP* knockout mouse.) A mouse with a hamster *PrP* gene is sensitive to infection by hamster PrP^{Sc}. This suggests that the conversion of cellular PrP^C protein into the Sc state requires that the PrP^{Sc} and PrP^C proteins have matched sequences.

There are different "strains" of PrP^{Sc}, which are distinguished by characteristic incubation periods upon inoculation into mice. This implies that the protein is not restricted solely to alternative states of PrP^C and PrP^{Sc}, but rather that there may be multiple Sc states. These differences must depend on some self-propagating property of the protein other than its sequence. If conformation is the feature that distinguishes PrP^{Sc} from PrP^C, then there must be multiple conformations, each of which has a self-templating property when it converts PrP^C.

The probability of conversion from PrP^C to PrP^{Sc} is affected by the sequence of PrP. Gerstmann–Straussler syndrome in humans is caused by a single amino acid change in PrP. This is inherited as a dominant trait. If the same change is made in the mouse PrP gene, mice develop the disease. This suggests that the mutant protein has an increased probability of spontaneous conversion into the Sc state. Similarly, the sequence of the PrP gene determines the susceptibility of sheep to develop the disease spontaneously; the combination of amino acids at three positions (codons 136, 154, and 171) determines susceptibility.

The prion offers an extreme case of epigenetic inheritance, in which the infectious agent is a protein that can adopt multiple conformations, each of which has a self-templating property. This property is likely to involve the state of aggregation of the protein.

29.13 Summary

The formation of heterochromatin occurs by proteins that bind to specific chromosomal regions (such as telomeres) and that interact with histones. The formation of an inactive structure may propagate along the chromatin thread from an initiation center. Similar events occur in silencing of the inactive yeast mating type loci. Repressive structures that are required to maintain the inactive states of particular genes are formed by the Pc-G protein complex in *Drosophila*. They share with heterochromatin the property of propagating from an initiation center.

Formation of heterochromatin may be initiated at certain sites and then propagated for a distance that is not precisely determined. When a heterochromatic state has been established, it is inherited through subsequent cell divisions. This gives rise to a pattern of epigenetic inheritance, in which two identical sequences of DNA may be associated with different protein structures, and therefore have different abilities to be expressed. This explains the occurrence of position effect variegation in *Drosophila*.

Modification of histone tails is a trigger for chromatin reorganization. Acetylation is generally associated with gene activation. Histone acetyltransferases are found in activating complexes, whereas histone deacetylases are found in inactivating complexes. Histone methylation is associated with gene inactivation or activation depending on the specific histone residues that are affected. Some histone modifications may be exclusive or synergistic with others.

Inactive chromatin at yeast telomeres and silent mating type loci appears to have a common cause, and involves the interaction of certain proteins with the N-terminal tails of histones H3 and H4. Formation of the inactive complex may be initiated by binding of one protein to a specific sequence of DNA; the other components may then polymerize in a cooperative manner along the chromosome.

Inactivation of one X chromosome in female (eutherian) mammals occurs at random. The *Xic* locus is necessary and sufficient to count the number of X chromosomes. The n−1 rule ensures that all but one X chromosome are inactivated. *Xic* contains the gene *Xist*, which codes for an RNA that is expressed only on the inactive X chromosome. Stabilization of *Xist* RNA is the mechanism by which the inactive X chromosome is distinguished; it is then inactivated by the activities of Poly-comb complexes, heterochromatin formation, and DNA methylation. The antisense RNA *Tsix* negatively regulates *Xist* on the future active X chromosome.

Methylation of DNA is inherited epigenetically. Replication of DNA creates hemimethylated products, and a maintenance methylase restores the fully methylated state. Epigenetic effects can be inherited during mitosis in somatic cells or they may be transmitted through organisms from one generation to another. Some methylation events depend on parental origin. Sperm and eggs contain specific and different patterns of methylation, with the result that paternal and maternal alleles are differently expressed in the embryo. This is responsible for imprinting, in which the unmethylated allele inherited from one parent is essential because it is the only active allele; the allele inherited from the other parent is silent. Patterns of methylation are reset during gamete formation in every generation after erasure in primordial germ cells, the cells that ultimately give rise to the germ line.

Prions are proteinaceous infectious agents that are responsible for the disease of scrapie in sheep and for related diseases in humans. The infectious agent is a variant of a normal cellular protein. The PrP^{Sc} form has an altered conformation that is self-templating: the normal PrP^{C} form does not usually take up this conformation, but does so in the presence of PrP^{Sc}. A similar effect is responsible for inheritance of the [*PSI*] element in yeast.

References

29.2 Heterochromatin Propagates from a Nucleation Event

Research

Ahmad, K. and Henikoff, S. (2001). Modulation of a transcription factor counteracts heterochromatic gene silencing in *Drosophila*. *Cell* 104, 839–847.

29.3 Heterochromatin Depends on Interactions with Histones

Reviews

Bühler, M. and Moazed, D. (2007). Transcription and RNAi in heterochromatic gene silencing. *Nat. Struct. Mol. Biol.* 14, 1041–1048.

Moazed, D. (2001). Common themes in mechanisms of gene silencing. *Mol. Cell* 8, 489–498.

Morris, C. A. and Moazed, D. (2007). Centromere assembly and propagation. *Cell* 128, 647–650.

Rusche, L. N., Kirchmaier, A. L., and Rine, J. (2003). The establishment, inheritance, and function of silenced chromatin in *Saccharomyces cerevisiae*. *Annu. Rev. Biochem.* 72, 481–516.

Zhang, Y. and Reinberg, D. (2001). Transcription regulation by histone methylation: interplay between different covalent modifications of the core histone tails. *Genes Dev.* 15, 2343–2360.

Research

Ahmad, K. and Henikoff, S. (2001). Modulation of a transcription factor counteracts heterochromatic gene silencing in *Drosophila*. *Cell* 104, 839–847.

Bannister, A. J., Zegerman, P., Partridge, J. F., Miska, E. A., Thomas, J. O., Allshire, R. C., and Kouzarides, T. (2001). Selective recognition of methylated lysine 9 on histone H3 by the HP1 chromo domain. *Nature* 410, 120–124.

Baum, M., Sanyal, K., Mishra, P. K., Thaler, N., and Carbon, J. (2006). Formation of functional centromeric chromatin is specified epigenetically in *Candida albicans*. *Proc. Natl. Acad. Sci. USA* 103, 14877–14882.

Bloom, K. S. and Carbon, J. (1982). Yeast centromere DNA is in a unique and highly ordered structure in chromosomes and small circular minichromosomes. *Cell* 29, 305–317.

Cheutin, T., McNairn, A. J., Jenuwein, T., Gilbert, D. M., Singh, P. B., and Misteli, T. (2003). Maintenance of stable heterochromatin domains by dynamic HP1 binding. *Science* 299, 721–725.

Eissenberg, J. C., Morris, G. D., Reuter, G., and Hartnett, T. (1992). The heterochromatin-associated protein HP-1 is an essential protein in *Drosophila* with dosage-dependent effects on position-effect variegation. *Genetics* 131, 345–352.

Folco, H. D., Pidoux, A. L., Urano, T., and Allshire, R. C. (2008). Heterochromatin and RNAi are required to establish CENP-A chromatin at centromeres. *Science* 319, 94–97.

Hecht, A., Laroche, T., Strahl-Bolsinger, S., Gasser, S. M., and Grunstein, M. (1995). Histone H3 and H4 N-termini interact with the silent information regulators SIR3 and SIR4: a molecular model for the formation of heterochromatin in yeast. *Cell* 80, 583–592.

Imai, S., Armstrong, C. M., Kaeberlein, M., and Guarente, L. (2000). Transcriptional silencing and longevity protein Sir2 is an NAD-dependent histone deacetylase. *Nature* 403, 795–800.

Kayne, P. S., Kim, U. J., Han, M., Mullen, R. J., Yoshizaki, F., and Grunstein, M. (1988). Extremely conserved histone H4 N terminus is dispensable for growth but essential for repressing the silent mating loci in yeast. *Cell* 55, 27–39.

Lachner, M., O'Carroll, D., Rea, S., Mechtler, K., and Jenuwein, T. (2001). Methylation of histone H3 lysine 9 creates a binding site for HP1 proteins. *Nature* 410, 116–120.

Landry, J., Sutton, A., Tafrov, S. T., Heller, R. C., Stebbins, J., Pillus, L., and Sternglanz, R. (2000). The silencing protein SIR2 and its homologs are NAD-dependent protein deacetylases. *Proc. Natl. Acad. Sci. USA* 97, 5807–5811.

Li, F., Huarte, M., Zaratiegui, M., Vaughn, M. W., Shi, Y., Martienssen, R., and Cande, W. Z. (2008). Lid2 is required for coordinating H3K4 and H3K9 methylation of heterochromatin and euchromatin. *Cell* 135, 272–283.

Manis, J. P., Gu, Y., Lansford, R., Sonoda, E., Ferrini, R., Davidson, L., Rajewsky, K., and Alt, F. W. (1998). Ku70 is required for late B cell development and immunoglobulin heavy chain class switching. *J. Exp. Med.* 187, 2081–2089.

Meluh, P. B. et al. (1998). Cse4p is a component of the core centromere of *S. cerevisiae*. *Cell* 94, 607–613.

Moretti, P., Freeman, K., Coodly, L., and Shore, D. (1994). Evidence that a complex of SIR proteins interacts with the silencer and telomere-binding protein RAP1. *Genes Dev.* 8, 2257–2269.

Nakagawa, H., Lee, J. K., Hurwitz, J., Allshire, R. C., Nakayama, J., Grewal, S. I., Tanaka, K., and Murakami, Y. (2002). Fission yeast CENP-B homologs nucleate centromeric heterochromatin by promoting heterochromatin-specific histone tail modifications. *Genes Dev.* 16, 1766–1778.

Nakayama, J., Rice, J. C., Strahl, B. D., Allis, C. D., and Grewal, S. I. (2001). Role of histone H3 lysine 9 methylation in epigenetic control of heterochromatin assembly. *Science* 292, 110–113.

Platero, J. S., Hartnett, T., and Eissenberg, J. C. (1995). Functional analysis of the chromodomain of HP1. *EMBO J.* 14, 3977–3986.

Schotta, G., Ebert, A., Krauss, V., Fischer, A., Hoffmann, J., Rea, S., Jenuwein, T., Dorn, R., and Reuter, G. (2002). Central role of *Drosophila* SU(VAR)3-9 in histone H3-K9 methylation and heterochromatic gene silencing. *EMBO J.* 21, 1121–1131.

Sekinger, E. A. and Gross, D. S. (2001). Silenced chromatin is permissive to activator binding and PIC recruitment. *Cell* 105, 403–414.

Smith, J. S., Brachmann, C. B., Celic, I., Kenna, M. A., Muhammad, S., Starai, V. J., Avalos, J. L., Escalante-Semerena, J. C., Grubmeyer, C., Wolberger, C., and Boeke, J. D. (2000). A phylogenetically conserved NAD+-dependent protein deacetylase activity in the Sir2 protein family. *Proc. Natl. Acad. Sci. USA* 97, 6658–6663.

Verdel, A., Jia, S., Gerber, S., Sugiyama, T., Gygi, S., Grewal, S. I., and Moazed, D. (2004). RNAi-mediated targeting of heterochromatin by the RITS complex. *Science* 303, 672–676.

29.4 Polycomb and Trithorax Are Antagonistic Repressors and Activators

Reviews

Henikoff, S. (2008). Nucleosome destabilization in the epigenetic regulation of gene expression. *Nat. Rev. Genet.* 9, 15–26.

Köhler, C and Villar, C. B. (2008). Programming of gene expression by Polycomb group proteins. *Trends Cell Biol.* 18, 236–243.

Ringrose, L. and Paro, R. (2004). Epigenetic regulation of cellular memory by the Polycomb and Trithorax group proteins. *Annu. Rev. Genet.* 38, 413–443.

Research

Brown, J. L., Fritsch, C., Mueller, J., and Kassis, J. A. (2003). The *Drosophila* pho-like gene encodes a YY1-related DNA binding protein that is redundant with pleiohomeotic in homeotic gene silencing. *Development* 130, 285–294.

Cao, R., Wang, L., Wang, H., Xia, L., Erdjument-Bromage, H., Tempst, P., Jones, R. S., and Zhang, Y. (2002). Role of histone H3 lysine 27 methylation in Polycomb-group silencing. *Science* 298, 1039–1043.

Chan, C. S., Rastelli, L., and Pirrotta, V. (1994). A Polycomb response element in the Ubx gene that determines an epigenetically inherited state of repression. *EMBO J.* 13, 2553–2564.

Cléard, F., Moshkin, Y., Karch, F., and Maeda, R. K. (2006). Probing long-distance regulatory interactions in the *Drosophila melanogaster bithorax* complex using Dam identification. *Nat. Genet.* 38, 931–935.

Czermin, B., Melfi, R., McCabe, D., Seitz, V., Imhof, A., and Pirrotta, V. (2002). *Drosophila* enhancer of Zeste/ESC complexes have a histone H3 methyltransferase activity that marks chromosomal Polycomb sites. *Cell* 111, 185–196.

Eissenberg, J. C., James, T. C., Fister-Hartnett, D. M., Hartnett, T., Ngan, V., and Elgin, S. C. R. (1990). Mutation in a heterochromatin-specific chromosomal protein is associated with suppression of position-effect variegation in *D. melanogaster. Proc. Natl. Acad. Sci. USA* 87, 9923–9927.

Fischle, W., Wang, Y., Jacobs, S. A., Kim, Y., Allis, C. D., and Khorasanizadeh, S. (2003). Molecular basis for the discrimination of repressive methyl-lysine marks in histone H3 by Polycomb and HP1 chromo domains. *Genes Dev.* 17, 1870–1881.

Francis, N. J., Kingston, R. E., and Woodcock, C. L. (2004). Chromatin compaction by a Polycomb group protein complex. *Science* 306, 1574–1577.

Franke, A., DeCamillis, M., Zink, D., Cheng, N., Brock, H. W., and Paro, R. (1992). Polycomb and polyhomeotic are constituents of a multimeric protein complex in chromatin of *D. melanogaster. EMBO J.* 11, 2941–2950.

Geyer, P. K. and Corces, V. G. (1992). DNA position-specific repression of transcription by a *Drosophila* zinc finger protein. *Genes Dev.* 6, 1865–1873.

Orlando, V. and Paro, R. (1993). Mapping Polycomb-repressed domains in the bithorax complex using *in vivo* formaldehyde cross-linked chromatin. *Cell* 75, 1187–1198.

Strutt, H., Cavalli, G., and Paro, R. (1997). Colocalization of Polycomb protein and GAGA factor on regulatory elements responsible for the maintenance of homeotic gene expression. *EMBO J.* 16, 3621–3632.

Wang, L., Brown, J. L., Cao, R., Zhang, Y., Kassis, J. A., and Jones, R. S. (2004). Hierarchical recruitment of Polycomb group silencing complexes. *Mol. Cell* 14, 637–646.

29.5 X Chromosomes Undergo Global Changes

Reviews

Plath, K., Mlynarczyk-Evans, S., Nusinow, D. A., and Panning, B. (2002). Xist RNA and the mechanism of X chromosome inactivation. *Annu. Rev. Genet.* 36, 233–278.

Wutz, A. (2007). Xist function: bridging chromatin and stem cells. *Trends Genet.* 23, 457–464.

Wutz, A. and Gribnau, J. (2007). X inactivation Xplained. *Curr. Opin. Genet. Dev.* 17, 387–393.

Research

Changolkar, L. N., Costanzi, C., Leu, N. A., Chen, D., McLaughlin, K. J., and Pehrson, J. R. (2007). Developmental changes in histone macroH2A1-mediated gene regulation. *Mol. Cell Biol.* 27, 2758–2764.

Erwin, J. A. and Lee, J. T. (2008). New twists in X-chromosome inactivation. *Curr. Opin. Cell Biol.* 20, 349–355.

Lee, J. T. et al. (1996). A 450 kb transgene displays properties of the mammalian X-inactivation center. *Cell* 86, 83–94.

Lyon, M. F. (1961). Gene action in the X chromosome of the mouse. *Nature* 190, 372–373.

Mlynarczyk-Evans, S., Royce-Tolland, M., Alexander, M. K., Andersen, A. A., Kalantry, S., Gribnau, J., and Panning, B. (2006). X chromosomes alternate between two states prior to random X-inactivation. *PLoS Biol.* 4, e159.

Penny, G. D. et al. (1996). Requirement for Xist in X chromosome inactivation. *Nature* 379, 131–137.

29.6 Chromosome Condensation Is Caused by Condensins

Reviews

Hirano, T. (2000). Chromosome cohesion, condensation, and separation. *Annu. Rev. Biochem.* 69, 115–144.

Hirano T. (2006). At the heart of the chromosome: SMC proteins in action. *Nat. Rev. Mol. Cell Biol.* 7, 311–322.

Jessberger, R. (2002). The many functions of SMC proteins in chromosome dynamics. *Nat. Rev. Mol. Cell Biol.* 3, 767–778.

Meyer, B. J. (2005). X-Chromosome dosage compensation. *WormBook*, ed. The *C. elegans* Research Community, WormBook, doi/10.1895/wormbook.1.8.1, http://www.wormbook.org.

Nasmyth, K. (2002). Segregating sister genomes: the molecular biology of chromosome separation. *Science* 297, 559–565.

Onn, I. et al. (2008). Sister chromatid cohesion: A simple concept with a complex reality. *Annu. Rev. Cell Dev. Biol.* 24, 105–129.

Peric-Hupkes, D. and van Steensel, B. (2008). Linking cohesin to gene regulation. *Cell* 132, 925–928.

Research

Blewitt, M. E., Gendrel, A. V., Pang, Z., Sparrow, D. B., Whitelaw, N., Craig, J. M., Apedaile, A., Hilton, D. J., Dunwoodie, S. L., Brockdorff, N., Kay, G. F., and Whitelaw E. (2008). SmcHD1, containing a structural-maintenance-of-chromosomes hinge domain, has a critical role in X inactivation. *Nat. Genet.* 40, 663–669.

Csankovszki, G., McDonel, P., and Meyer, B. J. (2004). Recruitment and spreading of the *C. elegans* dosage compensation complex along X chromosomes. *Science* 303, 1182–1185.

Ercan, S. et al. (2007). X chromosome repression by localization of the *C. elegans* dosage compensation machinery to sites of transcription initiation. *Nature Gen.* 39, 403–408.

Haering, C. H., Farcas, A. M., Arumugam, P., Metson, J., and Nasmyth, K. (2008). The cohesin ring concatenates sister DNA molecules. *Nature* 454, 297–301.

Kanno, T., Bucher, E., Daxinger, L., Huettel, B., Böhmdorfer, G, Gregor, W., Kreil, D. P., Matzke, M., and Matzke, A. J. (2008). A structural-maintenance-of-chromosomes hinge domain-containing protein is required for RNA-directed DNA methylation. *Nat. Genet.* 40, 670–675.

Kimura, K., Rybenkov, V. V., Crisona, N. J., Hirano, T., and Cozzarelli, N. R. (1999). 13S condensin actively reconfigures DNA by introducing global positive writhe: implications for chromosome condensation. *Cell* 98, 239–248.

29.7 CpG Islands Are Subject to Methylation

Reviews

Bird, A. (2002). DNA methylation patterns and epigenetic memory. *Genes Dev.* 16, 6–21.

Matzke, M., Matzke, A. J., and Kooter, J. M. (2001). RNA: guiding gene silencing. *Science* 293, 1080–1083.

Sharp, P. A. (2001). RNA interference—2001. *Genes Dev.* 15, 485–490.

Research

Amir, R. E., Van den Veyver, I. B., Wan, M., Tran, C. Q., Francke, U., and Zoghbi, H. Y. (1999). Rett syndrome is caused by mutations in X-linked MECP2, encoding methyl-CpG-binding protein 2. *Nat. Genet.* 23, 185–188.

Avvakumov, G. V., Walker, J. R., Xue, S., Li, Y., Duan, S., Bronner, C., Arrowsmith, C. H., and Dhe-Paganon, S. (2008). Structural basis for recognition of hemi-methylated DNA by the SRA domain of human UHRF1. *Nature* 455, 822–825.

Kangaspeska, S., Stride, B., Métivier, R., Polycarpou-Schwarz, M., Ibberson, D., Carmouche, R. P., Benes, V., Gannon, F., and Reid, G. (2008). Transient cyclical methylation of promoter DNA. *Nature* 452, 112–115.

Li, E., Bestor, T. H., and Jaenisch, R. (1992). Targeted mutation of the DNA methyltransferase gene results in embryonic lethality. *Cell* 69, 915–926.

Métivier, R., Gallais, R., Tiffoche, C., Le Péron, C., Jurkowska, R. Z., Carmouche, R. P., Ibberson, D., Barath, P., Demay, F., Reid, G., Benes, V., Jeltsch, A., Gannon, F., and Salbert, G. (2008). Cyclical DNA methylation of a transcriptionally active promoter. *Nature* 452, 45–50.

Morgan, H. D. et al. (2004). Activation-induced cytidine deaminase deaminates 5-methylcytosine in DNA and is expressed in pluripotent tissues: implications for epigenetic reprogramming. *J. Biol.Chem.* 279, 52353–52360.

Okano, M., Bell, D. W., Haber, D. A., and Li, E. (1999). DNA methyltransferases Dnmt3a and Dnmt3b are essential for *de novo* methylation and mammalian development. *Cell* 99, 247–257.

Penterman, J., Uzawa, R., and Fischer, R. L. (2007). Genetic interactions between DNA demethylation and methylation in Arabidopsis. *Plant Physiol.* 145, 1549–1557.

Xu, G. L., Bestor, T. H., Bourc'his, D., Hsieh, C. L., Tommerup, N., Bugge, M., Hulten, M., Qu, X., Russo, J. J., and Viegas-Paquinot, E. (1999). Chromosome instability and immunodeficiency syndrome caused by mutations in a DNA methyltransferase gene. *Nature* 402, 187–191.

29.8 DNA Methylation Is Responsible for Imprinting

Reviews

Horsthemke, B. and Wagstaff, J. (2008). Mechanisms of imprinting of the Prader-Willi/Angelman region. *Am. J. Med. Genet. A* 146A, 2041–2052.

McStay, B. (2006). Nucleolar dominance: a model for rRNA gene silencing. *Genes Dev.* 20, 1207–1214.

Wood, A. J. and Oakey, R. J. (2006). Genomic imprinting in mammals: emerging themes and established theories. *PLoS Genet.* Nov 24; 2:e147.

Research

Chaillet, J. R., Vogt, T. F., Beier, D. R., and Leder, P. (1991). Parental-specific methylation of an imprinted transgene is established during gametogenesis and progressively changes during embryogenesis. *Cell* 66, 77–83.

Lawrence, R. J., Earley, K., Pontes, O., Silva, M., Chen, Z. J., Neves, N., Viegas, W., and Pikaard, C. S. (2004). A concerted DNA methylation/histone methylation switch regulates rRNA gene dosage control and nucleolar dominance. *Mol. Cell* 13, 599–609.

Sahoo, T., del Gaudio, D., German, J. R., Shinawi, M., Peters, S. U., Person, R. E., Garnica, A., Cheung, S. W., and Beaudet, A. L. (2008). Prader-Willi phenotype caused by paternal deficiency for the HBII-85 C/D box small nucleolar RNA cluster. *Nat. Genet.* 40, 719–721.

Wood, A. J., Schulz, R., Woodfine, K., Koltowska, K., Beechey, C. V., Peters, J., Bourc'his, D., and Oakey, R. J. (2008). Regulation of alternative polyadenylation by genomic imprinting. *Genes Dev.* 22, 1141–1146.

29.9 Oppositely Imprinted Genes Can Be Controlled by a Single Center

Reviews

Edwards, C. A. and Ferguson-Smith, A. C. (2007). Mechanisms regulating imprinted genes in clusters. *Curr. Opin. Cell Biol.* 19, 281–289.

Research

Bell, A. C. and Felsenfeld, G. (2000). Methylation of a CTCF-dependent boundary controls imprinted expression of the Igf2 gene. *Nature* 405, 482–485.

Han, L., Lee, D. H., and Szabó, P. E. (2008). CTCF is the master organizer of domain-wide allele-specific chromatin at the H19/Igf2 imprinted region. *Mol. Cell. Biol.* 28, 1124–1135.

Hark, A. T., Schoenherr, C. J., Katz, D. J., Ingram, R. S., Levorse, J. M., and Tilghman, S. M. (2000). CTCF mediates methylation-sensitive enhancer-blocking activity at the H19/Igf2 locus. *Nature* 405, 486–489.

29.10 Epigenetic Effects Can Be Inherited

Reviews

Jirtle, R. L. and Skinner, M. K. (2007). Environmental epigenomics and disease susceptibility. *Nat. Rev. Genet.* 8, 253–262.

Morgan, D. K. and Whitelaw, E. (2008). The case for transgenerational epigenetic inheritance in humans. *Mamm. Genome* 19, 394–397.

Research

Bruder, C. E., Piotrowski, A., Gijsbers, A. A., Andersson, R., Erickson, S., de Ståhl, T. D., Menzel, U., Sandgren, J., von Tell, D., Poplawski, A., Crowley, M., Crasto, C., Partridge, E. C., Tiwari, H., Allison, D. B., Komorowski, J., van Ommen, G. J., Boomsma, D. I., Pedersen, N. L., den Dunnen, J. T., Wirdefeldt, K., and Dumanski, J. P. (2008). Phenotypically concordant and discordant monozygotic twins display different DNA copy-number-variation profiles. *Am. J. Hum. Genet.* 82, 763–771.

Mathieu, O., Reinders, J., Caikovski, M., Smathajitt, C., and Paszkowski, J. (2007). Transgenerational stability of the Arabidopsis epigenome is coordinated by CG methylation. *Cell* 130, 851–862.

29.11 Yeast Prions Show Unusual Inheritance

Reviews

Horwich, A. L. and Weissman, J. S. (1997). Deadly conformations: protein misfolding in prion disease. *Cell* 89, 499–510.

Lindquist, S. (1997). Mad cows meet psi-chotic yeast: the expansion of the prion hypothesis. *Cell* 89, 495–498.

Serio, T. R. and Lindquist, S. L. (1999). [PSI+]: an epigenetic modulator of translation termination efficiency. *Annu. Rev. Cell Dev. Biol.* 15, 661–703.

Wickner, R. B., Edskes, H. K., Roberts, B. T., Baxa, U., Pierce, M. M., Ross, E. D., and Brachmann, A. (2004). Prions: proteins as genes and infectious entities. *Genes Dev.* 18, 470–485.

Wickner, R. B., Shewmaker, F., Kryndushkin, D., and Edskes, H. K. (2008). Protein inheritance (prions) based on parallel in-register beta-sheet amyloid structures. *Bioessays* 30, 955–964.

Research

Derkatch, I. L., Bradley, M. E., Hong, J. Y., and Liebman, S. W. (2001). Prions affect the appearance of other prions: the story of [PIN(+)]. *Cell* 106, 171–182.

Derkatch, I. L., Bradley, M. E., Masse, S. V., Zadorsky, S. P., Polozkov, G. V., Inge-Vechtomov, S. G., and Liebman S. W. (2000). Dependence and independence of [PSI(+)] and [PIN(+)]: a two-prion system in yeast? *EMBO J.* 19, 1942–1952.

Du, Z., Park, K. W., Yu, H., Fan, Q., and Li, L. (2008). Newly identified prion linked to the chromatin-remodeling factor Swi1 in *Saccharomyces cerevisiae. Nat. Genet.* 40, 460–465.

Glover, J. R. et al. (1997). Self-seeded fibers formed by Sup35, the protein determinant of [PSI⁺], a heritable prion-like factor of *S. cerevisiae. Cell* 89, 811–819.

Osherovich, L. Z. and Weissman, J. S. (2001). Multiple gln/asn-rich prion domains confer susceptibility to induction of the yeast. *Cell* 106, 183–194.

Sparrer, H. E., Santoso, A., Szoka, F. C., and Weissman, J. S. (2000). Evidence for the prion hypothesis: induction of the yeast [PSI+] factor by *in vitro*-converted Sup35 protein. *Science* 289, 595–599.

Shorter, J. and Lindquist, S. (2008). Hsp104, Hsp70 and Hsp40 interplay regulates formation, growth and elimination of Sup35 prions. *EMBO J.* 27, 2712–2724.

29.12 Prions Cause Diseases in Mammals

Reviews

Chien, P., Weissman, J. S., and DePace, A. H. (2004). Emerging principles of conformation-based prion inheritance. *Annu. Rev. Biochem.* 73, 617–656.

Collinge, J. and Clarke, A. R. (2007). A general model of prion strains and their pathogenicity. *Science* 318, 930–936.

Prusiner, S. B. and Scott, M. R. (1997). Genetics of prions. *Annu. Rev. Genet.* 31, 139–175.

Harris, D. A. and True, H. L. (2006). New insights into prion structure and toxicity. *Neuron* 50, 353–357.

Research

Basler, K., Oesch, B., Scott, M., Westaway, D., Walchli, M., Groth, D. F., McKinley, M. P., Prusiner, S. B., and Weissmann, C. (1986). Scrapie and cellular PrP isoforms are encoded by the same chromosomal gene. *Cell* 46, 417–428.

Bueler, H. et al. (1993). Mice devoid of PrP are resistant to scrapie. *Cell* 73, 1339–1347.

Hsiao, K. et al. (1989). Linkage of a prion protein missense variant to Gerstmann-Straussler syndrome. *Nature* 338, 342–345.

McKinley, M. P., Bolton, D. C., and Prusiner, S. B. (1983). A protease-resistant protein is a structural component of the scrapie prion. *Cell* 35, 57–62.

Oesch, B. et al. (1985). A cellular gene encodes scrapie PrP27-30 protein. *Cell* 40, 735–746.

Scott, M. et al. (1993). Propagation of prions with artificial properties in transgenic mice expressing chimeric PrP genes. *Cell* 73, 979–988.

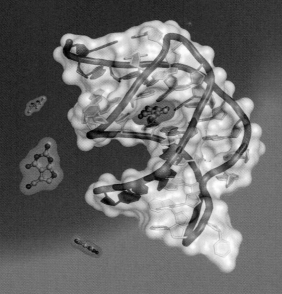

Regulatory RNA

Reproduced from R. C. Spitale, et al., *J. Biol. Chem.* 284 (2009): 11012–11016. © 2009, The American Society for Biochemistry and Molecular Biology. Photo courtesy of Joseph E. Wedekind, University of Rochester School of Medicine & Dentistry.

CHAPTER OUTLINE

30.1 Introduction

Key concepts

• RNA functions as a regulator by forming a region of secondary structure (either inter- or intramolecular) that changes the properties of a target sequence.

The basic principle of regulation is that gene expression is controlled by a regulator that interacts with a specific sequence or structure in DNA or mRNA at some stage prior to the synthesis of protein. The stage of expression that is controlled can be transcription, when the target for regulation is DNA, or it can be at translation, when the target for regulation is RNA. Control during transcription can be at initiation, elongation, or termination. The regulator can be a protein or an RNA. "Controlled" can mean that the regulator turns off (represses) or turns on (activates) the target. Expression of many genes can be coordinately controlled by a single regulator gene on the principle that each target contains a copy of the sequence or structure that the regulator recognizes. Regulators may themselves be regulated, most typically in response to small molecules whose supply responds to environmental conditions. Regulators may be controlled by other regulators to make complex circuits.

Let's compare the ways that different types of regulators work.

Many protein regulators work on the principle of allosteric changes. The protein has two binding sites—one for a nucleic acid target, the other for a small molecule. Binding of the small molecule to its site changes the conformation in such a way as to alter the affinity of the other site for the nucleic acid. The way in which this happens is known in detail for the *lac* Repressor in *E. coli* (see Chapter 26, *The Operon*). Protein regulators are often multimeric, with a symmetrical organization that allows two subunits to contact a palindromic or repeated target on DNA. This can generate cooperative binding effects that create a more sensitive response to regulation.

Regulation via RNA uses changes in secondary structure base pairing as the guiding principle. The ability of an RNA to shift between different conformations with regulatory consequences is the nucleic acid's alternative to the allosteric changes of protein conformation. The changes in structure may result from either intramolecular or intermolecular interactions.

The most common role for intramolecular changes is for an RNA molecule to assume

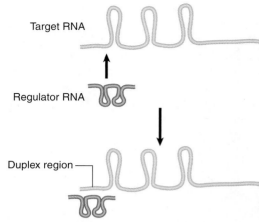

FIGURE 30.1 A regulator RNA is a small RNA with a single-stranded region that can pair with a single-stranded region in a target RNA.

alternative secondary structures by utilizing different schemes for base pairing. The properties of the alternative conformations may be different. Changes in secondary structure of an mRNA can result in a change in its ability to be translated. Secondary structure also is used to regulate the termination of transcription, when the alternative structures differ in whether they permit termination, or not (as we saw with attenuation in Chapter 26, *The Operon*).

In intermolecular interactions, an RNA regulator recognizes its target by the familiar principle of complementary base pairing. **FIGURE 30.1** shows that the regulator is usually a small RNA molecule with extensive secondary structure, but with a single-stranded region(s) that is complementary to a single-stranded region in its target. The formation of a double-helical region between regulator and target can have two types of consequence:

• Formation of the double-helical structure may itself be sufficient for regulatory purposes. In some cases, a protein can bind only to the single-stranded form of the target sequence and is therefore prevented from acting by duplex formation. In other cases, the duplex region becomes a target for binding—for example, by nucleases that degrade the RNA and therefore prevent its expression.

• Duplex formation may be important because it sequesters a region of the target RNA that would otherwise participate in some alternative secondary structure.

We once thought that RNA was merely structural: mRNA carried the blueprint for the synthesis of a protein, rRNA was the structural component of the ribosome, and tRNA shuttled amino acids to the ribosome. We now see a vast RNA world where RNAs have numerous functions, where mRNA can regulate its own translation (see *Section 26.13, The* trp *Operon Is Also Controlled by Attenuation*), where rRNA catalyzes peptide bond formation and where tRNAs participate in the mechanism of fidelity of translation (see *Section 24.3, Special Mechanisms Control the Accuracy of Translation*),

The RNA world extends far beyond the three major RNA types described above to include dozens of different RNAs. These RNAs can function as guide RNAs or splicing cofactors. In addition, there is a large and very heterogeneous class of RNAs with regulatory functions, to be described below. We have not yet uncovered all the mysteries in the RNA world.

30.2 A Riboswitch Can Alter Its Structure According to Its Environment

Key concepts

- A riboswitch is an RNA whose activity is controlled by the metabolite product or another small ligand (a ligand is any molecule that binds to another).
- A riboswitch may be a ribozyme.

As seen in *Section 26.13, The* trp *Operon Is Also Controlled by Attenuation,* and in *Section 26.15, Translation Can Be Regulated,* an mRNA is more than simply an open reading frame. We have seen that regions in the bacterial 5′ UTR (5′ untranslated region) contain elements that, due to coupled transcription/translation, can control transcription termination. We have also seen that the 5′ UTR sequence itself can make an mRNA into a "good" message, which supports a high level of translation, or a "poor" message, which does not. What we will see now is another type of element in a 5′ UTR that can control expression of the mRNA with a different mechanism, called a **riboswitch**. *A riboswitch is an RNA domain that contains a sequence that can change in secondary structure to control its activity.* This change can be mediated by small metabolites.

One type of riboswitch is an RNA element that can assume alternate base pairing configurations (controlled by metabolites in the environment) that can affect translation of the mRNA. **FIGURE 30.2** summarizes the regulation of the system that produces the metabolite

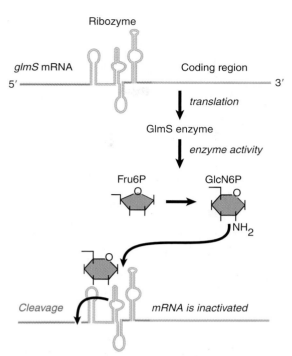

FIGURE 30.2 The 5′ untranslated region of the mRNA for the enzyme that synthesizes GlcN6P contains a ribozyme that is activated by the metabolic product. The ribozyme inactivates the mRNA by cleaving it.

GlcN6P. The gene *glmS* codes for an enzyme that synthesizes GlcN6P (Glucosamine-6-phosphate) from fructose-6-phosphate and glutamine. GlcN6P is a fundamental intermediate in bacterial cell wall biosynthesis. The mRNA contains a long 5′ UTR before the coding region of the mRNA. Within the 5′ UTR is a **ribozyme**—a sequence of RNA that has catalytic activity (see *Section 23.4, Ribozymes Have Various Catalytic Activities*). In this case, the catalytic activity is an endonuclease that cleaves its own RNA. It is activated by binding of the metabolite product, GlcN6P, to the **aptamer** region of the ribozyme. The aptamer is the RNA domain that binds the metabolite. The consequence is that accumulation of GlcN6P activates the ribozyme, which cleaves the mRNA, which in turn prevents further translation. This is an exact parallel to allosteric control of a repressor protein by the end product of a metabolic pathway. There are several examples of such riboswitches in bacteria.

Not all riboswitches encode a ribozyme that controls the mRNA stability. Other riboswitches have alternate configurations of the RNA that allow or prevent expression of the mRNA by affecting ribosome binding. Riboswitches are found predominantly in bacteria and less commonly in eukaryotes.

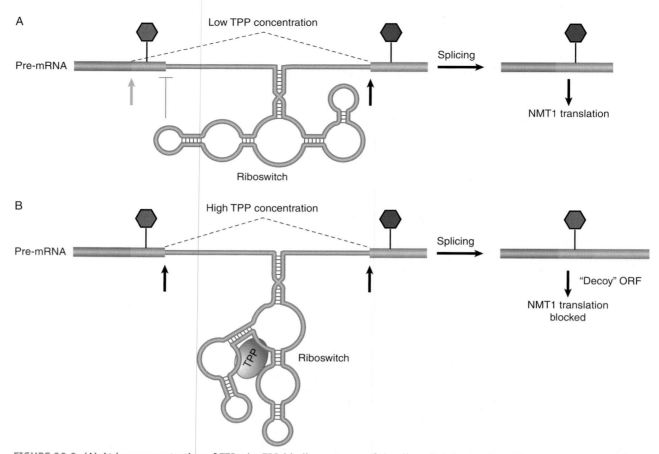

FIGURE 30.3 (A) At low concentration of TTP, the TPP-binding aptamer of the riboswitch base-pairs with sequences surrounding a splice site (red blocking line) in a nearby noncoding sequence, and prevents its selection by the splicing machinery. A distal splice site (green arrow) is selected, however, resulting in a shorter functional mRNA. (B) At high TTP levels, the aptamer undergoes a conformational rearrangement so that the region that was previously bound to the nearby splice site is now used to bind to TTP. This ultimately generates a longer, nonproductive splice variant, preventing gene expression. Reprinted by permission from Macmillan Publishers Ltd: *Nature*, B. J. Blencowe and M. Khanna, vol. 447, pp. 391–393, copyright 2007.

An interesting eukaryotic riboswitch has been described in the fungus *Neurospora* to control alternate splicing. The gene *NMT1* (involved with vitamin B1 synthesis) produces an mRNA precursor with a single intron that has two splice donor sites. Alternative use of these two sites can produce a functional or nonfunctional message depending on the concentration of a vitamin B1 metabolite, TTP (thiamine pyrophosphate). Thus, product concentration controls product formation, a form of repressible control. The selection of the splice site is controlled by a riboswitch in the intron. At a low concentration of TTP the proximal splice donor site is chosen and the distal splice donor site is blocked by the riboswitch, as seen in **FIGURE 30.3**. This splice produces a functional mRNA. At high TTP concentration, TTP binds the riboswitch to alter its configuration and prevents blocking the distal splice donor site to allow the alternate splice which produces a nonfunctional mRNA.

30.3 Noncoding RNAs Can Be Used to Regulate Gene Expression

Key concepts

- Vast tracts of the eukaryotic genome are transcribed.
- A regulator RNA can function by forming a duplex region with a target RNA.
- The duplex may block initiation of translation, cause termination of transcription, or create a target for an endonuclease.
- Transcriptional interference occurs when an overlapping transcript on the same or opposite strand prevents transcription of another gene.
- Noncoding RNAs (such as CUTs and PROMPTs) are often polyadenylated and very unstable.

Base pairing offers a powerful means for one RNA to control the activity of another. There are many cases in both prokaryotes and eukaryotes where a (usually rather short) single-

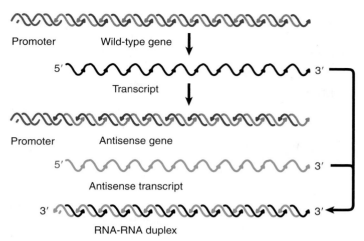

FIGURE 30.4 Antisense RNA can be generated by reversing the orientation of a gene with respect to its promoter and can anneal with the wild-type transcript to form duplex DNA.

stranded RNA base pairs with a complementary region of an mRNA, and as a result it prevents expression of the mRNA. One of the early illustrations of this effect was provided by an artificial situation in which **antisense genes** were introduced into eukaryotic cells.

Antisense genes are constructed by reversing the orientation of a gene with regard to its promoter, so that the "antisense" strand is transcribed into an antisense **noncoding RNA (ncRNA)**, as illustrated in **FIGURE 30.4**. Synthesis of **antisense RNA** can inactivate a target RNA in either prokaryotic or eukaryotic cells. An antisense RNA is in effect an RNA regulator. An antisense thymidine kinase gene inhibits synthesis of thymidine kinase from the endogenous gene. Quantitation of the effect is not entirely reliable, but it seems that an excess (perhaps a considerable excess) of the antisense RNA may be necessary.

At what level does the antisense RNA inhibit expression? It could in principle prevent transcription of the authentic gene, processing of its RNA product, or translation of the messenger. Results with different systems show that the inhibition depends on formation of RNA–RNA duplex molecules, but this can occur either in the nucleus or in the cytoplasm. In the case of an antisense gene stably carried by a cultured cell, sense–antisense RNA duplexes form in the nucleus, preventing normal processing and/or transport of the sense RNA. In another case, injection of antisense RNA into the cytoplasm inhibits translation by forming duplex RNA in the 5' region of the mRNA.

This technique offers a powerful approach for turning off genes at will; for example, the function of a regulatory gene can be investigated by introducing an antisense version. An extension of this technique is to place the antisense gene under the control of a promoter that is itself subject to regulation. The target gene can then be turned off and on by regulating the production of antisense RNA. This technique allows investigation of the importance of the timing of expression of the target gene.

Antisense RNA has been known for some time in eukaryotes. The first genome-sequencing projects demonstrated that **nested genes** (genes located within the introns of other genes) are widespread. They are more common than was first thought, comprising as much as 5%–10% of genes. If the nested gene is transcribed from the opposite strand, then antisense RNA is produced. This head-to-head arrangement of a nested gene will also lead to **transcriptional interference (TI)** because both genes cannot be transcribed simultaneously.

Transcriptional interference is emerging as a significant mechanism of transcriptional regulation, and it can actually occur both when an interfering RNA is produced in an antisense orientation, as described above, or in the sense orientation. For example, the yeast *SER3* gene (involved in serine biosynthesis) is normally repressed in the presence of serine and induced in its absence. It turns out that under serine-rich, repressive conditions, a noncoding RNA is expressed from the intergenic region upstream of the *SER3* promoter, and is transcribed from the same strand as *SER3*. This RNA (named *SER3 regulatory gene*, or *SRG1*) does not encode a protein, but its high expression serves to disrupt transcription initiation at the *SER3* promoter. *SRG1* is induced by serine, so in this case the end product of the biosynthetic pathway regulates

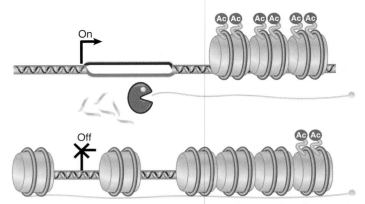

FIGURE 30.5 *PHO84* antisense RNA stabilization is paralleled by histone deacetylase recruitment, histone deacetylation and *PHO84* transcription repression. In wild-type cells, the RNA is rapidly degraded. In aging cells, antisense transcripts are stabilized and recruit the histone deacetylase to repress transcription. Adapted from J. Camblong, et al., *Cell* 131 (2007): 706–717.

SER3 by causing transcriptional interference by a *sense* transcript at the *SER3* promoter. It is important to note that in transcriptional interference, it can be transcription *per se*, rather than the RNA product, that is responsible for the regulatory effect.

Recent experiments using both whole genome tiling arrays (probing not just genes but whole genomes) and massive whole cell RNA sequencing experiments have shown that the vast majority of the eukaryotic genome is transcribed. This includes gene regions, of course, but surprisingly includes both the coding and noncoding strands. The estimate is that as much as 70% of human genes produce antisense RNA. This pattern varies with the cell type and is presumably regulated. Also transcribed are intergenic regions, previously assumed to house no information. Transcripts from the both the coding (sense) and noncoding (antisense) strands can result in noncoding RNAs with regulatory functions.

A direct role for antisense RNA in transcription control has recently been demonstrated. In the yeast *S. cerevisiae*, the gene *PHO84* is regulated in part by a class of noncoding RNAs called cryptic unstable transcripts, or **CUTs**. As shown in **FIGURE 30.5**, in addition to the promoter at the 5′ end of the gene, there is another promoter (which is unregulated) on the opposite strand. Transcription from this promoter on the opposite strand produces an antisense RNA. Under normal conditions, this RNA is rapidly degraded by the TRAMP and exosome complexes (see *Section 22.8, Newly Synthesized RNAs Are Checked for Defects via a Nuclear Surveillance System*) as it is produced. In the absence of degradation or

in aging cells, the antisense RNA persists. This antisense RNA, or CUT, recruits histone deacetylase enzymes that remove acetate groups from histones, thereby causing the chromatin over the gene region be remodeled and condensed so that the gene can no longer be transcribed (see *Section 28.9, Histone Acetylation Is Associated with Transcription Activation*). This is gene-specific remodeling directed by the antisense RNA and does not extend to the neighboring genes.

Since this discovery, similar examples of ncRNAs that result in alteration of local chromatin structure have been described, such as a long RNA transcribed from the *GAL1-10* locus (see *Section 28.14, Yeast GAL Genes: A Model for Activation and Repression*) that also results in histone deacetylation (as well as methylation) to promote *GAL* gene repression. ncRNAs also prevent Ty retrotransposition through changes in chromatin structure *in trans*; this is reminiscent of the role of piRNAs in *Drosophila* (discussed below in *Section 30.5, MicroRNAs Are Widespread Regulators in Eukaryotes*).

This phenomenon may be quite widespread. In human HeLa cells, when a component of the RNA degradation machinery is disabled, vast amounts of upstream transcripts are observed from active promoters, called **PROMPTs** (promoter upstream transcripts). Like CUTs in yeast, this RNA is polyadenylated and very unstable. It can occur in both directions and may by related to the fact that open chromatin is available.

30.4 Bacteria Contain Regulator RNAs

Key concepts

- Bacterial regulator RNAs are called sRNAs.
- Several of the sRNAs are bound by the protein Hfq, which increases their effectiveness.
- The *oxyS* sRNA activates or represses expression of >10 loci at the posttranscriptional level.
- Tandem repeats can be transcribed into powerful antiviral RNAs.

Bacteria contain many—up to hundreds—of genes that code for regulator RNAs. These are short RNA molecules, ranging from about 50 nucleotides to about 200 nucleotides, which are collectively known as **sRNAs**. Some of the sRNAs are general regulators that affect many target genes; others are specific for a single transcript. These sRNAs typically function as imperfect (meaning that only small regions within the sRNA are complementary to the target)

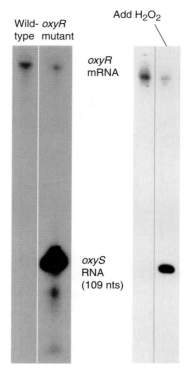

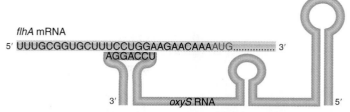

FIGURE 30.7 *oxyS* RNA inhibits translation of *flhA* mRNA by base pairing with a sequence just upstream of the AUG initiation codon.

FIGURE 30.6 The gels on the left show that *oxyS* RNA is induced in an *oxyR* constitutive mutant. The gels on the right show that *oxyS* RNA is induced within one minute of adding hydrogen peroxide to a wild-type culture. Reprinted from *Cell*, vol. 90, S. Altuvia, et al., A small stable RNA . . . , pp. 43–53. Copyright 1997, with permission from Elsevier [http://www.sciencedirect.com/science/journal/00928674]. Photo courtesy of Gisela Storz, National Institutes of Health.

antisense RNAs; that is, their sequences are complementary to their target RNAs.

At what level does the antisense RNA inhibit expression? As described for eukaryotic antisense RNAs, prokaryotic sRNAs could in principle (1) prevent transcription of the gene, (2) affect processing of its RNA product, (3) affect translation of the messenger, or (4) affect stability of the RNA. The action of sRNAs is primarily mediated by the formation of RNA–RNA duplex molecules.

Oxidative stress in *E. coli* provides an interesting example of a general control system in which an sRNA is the regulator. When exposed to reactive oxygen species, bacteria respond by inducing antioxidant defense genes. Hydrogen peroxide activates the transcription activator OxyR, which controls the expression of several inducible genes. One of these genes is *oxyS*, which codes for a small RNA.

FIGURE 30.6 shows two salient features of the control of *oxyS* expression. In a wild-type bacterium under normal conditions, it is not expressed. The pair of gels on the left side of the figure show that it is expressed at high levels in a mutant bacterium with a constitutively active *oxyR* gene. This identifies *oxyS* as a target for activation by *oxyR*. The pair of gels on the right side of the figure show that *oxyS* RNA is transcribed within one minute of exposure to hydrogen peroxide.

The *oxyS* RNA is a short sequence (109 nucleotides) that does not code for protein. It is a *trans*-acting regulator that affects gene expression at the level of translation. It has >10 target mRNAs; at some of them, it activates expression, and at others it represses expression. **FIGURE 30.7** shows the mechanism of repression of one target, the *flhA* mRNA. Three stem-loop double-stranded RNA structures protrude in the secondary structure of *oxyS* mRNA, and the loop closest to the 3′ terminus is complementary to a sequence just preceding the initiation codon of *flhA* mRNA. Base pairing between *oxyS* RNA and *flhA* RNA prevents the ribosome from binding to the initiation codon and therefore represses translation. There is also a second pairing interaction that involves a sequence within the coding region of *flhA*.

Another target for *oxyS* is *rpoS*, the gene coding for an alternative sigma factor (which activates a general stress response). By inhibiting production of the sigma factor, *oxyS* ensures that the specific response to oxidative stress does not trigger the response that is appropriate for other stress conditions. The *rpoS* gene is also regulated by two other sRNAs (*dsrA* and *rprA*), which activate it. These three sRNAs appear to be global regulators that coordinate responses to various environmental conditions.

The actions of all three sRNAs are assisted by an RNA-binding protein called Hfq. The Hfq protein was originally identified as a bacterial host factor needed for replication of the RNA bacteriophage Qβ. It is related to the Sm proteins of eukaryotes that bind to many of the snRNAs (small nuclear RNAs) that have regulatory roles in gene expression (see *Section 21.6, snRNAs Are Required for Splicing*). Mutations in its gene have many effects; this identifies it as

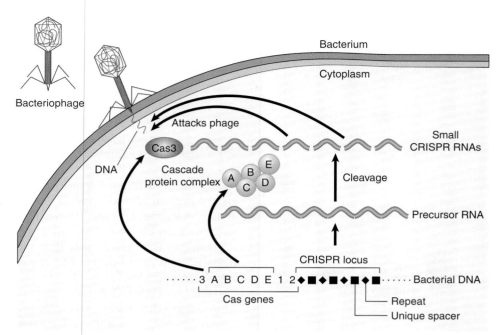

FIGURE 30.8 The CRISPR locus in *E. coli* is transcribed into a larger precursor RNA, which is processed by the Cascade protein complex into short fragments that contain unique spacers identical to sequences in the phage DNA. Assisted by the protein Cas3, these small CRISPR RNAs block the phage infection cycle. Reproduced from R. F. Young III, *Science* 321 (2008): 922–923 [http://www.sciencemag.org]. Reprinted with permission from AAAS.

a pleiotropic protein. Hfq binds to many of the sRNAs of *E. coli*, and it increases the effectiveness of *oxyS* RNA by enhancing its ability to bind to its target mRNAs. The effect of Hfq is probably mediated by causing a small change in the secondary structure of *oxyS* RNA that improves the exposure of the single-stranded sequences that pair with the target mRNAs.

We are just beginning to realize the vast potential that small RNAs possess in controlling so much of the life cycle of an organism. A system of bacterial defense against foreign invaders in the very well known bacterium *E. coli* provides an example of just how much we have yet to learn. This system is based upon clusters of short palindromic repeats called **CRISPRs** (clusters of regularly interspersed short palindromic repeats) and is widespread in both eubacteria and archaea. These sequences, probably phage derived, are used to provide the host bacteria with resistance to further phage infection, as seen in **FIGURE 30.8**.

The CRISPR defense system is used in conjunction with an RNA processing system of eight genes, called *cas* (CRISPR-associated) genes in *E. coli* K12. A complex of five Cas proteins can be identified and is called Cas-

cade (CRISPR-associated complex for antiviral defense). The CRISPR region is transcribed into a long RNA, pre-crRNA, which is processed into short CRISPR RNAs of about 57 nucleotides. The model proposed is that these RNAs, complementary to phage DNA, will base pair with and prevent expression of the phage genes.

These mechanisms offer powerful approaches for turning off genes at will. It is not, however, necessarily a one-way street where a regulatory RNA is produced and simply turns off expression of a message. This system can also be balanced by the production of a counter protein that can bind to and interfere with the sRNA. Thus, dynamic systems can exist that can change over time according to demands placed on the cell.

The function of a regulatory gene can be investigated by introducing an antisense version. An extension of this technique is to place the antisense gene under the control of a promoter that is itself subject to regulation. The target gene can then be turned off and on by regulating the production of antisense RNA. This technique allows investigation of the importance of the timing of expression of the target gene.

Key concepts

- Eukaryotic genomes code for many short (~22 base) RNA molecules called microRNAs.
- piRNAs regulate gene expression in germ cells and act to silence transposable elements.
- siRNA are complementary to viruses and transposable elements.

Eukaryotes, like bacteria, use RNAs to regulate gene expression. Noncoding RNAs are used to control gene expression in the nucleus at the level of DNA; in many cases the expression and function of these RNAs are inextricably linked to chromatin structure. Transcription of tandemly repeated simple sequence satellite heterochromatic DNA is required for the very formation of heterochromatin itself (see Chapter 28, *Eukaryotic Transcription Regulation*, and Chapter 29, *Epigenetic Effects Are Inherited*). We will focus here mainly on control in the cytoplasm at the level of the mRNA. As we will see, the eukaryote mechanisms, while related to the bacterial mechanisms, are very different.

Like bacteria, eukaryotes use RNA to regulate transcription. Note, though, that attenuation is not possible in eukaryotes (as it is in *E. coli*), because the nuclear membrane separates the processes of transcription and translation. Given that eukaryotic mRNA is so much more stable than bacterial mRNA, with an average half-life of hours as opposed to minutes, much more translation-level control is used in eukaryotes, both at the level of translation initiation and mRNA stability control itself (see Chapter 22, *mRNA Stability and Localization*).

There are numerous classes of small noncoding RNAs in eukaryotes. We have already seen some of these, such as the different classes of guide RNAs that are involved in RNA splicing, editing, and modification (see Chapter 21, *RNA Splicing and Processing*, and Chapter 23, *Catalytic RNA*).

Very small RNAs or microRNAs (**miRNA**s) are gene expression regulators found in most, if not all, eukaryotes. These bear some resemblance to their bacterial sRNA counterparts, but as we will see, they are typically smaller and their mechanism of action is different. The human genome has an estimated 1000 genes that code for miRNAs that participate in **RNA interference (RNAi)**, half from the introns of coding genes, and about half from large ncRNAs. Even more interesting, miRNAs can originate from pseudogenes, supposedly inactive gene-like regions that were thought to have no function. This is a general mechanism to repress gene expression, usually (but not always) at the level of translation. These miRNAs go by a number of names and are sometimes called short temporal RNA **(stRNA)**, because they are involved in development. Some miRNAs have also been shown to affect transcription initiation by binding to the gene's promoter. It is estimated that many hundreds of miRNAs control thousands of mRNAs, perhaps as much as 90% of the gene total, at all stages of development. Each miRNA may have hundreds of target mRNAs.

Piwi-associated RNAs, **piRNA**, are a special class of miRNA found in germ cells. Another type of very small RNA is **siRNA** (small interfering RNA), which is typically produced during a virus infection and both piRNAs are siRNAs that can be used to control the expression of transposable elements. These classes are summarized in **FIGURE 30.9**.

These RNAs have multiple origins and multiple mechanisms of synthesis and processing. Most are produced as larger precursor RNAs that are processed and cleaved to the correct size and then delivered to their target.

The miRNAs used in RNAi are produced as large RNA primary transcripts called pri-miRNA that are self complementary and can automatically fold into a double-strand hairpin structure, usually with some imperfect base pairing. The pri-miRNA is processed in a two-step reaction. The first step is catalyzed by **Drosha**, an RNase III superfamily member endonuclease, in the nucleus. Drosha reduces the pri-RNA to about a 70 bp precursor fragment, pre-miRNA. This cleavage determines the 5' and 3' ends of the precursor. After export from the nucleus to the cytoplasm, the second step is catalyzed by **Dicer** to produce a short double-stranded ~22 base pair segment with short, ~2 nucleotide single-stranded ends. Dicer has an N-terminal helicase activity, which enables it to unwind the double-stranded region, and two nuclease domains that are also related to the bacterial RNase III. Related enzymes are nearly universal in eukaryotes.

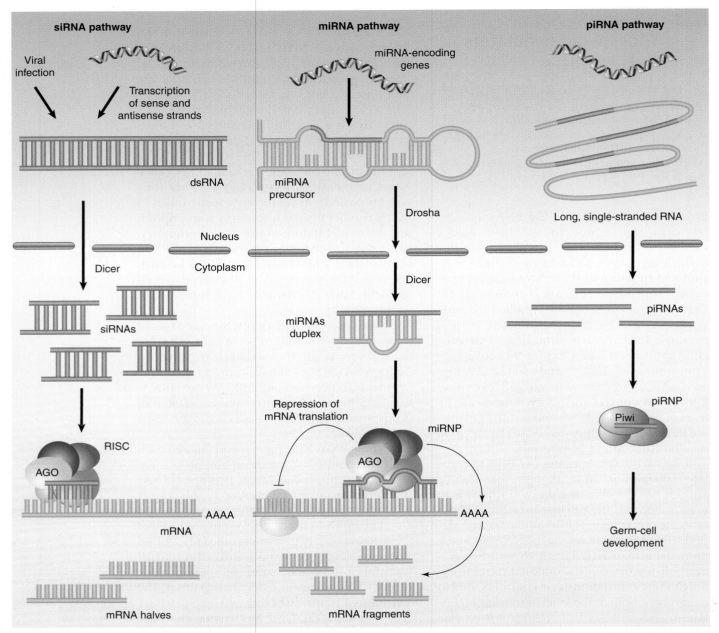

FIGURE 30.9 Small RNAs are generally produced by processing of longer precursors. Three separate but overlapping pathways exist for processing siRNAs, miRNAs and piRNAs. Reprinted by permission from Macmillan Publishers Ltd: *Nature*, H. Großhans and W. Filipowicz, vol. 451, pp. 414–416, copyright 2008.

These short double-stranded RNA fragments are delivered to, or loaded onto, a complex called **RISC** (RNA-induced silencing complex). Proteins in the Argonaute (Ago) family are components of this complex and are required for the final processing to a single strand, to be delivered to the 3′ UTR of its target mRNA. Humans have eight Ago family members, *Drosophila* has five, and *C. elegans* has 26. These proteins have an ancient origin and are found in bacteria, archaea, and eukaryotes (though this system is absent in the yeast *Saccharyomyces cerevisiae*). RISC has endonucle-

ase activity that cleaves the passenger strand, the one which will not be used, in the duplex miRNA.

The degree of base pairing and the sequence of the ends (determined by Dicer cleavage) of the duplex dictate which of the multiple Ago family members picks up the RNA duplex and which strand is selected as the passenger strand to be degraded, as shown in **FIGURE 30.10**. The RISC complex is now in a position to use the mature miRNA to guide it to its target mRNA.

A germline subset of miRNA is the recently discovered Piwi-interacting RNA or piRNA

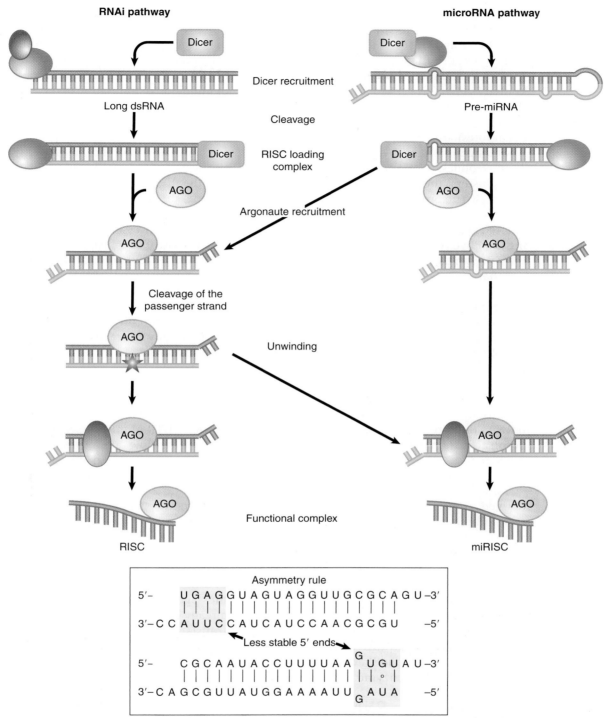

RNAi pathway

microRNA pathway

Dicer

Dicer recruitment

Long dsRNA

Dicer

Pre-miRNA

Cleavage

Dicer

RISC loading complex

Dicer

AGO

AGO

Argonaute recruitment

AGO

AGO

Cleavage of the passenger strand

AGO

Unwinding

AGO

AGO

AGO

Functional complex

AGO

AGO

RISC

miRISC

Asymmetry rule

```
5'-   U G A G G U A G U A G G U U G C G C A G U -3'
      | | | | | | | | | | | | | | | | | | | | |
3'-C C A U U C C A U C A U C C A A C G C G U    -5'
```

← Less stable 5' ends →

```
                                      G
5'-   C G C A A U A C C U U U U A A     U G U A U -3'
      | | | | | | | | | | | | | | | | | ∘ |
3'-C A G C G U U A U G G A A A A U U     A U A    -5'
                                      G
```

FIGURE 30.10 Assembly of the Argonaute–small RNA complex. Inside the cell, a double-stranded (ds)RNA duplex is bound by a recognition complex that contains a Dicer-family member and a dsRNA-binding protein (blue). In *Drosophila melanogaster*, the dsRNA-binding protein Loquacious forms the microRNA-induced silencing complex (miRISC) (in the microRNA pathway; right panel) with Dcr-1, whereas in the RNA interference (RNAi) pathway (left panel) Dcr-2 and R2D2 are important for recruiting the Argonaute (AGO) protein. Once Argonaute is associated with the small RNA duplex, the enzymatic activity conferred by the PIWI domain cleaves only the passenger strand (blue strand) of the small interfering (si)RNA duplex (RNAi pathway). Mismatches found in the microRNA (miRNA) duplex interfere with cleavage, although in some situations, the passenger strand might be cleaved if the RNA duplex is fully paired. RNA strand separation and incorporation into the Argonaute protein are guided by the strength of the base-pairing at the 5'-ends of the duplex; this is known as the asymmetry rule. In this example, the easiest 5'-end to unwind is highlighted in yellow. Once unwound, the siRNA or miRNA will associate with the Argonaute protein (and probably other cellular factors) to form the RNA-induced silencing complex (RISC) or miRISC, respectively. It has recently been demonstrated that the degree of complementarity between the two strands of the intermediate RNA duplex can define how miRNAs are sorted into AGO1 and/or AGO2 proteins in *D. melanogaster* (pathway indicated by the diagonal arrows in the center of the figure). The purple oval represents the unidentified 'unwindase' protein. The star represents an endonuclease event. Reprinted by permission from Macmillan Publishers Ltd: *Nat. Rev. Mol. Cell Biol.*, G. Hutvagner and M. J. Simard, vol. 9, pp. 22–32, copyright 2008.

(P-element induced wimpy testis). In *Drosophila*, these are sometimes called **rasiRNA**s for 'repeat-associated siRNAs.' These are so named because they interact with a different subfamily member of the Ago class proteins, known as Piwi (also called Miwi in mouse and Hiwi in humans). Piwi-class proteins are only found in metazoan organisms (multicellular eukaryotes). In addition, the piRNAs are somewhat longer than miRNAs, ranging from 24 to 31 nucleotides. piRNAs are found in giant tandem clusters; there can be tens of thousands of copies. The processing pathway has not yet been determined. They are delivered to different Ago family members than miRNAs, including the Piwi, Aubergine, and Ago3 proteins.

The function of the piRNAs is also different than miRNAs. Their primary function is nuclear, to repress the expression of transposable elements, preserve genome integrity, and control chromatin structure (see Chapter 17, *Transposons, Retroviruses and Retrotransposons* and Chapter 28, *Eukaryotic Transcription Regulation*). Only a small fraction of the piRNAs are complementary to transposable elements. Most map to single-copy DNA, both genes and intergenic regions. In *Drosophila*, it is maternally inherited piRNAs that provide protection against transposon activation to the female from P element-mediated hybrid dysgenesis (see *Section 17.10, P Elements Are Activated in the Germline*).

siRNAs have a different origin. These are derived from viral infections, which typically transcribe both genomic strands to produce complementary double-stranded RNAs. These large double-stranded RNAs are processed by Dicer in a manner similar to that of the miRNAs described above and are delivered to RISC. siRNAs are also derived from transcription of transposable elements and are used to silence them. This process can be amplified in plants and in *C. elegans* by an RNA-dependent RNA polymerase. Humans and *Drosophila* do not possess this polymerase enzyme.

30.6 How Does RNA Interference Work?

Key concepts

- MicroRNAs regulate gene expression by base pairing with complementary sequences in target mRNAs.
- RNA interference triggers degradation or translation inhibition of mRNAs complementary to miRNA or siRNA. It can also lead to mRNA activation.
- dsRNA may cause silencing of host genes.

RISC is the complex that carries out translational control, guided to its mRNA target in the cytoplasm by the associated miRNA. There are two primary mechanisms used to control mRNA expression: degradation of the mRNA or inhibition of translation of the mRNA. Plants use RNAi primarily for mRNA degradation, whereas animals primarily use translation inhibition. Both groups, however, do have both systems. The choice is primarily determined by the degree of base pairing between the miRNA and the mRNA. The higher the degree of base pairing, the more likely that the target mRNA will be degraded.

This is an essential mechanism for fine-tuned control of translation in eukaryotes. As noted earlier, eukaryotic mRNA is much more stable than bacterial mRNA, and because degradation of some mRNAs is stochastic, cells must be able to tightly control which mRNAs will be translated into protein. During development, it is especially critical to ensure rapid and complete turnover of key mRNAs, as we will see below.

RISC uses the miRNA as a guide to scan RNAs for small regions of homology. These regions are usually found in an AU-rich region in the 3' UTR of mRNAs. A given mRNA may contain multiple target sites and thus respond to different miRNAs. In binding to its target site on the mRNA, the 5'-end of the miRNA from nucleotide 2 to 8 is the most important—the *seed sequence*. These should have perfect base pairing.

Once binding has occurred, there are several different possible outcomes, as shown in **FIGURE 30.11**, ranging from various mechanisms of inhibiting translation to degradation of the message. RISC can interfere with translation already underway from a ribosome by blocking translation elongation (Figure 30.11a) or by inducing proteolysis of the nascent polypeptide being produced (Figure 30.11b).

RISC can also inhibit translation initiation in multiple ways, presumably by virtue of the fact that the central domain of the Ago polypeptide has homology to the cap-binding initiation factor, eIF4E (see *Section 24.9, Eukaryotes Use a Complex of Many Initiation Factors*). RISC can bind to the cap and inhibit eIF4E from joining (Figure 30.11c) or prevent the large 60S ribosomal subunit from joining (Figure 30.11d). RISC can also prevent the circularization of the mRNA by preventing cap binding to the polyA tail (Figure 30.11e). One way in which RISC can promote mRNA degradation is by promoting

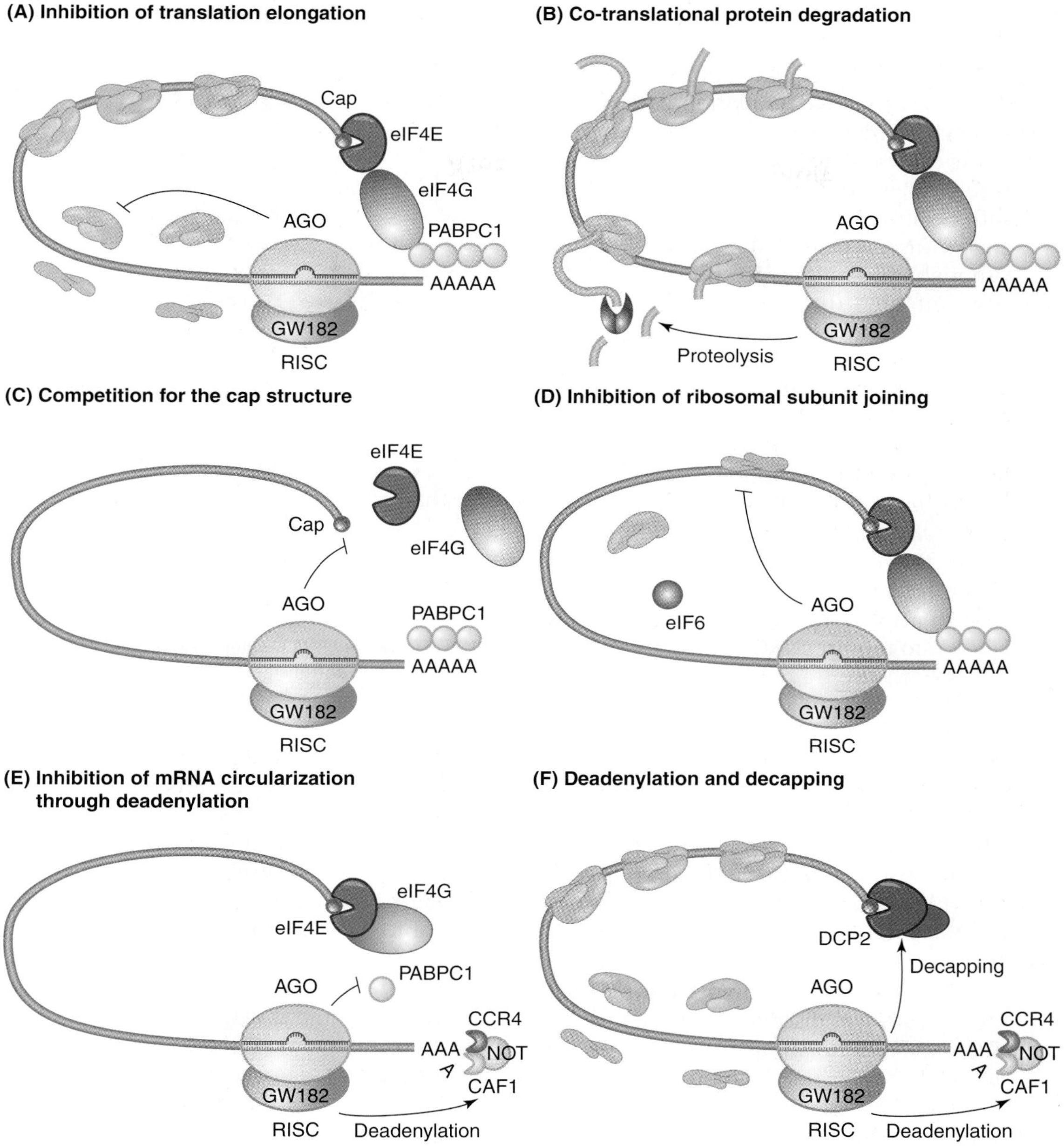

FIGURE 30.11 Mechanisms of miRNA-mediated gene silencing. (A) Postinitiation mechanisms. MicroRNAs (miRNAs; red) repress translation of target mRNAs by blocking translation elongation or by promoting premature dissociation of ribosomes (ribosome drop-off). (B) Cotranslational protein degradation. This model proposes that translation is not inhibited, but rather the nascent polypeptide chain is degraded cotranslationally. The putative protease is unknown. (C–E) Initiation mechanisms. MicroRNAs interfere with a very early step of translation, prior to elongation. (C) Argonaute proteins compete with eIF4E for binding to the cap structure (red dot). (D) Argonaute proteins recruit eIF6, which prevents the large ribosomal subunit from joining the small subunit. (E) Argonaute proteins prevent the formation of the closed loop mRNA configuration by an ill-defined mechanism that includes deadenylation. (F) MicroRNA-mediated mRNA decay. MicroRNAs trigger deadenylation and subsequent decapping of the mRNA target. Proteins required for this process are shown including components of the major deadenylase complex (CAF1, CCR4, and the NOT complex), the decapping enzyme DCP2, and several decapping activators (dark blue circles). (Note that mRNA decay could be an independent mechanism of silencing, or a consequence of translational repression, irrespective of whether repression occurs at the initiation or postinitiation levels of translation.) RISC is shown as a minimal complex including an Argonaute protein (yellow) and GW182 (blue). The mRNA is represented in a closed loop configuration achieved through interactions between the cytoplasmic poly (A) binding protein (PABPC1; bound to the 3' poly(A) tail) and eIF4G (bound to the cytoplasmic cap-binding protein eIF4E). Reprinted from *Cell,* vol. 132, A. Eulalio, E. Huntzinger, and E. Izaurralde, Getting to the root of miRNA . . . , pp. 9–14. Copyright 2008, with permission from Elsevier [http://www.sciencedirect.com/science/journal/00928674].

deadenylation and subsequent decapping of the message (Figure 30.11f). RISC can also indirectly facilitate mRNA degradation by targeting the mRNA to existing degradation pathways. RISC mediates the sequestering of mRNAs to processing centers called P bodies (cytoplasmic processing bodies). These are sites where mRNA can both be stored for future use and where decapped mRNA is degraded.

Although translation repression is the most common outcome (that we currently know about) for miRNA action, miRNAs can also lead to translation *activation*. The 3′ UTR of tumor necrosis factor-α (TNF-α) contains a regulatory RNA element called an ARE (AU-rich element). These are common elements that are usually involved in translation repression (see *Section 22.7, mRNA-Specific Half-Lives Are Controlled by Sequences or Structures Within the mRNA*). In this case, the ARE is involved in activation of translation of the mRNA upon serum starvation. This activation has now been shown to require RISC and its miRNA in a complex with the fragile X-related protein FXR1, an RNA-binding protein. The question of how the RISC complex is converted from its normal repression action to activation hinges on the exact makeup of the complex. Different protein partners in the complex will elicit different responses. Serum starvation leads to the recruitment of FXR1, which alters RISC action, perhaps because RISC is communicating between the 3′ UTR and the mRNA cap, where translation initiation is controlled.

One of the earliest known examples of RNAi in animals was discovered in the nematode *Caenorhabditis elegans* as the result of the interaction between the regulator gene *lin4* (lineage) and its target gene, *lin14*. **FIGURE 30.12** illustrates the behavior of this regulatory system. The *lin14* gene produces an mRNA that regulates larval developmental timing; it is a *heterochronic gene*. Lin14 is a critical protein for specifying the timing of mitotic divisions in a special group of cells. Both loss-of-function mutations and gain-of-function mutations result in embryos with severe defects. Expression of *lin14* is controlled by *lin4*, which codes for a miRNA. The *lin4* transcripts are complementary to a ten-base sequence that is imperfectly repeated seven times in the 3′ UTR of the *lin14* mRNA. *lin4* miRNA binds to these repeats both with a bulge (due to imperfect pairing) and without a bulge in the perfectly paired repeats.

As we described for bacterial sRNA, there can be a dynamic interplay between different elements that modulate the ultimate outcome.

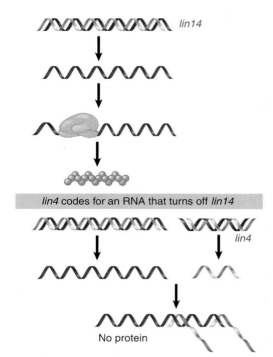

FIGURE 30.12 *lin4* RNA regulates expression of *lin14* by binding to the 3′ untranslated region.

There are multiple mechanisms to control the reaction between RISC and its target mRNA. Proteins can bind to mRNA target sequences to prevent their utilization by RISC, and the 3′ UTR of the mRNA itself may have alternate base-pairing structures that can influence the ability of RISC to identify and target a binding site. miRNA precursors can be edited by ADAR, an adenosine deaminase editing enzyme, which converts A to I and disrupts A:U base pairing. This can result in either activation or inactivation of a miRNA. *C. elegans* and some viruses can express an ncRNA, which can interfere with Dicer and alter the mRNA profile of a cell. Even more interesting is that some genes have alternate poly(A) cleavage sites and are able to produce two versions of the mRNA, differing in the length and therefore the makeup of the 3′ UTR, to either contain more or fewer miRNA target sites.

RNAi has become a powerful technique for ablating the expression of a specific target gene in invertebrates. The technique was initially more limited in mammalian cells, which have the more generalized response to dsRNA of shutting down protein synthesis and degrading mRNA. **FIGURE 30.13** shows that this happens as a result of two reactions. The dsRNA activates the enzyme PKR, which inactivates the translation initiation factor eIF2a by phosphorylating it. It also activates 2′,5′ oligoadenylate synthe-

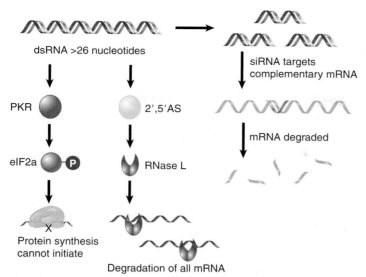

FIGURE 30.13 dsRNA inhibits protein synthesis and triggers degradation of all mRNA in mammalian cells, as well as having sequence-specific effects.

tase, whose product activates RNase L, which degrades all RNAs in the cell. It turns out, however, that these reactions require dsRNA that is longer than twenty-six nucleotides. If shorter dsRNA (twenty-one to twenty-three nucleotides) is introduced into mammalian cells, it triggers the specific degradation of complementary RNAs, just as with the RNAi technique in worms and flies. With this advance, RNAi has become the mechanism of choice for turning off the expression of a specific gene.

RNA interference is related to natural processes in which gene expression is silenced. Plants and fungi show **RNA silencing** (sometimes called *posttranscriptional gene silencing*), in which dsRNA inhibits expression of a gene. The most common sources of the RNA are a replicating virus or a transposable element. This mechanism may have evolved as a defense against these elements. When a virus infects a plant cell, the formation of dsRNA triggers the suppression of expression from the plant genome. Similarly, transposable elements also produce dsRNA. RNA silencing has the further remarkable feature that it is not limited to the cell in which the viral infection occurs: it can spread throughout the plant systemically. Presumably the propagation of the signal involves passage of RNA or fragments of RNA. It may require some of the same features that are involved in movement of the virus itself. RNA silencing in plants involves an amplification of the signal by an RNA-dependent RNA polymerase, which uses the siRNA as a primer to synthesize more RNA on a template of complementary RNA.

30.7 Heterochromatin Formation Requires MicroRNAs

Key concepts
• MicroRNAs can promote heterochromatin formation.

As we saw in the last chapter (see *Section 29.3, Heterochromatin Depends on Interaction with Histones*), heterochromatin is one of the major subdivisions that can be seen in chromosomes. It is visually different when stained because it is more condensed than euchromatin. It is late replicating and has few genes. The underlying DNA sequence is different from euchromatin in that it consists primarily of simple sequence satellite DNA organized in giant tandem blocks. Small islands of unique sequence DNA containing genes are found within heterochromatin. These simple sequence regions have been thought to be largely transcriptionally silent. We now understand that virtually the entire genome is transcribed, including the simple sequence satellite DNA that is often found surrounding centromeres. In fact, transcripts from these sequences are used to organize the heterochromatin structure and repress its transcription.

The centromeric heterochromatin of the fission yeast, *Schizosaccharomyces pombe,* has been a model for understanding heterochromatin formation. The outer region sequences of the heterochromatin are transcribed into ncRNAs

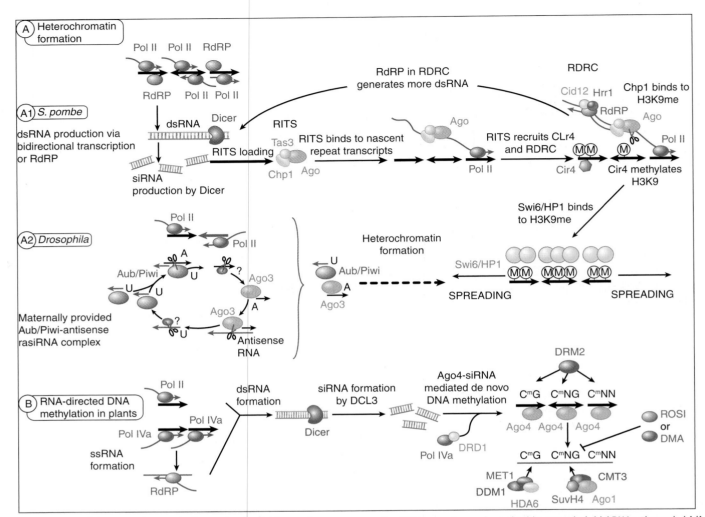

FIGURE 30.14 (A) Heterochromatin formation in *Schizosaccharomyces pombe*. DNA repeats produce double-stranded (ds)RNAs through bidirectional transcription or RNA-dependent RNA synthesis. dsRNAs are cut into small-interfering (si)RNAs that are loaded into an RNA-induced transcriptional silencing complex (RITS) that consists of Ago, Tas3, an *S. pombe* specific protein, and Chp1, a chromodomain containing protein. RITS finds the DNA repeats through siRNA base pairing with the nascent transcript and recruits the RNA-directed RNA polymerase complex (RDRC) and Clr4, a histone methyltransferase that methylates histone H3 at lysine 9 (H3K9me). RdRP in RDRC uses the Ago-cut nascent RNA as template to synthesize more dsRNA, which in turn will be cut into siRNAs to reinforce heterochromatin formation. Chp1 in the RITS complex binds to H3K9me, resulting in stable interaction of RITS and heterochromatic DNA. H3K9me also binds to another chromodomain protein, Swi6, an HP1 homolog, leading to the spreading of heterochromatin. (B) Heterochromatin Formation in *Drosophila*. Repeat associated small interfering RNAs (rasiRNAs) are produced in a Dicer independent, Aub/Piwi-Ago3 "pingpong" mechanism. Aub/Piwi associates with antisense rasiRNAs with a preference for a U at 5′ end, whereas Ago associates with sense-strand derived rasiRNA with a preference to an A at nucleotide 10. Aub/Piwi-rasiRNA complex binds to sense-strand RNA via a 10 nucleotide (nt) complementary sequence. Aub/Piwi cleaves sensestrand RNA, producing sense rasiRNA precursor. A yet-to-be-identified nuclease (?) generates the sense rasiRNAs that associate with Ago3. In turn, Ago3-sense siRNA binds to antisense RNA and generates more antisense rasiRNAs. In this ping-pong model, the initial Aub/Piwi-rasiRNA complex is maternally deposited. The resulting rasiRNA complexes initiate heterochromatin formation (dotted arrow line). As in yeast, H3K9me binds to a HP1 protein, leading to the spreading of heterochromatin. A similar mechanism has been reported in mammals. Reprinted from *Cell*, vol. 130, Y. Bei, S. Pressman, and R. Carthew, SnapShot: Small RNA-Mediated . . . , pp. 756.e1–756.e2. Copyright 2007, with permission from Elsevier [http://www.sciencedirect.com/science/journal/00928674].

by RNA polymerase II. This transcript is copied by an **RNA-dependent RNA polymerase (RDRP)** to give a double-stranded RNA, which is processed into siRNAs. Plants use a variation of the RNA polymerase called RNA polymerase IVb/V to amplify the ncRNA signal.

In a manner similar to what we saw in *Section 30.6, How Does RNA Interference Work?*, the RNA is processed by Dicer. An alternative processing pathway through the TRAMP exosome complex (Trf4-Air1-Mtr4 polyadenylation) also exists. The complex to which the fragments are delivered is called **RITS (RNA-induced transcriptional silencing)**. RITS contains an Argonaut subunit, Ago1. RITS and RDRP are in a complex together. Again, as we saw above, RITS uses the siRNA as a targeting mechanism back to its origin to begin the process of repressing transcription. This entails the recruitment of factors to begin chromatin modification, such as a histone H3K9 methyltransferase (see *Section 29.3, Heterochromatin Depends on Interaction with Histones*), as seen in FIGURE 30.14.

An analogous system is found in *Drosophila*, as described above for rasiRNAs that are targeted to the alternate RISC complex containing Piwi, Aubergine, and Ago3 proteins.

30.8   ## Summary

Gene expression can be regulated positively by factors that activate a gene or negatively by factors that repress a gene. Translation may be controlled by regulators that interact with mRNA. The regulatory products may be proteins, which often are controlled by allosteric interactions in response to the environment, or RNAs, which function by base pairing with the target nucleic acids to change its secondary structure or interfere with its function. Small metabolites can also bind to RNA aptamer domains and affect an alteration in secondary structure, as seen in riboswitches. Regulatory networks can be created by linking regulators so that the production or activity of one regulator is controlled by another.

ncRNAs such as antisense RNA are used in bacterial and eukaryotic cells as a powerful system to regulate gene expression. This regulation can be direct, at the level of interference with an RNA polymerase, or indirect, by affecting the chromatin configuration of the gene. Antisense transcripts can also function in the cytoplasm by giving rise to a host of small regulatory RNAs.

Small regulator RNAs are found in both bacteria and eukaryotes. *E. coli* has ~17 sRNA species. The *oxyS* sRNA controls about ten target loci at the posttranscriptional level; some of them are repressed, whereas others are activated. Repression is caused when the sRNA binds to a target mRNA to form a duplex region that includes the ribosome-binding site. MicroRNAs are ~22 bases long and are produced in most eukaryotes by Drosha and Dicer cleavage of a longer transcript, which is then delivered to RISC for delivery to its target mRNA. They function by base pairing with target mRNAs to form duplex regions that are susceptible to cleavage by endonucleases or inhibition of translation. These are dynamic systems, which themselves are controlled by both accessory protein and enzymes and by other RNAs. The technique of RNA interference is becoming the method of choice for inactivating eukaryotic genes. It uses the introduction of short dsRNA sequences with one strand complementary to the target RNA, and it works by inducing degradation of the targets. This may be related to a natural defense system in plants called RNA silencing.

References

30.2   A Riboswitch Can Alter Its Structure According to Its Environment

Research

Cheah, M. T., Wachter, A., Sudarsan, N., and Beaker, R. R. (2007). Control of alternate splicing and gene expression by eukaryote riboswitches. *Nature* 447, 497–500.

Winkler, W. C., Nahvi, A., Roth, A., Collins, J. A., and Breaker, R. R. (2004). Control of gene expression by a natural metabolite-responsive ribozyme. *Nature* 428, 281–286.

30.3   Noncoding RNA Can Be Used to Regulate Gene Expression

Research

Beretta, J., Pinskaya, M., and Morillon, A. (2008). A cryptic unstable transcript mediates transcriptional trans-silencing of the Ty1 retrotransposon in *S. cerevisiae*. *Genes Dev.* 22, 615–626.

Camblong, J., Iglesias, N., Fickentscher, C., Dieppois, G., and Stutz, F. (2007). Antisense RNA stabilization induces transcriptional gene silencing via histone deacetylation in *S. cerevisiae*. *Cell* 131, 706-717.

He, Y., Vogelstein, B. Velculescu, V. E., Papadopoulos, N., and Kinzler, K. W. (2008). The antisense transcriptomes of human cells. *Science* 322, 1855-1857.

Houseley, J., Rubbi, L., Grunstein, M., Tollervey, D., and Vogelauer, M. (2008). A ncRNA modulates histone modification and mRNA induction in the yeast *GAL* gene cluster. *Molecular Cell* 32, 685-695.

Izant, J. G. and Weintraub, H. (1984). Inhibition of thymidine kinase gene expression by antisense RNA: a molecular approach to genetic analysis. *Cell* 36, 1007–1015.

Martens, J. A., Laprade, L., and Winston, F. (2004). Intergenic transcription is required to repress the *Saccharomyces cerevisiae SER3* gene. *Nature* 429, 571–574.

Martens, J. A., Wu, P. Y., and Winston, F. (2005). Regulation of an intergenic transcript controls adjacent gene transcription in *Saccharomyces cerevisiae*. *Genes Dev.* 19, 2695–2704.

30.4 Bacteria Contain Regulator RNAs

Review

Gottesman, S. (2002). Stealth regulation: biological circuits with small RNA switches. *Genes Dev.* 16, 2829–2842.

Research

Altuvia, S., Weinstein-Fischer, D., Zhang, A., Postow, L., and Storz, G. (1997). A small, stable RNA induced by oxidative stress: role as a pleiotropic regulator and antimutator. *Cell* 90, 43–53.

Altuvia, S., Zhang, A., Argaman, L., Tiwari, A., and Storz, G. (1998). The E. coli OxyS regulatory RNA represses fhlA translation by blocking ribosome binding. *EMBO J.* 17, 6069–6075.

Brouns, S. J. J., Matthijs, M. J., Lundgren, M., Westra, E. R., Slijkhuis, R. J. K., Snijders, A. P. L., Dickman, M. J., Makarova, K. S., Koonin, E. V., and van der Oost, J. (2008). Small CRISPR RNAs guide antiviral defense in prokaryotes. *Science* 321, 960–964.

Maki, F., Uno, K., Morita, T., and Aiba, H. (2008). RNA, but not protein partners, is directly responsible for transcription silencing by a bacterial Hfq-binding small RNA. *Proc. Natl. Acad. Sciences* 105, 10332–10337

Massé, E., Escorcia, F. E., and Gottesman, S. (2003). Coupled degradation of a small regulatory RNA and its mRNA targets in *Escherichia coli*. *Genes Dev.* 17, 2374–2383.

Moller, T., Franch, T., Hojrup, P., Keene, D. R., Bachinger, H. P., Brennan, R. G., and Valentin-Hansen, P. (2002). Hfq: a bacterial Sm-like protein that mediates RNA-RNA interaction. *Mol. Cell* 9, 23–30.

Navarro, L., Jay, F., Nomura, K., He, S. Y., and Voinmet (2008). Suppression of the MicroRNA pathway by bacterial effector proteins. *Science* 321, 964–967.

Wassarman, K. M., Repoila, F., Rosenow, C., Storz, G., and Gottesman, S. (2001).

Identification of novel small RNAs using comparative genomics and microarrays. *Genes Dev.* 15, 1637–1651.

Zhang, A., Wassarman, K. M., Ortega, J., Steven, A. C., and Storz, G. (2002). The Sm-like Hfq protein increases OxyS RNA interaction with target mRNAs. *Mol. Cell* 9, 11–22.

30.5 MicroRNAs Are Widespread Regulators in Eukaryotes

Reviews

Eulalio, A., Huntzinger, E., and Izaurralde, E. (2008). Getting to the root of mi-mediated gene silencing. *Cell* 132, 9–14.

Großhans, H. and Filipowicz, W. (2008). The expanding world of small RNAs. *Nature* 451, 414–416.

Hutvagner, G. and Simard, M. J. (2008). Argonaute proteins: key players in RNA silencing. *Nature Rev. Mol. Cell Biol.* 9, 22–32.

Research

Bernstein, E., Caudy, A. A., Hammond, S. M., and Hannon, G. J. (2001). Role for a bidentate ribonuclease in the initiation step of RNA interference. *Nature* 409, 363–366.

Brennecke, J., Malone, C. D., Aravin, A. A., Sachidanandam, R., Stark, A., and Hannon, G. J. (2008). An epigenetic role for maternally inherited piRNAs in transposons silencing. *Science* 322, 1387–1392.

Ketting, R. F., Fischer, S. E., Bernstein, E., Sijen, T., Hannon, G. J., and Plasterk, R. H. (2001). Dicer functions in RNA interference and in synthesis of small RNA involved in developmental timing in *C. elegans*. *Genes Dev.* 15, 2654–2659.

Lau, N. C., Lim, l.e. E. P., Weinstein, E. G., and Bartel, d.a. V. P. (2001). An abundant class of tiny RNAs with probable regulatory roles in *C. elegans*. *Science* 294, 858–862.

Lee, R. C. and Ambros, V. (2001). An extensive class of small RNAs in *C. elegans*. *Science* 294, 862–864.

Lee, R. C., Feinbaum, R. L., and Ambros, V. (1993). The *C. elegans* heterochronic gene lin-4 encodes small RNAs with antisense complementarity to lin-14. *Cell* 75, 843–854.

Mourelatos, Z., Dostie, J., Paushkin, S., Sharma, A., Charroux, B., Abel, L., Rappsilber, J., Mann, M., and Dreyfuss, G. (2002). miRNPs: a novel class of ribonucleoproteins containing numerous microRNAs. *Genes Dev.* 16, 720–728.

Reinhart, B. J., Weinstein, E. G., Rhoades, M. W., Bartel, B., and Bartel, D. P. (2002). MicroRNAs in plants. *Genes Dev.* 16, 1616–1626.

Sullivan, C. S., Grundhoff, A. T., Tevethia, S., Pipas, J. M., and Ganem, D. (2005). SV40-

encoded microRNAs regulate viral gene expression and reduce susceptibility to cytotoxic T cells. *Nature* 435, 682–686.

Wightman, B., Ha, I., and Ruvkun, G. (1993). Posttranscriptional regulation of the heterochronic gene lin-14 by lin-4 mediates temporal pattern formation in *C. elegans. Cell* 75, 855–862.

Yu, B., Yang, Z., Li, J., Minakhina, S., Yang, M., Padgett, R. W., Steward, R., and Chen, X. (2005). Methylation as a crucial step in plant microRNA biogenesis. *Science* 307, 932–935.

Zamore, P. D., and Haley, B. (2005). Ribo-gnome: the big world of small RNAs. *Science* 309, 1519–1524.

30.6 How Does RNA Interference Work?

Reviews

Ahlquist, P. (2002). RNA-dependent RNA polymerases, viruses, and RNA silencing. *Science* 296, 1270–1273.

Matzke, M., Matzke, A. J., and Kooter, J. M. (2001). RNA: guiding gene silencing. *Science* 293, 1080–1083.

Schwartz, D. S. and Zamore, P. D. (2002). Why do miRNAs live in the miRNP? *Genes Dev.* 16, 1025–1031.

Sharp, P. A. (2001). RNA interference—2001. *Genes Dev.* 15, 485–490.

Tijsterman, M., Ketting, R. F., and Plasterk, R. H. (2002). The genetics of RNA silencing. *Annu. Rev. Genet.* 36, 489–519.

Research

Elbashir, S. M., Harborth, J., Lendeckel, W., Yalcin, A., Weber, K., and Tuschl, T. (2001). Duplexes of 21-nucleotide RNAs mediate RNA interference in cultured mammalian cells. *Nature* 411, 494–498.

Fire, A., Xu, S., Montgomery, M. K., Kostas, S. A., Driver, and Mello, C. C. (1998). Potent and specific genetic interference by double-stranded RNA in *Caenorhabditis elegans. Nature* 391, 806–811.

Hamilton, A. J. and Baulcombe, D. C. (1999). A species of small antisense RNA in posttranscriptional gene silencing in plants. *Science* 286, 950–952.

Kamath, R. S., Fraser, A. G., Dong, Y., Poulin, G., Durbin, R., Gotta, M., Kanapin, A., Le Bot, N., Moreno, S., Sohrmann, M., Welchman, D. P., Zipperlen, P., and Ahringer, J. (2003). Systematic functional analysis of the *C. elegans* genome using RNAi. *Nature* 421, 231–237.

Meister, G., Landthaler, M., Patkaniowska, A., Dorsett, Y., Teng, G., and Tuschl, T. (2004). Human argonaute2 mediates RNA cleavage targeted by miRNAs and siRNAs. *Mol. Cell* 15, 185–197.

Mette, M. F., Aufsatz, W., van der Winden, J., Matzke, M. A., and Matzke, A. J. (2000). Transcriptional silencing and promoter methylation triggered by double-stranded RNA. *EMBO J.* 19, 5194–5201.

Montgomery, M. K., Xu, S., and Fire, A. (1998). RNA as a target of double-stranded RNA-mediated genetic interference in *C. elegans. Proc. Natl. Acad. Sci. USA* 95, 15502–15507.

Ngo, H., Tschudi, C., Gull, K., and Ullu, E. (1998). Double-stranded RNA induces mRNA degradation in *Trypanosoma brucei. Proc. Natl. Acad. Sci. USA* 95, 14687–14692.

Ohta, H., Fujiwara, M., Ohshima, Y., and Ishihara, T. (2008). ADBP-1 regulates an ADAR RNA-editing enzyme to antagonize RNA-interference-mediated gene silencing in *Caenorhabditis elegans. Genetics* 180, 785–796.

Sandberg, R., Neilson, J. R., Sarma, A., Sharp, P. A., and Burge, C. B. (2008). Proliferating cells express mRNAs with shortened 3' untranslated regions and fewer microRNA target sites. *Science* 320, 1643–1647.

Schramke, V., Sheedy, D. M., Denli, A. M., Bonila, C., Ekwall, K., Hannon, G. J., and Allshire, R. C. (2005). RNA-interference-directed chromatin modification coupled to RNA polymerase II transcription. *Nature* 435, 1275–1279.

Vasudevan, S., Tong, Y., and Steitz, J. A. (2007). Switching from repression to activation: miRNAs can up-regulate translation. *Science* 318, 1931–1934.

Voinnet, O., Pinto, Y. M., and Baulcombe, D. C. (1999). Suppression of gene silencing: a general strategy used by diverse DNA and RNA viruses of plants. *Proc. Natl. Acad. Sci. USA* 96, 14147–14152.

Wassenegger, M., Heimes, S., Riedel, L., and Sanger, H. L. (1994). RNA-directed de novo methylation of genomic sequences in plants. *Cell* 76, 567–576.

Waterhouse, P. M., Graham, M. W., and Wang, M. B. (1998). Virus resistance and gene silencing in plants can be induced by simultaneous expression of sense and antisense RNA. *Proc. Natl. Acad. Sci. USA* 95, 13959–13964.

Yu, B., Yang, Z., Li, J., Minakhina, S., Yang, M., Padgett, R. W., Steward, R., and Chen, X. (2005). Methylation as a crucial step in plant microRNA biogenesis. *Science* 307, 932–935.

Zamore, P. D. and Haley, B. (2005). Ribo-gnome: the big world of small RNAs. *Science* 309, 1519–1524.

Zamore, P. D., Tuschl, T., Sharp, P. A., and Bartel, D. P. (2000). RNAi: double-stranded RNA directs the ATP-dependent cleavage of mRNA at 21 to 23 nucleotide intervals. *Cell* 101, 25–33.

30.7 Heterochromatin Formation Requires MicroRNAs

Review

Bei, Y., Pressman, S. and Carthew, R. (2007). Snapshot: small RNA-mediated epigenetic modifications. *Cell* 130, 756.

Grewel, S. I. S. and Elgin, S. C. R. (2007). Transcription and RNA interference in the formation of heterochromatin. *Nature* 447, 399–406.

Research

Bayne, E. H., Portoso, M., Kagansky, A., Kos-Braun, I. C., Urano, T., Ekwall, K., Alves, F., Rappsilber, J., and Allshire, R. C. (2008). Splicing factors facilitate RNA-directed silencing in fission yeast. *Science* 322, 602–606.

Buhler, M., Haas, W., Gygi, S. P., and Moazed, D. (2007). RNAi-dependent and -independent RNA turnover mechanisms contribute to heterochromatin gene silencing. *Cell* 129, 707–721.

Folco, H. D., Pidoux, A. L., Urano, T., and Allshire, R. C. (2008). Heterochromatin and RNAi are required to establish CENP-A chromatin at the centromeres. *Science* 319, 94–97.

Glossary

10 nm fiber   A linear array of nucleosomes, generated by unfolding from the natural condition of chromatin.

–10 element   The consensus sequence centered about 10 bp before the startpoint of a bacterial gene. It is involved in melting DNA during the initiation reaction.

2R hypothesis   The hypothesis that the early vertebrate genome underwent two rounds of duplication.

3′ UTR   The region in an mRNA between the termination codon and the end of the message.

30 nm fiber   A coil of nucleosomes. It is the basic level of organization of nucleosomes in chromatin.

–35 element   The consensus sequence centered about 35 bp before the startpoint of a bacterial gene. It is involved in initial recognition by RNA polymerase.

5′ end resection   The generation of 3′ overhanging single-stranded regions that occurs via exonucleolytic digestion of the 5′ ends at a double-strand break.

5′ UT   The region in an mRNA between the start of the message and the first codon.

A complex   The second splicing complex, formed by the binding of U2 snRNP to the E complex.

A domain   The conserved 11 bp sequence of A-T base pairs in the yeast ARS element that comprises the replication origin.

A site   The site of the ribosome that an aminoacyl-tRNA enters to base pair with the codon.

Abortive initiation   It describes a process in which RNA polymerase starts transcription but terminates before it has left the promoter. It then reinitiates. Several cycles may occur before the elongation stage begins.

Abundance   The average number of mRNA molecules per cell.

Abundant mRNA   Consists of a small number of individual species, each present in a large number of copies per cell.

***Ac* element**   Activator element; an autonomous transposable element in maize.

Acentric fragment   A fragment of a chromosome (generated by breakage) that lacks a centromere and is lost at cell division.

Acridines   Mutagens that act on DNA to cause the insertion or deletion of a single base pair. They were useful in defining the triplet nature of the genetic code.

Activator   A protein that stimulates the expression of a gene, typically by interacting with a promoter to stimulate RNA polymerase. In eukaryotes, the sequence to which it binds in the promoter is called an enhancer.

Adaptive (acquired) immunity   The response mediated by lymphocytes that are activated by their specific interaction with antigen. The response develops over several days as lymphocytes with antigen-specific receptors are stimulated to proliferate and become effector cells. It is responsible for immunological memory.

Addiction system   A survival mechanism used by plasmids. The mechanism kills the bacterium upon loss of the plasmid.

Agropine plasmids   Plasmids that carry genes coding for the synthesis of opines of the agropine type. The tumors usually die early.

Allele   One of several alternative forms of a gene occupying a given locus on a chromosome.

Allelic exclusion   The expression in any particular lymphocyte of only one allele coding for the expressed immunoglobulin. This is caused by feedback from the first immunoglobulin allele to be expressed that prevents activation of a copy on the other chromosome.

Allolactose   A byproduct of the LacZ enzyme, the true inducer of the *lac* operon.

Allopolyploidy   Polyploidization resulting from hybridization between two different but reproductively compatible species.

Allosteric control   The ability of a protein to change its conformation (and therefore activity) at one site as the result of binding a small molecule to a second site located elsewhere on the protein.

Alternative splicing   The production of different RNA products from a single product by changes in the usage of splicing junctions.

Alu element   One of a set of dispersed, related sequences, each ~300 bp long, in the human genome (members of the

SINE family). The individual members have Alu cleavage sites at each end.

Amber codon   The triplet UAG, one of the three termination codons that end polypeptide translation.

Amplicon   The precise, primer to primer double stranded nucleic acid product of a PCR or RT-PCR reaction.

Amyloid fibers   Insoluble fibrous protein polymers with a cross β-sheet structure, generated by prions or other dysfunctional protein aggregations (such as in Alzheimers).

Annealing   The renaturation of a duplex structure from single strands that were obtained by denaturing duplex DNA.

Anti-Sm   An autoimmune antiserum that defines the Sm domain that is common to a group of proteins found in snRNPs that are involved in RNA splicing.

Antibody   A protein that is produced by B lymphocytes and that binds a particular antigen. They are synthesized in membrane-bound and secreted forms. Those produced during an immune response recruit effector functions to help neutralize and eliminate the pathogen.

Antigen   A molecule that can bind specifically to an antigen receptor, such as an antibody.

Antiparallel   Strands of the double helix are organized in opposite orientation, so that the 5′ end of one strand is aligned with the 3′ end of the other strand.

Antirepressor   A positive regulator that functions in opening chromatin.

Antisense gene   A gene that codes for an (antisense) RNA that has a complementary sequence to an RNA that is its target.

Antisense RNA   RNA that has a complementary sequence to an RNA that is its target.

Antitermination   A mechanism of transcriptional control in which termination is prevented at a specific terminator site, allowing RNA polymerase to read into the genes beyond it.

Antitermination complex   Proteins that allow RNA polymerase to transcribe through certain terminator sites.

Anucleate cell   Bacteria that lack a nucleoid, but are of similar shape to wild-type bacteria.

Aptamer   An RNA domain that binds a small molecule; this can result in a conformation change in the RNA.

Architectural protein   A protein that when bound to DNA, can alter its structure, e.g., introduce a bend. They may have no other function.

ARS   An origin for replication in yeast. The common feature among different examples of these sequences is a conserved 11 bp sequence called the A domain.

Assembly factors   Proteins that are required for formation of a macromolecular structure but are not themselves part of that structure.

ATP-dependent chromatin remodeling complex   A complex of one or more proteins associated with an ATPase of the SWI2/SNF2 superfamily that uses the energy of ATP hydrolysis to alter or displace nucleosomes.

***att* sites**   The loci on a lambda phage and the bacterial chromosome at which recombination integrates the phage into, or excises it from, the bacterial chromosome.

Attenuation   The regulation of bacterial operons by controlling termination of transcription at a site located before the first structural gene.

Attenuator   A terminator sequence at which attenuation occurs.

Autoimmune disease   A pathological condition in which the immune response is directed to self antigen.

AU-rich element (ARE)   A eukaryotic mRNA cis sequence consisting largely of A and U ribonucleotides that acts as a destabilizing element.

Autonomous controlling element   An active transposon in maize with the ability to transpose.

Autopolyploidy   Polyploidization resulting from mitotic or meiotic errors within a species.

Autoradiography   A method of capturing an image of radioactive materials on film.

Autoregulation   A site or mutation that affects the properties only of its own molecule of DNA, often indicating that a site does not code for a diffusible product.

Autosplicing (self-splicing)   The ability of an intron to excise itself from an RNA by a catalytic action that depends only on the sequence of RNA in the intron.

Axial element   A proteinaceous structure around which the chromosomes condense at the start of synapsis.

B cell   A lymphocyte that produces antibodies. Development occurs primarily in bone marrow.

Back mutation   A mutation that reverses the effect of a mutation that had inactivated a gene; thus it restores the original sequence or function of the gene product.

Bacteriophage   A bacterial virus.

Bam islands   A series of short, repeated sequences found in the nontranscribed spacer of *Xenopus* rDNA genes.

Bands   Portions of polytene chromosomes visible as dense regions that contain the majority of DNA; they include active genes.

Basal apparatus   The complex of transcription factors that assembles at the promoter before RNA polymerase is bound.

Basal transcription factors   Transcription factors required by RNA polymerase II to form the initiation complex at all RNA polymerase II promoters. Factors are identified as $TF_{II}X$, where X is a letter.

Bidirectional replication A system in which an origin generates two replication forks that proceed away from the origin in opposite directions.

Bivalent The structure containing all four chromatids (two representing each homologue) at the start of meiosis.

Boundary (Insulator) element A DNA sequence element bound by proteins that prevent the spread of open or closed chromatin.

Branch migration The ability of a DNA strand partially paired with its complement in a duplex to extend its pairing by displacing the resident strand with which it is homologous.

Branch site A short sequence just before the end of an intron at which the lariat intermediate is formed in splicing by joining the 5′ nucleotide of the intron to the 2′ position of an adenosine.

Breakage and reunion The mode of genetic recombination in which two DNA duplex molecules are broken at corresponding points and then rejoined crosswise (involving formation of a length of heteroduplex DNA around the site of joining).

bZIP A bZIP (basic zipper) protein has a basic DNA-binding region adjacent to a leucine zipper dimerization motif.

C genes Genes that code for the constant regions of immunoglobulin protein chains.

C-value The total amount of DNA in the genome (per haploid set of chromosomes).

C-value paradox The lack of relationship between the DNA content (C-value) of an organism and its coding potential.

cAMP (cyclic AMP) The coregulator of CRP, it has an internal 3′-5′ phosphodiester bond. It concentration is inverse to the concentration of glucose.

Cap The structure at the 5′ end of eukaryotic mRNA, and is introduced after transcription by linking the terminal phosphate of 5′ GTP to the terminal base of the mRNA.

Capsid The external protein coat of a virus particle.

Carboxy terminal domain (CTD) The domain of eukaryotic RNA polymerase II that is phosphorylated at initiation and is involved in coordinating several activities with transcription.

Cascade A sequence of events, each of which is stimulated by the previous one. In transcriptional regulation, as seen in sporulation and phage lytic development, it means that regulation is divided into stages, and at each stage, one of the genes that is expressed codes for a regulator needed to express the genes of the next stage.

Catabolite regulation The ability of glucose to prevent the expression of a number of genes. In bacteria this is a positive control system; in eukaryotes, it is completely different.

Catabolite Repressor Protein (CRP) A positive regulator protein activated by cyclic AMP. It is needed for RNA polymerase to initiate transcription of many operons of *E. coli*.

Catenate To link together two circular molecules, as in a chain.

cDNA A single-stranded DNA complementary to an RNA, synthesized from it by reverse transcription *in vitro*.

Cell-mediated response The immune response that is mediated primarily by T lymphocytes. It is defined based on immunity that cannot be transferred from one organism to another by serum antibody.

Central dogma Information cannot be transferred from protein to protein or protein to nucleic acid, but can be transferred between nucleic acids and from nucleic acid to protein.

Central element A structure that lies in the middle of the synaptonemal complex, along which the lateral elements of homologous chromosomes align. It is formed from Zip proteins.

Centromere A constricted region of a chromosome that includes the site of attachment (the kinetochore) to the mitotic or meiotic spindle. It consists of unique DNA sequences and proteins not found anywhere else in the chromosome.

Checkpoint A biochemical control mechanism that prevents the cell from progressing from one stage to next unless specific goals and requirements have been met.

Chemical proofreading A proofreading mechanism in which the correction event occurs after the addition of an incorrect subunit to a polymeric chain, by means of reversing the addition reaction.

Chiasma (pl. chiasmata) A site at which two homologous chromosomes synapse during meiosis.

Chromatin The state of nuclear DNA and its associated proteins during the interphase (between mitoses) of the eukaryotic cell cycle.

Chromatin immunoprecipitation (ChIP) A method for detecting *in vivo* protein-DNA interactions that entails isolating proteins with an antibody and identifying DNA sequences that are associated with these proteins.

Chromatin remodeling The energy-dependent displacement or reorganization of nucleosomes that occurs in conjunction with activation of genes for transcription.

Chromocenter An aggregate of heterochromatin from different chromosomes.

Chromomeres Densely staining granules visible in chromosomes under certain conditions, especially early in meiosis, when a chromosome may appear to consist of a series of chromomeres.

Chromosomal walk A technique for locating a gene by using the mostly closely linked markers as a probe for a genetic library.

Chromosome A discrete unit of the genome carrying many genes. Each consists of a very long molecule of duplex DNA and an approximately equal mass of proteins. It is visible as a morphological entity only during cell division.

Chromosome pairing The coupling of the homologous chromosomes at the start of meiosis.

Chromosome scaffold A proteinaceous structure in the shape of a sister chromatid pair, generated when chromosomes are depleted of histones.

cis-acting A site that affects the activity only of sequences on its own molecule of DNA (or RNA); this property usually implies that the site does not code for protein.

cis-acting sequence A site that affects the activity only of sequences on its own molecule of DNA (or RNA); this property usually implies that the site does not code for protein

cis-dominant A site or mutation that affects the properties only of its own molecule of DNA, often indicating that a site does not code for a diffusible product.

Cistron The genetic unit defined by the complementation test; it is equivalent to a gene.

Clamp A protein complex that forms a circle around the DNA; by connecting to DNA polymerase, it ensures that the enzyme action is processive.

Clamp loader A 5-subunit protein complex that is responsible for loading the β clamp on to DNA at the replication fork.

Class switching A change in Ig gene organization in which the C region of the heavy chain is changed but the V region remains the same.

Clonal selection The theory proposed that each lymphocyte expresses a single antigen receptor specificity and that only those lymphocytes that bind to a given antigen are stimulated to proliferate and to function in eliminating that antigen. Thus, the antigen "selects" the lymphocytes to be activated. It is now an established principle in immunology.

Clone An exact replica or copy, whether it is Dolly the sheep or a fragment of DNA.

Cloning Propagation of a DNA sequence by incorporating it into a hybrid construct that can be replicated in a host cell.

Cloning vector DNA (often derived from a plasmid or a bacteriophage genome) that can be used to propagate an incorporated DNA sequence in a host cell; vectors contain selectable markers and replication origins to allow identification and maintenance of the vector in the host.

Closed (blocked) reading frame A reading frame that cannot be translated into protein because of the occurrence of termination codons.

Closed complex The stage of initiation of transcription before RNA polymerase causes the two strands of DNA to separate to form the "transcription bubble." The DNA is double stranded.

Coactivator Factors required for transcription that do not bind DNA, but are required for (DNA-binding) activators to interact with the basal transcription factors.

Coding end It is produced during recombination of immunoglobulin and T cell receptor genes. They are at the termini of the cleaved V and (D)J coding regions. Their subsequent joining yields a coding joint.

Coding region A part of a gene that codes for a polypeptide sequence.

Coding strand The DNA strand that has the same sequence as the mRNA and is related by the genetic code to the protein sequence that it represents.

Codon A triplet of nucleotides that codes for an amino acid, or a termination signal.

Codon bias A higher usage of one codon in genes to encode amino acids for which there are several synonymous codons.

Codon usage A description of the relative abundance of tRNAs for each codon.

Cognate tRNAs tRNAs recognized by a particular aminoacyl-tRNA synthetase. All are charged with the same amino acid.

Cointegrate A structure that is produced by fusion of two replicons, one originally possessing a transposon and the other lacking it; the cointegrate has copies of the transposon present at both junctions of the replicons, oriented as direct repeats.

Colinearity The relationship that describes the 1:1 correspondence of a sequence of triplet nucleotides to a sequence of amino acids.

Compatibility group A group of plasmids that contains members unable to coexist in the same bacterial cell.

Complement A set of ~20 proteins that function through a cascade of proteolytic actions to lyse infected target cells, or to attract macrophages.

Complementary Base pairs that match up in the pairing reactions in double helical nucleic acids (A with T in DNA or with U in RNA, and C with G).

Complementation test A test that determines whether two mutations are alleles of the same gene. It is accomplished by crossing two different recessive mutations that have the same phenotype and determining whether the wild-type phenotype can be produced. If so, the mutations are said to complement each other and are probably not mutations in the same gene.

Complex mRNA   see **Scarce mRNA**.

Composite elements Transposable elements consisting of two IS elements (can be the same or different) and the DNA sequences between the IS elements; the non-IS sequences often include gene(s) conferring antibiotic resistance.

Concerted evolution (coincidental evolution) The ability of two or more related genes to evolve together as though constituting a single locus.

Conditional lethal A mutation that is lethal under one set of conditions, but not lethal under a second set of conditions, such as temperature.

Conjugation A process in which two cells come in contact and transfer genetic material. In bacteria, DNA is transferred from a donor to a recipient cell. In protozoa, DNA passes from each cell to the other.

Consensus sequence An idealized sequence in which each position represents the base most often found when many actual sequences are compared.

Conservative transposition The movement of large elements that were originally classified as transposons but now are considered to be episomes. The mechanism of movement resembles that of phage excision and integration.

Conserved sequence Sequences in which many examples of a particular nucleic acid or protein are compared and the same individual bases or amino acids are always found at particular locations.

Constant region (C region) The part of an immunoglobulin or T cell receptor that varies least in amino acid sequence between different molecules. They are coded by C gene segments. The heavy chain regions identify the type of immunoglobulin and recruits effector functions.

Constitutive expression This describes a state in which a gene is expressed continuously.

Constitutive gene See **Housekeeping gene**.

Constitutive heterochromatin The inert state of permanently nonexpressed sequences, such as satellite DNA.

Context The fact that neighboring sequences may change the efficiency with which a codon is recognized by its aminoacyl-tRNA or is used to terminate polypeptide translation.

Controlling elements Transposable units in maize originally identified solely by their genetic properties. They may be autonomous (able to transpose independently) or nonautonomous (able to transpose only in the presence of an autonomous element).

Copy number The number of copies of a plasmid that is maintained in a bacterium (relative to the number of copies of the origin of the bacterial chromosome).

Core enzyme The complex of RNA polymerase subunits needed for elongation. It does not include additional subunits or factors that may be needed for initiation or termination.

Core histone One of the four types of histone (H2A, H2B, H3, and H4 and their variants) found in the core particle derived from the nucleosome. (This excludes linker histones.)

Core promoter The shortest sequence at which an RNA polymerase can initiate transcription (typically at a much lower level than that displayed by a promoter containing additional elements). For RNA polymerase II it is the minimal sequence at which the basal transcription apparatus can assemble, and it includes three sequence elements: the Inr, the TATA box and the DPE. It is typically ~40 bp long.

Corepressor A small molecule that triggers repression of transcription by binding to a regulator protein.

Core sequence The segment of DNA that is common to the attachment sites on both the phage lambda and bacterial genomes. It is the location of the recombination event that allows phage lambda to integrate.

Cosmid Cloning vector derived from a bacterial plasmid by incorporating the *cos* sites of phage lambda, which make the plasmid DNA a substrate for the lambda packaging system.

Countertranscript An RNA molecule that prevents an RNA primer from initiating transcription by base pairing with the primer.

Coupled transcription/translation The process in bacteria where a message is simultaneously being translated while it is still being transcribed.

CpG islands Stretches of 1–2 kb in mammalian genomes that are enriched in CpG dinucleotides; frequently found in promoter regions of genes.

CRISPR Clusters of Regularly Interspersed Short Palindromic Repeats in prokaryotes that are transcribed and processed into short RNAs that function in RNA interference.

Crossover fixation A possible consequence of unequal crossing over that allows a mutation in one member of a tandem cluster to spread through the whole cluster (or to be eliminated).

Crown gall disease A tumor that can be induced in many plants by infection with the bacterium *Agrobacterium tumefaciens*.

CRP A positive regulator protein activated by cyclic AMP. It is needed for RNA polymerase to initiate transcription of many operons of *E. coli*.

Cryptic satellite A satellite DNA sequence not identified as such by a separate peak on a density gradient; that is, it remains present in main band DNA.

Cryptic unstable transcripts (CUTs) Nonprotein-coding RNAs transcribed by RNA Pol II, frequently generated from the 3′ ends of genes (resulting in antisense transcripts) and rapidly degraded after synthesis.

CTD (C-terminal domain) The domain of RNA polymerase that is involved in stimulating transcription by contact with regulatory proteins.

ctDNA (cpDNA) Chloroplast DNA.

CUT Cryptic unstable transcripts, frequently generated by promoters located at the 3' end of genes (resulting in antisense transcripts)

Cytoplasmic domain The part of a transmembrane protein that is exposed to the cytosol.

Cytotoxic T cell A T lymphocyte (usually CD8+) that can be stimulated to kill cells containing intracellular pathogens, such as viruses.

Cytotype A cytoplasmic condition that affects P element activity. The effect of cytotype is due to the presence or absence of a repressor of transposition, which is provided by the mother to the egg.

D loop A region within mitochondrial DNA in which a short stretch of RNA is paired with one strand of DNA, displacing the original partner DNA strand in this region. The same term is used also to describe the displacement of a region of one strand of duplex DNA by a complementary single-stranded invader.

D segment An additional sequence that is found between the V and J regions of an immunoglobulin heavy chain.

de novo **methyltransferase** An enzyme that adds a methyl group to an unmethylated target sequence on DNA.

Deacylated tRNA tRNA that has no amino acid or polypeptide chain attached because it has completed its role in protein synthesis and is ready to be released from the ribosome.

Deadenylase (or poly(A) nuclease) An exoribonuclease that is specific for digesting poly(A) tails.

Decapping enzyme An enzyme that catalyzes the removal of the 7-methyl guanosine cap at the 5' end of eukaryotic mRNAs.

Degradosome A complex of bacterial enzymes, including RNAase and helicase activities, that is involved in degrading mRNA.

Delayed early genes Genes in phage lambda that are equivalent to the middle genes of other phages. They cannot be transcribed until regulator protein(s) coded by the immediate early genes have been synthesized.

Demethylase A casual name for an enzyme that removes a methyl group, typically from DNA, RNA, or protein.

Denaturation A molecule's conversion from the physiological conformation to some other (inactive) conformation. In DNA, this involves the separation of the two strands due to breaking of hydrogen bonds between bases.

Destabilizing element (DE) Any one of many different cis sequences, present in some mRNAs, that stimulates rapid decay of that mRNA.

Dicer An endonuclease that processes double stranded precursor RNA to 21 to 23 nucleotide RNAi molecules.

Dideoxynucleotide (dNTP) A chain-terminating nucleotide that lacks a 3'-OH group and therefore is not a substrate for DNA polymerization. Used in DNA sequencing.

Direct repeats Identical (or closely related) sequences present in two or more copies in the same orientation in the same molecule of DNA.

Distributive (nuclease) An enzyme that catalyzes the removal of only one or a few nucleotides before dissociating from the substrate.

Divergence The corrected percent difference in nucleotide sequence between two related DNA sequences or in amino acid sequences between two proteins.

DNA fingerprinting A technique for analyzing the differences between individuals of the fragments generated by using restriction enzymes to cleave regions that contain short repeated sequences or by PCR. The lengths of the repeated regions are unique to every individual, and as a result the presence of a particular subset in any two individuals can be used to define their common inheritance (e.g., a parent-child relationship).

DNA ligase The enzyme that makes a bond between an adjacent 3'–OH and 5'–phosphate end where there is a nick in one strand of duplex DNA.

DNA mutants Temperature-sensitive replication mutants in *E. coli* that identifies a set of loci called the *dna* genes.

DNA polymerase An enzyme that synthesizes a daughter strand(s) of DNA (under direction from a DNA template). Any particular enzyme may be involved in repair or replication (or both).

DNA repair The removal and replacement of damaged DNA by the correct sequence.

DNA replicase See **DNA polymerase**.

DNase An enzyme that degrades DNA.

Domain In reference to a chromosome, it may refer either to a discrete structural entity defined as a region within which supercoiling is independent of other regions or to an extensive region including an expressed gene that has heightened sensitivity to degradation by the enzyme DNase I. In a protein, it is a discrete continuous part of the amino acid sequence that can be equated with a particular function.

Dominant negative A mutation that results in a mutant gene product that prevents the function of the wild-type gene product, causing loss or reduction of gene activity in cells containing both the mutant and wild-type alleles. The most common cause is that the gene codes for a homomultimeric protein whose function is lost if only one of the subunits is a mutant.

Dosage compensation Mechanisms employed to compensate for the discrepancy between the presence of two X chromosomes in one sex but only one X chromosome in the other sex.

Down mutation A mutation in a promoter that decreases the rate of transcription.

Downstream Sequences proceeding farther in the direction of expression within the transcription unit.

Downstream promoter element (DPE) A common component of RNA polymerase II promoters that do not contain a TATA box.

Drosha An endonuclease that processes double stranded primary RNAs into short, ~70 base pair precursors for Dicer processing.

Ds element Dissociation element; a non-autonomous transposable element in maize, related to the autonomous Activator (*Ac*) element.

Double-strand breaks (DSB) Breaks that occur when both strands of a DNA duplex are cleaved at the same site. Genetic recombination is initiated by such breaks. The cell also has repair systems that act on breaks that are created at other times.

Doubling time The period (usually measured in minutes) that it takes for a bacterial cell to reproduce.

E complex The first complex to form at a splice site, consisting of U1 snRNP bound at the splice site together with factor ASF/SF2, U2AF bound at the branch site, and the bridging protein SF1/BBP.

Early genes Genes that are transcribed before the replication of phage DNA. They code for regulators and other proteins needed for later stages of infection.

Early infection The part of the phage lytic cycle between entry and replication of the phage DNA. During this time, the phage synthesizes the enzymes needed to replicate its DNA.

EF-Tu The elongation factor that binds aminoacyl-tRNA and places it into the A site of a bacterial ribosome.

EJC (exon junction complex) A protein complex that assembles at exon–exon junctions during splicing and assists in RNA transport, localization, and degradation.

Elongation The stage in a macromolecular synthesis reaction (replication, transcription, or translation) when the nucleotide or polypeptide chain is extended by the addition of individual subunits.

Elongation factors Proteins that associate with ribosomes cyclically during the addition of each amino acid to the polypeptide chain.

Endonuclease An enzyme that cleaves bonds within a nucleic acid chain; it may be specific for RNA or for single-stranded or double-stranded DNA.

Endoribonuclease A ribonuclease that cleaves an RNA at internal site(s).

Enhancer A *cis*-acting sequence that increases the utilization of (most) eukaryotic promoters, and can function in either orientation and in any location (upstream or downstream) relative to the promoter.

Error-prone polymerase A DNA polymerase that incorporates noncomplementary bases into the daughter strand.

Error-prone synthesis A repair process in which noncomplementary bases are incorporated into the daughter strand.

Epigenetic Changes that influence the phenotype without altering the genotype. They consist of changes in the properties of a cell that are inherited, but that do not represent a change in genetic information.

Episome A plasmid able to integrate into bacterial DNA.

Equilibrium density-gradient centrifugation A gradient method used to separate macromolecules on the basis of differences in their density. For DNA, it is prepared from a heavy soluble compound such as CsCl.

Euchromatin Regions that comprise most of the genome in the interphase nucleus, are less tightly coiled than heterochromatin, and contain most of the active or potentially active single copy genes.

Excision Release of phage or episome or other sequence from the host chromosome as an autonomous DNA molecule.

Excision repair A type of repair system in which one strand of DNA is directly excised and then replaced by resynthesis using the complementary strand as template.

Exon Any segment of an interrupted gene that is represented in the mature RNA product.

Exon definition The process in which a pair of splicing sites are recognized by interactions involving the 5′ site of the intron and also the 5′ site of the next intron downstream.

Exon junction complex (EJC) A protein complex that assembles at exon–exon junctions during splicing and assists in RNA transport, localization, and degradation.

Exon shuffling The hypothesis that genes have evolved by the recombination of various exons coding for functional protein domains.

Exon trapping Inserting a genomic fragment into a vector whose function depends on the provision of splicing junctions by the fragment.

Exonuclease An enzyme that cleaves nucleotides one at a time from the end of a polynucleotide chain; it may be specific for either the 5′ or 3′ end of DNA or RNA.

Exoribonuclease A ribonuclease that removes terminal ribonucleotides from RNA.

Exosome An exonuclease complex involved in nuclear processing and nuclear/cytoplasmic RNA degradation.

Expressed sequence tag (EST) A short sequenced fragment of a cDNA sequence that can be used to identify an actively expressed gene.

Extein   A sequence that remains in the mature protein that is produced by processing a precursor via protein splicing.

Extranuclear genes   Genes that reside outside the nucleus, in organelles such as mitochondria and chloroplasts.

F plasmid   An episome that can be free or integrated in *E. coli*, and that can sponsor conjugation in either form.

Facultative heterochromatin   The inert state of sequences that also exist in active copies, for example, one mammalian X chromosome in females.

Fixation   The process by which a new allele replaces the allele that was previously predominant in a population.

Fluorescence resonant energy transfer (FRET)   A process whereby the emission from an excited fluorophore is captured and reemitted at a longer wavelength by a nearby second fluorophore whose excitation spectrum matches the emission frequency of the first fluorophore.

Footprinting   A technique for identifying the site on DNA bound by some protein by virtue of the protection of bonds in this region against attack by nucleases.

Forward mutation   A mutation that inactivates a functional gene.

Frameshift   A mutation caused by deletions or insertions that are not a multiple of three base pairs. They change the frame in which triplets are translated into polypeptide.

Fully methylated   A site that is a palindromic sequence that is methylated on both strands of DNA.

G-bands   Bands generated on eukaryotic chromosomes by staining techniques that appear as a series of lateral striations. They are used for karyotyping (identifying chromosomes and chromosomal regions by the banding pattern).

Gain-of-function mutation   A mutation that causes an increase in the normal gene activity. It sometimes represents acquisition of certain abnormal properties. It is often, but not always, dominant.

Gap repair   A type of DNA repair in which one DNA duplex may act as a donor of genetic information that directly replaces the corresponding sequences in the recipient duplex by a process of gap generation, strand exchange, and gap filling.

Gene cluster   A group of adjacent genes that are identical or related.

Gene conversion   The alteration of one strand of a heteroduplex DNA to make it complementary with the other strand at any position(s) where there were mispaired bases, or the complete replacement of genetic material at one locus by a homologous sequence.

Gene expression   The process by which the information in a sequence of DNA in a gene is used to produce an RNA or polypeptide, involving transcription and (for polypeptides) translation.

Gene family   A set of genes within a genome that code for related or identical proteins or RNAs. The members were derived by duplication of an ancestral gene followed by accumulation of changes in sequence between the copies. Most often the members are related but not identical.

Gene knock-in   A process similar to a knockout, but more subtle mutations are made.

Gene knockout   A process in which a gene function is eliminated, usually by replacing most of the coding sequence with a selectable marker *in vitro* and transferring the altered gene to the genome by homologous recombination.

Genetic code   The correspondence between triplets in DNA (or RNA) and amino acids in polypeptide.

Genetic drift   The chance fluctuation (without selective pressure) of the frequencies of alleles in a population.

Genetic hitchhiking   The change in frequency of a genetic variant due to its linkage to a selected variant at another locus.

Genetic map   See **Linkage map**.

Genetic recombination   A process by which separate DNA molecules are joined into a single molecule, due to such processes as crossing-over or transposition.

Genome   The complete set of sequences in the genetic material of an organism. It includes the sequence of each chromosome plus any DNA in organelles.

Glycosylase   A repair enzyme that removes damaged bases by cleaving the bond between the base and the sugar.

GMP-PCP   An analog of GTP that cannot be hydrolyzed. It is used to test which stage in a reaction requires hydrolysis of GTP.

Gratuitous inducer   Inducers that resemble authentic inducers of transcription, but are not substrates for the induced enzymes.

Growing point   See **Replication fork**.

GU-AG rule   The rule that describes the presence of these constant dinucleotides at the first two and last two positions of introns of nuclear genes.

Guide RNA   A small RNA whose sequence is complementary to the sequence of an RNA that has been edited. It is used as a template for changing the sequence of the pre-edited RNA by inserting or deleting nucleotides.

Gyrase   An enzyme that changes the number of times the two strands in a closed DNA molecule cross each other. It does this by cutting the DNA, passing DNA through the break, and resealing the DNA.

Hairpin   An RNA sequence that can fold back on itself forming double stranded RNA.

Half-life (RNA) The time taken for the concentration of a given population of RNA molecules to decrease by half, in the absence of new synthesis.

Haplotype The particular combination of alleles in a defined region of some chromosome—in effect, the genotype in miniature. Originally used to described combinations of Major Histocompatibility Complex (MHC) alleles, it now may be used to describe particular combinations of RFLPs, SNPs, or other markers.

Hb anti-Lepore A fusion gene produced by unequal crossing over that has the N-terminal part of β globin and the C-terminal part of δ globin.

Hb Kenya A fusion gene produced by unequal crossing over between the ^Aγ and β globin genes.

Hb Lepore An unusual globin protein that results from unequal crossing over between the β and δ genes. The genes become fused together to produce a single β-like chain that consists of the N-terminal sequence of δ joined to the C-terminal sequence of β.

HbH disease A condition in which there is a disproportionate amount of the abnormal tetramer β4 relative to the amount of normal hemoglobin (α2β2).

Heatshock genes A set of loci activated in response to an increase in temperature (and other abuses to the cell). All organisms have them. Their products usually include chaperones that act on denatured proteins.

Helicase An enzyme that uses energy provided by ATP hydrolysis to separate the strands of a nucleic acid duplex.

Helix-loop-helix The motif that is responsible for dimerization of a class of transcription factors called HLH proteins. A bHLH protein has a basic DNA-binding sequence close to the dimerization motif.

Helix-turn-helix The motif that describes an arrangement of two α-helices that form a site that binds to DNA, one fitting into the major groove of DNA and other lying across it.

Helper T cell A T lymphocyte that activates macrophages and stimulates B cell proliferation and antibody production. They usually express cell surface CD4 but not CD8.

Helper virus A virus that provides functions absent from a defective virus, enabling the latter to complete the infective cycle during a mixed infection with the helper virus.

Hemimethylated DNA DNA that is methylated on one strand of a target sequence that has a cytosine on each strand.

Hemimethylated site A palindromic sequence that is methylated on only one strand of DNA.

Heterochromatin Regions of the genome that are highly condensed, are not transcribed, and are late-replicating. It is divided into two types: constitutive and facultative.

Heteroduplex DNA DNA that is generated by base pairing between complementary single strands derived from the different parental duplex molecules; it occurs during genetic recombination.

Heterogeneous nuclear RNA (hnRNA) RNA that comprises transcripts of nuclear genes made primarily by RNA polymerase II; it has a wide size distribution and variable stability.

Heteromultimer A molecular complex (such as a protein) composed of different subunits.

Heteroplasmy Having more than one mitochondrial allelic variant in a cell.

Hfr A bacterium that has an integrated F plasmid within its chromosome. Hfr stands for *high frequency recombination*, referring to the fact that chromosomal genes are transferred from an Hfr cell to an F⁻ cell much more frequently than from an F⁺ cell.

Histone acetyltransferase (HAT) An enzyme that modifies histones by addition of acetyl groups; some transcriptional coactivators have this activity. Also known as lysine acetyltransferase (KAT).

Histone code The hypothesis that combinations of specific modifications on specific histone residues act cooperatively to define chromatin function.

Histone deacetylase (HDAC) Enzyme that removes acetyl groups from histones; may be associated with repressors of transcription.

Histone fold A motif found in all four core histones in which three α-helices are connected by two loops.

Histone octamer The complex of 2 copies each of the four different core histones (H2A, H2B, H3 and H4); DNA wraps around the histone octamer to form the nucleosome.

Histone tails Flexible amino- or carboxy-terminal regions of the core histones that extend beyond the surface of the nucleosome; histone tails are sites of extensive post-translational modification.

Histone variant Any of a number of histones closely related to one of the core histones (H2A, H2B, H3 or H4) that can assemble into a nucleosome in the place of the related core histone; many histone variants have specialized functions or localization. There are also numerous linker histone variants.

Histones Conserved DNA-binding proteins that form the basic subunit of chromatin in eukaryotes. H2A, H2B, H3, and H4 form an octameric core around which DNA coils to form a nucleosome. Linker histones are external to the nucleosome.

hnRNP The ribonucleoprotein form of hnRNA (heterogeneous nuclear RNA), in which the hnRNA is complexed with proteins. Pre-mRNAs are not exported until processing is complete; thus they are found only in the nucleus.

Holliday junction A intermediate structure in homologous recombination, for which the two duplexes of DNA are connected by the genetic material exchanged between two of the four strands, one from each duplex. A joint molecule is said to be resolved when nicks in the structure restore two separate DNA duplexes.

Holoenzyme 1. The DNA polymerase complex that is competent to initiate replication. 2. The RNA polymerase form that is competent to initiate transcription. It consists of the five subunits of the core enzyme ($a2\beta\beta'\omega$) and σ factor.

Homeodomain A DNA-binding motif that typifies a class of transcription factors.

Homologous genes (homologs) Related genes in the same species, such as alleles on homologous chromosomes, or multiple genes in the same genome sharing common ancestry.

Homologous recombination Recombination involving a reciprocal exchange of sequences of DNA, e.g., between two chromosomes that carry the same genetic loci.

Homomultimer A molecular complex (such as a protein) in which the subunits are identical.

Horizontal transfer The transfer of DNA from one cell to another by a process other than cell division, such as bacterial conjugation.

Hotspots A site in the genome at which the frequency of mutation (or recombination) is very much increased, usually by at least an order of magnitude relative to neighboring sites.

Housekeeping gene A gene that is (theoretically) expressed in all cells because it provides basic functions needed for sustenance of all cell types.

Humoral response An immune response that is mediated primarily by antibodies. It is defined as immunity that can be transferred from one organism to another by serum antibody.

Hybrid dysgenesis The inability of certain strains of *D. melanogaster* to interbreed, because the hybrids are sterile (although otherwise they may be phenotypically normal).

Hybridization The pairing of complementary RNA and DNA strands to give an RNA-DNA hybrid.

Hydrops fetalis A fatal disease resulting from the absence of the hemoglobin a gene.

Hypermutation The introduction of somatic mutations in a rearranged immunoglobulin gene. The mutations can change the sequence of the corresponding antibody, especially in its antigen-binding site.

Hypersensitive site A short region of chromatin detected by its extreme sensitivity to cleavage by DNase I and other nucleases; it comprises an area from which nucleosomes are excluded.

IF-1 A bacterial initiation factor that stabilizes the initiation complex for polypeptide translation.

IF-2 A bacterial initiation factor that binds the initiator tRNA to the initiation complex for polypeptide translation.

IF-3 A bacterial initiation factor required for 30S ribosomal subunits to bind to initiation sites in mRNA. It also prevents 30S subunits from binding to 50S ribosomal subunits.

Immediate early genes Genes in phage lambda that are equivalent to the early class of other phages. They are transcribed immediately upon infection by the host RNA polymerase.

Immune response An organism's reaction, mediated by components of the immune system, to an antigen.

Immunity In phages, the ability of a prophage to prevent another phage of the same type from infecting a cell. In plasmids, the ability of a plasmid to prevent another of the same type from becoming established in a cell. It can also refer to the ability of certain transposons to prevent others of the same type from transposing to the same DNA molecule.

Immunity region A segment of the phage genome that enables a prophage to inhibit additional phage of the same type from infecting the bacterium. This region has a gene that encodes for the repressor, as well as the sites to which the repressor binds.

Immunoglobulin A protein that is produced by B cells and that binds to a particular antigen.

Immunoglobulin heavy chain One of two types of subunits in an antibody tetramer. Each antibody contains two of them. The N-terminus forms part of the antigen recognition site, whereas the C-terminus determines the subclass (isotype).

Immunoglobulin light chain (L) One of two types of subunits in an antibody tetramer. Each antibody contains two of them. The N-terminus forms part of the antigen recognition site.

Imprecise excision It occurs when the transposon removes itself from the original insertion site, but leaves behind some of its sequence.

Imprinting A change in a gene that occurs during passage through the sperm or egg with the result that the paternal and maternal alleles have different properties in the very early embryo. This is caused by methylation of DNA.

***In vitro* complementation** A functional assay used to identify components of a process. The reaction is reconstructed using extracts from a mutant cell. Fractions from wild-type cells are then tested for restoration of activity.

***In situ* hybridization** Hybridization performed by denaturing the DNA of cells squashed on a microscope slide so that reaction is possible with an added single-stranded RNA or DNA; the added preparation is radioactively labeled and its hybridization is followed by autoradiography.

Incision A step in a mismatch excision-repair system in which an endonuclease recognizes the damaged area in the DNA and isolates it by cutting the DNA strand on both sides of the damage.

Indirect end labeling A technique for examining the organization of DNA by making a cut at a specific site and identifying all fragments containing the sequence adjacent to one side of the cut; it reveals the distance from the cut to the next break(s) in DNA.

Induced mutations Mutations that result from the action of a mutagen. The mutagen may act directly on the bases in DNA or it may act indirectly to trigger a pathway that leads to a change in DNA sequence.

Inducer A small molecule that triggers gene transcription by binding to a regulator protein.

Inducible gene A gene that is turned on by the presence of its substrate.

Induction The ability to synthesize certain enzymes only when their substrates are present; applied to gene expression, it refers to switching on transcription as a result of interaction of the inducer with the regulator protein.

Induction of phage A phage's entry into the lytic (infective) cycle as a result of destruction of the lysogenic repressor, which leads to excision of free phage DNA from the bacterial chromosome.

Initiation The stages of transcription up to synthesis of the first bond in RNA. This includes binding of RNA polymerase to the promoter and melting a short region of DNA into single strands.

Initiation codon A special codon (usually AUG) used to start synthesis of a polypeptide.

Initiation factors (IFs) Proteins that associate with the small subunit of the ribosome specifically at the stage of initiation of polypeptide translation.

Initiator (Inr) The sequence of a pol II promoter between −3 and +5 and has the general sequence Py2CAPy5. It is the simplest possible pol II promoter.

Innate immunity A response triggered by receptors whose specificity is predefined for certain common motifs found in bacteria and other infective agents. The receptor that triggers the pathway is typically a member of the Toll-like class, and the pathway resembles the pathway triggered by Toll receptors during embryonic development. The pathway culminates in activation of transcription factors that cause genes to be expressed whose products inactivate the infective agent, typically by permeabilizing its membrane.

Insert The fragment of DNA that is to be cloned in a vector.

Insertion sequences (IS) A small bacterial transposon that carries only the genes needed for its own transposition.

Insulator A sequence that prevents an activating or inactivating effect passing from one side to the other.

Integrase An enzyme that is responsible for a site-specific recombination that inserts one molecule of DNA into another.

Integration Insertion of a viral or another DNA sequence into a host genome as a region covalently linked on either side to the host sequences.

Intein The part that is removed from a protein that is processed by protein splicing.

Interactome The complete set of protein complexes/protein-protein interactions present in a cell, tissue, or organism.

Interallelic complementation The change in the properties of a heteromultimeric protein brought about by the interaction of subunits coded by two different mutant alleles; the mixed protein may be more or less active than the protein consisting of subunits of only one or the other type.

Interbands The relatively dispersed regions of polytene chromosomes that lie between the bands.

Intercistronic region The distance between the termination codon of one gene and the initiation codon of the next gene.

Interrupted gene A gene in which the coding sequence is not continuous due to the presence of introns.

Intrinsic terminator Terminators that are able to terminate transcription by bacterial RNA polymerase in the absence of any additional factors.

Intron A segment of DNA that is transcribed, but later removed from within the transcript by splicing together the sequences (exons) on either side of it.

Intron definition The process in which a pair of splicing sites are recognized by interactions involving only the 5′ site and the branchpoint/3′ site.

Intron homing The ability of certain introns to insert themselves into a target DNA. The reaction is specific for a single target sequence.

Introns early model The hypothesis that the earliest genes contained introns and some genes subsequently lost them.

Introns late model The hypothesis that the earliest genes did not contain introns, and that introns were subsequently added to some genes.

Inverted terminal repeats The short related or identical sequences present in reverse orientation at the ends of some transposons.

IRES (internal ribosome entry site) A eukaryotic messenger RNA sequence that allows a ribosome to initiate polypeptide translation without migrating from the 5′ end.

Iron-response element (IRE)   A *cis* sequence found in certain mRNAs whose stability or translation is regulated by cellular iron concentration.

Isoaccepting tRNAs   See **Cognate tRNAs**.

J segments   Coding sequences in the immunoglobulin and T cell receptor loci. They are between the variable (V) and constant (C) gene segments.

Joint molecule   A pair of DNA duplexes that are connected together through a reciprocal exchange of genetic material.

Kinetic proofreading   A proofreading mechanism that depends on incorrect events proceeding more slowly than correct events, so that incorrect events are reversed before a subunit is added to a polymeric chain.

Kinetochore   A small organelle associated with the surface of the centromere that attaches a chromosome to the microtubules of the mitotic spindle. Each mitotic chromosome contains two "sisters" that are positioned on opposite sides of its centromere and face in opposite directions.

Kirromycin   An antibiotic that inhibits protein synthesis by acting on EF-Tu.

Kuru   A human neurological disease caused by prions. It may be caused by eating infected brains.

***lac* Repressor**   A negative gene regulator encoded by the *lacI* gene that turns off the *lac* operon

Lagging strand   The strand of DNA that must grow overall in the 3′ to 5′ direction and is synthesized discontinuously in the form of short fragments (5′–3′) that are later connected covalently.

Lampbrush chromosomes   The extremely extended meiotic bivalents of certain amphibian oocytes.

Lariat   An intermediate in RNA splicing in which a circular structure with a tail is created by a 5′ to 2′ bond.

Late genes   Genes transcribed when phage DNA is being replicated. They encode components of the phage particle.

Late infection   The part of the phage lytic cycle from DNA replication to lysis of the cell. During this time, the DNA is replicated and structural components of the phage particle are synthesized.

Lateral element   A structure in the synaptonemal complex that forms when a pair of sister chromatids condenses on to an axial element.

Leader (5′ UTR)   In mRNA, it is the untranslated sequence at the 5′ end that precedes the initiation codon.

Leader peptide   The product that would result from translation of a short coding sequence used to regulate transcription of an operon by controlling ribosome movement.

Leading strand   The strand of DNA that is synthesized continuously in the 5′ to 3′ direction.

Lesion bypass   Replication by an error-prone DNA polymerase on a template that contains a damaged base. The polymerase can incorporate a noncomplementary base into the daughter strand.

Leucine-rich region   A motif found in the extracellular domains of some surface receptor proteins in animal and plant cells.

Leucine zipper   A dimerization motif that is found in a class of transcription factors.

Licensing factor   A factor located in the nucleus and necessary for replication; it is inactivated or destroyed after one round of replication. New factors must be provided for further rounds of replication to occur.

Linkage disequilibrium   A nonrandom association between alleles at two different loci, often as a result of linkage.

Linkage map   A map of the positions of loci or other genetic markers on a chromosome obtained by measuring recombination frequencies between markers.

Linker DNA   Non-nucleosomal DNA present between nucleosomes.

Linker histones   A family of histones (such as histone H1) that are not components of the nucleosome core; linker histone bind nucleosomes and/or linker DNA and promote 30 nm fiber formation.

Locus   The position on a chromosome at which the gene for a particular trait resides; it may be occupied by any one of the alleles for the gene.

Locus control region (LCR)   The region that is required for the expression of several genes in a domain.

Long interspersed elements (LINEs)   Long interspersed nuclear elements; a major class of retrotransposons that occupy ~21% of the human genome (see **Retrotransposon**).

Long terminal repeat (LTR)   The sequence that is repeated at each end of the provirus (integrated retroviral sequence).

Loss-of-function mutation   A mutation that eliminates or reduces the activity of a gene. It is often, but not always, recessive.

Luxury gene   A gene coding for a specialized function, synthesized (usually) in large amounts in particular cell types.

Lyase   A repair enzyme (usually also a glycosylase) that opens the sugar ring at the site of a damaged base.

Lysine (K) acetyltransferase (KAT)   An enzyme (typically present in large complexes) that acetylates lysine residues in histones (or other proteins). Previously known as histone acetyltransferase (HAT).

Lysis   The death of bacteria at the end of a phage infective cycle when they burst open to release the progeny of an infecting phage (because phage enzymes disrupt the bacterium's cytoplasmic membrane or cell wall). The same term also applies to eukaryotic cells; for example, when infected cells are attacked by the immune system.

Lysogenic   The ability of a phage to survive in a bacterium as a stable prophage component of the bacterial genome.

Lysogeny   The ability of a phage to survive in a bacterium as a stable prophage component of the bacterial genome.

Lytic infection   Infection of a bacterium by a phage that ends in the destruction of the bacterium with release of progeny phage.

Maintenance methyltransferase   An enzyme that adds a methyl group to a target site that is already hemimethylated.

Major groove   A fissure running the length of the DNA double helix that is 22 Å across.

Major histocompatibility complex (MHC)   A chromosomal region containing genes that are involved in the immune response. The genes encode proteins for antigen presentation, cytokines, and complement, as well as other functions. It is highly polymorphic. Its genes and proteins are divided into three classes.

Maternal inheritance   The preferential survival in the progeny of genetic markers provided by one parent.

Maternal mRNA granules   Oocyte particles containing translationally repressed mRNAs awaiting activation later in development.

Mating type cassette   Yeast mating type is determined by a single active locus (the active cassette) and two inactive copies of the locus (the silent cassettes). Mating type is changed when an active cassette of one type is replaced by a silent cassette of the other type.

Maturase   A protein encoded by a group I or group II intron that is needed to assist the RNA to form the active conformation that is required for self-splicing.

Mature transcript   A modified RNA transcript. Modification may include the removal of intron sequences and alterations to the 5′ and 3′ ends.

Matrix attachment region (MAR)   A region of DNA that attaches to the nuclear matrix. It is also known as a scaffold attachment site (SAR).

MCS (multiple cloning site)   A sequence of DNA containing a series of tandem restriction endonuclease sites, used in cloning vectors for creating recombinant molecules.

Mediator   A large protein complex associated with yeast bacterial RNA polymerase II. It contains factors that are necessary for transcription from many or most promoters.

Melting temperature   The midpoint of the temperature range over which the strands of DNA separate.

Messenger RNA (mRNA)   The intermediate that represents one strand of a gene coding for polypeptide. Its coding region is related to the polypeptide sequence by the triplet genetic code.

Metaphase (or mitotic) scaffold   A proteinaceous structure in the shape of a sister chromatid pair, generated when chromosomes are depleted of histones.

Methyltransferase   An enzyme that adds a methyl group to a substrate, which can be a small molecule, a protein, or a nucleic acid.

Microarray   An arrayed series of thousands of tiny DNA oligonucleotide samples imprinted on a small chip. mRNAs can be hybridized to microarrays to assess the amount and level of gene expression.

Micrococcal nuclease (MNase)   An endonuclease that cleaves DNA; in chromatin, DNA is cleaved preferentially between nucleosomes.

microRNA (miRNA)   Very short RNAs that may regulate gene expression.

Microsatellite   DNAs consisting of tandem repetitions of very short (typically <10 bp) units repeated a small number of times.

Microtubule organizing center (MTOC)   A region from which microtubules emanate. In animal cells the centrosome is the major microtubule organizing center.

Middle genes   Phage genes that are regulated by the proteins encoded by early genes. Some proteins coded by them catalyze replication of the phage DNA; others regulate the expression of a later set of genes.

Minisatellite   DNAs consisting of tandemly repeated copies of a short repeating sequence, with more repeat copies than a microsatellite but fewer than a satellite. The length of the repeating unit is measured in tens of base pairs. The number of repeats varies between individual genomes.

Minicell   An anucleate bacterial (*E. coli*) cell produced by a division that generates a cytoplasm without a nucleus.

Minor groove   A fissure running the length of the DNA double helix that is 12 Å across.

Minus strand DNA   The single-stranded DNA sequence that is complementary to the viral RNA genome of a plus strand virus.

miRNA   Very short RNAs that may regulate gene expression.

Mismatch repair   Repair that corrects recently inserted bases that do not pair properly. The process preferentially corrects the sequence of the daughter strand by distinguishing the daughter strand and parental strand, sometimes on the basis of their states of methylation.

Missense suppressor   A suppressor that codes for a tRNA that has been mutated to recognize a different codon. By inserting a different amino acid at a mutant codon, the tRNA suppresses the effect of the original mutation.

Modification   All changes made to the nucleotides of DNA or RNA after their initial incorporation into the polynucleotide chain.

Molecular clock   An approximately constant rate of evolution that occurs in DNA sequences, such as by the genetic drift of neutral mutations.

Monocistronic mRNA   mRNA that codes for one polypeptide.

mRNA decay   mRNA degradation, assuming that the degradation process is stochastic.

mtDNA   Mitochondrial DNA.

Multicopy replication control   Occurs when the control system allows the plasmid to exist in more than one copy per individual bacterial cell.

Multiforked chromosome   A bacterial chromosome that has more than one set of replication forks, because a second initiation has occurred before the first cycle of replication has been completed.

Mutagens   Substances that increase the rate of mutation by inducing changes in DNA sequence, directly or indirectly.

Mutation hotspot   A site in the genome at which the frequency of mutation (or recombination) is very much increased, usually by at least an order of magnitude relative to neighboring sites.

Mutator   A mutation or a mutated gene that increases the basal level of mutation. Such genes often code for proteins that are involved in repairing damaged DNA.

N nucleotide   A short nontemplated sequence that is added randomly by the enzyme at coding joints during rearrangement of immunoglobulin and T cell receptor genes. They augment the diversity of antigen receptors.

n-1 rule   The rule that states that only one X chromosome is active in female mammalian cells; any others are inactivated.

N-formyl-methionyl-tRNA   The aminoacyl-tRNA that initiates bacterial polypeptide translation. The amino group of the methionine is formylated.

Nascent polypeptide   A protein that has not yet completed its synthesis; the polypeptide chain is still attached to the ribosome via a tRNA.

Nascent RNA   A ribonucleotide chain that is still being synthesized, so that its 3′ end is paired with DNA where RNA polymerase is elongating.

ncRNA   Noncoding RNA which does not contain an open reading frame.

Negative complementation   This occurs when interallelic complementation allows a mutant subunit to suppress the activity of a wild-type subunit in a multimeric protein.

Negative control   This describes a mechanism of gene regulation in which a regulator is required to turn the gene off.

Negative inducible   A control circuit in which an active repressor is inactivated by the substrate of the operon.

Negative repressible   A control circuit in which an inactive repressor is activated by the product of the operon.

Nested gene   A gene located within an intron of another gene.

Neuronal granules   Particles containing translationally repressed mRNAs in transit to final cell destinations.

Neutral mutation   A mutation that has no significant effect on evolutionary fitness and usually has no effect on the phenotype.

Neutral substitutions   Substitutions in a protein that cause changes in amino acids that do not affect activity.

Nick translation   The ability of *E. coli* DNA polymerase I to use a nick as a starting point from which one strand of a duplex DNA can be degraded and replaced by resynthesis of new material; is used to introduce radioactively labeled nucleotides into DNA *in vitro*.

No-go decay (NGD)   A pathway that rapidly degrades an mRNA with ribosomes stalled in its coding region.

Non-Mendelian inheritance   A pattern of inheritance that does not follow that expected by Mendelian principles (each parent contributing a single allele to offspring). Extranuclear genes show a non-Mendelian inheritance pattern.

Non-template strand   See **Coding strand**.

Nonallelic genes   Two (or more) copies of the same gene that are present at different locations in the genome (contrasted with alleles, which are copies of the same gene derived from different parents and present at the same location on the homologous chromosomes).

Nonautonomous controlling element   A transposon in maize that encodes a nonfunctional transposase; it can transpose only in the presence of a *trans*-acting autonomous member of the same family.

Nonhistone   Any structural protein found in a chromosome except one of the histones.

Nonhomologous end-joining (NHEJ)   The process that ligates blunt ends. It is common to many repair pathways and to certain recombination pathways (such as immunoglobulin recombination).

Nonprocessed pseudogene   An inactive gene copy that arises by incomplete gene duplication or duplication followed by inactivating mutations.

Nonproductive rearrangement   This occurs as a result of the recombination of V, (D), J gene segments if the rearranged gene segments are not in the correct reading frame. It occurs when nucleotide addition or subtraction disrupts the reading frame or when a functional protein is not produced.

Nonrepetitive DNA DNA that is unique (present only once) in a genome.

Nonreplicative transposition The movement of a transposon that leaves a donor site (usually generating a double-strand break) and moves to a new site.

Nonsense-mediated mRNA decay (NMD) A pathway that degrades an mRNA that has a nonsense mutation prior to the last exon.

Nonsense suppressor A gene coding for a mutant tRNA that is able to respond to one or more of the termination codons and insert an amino acid at that site.

Nonstop decay (NSD) A pathway that rapidly degrades an mRNA that lacks an in-frame termination codon.

Nonsynonymous sites Sites in a coding region at which mutations have altered the amino acid that is encoded.

Nontranscribed spacer The region between transcription units in a tandem gene cluster.

Nonviral superfamily Transposons originated independently of retroviruses.

Nopaline plasmids Ti plasmids of *Agrobacterium tumefaciens* that carry genes for synthesizing the opine, nopaline. They retain the ability to differentiate into early embryonic structures.

Nuclease An enzyme that can break a phosphodiester bond.

Nucleation center A duplex hairpin in TMV (tobacco mosaic virus) in which assembly of coat protein with RNA is initiated.

Nucleoid The structure in a prokaryotic cell that contains the genome. The DNA is bound to proteins and is not enclosed by a membrane.

Nucleolar organizer The region of a chromosome carrying genes coding for rRNA.

Nucleolus A discrete region of the nucleus where ribosomes are produced.

Nucleoside A molecule consisting of a purine or pyrimidine base linked to the 1′ carbon of a pentose sugar.

Nucleosome The basic structural subunit of chromatin, consisting of ~200 bp of DNA and an octamer of histone proteins.

Nucleosome positioning The placement of nucleosomes at defined sequences of DNA instead of at random locations with regard to sequence.

Nucleotide A molecule consisting of a purine or pyrimidine base linked to the 1′ carbon of a pentose sugar and a phosphate group linked to either the 5′ or 3′ carbon of the sugar.

Null mutation A mutation that completely eliminates the function of a gene.

Nut An acronym for N utilization site, the sequence of DNA that is recognized by the N antitermination factor.

Ochre codon The triplet UAA, one of the three termination codons that end polypeptide translation.

Octopine plasmids Plasmids of *Agrobacterium tumefaciens* that carry genes coding the synthesis of opines of the octopine type. The tumors are undifferentiated.

Okazaki fragment Short stretches of 1000 to 2000 bases produced during discontinuous replication; they are later joined into a covalently intact strand.

Oligo(A) tail A short poly(A) tail, generally referring to a stretch of less than 15 adenylates.

One gene : one enzyme hypothesis Beadle and Tatum's hypothesis that a gene is responsible for the production of a single enzyme.

One gene : one polypeptide hypothesis A modified version of the not generally correct one gene : one enzyme hypothesis; the hypothesis that a gene is responsible for the production of a single polypeptide.

Opal codon The triplet UGA, one of the three termination codons that end polypeptide translation. It has evolved to code for an amino acid in a small number of organisms or organelles.

Open complex The stage of initiation of transcription when RNA polymerase causes the two strands of DNA to separate to form the "transcription bubble."

Open reading frame (ORF) A sequence of DNA consisting of triplets that can be translated into amino acids starting with an initiation codon and ending with a termination codon.

Operon A unit of bacterial gene expression and regulation, including structural genes and control elements in DNA recognized by regulator gene product(s).

Operator The site on DNA at which a repressor protein binds to prevent transcription from initiating at the adjacent promoter.

Opine A derivative of arginine that is synthesized by plant cells infected with crown gall disease.

ORC Origin recognition complex, found in eukaryotes, a multiprotein complex that binds to the replication origin, the ARS, and remains associated with it throughout the cell cycle.

Origin A sequence of DNA at which replication is initiated.

Orthologous genes (orthologs) Related genes in different species.

Overlapping gene A gene in which part of the sequence is found within part of the sequence of another gene.

Overwound B-form DNA that has more than 10.5 base pairs per turn of the helix.

P element A type of transposon in *D. melanogaster*.

P nucleotide A short palindromic (inverted repeat) sequence that is generated during rearrangement of immunoglobulin and T cell receptor V, (D), J gene segments. They are generated at coding joints when RAG proteins cleave the hairpin ends generated during rearrangement.

P site The site in the ribosome that is occupied by peptidyl-tRNA, the tRNA carrying the nascent polypeptide chain, still paired with the codon to which it bound in the A site.

Packing ratio The ratio of the length of DNA to the unit length of the fiber containing it.

Palindrome A symmetrical sequence that reads the same forward and backward.

Patch recombinant DNA that results from a Holliday junction being resolved by cutting the exchanged strands. The duplex is largely unchanged, except for a DNA sequence on one strand that came from the homologous chromosome.

Pathogenicity islands DNA segments that are present in pathogenic bacterial genomes but absent in their non-pathogenic relatives.

Peptidyl transferase The activity of the large ribosomal subunit that synthesizes a peptide bond when an amino acid is added to a growing polypeptide chain. The actual catalytic activity is a property of the rRNA.

Peptidyl-tRNA The tRNA to which the nascent polypeptide chain has been transferred following peptide bond synthesis during polypeptide translation.

Phage An abbreviation of bacteriophage or bacterial virus.

Phosphatase An enzyme that can break a phosphomono-ester bond, cleaving a terminal phosphate.

Phosphorelay A pathway in which a phosphate group is passed along a series of proteins.

Photoreactivation A repair mechanism that uses a white light-dependent enzyme to split cyclobutane pyrimidine dimers formed by ultraviolet light.

Pili A surface appendage on a bacterium that allows the bacterium to attach to other bacterial cells. It appears as a short, thin, flexible rod. During conjugation, it is used to transfer DNA from one bacterium to another.

Pilin The subunit that is polymerized into the pilus in bacteria.

Pioneer round of translation The first translation event for a newly synthesized and exported mRNA.

piRNA Piwi RNA, a special form of miRNA found in germ cells.

Plasmid Circular, extrachromosomal DNA. It is autonomous and can replicate itself.

Plus strand DNA The strand of the duplex sequence representing a retrovirus that has the same sequence as that of the RNA.

Plus strand virus A virus with a single-stranded nucleic acid genome whose sequence directly codes for the protein products.

Point mutation A change in the sequence of DNA involving a single base pair.

Polarity The effect of a mutation in one gene in influencing the expression (at transcription or translation) of subsequent genes in the same transcription unit.

Poly(A) A stretch of adenylic acid that is added to the 3′ end of mRNA following its synthesis.

Poly(A)⁺ mRNA mRNA that has a 3′ terminal stretch of poly(A).

Poly(A) binding protein (PABP) The protein that binds to the 3′ stretch of poly(A) on a eukaryotic mRNA.

Poly(A) nuclease (or deadenylase) An exoribonuclease that is specific for digesting poly(A) tails.

Poly(A) polymerase (PAP) The enzyme that adds the stretch of polyadenylic acid to the 3′ end of eukaryotic mRNA. It does not use a template.

Polycistronic mRNA mRNA that includes coding regions representing more than one gene.

Polymerase Chain Reaction (PCR) A process for the amplification of a defined nucleic acid section through repeated thermal cycles of denaturation, annealing, and polymerase extension.

Polymorphism The simultaneous occurrence in the population of alleles showing variations at a given position.

Polynucleotide A chain of nucleotides, such as DNA or RNA.

Polyploidization An event that results in an increase in the number of haploid chromosome sets in the cell, typically from diploid to tetraploid and usually as a result of fertilization of unreduced gametes.

Polyribosome (or polysome) An mRNA that is simultaneously being translated by multiple ribosomes.

Polytene chromosomes Chromosomes that are generated by successive replications of a chromosome set without separation of the replicas.

Position effect variegation (PEV) Silencing of gene expression that occurs as the result of proximity to heterochromatin.

Positive control This describes a system in which a gene is not expressed unless some action turns it on.

Positive inducible A control circuit in which an inactive positive regulator is converted into an active regulator by the substrate of the operon.

Positive repressible A control circuit in which an active positive regulator is inactivated by the product of the operon.

Postreplication complex A protein-DNA-complex in *S. cerevisiae* that consists of the ORC complex bound to the origin.

pre-mRNA The nuclear transcript that is processed by modification and splicing to give an mRNA.

Precise excision The removal of a transposon plus one of the duplicated target sequences from the chromosome. Such an event can restore function at the site where the transposon inserted.

Preinitiation complex The assembly of transcription factors at the promoter before RNA polymerase binds in eukaryotic transcription.

Premature termination The termination of protein or of RNA synthesis before the chain has been completed. In translation it can be caused by mutations that create stop codons within the coding region. In RNA synthesis it is caused by various events that act on RNA polymerase.

Prereplication complex A protein-DNA complex at the origin in *S. cerevisiae* that is required for DNA replication. The complex contains the ORC complex, Cdc6, and the MCM proteins.

Presynaptic filaments Single-stranded DNA bound in a helical nucleoprotein filament with a strand transfer protein such as Rad51 or RecA.

Primary (RNA) transcript The original unmodified RNA product corresponding to a transcription unit.

Primase A type of RNA polymerase that synthesizes short segments of RNA that will be used as primers for DNA replication.

Primer A short sequence (often of RNA) that is paired with one strand of DNA and provides a free 3′–OH end at which a DNA polymerase starts synthesis of a deoxyribonucleotide chain.

Prion A proteinaceous infectious agent that behaves as an inheritable trait, although it contains no nucleic acid. Examples are PrPSc, the agent of scrapie in sheep and bovine spongiform encephalopathy, and Psi, which confers an inherited state in yeast.

Probe A radioactive nucleic acid, DNA or RNA, used to identify a complementary fragment.

Processed pseudogene An inactive gene copy that lacks introns, contrasted with the interrupted structure of the active gene. Such genes originate by reverse transcription of mRNA and insertion of a duplex copy into the genome.

Processing body (PB) A particle containing multiple mRNAs and proteins involved in mRNA degradation and translational repression, occurring in many copies in the cytoplasm of eukaryotes.

Processive (nuclease) An enzyme that remains associated with the substrate while catalyzing the sequential removal of nucleotides.

Processivity The ability of an enzyme to perform multiple catalytic cycles with a single template instead of dissociating after each cycle.

Productive rearrangement This occurs as a result of the recombination of V, (D), J gene segments if all the rearranged gene segments are in the correct reading frame.

Programmed frameshifting Frameshifting that is required for expression of the polypeptide sequences encoded beyond a specific site at which a +1 or −1 frameshift occurs at some typical frequency.

Promoter A region of DNA where RNA polymerase binds to initiate transcription.

PROMPT Promoter upstream transcripts, short RNAs produced from both strands of DNA from active promoters.

Proofreading A mechanism for correcting errors in DNA synthesis that involves scrutiny of individual units after they have been added to the chain.

Prophage A phage genome covalently integrated as a linear part of the bacterial chromosome.

Protein splicing The autocatalytic process by which an intein is removed from a protein and the exteins on either side become connected by a standard peptide bond.

Proteome The complete set of proteins that is expressed by the entire genome. Sometimes the term is used to describe the complement of proteins expressed by a cell at any one time.

Provirus A duplex sequence of DNA integrated into a eukaryotic genome that represents the sequence of the RNA genome of a retrovirus.

Pseudogenes Inactive but stable components of the genome derived by mutation of an ancestral active gene. Usually they are inactive because of mutations that block transcription or translation or both.

Puff An expansion of a band of a polytene chromosome associated with the synthesis of RNA at some locus in the band.

Purine A double-ringed nitrogenous base, such as adenine or guanine.

Puromycin An antibiotic that terminates protein synthesis by mimicking a tRNA and becoming linked to the nascent protein chain.

Pyrimidine A single-ringed nitrogenous base, such as cytosine, thymine, or uracil.

Pyrimidine dimer A dimer that forms when ultraviolet irradiation generates a covalent link directly between two adjacent pyrimidine bases in DNA. It blocks DNA replication and transcription.

Quick stop mutant Temperature sensitive replication mutants that are defective in replication elongation during synthesis of DNA.

R segments The sequences that are repeated at the ends of a retroviral RNA. They are called R-U5 and U3-R.

rasiRNA Repeat associated silencer RNA is a germline subset of miRNA transcribed from transposable elements and other repeated elements that is used to silence them.

rDNA Genes encoding ribosomal RNA (rRNA).

Reading frame One of three possible ways of reading a nucleotide sequence. Each divides the sequence into a series of successive triplets.

Readthrough It occurs at transcription or translation when RNA polymerase or the ribosome, respectively, ignores a termination signal because of a mutation of the template or the behavior of an accessory factor.

Real-time PCR or RT-PCR technique with continuous monitoring of product formation as process proceeds, usually through fluorometric methods.

Recoding Events that occur when the meaning of a codon or series of codons is changed from that predicted by the genetic code. It may involve altered interactions between aminoacyl-tRNA and mRNA that are influenced by the ribosome.

Recognition helix One of the two helices of the helix-turn-helix motif that makes contacts with DNA that are specific for particular bases. This determines the specificity of the DNA sequence that is bound.

Recombinant joint The point at which two recombining molecules of duplex DNA are connected (the edge of the heteroduplex region).

Recombination nodules (nodes) Dense objects present on the synaptonemal complex; they may represent protein complexes involved in crossing-over.

Recombination-repair A mode of filling a gap in one strand of duplex DNA by retrieving a homologous single strand from another duplex.

Recombinase Enzyme that catalyzes site-specific recombination.

Redundancy The concept that two or more genes may fulfill the same function, so that no single one of them is essential.

Regulator gene A gene that codes for a product (typically protein) that controls the expression of other genes (usually at the level of transcription).

Relaxase An enzyme that cuts one strand of DNA and binds to the free 5′ end.

Relaxed mutants In *E. coli*, these do not display the stringent response to starvation for amino acids (or other nutritional deprivation).

Release factor (RF) A protein required to terminate polypeptide translation to cause release of the completed polypeptide chain and the ribosome from mRNA.

Renaturation The reassociation of denatured complementary single strands of a DNA double helix.

Repetitive DNA DNA that is present in many (related or identical) copies in a genome.

Replication bubble A region in which DNA has been replicated within a longer, unreplicated region.

Replication-defective virus A virus that cannot perpetuate an infective cycle because some of the necessary genes are absent (replaced by host DNA in a transducing virus) or mutated.

Replication fork The point at which strands of parental duplex DNA are separated so that replication can proceed. A complex of proteins including DNA polymerase is found there.

Replicative transposition The movement of a transposon by a mechanism in which first it is replicated, and then one copy is transferred to a new site.

Replicon A unit of the genome in which DNA is replicated. Each contains an origin for initiation of replication.

Replisome The multiprotein structure that assembles at the bacterial replication fork to undertake synthesis of DNA. It contains DNA polymerase and other enzymes.

Repressible gene A gene that is turned off by its product.

Repression The ability to prevent synthesis of certain enzymes when their products are present; more generally, it refers to inhibition of transcription (or translation) by binding of repressor protein to a specific site on DNA (or mRNA).

Repressor A protein that inhibits expression of a gene. It may act to prevent transcription by binding to an enhancer or silencer.

Resolution Resolution occurs by a homologous recombination reaction between the two copies of the transposon in a cointegrate. The reaction generates the donor and target replicons, each with a copy of the transposon.

Resolvase The enzyme activity involved in site-specific recombination between two copies of a transposon that has been duplicated.

Restriction endonuclease An enzyme that recognizes specific short sequences of DNA and cleaves the duplex (sometimes at the target site, sometimes elsewhere, depending on type).

Restriction fragment length polymorphism (RFLP) Inherited differences in sites for restriction enzymes (for example, caused by base changes in the target site) that result in differences in the lengths of the fragments produced by cleavage with the relevant restriction enzyme. They are used for genetic mapping to link the genome directly to a conventional genetic marker.

Restriction mapping Determination of a linear array of sites on DNA cleaved by various restriction endonucleases.

Retrotransposon (retroposon) A transposon that mobilizes via an RNA form; the DNA element is transcribed into RNA, and then reverse-transcribed into DNA, which is inserted at a new site in the genome. It does not have an infective (viral) form.

Retrovirus An RNA virus with the ability to convert its sequence into DNA by reverse transcription.

Reverse gyrase Enzyme that introduces positive supercoils into DNA.

Reverse transcriptase An enzyme that uses single stranded RNA as a template to synthesize a complementary DNA strand.

Reverse transcription Synthesis of DNA on a template of RNA. It is accomplished by the enzyme reverse transcriptase.

Reverse transcription polymerase chain reaction (RT-PCR) A technique for the detection and quantification of expression of a gene by reverse transcription and amplification of RNAs from a cell sample.

Revertants Reversions of a mutant cell or organism to the wild-type phenotype.

RF1 The bacterial release factor that recognizes UAA and UAG as signals to terminate polypeptide translation.

RF2 The bacterial release factor that recognizes UAA and UGA as signals to terminate polypeptide translation.

RF3 A polypeptide translation termination factor related to the elongation factor EF-G. It functions to release the factors RF1 or RF2 from the ribosome when they act to terminate polypeptide translation.

Rho dependent termination Transcriptional termination by bacterial RNA polymerase in the presence of the rho factor.

Rho factor A protein involved in assisting *E. coli* RNA polymerase to terminate transcription at certain terminators (called rho-dependent terminators).

Ri plasmid Plasmids found in *Agrobacterium tumefaciens*. Like Ti plasmids, they carry genes that cause disease in infected plants. The disease may take the form of either hairy root disease or crown gall disease.

Ribonuclease An enzyme that cleaves phosphodiester linkages between RNA ribonucleotides.

Ribonucleoprotein (RNP) A complex of RNA and proteins. Larger complexes are sometimes called ribonucleoprotein particles.

Ribosomal RNAs (rRNAs) A major component of the ribosome.

Ribosome A large assembly of RNA and proteins that synthesizes proteins under direction from an mRNA template.

Ribosome-binding site A sequence on bacterial mRNA that includes an initiation codon that is bound by a 30S subunit in the initiation phase of polypeptide translation.

Ribosome stalling The inhibition of movement that occurs when a ribosome reaches a codon for which there is no corresponding charged aminoacyl-tRNA.

Riboswitch A catalytic RNA whose activity responds to a small ligand.

Ribozyme An RNA that has catalytic activity.

RISC RNA-induced silencing complex, a ribonucleoprotein particle composed of a short single-stranded siRNA and a nuclease that cleaves mRNAs complementary to the siRNA. It receives siRNA from Dicer and delivers it to the mRNA.

RNA-binding protein (RBP) a protein containing one or more domains that confer an affinity for RNA, usually in an RNA sequence- or structure-specific manner.

RNA editing A change of sequence at the level of RNA following transcription.

RNA interference (RNAi) A process by which short 21 to 23 nucleotide antisense RNAs, derived from longer double-stranded RNAs, can modulate expression of mRNA by translation inhibition or degradation.

RNA ligase An enzyme that functions in tRNA splicing to make a phosphodiester bond between the two exon sequences that are generated by cleavage of the intron.

RNA polymerase An enzyme that synthesizes RNA using a DNA template (formally described as DNA-dependent RNA polymerases).

RNA processing Modifications to RNA transcripts of genes. This may include alterations to the 3' and 5' ends and the removal of introns.

RNA silencing The ability of an RNA, especially ncRNA, to alter chromatin structure in order to prevent gene transcription.

RNA splicing The process of excising introns from RNA and connecting the exons into a continuous mRNA.

RNA surveillance systems Systems that check RNAs (or RNPs) for errors. The system recognizes an invalid sequence or structure and triggers a response.

RNase An enzyme that degrades RNA.

Rolling circle A mode of replication in which a replication fork proceeds around a circular template for an indefinite number of revolutions; the DNA strand newly synthesized in each revolution displaces the strand synthesized in the previous revolution, giving a tail containing a linear series of sequences complementary to the circular template strand.

Rotational positioning The location of the histone octamer relative to turns of the double helix, which determines which face of DNA is exposed on the nucleosome surface.

Rut An acronym for rho utilization site, the sequence of RNA that is recognized by the rho termination factor.

S phase The restricted part of the eukaryotic cell cycle during which synthesis of DNA occurs.

S region   A sequence involved in immunoglobulin class switching. They consist of repetitive sequences at the 5′ ends of gene segments encoding the heavy chain constant regions.

Satellite DNA   DNA that consists of many tandem repeats (identical or related) of a short basic repeating unit.

Scarce mRNA   mRNA that consists of a large number of individual mRNA species, each present in very few copies per cell. This accounts for most of the sequence complexity in RNA.

Scrapie   A disease caused by an infective agent made of protein (a prion).

Second-site reversion   A second mutation suppressing the effect of a first mutation.

Selfish DNA   DNA sequences that do not contribute to the phenotype of the organism but have self-perpetuation within the genome as their sole function.

Semiconservative replication   DNA replication accomplished by separation of the strands of a parental duplex, each strand then acting as a template for synthesis of a complementary strand.

Semidiscontinuous replication   The mode of replication in which one new strand is synthesized continuously while the other is synthesized discontinuously.

Septal ring   A complex of several proteins coded by *fts* genes of *E. coli* that forms at the mid-point of the cell. It gives rise to the septum at cell division. The first of the proteins to be incorporated is FtsZ, which gave rise to the original name of the Z-ring.

Septum   The structure that forms in the center of a dividing bacterium, providing the site at which the daughter bacteria will separate. The same term is used to describe the cell wall that forms between plant cells at the end of mitosis.

Sequence context   The sequence surrounding a consensus sequence. It may modulate the activity of the consensus sequence.

Shine-Dalgarno sequence   The polypurine sequence AGGAGG centered about 10 bp before the AUG initiation codon on bacterial mRNA. It is complementary to the sequence at the 3′ end of 16S rRNA.

Short interspersed elements (SINEs)   Short interspersed nuclear elements; a major class of short (<500 bp) nonautonomous retrotransposons that occupy ~13% of the human genome (see **Retrotransposon**).

Sigma factor   The subunit of bacterial RNA polymerase needed for initiation; it is the major influence on selection of promoters.

Signal end   It is produced during recombination of immunoglobulin and T cell receptor genes. Signal ends are at the termini of the cleaved fragment containing the recombination signal sequences. Their subsequent joining yields a signal joint.

Silencer   A short sequence of DNA that can inactivate expression of a gene in its vicinity.

Silent mutation   A mutation that does not change the sequence of a polypeptide because it produces synonymous codons.

Simple sequence DNA   Short repeating units of DNA sequence.

Single-copy replication control   A control system in which there is only one copy of a replicon per unit bacterium. The bacterial chromosome and some plasmids have this type of regulation.

Single nucleotide polymorphism (SNP)   A polymorphism (variation in sequence between individuals) caused by a change in a single nucleotide. This is responsible for most of the genetic variation between individuals.

Single-strand binding protein (SSB)   The protein that attaches to single-stranded DNA, thereby preventing the DNA from forming a duplex.

Single-strand exchange   A reaction in which one of the strands of a duplex of DNA leaves its former partner and instead pairs with the complementary strand in another molecule, displacing its homologue in the second duplex.

Single-strand invasion (or single-strand assimilation)   The process in which a single strand of DNA displaces its homologous strand in a duplex.

Single X hypothesis   The theory that describes the inactivation of one X chromosome in female mammals.

Sister chromatid   Each of two identical copies of a replicated chromosome; this term is used as long as the two copies remain linked at the centromere. Sister chromatids separate during anaphase in mitosis or anaphase II in meiosis.

Site-specific recombination   Recombination that occurs between two specific sequences, as in phage integration/ excision or resolution of cointegrate structures during transposition.

SKI proteins   A set of protein factors that target nonstop decay (NSD) substrates for degradation.

Slow stop mutant   Temperature sensitive replication mutants that are defective in initiation of replication.

SL RNA (spliced leader RNA)   A small RNA that donates an exon in the *trans*-splicing reaction of trypanosomes and nematodes.

Small cytoplasmic RNAs (scRNA; scyrps)   RNAs that are present in the cytoplasm (and sometimes are also found in the nucleus).

Small nuclear RNA (snRNA; snurps)   One of many small RNA species confined to the nucleus; several of them are involved in splicing or other RNA processing reactions.

Small nucleolar RNA (snoRNA) A small nuclear RNA that is localized in the nucleolus.

Somatic mutation A mutation occurring in a somatic cell, therefore affecting only its daughter cells; it is not inherited by descendants of the organism.

Somatic recombination Recombination that occurs in non-germ cells (i.e., it does not occur during meiosis); most commonly used to refer to recombination in the immune system.

Southern blotting A process for the transfer of DNA bands separated by gel electrophoresis from the gel matrix to a solid support matrix such as a nylon membrane for subsequent probing and detection.

Spindle A structure made up of microtubules that guides the movements of the chromosomes during mitosis.

Splice recombinant DNA that results from a Holliday junction being resolved by cutting the nonexchanged strands. Both strands of DNA before the exchange point come from one chromosome; the DNA after the exchange point come from the homologous chromosome.

Spliceosome A complex formed by snRNPs and additional protein factors that is required for RNA splicing.

Splicing The process of excising introns from RNA and connecting the exons into a continuous mRNA.

Splicing factor A protein component of the spliceosome that is not part of one of the snRNPs.

Spontaneous mutations Mutations occurring in the absence of any added reagent to increase the mutation rate, as the result of errors in replication (or other events involved in the reproduction of DNA) or by random changes to the chemical structure of bases.

Sporulation The generation of a spore by a bacterium (by morphological conversion) or by a yeast (as the product of meiosis).

SR protein A protein that has a variable length of a Ser-Arg-rich region and is involved in splicing.

sRNA A small bacterial RNA that functions as a regulator of gene expression.

siRNA Short interfering RNA, a miRNA that prevents gene expression.

Stabilizing element One of a variety of cis sequences present in some mRNAs that confers a long half-life on that mRNA.

Startpoint The position on DNA corresponding to the first base incorporated into RNA.

Steady state (molecular concentration) The concentration of population of molecules when the rates of synthesis and degradation are constant.

Stem-loop A secondary structure that appears in RNAs consisting of a base-paired region (stem) and a terminal loop of single-stranded RNA. Both are variable in size.

Steroid receptor Transcription factors that are activated by binding of a steroid ligand.

Stop codon One of three triplets (UAG, UAA, or UGA) that cause polypeptide translation to terminate. They are also known historically as nonsense codons. The UAA codon is called ochre and the UAG codon is called amber, after the names of the nonsense mutations by which they were originally identified.

Strand displacement A mode of replication of some viruses in which a new DNA strand grows by displacing the previous (homologous) strand of the duplex.

Stress granules Cytoplasmic particles, containing translationally inactive mRNAs, that form in response to a general inhibition of translation initiation.

Stringency A measure of the exactness of complementarity required between two DNA strands to allow them to hybridize. Stringency is related to buffer ionic strength and reaction temperature above or below TM with lower ionic strengths and higher temperatures giving higher stringencies (greater exactness required).

stRNA Short temporal RNA, a form of miRNA in eukaryotes that modulates mRNA expression during development.

Structural gene A gene that codes for any RNA or polypeptide product other than a regulator.

Subclone The process of breaking a cloned fragment into smaller fragments for further cloning.

Supercoiling The coiling of a closed duplex DNA in space so that it crosses over its own axis.

Superfamily A set of genes all related by presumed descent from a common ancestor, but now showing considerable variation.

Suppression mutation A second event eliminates the effects of a mutation without reversing the original change in DNA.

Suppressor A second mutation that compensates for or alters the effects of a primary mutation.

Synapsis The association of the two pairs of sister chromatids (representing homologous chromosomes) that occurs at the start of meiosis; the resulting structure is called a bivalent.

Synaptonemal complex The morphological structure of synapsed chromosomes.

Synonymous codons Codons that have the same meaning (specifying the same amino acid, or specifying termination of translation) in the genetic code.

Synonymous sites Sites in a coding region at which mutations have not changed the amino acid that is encoded.

Synteny A relationship between chromosomal regions of different species where homologous genes occur in the same order.

Synthetic genetic array analysis (SGA) An automated technique in budding yeast whereby a mutant is crossed to an array of approximately 5000 deletion mutants to determine if the mutations interact to cause a synthetic lethal phenotype.

Synthetic lethality This occurs when two mutations that are viable by themselves cause lethality when combined.

T cell receptor (TCR) The antigen receptor on T lymphocytes. It is clonally expressed and binds to a complex of MHC class I or class II protein and antigen-derived peptide.

T cells Lymphocytes of the T (thymic) lineage; they may be subdivided into several functional types. They carry TcR and are involved in the cell-mediated immune response.

TAFs The subunits of TFIID that assist TBP in binding to DNA. They also provide points of contact for other components of the transcription apparatus.

TATA-binding protein (TBP) The subunit of transcription factor TFIID that binds to the TAT box in the promoter and is positioned at the promoters that do not contain a TATA box by other factors.

TATA box A conserved A-T-rich octamer found about 25 bp before the startpoint of each eukaryotic RNA polymerase II transcription unit; it is involved in positioning the enzyme for correct initiation.

TATA-less promoter It does not have a TATA box in the sequence upstream of its startpoint.

T-DNA The ribonucleoprotein enzyme that creates repeating units of one strand at the telomere by adding individual bases to the DNA 3′ end, as directed by an RNA sequence in the RNA component of the enzyme.

Telomerase The ribonucleoprotein enzyme that creates repeating units of one strand at the telomere by adding individual bases to the DNA 3′ end, as directed by an RNA sequence in the RNA component of the enzyme.

Telomere The natural end of a chromosome; the DNA sequence consists of a simple repeating unit with a protruding single-stranded end.

Telomeric silencing The repression of gene activity that occurs in the vicinity of a telomere.

Temperate phage A bacteriophage that can follow the lytic or lysogenic pathway.

Template strand The DNA strand that is copied by the polymerase.

Teratoma A growth in which many differentiated cell types—including skin, teeth, bone, and others—grow in a disorganized manner after an early embryo is transplanted into one of the tissues of an adult animal.

Terminal protein A protein that allows replication of a linear phage genome to start at the very end. It attaches to the 5′ end of the genome through a covalent bond, is associated with a DNA polymerase, and contains a cytosine residue that serves as a primer.

Terminase An enzyme cleaves multimers of a viral genome and then uses hydrolysis of ATP to provide the energy to translocate the DNA into an empty viral capsid starting with the cleaved end.

Termination A separate reaction that ends a macromolecular synthesis reaction (replication, transcription, or translation) by stopping the addition of subunits and (typically) causing disassembly of the synthetic apparatus.

Termination codon One of the three codons (UAA, UAG, UGA) that signal the termination of translation of a polypeptide.

Terminator A sequence of DNA that causes RNA polymerase to terminate transcription.

Terminus A segment of DNA at which replication ends.

Ternary complex The complex in initiation of transcription that consists of RNA polymerase and DNA as well as a dinucleotide that represents the first two bases in the RNA product.

TF$_{II}$D The transcription factor that binds to the TATA sequence upstream of the startpoint of promoters for RNA polymerase II. It consists of TBP (TATA binding protein) and the TAF subunits that bind to TBP.

Thalassemia A disease of red blood cells resulting from lack of either α or β globin.

Third-base degeneracy The lesser effect on codon meaning of the nucleotide present in the third (3′) codon position.

Threshold cycle (C$_T$) The thermocycle number in a real-time PCR or RT-PCR reaction at which the product signal rises above a specified cutoff value to indicate amplicon production is occurring.

Ti plasmid An episome of the bacterium *Agrobacterium tumefaciens* that carries the genes responsible for the induction of crown gall disease in infected plants.

Tiling array An array of immobilized nucleic acid sequences which together represent the entire genome of an organism. The shorter each array spot is, the larger the total number of spots is required but the greater the genetic resolution of the array.

T$_M$ The theoretical melting temperature of a duplex nucleic acid segment into separate strands. TM is dependant on parameters including sequence composition, duplex length, and buffer ionic strength.

tmRNA A mRNA-tRNA hybrid that allows recycling of stalled ribosomes.

Tn Followed by a number, it denotes bacterial transposons carrying markers that are not related to their function, e.g., drug resistance.

Topoisomerase An enzyme that changes the number of times the two strands in a closed DNA molecule cross each other. It does this by cutting the DNA, passing DNA through the break, and resealing the DNA.

Trailer (3′ UTR) An untranslated sequence at the 3′ end of an mRNA following the termination codon.

TRAMP A protein complex that identifies and polyadenylates aberrant nuclear RNAs in yeast, recruiting the nuclear exosome for degradation.

***trans*-acting** A product that can function on any copy of its target DNA. This implies that it is a diffusible protein or RNA.

***trans*-acting sequence** DNA sequence coding for a product that can function on any copy of its target DNA. This implies that it is a diffusible protein or RNA.

Transcription Synthesis of RNA on a DNA template.

Transcription unit The sequence between sites of initiation and termination by RNA polymerase; it may include more than one gene.

Transcriptional interference (TI) The phenomenon in which transcription from one promoter interferes directly with transcription from a second, linked promoter.

Transcriptome The complete set of RNAs present in a cell, tissue, or organism. Its complexity is due mostly to mRNAs, but it also includes noncoding RNAs.

Transducing virus A virus that carries part of the host genome in place of part of its own sequence. The best known examples are retroviruses in eukaryotes and DNA phages in *E. coli*.

Transesterification A reaction that breaks and makes chemical bonds in a coordinated transfer so that no energy is required.

Transfection In eukaryotic cells, it is the acquisition of new genetic markers by incorporation of added DNA.

Transfer region A segment on the F plasmid that is required for bacterial conjugation.

Transfer RNA (tRNA) The intermediate in protein synthesis that interprets the genetic code. Each molecule can be linked to an amino acid. It has an anticodon sequence that is complementary to a triplet codon representing the amino acid.

Transformation In bacteria, it is the acquisition of new genetic material by incorporation of added DNA.

Transforming principle DNA that is taken up by a bacterium and whose expression then changes the properties of the recipient cell.

Transgenic animals Animals created by introducing DNA prepared in test tubes into the germline. The DNA may be inserted into the genome or exist in an extrachromosomal structure.

Transition A mutation in which one pyrimidine is replaced by the other, or in which one purine is replaced by the other.

Translation Synthesis of protein on an mRNA template.

Translational positioning The location of a histone octamer at successive turns of the double helix, which determines which sequences are located in linker regions.

Translocation The reciprocal or nonreciprocal exchange of chromosomal material between nonhomologous chromosomes.

Transmembrane region (domain) The part of a protein that spans the membrane bilayer. It is hydrophobic and in many cases contains approximately 20 amino acids that form an α-helix.

Transposase The enzyme activity involved in insertion of transposon at a new site.

Transposition The movement of a transposon to a new site in the genome.

Transposon A DNA sequence able to insert itself (or a copy of itself) at a new location in the genome without having any sequence relationship with the target locus.

Transversion A mutation in which a purine is replaced by a pyrimidine or vice versa.

tRNA$_f^{Met}$ The special RNA used to initiate polypeptide translation in bacteria. It mostly uses AUG, but can also respond to GUG and CUG.

tRNA$_m^{Met}$ The bacterial tRNA that inserts methionine at internal AUG codons.

True activator A positive transcription faction that functions by making contact, direct or indirect, with the basal apparatus to activate transcription.

True reversion A mutation that restores the original sequence of the DNA.

U3 The repeated sequence at the 3′ end of a retroviral RNA.

U5 The repeated sequence at the 5′ end of a retroviral RNA.

UAS (upstream activating sequence) The equivalent in yeast of the enhancer in higher eukaryotes and is bound by transcriptional activator proteins.

Underwound B-form DNA that has fewer than 10.5 base pairs per turn of the helix.

Unequal crossing-over (nonreciprocal recombination) It results from an error in pairing and crossing-over in which nonequivalent sites are involved in a recombination event. It produces one recombinant with a deletion of material and one with a duplication.

Unidentified reading frame (URF) An open reading frame with an as yet undetermined function.

Unidirectional replication The movement of a single replication fork from a given origin.

Uninducible A mutant in which the affected gene(s) cannot be expressed.

UP element A sequence in bacteria adjacent to the promoter, upstream of the −35 element, that enhances transcription.

UPF proteins A set of protein factors that target nonsense-mediated decay (NMD) substrates for degradation.

Upstream Sequences in the opposite direction from expression.

Up mutation A mutation in a promoter that increases the rate of transcription.

Upstream activating sequence (UAS) The equivalent in yeast of the enhancer in higher eukaryotes; a UAS cannot function downstream of the promoter.

V gene A sequence coding for the major part of the variable (N-terminal) region of an immunoglobulin chain.

Variable number tandem repeat (VNTR) Very short repeated sequences, including microsatellites and minisatellites.

Variable region (V region) An antigen-binding site of an immunoglobulin or T cell receptor molecule. They are composed of the variable domains of the component chains. They are coded by V gene segments and vary extensively among antigen receptors as the result of multiple, different genomic copies and of changes introduced during synthesis.

Vector An engineered DNA molecule used to transfer and propagate various insert DNAs.

Vegetative phase The period of normal growth and division of a bacterium. For a bacterium that can sporulate, this contrasts with the sporulation phase, when spores are being formed.

Viral superfamily Transposons that are related to retroviruses. They are defined by sequences that code for reverse transcriptase or integrase.

Viroid A small infectious nucleic acid that does not have a protein coat.

Virulent mutations Phage mutants that are unable to establish lysogeny.

Virulent phage A bacteriophage that can only follow the lytic cycle.

Virusoid (satellite RNA) A small infectious nucleic acid that is encapsidated by a plant virus together with its own genome.

Wobble hypothesis The ability of a tRNA to recognize more than one codon by unusual (non-G-C, non-A-T) pairing with the third base of a codon.

Xeroderma pigmentosum (XP) A disease caused by mutation in one of the *XP* genes, which results in hypersensitivity to sunlight (particularly ultraviolet light), skin disorders and cancer predisposition.

Z-ring See **Septal ring**.

Zinc finger A DNA-binding motif that typifies a class of transcription factor.

Zipcode (or localization signal) Any of the number of mRNA cis elements involved in directing cellular localization.

Zoo blot The use of Southern blotting to test the ability of a DNA probe from one species to hybridize with the DNA from the genomes of a variety of other species.

Index

synonymous, 706
  termination, 36–37, 37f, 686–689, 688f, 689f,
    706, 706f
  translation of, 705–706
  in translation regulation, 760
  usage biases in, 185–186
Codon usage, 760
Cognate tRNAs, 717
Cohesins, 361–362
  chromosome condensation and, 840–843,
    841f, 842f
Coincidental evolution, 148
Cointegrates, 427–428, 427f, 428f
ColE1 plasmid, 313–315, 313f, 314f, 315f
Colinearity of gene and protein, 37, 37f
Colorimetric detection, 67–68
Commitment complex, 584, 585f
Comparative genomics, 161
Compatibility groups, 312–315, 313f, 314f, 315f
Complement, 464–465
  major histocompatibility, 494
Complementarity
  in guide RNA, 658
  hybridization and, 17
  in recombination, 33
  in single-strand annealing, 359
Complementary base pairing, 10–11, 11f,
    14–15, 15f
Complementary sequences, 52–53, 54f
Complementation
  interallelic, 745
  negative, 745
  in vitro, 322
Complementation test, 29–30, 30f
Complex mRNA, 134
Composite transposons (Tn), 425, 425f
Concerted evolution, 148
Condensation, DNA
  condensins in, 840–843, 841f, 842f
  histone phosphorylation and, 816–818, 818f
  partition and, 309–310, 309f
  in phage heads, 192–193, 193f
Condensins, 840–843, 841f, 842f
Conditional knockouts, 77
Conditional lethals, 322
Conjugation, bacterial
  F plasmid transfer in, 289–290, 289f, 290f
  single-stranded DNA transfer in, 290–292, 291f
  T-DNA transfer and, 295–297, 296f
Connecting domain, 801–802, 802f
Consensus sequences, 154, 154f
  in immune recombination, 473, 473f
  in phages, 786
  promoter mutations and, 517
  in promoters, 515
  sigma factor recognition of, 517–518, 518f
Conservation
  in matrix attachment regions, 198f, 199
  promoter sequence, 515
Constitutive expression, 741–742
  cis-acting mutations and, 744, 744f
Constitutive genes, 134
Constitutive heterochromatin, 200, 838
Constitutive mutants, 744, 744f
Constitutive mutations, 745
Constrained supercoiling, 196, 196f
Context, in translation, 675–676
Control, allosteric, 743, 862
Controlling elements, 430, 431f
Conventional phenotype, 94

Cooperative binding, 782–783, 782f, 783f
Copia, 448, 450
Copy choice, 441–442, 442f
Copy number, 310
Cordycepin, 601
Core DNA, 223f, 224, 234–235
Core enzymes, 511, 511f
  movement of in transcription, 523–525,
    523f, 524f
  in promoter escape, 522–523, 523f
  in RNA polymerase contact, 517–520, 518f,
    519f, 520f
  sigma factor competition and, 531–533, 532f
Core histones, 225–227, 226f
Corepressors, 739
Core promoters, 548, 551
Core sequences, 375
Cotranslational protein degradation, 872, 873f
Counterselectable markers, 76, 76f
Countertranscripts, 314
Coupled transcription/translation, 538–540, 539f,
    540f, 737
Covalent modification, nucleosome, 228–231,
    229f, 230f, 231f
CoxII gene, 657–658, 658f
CpDNA, 112
CpG islands, 567–569, 568f
  methylation of, 843–846, 844f, 845f
CPSF75, 604f, 605
CREB-binding protein (CBP), 812–813, 813f
C (constant) region, 468–469, 468f
  in heavy and light chains, 472
Cre/lox system, 76–77, 77f, 78f
  in lambda integration, 379
  in site-specific recombination, 307–308,
    307f, 308f
  targeted recombination and knockout in,
    383–386
Creutzfeldt–Jakob disease (CJD), 23, 853
Crick, Francis, 10, 11
Criminal forensics
  polymerase chain reaction in, 64
CRISPRs (clusters of regularly interspersed short
    palindromic repeats), 868, 868f
Cro genes, 775–776, 775f
  in antitermination, 776–777, 777f
Cro repressor
  in lysogeny/lytic cycle balance, 790–792, 791f
  in lytic infection, 789–790, 790f
Crossovers, 33–34, 33f, 34f
  branch migration in, 354–355, 354f
  chiasmata in, 353
  cointegrates and, 428, 428f
  double-strand breaks and, 363
  fixation for identical repeats in, 147–150,
    148f, 149f
  gene conversion and, 356–357
  in minisatellites, 157
  in nonreplicative transposition, 428–430
  recombinant DNA, 355
  in telomeric regions, 215, 215f
  unequal, 141, 141f, 142–144, 142f, 147
Crown gall disease, 292–293, 292f
CRP (catabolite repressor protein), 752–755, 753f
Cryptic satellites, 151
Cryptic unstable transcripts (CUTs), 632
  in transcription control, 866, 866f
Cse4, 210
CSR. See Class switch DNA recombination (CSR)
CstF, 600

CTD. See Carboxy-terminal domains (CTDs)
CtDNA, 112
C-terminal domains (CTDs), 510
  lambda repressor and, 778–779, 779f
  sigma factor and, 520
CTLs, antigen-presenting cells and, 490
CUCU sequence, 648–649, 649f
Cut-and-paste, 425, 427
CUTs. See Cryptic unstable transcripts (CUTs)
CUUAGGC, 729, 729f
C-value, 175, 175f
C-value paradox, 176
Cycle-dependent elements (CDEs), 209–210, 209f
Cyclic AMP (cAMP), 752–755, 753f
Cyclic phosphodiesterase, 607, 608
CyL mutations, 786–787, 787f
CyR mutations, 786–787, 787f
Cys$_2$/Cys$_2$ fingers, 805, 805f
Cysteine, 715–716
Cystic fibrosis, 102–103
Cytidine
  deamination of, 482
  in modified bases, 711, 711f
Cytidine deaminases, 845
Cytological maps, 204
Cytoplasm, mRNA translation in, 593, 593f
Cytoplasmic cap-binding protein, 626–627
Cytoplasmic poly(A) binding protein
    (PABP I), 602
Cytoplasmic surveillance systems, 633–635, 633f,
    634f, 635f
Cytosine (C), 8, 10
  in anticodon–codon pairing, 712–713, 712f
  melting temperature and, 16
  spontaneous deamination of, 21–22, 21f, 22f
Cytosol, 667
Cytotypes, 437–438, 437f

D

Dat, DnaA binding to, 271–272
Daughter cells
  compatibility groups and, 312–315, 313f,
    314f, 315f
  mitochondria assignment to, 316, 316f
  partition of, 308–312
  recombination-repair systems in, 405–406, 405f
  septum formation in, 302–303, 303f
Daughter DNA strands, 12–13, 12f
DdNTPS (dideoxynucleotides), 58, 58f
D-D rearrangements, 490
Deacylated tRNA, 669
  in translocation, 684–685, 684f
Deadenylases, 625–626
Deadenylation, 625–627, 626f, 629f
  microRNA inhibition of, 872, 873f
Deadenylation-independent decapping, 628,
    628f, 629f
Deamination, 21–22, 21f, 22f
  in class switching, 482
  mismatch repair and, 403
  repair systems for, 395, 395f
  in RNA editing, 657, 657f
Decapping
  deadenylation-independent, 628, 628f, 629f
  enzymes in, 626–627
  microRNA inhibition of, 872, 873f
Decay, mRNA, 621–622, 622f
  no-go, 633f, 635
  nonsense-mediated, 633–634, 633f, 693, 693f
  nonstop, 633f, 635

Enzymes. *See also specific enzymes*
  in catalytic RNA, 643
  core, in holoenzymes, 511, 511f
  Cre, 76–77, 77f
  deamination, 657
  decapping, 626–627
  DNA polymerases, 324–326, 325f
  genetic information on, 94
  linking number and, 10
  in modified bases, 710–711
  movement of in transcription, 523–525, 523f, 524f
  one gene : one enzyme hypothesis of, 28–29
  processivity of, 327
  in replication, 13–14, 13f, 14f
  restriction, 45–46, 45f
    satellite DNA and, 154–155, 154f, 155f
  in septum formation, 303
  in western blotting, 67–68
Enzyme units, 333
Epigenetic effects, 797–798, 828–860
  condensins in, 840–843, 841f, 842f
  CpG island methylation in, 843–846, 844f, 845f
  definition of, 829
  in disease inheritance, 853–854, 854f
  heterochromatin in
    histone interactions with, 832–835, 833f, 834f
    propagation of, 831—832, 831f
  imprinting, 846–848, 847f, 848f
  inheritance of, 848–851, 850f, 859f
  polycomb group proteins in, 835–837, 836f
  in prions, 851–854, 851f, 852f, 854f
  transgenerational, 850–851
  X chromosome global changes in, 837–840, 838f, 839f, 840f
Epimutation, 850–851
Episomes, 283, 770
Epitopes, 467
Epitope tags, 67
ε DNA polymerase, 339, 340, 340f
Error-prone polymerases, 325, 339
Error-prone repair, 402
Error-prone synthesis, 402
Error-prone systems, 394
Esc-E(z), 837
ESEs (exonic splicing enhancers), 596–598, 597f
ESSs (exonic splicing silencers), 697–698, 697f
Ester bonds, 44, 44f
ESTs (expressed sequence tags), 110
ETS (external transcribed), 609–612, 610f, 611f
Euchromatin, 126, 151, 199–200
  initiation in, 339
Eukaryotes, 3
  activators in, 798–801, 799f, 800f, 801f
  alternative splicing in, 594–596, 594f, 595f, 596f
  basal apparatus in, 803–804, 803f, 804f
  bioswitches in, 863–864, 864f
  chromatids in, 199, 199f
  chromatin remodeling in, 806–809, 807f, 808f
  chromosome segregation in, 205–206
  in cloning vectors, 49–51, 49f, 50f, 51f
  condensins in, 840–843, 841f, 842f
  conservation of organization in, 82–84, 83f, 84f
  CpG island methylation in, 843–846, 844f, 845f
  DNA in, 6–7, 7f
    repair, 410–412, 411f, 412f
    scaffolds, 197–198
  DNA polymerases in, 325–326, 338–340
  excision repair systems in, 397, 398f, 399
  gene activation in, 818–819, 819f
  gene expression in, 38–39, 133–134

  gene number of, 121–123, 121f
  gene organization in, 739–740
  gene size in, 88
  genomes of, 104–105, 104f
    size of, 176–178, 177f
  histones in
    acetylation, 811–814, 812f, 813f, 814f
    methylation, 814–815, 815f
    phosphorylation, 816–818, 818f
  homologous recombination in, 352, 371–374, 372f
  imprinting in, 846–848, 847f, 848f
  initiation factors in, 678–681, 679f, 680f, 681f
  interrupted genes in, 28, 80
  licensing factor in, 275–278, 276f, 277f
  microRNA regulation in, 869–872, 870f, 871f
  minimum number of genes in, 119, 119f, 120
  mismatch repair in, 403–405, 405f
  mRNA in, 620, 620f
    cell localization of, 636–638, 636f, 638f
    degradation of, 625–627, 626f, 627f
    mRNPs, 622–623, 623f
  mutation rates in, 18
  nucleic acid length in, 190–191, 191f
  nucleosome organization in, 809–811, 810f, 811f
  polycomb group proteins in, 835–837, 836f
  promoter activation in, 815–816, 816f
  promoter clearance and elongation in, 560–562, 561f, 562f
  release factors in, 688
  repair systems in, 392
  replication and cell cycle in, 300
  replication enzymes in, 338–340, 339f
  replicons in, 272–275, 274f, 275f
  repressors in, 798–801, 799f, 800f, 801f
  RNA polymerases in, 245–246, 246f
    subunits, 549–550, 550f
  rRNA production in, 609–612, 610f, 611f
  satellite DNA in, 150–151
  SOS response in, 413–414, 413f
  transcription in, 546–572
    regulation of, 795–827
  translation initiation site scanning in, 677–678, 677f
  X chromosome global changes in, 837–840, 838f, 839f, 840f
Euplotes crassus, 716
Euplotes octacarinatus, 714, 714f
Evolution, genome, 159–188
  biases in, 185–186
  coincidental, 148
  concerted, 148
  gene duplication in, 178–179, 178f
  genome duplication in, 182–184, 183f
  genome size and, 175–176, 175f, 176f
  globin clusters in, 179–181, 179f
  of interrupted genes, 172–175
  morphological complexity and, 176–178, 177f
  mutation and sorting mechanisms in, 161–163, 162f, 163f
  neutral substitution measurement and, 170–171
  pseudogenes in, 181–182, 181f, 182f
  RNA in, 575
  selection measurement and, 163–167, 165f, 166f, 167f
  sequence divergence in, 167–170, 168f
  transposable elements in, 184
Evolutionary biology
  conservation of gene organization and, 91–93
  polymerase chain reaction in, 64
  synteny and, 109–110

Excision, 375
  imprecise, 427
  in lambda integration, 379
  in phages, 770
  precise, 427
  repair systems, 393–394, 393f
    base, 393
      in *E. coli*, 396–397, 396f
      in eukaryotes, 397, 398f, 399
      nucleotide, 393–394
    transposons and, 426–427, 426f
Exclusion, allelic, 474–476, 475f
Exonic splicing enhancers (ESEs), 596–598, 597f
Exonic splicing silencers (ESSs), 597–598, 597f
Exon junction complex (EJC), 592–593, 592f
  in mRNA regulation, 634, 634f
Exons, 39, 80
  alternative splicing of, 89, 89f, 594–596, 594f, 595f, 596f
  base composition of, 82
  definition of, 585f, 586
  duplication of in genome evolution, 178–179
  evolutionary role of, 93
    in interrupted genes, 172–175, 172f
  in the human genome, 126, 128, 128f
  in interrupted genes, 81–82, 81f
  negative selection and, 84–85
  in P elements, 436–437, 436f
  positive selection and, 85–86
  protein-coding genes and conservation of, 105–108, 106f, 107f, 108f
  protein functional domains and, 90–91, 90f
  shuffling, 172–173
  size distribution of, 87–88, 87f
  splice site, 578
  in T cell receptors, 489–490, 489f
  trapping, 107–108, 108f, 173
  in tRNA splicing, 605–608, 605f, 606f, 607f
Exonucleases, 44, 44f
  DNA polymerases in, 326, 326f
  in mispaired base removal, 329
  in replication, 13–14, 13f, 14f
  in RNA editing, 657–660, 658f, 659f
Exoribonucleases, 621, 621f
Exosomes, 627
  in RNA surveillance systems, 631–632, 631f, 632f
Expressed sequence tags (ESTs), 110
Expression vectors, 49
Exteins, 660–661, 660f, 661f
Extended −10 elements, 515, 516, 516f
External transcribed spacers (ETS), 609–612, 610f, 611f
Extracellular defenses, 95
Extrachromosomal replicons, 282–298
  in crown gall disease, 292–293, 292f
  F plasmid transfer in, 289–290, 289f, 290f
  rolling circles in, 286–289, 287f, 288–289, 288f
  terminal proteins in, 285–286, 285f, 286f
Extranuclear genes, 111

F

Fab-7 element, 254, 254f
FACT (facilitates chromatin transcription), 247–248, 248f, 821
Facultative heterochromatin, 200, 838
Fast-clock effects, 166
FEN1 (flap endonuclease 1), 338, 338f, 340
Ferments, 94
F (fertility) factor, 292

Helix-loop-helix (HLH) motif, 806, 806f
Helix-turn-helix (HTH) model, 780–782, 780f, 781f, 782f
  Cro repressor in, 789, 790f
  in DNA-binding domains, 805–806, 806f
Helper T (T$_h$) cells, 464–465, 465f
Helper viruses, 444–445, 444f, 445f
Hemimethylated DNA, 270, 270f
  in epigenetic effects, 830, 830f
Hemimethylated sites, 843–844, 844f
Hemizygous embryos, 847
Hereditary nonpolyposis colorectal cancer (HNPCC), 405
Hershey, Alfred, 6
Heterochromatin, 199–200, 200f
  around centromeres, 206
  constitutive, 200, 838
  in epigenetic effects, 830, 830f
  facultative, 200, 838
  histone acetylation in, 814
  histone interactions with, 832–835, 833f, 834f
  initiation in, 339
  insulators and, 251, 251f
  microRNAs in formation of, 875, 876f, 877
  satellite DNA in, 151
Heterochronic genes, 874, 874f
Heteroduplex DNA, 354, 354f
  extension of, 373
  in interallelic recombination, 355–357, 356f
  sigma factor and, 519
  single-strand assimilation, 368, 368f
Heterogeneous nuclear RNA (hnRNA), 576, 597–598
Heteroplasmy, 315
Heterozygotes, 31
HflA protein, 786, 786f
Hfq protein, 867–868
Highly repetitive DNA, 150
High-mobility group (HMG) proteins, 222
HIL-1 kinase, 818, 818f
Him genes, 378
Hinge helix, 749
Histone acetyltransferases (HATs), 812–813, 813f
Histone code hypothesis, 229–231
Histone deacetylases (HDACs), 812–813, 813f
Histone demethylases, 837
Histone downstream elements (HDEs), 604–605, 604f
Histone fold, 226–227, 227f, 228f
Histone H3, 207–208, 207f
Histones
  acetylation of, 811–814, 812f, 813f, 814f
  biogenesis of, 604–605, 604f
  in chromatin remodeling, 807–809, 807f, 808f
  in chromatin replication, 239–241
  core, 225–227, 226f
  covalent modification of, 228–231, 229f, 230f, 231f
  in CpG islands, 567–569
  demethylation and, 435
  exchange of, 809
  in gene activation, 797
  H2B, 562
  heterochromatin interactions with, 832–835, 833f, 834f
  linker, 222, 227–228, 237
  methylation of, 814–815, 815f
  modification of in chromatin repair, 410–412, 411f, 412f
  in mRNA degradation, 629, 629f

in nucleosomes, 225–228, 226f, 227f, 228f
  octamers
    DNA wrapping around, 234–235, 234f
    nucleosome positioning and, 243–244
    in nucleosomes, 224, 225
    in transcription, 246, 246f
  phosphorylation of, 816–818, 818f
  tails of, 227–228, 228f
  3′ end formation, 604–605, 604f
  in 30 nm fiber, 237
Histone tails, 222
  in 30 nm fiber, 237
  acetylation of, 849–850, 849f
Histone variants, 231–234, 232f, 233f
HIV
  budding in, 440f
  recombination in, 442
HLA-DP, 492, 494
HLA-DQ, 492, 494
HLA-DR, 492, 494
HLH. *See* Helix-loop-helix (HLH) motif
HMG (high-mobility group) proteins, 222
HML, 380–382, 381f, 382f
  heterochromatin formation and, 832f, 833–834, 833f
HMLH1, 410
HMR, 380–382, 381f, 382f
  heterochromatin formation and, 832f, 833–834, 833f
HnRNA, 576, 597–598
HnRNP, 576
HO endonuclease, in mating type switching, 381–382, 382f
Holliday junctions, 308, 308f
  dissolution of, 373–374, 374f
  double-strand breaks and, 363
  in lambda integration, 378
  migration of, 330
  in recombination, 355
  resolution of, 369–371, 370f
  in single-strand assimilation, 369
Holoenzymes, 325, 327
  in basal apparatus interactions, 804, 804f
  clamps in, 334–337, 334f, 335f, 336f
  core enzyme in, 511, 511f
  in promoter escape, 522–523, 523f
  promoter recognition in, 512–514, 513f
  RNA polymerase subunits in, 510
  sigma factor in, 511, 511f
  subcomplexes in, 333–334, 334f
Homing, intron, 652
Homing endonuclease genes (HEGs), 651–652, 651f
Homing introns, 652, 653, 653f
  in inteins, 661, 661f
Homologous genes (homologs), 92, 92f
Homologous recombination, 343–344, 350–351
  in antigenic variation, 383, 383f
  definition of, 351
  double-strand break model of, 353–355, 354f
  eukaryotic genes in, 371–374, 372f
  experimental adaptations of, 384–386, 384f, 385f
  recombination repair in, 408–409, 408f
  single-strand annealing model of, 359, 359f
  synaptonemal complex in, 360–365
  transposons in, 426–427, 426f
Homologs. *See* Homologous genes (homologs)
Homozygotes, 31, 31f
HOP2 gene, 364

Horizontal transfer, 95, 121
Hormone receptors (HRs), 811, 811f
Host defenses, 95
Hotspots, 20–23, 21f, 22f
  gene conversion and, 357
  replication fidelity control and, 327
  somatic hypermutation, 483–484
Hot start techniques, 60–61
Housekeeping genes, 134
  CpG islands and, 568
  in transcription, 549
HP1 (heterochromatin protein 1), 832–834, 832f, 833f
Hpg mice, 73–74, 74f
Hsp (heat-shock protein) 70, 252, 252f
HTH (helix-turn-helix) model, 746, 746f
Human genome
  CpG islands in, 567–569, 568f
  gene and sequence distribution in, 127–128
  gene duplication in evolution of, 178–179, 178f
  minisatellites in, 157
  neutral substitution in, 170–171
  number of essential genes in, 132
  number of genes in, 125–127, 126f, 127f
  pseudogenes in, 125
  repair systems in, 392–393, 393f
  ribosomal protein pseudogenes in, 182, 182f
  size of, 176–178, 176f, 177f
  Y chromosome in, 129–130, 129f
Humoral response, 461
HU protein, 195
HU protein n replication initiation, 323
Hybrid dysgenesis, 435–436, 435f, 437f
Hybridization
  definition of, 17
  filter, 17, 17f
  FISH, 53, 54f
  microarrays in, 70
  nucleic acid, 16–17, 16f, 17f, 52–53, 54f
  polytene chromosome, 204
  satellite DNA, 151
  in situ, 53, 54f, 204
Hybridomas, 483–484
Hybrid state model, 684–685, 684f, 694
Hydrogen bonds, 10–11, 11f
Hydrolysis
  ATP, 330
    spliceosome assembly pathway, 586–588
  GTP, in translation accuracy, 726–728, 727f
Hydrops fetalis, 143
Hydroxyls, 8
Hypersensitive sites, 248–251, 249f, 250f, 251f
  in chromatin remodeling, 808
Hypogonadal mice, 73–74
Hypoxanthine, 401

I

Iab-6 element, 254
Iab-7 element, 254
Icosahedral symmetry, 192
ICRs. *See* Internal control regions (ICRs)
ICRs (imprinting control regions), 848, 848f
Identity set, tRNA, 717–718
IFs. *See* Initiation factors (IFs)
Igf2, 848, 848f
IGF-II (insulin-like growth factor II), 846
Ig genes, avian, 485–486, 486f
IgH
  chain expression, 479–480, 479f
  consensus sequences and, 473, 473f

for single-base changes, 394, 395f
single-strand annealing, 359
SOS system, 413–414, 413f
for structural distortions, 395, 395f
Repeats
crossover fixation for identical repeats in, 147–150, 148f, 149f
direct, 424, 441, 441f
function of, 208
in heterochromatin, 200
inverted terminal, 423, 424
in minisatellites, 156–157, 156f
rRNA, 145–147, 145f, 146f
in satellite DNA, 150–156
arthropod, 152, 152f
mammalian, 152–156
sequence divergence in, 170–171, 171f
telomerase synthesis of, 213–214
in telomeres, 210–211, 210f
Repetitive DNA sequences, 104–105, 104f
Replicases, DNA, 325
error-prone repair and, 402
replication fidelity and, 328
Replication, 320–347. *See also* Repair systems, DNA
acetylation in, 231f
bacterial, 299–319
base pairing in, 12, 12f
bidirectional, 266
break-induced, 359–360, 360f
cell cycle linkage with, 264–265, 299–319
central dogma on, 14–15, 14f, 15f
chromosome segregation in, 307–308, 307f, 308f
clamps in, 334–337, 334f, 335f, 336f
D loops in, 278–279, 279f
doubling time and, 301–302
elongation in, 321–322
error repair in, 406–407, 407f
FACT in, 247–248, 248f
fidelity control in, 326–328, 327f
FtsZ in, 304–306, 305f
gene activation and, 818–819
helicases in, 330–331, 330f
initiation of, 300–301, 300f, 321
DNA polymerases in, 338–340
lesion bypass in, 342–344
licensing factor in, 275–278, 276f, 277f
linear DNA ends in, 284–286, 284f, 285f, 286f
mitochondrial, 315–316, 316f
multicopy control of, 265
mutations and cell shape in, 304, 304f
Okazaki fragment linkage in, 337–338
origin sequestering in, 271–272
partition in, 308–312, 311f
phage T4, 340–342, 341f
polymerases in, 13–14, 13f, 14f
priming in, 331–332, 331f
repair systems for, 395, 395f
replicons in, 263–281
extrachromosomal, 282–298
rolling circle, 279, 286–289, 287f, 288f
semiconservative, 12–13, 12f, 265–267, 324
semidiscontinuous, 330
septum formation in, 302–303, 303f
septum location in, 306–307, 306f
single-copy control of, 265
in single-copy plasmids, 310–312, 311f
single-strand binding proteins in, 330–331, 330f
strand synthesis modes in, 329–330, 329f
synthesis coordination in, 33f, 332–333
telomeres in, 211

unidirectional, 266
viroid, 23, 24
Replication bubbles, 265–266, 266f, 267f
mapping movement of, 267–268, 267f, 268f
Replication-coupled (RC) pathway, 240
Replication-defective viruses, 444–445, 444f, 445f
Replication forks, 13–14, 13f, 14f
in bacterial replicons, 268–270, 268f, 269f
in chromatin replication, 239–240, 241f
creation of, 322–324, 323f
DNA polymerases in, 338–340, 340f
doubling time and, 301–302, 302f
error repair in, 406–407, 407f
gene activation and, 818–819, 819f
lesion bypass in, 342–344
mismatch repair and, 404, 404f
movement of, 266
mapping, 267–268, 267f, 268f
in multiforked chromosomes, 302
in phage rolling circle replication, 288
reactivation of, 343–344
stalled, 406–407, 407f, 408f
traps, 269
Replication-independent (RI) pathway, 240, 241
Replication slippage, 157, 157f
repair of, 404–405
Replicative transposition, 425, 425f
cointegrates in, 427–428, 427f, 428f
Replicons, 263–281
archaeal, multiple, 272
attachment of, 308–309
bacterial, 268–270
circular bacterial, 268–269, 268f, 269f
compatibility groups in, 312–315, 313f, 314f, 315f
eukaryotic, 272–274, 273f, 274f, 339
extrachromosomal, 282–298
licensing factor and, 275–278, 276f, 277f
linear DNA ends in, 284–286, 284f, 285f, 286f
linear vs. circular, 265–267, 266f, 267f
multiple, 272–274, 273f, 274f
origins of, 264
isolation in yeast, 274–275, 274f
mapping, 267–268, 267f, 268f
methylation in, 270, 270f
sequestering in, 271–272
rolling circle replication of, 286–289, 287f, 288f
terminus of, 264
Replisome, 321–322, 341
Reporter genes, 49–51, 50f, 51f
in fluorescence resonant energy transfer, 62–64, 63f
in two-hybrid assay, 802–803, 802f
Repressible genes, 739
Repressible operons, 755–756, 755f
Repression
attenuation and, 756–757
catabolite, 752–755, 753f, 754f
CpG islands in, 567–569, 568f
deacetylation in, 813–814
definition of, 739
glucose, 822
low-affinity sites and, 751–752, 751f, 752f
in yeasts, 813–814, 814f
Repressors, 564
antirepressors and, 800
competition with, 801, 801f
in eukaryotes, 798–801, 799f, 800f, 801f
helix-turn-helix model in, 780–782, 780f, 781f, 782f
lac, 741, 742–743, 743f

in lambda phages, 777–778, 777f, 778f
cooperative binding in, 782–783, 782f, 783f
immunity region definition by, 778–779, 779f
in lysogeny initiation, 785–787, 785f, 787–788, 787f, 788f
low-affinity sites and, 751–752, 751f, 752f
in lytic infection, 789, 790f
Mad:Max, 814
masked, 801, 801f
polycomb group proteins, 835–837, 836f
in translation regulation, 760–761, 761f
trp, 755–756, 755f
Resolution, recombination, 355, 428
Resolvases, 373–374
cointegrates and, 428
in mismatch repair, 403–405
in repair systems, 394
in replicative transposition, 425–426, 425f
in stalled replication forks, 407, 407f
Resolvasome complexes, 370–371
Restriction endonucleases, 43, 45–46, 45f
Restriction enzymes
digestion, 256–257, 256f
satellite DNA and, 154–155, 154f, 155f
Restriction fragment length polymorphism (RFLP), 102, 102f
genetic mapping with, 102–104, 103f
Restriction maps, 46, 46f, 100–101
chromosomal walk and, 107
Restriction markers, 102–104, 103f
Restriction polymorphisms, 102–104, 103f
Restriction sites, nucleosome positioning and, 242–245
Retroposons, non-LTR, 449–450, 449f, 450f
Retrotransposons, 422, 422f
classes of, 449–451, 449f, 450f
copia, 448
LINEs, 449, 449f, 450, 450f, 451–453, 452f, 453f
LTR, 445, 449–450, 449f, 450f
non-LTR, 449, 449f, 450, 450f
SINEs, 449–451, 449f, 450f
Ty elements, 445–447, 446f, 447f
Retroviruses, 422, 422f, 438–447
budding in, 439, 440f
life cycle of, 438–439
polyprotein coding in, 439, 440f
reverse transcription in, 440–443, 441f, 442f, 443f
transducing, 444–445, 444f, 445f
Ty elements and, 445–447, 446f, 447f
Reverse transcriptase, 15
Group II intron multifunction proteins and, 652, 652f
in retroviruses, 438, 439
telomerase, 213–214
Reverse transcription, 15, 78. *See also* Retrotransposons; Retroviruses
in LINEs, 451–453, 452f, 453f
in polymerase chain reaction, 61–62
priming for, 331
processed pseudogenes from, 181
Reverse transcription polymerase chain reaction (RT-PCR), 109
Reversion mutations, 20, 20f
Revertants, 20, 20f
RFLP (restriction fragment length polymorphism), 102, 102f
Rho-dependent terminators, 525, 526, 529
Rho factor, 526, 527–529, 527f, 528f
Ribonucleases, 621, 621f

Ribonucleoprotein particles (RNPs), 622–623, 623f
  in ribosomes, 667, 667f
  as tRNA precursors, 709–710, 709f
  tRNA synthetases, 716–718
Ribonucleoproteins, 213–214
Ribosomal protein (RP) pseudogenes, 182, 182f
Ribosomal proteins, 667, 667f, 761–763,
    762f, 763f
  synthesis of, 761–763, 762f, 763f
Ribosomal RNAs (rRNAs), 4, 39, 667–668, 667f
  in antitermination, 538
  in catalysis, 697–698, 698s
  chloroplast genome and, 114, 114f
  domains in, 689–690
  ribosomal structure changes in, 698–699
  in ribosomal subunits, 689–692
  ribosome active centers in, 692–695, 693f, 694f
  r-protein binding to, 762–763, 763f
  16S, 695–697, 695f, 696f
  small RNAs in production of, 609–612, 610f, 611f
  tandem repeats of, 145–147, 145f, 146f
  in translation, 695–697, 695f
  23S, 697–698, 698s
Ribosome-binding sites, 671–672, 671f, 672f
Ribosome recycling factor (RRF), 689, 689f
Ribosomes, 39
  active centers in, 692–695, 693f, 694f
  in attenuation, 756–757
  bypassing by, 730–731, 730f
  in coupled translation/transcription, 539–540
  elongation factor binding to, 685–686,
    685f, 686f
  fMET-rTNA_f controlled by, 675—677, 676f
  40S subunits in, 677–678, 677f
  rRNA in subunits of, 689–692
  16S rRNA and, 695–697
  stalled, 731, 758, 758f
  subunit structure changes in, 698–699
  in translation, 667–668, 667f
    accuracy influences of, 670–671, 670f,
      726–728, 727f
  in translocation, 669, 669f
  translocation of, 684–685, 684f
Riboswitches, 650, 650f, 863–864, 864f
Ribothymidine, 711, 711f
Ribozymes, 621
  in catalytic RNA, 643, 648–651, 648f, 649f, 650f
  5′ UTR, 863
  hammerhead, 654–656, 655f
  riboswitches and, 863–864, 864f
Rickettsia, 115, 120f
Rifampicin, 523
Rif operon, 762, 762f
Ri plasmids, 293
RISC (RNA-induced silencing) complex, 629, 870
  RNA interference and, 872–874, 873f
RITS (RNA-induced transcriptional silencing),
    867f, 877
RNA. *See also specific types*
  antisense, 314
  catalytic, 642–664
  central dogma on, 14–15
  defect surveillance system in, 631–632,
    631f, 632f
  editing, 644
  in evolution of interrupted genes, 173–174
  5.8S, 690
  in gene expression, 575
  genes for, 4, 4f
  guide, 657–660, 657f, 658f, 659f
  heterogeneous nuclear, 576

I, in compatibility regulation, 314–315,
    314f, 315f
  in IgH chain processing, 479–480, 479f
  nascent, 540, 540f
  nucleic acid detection, 52–53, 54f
  orphan, 612
  polymerases, 245, 246f
  primer, 314, 331
  regulatory, 861–880
  spliced leader, 599–600, 599f, 600f
  sugars in, 7–8, 7f
  telomerase and, 213–214, 214f
  in transposition, 421–422
  viral, 192, 192f, 194
  viroid, 23, 24
RNA-binding proteins (RBPs), 623, 867–868
RNA-dependent RNA polymerase (RDRP),
    867f, 877
RNAi. *See* RNA interference (RNAi)
RNA-induced transcriptional silencing (RITS),
    867f, 877
RNA interference (RNAi), 869
  in gene expression, 874–875
  gene knockouts with, 74–75
  pathways
    heterochromatin formation and, 832f,
      833–834, 833f
    RNA interference in, 872
RNA ligase, 606
RNA polymerases
  in antitermination, 537–538, 537f, 538f
  attenuation of, 756–757
  bacterial
    subunits in, 509–511, 510f, 511f
    termination in, 525–527, 526f
  in basal apparatus interactions, 804
  binding to promoter site, 508–509, 509f
  catabolite repression and, 754–755, 754f
  chromatin in binding to, 547–548
  crystal structure of, 523–525, 523f, 524f
  DNA promoter contact by, 517–520, 518f,
    519f, 520f
  enzyme movement model in, 523–525,
    523f, 524f
  footprinting, 520–522, 521f, 522f
  in heterochromatin formation, 875, 876f, 877
  I, 549–550, 550f
    promoter for, 551–552, 551f
  II, 135, 135f, 549–550, 550f
    basal assembly for, 558–559, 558f
    in chromatin immunoprecipitation, 72
    nucleosome-free regions in, 810–811, 810f
    regulation of, 549
    startpoint for, 554–555, 555f
    stuttering by, 561
  III, 549–550, 550f
    Alu family and, 451
    promoters for, 552–554, 552f, 553f
  lac repressor interaction with, 750–751, 750f
  mRNA half-life and, 621–622, 622f
  in phage immunity region, 778–779, 779f
  in promoter escape, 522–523, 523f
  in promoter recognition, 512–514, 513f, 514f,
    515–516, 516f
  in replication forks, 269
  restarts of in transcription, 525
  rho factor and, 526, 527–529, 527f, 528f
  in sporulation, 535–536
  subunits in, 549–550, 550f
  T7, 530–531, 531f
  in transcription, 506–507

  bubble creation by, 507–508, 508f
  termination in, 602–604, 603f
RNA processing, 38–39, 39f
RNA regulons, 623, 623f
RNases (ribonucleases), 13–14, 13f, 14f
  E, 624–625
  H, in ColE1, 313–314
  in mRNA degradation, 624–625
  MRP, 629, 629f
  P
    in catalytic RNA, 643, 653–654
    as tRNA precursor, 709–710, 709f
RNA silencing, 875
RNA splicing, 81
RNPs. *See* Ribonucleoprotein particles (RNPs)
RodA, in septum formation, 303
Rolling circle replication, 279, 286–289, 287f, 288f
Rotational positioning, 244, 244f
RPD3, 813–814, 814f
RpoB, 510, 529
RpoC, 510
RpoH, 532, 532f
RpoS gene, 867
R-proteins. *See* Ribosomal proteins
RRF. *See* Ribosome recycling factor (RRF)
RRNAs. *See* Ribosomal RNAs (rRNAs)
Rrn operons, 538
RSC complex, 412, 412f
RseA, 532–533, 533f
R segments, 441–442, 442f
RSS. *See* Recombination signal sequences (RSSs)
RTNA, Trp, 757–760
RT-PCR. *See* Reverse transcription polymerase
    chain reaction (RT-PCR)
Rtt106, 248
R-U5, 441, 441f, 442f
Rut sites, 527–529, 527f, 528f
RuvAB complex, 370, 370f

S

S10 operon, 762, 762f
Saccharomyces cerevisiae, 82
  centromeres in, 208
  GAL genes in, 819–822
  gene expression in, 135
  gene number of, 121–122, 122f
  genome size of, 112, 113f
  Holliday junction resolution in, 370
  homologous chromosomes in, 364
  interrupted genes in, 86, 86f
  licensing factor in, 278
  methylation in, 489
  nucleosome positioning in, 245
  number of essential genes in, 130–131, 131f
  point centromeres in, 208–209
  PTC detection in, 634
  recombination pathway adaptation experiments
    on, 385
  recombination repair in, 408, 408f
  replication origins in, 274–275, 274f
  replicons in, 273
  RNA-binding proteins in, 623
  RNA polymerase II in, 550, 550f
  shuttle vectors, 49, 50f
  silent and active loci in, 380–382, 381f, 382f
  snoRNAs in, 610, 610f
  SR proteins in, 584–586, 585f
  synaptonemal complex formation in, 362, 363f
  tRNA splicing in, 605–608, 605f, 606f, 607f
  Ty elements in, 445–447, 446f, 447f
Saccharomyces pombe, 834–835

Specificity loops, 531, 531f
SpH2B variant, 233–234
S phase, 272–273, 811–812
Spindle, mitotic, 206
Spliced leader RNA. *See* SL RNA (spliced leader RNA)
Splice junctions
  in alternative splicing, 594–596, 594f, 595f, 596f
  nuclear, 578
  reading, 578–580, 579f
Spliceosomes
  alternative, 589
  assembly pathway in, 586–588, 587f, 588f
  definition of, 582
  snRNAs in, 581–583, 582f
Splice recombinants, 370
Splicing, 39
  alternative, 89–90, 89f
  in interrupted genes, 81
  junctions in, 173
  protein, 99, 660–661, 660f, 661f
  splice junctions in, 578–580, 579f
Splicing, RNA, 575–617
  5′ cap in, 576–577, 577f
  alternative, in eukaryotes, 594–596, 594f, 595f, 596f
  alternative spliceosomes in, 589
  auto-, 590–591
  cleavage and polyadenylation in, 601–602, 601f
  definition of, 576
  enhancers and silencers in, 596–598, 597f
  gene expression and, 591–593, 592f, 593f
  nuclear, 590–591, 591f
  pre-mRNA, 580–581, 581f, 589–591, 590f, 591f
  pre-mRNA pathway in, 583–586, 583f, 584f, 585f
  snRNAs in, 581–583, 582f
  spliceosome assembly pathway in, 586–588, 587f, 588f
  trans-, 598–600, 598f, 599f, 600f
  tRNA, 605–608, 605f, 606f, 607f
  unfolded protein response in, 608–609
Splicing factors, 582
SPO1 phage, 533–534, 534f
SpoIIIE, 308
S. pombe fbp1 gene promoter, 558
Spontaneous mutations, 17, 18
  hotspots for, 21, 21f
  modified bases and, 21
SpoOA, 535
SpoOJ, 311–312
Sporulation, 310, 534–536, 535f, 536f
Spt6 protein, 248
S regions, in class switching, 481–482, 481f, 482f
SRG1, 865–866, 866f
SRNAs, 866–868, 867f, 868f
SR proteins, 584–586, 585f
  in splice site recognition, 597–598
SSA. *See* Single-strand annealing (SSA)
SSB. *See* Single-strand binding proteins (SSBs)
Stabilizing elements (SEs), 630, 630f
Stahl, Franklin, 13
Stalling, ribosome, 731, 758, 758f
Startpoints, transcription, 506, 515–516, 516f
  polymerase binding to, 548
  RNA polymerase II, 554–555, 555f
Steady state, 622
Stem-loop binding protein (SLBP), 604–605, 604f

Stem-loop structures
  in hammerhead ribozymes, 655, 655f
  in modified bases, 711, 711f
  mRNA, 620, 620f
Steroid receptors, 805, 805f
Stop codons. *See* Termination codons
Strand displacement, 285, 285f
Streptococcus pneumoniae
  antibody response to, 461
  transformation of, 5–6, 5f
Stress granules, 627
Stringency, 52
StRNA (short temporal RNA), 869
Strong-stop minus DNA, 441, 441f
Strong-stop plus DNA, 442
Str operon, 762, 762f
Structural distortions, 395, 395f
Structural genes, 4, 737
Structural maintenance of chromosomes (SMC) proteins, 309
Subcloned fragments, 46
Subfamilies, protein, 808
Substitutions, replication fidelity control and, 327–328, 327f
Substrate binding sites, in catalytic RNA, 648, 648f
Su(Hw)/mod(mdg4) complex, 253–254, 254f
Suicide substrates, 376, 378
Sulfolobus, replicons in, 272
Sumoylation, of histone tails, 229, 229f
SUP35, 851–853, 851f, 852f
Supercoiling, DNA, 8–10, 8f
  bacterial nucleoid, 195–197, 196f
  constrained, 196, 196f
  eukaryotic, 197
  gel electrophoresis and, 55, 56f
  in lambda integration, 378, 379
  negative, 8–9, 196, 530, 530f
  nucleosomal, 235–236, 236f
  in phage rolling circle replication, 288
  positive, 8, 530, 530f
  relaxation of in replication, 321
  in site-specific recombination, 351
  in transcription, 530, 530f
  unconstrained, 196, 196f
Superfamilies, gene, 91, 466
Superfamilies, protein, 808
Suppression mutations, 20
  in tRNA, 722–723, 723f
Suppressors, 722–723, 723f
  missense, 722, 725
  nonsense, 722
  partial, 725–726
Surveillance systems, RNA, 631–632, 631f, 632f
  cytoplasmic, 633–635, 633f, 634f, 635f
Su(var) proteins, 832–834, 832f, 833f
SV5, 658
SV40 minichromosomes, 245, 246f
SWI/SNF (switch sniff) complex, 808–809
  nucleosome organization and, 809
  in prions, 853
SWI2/SNF2
  in double-strand break repair, 412, 412f
  in recombination repair, 409
Switch sniff complex. *See* SWI/SNF (switch sniff) complex
Symplekin, 604f, 605
Synapsis, chromosome, 353
  homologous recombination and, 373
  recombination and, 363

Synaptonemal complex, 353
  chromosome pairing and, 364–365
  formation of, 362–363, 364f
  meiotic chromosome connection in, 360–362, 361f
Synonymous codons, 706
Synonymous mutations, 162
  neutral, 164
  in sequence divergence, 168–169, 168–170, 168f
Synteny, 109–110, 171, 171f
Synthesis
  abortive, 531
  error-prone, 402
  in excision repair, 396, 396f
  translesion, 402
Synthesis-dependent strand-annealing model (SDSA), 357–358, 357f
  double-strand breaks in, 363
Synthetases, 670–671, 670f
Synthetic genetic array analysis (SGA), 132
Synthetic lethals, 132

T

T4 phage
  functional clustering in, 773–775, 773f, 774f
T7 phage
  functional clustering in, 773–775, 773f, 774f
T7 RNA polymerases, 530–531, 531f
TAFs. *See* TBP-associated factors (TAFs)
Tandem gene clusters, 145–147, 145f, 146f
  crossover fixation in, 147–150
TAP, 494
TATA-binding protein (TBP), 551, 551f
TATA box, 515. *See also* TBP (TATA binding protein)
  in basal apparatus interactions, 803–804, 803f, 804f
  in basal assemblies, 559–560
  bending of, 556–557
  in RNA polymerase II, 554–555
TATA-less promoters, 555
TAZ1 gene, 215
TBP (TATA binding protein), 555–557, 556f, 557f
  in TF$_{III}$C, 553
TBP-associated factors (TAFs), 555–556
  in basal apparatus interactions, 803–804, 803f, 804f
  histone acetyltransferases and, 812
T cell receptors (TCRs), 461
  in adaptive immunity, 464–466, 465f
  antigen-presenting cells and, 490–491, 491f
  clonal selection in, 466–467, 467f
  expression of, 488–490, 489f, 490f
  gene encoding of, 469
  major histocompatibility complex and, 465
T cells, 460, 461. *See also* Immune system
  in cell-mediated response, 465
  clonal selection in, 466–467, 467f
  cytotoxic (killer), 465
  helper, 464–465, 465f
  TCR expression in, 488–490, 489f, 490f
T-cells, locus control regions and, 257
TCRs. *See* T cell receptors (TCRs)
T-DNA
  infection genes in, 292–293, 293f, 294f, 295f
  transfer of, 295–297, 296f
TdT. *See* Terminal deoxynucleotdyl transferase (TdT)
Telomerase, 213–214
Telomeres, 210–215
  definition of, 210
  exchange of in antigenic variation, 383